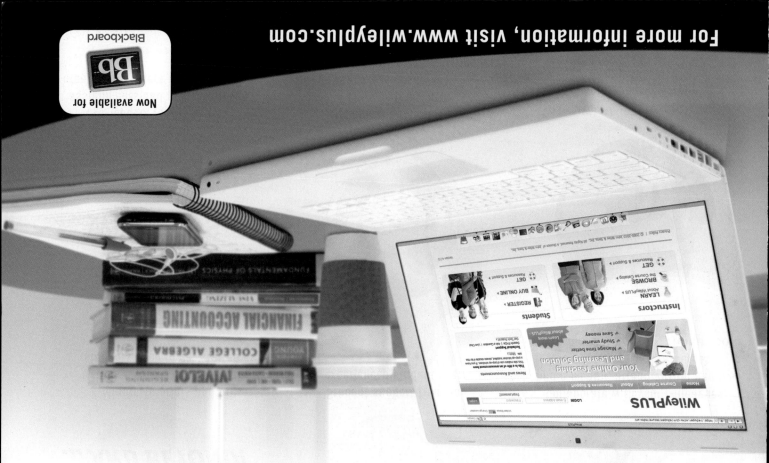

WileyPLUS

ALL THE HELP, RESOURCES, AND PERSONAL SUPPORT YOU AND YOUR STUDENTS NEED!

www.wileyplus.com/resources

1st DAY OF CLASS ... AND BEYOND!

2-Minute Tutorials and all of the resources you and your students need to get started

WileyPLUS Student Partner Program

Student support from an experienced student user

Wiley Faculty Network

Collaborate with your colleagues, find a mentor, attend virtual and live events, and view resources
www.WhereFacultyConnect.com

WileyPLUS Quick Start

Pre-loaded, ready-to-use assignments and presentations created by subject matter experts

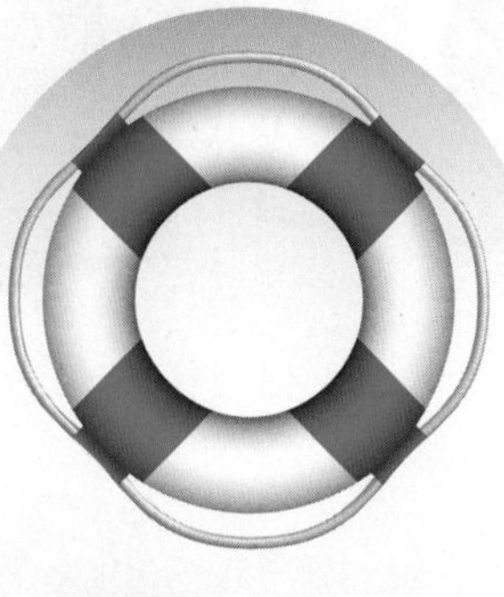

Technical Support 24/7
FAQs, online chat, and phone support
www.wileyplus.com/support

© Courtney Keating/iStockphoto

Your *WileyPLUS* Account Manager, providing personal training and support

NINTH EDITION

Environmental Science

Earth as a Living Planet

Daniel B. Botkin

Professor (Adjunct)
Department of Biology
University of Miami
Coral Gables, FL

Professor Emeritus
Department of Ecology, Evolution, and Marine Biology
University of California, Santa Barbara

Edward A. Keller

Professor of Environmental Studies and Earth Science
University of California, Santa Barbara

WILEY

VICE PRESIDENT AND PUBLISHER Petra Recter
EXECUTIVE EDITOR Ryan Flahive
ASSISTANT EDITOR Elizabeth Baird
EDITORIAL ASSISTANT Chloe Moffett
CONTENT MANAGER Juanita Thompson
MARKETING MANAGER Clay Stone
CREATIVE DIRECTOR Harry Nolan
SENIOR PRODUCTION EDITOR Trish McFadden
DESIGNER Wendy Lai
SENIOR PHOTO EDITOR Billy Ray
COVER PHOTOS: Main photo: ©Hal Bergman Photography/Flickr/Getty Images
Inset photos (from left to right): Stock Image ©Anton Balazh 2011/istock; ©ssguy/ShutterStock; ©Sean Randall/Getty Images, Inc.; ©George Doyle/Getty Images, Inc.

This book was set in Adobe Garamond by cMPreparé and printed and bound by Courier/Kendallville. The cover was printed by Courier/Kendallville.

This book is printed on acid-free paper.

Founded in 1807, John Wiley & Sons, Inc. has been a valued source of knowledge and understanding for more than 200 years, helping people around the world meet their needs and fulfill their aspirations. Our company is built on a foundation of principles that include responsibility to the communities we serve and where we live and work. In 2008, we launched a Corporate Citizenship Initiative, a global effort to address the environmental, social, economic, and ethical challenges we face in our business. Among the issues we are addressing are carbon impact, paper specifications and procurement, ethical conduct within our business and among our vendors, and community and charitable support. For more information, please visit our website: *www.wiley.com/go/citizenship*.

Library of Congress Cataloging-in-Publication Data:

ISBN-13: 978-1-118-42732-3
BRV ISBN: 978-1-118-29197-9

Printed in the United States of America

10 9 8 7 6 5 4 3 2

Preface

What Is Environmental Science?

Environmental science is a group of sciences that attempt to explain how life on the Earth is sustained, the causes of environmental problems, and how these problems can be solved.

Why Is This Study Important?

- We depend on our environment. We enjoy it and the condition of our environment has a large effect on the quality of our lives. People can only live in an environment with certain kinds of characteristics and within certain ranges of availability of resources. Because modern science and technology give us the power to affect the environment, we have to understand how the environment works, so that we can live within its constraints.
- People have always been fascinated with nature, which, in its broadest view, is our environment. As long as people have written, they have asked three questions about ourselves and nature:

  What is nature like when it is undisturbed by people?

  What are the effects of people on nature?

  What are the effects of nature on people?

Environmental science is our modern way of seeking answers to these questions.

What Is the "Science" in Environmental Science?

Many sciences are important to environmental science. These include biology (especially ecology, the part of biology that deals with the relationships among living things and their environment), geology, hydrology, climatology, meteorology, oceanography, and soil science.

How Is Environmental Science Different from other Sciences?

It involves many sciences.

It includes sciences, but also involves related nonscientific fields that have to do with how we value the environment, from environmental philosophy to environmental economics.

It deals with many topics that have great emotional effect on people, and therefore are subject to political debate and to strong feelings that often ignore scientific information.

What Is Your Role as a Student and as a Citizen?

Your role is to understand how to think through environmental issues so that you can arrive at your own conclusions.

What Are the Professions That Grow Out of Environmental Science?

Many professions have grown out of the modern concern with environment, or have been extended and augmented by modern environmental sciences. These include park, wildlife, and wilderness management; urban planning and design; landscape planning and design; conservation and sustainable use of our natural resources; pollution control; environmental energy engineering.

Goals of This Book

Environmental Science: Earth as a Living Planet provides an up-to-date introduction to the study of the environment. Information is presented in an interdisciplinary perspective necessary to deal successfully with environmental problems. The goal is to teach you, the student, how to think through environmental issues.

Critical Thinking

We must do more than simply identify and discuss environmental problems and solutions. To be effective, we must know what science is and is not. Then, we need to develop critical thinking skills. Critical thinking is so important that we have made it the focus of its own chapter, Chapter 2. With this need in mind, we have also developed *Environmental Science* to present the material in a factual and unbiased format. Our goal is to help you think through the issues, not tell you what to think. To this purpose, at the end of each chapter, we present "Critical Thinking Issues." Critical thinking is further emphasized throughout the text in analytical discussions of topics, evaluation of perspectives, and integration of important themes, which are described in detail later.

Interdisciplinary Approach

The approach of *Environmental Science* is interdisciplinary in nature. Environmental science integrates many disciplines, including the natural sciences, in addition to fields such as anthropology, economics, history, sociology, and philosophy of the environment. Not only do we need the best ideas and information to deal successfully with our environmental problems, but we also must be aware of the cultural and historical contexts in which we make decisions about the environment. Thus, the field of environmental science also integrates the natural sciences with environmental law, environmental impact, and environmental planning.

Themes

Our book is based on the philosophy that six threads of inquiry are of particular importance to environmental science. These key themes—human population, sustainability, global perspective, urban world, people and nature, and science and values—are woven throughout the book.

These six key themes are discussed in more detail in Chapter 1. They are also revisited at the end of each chapter and are emphasized in the Closer Look boxes, each of which is highlighted by an icon suggesting the major underlying theme of the discussion. In many cases, more than one theme is relevant.

Human Population

Underlying nearly all environmental problems is the rapidly increasing human population. Ultimately, we cannot expect to solve environmental problems unless the total number of people on Earth is an amount the environment can sustain. There is considerable debate about what are practical limits to what can be sustained.

Sustainability

Sustainability means that a resource is used in a way that it continues to be available. However, the term is used vaguely, and it is a concept experts continually struggle to clarify. Some would define it as ensuring that future generations have equal opportunities to access the resources that our planet offers. Others would argue that sustainability refers to types of developments that are economically viable, do not harm the environment, and are socially just. We all agree that we must learn how to sustain our environmental resources so that they continue to provide benefits for people and other living things on our planet.

A Global Perspective

Until recently, it was common to believe that human activity caused only local, or at most regional, environmental change. We now know that human activities can affect the environment globally. An emerging science known as Earth System Science seeks a basic understanding of how our planet's environment works as a global system. This understanding can then be applied to help solve global environmental problems. The emergence of Earth System Science has opened up a new area of inquiry for faculty and students.

The Urban World

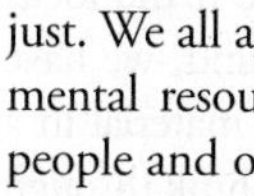

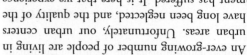

An ever-growing number of people are living in urban areas. Unfortunately, our urban centers have long been neglected, and the quality of the urban environment has suffered. It is here that we experience the worst of air pollution, waste-disposal problems, and other stresses on the environment. In the past our studies of the environment have focused more on wilderness than on the urban environment. In the future we must place greater emphasis on towns and cities as livable environments.

People and Nature

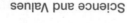

People seem to be always interested—amazed, fascinated, pleased, curious—in our environment. Why is it suitable for us? How can we keep it that way? We know that people and our civilizations are having major effects on the environment, from local effects (the street where you live) to the entire planet (we have created a hole in the Earth's ozone layer), which can affect us and many forms of life.

Science and Values

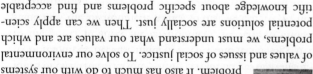

Finding solutions to environmental problems involves more than simply gathering facts and understanding the scientific issues of a particular problem. It also has much to do with our systems of values and issues of social justice. To solve our environmental problems, we must understand what our values are and which potential solutions are socially just. Then we can apply scientific knowledge about specific problems and find acceptable solutions.

Special Features

In writing *Environmental Science* we have designed a text that incorporates a number of special features that we believe will help teachers to teach and students to learn. These include the following:

A **Case Study** introduces each chapter. The purpose of this feature is to interest students in the chapter's subject and to raise important questions on the subject matter For example, in Chapter 11, Agriculture, Aquaculture, and Environment, the opening case study tells about a farmer feeding his pigs trail mix, banana chips, yogurt-covered raisins, dried papaya, and cashews because growing corn for biofuels is raising the costs of animal feed so much.

Learning Objectives at the beginning of each chapter challenge students to apply the knowledge in the chapter to solve some problems and to integrate the chapter's materials.

A **Closer Look** is the name of special learning modules that present more detailed information concerning a particular concept or issue. For example, A Closer Look 13.2 discusses the reasons for conserving endangered species.

Many of these special features contain **figures** and **data** designed to enrich the reader's understanding and relate back to the book themes.

Near the end of each chapter, a **Critical Thinking Issue** encourages critical thinking about the environment, helping students understand how the issue may be studied and evaluated. For example, Chapter 22 presents a critical thinking issue title "How Can Urban Sprawl Be Controlled?"

Following the Summary, a special section, **Reexamining Themes and Issues**, reinforces the six major themes of the textbook.

Brief Contents

Contents

Acknowledgments

Completion of this book was possible only through the cooperation and work of many people. To all those who so freely offered their advice and encouragement in this endeavor, we offer our most sincere appreciation. We are indebted to our colleagues who made contributions.

We greatly appreciate the work of our editors at John Wiley & Sons, Rachel Falk and Ryan Flahive, for support, encouragement, assistance, and professional work. We extend thanks to our production editor Trish McFadden, who did a great job and made important contributions in many areas; to Wendy Lai for a beautiful interior design; and to Billy Ray for photo research. Thanks go out also for editorial assistance from Elizabeth Baird and Chloe Moffett.

Reviewers of This Edition

Renée E. Bishop Pierce, Penn State Worthington Scranton
Ron Cisar, Iowa Western Community College
James Kubicki, Penn State
Bryan Mark, Ohio State University
Stephen Overmann, Southeast Missouri State University
Dork Sahagian, Lehigh University
Santhosh Seelan, University of North Dakota
Don Williams, Park University

Reviewers of Previous Editions

David Aborn, University of Tennessee, Chattanooga
Marc Abrams, Pennsylvania State University
John All, Western Kentucky University
Diana Anderson, Northern Arizona University
Mark Anderson, University of Maine
Robert J. Andres, University of North Dakota
Walter Arenstein, Ohlone College
Daphne Babcock, Collin County Community College
Marvin Baker, University of Oklahoma
Kevin Baldwin, Monmouth College
Michele Barker-Bridges, Pembroke State University (NC)
James W. Bartolome, University of California, Berkeley
Colleen Baxter, Georgia Military College
Laura Beaton, York College
Susan Beatty, University of Colorado, Boulder
David Beckett, University of Southern Mississippi
Brian Beeder, Morehead State University

About the Authors

Photo by Maguire Neblet

Daniel B. Botkin is Professor (Adjunct), Department of Biology, University of Miami, Coral Gables, Florida, and Professor Emeritus of Ecology, Evolution, and Marine Biology, University of California, Santa Barbara, where he joined the faculty in 1978, serving as chairman of the Environmental Studies Program from 1978 to 1985. For more than four decades, Professor Botkin has been active in the application of ecological science to environmental management. He is winner of Santa Barbara Botanic Garden's John C. Pritzlaff Conservation Award in 2012, in recognition of his career contributions to conservation of biodiversity. Other awards include Oxford University Astor Lectureship, 2007; the Mitchell International Prize for Sustainable Development, and the Fernow Prize for International Forestry; he has also been elected to the California Environmental Hall of Fame.

Trained in physics and biology, Professor Botkin is a leader in the application of advanced technology to the study of the environment. The originator of widely used forest gap-models, he has conducted research on endangered species, characteristics of natural wilderness areas, the biosphere, and global environmental problems, including possible ecological effects of global warming. During his career, Professor Botkin has advised the World Bank about tropical forests, biological diversity and sustainability; the Rockefeller Foundation about global environmental issues; the government of Taiwan about approaches to solving environmental problems; and the state of California on the environmental effects of water diversion on Mono Lake. He served as the primary advisor to the National Geographic Society for its centennial edition map on "The Endangered Earth." He directed a study for the states of Oregon and California concerning salmon and their forested habitats.

He has published many articles and books about environmental issues. His latest books are: *The Moon in the Nautilus Shell: Discordant Harmonies Reconsidered (Oxford University Press, 2012*) and *Powering the Future: A Scientist's Guide to Energy Independence (FT Press, 2010).* His other books are: *Beyond the Stoney Mountains: Nature in the American West from Lewis and Clark to Today* (Oxford University Press) and available as an ebook (Croton River Publishers); *Strange Encounters: Adventures of a Renegade Naturalist* (Penguin/Tarcher), *The Blue Planet* (Wiley), available as an ebook (Croton River Publishers); *No Man's Garden: Thoreau and a New Vision for Civilization and Nature* (Island Press; now available as an ebook (Croton River Publishers); *Passage of Discovery: The American Rivers Guide to The Missouri River of Lewis and Clark* (Perigee Books (a Division of Penguin-Putnam, now available as an ebook (Croton River Publisher); *Our Natural History: The Lessons of Lewis and Clark* (Oxford University Press), *Discordant Harmonies: A New Ecology for the 21st Century* (Oxford University Press), and *Forest Dynamics: An Ecological Model* (Oxford University Press).

Professor Botkin was on the faculty of the Yale School of Forestry and Environmental Studies (1968–1974) and was a member of the staff of the Ecosystems Center at the Marine Biological Laboratory, Woods Hole, MA (1975–1977). He received a B.A. from the University of Rochester, an M.A. from the University of Wisconsin, and a Ph.D. from Rutgers University.

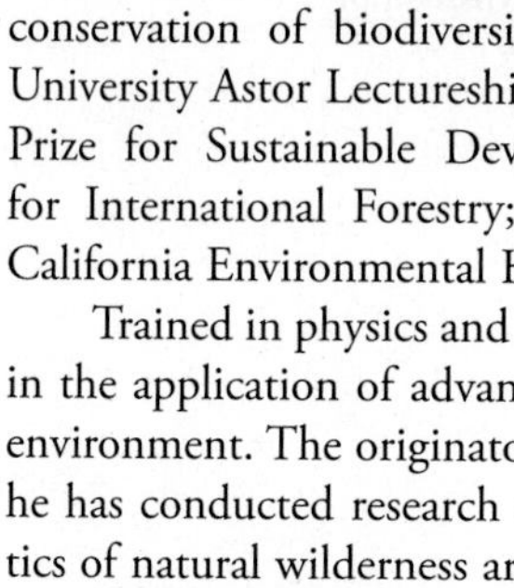

Edward A. Keller was chair of the Environmental Studies and Hydrologic Sciences Programs from 1993 to 1997 and is Professor of Earth Science at the University of California, Santa Barbara, where he teaches earth surface processes, environmental geology, environmental science, river processes, and engineering geology. Prior to joining the faculty at Santa Barbara, he taught geomorphology, environmental studies, and earth science at the University of North Carolina, Charlotte. He was the 1982–1983 Hartley Visiting Professor at the University of Southampton, a Visiting Fellow in 2000 at Emmanuel College of Cambridge University, England, and recipient of the Easterbrook Distinguished Scientist award from the Geological Society of America in 2004.

Professor Keller has focused his research efforts into three areas: studies of Quaternary stratigraphy and tectonics as they relate to earthquakes, active folding, and mountain building processes; hydrologic process and wildfire in the chaparral environment of Southern California; and physical habitat requirements for the endangered Southern California steelhead trout. He is the recipient of various Water Resources Research Center grants to study fluvial processes and U.S. Geological Survey and Southern California Earthquake Center grants to study earthquake hazards.

Professor Keller has published numerous papers and is the author of the textbooks *Environmental Geology*; *Introduction to Environmental Geology*; *Natural Hazards* (with Duane DeVecchio); and *Active Tectonics* (with Nicholas Pinter). He holds bachelor's degrees in both geology and mathematics from California State University, Fresno, an M.S. in geology from the University of California, and a Ph.D. in geology from Purdue University.

Steven Kazlowski/NG Image Collection

FIGURE 1.2 Sockeye salmon spawning in an Alaskan River.

swim through very fast-running water. Because salmon tend to return to the same stream where they were born, each stream's population can be genetically isolated from others long enough so that different populations differ genetically from each other. The American Fisheries Society has listed up to 500 genetically distinct salmon populations among the six species. Unfortunately, numerous populations of salmon have declined in many rivers in northern California, Oregon, Washington, British Columbia, and Alaska.

Salmon are important to various ecosystem food chains. They have been long vital to the cultures of Native Americans, as we see in the Coyote legend. Salmon fishing is also a major recreational activity benefiting many towns along the Pacific coast. Moreover, there is a worldwide market for salmon. As a result, many attempts have been made to increase the size of salmon populations. The four culprits usually listed as causes of decline in salmon populations are overfishing by people, forest logging that destroys spawning and the breeding habitat, construction of dams that interfere with salmon migration, and harvests of salmon by sea lions.

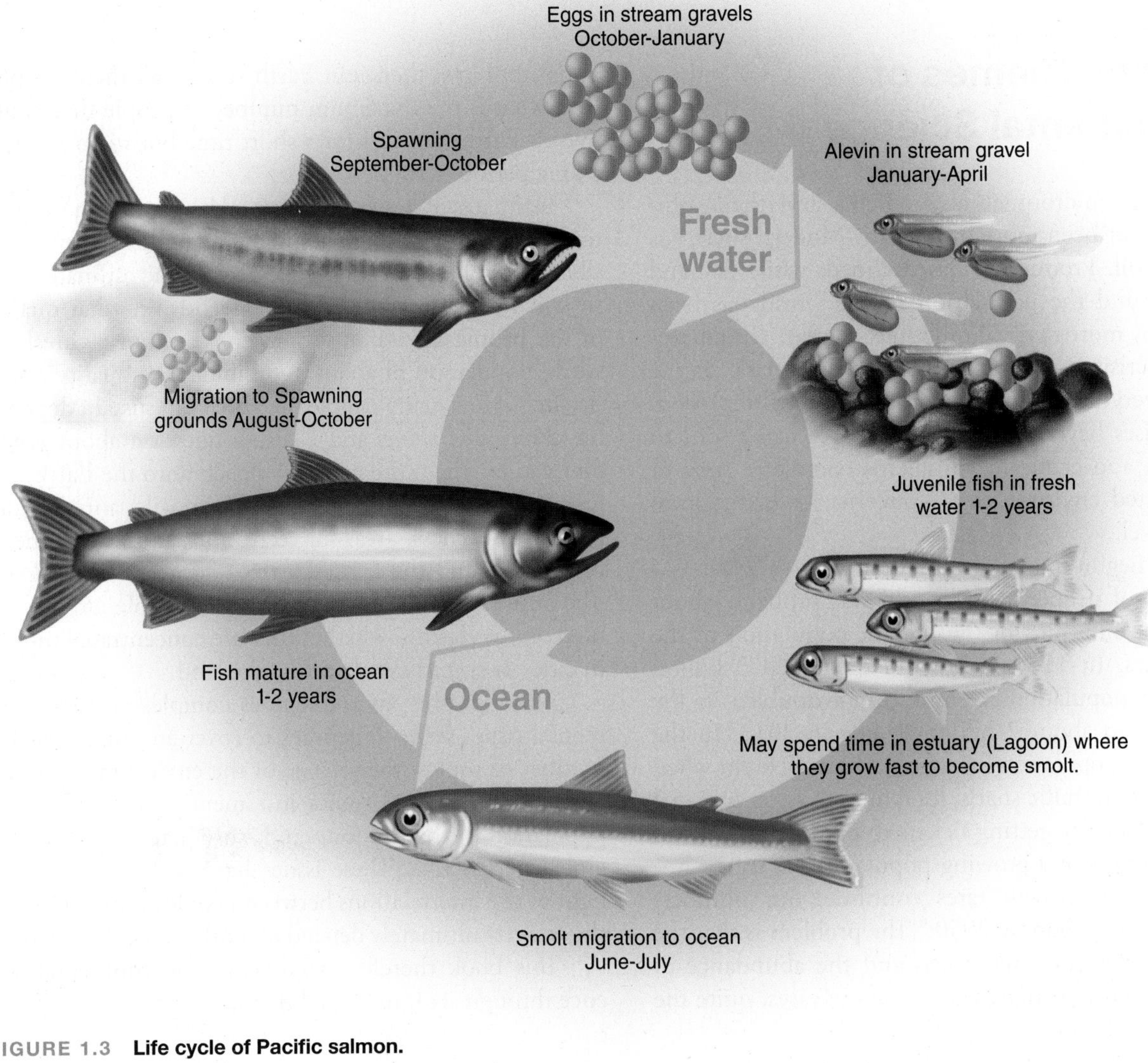

(*Source:* Based on figures from Alaskan Department of Fish and Game and the US Fish and Wildlife Service).

FIGURE 1.3 **Life cycle of Pacific salmon.**

But other factors, which receive less attention, may also be important. These include problems arising from hatchery-bred fish intermixing with wild fish, thereby mixing genotypes that may be less well adapted to a stream; pollution of streams and therefore salmon habitat from agricultural chemical runoff; and introduction of exotic species, such as parasites, predators, and competitors with native species and populations.[2] Deforestation also leads to warmer waters and the addition of fine sediment that plugs stream gravel, reducing oxygen and egg vitality.

So what is the future of salmon, these iconic fish of the Pacific Northwest, often viewed as an indicator of the health of our rivers? Large amounts of money have been spent to study the life cycle of salmon in various rivers and to develop management plans. In general, these efforts have varied from unsuccessful to moderately successful.

The history of salmon exploitation around the world, from Europe to the United States, as it is with most fish species, seems to follow a similar pattern: discovery of a very abundant source; habitat degradation; and then overfishing that leads to great declines, sometimes leaving populations so low as to make harvests impractical. The good news is that it is very likely that many salmon populations will recover, although probably not to pre-European settlement levels.[3]

The story of Pacific Northwest salmon emphasizes the six major themes of *Environmental Science:*

1. Human population increase has impacted salmon habitat.
2. We are concerned for the future (sustainability) of salmon.
3. Climate change, with warming river water and changing flows, is impacting the future of salmon.
4. Urbanization of coastal areas impacts some rivers where salmon are present.
5. People and salmon in the Pacific Northwest have shared a common history for thousands of years.
6. While science can provide choices to manage salmon, the choices we make will reflect our values.

1.1 Major Themes of Environmental Science

The study of environmental problems and their solutions has never been more important. Modern society is hooked on oil. Production has declined, while demand has grown, and the population of the world has been increasing by more than 70 million each year. The emerging energy crisis is producing an economic crisis, as the prices of everything produced from oil (fertilizer, food, and fuel) rises beyond what some people can afford to pay. Energy and economic problems come at a time of unprecedented environmental concerns, from the local to global level.

At the beginning of the modern era—in A.D. 1—the number of people in the world was probably about 100 million, one-third of the present population of the United States. In 1960, the world contained 3 billion people. Our population has more than doubled in the last 50 years to about 7 billion people in 2013. In the United States, population increase is often apparent when we travel. Urban traffic snarls, long lines to enter national parks, and difficulty getting tickets to popular attractions are all symptoms of a growing population. If recent human population growth rates continue, our numbers could reach 9.4 billion by 2050.[4] The problem is that the Earth has not grown any larger, and the abundance of its resources has not increased—in many cases, quite the opposite. How, then, can Earth sustain all these people? And what is the maximum number of people that could live on Earth, not just for a short time but *sustained* over a long period?

Estimates of how many people the planet can support range from 2.5 billion to 40 billion (a population not possible with today's technology). Why do the estimates vary so widely? Because the answer depends on what quality of life people are willing to accept. Beyond a threshold world population of about 4–6 billion, the quality of life declines. How many people the Earth can sustain depends on *science and values* and is also a question about *people and nature*. The more people we pack onto the Earth, the less room and resources there are for wild animals and plants, wilderness, areas for recreation, and other aspects of nature—and the faster Earth's resources will be used. The answer also depends on how the people are distributed on the Earth—whether they are concentrated mostly in cities or spread evenly across the land.

Although the environment is complex and environmental issues seem sometimes to cover an unmanageable number of topics, the science of the environment comes down to the central topics just mentioned: the human population, urbanization, and sustainability within a global perspective. These issues have to be evaluated in light of the interrelations between people and nature, and the answers ultimately depend on both science and nature.

This book therefore approaches environmental science through six interrelated themes:

1. *Human population growth* (the environmental problem)
2. *Sustainability* (the environmental goal)
3. *A global perspective* (many environmental problems require a global solution)
4. *An urbanizing world* (most of us live and work in urban areas)
5. *People and nature* (we share a common history with nature)
6. *Science and values* (science provides solutions; which ones we choose are in part value judgments)

You may ask, "If this is all there is to it, what is in the rest of this book?" (See A Closer Look 1.1.) The answer lies with the old saying "The devil is in the details." The solution to specific environmental problems requires specific knowledge. The six themes listed above help us see the big picture and provide a valuable background. The opening case study illustrates linkages among the themes, as well as the importance of details.

In this chapter we introduce the six themes with brief examples, showing the linkages among them and touching on the importance of specific knowledge that will be the concern of the rest of the book. We start with human population growth.

1.2 Human Population Growth

Our Rapid Population Growth

The most dramatic increase in the history of the human population occurred in the last part of the 20th century and continues today into the early 21st century. As mentioned, in merely the past 50 or so years, the human population of the world has more than doubled—to more than 7 billion. Figure 1.4 illustrates this population explosion, sometimes referred to as the "population bomb." The figure shows that the expected decrease in population in the developed regions (for example, the United States and western Europe) is more than offset by rapid population growth in the developing regions (for example, Africa, India, and South America).

A CLOSER LOOK 1.1

A Little Environmental History

A brief historical explanation will help clarify what we seek to accomplish. Before 1960, few people had ever heard the word *ecology*, and the word *environment* meant little as a political or social issue. Then came the publication of Rachel Carson's landmark book, *Silent Spring* (Boston: Houghton Mifflin, 1960, 1962). At about the same time, several major environmental events occurred, such as oil spills along the coasts of Massachusetts and southern California and highly publicized threats of extinction of many species, including whales, elephants, and songbirds. The environment became a popular issue.

As is true of any new social or political issue, at first relatively few people recognized its importance. Those who did found it necessary to stress the problems—to emphasize the negative—in order to bring public attention to environmental concerns. Adding to the limitations of the early approach to environmental issues was a lack of scientific knowledge and practical know-how. Environmental sciences were in their infancy. Some people even saw science as part of the problem.

The early days of modern environmentalism were dominated by confrontations between those labeled environmentalists and those labeled anti-environmentalists. Stated in the simplest terms, environmentalists believed that the world was in peril. To them, economic and social development meant destruction of the environment and ultimately the end of civilization, the extinction of many species, and perhaps the extinction of human beings. Their solution was a new worldview that depended only secondarily on facts, understanding, and science. In contrast, again in simplest terms, the anti-environmentalists believed that whatever the environmental effects, social and economic health and progress were necessary for people and civilization to prosper. From their perspective, environmentalists represented a dangerous and extreme view with a focus on the environment to the detriment of people, a focus they thought would destroy the very basis of civilization and lead to the ruin of our modern way of life.

Today, the situation has changed. Public-opinion polls now show that people around the world rank the environment among the most important social and political issues. There is no longer a need to prove that environmental problems are serious.

We have made significant progress in many areas of environmental science (although our scientific understanding of the environment still lags behind our need to know). We have also begun to create legal frameworks for managing the environment, thus providing a new basis for addressing environmental issues. The time is now ripe to seek truly lasting, more rational solutions to environmental problems.

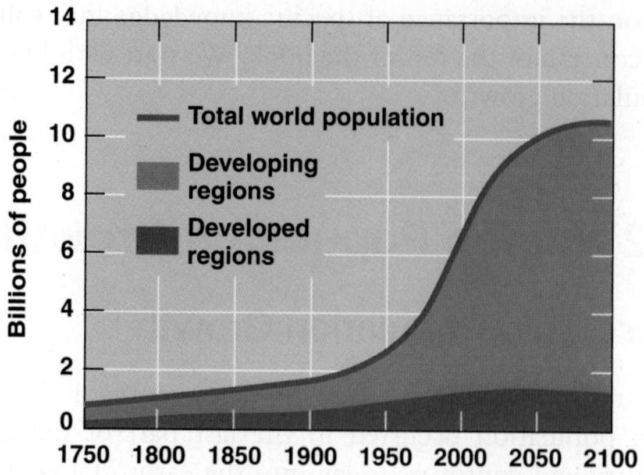

FIGURE 1.4 Population growth in developed and developing nations. 1750 projected to 2100.

Human population growth is, in some important ways, *the* underlying issue of the environment. Much current environmental damage is directly or indirectly the result of the very large number of people on Earth and our rate of increase. As you will see in Chapter 5, where we consider the human population in more detail, for most of human history, the total population was small and the average long-term rate of increase was low relative to today's growth rate.[4]

Although it is customary to think of the population as increasing continuously without declines or fluctuations, the growth of the human population has not been a steady march. For example, great declines occurred during the time of the Black Death in the 14th century. At that time, entire towns were abandoned, food production declined, and in England one-third of the population died within a single decade.[5]

Famine and Food Crisis

Famine is one of the phenomena that happen when a human population exceeds its environmental resources. Famines have occurred in recent decades in Africa. In the mid-1970s, following a drought in the Sahel region, 500,000 Africans starved to death and several million more were permanently affected by malnutrition.[6] Starvation in African nations gained worldwide attention some ten years later, in the 1980s.[7, 8]

Famine in Africa has had multiple interrelated causes. One, as suggested, is drought. Although drought is not new to Africa, the size of the population affected by drought is unprecedented. In addition, deserts in Africa appear to be spreading, in part because of changing climate but also because of human activities. Poor farming practices have increased erosion, and deforestation may be helping to make the environment drier. In addition, the control and destruction of food have sometimes been used as a weapon in political disruptions (Figure 1.5). Today, malnutrition contributes to the death of about 6 million children per year. Low- and middle-income countries suffer the most from malnutrition, as measured by low weight for age (underweight, as shown in Figure 1.6).[9]

The emerging global food crisis in the first decade of the 21st century has not been caused by war or drought but by rising food costs. The cost of basic items, such as rice, corn, and wheat, has risen to the point where

Peter Turnley/© Corbis

dbimages/Alamy

FIGURE 1.5 Science and values. Social conditions affect the environment, and the environment affects social conditions. Political disruption in Somalia (illustrated by a Somalian boy with a gun, left photo) interrupted farming and food distribution, leading to starvation. Overpopulation, climate change, and poor farming methods also lead to starvation, which in turn promotes social disruption. Famine has been common in parts of Africa since the 1980s, as illustrated by gifts of food from aid agencies.

(*Source:* World Population Data Sheet [Washington, DC: Population Reference Bureau, 2007. Accessed 5/19/08 @www.prb.org].)

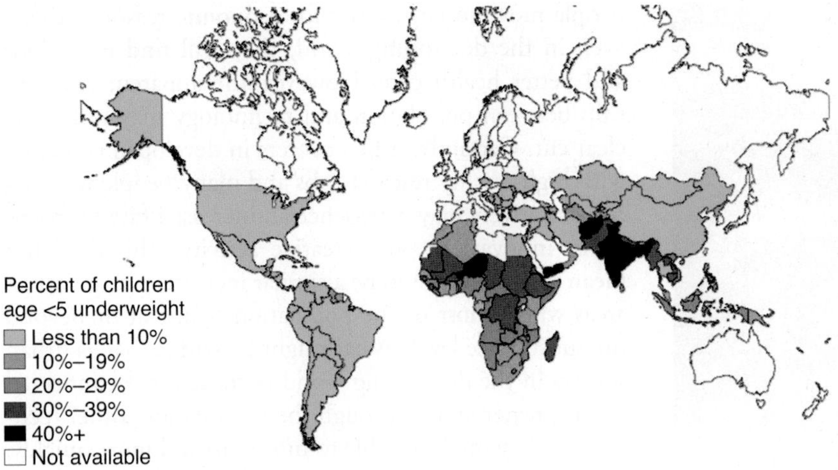

FIGURE 1.6 **Underweight children under the age of five by region.** Most are in low- and middle-income countries.

low- and moderate-income countries are experiencing a serious crisis. In 2007 and 2008, food riots occurred in many locations, including Mexico, Haiti, Egypt, Yemen, Bangladesh, India, and Sudan (Figure 1.7). The rising cost of oil used to produce food (in fertilizer, transportation, working fields, etc.) and the conversion of some corn production to biofuels have been blamed. This situation involves yet another key theme: science and values. Scientific knowledge has led to increased agricultural production and to a better understanding of population growth and what is required to conserve natural resources. With this knowledge, we are forced to confront a choice: which is more important, the survival of people alive today or conservation of the environment on which future food production and human life depend?[10]

Answering this question requires both *value judgments* and the information and knowledge with which to make such judgments. For example, we must determine whether we can continue to increase agricultural production without destroying the very environment on which agriculture and, indeed, the persistence of life on Earth depend. Put another way, a technical, scientific investigation provides a basis for a value judgment.

Human Population and the Incidence of Natural Disasters

It is difficult to pick up a paper without hearing about a natural disaster striking some part of the world. It might be tornadoes in the midwestern United States, flooding in China or Indonesia, a tsunami in Japan, or hurricanes and typhoons around various locations in the Pacific and Atlantic oceans. What is clear is that the number of natural events has not dramatically increased, but the impacts certainly have. Going back a few hundred years, human population on Earth was much less, and when hazardous events struck an area, there was less chance of loss of structures and lives. As human population has grown, more people are in harm's way. A disaster that occurred before the planet was widely populated might have affected a few people in a small, isolated community. Today, a similar natural event is likely to be a catastrophe in which damages may exceed billions of dollars and take hundreds of lives (Figure 1.8). And with growing human population, more people are forced to live on marginal lands where hazardous events are more likely to occur. Thus, increasing population and poor land-use choices (some of which

Thony Belizaire/AFP/Getty Images, Inc.

(a)

STRDEL/AFP/Getty Images, Inc.

(b)

FIGURE 1.7 **Food riots over the rising cost of food in 2007. (a)** Haiti and **(b)** Bangladesh.

REUTERS/MAINCHI SHIMBUN LANDOV

FIGURE 1.8 Tsunami waves in 2011 produced by a very large offshore earthquake inundating part of east coastal Japan (about 400 km or 250 miles north of Tokyo). More than 20,000 people died, and entire coastal communities were destroyed. High population density of about 300 people per square kilometer (800 per square mile) was achieved during the 19th and early 20th centuries, contributing to the large loss of life in the hazardous coastal zone where very large tsunamis happen about every 1,000 years.

are difficult to avoid) are leading to greater property damage and loss of life.

The Age of Abundance and Human Population Increase

Human population on Earth is increasing exponentially (see Working it Out 4.2 on exponential growth in Chapter 4), and on one hand, the general consensus is that this growth is not sustainable from a resource perspective. On the other hand, it has been argued that, along with the exponential growth of population, there has been an exponential growth of technology that has resulted in a much improved environment for people. The idea is that, as population continues to increase, people have a greater chance of making important innovations that will eventually provide clean water and energy on Earth, with affordable housing and education and good health care for all citizens.[11]

Let's explore this hypothesis in more detail. A lot of research has been done on innovation and what makes for creative environments. It certainly appears that large numbers of people working collaboratively in urban centers are more productive, and innovation and creativity occur at higher rates in cities. Thus, it appears that, as the urban regions have grown, so have technological advances and creativity. However, this is not true across the board, particularly for developing countries where the increase in population is most apparent and where large urban regions are developing.

Our large urban centers must provide a quality environment for people to be truly successful. Certainly, people move to large cities for economic reasons. There, even in the developing world, they will find more jobs and better health care. However, the apparent relationship between population and technology may not be so clear cut. Certainly, urban centers in developed countries with high-quality research labs and many people working collaboratively may experience an increased rate of inventions, innovations, and creative activity. This does not mean that there would be a similar increase in large urban areas where most of the population is just at or beyond the subsistence level. What might be said is that as urban centers in the developing world become more connected to the greater world through the Internet and other technology, they might be able to move forward more quickly. However, an ever-growing human population that exceeds the ability of the planet to support it will eventually lead to a failure of ecosystems and cause problems for humans. Later in this chapter and in Chapter 22 on urban environments, we will return to this discussion when we consider the principle that we live in an urban world.

1.3 Sustainability and Carrying Capacity

The story of recent famines and food crises brings up one of the central environmental questions of our day: What is the maximum number of people the Earth can sustain? That is, what is the sustainable human carrying capacity of the Earth? Much of this book will deal with information that helps answer this question. However, there is little doubt that we are using many renewable environmental resources faster than they can be replenished—in other words, we are using them *unsustainably*. In general, we are using forests and fish faster than they can regrow, and we are eliminating habitats of endangered species and other wildlife faster than they can be replenished. We are also extracting minerals, petroleum, and groundwater without sufficient concern for their limits or the need to recycle them. As a result, there is a shortage of some resources and a probability of more shortages in the future. Clearly, we must learn how to sustain our environmental resources so that they continue to provide benefits for people and other living things on our planet.[3]

Sustainability: The Environmental Objective

The environmental catchphrase of the 1990s was "saving our planet." Are all life and the environments on which life depends really in danger? Will we leave behind a dead planet?

In the long view of planetary evolution, it is certain that planet Earth will survive us. Our sun is likely to last another several billion years, and if all humans became extinct in the next few years, life would still flourish here on Earth. The changes we have made—in the landscape, the atmosphere, the waters—would last for a few hundred or thousands of years but in a modest length of time would be erased by natural processes. What we are concerned with, as environmentalists, is the quality of the *human* environment on Earth, for us today and for our children.

Environmentalists agree that sustainability must be achieved, but we are unclear about how to achieve it, in part because the word is used to mean different things, often leading to confusion that causes people to work at cross-purposes. **Sustainability** has two formal scientific meanings with respect to environment: (1) *sustainability of resources*, such as a species of fish from the ocean, a kind of tree from a forest, or coal from mines; and (2) *sustainability of an ecosystem*. Strictly speaking, harvesting a resource at a certain rate is sustainable if we can continue to harvest that resource at that same rate for some specified time well into the future. An ecosystem is sustainable if it can continue its primary functions for a specified time in the future. (Economists refer to the specified time in the future as a "planning time horizon.") Commonly, in discussions about environmental problems, the time period is not specified and is assumed to be very long—mathematically an infinite planning time, but in reality as long as it could possibly matter to us. For conservation of the environment and its resources to be based on quantitative science, both a rate of removal and a planning time horizon must be specified. However, ecosystems and species are always undergoing change, and a completely operational definition of *sustainability* will have to include such variation over time.

Economists, political scientists, and others also use the term *sustainability* in reference to types of development that are economically viable, do not harm the environment, and are socially just (fair to all people). We should also point out that the term *sustainable growth* is an oxymoron (i.e., a contradictory term) because any steady growth (fixed-percentage growth per year) produces large numbers in modest periods of time (see Working it Out 4.2 on exponential growth in Chapter 4).

One of the environmental paradigms of the 21st century will be sustainability, but how will it be attained? Economists have begun to consider what is known as the *sustainable global economy:* the careful management and wise use of the planet and its resources, analogous to the management of money and goods. Those focusing on a sustainable global economy generally agree that under present conditions the global economy is *not* sustainable. Increasing numbers of people have resulted in so much pollution of the land, air, and water that the ecosystems that people depend on are in danger of collapse. What, then, are the attributes of a sustainable economy in the information age?[12]

- Populations of humans and other organisms living in harmony with the natural support systems, such as air, water, and land (including ecosystems)
- An energy policy that does not pollute the atmosphere, cause climate change (such as global warming), or pose unacceptable risk (a political or social decision)
- A plan for renewable resources—such as water, forests, grasslands, agricultural lands, and fisheries—that will not deplete the resources or damage ecosystems
- A plan for nonrenewable resources that does not damage the environment, either locally or globally, and ensures that a share of our nonrenewable resources will be left to future generations
- A social, legal, and political system that is dedicated to sustainability, equity, and justice for all people, with a democratic mandate to produce such an economy

Recognizing that population is *the* environmental problem, we should keep in mind that a sustainable global economy will not be constructed around a completely stable global population. Rather, such an economy will take into account the fact that the size of the human population will fluctuate within some stable range necessary to maintain healthy relationships with other components of the environment. To achieve a sustainable global economy, we need to do the following:[12]

- Develop an effective population-control strategy. This will at least require more education of people, since literacy and population growth are inversely related.
- Completely restructure our energy programs. A sustainable global economy is probably impossible if it is based on the use of fossil fuels. New energy plans will be based on an integrated energy policy, with more emphasis on renewable energy sources (such as solar and wind) and on energy conservation.
- Institute economic planning, including a tax structure that will encourage population control and wise use of resources. Financial aid for developing countries is absolutely necessary to narrow the gap between rich and poor nations.
- Implement social, legal, political, and educational changes that help to maintain a quality local, regional, and global environment. This must be a serious commitment that all the people of the world will cooperate with.

Moving Toward Sustainability: Some Criteria

Stating that we wish to develop a sustainable future acknowledges that our present practices are not sustainable. Indeed, continuing on our present paths of overpopulation, resource consumption, and pollution will not lead to sustainability. We will need to develop new concepts that will mold industrial, social, and environmental interests into an integrated, harmonious system. In other words, we need to develop a new paradigm, an alternative to our present model for running society and creating wealth.[13] The new paradigm might be described as follows.[14]

- *Evolutionary rather than revolutionary.* Developing a sustainable future will require an evolution in our values that involves our lifestyles as well as social, economic, and environmental justice.
- *Inclusive, not exclusive.* All peoples of Earth must be included. This means bringing all people to a higher standard of living in a sustainable way that will not compromise our environment.
- *Proactive, not reactive.* We must plan for change and for events, such as human population problems, resource shortages, and natural hazards, rather than waiting for them to surprise us and then reacting. This may sometimes require us to apply the Precautionary Principle, which we discuss with science and values (Section 1.7).
- *Attracting, not attacking.* People must be attracted to the new paradigm because it is right and just. Those who speak for our environment should not take a hostile stand but should attract people to the path of sustainability through sound scientific argument and appropriate values.
- *Assisting the disadvantaged, not taking advantage.* This involves issues of environmental justice. All people have the right to live and work in a safe, clean environment. Working people around the globe need to receive a living wage—wages sufficient to support their families. Exploitation of workers to reduce the costs of manufacturing goods or growing food diminishes us all.

The Carrying Capacity of Earth

Carrying capacity is a concept related to sustainability. It is usually defined as the maximum number of individuals of a species that can be sustained by an environment without decreasing the capacity of the environment to sustain that same number in the future.

There are limits to Earth's potential to support humans. If we used Earth's total photosynthetic potential with present technology and efficiency to support 7 billion people, Earth could support a human population of about 15 billion. However, in doing this, we would share our land with very little else.[15,16] When we ask "What is the maximum number of people that Earth can sustain?" we are asking not just about Earth's carrying capacity but also about sustainability.

As we pointed out, what we consider a "desirable human carrying capacity" depends in part on our values (Figure 1.9). Do we want those who follow us to live short lives in crowded conditions, without a chance to enjoy Earth's scenery and diversity of life? Or do we hope that our descendants will have a life of high quality and good health? Once we choose a goal regarding the quality of life, we can use scientific information to understand what the sustainable carrying capacity might be and how we might achieve it.

1.4 A Global Perspective

Our actions today are experienced worldwide. Because human actions have begun to change the environment all over the world, the next generation, more than the present generation, will have to take a global perspective on environmental issues (Figure 1.10).

Recognition that civilization can change the environment at a global level is relatively recent. As we discuss in detail in later chapters, scientists now believe that emissions of modern chemicals are changing the ozone layer high in the atmosphere. Scientists also believe that burning fossil fuels increases the concentration of greenhouse gases in the atmosphere, which may change Earth's climate. These atmospheric changes suggest that the actions of many groups of people, at many locations, affect

Mira/Alamy

Comstock/Jupiter Images

(a) (b)

FIGURE 1.9 **How many people do we want on Earth? (a)** Streets of Calcutta; **(b)** Davis, California.

FIGURE 1.10 **Earth from space.** Isolated from other planets, Earth is "home," the only habitat we have.

the environment of the entire world.[17] Another new idea explored in later chapters is that not only human life but also nonhuman life affects the environment of our whole planet and has changed it over the course of several billion years. These two new ideas have profoundly affected our approach to environmental issues.

Awareness of the global interactions between life and the environment has led to the development of the **Gaia hypothesis**. Originated by British chemist James Lovelock and American biologist Lynn Margulis, the Gaia hypothesis (discussed in Chapter 4) proposes that over the history of life on Earth, life has profoundly changed the global environment, and that these changes have tended to improve the chances for the continuation of life. Because life affects the environment at a global level, the environment of our planet is different from that of a lifeless one.

1.5 An Urban World

In part because of the rapid growth of the human population and in part because of changes in technology, we are becoming an urban species, and our effects on the environment are more and more the effects of urban life (Figure 1.11a). Economic development leads to urbanization; people move from farms to cities and then perhaps to suburbs. Cities and towns get larger, and because they are commonly located near rivers and along coastlines, urban sprawl often overtakes the agricultural land of river floodplains, as well as the coastal wetlands, which are important habitats for many rare and endangered species. As urban areas expand, wetlands are filled in, forests cut down, and soils covered over with pavement and buildings.

In developed countries, about 75% of the population lives in urban areas and 25% in rural areas, but in developing countries only 40% of the people are city dwellers. By 2008, for the first time, more than half of the people on Earth lived in urban areas, and it is estimated that by 2025 almost two-thirds of the population—5 billion people—will live in cities. Only a few urban areas had populations over 4 million in 1950. In 1999 Tokyo, Japan, was the world's largest city, with a population of about 12 million, and by 2015 Tokyo will likely still be the world's largest city, with a projected population of 28.9 million. The number of **megacities**—urban areas with at least 10 million inhabitants—increased from 2 (New York City and London) in 1950 to 22 (including Los Angeles and New York City) in 2005 (Figure 1.11b). Most megacities are in the developing world, and it is estimated that by 2015 most megacities will be in Asia.[18, 19]

In the past, environmental organizations often focused on nonurban issues—wilderness, endangered species, and natural resources, including forests, fisheries, and wildlife. Although these will remain important issues, in the future we must place more emphasis on urban environments and their effects on the rest of the planet.

Cities Can Help with Some of Our Environmental Problems

Cities are often criticized by environmentalists because of the adverse environmental effects they have had over the past hundred years or more. For example, London in the 1880s was a booming city but filled with slums and desperate people crowded into small areas; it had poor sanitation and little medical support. At that time, some of the leaders of London described the city as a disease that was sucking the life and blood out of the rural districts. What was happening, of course, was that people were moving to cities because they perceived them to have expanded opportunities. One of the great movements of people in human history is going on now with large numbers of people in Asia and the Americas moving to large cities. The main reason identified by economists is that urbanized centers tend to be the richer and rural areas poorer. But is that the entire story?[20]

From an environmental perspective, it is becoming clear that large cities with a good environment for people are preferable environmentally to people dispersed over the land in numerous small communities. People in cities have a much reduced cost of transportation, partly because many people in cities rely entirely on mass transit. People in cities tend to live in smaller places, such as apartments or townhouses, which are easier to heat and cool than larger country homes. Also, because cities are where the money tends to be, there is a lot of economic activity, with goods being produced and sold without having to make long-distance travel from producers to consumers.

In conclusion, when comparing urban city dwellers to those in the countryside, it becomes clear that urban people tread lighter on the environment. The number of roads and length of roads, along with sewers and power lines, all tend to be shorter. There is more efficient use of energy, and,

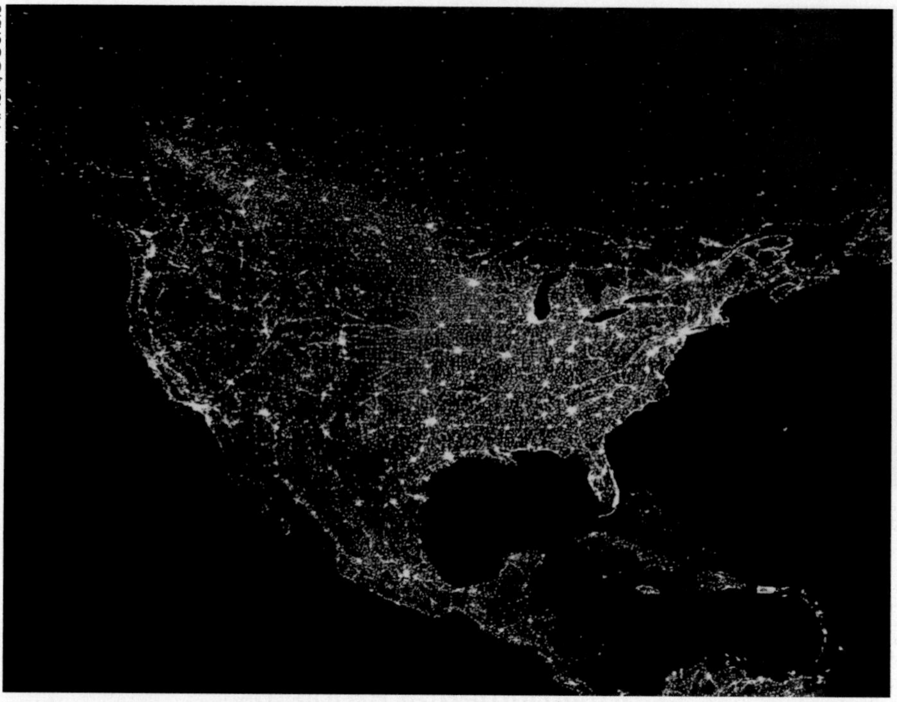

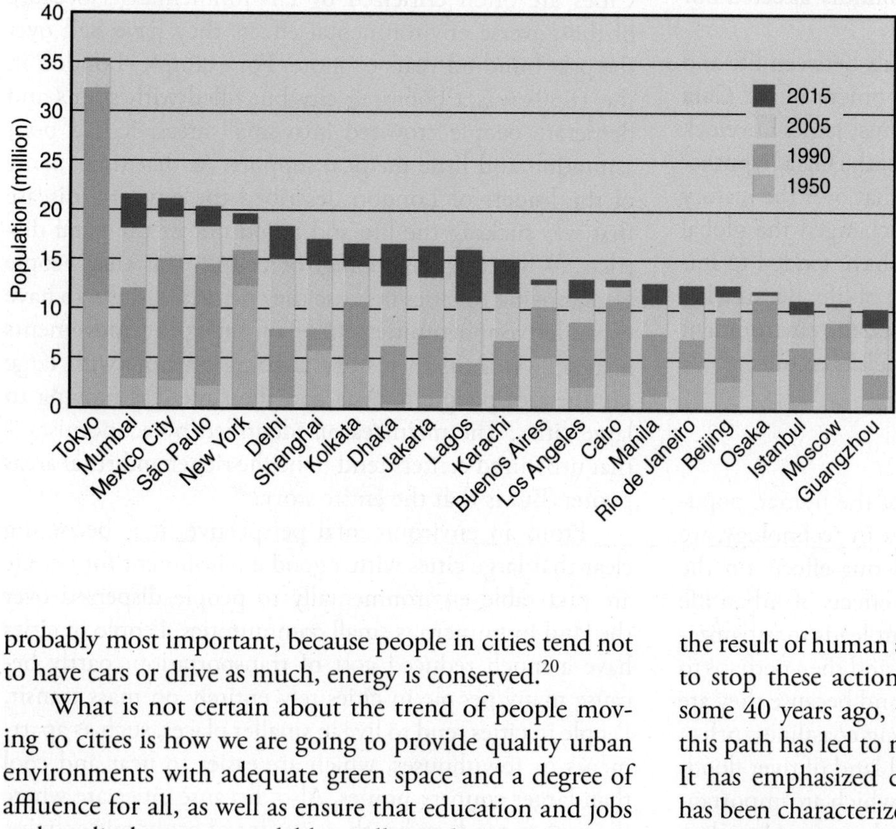

(*Source:* Data from United Nations Population Division, World Urbanization 2005, and State of the World 2007. Worldwatch Institute.)

FIGURE 1.11 **(a) An urban world and a global perspective.** When the United States is viewed at night from space, the urban areas show up as bright lights. The number of urban areas reflects the urbanization of our nation. **(b) Megacities by 2015.**

probably most important, because people in cities tend not to have cars or drive as much, energy is conserved.[20]

What is not certain about the trend of people moving to cities is how we are going to provide quality urban environments with adequate green space and a degree of affluence for all, as well as ensure that education and jobs and medical care are available to all people.

1.6 People and Nature

Today we stand at the threshold of a major change in our approach to environmental issues. Two paths lie before us. One path is to assume that environmental problems are the result of human actions and that the solution is simply to stop these actions. Based on the notion, popularized some 40 years ago, that people are separate from nature, this path has led to many advances but also many failures. It has emphasized confrontation and emotionalism and has been characterized by a lack of understanding of basic facts about the environment and how natural ecological systems function, often basing solutions on political ideologies and ancient myths about nature.

The second path begins with a scientific analysis of an environmental controversy and leads from there to cooperative problem solving. It accepts the connection between people and nature and offers the potential for long-lasting, successful solutions to environmental problems. One purpose of this book is to take the student down the second pathway.

People and nature are intimately integrated. Each affects the other. We depend on nature in countless ways. We depend on nature directly for many material resources, such as wood, water, and oxygen. We depend on nature indirectly through what are called public-service functions. For example, soil is necessary for plants and therefore for us; the atmosphere provides a climate in which we can live; the ozone layer high in the atmosphere protects us from ultraviolet radiation; trees absorb some air pollutants; and wetlands can cleanse water. We also depend on nature for beauty and recreation—the needs of our inner selves—as people always have.

We in turn affect nature. For as long as we have had tools, including fire, we have changed nature, often in ways that we like and have considered "natural." One can argue that it is natural for organisms to change their environment. Elephants topple trees, changing forests to grasslands, and people cut down trees and plant crops (Figure 1.12). Who is to say which is more natural? In fact, few organisms do *not* change their environment.

People have known this for a long time, but the idea that people might change nature to their advantage was unpopular in the last decades of the 20th century. At that time, the word *environment* suggested something separate—"out there"—implying that people were not part of nature. Today, environmental sciences are showing us how people and nature connect and in what ways this is beneficial to both.

With growing recognition of the environment's importance, we are becoming more Earth centered. We seek to spend more time in nature for recreation and spiritual activities. We accept the idea that we have evolved on and with the Earth and are not separate from it. Although we are evolving fast, we remain genetically similar to people who lived more than 100,000 years ago. Do you ever wonder why we like to go camping, to sit around a fire at night roasting marshmallows and singing, or exchanging scary stories about bears and mountain lions (Figure 1.13)?

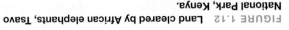

FIGURE 1.12 **Land cleared by African elephants, Tsavo National Park, Kenya.**

Courtesy Dan Botkin

Exactostock/SuperStock

FIGURE 1.13 **People and nature.** We feel safe around a campfire—a legacy from our Pleistocene ancestors?

More than ever, we understand and celebrate our union with nature as we work toward sustainability.

Most people recognize that we must seek sustainability not only of the environment but also of our economic activities, so that humanity and the environment can persist together. The dichotomy of the 20th century is giving way to a new unity: the idea that a sustainable environment and a sustainable economy may be compatible, that people and nature are intertwined, and that success for one involves success for the other.

1.7 Science and Values

Deciding what to do about an environmental problem involves both values and science, as we have already seen. We must choose what we want the environment to be. But to make this choice, we must first know what is possible. That requires knowing the scientific data and understanding its implications. Scientists rely on critical thinking. Critical scientific thinking is disciplined, using intellectual standards, effective communication, clarity, and commitment to developing scientific knowledge and skills. It leads to conclusions, generalizations, and, sometimes, scientific theories and even scientific laws. Taken together, these comprise a body of beliefs that, at the present time, account for all known observations about a particular phenomenon. Some of the intellectual standards are as follows:

Selected Intellectual Standards

- ***Clarity:*** If a statement is unclear, you can't tell whether it is relevant or accurate.
- ***Accuracy:*** Is a statement true? Can it be checked? To what extent does a measurement agree with the accepted value?
- ***Precision:*** The degree of exactness to which something is measured. Can a statement be more specific, detailed, and exact?
- ***Relevance:*** How well is a statement connected to the problem at hand?

- ***Depth:*** Did you deal with the complexities of a question?
- ***Breadth:*** Did you consider other points of view or look at it from a different perspective?
- ***Logic:*** Does a conclusion make sense and follow from the evidence?
- ***Significance:*** Is the problem an important one? Why?
- ***Fairness:*** Are there any vested interests, and have other points of view received attention?

Modified after R. Paul and L. Elder. 2003. *Critical Thinking* (Dillon Beach, CA: The Foundation for Critical Thinking.)

Once we know our options, we can select from among them. What we choose is determined by our values. An example of a value judgment regarding the world's human environmental problem is the choice between the desire of an individual to have many children and the need to find a way to limit the human population worldwide.

After we have chosen a goal based on knowledge and values, we have to find a way to attain that goal. This step also requires knowledge. And the more technologically advanced and powerful our civilization, the more knowledge is required.

Learning and Discovery in Environmental Science

It may be helpful as you progress with your understanding of environmental science to be aware of learning methods—for your understanding as well as for an awareness of the broader context of how scientists approach theories and discovery.

Learning

Each of us studies, learns, and thinks in slightly different ways, but there are some common denominators. In this book, we present learning objectives at the beginning of each chapter; these are statements about what students should be able to do when they have completed a particular segment of instruction. When you are thinking about the learning objectives for a specific chapter and wonder how to approach achieving your best results, it is worthwhile to think about some learning theory. In general, the learning objectives are in the cognitive (thinking) domain, as well as in the affective domain (motivation, attitude, perception, and values).[21-23] Cognitive learning pertains to mental processes including, but not limited to, memory, judgment, synthesis, and reasoning. The cognitive domain includes those objectives that are related to information or knowledge that can be arranged in a hierarchical classification known as levels of Bloom's Taxonomy.[22]

- Knowledge: Recall of basic information.
- Comprehension: The lowest level of understanding; involves interpretation of information in your own words.
- Application: Application of knowledge or generalization of knowledge in a new situation.
- Analysis: Breakup of knowledge into parts and focus on relationships among parts.
- Synthesis: The bringing together of various parts of knowledge to form a new understanding based on relationships in new situations.
- Evaluation: Judgment about various materials and methods, based on given criteria.

When we acquire new knowledge and are able to define it, label it, and list it, we are operating on the beginning level of cognitive abilities. When we are able to classify, describe, and explain it, we are on a slightly higher level. If we practice application, we are concerned with demonstrating, choosing, and applying our knowledge to new conditions or ideas that we generate. We begin the higher cognitive processes when we apply the steps of analysis, synthesis, and evaluation. When we do analysis, we may be involved with calculations, comparing, and contrasting. Synthesis is a higher level of cognitive activity that involves producing a whole new set of ideas that are assembled and collected and composed as a careful, constructed understanding of new situations. Finally, the evaluation phase, the highest level of cognitive thinking, involves construction of arguments, choosing, comparing, and appraising, with the goal of making judgments about the materials, methods, and criteria used in our study.

In working through our cognitive processes and levels of learning, we don't want to forget the affective domain, which is concerned with values that enhance our deeper appreciation and understanding of a given environmental problem. The affective domain is also important in providing motivation and forming a plan of action to help guide our cognitive processes. For example, we may ask, "what is our motivation for studying environmental science?" Is it, in part, based on our respect for the environment or our obligations to future generations? We might also be motivated because we want to become informed citizens, better able to make decisions in a complex, changing world.

Discovery

The joy of discovery is a driving force for many scientists. Some of you studying environmental science may move forward with your studies and eventually become scientists. This involves creativity, imagination, and a lot of hard work. Many people come up with ideas, some of which are original, but do not work them through to fruition. Moving beyond the idea stage involves mostly very hard work and persistence and not letting go of an idea. Many ideas that are original may take a lot of time to develop. If those ideas lead to a paradigm change in how we view the Earth, then we must be ready for criticism. New ideas are seldom brought forward without criticism because we are "shaking the tree" of an older, prevailing paradigm that is comfortable and tightly held by other scientists.

Although there are many ways that creativity and imagination may move forward, sometimes it is worthwhile to consider a framework of inquiry. For example, a simple way of approaching research is to consider the various levels of scientific inquiry.[24] We can organize this way of thinking into three basic levels:

- **Level 1:** Studying elements and processes of a particular study, independent of each other and of time.
- **Level 2:** Obtaining relationships between independent variables (causes) and dependent variables (effects), independent of time.
- **Level 3:** Evaluating relations between variables over time. In terms of mathematics, this requires integrating the relationships and naturally leads to the study of evolution.

As a simple example, in the physical science Earth Surface Process (part of geology and physical geography), a scientist may study materials such as rocks and soils, along with individual processes such as those associated with running water, weathering of rock and soil, landslides, or tectonic processes, including earthquakes. Studying those processes independent of each other and of time is the first level. Level 2 begins when we start to derive relationships between materials and processes, independent of time—for example, the relationship between rock strength and landslides (the strongest rocks generally have the fewest landslides). The final stage, Level 3, is the integration of the materials and processes and their relationships through time, and this is the study of landscape evolution.[25]

Perhaps the most famous example in the biological sciences that pertains to discovery is the story of Charles Darwin that led to his famous book *The Origin of Species*.[26] Darwin began his studies of plants and animals at a young age and took the famous adventure of scientific inquiry aboard the *H.M.S. Beagle* in 1831. Darwin was very cautious for many years, prior to publishing his ideas concerning natural selection. His work moved through levels 1–3, ending with biologic evolution through natural selection. For example, he did extensive work on pigeons. Darwin raised pigeons and compared anatomy (bone structure and function) between what he called variation between breeds of pigeons. At level 2, he developed relations between variations of the same species with other similar species. At level 3, Darwin integrated his work with the known fossil record of the succession of life through geologic time to develop a more general evolutionary view of life.

Darwin wrote his book, *The Origin of Species*, years after the voyage of discovery to the Galápagos Islands off the coast of South America and many other places. One of the important things to take away from his book is how he prepared his argument on natural selection for a world where many were not necessarily willing to accept it. He began by discussing the past work of others, giving credit to other scientists who had talked about variation of species. He gave extensive attention in the early chapters of his book to how humans select animals such as horses, dogs, and pigeons for particular characteristics and, through selective breeding, create subspecies. He wrote this part of his book in a very logical way, and, as a result, it is not such a big jump for biologists to consider natural selection, where nature, not people, does the selecting. Darwin ends his discussion of natural selection and how new species might evolve with a famous modest statement for the time—that his work would shed light on the origin of man.

The Precautionary Principle

Science and values come to the forefront when we think about what action to take about a perceived environmental problem for which the science is only partially known. This is often the case because all science is preliminary and subject to analysis of new data, ideas, and tests of hypotheses. Even with careful scientific research, it can be difficult, even impossible, to prove with absolute certainty how relationships between human activities and other physical and biological processes lead to local and global environmental problems, such as global warming, depletion of ozone in the upper atmosphere, loss of biodiversity, and declining resources. For this reason, in 1992 the Rio Earth Summit on Sustainable Development listed as one of its principles what we now call the **Precautionary Principle**. Basically, it says that when there is a threat of serious, perhaps even irreversible, environmental damage, we should not wait for scientific proof before taking precautionary steps to prevent potential harm to the environment.

The Precautionary Principle requires critical thinking about a variety of environmental concerns, such as the manufacture and use of chemicals, including pesticides, herbicides, and drugs; the use of fossil fuels and nuclear energy; the conversion of land from one use to another (for example, from rural to urban); and the management of wildlife, fisheries, and forests.[27]

One important question in applying the Precautionary Principle is how much scientific evidence we should have before taking action on a particular environmental problem. The principle recognizes the need to evaluate all the scientific evidence we have and to draw provisional conclusions while continuing our scientific investigation, which may provide additional or more reliable data. For example, when considering environmental health issues related to the use of a pesticide, we may have a lot of scientific data, but with gaps, inconsistencies, and other scientific uncertainties. Those in favor of continuing to use that pesticide may argue that there isn't enough proof of its danger to ban it. Others may argue that absolute proof of safety is necessary before a new pesticide is used. Those advocating the Precautionary Principle would argue that we should continue to investigate but, to be on the safe side, should not wait to take cost-effective precautionary measures to prevent environmental damage or health problems. What constitutes a cost-effective measure? Certainly we would need to examine the benefits and costs of taking a particular action versus taking no action. Other economic analyses may also be appropriate.[27, 28]

The Precautionary Principle is emerging as a new tool for environmental management and has been adopted by the city of San Francisco and the European Union. There will always be arguments over what constitutes sufficient scientific knowledge for decision making. Nevertheless, the Precautionary Principle, even though it may be difficult to apply, is becoming a common part of environmental analysis with respect to environmental protection and environmental health issues. It requires us to think ahead and predict potential consequences before they occur. As a result, the Precautionary Principle is a *proactive*, rather than a *reactive*, tool—that is, we can use it when we see real trouble coming, rather than reacting after the trouble arises.

Placing a Value on the Environment

How do we place a value on any aspect of our environment? How do we choose between two different concerns? The value of the environment is based on eight justifications: utilitarian (materialistic), ecological, aesthetic, recreational, inspirational, creative, moral, and cultural.

The **utilitarian justification** is that some aspect of the environment is valuable because it benefits individuals economically or is directly necessary to human survival. For example, conserving salmon in the Pacific Northwest provides a livelihood for local people through tourism and recreational fishing.

The **ecological justification** is that an ecosystem is necessary for the survival of some species of interest to us, or that the system itself provides some benefit. For example, a mangrove swamp (a type of coastal wetland) provides habitat for marine fish, and although we do not eat mangrove trees, we may eat the fish that depend on them. Also, the mangroves are habitat for many noncommercial species, some endangered. Therefore, conservation of the mangrove is important ecologically. Here is another example: Burning coal and oil adds greenhouse gases to the atmosphere, which may lead to a climate change that could affect the entire Earth. Such ecological reasons form a basis for the conservation of nature that is essentially enlightened self-interest.

Aesthetic and **recreational justifications** have to do with our appreciation of the beauty of nature and our desire to get out and enjoy it. For example, many people find wilderness scenery beautiful and would rather live in a world with wilderness than without it. One way we enjoy nature's beauty is to seek recreation in the outdoors.

The aesthetic and recreational justifications are gaining a legal basis. The state of Alaska acknowledges that sea otters have an important recreational role in that people enjoy watching and photographing them in a wilderness setting. And there are many other examples of the aesthetic importance of the environment. When people mourn the death of a loved one, they typically seek out places with grass, trees, and flowers; thus we use these to beautify our graveyards. Conservation of nature can be based on its benefits to the human spirit, our "inner selves" (*inspirational justification*). Nature is also often an aid to human creativity (the *creative justification*). The creativity of artists and poets, among others, is often inspired by their contact with nature. But while nature's aesthetic, recreational, and inspirational value is a widespread reason that people enjoy nature, it is rarely used in formal environmental arguments, perhaps in the belief that they might seem like superficial justifications for conserving nature. In fact, however, beauty in their surroundings is of profound importance to people. Frederick Law Olmsted, the great American landscape planner, argued that plantings of vegetation provide medical, psychological, and social benefits and are essential to city life.[15]

Moral justification has to do with the belief that various aspects of the environment have a right to exist and that it is our moral obligation to help them, or at least allow them, to persist. Moral arguments have been extended to many nonhuman organisms, to entire ecosystems, and even to inanimate objects. The historian Roderick Nash, for example, wrote an article entitled "Do Rocks Have Rights?" that discusses such moral justification, and the United Nations General Assembly World Charter for Nature, signed in 1982, states that species have a moral right to exist.

Cultural justification refers to the fact that different cultures have many of the same values but also some different values with respect to the environment. This may also be in terms of specifics of a particular value. All cultures may value nature, but, depending on their religious beliefs, may value it in different degrees of intensity. For example, Buddhist monks when preparing ground for a building may pick up and move disturbed earthworms, something few others would do. Different cultures integrate nature into their towns, cities, and homes in different ways, depending on their view of nature.

Analysis of environmental values is the focus of a new discipline, known as environmental ethics. Another concern of environmental ethics is our obligation to future generations: Do we have a moral obligation to leave the environment in good condition for our descendants, or are we at liberty to use environmental resources to the point of depletion within our own lifetimes?

CRITICAL THINKING ISSUE

Easter Island

The story of Easter Island has been used as an example of how people may degrade the environment as they grow in number, until eventually their overuse of the environment results in the collapse of the society. This story has been challenged by recent work. We will present what is known, and you should examine the case history critically. To help with this issue, look back to the list of intellectual standards useful in critical thinking.

Easter Island's history illustrates the importance of science and the sometimes irreversible consequences of human population growth and the introduction of a damaging exotic species, accompanied by depletion of resources necessary for survival. Evidence of the island's history is based on detailed studies by earth scientists and social scientists who investigated the anthropological record left in the soil where people lived and the sediment in ponds where pollen from plants that lived at different times was deposited. The goals of the studies were to estimate the number of people, their diet, and their use of resources. This was linked to studies of changes in vegetation, soils, and land productivity.

Easter Island lies about 3,700 km (2,300 mi) west of South America and 4,000 km from Tahiti (Figure 1.14a), where the people may have come from. The island is small, about 170 km^2, with a rough triangular shape and an inactive volcano at each corner. The elevation is less than about 500 m (1,640 ft) (Figure 1.14b), too low to hold clouds like those in Hawaii that bring rain. As a result, water resources are limited. When Polynesian people first reached it about 800–1,500 years ago, they colonized a green island covered with rich soils and forest. The small group of settlers grew rapidly, to perhaps over 10,000 people, who eventually established a complex society that was spread among a number of small villages. They raised crops and chickens, supplementing their diet with fish from the sea. They used the island's trees to build their homes and to build boats. They also carved massive 8-meter-high statues from volcanic rock and moved them into place at various parts of the island, using tree trunks as rollers (Figure 1.14b, c).

When Europeans first reached Easter Island in 1722, the only symbols of the once-robust society were the statues. A study suggested that the island's population had collapsed in just a few decades to about 2,000 people because they had used up (degraded) the isolated island's limited resource base.[29, 30]

At first there were abundant resources, and the human population grew fast. To support their growing population, they cleared more and more land for agriculture and cut more trees for fuel, homes, and boats—and for moving the statues into place. Some of the food plants they brought to the island didn't survive, possibly because the voyage was too long or the climate unsuitable for them. In particular, they did not have the breadfruit tree, a nutritious starchy food source, so they relied more heavily on other crops, which required clearing more land for planting. The island was also relatively dry, so it is likely that fires for clearing land got out of control sometimes and destroyed even more forest than intended.[29, 30]

The cards were stacked against the settlers to some extent—but they didn't know this until it was too late. Other islands of similar size that the Polynesians had settled did not suffer forest depletion and fall into ruin.[29, 30] This isolated island, however, was more sensitive to change. As the forests were cut down, the soils, no longer protected by forest cover, were lost to erosion. Loss of the soils reduced agricultural productivity, but the biggest loss was the trees. Without wood to build homes and boats, the people were forced to live in caves and could no longer venture out into the ocean for fish.[29]

These changes did not happen overnight—it took more than 1,000 years for the expanding population to deplete its resources. Loss of the forest was irreversible: Because it led to loss of soil, new trees could not grow to replace the forests. As resources grew scarcer, wars between the villages became common, as did slavery, and perhaps even cannibalism.

Easter Island is small, but its story is a dark one that suggests what can happen when people use up the resources of an isolated area. We note, however, that some aspects of the above history of Easter Island have been challenged recently. New data suggest that people first arrived about 800 years ago, not 1,500; thus, much less time was available for people to degrade the land.[31, 32] Deforestation certainly played a role in the loss of trees, and the rats that arrived with the Polynesians were evidently responsible for eating seeds of the palm trees, preventing regeneration. According to the alternative explanation of the island's demise, the Polynesian people on the island at the time of European contact in 1722 numbered about 3,000; this may have been close to the maximum reached around the year 1350. Contact with Europeans introduced new diseases and enslavement, which reduced the population to about 100 by the late 1870s.[31]

Easter Island, also called Rapa Nui, was annexed by Chile in 1888. Today, about 3,000 people live on the island. Tourism is the main source of income; about 90% of the island is grassland, and thin, rocky soil is common. There have been reforestation projects, and about 5% of the island is now forested, mostly by eucalyptus plantations in the central part of the island. There are also fruit trees in some areas.

As more of the story of Easter Island emerges from scientific and social studies, the effects of resource exploitation, invasive rats, and European contact will become clearer, and the environmental lessons of the collapse will lead to a better understanding of how we can sustain our global human culture. However, the primary lesson is that *limited resources can support only a limited human population.*

Like Easter Island, our planet Earth is isolated in our solar system and universe and has limited resources. As a result, the world's growing population is facing the problem of how to conserve those resources. We know it takes a while before environmental damage begins to show, and we know that some environmental damage may be irreversible. We are striving to develop plans to ensure that our natural resources, as well as the other living things we share our planet with, will not be damaged beyond recovery.[33]

Critical Thinking Questions

1. Assuming that an increasing human population, introduction of invasive rats, loss of trees, the resulting soil erosion, and, later, introduced European diseases led to the collapse of the society, can Easter Island be used as a model for what could happen to Earth? Why? Why not?
2. People may have arrived at Easter Island 1,500 years ago or later, perhaps 800 years ago. Does the timing make a significant difference in the story? How?

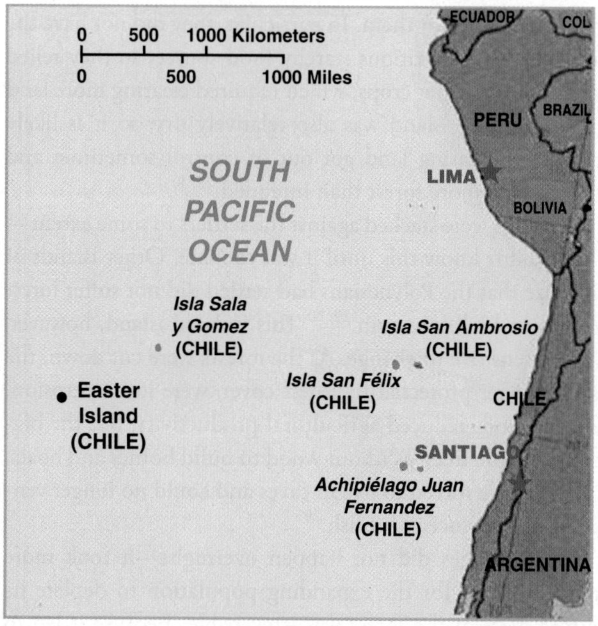

FIGURE 1.14 **Easter Island, collapse of a society. (a)** Location of Easter Island in the Pacific Ocean, several thousand kilometers west of South America; **(b)** map of Easter Island showing the three major volcanoes that anchor the three corners of the small island; and **(c)** large statues carved from volcanic rock before the collapse of a society with several thousand people.

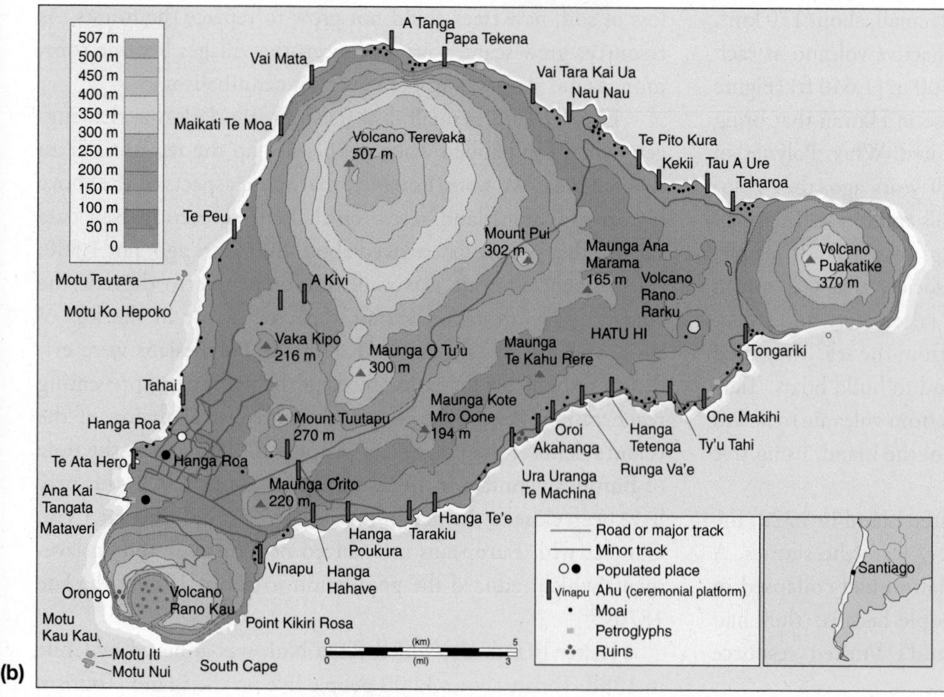

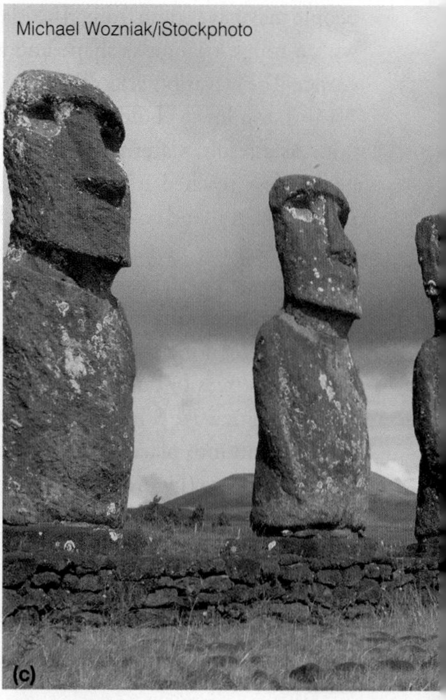

SUMMARY

- Six themes run through this text: the urgency of the population issue; the importance of urban environments; the need for sustainability of resources; the importance of a global perspective; people and nature; and the role of science and values in the decisions we face.
- People and nature are intertwined. Each affects the other.
- The human population grew at a rate unprecedented in history in the 20th century. Population growth is the underlying environmental problem.
- Achieving sustainability, the environmental goal, is a long-term process to maintain a quality environment for future generations. Sustainability is becoming an important environmental paradigm for the 21st century.
- The combined impact of technology and population multiplies the impact on the environment.
- In an increasingly urban world, we must focus much of our attention on the environments of cities and the effects of cities on the rest of the environment.
- Determining Earth's carrying capacity for people and levels of sustainable harvests of resources is difficult but crucial if we are to plan effectively to meet our needs in the future. Estimates of Earth's carrying capacity for people range from 2.5 to 40 billion, but about 15 billion is the upper limit with today's technology. The differences in capacity have to do with the quality of life projected for people—the poorer the quality of life, the more people can be packed onto the Earth.
- Awareness of how people at a local level affect the environment globally gives credence to the Gaia hypothesis. Future generations will need a global perspective on environmental issues.
- Placing a value on various aspects of the environment requires knowledge and understanding of the science, but it also depends on our judgments about the uses and aesthetics of the environment and on our moral commitments to other living things and to future generations.
- The Precautionary Principle is emerging as a powerful new tool for environmental management.

REEXAMINING THEMES AND ISSUES

Sean Randall/Getty Images, Inc.

HUMAN POPULATION

What is more important: the quality of life of people alive today or the quality of life of future generations?

© Biletskiy_Evgeniy/iStockphoto

SUSTAINABILITY

What is more important: abundant resources today—as much as we want and can obtain—or the availability of these resources for future generations?

© Anton Balazh 2011/iStockphoto

GLOBAL PERSPECTIVE

What is more important: the quality of your local environment or the quality of the global environment—the environment of the entire planet?

ssguy/ShutterStock

URBAN WORLD

What is more important: human creativity and innovation, including arts, humanities, and science, or the persistence of certain endangered species? Must this always be a trade-off, or are there ways to have both?

B2M Productions/Getty Images, Inc.

PEOPLE AND NATURE

If people have altered the environment for much of the time our species has been on Earth, what then is "natural"?

George Doyle/Getty Images, Inc.

SCIENCE AND VALUES

Does nature know best, so that we never have to ask what environmental goal we should seek, or do we need knowledge about our environment, so that we can make the best judgments given available information?

KEY TERMS

aesthetic justification 16
carrying capacity 10
cultural justification 16
ecological justification 16
Gaia hypothesis 11
megacities 11
moral justification 16
Precautionary Principle 15
recreational justification 16
sustainability 9
utilitarian justification 16

STUDY QUESTIONS

1. Support or criticize the idea that in the past few decades a convergence between energy, economics, and environment has become increasingly more apparent.
2. Compare and contrast the effects on the environment of a resident of a large city with the effects of someone living on a farm. In what ways are the effects similar?
3. Programs have been established to supply food from Western nations to starving people in Africa. Some people argue that such programs, which may have short-term benefits, actually increase the threat of starvation in the future. What are the pros and cons of international food relief programs?
4. Why is there an emerging food crisis that is different from any in the past?
5. Which of the following are global environmental problems? Why?
   (a) Growth of the human population.
   (b) Furbish's lousewort, a small flowering plant found in the state of Maine and in New Brunswick, Canada. It is so rare that it has been seen by few people and is considered endangered.
   (c) The blue whale, listed as an endangered species under the U.S. Marine Mammal Protection Act.
   (d) A car that has air-conditioning.
   (e) Seriously polluted harbors and coastlines in major ocean ports.
6. Construct a research plan to estimate the carrying capacity of Earth.
7. Is it possible that sometime in the future all the land on Earth will become one big city? If not, why not? To what extent does the answer depend on the following:
   (a) global environmental considerations
   (b) scientific information
   (c) values
8. Explain why knowing about both the affective and cognitive domains of learning will help motivate you to find solutions to environmental problems.

FURTHER READING

Botkin, D.B., ***No Man's Garden: Thoreau and a New Vision for Civilization and Nature*** (Washington, DC: Island Press, 2000). A discussion of many of the central themes of this textbook, with special emphasis on values and science and on an urban world. Henry David Thoreau's life and works illustrate approaches that can help us deal with modern environmental issues.

Botkin, D.B., ***Discordant Harmonies: A New Ecology for the 21st Century*** (New York: Oxford University Press, 1990). An analysis of the myths that underlie attempts to solve environmental issues.

Erlich, P.R., Kareiva, P.M., and Daily, G.C., Securing natural capital and expanding equity to rescale civilization (2012.

chapter). A statement can be termed "scientific" if someone can state a method of disproving it. If no one can think of such a test, then the statement is said to be nonscientific. Consider, for example, the crop circles discussed in A Closer Look 2.1. One Web site says that some people believe the crop circles are a "spiritual nudge . . . designed to awaken us to our larger context and milieu, which is none other than our collective earth soul." Whether or not this is true, it does not seem open to disproof.

Science is a process of discovery—a continuing process whose essence is change in ideas. The fact that scientific ideas change is frustrating. Why can't scientists agree on what is the best diet for people? Why is a chemical considered dangerous in the environment for a while and then determined not to be? Why do scientists in one decade consider forest fires undesirable disturbances and in a later decade decide forest fires are natural and in fact important? Are we causing global warming or not? And on and on. Can't scientists just find out the truth and give us the final word on all these questions once and for all, and agree on it?

The answer is no—because science is a continuing adventure during which scientists make better and better approximations of how the world works. Sometimes changes in ideas are small, and the major context remains the same. Sometimes a science undergoes a fundamental revolution in ideas.

Science makes certain assumptions about the natural world: that events in the natural world follow patterns that can be understood through careful observation and scientific analysis, which we will describe later; and that these basic patterns and the rules that describe them are the same throughout the universe.

A CLOSER LOOK 2.1

The Case of the Mysterious Crop Circles

For 13 years, circular patterns appeared "mysteriously" in grain fields in southern England. Proposed explanations included aliens, electromagnetic forces, whirlwinds, and pranksters. The mystery generated a journal and a research organization headed by a scientist, as well as a number of books, magazines, and clubs devoted solely to crop circles. Scientists from Great Britain and Japan brought in scientific equipment to study the strange patterns. Then, in September 1991, two men confessed to having created the circles by entering the fields along paths made by tractors (to disguise their footprints) and dragging planks through the fields. When they made their confession, they demonstrated their technique to reporters and some crop-circle experts (Figure 2.5).[a]

Despite their confession, some people still believe that the crop circles were caused by something else. Books have been written about these circles, some published as recently as 2012. One report published on the Internet in 2003 stated that "strange orange lightning" was seen one evening and that crop circles appeared the next day.

How is it that so many people, including some scientists, still take those English crop circles seriously? Probably some of these people misunderstood the scientific method and used it incorrectly—and some simply wanted to believe in a mysterious cause and therefore chose to reject scientific information. We run into this way of thinking frequently with environmental issues. People often believe that some conclusions or some action is good, based on their values. They wish it were so and decide therefore that it must be so. The false logic here can be phrased: *If it sounds good, it must be good, and if it must be good, we must make it happen.*

(a)

© Trinity Mirror / Mirrorpix / Alamy

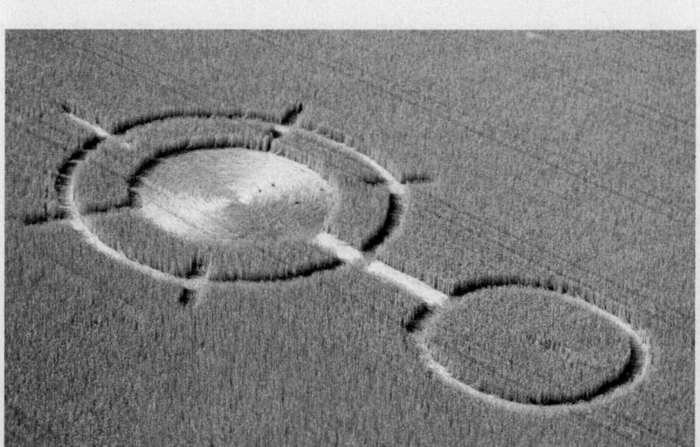

(b)

© Adam Woolfitt/Age Fotostock America, Inc.

FIGURE 2.5 **(a)** A crop circle close up at the Vale of Pewsey in southern England in July 1990. **(b)** Crop circles seen from the air make distinctive patterns.

2.2 Observations, Facts, Inferences, and Hypotheses

I have no data yet. It is a capital mistake to theorize before one has data. Insensibly one begins to twist facts to suit theories, instead of theories to suit facts.

—Sherlock Holmes,
in Sir Arthur Conan Doyle's
A Scandal in Bohemia

Now we can turn to the specific characteristics of the scientific method. (The steps in the scientific method are shown in Table 2.2.) It is important to distinguish between observations and inferences. **Observations**, the basis of science, may be made through any of the five senses or by instruments that measure beyond what we can sense. **Inferences** are generalizations that arise from a set of observations. When everyone or almost everyone agrees with what is observed about a particular thing, the inference is often called a **fact**.

We might *observe* that a substance is a white, crystalline material with a sweet taste. We might *infer* from these observations alone that the substance is sugar. Before this inference can be accepted as fact, however, it must be subjected to further tests. Confusing observations with inferences and accepting untested inferences as facts are kinds of sloppy thinking described as "Thinking makes it so." When scientists wish to test an inference, they convert it into a **hypothesis**, which is a statement that can be disproved. The hypothesis continues to be accepted until it is disproved.

For example, a scientist is trying to understand how a plant's growth will change with the amount of light it receives. She proposed a hypothesis that a plant can use only so much light and no more—it can be "saturated" by an abundance of light. She measures the rate of photosynthesis at a variety of light intensities. The rate of photosynthesis is called the **dependent variable** because it is affected by, and in this sense depends on, the amount of light, which is called the **independent variable**. The independent variable is also sometimes called a **manipulated variable** because it is deliberately changed, or manipulated, by the scientist. The dependent variable is then referred to as a **responding variable**—one that responds to changes in the manipulated variable. These values are referred to as *data* (singular: *datum*). They may be numerical, **quantitative data**, or nonnumerical, **qualitative data**. In our example, qualitative data would be the species of a plant; quantitative data would be the tree's mass in grams or the diameter in centimeters. The result of the scientist's observations: The hypothesis is confirmed—the rate of photosynthesis increases to a certain level and does not go higher at higher light intensities (Figure 2.6).

Controlling Variables

In testing a hypothesis, a scientist tries to keep all relevant **variables** constant except for the independent and dependent variables. This practice is known as *controlling variables*. In a **controlled experiment**, the experiment is compared to a standard, or control—an exact duplicate of the experiment except for the one variable being tested (the independent variable). Any difference in outcome (dependent variable) between the experiment and the control can be attributed to the effect of the independent variable.

An important aspect of science, but one frequently overlooked in descriptions of the scientific method, is the need to define or describe variables in exact terms

Table 2.2 STEPS IN THE SCIENTIFIC METHOD (TERMS USED HERE ARE DEFINED IN THE TEXT.)

1. Make observations and develop a question about the observations.
2. Develop a tentative answer to the question—a hypothesis.
3. Design a controlled experiment to test the hypothesis (implies identifying and defining independent and dependent variables).
4. Collect data in an organized form, such as a table.
5. Interpret the data visually (through graphs), quantitatively (using statistical analysis), and/or by other means.
6. Draw a conclusion from the data.
7. Compare the conclusion with the hypothesis and determine whether the results support or disprove the hypothesis.
8. If the hypothesis is consistent with observations in some limited experiments, conduct additional experiments to test it further. If the hypothesis is rejected, make additional observations and construct a new hypothesis.

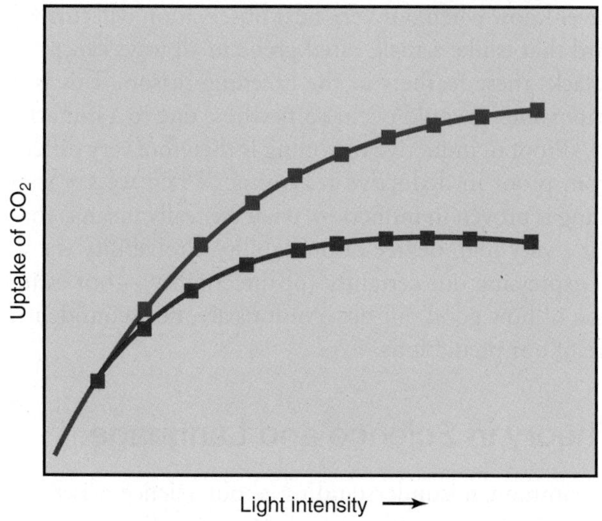

FIGURE 2.6 **Dependent and independent variables: Photosynthesis as affected by light.** In this diagram, photosynthesis is represented by carbon dioxide (CO_2) uptake. Light is the independent variable, uptake is the dependent variable. The blue and red lines represent two plants with different responses to light.

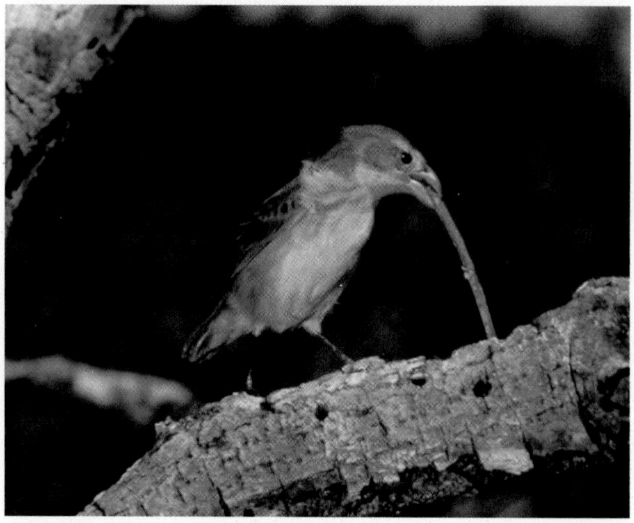

Alan Root/Photo Researchers, Inc

FIGURE 2.7 **A woodpecker finch in the Galápagos Islands** uses a twig to remove insects from a hole in a tree, demonstrating tool use by nonhuman animals. Because science is based on observations, its conclusions are only as true as the premises from which they are deduced.

that all scientists can understand. The least ambiguous way to define or describe a variable is in terms of what one would have to do to duplicate the measurement of that variable. Such definitions are called **operational definitions**. Before carrying out an experiment, both the independent and dependent variables must be defined operationally. Operational definitions allow other scientists to repeat experiments exactly and to check on the results reported.

Science is based on **inductive reasoning**, also called *induction*: It begins with specific observations and then extends to generalizations, which may be disproved by testing them. If such a test cannot be devised, then we cannot treat the generalization as a scientific statement. Although new evidence can disprove existing scientific theories, science can never provide absolute proof of the truth of its theories.

The Nature of Scientific Proof

One source of serious misunderstanding about science is the use of the word *proof*, which most students encounter in mathematics, particularly in geometry. Proof in mathematics and logic involves reasoning from initial definitions and assumptions. If a conclusion follows logically from these assumptions, or premises, we say it is proven. This process is known as **deductive reasoning**. An example of deductive reasoning is the following syllogism, or series of logically connected statements:

Premise: A straight line is the shortest distance between two points.

Premise: The line from A to B is the shortest distance between points A and B.

Conclusion: Therefore, the line from A to B is a straight line.

Note that the conclusion in this syllogism follows directly from the premises.

Deductive proof does not require that the premises be true, only that the reasoning be foolproof. Statements that are logically valid but untrue can result from false premises, as in the following example (Figure 2.7):

Premise: Humans are the only toolmaking organisms.

Premise: The woodpecker finch uses tools.

Conclusion: Therefore, the woodpecker finch is a human being.

In this case, the concluding statement must be true if both of the preceding statements are true. However, we know that the conclusion is not only false but ridiculous. If the second statement is true (which it is), then the first cannot be true.

The rules of deductive reasoning govern only the process of moving from premises to conclusion. *Science, in contrast, requires not only logical reasoning but also correct premises.* Returning to the example of the woodpecker finch, to be scientific the three statements should be expressed conditionally (that is, with reservation):

If humans are the only toolmaking organisms
and
the woodpecker finch is a toolmaker,
then
the woodpecker finch is a human being.

When we formulate generalizations based on a number of observations, we are engaging in inductive reasoning. To illustrate: One of the birds that feeds at Mono Lake is the eared grebe. The "ears" are a fan of golden feathers that occur behind the eyes of males during the breeding season. Let us define birds with these golden feather fans as eared grebes (Figure 2.8). If we always observe that the breeding male grebes have this feather fan, we may make the inductive statement "All male eared grebes have golden feathers during the breeding season." What we really mean is "All of the male eared grebes *we have seen* in the breeding season have golden feathers." We never know when our very next observation will turn up a bird that is like a male eared grebe in all ways except that it lacks these feathers in the breeding season. This is not impossible; it could occur somewhere due to a mutation.

All Canada Photos/SuperStock

FIGURE 2.8 **Male eared grebe in breeding season.**

Proof in inductive reasoning is therefore very different from proof in deductive reasoning. When we say something is proven in induction, what we really mean is that it has a very high degree of probability. Probability is a way of expressing our certainty (or uncertainty)—our estimation of how good our observations are, how confident we are of our predictions.

Theory in Science and Language

A common misunderstanding about science arises from confusion between the use of the word *theory* in science and its use in everyday language. A **scientific theory** is a grand scheme that relates and explains many observations and is supported by a great deal of evidence. In contrast, in everyday usage a theory can be a guess, a hypothesis, a prediction, a notion, a belief. We often hear the phrase "It's just a theory." That may make sense in everyday conversation but not in the language of science. In fact, theories have tremendous prestige and are considered the greatest achievements of science.[3]

Further misunderstanding arises when scientists use the word *theory* in several different senses. For example, we may encounter references to a currently accepted, widely supported theory, such as the theory of evolution by natural selection; a discarded theory, such as the theory of inheritance of acquired characteristics; a new theory, such as the theory of evolution of multicellular organisms by symbiosis; and a model dealing with a specific or narrow area of science, such as the theory of enzyme action.[5]

One of the most important misunderstandings about the scientific method pertains to the relationship between research and theory. Theory is usually presented as growing out of research, but in fact theories also guide research. When a scientist makes observations, he or she does so in the context of existing theories. At times, discrepancies between observations and accepted theories become so great that a scientific revolution occurs: The old theories are discarded and are replaced with new or significantly revised theories.[6]

Knowledge in an area of science grows as more hypotheses are supported. Ideally, scientific hypotheses are continually tested and evaluated by other scientists, and this provides science with a built-in self-correcting feedback system. This is an important, fundamental feature of the scientific method. If you are told that scientists have reached a consensus about something, you want to check carefully to see if this feedback process has been used correctly and is still possible. If not, what began as science can be converted to ideology—a way that certain individuals, groups, or cultures may think despite evidence to the contrary.

Models and Theory

Scientists use accumulated knowledge to develop explanations that are consistent with currently accepted hypotheses. Sometimes an explanation is presented as a model. A **model** is "a deliberately simplified construct of nature."[7] It may be a physical working model, a pictorial model, a set of mathematical equations, or a computer simulation. For example, the U.S. Army Corps of Engineers has a physical model of San Francisco Bay. Open to the public to view, it is a miniature in a large aquarium with the topography of the bay reproduced to scale and with water flowing into it in accordance with tidal patterns. Elsewhere, the Army Corps develops mathematical equations and computer simulations, which are models designed to explain some aspects of such water flow.

As new knowledge accumulates, models may no longer be consistent with observations and may have to be revised or replaced, with the goal of finding models more consistent with nature.[6] Computer simulation of the atmosphere has become important in scientific analysis of the possibility of global warming. Computer simulation is becoming important for biological systems as well, such as simulations of forest growth (Figure 2.9).

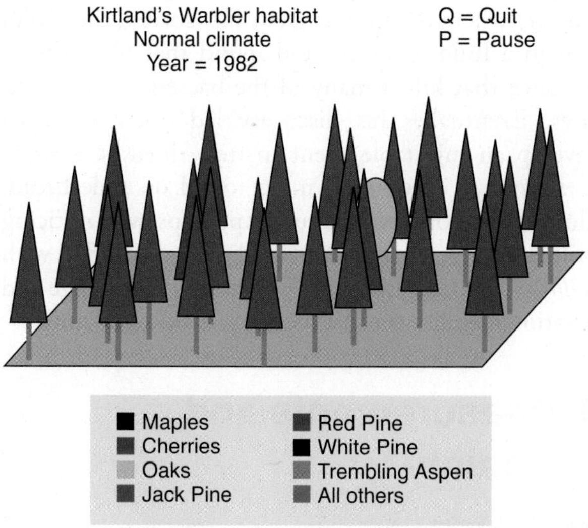

(Source: *JABOWA-II* by D.B. Botkin. Copyright © 1993, 2009 by D.B. Botkin.)

FIGURE 2.9 **A computer simulation of forest growth.** Shown here is a screen display of individual trees whose growth is forecast year by year, depending on environmental conditions. In this computer run, only three types of trees are present. This kind of model is becoming increasingly important in environmental sciences.

Some Alternatives to Direct Experimentation

Environmental scientists have tried to answer difficult questions using several approaches, including historical records and observations of modern catastrophes and disturbances.

Historical Evidence

Ecologists have made use of both human and ecological historical records. A classic example is a study of the history of fire in the Boundary Waters Canoe Area (BWCA) of Minnesota, 1 million acres of boreal forests, streams, and lakes well known for recreational canoeing.

Murray ("Bud") Heinselman had lived near the BWCA for much of his life and was instrumental in having it declared a wilderness area. A forest ecological scientist, Heinselman set out to determine the past patterns of fires in this wilderness. Those patterns are important in maintaining the wilderness. If the wilderness has been characterized by fires of a specific frequency, then one can argue that this frequency is necessary to maintain the area in its most "natural" state.

Heinselman used three kinds of historical data: written records, tree-ring records, and buried records (fossil and pre-fossil organic deposits). Trees of the boreal forests, like most trees that are conifers or angiosperms (flowering plants), produce annual growth rings. If a fire burns through the bark of a tree, it leaves a scar, just as a serious burn leaves a scar on human skin. The tree grows over the scar, depositing a new growth ring for each year. (Figure 2.10 shows fire scars and tree rings on a cross section of a tree.) By examining cross sections of trees, it is possible to determine the date of each fire and the number of years between fires. From written and tree-ring records, Heinselman found that the frequency of fires had varied over time but that since the 17th century the BWCA forests had burned, on average, once per century. Furthermore, buried charcoal dated using carbon-14 revealed that fires could be traced back more than 30,000 years.[8]

The three kinds of historical records provided important evidence about fire in the history of the BWCA. At the time Heinselman did his study, the standard hypothesis was that fires were bad for forests and should be suppressed. The historical evidence provided a disproof of this hypothesis. It showed that fires were a natural and an integral part of the forest and that the forest had persisted with fire for a very long time. Thus, the use of historical information meets the primary requirement of the scientific method—the ability to disprove a statement. Historical evidence is a major source of data that can be used to test scientific hypotheses in ecology.

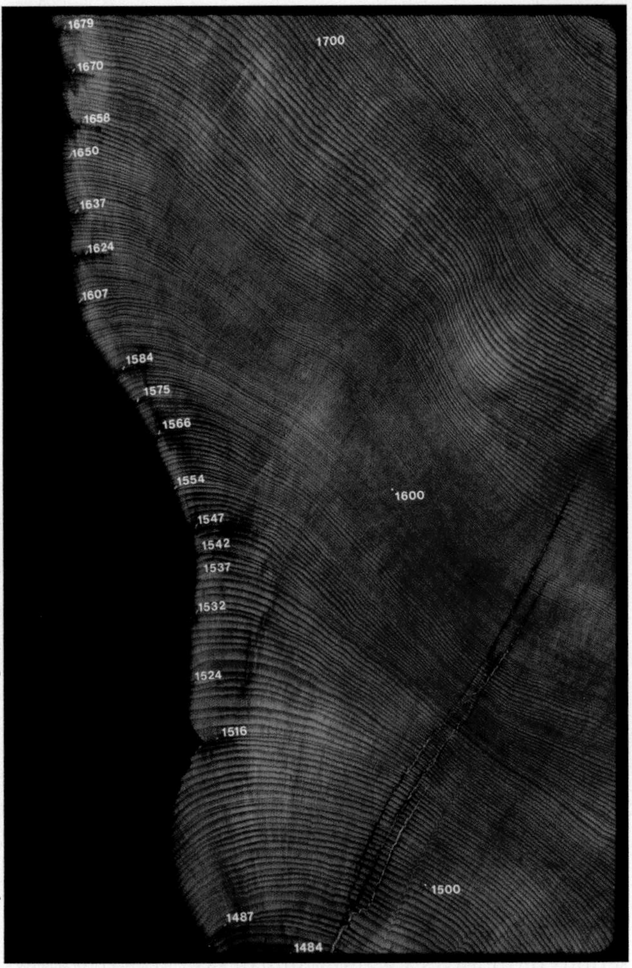

Photo by Bruce Dale/National Geographic/Getty Images

FIGURE 2.10 **Cross section of a tree showing fire scars and tree rings.** Together these allow scientists to date fires and to average the time between fires.

Modern Catastrophes and Disturbances as Experiments

Sometimes a large-scale catastrophe provides a kind of modern ecological experiment. The volcanic eruption of Mount St. Helens in 1980 supplied such an experiment, destroying vegetation and wildlife over a wide area. The recovery of plants, animals, and ecosystems following this explosion gave scientists insights into the dynamics of ecological systems and provided some surprises. The main surprise was how quickly vegetation recovered and wildlife returned to parts of the mountain. In other ways, the recovery followed expected patterns in ecological succession (see Chapters 6 and 12).

It is important to point out that the greater the quantity and the better the quality of ecological data prior to such a catastrophe, the more we can learn from the response of ecological systems to the event. This calls for careful monitoring of the environment.

Uncertainty in Science

In science, when we have a fairly high degree of confidence in our conclusions, we often forget to state the degree of certainty or uncertainty. Instead of saying, "There is a 99.9% probability that . . ." we say, "It has been proved that . . ." Unfortunately, many people interpret this as a deductive statement, meaning the conclusion is absolutely true, which has led to much misunderstanding about science. Although science begins with observations and therefore inductive reasoning, deductive reasoning is useful in helping scientists analyze whether conclusions based on inductions are logically valid. *Scientific reasoning combines induction and deduction*—different but complementary ways of thinking.

Leaps of Imagination and Other Nontraditional Aspects of the Scientific Method

What we have described so far is the classic scientific method. Scientific advances, however, often happen somewhat differently. They begin with instances of insight—leaps of imagination that are then subjected to the stepwise inductive process. And some scientists have made major advances by being in the right place at the right time, noticing interesting oddities, and knowing how to put these clues together. For example, penicillin was discovered "by accident" in 1928 when Sir Alexander Fleming was studying the pus-producing bacterium *Staphylococcus aureus*. When a culture of these bacteria was accidentally contaminated by the green fungus *Penicillium notatum*, Fleming noticed that the bacteria did not grow in areas of the culture where the fungus grew. He isolated the mold, grew it in a fluid medium, and found that it produced a substance that killed many of the bacteria that caused diseases. Eventually this discovery led other scientists to develop an injectable agent to treat diseases. *Penicillium notatum* is a common mold found on stale bread. No doubt many others had seen it, perhaps even noticing that other strange growths on bread did not overlap with *Penicillium notatum*. But it took Fleming's knowledge and observational ability for this piece of "luck" to occur.

2.3 Measurements and Uncertainty

A Word about Numbers in Science

We communicate scientific information in several ways. The written word is used for conveying synthesis, analysis, and conclusions. When we add numbers to our analysis, we obtain another dimension of understanding that goes beyond qualitative understanding and synthesis of a problem. Using numbers and statistical analysis allows us to visualize relationships in graphs and make predictions. It also allows us to analyze the strength of a relationship and in some cases discover a new relationship.

People in general put more faith in the accuracy of measurements than do scientists. Scientists realize that all measurements are only approximations, limited by the accuracy of the instruments used and the people who use them. Measurement uncertainties are inevitable; they can be reduced but never completely eliminated. For this reason, *a measurement is meaningless unless it is accompanied by an estimate of its uncertainty.*

Consider the loss of the *Challenger* space shuttle in 1986, the first major space shuttle accident, which appeared to be the result of the failure of rubber O-rings that were supposed to hold sections of rockets together. Imagine a simplified scenario in which an engineer is given a rubber O-ring used to seal fuel gases in a space shuttle. The engineer is asked to determine the flexibility of the O-rings under different temperature conditions to help answer two questions: At what temperature do the O-rings become brittle and subject to failure? And at what temperature(s) is it unsafe to launch the shuttle? After doing some tests, the engineer says that the rubber becomes brittle at $-1°C$ (30°F). So, can you assume it is safe to launch the shuttle at 0°C (32°F)?

At this point, you do not have enough information to answer the question. You assume that the temperature data may have some degree of uncertainty, but you have no idea how great a degree. Is the uncertainty $\pm 5°C$, $\pm 2°C$, or $\pm 0.5°C$? To make a reasonably safe and economically sound decision about whether to launch the shuttle, you must know the amount of uncertainty of the measurement.

Dealing with Uncertainties

There are two sources of uncertainty. One is the real variability of nature. The other is the fact that every measurement has some error. Measurement uncertainties and other errors that occur in experiments are called **experimental errors**. Errors that occur consistently, such as those resulting from incorrectly calibrated instruments, are **systematic errors**.

Scientists traditionally include a discussion of experimental errors when they report results. Error analysis often leads to greater understanding and sometimes even to important discoveries. For example, scientists discovered the eighth planet in our solar system, Neptune, when they investigated apparent inconsistencies—observed "errors"—in the orbit of the seventh planet, Uranus.

We can reduce measurement uncertainties by improving our measurement instruments, standardizing measurement procedures, and using carefully designed experiments and appropriate statistical procedures. Even then, however, uncertainties can never be completely eliminated. Difficult as it is for us to live with uncertainty, that is the nature of nature, as well as the nature of measurement and of science. Our awareness of these uncertainties should lead us to read reports of scientific studies critically, whether they appear in science journals or in popular magazines and newspapers. (See A Closer Look 2.2.)

Accuracy and Precision

A friend inherited some land on an island off the coast of Maine. However, the historical records were unclear about the land's boundaries, and to sell any portion of the land, he first had to determine where his neighbor's land ended and his began. There were differences of opinion about this. In fact, some people said one boundary went right through the house, which would have caused a lot of problems! Clearly, what was needed was a good map that everybody could agree on, so our friend hired a surveyor to determine exactly where the boundaries were.

The original surveyor's notes from the early 19th century had vague guidelines, such as "beginning at the mouth of Marsh brook on the Eastern side of the bars at a stake and stones. . . thence running South twenty six rods to a stake & stones. . . ." Over time, of course, the shore, the brook, its mouth, and the stones had moved and the stakes had disappeared. The surveyor was clear about the total distance (a rod, by the way, is an old English measure equal to 16.5 feet or 5.02 meters), but "South" wasn't very specific. So where and in exactly which direction was the true boundary? (This surveyor's method was common in early-19th-century New England. One New Hampshire survey during that time began with "Where you and I were standing yesterday . . ." Another began, "Starting at the hole in the ice [on the pond] . . .").

The 21st-century surveyor who was asked to find the real boundary used the most modern equipment—laser and microwave surveying transits, GPS devices—so he knew where the line he measured went to in millimeters. He could remeasure his line and come within millimeters of his previous location. But because the original starting point couldn't be determined within many meters, the surveyor didn't know where the true boundary line went; it was just somewhere within 10 meters or so of the line he had surveyed. So the end result was that even after this careful, modern, hi-tech survey, nobody really knew where the original boundary lines went. Scientists would say that the modern surveyor's work was precise but not accurate. *Accuracy* refers to what we know; *precision,* to how well we measure. With such things as this land survey, this is an important difference.

Accuracy also has another, slightly different scientific meaning. In some cases, certain measurements have been

A CLOSER LOOK 2.2

Measurement of Carbon Stored in Vegetation

A number of people have suggested that a partial solution to global warming might be a massive worldwide program of tree planting. Trees take carbon dioxide (an important greenhouse gas) out of the air in the process of photosynthesis. And because trees live a long time, they can store carbon for decades, even centuries. But how much carbon can be stored in trees and in all perennial vegetation? Many books and reports published during the past 20 years contained numbers representing the total stored carbon in Earth's vegetation, but all were presented without any estimate of error (Table 2.3). Without an estimate of that uncertainty, the figures are meaningless, yet important environmental decisions have been based on them.

Recent studies have reduced error by replacing guesses and extrapolations with scientific sampling techniques similar to those used to predict the outcomes of elections. Even these improved data would be meaningless, however, without an estimate of error. The new figures show that the earlier estimates were three to four times too large, grossly overestimating the storage of carbon in vegetation and therefore the contribution that tree planting could make in offsetting global warming.

Table 2.3 ESTIMATES OF ABOVEGROUND BIOMASS IN NORTH AMERICAN BOREAL FOREST

SOURCE	BIOMASS[a] (kg/m^2)	CARBON[b] (kg/m^2)	TOTAL BIOMASS[c] (10^9 metric tons)	TOTAL CARBON[c] (10^9 metric tons)
This study[d]	4.2 ± 1.0	1.9 ± 0.4	22 ± 5	9.7 ± 2
Previous estimates[e]				
1	17.5	7.9	90	40
2	15.4	6.9	79	35
3	14.8	6.7	76	34
4	12.4	5.6	64	29
5	5.9	2.7	30	13.8

Source: D.B. Botkin and L. Simpson, "The First Statistically Valid Estimate of Biomass for a Large Region," *Biogeochemistry* 9 (1990): 161–274. Reprinted by permission of Klumer Academic, Dordrecht, The Netherlands.

[a]Values in this column are for total aboveground biomass. Data from previous studies giving total biomass have been adjusted using the assumption that 23% of the total biomass is in below-ground roots. Most references use this percentage; Leith and Whittaker use 17%. We have chosen to use the larger value to give a more conservative comparison.

[b]Carbon is assumed to be 45% of total biomass following R.H. Whittaker, *Communities and Ecosystems* (New York: Macmillan, 1974).

[c]Assuming our estimate of the geographic extent of the North American boreal forest: 5,126,427 km^2 (324,166 mi^2).

[d]Based on a statistically valid survey; aboveground woodplants only.

[e]Lacking estimates of error: Sources of previous estimates by number (1) G.J. Ajtay, P. Ketner, and P. Duvigneaud, "Terrestrial Primary Production and Phytomass," in B. Bolin, E.T. Degens, S. Kempe, and P. Ketner, eds., *The Global Carbon Cycle* (New York: Wiley, 1979), pp. 129–182. (2) R.H. Whittaker and G.E. Likens, "Carbon in the Biota," in G.M. Woodwell and E.V. Pecam, eds., *Carbon and the Biosphere* (Springfield, VA: National Technical Information Center, 1973), pp. 281–300. (3) J.S. Olson, H.A. Pfuderer, and Y.H. Chan, *Changes in the Global Carbon Cycle and the Biosphere,* ORNL/EIS-109 (Oak Ridge, TN: Oak Ridge National Laboratory, 1978). (4) J.S. Olson, I.A. Watts, and L.I. Allison, *Carbon in Live Vegetation of Major World Ecosystems,* ORNL-5862 (Oak Ridge, TN: Oak Ridge National Laboratory, 1983). (5) G.M. Bonnor, *Inventory of Forest Biomass in Canada* (Petawawa, Ontario: Canadian Forest Service, Petawawa National Forest Institute, 1985).

Additional references: Lieth, H.,and R.H. Whittaker (eds) 1975, *Primary Productivity of the Biosphere*, N.Y., Springer-Verlag.

made very carefully by many people over a long period, and accepted values have been determined. In that kind of situation, *accuracy* means the extent to which a measurement agrees with the accepted value. But, as before, *precision* retains its original meaning—the degree of exactness with which a quantity is measured. In the case of the land in Maine, we can say that the new measurement had no accuracy in regard to the previous ("accepted") value.

Although a scientist should make measurements as precisely as possible, this friend's experience with surveying his land shows us that it is equally important not to report measurements with more precision than they warrant. Doing so conveys a misleading sense of both precision and accuracy.

2.4 Misunderstandings about Science and Society

Science and Decision Making

Like the scientific method, the process of making decisions is sometimes presented as a series of steps

1. Formulate a clear statement of the issue to be decided.
2. Gather the scientific information related to the issue.
3. List all alternative courses of action.
4. Predict the positive and negative consequences of each course of action and the probability that each consequence will occur.
5. Weigh the alternatives and choose the best solution.

Such a procedure is a good guide to rational decision making, but it assumes a simplicity not often found in real-world issues. It is difficult to anticipate all the potential consequences of a course of action, and unintended consequences are at the root of many environmental problems. Often the scientific information is incomplete and even controversial. For example, the insecticide DDT causes eggshells of birds that feed on insects to be so thin that unhatched birds die. When DDT first came into use, this consequence was not predicted. Only when populations of species such as the brown pelican became seriously endangered did people become aware of it.

In the face of incomplete information, scientific controversies, conflicting interests, and emotionalism, how can we make sound environmental decisions? We need to begin with the scientific evidence from all relevant sources and with estimates of the uncertainties in each. Avoiding emotionalism and resisting slogans and propaganda are essential to developing sound approaches to environmental issues. Ultimately, however, environmental decisions are policy decisions negotiated through the political process. Policymakers are rarely professional scientists; generally, they are political leaders and ordinary citizens. Therefore, the scientific education of those in government and business, as well as of all citizens, is crucial.

Science and Technology

Science is often confused with technology. As noted earlier, science is a search for understanding of the natural world, whereas technology is the application of scientific knowledge in an attempt to benefit people. Science often leads to technological developments, just as new technologies lead to scientific discoveries. The telescope began as a technological device, such as an aid to sailors, but when Galileo used it to study the heavens, it became a source of new scientific knowledge. That knowledge stimulated the technology of telescope-making, leading to the production of better telescopes, which in turn led to further advances in the science of astronomy.

Science is limited by the technology available. Before the invention of the electron microscope, scientists were limited to magnifications of 1,000 times and to studying objects about the size of one-tenth of a micrometer. (A micrometer is 1/1,000,000 of a meter, or 1/1,000 of a millimeter.) The electron microscope enabled scientists to view objects far smaller by magnifying more than 100,000 times. The electron microscope, a basis for new science, was also the product of science. Without prior scientific knowledge about electron beams and how to focus them, the electron microscope could not have been developed.

Most of us do not come into direct contact with science in our daily lives; instead, we come into contact with the products of science—technological devices such as computers, iPods, and microwave ovens. Thus, people tend to confuse the products of science with science itself. As you study science, it will help if you keep in mind the distinction between science and technology.

Science and Objectivity

One myth about science is the myth of objectivity, or value-free science—the notion that scientists are capable of complete objectivity independent of their personal values and the culture in which they live, and that science deals only with objective facts. Objectivity is certainly a goal of scientists, but it is unrealistic to think they can be totally free of influence by their social environments and personal values. It would be more realistic to admit that scientists do have biases and to try to identify these biases rather than deny or ignore them. In some ways, this situation is similar to that of measurement error: It is inescapable, and we can best deal with it by recognizing it and estimating its effects.

To find examples of how personal and social values affect science, we have only to look at recent controversies about environmental issues, such as whether or not to adopt more stringent automobile emission standards. Genetic engineering, nuclear power, global warming, and the preservation of threatened or endangered species involve conflicts among science, technology, and society. When we function as *scientists* in society, we want to explain the results of science objectively. As citizens who are not scientists, we want scientists to always be objective and tell us the truth about their scientific research.

That science is not entirely value-free should not be taken to mean that fuzzy thinking is acceptable in science. It is still important to think critically and logically about science and related social issues. Without the high standards of evidence held up as the norm for science, we run the risk of accepting unfounded ideas about the world. When we confuse what we would like to believe with what we have the evidence to believe, we have a weak basis for making critical environmental decisions that could have far-reaching and serious consequences.

The great successes of science, especially as the foundation for so many things that benefit us in modern technological societies—from cell phones to CAT scans to space exploration—give science and scientists a societal authority that makes it all the more difficult to know when a scientist might be exceeding the bounds of his or her scientific knowledge. It may be helpful to realize that scientists play three roles in our society: first, as researchers simply explaining the results of their work; second, as almost priest-like authorities who often seem to speak in tongues the rest of us can't understand; and third, as what we could call expert witnesses. In this third role, they will discuss broad areas of research that they are familiar with and that are within their field of study, but about which they may not have done research themselves. Like an expert testifying in court, they are basically saying to us, "Although I haven't done this particular research myself, my experience and knowledge suggest to me that . . ."

The roles of researcher and expert witness are legitimate as long as it is clear to everybody which role a scientist is playing. Whether you want a scientist to be your authority about everything, within science and outside of science, is a personal and value choice.

In the modern world, there is another problem about the role of scientists and science in our society. Science has been so potent that it has become fundamental to political policies. As a result, science can become politicized, which means that rather than beginning with objective inquiry, people begin with a belief about something and pick and choose only the scientific evidence that supports that belief. This can even be carried to the next step, where research is funded only if it fits within a political or an ethical point of view.

Scientists themselves, even acting as best they can *as* scientists, can be caught up in one way of thinking when the evidence points to another. These scientists are said to be working under a certain paradigm, a particular theoretical framework. Sometimes their science undergoes a paradigm shift: New scientific information reveals a great departure from previous ways of thinking and from previous scientific theories, and it is difficult, after working within one way of thinking, to recognize that some or all of their fundamentals must change. Paradigm shifts happen over and over again in science and lead to exciting and often life-changing results for us. The discovery and understanding of electricity are examples, as is the development of quantum mechanics in physics in the early decades of the 20th century.

We can never completely escape biases, intentional and unintentional, in fundamental science, its interpretation, and its application to practical problems, but understanding the nature of the problems that can arise can help us limit this misuse of science. The situation is complicated by legitimate scientific uncertainties and differences in scientific theories. It is hard for us, as citizens, to know when scientists are having a legitimate debate about findings and theories, and when they are disagreeing over personal beliefs and convictions that are outside of science. Because environmental sciences touch our lives in so many ways, because they affect things that are involved with choices and values, and because these sciences deal with phenomena of great complexity, the need to understand where science can go astray is especially important.

Consensus Science

It has become common in recent years for people to talk about consensus science. This happens especially with complex environmental problems that are yet only poorly understood. One hears that you can believe that a statement is true because the "scientific consensus"—meaning a group of scientists, or sometimes the majority of scientists in a field—is that a statement is scientifically correct. But there is a fundamental difference between a scientific finding and what a scientist says. And a scientific consensus is basically what a group of scientists are saying.

There are famous stories in the past where a scientific consensus proved to be wrong. One of the most well known is the theory of plate tectonics. Continental drift was first proposed in 1915 by Alfred Wegener in his book *The Origins of Continents and Oceans*. He based his idea on the shape of the west coast of Africa and the east coast of South America, which looked as if they would fit together like a jigsaw puzzle, as though they had once been together but had drifted apart, and also on the fact that fossils of identical animals and plants had been found on these two coasts. Since nobody could come up with a mechanism that would make the continents move, the idea was rejected pretty much out of hand. But after World War II, geological research revealed forces that moved continents, and Wegener's theory became the accepted norm.

This kind of scientific revolution is common enough to have its own name—a paradigm shift. When a science is changing rapidly, the conventional wisdom becomes less and less trustworthy.

Citizens and Science

The question comes up often these days: How can an ordinary citizen who is not trained as a scientist decide when a scientist is staying true to the scientific information and when he or she is being persuaded by political biases? There are several questions to ask of a scientist or of a scientist's written report.

For quantitative information, one question to ask is: What are the statistical errors of the results? Especially in environmental scientific research, there are often many measurements required, as we learned from Mono Lake. A statistically valid estimate will be the result of a number of measurements of the same thing. In cases where what is being measured varies over time or space (such as the number of gulls nesting at Mono Lake), more multiple measurements must be taken to represent those variations. These variations can then be analyzed to give an average—the statistical mean—and a measure of how variable and therefore how reliable that estimate is; this is known as the statistical error. A quantitative measurement with no statistical estimate of error is not a sufficient scientific measure.

Another question is: Were different methods used to measure the same thing, and then were these compared? This is always a mark of careful science, even though it is not always possible.

In this chapter's Critical Thinking feature, we explore this question further. As you read the other chapters in this book, think about what other questions you could ask a scientist to decide whether what he or she was saying is the result of good science.

Science, Pseudoscience, and Frontier Science

Some ideas presented as scientific are in fact not scientific because they are inherently untestable, lack empirical support, or are based on faulty reasoning or poor scientific methodology, as illustrated by the case of the mysterious crop circles (A Closer Look 2.1). Such ideas are referred to as pseudoscientific (the prefix *pseudo-* means false).

Pseudoscientific ideas arise from various sources. With more research, however, some of the frontier ideas may move into the realm of accepted science, and new ideas will take their place at the advancing frontier (Figure 2.11).[9] Research may not support other hypotheses at the frontier, and these will be discarded. Accepted science may merge into frontier science, which in turn may merge into farther-out ideas, or fringe science. Really wild ideas may be considered beyond the fringe.

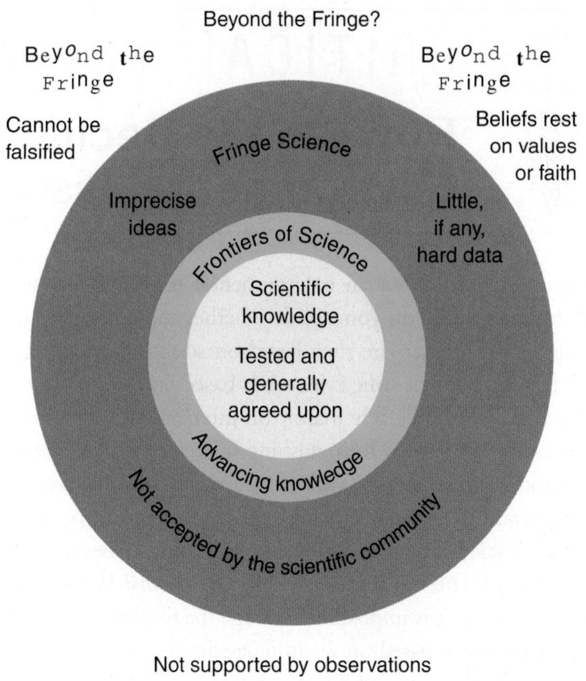

FIGURE 2.11 Beyond the fringe? A diagrammatic view of different kinds of knowledge and ideas.

2.5 Environmental Questions and the Scientific Method

Environmental sciences deal with especially complex systems and include a relatively new set of sciences. Therefore, the process of scientific study has not always neatly followed the formal scientific method discussed earlier in this chapter. Often, observations are not used to develop formal hypotheses. Controlled laboratory experiments have been the exception rather than the rule. Much environmental research has been limited to field observations of processes and events that have been difficult to subject to controlled experiments.

Environmental research presents several obstacles to following the classic scientific method. The long time frame of many ecological processes relative to human lifetimes, professional lifetimes, and lengths of research grants poses problems for establishing statements that can in practice be subject to disproof. What do we do if a theoretical disproof through direct observation would take a century or more? Other obstacles include difficulties in setting up adequate experimental controls for field studies, in developing laboratory experiments of sufficient complexity, and in developing theory and models for complex systems. Throughout this text, we present differences between the "standard" scientific method and the actual approach that has been used in environmental sciences.

CRITICAL THINKING ISSUE

How Do We Decide What to Believe about Environmental Issues?

When you read about an environmental issue in a newspaper or magazine, how do you decide whether to accept the claims made in the article? Are they based on scientific evidence, and are they logical? Scientific evidence is based on observations, but media accounts often rely mainly on inferences (interpretations) rather than evidence. Distinguishing inferences from evidence is an important first step in evaluating articles critically. Second, it is important to consider the source of a statement. Is the source a reputable scientific organization or publication? Does the source have a vested interest that might bias the claims? When sources are not named, it is impossible to judge the reliability of claims.

If a claim is based on scientific evidence presented logically from a reliable, unbiased source, it is appropriate to accept the claim tentatively, pending further information. Practice your critical evaluation skills by reading the material below and answering the critical thinking questions. The material is taken from an article published in a Canadian newspaper in 1996. We present the essence of the article here as an exercise to practice your critical thinking skills.

Mystery of Deformed Frogs: A Clue Is Found

Note: Much of the material below was apparently taken during an interview (for a newspaper article) with a scientist in the Midwestern United States who was a member of the U.S. section of the Worldwide Declining Amphibian Population Taskforce.

The scientist pointed out that limb deformities found in frogs have been noticed and reported for over 250 years. However, the rates of deformities that are being observed today are thought to be unprecedented in some species of frogs. Some of these deformities have been observed in particular frog species and in one case have affected more than half the population of a species of frog living in a particular area. He mentioned that, while deformities in California frogs may be due to a parasite, those same deformities in frogs in the Midwestern United States are probably not the result of parasites. The deformities mentioned include misshapen legs, extra limbs, and missing or misaligned eyes.[10]

Today a large number of chemicals are being applied to our environment, and we do not completely understand their breakdown products and their potential effects on frogs. The scientist speculated that the chemical methoprene, which is an insecticide used in mosquito control, might be linked to the deformities showing up in frogs. The scientist pointed out that a breakdown product of methoprene resembles retinoic acid, known to be important in the development of frogs. He went on to point out that exposure of frogs in the laboratory to retinoic acid can produce all the limb deformities that have been observed in nature. Having said that, he admitted that it is not necessarily the retinoic acid that is causing the problem in nature. However, he stated that it is the best guess as to what is happening.[10]

The deformities and decline of amphibians in the United States and other areas are of concern because amphibians, and, in particular, frogs, are species that may provide early warnings of potential risk of chemicals to people. Frogs are particularly susceptible to chemicals because their skin is permeable, putting them at special risk to chemicals found in the water.

As a final note, a paper published in 2003 concluded that while laboratory tests of exposure to high-concentration retinoic acid could cause deformities of legs in frogs, methoprene did not result in such deformities. Field measurements led to the conclusion that frogs exposed to methoprene could not be the sole cause of abnormalities in frogs. Methoprene breaks down quickly and was not found to persist in the environment where the number of deformities was high.[11]

Critical Thinking Questions

1. What is the major claim made in the 1996 newspaper article?
2. What evidence is presented to support the claim?
3. Is the evidence based on observations, and is the source of the evidence reputable and unbiased?
4. Is the argument for the claim, whether or not based on evidence, logical?
5. Would you accept or reject the claim?
6. Even if the claim were well supported by evidence based on good authority, why might your acceptance be only tentative?
7. Your Environmental Science professor has asked you to write a term paper about deformities in Midwestern U.S. frogs and you are able to find the 1996 newspaper article on methoprene and frog deformities. You decide to do an Internet search (try EPA methoprene exposure to frogs or methoprene and deformities in frogs) to update what the article states. Do this search and report your results. Does your research support the contention that newspaper articles can be misleading? Or does it not in this case? Would you agree that newspaper articles are not acceptable references for term papers, because they may not correctly report scientific information, particularly if the source (a refereed paper or reliable government report) is not cited?
8. Suppose your search found the 2003 Scientific American paper (discussed above—Blaustein, A.R., and Johnson, P.J. 2003, February. Explaining frog deformities. *Scientific American*), which presented a different point of view from that in the 1996 newspaper article. Would your position on methoprene change after reading the 2003 paper?

SUMMARY

- Science is one path to critical thinking about the natural world. Its goal is to gain an understanding of how nature works. Decisions on environmental issues must begin with an examination of the relevant scientific evidence. However, environmental decisions also require careful analysis of economic, social, and political consequences. Solutions will reflect religious, aesthetic, and ethical values as well.
- Science is an open-ended process of finding out about the natural world. In contrast, science lectures and texts are usually summaries of the answers arrived at through this process, and science homework and tests are exercises in finding the right answer. Therefore, students often perceive science as a body of facts to be memorized, and they view lectures and texts as authoritative sources of absolute truths about the world.
- Science begins with careful observations of the natural world, from which scientists formulate hypotheses. Whenever possible, scientists test hypotheses with controlled experiments.
- Although the scientific method is often taught as a prescribed series of steps, it is better to think of it as a general guide to scientific thinking, with many variations.
- We acquire scientific knowledge through inductive reasoning, basing general conclusions on specific observations. Conclusions arrived at through induction can never be proved with certainty. Thus, because of the inductive nature of science, it is possible to disprove hypotheses but not possible to prove them with 100% certainty.
- Measurements are approximations that may be more or less exact, depending on the measuring instruments and the people who use them. A measurement is meaningful when accompanied by an estimate of the degree of uncertainty, or error.
- Accuracy in measurement is the extent to which the measurement agrees with an accepted value. Precision is the degree of exactness with which a measurement is made. A precise measurement may not be accurate. The estimate of uncertainty provides information on the precision of a measurement.
- A general statement that relates and explains a great many hypotheses is called a theory. Theories are the greatest achievements of science.
- Critical thinking can help us distinguish science from pseudoscience. It can also help us recognize possible bias on the part of scientists and the media. Critical thinking involves questioning and synthesizing information rather than merely acquiring information.

REEXAMINING THEMES AND ISSUES

© Anton Balazh 2011/iStockphoto

GLOBAL PERSPECTIVE

The global perspective on environment arises out of new findings in environmental science.

ssguy/ShutterStock

URBAN WORLD

Our increasingly urbanized world is best understood with the assistance of scientific investigation.

B2M Productions/Getty Images, Inc.

PEOPLE AND NATURE

Solutions to environmental problems require both values and knowledge. Understanding the scientific method is especially important if we are going to understand the connection between values and knowledge, and the relationship between people and nature. Ultimately, environmental decisions are policy decisions, negotiated through the political process. Policymakers often lack sufficient understanding of the scientific method, leading to false conclusions. Uncertainty is part of the nature of measurement and science. We must learn to accept uncertainty as part of our attempt to conserve and use our natural resources.

George Doyle/Getty Images, Inc.

SCIENCE AND VALUES

This chapter summarizes the scientific method, which is essential to analyzing and solving environmental problems and to developing sound approaches to sustainability.

KEY TERMS

controlled experiment 28
deductive reasoning 29
dependent variable 28
disprovability 26
experimental errors 33
fact 28
hypothesis 28
independent variable 28
inductive reasoning 29
inferences 28
manipulated variable 28
model 31
observations 28
operational definitions 29
qualitative data 28
quantitative data 28
responding variable 28
scientific method 25
scientific theory 30
systematic errors 33
variables 28

STUDY QUESTIONS

1. Which of the following are scientific statements and which are not? What is the basis for your decision in each case?
   (a) The amount of carbon dioxide in the atmosphere is increasing.
   (b) Condors are ugly.
   (c) Condors are endangered.
   (d) Today there are 280 condors.
   (e) Crop circles are a sign from Earth to us that we should act better.
   (f) Crop circles can be made by people.
   (g) The fate of Mono Lake is the same as the fate of the Aral Sea.
2. What is the logical conclusion of each of the following syllogisms? Which conclusions correspond to observed reality?
   (a) All men are mortal. Socrates is a man. Therefore____________________
   (b) All sheep are black. Mary's lamb is white. Therefore____________________
   (c) All elephants are animals. All animals are living beings. Therefore____________________
3. Which of the following statements are supported by deductive reasoning and which by inductive reasoning?
   (a) The sun will rise tomorrow.
   (b) The square of the hypotenuse of a right triangle is equal to the sum of the squares of the other two sides.
   (c) Only male deer have antlers.
   (d) If $A = B$ and $B = C$, then $A = C$.
   (e) The net force acting on a body equals its mass times its acceleration.

4. The accepted value for the number of inches in a centimeter is 0.3937. Two students mark off a centimeter on a piece of paper and then measure the distance using a ruler (in inches). Student A finds the distance equal to 0.3827 in., and student B finds it equal to 0.39 in. Which measurement is more accurate? Which is more precise? If student B measured the distance as 0.3900 in., what would be your answer?
5. (a) A teacher gives five students each a metal bar and asks them to measure the length. The measurements obtained are 5.03, 4.99, 5.02, 4.96, and 5.00 cm. How can you explain the variability in the measurements? Are these systematic or random errors?
   (b) The next day, the teacher gives the students the same bars but tells them that the bars have contracted because they have been in the refrigerator. In fact, the temperature difference would be too small to have any measurable effect on the length of the bars. The students' measurements, in the same order as in part (a), are 5.01, 4.95, 5.00, 4.90, and 4.95 cm. Why are the students' measurements different from those of the day before? What does this illustrate about science?
6. Identify the independent and dependent variables in each of the following:
   (a) Change in the rate of breathing in response to exercise.
   (b) The effect of study time on grades.
   (c) The likelihood that people exposed to smoke from other people's cigarettes will contract lung cancer.
7. (a) Identify a technological advance that resulted from a scientific discovery.
   (b) Identify a scientific discovery that resulted from a technological advance.
   (c) Identify a technological device you used today. What scientific discoveries were necessary before the device could be developed?
8. What is fallacious about each of the following conclusions?
   (a) A fortune cookie contains the statement "A happy event will occur in your life." Four months later, you find a $100 bill. You conclude that the prediction was correct.
   (b) A person claims that aliens visited Earth in prehistoric times and influenced the cultural development of humans. As evidence, the person points to ideas among many groups of people about beings who came from the sky and performed amazing feats.
   (c) A person observes that light-colored animals almost always live on light-colored surfaces, whereas dark forms of the same species live on dark surfaces. The person concludes that the light surface causes the light color of the animals.
   (d) A person knows three people who have had fewer colds since they began taking vitamin C on a regular basis. The person concludes that vitamin C prevents colds.
9. Find a newspaper article on a controversial topic. Identify some loaded words in the article—that is, words that convey an emotional reaction or a value judgment.
10. Identify some social, economic, aesthetic, and ethical issues involved in a current environmental controversy.

FURTHER READING

American Association for the Advancement of Science (AAAS), ***Science for All Americans*** (Washington, DC: AAAS, 1989). This report focuses on the knowledge, skills, and attitudes a student needs in order to be scientifically literate.

Botkin, D.B., ***No Man's Garden: Thoreau and a New Vision for Civilization and Nature*** (Washington, DC: Island Press, 2001). The author discusses how science can be applied to the study of nature and to problems associated with people and nature. He also discusses science and values.

Grinnell, F., ***The Scientific Attitude*** (New York: Guilford, 1992). The author uses examples from biomedical research to illustrate the processes of science (observing, hypothesizing, experimenting) and how scientists interact with each other and with society.

Kuhn, Thomas S., ***The Structure of Scientific Revolutions*** (Chicago: University of Chicago Press, 1996). This is a modern classic in the discussion of the scientific method, especially regarding major transitions in new sciences, such as environmental sciences.

McCain, G., and E.M. Segal, ***The Game of Science*** (Monterey, CA: Brooks/Cole, 1982). The authors present a lively look into the subculture of science.

Sagan, C., ***The Demon-Haunted World*** (New York: Random House, 1995). The author argues that irrational thinking and superstition threaten democratic institutions and discusses the importance of scientific thinking to our global civilization.

NOTES

1. Wiens, J.A., D.T. Pattern, and D.B. Botkin. 1993. Assessing ecological impact assessment: Lessons from Mono Lake, California. *Ecological Applications* 3(4):595–609.
2. Botkin, D.B., W.S. Broecker, L.G. Everett, J. Shapiro, and J.A. Wiens. 1988. *The Future of Mono Lake* (Report No. 68). Riverside: California Water Resources Center, University of California.
3. Lerner, L.S., and W.J. Bennetta. 1988 (April). The treatment of theory in textbooks. *The Science Teacher*, pp. 37–41.
4. Pullis La Rouche, G. 2003. (PDF). Birding in the United States: A demographic and economic analysis. Addendum to the 2001 National Survey of Fishing, Hunting and Wildlife-Associated Recreation. Report 2001-1. U.S. Fish and Wildlife Service, Arlington, Virginia.
5. Vickers, B., ed. 1987. *English Science: Bacon to Newton* (Cambridge English Prose Texts).
6. Kuhn, T.S. 1970. *The Structure of Scientific Revolutions*. Chicago: University of Chicago Press.
7. Pease, C.M., and J.J. Bull. 1992. Is science logical? *Bioscience* 42:293–298.
8. Heinselman, H.M. 1973. Fire in the virgin forests of the Boundary Waters Canoe Area, Minnesota. *Journal of Quaternary Research* 3:329–382.
9. Trefil, J.S. 1978. A consumer's guide to pseudoscience. *Saturday Review* 4:16–21.
10. Conlon, M. 1996. Clue found in deformed frog mystery. *Toronto Star*, November 6.
11. Blaustein, A.R., and Johnson, P.J. 2003, February. Explaining frog deformities. *Scientific American*.

A CLOSER LOOK 2.1 NOTES

a. Schmidt, W.E. 1991 (September 10). "Jovial con men" take credit(?) for crop circles. *New York Times*, p. 81.

CHAPTER 3

Dollars and Environmental Sense: Economics of Environmental Issues

LEARNING OBJECTIVES

Why do people value environmental resources? To what extent are environmental decisions based on economics? Other chapters in this text explain the causes of environmental problems and discuss technical solutions. The scientific solutions, however, are only part of the answer. This chapter introduces some basic concepts of environmental economics and shows how these concepts help us to understand environmental issues. After studying this chapter, you should be able to . . .

- **Calculate the future value of an environmental benefit and determine how it affects willingness to pay for it now**
- **Explain why the Arctic Ocean is a commons and present a plan for how parts of it might be removed as a commons**
- **Analyze why decision making about environmental issues involves economics as well as scientific information**
- **Support or argue against the statement: Economic factors are more important than scientific findings in resolving an environmental issue**
- **Consider the conservation of groundwater where you live; explain how its perceived future value affects willingness to pay for it now**
- **Compare the risk of bicycling versus driving a car and explain why some people might accept a higher risk for bicycling than they would for driving a car**
- **Determine whether environmental intangibles affect the housing prices where you live**
- **Decide whether the scenery at the Grand Canyon is an economic externality or has intrinsic economic value**

A lion like those found in the Mbirikani in Kenya described in this chapter's Case Study.

John Lindsay-Smith/Shutterstock

CASE STUDY

How Conservation Can Be an Economic Benefit

Saving tropical forests and other endangered habitats has not been easy despite widespread support. One approach tried by environmental organizations is to establish national parks in nations where these habitats exist. But that is difficult. It can require new laws and international support. In the past, some parks have excluded people as long-term residents, causing conflicts with local indigenous groups and making new parks controversial. It becomes a people versus nature issue.

Parks also mean no development, which is politically unpopular, even where the revenues from development are low. To Richard Rice, former chief economist at Conservation International (CI), this lack of tangible economic value is one of the most serious threats facing the world's dwindling stock of biodiversity. Though a seemingly daunting dilemma, the solution, according to Rice, is surprisingly straightforward: Make conservation itself a source of economic benefit.

In recent years, Rice and his colleagues have taken this prescription to the ground in a dozen different countries in a series of efforts that they call, simply, conservation agreements. As the name suggests, conservation agreements are essentially business deals in which governments or local communities agree to protect natural ecosystems in exchange for a steady stream of compensation from conservationists or other investors. Rice and his colleagues have found that this approach holds much promise in a wide variety of different contexts, including areas such as private or indigenous lands where traditional protected areas might be difficult or impossible to establish.

Experience has also shown that effective incentives in many situations, particularly those involving rural communities in remote areas, can take the form of funding for social services such as health and education, rather than only direct cash payments. This flexibility, which is one of the most important strengths of the approach, means that conservation agreements can be tailored to fit the needs of a given situation. An interesting example includes a lion conservation project in Kenya. Prior to European settlement, 1 million lions lived in Africa. But today, due to overhunting and human encroachment, fewer than 30,000 lions exist in the wild, mostly in parks and protected areas that are too small to maintain viable populations.

The situation is particularly dire in Kenya where fewer than 2,000 lions remain. To address this issue, in 2003, the Maasailand Preservation Trust (MPT) entered into an agreement with indigenous Maasai landowners in southeastern Kenya to keep lions and other predators safe on the community-owned Mbirikani Group Ranch. Mbirikani (pronounced em-beer-e-con-e) is one of six contiguous Masaai-owned grazing areas that together form a natural corridor essential to the ecological integrity of four surrounding national parks. (See Figure 3.1.)

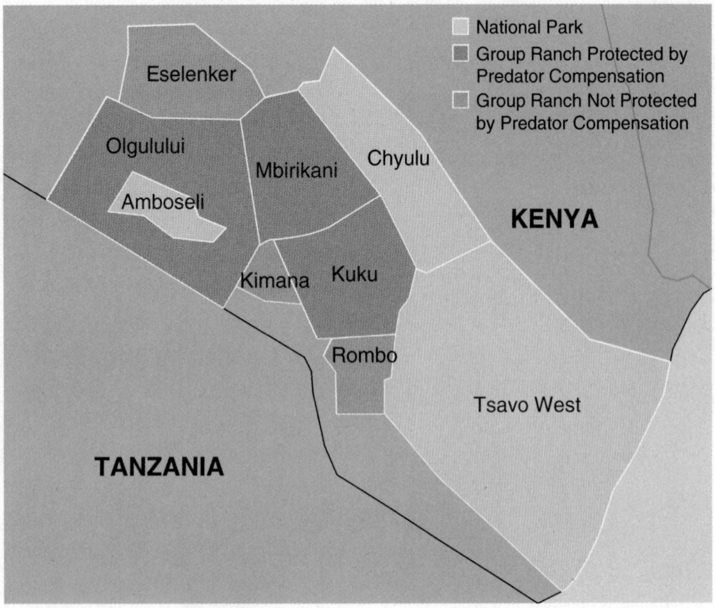

FIGURE 3.1 **Maasai Group Ranches participating in the Predator Compensation Program (PCF) and surrounding national parks.**

At its heart, the Mbirikani program (called the Predator Compensation Fund—PCF) is just a business agreement with Masaai herdsman. In exchange for their commitment to not kill lions, PCF agreed to compensate the Masaai for all verified kills of livestock by lions and other predators. In a region where lions once thrived but are now on the brink of local extinction, the agreement, which has since been expanded to adjacent ranches, has virtually stopped the killing of predators across more than 1 million acres. As a result, for the first time in many decades, the number of lions in an important part of their remaining range is actually on the rise.

Based on his work with conservation agreements in Kenya and other countries, Richard Rice established his own nonprofit corporation, the Conservation Agreement Fund, to use similar methods elsewhere in the world. He and others have been able to extend this and related economic methods to help save coral reefs in the South Pacific among other conservation goals. As the story of conservation agreements shows, economics and economic analyses have important roles to play in solving environmental problems. Because of this, we introduce the concepts of environmental economics near the beginning of our book.

3.1 Overview of Environmental Economics

Modern environmental law began in the 1960s with the start of the social and political movement we know as environmentalism. Its foundation is the "three E's": ecology, engineering, and economics.[1] Economics plays an important role in environmental decision making. It is a factor in finding solutions that work, are efficient, and are fair. This is why we are devoting one of our early overview chapters to environmental economics.

Environmental economics is not simply about money; it is about how to persuade people, organizations, and society at large to act in a way that benefits the environment, keeping it as free as possible of pollution and other damage, keeping our resources sustainable, and accomplishing these goals within a democratic framework. Put most simply, environmental economics focuses on two broad areas: controlling pollution and environmental damage in general, and sustaining renewable resources—forests, fisheries, recreational lands, and so forth. Environmental economists also explore the reasons why people don't act in their own best interests when it comes to the environment. Are there rational explanations for what seem to be irrational choices? If so, and if we can understand them, perhaps we can do something about them. What we do, what we can do, and how we do it are known collectively as **policy instruments**.

Environmental decision making often, perhaps even usually, involves analysis of tangible and intangible factors. In the language of economics, a **tangible factor** is one you can touch, buy, and sell. A house lost in a mudslide due to altering the slope of the land is an example of a tangible factor. For economists, an **intangible factor** is one you can't touch directly, but you value it, as with the beauty of the slope before the mudslide. Of the two, the intangibles are obviously more difficult to deal with because they are harder to measure and to value economically. Nonetheless, evaluation of intangibles is becoming more important. As you will see in later chapters, huge amounts of money and resources are involved in economic decisions about both tangible and intangible aspects of the environment: There are the costs of pollution and the loss of renewable resources, and there are the costs of doing something about these problems.

In addition, environmental economics deals with the question: Who should pay for improving or protecting the environment? Those who produce the problem? Those who suffer from it? The government, through subsidies and direct financing of programs? Should the free market determine what aspect of the environment is to be valued, or should the voters in a democracy decide? Should decisions be left to those who know the most about science, economics, and other fields that influence public policy?

These are difficult questions to answer, and it is unlikely that the answer will be universal—that one approach will work for every environmental problem.

Put another way, in every environmental matter, there is a desire on the one hand to maintain individual freedom of choice and on the other to achieve a specific social goal. In ocean fishing, for example, we want to allow every individual to choose whether or not to fish, but we want to prevent everyone from fishing at the same time and bringing fish species to extinction. This interplay between private good and public good is at the heart of environmental issues.

3.2 Public-Service Functions of Nature

A complicating factor in maintaining clean air, soils, and water, and sustaining our renewable resources is that ecosystems do some of this without our help. Forests absorb particulates, salt marshes convert toxic compounds to nontoxic forms, wetlands and organic soils treat sewage. These are called the **public-service functions** of nature. Economists refer to the ecological systems that provide these benefits as **natural capital**.

The atmosphere performs a public service by acting as a large disposal site for toxic gases. For example, carbon monoxide is eventually converted to nontoxic carbon dioxide either by inorganic chemical reactions or by soil bacteria. Bacteria also clean water in the soil by decomposing toxic chemicals, and bacteria fix nitrogen in the oceans, lakes, rivers, and soils. Since the Industrial Revolution, nitrogen fertilizers have been made artificially. But if we replaced natural bacterial production everywhere by producing nitrogen fertilizers artificially and transporting them ourselves, the cost would be immense—but, again, we rarely think about this activity of bacteria.

Among the most important public-service providers are the pollinators, which include birds, bats, ants, bees, wasps, beetles, butterflies, moths, flies, mosquitoes, and midges (Figure 3.2.) It is estimated that pollinating animals pollinate about $15 billion worth of crops grown on 2 million acres in the United States,[2, 3] and about one bite in three of the food you eat depends on pollinators. Their total economic impact can reach $40 billion a year.[4] The cost of pollinating these crops by hand would be exorbitant, so a pollutant that eliminated bees would have large indirect economic consequences. We rarely think of this benefit of bees, but it has received wide attention in recent years because of what is called Colony Collapse Disorder (CCD)—a situation where many of a hive's worker bees disappear, the causes of which continue to be controversial. CCD affects agricultural practices, including the bottom line of many companies that provide bees to pollinate crops. As a result, food prices go up.[5]

Dr. Merlin D. Tuttle/Bat Conservation International/Science Source

FIGURE 3.2 **Public-service function of living things.** Among the animals that pollinate flowers and thereby play a public-service function are bats, such as this lesser long-nosed bat (*Leptonycteris-yerbabuenae*) found in the mountains of Central America, including Mexico, and some parts of North America. Long-nosed bats of various species pollinate the century plant (agave) of American deserts—important food for many animals and an important plant over the centuries to Native Americans of Mexico.

Public-service functions of living things are estimated to provide between $3 trillion and $33 trillion in benefits to human beings and other forms of life per year.[6] However, current estimates are only rough approximations because the value is difficult to measure.

3.3 The Environment as a Commons

Often people use a natural resource without regard for maintaining that resource and its environment in a renewable state—that is, they don't concern themselves with that resource's sustainability. At first glance, this seems puzzling, but economic analysis suggests that the profit motive, by itself, will not always lead a person to act in the best interests of the environment. This chapter's critical thinking exercise discusses the problem of making one of the world's major fisheries, Georges Bank, in the Atlantic Ocean, sustainable.

In some situations, what a group of people from one time and one culture value and wish to have sustained another culture considers to be without value or undesirable. For example, many of the early settlers to the northeastern United States found themselves in vast areas of forests. They needed to clear land for farming, and, for them, forests were simply in the way—something to be gotten rid of. Large areas of forests were cleared, sometimes with fire, sometimes by cutting or just girdling the trees. (Girdling means cutting all the way through the bark of a tree around the entire trunk. This prevents water and minerals from going up to the leaves and sugars and other products of photosynthesis from descending to feed the lower stem and roots.) To these settlers, there were plenty of forests for firewood and construction. So much in fact that it was commonly believed that America could never run out of forests—by the time one area of forest was cleared, another area would regrow. This, of course, is a very different valuation of forests than the one that predominates in 21st-century America.

A modern example of the difference in valuation is between those who want to drill for oil in or near a national park, because the value and utility of the energy seem more important than the park, while environmentally oriented people value the national park more than an additional source of petroleum. Resolving this conflict in valuation is a major problem in a modern democracy, one in which an understanding of both economics and ecology are essential.

Another reason natural resources are not sustained has to do with what the ecologist Garrett Hardin called the tragedy of the commons.[7] When a resource is shared, an individual's personal share of profit from its exploitation is usually greater than his or her share of the resulting loss. A second reason has to do with the low growth rate, and therefore low productivity, of a resource.

A **commons** is land (or another resource) owned publicly, with public access for private uses. The term *commons* originated from land owned publicly in English and New England towns and set aside so that all the farmers of the town could graze their cattle. Sharing the grazing area worked as long as the number of cattle was low enough to prevent overgrazing. It would seem that people of goodwill would understand the limits of a commons. But take a dispassionate view and think about the benefits and costs to each farmer as if it were a game. Phrased simply, each farmer tries to maximize personal gain and must periodically consider whether to add more cattle to the herd on the commons. The addition of one cow has both a positive and a negative value. The positive value is the benefit when the farmer sells that cow. The negative value is the additional grazing by the cow. The personal profit from selling a cow is greater than the farmer's share of the loss caused by the degradation of the commons. Therefore, the short-term successful game plan is always to add another cow.

Since individuals will act to increase use of the common resource, eventually the common grazing land is so crowded with cattle that none can get adequate food and the pasture is destroyed. In the short run, everyone seems to gain, but in the long run, everyone loses. This principle applies generally: Complete freedom of action in a commons inevitably brings ruin to all. The implication seems clear: Without some management or control, all natural resources treated like a commons will inevitably be destroyed.

How can we deal with the tragedy of the commons? It is only a partially solved problem. Still, in trying to solve this puzzle, economic analysis can be helpful.

There are many examples of commons, both past and present. In the United States, 38% of forests are on publicly owned lands; as such, these forests are commons. Resources in international regions, such as ocean fisheries away from coastlines, and the deep-ocean seabed, where valuable mineral deposits lie, are international commons not controlled by any single nation.

The Arctic sea ice and the Arctic seas are commons (Figure 3.3), as is most of the continent of Antarctica, although there are some national territorial claims, and international negotiations have continued for years about conserving Antarctica and about the possible use of its resources. The atmosphere, too, is a commons, nationally and internationally, both in its use and as a place where people dump wastes.

In the 19th century, burning wood in fireplaces was the major source of heating in the United States (and fuel wood is still the major source of heat in many nations). Until the 1980s, a wood fire in a fireplace or woodstove was considered a simple good, providing warmth and beauty. People enjoyed sitting around a fire and watching the flames—an activity with a long history in human societies. But in the 1980s, with increases in populations and vacation homes in states such as Vermont and Colorado, home burning of wood began to pollute air locally. Especially in valley towns surrounded by mountains, the air became fouled, visibility declined, and there was a potential for ill effects on human health and the environment. Several states, including Vermont, have had programs offering rebates to buyers of newer, lower-polluting woodstoves.[8] The local air is a commons, and its overuse required a societal change.

Bryan and Cherry Alexander/Science Source

FIGURE 3.3 **Arctic sea ice and polar bears, which live in many areas of the Arctic, are part of a commons.**

Recreation is a problem of the commons—overcrowding of national parks, wilderness areas, and other nature–recreation areas. An example is Voyageurs National Park in northern Minnesota. The park, within North America's boreal-forest biome, includes many lakes and islands and is an excellent place for fishing, hiking, canoeing, and viewing wildlife. Before the area became a national park, it was used for motorboating, snowmobiling, and hunting; a number of people in the region made their living from tourism based on these kinds of recreation. Some environmental groups argue that Voyageurs National Park is ecologically fragile and needs to be legally designated a U.S. wilderness area to protect it from overuse and from the adverse effects of motorized vehicles. Others argue that the nearby million-acre Boundary Waters Canoe Area provides ample wilderness, that Voyageurs can withstand a moderate level of hunting and motorized transportation, and that these uses should be allowed. At the heart of this conflict is the problem of the commons, which in this case can be summed up as follows:

- What is the appropriate public use of public lands?
- Should all public lands be open to all public uses?
- Should some public lands be protected from people?

FIGURE 3.4 **Voyageurs National Park** in northern Minnesota has many lakes well suited to recreational boating. But what kind of boating—what kinds of motors, what size boats—is a long-running controversy. In a commons such as a national park, these are the kinds of conflicts that arise over intangible value (such as scenic beauty) and tangible value (such as the opportunity for boat owners and guides to make a living).

At present, the United States has a policy of different uses for different lands. In general, national parks are open to the public for many kinds of recreation, whereas designated wildernesses have restricted visitorship and kinds of uses (Figure 3.4).

3.4 Low Growth Rate and Therefore Low Profit as a Factor in Exploitation

Another reason individuals tend to overexploit natural resources held in common is the low growth rate of many biological resources.[9] For example, one way to view whales economically is to consider them solely in terms of whale oil. Whale oil, a marketable product, and the whales alive in the ocean, can be thought of as the capital investment of the industry.

From an economic point of view, how can whalers get the best return on their capital? Keeping in mind that whale populations, like other populations, increase only if there are more births than deaths, we will examine two approaches: *resource sustainability* and *maximum profit*. If whalers adopt a simple, one-factor resource-sustainability policy, they will harvest only the net biological productivity each year (the number by which the population increased). Barring disease or disaster, this will maintain the total abundance of whales at its current level and keep the whalers in business indefinitely. If, on the other hand, they choose to simply maximize immediate profit, they will harvest all the whales now, sell the oil, get out of the whaling business, and invest their profits.

Suppose they adopt the first policy. What is the maximum gain they can expect? Whales, like other large, long-lived creatures, reproduce slowly, with each female typically giving birth to a calf every three or four years. Thus, the total net growth of a whale population is likely to be no more than 5% per year and probably more like 3%. This means that if all the oil in the whales in the oceans today represented a value of $100 million, then the most the whalers could expect to take in each year would be no more than 5% of this amount, or $5 million. Until the 2008 economic recession, 5% interest was considered a modest, even poor, rate of return on one's money. And meanwhile the whalers would have to pay for the upkeep of ships and other equipment, salaries of employees, and interest on loans—all of which would decrease profit.

However, if whalers opted for the second policy and harvested all the whales, they could invest the money from the oil. Although investment income varies, even a conservative return on their investment of $100 million would likely yield millions of dollars annually, and since they would no longer be hunting whales, this would be clear profit, without the costs of paying a crew, maintaining ships, buying fuel, marketing the oil, and so on.

Clearly, if one considers only direct profit, it makes sense to adopt the second policy: Harvest all the whales, invest the money, and relax. And this seems to have been the case for those who hunted bowhead whales in the 19th and early 20th centuries (Figures 3.5 and 3.6).[10] Whales simply are not a highly profitable long-term investment under the resource-sustainability policy. From a tangible economic perspective, without even getting into the intangible ethical and environmental concerns, it is no wonder that there are fewer and fewer whaling companies and that companies left the whaling business when their ships became old and inefficient. Few nations support whaling; those that do have stayed with whaling for cultural reasons. For example, whaling is important to the

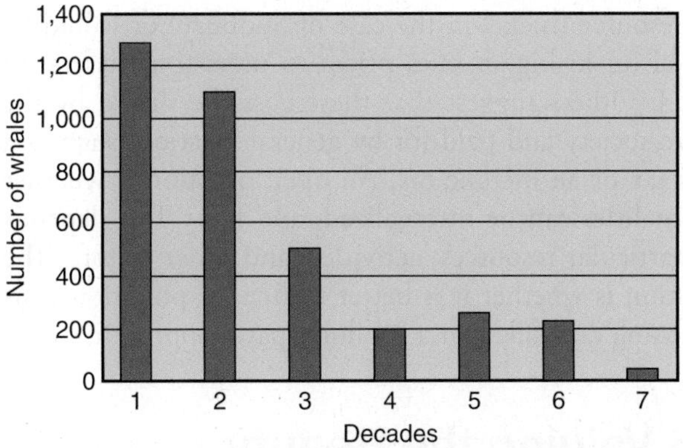

FIGURE 3.5 Bowhead whales caught and killed by Yankee whalers from 1849 to 1914. The number killed, shown for each decade, declined rapidly, indicating that the whale population was unable to reproduce at a rate that could replace the initial large catches, yet the whalers kept killing at a nonsustainable rate, as economics would predict. (*Source:* Redrawn from J.R. Bockstoce and D.B. Botkin, *The Historical Status and Reduction of the Western Arctic Bowhead Whale* (Balaenamysticetus) *Population by the Pelagic Whaling Industry, 1849–1914.* First report to the U.S. National Marine Fisheries Service by the Old Dartmouth Historical Society, 1980; and J.R. Bockstoce, D.B. Botkin, A. Philip, B.W. Collins, and J.C. George. The geographic distribution of bowhead whales in the Bering, Churchi, and Beaufort seas: Evidence from whaleship records, 1849–1914, *Marine Fisheries Review* 67(3) [2007]:1–43.)

Courtesy of the New Bedford Whaling Museum

FIGURE 3.6 Bowhead whale baleen (their modified teeth) on a dock in San Francisco in the late 19th century, when these flexible plates were used for women's corsets and other uses where strength and flexibility were important. Baleen and whale oil were the two commercial products obtained from bowheads. When baleen was replaced by new forms of steel, the baleen market disappeared. Commercial bowhead whale hunting ended with the beginning of World War I. However, whale oil remained listed as a strategic material by the U.S. Department of Defense for decades afterward because of its lubricating qualities.

Eskimo culture, so some harvest of bowheads takes place in Alaska; and whale meat is a traditional Japanese and Norwegian food, so these countries continue to harvest whales for this reason.

Scarcity Affects Economic Value

The relative scarcity of a necessary resource is another factor to consider in resource use, because this affects its value and therefore its price. For example, if a whaler lived on an isolated island where whales were the only food and he had no communication with other people, then his primary interest in whales would be as a way for him to stay alive. He couldn't choose to sell off all whales to maximize profit, since he would have no one to sell them to. He might harvest at a rate that would maintain the whale population. Or, if he estimated that his own life expectancy was only about ten years, he might decide that he could take a chance on consuming whales beyond their ability to reproduce. Cutting it close to the line, he might try to harvest whales at a rate that would cause them to become extinct at the same time that he would. "You can't take it with you" would be his attitude.

If ships began to land regularly at this island, he could leave, or he could trade and begin to benefit from some of the future value of whales. If ocean property rights existed, so that he could "own" the whales that lived within a certain distance of his island, then he might consider the economic value of owning this right to the whales. He could sell rights to future whales, or mortgage against them, and thus reap the benefits during his lifetime from whales that could be caught after his death. Causing the extinction of whales would not be necessary.

From this example, we see that policies that seem ethically good may not be the most profitable for an individual. We must think beyond the immediate, direct economic advantages of harvesting a resource. Economic analysis clarifies how an environmental resource can be and even perhaps should be used—what is perceived as its intrinsic value and therefore its price. And this brings us to the question of externalities.

3.5 Externalities

One gap in our thinking about whales, an environmental economist would say, is that we must be concerned with externalities in whaling. An **externality**, also called an **indirect cost**, is often not recognized by producers as part of their costs and benefits, and therefore not normally accounted for in their cost-revenue analyses.[9] Put simply, externalities are costs or benefits that don't show up in the price tag.[11] In the case of whaling, externalities include the loss of revenue to whale-watching tourist boats and the loss of the ecological role that whales play in marine ecosystems. Classically, economists agree that the only way for a consumer to make a rational decision is by comparing the true costs—including externalities—against the benefits the consumer seeks.

Air and water pollution provide good examples of externalities. For many years large smelters have produced nickel from ore at Sudbury, Ontario. The smelters have had serious environmental effects, including destroying forests and damaging lakes over a surprisingly large area. According to the Ontario Ministry of the Environment and Energy, 7,000 lakes encompassing more than 17,000 km^2 have been affected.[12] And this activity continues. In 2012 a new $1.8 billion (Canadian) smelter was announced.

Traditionally, the economic costs associated with producing commercially usable nickel from an ore were only the **direct costs**—that is, those borne by the producer in obtaining, processing, and distributing a product—passed directly on to the user or purchaser. In this case, direct costs include purchasing the ore, buying energy to run the smelter, building the plant, and paying employees. The externalities, however, include costs associated with degradation of the environment from the plant's emissions. For example, prior to implementation of pollution control, the Sudbury smelter destroyed vegetation over a wide area, which led to increased erosion. Although air emissions from smelters have been substantially reduced and restoration efforts have initiated a slow recovery of the area, pollution remains a problem, and total recovery of the local ecosystem may take a century or more.[12] There are costs associated with the value of trees and soil, with restoring vegetation and land to a productive state.

Problem number one: What is the true cost of clean air over Sudbury? Economists say that there is plenty of disagreement about the cost, but that everyone agrees that it is larger than zero. In spite of this, clean air and water are in the large traded and dealt with in today's world as if their value were zero. (As we will see in Chapter 20, this is no longer the case in all situations for greenhouse gases.) How do we get the value of clean air and water and other environmental benefits to be recognized socially as greater than zero? In some cases, we can determine the dollar value. We can evaluate water resources for power or other uses based on the amount of flow of the rivers and the quantity of water storage in rivers and lakes. We can evaluate forest resources based on the number, types, and sizes of trees and their subsequent yield of lumber. We can evaluate mineral resources by estimating how many metric tons of economically valuable mineral material exist at particular locations. Quantitative evaluation of the tangible natural resources—such as air, water, forests, and minerals—prior to development or management of a particular area is now standard procedure.

Problem number two: Who should bear the burden of these costs? Some suggest that environmental and ecological costs should be included in costs of production through taxation or fees. The expense would be borne by the corporation that benefits directly from the sale of the resource (nickel in the case of Sudbury) or would be passed on in higher sales prices to users (purchasers) of nickel. Others suggest that these costs be shared by the entire society and paid for by general taxation, such as a sales tax or an income tax. An open question is whether externalities can be internalized, and how. This depends on particular resources, activities, and other factors. The question is whether it is better to finance pollution control using tax dollars or a "polluter pays" approach.

3.6 Valuing the Beauty of Nature

The beauty of nature—technically termed *landscape aesthetics*—is an environmental intangible that has probably been important to people as long as our species has existed. We know it has been important since people have written, because the beauty of nature is a continuous theme in literature and art. Once again, as with forests cleaning the air, we face the difficult question: How do we arrive at a price for the beauty of nature? The problem is even more complicated because among the kinds of scenery we enjoy are many modified by people. For example, the open farm fields in Vermont improved the view of the mountains and forests in the distance, so when farming declined in the 1960s, the state began to provide tax incentives for farmers to keep their fields open and thereby help the tourism economy. As another example, about 5 million people visit the Grand Canyon every year, to see the view, hike through it, or take mule rides down into it. Would you consider the scenery of the Grand Canyon an externality? (Figure 3.7).

One of the perplexing problems of aesthetic evaluation is personal preference. One person may appreciate a high mountain meadow far removed from civilization; a second

Photo courtesy of D.B. Botkin

FIGURE 3.7 How much is a beautiful scene worth? Consider, for example, this view of the Grand Canyon from the north rim. Is landscape beauty an externality? People spend a good deal of money to reach this location and see this view.

person may prefer visiting with others on a patio at a trailhead lodge; a third may most want to visit a city park; and a fourth may seek the austere beauty of a desert. If we are going to consider aesthetic factors in environmental analysis, we must develop a method of aesthetic evaluation that allows for individual differences—another yet unsolved topic.

One way the intangible value of landscape beauty is determined is by how much people are willing to pay for it and by how high a price people will pay for land with a beautiful view, compared with the price of land without a view. As apartment dwellers in any big city will tell you, the view makes a big difference in the price of their unit. For example, in mid-2009, *The New York Times* listed two apartments, both with two bedrooms, for sale in the same section of Manhattan, one without a view for $850,000 and one with a wonderful view of the Hudson River estuary for $1,315,000 (Figure 3.8).

Some philosophers suggest that there are specific characteristics of landscape beauty and that we can use these characteristics to help us set the value of intangibles.

Photonica/Getty Images, Inc.

FIGURE 3.8 **A view from a New York City apartment greatly increases its price compared with similar apartments without a view.**

Some suggest that the three key elements of landscape beauty are coherence, complexity, and mystery—mystery in the form of something seen in part but not completely, or not completely explained. Other philosophers suggest that the primary aesthetic qualities are unity, vividness, and variety.[13] *Unity* refers to the quality or wholeness of the perceived landscape—not as an assemblage but as a single, harmonious unit. *Vividness* refers to that quality of a landscape that makes a scene visually striking; it is related to intensity, novelty, and clarity. Variety refers to the amount of different things illustrated. For example, forest trees only will appear to most people less fascinating than trees with an animal in it, or trees around a pond. People differ in what they believe are the key qualities of landscape beauty, but again, almost everyone would agree that the value is greater than zero.

3.7 How Is the Future Valued?

The earlier discussion of whaling—explaining why whalers may not find it advantageous to conserve whales—reminds us of the old saying "A bird in the hand is worth two in the bush." In economic terms, a profit now is worth much more than a profit in the future. This brings up the economic concept important to environmental issues: the future value of something compared with its present value.

Suppose you are dying of thirst in a desert and meet two people. One offers to sell you a glass of water now, and the other offers to sell you a glass of water if you can be at the well tomorrow. How much is each glass worth? If you believe you will die today without water, the glass of water today is worth all your money, and the glass tomorrow is worth nothing. If you believe you can live another day without water, but will die in two days, you might place more value on tomorrow's glass than on today's, since it will gain you an extra day—three rather than two.

In practice, things are rarely so simple and distinct. We know we aren't going to live forever, so we tend to value personal wealth and goods more if they are available now than if they are promised in the future. This evaluation is made more complex, however, because we are accustomed to thinking of the future—to planning a nest egg for retirement or for our children. Indeed, many people today argue that we have a debt to future generations and must leave the environment in at least as good a condition as we found it. These people would argue that the future environment is not to be valued less than the present one (Figure 3.9).

Since the future existence of whales and other endangered species has value to those interested in biological conservation, the question arises: Can we place a dollar value on the *future* existence of anything? The future value

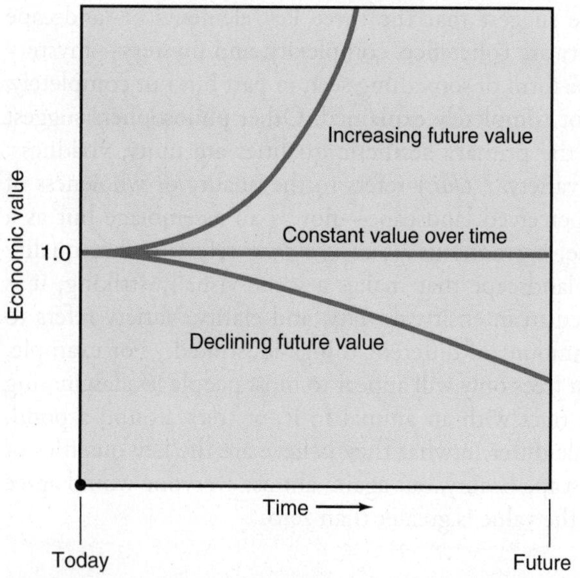

FIGURE 3.9 Economic value as a function of time—a way of comparing the value of having something now with the value of having it in the future. A negative value means that there is more value attached to having something in the present than having it in the future. A positive value means that there is more value attached to having something in the future than having it today.

depends on how far into the future you are talking about. The future times associated with some important global environmental topics, such as stratospheric ozone depletion and global warming, extend longer than a century. This is because chlorofluorocarbons (CFCs) have such a long residence time in the atmosphere and because of the time necessary to realize benefits from changing energy policy to offset global climate change.

Another aspect of future versus present value is that spending on the environment can be viewed as diverting resources from alternative forms of productive investment that will be of benefit to future generations. (This assumes that spending on the environment is not itself a productive investment.)

A further issue is that as we get wealthier, the value we place on many environmental assets (such as wilderness areas) increases dramatically. Thus, if society continues to grow in wealth over the next century as it has over the past century, the environment will be worth far more to our great-grandchildren than it was to our great-grandparents, at least in terms of willingness to pay to protect it. The implication—which complicates this topic even more—is that conserving resources and environment for the future is tantamount to taking from the poor today and giving to the possibly rich in the future. To what extent should we ask the average American today to sacrifice now for richer great-great-grandchildren? How can we know the future usefulness of today's sacrifices? Put another way, what would you have liked your ancestors in 1900 to have sacrificed for our benefit today? Should they have increased research and development on electric transportation? Should they have saved more tall-grass prairie or restricted whaling?

Economists observe that it is an open question whether something promised in the future will have more value in the years to come than it does today. Future economic value is difficult enough to predict because it is affected by how future consumers view consumption. But if, in addition, something has greater value in the future than it does today, then that leads to the mathematical conclusion that in the very long run, the future value will become infinite, which of course is impossible. So in terms of the future, the basic issues are (1) that since we are so much richer and better off than our ancestors, their sacrificing for us might have been inappropriate; and (2) even if they had wanted to sacrifice, how would they have known what sacrifices would be important to us?

As a general rule, one answer to the thorny questions about future value is: Do not throw away or destroy something that cannot be replaced if you are not sure of its future value. For example, if we do not fully understand the value of the wild relatives of potatoes that grow in Peru but do know that their genetic diversity might be helpful in developing future strains of potatoes, then we ought to preserve those wild strains.

3.8 Risk-Benefit Analysis

Death is the fate of all individuals, and almost every activity in life involves some risk of death or injury. How, then, do we place a value on saving a life by reducing the level of a pollutant? This question raises another important area of environmental economics: **risk-benefit analysis**, in which the riskiness of a present action in terms of its possible outcomes is weighed against the benefit, or value, of the action. Here, too, difficulties arise.

With some activities, the relative risk is clear. It is much more dangerous to stand in the middle of a busy highway than to stand on the sidewalk, and hang gliding has a much higher mortality rate than hiking. The effects of pollutants are often more subtle, so the risks are harder to pinpoint and quantify. Table 3.1 gives the lifetime risk of death associated with a variety of activities and some forms of pollution. In looking at the table, remember that since the ultimate fate of everyone is death, the total lifetime risk of death from all causes must be 100%. So if you are going to die of something and you smoke a pack of cigarettes a day, you have 8 chances in 100 that your death will be a result of smoking. At the same time, your risk of death from driving an automobile is 1 in 100. Risk tells you the chance of an event but not its timing. So you

Table 3.1 RISK OF DEATH FROM VARIOUS CAUSES

CAUSE	RESULT	RISK OF DEATH (PER LIFETIME)	LIFETIME RISK OF DEATH (%)	COMMENT
Cigarette smoking (pack a day)	Cancer, effect on heart, lungs, etc.	8 in 100	8.0%	
Breathing radon-containing air in the home	Cancer	1 in 100	1.0%	Naturally occurring
Automobile driving		1 in 100	1.0%	
Death from a fall		4 in 1,000	0.4%	
Drowning		3 in 1,000	0.3%	
Fire		3 in 1,000	0.3%	
Artificial chemicals in the home	Cancer	2 in 1,000	0.2%	Paints, cleaning agents, pesticides
Sunlight exposure	Melanoma	2 in 1,000	0.2%	Of those exposed to sunlight
Electrocution		4 in 10,000	0.04%	
Air outdoors in an industrial area		1 in 10,000	0.01%	
Artificial chemicals in water		1 in 100,000	0.001%	
Artificial chemicals in foods		less than 1 in 100,000	0.001%	
Airplane passenger (commercial airline)		less than 1 in 1,000,000	0.00010%	

Source: From *Guide to Environmental Risk* (1991), U.S. EPA Region 5 Publication Number 905/91/017.

might smoke all you want and die from the automobile accident first.

One of the striking things about Table 3.1 is that death from outdoor environmental pollution is comparatively low—even compared to the risks of drowning or of dying in a fire. This suggests that the primary reason we value lowering air pollution is not to lengthen our lives but to improve the quality of our lives. Considering people's great interest in air pollution today, the quality of life must be much more important than is generally recognized. We are willing to spend money on improving that quality rather than just extending our lives. Another striking observation in this table is that natural *indoor* air pollution is much more deadly than most outdoor air pollution—unless, of course, you live at a toxic-waste facility.

It is commonly believed that future discoveries will help to decrease various risks, perhaps eventually allowing us to approach a zero-risk environment. But complete elimination of risk is generally either technologically impossible or prohibitively expensive. Societies differ in their views of what constitutes socially, psychologically, and ethically acceptable levels of risk for any cause of death or injury, but we can make some generalizations about the acceptability of various risks. One factor is the number of people affected. Risks that affect a small population (such as employees at nuclear power plants) are usually more acceptable than those that involve all members of a society (such as risk from radioactive fallout).

In addition, novel risks appear to be less acceptable than long-established or natural risks, and society tends to be willing to pay more to reduce such risks. For example, in the late part of the 20th century, France spent about $1 million each year to reduce the likelihood of one air-traffic death but only $30,000 for the same reduction in automobile deaths.[14] Some argue that the greater safety of commercial air travel versus automobile travel is in part due to the relatively novel fear of flying compared with the more ordinary fear of death from a road accident. That is, because the risk is newer to us and thus less acceptable, we are willing to spend more per life to reduce the risk from flying than to reduce the risk from driving.

People's willingness to pay for reducing a risk also varies with how essential and desirable the activity associated with the risk is. For example, many people accept much higher risks for athletic or recreational activities than they would for transportation or employment-related activities (see Table 3.1). People volunteer to climb Mount Everest even though many who have attempted it have died, but the same people could be highly averse to risking death in a train wreck or commercial airplane crash. The risks associated with playing a sport or using transportation are assumed to be inherent in the activity. The risks to human health from pollution may be widespread and linked to a large number of deaths. But although risks from pollution are often unavoidable and unseen, people want a lesser risk from pollution than from, say, driving a car or playing a sport.

In an ethical sense, it is impossible to put a value on a human life. However, it is possible to determine how much people are willing to pay for a certain amount of risk reduction or a certain probability of increased longevity. For example, a study by the Rand Corporation considered measures that would save the lives of heart-attack victims, including increasing ambulance services and initiating pretreatment screening programs. According to the study, which identified the likely cost per life saved and people's willingness to pay, people favored government spending of about $32,000 per life saved, or $1,600 per year of longevity. Although information is incomplete, it is possible to estimate the cost of extending lives in terms of dollars per person per year for various actions (Table 3.1). For example, on the basis of direct effects on human health, it costs more to increase longevity by reducing air pollution than to directly reduce deaths by adding a coronary-ambulance system. Also, a general trend is that costs to reduce risks of death increase very rapidly as the risk is greatly reduced—generally speaking, costs increase exponentially as the average risk of death is decreased. For example, one study showed that reducing carbon monoxide to 15 parts per million in automobile exhaust would cost in the tens of thousands of dollars to increase average longevity by a few days; reducing it further to 3.4 parts per million would cost in the tens of millions of dollars and increase average longevity by just a few minutes.

Such a comparison is useful as a basis for decision making. Clearly, though, when a society chooses to reduce air pollution, many factors beyond the direct, measurable health benefits are considered. Pollution not only directly affects our health but also causes ecological and aesthetic damage, which can indirectly affect human health (see Section 3.4). We might want to choose a slightly higher risk of death in a more pleasant environment rather than increase the chances of living longer in a poor environment—spend money to clean up the air rather than increase ambulance services to reduce deaths from heart attacks.

Comparisons like these may make you uncomfortable. But like it or not, we cannot avoid making choices of this kind. The issue boils down to whether we should improve the quality of life for the living or extend life expectancy regardless of the quality of life.[15]

The degree of risk is an important concept in our legal processes. For example, the U.S. Toxic Substances Control act states that no one may manufacture a new chemical substance or process a chemical substance for a new use without obtaining clearance from the EPA. The act establishes procedures for estimating the hazard to the environment and to human health of any new chemical before its use becomes widespread. The EPA examines the data provided and judges the degree of risk associated with all aspects of the production of the new chemical or process, including extraction of raw materials, manufacturing, distribution, processing, use, and disposal. The chemical can be banned or restricted in either manufacturing or use if the evidence suggests that it will pose an unreasonable risk to human health or to the environment.

But what is unreasonable?[16] This question brings us back to Table 3.1 and makes us realize that deciding what is "unreasonable" involves judgments about the quality of life as well as the risk of death. The level of acceptable pollution (and thus risk) is a social-economic-environmental trade-off. Moreover, the level of acceptable risk changes over time in society, depending on changes in scientific knowledge, comparison with risks from other causes, the expense of decreasing the risk, and the social and psychological acceptability of the risk.

When adequate data are available, it is possible to take scientific and technological steps to estimate the level of risk and, from this, to estimate the cost of reducing risk and compare the cost with the benefit. However, what constitutes an acceptable risk is more than a scientific or technical issue. The acceptability of a risk involves ethical and psychological attitudes of individuals and society. We must therefore ask several questions: What risk from a particular pollutant is acceptable? How much is a given reduction in risk from that pollutant worth to us? How much will each of us, as individuals or collectively as a society, be willing to pay for a given reduction in that risk?

The answers depend not only on facts but also on societal and personal values. What must also be factored into the equation is that the costs of cleaning up pollutants and polluted areas and the costs of restoration programs can be minimized, or even eliminated, if a recognized pollutant is controlled initially.

CRITICAL THINKING ISSUE

Georges Bank: How Can U.S. Fisheries Be Made Sustainable?

Ocean fishing in Georges Bank—a large, shallow area between Cape Cod, Massachusetts, and Cape Sable Island in Nova Scotia, Canada—illustrates different ways of making a policy work. Both overfishing and pollution have been blamed for the alarming decline in groundfish (cod, haddock, flounder, redfish, pollack, hake) off the northeastern coast of the United States and Canada. Governments' attempts to regulate fishing have generated bitter disputes with fishermen, many of whom contend that restrictions on fishing make them scapegoats for pollution problems. The controversy has become a classic battle between short-term economic interests and long-term environmental concerns.

The oceans outside of national territorial waters are *commons*—open to free use by all—and thus the fish and mammals that live in them are common resources. What is a common resource may change over time, however. The move by many nations to define international waters as beginning 325 km(200 mi) from their coasts has turned some fisheries that used to be completely open common resources into national resources open only to domestic fishermen.

In fisheries, there have been four main management options:

1. Establish total-catch quotas for the entire fishery and allow anybody to fish until the total is reached.
2. Issue a restricted number of licenses but allow each licensed fisherman to catch many fish.
3. Tax the catch (the fish brought in) or the effort (the cost of ships, fuel, and other essential items).
4. Allocate fishing rights—that is, assign each fisherman a transferable and salable quota.

With total-catch quotas, the fishery is closed when the quota is reached. Whales, Pacific halibut, tropical tuna, and anchovies have been regulated in this way. Although the total catch can be regulated in a way that helps the fish, total-catch quotas tend to increase the number of fishermen and encourage them to buy larger and larger boats. The end result is a hardship for fishermen—huge boats usable for only a brief time each year. When Alaska tried this, all of the halibut were caught in a few days, with the result that restaurants no longer had halibut available for most of the year. This undesirable result led to a change in policy: The total-catch approach was replaced by the sale of licenses.

Issues relating to U.S. fisheries are hardly new. In the early 1970s, fishing was pretty open, but in 1977, in response to concerns about overfishing in U.S. waters by foreign factory ships, the U.S. government extended the nation's coastal waters from 19 to 322 km (from 12 to 200 mi). To encourage domestic fishermen, the National Marine Fisheries Service provided loan guarantees for replacing older vessels and equipment with newer boats carrying high-tech equipment for locating fish. During this same period, demand for fish increased as Americans became more concerned about cholesterol in red meat. Consequently, the number of fishing boats, the number of days at sea, and fishing efficiency increased sharply, and 50–60% of the populations of some species were landed each year.

The international battle over Georges Bank led to a consideration by the International Court of Justice in The Hague. This court's 1984 decision intensified competition. Overfishing continued, and in 1992 Canada was forced to suspend all cod fishing to save the stock from complete annihilation. Later that year, Canada prohibited fishing at certain times and in certain areas on Georges Bank, mandated minimum net sizes, and set quotas on the catch.

These measures were intended to cut the fishing effort in half by 1997. A limited number of fishing permits were issued, limiting the number of days at sea and the number of trips for harvesting certain species. High-tech monitoring equipment ensured compliance. Still, things got worse. Recently, portions of Georges Bank were closed indefinitely to fishing for some fish species, including yellowtail, cod, and haddock.

In the spring of 2009, fishermen suggested that the limit on individual fishermen be replaced by a group quota, a variation on Management Option 4 (above). Fishermen would work together in groups called "sectors," and each sector could take a set percentage of the annual catch of one species. This approach is being used elsewhere in U.S. waters. It places fewer restrictions on individual fishermen, such as limiting each one's number of trips or days at sea.

In 2011, the U.S. quota for yellowtail flounder was 18% more than the 2010 quota, suggesting that the National Oceanographic and Atmospheric Administration believes that the abundance of this species has improved.

Recent economic analysis suggests that taxes taking into account the cost of externalities (such as water pollution from motorboat oil) can work to the best advantage of fishermen and fish. Allocating a transferable and salable quota (referred to as Individual Transferable Quotas (ITQs) to each fisherman produces similar results. However, after decades of trying to find a way to regulate fishing so that Georges Bank becomes a sustainable fishery, nothing has worked well. The fisheries remain in trouble.

Critical Thinking Questions

1. Which of the policy options described above attempt to convert the fishing industry from a commons system to private ownership? How might these measures help prevent overfishing? Is it right to institute private ownership of public resources?
2. Think over the choices discussed in this chapter; what policy option do you think has the best chance of sustaining the fisheries on Georges Bank? Explain your answer.
3. What approach to future value (approximately) do each of the following people assume for fish?

   Fisherman: If you don't get it now, someone else will.

   Fisheries manager: By sacrificing now, we can do something to protect fish stocks.
4. Develop a list of the environmental and economic advantages and disadvantages of Individual Transferable Quotas (ITQs). Would you support instituting ITQs in New England? Explain why or why not.
5. Do you think it is possible to reconcile economic and environmental interests in the case of the New England fishing industry? If so, how? If not, why not?

SUMMARY

- Economic analysis can help us understand why environmental resources have been poorly conserved in the past and how we might more effectively achieve conservation in the future.
- Economic analysis applies to three different kinds of environmental issues: the use of desirable resources (fish in the ocean, oil in the ground, forests on the land); the minimization of pollution; and ownership of land, resources, and rights to action.
- Resources may be common property or privately controlled. The kind of ownership affects the methods available to achieve an environmental goal. There is a tendency to overexploit a common-property resource and to harvest to extinction nonessential resources whose innate growth rate is low, as suggested in Hardin's tragedy of the commons.
- Future worth compared with present worth can be an important determinant of the level of exploitation.
- The relation between risk and benefit affects our willingness to pay for an environmental good.
- Evaluation of environmental intangibles, such as landscape aesthetics, is becoming more common in environmental analysis. Such evaluation can be used to balance the more traditional economic evaluation and to help separate facts from emotion in complex environmental problems.
- Societal methods to achieve an environmental goal include moral suasion, direct controls, market processes, and government investment. Many kinds of controls have been applied to pollution and the use of desirable resources.
- People sometimes are not interested in sustaining an environmental resource from which they make a living. When the goal is simply to maximize profits, it is sometimes a rational decision to liquidate an environmental resource and put the money gained into a bank or another investment. To avoid such liquidation, we need to understand economic externalities and intangible values.
- How do we value the environment, and when can we attach a monetary value to the benefits and costs of environmental actions? People are intimately involved with nature. While we seek rational methods to put a value on nature, the values we choose often derive from intangible benefits, such as an appreciation of the beauty of nature.
- One of the central questions of environmental economics concerns how to develop equivalent economic valuation for tangible and intangible factors. For example, how can we compare the value of timber with the beauty people attach to the scenery, with trees intact? How can we compare the value of a dam that provides irrigation water and electrical power on the Columbia River with the scenery without the dam and the salmon that could inhabit that river?

REEXAMINING THEMES AND ISSUES

Sean Randall/Getty Images, Inc.

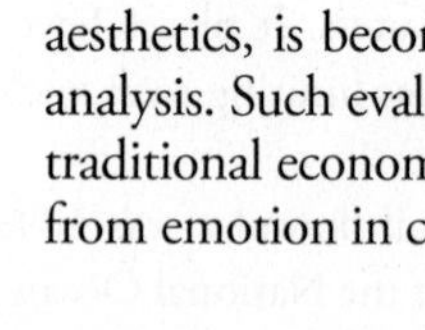

HUMAN POPULATION

The tragedy of the commons will worsen as human population density increases because more and more individuals will seek personal gain at the expense of community values. For example, more and more individuals will try to make a living from harvesting natural resources. How people can use resources while at the same time conserving them requires an understanding of environmental economics.

© Biletskiy_Evgeniy/iStockphoto

SUSTAINABILITY

From this chapter, we learn why people sometimes are not interested in sustaining an environmental resource from which they make a living. When the goal is simply to maximize profits, it is sometimes a rational decision to liquidate an environmental resource and put the money gained into a bank or another investment, to avoid such liquidation, we need to understand economic externalities and intangible values.

© Anton Balazh 2011/iStockphoto

GLOBAL PERSPECTIVE

Solutions to global environmental issues, such as global warming, require that we understand the different economic interests of developed and developing nations. These can lead to different economic policies and different valuation of global environmental issues.

ssguy/ShutterStock

URBAN WORLD

The tragedy of the commons began with grazing rights in small villages. As the world becomes increasingly urbanized, the pressure to use public lands for private economic gain is likely to increase. An understanding of environmental economics can help us find solutions to urban environmental problems.

B2M Productions/Getty Images, Inc.

PEOPLE AND NATURE

This chapter brings us to the heart of the matter: How do we value the environment, and when can we attach a monetary value to the benefits and costs of environmental actions? People are intimately involved with nature. While we seek rational methods to put a value on nature, the values we choose often derive from intangible benefits, such as an appreciation of the beauty of nature.

George Doyle/Getty Images, Inc.

SCIENCE AND VALUES

One of the central questions of environmental economics concerns how to develop equivalent economic valuation for tangible and intangible factors. For example, how can we compare the value of timber with the beauty people attach to the scenery, trees intact? How can we compare the value of a dam that provides irrigation water and electrical power on the Columbia River with the scenery without the dam, and the salmon that could inhabit that river?

KEY TERMS

commons 47
direct costs 50
environmental economics 45
externality 49
indirect cost 49
intangible factor 45
natural capital 46
policy instruments 45
public-service functions 46
risk-benefit analysis 52
tangible factor 45

STUDY QUESTIONS

1. Why is an understanding of economics important in environmental science?

2. What is meant by the term *the tragedy of the commons?* Which of the following are results of this tragedy?

   (a) The fate of the California condor
   (b) The fate of the gray whale
   (c) The high price of walnut wood used in furniture

3. Describe what is meant by risk-benefit analysis.

4. Cherry and walnut are valuable woods used to make fine furniture. Basing your decision on the information in the following table, which would you invest in? (*Hint:* Refer to the discussion of whales in this chapter.)
   (a) A cherry plantation
   (b) A walnut plantation
   (c) A mixed stand of both species
   (d) An unmanaged woodland where you see some cherry and walnut growing

Species	*Longevity*	*Maximum Size*	*Maximum Value*
Walnut	400 years	1 m	$15,000/tree
Cherry	100 years	1 m	$10,000/tree

5. Bird flu is spread in part by migrating wild birds. How would you put a value on (a) the continued existence of one species of these wild birds; (b) domestic chickens important for food but also a major source of the disease; (c) control of the disease for human health? What relative value would you place on each (that is, which is most important and which least)? To what extent would an economic analysis enter into your valuation?

6. Which of the following are intangible resources? Which are tangible?
   (a) The view of Mount Wilson in California
   (b) A road to the top of Mount Wilson
   (c) Porpoises in the ocean
   (d) Tuna in the ocean
   (e) Clean air

7. What kind of future value is implied by the statement "Extinction is forever"? Discuss how we might approach providing an economic analysis for extinction.

8. Which of the following can be thought of as commons in the sense meant by Garrett Hardin? Explain your choice.
   (a) Tuna fisheries in the open ocean
   (b) Catfish in artificial freshwater ponds
   (c) Grizzly bears in Yellowstone National Park
   (d) A view of Central Park in New York City
   (e) Air over Central Park in New York City

FURTHER READING

Daly, H.E., and J. Farley, *Ecological Economics: Principles and Applications* (Washington, DC: Island Press, 2003). A discussion of an interdisciplinary approach to the economics of environment.

Goodstein, E.S., *Economics and the Environment*, 6th ed. (New York: Wiley, 2010). A modern summary of the topic.

Hanley, N. J. Shogren, and B. White, *Environmental Economics in Theory and Practice*, 2nd ed., paperback (New York: Macmillan, 2007). Another modern summary of the topic.

Hardin, G., Tragedy of the commons, *Science* 162:1243–1248, 1968. One of the most cited papers in science and social science; this classic work outlines the differences between individual interest and the common good.

Samuelson, P. and W. Nordhaus. ***Economics***, 19th ed. (New York,: McGraw-Hill/Irwin, 2009). Probably the most famous textbook on economics.

NOTES

1. Tarlock, D. 1994. The nonequilibrium paradigm in ecology and the partial unraveling of environmental law. *Loyola of Los Angeles Law Review* 2(1121):1–25.
2. Senate Resolution 580 2007. Recognizing the importance of pollinators to ecosystem health and agriculture in the United States and the value of partnership efforts to increase awareness about pollinators and support (Agreed to by Senate). http://www.fws.gov/pollinators/pdfs/pollinatorweekres580.pdf.
3. Morse, R. A., and N. W. Calderone. 2000 The value of honey bees as pollinators of U.S. crops in 2000. *Pollination 2000.* Ithaca, NY: Cornell University Press.
4. Marks, R. (2005). Native Pollinators: Fish and Wildlife Habitat Management Leaflet 34. U.S. Department of Agriculture. Washington, DC: USDA .
5. Hills, S. 2008. Study sheds light on bee decline threatening crops. Food Navigator—USA.com, August 20, 2008. http://www.foodnavigator-usa.com/Business/Study-sheds-light-on-bee-decline-threatening-crops. Accessed August 27, 2012.
6. Costanza, R., et al. 1997. The value of the world's ecosystem services and natural capital. *Nature* 387:253–260.
7. Hardin, G. 1968. The tragedy of the commons. *Science* 162:1243–1248.

8. State of Vermont. http://www.anr.state.vt.us/air/htm/woodstoverebate.htm. Accessed August 25, 2012.

9. Clark, C.W. 1973. The economics of overexploitation. *Science* 181:630–634.

10. Bockstoce, J.R., D.B. Botkin, A. Philp, B.W. Collins, and J.C. George. 2007. The geographic distribution of bowhead whales in the Bering, Chukchi, and Beaufort seas: Evidence from whaleship records, 1849–1914. 2007 *Marine Fisheries Review* 67(3):1–43.

11. Freudenburg, W.R. 2004. Personal communication.

12. Gunn, J.M., ed. 1995. *Restoration and Recovery of an Industrial Region: Progress in Restoring the Smelter-damaged Landscape near Sudbury*, Canada. New York: Springer-Verlag.

13. Litton, R.B. 1972. Aesthetic dimensions of the landscape. In J.V. Krutilla, ed. *Natural Environments*. Baltimore, MD: Johns Hopkins University Press.

14. Schwing, R.C. 1979. Longevity and benefits and costs of reducing various risks. *Technological Forecasting and Social Change* 13:333–345.

15. Gori, G.B. 1980. The regulation of carcinogenic hazards. *Science* 208:256–261.

16. Cairns, J., Jr. 1980. Estimating hazard. *BioScience* 20: 101–107.

CHAPTER 4

The Big Picture: Systems of Change

LEARNING OBJECTIVES

In this book we discuss a wide range of phenomena. One thing that links them is that they are all part of complex systems. Systems have well-defined properties. Understanding these properties, common to so much of the environment, guides our way to achieving an understanding of all aspects of environmental science. Changes in systems may occur naturally or may be induced by people, but a key to understanding these systems is that change in them is natural. After reading this chapter, you should be able to . . .

- **Summarize how the study of systems and rates of change is linked to solutions of environmental problems**
- **Compare and contrast positive and negative feedback and discuss how they are important to systems**
- **Differentiate between open and closed systems, static systems, and dynamic systems**
- **Delineate potential problems associated with the concept of average residence time as calculated in this chapter**
- **Use the principle of uniformitarianism in a discussion of how we might anticipate future environmental changes**
- **Analyze the principle of environmental unity in terms of why it is important in studying environmental problems**
- **Synthesize helpful ways to think about systems when trying to solve environmental problems that arise from complex natural systems**
- **Discuss what a stable system is and how this idea relates to the prescientific idea of a balance of nature**

Shot of elephants on a plain with Mt. Kilimanjaro in the background.

Danita Delimont/Gallo Images/Getty Images

CASE STUDY

Amboseli National Reserve: A Story of Change

Amboseli National Reserve in southern Kenya is home to the Maasai people who are nomadic some of the time and raise cattle. The reserve is also a major tourist destination where people from around the world can experience Africa and wild animals, such as lions and elephants. Today, environmental change and the future of tourism are being threatened in the area. We will consider long-term change and the more recent management of lions that may result in their local extinction.

Environmental change is often caused by a complex web of interactions among living things and between living things and their environment. In seeking to determine what caused a particular change, the most obvious answer may not be the right answer. Amboseli National Reserve is a case in point. In the short span of a few decades, this reserve, located at the foot of Mount Kilimanjaro (Figure 4.1), underwent a significant environmental change. An understanding of physical, biological, and human-use factors—and how these factors are linked—is needed to explain what happened.

Before the mid-1950s, fever-tree woodlands—mostly acacia trees and associated grasses and shrubs—dominated the land and provided habitat for mammals that lived in these open woodlands, such as kudu, baboons, vervet monkeys, leopards, and impalas. Then, beginning in the 1950s and accelerating in the 1960s, these woodlands disappeared and were replaced by short grass and brush, which provided habitat for typical plains animals, such as zebras and wildebeest. Since the mid-1970s, Amboseli has remained a grassland with scattered brush and few trees.

Loss of the woodland habitat was initially blamed on overgrazing of cattle by the Maasai people (Figure 4.2) and damage to the trees from elephants (Figure 4.3). Environmental scientists eventually rejected these hypotheses as the main causes of the environmental change. Their careful work showed that changes in rainfall and soils were the primary culprits, rather than people or elephants.[1, 2] How did they arrive at this explanation?

During recent decades, the mean daily temperature rose dramatically; annual rainfall increased but continued to vary from year to year by a factor of four, though with no regular pattern.[1, 2] Increased rainfall is generally associated with an increased abundance of trees, unlike what happened at Amboseli.

Why did scientists reject the overgrazing and elephant-damage hypothesis as the sole explanation for changes in

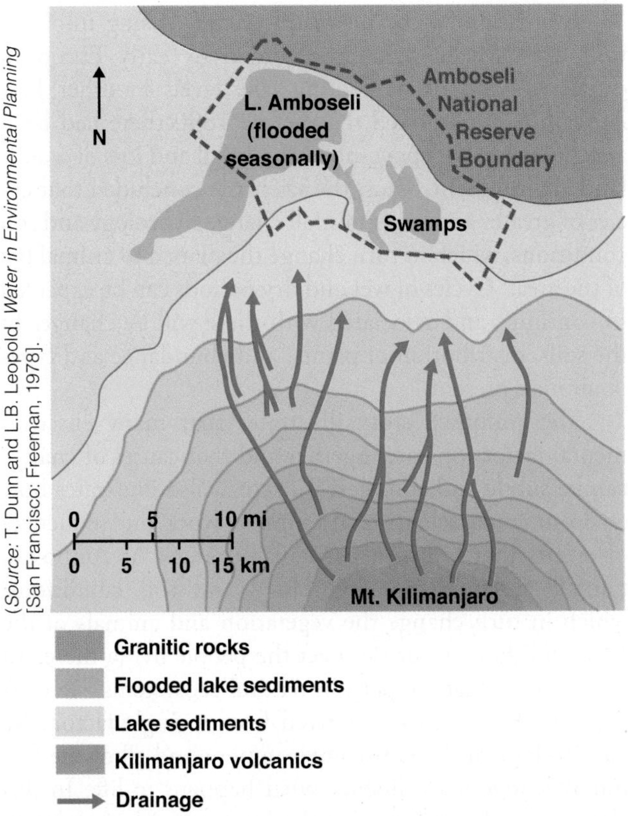

FIGURE 4.1 Generalized geology and landforms of Amboseli National Reserve, southern Kenya, Africa, and Mount Kilimanjaro.

FIGURE 4.2 Maasai people grazing cattle in Amboseli National Reserve, Kenya. Grazing was prematurely blamed for loss of fever-tree woodlands.

FIGURE 4.3 **Elephant feeding on a yellow-bark acacia tree.** Elephant damage to trees is considered a factor in loss of woodland habitat in Amboseli National Reserve. However, elephants probably play a relatively minor role compared with oscillations in climate and groundwater conditions.

Amboseli? Investigators were surprised to note that most dead trees were in an area that had been free of cattle since 1961, which was before the major decline in the woodland environment. Furthermore, some of the woodlands that suffered the least decline had the highest density of people and cattle. These observations suggested that overgrazing by cattle was not responsible for loss of the trees.

Elephant damage was thought to be a major factor because elephants had stripped bark from more than 83% of the trees in some areas and had pushed over some younger, smaller trees. However, researchers concluded that elephants played only a secondary role in changing the habitat. As the density of fever trees and other woodland plants decreased, the incidence of damage caused by elephants increased. In other words, elephant damage interacted with some other, primary factor in changing the habitat.[1]

Figure 4.1 shows the boundary of the reserve and the major geologic units. The park is centered on an ancient lakebed, remnants of which include the seasonally flooded Lake Amboseli and some swampland. Mount Kilimanjaro is a well-known volcano, composed of alternating layers of volcanic rock and ash deposits. Rainfall that reaches the slopes of Mount Kilimanjaro infiltrates the volcanic material (becomes groundwater) and moves slowly down the slopes to saturate the ancient lakebed, eventually emerging at springs in the swampy, seasonally flooded land. The groundwater becomes saline (salty) as it percolates through the lakebed, since the salt stored in the lakebed sediments dissolves easily when the sediments are wet.

Because a lot of land has been transformed to agricultural uses, the slopes of Mount Kilimanjaro above Amboseli have less forest cover than they did 25 years ago. The loss of trees exposed dark soils that absorb solar energy, and this could cause local warming and drier conditions. In addition, there had been a significant decrease in snow and ice cover on the high slopes and summit of the mountain. Snow and ice reflect sunlight. As snow and ice decrease and dark rock is exposed, more solar energy is absorbed at the surface, warming it. Therefore, decreased snow and ice might cause some local warming.[3]

Research on rainfall, groundwater history, and soils suggested that the area is very sensitive to changing amounts of rainfall. During dry periods, the salty groundwater sinks lower into the earth, and the soil near the surface has a relatively low salt content. The fever trees grow well in the non-salty soil. During wet periods, the groundwater rises closer to the surface, bringing with it salt, which invades the root zones of trees and kills them. The groundwater level rose as much as 3.5 m (11.4 ft) in response to unusually wet years in the 1960s. Analysis of the soils confirmed that the tree stands that suffered the most damage were those growing in highly saline soils. As the trees died, they were replaced by salt-tolerant grasses and low brush.[1, 2]

Evaluation of the historical record—using information from Maasai herders recorded by early European explorers—and of fluctuating lake levels in other East African lakes suggested that before 1890 there had been another period of above-normal rainfall and loss of woodland environment. Thus, the scientists concluded that cycles of greater and lesser rainfall change hydrology and soil conditions, which in turn change the plant and animal life of the area.[1] Cycles of wet and dry periods can be expected to continue, and associated with these will be changes in the soils, distribution of plants, and abundance and types of animals present.[1]

The Amboseli story illustrates that many environmental factors operate together and that causes of change can be subtle and complex. The story also illustrates how environmental scientists attempt to work out sequences of events that follow a particular change. At Amboseli, rainfall cycles change hydrology and soil conditions, which in turn change the vegetation and animals of the area, and these in turn impact the people living there. To understand what happens in natural ecosystems, we can't just look for an answer derived from a single factor. We have to look at the entire environment and all of the factors that together influence what happens to life. In this chapter, we discuss some of the fundamental concepts of studying environmental change and systems.

4.1 Basic Systems Concepts

A **system** is a set of components, or parts, that function together as a whole. A single organism, such as your body, is a system, as are a sewage-treatment plant, a city, and a river. On a much different scale, the entire Earth is a system. In a broader sense, a system is any part of the universe you can isolate in thought (in your brain or on your computer) or, indeed, physically, for the purpose of study. Key systems concepts that we will explain are:

- how a system is connected to the rest of the environment;
- how matter and energy flow between parts of a system;
- whether a system is static or dynamic—whether it changes over time;
- average residence time—how long something stays within a system or part of a system;
- feedback—how the output from a system can affect its inputs; and
- linear and nonlinear flows.

In its relation to the rest of the environment, a system can be open or closed. In an **open system**, some energy or material (solid, liquid, or gas) moves into or out of the system. The ocean is an open system with regard to water because water moves into the ocean from the atmosphere and out of the ocean into the atmosphere. In a **closed system**, no such transfers take place. For our purposes, a **materially closed system** is one in which no matter moves in and out of the system, although energy and information can move across the system's boundaries. Earth is a materially closed system (for all practical purposes).

Systems respond to **inputs** and have **outputs**. For example, think of your body as a complex system and imagine you are hiking in Yellowstone National Park and see a grizzly bear. The sight of the bear is an input. Your body reacts to that input: The adrenaline level in your blood goes up, your heart rate increases, and the hair on your head and arms may rise. Your response—perhaps to move slowly away from the bear—is an output.

Static and Dynamic Systems

A **static system** has a fixed condition and tends to remain in that exact condition. A **dynamic system** changes, often continually, over time. A birthday balloon attached to a pole is a static system in terms of space—it stays in one place. A hot-air balloon is a simple dynamic system in terms of space—it moves in response to the winds, air density, and controls exerted by a pilot (Figure 4.4a and 4.4b).

An important kind of static system is one with **classical stability**. Such a system has a constant condition, and if it is disturbed from that condition, it returns to it once the disturbing factor is removed. The pendulum of an old-fashioned grandfather clock is an example of classical stability. If you push it, the pendulum moves back and forth for a while, but then friction gradually dissipates the energy you just gave it and the pendulum comes to rest exactly where it began. This rest point is known as the **equilibrium** (Figure 4.4c).

We will see that the classic interpretation of populations, species, ecosystems, and Earth's entire biosphere has been to assume that each is a stable, static system. But the more these ecological systems are studied scientifically, the clearer it becomes that these are dynamic systems, always changing *and always requiring change.* An important practical question that keeps arising in many environmental controversies is whether we want to, and should, force ecological systems to be static if and when they are naturally dynamic. You will find this question arising in many of the chapters in this book.

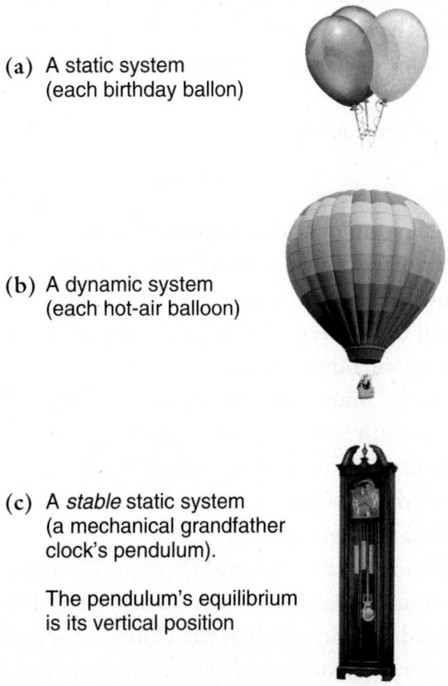

FIGURE 4.4 Static and dynamic systems. (a) *A static system* (each birthday balloon). Balloons are tied down and can't move vertically. **(b)** *A dynamic system* (each hot-air balloon). Hot air generated by a heater fills the balloon with warm air, which is lighter than outside air, so it rises; as air in the balloon cools, the balloon sinks, and winds may move it in any direction. **(c)** *A classical stable static system* (the pendulum on a mechanical grandfather clock). The pendulum's equilibrium is its vertical position. The pendulum will move if you push it or if the clock's mechanism is working. When the source of energy is no longer active (you forgot to wind the clock), the pendulum will come to rest exactly where it started.

Open Systems

With few exceptions, all real systems that we deal with in the environment are open to the flow of matter, energy, and information. (For all practical purposes, as we noted earlier, Earth as a planet is a materially closed system.) An important distinction for open systems is whether they are steady-state or nonsteady-state. In a **steady-state system**, the inputs (of anything of interest) are equal to the outputs, so the amount stored within the system is constant. An idealized example of a steady-state system is a dam and lake into which water enters from a river and out of which water flows. If the water input equals the water output and evaporation is not considered, the water level in the lake does not change, and so, in regard to water, the lake is in a steady state. (Additional characteristics of systems are discussed in A Closer Look 4.1.)

A CLOSER LOOK 4.1

Simple Systems

A simple way to think about a system is to view it as a series of compartments (also called reservoirs, and we will use these terms interchangeably), each of which can store a certain amount of something you are interested in and each of which receives input from other compartments and transfers some of its stored material to other compartments (Figure 4.5a).

The general equation is

$$I = O \pm \Delta S$$

where I is input into a compartment; O is output, and ΔS is change in storage. This equation defines a budget for what is being considered. For example, if your checking account has \$1,000 in it (no interest rate) and you earn \$500 per month at the bookstore, input is \$500 per month. If you spend \$500 per month, the amount in your account will be \$1,000 at the end of the month (no change in storage). If you spend less than \$500 per month, your account will grow ($+\Delta S$). If you spend more than \$500 per month, the amount of money in your account will decrease ($-\Delta S$).

The history of the Missouri River provides an example of input–output analysis for river management. The Missouri River drains one-sixth of the United States (excluding Alaska and Hawaii) and flows for more than 3,200 km (2,000 mi). After large towns and cities were built on the land near the river, flooding of the Missouri became a major problem. The "wild Missouri" became famous in history and folklore for its great fluctuations, its flows and droughts, and as the epitome of unpredictability in nature. One settler said that the Missouri "makes farming as fascinating as gambling. You never know whether you are going to harvest corn or catfish."[a, b]

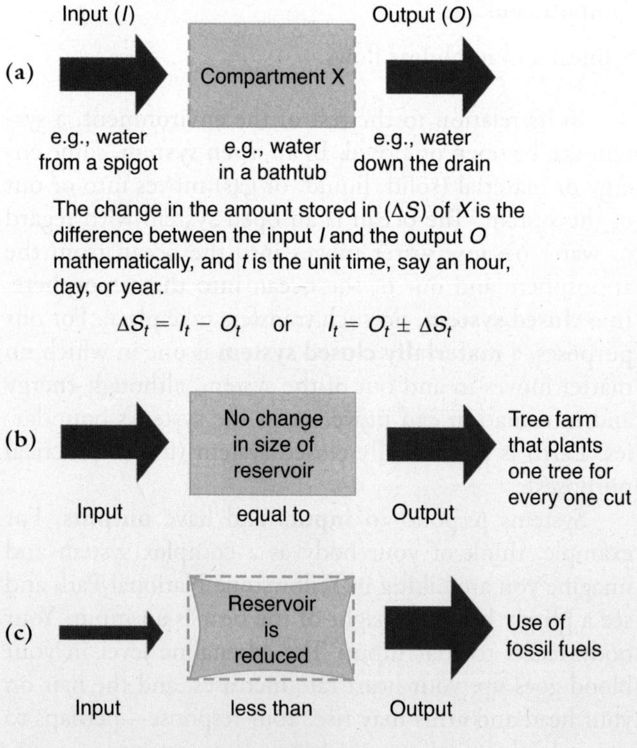

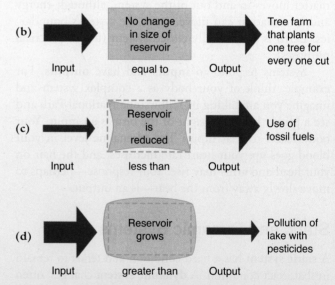

FIGURE 4.5 **(a)** General equation for ways in which a compartment of some material can change Row **(b)** represents steady-state conditions; rows **(c)** and **(d)** are examples of negative and positive changes in storage.

(*Source:* Modified from P.R. Ehrlich, A.H. Ehrlich, and J.P. Holvren, *Ecoscience: Population, Resources, Environment*, 3rd ed. [San Francisco: W.H. Freeman, 1977].)

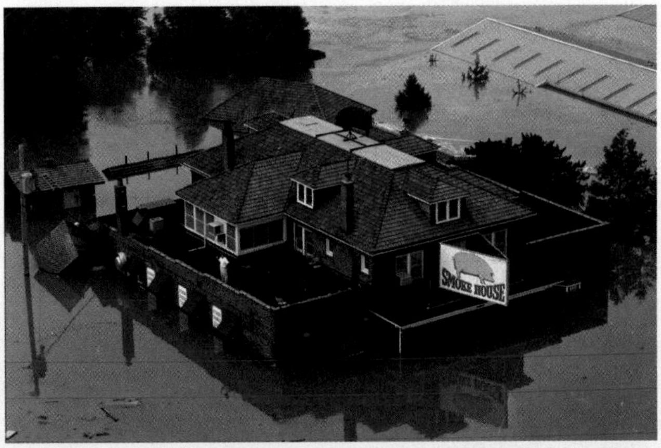

©AP/Wide World Photos

FIGURE 4.6 **St. Louis, Missouri, during the 1993 flood of the Missouri River.** No matter how hard we try to keep this huge river flowing at a fixed rate we always seem to fail. So it is when we try to tame most natural ecological and environmental systems that are naturally dynamic and always changing.

(Based on a drawing by Gary Pound from Daniel B. Botkin, *Passage of Discovery: The American Rivers Guide to the Missouri River of Lewis and Clark* [New York: Perigee Books, a division of Penguin-Putnam, 1999].)

FIGURE 4.7 **The six major dams on the Missouri River.**

Two of the river's great floods were in 1927 and 1993 (Figure 4.6). After the 1927 flood, the federal government commissioned the Army Corps of Engineers to build six major dams on the river (Figure 4.7). (The attempt to control the river's flow also included many other alterations of the river, such as straightening the channel and building levees.) Of the six dams, the three largest were built upstream, and each of their reservoirs was supposed to hold the equivalent of an entire year's average flow. The three smaller, downstream dams were meant to serve as safety valves to control the flow more precisely.

The underlying idea was to view the Missouri as a large plumbing system that needed management. When rainfall was sparse in the huge watershed of the river, the three upstream dams were supposed to be able to augment the flow for up to three years, ensuring a constant and adequate supply of water for irrigation and personal use. In flood years, the six dams were supposed to be able to store the dangerous flow so that the water could be released slowly, the floods controlled, and the flow once again constant. In addition, levees—narrow ridges of higher ground—were built along the river and into it to protect the settled land along the river from floodwaters not otherwise contained. But these idealistic plans did not stop the Missouri from flooding in 1993 (Figures 4.6 and 4.8).

NASA images created by Jesse Allen, Earth Observatory, using data provided courtesy of the Landsat Project Science Office.

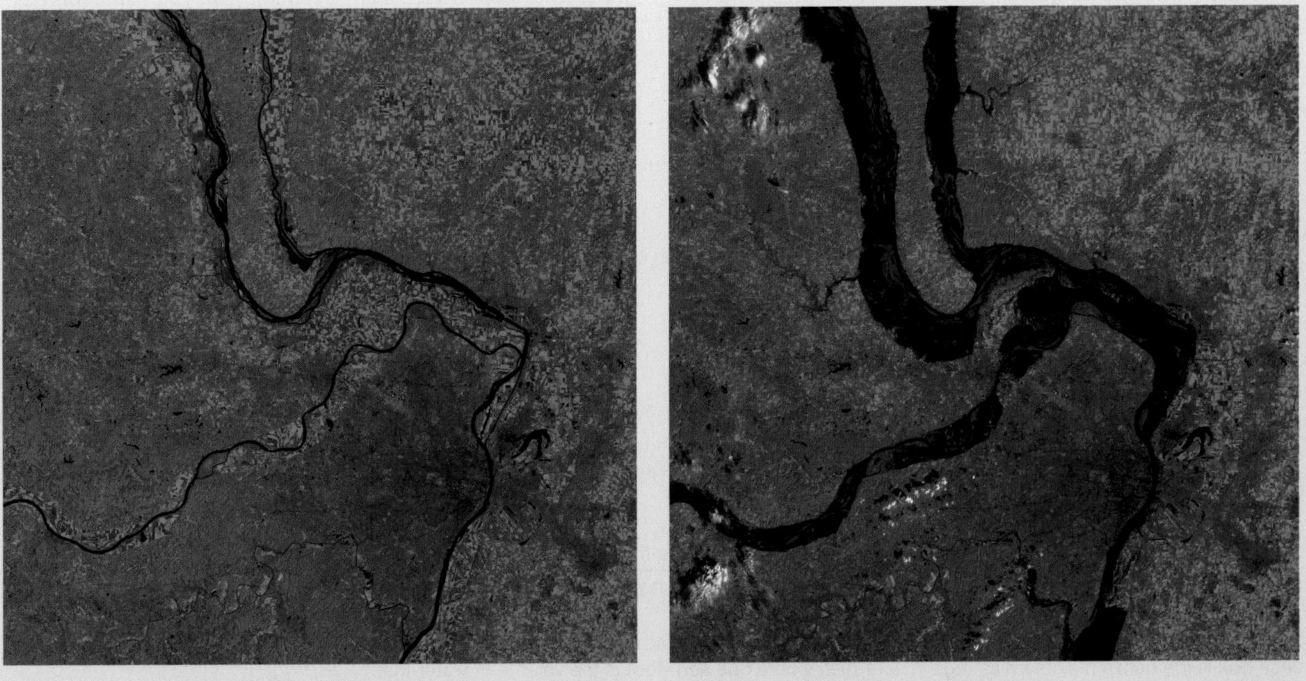

(a) (b)

NASA images created by Jesse Allen, Earth Observatory, using data provided courtesy of the Landsat Project Science Office.

FIGURE 4.8 **Satellite image of the Missouri River at St. Louis before the flood in 1991 (left) and during the 1993 flood.** The dark area is water.

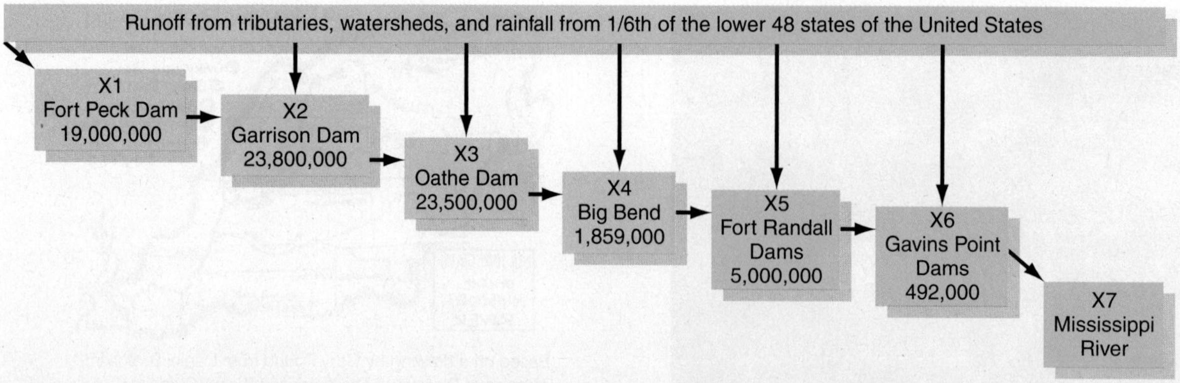

FIGURE 4.9 Imagine the Missouri River as one large lake (composed of the series of dams whose water level is controlled. The water level remains constant as water flows into the lake (Fort Peck Dam) at the same rate as water flows out. If more water comes in, more leaves (Gavins Point Dam); if less water comes in, less flows out, and the water level remains at the spillway level. The number inside each box is the dam's maximum storage in acre-feet. The average annual water flow for the Missouri River is 25 million acre-feet (the amount reaching its mouth where it meets the Mississippi at St. Louis, MO).

Taking the large view, standing way back from the river, led to a perception of the Missouri River as one huge lake (Figure 4.9) into which water flowed, then drained downstream and out at its mouth at St. Louis, Missouri, into the Mississippi, which carried the waters to New Orleans, Louisiana, and out into the Gulf of Mexico. The Army Corps of Engineers' hope was that the Missouri River could be managed the way we manage our bathwater—keeping it at a constant level by always matching the outflow down the drain with inflow from the spigot. This is a perception of the river as a system held in *steady state,* a term we defined earlier.

An environmental water engineer could use this kind of systems diagram (Figure 4.10) to plan the size of the various dams to be built on the Missouri River, taking into account the desired total storage among the dams and the role of each dam in managing the river's flow. In Figure 4.10, the amount stored in a dam's reservoir is listed as Xn, where X is the amount of water stored and n is the number of the compartment. (In this case the dams are numbered in order from upstream to downstream.) Water flows from the environment—tributaries, watersheds, and direct rainfall—into each of the reservoirs, and each is connected to the adjacent reservoirs by the river. Finally, all of the Missouri's water flows into the Mississippi, which carries it to the Gulf of Mexico.

Looking at Figure 4.10, can you think of problems associated with the input–output of the river and managing water reservoirs in such a large system? Can you name some consequences likely to arise when attempting to keep a river in a steady state?

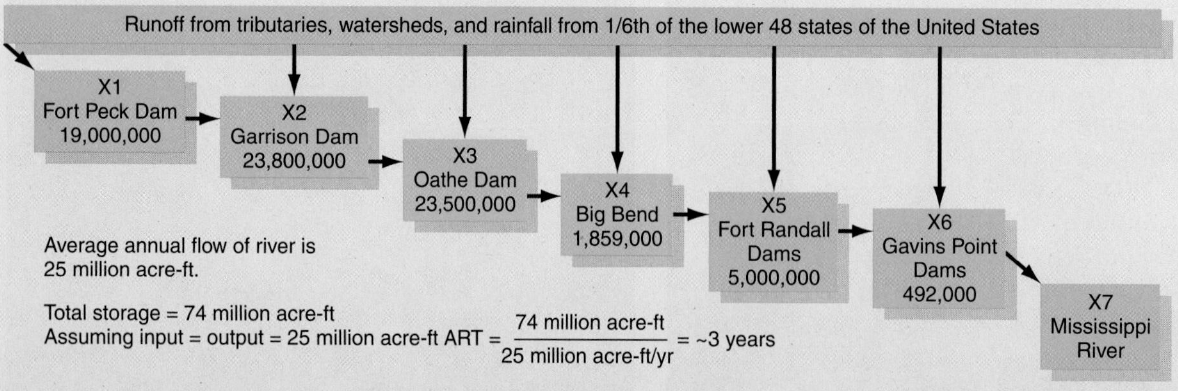

FIGURE 4.10 The Missouri River and its dams viewed as a systems flow chart. The number inside each box is the dam's maximum storage in acre-feet, where one acre-foot is the volume of water that would cover one acre to a depth of 1,233 m^3). (1 ft). The average annual water flow for the Missouri River is 25 million acre-feet (the amount reaching its mouth where it meets the Mississippi at St. Louis, Missouri).

WORKING IT OUT 4.1 Average Residence Time (ART)

The average residence time (ART) is the ratio of the size of a reservoir of some material—say, the amount of water in a reservoir—to the rate of its transfer through the reservoir. The equation is

$$ART = S/F$$

where S is the size of the reservoir and F is the rate of transfer.

For example, we can calculate the average residence time for water in the Gavins Point Dam (see Figure 4.10), the farthest downstream of all the dams on the Missouri River, by realizing that the average flow into and out of the dam is about 25 million acre-feet (31 km^3) a year, and that the dam stores about 492,000 acre-feet (0.6 km^3). This suggests that the average residence time in the dam is only about seven days:

$$ART = S/F = 0.6\ km^3/31\ km^3\ per\ year$$

$$S/F = 0.019\ year\ (about\ 7\ days)$$

If the total flow were to go through Garrison Dam, the largest of the dams, the residence time would be 347 days, almost a year.

The ART for a chemical element or compound is important in evaluating many environmental problems. For example, knowing the ART of a pollutant in the air, water, or soil gives us a more quantitative understanding of that pollutant, allows us to evaluate the extent to which the pollutant acts in time and space, and helps us to develop strategies to reduce or eliminate the pollutant.

Figure 4.11 shows a map of Big Lake, a hypothetical reservoir impounded by a dam. Three rivers feed a combined 10 m^3/sec (2,640 gal/sec) of water into the lake, and the outlet structure releases an equal 10 m^3/sec. In this simplified example, we will assume that evaporation of water from the lake is negligible. A water pollutant, MTBE (methyl tertiary—butyl ether), is also present in the lake. MTBE is added to gasoline to help reduce emissions of carbon monoxide. MTBE readily dissolves in water and so travels with it. It is toxic; in small concentrations of 20–40 μg/l (millionths of grams per liter) in water, it smells like turpentine and is nauseating to some people. Concern over MTBE in California led to a decision to stop adding it to gasoline. The sources of MTBE in "Big Lake" are urban runoff from Bear City gasoline stations, gasoline spills on land or in the lake, and gasoline engines used by boats on the lake.

We can ask several questions concerning the water and MTBE in Big Lake.

1. What is the ART of water in the lake?
2. What is the amount of MTBE in the lake, the rate (amount per time) at which MTBE is being put into the lake, and the ART of MTBE in the lake? Because the water and MTBE move together, their ARTs should be the same. We can test this.

ART of Water in Big Lake

For these calculations, use multiplication factors and conversions in Appendixes B and C at the end of this book.

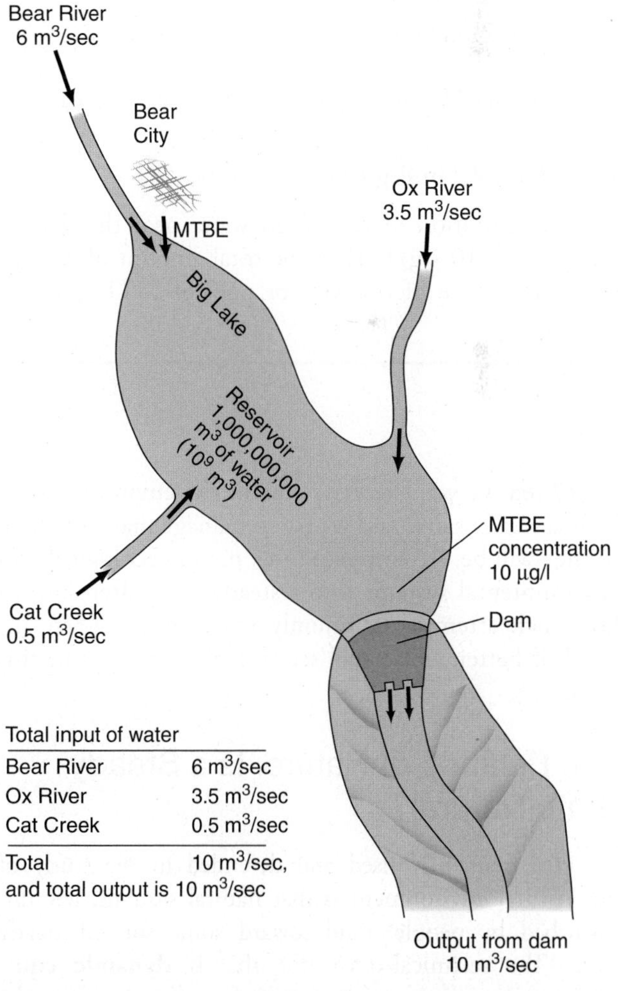

FIGURE 4.11 Idealized diagram of a lake system with MTBE contamination.

$$ART_{water} = \frac{S}{F} = ART_{water} = \frac{1{,}000{,}000{,}000\ m^3}{10\ m^3/sec}$$

$$or\ \frac{10^9 m^3}{10\ m^3/sec}$$

The units m^3 cancel out and

$$ART = 100{,}000{,}000\ sec\ or\ 10^8\ sec$$

Convert 10^8 sec to years:

$$\frac{seconds}{year} = \frac{60\ sec}{1\ minute} \times \frac{60\ minute}{1\ hour} \times \frac{24\ hours}{1\ day} \times \frac{365\ days}{1\ year}$$

Canceling units and multiplying, there are 31,536,000 sec/year, which is

$$3.1536 \times 10^7\ sec/year$$

Then the ART for Big Lake is

$$\frac{100{,}000{,}000\ sec}{31{,}536{,}000\ sec/yr}\ or\ \frac{10^8\ sec}{3.1536 \times 10^7\ sec/yr}$$

Therefore the ART for water in Big Lake is 3.17 years.

ART of MTBE in Big Lake

The concentration of MTBE in water near the dam is measured as 10 μg/l. Then the total amount of MTBE in the lake (size of reservoir or pool of MTBE) is the product of volume of water in the lake and concentration of MTBE:

$$10^9\ m^3 \times \frac{10^3 l}{m^3} \times \frac{10\ \mu g}{l} = 10^{13}\ \mu g\ or\ 10^7\ g$$

which is 10^4 kg, or 10 metric tons, of MTBE.

The output of water from Big Lake is 10 m^3/sec, and this contains 10 μg/l of MTBE; the transfer rate of MTBE (g/sec) is

$$MTBE/sec = \frac{10\ m^3}{sec} \times \frac{10^3 l}{m^3} \times \frac{10\ \mu g}{l} \times \frac{10^{-6} g}{\mu g}$$

$$= 0.1\ g/sec$$

Because we assume that input and output of MTBE are equal, the input is also 0.1 *g/sec.*

$$ART_{MTBE} = \frac{S}{F} = \frac{10^7\ g}{0.1\ g/sec} = 10^8\ sec,\ or\ 3.17\ years$$

Thus, as we suspected, the ARTs of the water and MTBE are the same. This is because MTBE is dissolved in the water. If it attached to the sediment in the lake, the ART of the MTBE would be much longer. Chemicals with large reservoirs or small rates of transfer tend to have long ARTs. In this exercise we have calculated the ART of water in Big Lake as well as the input, total amount, and ART of MTBE.[a]

Often we want real systems in the environment to be in a steady state, and we try to manage many of them so they will be. Attempts to force natural ecological and environmental systems into a steady state often fail. In fact, such attempts commonly make things worse instead of better, as we will see in many chapters in this book.

The Balance of Nature: Is a Steady State Natural?

An idea frequently used and defended in the study of our natural environment is that natural systems, left undisturbed by people, tend toward some sort of steady state. The technical term for this is **dynamic equilibrium**, but it is more familiarly referred to as the **balance of nature** (see Figure 4.12). Certainly, negative

Morning in the Tropics, c.1858 (oil on 207151 canvas), Church, Frederic Edwin (1826-1900)/© Walters Art Museum, Baltimore, USA/The Bridgeman Art Library

FIGURE 4.12 **The balance of nature.** This painting, *Morning in the Tropics* by Frederic Edwin Church, illustrates the idea of the balance of nature and a dynamic steady state, with everything stationary and still, unchanging.

feedback operates in many natural systems and may tend to hold a system at equilibrium. Nevertheless, we need to ask how often the equilibrium model really applies.[4]

If we examine natural ecological systems or **ecosystems** (simply defined here as communities of organisms and their nonliving environment in which nutrients and other chemicals cycle and energy flows) in detail and over a variety of time frames, it is evident that a steady state is seldom attained or maintained for very long. Rather, systems are characterized not only by human-induced disturbances but also by natural disturbances (sometimes large-scale ones called natural disasters, such as floods and wildfires). Thus, changes over time can be expected. In fact, studies of such diverse systems as forests, rivers, and coral reefs suggest that disturbances due to natural events, such as storms, floods, and fires, are necessary for the maintenance of those systems, as we will see in later chapters. The environmental lesson is that systems change naturally. If we are going to manage systems for the betterment of the environment, we need to gain a better understanding of how they change.[4, 5]

Residence Time

By using rates of change or input–output analysis of systems, we can derive an **average residence time**—how long, on average, a unit of something of interest to us will remain in a reservoir. This is obviously important, as in the case of how much water can be stored for how long in a reservoir. To compute the average residence time (assuming input is equal to output), we divide the total volume of stored water in the reservoir by the average rate of transfer through the system. For example, suppose a university has 10,000 students, and each year 2,500 freshmen start and 2,500 seniors graduate. The average residence time for students is 10,000 divided by 2,500, or four years.

Average residence time has important implications for environmental systems. A system such as a small lake with an inlet and an outlet and a high transfer rate of water has a short residence time for water. On the one hand, from our point of view, that makes the lake especially vulnerable to change because change can happen quickly. On the other hand, any pollutants soon leave the lake.

In large systems with a slow rate of transfer of water, such as oceans, water has a long residence time, and such systems are thus much less vulnerable to quick change. However, once they are polluted, large systems with slow transfer rates are difficult to clean up. (See Working It Out 4.1.)

Feedback

Feedback occurs when the output of a system (or a compartment in a system) affects its input. Changes in the output "feed back" on the input. There are two kinds of feedback: negative and positive. A good example of feedback is human temperature regulation. If you go out in the sun and get hot, the increase in temperature affects your sensory perceptions (input). If you stay in the sun, your body responds physiologically: Your pores open, and you are cooled by evaporating water (you sweat). The cooling is output, and it is also input to your sensory perceptions. You may respond behaviorally as well: Because you feel hot (input), you walk into the shade (output) and your temperature returns to normal. In this example, an increase in temperature is followed by a response that leads to a decrease in temperature. This is an example of negative feedback, in which an increase in output leads to a further *decrease* in output. **Negative feedback** is self-regulating or stabilizing. It is the way that steady-state systems can remain in a constant condition.

Positive feedback occurs when an increase in output leads to a further *increase* in output. A fire starting in a forest provides an example of positive feedback. The wood may be slightly damp at the beginning and so may not burn readily. Once a fire starts, wood near the flame dries out and begins to burn, which in turn dries out a greater quantity of wood and leads to a larger fire. The larger the fire, the faster more wood becomes dry and the more rapidly the fire grows. Positive feedback, sometimes called a "vicious cycle," is destabilizing.

Environmental damage can be especially serious when people's use of the environment leads to positive feedback. For example, off-road vehicles—including bicycles—may cause positive feedback to soil erosion (Figure 4.13). The vehicles' churning tires are designed to grip the earth, but they also erode the soil and uproot plants. Without vegetation, the soil erodes faster, exposing even more soil (positive feedback). As more soil is exposed, rainwater more easily carves out ruts and gullies (more positive feedback). Drivers of off-road vehicles then avoid the ruts and gullies by driving on adjacent sections that are not as eroded, thus widening paths and further increasing erosion (more positive feedback). The gullies themselves increase erosion because they concentrate runoff and have steep side slopes. Once formed, gullies tend to get longer, wider, and deeper, causing additional erosion (even more positive feedback). Eventually, an area of intensive off-road vehicle use may become a wasteland of eroded paths and gullies. Positive feedback has made the situation increasingly worse.

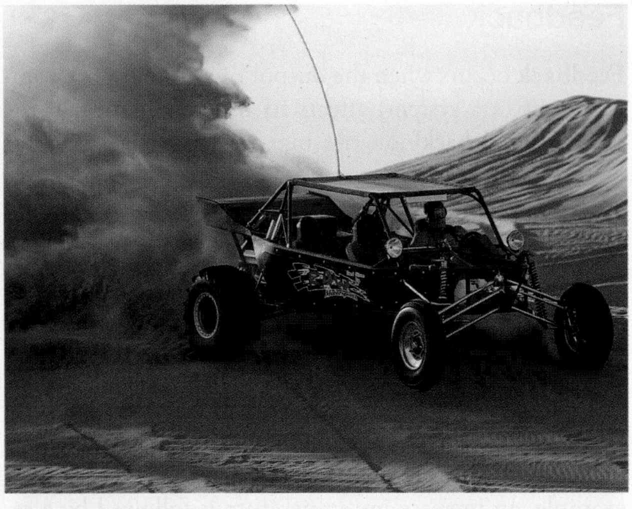

Tim Tadder/©AP/Wide World Photos

(a)

Photo by Ed Keller

(c)

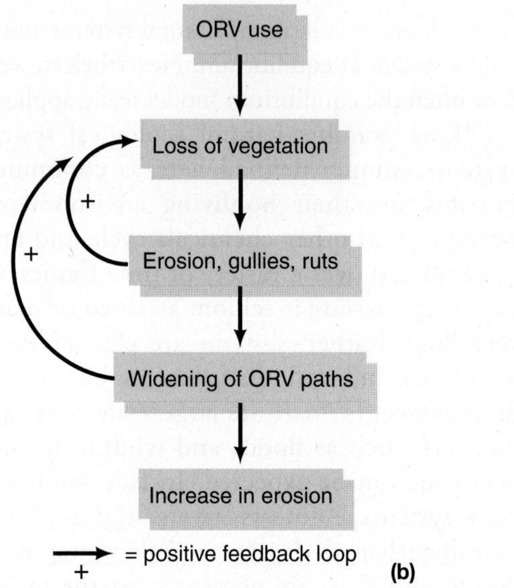

(b)

FIGURE 4.13 How off-road vehicles (a) create positive feedback on soil erosion **(b)** and **(c)**.

Some systems have both positive and negative feedbacks, as can occur, for example, for the human population in large cities (Figure 4.14). Positive feedback on the population size may occur when people perceive greater opportunities in cities and move there, hoping for a higher standard of living. As more people move to cities, opportunities may increase, leading to even more migration to cities. Negative feedback can then occur when crowding increases air and water pollution, disease, crime, and discomfort. These negatives encourage some people to migrate from the cities to rural areas, reducing the city's population.

Practicing your critical thinking skills, you may ask, "Is negative feedback generally desirable, and is positive feedback generally undesirable?" Reflecting on this question, we can see that, although negative feedback is self-regulating, it may in some instances not be desirable. The period over which the positive or negative feedback occurs is the important factor. For example, suppose we are interested in restoring wolves to Yellowstone National Park. We will expect positive feedback in the wolf population for a time as the number of wolves grows. (The more wolves, the greater their population growth, through exponential growth.) Positive feedback, for a time, is desirable because it produces a change we want.

We can see that whether we view positive or negative feedback as desirable depends on the system and potential changes. Nevertheless, some of the major environmental problems we face today result from positive feedback mechanisms. These include resource use and growth of the human population.

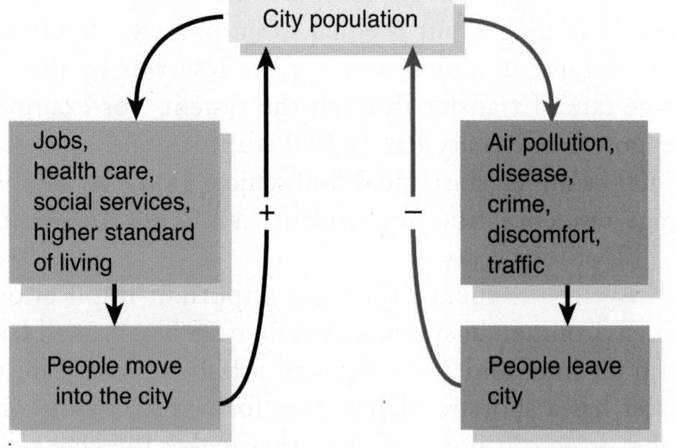

(*Source:* Modified from M. Maruyama, The second cybernetics: Deviation-amplifying mutual causal processes, *American Scientist* 51 [1963]:164–670. Reprinted by permission of *American Scientist* magazine of Sigma Xi, The Scientific Research Society.)

FIGURE 4.14 Potential positive and negative feedback loops for changes of human population in large cities. The left side of the figure shows that as jobs increase and health care and the standard of living improve, migration and the city population increase. Conversely, the right side of the figure shows that increased air pollution, disease, crime, discomfort, and traffic tend to reduce the city population.

4.2 System Responses: Some Important Kinds of Flows[5]

Within systems, there are certain kinds of flows that we come across over and over in environmental science. (Note that **flow** is an amount transferred; we also refer to the **flux**, which is the rate of transfer per unit time.) Because these are so common, we will explain a few of them here.

Linear and Nonlinear Flows

An important distinction among environmental and ecological systems is whether they are characterized by linear or nonlinear processes. Put most simply, in a **linear process**, if you add the same amount of anything to a compartment in a system, the change will always be the same, no matter how much you have added before and no matter what else has changed about the system and its environment. If you harvest one apple and weigh it, then you can estimate how much 10 or 100 or 1,000 or more of the apples will weigh—adding another apple to a scale does not change the amount by which the scale shows an increase. One apple's effect on a scale is the same, no matter how many apples were on the scale before. This is a linear effect.

Many important processes are **nonlinear**, which means that the effect of adding a specific amount of something changes, depending on how much has been added before. If you are very thirsty, one glass of water makes you feel good and is good for your health. Two glasses may also be helpful. But what about 100 glasses? Drinking more and more glasses of water leads quickly to diminishing returns and eventually to water becoming a poison.

Lag Time

Many responses to environmental inputs (including human population change; pollution of land, water, and air; and use of resources) are nonlinear and may involve delays, which we need to recognize if we are to understand and solve environmental problems. For example, when you add fertilizer to help a tree grow, it takes time for it to enter the soil and be used by the tree.

Lag time is the delay between a cause and the appearance of its effect. (This is also referred to as the time between a stimulus and the appearance of a response.) If the lag time is long, especially compared to human lifetimes (or attention spans or our ability to continue measuring and monitoring), we can fail to recognize the change and know what is the cause and what is the effect. We can also come to believe that a possible cause is not having a detrimental effect, when in reality the effect is only delayed. For example, logging on steep slopes can increase the likelihood and rate of erosion, but in comparatively dry environments this may not become apparent until there is heavy rain, which might not occur until a number of years afterward. If the lag time is short, cause and effect are easier to identify. For example, highly toxic gas released from a chemical plant will likely have rapid effects on the health of people living nearby.

With an understanding of input and output, positive and negative feedback, stable and unstable systems, and systems at steady state, we have a framework for interpreting some of the changes that may affect systems.

Selected Examples of System Responses

Although environmental science deals with very complex phenomena, there are recurring relationships that we can represent with a small number of graphs that show how one part of a system responds to inputs from another part. These graphs include responses of individual organisms, responses of populations and species, responses of entire ecosystems and then large units of the **biosphere**, the planetary system that includes and sustains life, such as how the atmosphere responds to the burning of fossil fuels. Each of these graphs has a mathematical equation that can explain the curve, but it is the shape of the graph and what that shape represents that are the keys to understanding environmental systems. These curves represent, in one manifestation or another, the fundamental dynamics found in these systems. The graphs show (1) a straight line (linear), (2) the positive exponential, (3) the negative exponential, and (4) the saturation (Michaelis-Menton) curve. An example of each is shown in Figures 4.15 to 4.17.

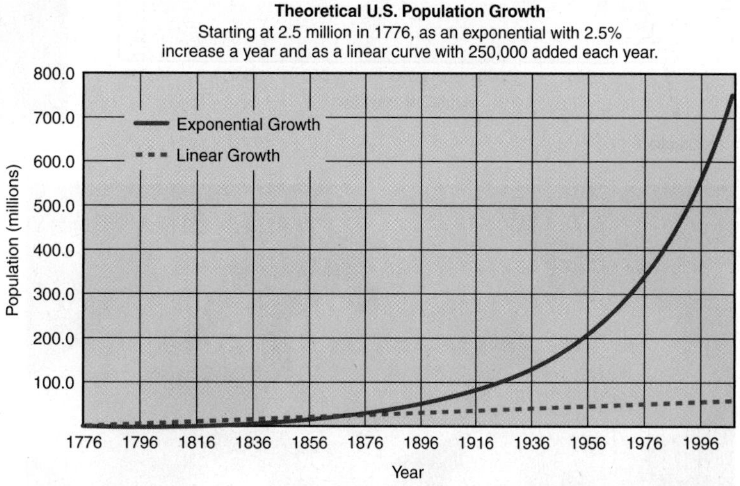

FIGURE 4.15 Curves 1 and 2: linear and positive exponential. This graph shows theoretical growth of the population of the United States, starting with the 2.5 million people estimated to have been here in 1776 and growing as an exponential and a linear curve. Even though the linear curve adds 250,000 people a year—10% of the 1776 population—it greatly lags the exponential by the beginning of the 20th century, reaching fewer than 100 million people today, while the exponential would have exceeded our current population.

(Source: D.B. Botkin and R.S. Miller, 1974, Mortality rates and survival of birds, *American Nat.* 108:181–192.)

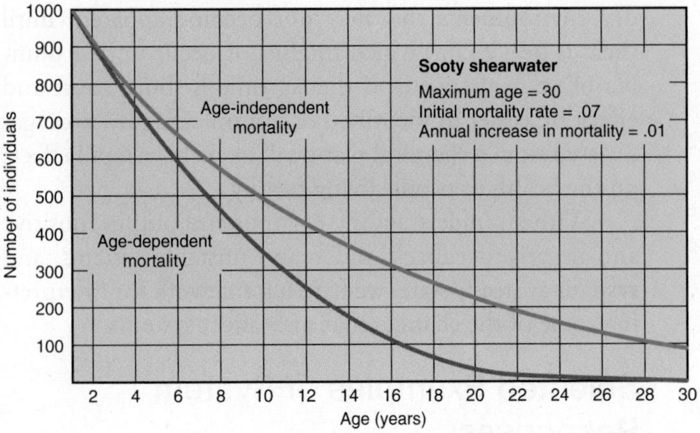

(a)

All Canada Photos/Alamy

(b)

FIGURE 4.16 Negative exponential. Example: the decline in a population of a species of birds when there are no births and the mortality rate is 7% per year. The upper curve is a pure negative exponential.

Figure 4.15 shows both a linear relation and a positive exponential relation. A linear relation is of the form $y = a + bx$, where a is the y intercept (in this case, o) and b is the slope of the line (change in y to change in x, where y is the vertical axis and x the horizontal). The form of the positive exponential curve is $y = ax^b$, where a is the y intercept (in this case o) and b is the slope. However, b is a positive exponent (power).

(*Source:* F.B. Salisbury and C. Ross, *Plant Physiology* [Belmont, CA: Wadsworth, 1969, p. 292, Figure 14-9.] Data from R. Bohning and C. Burnside, 1956, *American Journal of Botany* 43:557].)

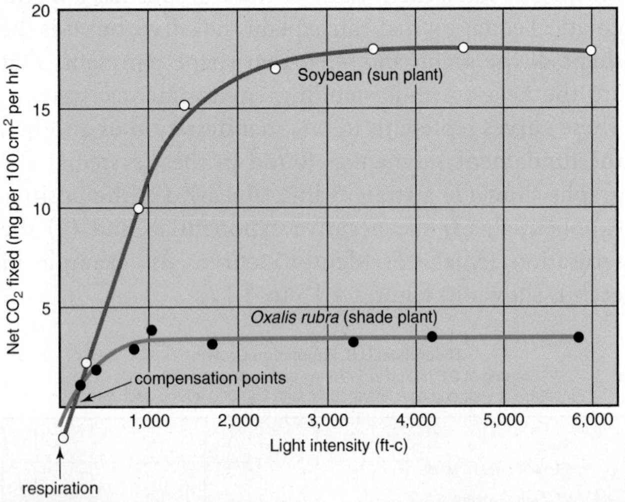

(a)

AGStock USA/Alamy

(b)

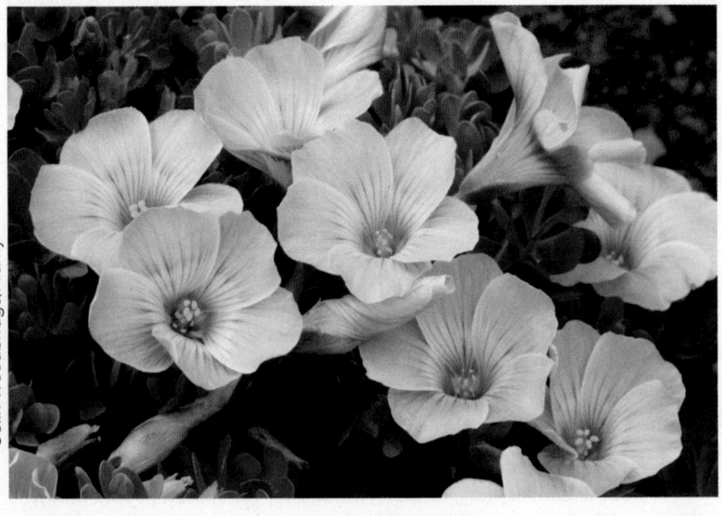

Colin Woodbridge/Alamy

(c)

FIGURE 4.17 (a) The saturation (Michaelis-Menton) curve; **(b)** *Glycine max* (soybeans); **(c)** *Oxalis rubra* (shade plant).

WORKING IT OUT 4.2 Exponential Growth

If the quantity of something (say, the number of people on Earth) increases or decreases at a fixed fraction per unit of time, whose symbol is k (for example, $k = +0.02$ per year), then the quantity is changing exponentially. With positive k, we have exponential growth. With negative k, we have exponential decay.

The growth rate R is defined as the percent change per unit of time—that is, $k = R/100$. Thus, if $R = 2\%$ per year, then $k = +0.02$ per year. The equation to describe exponential growth is

$$N = N_0 e^{kt}$$

where N is the future value of whatever is being evaluated; N_0 is present value; e, the base of natural logarithms, is a constant 2.71828; k is as defined above; and t is the number of years over which the growth is to be calculated.

This equation can be solved using a simple hand calculator, and a number of interesting environmental questions can then be answered. For example, assume that we want to know what the world population is going to be in the year 2020, given that the population in early 2012 was 7.0 billion and the population is growing at a constant rate of 1.09% per year ($k = 0.0109$). We can estimate N, the world population for the year 2020, by applying the preceding equation:

$$\begin{aligned} N &= (7.0 \times 10^9) \times e^{(0.0109 \times 8)} \\ &= 7.0 \times 10^9 \times e^{0.0872} \\ &= 7.0 \times 10^9 \times 2.71828^{\,0.0872} \\ &= 7.64 \times 10^9, \text{ or } 7.64 \text{ billion people} \end{aligned}$$

The doubling time for a quantity undergoing exponential growth (i.e., increasing by 100%) can be calculated by the following equation:

$$2N_0 = N_0 e^{kTd}$$

where T_d is the doubling time.

Take the natural logarithm of both sides.

$$\ln 2 = kT_d \text{ and } T_d = \ln 2/k$$

Then, remembering that $k = R/100$,

$$\begin{aligned} T_d &= 0.693/(R/100) \\ &= 100(0.693)/R \\ &= 69.3/R, \textit{ or about } 70/R \end{aligned}$$

This result is our general rule—that the doubling time is approximately 70 divided by the growth rate. For example, if $R = 10\%$ per year, then $T = 7$ years.

Exponential growth is a particularly important kind of positive feedback. Change is exponential when it increases or decreases at a constant rate per time period, rather than by a constant amount. For instance, suppose you have $1,000 in the bank and it grows at 10% per year. The first year, $100 in interest is added to your account. The second year, you earn more, $110, because you earn 10% on a higher total amount of $1,100. The greater the amount, the greater the interest earned, so the money increases by larger and larger amounts. When we plot data in which exponential growth is occurring, the curve we obtain is J-shaped. It looks like a skateboard ramp, starting out nearly flat and then rising steeply.

Two important qualities of exponential growth are (1) the rate of growth measured as a percentage and (2) the doubling time in years. The **doubling time** is the time necessary for the quantity being measured to double. A useful rule is that the doubling time is approximately equal to 70 divided by the annual percentage growth rate. Working It Out 4.2 describes exponential growth calculations and explains why 70 divided by the annual growth rate is the doubling time.

Figure 4.16 shows two examples of negative exponential relations. The saturation (Michaelis-Menton) curve (Figure 4.17) shows initial fast change, followed by a leveling off at saturation. At the point of saturation, the net CO_2 fixed (for soybean) is at a light-intensity value of about 3,000 (Figure 4.17**a**). As light intensity increases above about 3,000, net fixed CO_2 is nearly constant (that is, fixed CO_2 saturates at light intensity of 3,000 and does not change if intensity increases).

4.3 Overshoot and Collapse

Figure 4.18 shows the relationship between carrying capacity (maximum population possible without degrading the environment necessary to support the population) and the human population. The carrying capacity starts out being much higher than the human population, but if a population grows exponentially (see Working It Out

(*Source:* Modified after D.H. Meadows and others, 1992.)

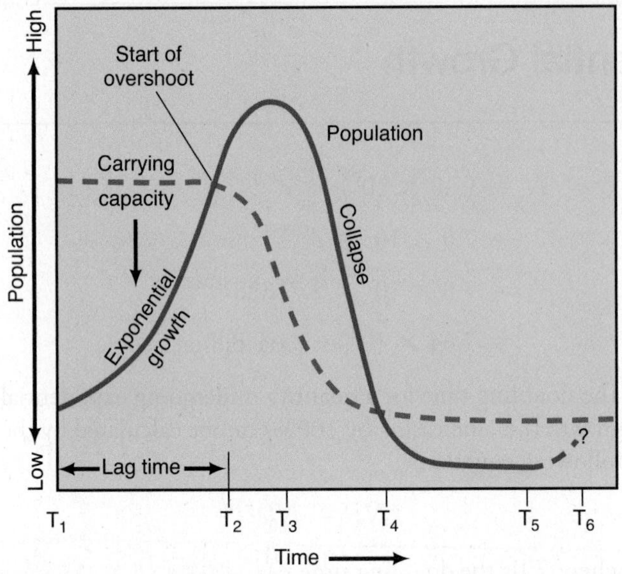

FIGURE 4.18 **The concept of overshoot.** A population starts out growing exponentially, but as this growth cannot continue indefinitely, it reaches a peak, then declines sharply. Sometimes the population is assumed to have a carrying capacity, which is the maximum number possible, and if the population's habitat is damaged by too great an abundance, the carrying capacity also decreases.

FIGURE 4.19 **Timber harvest (clear-cut) can result in soil erosion.** Once soil is removed, it can take such a long time for it to rebuild that the damage may be viewed as irreversible on a human time scale.

4.2), it eventually exceeds—**overshoots**—the carrying capacity. This ultimately results in the **collapse** of a population to some lower level, and the carrying capacity may be reduced as well. In this case, the lag time is the period of exponential growth of a population before it exceeds the carrying capacity. A similar scenario may be posited for harvesting species of fish or trees.

4.4 Irreversible Consequences

The adverse consequences of environmental change do not necessarily lead to irreversible consequences. Some do, however, and these lead to particular problems. When we talk about irreversible consequences, we mean consequences that may not be easily rectified on a human scale of decades or a few hundred years.

Good examples of this are soil erosion and the harvesting of old-growth forest (Figure 4.19). With soil erosion, there may be a long lag time until the soil erodes to the point where crops no longer have their roots in active soil that has the nutrients necessary to produce a successful crop. But once the soil is eroded, it may take hundreds or thousands of years for new soil to form, and so the consequences are irreversible in terms of human planning. Similarly, when old-growth forests are harvested, it may take hundreds of years for them to be restored. Lag times may be even longer if the soils have been damaged or eroded by timber harvesting.

4.5 Environmental Unity

Our discussion of positive and negative feedback sets the stage for another fundamental concept in environmental science: **environmental unity**—the idea that it is impossible to change only one thing; everything affects everything else. Of course, this is something of an overstatement; the extinction of a species of snails in North America, for instance, is hardly likely to change the flow characteristics of the Amazon River. However, many aspects of the natural environment are in fact closely linked, and thus changes in one part of a system often have secondary and tertiary effects within the system and on adjacent systems as well. Earth and its ecosystems are complex entities in which any action may have many effects.

We will find many examples of environmental unity throughout this book. Urbanization illustrates it. When cities, such as Chicago and Indianapolis, were developed in the eastern and midwestern United States, the clearing of forests and prairies and the construction of buildings and paved streets increased surface-water runoff and soil erosion, which in turn affected the shape of river channels—some eroded soil was deposited on the bottom of the channel, reducing channel depth and increasing flood hazard. Increased fine sediment made the water muddy, and chemicals from street and yard runoff polluted streams.[6,7] These changes affected fish and other life in the river, as well as terrestrial wildlife that depended on the river. The point here is that land-use conversion can set off a series of changes in the environment, and each change is likely to trigger additional changes.

4.6 Uniformitarianism

Uniformitarianism is the idea that geological and biological processes that occur today are the same kinds of processes that occurred in the past and vice versa. Thus, the present is the key to the past, and the past the key to the future. For example, we use measurements of the current rate of erosion of soils and bedrock by rivers and streams to calculate the rate at which this happened in the past and to estimate how long it took for certain kinds of deposits to develop. If a deposit of gravel and sand found at the top of a mountain is similar to stream gravels found today in an adjacent valley, we may infer by uniformitarianism that a stream once flowed in a valley where the mountaintop is now. The concept of uniformitarianism helps explain the geologic and evolutionary history of Earth.

Uniformitarianism was first suggested in 1785 by the Scottish scientist James Hutton, known as the father of geology. Charles Darwin was impressed by the concept, and it pervades his ideas on biological evolution. Today, uniformitarianism is considered one of the fundamental principles of the biological and Earth sciences.

Uniformitarianism does not demand or even suggest that the magnitude and frequency of natural processes remain constant, only that the processes themselves continue. For the past several billion years, the continents, oceans, and atmosphere have been similar to those of today. We assume that the physical and biological processes that form and modify the Earth's surface have not changed significantly over this period. To be useful from an environmental standpoint, the principle of uniformitarianism has to be more than a key to the past; we must turn it around and say that a study of past and present processes is the key to the future. That is, we can assume that in the future the same physical and biological processes will operate, although the rates will vary as the environment is influenced by natural change and human activity. Geologically short-lived landforms, such as beaches (Figure 4.20) and lakes, will continue to appear and disappear in response to storms, fires, volcanic eruptions, and earthquakes. Extinctions of animals and plants will continue, in spite of, as well as because of, human activity.

Obviously, some processes do not extend back through all of geologic time. For example, the early Earth atmosphere did not contain free oxygen. Early photosynthetic bacteria converted carbon dioxide in the atmosphere to hydrocarbons and released free oxygen; before life, this process did not occur. But the process began a long time ago—3.5 billion years ago—and as long as there are photosynthetic organisms, this process of carbon dioxide uptake and oxygen release will continue.

Knowledge of uniformitarianism is one way that we can decide what is "natural" and ascertain the characteristics of nature undisturbed by people. One of the environmental questions we ask repeatedly, in many contexts, is whether human actions are consistent with the processes of the past. If not, we are often concerned that these actions will be harmful. We want to improve our ability to predict what the future may bring, and uniformitarianism can assist in this task.

Diane Cook and Len Jenshel/The Image Bank/Getty Images

FIGURE 4.20 **This beach on the island of Bora Bora, French Polynesia,** is an example of a geologically short-lived landform, vulnerable to rapid change from storms and other natural processes.

4.7 Earth as a System

The discussion in this chapter sets the stage for a relatively new way of looking at life and the environment—a global perspective, or thinking about our entire planet's life-supporting and life-containing system. This is known as Earth Systems Science, and it has become especially important in recent years, with concerns about climate change (see Chapter 20).

Our discussion of Earth as a system—life in its environment, the biosphere, and ecosystems—leads us to the question of how much life on Earth has affected our planet. In recent years, the **Gaia hypothesis**—named for Gaia, the Greek goddess Mother Earth—has become a hotly debated subject.[8] The hypothesis states that life manipulates the environment for the maintenance of life. For example, scientists have evidence that algae floating near the surface of the ocean influence rainfall at sea and the carbon dioxide content of the atmosphere, thereby significantly affecting the global climate. It follows, then, that the planet Earth is capable of physiological self-regulation.

The idea of a living Earth can be traced back at least to Roman times in the writing of Lucretius.[4] James Hutton, whose theory of uniformitarianism was discussed earlier, stated in 1785 that he believed Earth to be a superorganism, and he compared the cycling of nutrients from soils and rocks in streams and rivers to the circulation of blood in an animal.[8] In this metaphor, the rivers are the arteries and veins, the forests are the lungs, and the oceans are the heart of Earth.

The Gaia hypothesis is really a series of hypotheses. The first is that life, since its inception, has greatly affected the planetary environment. Few scientists would disagree. The second hypothesis asserts that life has altered Earth's environment in ways that have allowed life to persist. Certainly, there is some evidence that life has had such an effect on Earth's climate. A popularized extension of the Gaia hypothesis is that life *deliberately* (consciously) controls the global environment. Few scientists accept this idea.

Since the Gaia hypothesis was introduced 40 years ago, there has been heated debate over it.[9-11] The hypothesis, like previous paradigm shifts on how we view Earth history, is moving through a path that is as yet incomplete. When new ideas are suggested, they often are not believed because they threaten our previous ideas. After a while, scientific evidence is gathered to test the hypothesis, and attempts are made to negate or support it in scientific journals. Generally, the original hypothesis is refuted, modified, or accepted. If after years (sometimes decades or longer) a modified hypothesis is accepted, scientists often proclaim that they knew it was basically right all along.

Scientific evaluation of the Gaia hypothesis generally accepts that:[9-11]

- Through biogeochemical cycles, life has and is playing a significant role in producing Earth's physical and chemical environment.
- There may be mechanisms through which life plays a role of particular importance in modulating Earth's climate.
- It produced an important metaphor to be used in scientific exploration.
- Biologic evolution is an important factor in the changing thoughts about the original Gaia hypothesis.
- The hypothesis has encouraged the study of Earth as a single, unified system, rather than a set of components.

Criticism of the Gaia hypothesis includes the following:[9-11]

- Gaia is limited by its broad generality in defining a description of the role of life in Earth history.
- Attempts to test a metaphor will likely be difficult and perhaps futile.
- Global change (the temperature of the atmosphere, land, and ocean is accelerating) is making a stable environment controlled by life less likely.
- Gaia has metaphorical and religious connotations.
- The Gaia hypothesis predicts that biological by-products in the atmosphere should act to regulate Earth's climate. However, available evidence suggests that biological by-products such as carbon dioxide and methane make the Earth warmer when it is warm and colder when it is cold. The Gaia hypothesis implies that negative feedbacks linked to life should regulate Earth's climate over geologic time. However, over the past 300 million years of Earth history peaks in past temperature correspond to peaks in past levels of carbon dioxide.
- Feedback between life and the environment does not necessarily enhance the environment, although this may appear to an observer to be the case. Biologic evolution through natural selection will favor organisms that do well in their environment at the time they are living. Gaia feedbacks can evolve by natural selection, but so can anti-Gaia feedback. Natural selection will favor a trait that gives a particular life-form a reproductive advantage, whether or not that trait improves the environment.

Countering all this criticism are the following:[9]

- Several interpretations of the original Gaia hypothesis have been discarded. For example, referring to Gaia as a "superorganism" has been largely abandoned in favor of an interpretation that involves a tightly coupled system of life with its nonliving environment.
- Gaia thinking has evolved over recent decades and will continue to evolve.
- The Gaia hypothesis assumes that regulatory feedbacks that control the environment are a probable outcome of planets with abundant life. If this assumption is proven wrong, the assumption will have served to encourage an important research agenda. It seems unlikely (a value judgment) that our Earth system has evolved by random processes resulting in the good luck of abundant, sustained life on Earth.

One positive note associated with the Gaia hypothesis is that it may have made us more conscious of our effects on the planet, leading us to understand that we can make a difference in the future of our planet. The future status of the human environment may depend in part on actions we take now and in coming years. This aspect of the Gaia hypothesis exemplifies the key theme of thinking globally, which was introduced in Chapter 1.

4.8 Types of Change

Change comes in several forms. Some changes brought on by human activities involve rather slow processes—at least from our point of view—with cumulative effects. For example, in the middle of the 19th century, people began to clear-cut patches of the Michigan forests. It was commonly believed that the forests were so large that it would be impossible to cut them all down before they grew back just as they were. But with many people logging in different, often isolated areas, it took less than 100 years for all but about 100 hectares to be clear-cut.

Another example: With the beginning of the Industrial Revolution, people in many regions began to burn fossil fuels, but only since the second half of the 20th century have the possible global effects become widely evident. Many fisheries appear capable of high harvests for many years. But then suddenly, at least from our perspective—sometimes within a year or a few years—an entire species of fish suffers a drastic decline. In such cases, long-term damage can be done. It has been difficult to recognize when harvesting fisheries is overharvesting and, once it has started, figuring out what can be done to enable a fishery to recover in time for fishermen to continue making a living. A famous example of this was the harvesting of anchovies off the coast of Peru. Once the largest fish catch in the world, within a few years the fish numbers declined so greatly that commercial harvest was threatened. The same thing has happened with the fisheries of Georges Banks and the Grand Banks in the Atlantic Ocean.

You can see from these few examples that environmental problems are often complex, involving a variety of linkages among the major components and within each component, as well as linear and exponential change, lag times, and the possibility of irreversible consequences.

As stated, one of our goals in understanding the role of human processes in environmental change is to help manage our global environment. To accomplish this goal, we need to be able to predict changes, but as the examples above demonstrate, prediction poses great challenges. Although some changes are anticipated, others come as a surprise. As we learn to apply the principles of environmental unity and uniformitarianism more skillfully, we will be better able to anticipate changes that would otherwise have been surprises.

CRITICAL THINKING ISSUE

Is the Gaia Hypothesis Science?

According to the Gaia hypothesis, Earth and all living things form a single system with interdependent parts, communication among these parts, and the ability to self-regulate. Are the Gaia hypothesis and its component hypotheses science, fringe science, or pseudoscience? Is the Gaia hypothesis anything more than an attractive metaphor? Does it have religious overtones? Answering these questions is more difficult than answering similar questions about, say, crop circles, described in Chapter 2. Analyzing the Gaia hypothesis forces us to deal with some of our most fundamental ideas about science and life.

Critical Thinking Questions

Before attempting the questions, revisit the discussion of the Gaia hypothesis in Section 4.7.

1. Compare and contrast the main hypotheses included in the Gaia hypothesis.
2. Discuss what kind of evidence is necessary to support each hypothesis.
3. Summarize how each hypothesis can be tested.
4. How could you determine if each hypothesis is science, fringe science, or pseudoscience?
5. Compare the strengths and weaknesses of the Gaia hypothesis.

SUMMARY

- A system is a set of components or parts that function together as a whole. Environmental studies deal with complex systems, and solutions to environmental problems often involve understanding systems and their rates of change.
- Systems respond to inputs and have outputs. Feedback is a special kind of system response, where the output affects the input. Positive feedback, in which increases in output lead to increases in input, is destabilizing, whereas negative feedback, in which

increases in output lead to decreases in input, tends to stabilize or encourage more constant conditions in a system.

- Relationships between the input (cause) and output (effect) of systems may be linear, exponential, or represented by a logistic curve or a saturation curve.
- The principle of environmental unity, simply stated, holds that everything affects everything else. It emphasizes linkages among parts of systems.
- The principle of uniformitarianism can help predict future environmental conditions on the basis of the past and the present.
- Although environmental and ecological systems are complex, much of what happens with them can be characterized by just a few response curves or equations: the straight line and the exponential, the logistic, and the saturation curves.
- Exponential growth, long lag times, and the possibility of irreversible change can combine to make solving environmental problems difficult.
- Change may be slow, fast, expected, unexpected, or chaotic. One of our goals is to learn to better recognize change and its consequences in order to better manage the environment.

REEXAMINING THEMES AND ISSUES

Sean Randall/Getty Images, Inc.

HUMAN POPULATION

Due partly to a variety of positive-feedback mechanisms, Earth's human population is increasing. Of particular concern are local or regional increases in population density (the number of people per unit area), which strain resources and lead to human suffering.

© Biletskiy_Evgeniy/iStockphoto

SUSTAINABILITY

Negative feedback is stabilizing. If we are to have a sustainable human population and use our resources sustainably, then we need to put in place a series of negative feedbacks within our agricultural, urban, and industrial systems.

© Anton Balazh 2011/iStockphoto

GLOBAL PERSPECTIVE

This chapter introduced Earth as a system. One of the most fruitful areas for environmental research remains the investigation of relationships between physical and biological processes on a global scale. More of these relationships must be discovered if we are to solve environmental problems related to such issues as potential global warming, ozone depletion, and disposal of toxic waste.

ssguy/ShutterStock

URBAN WORLD

The concepts of environmental unity and uniformitarianism are particularly applicable to urban environments, where land-use changes result in a variety of changes that affect physical and biochemical processes.

B2M Productions/Getty Images, Inc.

PEOPLE AND NATURE

People and nature are linked in complex ways in systems that are constantly changing. Some changes are not related to human activity, but many are—and human-caused changes from local to global in scale are accelerating.

George Doyle/Getty Images, Inc.

SCIENCE AND VALUES

Our discussion of the Gaia hypothesis reminds us that we still know very little about how our planet works and how physical, biological, and chemical systems are linked. What we do know is that we need more scientific understanding. This understanding will be driven in part by the value we place on our environment and on the well-being of other living things.

KEY TERMS

average residence time 69
balance of nature 68
biosphere 71
classical stability 63
closed system 63
doubling time 73
dynamic equilibrium 68
dynamic system 63
ecosystem 69
environmental unity 74
equilibrium 63
exponential growth 73
feedback 69
flow 71
flux 71
Gaia hypothesis 75
input 63
lag time 71
linear process 71
materially closed system 63
negative feedback 69
nonlinear process 71
open system 63
output 63
overshoot and collapse 74
positive feedback 69
static system 63
steady-state system 64
system 63
uniformitarianism 75

STUDY QUESTIONS

1. What is the difference between positive and negative feedback in systems? Provide an example of each.
2. What is the main point concerning exponential growth? Is exponential growth good or bad?
3. Why is the idea of equilibrium in systems somewhat misleading in regard to environmental questions? Is it ever possible to establish a balance of nature?
4. Why is the average residence time important in the study of the environment?
5. Consider the six dams on the Missouri River in Figures 4.9 and 4.10. Is the average residence time (ART) of 3 years for the system a reasonable value? What are the assumptions and how and why could the value change?
6. How might you use the principle of uniformitarianism to help evaluate environmental problems? Is it possible to use this principle to help evaluate the potential consequences of too many people on Earth?
7. Why does overshoot occur, and what could be done to anticipate and avoid it?

FURTHER READING

Botkin, D.B., M. Caswell, J.E. Estes, and A. Orio, eds., *Changing the Global Environment: Perspectives on Human Involvement* (New York: Academic Press, 1989). One of the first books to summarize the effects of people on nature; it includes global aspects and uses satellite remote sensing and advanced computer technologies.

Bunyard, P., ed., *Gaia in Action: Science of the Living Earth* (Edinburgh: Floris Books, 1996). This book presents investigations into implications of the Gaia hypothesis.

Lovelock, J., *The Ages of Gaia: A Biography of Our Living Earth* (New York: Norton, 1995). This small book explains the Gaia hypothesis, presenting the case that life very much affects our planet and in fact may regulate it for the benefit of life.

NOTES

1. Western, D., and C. Van Prat. 1973. Cyclical changes in habitat and climate of an East African ecosystem. *Nature* 241(549):104–106.
2. Dunne, T., and L.B. Leopold. 1978. *Water in Environmental Planning.* San Francisco: Freeman.
3. Altmann, J., S.C. Alberts, S.A. Altman, and S.B. Roy. 2002. Dramatic change in local climate patterns in the Amboseli basin, Kenya. *African Journal of Ecology* 40:248–251.
4. Botkin, D.B. 1990. *Discordant Harmonies: A New Ecology for the 21st Century.* New York: Oxford University Press.
5. The discussion of basic systems responses in Botkin, D.B., and K. Woods, *Fundamentals of Ecology* (in press).
6. Dunne, T., and L.B. Leopold. 1978. *Water in Environmental Planning.* San Francisco: Freeman.
7. Leach, M.K., and T.J. Givnich. 1996. Ecological determinants of species loss in remnant prairies. *Science* 273:1555–1558.
8. Lovelock, J. 1995. *The Ages of Gaia: A Biography of Our Living Earth*, rev. ed. New York: Norton.
9. Lenton, T.M., and Wilkinson, D.M. 2002. Developing the Gaia theory: A response to the criticisms of Kirchner and Volk. *Climate Change* 58:1–12.
10. Kirchner, J.W. 2002. The Gaia hypothesis—Fact, theory, and wishful thinking. *Climate Change* 52:391–408.
11. Kirchner, J.W. 2003. The Gaia hypothesis: Conjectures and refutations. *Climate Change* 58:21–45.

A CLOSER LOOK 4.1 NOTES

a. Botkin, D.B. 1999. *Passage of Discovery: The American Rivers Guide to the Missouri River of Lewis and Clark.* New York: Perigee Books (a Division of Penguin-Putnam).

b. Botkin, D.B. 2004. *Beyond the Stony Mountains: Nature in the American West from Lewis and Clark to Today.* New York: Oxford University Press.

WORKING IT OUT 4.1 NOTE

a. Bartlett, A.A. 1993. The arithmetic of growth: Methods of calculation. *Population and Environment* 14(4): 359–387.

CHAPTER 5

The Human Population and the Environment

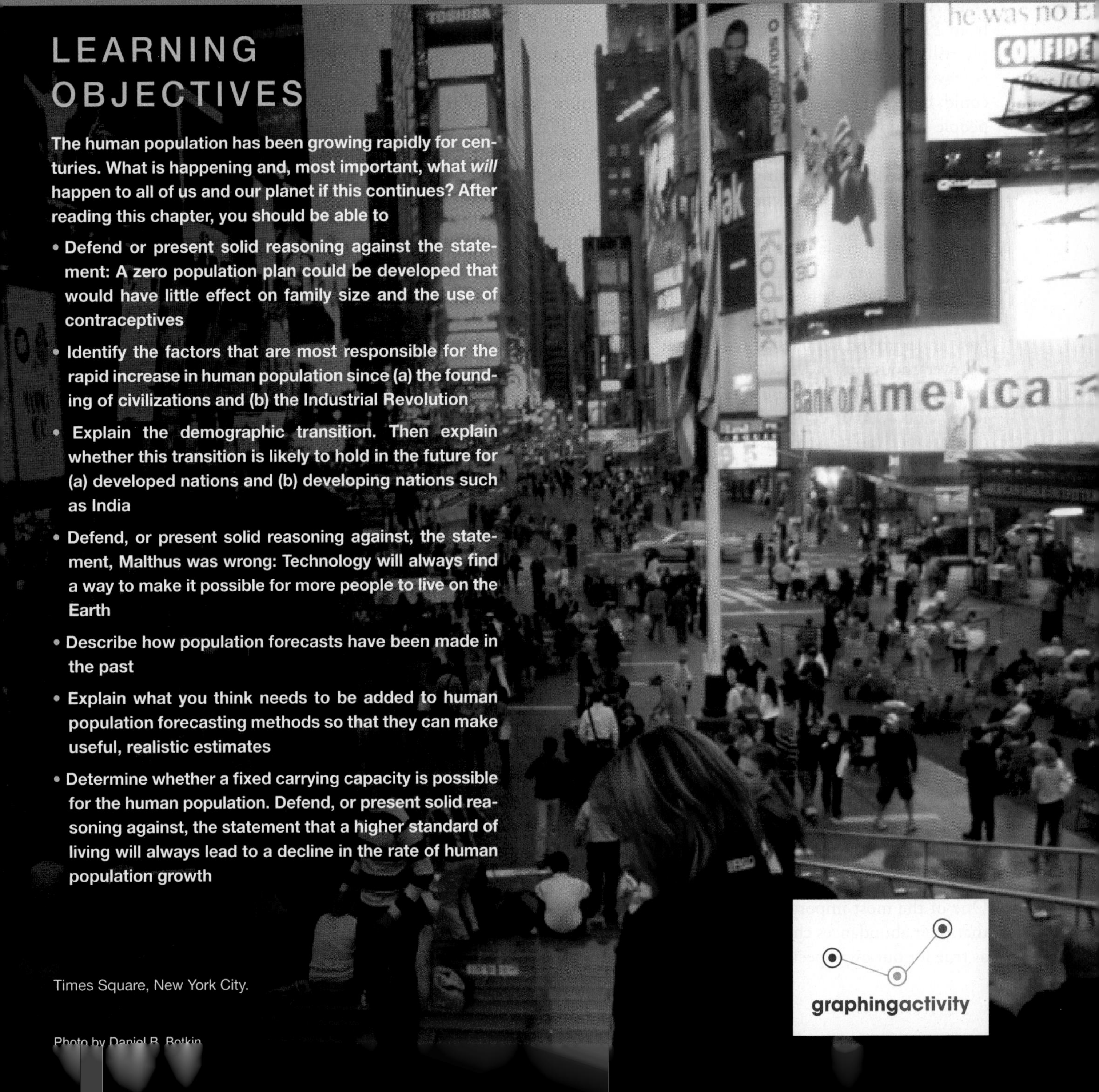

LEARNING OBJECTIVES

The human population has been growing rapidly for centuries. What is happening and, most important, what *will* happen to all of us and our planet if this continues? After reading this chapter, you should be able to

- Defend or present solid reasoning against the statement: A zero population plan could be developed that would have little effect on family size and the use of contraceptives
- Identify the factors that are most responsible for the rapid increase in human population since (a) the founding of civilizations and (b) the Industrial Revolution
- Explain the demographic transition. Then explain whether this transition is likely to hold in the future for (a) developed nations and (b) developing nations such as India
- Defend, or present solid reasoning against, the statement, Malthus was wrong: Technology will always find a way to make it possible for more people to live on the Earth
- Describe how population forecasts have been made in the past
- Explain what you think needs to be added to human population forecasting methods so that they can make useful, realistic estimates
- Determine whether a fixed carrying capacity is possible for the human population. Defend, or present solid reasoning against, the statement that a higher standard of living will always lead to a decline in the rate of human population growth

graphingactivity

Times Square, New York City.

Photo by Daniel B. Botkin

CASE STUDY

Your Future

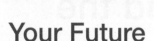

By 2050, when a college student about 20 years old today is 58 and beginning to think about retirement, the U.S. population will have reached 420 million—120 million more than today, a 40% increase, and life will have to be different. If each of the 420 million people, on average, continues to use the same amount of energy that the average American uses today, total energy use will increase from 29 to 40 trillion kilowatt hours and the energy supply will have to increase by 40%. It is unlikely that our energy sources can increase by that amount. And if they could, one could expect many environmental effects that people would at least complain about and not like, and many would be harmful to people and to other life-forms. This means that your future includes a considerable reduction in the amount of energy each of you will be able to use.

Currently, people in the United States use more fresh water per capita than people in any other nation, and we are reaching the limit of an easily renewable freshwater supply. As you will learn in Chapter 18, we are depleting deep aquifers, underground supplies of fresh water that have built up over thousands of years. You and your 419,999,999 companions will have to use a lot less water than you do today. You will, of course, have choices as to where to decrease per capita water use. Today, 49% of the fresh water in the United States is used to cool steam-generating, steam-turbine-running power plants, which includes all those that today use fossil fuels and nuclear power.

Want to go the nuclear power route? Today, our 104 nuclear power plants provide 8% of our electricity. That's 13 nuclear power plants for each 1% increase in energy. To increase America's total energy by 40% only from a nuclear source would require 520 more conventional such plants. That's an average of more than 10 per state. Just about everybody would be living within a zone that could be dangerously damaged by a nuclear power plant accident, like the one that happened in 2011 in Japan (see Chapter 17).

If you are hoping that the recently discovered new sources of natural gas and oil—shale gas and tar sands oil—will let us keep using as much electricity as you and your companions would like, then the water needed to cool all those new steam-generating power plants would increase way beyond our nation's environmental capacity to provide the water. You could take a lot fewer showers—use nonwater-consuming toilets, as starters—but since domestic water use is only about 11% of the total U.S. water use, that won't get you far in water conservation.

Wind, solar, ocean energy, and low-density geothermal could provide that 40%, but the land area and near-shore area required would make a lot of you unhappy and might not be the best for biodiversity.

Of course, these changes aren't going to happen all of a sudden in 2050. They are happening now. Each year you live is going to require that you use less energy and less water. Critics of this forecast of your future will tell you this is just another one of those Malthusian projections that have never yet come true, and that in a dynamic, inventive nation like ours technology will always find a way. In this chapter we will explore what that Malthusian projection is and its criticisms. In reading this chapter, you should be better prepared to plan for your own future and that of your nation. This isn't a case history about some far away little island of people you'll never meet and don't care about; it's about you, sitting right there now and beginning to read this chapter. So what are you going to do about it? First, you will need to understand basic concepts of population growth. Then you need to keep this population problem in mind as you study the rest of the chapters in this book, and think about how you and the other 419,999,999 people living on Earth in 2050 will deal with each of those environmental problems.

5.1 Basic Concepts of Population Dynamics

One of the most important properties of living things is that their abundances change over time and space. This is as true for our own species as it is for all others, including species that directly or indirectly affect our lives—for example, by providing our food, or materials for our shelter, or causing diseases and other problems—and those that we just like having around us or knowing that they exist.

In this chapter we focus on the human population because it is so important to all environmental

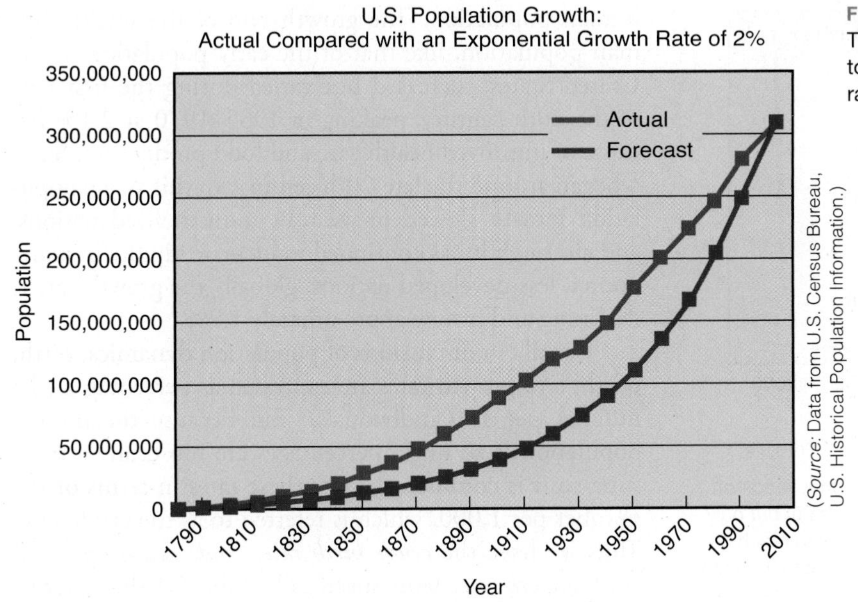

FIGURE 5.1 **U.S. population, 1790 to 2010.** The actual population growth is shown compared to an exponential curve with an annual growth rate of 2%.

problems, but the concepts we discuss here are useful for the populations of all species, and we will use these concepts throughout this book. You should also familiarize yourself with the following definitions and ideas:

- **Population dynamics** is the general study of population changes.
- A **population** is a group of individuals of the same species living in the same area or interbreeding and sharing genetic information.
- A **species** is all individuals that are capable of interbreeding, and so a species is composed of one or more populations.
- **Demography** is the statistical study of human populations, and people who study the human population include demographers.
- Five key properties of any population are **abundance**, which is the size of a population; **birth rates**, **death rates**, **growth rates**, and **age structure.** How rapidly a population's abundance changes over time depends on its growth rate, which is the difference between the birth rate and the death rate.
- The three rates—birth, death, and growth—are usually expressed as a percentage of a population per unit of time. For people, the unit of time is typically a year or greater. Sometimes these rates are expressed as actual numbers within a population during a specified time. (See Useful Human Population Terms in Table 5.1.)
- Let us begin with the population of the United States, which has grown rapidly since European settlement (Figure 5.1).

The Human Population as an Exponential Growth Curve

It is common to say that human populations, like that of the United States, grow at an **exponential rate,** which means that the annual growth rate is a constant percentage of the population (see Chapter 4). But Figure 5.1 shows that for much of the nation's history the population has grown at a rate that exceeds an exponential. The annual growth rate has changed over time, increasing in the early years, in part because of large immigrations to North America, and decreasing later. An exponential curve growing at 2% per year lags the actual increase in the U.S. population for most of the nation's history but catches up with it today. That is because the growth rate has slowed considerably. It is now 0.6%; in contrast, between 1790 and 1860, the year the Civil War began, the population increased more than 30% per year! (This is a rate that for a human population can be sustained only by immigration.)

Like that of the U.S. population, the world's human population growth is typically also shown as an exponential (Figure 5.2), although we know very little about the variation in the number of people during the early history of our species.

We can divide the history of our species' population into four phases (see A Closer Look 5.1 for more about this history). In Stage 1, the early period of hunters and gatherers, the world's total human population was probably less than a few million. Stage 2 began with the rise of agriculture, which allowed a much greater density of people and the first major increase in the human population. Stage 3, the Industrial Revolution in the late-18th and

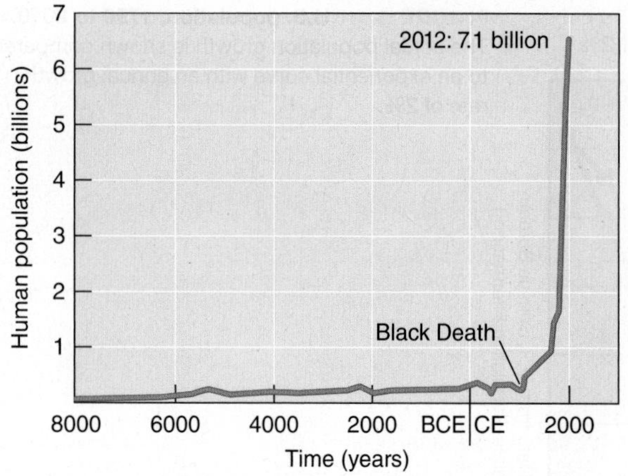

(*Source:* Reprinted with permission of John Wiley & Sons, Inc.)

FIGURE 5.2 **Human population growth.** It took thousands of years for the human population to reach 1 billion (in 1800) but only 30 years to reach 2 billion (1930). It only took 30 years to reach 3 billion (1960), 15 years to reach 4 billion (1975), 12 years to reach 5 billion (1987), 12 years to reach 6 billion (1999), and 13 years to reach 7 billion (2012).

early-19th centuries, saw improvements in health care and the food supply, which led to a rapid increase in the human population. The growth rate of the world's human population, like that of the early population of the United States, increased but varied during the first part of the 20th century, peaking in 1965–1970 at 2.1% because of improved health care and food production. Stage 4 began around the late-20th century. In this stage, population growth slowed in wealthy, industrialized nations, and although it has continued to increase rapidly in many poorer, less-developed nations, globally the growth rate is declining and is now approximately 1.2%.[1]

Usually in discussions of population dynamics, birth, death, and growth rates are expressed as percentages (the number per 100 individuals). But because the human population is so huge, percentages are too crude a measure; so it is common to state these rates in terms of the number per 1,000, which is referred to as the crude rate. Thus we have the *crude birth rate, crude death rate*, and *crude growth rate*. More specifically, Table 5.1 shows terms that are used frequently in discussions of human population change. Most of these terms will be explained in this chapter and will be useful to us in this book from time to time.

Table 5.1 USEFUL HUMAN POPULATION TERMS

Crude birth rate: number of births per 1,000 individuals per year; called "crude" because the population age structure is not taken into account.

Crude death rate: number of deaths per 1,000 individuals per year.

Crude growth rate: net number added per 1,000 individuals per year; also equal to the crude birth rate minus crude death rate.

Fertility: pregnancy or the capacity to become pregnant or to have children.

General fertility rate: number of live births expected in a year per 1,000 women aged 15–49, considered the childbearing years.

Total fertility rate (TFR): the average number of children expected to be born to a woman throughout her childbearing years.

Age-specific birth rate: number of births expected per year among a fertility-specific age group of women in a population. The fertility-specific age group is, in theory, all ages of women that could have children. In practice, it is typically assumed to be all women between 15 and 49 years of age.

Cause-specific death rate: the number of deaths from one cause per 100,000 total deaths.

Morbidity: a general term meaning the occurrence of disease and illness in a population.

Incidence: with respect to disease, the number of people contracting a disease during a specific time period, usually measured per 100 people.

Prevalence: with respect to a disease, the number of people afflicted at a particular time.

Case fatality rate: the percentage of people who die once they contract a disease.

Rate of natural increase (RNI): the birth rate minus the death rate, implying an annual rate of population growth not including migration.

Doubling time: the number of years it takes for a population to double, assuming a constant rate of natural increase.

Infant mortality rate: the annual number of deaths of infants under age 1 per 1,000 live births.

Life expectancy at birth: the average number of years a newborn infant can expect to live given current mortality rates.

GNP per capita: gross national product (GNP), which includes the value of all domestic and foreign output.

(*Source:* C. Haub and D. Cornelius, *World Population Data Sheet* [Washington, DC: Population Reference Bureau, 1998].)

A CLOSER LOOK 5.1

A Brief History of Human Population Growth

STAGE 1. **Hunters and Gatherers: From the first evolution of humans to the beginning of agriculture.**[a]

Population density: About 1 person per 130–260 km^2 in the most habitable areas.

Total human population: As low as one-quarter million, less than the population of modern small cities such as Hartford, Connecticut, and certainly fewer than the number of people—commonly a few million—who now live in many of our largest cities.

Average annual rate of growth: Over the entire history of human population, less than 0.00011% per year.

STAGE 2. **Early, Preindustrial Agriculture: Beginning sometime between 10,000 B.C. and 6000 B.C. and lasting until approximately the 16th century.**

Population density: With the domestication of plants and animals and the rise of settled villages, human population density increased greatly, to about 1 or 2 people/km^2 or more, beginning a second period in human population history.

Total human population: About 100 million by A.D. 1 and 500 million by A.D. 1600.

Average rate of growth: Perhaps about 0.03%, which was high enough to increase the human population from 5 million in 10,000 B.C. to about 100 million in A.D. 1. The Roman Empire accounted for about 54 million. From A.D. 1 to A.D. 1000, the population increased to 200–300 million.

STAGE 3. **The Machine Age: Beginning in the 16th century.**

Some experts say that this period marked the transition from agricultural to literate societies, when better medical care and sanitation were factors in lowering the death rate.

Total human population: About 900 million in 1800, almost doubling in the next century and doubling again (to 3 billion) by 1960.

Average rate of growth: By 1600, about 0.1% per year, with rate increases of about 0.1% every 50 years until 1950. This rapid increase occurred because of the discovery of causes of diseases, invention of vaccines, improvements in sanitation, other advances in medicine and health, and advances in agriculture that led to a great increase in the production of food, shelter, and clothing.

STAGE 4. **The Modern Era: Beginning in the mid-20th century.**

Total human population: Reaching and exceeding 6.6 billion.

Average rate of growth: The growth rate of the human population reached 2% in the middle of the 20th century and has declined to 1.2%.[b]

How Many People Have Lived on Earth?

Before written history, there were no censuses. The first estimates of population in Western civilization were attempted in the Roman era. During the Middle Ages and the Renaissance, scholars occasionally estimated the number of people. The first modern census was taken in 1655 in the Canadian colonies by the French and the British.[c] The first series of regular censuses by a country began in Sweden in 1750, and the United States has taken a census every decade since 1790. Most countries began counting their populations much later. The first Russian census, for example, was not taken until 1870. Even today, many countries do not take censuses or do not do so regularly. The population of China has only recently begun to be known with any accuracy. However, studying modern primitive peoples and applying principles of ecology can give us a rough idea of the total number of people who may have lived on Earth.

Summing all the values, including those since the beginning of written history, about 50 billion people are estimated to have lived on Earth.[d] If so, then, surprisingly, the more than 7 billion people alive today represent more than 10% of all of the people who have ever lived.

5.2 Analyzing and Estimating Future Population Growth

With human population growth a central issue, it is important that we develop ways to forecast what will happen to our population in the future. One of the simplest approaches is to calculate the doubling time.

Exponential Growth and Doubling Time

Recall from Chapter 4 and from Table 5.1 that **doubling time**, a concept used frequently in discussing human population growth, is the time required for a population to double in size (see Working It Out 5.1). The standard way to estimate doubling time is to assume that the population is growing exponentially and then divide 70 by the annual growth rate stated as a percentage. (Dividing into 70 is a consequence of the mathematics of exponential growth.)

The doubling time based on exponential growth is very sensitive to the growth rate—it changes quickly as the growth rate changes (Figure 5.3). A few examples demonstrate this sensitivity. With a current population growth of 1.2%, the world has a doubling time of 58 years. The current growth rate of Nicaragua is 2.5% (an increase from 2.0% in the previous estimate), giving that nation a doubling time of 28 years (so that the doubling time decreased from 35 years in just a few years of observations).[2] In contrast, Sweden, with an annual rate of about 0.2%, has a doubling time of 350 years. The world's most populous country, China, has a growth rate of 0.5% and a 140-year doubling time. The sensitivity of population growth to small changes in growth rate is well illustrated by China, whose growth rate just a few years ago was 0.6%, resulting in a 117-year doubling time.[2] The world's population growth rate peaked in the 1960s at about 2.2% and is now about 1.2% (Figure 5.4). If the growth rate had continued indefinitely at the 1960s peak, the world population would have doubled in 32 years. At today's rate, it will double in 58 years.

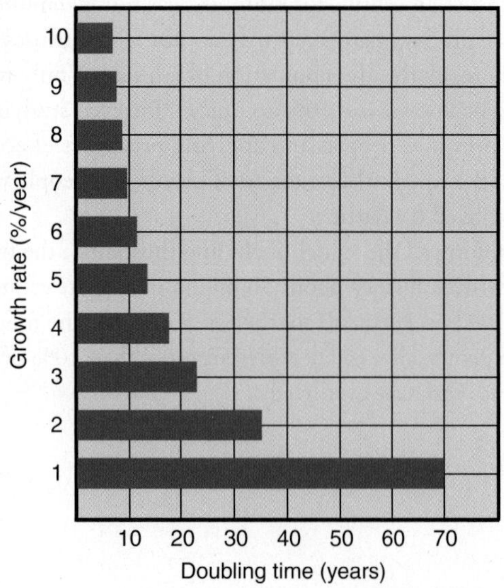

FIGURE 5.3 **Doubling time changes rapidly** with the population growth rate. Because the world's population is increasing at a rate between 1 and 2%, we expect it to double within the next 35 to 70 years.

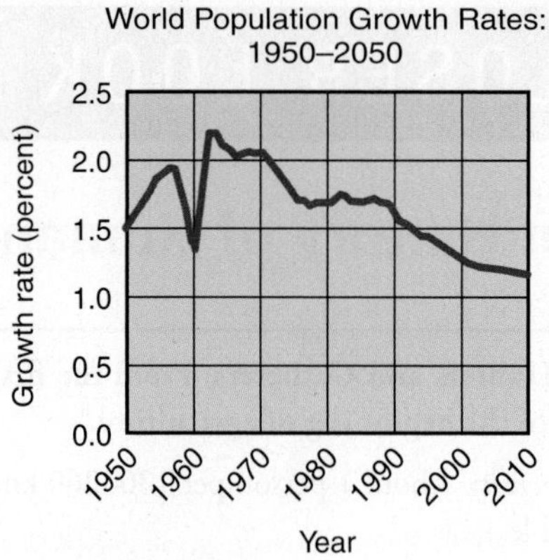

FIGURE 5.4 The annual growth rate of the world's population has been declining since the 1960s.

Human Population as a Logistic Growth Curve

An exponentially growing population theoretically increases forever. However, on Earth, which is limited in size, this is not possible, as Thomas Henry Malthus pointed out in the 18th century (see A Closer Look 5.2). Eventually, the population would run out of food and space and large numbers of people would become increasingly vulnerable to catastrophes, as we are already beginning to observe (see A Closer Look 5.3). Consider: A population of 100 increasing at 5% per year would grow to 1 billion in less than 325 years. *If the human population had increased at this rate since the beginning of recorded history, it would now exceed all the known matter in the universe.*

If a population cannot increase forever, what changes in the population can we expect over time? One of the first suggestions made about population growth is that it would follow a smooth S-shaped curve known as the **logistic growth curve**. This concept was first suggested in 1838 by a European scientist, P. F. Verhulst, as a theory for the growth of animal populations. It has been applied widely to the growth of many animal populations, including those important in wildlife management, endangered species and those in fisheries (see Chapter 13), as well as the human population.

A logistic population would increase exponentially only temporarily. After that, the rate of growth would gradually decline (i.e., the population would increase more slowly) until an upper population limit, called the

WORKING IT OUT 5.1 Forecasting Population Change

Populations change in size through births, deaths, immigration (arrivals), and emigration (departures). We can write a formula to represent population change in *terms of actual numbers in a population*:

$$P_2 = P_1 + (B - D) + (I - E)$$

where P_1 is the number of individuals in a population at time 1, P_2 is the number of individuals in that population at some later time 2, B is the number of births in the period from time 1 to time 2, D is the number of deaths from time 1 to time 2, I is the number entering as immigrants, and E is the number leaving as emigrants.

So far we have expressed population change in terms of total numbers in the population. We can also express these numbers as rates, including the birth rate (number born divided by the total number in the population), death rate (number dying divided by the total number in the population), and growth rate (change in the population divided by the total number in the population). (In this section, we will use lowercase letters to represent a rate and uppercase letters to represent total amounts.)

Ignoring for the moment immigration and emigration, we can state that how rapidly a population changes depends on the growth rate, g, which is the difference between the birth rate and the death rate (see the earlier list of useful terms). For example, in 1999 the crude death rate, d, in the United States was 9, meaning that 9 of every 1,000 people died each year. (The same information expressed as a percentage is a rate of 0.9%.) In 1999 the crude birth rate, b, in the United States was 15.[a] The crude growth rate is the net change—the birth rate minus the death rate. Thus the crude growth rate, g, in the United States in 1999 was 6. For every 1,000 people at the beginning of 1999, there were 1,006 at the end of the year.

Continuing for the moment to ignore immigration and emigration, we can state that how rapidly a population grows also depends on the difference between the birth rate and the death rate. The growth rate of a population is then

$$g = (B - D)/N \text{ or } g = G/N$$

Note that in all these cases, the units are numbers per unit of time.

It is important to be consistent in using the population at the beginning, middle, or end of the period. Usually, the number at the beginning or the middle is used. Consider an example: There were 19,700,000 people in Australia in mid-2002, and 394,000 births from 2002 to 2003. The birth rate, b, calculated against the mid-2002 population was 394,000/19,700,000, or 2%. During the same period, there were 137,900 deaths; the death rate, d, was 137,900/19,700,000, or 0.7%. The growth rate, g, was (394,000 – 137,900)/19,700,000, or 1.3%.[a]

Recall from Chapter 4 that doubling time—the time it takes a population to reach twice its present size—can be estimated by the formula

$$T = 70/\text{annual growth rate}$$

where T is the doubling time and the annual growth rate is expressed as a percentage. For example, a population growing 2% per year would double in approximately 35 years.

---

logistic carrying capacity, was reached. Once that had been reached, the population would remain at that number.

Although the logistic growth curve is an improvement over the exponential, it too involves assumptions that are unrealistic for humans and other mammals. Both the exponential and logistic assume a constant environment and a homogeneous population—one in which all individuals are identical in their effects on each other. In addition to these two assumptions, the logistic assumes a constant carrying capacity, which is also unrealistic in most cases, as we will discuss later. There is, in short, little evidence that human populations—or any animal populations, for that matter—actually follow this growth curve, for reasons that are pretty obvious if you think about all the things that can affect a population.[3]

Nevertheless, the logistic curve has been used—and continues to be used—for most long-term forecasts of the size of human populations in specific nations. As we said, this S-shaped curve first rises steeply upward and then changes slope, curving toward the horizontal carrying capacity. The point at which the curve changes is the **inflection point**, and until a population has reached this point, we cannot project its final logistic size. Even before the human population had not yet made the bend around the inflection point, forecasters typically dealt with this problem by assuming that the population was just reaching the inflection point at the time the forecast was made. This standard practice inevitably led to a great underestimate of the maximum population. For example, one of the first projections of the upper limit of the

A CLOSER LOOK 5.2

The Prophecy of Malthus

Almost 200 years ago, the English economist Thomas Malthus eloquently stated the human population problem. His writings have gone in and out of fashion, and some people think his views may be out of date, but in 2008 Malthus was suddenly back on the front page, the focus of major articles in *The New York Times*[a] and *The Wall Street Journal*, among other places. Perhaps this is because recent events—from natural catastrophes in Asia to rising prices for oil, food, and goods in general—suggest that the human population problem really is a problem.

Malthus based his argument on three simple premises:[b]

- **Food is necessary for people to survive.**
- **"Passion between the sexes is necessary and will remain nearly in its present state"—so children will continue to be born.**
- **The power of population growth is infinitely greater than the power of Earth to produce subsistence.**

Malthus reasoned that it would be impossible to maintain a rapidly multiplying human population on a finite resource base. His projections of the ultimate fate of humankind were dire, as dismal a picture as that painted by today's most extreme pessimists. The power of population growth is so great, he wrote, that "premature death must in some shape or other visit the human race. The vices of mankind are active and able ministers of depopulation, but should they fail, sickly seasons, epidemics, pestilence and plague, advance in terrific array, and sweep off their thousands and ten thousands." Should even these fail, he said, "gigantic famine stalks in the rear, and with one mighty blow, levels the population with the food of the world."

Malthus's statements are quite straightforward. From the perspective of modern science, they simply point out that in a finite world nothing can grow or expand forever, not even the population of the smartest species ever to live on Earth. Critics of Malthus continue to point out that his predictions have yet to come true, that whenever things have looked bleak, technology has provided a way out, allowing us to live at greater densities. Our technologies, they insist, will continue to save us from a Malthusian fate, so we needn't worry about human population growth. Supporters of Malthus respond by reminding them of the limits of a finite world.

Who is correct? Ultimately, in a finite world, Malthus must be correct about the final outcome of unchecked growth. He may have been wrong about the timing; he did not anticipate the capability of technological changes to delay the inevitable. But although some people believe that Earth can support many more people than it does now, in the long run there must be an upper limit. The basic issue that confronts us is this: How can we achieve a constant world population, or at least halt the increase in population, in a way most beneficial to most people? This is undoubtedly one of the most important questions that has ever faced humanity, and it is coming home to roost now.

Recent medical advances in our understanding of aging, along with the potential of new biotechnology to increase both the average longevity and maximum lifetime of human beings, have major implications for the growth of the human population. As medical advances continue to take place, the death rate will drop and the growth rate will rise even more. Thus, a prospect that is positive from the individual's point of view—a longer, healthier, and more active life—could have negative effects on the environment. We will therefore ultimately face the following choices: Stop medical research into chronic diseases of old age and other attempts to increase people's maximum lifetime; reduce the birth rate; or do neither and wait for Malthus's projections to come true—for famine, environmental catastrophes, and epidemic diseases to cause large and sporadic episodes of human death. The first choice seems inhumane, but the second is highly controversial, so doing nothing and waiting for Malthus's projections may be what actually happens, a future that nobody wants. For the people of the world, this is one of the most important issues concerning science and values, and people and nature.

U.S. population, made in the 1930s, assumed that the inflection point had been reached then. That assumption resulted in an estimate that the final population of the United States would be approximately 200 million.[4]

Fortunately for us, Figure 5.4 suggests that our species' growth rate has declined consistently since the 1960s, as we noted before, and therefore we can make projections using the logistic method, assuming that we have passed the inflection point. The United Nations has made a series of projections based on current birth rates and death rates and assumptions about how these rates will change. These projections form the basis for the curves presented in Figure 5.5. The logistic projections assume that (1) mortality will fall everywhere and level off when female life expectancy reaches 82 years; (2) fertility will reach replacement levels everywhere between 2005 and 2060; and (3) there will be no worldwide catastrophe. This approach projects an equilibrium world population of 10.1–12.5 billion.[5] Developed nations would experience population growth from 1.2 billion today to 1.9

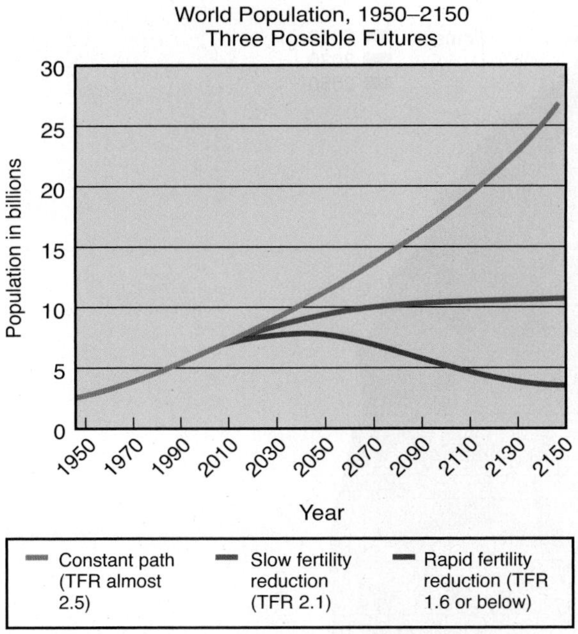

(*Source:* U.S. Census Bureau, *Global Population Profile: 2002*, Table A-10, p. 1. http://www.census.gov.)

FIGURE 5.5 **UN projections of world population growth based on the logistic curve and using different total fertility rates (the expected number of children a woman will have during her life).** The constant path assumes the 1998 growth rate will continue unchanged, resulting in an exponential increase. The slow-fertility-reduction path assumes the world's fertility declines to replacement level by 2050 and the world's population stabilizes at about 11 billion by the 22nd century. The rapid-fertility-reduction path assumes that the total fertility rate will decline in the 21st century, with population peaking at 7.7 billion in 2050 and dropping to 3.6 billion by 2150. These are theoretical curves.

billion, but populations in developing nations would increase from 4-5 billion to 9.6 billion. Bangladesh (an area the size of Wisconsin) would reach 257 million; Nigeria, 453 million; and India, 1.86 billion. In these projections, the developing countries contribute 95% of the increase.[6]

5.3 Age Structure

As we noted earlier, the two standard methods for forecasting human population growth—the exponential and the logistic—ignore all characteristics of the environment and in that way are seriously incomplete. A more comprehensive approach would take into account the effects of the supply of food, water, and shelter; the prevalence of diseases; and other factors that can affect birth and death rates. But with long-lived organisms like ourselves, these environmental factors have different effects on different age groups, and so the next step is to find a way to express how a population is divided among ages. This is known as the population age structure, which is the proportion of the population of each age group. The age structure of a population affects current and future birth rates, death rates, and growth rates; has an impact on the environment; and has implications for current and future social and economic conditions.

We can picture a population's age structure as a pile of blocks, one for each age group, with the size of each block representing the number of people in that group (Figure 5.6). Although age structures can take many shapes, four general types are most important to our discussion: a pyramid, a column, an inverted pyramid (top-heavy), and a column with a bulge. The pyramid age structure occurs in a population that has many young people and a high death rate at each age—and therefore

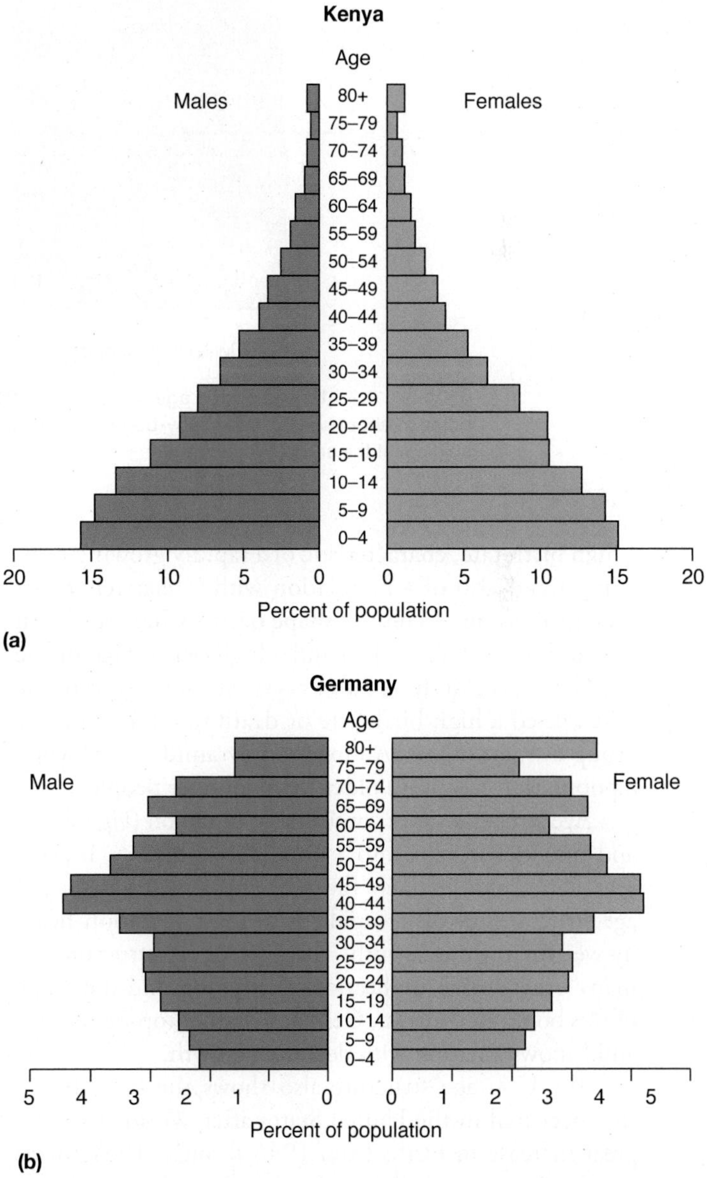

FIGURE 5.6 **(a) Age structure of Kenya and Germany** In 2011 43% of the total population of Kenya is under age 15. Kenya's population is growing rapidly, more than tripling from 10.9 million people in 1969 to 38.6 million people in 2009. **(b)** In Germany in 2009, there are much fewer young people.

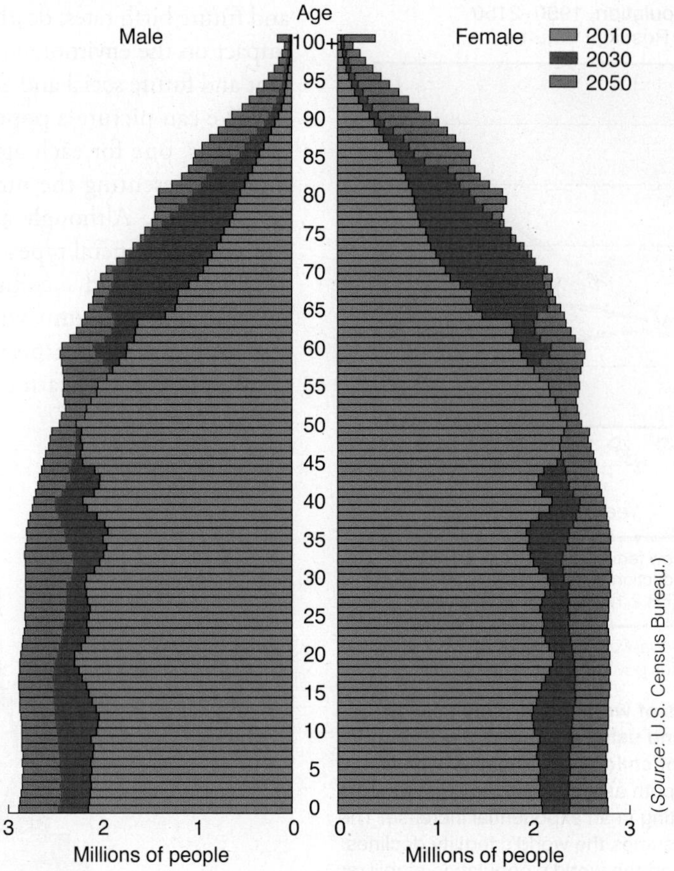

FIGURE 5.7 **United States age structure in 2010 and as projected for 2030 and 2050.** The current bulge at about age 50 is the baby-boom generation—the result of a spurt of births following the return of soldiers at the end of World War II.

a high birth rate, characteristic of a rapidly growing population and also of a population with a relatively short average lifetime. A column shape occurs where the birth rate and death rate are low and a high percentage of the population is elderly. A bulge occurs if some event in the past caused a high birth rate or death rate for some age group but not others. An inverted pyramid occurs when a population has more older than younger people.

Age structure varies considerably by nation (Figure 5.6) and provides insight into a population's history, its current status, and its likely future. Kenya's pyramid-shaped age structure reveals a rapidly growing population heavily weighted toward youth. This type of age structure has many other social implications that go beyond the scope of this book. In contrast, Germany's rather top-heavy pyramid shows a nation with declining growth.

The U.S. age structure also shows the baby boom that occurred in the United States after World War II; a great increase in births from 1946 through 1964 forms a pulse in the population that can be seen as a bulge in the age structure, especially of those aged 45–55 in 2010 (Figure 5.7). At each age, baby boomers increased demand for social and economic resources; for example, schools were crowded when the baby boomers were of primary- and secondary-school age. You can see a secondary bulge from the children of the baby-boom generation, around ages 20 and 25. In the forecast for 2030, the baby-boom generation moves upward, leading to changes in demands for social resources such as retirement homes, and declines in certain kinds of purchases, characteristic of younger people still buying things for their home and their activities.

5.4 The Demographic Transition

The **demographic transition** is a multistage pattern of change in birth rates and death rates that has occurred during the process of industrial and economic development in Western nations. It leads to a decline in population growth.

A decline in the death rate is the first stage of the demographic transition (Figure 5.8).[7] In a nonindustrial country, birth rates are high, but the growth rate is low because death rates are also high, canceling out the effect of the high birth rates. With industrialization, health and sanitation improve and the death rate drops rapidly.

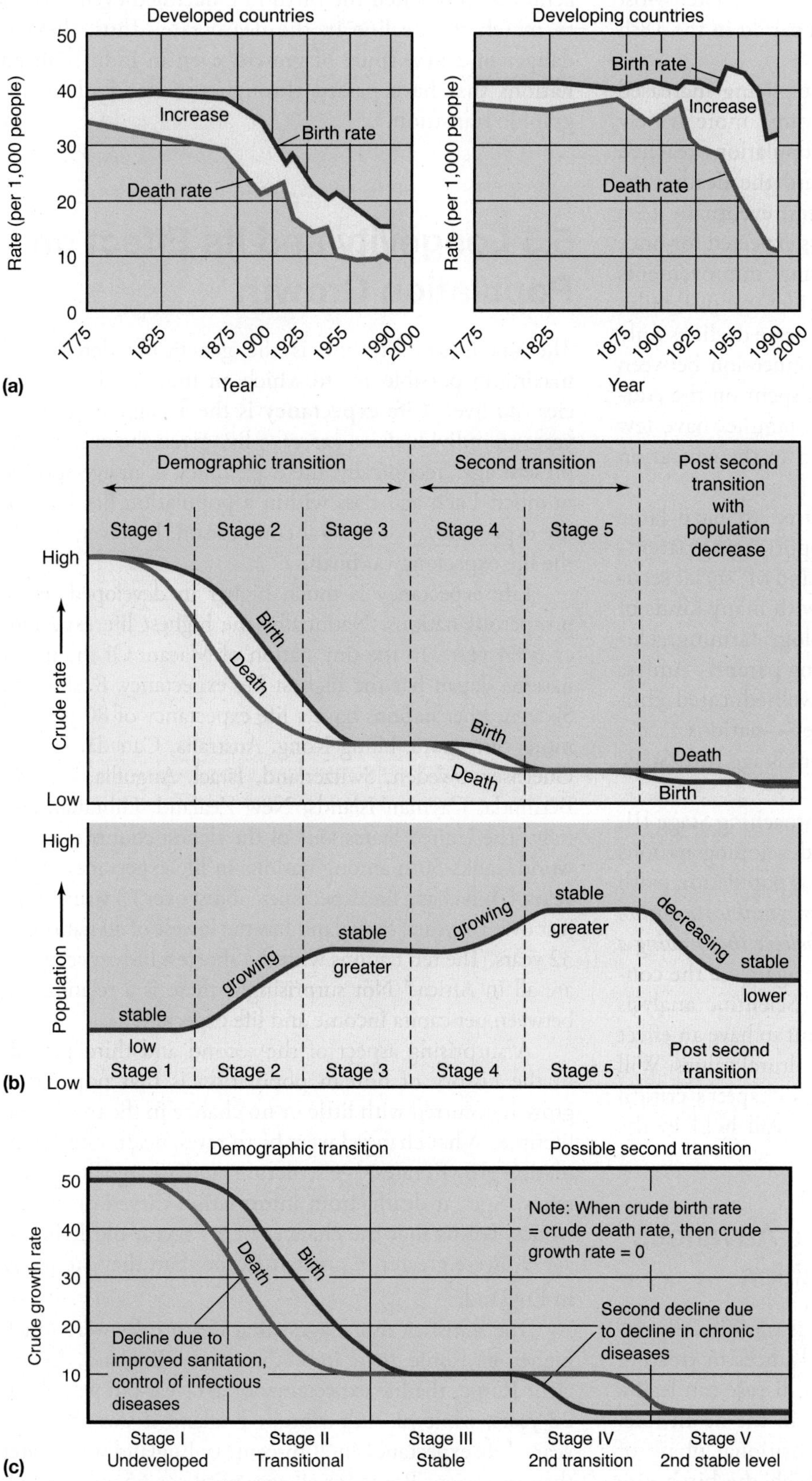

(*Source:* M.M. Kent and K.A. Crews, *World Population: Fundamentals of Growth* [Washington, DC: Population Reference Bureau, 1990]. Copyright 1990 by the Population Reference Bureau, Inc. Reprinted by permission.)

FIGURE 5.8 **The demographic transition: (a)** Theoretical, including possible fourth and fifth stages that might take place in the future; **(b)** the resulting relative change in population; **(c)** the change in birth rates and death rates from 1775 to 2000 in developed and developing countries.

The birth rate remains high, however, and the population enters Stage II, a period with a high growth rate. Most European nations passed through this period in the 18th and 19th centuries.

As education and the standard of living increased and as family-planning methods became more widely used in industrialized nations, the populations reached Stage III. The birth rate drops toward the death rate, and the growth rate therefore declines, eventually to a low or zero growth rate. This change is expected for once undeveloped nations now undergoing improvements in education, living standards, and wider use of family-planning methods. However, the birth rate declines only if families believe there is a direct connection between future economic well-being and funds spent on the education and care of their young. Such families have few children and devote all their resources to the education and well-being of those few.

Historically, parents have preferred to have large families. Without other means of support, aging parents can depend on grown children for a kind of "social security," and even young children help with many kinds of hunting, gathering, and low-technology farming. Unless there is a change in attitude among parents—unless they see more benefits from a few well-educated children than from many poorer children—nations face a problem in making the transition from Stage II to Stage III (see Figure 5.8).

Some developed countries are approaching Stage III, but it is an open question whether developing nations will make the transition before a serious population crash occurs. *The key point here is that the demographic transition will take place only if parents come to believe that having a small family is to their benefit.* Here we again see the connection between science and values. Scientific analysis can show the value of small families, but to have an effect this knowledge must become part of cultural values. Will the demographic transition hold? This chapter's critical thinking exercise considers whether it will hold in the United States.

Potential Effects of Medical Advances on the Demographic Transition

Although the demographic transition is traditionally defined as consisting of three stages, advances in treating chronic health problems such as heart disease can lead a Stage III country to a second decline in the death rate. This could bring about a second transitional phase of population growth (Stage IV), in which the birth rate would remain the same while the death rate fell. A second stable phase of low or zero growth (Stage V) would be achieved only when the birth rate declined even further to match the decline in the death rate. Thus, there is danger of a new spurt of growth even in industrialized nations that have passed through the standard demographic transition.

5.5 Longevity and Its Effect on Population Growth

The **maximum lifetime** is the genetically determined maximum possible age to which an individual of a species *can* live. **Life expectancy** is the average number of years an individual *can expect* to live given the individual's present age. Technically, life expectancy is an age-specific number: Each age class within a population has its own life expectancy. For general comparison, however, we use the life expectancy at birth.

Life expectancy is much higher in developed, more prosperous nations. Nationally, the highest life expectancy is 84 years, in the tiny nation of Macau. Of the major nations, Japan has the highest life expectancy, 82.1 years. Sixteen other nations have a life expectancy of 80 years or more: Singapore, Hong Kong, Australia, Canada, France, Guernsey, Sweden, Switzerland, Israel, Anguilla, Iceland, Bermuda, Cayman Islands, New Zealand, Gibraltar, and Italy. The United States, one of the richest countries in the world, ranks 50th among nations in life expectancy, at 78 years. China has a life expectancy of just over 73 years; India just over 69 years. Swaziland has the lowest of all nations at 32 years. The ten nations with the shortest life expectancies are all in Africa.[8] Not surprisingly, there is a relationship between per capita income and life expectancy.

A surprising aspect of the second and third periods in the history of human population is that population growth occurred with little or no change in the maximum lifetime. What changed were birth rates, death rates, population growth rates, age structure, and average life expectancy. Ages at death, from information carved on tombstones, tell us that the chances of a 75-year-old living to age 90 were greater in ancient Rome than they are today in England.

The statistics also suggest that death rates were much higher in Rome than in 20th-century England. In ancient Rome, the life expectancy of a 1-year-old was about 22 years, while in 20th-century England it was about 50 years. Life expectancy in 20th-century England was greater than in ancient Rome for all ages until age 55, after which it appears to have been higher for ancient Romans than for 20th-century Britons. This suggests that many hazards of modern life may be concentrated more on the aged. Pollution-induced diseases are one factor in this change.

Human Death Rates and the Rise of Industrial Societies

We return now to further consideration of the first stage in the demographic transition. We can get an idea of the first stage by comparing a modern industrialized country, such as Switzerland, which has a crude death rate of 8.8 per 1,000, with a developing nation, such as Central African Republic, which has a crude death rate of 14.71.[2] Modern medicine has greatly reduced death rates from disease in countries such as Switzerland, particularly with respect to death from acute or epidemic diseases, such as flu, severe acute respiratory syndrome (SARS), and West Nile virus, which we will discuss in A Closer Look 5.3.

A CLOSER LOOK 5.3

Pandemics And World Population Growth: West Nile Virus

Before 1999, West Nile virus occurred in Africa, West Asia, and the Middle East, but not in the New World. Related to encephalitis, West Nile virus is spread by mosquitoes, which bite infected birds, ingest the virus, and then bite people. It reached the Western Hemisphere through infected birds and has now been found in more than 25 species of birds native to the United States, including crows, the bald eagle, and the black-capped chickadee—a common visitor to bird feeders in the U.S. Northeast. Fortunately, in human beings this disease has lasted only a few days and has rarely caused severe symptoms.[a] By 2007, more than 3,600 people in the United States had contracted this disease, most in California and Colorado, with 124 fatalities.[b] But the speed with which it spread led to concerns about other possible new pandemics.

Four years earlier, in February 2003, the sudden occurrence of a new disease, severe acute respiratory syndrome (SARS), had demonstrated that *modern transportation and the world's huge human population could lead to the rapid spread of epidemic diseases.* Jet airliners daily carry vast numbers of people and goods around the world. The disease began in China, perhaps spread from some wild animal to human beings. China had become much more open to foreign travelers, with more than 90 million visitors in a recent year.[c] By late spring 2003, SARS had spread to two dozen countries; more than 8,000 people were affected and 774 died. Quick action, led by the World Health Organization (WHO), contained the disease.[d]

And behind all of this is the knowledge of the 1918 world flu virus, which is estimated to have killed as many as 50 million people in one year, probably more than any other single epidemic in human history. It spread around the world in the autumn, striking otherwise healthy young adults in particular. Many died within hours! By the spring of 1919, the virus had virtually disappeared.[e]

Although outbreaks of the well-known traditional epidemic diseases have declined greatly during the past century in industrialized nations, there is now concern that the incidence of **pandemics** may increase due to several factors. One is that as the human population grows, people live in new habitats, where previously unknown diseases occur. Another is that strains of disease organisms have developed resistance to antibiotics and other modern methods of control.

A broader view of why diseases are likely to increase comes from an ecological and evolutionary perspective (which will be explained in later chapters). Stated simply, the more than 6.6 billion people on Earth constitute a great resource and opportunity for other species; it is naive to think that other species will not take advantage of this huge and easily accessible host. From this perspective, the future promises more diseases rather than fewer. This is a new perspective. In the mid-20th century it was easy to believe that modern medicine would eventually cure all diseases and that most people would live the maximum human life span. It is generally believed, and often forecast, that the human population will simply continue increasing, without any decline. But with increased crowding and its many effects on the environment, there is also concern that the opposite might happen, that our species might suffer a large, if temporary, dieback. This leads us to consider how populations change over time and space, especially our own populations.

An **acute disease** or **epidemic disease** appears rapidly in the population, affects a comparatively large percentage of it, and then declines or almost disappears for a while, only to reappear later. Epidemic diseases typically are rare but have occasional outbreaks during which a large proportion of the population is infected. A **chronic disease**, in contrast, is always present in a population, typically occurring in a relatively small but relatively constant percentage of the population. Heart disease, cancer, and stroke are examples.

The great decrease in the percentage of deaths due to acute or epidemic diseases can be seen in a comparison of causes of deaths in Ecuador in 1987 and in the United States in 1900, 1987, and 1998 (Figure 5.9).[9] In Ecuador, a developing nation, acute diseases and those listed as "all others" accounted for about 60% of mortality in 1987. In the United States in 1987, these accounted for only 20% of mortality. Chronic diseases account for about 70% of mortality in the modern United States. In contrast, chronic diseases accounted for less than 20% of the deaths in the United States in 1900 and about 33% in Ecuador in 1987. Ecuador in 1987, then, resembled the United States of 1900 more than it resembled the United States of either 1987 or 1998.

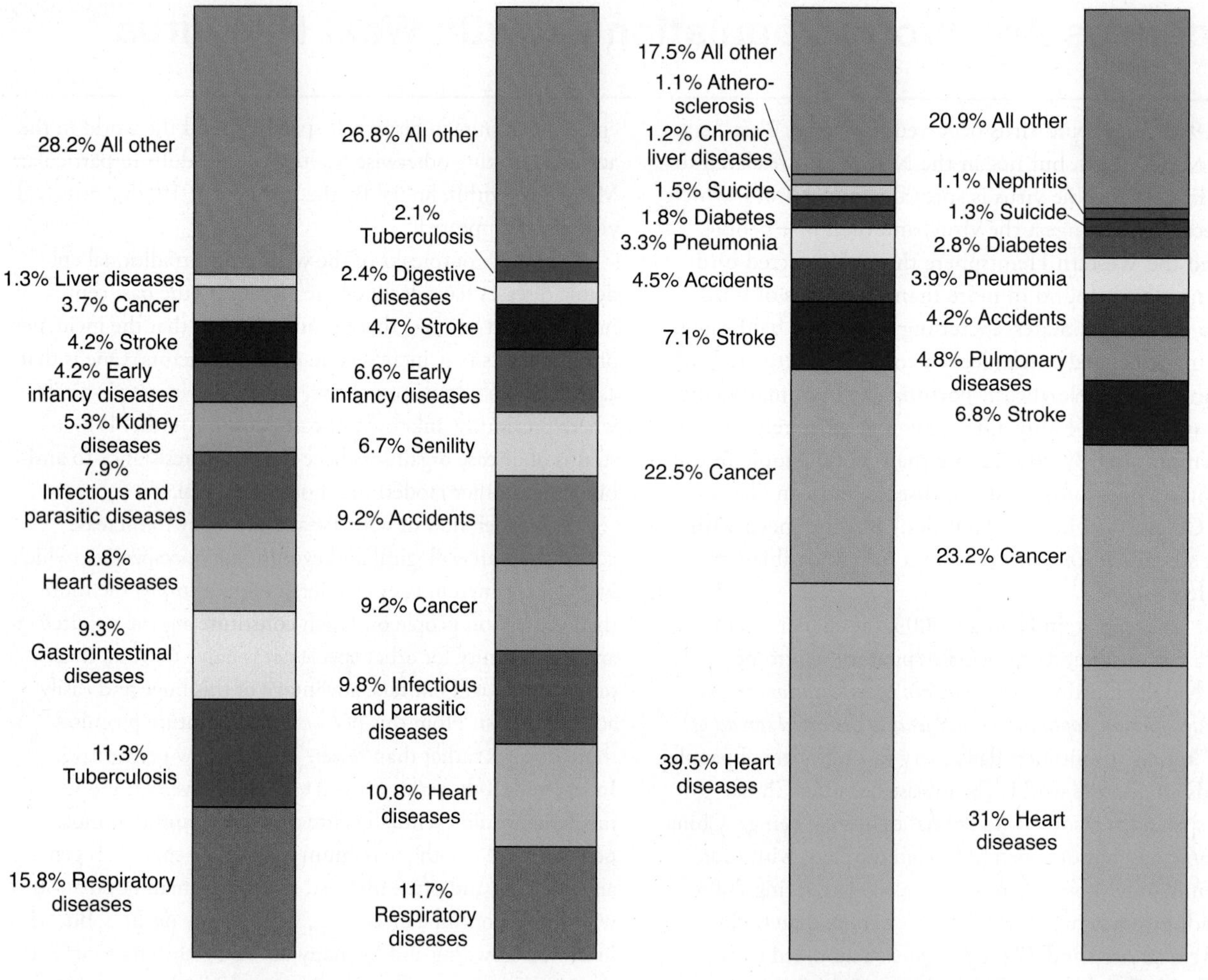

(*Sources:* U.S. 1900, Ecuador 1987, and U.S. 1987 data from M.M. Kent and K. A. Crews, *World Population: Fundamentals of Growth* [Washington, DC: Population Reference Bureau, 1990]. Copyright 1990 by the Population Reference Bureau, Inc. Reprinted by permission. *National Vital Statistics Report* 48 [11], July 24, 2000.)

FIGURE 5.9 **Causes of mortality in industrializing, developing, and industrialized nations.**

5.6 The Human Population's Effects on the Earth

The danger that the human population poses to the environment is the result of two factors: the number of people and the environmental impact of each person. When there were fewer people on Earth and limited technology, the human impact was less than in today's industrialized, global, society. Even so, people have affected the environment for a surprisingly long time. It started with the use of fire to clear land. There is some evidence that early human ancestors used fire, from burned animal bones in *Homo erectus* sites as early as 500,000 years ago,[3] but it is unclear how early our ancestors started fires that burned and cleared large areas. Effects on the environment increased with the rise of early civilizations. For example, the Mayan temples in South America, standing now in the midst of what were recently believed to be ancient rain forests, actually stood in large areas of farmed land cleared by the Maya. Large areas of North America were modified by American Indians, who used fire for a variety of reasons and modified the forests of the eastern United States.[3] The problem now is that there are so many people and our technologies are so powerful that our effects on the environment are even more global and significant. This could cause a negative feedback—the more people, the worse the environment; the worse the environment, the fewer people.

The simplest way to characterize the total impact of the human population on the environment is to multiply the average impact of an individual by the total number of individuals,[10] or

$$T = P \times I$$

where P is the population size—the number of people—and I is the average environmental impact per person. Of course, the impact per person varies widely, within the same nation and also among nations. The average impact of a person who lives in the United States is much greater than the impact of a person who lives in a low-technology society. But even in a poor, low-technology nation like Bangladesh, the sheer number of people leads to large-scale environmental effects.

Modern technology increases the use of resources and enables us to affect the environment in many new ways, compared with hunters and gatherers or people who farmed with simple wooden and stone tools. For example, before the invention of chlorofluorocarbons (CFCs), which are used as propellants in spray cans and as coolants in refrigerators and air conditioners, we were not causing depletion of the ozone layer in the upper atmosphere. Similarly, before we started driving automobiles, there was much less demand for steel, little demand for oil, and much less air pollution. These linkages between people and the global environment illustrate the global theme and the people-and-nature theme of this book.

The population-times-technology equation reveals a great irony involving two standard goals of international aid: improving the standard of living and slowing overall human population growth. Improving the standard of living increases the total environmental impact, countering the environmental benefits of a decline in population growth.

5.7 Estimating How Many People Could Live on Earth

How many people can live on Earth at the same time? Of course, the answer will vary with environmental conditions, including climate. The maximum number of people that Earth—often referred to as the **human carrying capacity**—could support at the height of an Ice Age would be very different from what could be supported during an interglacial. But the answer depends on what quality of life people desire and are willing to accept.

On our finite planet, the human population will eventually be limited by some factor or combination of factors—there can't be an infinite number of people living on a finite planet. The maximum number also depends in important ways on the technology available. The invention of farming allowed for a much higher density of people in an area than was possible before, and modern advances in agriculture have greatly increased the production of food per land area, so that many more people can be supported on a smaller agricultural land base.

The role that technological advances will play in the future is at the heart of the debate that has gone on for several centuries and continues today. On the one hand is the prophecy of Malthus (A Closer Look 5.2). Malthus basically pointed out that at some population level resources could not increase at the rate that a human population could grow, and therefore there had to be an upper limit to how many people could live on Earth. Critics of Malthus's prediction assert that technological advances have always found a way for more people to live on less and less land area. This has been the case, but it is obvious that in a finite world this cannot go on forever. The debate between the pro- and anti-population growth people is how many more people (if any more) the Earth can sustain. It is a question that will occupy us and be with us indirectly, if not directly, throughout this book.

On our finite planet the human population will eventually be limited by some factor or combination of factors.

We can group limiting factors into those that affect a population during the year in which they become limiting (short-term factors), those whose effects are apparent after one year but before ten years (intermediate-term factors), and those whose effects are not apparent for ten years (long-term factors). Some factors fit into more than one category, having, say, both short-term and intermediate-term effects.

Important *short-term* factors include the disruption of food distribution in a country, commonly caused by drought; introductions of new crop and domestic animal diseases; shortages of energy for transporting food; and societal actions including war.

Intermediate-term factors include desertification; dispersal of certain pollutants, such as toxic metals, into waters and fisheries; disruption in the supply of nonrenewable resources, such as rare metals used in making steel alloys for transportation machinery; and a decrease in the supply of firewood or other fuels for heating and cooking.

Long-term factors include soil erosion, a decline in groundwater supplies, and climate change. A decline in resources available per person suggests that we may already have exceeded Earth's long-term maximum population size under present environmental conditions. For example, food supply per person peaked in the late 1980s and has declined since then (Figure 5.10). Malthusians will argue that this is evidence that Earth is reaching its limits to supply people. Anti-Malthusians will argue just to wait for the next technological advance in agriculture.

Since the rise of the modern environmental movement in the second half of the 20th century, much attention has focused on estimating what is commonly referred to as the human carrying capacity of Earth—the total number of people that our planet could support indefinitely. This estimation has typically involved three methods. One method, which we have already discussed, is to simply extrapolate from past growth, assuming that the population will follow an S-shaped logistic growth curve and gradually level off (Figure 5.5).

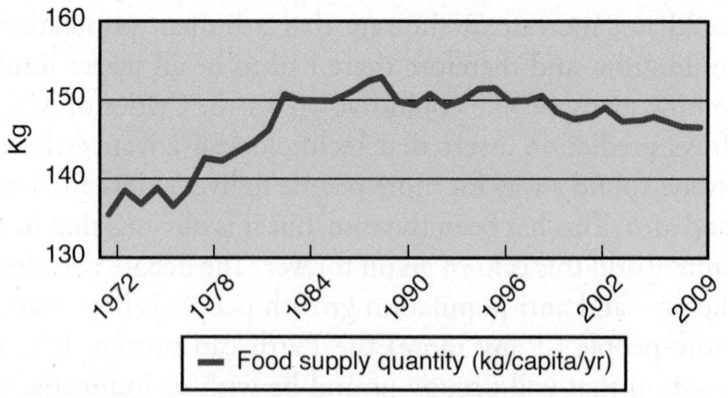

FIGURE 5.10 World food supply per person per year. (*Source:* UN Food and Agriculture Organization Statistics http://faostat3.fao.org/home/index.html#VISUALIZE.)

The second method can be referred to as the packing-problem approach. This method simply considers how many people might be packed onto Earth, not taking into sufficient account the need for land and oceans to provide food, water, energy, and construction materials, as well as the need to maintain biological diversity and the human need for scenic beauty. This approach, which could also be called the standing-room-only approach, has led to very high estimates of the total number of people that might occupy Earth—as many as 50 billion.

In the second half of the 20th century, a philosophical movement has developed at the other extreme. Known as deep ecology, this third method makes sustaining the biosphere the primary moral imperative. Its proponents argue that the whole Earth is necessary to sustain life, and therefore everything else must be sacrificed to the goal of sustaining the biosphere. People are considered active agents of destruction of the biosphere, and therefore the total number of people should be greatly reduced.[11] Estimates based on this rationale for the desirable number of people vary greatly, from a few million up.

Between the packing-problem approach and the deep-ecology approach are many options. It is possible to set goals in between these extremes, but each of these goals is a value judgment, again reminding us of one of this book's themes: *science and values.* What constitutes a desirable quality of life is a value judgment. The perception of what is desirable will depend in part on what we are used to, and this varies greatly. For example, in 2012, New Jersey had only a half acre (0.22 ha) per person, while Wyoming, the most sparsely populated of the lower 48 states, had 110 acres (44 ha) per person. For comparison, that year New York City's Manhattan Island had 71,000 people per square mile, which works out to an area of about 20 x 20 feet per person. Manhattanites manage to live comfortably by using not just the land area but also the airspace to a considerable height. Still, it's clear that people used to living in Wyoming and people living in New Jersey or in Manhattan skyscrapers are likely to have very different views on what is a desirable population density.

If all the people of the world were to live at the same level as those of the United States, with our high resource use, then the carrying capacity would be comparatively low. If all the people of the world were to live at the level of those in Bangladesh, with all of its risks as well as its poverty and its heavy drain on biological diversity and scenic beauty, the carrying capacity would be much higher.

In summary, the acceptable carrying capacity is not simply a scientific issue; it is an issue combining science and values, within which science plays two roles. First, by leading to new knowledge, which in turn leads to new technology, it makes possible a higher density of human beings. Second, scientific methods can be used to forecast a probable range of carrying capacities once a goal for the average quality of life, in terms of human values, is chosen. In this second use, science can tell us the implications of our value judgments, but it cannot provide those value judgments.

5.8 Can We Achieve Zero Population Growth?

We have surveyed several aspects of population dynamics. The underlying question is: Can we achieve **zero population growth**—a condition in which the human population, on average, neither increases nor decreases? Much of environmental concern has focused on how to lower the human birth rate and decrease our population growth. As with any long-lived animal population, our species could take several possible approaches to achieving zero population growth. Here are a few.

Age of First Childbearing

The simplest and one of the most effective means of slowing population growth is to delay the age of first childbearing.[12] As more women enter the workforce and as education levels and standards of living rise, this delay occurs naturally. Social pressures that lead to deferred marriage and childbearing can also be effective, or they can work the other way, if having children early becomes fashionable. The median age for women when they first give birth is 19.2 years in India and 29 years in New Zealand, so even if the birth rates were the same in those two countries, India's population would grow much faster.[12] In the United States, the average age at first child increased from 21.4 years in 1970 to 25.2 in 2009.[12]

Typically, countries where early marriage is common have high population growth rates, because this also decreases the age of first childbearing. In South Asia and in Sub-Saharan Africa, about 50% of women marry between the ages of 15 and 19, and in Bangladesh women marry on average at age 16. In Sri Lanka, however, the average age for marriage is 25. The World Bank estimates that if Bangladesh adopted Sri Lanka's marriage pattern, families could average 2.2 fewer children.[13] For many countries, raising the marriage age could account for 40–50% of the drop in fertility required to achieve zero population growth.

Birth Control: Biological and Societal

Another simple way to lower the birth rate is breast feeding, which can delay resumption of ovulation after childbirth.[14] Women in a number of countries use this deliberately as a birth-control method—in fact, according to the World Bank, in the mid-1970s breast feeding provided more protection against conception in developing countries than did family-planning programs.[13]

Family planning is still emphasized, however.[5] Traditional methods range from abstinence to the use of natural agents to induced sterility. Modern methods include the birth-control pill, which prevents ovulation through control of hormone levels; surgical techniques for permanent sterility; and mechanical devices.

The world's average total fertility rate has dropped dramatically since 1950, when the average was 5 children per woman. Today the average is 2.6 children per woman. According to the United Nations World Health Organization, this change is largely due to widespread use of modern contraceptives, especially in the developing world. Income and national development affect the percentage of women using contraceptives. In developed nations, an average of 62% of women use them, but in less-developed countries, on average of only 43% of women use contraceptives.[15]

National Programs to Reduce Birth Rates

Reducing birth rates requires a change in attitude, knowledge of the means of birth control, and the ability to afford these means. As we have seen, a change in attitude can occur simply with a rise in the standard of living. In many countries, however, it has been necessary to provide formal family-planning programs to explain the problems arising from rapid population growth and to describe the ways that individuals will benefit from reduced population growth. These programs also provide information about birth-control methods and provide access to these methods.[15] Which methods to promote and use involves social, moral, and religious beliefs, which vary from country to country.

The first country to adopt an official population policy was India in 1952. Few developing countries had official family-planning programs before 1965. Since 1965, many such programs have been introduced, and the World Bank has lent $4.2 billion to more than 80 countries to support "reproductive" health projects.[5, 13] Although most countries now have some kind of family-planning program, effectiveness varies greatly.

A wide range of approaches have been used, from simply providing more information to promoting and providing means for birth control, offering rewards, and imposing penalties. Penalties usually take the form of taxes. Ghana, Malaysia, Pakistan, Singapore, and the Philippines have used a combination of methods, including limits on tax allowances for children and on maternity benefits. Tanzania has restricted paid maternity leave for women to a frequency of once in three years. Singapore does not take family size into account in allocating government-built housing, so larger families are more crowded. Singapore also gives higher priority in school admission to children from smaller families. Some countries, including Bangladesh, India, and Sri Lanka, have paid people to be voluntarily sterilized.

CRITICAL THINKING ISSUE

Will the Demographic Transition Hold in the United States?

Earlier in this chapter, we presented the idea of the demographic transition and suggested that it has occurred in developed nations and may continue in the future. But we also noted that improvements in health care can further decrease death rates, which is something everybody wants to see happen but which will increase the human population growth rate, even in technologically developed nations. In 2008 Robert Engelman, a vice president of the Worldwatch Institute of Washington, D.C., proposed another problem for the demographic transition—an increase in birth rates in nations such as the United States.[16] The accompanying text box presents selections from Engelman's article. Using the material in the chapter, the quotes from Engelman here, and any other information you would like to introduce, present an argument either for or against the following: Growth rates will continue to decline in technologically developed nations, leading toward zero population growth.

Robert Engelman, Vice President of the Worldwatch Institute, "World Population Growth: Fertile Ground for Uncertainty," 2008.

Although the average woman worldwide is giving birth to fewer children than ever before, an estimated 136 million babies were born in 2007. Global data do not allow demographers to be certain that any specific year sets a record for births, but this one certainly came close. The year's cohort of babies propelled global population to an estimated 6.7 billion by the end of 2007.

The seeming contradiction between smaller-than-ever families and near-record births is easily explained. The number of women of childbearing age keeps growing, and global life expectancy at birth continues to rise. These two trends explain why population continues growing despite declines in family size. There were 1.7 billion women aged 15 to 49 in late 2007, compared with 856 million in 1970. The average human being born today can expect to live 67 years, a full decade longer than the average newborn could expect in 1970.

Only the future growth of the reproductive-age population is readily predictable, however: All but the youngest of the women who will be in this age group in two decades are already alive today. But sustaining further declines in childbearing and increases in life expectancy will require continued efforts by governments to improve access to good health care, and both trends could be threatened by environmental or social deterioration. The uncertain future of these factors makes population growth harder to predict than most people realize.

SUMMARY

- The human population is often referred to as the underlying environmental issue because much current environmental damage results from the very high number of people on Earth and their great power to change the environment.
- Throughout most of our history, the human population and its average growth rate were small. The growth of the human population can be divided into four major phases. Although the population has increased in each phase, the current situation is unprecedented.
- Countries whose birth rates have declined have experienced a demographic transition marked by a decline in death rates followed by a decline in birth rates. In contrast, many developing nations have undergone a great decline in their death rates but still have very high birth rates. It remains an open question whether some of these nations will be able to achieve a lower birth rate before reaching disastrously high population levels.
- The maximum population Earth can sustain and the population size human beings will ultimately attain are controversial questions. Standard estimates suggest that the human population will reach 10–16 billion before stabilizing.
- How the human population might stabilize, or be stabilized, raises questions concerning science, values, people, and nature.
- One of the most effective ways to lower a population's growth rate is to delay the age of first childbearing. This approach also involves relatively few societal and value issues.

REEXAMINING THEMES AND ISSUES

Sean Randall/Getty Images, Inc.

HUMAN POPULATION

Our discussion in this chapter reemphasizes the point that there can be no long-term solution to our environmental problems unless the human population stops growing at its present rate. This makes the problem of human population a top priority.

© Biletskiy_Evgeniy/iStockphoto

SUSTAINABILITY

As long as the human population continues to grow, it is doubtful that our other environmental resources can be made sustainable.

© Anton Balazh 2011/iStockphoto

GLOBAL PERSPECTIVE

Although the growth rate of the human population varies from nation to nation, the overall environmental effects of the rapidly growing human population are global. For example, the increased use of fossil fuels in Western nations since the beginning of the Industrial Revolution has affected the entire world. The growing demand for fossil fuels and their increasing use in developing nations are also having a global effect.

ssguy/ShutterStock

URBAN WORLD

One of the major patterns in the growth of the human population is the increasing urbanization of the world. Cities are not self-contained but are linked to the surrounding environment, depending on it for resources and affecting environments elsewhere.

B2M Productions/Getty Images, Inc.

PEOPLE AND NATURE

As with any species, the growth rate of the human population is governed by fundamental laws of population dynamics. We cannot escape these basic rules of nature. People greatly affect the environment, and the idea that human population growth is the underlying environmental issue illustrates the deep connection between people and nature.

George Doyle/Getty Images, Inc.

SCIENCE AND VALUES

The problem of human population exemplifies the connection between values and knowledge. Scientific and technological knowledge has helped us cure diseases, reduce death rates, and thereby increase growth of the human population. Our ability today to forecast human population growth provides a great deal of useful knowledge, but what we do with this knowledge is hotly debated around the world because values are so important in relation to birth control and family size.

KEY TERMS

abundance 83
acute disease 94
age structure 83
birth rate 83
chronic disease 94
death rate 83
demographic transition 90
demography 83
doubling time 86
epidemic disease 94
exponential rate 83
growth rate 83
human carrying capacity 95
inflection point 87
life expectancy 92
logistic carrying capacity 87
logistic growth curve 86
maximum lifetime 92
pandemic 93
population 83
population dynamics 83
species 83
zero population growth 97

STUDY QUESTIONS

1. Refer to three forecasts for the future of the world's human population in Figure 5.5. Each forecast makes a different assumption about the future total fertility rate: that the rate remains constant; that it decreases slowly and smoothly; and that it decreases rapidly and smoothly. Which of these do you think is realistic? Explain why.
2. Why is it important to consider the age structure of a human population?
3. Three characteristics of a population are the birth rate, growth rate, and death rate. How has each been affected by (a) modern medicine, (b) modern agriculture, and (c) modern industry?
4. What is meant by the statement "What is good for an individual is not always good for a population"?
5. Strictly from a biological point of view, why is it difficult for a human population to achieve a constant size?
6. What environmental factors are likely to increase the chances of an outbreak of an epidemic disease?
7. To which of the following can we attribute the great increase in human population since the beginning of the Industrial Revolution: changes in human (a) birth rates, (b) death rates, (c) longevity, or (d) death rates among the very old? Explain.
8. What is the demographic transition? When would one expect replacement-level fertility to be achieved—before, during, or after the demographic transition?
9. Based on the history of human populations in various countries, how would you expect the following to change as per capita income increased: (a) birth rates, (b) death rates, (c) average family size, and (d) age structure of the population? Explain.

FURTHER READING

Barry, J.M., *The Great Influenza: The Story of the Deadliest Pandemic in History* (New York: Penguin Books, paperback, 2005). Written for the general reader but praised by such authorities as the *New England Journal of Medicine;* discusses the connection between politics, public health, and pandemics.

Cohen, J.E., *How Many People Can the Earth Support?* (New York: Norton, 1995). A detailed discussion of world population growth, Earth's human carrying capacity, and factors affecting both.

Livi-Bacci, M. A *Concise History of World Population* (Hoboken, NJ: Wiley-Blackwell, paperback, 2001). A well-written introduction to the field of human demography.

McKee, J.K., *Sparing Nature: The Conflict between Human Population Growth and Earth's Biodiversity* (New Brunswick, NJ: Rutgers University Press, 2003). One of the few recent books about human populations.

NOTES

1. U.S. Census Bureau. International Data Base, http://www.census.gov/ipc/www/idb/idbsprd.html. Accessed May 14, 2009.
2. All population growth rate data are from U.S. Census Bureau, http://www.census.gov/ipc/www/idb/country/usportal.html. Accessed May 12, 2009.
3. Botkin, D.B. 1990. *Discordant Harmonies: A New Ecology for the 21st Century*. New York: Oxford University Press.
4. Zero Population Growth. 2000. *U.S. population.* Washington, DC: Zero Population Growth.
5. World Bank. 1984. *World Development Report 1984.* New York: Oxford University Press.
6. Fathalla, M.F. 1992. Family planning: Future needs, *AMBIO* 21:84–87.
7. Keyfitz, N. 1992. Completing the worldwide demographic transition: The relevance of past experience. *Ambio* 21:26–30.
8. U.S. Census Bureau, International Data Base. http://www.census.gov/cgi-bin/ipc/idbagg.
9. Bureau of the Census, 1990. *Statistical Abstract of the United States* 1990. Washington, DC: U.S. Department of Commerce.
10. Erhlich, P.R. 1971. *The Population Bomb*, rev. ed. New York: Ballantine.
11. Naess, A. 1989. *Ecology, Community and Lifestyle: Outline of an Ecosophy*. New York: Cambridge University Press.
12. National Master Statistics. http://www.nationmaster.com/graph/hea_age_of_wom_at_fir_chi-health-age-women-first-childbirth
13. World Bank. 1992. World Development Report. *The Relevance of Past Experience*. Washington, DC: World Bank.
14. Guz, D., and J. Hobcraft. 1991. Breastfeeding and fertility: A comparative analysis. *Population Studies* 45:91–108.
15. Creanga, A. A., D. Gillespie, S. Karklins, and A. O. Tsui. 2011. Low use of contraception among poor women in Africa: An equity issue, *Bulletin of the World Health Organization* 2011;89:258–266. doi: 10.2471/BLT.10.083329.
16. Revkin, A.C. 2008 (March 14). *Earth 2050: Population Unknowable?* Dot Earth (a Web page of the New York Times). http://dotearth.blogs.nytimes.com/2008/03/14/earth-2050-population-unknowable/?scp=2-b&sq =human+population+growth&st=nyt.

A CLOSER LOOK 5.1 NOTES

a. Keyfitz, N. 1992. Completing the worldwide demographic transition: The relevance of past experience. *Ambio* 21:26–30.

b. U.S. Census Bureau. International Data Base, http://www.census.gov/ipc/www/idb/idbsprd.html. Accessed May 14, 2009.

c. Graunt, J. *Natural and Political Observations Made upon the Bill of Mortality*.

d. Dumond, D.E. 1975. The limitation of human population: A natural history. *Science* 187:713–721.

A CLOSER LOOK 5.2 NOTES

a. McNeil, Donald G., Jr. 2008 (June 15). Malthus redux: Is doomsday upon us, again? *The New York Times.*

b. Malthus, T.R. (1803). 1992. *An Essay on the Principle of Population.* Selected and introduced by Donald Winch. Cambridge, England: Cambridge University Press.

A CLOSER LOOK 5.3 NOTES

a. U.S. Centers for Disease Control. West Nile virus statistics, surveilance, and control. http://www.cdc.gov/ncidod/dvbid/westnile/surv&controlCaseCount07_detailed.htm.

b. U.S. Centers for Disease Control Web site, available at http://www.cdc.gov/ncidod/sars/factsheet.htm and http://www.cdc.gov/ncidod/dvbid/westnile/qa/overview.htm.

c. Xinhua News Agency, 2002 (December 31). China's cross-border tourism prospers in 2002. From the Population Reference Bureau Web site, http://www.prb.org/Articles/2003/EmergingEpidemicsinChina.aspx.

d. Graunt, J. (1662). 1973. *Natural and Political Observations Made upon the Bill of Mortality.* London, originally published in 1662.

e. Barry, J.M. 2005. *The Great Influenza: The Story of the Deadliest Pandemic in History.* New York: Penguin Books.

WORKING IT OUT 5.1 NOTE

a. Population Reference Bureau. 2005. World Population Data Sheet.

CHAPTER 6

Ecosystems: Concepts and Fundamentals

Photo by Daniel B. Botkin

LEARNING OBJECTIVES

After reading this chapter, you should be able to . . .

- **Explain why the ecosystem is the basic system that supports life and allows it to persist**
- **Identify an ecosystem near your classroom and its boundaries**
- **Explain how the laws of thermodynamics ultimately limit the abundance of life**
- **Defend, or present solid reasoning against, the statement: Ecological succession makes it possible for an ecosystem to develop to a single, permanent, mature stage**
- **Describe a city as an ecosystem, or explain why the ecosystem concept can't be applied to a city**
- **Determine whether mice and/or deer in the chapter opening case study are keystone species**
- **Determine what trophic level the following are on:**
  - **people**
  - **a pet cat**

A view of southern New Hampshire in the autumn, when fall colors show many areas in different stages of forest succession. This mixture of successional stages provides variety and scenic beauty to New Hampshire.

graphingactivity

CASE STUDY

The Acorn Connection

As young children, most people learn that acorns grow into oak trees. In fact, most acorns do *not* grow into trees but become food for mice, chipmunks, squirrels, and deer.

In the woodlands of the northeastern United States, where oak trees abound, large crops of acorns (Figure 6.1a) are produced every three to four years. Acorns are rich in proteins and fats and are an excellent source of nutrition. A steady supply of acorns would be an excellent food base for the woodland animals.

The production of acorns is affected by the amount of light and rain, the temperature patterns over the year, and the quality of the soil. Scientists reason that if oaks produced the same number of acorns each year, the populations of animals that feed on them would grow so large that very few acorns would survive.

In reality, the number of acorns produced varies from year to year, with "mast" years—years of high production—occurring occasionally. In the years between bumper crops of acorns, mice populations decline. With the next bumper crop, there are more acorns than can be eaten by the consumers of acorns, so many acorns survive to become oaks. Also, because of the abundance of food, the mice populations increase.

White-footed mice **(b)**, feeders on acorns, also carry tick larvae **(c)**. When the ticks feed on the blood of the mice, they inject the microorganisms responsible for Lyme disease into the mice. Mice populations are highest during the summer following a bumper crop of acorns, and so are tick larvae.

In later stages of their life cycle, ticks attach to other animals, including deer **(d)**. Deer feed on oak leaves **(e)**

FIGURE 6.1 The tick that carries Lyme disease **(c)** feeds on both the white-footed mouse **(b)** and the white-tailed deer **(d)**. Oak leaves **(e)** are an important food for the deer **(d)** and for gypsy moths **(f)**, while oak acorns **(a)** are important food for the mouse. But the mouse also eats the moths. The more mice, the fewer gypsy moths, but the more ticks.

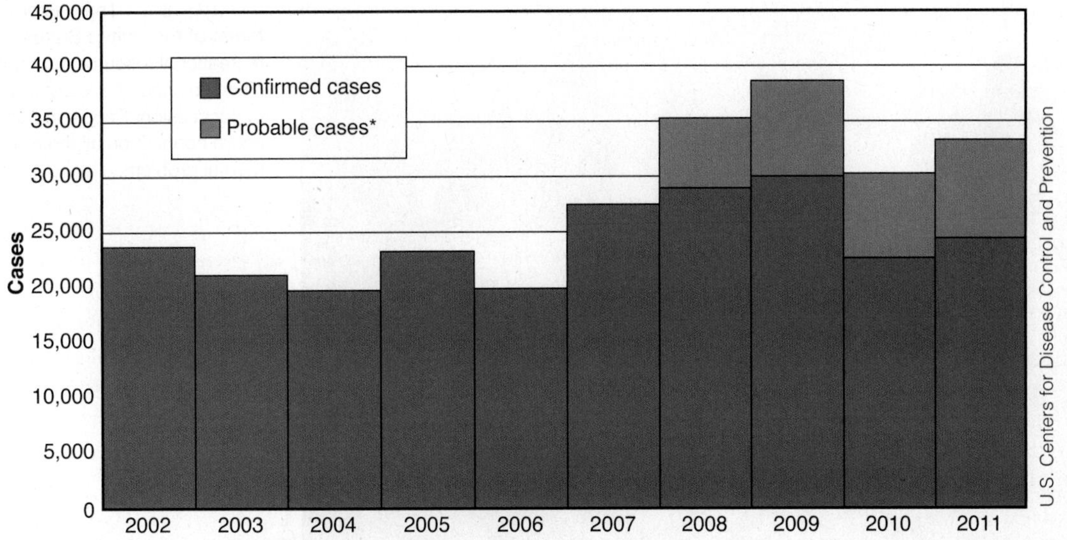

FIGURE 6.2 **Total U.S. Lyme Disease Cases, 2002–2011**

and acorns, and as deer brush against plants, ticks are deposited. The ticks can be picked up by people brushing against the plants as they walk past. If an infected tick bites a person, the person may contract Lyme disease. As second-growth forest area has increased and deer populations have soared, Lyme disease has become the most common tick-borne disease in the United States.[1,2] Between 1993 and 2005, the number of cases of Lyme disease diagnosed in a year almost tripled, reaching more than 23,000 nationwide and climbing since then to almost 30,000 cases in 2009, dropping in 2010 to about the 2005 level and then increasing slightly in 2011[3] (Figure 6.2). Interestingly, seven states—Pennsylvania, New Jersey, New York, Massachusetts, Connecticut, Maryland, and Minnesota—account for 80% of the 2009 cases. These are states with large areas of eastern deciduous forests.

Why has forest area increased? Beginning in colonial times, the forests of the northeastern United States were cleared to make space for farming and settlements, to provide fuel, and to supply timber for commercial uses. As coal, oil, and gas replaced wood as a primary fuel, and as farming moved westward to the more fertile Great Plains, fields that had been cleared were abandoned. In many areas, the maximum clearing occurred around 1900. Since then, forests have grown back.

But let's return to the mice. In addition to feeding on acorns (and other grains), mice feed on insects, including larvae of the gypsy moth. Gypsy moth larvae **(f)** feed on leaves of trees and are particularly fond of oak leaves. Studies suggest that in years when mice populations are low—the years between bumper crops of acorns—gypsy moth populations can increase dramatically. During these periodic outbreaks, gypsy moth larvae can virtually denude an area, stripping the leaves from the trees. Oaks that have lost most or all of their leaves may not produce bumper crops of acorns.

Once the leaves are off the trees, more light reaches the ground, and seedlings of many plants that could not do well in deep forest shade begin to grow. As a result, other species of trees may gain a foothold in the forest and change its species profile. Of course, the next generation of gypsy moth larvae find little to eat, and the population of gypsy moths begins to decline again.

Abundant acorns draw deer into the woods, where they browse on small plants and tree seedlings. Ticks drop off the deer and lay eggs in the leaf litter. When the eggs hatch, the larvae attach to mice, and the cycle of Lyme disease continues. Deer do not eat ferns, however, and in areas where deer populations are dense, many ferns but few wildflowers and tree seedlings are found.

Predators are also affected by the periodic nature of acorn crops. For example, birds that feed on gypsy moth larvae lose a food source when moth populations are low. When moth populations are high, however, bird nests are more exposed to predation because trees lose so many leaves.[2] The acorn connection illustrates many of the basic characteristics of ecosystems and ecological communities. First, all of the living parts of the oak forest community depend on the nonliving parts of the ecosystem for their survival: water, soil, air, and the light that provides energy for photosynthesis. Second, the members of the ecological community affect the nonliving parts of the ecosystem. When gypsy moths denude an area, for example, more sunlight can reach the forest floor. Third, the living organisms in the ecosystem are connected in complex relationships that make it difficult to change one thing without changing many

FIGURE 6.3 The Eastern deciduous forest of the United States is a major recreational resource, as illustrated by these hikers looking at a forest in various stages of succession. People in nature make the acorn connection all the more an environmental problem.

others. Fourth, the relationships among the members of the ecological community are dynamic and constantly changing. Many species are adapted to and benefit from a changing environment, as shown by the advantages provided oaks by varying acorn production. Fifth, the implication for human management of ecosystems is that any management practice involves trade-offs (Figure 6.3). In this case, managing the forest to protect people against Lyme disease only results in more potential for gypsy moth damage.

6.1 The Ecosystem: Sustaining Life on Earth

We tend to associate life with individual organisms, for the obvious reason that it is individuals that are alive. But sustaining life on Earth requires more than individuals or even single populations or species. Life is sustained by the interactions of many organisms functioning together, interacting through their physical and chemical environments. We call this an **ecosystem**. Sustained life on Earth, then, is a characteristic of ecosystems, not of individual organisms or populations. As the opening case study about Lyme disease illustrates, to understand important environmental issues—such as controlling undesirable species; conserving endangered species; sustaining renewable resources; and minimizing the effects of toxic substances—we must understand the basic characteristics of ecosystems.

Basic Characteristics of Ecosystems

Ecosystems have several fundamental characteristics, which we can group as *structure* and *processes*.

Ecosystem Structure

An ecosystem has two major parts: nonliving and living. The nonliving part is the physical-chemical environment, including the local atmosphere, water, and mineral soil (on land) or other substrate (in water). The living part, called the **ecological community**, is the set of species interacting within the ecosystem.

Ecosystem Functions and Processes

Two basic kinds of processes (sometimes referred to as ecosystem functions) must occur in an ecosystem: a cycling of chemical elements and a flow of energy. These processes are necessary for all life, but no single species can carry out all necessary chemical cycling and energy flow alone. That is why we said that sustained life on Earth is a characteristic of ecosystems, not of individuals or populations. At its most basic, an ecosystem consists of several species and a fluid medium—air, water, or both (Figure 6.4). Ecosystem energy flow places a fundamental limit on the abundance of life. Energy flow is a difficult subject, which we will discuss in Section 6.4.

Ecosystem chemical cycling is complex as well, and for that reason we have devoted a separate chapter (Chapter 7) to chemical cycling within ecosystems and throughout

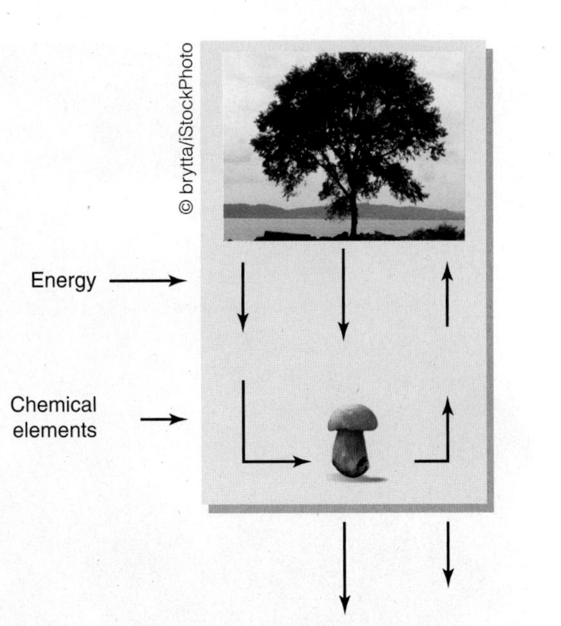

FIGURE 6.4 **An idealized minimum ecosystem.** Energy flows through an ecosystem one way. A small amount is stored within the system. As chemical elements cycle, there is some small loss, depending on the characteristics of the ecosystem. The size of the arrows in the figure approximates the amount of flow.

the entire Earth's biosphere. Briefly, 21 chemical elements are required by at least some form of life, and each chemical element required for growth and reproduction must be available to each organism at the right time, in the right amount, and in the right ratio relative to other elements. These chemical elements must also be recycled—converted to a reusable form: Wastes are converted into food, which is converted into wastes, which must be converted once again into food, with the cycling going on indefinitely if the ecosystem is to remain viable.

For recycling of chemical elements to take place, several species must interact. In the presence of light, green plants, algae, and photosynthetic bacteria produce sugar from carbon dioxide and water. From sugar and inorganic compounds, they make many organic compounds, including proteins and woody tissue. But no green plant, algae, or photosynthetic bacteria can decompose woody tissue back to its original inorganic compounds. Other forms of life—primarily bacteria and fungi—can decompose organic matter. But they cannot produce their own food; instead, they obtain energy and chemical nutrition from the dead tissues on which they feed. In an ecosystem, chemical elements recycle, but energy flows one way, into and out of the system, with a small fraction of it stored, as we will discuss later in this chapter.

To repeat: Theoretically, at its simplest, an ecosystem consists of at least one species that produces its own food from inorganic compounds in its environment and another species that decomposes the wastes of the first species, plus a fluid medium—air, water, or both (Figure 6.4). But the reality is never as simple as that.

6.2 Ecological Communities and Food Chains

In practice, ecologists define the term *ecological community* in two ways. One method defines the community as a set of *interacting* species found in the same place and functioning together, thus enabling life to persist. That is essentially the definition we used earlier. A problem with this definition is that it is often difficult in practice to know the entire set of interacting species. Ecologists therefore may use a practical or an operational definition, in which the community consists of all the species found in an area, whether or not they are known to interact. Animals in different cages in a zoo could be called a community according to this definition.

One way that individuals in a community interact is by feeding on one another. Energy, chemical elements, and some compounds are transferred from creature to creature along **food chains**, the linkage of who feeds on whom. The more complex linkages are called **food webs**. Ecologists group the organisms in a food web into trophic levels. A **trophic level** (from the Greek word *trephein*, meaning to nourish, thus the "nourishing level") consists of all organisms in a food web that are the same number of feeding levels away from the original energy source. The original source of energy in most ecosystems is the sun. In other cases, it is the energy in certain inorganic compounds.

Green plants, algae, and certain bacteria produce sugars through the process of **photosynthesis**, using only energy from the sun, water (H_2O), and carbon dioxide (CO_2) from the local environment. They are called **autotrophs**, from the words *auto* (self) and *trephein* (to nourish), thus "self-nourishing," and are grouped into the first trophic level. All other organisms are called **heterotrophs**. Of these, **herbivores**—organisms that feed on plants, algae, or photosynthetic bacteria—are members of the second trophic level. **Carnivores**, or meat-eaters, that feed directly on herbivores make up the third trophic level. Carnivores that feed on third-level carnivores are in the fourth trophic level, and so on. **Decomposers**, those that feed on dead organic material, are classified in the highest trophic level in an ecosystem.

Food chains and food webs are often quite complicated and thus not easy to analyze. For starters, the number of trophic levels differs among ecosystems.

A Simple Ecosystem

One of the simplest natural ecosystems is a hot spring, such as those found in geyser basins in Yellowstone National Park, Wyoming.[4] They are simple because few organisms can live in these severe environments. In and near the center of a spring, water is close to the boiling point, while at the edges, next to soil and winter snow, water is much cooler. In addition, some springs are very acidic and others are very alkaline; either extreme makes a harsh environment.

Photosynthetic bacteria and algae make up the spring's first trophic level. In a typical alkaline hot spring, the hottest waters, between 70° and 80°C (158–176°F), are colored bright yellow-green by photosynthetic blue-green bacteria. One of the few kinds of photosynthetic organisms that can survive at those temperatures, these give the springs the striking appearance for which they are famous (Figure 6.5). In slightly cooler waters, 50° to 60°C (122–140°F), thick mats of other kinds of bacteria and algae accumulate, some becoming 5 cm thick (Figures 6.5, 6.6).

The bacteria living in these hot springs became of special interest in 1985 when an enzyme was isolated from one species that allows "DNA fingerprinting," a now major method in molecular biology.[5]

Ephydrid flies make up the second (herbivore) trophic level. Note that because the environment is so stressful, they are the only genus on that entire trophic level,

FIGURE 6.5 One of the many hot springs in Yellowstone National Park. The bright yellowish-green color comes from photosynthetic bacteria, one of the few kinds of organisms that can survive in the hot temperatures and chemical conditions of the springs.

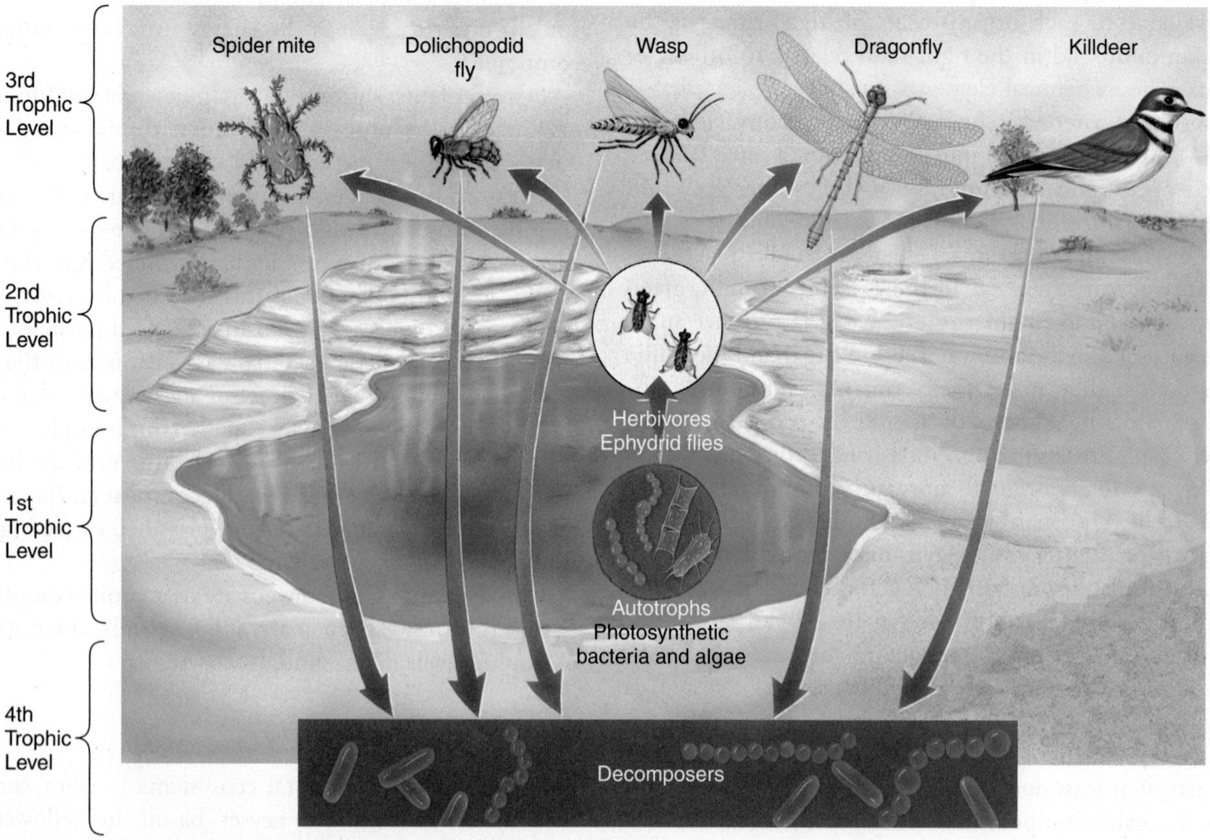

FIGURE 6.6 Food web of a Yellowstone National Park hot spring. Even though this is one of the simplest ecological communities in terms of the numbers of species, a fair number are found. About 20 species in all are important in this ecosystem.

and they live only in the cooler areas of the springs. One of these species, *Ephydra bruesi*, lays bright orange-pink egg masses on stones and twigs that project above the mat. These larvae feed on the bacteria and algae.

The third (carnivore) trophic level is made up of a dolichopodid fly, which feeds on the eggs and larvae of the herbivorous flies, and dragonflies, wasps, spiders, tiger beetles, and one species of bird, the killdeer, that feeds on the ephydrid flies. (Note that the killdeer is a carnivore of the hot springs but also feeds widely in other ecosystems. An interesting question, little addressed in the ecological scientific literature, is how we should list this partial member of the food web: Should there be a separate category of "casual" members? What do you think?)

In addition to their other predators, the herbivorous ephydrid flies have parasites. One is a red mite that feeds on the flies' eggs and travels by attaching itself to the adult flies. Another is a small wasp that lays its eggs within the fly larvae. These are also on the third trophic level.

Wastes and dead organisms of all trophic levels are fed on by decomposers, which in the hot springs are primarily bacteria. These form the fourth trophic level.

The entire hot-springs community of organisms—photosynthetic bacteria and algae, herbivorous flies, carnivores, and decomposers—is maintained by two factors: (1) sunlight, which provides usable energy for the organisms; and (2) a constant flow of hot water, which provides a continual new supply of chemical elements required for life and a habitat in which the bacteria and algae can persist. (See A Closer Look 6.1 and Figure 6.7 for another example of trophic levels and keystone species.)

A CLOSER LOOK 6.1

Sea Otters, Sea Urchins, and Kelp: Indirect Effects of Species on One Another

Within an ecological community, species often have indirect and important effects on one another. A classic example is sea otters, the lovable animals often shown lying face up among kelp as they eat shellfish (Figure 6.7a). Although they feed on a variety of shellfish, sea otters especially like sea urchins. Sea urchins, in turn, feed on kelp, large brown algae that form undersea "forests" and provide important habitat for many species that require kelp beds for reproduction, places to feed, or havens from predators. Sea urchins graze along the bottoms of the beds, feeding on the base of kelp, called *holdfasts*, which attach the kelp to the bottom. When holdfasts are eaten through, the kelp floats free and dies. Sea urchins thus can clear cut kelp beds, so to speak. While sea otters affect the abundance of kelp, their influence is indirect (Figure 6.7b) because they neither feed on kelp nor protect individual kelp plants from attack by sea urchins. But sea otters reduce the number of sea urchins. With fewer sea urchins, less kelp is destroyed. With more kelp, there is more habitat for many other species; so sea otters indirectly increase the diversity of species.[a,b,c,d] This is called a **community effect**, and the otters are referred to as **keystone species** in their ecological community and ecosystem.

Sea otters originally occurred throughout a large area of the Pacific coasts, from northern Japan northeastward along the Russian and Alaskan coasts, and southward along the coast of North America to Morro Hermoso in Baja California and to Mexico.[e] Sea otters like to eat abalone, and this brings them into direct conflict with people, since abalone is a prized seafood for humans, too. They also have one of the finest furs in the world and were brought almost to extinction by commercial hunting for their fur during the 18th and 19th centuries.

Several small populations survived and have increased since then, so that today sea otters number in the hundreds of thousands—mainly found in North America off the coasts of California, Alaska, British Columbia, and Washington state.[f]

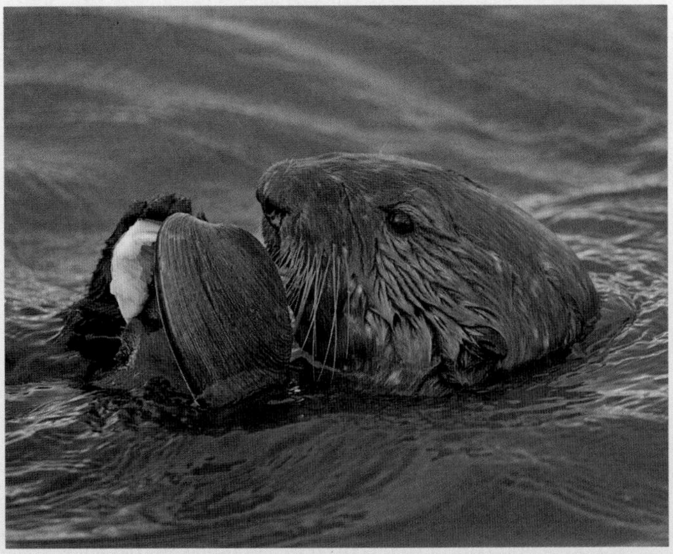

Enrique R Aguirre Aves / Getty Images

FIGURE 6.7a **A sea otter with prey.**

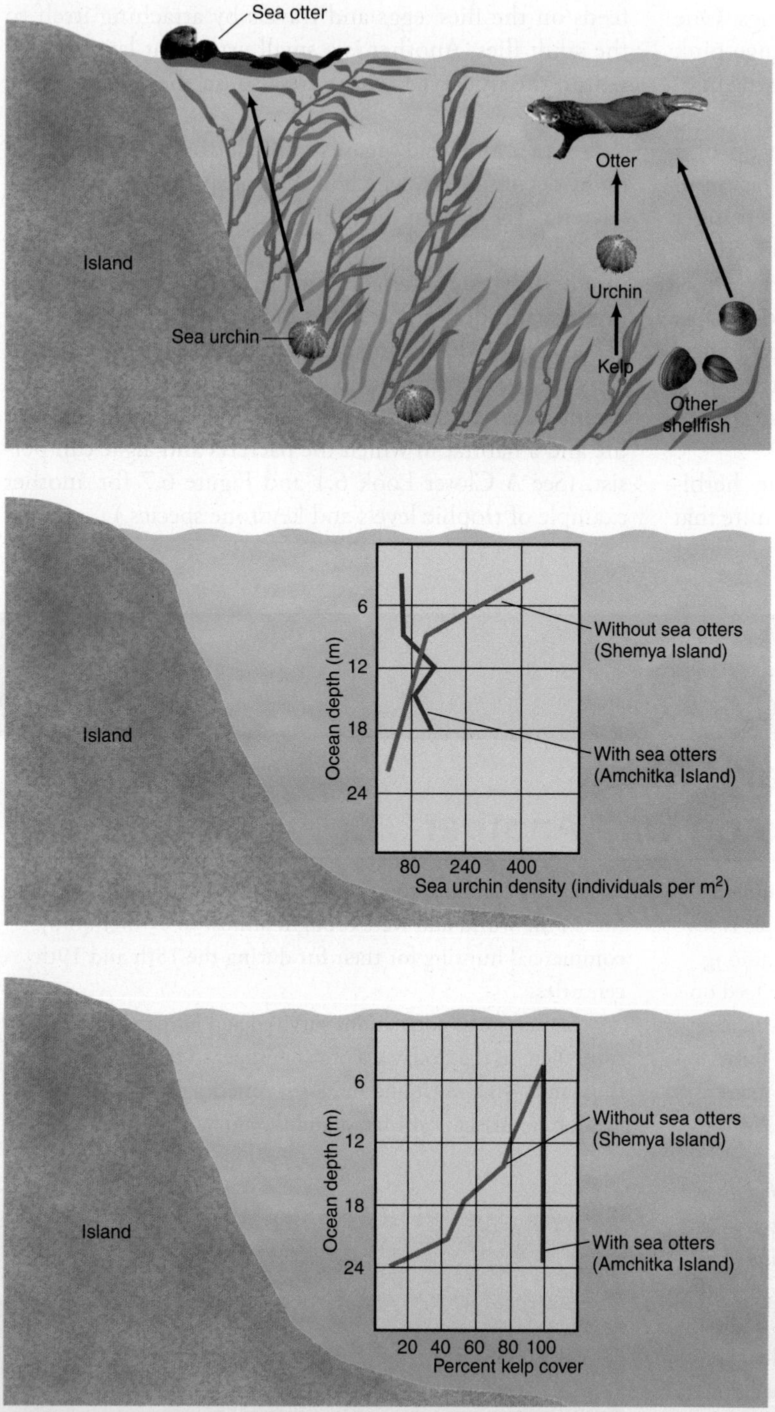

FIGURE 6.7b **The effect of sea otters on kelp.** Sea otters feed on shellfish, including sea urchins. Sea urchins feed on kelp. Where sea otters are abundant, as on Amchitka Island in the Aleutian Islands, there are few sea urchins and kelp beds are abundant. At nearby Shemya Island, which lacks sea otters, sea urchins are abundant and there is little kelp.[A1] Experimental removal of sea urchins has led to an increase in kelp.[A2]

[A1] Kvitek, R.G., J.S. Oliver, A.R. DeGange, and B.S. Anderson. 1992. Changes in Alaskan soft-bottom prey communities along a gradient in sea otter predation. *Ecology*
[A2] Duggins, D.O. 1980. Kelp beds and sea otters: An experimental approach. *Ecology* 61:447–453.

Sea otter populations were growing rapidly at the end of the 20th century, but that growth rate has slowed considerably in the past decade, for reasons so far not understood.[f]

Legal protection of the sea otter by the U.S. government began in 1911 and continues under the U.S. Marine Mammal Protection Act of 1972 and the Endangered Species Act of 1973. Today, however, the sea otter continues to be a focus of controversy. On the one hand, fishermen argue that the sea otter population has recovered—so much so that they now interfere with commercial fishing because they take large numbers of abalone.[b] On the other hand, conservationists argue that the community and ecosystem effects of sea otters make them necessary for the persistence of many oceanic species, and that there are still not enough sea otters to maintain this role at a satisfactory level. Thus, sea otters' indirect effects on many other species have practical consequences. They also demonstrate certain properties of ecosystems and ecological communities that are important to us in understanding how life persists—and how we may be able to help solve certain environmental problems.

common to all ecosystems. Energy enters an ecosystem by two pathways: Energy that is fixed by organisms and moving through food webs within an ecosystem; and heat energy that is transferred by air or water currents or by convection through soils and sediments and warms living things. For instance, when a warm air mass passes over a forest, heat energy is transferred from the air to the land and to the organisms.

Energy is a difficult and an abstract concept. When we buy electricity, what are we buying? We cannot see it or feel it, even if we have to pay for it. At first glance, and as we think about it with our own diets, energy flow seems simple enough: We take energy in and use it, just like machines do—our automobiles, cell phones, and so on. But if we dig a little deeper into this subject, we discover a philosophical importance: We learn what distinguishes Earth's life and life-containing systems from the rest of the universe.

Although most of the time energy is invisible to us, with infrared film we can see the differences between warm and cold objects, and we can see some things about energy flow that affect life. With infrared film, warm objects appear red and cool objects blue. Figure 6.13 shows birch trees in a New Hampshire forest, both as we see them, using standard film, and with infrared film. The infrared film shows tree leaves bright red, indicating that they have been warmed by the sun and are absorbing and reflecting energy,

Photo by Daniel B. Botkin (a)
Photo by Daniel B. Botkin (b)
Photo by Daniel B. Botkin (c)
Photo by Daniel B. Botkin (d)

FIGURE 6.13 **Making energy visible.** Top: A birch forest in New Hampshire as we see it, using normal photographic film **(a)** and the same forest photographed with infrared film **(b)**. Red color means warmer temperatures; the leaves are warmer than the surroundings because they are heated by sunlight. Bottom: A nearby rocky outcrop as we see it, using normal photographic film **(c)** and the same rocky outcrop photographed with infrared film **(d)**. Blue means that a surface is cool. The rocks appear deep blue, indicating that they are much cooler than the surrounding trees.

whereas the white birch bark remains cooler. The ability of tree leaves to absorb energy is essential; it is this source of energy that ultimately supports all life in a forest. Energy flows through life, and energy flow is a key concept.

The Laws of Thermodynamics and the Ultimate Limit on the Abundance of Life

When we discuss ecosystems, we are talking about some of the fundamental properties of life and of the ecological systems that keep life going. A question that frequently arises both in basic science and when we want to produce a lot of some kind of life—a crop, biofuels, pets—is: *What ultimately limits the amount of organic matter in living things that can be produced anywhere, at any time, forever on the Earth or anywhere in the universe?*

We ask this question when we are trying to improve the production of some form of life. We want to know: How closely do ecosystems, species, populations, and individuals approach this limit? Are any of these near to being as productive as possible?

The answers to these questions, which are at the same time practical, scientifically fundamental, and philosophical, lie in the laws of thermodynamics. The **first law of thermodynamics**, known as the *law of conservation of energy* states that in any physical or chemical change, energy is neither created nor destroyed but merely changed from one form to another. (See Working It Out 6.1.) This seems to lead us to a confusing, contradictory answer—it seems to say that we don't need to take in any energy at all! If the total amount of energy is always conserved—if it remains constant—then why can't we just recycle energy inside our bodies? The famous 20th-century physicist Erwin Schrödinger asked this question in a wonderful book entitled *What Is Life?* He wrote:

> *In some very advanced country (I don't remember whether it was Germany or the U.S.A. or both) you could find menu cards in restaurants indicating, in addition to the price, the energy content of every dish. Needless to say, taken literally, this is . . . absurd. For an adult organism the energy content is as stationary as the material content. Since, surely, any calorie is worth as much as any other calorie, one cannot see how a mere exchange could help.*[8]

WORKING IT OUT 6.1 Some Chemistry of Energy Flow

For Those Who Make Their Own Food (autotrophs)

Photosynthesis—the process by which autotrophs make sugar from sunlight, carbon dioxide, and water—is:

$$6CO_2 + 6H_2O + \text{energy} = C_6H_{12}O_6 + 6O_2$$

Chemosynthesis takes place in certain environments. In chemosynthesis, the energy in hydrogen sulfide (H_2S) is used by certain bacteria to make simple organic compounds. The reactions differ among species and depend on characteristics of the environment (Figure 6.14).

Net production for autotrophs is given as

$$NPP = GPP - R_a$$

where *NPP* is net primary production, *GPP* is gross primary production, and R_a is the respiration of autotrophs.

For Those Who Do Not Make Their Own Food (heterotrophs)

Secondary production of a population is given as

$$NSP = B_2 - B_1$$

where *NSP* is net secondary production, B_2 is the biomass (quantity of organic matter) at time 2, and B_1 is the biomass at time 1. (See the discussion of biomass in Section 6.5.) The change in biomass is the result of the addition of weight of living individuals, the addition of newborns and immigrants, and loss through death and emigration. The biological use of energy occurs through respiration, which is most simply expressed as

$$C_6H_{12}O_6 + 6O_2 = 6CO_2 + 6H_2O + \text{Energy}$$

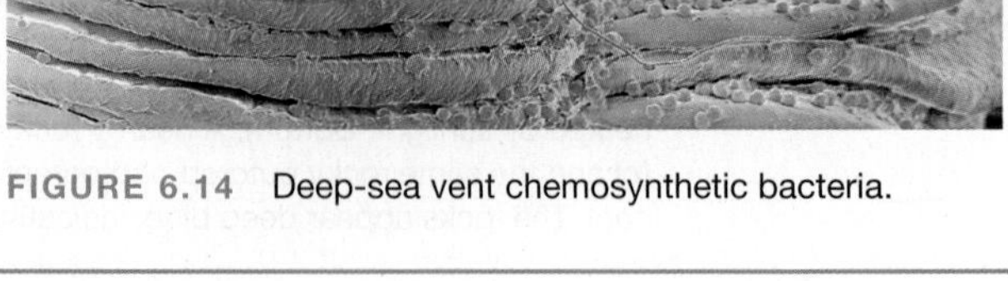

Mona Lisa Productions/Science Photo Library / Photo Researchers, Inc.

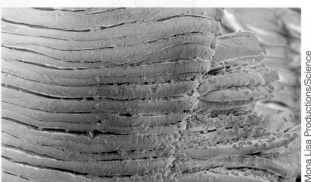

FIGURE 6.14 Deep-sea vent chemosynthetic bacteria.

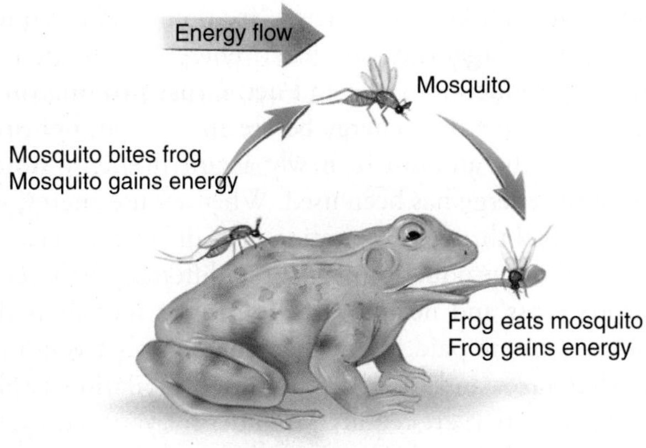

FIGURE 6.15 **An impossible ecosystem.** Energy always changes from a more useful, more highly organized form to a less useful, disorganized form. That is, energy cannot be completely recycled to its original state of organized, high-quality usefulness. For this reason, the mosquito–frog system will eventually stop when not enough useful energy is left. (There is also a more mundane reason: Only female mosquitoes require blood, and then only in order to reproduce. Mosquitoes are otherwise herbivorous.)

Schrödinger was saying that, according to the first law of thermodynamics, we should be able to recycle energy in our bodies and never have to eat anything. Similarly, we can ask: Why can't energy be recycled in ecosystems and in the biosphere?

Let us imagine how that might work, say, with frogs and mosquitoes. Frogs eat insects, including mosquitoes. Mosquitoes suck blood from vertebrates, including frogs. Consider an imaginary closed ecosystem consisting of water, air, a rock for frogs to sit on, frogs, and mosquitoes. In this system, the frogs get their energy from eating the mosquitoes, and the mosquitoes get their energy from biting the frogs (Figure 6.15). Such a closed system would be a biological perpetual-motion machine. It could continue indefinitely without an input of any new material or energy. This sounds nice, but unfortunately it is impossible. Why? The general answer is found in the *second* law of thermodynamics, which addresses how energy changes in form.

To understand why we cannot recycle energy, imagine a closed system (a system that receives no input after the initial input) containing a pile of coal, a tank of water, air, a steam engine, and an engineer (Figure 6.16). Suppose the engine runs a lathe that makes furniture. The engineer lights a fire to boil the water, creating steam to run the engine. As the engine runs, the heat from the fire gradually warms the entire system.

When all the coal is completely burned, the engineer will not be able to boil any more water, and the engine will stop. The average temperature of the system is now higher than the starting temperature. The energy that was in the coal is dispersed throughout the entire system, much of it as heat in the air. Why can't the engineer recover all that energy, recompact it, put it under the boiler, and run the engine? The answer is in the **second law of thermodynamics**. Physicists have discovered that *no use of energy in the real (not theoretical) world can ever be 100% efficient.* Whenever useful work is done, some energy is inevitably converted to heat. Collecting all the energy dispersed in this closed system would require more energy than could be recovered.

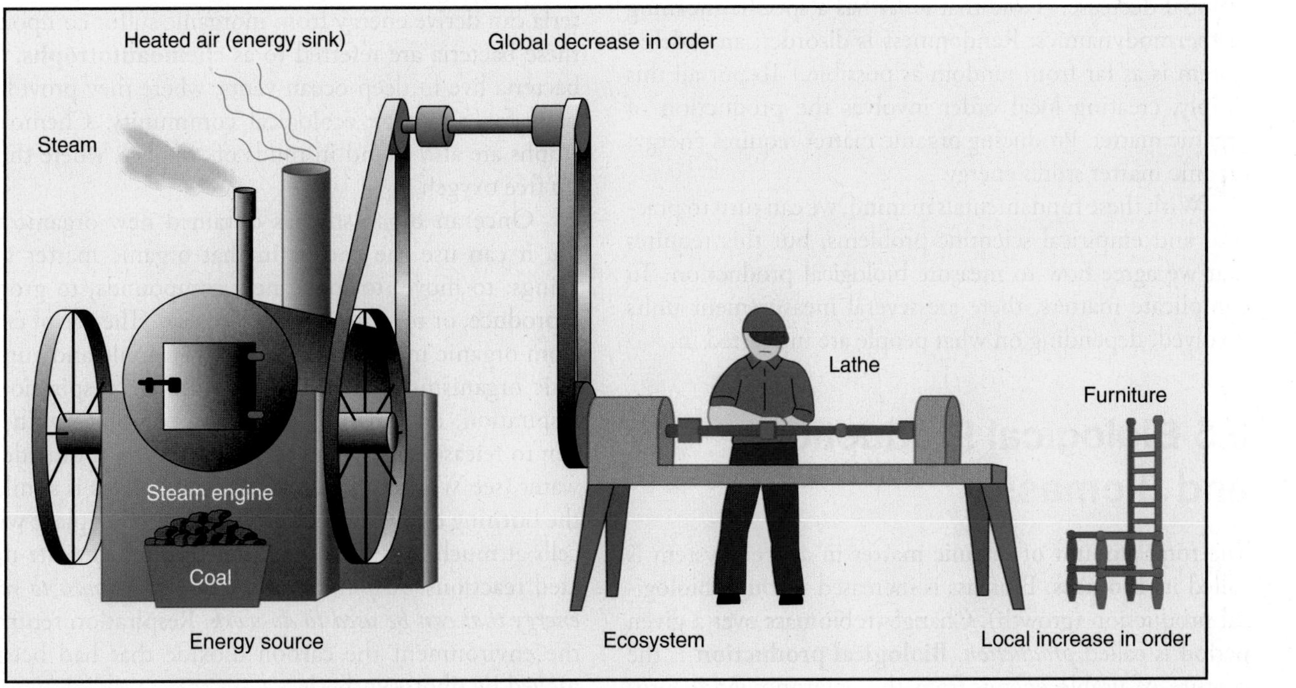

FIGURE 6.16 **A system closed to the flow of energy.**

Our imaginary system begins in a highly organized state, with energy compacted in the coal. It ends in a less organized state, with the energy dispersed throughout the system as heat. The energy has been degraded, and the system is said to have undergone a decrease in order. The measure of the decrease in order (the disorganization of energy) is called **entropy**. The engineer did produce some furniture, converting a pile of lumber into nicely ordered tables and chairs. The system had a local increase of order (the furniture) at the cost of a general increase in disorder (the state of the entire system). All energy of all systems tends to flow toward states of increasing entropy.

The second law of thermodynamics gives us a new understanding of a basic quality of life. *It is the ability to create order on a local scale that distinguishes life from its nonliving environment.* This ability requires obtaining energy in a usable form, and that is why we eat. This principle is true for every ecological level: individual, population, community, ecosystem, and biosphere. Energy must continually be added to an ecological system in a usable form. Energy is inevitably degraded into heat, and this heat must be released from the system. If it is not released, the temperature of the system will increase indefinitely. The net flow of energy through an ecosystem, then, is a one-way flow.

Based on what we have said about the energy flow through an ecosystem, we can see that an ecosystem must lie between a source of usable energy and a sink for degraded (heat) energy. The ecosystem is said to be an *intermediate system* between the energy source and the energy sink. The energy source, ecosystem, and energy sink together form a thermodynamic system. The ecosystem can undergo an increase in order, called a *local increase*, as long as the entire system undergoes a decrease in order, called a global decrease. (Note that *order* has a specific meaning in thermodynamics: Randomness is disorder; an ordered system is as far from random as possible.) To put all this simply, creating local order involves the production of organic matter. Producing organic matter requires energy; organic matter stores energy.

With these fundamentals in mind, we can turn to practical and empirical scientific problems, but this requires that we agree how to measure biological production. To complicate matters, there are several measurement units involved, depending on what people are interested in.

6.5 Biological Production and Biomass

The total amount of organic matter in any ecosystem is called its **biomass**. Biomass is increased through biological production (growth). Change in biomass over a given period is called *production*. **Biological production** is the capture of usable energy from the environment to produce organic matter (or organic compounds). This capture is often referred to as energy "fixation," and it is often said that the organism has "fixed" energy. There are two kinds of production, gross and net. **Gross production** is the increase in stored energy before any is used; **net production** is the amount of newly acquired energy stored after some energy has been used. When we use energy, we "burn" a fuel through respiration. The difference between gross and net production is like the difference between a person's gross and net income. Your gross income is the amount you are paid. Your net income is what you have left after taxes and other fixed costs. Respiration is like the expenses that are required in order for you to do your work. (See Working It Out 6.2.)

Measuring Biomass and Production

Three measures are used for biomass and biological production: The quantity of organic material (biomass), energy stored, and carbon stored. We can think of these measures as the currencies of production. Biomass is usually measured as the amount per unit surface area—for example, as grams per square meter (g/m^2) or metric tons per hectare (MT/ha). Production, a rate, is the change per unit area in a unit of time—for example, grams per square meter per year. (Common units of measure of production are given in the Appendix.)

The production carried out by autotrophs is called **primary production**; that of heterotrophs is called **secondary production**. As we have said, most autotrophs make sugar from sunlight, carbon dioxide, and water in a process called photosynthesis, which releases free oxygen (see Working It Out 6.1 and 6.2). Some autotrophic bacteria can derive energy from inorganic sulfur compounds; these bacteria are referred to as **chemoautotrophs**. Such bacteria live in deep-ocean vents, where they provide the basis for a strange ecological community. Chemoautotrophs are also found in muds of marshes, where there is no free oxygen.

Once an organism has obtained new organic matter, it can use the energy in that organic matter to do things: to move, to make new compounds, to grow, to reproduce, or to store it for future uses. The use of energy from organic matter by most heterotrophic and autotrophic organisms is accomplished through respiration. In respiration, an organic compound combines with oxygen to release energy and to produce carbon dioxide and water (see Working It Out 6.2). The process is similar to the burning of organic compounds but takes place within cells at much lower temperatures through enzyme-mediated reactions. *Respiration is the use of biomass to release energy that can be used to do work.* Respiration returns to the environment the carbon dioxide that had been removed by photosynthesis.

WORKING IT OUT 6.2 Gross and Net Production

The production of biomass and its use as a source of energy by autotrophs include three steps:

1. An organism produces organic matter within its body.
2. It uses some of this new organic matter as a fuel in respiration.
3. It stores some of the newly produced organic matter for future use.

The first step, production of organic matter before use, is called *gross production.* The amount left after utilization is called *net production.*

Net production = Gross production − Respiration

The gross production of a tree—or any other plant—is the total amount of sugar it produces by photosynthesis before any is used. Within living cells in a green plant, some of the sugar is oxidized in respiration. Energy is used to convert sugars to other carbohydrates; those carbohydrates to amino acids; amino acids to proteins and to other tissues, such as cell walls and new leaf tissue. Energy is also used to transport material within the plant to roots, stems, flowers, and fruits. Some energy is lost as heat in the transfer. Some is stored in these other parts of the plant for later use. For woody plants like trees, some of this storage includes new wood laid down in the trunk, new buds that will develop into leaves and flowers the next year, and new roots.

Equations for Production, Biomass, and Energy Flow

We can write a general relation between biomass (*B*) and net production (*NP*):

$$B_2 = B_1 + NP$$

where B_2 is the biomass at the end of the time period, B_1 is the amount of biomass at the beginning of the time period, and *NP* is the change in biomass during the time period.

Thus,

$$NP = B_2 - B_1$$

General production equations are given as

$$GP = NP + R$$

$$NP = GP - R$$

where *GP* is gross production, *NP* is net production, and *R* is respiration.

Several units of measure are used in the discussion of biological production and energy flow: calories when people talk about food content; watt-hours when people talk about biofuels; and kilojoules in standard international scientific notation. To make matters even more complicated, there are two kinds of calories: the standard calorie, which is the heat required to heat one gram of water from 15.5°C to 16.5°C, and the kilocalorie, which is 1,000 of the little calories. Even more confusing, when people discuss the energy content of food and diets, they use *calorie* to mean the kilocalorie. Just remember: The calorie you see on the food package is the big calorie, the kilocalorie, equal to 1,000 little calories. Almost nobody uses the "little" calorie, regardless of what they call it. To compare, an average apple contains about 100 Kcal or 116 watt-hours. This means that an apple contains enough energy to run a 100-watt bulb for 1 hour and 9 minutes. For those of you interested in your diet and weight, a Big Mac contains 576 Kcal, which is 669 watt-hours, enough to keep that 100-watt bulb burning for 6 hours and 41 minutes.

The average energy stored in vegetation is approximately 5 Kcal/gram (21 kilojoules per gram [kJ/g]). The energy content of organic matter varies. Ignoring bone and shells, woody tissue contains the least energy per gram, about 4 Kcal/g (17 kJ/g); fat contains the most, about 9 Kcal/g (38 kJ/g); and muscle contains approximately 5–6 Kcal/g (21–25 kJ/g). Leaves and shoots of green plants have about 5 Kcal/g (21–23 kJ/g); roots have about 4.6 Kcal/g (19 kJ/g).

6.6 Energy Efficiency and Transfer Efficiency

How efficiently do living things use energy? This is an important question for the management and conservation of all biological resources. We would like biological resources to use energy efficiently—to produce a lot of biomass from a given amount of energy. This is also important for attempts to sequester carbon by growing trees and other perennial vegetation to remove carbon dioxide from the atmosphere and store it in living and dead organic matter (see Chapter 20).

As you learned from the second law of thermodynamics, no system can be 100% efficient. As energy flows through a food web, it is degraded, and less and less is usable. Generally, the more energy an organism gets, the more it has for its own use. However, organisms differ in how efficiently they use the energy they obtain. A more efficient organism has an advantage over a less efficient one.

Efficiency can be defined for both artificial and natural systems: machines, individual organisms, populations, trophic levels, ecosystems, and the biosphere. **Energy efficiency** is defined as the ratio of output to input, and it is usually further defined as the amount of useful work obtained from some amount of available energy. *Efficiency* has different meanings to different users. From the point of view of a farmer, an efficient corn crop is one that converts a great deal of solar energy to sugar and uses little of that sugar to produce stems, roots, and leaves. In other words, the most efficient crop is the one that has the most harvestable energy left at the end of the season. A truck driver views an efficient truck as just the opposite: For him, an efficient truck uses as much energy as possible from its fuel and stores as little energy as possible (in its exhaust). When we view organisms as food, we define *efficiency* as the farmer does, in terms of energy storage (net production from available energy). When we are energy users, we define *efficiency* as the truck driver does, in terms of how much useful work we accomplish with the available energy.

Consider the use of energy by a wolf and by one of its principal prey, moose. The wolf needs energy to travel long distances and hunt, and therefore it will do best if it uses as much of the energy in its food as it can. For itself, a highly energy-efficient wolf stores almost nothing. But from its point of view, the best moose would be one that used little of the energy it took in, storing most of it as muscle and fat, which the wolf can eat. Thus what is efficient depends on your perspective.

A common ecological measure of energy efficiency is called *food-chain efficiency*, or *trophic-level efficiency*, which is the ratio of production of one trophic level to the production of the next-lower trophic level. This efficiency is never very high. Green plants convert only 1–3% of the energy they receive from the sun during the year to new plant tissue. The efficiency with which herbivores convert the potentially available plant energy into herbivorous energy is usually less than 1%, as is the efficiency with which carnivores convert herbivores into carnivorous energy. In natural ecosystems, the organisms in one trophic level tend to take in much less energy than the potential maximum available to them, and they use more energy than they store for the next trophic level. At Isle Royale National Park, an island in Lake Superior, wolves feed on moose in a natural wilderness. In one study, a pack of 18 wolves killed an average of one moose approximately every 2.5 days,[9] which gives wolves a trophic-level efficiency of about 0.01%. (In 2012, the wolf population had dropped to nine individuals on the island, for reasons not yet understood.[10])

The rule of thumb for ecological trophic energy efficiency is that more than 90% (usually much more) of all energy transferred between trophic levels is lost as heat. Less than 10% (approximately 1% in natural ecosystems) is fixed as new tissue. In highly managed ecosystems, such as ranches, the efficiency may be greater. But even in such systems, it takes an average of 3.2 kg (7 lb) of vegetable matter to produce 0.45 kg (1 lb) of edible meat. Cattle are among the least efficient producers, requiring around 7.2 kg (16 lb) of vegetable matter to produce 0.45 kg (1 lb) of edible meat. Chickens are much more efficient, using approximately 1.4 kg (3 lb) of vegetable matter to produce 0.45 kg (1 lb) of eggs or meat. Much attention has been paid to the idea that humans should eat at a lower trophic level in order to use resources more efficiently. (See Critical Thinking Issue: Should People Eat Lower on the Food Chain?)

6.7 Ecological Stability and Succession

Ecosystems are dynamic: They change over time both from external (environmental) forces and from their internal processes. It is worth repeating the point we made in Chapter 4 about dynamic systems: The classic interpretation of populations, species, ecosystems, and Earth's entire biosphere has been to assume that each is a stable, static system. But the more we study these ecological systems, the clearer it becomes that these are dynamic systems, always changing *and always requiring change.* Curiously, they persist while undergoing change.[11] We say "curiously" because in our modern technological society we are surrounded by mechanical and electronic systems that stay the same in most characteristics and are designed to do so. We don't expect our car or television or cell phone to shrink or get larger and then smaller again; we don't expect that one component will get bigger or smaller over time. If anything like this were to happen, those systems would break.

Ecosystems, however, not only change but also then recover and overcome these changes, and life continues on. It takes some adjustment in our thinking to accept and understand such systems.

When disturbed, ecosystems can recover through **ecological succession** if the damage is not too great. We can classify ecological succession as either primary or secondary. **Primary succession** is the establishment and development of an ecosystem where one did not exist previously. Coral reefs that form on lava emitted from a volcano and cooled in shallow ocean waters are examples of primary succession. So are forests that develop on new lava flows, like those released by the volcano on the big island of Hawaii (Figure 6.17a), and forests that develop at the edges of retreating glaciers (Figure 6.17b).

Masha Nordbye/Bruce Coleman, Inc. / Photoshot Holdings Ltd.
Robert Harding Picture Library Ltd/Alamy Limited
(a) (b)

FIGURE 6.17 **Primary succession. (a)** Forests developing on new lava flows in Hawaii and **(b)** at the edge of a retreating glacier.

Secondary succession is reestablishment of an ecosystem after disturbances. In secondary succession, there are remnants of a previous biological community, including such things as organic matter and seeds. A coral reef that has been killed by poor fishing practices, pollution, climate change, or predation, and then recovers, is an example of secondary succession. Forests that develop on abandoned pastures or after hurricanes, floods, or fires are also examples of secondary succession.

Succession is one of the most important ecological processes, and the patterns of succession have many management implications (discussed in detail in Chapter 12). We see examples of succession all around us. When a house lot is abandoned in a city, weeds begin to grow. After a few years, shrubs and trees can be found; secondary succession is taking place. A farmer weeding a crop and a homeowner weeding a lawn are both fighting against the natural processes of secondary succession.

Patterns in Succession

Succession follows certain general patterns. When ecologists first began to study succession, they focused on three cases involving forests: (1) on dry sand dunes along the shores of the Great Lakes in North America; (2) in a northern freshwater bog; and (3) in an abandoned farm field. These were particularly interesting because each demonstrated a repeatable pattern of recovery and each tended to produce a late stage that was similar to the late stages of the others.

Dune Succession

Sand dunes are continually being formed along sandy shores and then breached and destroyed by storms. In any of the Great Lakes states, soon after a dune is formed on the shores of one of the Great Lakes, dune grass invades. This grass has special adaptations to the unstable dune. Just under the surface, it puts out runners with sharp ends (if you step on one, it will hurt). The dune grass rapidly forms a complex network of underground runners, crisscrossing almost like a coarsely woven mat. Above the ground, the green stems carry out photosynthesis, and the grasses grow. Once the dune grass is established, its runners stabilize the sand, and seeds of other plants have a better chance of germinating. The seeds germinate, the new plants grow, and an ecological community of many species begins to develop. The plants of this early stage tend to be small, grow well in bright light, and withstand the harsh environment—high temperatures in summer, low temperatures in winter, and intense storms.

Slowly, larger plants, such as eastern red cedar and eastern white pine, are able to grow on the dunes. Eventually, a forest develops, which may include species such as beech and maple. A forest of this type can persist for many years, but at some point a severe storm breaches even these heavily vegetated dunes, and the process begins again (Figure 6.18).

Bog Succession

A **bog** is an open body of water with surface inlets—usually small streams—but no surface outlet. As a result, the waters of a bog are quiet, flowing slowly if at all. Many bogs that exist today originated as lakes that filled depressions in the land, which in turn were created by glaciers during the Pleistocene ice age. Succession in a northern bog, such as the Livingston Bog in Michigan (Figure 6.19), begins when a sedge (a grasslike herb) puts out floating runners (Figure 6.20a, b). These runners form a complex, matlike network similar to that formed by dune grass. The stems of the sedge grow on the runners and

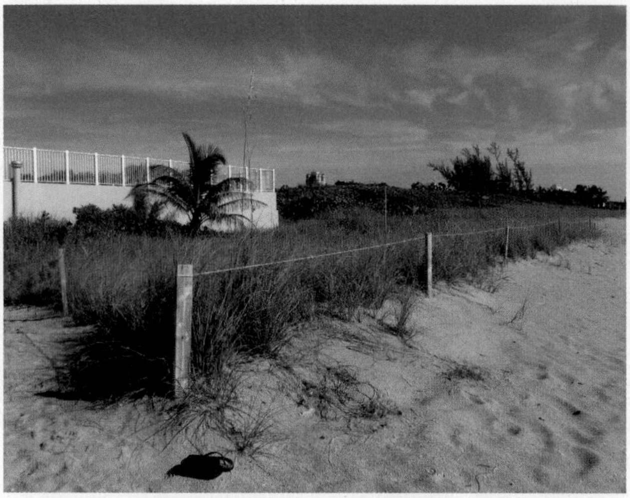

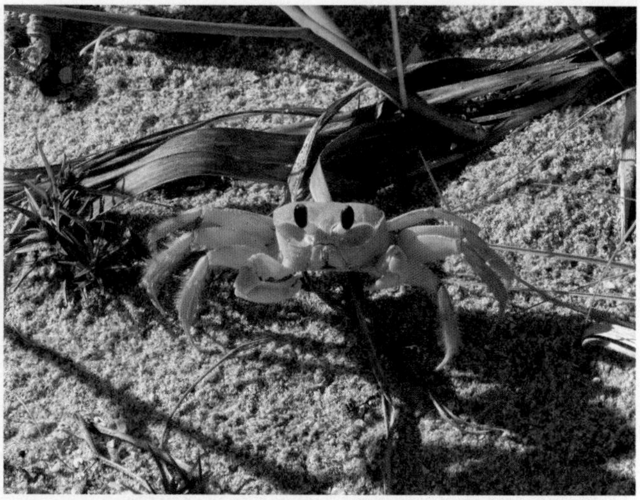

FIGURE 6.18 **Dune Succession** A section of the Atlantic Ocean coastal dunes in south Florida was partitioned off to let vegetation grow back and stabilize the dune **(a)** to help prevent storm damage. After about five years, the vegetation of many species has become well established. Many small animals, including this Atlantic ghost crab (*Ocypode quadrata*) **(b)**, benefit from these dune successional stages, which provide habitat.

FIGURE 6.19 **Livingston Bog,** a famous bog in the northern part of Michigan's lower peninsula.

carry out photosynthesis. Wind blows particles onto the mat, and soil, of a kind, develops. Seeds of other plants, instead of falling into the water, land on the mat and can germinate. The floating mat becomes thicker as small shrubs and trees, adapted to wet environments, grow. In the North, these include species of the blueberry family.

The bog also fills in from the bottom as streams carry fine particles of clay into it (Figure 6.20b, c). At the shore, the floating mat and the bottom sediments meet, forming a solid surface. But farther out, a "quaking bog" occurs. You can walk on this mat; and if you jump up and down, all the plants around you bounce and shake because the mat is really floating. Eventually, as the bog fills in from the top and the bottom, trees grow that can withstand wetter conditions—such as northern cedar, black spruce, and balsam fir. The formerly open-water bog becomes a wetland forest. If the bog is farther south, it may eventually be dominated by beech and maple, the same species that dominate the late stages of the dunes.

Old-Field Succession

In the northeastern United States, a great deal of land was cleared and farmed in the 18th and 19th centuries. Today, much of this land has been abandoned for farming and allowed to grow back to forest (Figure 6.21). The first plants to enter the abandoned farmlands are small plants adapted to the harsh and highly variable conditions of a clearing—a wide range of temperatures and precipitation. As these plants become established, other, larger plants enter. Eventually, large trees grow, such as sugar maple, beech, yellow birch, and white pine, forming a dense forest.

Since these three different habitats—one dry (the dunes), one wet (the bog), and one in between (the old field)—tend to develop into similar forests, early ecologists believed that this late stage was in fact a steady-state condition. They referred to it as the "climatic climax," meaning that it was the final, ultimate, and permanent stage to which all land habitats would proceed if undisturbed by people. Thus, the examples of succession were among the major arguments in the early 20th century that nature did in fact achieve a constant condition, a steady state, and there actually was a balance of nature. We know today that this is not true, a point we will return to later.

Coral Reef Succession

Coral reefs (Figure 6.22) are formed in shallow warm waters by corals—small marine animals that live in colonies and are members of the phylum Coelenterata, which also includes sea anemones and jellyfishes. Corals have a whorl of tentacles surrounding the mouth, and feed by catching prey, including planktonic algae, as it passes by. The corals settle on a solid surface and produce a hard

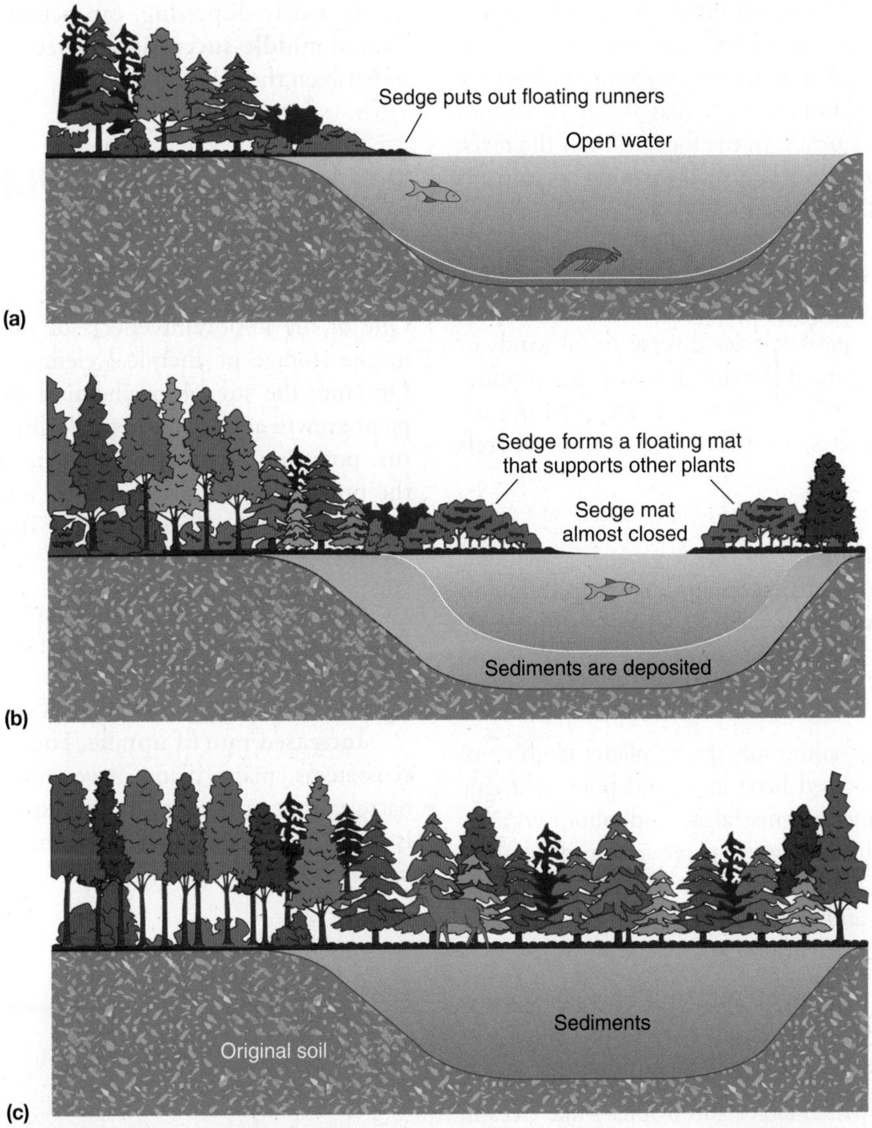

FIGURE 6.20 **Diagram of bog succession.** Open water **(a)** is transformed through formation of a floating mat of sedge and deposition of sediments **(b)** into wetland forest **(c)**.

Photo by Daniel B. Botkin

FIGURE 6.21 This woodland in New Hampshire has grown up after a field cleared for farming (and used many decades for that purpose) had been abandoned for more than 30 years. The trees include white pine and white birch, which are early-successional species characteristic of this part of New England.

Photo by Daniel B. Botkin; from Pennekamp State Park, FL, coral reef aquarium

FIGURE 6.22 The coral reefs, animals, and algae shown here occur in the world's third largest coral reef, extending along the southern Atlantic Ocean coast of Florida down into and beyond the Florida Keys.

polyp of calcium carbonate (in other words, limestone). As old individuals die, this hard material becomes the surface on which new individuals establish themselves. Coralline algae, a kind of red algae that also uses calcium carbonate, is also important in the formation of the reefs. In addition, other limestone-shell-forming organisms such as certain snails and sea urchins live and die on the reef and are glued together by coralline algae. Eventually a large and complex structure results involving many other species, including autotrophs and heterotrophs, creating one of the most species-diverse of all kinds of ecosystems. Highly valued for this diversity, for production of many edible fish, for the coral itself (used in various handicrafts and arts), and for recreation, coral reefs attract lots of attention.

Succession, in Sum

Even though the environments are very different, these four examples of ecological succession—dune, bog, old field, and coral reef—have common elements found in most ecosystems:

1. An initial kind of autotroph (green plants in three of the examples discussed here; algae and photosynthetic bacteria in marine systems; algae and photosynthetic bacteria, along with some green plants in some freshwater and near-shore marine systems). These are typically small in stature and specially adapted to the unstable conditions of their environment.
2. A second stage with autotrophs still of small stature, rapidly growing, with seeds or other kinds of reproductive structures that spread rapidly.
3. A third stage in which larger autotrophs—like trees in forest succession—enter and begin to dominate the site.
4. A fourth stage in which a mature ecosystem develops.

Although we list four stages, it is common practice to combine the first two and speak of early-, middle-, and late-successional stages. The stages of succession are described here in terms of autotrophs, but similarly adapted animals and other life forms are associated with each stage. We discuss other general properties of succession later in this chapter.

Species characteristic of the early stages of succession are called pioneers, or **early-successional species**. They have evolved and are adapted to the environmental conditions in early stages of succession. In terrestrial ecosystems, vegetation that dominates late stages of succession, called **late-successional species**, tends to be slower growing and longer lived, and can persist under intense competition with other species. For example, in terrestrial ecosystems, late-successional vegetation tends to grow well in shade and have seeds that, though not as widely dispersing, can persist a rather long time. Typical **middle-successional species** have characteristics in between the other two types.

6.8 Chemical Cycling and Succession

In Chapter 4, we introduced the biogeochemical cycle. One of the important effects of succession is a change in the storage of chemical elements necessary for life. On land, the storage of chemical elements essential for plant growth and function (including nitrogen, phosphorus, potassium, and calcium) generally increases during the progression from the earliest stages of succession to middle succession (Figure 6.23). There are three reasons for this:

Increased storage. Organic matter, living or dead, stores chemical elements. As long as there is an increase in organic matter within the ecosystem, there will be an increase in the storage of chemical elements.

Increased rate of uptake. For example, in terrestrial ecosystems, many plants have root nodules containing bacteria that can assimilate atmospheric nitrogen, which is then used by the plant in a process known as *nitrogen fixation*.

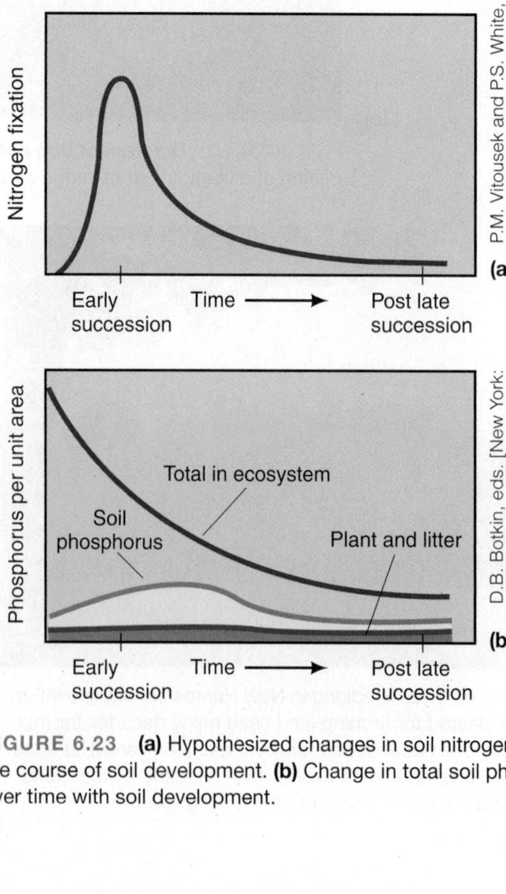

P.M. Vitousek and P.S. White, *Process Studies in Forest Succession*, in D.C. West, H.H. Shugart D.B. Botkin, eds. [New York: Springer-Verlag, 1981], Figure 17.1, p. 269

FIGURE 6.23 **(a)** Hypothesized changes in soil nitrogen during the course of soil development. **(b)** Change in total soil phosphorus over time with soil development.

Decreased rate of loss. The presence of live and dead organic matter helps retard erosion. Both organic and inorganic soil can be lost to erosion by wind and water. Vegetation and, in certain marine and freshwater ecosystems, large forms of algae tend to prevent such losses and therefore increase total stored material.

Ideally, chemical elements could be cycled indefinitely in ecosystems, but in the real world there is always some loss as materials are moved out of the system by wind and water. As a result, ecosystems that have persisted continuously for the longest time are less fertile than those in earlier stages. For example, where glaciers melted back thousands of years ago in New Zealand, forests developed, but the oldest areas have lost much of their fertility and have become shrublands with less diversity and biomass. The same thing happened to ancient sand dune vegetation in Australia.[11]

6.9 How Species Change Succession

Early-successional species can affect what happens later in succession in three ways: through (1) facilitation, (2) interference, or (3) life history differences (Figure 6.24).

Facilitation

In facilitation, an earlier-successional species changes the local environment in ways that make it suitable for another species that is characteristic of a later successional stage. Dune and bog succession illustrate facilitation. The first plant species—dune grass and floating sedge—prepare the way for other species to grow. Facilitation is common in tropical rain forests, where early-successional species speed the reappearance of the microclimatic conditions that occur in a mature forest. Because of the rapid growth of early-successional plants, after only 14 years the temperature, relative humidity, and light intensity at the soil surface in tropical forests can approximate those of a mature rain forest. Once these conditions are established, species adapted to deep forest shade can germinate and persist.

Facilitation also occurs in coral reefs, mangrove swamps along ocean shores, kelp beds along cold ocean shores such as the Pacific coast of the United States and Canada's Pacific Northwest, and in shallow marine benthic areas where the water is relatively calm and large algae can become established.

Knowing the role of facilitation can be useful in the restoration of damaged areas. Plants that facilitate the presence of others should be planted first. On sandy areas, for example, dune grasses can help hold the soil before we attempt to plant shrubs or trees.

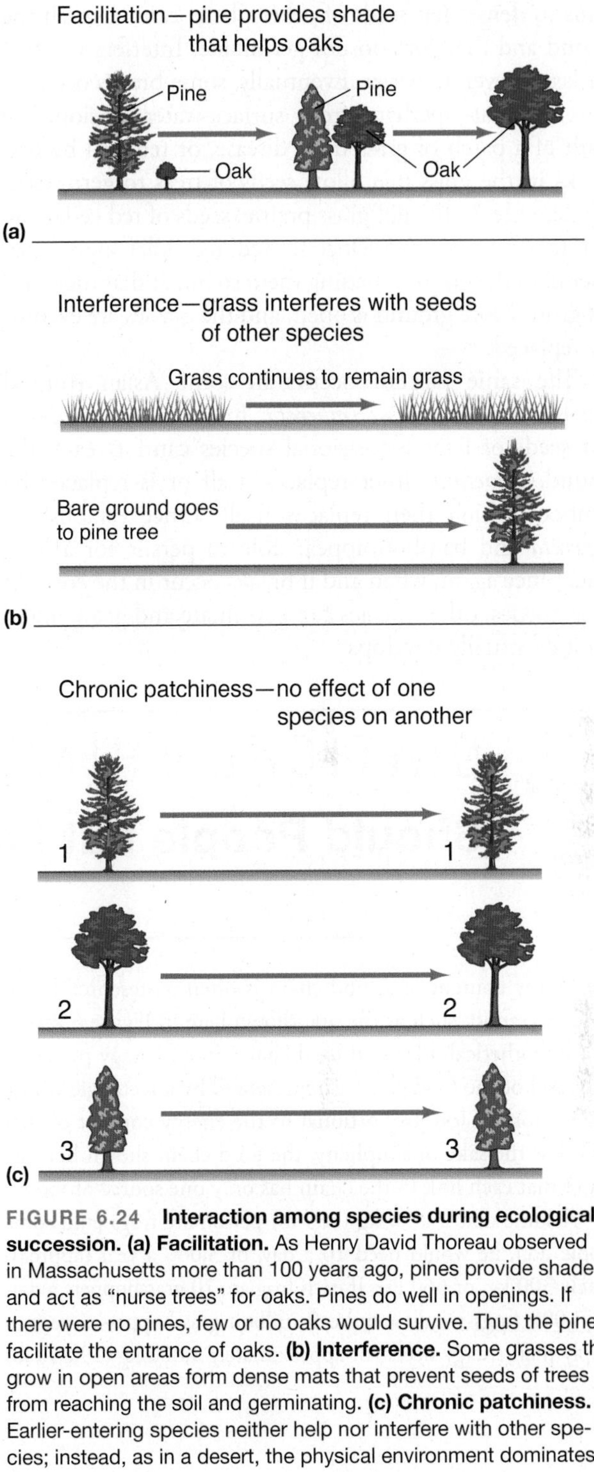

FIGURE 6.24 Interaction among species during ecological succession. (a) Facilitation. As Henry David Thoreau observed in Massachusetts more than 100 years ago, pines provide shade and act as "nurse trees" for oaks. Pines do well in openings. If there were no pines, few or no oaks would survive. Thus the pines facilitate the entrance of oaks. **(b) Interference.** Some grasses that grow in open areas form dense mats that prevent seeds of trees from reaching the soil and germinating. **(c) Chronic patchiness.** Earlier-entering species neither help nor interfere with other species; instead, as in a desert, the physical environment dominates.

Interference

In contrast to facilitation, interference refers to situations where an earlier-successional species changes the local environment so that it is *unsuitable* to another species characteristic of a later-successional stage. Interference is common, for example, in American tall-grass prairies, where prairie grasses like little bluestem form a mat of living and dead stems so dense that seeds of other plants cannot reach the ground and therefore do not germinate. Interference does not last forever, however. Eventually, some breaks occur in the grass mat—perhaps from surface-water erosion, the death of a patch of grass from disease, or removal by fire. Breaks in the grass mat allow seeds of trees to germinate. For example, in the tall-grass prairie, seeds of red cedar can then reach the ground. Once started, red cedar soon grows taller than the grasses, shading them so much that they cannot grow. More ground is open, and the grasses are eventually replaced.

The same pattern occurs in some Asian tropical rain forests. The grass, *Imperata*, forms stands so dense that seeds of later-successional species cannot reach the ground. *Imperata* either replaces itself or is replaced by bamboo, which then replaces itself. Once established, *Imperata* and bamboo appear able to persist for a long time. Once again, when and if breaks occur in the cover of these grasses, other species can germinate and grow, and a forest eventually develops.

Life History Differences

In this case, changes in the time it takes different species to establish themselves give the appearance of a succession caused by species interactions, but it is not. In cases where no species interact through succession, the result is termed **chronic patchiness**. Chronic patchiness is characteristic of highly disturbed environments and highly stressful ones in terms of temperature, precipitation, or chemical availability, such as deserts. For example, in the warm deserts of California, Arizona, and Mexico, the major shrub species grow in patches, often consisting of mature individuals with few seedlings. These patches tend to persist for long periods until there is a disturbance. Similarly, in highly polluted environments, a sequence of species replacement may not occur. Chronic patchiness also describes planktonic ecological communities and their ecosystems, which occur in the constantly moving waters of the upper ocean and the upper waters of ponds, lakes, rivers, and streams.

CRITICAL THINKING ISSUE

Should People Eat Lower on the Food Chain?

The energy content of a food chain is often represented by an energy pyramid, such as the one shown here in Figure 6.25a—for a hypothetical, idealized food chain. In an energy pyramid, each level of the food chain is represented by a rectangle whose area is more or less proportional to the energy content of that level. For the sake of simplicity, the food chain shown here assumes that each link in the chain has only one source of food.

Assume that if a 75 kg (165 lb) person ate frogs (and some people do!), he would need 10 a day, or 3,000 a year (approximately 300 kg, or 660 lb). If each frog ate 10 grasshoppers a day, the 3,000 frogs would require 9 million grasshoppers a year to supply their energy needs, or approximately 9,000 kg (19,800 lb) of grasshoppers. A horde of grasshoppers of that size would require 333,000 kg (732,600 lb) of wheat to sustain them for a year.

As the pyramid illustrates, the energy content decreases at each higher level of the food chain. The result is that the amount of energy at the top of a pyramid is related to the number of layers the pyramid has. For example, if people fed on grasshoppers rather than frogs, each person could probably get by on 100 grasshoppers a day. The 9 million grasshoppers could support 300 people for a year, rather than only one. If, instead of grasshoppers, people ate wheat, then 333,000 kg of wheat could support 666 people for a year.

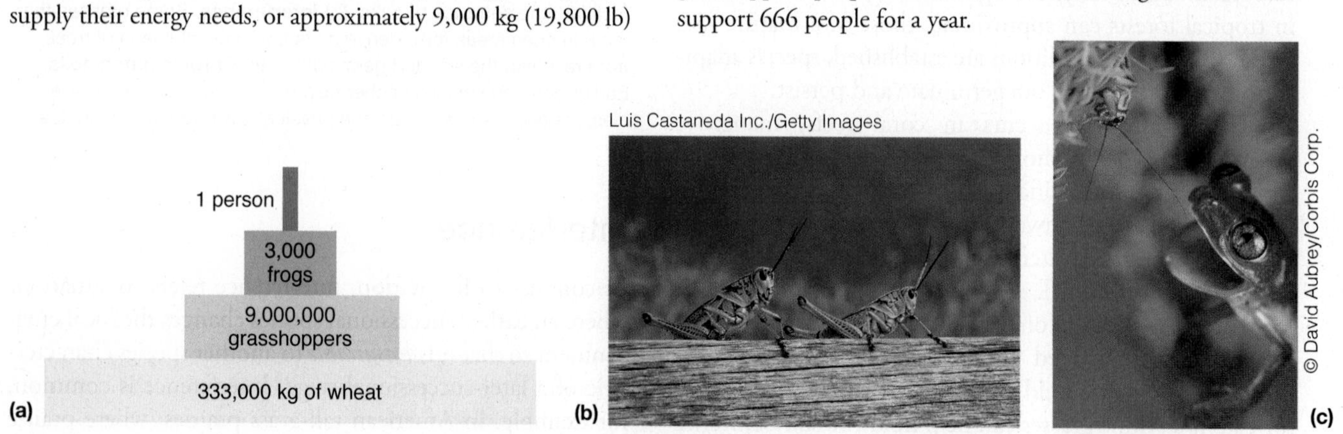

FIGURE 6.25 (a) Energy pyramid. (b) Grasshoppers. (c) Frogs that eat grasshoppers.

This argument is often extended to suggest that people should become herbivores (vegetarians) and eat directly from the lowest level of all food chains, the autotrophs. Consider, however, that humans can eat only parts of some plants. Herbivores can eat some parts of plants that humans cannot eat and some plants that humans cannot eat at all. When people eat these herbivores, more of the energy stored in plants becomes available for human consumption.

The most dramatic example of this is in aquatic food chains. Because people cannot digest most kinds of algae, which are the base of most aquatic food chains, they depend on eating fish that eat algae and fish that eat other fish. So if people were to become entirely herbivorous, they would be excluded from many food chains. In addition, there are major areas of Earth where crop production damages the land but grazing by herbivores does not. In those cases, conservation of soil and biological diversity lead to arguments that support the use of grazing animals for human food. This creates an environmental issue: How low on the food chain should people eat?

Critical Thinking Questions

1. Why does the energy content decrease at each higher level of a food chain? What happens to the energy lost at each level?
2. The pyramid diagram uses mass as an indirect measure of the energy value for each level of the pyramid. Why is it appropriate to use mass to represent energy content?
3. Using the average of 21 kilojoules (kJ) of energy to equal 1 g of completely dried vegetation (see Working It Out 6.2) and assuming that wheat is 80% water, what is the energy content of the 333,000 kg of wheat shown in the pyramid?
4. Make a list of the environmental arguments for and against an entirely vegetarian diet for people. What might be the consequences for U.S. agriculture if everyone in the country began to eat lower on the food chain?
5. How low do you eat on the food chain? Would you be willing to eat lower? Explain.

SUMMARY

- An ecosystem is the simplest entity that can sustain life. At its most basic, an ecosystem consists of several species and a fluid medium (air, water, or both). The ecosystem must sustain two processes—the cycling of chemical elements and the flow of energy.
- The living part of an ecosystem is the ecological community, a set of species connected by food webs and trophic levels. A food web or food chain describes who feeds on whom. A trophic level consists of all the organisms that are the same number of feeding steps from the initial source of energy.
- Community-level effects result from indirect interactions among species, such as those that occur when sea otters influence the abundance of sea urchins.
- Ecosystems are real and important, but it is often difficult to define the limits of a system or to pinpoint all the interactions that take place. Ecosystem management is considered key to the successful conservation of life on Earth.
- Energy flows one way through an ecosystem; the second law of thermodynamics places a limit on the abundance of productivity of life and requires the one-way flow.
- Chemical elements cycle, and in theory could cycle forever, but in the real world there is always some loss.
- Ecosystems recover from changes through ecological succession, which has repeatable patterns.
- Ecosystems are nonsteady-state systems, undergoing changes all the time and requiring change.

REEXAMINING THEMES AND ISSUES

Sean Randall/Getty Images, Inc.

HUMAN POPULATION

The human population depends on many ecosystems that are widely dispersed around the globe. Modern technology may appear to make us independent of these natural systems. In fact, though, the more connections we establish through modern transportation and communication, the more kinds of ecosystems we depend on. Therefore, the ecosystem concept is one of the most important we will learn about in this book.

© Biletskiy_Evgeniy/iStockphoto

SUSTAINABILITY

The ecosystem concept is at the heart of managing for sustainability. When we try to conserve species or manage living resources so that they are sustainable, we must focus on their ecosystem and make sure that it continues to function.

© Anton Balazh 2011/iStockphoto

GLOBAL PERSPECTIVE

Our planet has sustained life for approximately 3.5 billion years. To understand how Earth as a whole has sustained life for such a long time, we must understand the ecosystem concept because the environment at a global level must meet the same basic requirements as those of any local ecosystem.

ssguy/ShutterStock

URBAN WORLD

Cities are embedded in larger ecosystems. But like any life-supporting system, a city must meet basic ecosystem needs. This is accomplished through connections between cities and surrounding environments. Together, these function as ecosystems or sets of ecosystems. To understand how we can create pleasant and sustainable cities, we must understand the ecosystem concept.

B2M Productions/Getty Images, Inc.

PEOPLE AND NATURE

The feelings we get when we hike through a park or near a beautiful lake are as much a response to an ecosystem as to individual species. This illustrates the deep connection between people and ecosystems. Also, many effects we have on nature are at the level of an ecosystem, not just on an individual species.

George Doyle/Getty Images, Inc.

SCIENCE AND VALUES

The introductory case study about acorns, white-footed mice, gypsy moths, and Lyme disease illustrates the interactions between values and scientific knowledge about ecosystems. Science can tell us how organisms interact. This knowledge confronts us with choices. Do we want to manage forests to protect people from Lyme disease, which could result in gypsy moths denuding oak trees of foliage and possibly changing the ecosystem dynamic? The choices we make depend on our values.

KEY TERMS

autotrophs 107
biological production 118
biomass 118
bog 121
carnivores 107
chemoautotrophs 118
chronic patchiness 126
community effect 109
decomposers 107
early-successional species 124
ecological community 106
ecological succession 120
ecosystem 106
ecosystem energy flow 114
energy efficiency 120
entropy 118
first law of thermodynamics 116
food chains 107
food webs 107
gross production 118
herbivores 107
heterotrophs 107
keystone species 109
late-successional species 124
middle-successional species 124
net production 118
omnivores 113
pelagic ecosystem 111
photosynthesis 107
primary production 118
primary succession 120
secondary production 118
secondary succession 121
second law of thermodynamics 117
succession 121
trophic level 107
watershed 112

STUDY QUESTIONS

1. Farming has been described as managing land to keep it in an early stage of succession. What does this mean, and how is it achieved?
2. Redwood trees reproduce successfully only after disturbances (including fire and floods), yet individual redwood trees may live more than 1,000 years. Is redwood an early- or late-successional species?
3. What is the difference between an ecosystem and an ecological community?
4. What is the difference between the way energy puts limits on life and the way phosphorus does so?
5. Based on the discussion in this chapter, would you expect a highly polluted ecosystem to have many species or few species? Is our species a keystone species? Explain.
6. Keep track of the food you eat during one day and make a food chain linking yourself with the sources of those foods. Determine the biomass (grams) and energy (kilocalories) you have eaten. Using an average of 5 kcal/g, then using the information on food packaging or assuming that your net production is 10% efficient in terms of the energy intake, how much additional energy might you have stored during the day? What is your weight gain from the food you have eaten?
7. Which of the following are ecosystems? Which are ecological communities? Which are neither?
   (a) Chicago
   (b) A 1,000-ha farm in Illinois
   (c) A sewage-treatment plant
   (d) The Illinois River
   (e) Lake Michigan

FURTHER READING

Modern Studies

Botkin, D.B., ***No Man's Garden: Thoreau and a New Vision for Civilization and Nature*** (Washington, DC: Island Press, 2001).

Chapin, F. Stuart, III, Harold A. Mooney, Melissa C. Chapin, and Pamela Matson, ***Principles of Terrestrial Ecosystem Ecology*** (New York: Springer, 2004, paperback).

Kaiser, Michel J., Martin J. Attrill, Simon Jennings and David N. Thomas, *Marine Ecology: Processes, Systems, and Impacts* (New York: Oxford University Press, 2011).

Some Classic Studies and Books

Blum, H.F., ***Time's Arrow and Evolution*** (New York: Harper & Row, 1962). A very readable book; the author discusses how life is connected to the laws of thermodynamics and why this matters.

Bormann, F.H., and G.E. Likens, ***Pattern and Process in a Forested Ecosystem,*** **2nd ed.** (New York: Springer-Verlag, 1994). A synthetic view of the northern hardwood ecosystem, including its structure, function, development, and relationship to disturbance.

Gates, D.M., ***Biophysical Ecology*** (New York: Springer-Verlag, 1980). A discussion about how energy in the environment affects life.

Morowitz, H.J., ***Energy Flow in Biology*** (Woodbridge, CT: Oxbow, 1979). The most thorough and complete discussion available about the connection between energy and life, at all levels, from cells to ecosystems to the biosphere.

Odum, Eugene, and G.W. Barrett, ***Fundamentals of Ecology*** (Duxbury, MA: Brooks/Cole, 2004). Odum's original textbook was a classic, especially in providing one of the first serious introductions to ecosystem ecology. This is the latest update of the late author's work, done with his protégé.

Schrödinger, E. (ed. Roger Penrose), ***What Is Life?: With Mind and Matter and Autobiographical Sketches (Canto)*** (Cambridge: Cambridge University Press, 1992). The original statement about how the use of energy differentiates life from other phenomena in the universe. Easy to read and a classic.

NOTES

1. U. S. Centers for Disease Control. *Reported cases of Lyme disease by year, United States, 2002–2011*. www.cdc.gov/lyme/stats/chartstables/casesbyyear.html. Accessed October 18. 2012.
2. Ostfield, R.S., C.G. Jones, and J.O. Wolff. 1996 (May). Of mice and mast: Ecological connections in eastern deciduous forests. *BioScience* 46(5):323–330.
3. U. S. Centers for Disease Control. Reported Lyme disease cases by state, 2002–2011. http://www.cdc.gov/lyme/stats/chartstables/reportedcases_statelocality.html. Accessed October 18, 2012.
4. Brock, T.D. 1967. Life at high temperatures. *Science* 158:1012–1019.

5. NPS Evaluates Benefits-Sharing with Researchers Draft Environmental Impact Statement Available for Review September 2006. National Park Service. http://nature.nps.gov/benefitssharing/pdf/92206finalBenSharNewsletter.pdf. See also NPS Releases Final Benefits-Sharing EIS. http://www.nps.gov/yell/parknews/09115.htm.

6. National Oceanic and Atmospheric Administration (NOAA), Office of Protected Resources. http://www.nmfs.noaa.gov/pr/species/mammals/pinnipeds/harpseal.htm#population. Accessed August 28, 2012.

7. Lavigne, D.M., W. Barchard, S. Innes, and N.A. Oritsland. 1976. *Pinniped bioenergetics*. ACMRR/MM/SC/12. Rome: United Nations Food and Agriculture Organization.

8. Schrödinger, Erwin, 1944. *What Is Life? The Physical Aspect of the Living Cell*. Cambridge: The University Press. Based on lectures delivered under the auspices of the Institute of Trinity College, Dublin, in February 1943. Widely reprinted since then and available as a PDF file on the Internet.

9. Peterson, R.O. 1995. *The Wolves of Isle Royale: A Broken Balance*. Minocqua, WI: Willow Creek Press.

10. Vucetich, J.A., Michael P. Nelson, et al. 2012. Should Isle Royale wolves be reintroduced? A case study on wilderness management. *The George Wright Forum* 29(1):126–147; and personal communication with Rolf Peterson, May 2012

11. Botkin, D.B. 2012. *The Moon in the Nautilus Shell: Discordant Harmonies Reconsidered*. New York: Oxford University Press.

A CLOSER LOOK 6.1 NOTES

a. Estes, J.A., and J.F. Palmisano. 1974. Sea otters: Their role in structuring nearshore communities. *Science* 185:1058–1060.

b. Kvitek, R.G., J.S. Oliver, A.R. DeGange, and B.S. Anderson. 1992. Changes in Alaskan soft-bottom prey communities along a gradient in sea otter predation. *Ecology* 73:413–428.

c. Paine, R.T. 1969. A note on trophic complexity and community stability. *American Naturalist* 100:65–75.

d. Duggins, D.O. 1980. Kelp beds and sea otters: An experimental approach. *Ecology* 61:447–453.

e. Kenyon, K.W. 1969. The sea otter in the eastern Pacific Ocean. *North American Fauna*, no. 68. Washington, DC: Bureau of Sports Fisheries and Wildlife, U.S. Department of the Interior.

f. Sea otter 2008 population size from the following: U.S. Geological Survey, Western Ecological Research Center, Santa Cruz Field Station, Spring 2008 Mainland California Sea Otter Survey Results. Brian Hatfield and Tim Tinker, brian_hatfield@usgs.gov, ttinker@usgs.gov June, 24 2008, U.S. Geological Survey http://www.werc.usgs.gov/project.aspx?projectid=91; U.S. Geological Survey Alaska Science Center. IUCN otter specialist group. 2006. http://www.usgs.gov/newsroom/article.asp?ID=3369#.UGClhxgxfXd. The current worldwide population estimate for *E. lutris* is 108,000.

surface.[4] It's a controversial idea, but it suggests how close the present atmosphere is to a violent disequilibrium.[5]

The Fitness of the Environment[6]

Early in the 20th century, a scientist named Lawrence Henderson wrote a book with a curious title: *The Fitness of the Environment.*[7] In this book, Henderson observed that the environment on Earth was peculiarly suited to life. The question was, how did this come about? Henderson sought to answer this question in two ways: first, by examining the cosmos and seeking an answer in the history of the universe and in fundamental characteristics of the universe; second, by examining the properties of Earth and trying to understand how these may have come about.

"In the end there stands out a perfectly simple problem which is undoubtedly soluble," Henderson wrote. "In what degree are the physical, chemical, and general meteorological characteristics of water and carbon dioxide and of the compounds of carbon, hydrogen, and oxygen favorable to a mechanism which must be physically, chemically, and physiologically complex, which must be itself well regulated in a well-regulated environment, and which must carry on an active exchange of matter and energy with that environment?" In other words, to what extent are the nonbiological properties of the global environment favorable to life? And why is Earth so fit for life?

Today, we can give partial answers to Henderson's question. The answers involve recognizing that "environmental fitness" is the result of a two-way process. Life evolved in an environment conducive for that to occur, and then, over time, life altered the environment at a global level. These global alterations were originally problems for existing organisms, but they also created opportunities for the evolution of new life-forms adapted to the new conditions.

The Rise of Oxygen

The fossil record provides evidence that before about 2.3 billion years ago, Earth's atmosphere was very low in oxygen (anoxic), much closer to the atmospheres of Mars and Venus. The evidence for this exists in water-worn grains of pyrite (iron sulfide, FeS_2), which appear in sedimentary rocks formed more than 2.3 billion years ago. Today, when pure iron gets into streams, it is rapidly oxidized because there is so much oxygen in the atmosphere, and the iron forms sediments of iron oxides (what we know familiarly as rusted iron). If there were similar amounts of oxygen in the ancient waters, these ancient deposits would not have been pyrite—iron combined with sulfur—but would have been oxidized, just as they are today. This tells us that Earth's ancient Precambrian atmosphere and oceans were low in oxygen.

The ancient oceans had a vast amount of dissolved iron, which is much more soluble in water in its unoxidized state. Oxygen released into the oceans combined with the dissolved iron, changing it from a more soluble to a less soluble form. No longer dissolved in the water, the iron settled (precipitated) to the bottom of the oceans and became part of deposits that slowly were turned into rock. Over millions of years, these deposits formed the thick bands of iron ore that are mined today all around Earth, with notable deposits found today from Minnesota to Australia. That was the major time when the great iron ore deposits, now mined, were formed. It is intriguing to realize that very ancient Earth history affects our economic and environmental lives today.

The most convincing evidence of an oxygen-deficient atmosphere is found in ancient chemical sediments called *banded-iron formations*. These sediments were laid down in the sea, a sea that must have been able to carry dissolved iron—something it can't do now because oxygen precipitates the iron. If the ancient sea lacked free oxygen, then oxygen must also have been lacking in the atmosphere; otherwise, simple diffusion would have brought free oxygen into the ocean from the atmosphere.

How did the atmosphere become high enough in oxygen to change iron deposits from unoxidized to oxidized? The answer is that life changed the environment at a global level, adding free oxygen and removing carbon dioxide, first within the oceans, then in the atmosphere. This came about as a result of the evolution of photosynthesis. In photosynthesis, you will recall, carbon dioxide and water are combined, in the presence of light, to form sugar and free oxygen. But in early life, oxygen was a toxic waste that was eliminated from the cell and emitted into the surrounding environment. Scientists calculate that before oxygen started to accumulate in the air, 25 times the present-day amount of atmospheric oxygen had been neutralized by reducing agents such as dissolved iron. It took about 2 billion years for the unoxidized iron in Earth's oceans to be used up.

Life Responds to an Oxygen Environment

The oxygen-deficient phase of life's history occurred early in Earth's history, from 4.6 billion years ago until about 2.3 billion years ago. Some of the earliest photosynthetic organisms (3.4 billion years ago) were ancestors of bacteria that formed mats in shallow water. Photosynthesis became well established by about 1.9 billion years ago, but until sufficient free oxygen was in the atmosphere, organisms could get their energy only by fermentation, and the only kinds of organisms were bacteria and their relatives called **prokaryotes**. These have a simpler

Dr. Patricia Schulz

(a)

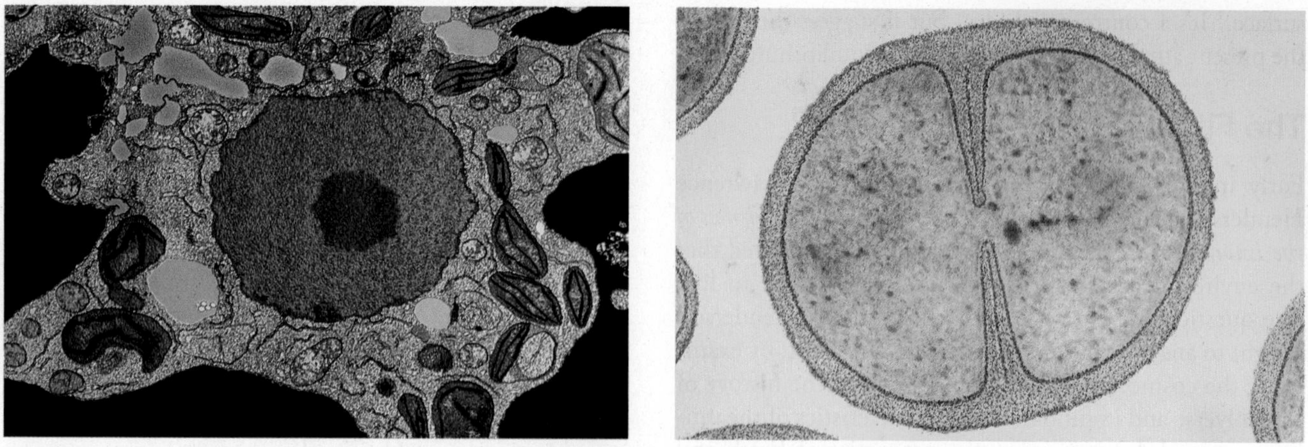

Kari Lounatmaa / Photo Researchers Inc

(b)

FIGURE 7.4 **Prokaryotes and eukaryotes.** Photomicrograph of **(a)** a eukaryotic cell—such cells are found in animals, plants, and fungi. These have distinct organelles, some structures within the cell that perform specific functions, and **(b)** a bacterial (prokaryote) cell.

internal cell structure than that of the cells of our bodies and other familiar forms of life, known as **eukaryotes** (Figure 7.4).

Without oxygen, organisms cannot completely "burn" organic compounds. Instead, they can get some energy from what we call fermentation, whose waste products are carbon dioxide and alcohol. Alcohol is a high-energy compound that, for the cell in an oxygenless atmosphere, is a waste product that has to be gotten rid of. Fermentation's low-energy yield to the organism puts limitations on the anaerobic cell. For example:

- Anaerobic cells must be small because a large surface-to-volume ratio is required to allow rapid diffusion of food in and waste out.
- Anaerobic cells have trouble keeping themselves supplied with energy. They cannot afford to use energy to maintain **organelles**—specialized cell parts that function like the organs of multicelled organisms. This means that all anaerobic bacteria were prokaryotes lacking specialized organelles, including a nucleus.
- Prokaryotes need free space around them; crowding interferes with the movement of nutrients and water into and out of the cell. Therefore, they live singly or strung end-to-end in chains. They cannot form three-dimensional structures. They are life restricted to a plane.

For other forms of life to evolve and persist, bacteria had to convert Earth's atmosphere to one high in oxygen. Once the atmosphere became high in oxygen, complete respiration was possible, and organisms could have much more complex structures, with three-dimensional bodies, and could use energy more efficiently and rapidly. Thus, the presence of free oxygen, a biological product, made possible the evolution of eukaryotes, organisms with more structurally complex cells and bodies, including the familiar animals and plants. Eukaryotes can do the following:

- Use oxygen for respiration, and because oxidative respiration is much more efficient (providing more energy) than fermentation, eukaryotes do not require as large a surface-to-volume ratio as anaerobic cells do, so eukaryote cells are larger.
- Maintain a nucleus and other organelles because of their superior metabolic efficiency.
- Form three-dimensional colonies of cells. Unlike prokaryotes, aerobic eukaryotes are not inhibited by crowding, so they can exist close to each other. This made possible the complex, multicellular body structures of animals, plants, and fungi.

Animals, plants, and fungi first evolved about 700 million to 500 million years ago. Inside their cells, DNA, the genetic material, is concentrated in a nucleus rather than distributed throughout the cell, and the cell contains organelles such as mitochondria which process energy. With the appearance of eukaryotes and the growth of an oxygenated atmosphere, the **biosphere**—our planet's system that includes and sustains all life—started to change rapidly and to influence more processes on Earth.

In sum, life affected Earth's surface—not just the atmosphere, but the oceans, rocks, and soils—and it still does so today. Billions of years ago, Earth presented a habitat where life could originate and flourish. Life, in turn, fundamentally changed the characteristics of the planet's surface, providing new opportunities for life and ultimately leading to the evolution of us.

7.2 Life and Global Chemical Cycles

All living things are made up of chemical elements, but of the more than 103 known chemical elements, only 24 are required by organisms (see the Periodic Table of Elements, Figure 7.5). These 24 are divided into the **macronutrients**, elements required in large amounts by all life, and **micronutrients**, elements required either in small amounts by all life, or in moderate amounts by some forms of life, and not at all by others.

The macronutrients, in turn, include the "big six" elements that are the fundamental building blocks of life: carbon, hydrogen, nitrogen, oxygen, phosphorus, and sulfur. Each one plays a special role in organisms. Carbon is the basic building block of organic compounds; along with oxygen and hydrogen, carbon forms carbohydrates. Nitrogen, along with these other three, makes proteins. Phosphorus is required in large quantities by all forms of life and is often a limiting factor for plant and algae growth. We call phosphorus the energy element because it is fundamental to a cell's use of energy. Phosphorus is in DNA, which carries the genetic material of life and is an important ingredient in cell membranes. Phosphorus occurs in compounds called ATP and ADP, which are important in the transfer and use of energy within cells.

Other macronutrients also play specific roles. Calcium, for example, is the structure element, occurring in bones and teeth of vertebrates, shells of shellfish, and wood-forming cell walls of vegetation. Sodium and potassium are important to nerve-signal transmission. Many of the metals required by living things are necessary for specific enzymes. (An enzyme is a complex organic compound that acts as a catalyst—it causes or speeds up chemical reactions, such as digestion.)

For any form of life to persist, chemical elements must be available at the right times, in the right amounts, and in the right concentrations. When this does not happen, a chemical can become a **limiting factor**, preventing the growth of an individual, a population, or a species, or even causing its local extinction.

Chemical elements may also be toxic to some life-forms and ecosystems. Mercury, for example, is toxic, even in low concentrations. Copper and some other elements are required in low concentrations for life processes but are toxic in high concentrations.

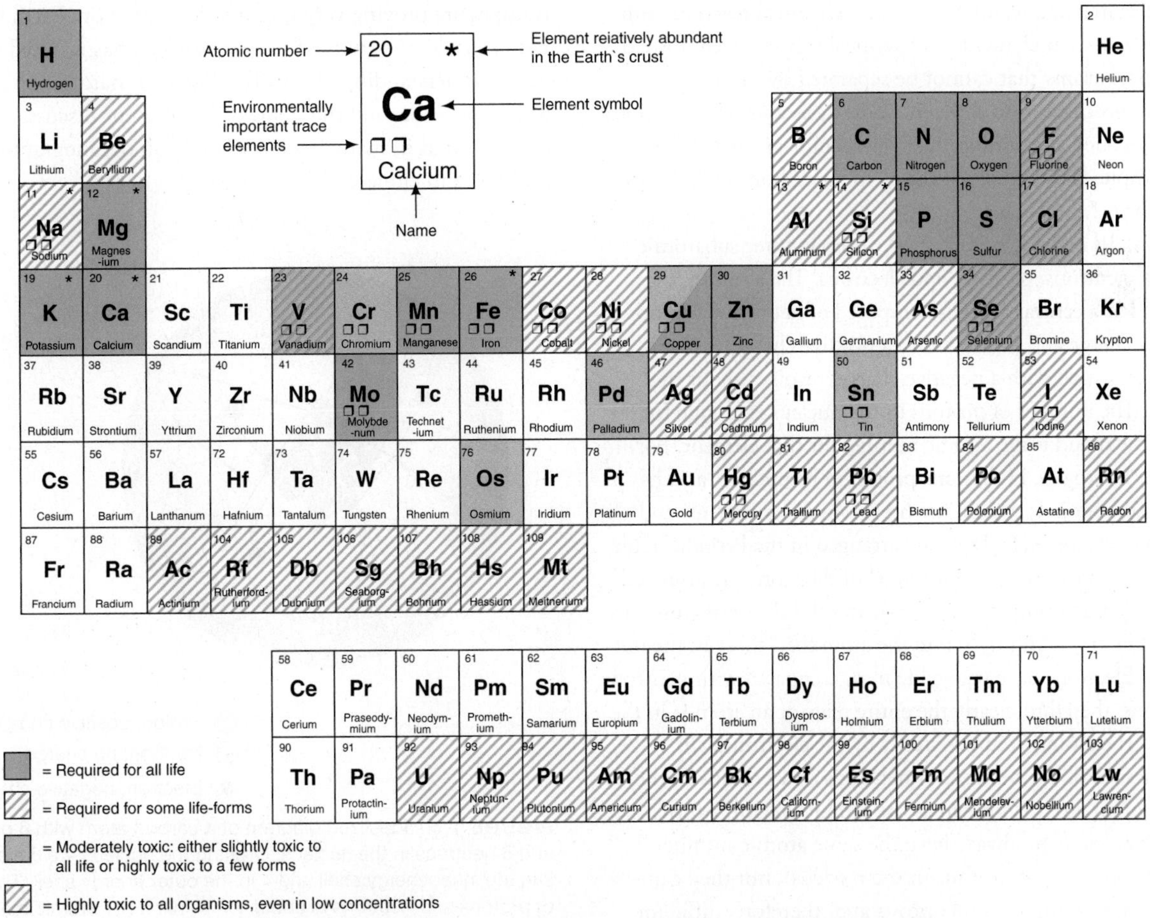

FIGURE 7.5 The Periodic Table of the Elements. The elements in green are required by all life; those in hatched green are micronutrients—required in very small amounts by all life-forms or required by only some forms of life. Those that are moderately toxic are in hatched red, and those that are highly toxic are solid red.

Finally, some elements are neutral for life. Either they are chemically inert, such as the noble gases (for example, argon and neon), which do not react with other elements, or they are present on Earth in very low concentrations.

7.3 General Aspects of Biogeochemical Cycles

A **biogeochemical cycle** is the complete path a chemical takes through the four major components, or reservoirs, of Earth's system: atmosphere, hydrosphere (oceans, rivers, lakes, groundwaters, and glaciers), lithosphere (rocks and soils), and biosphere (plants and animals). (Refer to A Closer Look 7.1 for a primer on environmental chemistry.) A biogeochemical cycle is *chemical* because it is chemicals that are cycled, *bio-* because the cycle involves life, and *geo-* because a cycle may include atmosphere, water, rocks, and soils. Although there are as many biogeochemical cycles as there are chemicals, certain general concepts hold true for these cycles:

A CLOSER LOOK 7.1

Basic Chemistry for Environmental Science

An understanding of basic chemistry will help you in understanding biogeochemical cycles in the context of environmental problems and solutions. An atom is the smallest piece of matter that can take part in a chemical reaction with another atom. An element is a chemical substance composed of identical atoms that cannot be separated by ordinary chemical processes into different substances and share chemical properties such as bonding. Each element is given a symbol. For example, the symbol for the element carbon is C, and for phosphorus P.

A model of an atom (Figure 7.6) shows three subatomic particles: neutrons, protons, and electrons. The atom is visualized as a central nucleus composed of neutrons with no electrical charge and protons with a positive charge. A cloud of electrons, each with a negative charge, revolves about the nucleus. The number of protons in the nucleus is unique for each element and is the atomic number for that element. For example, hydrogen, H, has one proton in its nucleus, and its atomic number is 1; carbon has 6 protons in its nucleus, and its atomic number is 6. These are arranged in the Periodic Table (Figure 7.5). Electrons, in our model of the atom, are arranged in shells (representing energy levels), and the electrons closest to the nucleus are bound tighter to the atom than those in the outer shells. Electrons have negligible mass compared with neutrons or protons; therefore, nearly the entire mass of an atom is in the nucleus. The atoms' volumes are set by the electron shells.

The sum of the number of neutrons and protons in the nucleus of an atom is known as the atomic weight. Atoms of the same element always have the same atomic number (the same number of protons in the nucleus), but they can have different numbers of neutrons and, therefore, different atomic weights. Two atoms of the same element with different numbers of neutrons in their nuclei and different atomic weights are known as isotopes of that element. For example, two isotopes of oxygen are ^{16}O and ^{18}O, where 16 and 18 are the atomic weights. Both isotopes have an atomic number of 8, but ^{18}O has two more neutrons than ^{16}O. Investigations of isotopes are proving very useful in learning how Earth works. For example, the study of oxygen isotopes has resulted in a better understanding of how the global climate has changed. This topic is beyond the scope of our present discussion but can be found in many basic textbooks on oceanography, geology, and climate.

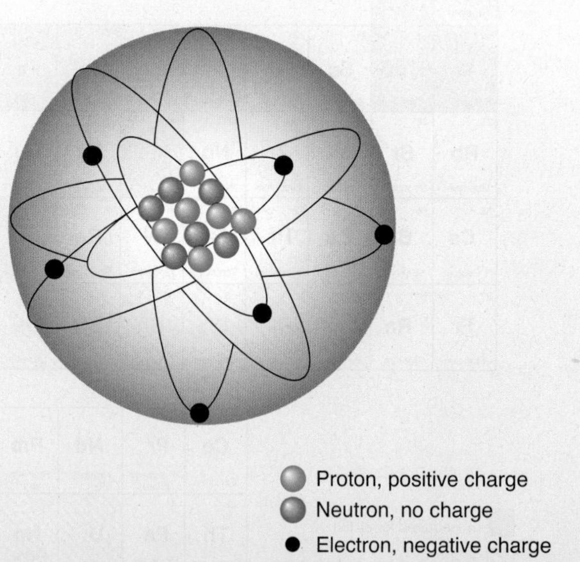

FIGURE 7.6 Idealized diagram of a carbon atom with 6 protons and 6 neutrons in the nucleus. Orbiting the nucleus are 6 electrons, 2 in the inner energy shell and 4 in the outer energy shell. The mass of protons and neutrons is about 1.7×10^{-24} g (1 atomic mass unit, amu). That of the electron is about 9.1×10^{-28} g (0.0005 amu). Thus the mass of an electron is about 0.05% of the mass of a proton or neutron. The diameter of a carbon atom is about 2.2×10^{-9} m, and that of human hair is about 300,000 times wider.

An atom is chemically balanced in terms of electric charge when the number of protons in the nucleus is equal to the number of electrons. However, an atom may lose or gain electrons, changing the balance in the electrical charge. An atom that has lost or gained electrons is called an ion, and its charge is represented by a superscripted number to the right of the symbol. An atom that has lost one or more electrons has a net positive charge and is called a cation. For example, the potassium ion K^+ has lost one electron, and the calcium ion Ca^{2+} has lost two electrons. An atom that has gained electrons has a net negative charge and is called an anion. For example, O^{2-} is an anion of oxygen that has gained two electrons. Other important ions in environmental science include the hydrogen (H^+), the carbonate ion (CO_3^{2-}), and the bicarbonate ion (HCO_3^-). These three ions are important in explaining seawater chemistry and formation or dissolution of calcium carbonate, the chemical most marine shells are composed of (discussed in the opening case study).

A compound is a chemical substance composed of two or more atoms of the same or different elements. The smallest unit of a compound is a molecule. For example, each molecule of water, H_2O, contains two atoms of hydrogen and one atom of oxygen, held together by chemical bonds. An important compound to the carbon cycle is CO_2 or carbon dioxide, which contains one atom of carbon and two atoms of oxygen.

The atoms that constitute a compound are held together by *chemical bonding*. The three main types of chemical bonds are covalent, ionic, and metallic. Chemical bonds describe how the electrons of elements interact. It is important to recognize that, when talking about chemical bonding of compounds, we are dealing with a complex subject. Although some compounds have a particular type of bond, many other compounds have more than one type of bond. With this caveat in mind, let's define each type of bond.

Covalent bonds result when atoms share electrons. This sharing takes place in the region between the atoms, and the strength of the bond is related to the number of pairs of electrons that are shared. Some important environmental compounds are held together solely by covalent bonds. These include carbon dioxide (CO_2) and water (H_2O). Covalent bonds are stronger than *ionic bonds*, which form as a result of attraction between cations and anions. An example of an environmentally important compound with ionic bonds is sea salt (mineral halite), or sodium chloride (NaCL). Compounds with ionic bonds such as sodium chloride tend to be soluble in water and, thus, dissolve easily.

Metallic bonds are those in which electrons are shared, as with covalent bonds. However, they differ because, in metallic bonding,the electrons are shared by all atoms of the solid rather than by specific atoms. As a result, the electrons can flow. For example, the mineral and element gold is an excellent conductor of electricity and can be pounded easily into thin sheets because the electrons have the freedom of movement characteristic of metallic bonding.

There are also weak forces of attraction known as Van der Waals force between molecules that are much weaker than either covalent or ionic bonding. For example, the mineral graphite (which is the "lead" in pencils) is black and consists of sheets of carbon atoms that easily part or break from one another because they are weak van der Waals forces.

In summary, in the study of the environment, we are concerned with matter (chemicals) that moves in and between the major components of the Earth system. An example is the element carbon, which moves through the atmosphere, hydrosphere, lithosphere, and biosphere in a large variety of compounds. These include carbon dioxide (CO_2) and methane (CH_4), which are gases in the atmosphere; sugar ($C_6H_{12}O_6$) in plants and animals; and complex hydrocarbons (compounds of hydrogen and carbon) in coal and oil deposits.

Chemical Reactions

It is important to acknowledge in our discussion of how chemical cycles work that the emphasis is on chemistry. Many chemical reactions occur within and between the living and nonliving portions of ecosystems. A **chemical reaction** is a process in which new chemicals (products) are formed from elements and compounds (reactants) that undergo a chemical change. For example, a simple reaction between seawater (H_2O) and carbon dioxide (CO_2) dissolved in the seawater produces carbonic acid:

$$H_2O + CO_2 \rightarrow H_2CO_3$$

Carbonic acid (H_2CO_3) is a weak acid that dissociates in seawater to H^+ (hydrogen cation) and HCO_3^- (bicarbonate anion). Bicarbonate further dissociates to H^+ and the CO_3^{2-} ion. Therefore, each molecule of CO_2 that enters seawater can end up in one of four dissolved pools: aqueous CO_2, H_2CO_3, HCO_3^-, or CO_3^{2-}. Biogenic shells, calcium carbonate ($CaCO_3$), are formed when dissolved Ca^{2+} and CO_3^{2-} combine.

As we saw in this chapter's case study, acidity is an important characteristic of a solution such as seawater because more acidic solutions have faster reactions and cause more environmental change. Acidity is the concentration of hydrogen ions (H^+) in a solution such as seawater. Acidity is commonly expressed as pH, where $pH = -\log(H^+)$. Relatively low values of pH are acids (for example, stomach acid, pH of 2), and relatively high values are basic (for example, household bleach, pH of 13). Changing the value of pH by one unit corresponds to a tenfold change in acidity. For example, if pH is reduced from about 8 (seawater) to about 5 (coffee), acidity increases by about 1,000 times. Figure 7.7 lists values of pH for several common solutions. The pH of seawater in recent years has decreased by about 0.11m pH unit, which equates to an increase in acidity of about 30%. This is the ocean acidification problem.

Chemical reactions determine when, where, and whether chemicals are available to life. For example, as we have discussed in this chapter, photosynthesis is a series of chemical reactions by which living green plants, with sunlight as an energy source, convert carbon dioxide (CO_2) and water (H_2O)

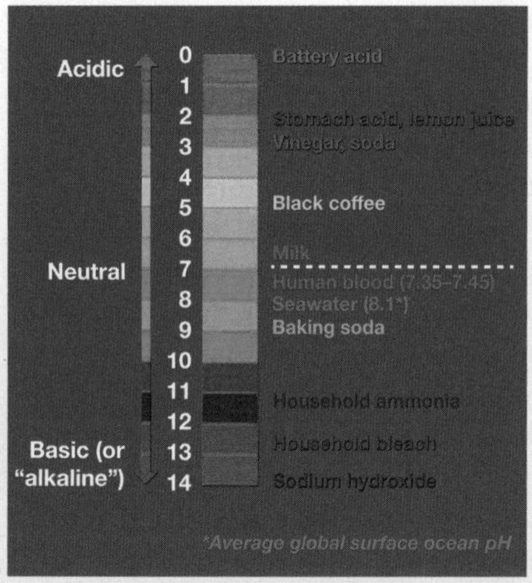

Source: National Oceanic and Atmospheric Administration. A Primer on pH. www.pmel.noaa.gov. Accessed 7/14/12

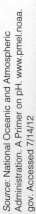

FIGURE 7.7 Diagram illustrating the pH scale from basic to acidic solutions.

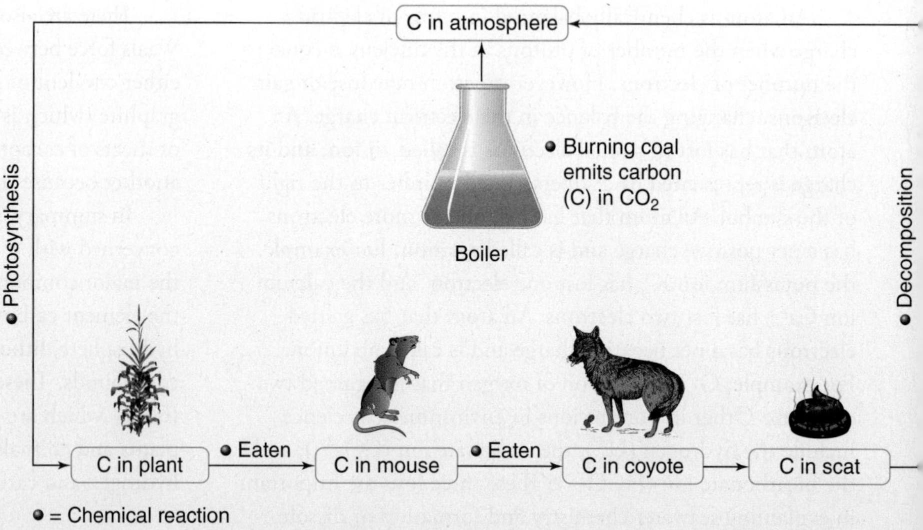

FIGURE 7.8 Idealized diagram illustrating several possible chemical reactions involving an atom of carbon from release into the atmosphere as a result of burning coal to uptake in a plant, which is consumed by a mouse, which is then eaten by a coyote and returned to the atmosphere through decomposition.

to sugar ($C_6H_{12}O_6$) and oxygen (O_2). The general chemical reaction for photosynthesis is:

$$6CO_2 + 6H_2O \xrightarrow{\text{sunlight}} C_6H_{12}O_6 + 6O_2$$

Photosynthesis produces oxygen as a by-product, and that is why we have free oxygen in our atmosphere. Many chemical reactions can occur in ecosystems. Figure 7.8 illustrates several reactions that start with burning of coal that releases carbon dioxide into the atmosphere; a plant takes up carbon through photosynthesis; a mouse eats the plant and is eaten by a coyote; its body is eventually released to the atmosphere through decomposition.

- Some chemical elements, such as oxygen and nitrogen, cycle quickly and are readily regenerated for biological activity. Typically, these elements have a gas phase and are present in the atmosphere and/or easily dissolved in water and carried by the hydrologic cycle (discussed later in the chapter).
- Other chemical elements are easily tied up in relatively immobile forms and are returned slowly, by geologic processes, to where they can be reused by life. Typically, they lack a gas phase and are not found in significant concentrations in the atmosphere. They also are relatively insoluble in water. Phosphorus is an example.
- Most required nutrient elements have a light atomic weight. The heaviest required micronutrient is iodine, element 53.
- Since life evolved, it has greatly altered biogeochemical cycles, and this alteration has changed our planet in many ways.
- The continuation of processes that control biogeochemical cycles is essential to the long-term maintenance of life on Earth.

Through modern technology, we have begun to transfer chemical elements among air, water, and soil, in some cases at rates comparable to natural processes. These transfers can benefit society, as when they improve crop production, but they can also pose environmental dangers, as illustrated by the opening case study. To live wisely with our environment, we must recognize the positive and negative consequences of altering biogeochemical cycles.

The simplest way to visualize a biogeochemical cycle is as a box-and-arrow diagram of a system (see the discussion of systems in Chapter 4), with the boxes representing places where a chemical is stored (*storage compartments*) and the arrows representing pathways of transfer (Figure 7.9**a**). In this kind of diagram, the **flow** is the amount moving from

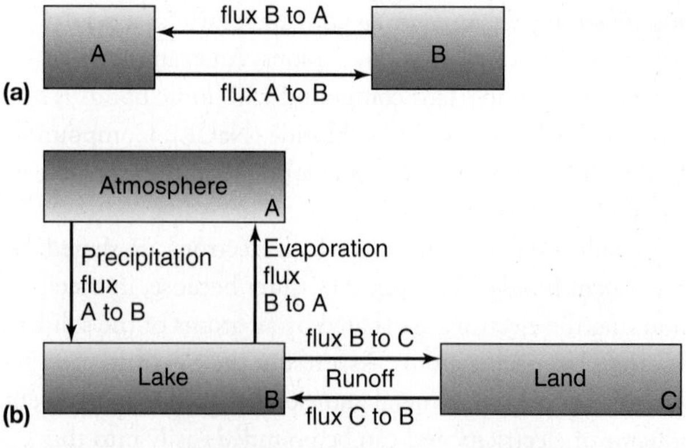

FIGURE 7.9 **(a)** A unit of a biogeochemical cycle viewed as a systems diagram; **(b)** a highly simplified systems diagram of the water cycle.

one compartment to another, whereas the **flux** is the rate of transfer—the amount per unit time—of a chemical that enters or leaves a storage compartment. The **residence time** is the average time that an atom is stored in a compartment. The donating compartment is a **source**, and the receiving compartment is a **sink**.

A biogeochemical cycle is generally drawn for a single chemical element, but sometimes it is drawn for a compound—for example, water (H_2O). Figure 7.9b shows the basic elements of a biogeochemical cycle for water, represented about as simply as it can be, as three compartments: water stored temporarily in a lake (compartment B); entering the lake from the atmosphere (compartment A) as precipitation; and from the land around the lake as runoff (compartment C). It leaves the lake through evaporation to the atmosphere or as runoff via a surface stream or subsurface flows. We diagrammed the Missouri River in Chapter 4 in this way.

As an example, consider a salt lake with no transfer out except by evaporation. Assume that the lake contains 3,000,000 m^3 (106 million ft^3) of water and the evaporation is 3,000 m^3/day (106,000 ft^3/day). Surface runoff into the lake is also 3,000 m^3/day, so the volume of water in the lake remains constant (input = output). We can calculate the average residence time of the water in the lake as the volume of the lake divided by the evaporation rate (rate of transfer), or 3,000,000 m^3 divided by 3,000 m^3/day, which is 1,000 days (or 2.7 years).

7.4 The Geologic Cycle

Throughout the 4.6 billion years of Earth's history, rocks and soils have been continuously created, maintained, changed, and destroyed by physical, chemical, and biological processes. This is another illustration that the biosphere is a dynamic system, not in steady state. Collectively, the processes responsible for formation and change of Earth materials are referred to as the **geologic cycle** (Figure 7.10). The geologic cycle is best described

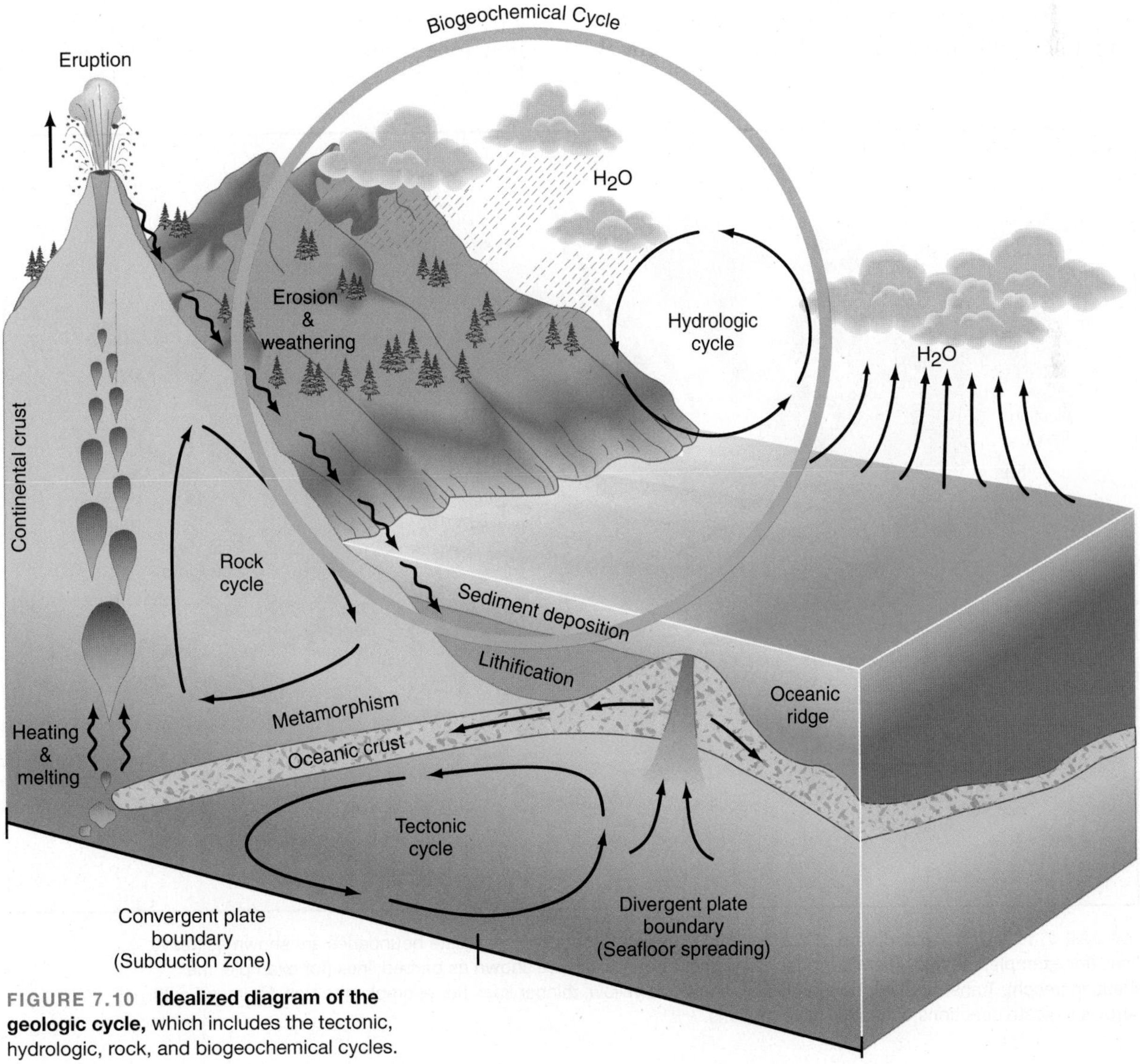

FIGURE 7.10 **Idealized diagram of the geologic cycle,** which includes the tectonic, hydrologic, rock, and biogeochemical cycles.

as a group of cycles: tectonic, hydrologic, rock, and biogeochemical. (We discuss the last cycle separately because it requires lengthier examination.)

The Tectonic Cycle

The **tectonic cycle** involves the creation and destruction of Earth's solid outer layer, the *lithosphere*. The lithosphere is about 100 km (60 mi) thick on average and is broken into several large segments called *plates*, which are moving relative to one another (Figure 7.11). The slow movement of these large segments of Earth's outermost rock shell is referred to as **plate tectonics**. The plates "float" on denser material and move at rates of 2 to 15 cm/year (0.8 to 6.9 in./year), about as fast as your fingernails grow. The tectonic cycle is driven by forces originating deep within the earth. Closer to the surface, rocks are deformed by spreading plates, which produce ocean basins, and by collisions of plates, which produce mountain ranges and island-arc volcanoes.

Plate tectonics has important environmental effects. Moving plates change the location and size of continents, altering atmospheric and ocean circulation and thereby altering climate. Plate movement has also created ecological islands by breaking up continental areas. When this happens, closely related life-forms are isolated from one another for millions of years, leading to the evolution of new species. Finally, boundaries between plates are geologically active areas, and most volcanic activity and earthquakes occur there. Earthquakes occur when the brittle upper lithosphere fractures along faults (fractures in rock within the Earth's crust). Movement of several meters between plates can occur within a few seconds or minutes, in contrast to the slow, deeper plate movement described above.

Three types of plate boundaries occur: divergent, convergent, and transform faults.

A divergent plate boundary occurs at a spreading ocean ridge, where plates are moving away from one another and new lithosphere is produced. This process, known as *seafloor spreading*, produces ocean basins.

A convergent plate boundary occurs when plates collide. When a plate composed of relatively heavy ocean-basin rocks dives (subducts) beneath the leading edge of a plate composed of lighter continental rocks, a subduction zone is present. Such a convergence may produce

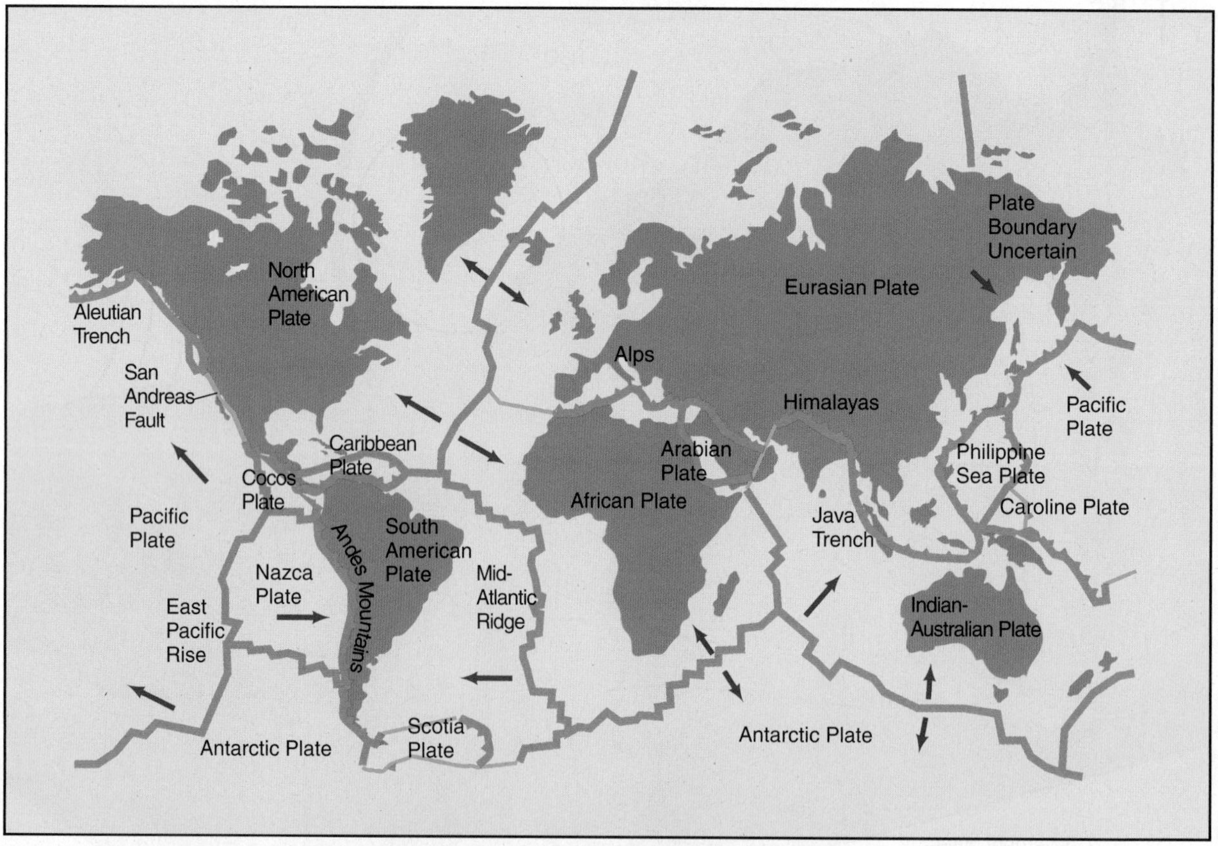

(*Source*: Modified from B.C. Burchfiel, R.J. Foster, E.A. Keller, W.N. Melhorn, D.G. Brookins, L.W. Mintz, and H.V. Thurman, *Physical Geology: The Structures and Processes of the Earth* [Columbus, Ohio: Merrill, 1982].)

FIGURE 7.11 **Generalized map of Earth's lithospheric plates.** Divergent plate boundaries are shown as heavy lines (for example, the Mid-Atlantic Ridge). Convergent boundaries are shown as barbed lines (for example, the Aleutian trench). Transform fault boundaries are shown as yellow, thinner lines (for example, the San Andreas Fault). Arrows indicate directions of relative plate motions.

Jeff Foott/Getty Images

FIGURE 7.12 **Plate tectonics and landscapes.** In response to slow tectonic uplift of the region, the Colorado River has eroded through the sedimentary rocks of the Colorado plateau to produce the spectacular Grand Canyon. In recent years the river has been greatly modified by dams and reservoirs above and below the canyon. Sediment once carried to the Gulf of California is now deposited in reservoirs. The dam stores sediments, and some of the water released is from the deeper and thus cooler parts of the reservoir, so water flowing out of the dam and down through the Grand Canyon is clearer and colder than it used to be. Fewer sandbars are created; this and the cooler water change which species of fish are favored. Thus this upstream dam has changed the hydrology and environment of the Colorado River in the Grand Canyon.

linear coastal mountain ranges, such as the Andes in South America. When two plates that are both composed of lighter continental rocks collide, a continental mountain range may form, such as the Himalayas in Asia.

A transform fault boundary occurs where one plate slides past another. An example is the San Andreas Fault in California, which is the boundary between the North American and Pacific plates. The Pacific plate is moving north, relative to the North American plate, at about 5 cm/year (2 in./year). As a result, Los Angeles is moving slowly toward San Francisco, about 500 km (300 mi) north. If this continues, in about 10 million years San Francisco will be a suburb of Los Angeles.

Uplift and subsidence of rocks, along with erosion, produce Earth's varied topography. The spectacular Grand Canyon of the Colorado River in Arizona (Figure 7.12), sculpted from mostly sedimentary rocks, is one example.

The Hydrologic Cycle

The **hydrologic cycle** (Figure 7.13) is the transfer of water from the oceans to the atmosphere to the land and back to the oceans. It includes evaporation of water from the oceans; precipitation on land; evaporation from land; transpiration of water by plants; and runoff from streams, rivers, and subsurface groundwater. Solar energy drives the hydrologic cycle by evaporating water from oceans, freshwater bodies, soils, and vegetation. Of the total 1.3 billion km^3 of water on Earth, about 97% is in oceans

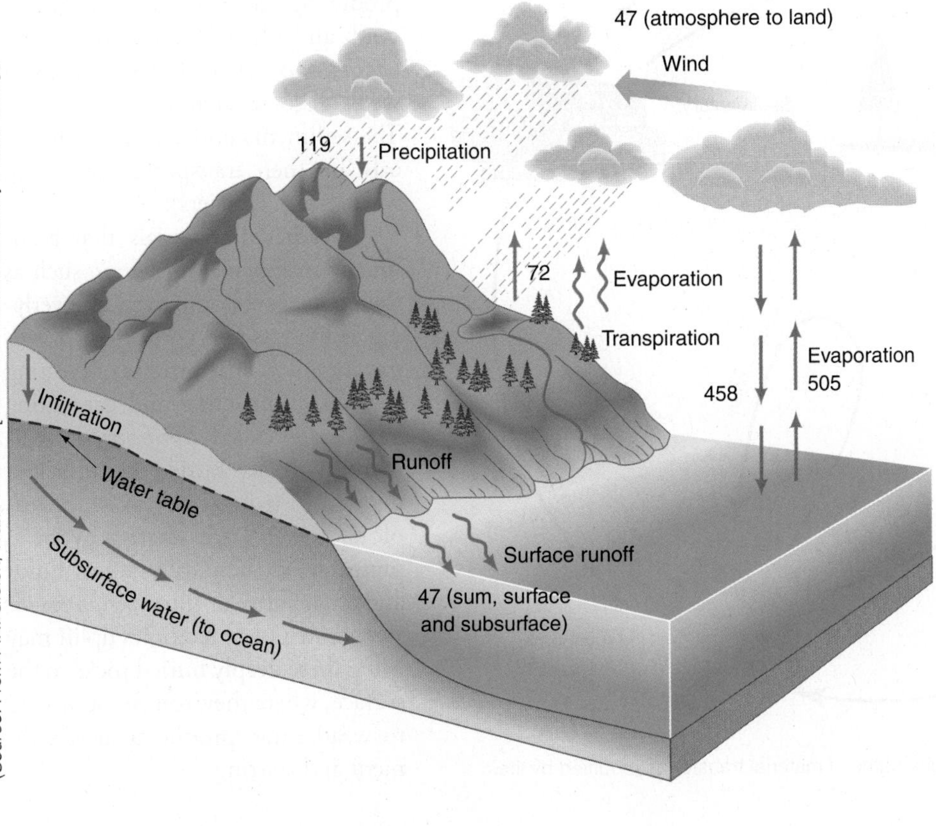

(*Source:* From P.H. Gleick, *Water in Crisis* [New York: Oxford University Press, 1993].)

FIGURE 7.13 **The hydrologic cycle**, showing the transfer of water (thousands of km^3/yr) from the oceans to the atmosphere to the continents and back to the oceans again.

and about 2% is in glaciers and ice caps; 0.76% is shallow groundwater; 0.013% is in lakes and rivers; and only 0.001% is in the atmosphere. Although water on land and in the atmosphere accounts for only a small fraction of the water on Earth, this water is important in moving chemicals, sculpting landscape, weathering rocks, transporting sediments, and providing our water resources.

The rates of transfer of water from land to the ocean are relatively low, and the land and oceans are somewhat independent in the water cycle because most of the water that evaporates from the ocean falls back into the ocean as precipitation, and most of the water that falls as precipitation on land comes from evaporation of water from land, as shown in Figure 7.13. Approximately 60% of precipitation on land evaporates each year back to the atmosphere, while the rest, about 40%, returns to the ocean as surface and subsurface runoff. The distribution of water is far from uniform on the land, and this has many environmental and ecological effects, which we discuss in Chapter 8 (on biological diversity), Chapters 18 and 20 (on water and climate), and Chapter 22 (urban environments).

At the regional and local levels, the fundamental hydrologic unit of the landscape is the *drainage basin* (also called a *watershed* or *catchment*). As explained in Chapter 6, a watershed is the area that contributes surface runoff to a particular stream or river. The term is used in evaluating the hydrology of an area (such as the stream flow or runoff from slopes) and in ecological research and biological conservation. Watersheds are best categorized by drainage basin area, and further by how many streams flow into the final, main channel. A first-order watershed is drained by a single small stream; a second-order watershed includes streams from first-order watersheds, and so on. Drainage basins vary greatly in size, from less than a hectare (2.5 acres) for a first-order watershed to millions of square kilometers for major rivers like the Missouri, Amazon, and Congo. A watershed is usually named for its main stream or river, such as the Mississippi River drainage basin.

The Rock Cycle

The **rock cycle** consists of numerous processes that produce rocks and soils. The rock cycle depends on the tectonic cycle for energy and on the hydrologic cycle for water. As shown in Figure 7.14, rock is classified as igneous, sedimentary, or metamorphic. These three types of rock are involved in a worldwide recycling process. Internal heat from the tectonic cycle produces *igneous rocks* from molten material (magma) near the surface, such as lava from volcanoes. When magma crystallized deep in the earth, the igneous rock granite was formed. These new rocks weather when exposed at the surface. Water in cracks of rocks expands when it freezes, breaking the rocks apart. This physical weathering makes smaller particles of rock from bigger ones, producing sediment, such as gravel, sand, and silt. Chemical weathering occurs, too, when the weak acids in water dissolve chemicals from rocks. The sediments and dissolved chemicals are then transported by water, wind, or ice (glaciers).

Weathered materials that accumulate in *depositional basins*, such as the oceans, are compacted by overlying sediments and converted to *sedimentary rocks.* The process of creating rock by compacting and cementing particles is called *lithification.* Sedimentary rocks buried at sufficient depths (usually tens to hundreds of kilometers) are altered by heat, pressure, or chemically active fluids and transformed into *metamorphic rocks.* Later, plate tectonics uplift may bring these deeply buried rocks to the surface, where they, too, are subjected to weathering, producing new sediment and starting the cycle again.

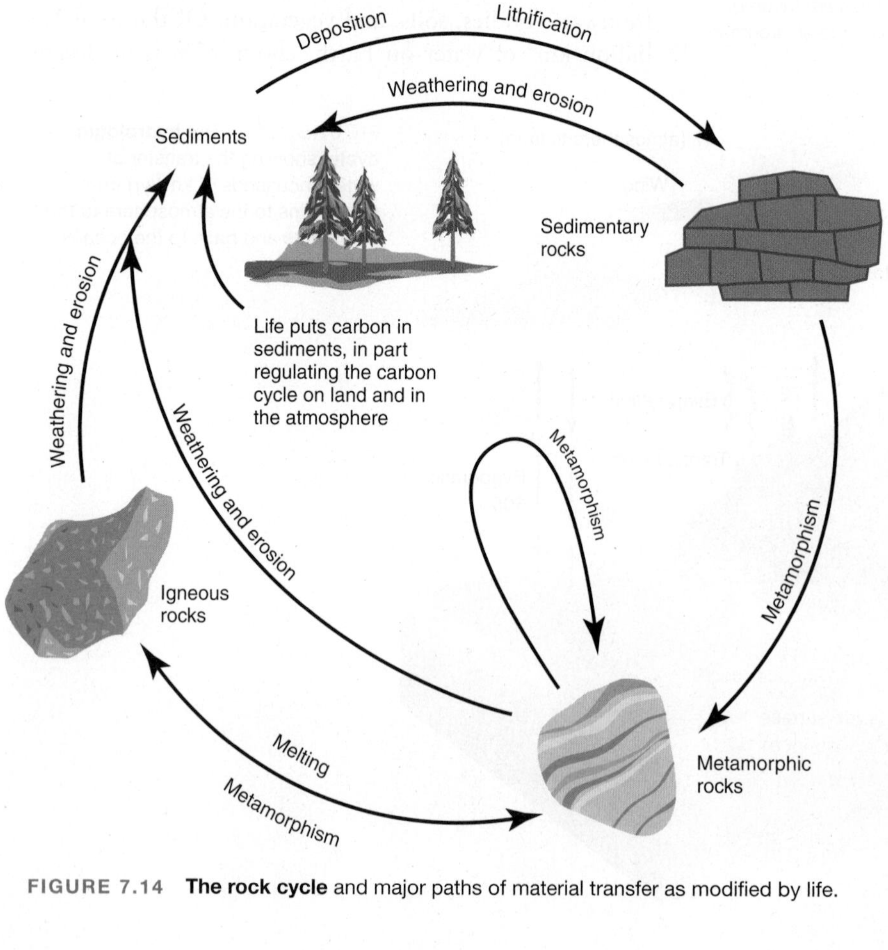

FIGURE 7.14 **The rock cycle** and major paths of material transfer as modified by life.

You can see in Figure 7.14 that life processes play an important role in the rock cycle by adding organic carbon to rocks. The addition of organic carbon produces rocks such as limestone, which is mostly calcium carbonate (the material of seashells and bones), as well as fossil fuels, such as coal.

Our discussion of geologic cycles has emphasized tectonic, hydrologic, and rock-forming processes. We can now begin to integrate biogeochemical processes into the picture.

7.5 Some Major Global Biogeochemical Cycles

With Figure 7.14's basic diagram in mind, we can now consider complete chemical cycles, though still quite simplified. Each chemical element has its own specific cycle, but all the cycles have certain features in common (Figure 7.15).

The Carbon Cycle

Carbon is the basic building block of life and the element that anchors all organic substances, from coal and oil to DNA (deoxyribonucleic acid), the compound that carries genetic information. Although of central importance to life, carbon is not one of the most abundant elements in Earth's crust. It contributes only 0.032% of the weight of the crust, ranking far behind oxygen (45.2%), silicon (29.5%), aluminum (8.0%), iron (5.8%), calcium (5.1%), and magnesium (2.8%).[8, 9]

The major pathways and storage reservoirs of the **carbon cycle** are shown in Figure 7.16. This diagram is simplified to show the big picture of the carbon cycle. Details are much more complex.[10]

- **Oceans and land ecosystems:** In the past half-century, ocean and land ecosystems have removed about 3.1 ± 0.5 GtC/yr, which is approximately 45% of the carbon emitted from burning fossil fuels during that period.
- **Land-use change:** Deforestation and decomposition of what is cut and left, as well as burning of forests to make room for agriculture in the tropics, are the main reasons 2.2 ± 0.8 GtC/yr is added to the atmosphere. A small flux of carbon (0.2 ± 0.5 GtC/yr) from hot tropical areas is pulled from the atmosphere by growing forests. In other words, when considering land-use change, deforestation is by far the dominant process.
- **Residual land sink:** The observed net uptake of CO_2 from the atmosphere (see Figure 7.16) by land ecosystems suggests there must be a sink for carbon in land ecosystems that has not been adequately identified.

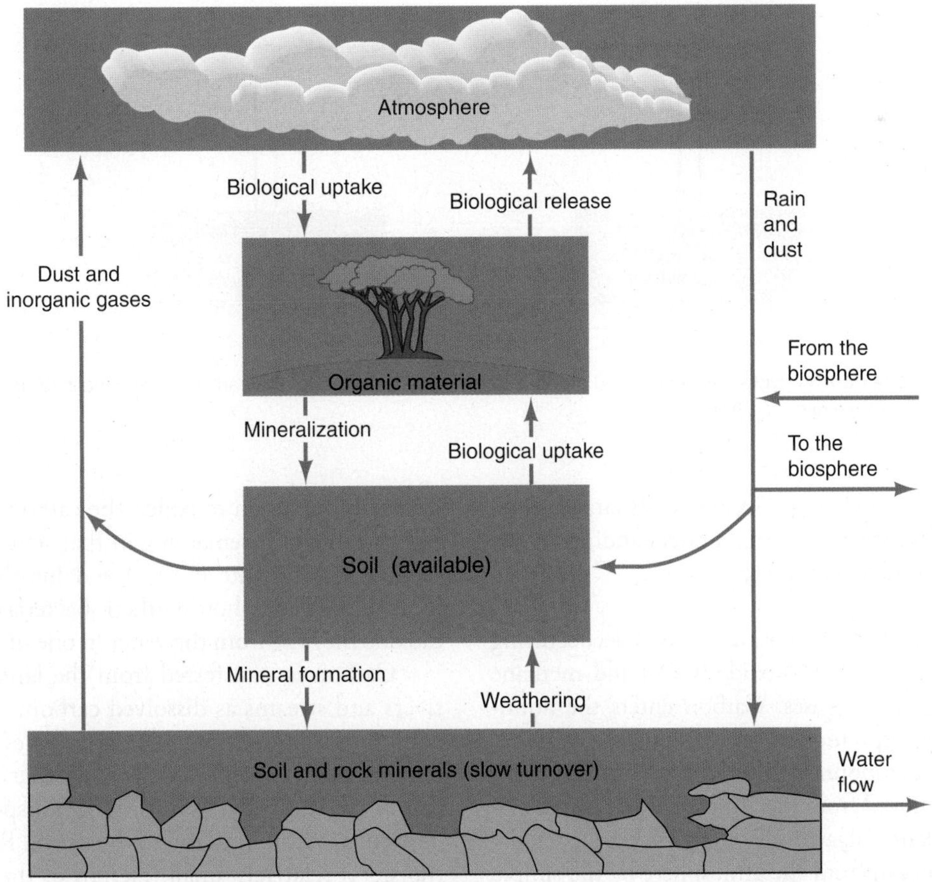

FIGURE 7.15 **Basic biogeochemical cycle.**

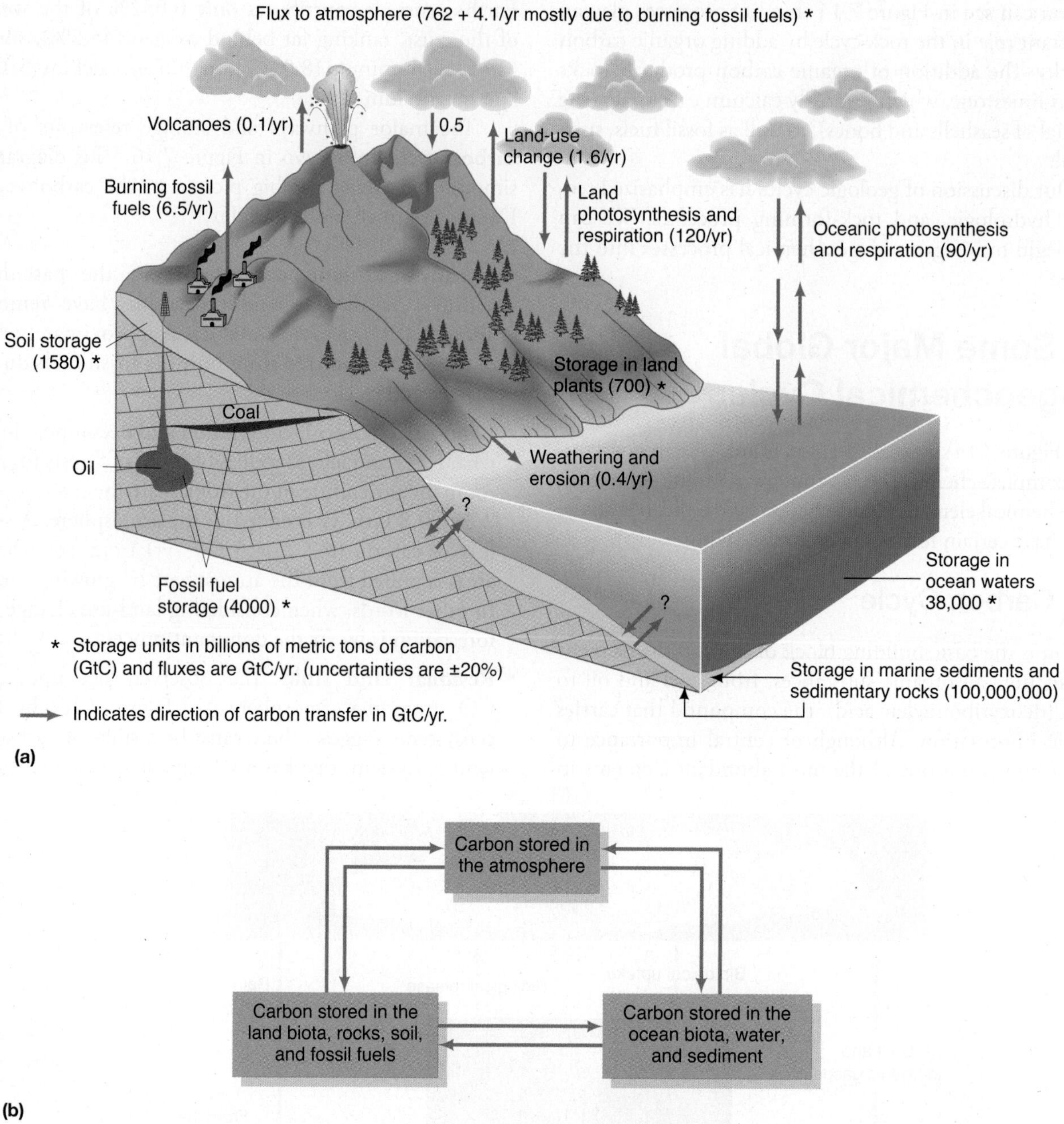

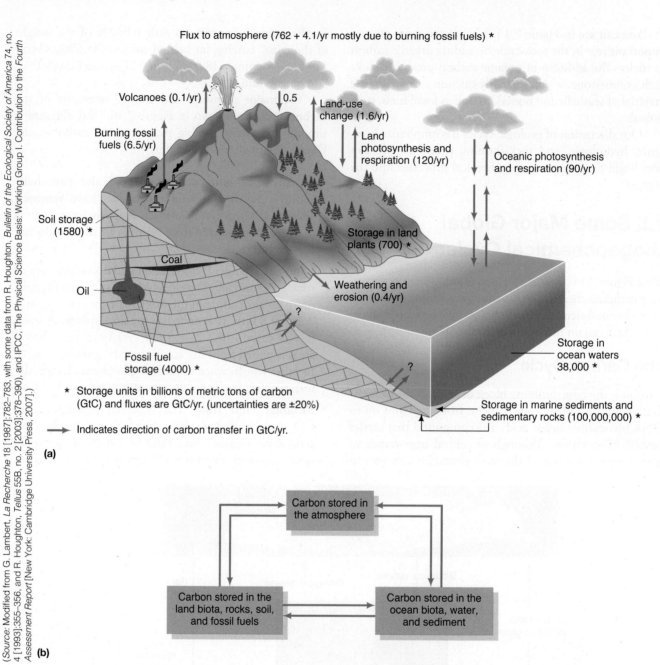

(*Source:* Modified from G. Lambert, *La Recherche* 18 [1987]:782–783, with some data from R. Houghton, *Bulletin of the Ecological Society of America* 74, no. 4 [1993]:355–356, and R. Houghton, *Tellus* 55B, no. 2 [2003]:378–390), and IPCC, The Physical Science Basis: Working Group I. Contribution to the *Fourth Assessment Report* [New York: Cambridge University Press, 2007].)

FIGURE 7.16 The carbon cycle. (a) Generalized global carbon cycle. **(b)** Parts of the carbon cycle simplified to illustrate the cyclic nature of the movement of carbon.

The sink is large, at 2 to 3 GtC/yr, with large uncertainty (±1.7GtC/yr). Thus, our understanding of the carbon cycle is not yet complete.

Carbon has a gaseous phase as part of its cycle, occurring in the atmosphere as carbon dioxide (CO_2) and methane (CH_4), both greenhouse gases. Carbon enters the atmosphere through the respiration of living things, through fires that burn organic compounds, and through diffusion from the ocean. It is removed from the atmosphere by photosynthesis of green plants, algae, and photosynthetic bacteria, and it enters the ocean from the atmosphere by the simple diffusion of carbon dioxide. The carbon dioxide then dissolves, some of it remaining in that state and the rest converting to carbonate (CO_3^-) and bicarbonate (HCO_3^-). Marine algae and photosynthetic bacteria obtain the carbon dioxide they use from the water in one of these three forms.

Carbon is transferred from the land to the ocean in rivers and streams as dissolved carbon, including organic compounds, and as organic particulates (fine particles of organic matter) and seashells and other forms of calcium carbonate ($CaCO_3$). Winds, too, transport small organic particulates from the land to the ocean. Rivers and streams transfer a relatively small fraction of the total global carbon flux to the oceans. However, on the local and regional scale, input of carbon from rivers to nearshore areas, such as deltas and salt marshes, which are often highly biologically productive, is important.

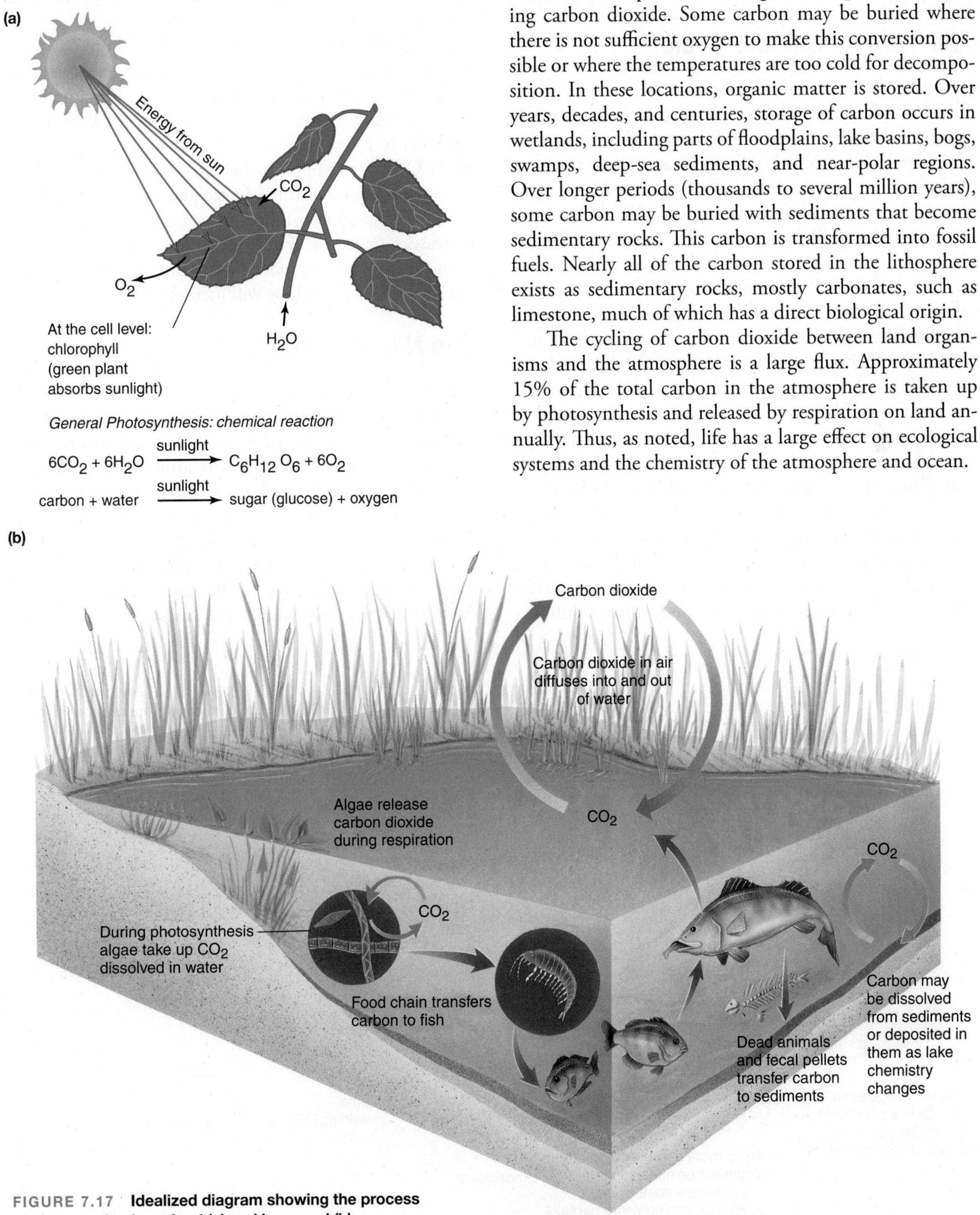
(a)
Energy from sun
CO2
O2
H2O
At the cell level: chlorophyll (green plant absorbs sunlight)
General Photosynthesis: chemical reaction
sunlight
6CO2 + 6H2O → C6H12 O6 + 6O2
sunlight
carbon + water → sugar (glucose) + oxygen
(b)
Carbon dioxide
Carbon dioxide in air diffuses into and out of water
CO2
Algae release carbon dioxide during respiration
CO2
CO2
During photosynthesis algae take up CO2 dissolved in water
Food chain transfers carbon to fish
Dead animals and fecal pellets transfer carbon to sediments
Carbon may be dissolved from sediments or deposited in them as lake chemistry changes

FIGURE 7.17 **Idealized diagram showing the process of photosynthesis on land (a) and in a pond (b).**

Carbon enters the **biota**—the term for all life in a region—through photosynthesis and is returned to the atmosphere or waters by respiration or by wildfire (Figure 7.17). When an organism dies, most of its organic material decomposes into inorganic compounds, including carbon dioxide. Some carbon may be buried where there is not sufficient oxygen to make this conversion possible or where the temperatures are too cold for decomposition. In these locations, organic matter is stored. Over years, decades, and centuries, storage of carbon occurs in wetlands, including parts of floodplains, lake basins, bogs, swamps, deep-sea sediments, and near-polar regions. Over longer periods (thousands to several million years), some carbon may be buried with sediments that become sedimentary rocks. This carbon is transformed into fossil fuels. Nearly all of the carbon stored in the lithosphere exists as sedimentary rocks, mostly carbonates, such as limestone, much of which has a direct biological origin.

The cycling of carbon dioxide between land organisms and the atmosphere is a large flux. Approximately 15% of the total carbon in the atmosphere is taken up by photosynthesis and released by respiration on land annually. Thus, as noted, life has a large effect on ecological systems and the chemistry of the atmosphere and ocean.

Because carbon forms two of the most important greenhouse gases—carbon dioxide and methane—much research has been devoted to understanding the carbon cycle, which will be discussed in Chapter 20 about the atmosphere and climate change.

The Carbon–Silicate Cycle

Carbon cycles rapidly among the atmosphere, oceans, and life. However, over geologically long periods, the cycling of carbon becomes intimately involved with the cycling of silicon. The combined carbon–silicate cycle is therefore of geologic importance to the long-term stability of the biosphere over periods that exceed half a billion years.[10]

The **carbon–silicate cycle** begins when carbon dioxide in the atmosphere dissolves in the water to form weak carbonic acid (H_2CO_3) that falls as rain (Figure 7.18). As the mildly acidic water migrates through the ground, it chemically weathers (dissolves) rocks and facilitates the erosion of Earth's abundant silicate-rich rocks. Among other products, weathering and erosion release calcium ions (Ca^{++}) and bicarbonate ions (HCO_3^-). These ions enter the groundwater and surface waters and eventually are transported to the ocean. Calcium and bicarbonate ions make up a major portion of the chemical load that rivers deliver to the oceans.

Tiny floating marine organisms use the calcium and bicarbonate to construct their shells. When these organisms die, the shells sink to the bottom of the ocean, where they accumulate as carbonate-rich sediments. Eventually, carried by moving tectonic plates, they enter a subduction zone (where the edge of one continental plate slips under the edge of another). There they are subjected to increased heat, pressure, and partial melting. The resulting magma releases carbon dioxide, which rises in volcanoes and is released into the atmosphere. This process provides a lithosphere-to-atmosphere flux of carbon.

The long-term carbon–silicate cycle (Figure 7.18) and the short-term carbon cycle (Figure 7.16) interact to affect levels of CO_2 and O_2 in the atmosphere. For example, the burial of organic material in an oxygen-poor environment amounts to a net increase of photosynthesis (which produces O_2) over respiration (which produces CO_2). Thus, if burial of organic carbon in oxygen-poor environments increases, the concentration of atmospheric oxygen will increase. Conversely, if more organic carbon escapes burial and is oxidized to produce CO_2, then the CO_2 concentration in the atmosphere will increase.[11]

The Nitrogen Cycle

Nitrogen is essential to life in proteins and DNA. As we discussed at the beginning of this chapter, free or diatonic nitrogen (N_2 uncombined with any other element) makes up approximately 78% of Earth's atmosphere. However, no organism can use molecular nitrogen directly. Some organisms, such as animals, require nitrogen in an organic compound. Others, including plants, algae, and bacteria, can take up nitrogen either as the nitrate ion (NO_3^-) or the ammonium ion (NH_4^+). Because nitrogen is a relatively unreactive element, few processes convert molecular

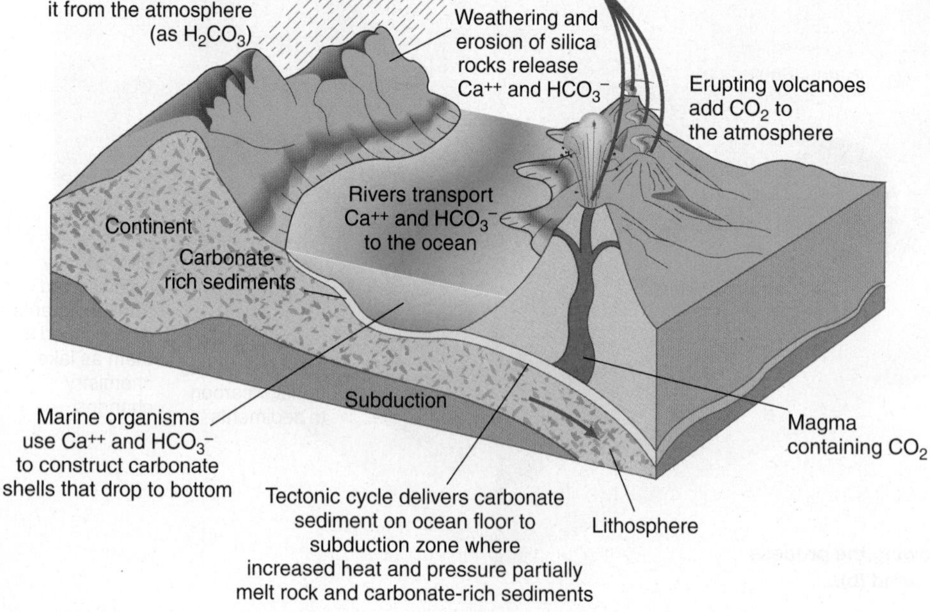

(*Source:* Modified from J.E. Kasting, O.B. Toon, and J.B. Pollack, How climate evolved on the terrestrial planets, *Scientific American* 258 [1988]:2.)

FIGURE 7.18 An idealized diagram showing the carbon–silicate cycle.

nitrogen to one of these compounds. Lightning oxidizes nitrogen, producing nitric oxide. In nature, essentially all other conversions of molecular nitrogen to biologically useful forms are conducted by bacteria.

The **nitrogen cycle** is one of the most important and most complex of the global cycles (Figure 7.19). The process of converting inorganic, molecular nitrogen in the atmosphere to ammonia or nitrate is called **nitrogen fixation**. Once in these forms, nitrogen can be used on land by plants and in the oceans by algae. Bacteria, plants, and algae then convert these inorganic nitrogen compounds into organic ones through chemical reactions, and the nitrogen becomes available in ecological food chains. When organisms die, bacteria convert the organic compounds containing nitrogen back to ammonia, nitrate, or molecular nitrogen, which enters the atmosphere. The process of releasing fixed nitrogen back to molecular nitrogen is called **denitrification**.

Thus, all organisms depend on nitrogen-converting bacteria. Some organisms, including termites and ruminant (cud-chewing) mammals, such as cows, goats, deer, and bison, have evolved symbiotic relationships with these bacteria. For example, the roots of the pea family have nodules that provide a habitat for the bacteria. The bacteria obtain organic compounds for food from the plants, and the plants obtain usable nitrogen. Such plants can grow in otherwise nitrogen-poor environments. When these plants die, they contribute nitrogen-rich organic matter to the soil, improving the soil's fertility. Alder trees, too, have nitrogen-fixing bacteria in their roots. These trees grow along streams, and their nitrogen-rich leaves fall into the streams and increase the supply of organic nitrogen to freshwater organisms.

In terms of availability for life, nitrogen lies somewhere between carbon and phosphorus. Like carbon, nitrogen has a gaseous phase and is a major component of Earth's atmosphere. Unlike carbon, however, it is not very

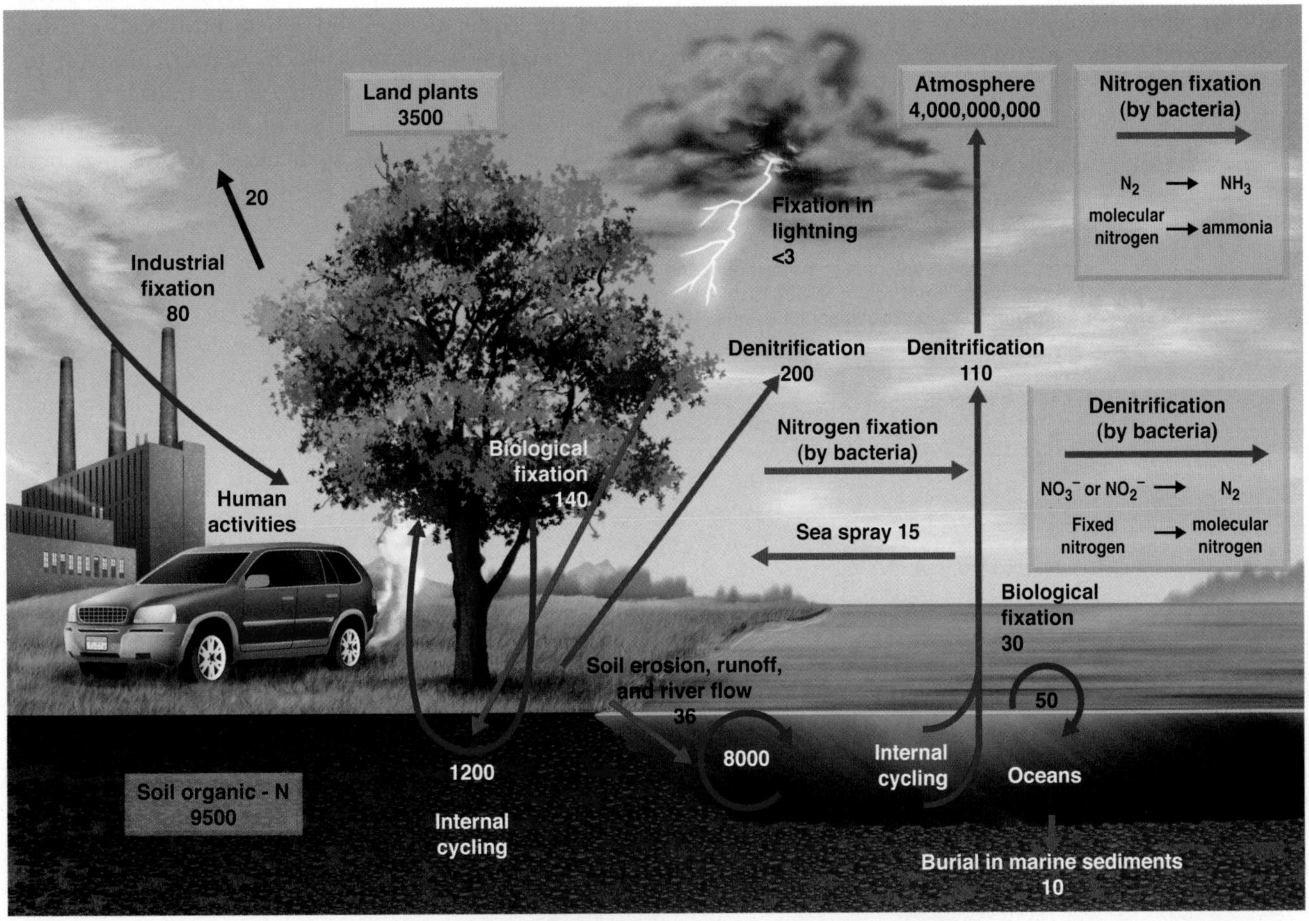

(*Source:* Data from R. Söderlund and T. Rosswall, in *The Handbook of Environmental Chemistry*, Vol. 1, Pt. B, O. Hutzinger, ed. [New York: Springer-Verlag, 1982]; W.H. Schlosinger, *Biogeochemistry: An Analysis of Global Change* [San Diego: Academic Press, 1997], p. 386; and Peter M. Vitousek, Chair, John Aber, Robert W. Howarth, Gene E. Likens, Pamela A. Matson, David W. Schindler, William H. Schlesinger, and G. David Tilman, Human alteration of the global nitrogen cycle: Causes and consequences, *Issues in Ecology—Human Alteration of the Global Nitrogen Cycle*, Ecological Society of America publication.)

FIGURE 7.19 **The global nitrogen cycle.** Numbers in boxes indicate amounts stored, and numbers with arrows indicate annual flux, in millions of metric tons of nitrogen. Note that the industrial fixation of nitrogen is nearly equal to the global biological fixation.

reactive, and its conversion depends heavily on biological activity. Thus, the nitrogen cycle is not only essential to life but is also primarily driven by life.

In the early part of the 20th century, scientists invented industrial processes that could convert molecular nitrogen into compounds usable by plants. This greatly increased the availability of nitrogen in fertilizers. Today, industrial fixed nitrogen is about 60% of the amount fixed in the biosphere and is a major source of commercial nitrogen fertilizer.[12]

Although nitrogen is required for all life, and its compounds are used in many technological processes and in modern agriculture, nitrogen in agricultural runoff can pollute water, and many industrial combustion processes and automobiles that burn fossil fuels produce nitrogen oxides that pollute the air and play a significant role in urban smog (see Chapter 21).

The Phosphorus Cycle

Phosphorus, one of the "big six" elements required in large quantities by all forms of life, is often a limiting nutrient for plant and algae growth. The **phosphorus cycle** is significantly different from the carbon and nitrogen cycles. Unlike carbon and nitrogen, phosphorus does not have a gaseous phase on Earth; it is found in the atmosphere only in small particles of dust (Figure 7.20). In addition, phosphorus tends to form compounds that are relatively insoluble in water, so phosphorus is not readily weathered chemically. It does occur commonly in an oxidized state as phosphate, which combines with calcium, potassium, magnesium, or iron to form minerals. All told, however, the rate of transfer of phosphorus in Earth's system is slow compared with that of carbon or nitrogen.

Phosphorus enters the biota through uptake as phosphate by plants, algae, and photosynthetic bacteria. It is recycled locally in life on land nearly 50 times before being transported by weathering and runoff. Some phosphorus is inevitably lost to ecosystems on the land. It is transported by rivers to the oceans, either in a water-soluble form or as suspended particles. When it finally reaches the ocean, it may be recycled about 800 times before entering marine sediments to become part of the rock cycle. Over tens to hundreds of millions of years, the sediment is transformed into sedimentary rocks, after which it may eventually be returned to the land by uplift, weathering, and erosion.[13]

Ocean-feeding birds, such as the brown pelican, provide an important pathway in returning phosphorus from the ocean to the land. These birds feed on small fish, especially anchovies, which, in turn, feed on tiny ocean plankton. Plankton thrive where nutrients, such as phosphorus, are present. Areas of rising oceanic currents known as upwellings are such places. Upwellings occur

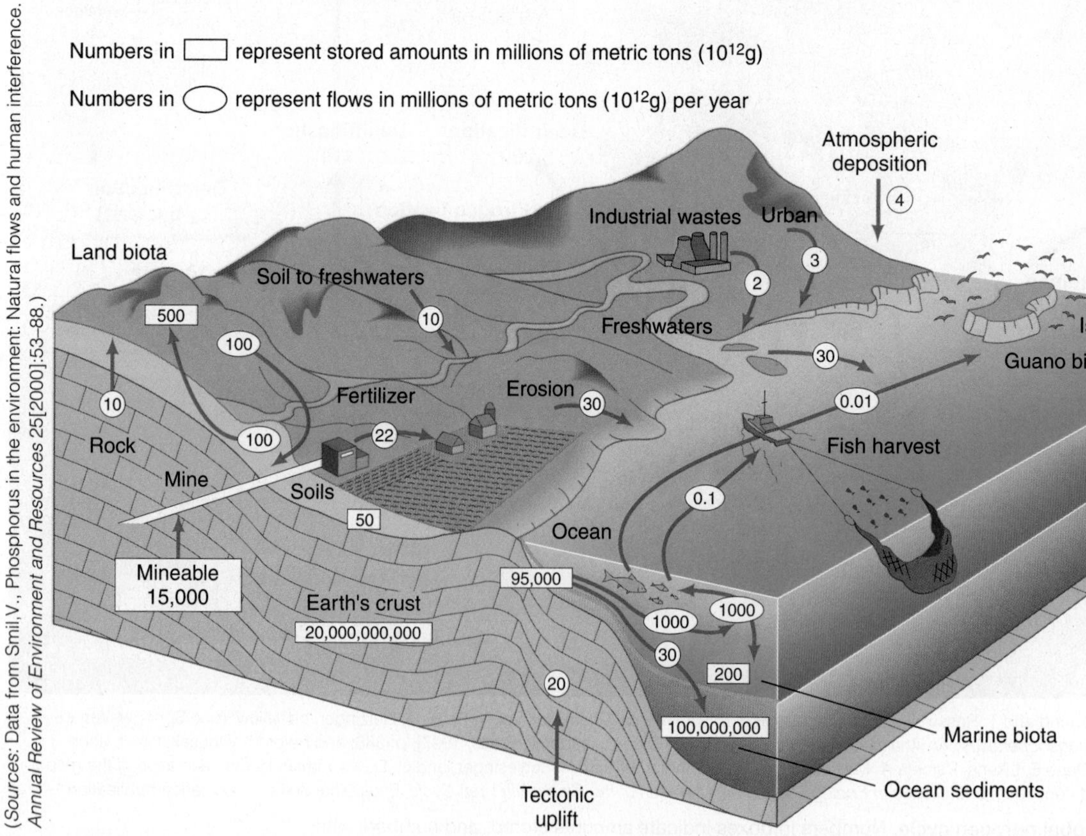

FIGURE 7.20 **Global phosphorus cycle.** Values are approximated with errors ±20%. Note that the amount mined (22) is about equivalent to the amount that is eroded from the land and enters the oceans by runoff (25).

near continents where the prevailing winds blow offshore, pushing surface waters away from the land and allowing deeper waters to rise and replace them. Upwellings carry nutrients, including phosphorus, from the depths of the oceans to the surface.

The fish-eating birds nest on offshore islands, where they are protected from predators. Over time, their nesting sites become covered with their phosphorus-laden excrement, called guano. The birds nest by the thousands, and deposits of guano accumulate over centuries. In relatively dry climates, guano hardens into a rocklike mass that may be up to 40 m (130 ft) thick. The guano results from a combination of biological and nonbiological processes. Without the plankton, fish, and birds, the phosphorus would have remained in the ocean. Without the upwellings, the phosphorus would not have been available.

Guano deposits were once major sources of phosphorus for fertilizers. In the mid-1800s, as much as 9 million metric tons per year of guano deposits were shipped to London from islands near Peru (Figure 7.21). Today, most phosphorus fertilizers come from the mining of phosphate-rich sedimentary rocks containing fossils of marine animals. The richest phosphate mine in the world is Bone Valley, 40 km east of Tampa, Florida. But 10–15 million years ago Bone Valley was the bottom of a shallow sea where marine invertebrates lived and died.[14] Through tectonic processes, the valley was slowly uplifted, and in the 1880s and 1890s phosphate ore was discovered there. Today, Bone Valley provides about 20% of the world's phosphate (Figure 7.22).

About 80% of phosphorus is produced in four countries: the United States, China, South Africa, and Morocco.[15,16] The global supply of phosphorus that can be extracted economically is about 15 billion tons (15,000 million tons). Total U.S. reserves are estimated at 1.2 billion metric tons. Most of the U.S. phosphorus, about 85%, is from Florida and North Carolina; the rest is from Utah and Idaho.[17] All of our industrialized agriculture—most of the food produced in the United States—depends on phosphorus for fertilizers that comes from just four states!

Phosphorus may become much more difficult to obtain in the next few decades. According to the U.S. Geological Survey, in recent years, the price of phosphate rock has "jumped dramatically worldwide owing to increased agricultural demand and tight supplies." The average U.S. price has more than doubled since 2007.

One fact is clear: Without phosphorus, we cannot produce food. Thus, declining phosphorus resources will harm the global food supply and affect all of the world's economies. Extraction continues to increase as the expanding human population demands more food and as we grow more corn for biofuel. However, if the price of phosphorus rises as high-grade deposits dwindle, phosphorus from lower-grade deposits can be mined at a profit. Florida is thought to have as much as 8 billion metric tons of phosphorus that might eventually be recovered if the price is right.

Mining, of course, may have negative effects on the land and ecosystems. For example, in some phosphorus mines, huge pits and waste ponds have scarred the landscape, damaging biologic and hydrologic resources. Balancing the need for phosphorus with the adverse environmental impacts of mining is a major environmental

(a)

(b)

FIGURE 7.21 **Guano Island, Peru.** For centuries the principal source of phosphorus fertilizer was guano deposits from seabirds. The birds feed on fish and nest on small islands. Their guano accumulates in this dry climate over centuries, forming rocklike deposits that continue to be mined commercially for phosphate fertilizers. On the Peruvian Ballestas Islands **(a)** seabirds (in this case, Incan terns) nest, providing some of the guano, and **(b)** sea lions haul out and rest on the rocklike guano.

Melissa Farlow/National Geographic Society

FIGURE 7.22 A large open-pit phosphate mine in Florida (similar to Bone Valley), with piles of waste material. The land in the upper part of the photograph has been reclaimed and is being used for pasture.

issue. Following phosphate extraction, land disrupted by open-pit phosphate mining, shown in Figure 7.22, is reclaimed to pastureland, as mandated by law.

As with nitrogen, an overabundance of phosphorus causes environmental problems. In bodies of water, from ponds to lakes and the ocean, phosphorus can promote unwanted growth of photosynthetic bacteria. As the algae proliferate, oxygen in the water may be depleted. In oceans, dumping of organic materials high in nitrogen and phosphorus has produced several hundred "dead zones," collectively covering about 250,000 km^2. Although this is an area almost as large as Texas, it represents less than 1% of the area of the Earth's oceans (335,258,000 km^2).

What might we do to maintain our high agriculture production but reduce our need for newly mined phosphate? Among the possibilities:

- Recycle human waste in the urban environment to reclaim phosphorus and nitrogen.
- Use wastewater as a source of fertilizer, rather than letting it end up in waterways.
- Recycle phosphorus-rich animal waste and bones for use in fertilizer.
- Further reduce soil erosion from agricultural lands so that more phosphorus is retained in the fields for crops.
- Apply fertilizer more efficiently so less is immediately lost to wind and water erosion.
- Find new phosphorus sources and more efficient and less expensive ways to mine it.
- Use phosphorus to grow food crops rather than biofuel crops.
- Encourage organic farming methods.
- Encourage cultivation of crops that use less phosphorus, such as a genetically modified rice with a root system that takes up phosphorus from the soil.[18]

We have focused on the biogeochemical cycles of three of the macronutrients, illustrating the major kinds of biogeochemical cycles—those with and those without an atmospheric component—but, obviously, this is just an introduction about methods that can be applied to all elements required for life and especially in agriculture.

CRITICAL THINKING ISSUE

How are Human Activities Linked to the Phosphorus and Nitrogen Cycles?

Scientists estimate that nitrogen deposition to Earth's surface will double in the next 25 years and that the demand for phosphorus will also increase greatly as we attempt to feed a few billion more people in coming decades. The natural rate of nitrogen fixation is estimated to be 140 teragrams (Tg) of nitrogen a year (1 teragram = 1 million metric tons). Human activities—such as the use of fertilizers, draining of wetlands, clearing of land for agriculture, and burning of fossil fuels—are causing additional nitrogen to enter the environment. Currently, human activities are responsible for more than half of the fixed nitrogen that is deposited on land. Before the 20th century, fixed nitrogen was recycled by bacteria, with no net accumulation. Since 1900, however, the use of commercial fertilizers has increased exponentially (Figure 7.23). Nitrates and ammonia from burning fossil fuels have increased about 20% in the last decade or so. These inputs have overwhelmed the denitrifying part of the nitrogen cycle and the ability of plants to use fixed nitrogen.

Nitrate ions, in the presence of soil or water, may form nitric acid. With other acids in the soil, nitric acid can leach out chemicals important to plant growth, such as magnesium

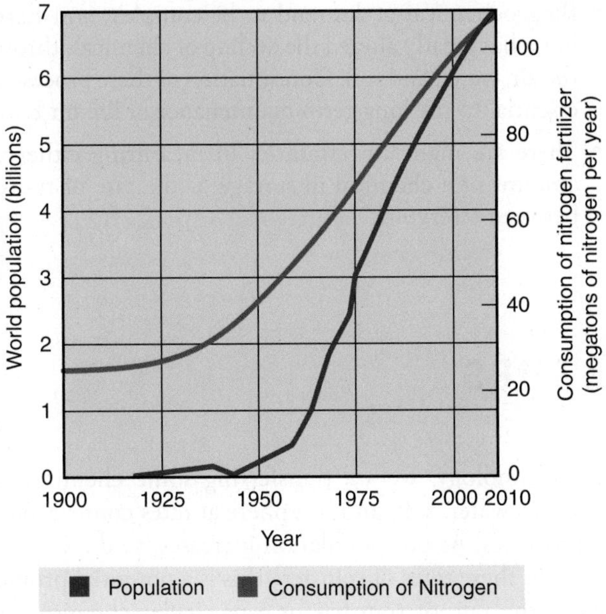

(*Source:* Modified from Rhodes, D. 2009. Purdue University Department of Horticulture & Landscape Architecture.)

FIGURE 7.23 The use of nitrogen fertilizers has increased greatly.

and potassium. When these chemicals are depleted, more toxic ones, such as aluminum, may be released, damaging tree roots. Acidification of soil by nitrate ions is also harmful to organisms. When toxic chemicals wash into streams, they can kill fish. Excess nitrates in rivers and along coasts can cause algae to overgrow, damaging ecosystems. High levels of nitrates in drinking water from streams or groundwater contaminated by fertilizers are a health hazard.[12, 19-21]

The nitrogen, phosphorus, and carbon cycles are linked because nitrogen is a component of chlorophyll, the molecule that plants use in photosynthesis. Phosphorus taken up by plants enters the food chain and, thus, the carbon cycle. It is an irreplaceable ingredient in life. Because nitrogen is a limiting factor on land, it has been predicted that rising levels of global nitrogen may increase plant growth. Recent studies have suggested, however, that a beneficial effect from increased nitrogen would be short-lived. As plants use additional nitrogen, some other factor, such as phosphorus, will become limiting. When that occurs, plant growth will slow, and so will the uptake of carbon dioxide. More research is needed to understand the interactions between carbon and the phosphorus and nitrogen cycles and to be able to predict the long-term effects of human activities.

Critical Thinking Questions

1. The supply of phosphorus from mining is a limited resource. In the United States, extraction is decreasing, and the price is rising dramatically. Speculate how phosphorus can be used sustainably? If it can't, what are the potential consequences for agriculture?
2. Do you think phosphorus use should be governed by an international body? Why? Why not?
3. Compare the rate of human contributions to nitrogen fixation with the natural rate.
4. Defend or criticize the statement that the increase in human population must be accompanied by an increase in mining of phosphorus.
5. Develop a diagram to illustrate the links between the phosphorus, nitrogen, and carbon cycles.
6. Discuss ways in which we could modify our activities to reduce our contributions to the phosphorus and nitrogen cycles.
7. Should phosphorus and nitrogen be used to produce corn as a biofuel (alcohol)? Why? Why not?

SUMMARY

- Biogeochemical cycles are the major way that elements important to Earth processes and life are moved through the atmosphere, hydrosphere, lithosphere, and biosphere.
- Biogeochemical cycles can be described as a series of reservoirs, or storage compartments, and pathways, or fluxes, between reservoirs.
- In general, some chemical elements cycle quickly and are readily regenerated for biological activity. Elements whose biogeochemical cycles include a gaseous phase in the atmosphere tend to cycle more rapidly.
- Life on Earth has greatly altered biogeochemical cycles, creating a planet with an atmosphere unlike that of any other known and especially suited to sustain life.
- Every living thing, plant or animal, requires a number of chemical elements. These chemicals must be available at the appropriate time and in the appropriate form and amount.
- Chemicals can be reused and recycled, but in any real ecosystem, some elements are lost over time and must be replenished if life in the ecosystem is to persist. Change and disturbance of natural ecosystems are the norm. A steady state, in which the net storage of chemicals in an ecosystem does not change with time, cannot be maintained.
- Our modern technology has begun to alter and transfer chemical elements in biogeochemical cycles at rates comparable to those of natural processes.

Some of these activities are beneficial to society, but others create problems, such as pollution by nitrogen and phosphorus

- To be better prepared to manage our environment, we must recognize both the positive and the negative consequences of activities that transfer chemical elements, and we must deal with them appropriately.
- Biogeochemical cycles tend to be complex, and Earth's biota has greatly altered the cycling of chemicals through the air, water, and soil. Continuation of these processes is essential to the long-term maintenance of life on Earth.
- There are many uncertainties in measuring either the amount of a chemical in storage or the rate of transfer between reservoirs.

REEXAMINING THEMES AND ISSUES

Sean Randall/Getty Images, Inc.

HUMAN POPULATION

Through modern technology, we are transferring some chemical elements through the air, water, soil, and biosphere at rates comparable to those of natural processes. As our population increases, so does our use of resources and so do these rates of transfer. This is a potential problem because, eventually, the rate of transfer for a particular chemical may become so large that pollution of the environment results.

© Biletskiy_Evgeniy/iStockphoto

SUSTAINABILITY

If we are to sustain a high-quality environment, the major biogeochemical cycles must transfer and store the chemicals necessary to maintain healthy ecosystems. That is one reason understanding biogeochemical cycles is so important. For example, the release of sulfur into the atmosphere is degrading air quality at local to global levels. As a result, the United States is striving to control these emissions.

© Anton Balazh 2011/iStockphoto

GLOBAL PERSPECTIVE

The major biogeochemical cycles discussed in this chapter are presented from a global perspective. Through ongoing research, scientists are trying to better understand how major biogeochemical cycles work. For example, the carbon cycle and its relationship to the burning of fossil fuels and the storage of carbon in the biosphere and oceans are being intensely investigated. Results of these studies are helping us to develop strategies for reducing carbon emissions. These strategies are implemented at the local level, at power plants, and in cars and trucks that burn fossil fuels.

ssguy/ShutterStock

URBAN WORLD

Our society has concentrated the use of resources in urban regions. As a result, the release of various chemicals into the biosphere, soil, water, and atmosphere is often greater in urban centers, resulting in biogeochemical cycles that cause pollution problems.

B2M Productions/Getty Images, Inc.

PEOPLE AND NATURE

Humans, like other animals, are linked to natural processes and nature in complex ways. We change ecosystems through land-use changes and the burning of fossil fuels, both of which change biogeochemical cycles, especially the carbon cycle that anchors life and affects Earth's climate.

George Doyle/Getty Images, Inc.

SCIENCE AND VALUES

Our understanding of biogeochemical cycles is far from complete. There are large uncertainties in the measurement of fluxes of chemical elements—nitrogen, carbon, phosphorus, and others. We are studying biogeochemical cycles because understanding them will help us to solve environmental problems. Which problems we address first will reflect the values of our society.

KEY TERMS

biogeochemical cycle 138
biosphere 136
biota 147
carbon cycle 145
carbon–silicate cycle 148
chemical reaction 139
denitrification 149
eukaryote 136
flow 140
flux 140
geologic cycle 141
hydrologic cycle 143
limiting factor 137
macronutrients 137
micronutrients 137
nitrogen cycle 149
nitrogen fixation 149
ocean acidification 132
organelle 136
phosphorus cycle 150
plate tectonics 142
prokaryote 135
residence time 141
rock cycle 144
sink 141
source 141
tectonic cycle 142
thermodynamic equilibrium 134

STUDY QUESTIONS

1. Why is an understanding of biogeochemical cycles important in environmental science? Explain your answer, using two examples.
2. Summarize some of the general rules that govern biogeochemical cycles, especially the transfer of material.
3. Synthesize the major aspects of the carbon cycle and the environmental concerns associated with it.
4. Contrast the geochemical cycles for phosphorus and nitrogen and explain why the differences are important in environmental science.
5. What are the major ways that people have altered biogeochemical cycles?
6. Speculate on what would happen if all life on Earth ceased. Would the atmosphere become like that of Venus and Mars? Why or why not.

FURTHER READING

Lane, Nick, ***Oxygen: The Molecule That Made the World*** (Oxford: Oxford University Press, 2009).

Lovelock, J., ***The Ages of Gaia: A Biography of Our Living Earth*** (Oxford: Oxford University Press, 2000).

Schlesinger, W.H., ***Biogeochemistry: An Analysis of Global Change,*** **2nd ed.** (San Diego: Academic Press, 1997). This book provides a comprehensive and up-to-date overview of the chemical reactions on land, in the oceans, and in the atmosphere of Earth.

Wollast, R., McKenzie, E.T., and Chun, L., ***Interactions of C, N, P, and S Biogeochemical Cycles and Global Change.*** NATO AST Series. Series 1: Global Environmental Change, v4, (New York: Springer, 2012).

NOTES

1. Barton, A., B. Hales, G.G. Waldbusser, C. Langdon, and R. Feely. 2012. The Pacific oyster, *Crassostrea gigas*, shows negative correlation to naturally elevated carbon dioxide levels: Implications for near-term ocean acidification effects. *Limnology and Oceanography* 57 (3): 698 DOI: 10.4319/lo.2012.57.3.0698
2. Christian, D. 2004. *Maps of Time.* Berkeley: University of California Press.
3. Gaddis, J.L. 2002. *The Landscape of History.* New York: Oxford University Press.
4. Lovelock, J. 1995. *The Ages of Gaia: A Biography of the Earth.* Oxford: Oxford University Press.
5. Lane, Nick. 2009. *Oxygen: The Molecule That Made the World.* Oxford: Oxford University Press.
6. Botkin, D.B. 2012. *The Moon in the Nautilus Shell: Discordant Harmonies Reconsidered.* New York: Oxford University Press.
7. Henderson, Lawrence J. 1913; reprinted 1958 by the same publisher. *The Fitness of the Environment.* Boston: Beacon Press.
8. Ehrlich, P.R., A.H. Ehrlich, and J.P. Holdren. 1970. *Ecoscience: Population, Resources, Environment.* San Francisco: W.H. Freeman, p. 1051.
9. Post, W.M., T. Peng, W.R. Emanuel, A.W. King, V.H. Dale, and D.L. De Angelis. 1990. The global carbon cycle. *American Scientist* 78:310–326.
10. Kasting, J.F., O.B. Toon, and J.B. Pollack. 1988. How climate evolved on the terrestrial planets. *Scientific American* 258:90–97.
11. Berner, R.A. 1999. A new look at the long-term carbon cycle. *GSA Today* 9(11):2–6.
12. Vitousek, P.M., J. Aber, R.W. Howarth, G.E. Likens, P.A. Matson, D.W. Schindler, W.H. Schlesinger, and G.D. Tilman. 1997. Human alteration of the global nitrogen cycle: Causes and consequences. *Issues in Ecology—Human Alteration of the Global Nitrogen Cycle.* ESA publication.
13. Chameides, W.L., and E.M. Perdue. 1997. *Biogeochemical Cycles.* New York: Oxford University Press.
14. Carter, L.J. 1980. Phosphate: Debate over an essential resource. *Science* 209:4454.
15. Vaccari, D.A. 2009. Phosphorus: A looming crisis. *Scientific American* 300(6):54–59.
16. Smil, V. 2000. Phosphorus in the environment: Natural flows and human interference. *Annual Review of Environment and Resources* 25:53–88.
17. U.S. Geological Survey. 2012. http://minerals.usgs.gov/minerals/pubs/commodity/phosphate_rock.
18. Gamuvao, Rico, et al. 2012 (August 23). The protein kinase Pstol1 from traditional rice confers tolerance of phosphorus deficiency. *Nature* 488:535–539.
19. Asner, G.P., T.R. Seastedt, and A.R. Townsend. 1997 (April). The decoupling of terrestrial carbon and nitrogen cycles. *Bioscience* 47 (4):226–234.
20. Hellemans, A. 1998 (February 13). Global nitrogen overload problem grows critical. *Science* 279:988–989.
21. Smil, V.1997 (July). Global populations and the nitrogen cycle. *Scientific American* 76–81.

CHAPTER 8

Environmental Health, Pollution, and Toxicology

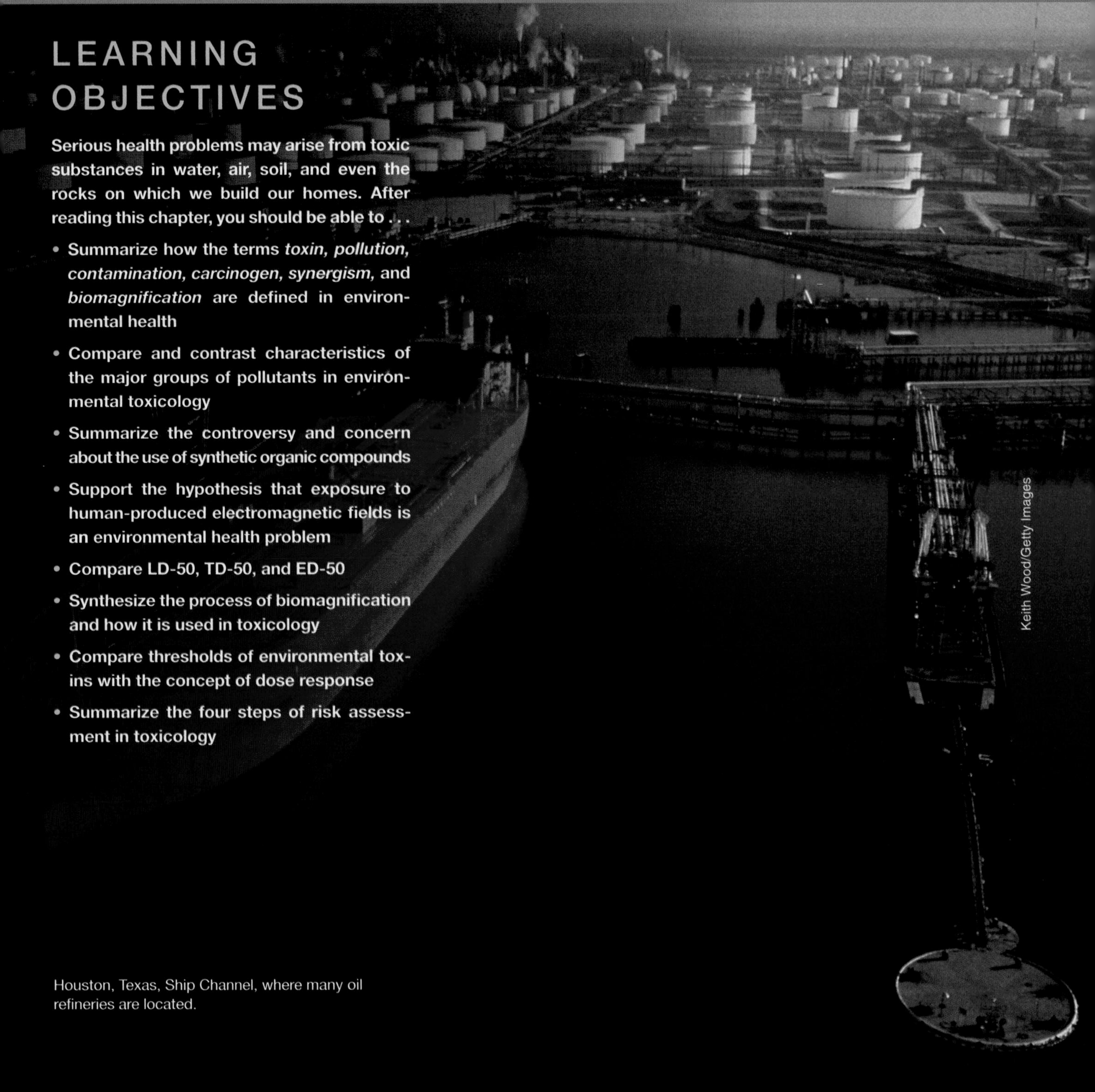

LEARNING OBJECTIVES

Serious health problems may arise from toxic substances in water, air, soil, and even the rocks on which we build our homes. After reading this chapter, you should be able to . . .

- Summarize how the terms *toxin, pollution, contamination, carcinogen, synergism,* and *biomagnification* are defined in environmental health
- Compare and contrast characteristics of the major groups of pollutants in environmental toxicology
- Summarize the controversy and concern about the use of synthetic organic compounds
- Support the hypothesis that exposure to human-produced electromagnetic fields is an environmental health problem
- Compare LD-50, TD-50, and ED-50
- Synthesize the process of biomagnification and how it is used in toxicology
- Compare thresholds of environmental toxins with the concept of dose response
- Summarize the four steps of risk assessment in toxicology

Keith Wood/Getty Images

Houston, Texas, Ship Channel, where many oil refineries are located.

CASE STUDY

Toxic Air Pollution and Human Health: Story of a Southeast Houston Neighborhood

Manchester is a neighborhood in southeast Houston, Texas, that is nearly surrounded by oil refineries and petrochemical plants. Residents and others have long noted the peculiar and not so pleasant smells of the area, but only recently have health concerns been raised. The neighborhood is close to downtown Houston, the houses are relatively inexpensive, and the streets are safe. It was a generally positive neighborhood except for the occasional complaints about nosebleeds, coughing, and acidic smoke smells. Over a period of years, the number of oil refineries, petrochemical plants, and waste-disposal sites grew along what is known as the Houston Ship Channel (see opening photograph).[1]

Cancer is the second leading cause of death of U.S. children who are not linked to a known health risk before being stricken by the disease. Investigations into childhood cancer from air pollution are few in number but now include exposure to benzene, 1,3-butadiene (commonly referred to simply as butadiene), and chromium.[1-3]

Benzene is a colorless toxic liquid that evaporates into the air. Exposure to benzene has a whole spectrum of possible consequences for people, such as drowsiness, dizziness, and headaches; irritation to eyes, skin, and respiratory tract; and loss of consciousness at high levels of exposure. Long-term (chronic) exposure through inhalation can cause blood disorders, including reduced numbers of red blood cells (anemia), in industrial settings. Inhalation has reportedly resulted in reproductive problems for women and, in tests on animals, adverse effects on the developing fetus. In humans, occupational exposure to benzene is linked to increased incidence of leukemia (a cancer of the tissues that form white blood cells). The many potential sources of exposure to benzene include tobacco smoke and evaporating gasoline at service stations. Of particular concern are industrial sources; for example, the chemical is released when gasoline is refined from oil.

The chemical 1,3-butadiene is a colorless gas with a mild gasoline-like odor. One way it is produced is as a by-product of refining oil. Health effects from this toxin are fairly well-known and include both acute and chronic problems. Some of the acute problems are irritation of the eyes, throat, nose, and lungs. Possible chronic health effects of exposure to 1,3-butadiene include cancer, disorders of the central nervous system, damage to kidneys and liver, birth defects, fatigue, lowered blood pressure, headache, nausea, and cancer.[1,2] While there is controversy as to whether exposure to 1,3-butadiene causes cancer in people, more definitive studies of animals (rats and mice) exposed to the toxin have prompted the Environmental Protection Agency to classify 1,3-butadiene as a known human carcinogen.[1,2]

Solving problems related to air toxins in the Houston area has not been easy. First of all, the petrochemical facilities along the Houston Ship Channel were first established decades ago, during World War II, when the area was nearly unpopulated; since then, communities such as Manchester have grown up near the facilities. Second, the chemical plants at present are not breaking state or federal pollution laws. Texas is one of the states that have not established air standards for toxins emitted by the petrochemical industry. Advocates of clean air argue that the chemical industry doesn't own the air and doesn't have the right to contaminate it. People in the petrochemical industry say they are voluntarily reducing emissions of some of the chemicals known to cause cancer. Butadiene emissions have in fact decreased significantly in the last several years, but this is not much comfort to parents who believe their child contracted leukemia as a result of exposure to air toxins. Some people examining the air toxins released along Houston's Ship Channel have concluded that although further reducing emissions would be expensive, the technology to do it is available. Petrochemical companies are taking steps to reduce the emissions and the potential health risks associated with them, but more may be necessary.

A recent study set out to study neighborhoods (census tracts near the ship channel) with the highest levels of benzene and 1,3-butadiene in the air and to evaluate whether these neighborhoods had a higher incidence of childhood lymphohematopoietic cancer. After adjusting for sex, ethnicity, and socioeconomic status, the study found that census tracts with the highest exposure to benzene had higher rates of leukemia.[1] The study concluded that elevated exposure to benzene and 1,3-butadiene may contribute to increased rates of childhood leukemia, but the possible link between the air pollution and disease needs further exploration.

Another study that examined the cancer risk near the Ship Channel concluded that, in addition to benzene, chromium IV (a combustion waste product produced in industrial processes), when inhaled, is a known human carcinogen. Chromium IV is thought to be responsible for up to 75% of the cancer risk along the Ship Channel.[3]

The case history of the Houston Ship Channel, oil refineries, and disease is a complex problem for several reasons:

1. Disease may have more than a one-cause/one-effect relationship.
2. Data on air-pollution exposure are difficult to collect and link to a population that is moving around and has different responses to exposure to chemicals.
3. It is difficult to definitively link health problems to toxic air pollutants.
4. There have been few other studies with which the Houston study can be compared.

In this chapter we will explore selected aspects of exposure to toxins in the environment and real and potential health consequences to people and ecosystems.

8.1 Some Basics

Environmental Health

The World Health Organization broadly defines **environment health** as human health and disease that is determined by or related to environmental factors such as toxic chemicals, toxic biological agents, or radiation. Also included in the realm of environmental health are direct and more often indirect adverse health effects resulting from housing (such as indoor air pollution from formaldehyde); urban development (including construction materials such as some paints); land use (including exposure to harmful trace metals such as arsenic from mining); or pesticides from agriculture and transport systems (including exposure to gasoline or diesel fuel).[4]

Disease may be defined as an impairment of an individual's health. The incidence of disease depends on several factors, including the physical environment, biological environment, human-made environment, and lifestyle. Linkages between these factors are often related to other factors such as cultural customs and levels of urbanization and industrialization. Modern medicine in highly developed societies as found in the United States and western Europe has greatly reduced infectious diseases such as polio, cholera, dysentry, and typhoid fever. However, through modern agriculture, urbanization, and industrialization, we are exposed to more chemicals directly or indirectly released into the environment that may cause acute or, more often, chronic health problems. In this chapter, we will explore some of the factors and linkages between factors that influence our health.

Terminology

What do we mean when we use the terms *pollution, contamination, toxin, and carcinogen?* A polluted environment is one that is impure, dirty, or otherwise unclean. The term **pollution** refers to an unwanted change in the environment caused by the introduction of harmful materials or the production of harmful conditions (heat, cold, sound). **Contamination** has a meaning similar to that of *pollution* and implies making something unfit for a particular use through the introduction of undesirable materials—for example, the contamination of water by hazardous waste. The term **toxin** refers to substances (pollutants) that are poisonous to living things. **Toxicology** is the science that studies toxins or suspected toxins, and toxicologists are scientists in this field. A **carcinogen** is a toxin that increases the risk of cancer. Carcinogens are among the most feared and regulated toxins in our society.

Synergism

An important concept we need to learn in considering pollution problems is **synergism**—the interaction of different substances, resulting in a total effect that is greater than the sum of the effects of the separate substances. For example, both sulfur dioxide (SO_2) and coal dust particulates are air pollutants. Either one taken separately may cause adverse health effects, but when they combine, as when SO_2 adheres to the coal dust, the dust with SO_2 is inhaled deeper than SO_2 alone and causes greater damage to lungs. Another aspect of synergistic effects is that the body may be more sensitive to a toxin if it is simultaneously subjected to other toxins.

Point, Area, and Mobile Sources

Pollutants are commonly introduced into the environment by way of **point sources**, such as smokestacks, pipes discharging into waterways, a small stream entering the ocean (Figure 8.1), or accidental spills **Area sources**, also called *nonpoint sources*, are more diffused over the land and include urban runoff and **mobile sources**, such as automobile exhaust. Area sources are difficult to isolate and correct because the problem is often widely dispersed over a region, as in agricultural runoff that contains pesticides.

Prof. Ed Keller

FIGURE 8.1 **This southern California urban stream flows into the Pacific Ocean at a coastal park.** The stream water often carries high counts of fecal coliform bacteria. As a result, the stream is a point source of pollution for the beach, which is sometimes closed to swimming following runoff events.

Sudbury, Ontario

A famous example of a point source of pollution is provided by the smelters that refine nickel and copper ores at Sudbury, Ontario. Sudbury contains one of the world's major nickel and copper ore deposits. A number of mines, smelters, and refineries lie within a small area. The smelter stacks used to release large amounts of particulates containing toxic metals—including arsenic, chromium, copper, nickel, and lead—into the atmosphere, much of which was then deposited locally in the soil. In addition, because the areas contained a high percentage of sulfur, the emissions included large amounts of sulfur dioxide (SO_2). During its peak output in the 1960s, this complex was the largest single source of SO_2 emissions in North America, emitting 2 million metric tons per year.

As a result of the pollution, nickel contaminated soils up to 50 km (about 31 mi) from the stacks. The forests that once surrounded Sudbury were devastated by decades of acid rain (produced from SO_2 emissions) and the deposition of particulates containing heavy metals. An area of approximately 250 km^2 (96 mi^2) was nearly devoid of vegetation, and damage to forests in the region has been visible over an area of approximately 3,500 km^2 (1,350 mi^2); see Figure 8.2**a**. To control emissions from Sudbury, the Ontario government set standards to reduce emissions.[5] During the past few decades, emissions of SO_2 and metal particulates have been reduced by about 90%.[6]

Reducing emissions from Sudbury has allowed surrounding areas to begin to recover from the pollution (Figure 8.2**b**). Species of trees once eradicated from some areas have begun to grow again. Recent restoration efforts have included planting over 7 million trees and 75 species of herbs, moss, and lichens—all of which have contributed to the increase of biodiversity. Lakes damaged by acid precipitation in the area are rebounding and now support populations of plankton and fish.[5] However, recovery of land aquatic ecosystems is a slow process.[6]

(a)

(b)

Bill Brooks/Masterfile

© Greg Taylor/Alamy

FIGURE 8.2 **(a) Lake St. Charles, Sudbury, Ontario, prior to restoration.** Note high stacks (smelters) in the background and lack of vegetation in the foreground, resulting from air pollution (acid and heavy-metal deposition). **(b) Recent photo showing regrowth and restoration.**

The case of the Sudbury smelters provides a positive example of emphasizing the key theme of thinking globally but acting locally to reduce air pollution. It also illustrates the theme of science and values: Scientists and engineers can design pollution-abatement equipment, but spending the money to purchase the equipment reflects what value we place on clean air.

Toxic Pathways

Chemical elements released from rocks or human processes can become concentrated in people (see Chapter 7) through many pathways (Figure 8.3). These pathways may involve what is known as **biomagnification**—the accumulation or increasing concentration of a substance in living tissue as it moves through a food web (also known as *bioaccumulation*). For example, cadmium, which increases the risk of heart disease, may enter the environment via ash from burning coal. The cadmium in coal is in very low concentrations. However, after coal is burned in a power plant, the ash is collected in a solid form and disposed of in a landfill. The landfill is covered with soil and revegetated. The low concentration of cadmium in the ash and soil is taken into the plants as they grow, but the concentration of cadmium in the plants is three to five times greater than the concentration in the ash. As the cadmium moves through the food chain, it becomes more and more concentrated. By the time it is incorporated into the tissue of people and other carnivores, the concentration is approximately 50 to 60 times the original concentration in the coal.

8.2 Categories of Pollutants and Toxins

A partial classification of pollutants by arbitrary categories is presented below. We discuss examples of other pollutants in other parts of the book.

Infectious Agents

Infectious diseases—spread by interactions between individuals and by the food, water, air, soil, and animals we come in contact with—constitute some of the oldest health problems that people face. Infectious disease may be caused by bacteria, virus, or fungus (but most of these organisms do not cause disease).

Bacteria compose a large group of unicellular microorganisms lacking organelles with an organized nucleus (see Chapter 7). Some bacteria cause disease, but most do not. Viruses are very tiny organisms (particles smaller than bacteria) that consist of a combination of long molecules that carry genetic information (genes) and a protein coat that protects the genes. Some viruses cause mild to severe illnesses in hosts (humans, animals, and plants). Because the original host may either die or eliminate the infection, a virus needs to spread to another host to survive. Some transfer mechanisms for the human host include coughing (colds and influenza); fecal or oral emanations (hepatitis A, polio); sexual activity (HIV, hepatitis B); insect or rodent vector (yellow fever, dengue fever, West Nile, hantavirus); and animal bites (rabies).

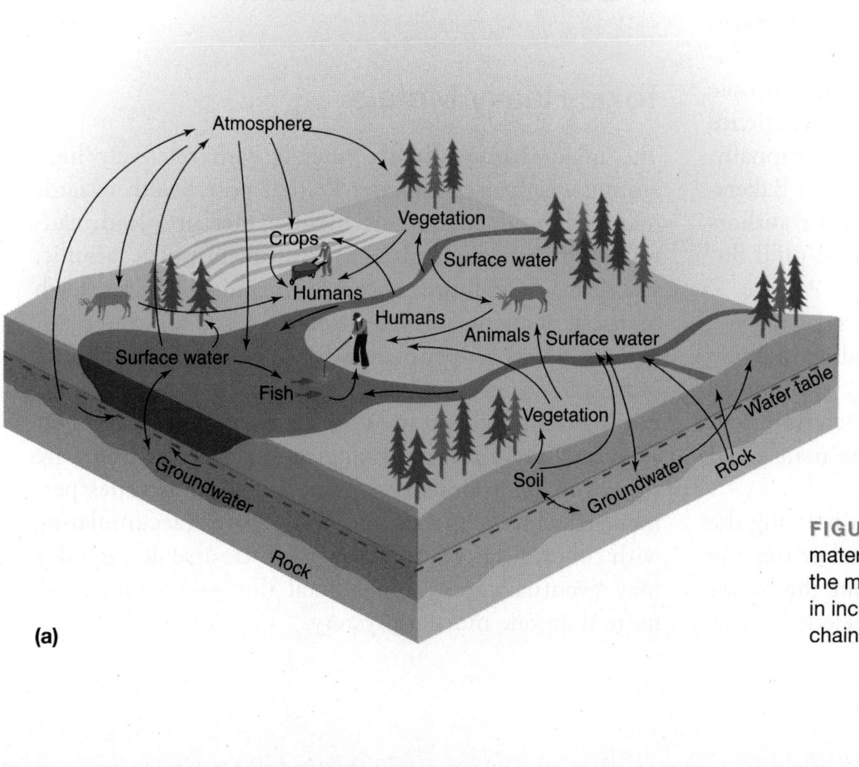

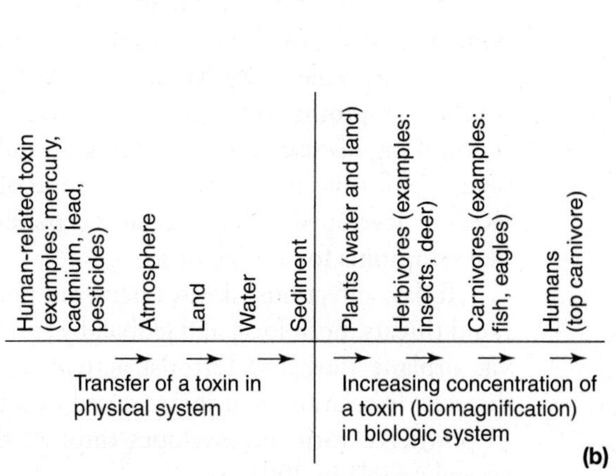

FIGURE 8.3 **Potential complex pathways. (a)** for toxic materials through the living and nonliving environment. Note the many arrows into humans and other animals, sometimes in increasing concentrations as they move through the food chain **(b)**.

Hantavirus is potentially a life-threatening disease that is spread from the feces and urine of rodents (especially deer mice). Recently, there was a small outbreak in the Sierra Nevada (Yosemite National Park) among tent–cabin campers. Nine cases were confirmed, with three deaths. Most infections result from persons exposed to rodent droppings in their own homes or when cleaning places such as sheds that have been empty for some time and inhabited by mice.[7]

Fungi are a member of a large group of eukaryotic organisms that includes microorganisms such as molds as well as mushrooms. Fungal infections, especially but not limited to mold, are often carried by mold spores that are the asexual reproductive structures of fungi. In our homes and other buildings, mold may spread due to moisture in walls and foundation areas. Molds often become a problem following floods but grow whenever there is sufficient moisture and heat. For example, following Hurricane Katrina in 2005 (which caused widespread flooding), there was a significant rise in reports of allergies and asthma in children in New Orleans—about three times the U.S. average. When people sensitive to mold inhale airborne mold spores, they may experience allergic reactions with asthma symptoms. Under specific growing conditions, molds may excrete toxic compounds (miotoxins). One miotoxin is called aflatoxin (a carcinogenic miotoxin), which is widespread in nature and has been known to contaminate peanuts, peanut butter, and milk. Other fungal infections are found in arid regions.

Valley fever or coccidioidomycosis is caused by breathing fungal spores that are naturally found in arid soil. It commonly occurs in the southern San Joaquin Valley of California. Agriculture that disturbs soil can expose people to valley fever as soil erodes. Dust storms are another source of exposure because they erode soils. And dust storms become more severe when land is denuded of natural vegetation by overgrazing, wildfire, or development that exposes soil to wind erosion. Most people infected with valley fever have no symptoms or experience flu-like symptoms (often in the summer). Occasionally, serious lung disease occurs. Working on soils near Bakersfield, California, in the mid 1980s, one of your authors (Keller) developed valley fever and experienced mild flu-like symptoms for a week or so.

Today, infectious diseases have the potential to pose rapid threats, both local and global, by spreading in hours via airplane travelers. Terrorist activity may also spread diseases. Inhalation of anthrax caused by a bacterium sent in powdered form in envelopes through the mail killed several people in 2001.

Some diseases can be controlled by manipulating the environment, such as by improving sanitation or treating water. Although there is great concern about the toxins and carcinogens produced in industrial societies today, the greatest mortality in developing countries is caused by environmentally transmitted infectious disease. In the United States, thousands of cases of waterborne illness and food poisoning occur each year. These diseases can be spread by people; by mosquitoes and fleas; or by contact with contaminated food, water, or soil. They can also be transmitted through ventilation systems in buildings. The following are selected examples of environmentally transmitted infectious diseases:

- Legionellosis, or Legionnaires' disease, which often occurs where air-conditioning systems have been contaminated by disease-causing organisms (bacteria).
- Giardiasis, a protozoan infection of the small intestine, spread via food, water, or person-to-person contact.
- Salmonella, a food-poisoning bacterial infection that is spread via water or food.
- Malaria, a protozoan infection transmitted by mosquitoes.
- Lyme borreliosis (Lyme disease), transmitted by ticks (caused by at least three species of bacteria).
- Cryptosporidiosis, a protozoan infection transmitted via water or person-to-person contact (see Chapter 19).
- Anthrax, spread by terrorist activity (caused by a bacteria).

We sometimes hear about epidemics in developing nations. An example is the highly contagious Ebola virus in Africa, which causes external and internal bleeding and kills 80% of those infected. We may tend to think of such epidemics as problems only for developing nations, but such thinking may give us a false sense of security. True, monkeys and bats spread Ebola, but the origin of the virus in the tropical forest remains unknown.

Toxic Heavy Metals

The major **heavy metals** (metals with relatively high atomic weight; see Chapter 7) that pose health hazards to people and ecosystems include mercury, lead, cadmium, nickel, gold, platinum, silver, bismuth, arsenic, selenium, vanadium, chromium, and thallium. Each of these elements may be found in soil or water not contaminated by people; each has uses in our modern industrial society; and each is also a by-product of the mining, refining, and use of other elements. Heavy metals often have direct physiological toxic effects. Some are stored or incorporated in living tissue, sometimes permanently. Heavy metals tend to be stored (accumulating with time) in fatty body tissue. A little arsenic each day may eventually result in a fatal dose—the subject of more than one murder mystery.

The quantity of heavy metals in our bodies is referred to as the *body burden*. The body burden of toxic heavy elements for an average human body (70 kg) is about 8 mg of antimony, 13 mg of mercury, 18 mg of arsenic, 30 mg of cadmium, and 150 mg of lead. The average body burden of lead (for which we apparently have no biological need) is about twice that of the others combined, reflecting our heavy use of this potentially toxic metal.

Mercury, thallium, and lead are very toxic to people. They have long been mined and used, and their toxic properties are well known. Mercury, for example, is the "Mad Hatter" element. At one time, mercury was used to stiffen felt hats, and because mercury damages the brain, hatters in Victorian England were known to act peculiarly. Thus, the Mad Hatter in Lewis Carroll's *Alice in Wonderland* had real antecedents in history.

Mercury in aquatic ecosystems offers an example of biomagnification. Mercury is a potentially serious pollutant of aquatic ecosystems such as ponds, lakes, rivers, and the ocean. Natural sources of mercury in the environment include volcanic eruptions and erosion of natural mercury deposits, but we are most concerned with human input of mercury into the environment by, for example, burning coal in power plants, incinerating waste, and processing metals such as gold. Rates of input of mercury into the environment through human processes are poorly understood. However, it is believed that human activities have doubled or tripled the amount of mercury in the atmosphere, and it is increasing at about 1.5% per year.[8]

A major source of mercury in many aquatic ecosystems is deposition from the atmosphere through precipitation. Most of the deposition is of inorganic mercury (Hg^{++}, ionic mercury). Once this mercury is in surface water, it enters into complex biogeochemical cycles, and a process known as *methylation* may occur. Methylation changes inorganic mercury to methyl mercury $[CH_3Hg]^+$ through bacterial activity. Methyl mercury is much more toxic than inorganic mercury, and it is eliminated more slowly from animals' systems. As the methyl mercury works its way through food chains, biomagnification occurs, resulting in higher concentrations of methyl mercury farther up the food chain. In short, big fish that eat little fish contain higher concentrations of mercury than do smaller fish and the aquatic insects on which the fish feed.

Selected aspects of the mercury cycle in aquatic ecosystems are shown in Figure 8.4. The figure emphasizes the input side of the cycle, from deposition of inorganic mercury through formation of methyl mercury, biomagnification, and sedimentation of mercury at the bottom of a pond. On the output side of the cycle, the mercury that enters fish may be taken up by animals that eat the fish; and sediment may release mercury by a variety of processes, including resuspension in the water, where, eventually, the mercury enters the food chain or is released into the atmosphere through volatilization (conversion of liquid mercury to a vapor form).

Biomagnification also occurs in the ocean. Because large fish, such as tuna and swordfish, have elevated mercury levels, we are advised to limit our consumption

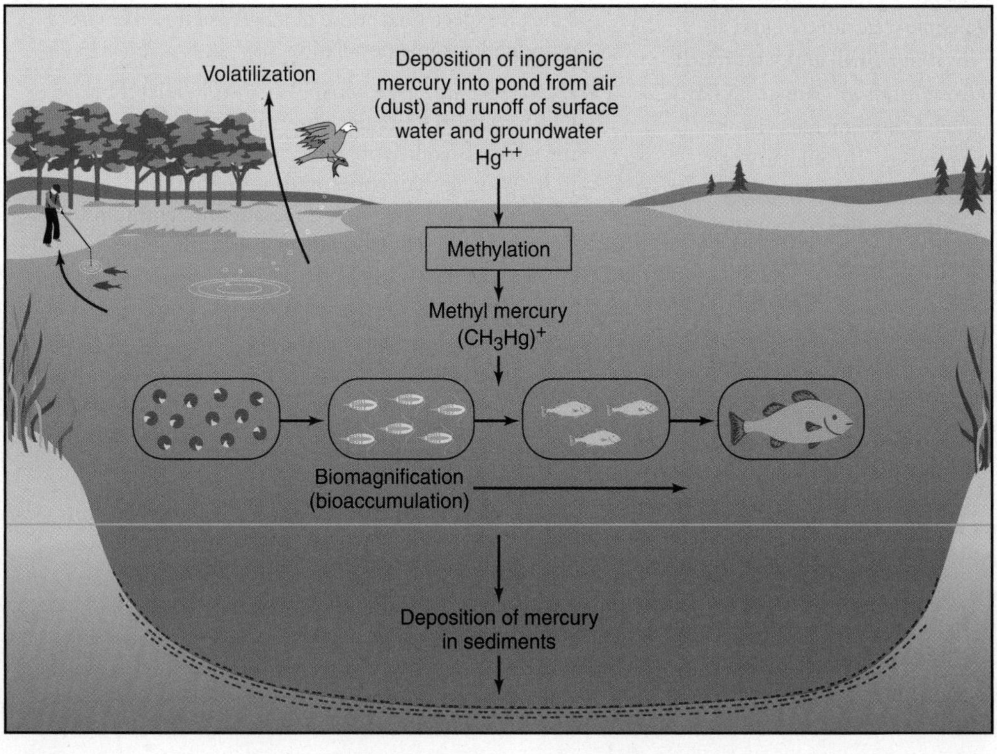

FIGURE 8.4 Idealized diagram showing selected pathways for movement of mercury into and through an aquatic ecosystem.

Modified from G.L. Waldbott, *Health Effects of Environmental Pollutants*, 2nd ed. [Saint Louis, MO: C.V. Mosby, 1978].)

of these fish, and pregnant women are advised not to eat them at all.

The threat of mercury poisoning is widespread. Millions of young children in Europe, the United States, and other industrial countries have mercury levels that exceed health standards.[9] Even children in remote areas of the far north are exposed to mercury through their food chain.

During the 20th century, several significant incidents of methyl mercury poisoning were recorded. One, in Minamata Bay, Japan, involved the industrial release of methyl mercury.[8] (See A Closer Look 8.1.)

Minamata Bay involved local exposure to mercury. Another example of mercury poisoning is being reported in the Arctic. However, in this case the contamination would be classified as mobile, occurring at the global level, in a

A CLOSER LOOK 8.1

Mercury and Minamata, Japan

In the Japanese coastal town of Minamata, on the island of Kyushu, a strange illness began to occur in the middle of the 20th century. It was first recognized in birds that lost their coordination and fell to the ground or flew into buildings and in cats that went mad, running in circles and foaming at the mouth.[a] The affliction, known by local fishermen as the "disease of the dancing cats," subsequently affected people, particularly families of fishermen. The first symptoms were subtle: fatigue, irritability, headaches, numbness in arms and legs, and difficulty in swallowing. More severe symptoms involved the sensory organs; vision was blurred, and the visual field was restricted. Afflicted people became hard of hearing and lost muscular coordination. Some complained of a metallic taste in their mouths; their gums became inflamed, and they suffered from diarrhea. Lawsuits were brought, and approximately 20,000 people claimed to be affected. In the end, according to the Japanese government, an estimated 3,000 people were affected and almost 1,800 died. Those affected lived in a small area, and much of the protein in their diet came from fish from Minamata Bay.

A vinyl chloride factory on the bay used mercury in an inorganic form in its production processes. The mercury was released in waste that was discharged into the bay. Mercury forms few organic compounds, and it was believed that the mercury, though poisonous, would not get into food chains. But the inorganic mercury released by the factory was converted by bacterial activity in the bay into methyl mercury, an organic compound that turned out to be much more harmful. Unlike inorganic mercury, methyl mercury readily passes through cell membranes. It is transported by the red blood cells throughout the body, and it enters and damages brain cells.[b] Fish absorb methyl mercury from water 100 times faster than they absorb inorganic mercury. (This was not known before the epidemic in Japan.) And once absorbed, methyl mercury is retained two to five times longer than is inorganic mercury.

In 1982, lawsuits were filed by plaintiffs affected by the mercury. Twenty-two years later, in 2004—almost 50 years after the initial poisonings—the government of Japan agreed to a settlement of $700,000.

Harmful effects of methyl mercury depend on a variety of factors, including the amount and route of intake, the duration of exposure, and the species affected. The effects of the mercury are delayed from three weeks to two months from the time of ingestion. If mercury intake ceases, some symptoms may gradually disappear, but others are difficult to reverse.[b]

The mercury episode at Minamata illustrates four major factors that must be considered in evaluating and treating toxic environmental pollutants:

- *Individuals vary in their response to exposure to the same dose, or amount, of a pollutant.* Not everyone in Minamata responded in the same way; there were variations even among those most heavily exposed. Because we cannot predict exactly how any single individual will respond, we need to find a way to state an expected response of a particular percentage of individuals in a population.
- *Pollutants may have a threshold*—that is, a level below which the effects are not observable and above which the effects become apparent. Symptoms appeared in individuals with concentrations of 500 ppb of mercury in their bodies; no measurable symptoms appeared in individuals with significantly lower concentrations.
- *Some effects are reversible.* Some people recovered when the mercury-filled seafood was eliminated from their diet.
- *The chemical form of a pollutant, its activity, and its potential to cause health problems may be changed markedly by ecological and biological processes.* In the case of mercury, its chemical form and concentration changed as the mercury moved through the food webs.

Source: Mary Kugler, R.N. *Thousands poisoned, disabled, and killed.* About.com Created October 23, 2004. About.com Health's Disease and Condition content is reviewed by our Medical Review Board. Also, BBC News, "Japan remembers mercury victims." http://news.bbc.co.uk/2/hi/asia-pacific/4959562.stm Published 2006/05/01 GMT © BBC MM VIII.

region far from emission sources of the toxic metal. The Inuit people in Quanea, Greenland, live above the Arctic Circle, far from any roads and 45 minutes by helicopter from the nearest outpost of modern society. Nevertheless, they are some of the most chemically contaminated people on Earth, with as much as 12 times more mercury in their blood than is recommended in U.S. guidelines. The mercury gets to the Inuit from the industrialized world by way of what they eat. The whale, seal, and fish they eat contain mercury that is further concentrated in their tissues and blood. The process of increasing concentrations of mercury farther up the food chain is an example of biomagnification.[9] A recent series of studies by a group of scientists working in conjunction with the U.S. National Park Service found surprising amounts of toxic substances in nine national parks, including mercury, dieldrin, and DDT; they also found serious indications of biomagnification through study of fish in the parks. Wind tends to carry contaminants upslope and into higher elevations where they vaporize and condense in cold weather. Hence high, more pristine areas may have greater pollution levels than lower elevations.[10]

What needs to be done to stop mercury toxicity from the local to the global level is straightforward. The answer is to reduce emissions of mercury by capturing it before emission or by using alternatives to mercury in industry. Success will require international cooperation and technology transfer to countries such as China and India, which, with their tremendous increases in manufacturing, are the world's largest emitters of mercury today.[9]

Organic Compounds

Organic compounds are carbon compounds produced naturally by living organisms or synthetically by industrial processes. It is difficult to generalize about the environmental and health effects of artificially produced organic compounds because there are so many of them, they have so many uses, and they can produce so many different kinds of effects.

Synthetic organic compounds are used in industrial processes, pest control, pharmaceuticals, and food additives. We have produced over 20 million synthetic chemicals, and new ones are appearing at a rate of about 1 million per year! Most are not produced commercially, but up to 100,000 chemicals are now being used or have been used in the past. Once used and dispersed in the environment, they may become a hazard for decades or even hundreds of years. Some synthetic compounds are called **persistent organic pollutants** or **POPs**. Many were first produced decades ago, when their harm to the environment was not known, and they are now banned or restricted (see Table 8.1). POPs are defined by several properties:[11]

- They have a carbon-based molecular structure, often containing highly reactive chlorine.
- Most are manufactured by people—that is, they are synthetic chemicals.
- They are persistent in the environment—they do not easily break down in the environment.
- They are polluting and toxic.
- They are soluble in fat and likely to accumulate in living tissue.
- They occur in forms that allow them to be transported by wind, water, and sediments for long distances.

For example, consider polychlorinated biphenyls (PCBs), which are heat-stable oils originally used as an insulator in electric transformers.[11] A factory in Alabama manufactured PCBs in the 1940s, shipping them to a General Electric factory in Massachusetts. They were put in insulators and mounted on poles in thousands of locations. The transformers deteriorated over time. Some were damaged by lightning, and others were damaged or destroyed during demolition. The PCBs leaked into the soil or were carried by surface runoff into streams and rivers. Others combined with dust, were transported by wind around the world, and were deposited in ponds, lakes, or rivers, where they entered the food chain. First, the PCBs entered algae. Insects ate the algae and were in turn eaten by shrimp and fish. In each stage up the food web, the concentration of PCBs increased. Fish are caught

Table 8.1 SELECTED COMMON PERSISTENT ORGANIC POLLUTANTS (POPs)

CHEMICAL	EXAMPLE OF USE
Aldrin[a]	Insecticide
Atrazine[b]	Herbicide
DDT[a]	Insecticide
Dieldrin[a]	Insecticide
Endrin[c]	Insecticide
PCBs[a]	Liquid insulators in electric transformers
Dioxins	By-product of herbicide production

[a] Banned in the United States and many other countries.

[b] Degrades in the environment. It is persistent when reapplied often.

[c] Restricted or banned in many countries.

Source: Data in part from Anne Platt McGinn, "Phasing Out Persistent Organic Pollutants," in Lester R. Brown et al., *State of the World 2000* (New York: Norton, 2000).

and eaten, passing the PCBs on to people, where they are concentrated in fatty tissue and mother's milk.

Dioxin

Dioxin, a persistent organic pollutant, or POP, may be one of the most toxic human-made chemicals in the environment. The history of the scientific study of dioxin and its regulation illustrates the interplay of science and values.

Dioxin is a colorless crystal made up of oxygen, hydrogen, carbon, and chlorine. It is classified as an organic compound because it contains carbon. About 75 types of dioxin and dioxin-like compounds are known; they are distinguished from one another by the arrangement and number of chlorine atoms in the molecule.

Dioxin is not normally manufactured intentionally but is a by-product of chemical reactions, including the combustion of compounds containing chlorine in the production of herbicides.[12] Dioxins are emitted into the air through such processes as incineration of municipal waste (the major source), incineration of medical waste, burning of gasoline and diesel fuels in vehicles, burning of wood as a fuel, and refining of metals such as copper. Releases of dioxins have decreased about 75% since 1987. However, we are only beginning to understand the many sources of dioxin emissions into the air, water, and land and the linkages and rates of transfer from dominant airborne transport to deposition in water, soil, and the biosphere.[13]

Studies of animals exposed to dioxin suggest that some fish, birds, and other animals are sensitive to even small amounts. As a result, it can cause widespread damage to wildlife, including birth defects and death. However, the concentration at which it poses a hazard to human health is still controversial. Studies suggest that workers exposed to high concentrations of dioxin for longer than a year have an increased risk of dying of cancer.[14]

The Environmental Protection Agency (EPA) has classified dioxin as a known human carcinogen, but its decision is controversial. For most of the exposed people, such as those eating a diet high in animal fat, the EPA puts the risk of developing cancer between 1 in 1,000 and 1 in 100. This estimate represents the highest possible risk for individuals who have had the greatest exposure. For most people, the risk will likely be much lower.[15]

The dioxin problem became well known in 1983 when Times Beach, Missouri, a river town just west of Saint Louis with a population of 2,400, was evacuated and purchased for $36 million by the government. The evacuation and purchase occurred after the discovery that oil sprayed on the town's roads to control dust contained dioxin, and that the entire area had been contaminated. Times Beach was labeled a dioxin ghost town (Figure 8.5). The buildings were bulldozed, and all that was left was a grassy and woody area enclosed by a barbed-wire-topped chain-link fence. The evacuation has since been viewed by some scientists (including the person who ordered the evacuation) as a government overreaction to a perceived dioxin hazard. Following cleanup, trees were planted, and today Times Beach is part of Route 66 State Park and a bird refuge.

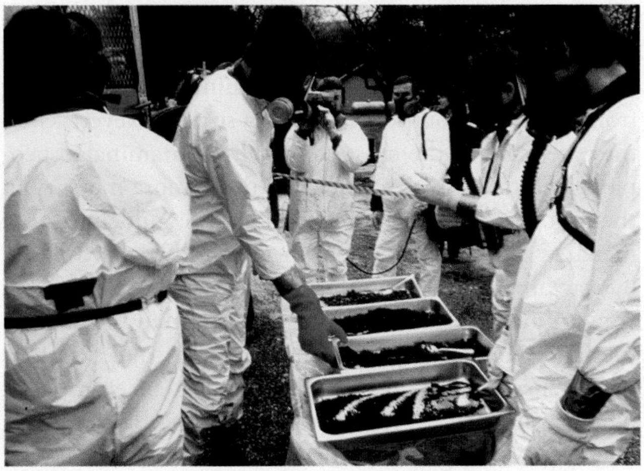
O. Franken/©Corbis

FIGURE 8.5 **Soil samples from Times Beach, Missouri,** thought to be contaminated by dioxin.

The controversy about the toxicity of dioxin is not over.[16-19] Some environmental scientists argue that the regulation of dioxin must be tougher, whereas the industries producing the chemical insist that the dangers of exposure are exaggerated.

Hormonally Active Agents (HAAs)

Persistent organic chemicals that interact with the hormone systems of an organism, whether or not they are linked to disease or abnormalities, are known as **hormonally active agents (HAAs)**. What happens when HAAs—in particular, hormone disrupters (such as pesticides and herbicides)—are introduced into the system is shown in Figure 8.6. Natural hormones produced by the body send chemical messages to cells, where receptors for the hormone molecules are found on the outside and inside of cells. These natural hormones then transmit instructions to the cells' DNA, eventually directing development and growth. We now know that chemicals, such as some pesticides and herbicides, can also bind to the receptor molecules and either mimic or obstruct the role of the natural hormones. Thus, hormonal disrupters may also be known as HAAs.[20-25]

An increasing body of scientific evidence indicates that certain HAAs in the environment may cause developmental and reproductive abnormalities in animals, including humans. (See A Closer Look 8.2.) HAAs include a wide variety of chemicals, such as some herbicides, pesticides, phthalates (compounds found in many chlorine-based plastics), and PCBs. Evidence in support of

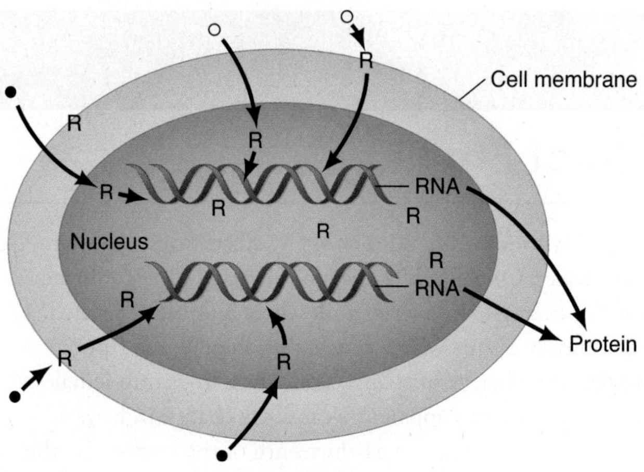

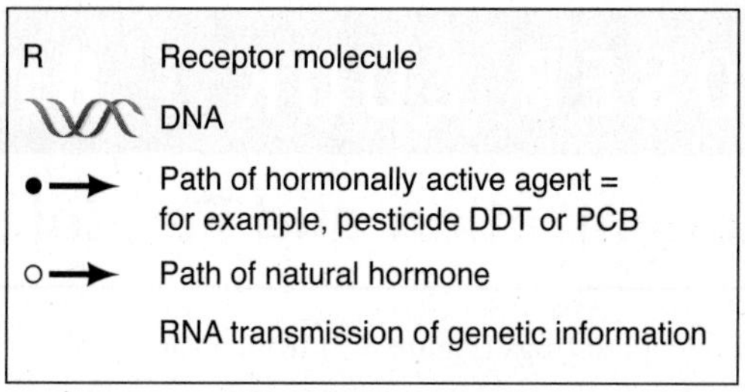

FIGURE 8.6 Idealized diagram of hormonally active agents (HAAs) binding to receptors on the surface of and inside a cell. When HAAs, along with natural hormones, transmit information to the cells' DNA, the HAAs may obstruct the role of the natural hormones that produce proteins that in turn regulate the growth and development of an organism.

the hypothesis that HAAs are interfering with the growth and development of organisms comes from studies of wildlife in the field and laboratory studies of human diseases, such as breast, prostate, and ovarian cancer, as well as abnormal testicular development and thyroid-related abnormalities.[20]

Studies of wildlife include evidence that alligator populations in Florida that were exposed to pesticides, such as DDT, have genital abnormalities and low egg production. Pesticides have also been linked to reproductive problems in several species of birds, including gulls, cormorants, brown pelicans, falcons, and eagles. Studies are ongoing on Florida panthers; they apparently have abnormal ratios of sex hormones, and this may be affecting their reproductive capability. In sum, the studies of major disorders in wildlife have centered on abnormalities, including thinning of birds' eggshells, decline in populations of various animals and birds, reduced viability of offspring, and changes in sexual behavior.[21]

With respect to human diseases, much research has been done on linkages between HAAs and breast cancer by exploring relationships between environmental estrogens and cancer. Other studies are ongoing to understand relationships between PCBs and neurological behavior that result in poor performance on standard intelligence tests. Finally, there is concern that exposure of people to phthalates that are found in plastics containing chlorine is also causing problems. Consumption of phthalates in the United States is considerable, with the highest exposure in women of childbearing age. The products being tested as the source of contamination include perfumes and other cosmetics, such as nail polish and hairspray.[21]

In sum, there is good scientific evidence that some chemical agents, in sufficient concentrations, will affect human reproduction through endocrine and hormonal disruption. The human endocrine system is of primary importance because it is one of the two main systems (the other is the nervous system) that regulate and control growth, development, and reproduction. The human endocrine system consists of a group of hormone-secreting glands, including the thyroid, pancreas, pituitary, ovaries (in women), and testes (in men). The bloodstream transports the hormones to virtually all parts of the body, where they act as chemical messengers to control growth and development of the body.[20]

The National Academy of Sciences completed a review of the available scientific evidence concerning HAAs and recommends continued monitoring of wildlife and human populations for abnormal development and reproduction. Furthermore, where wildlife species are known to be experiencing declines in population associated with abnormalities, experiments should be continued to study the phenomena with respect to chemical contamination. For people, the recommendation is for additional studies to document the presence or absence of associations between HAAs and human cancers. When associations are discovered, the causality is investigated in the relationship between exposure and disease and indicators of susceptibility to disease of certain groups of people by age and sex.[21]

As an example, consider the exposure of frogs to a common herbicide in A Closer Look 8.2.

Nuclear Radiation

Nuclear radiation is introduced here as a category of pollution. We discuss it in detail in Chapter 17, in conjunction with nuclear energy. We are concerned about nuclear radiation because excessive exposure is linked to serious health problems, including cancer. (See also Chapter 21 for a discussion of radon gas as an indoor air pollutant.)

Thermal Pollution

Thermal pollution, also called *heat pollution*, occurs when heat released into water or air produces undesirable effects. Heat pollution can occur as a sudden, acute event or as a long-term, chronic release. Sudden heat releases may result from natural events, such as brush or forest fires and volcanic eruptions, or from human activities, such as agricultural burning.

A CLOSER LOOK 8.2

Demasculinization and Feminization of Frogs

The story of wild leopard frogs (Figure 8.7) from a variety of areas in the midwestern United States sounds something like a science fiction horror story. In affected areas, between 10 and 92% of male frogs exhibit gonadal abnormalities, including retarded development and hermaphroditism, meaning they have both male and female reproductive organs. Other frogs have vocal sacs with retarded growth. Since their vocal sacs are used to attract female frogs, these frogs are less likely to mate.[a-d]

What is apparently causing some of the changes in male frogs is exposure to atrazine, the most widely used herbicide in the United States today. The chemical is a weed killer (and HAA) used primarily in agricultural areas. The region of the United States with the highest frequency (92%) of sex reversal of male frogs is in Wyoming, along the North Platte River. Although the region is not near any large agricultural activity, and the use of atrazine there is not particularly significant, hermaphrodite frogs are common there because the North Platte River flows from areas in Colorado where atrazine is commonly used.[a-d]

The amount of atrazine released into the environment of the United States is estimated at approximately 7.3 million kg (16 million lb) per year. The chemical degrades in the environment, but the degradation process is longer than the application cycle. Because of its continual application every year, the waters of the Mississippi River basin, which drains about 40% of the lower United States, discharge approximately 0.5 million kg (1.2 million lb) of atrazine per year to the Gulf of Mexico. Atrazine easily attaches to dust particles and has been found in rain, fog, and snow. As a result, it has contaminated groundwater and surface water in regions where it isn't used. The EPA states that up to 3 parts per billion (ppb) of atrazine in drinking water is acceptable, but at this concentration it definitely affects frogs that swim in the water. Other studies around the world have confirmed this. For example, in Switzerland, where atrazine is banned, it commonly occurs with a concentration of about 1 ppb, and that is sufficient to change some male frogs into females. In fact, atrazine can apparently cause sex change in frogs at concentrations as low as one-thirteenth of the level set by the EPA for drinking water.[a-d]

Mother Daughter Press/Getty Images

FIGURE 8.7 Wild leopard frogs in America have been affected by human-made chemicals (the herbicide atrazine) in the environment.

Of particular interest and importance is the process that causes the changes in leopard frogs. We begin the discussion with the endocrine system, composed of glands that secrete hormones such as testosterone and estrogen directly into the bloodstream, which carries them to parts of the body where they regulate and control growth and sexual development. Testosterone in male frogs is partly responsible for development of male characteristics. The atrazine is believed to switch on a gene that turns testosterone into estrogen, a female sex hormone. It's the hormones, not the genes, that actually regulate the development and structure of reproductive organs.

Frogs are particularly vulnerable during their early development, before and as they metamorphose from tadpoles into adult frogs. This change occurs in the spring, when atrazine levels are often at a maximum in surface water. Apparently, a single exposure to the chemical may affect the frog's development. Thus, the herbicide is known as a hormone disrupter.

Frogs have also been studied in the laboratory. All male African clawed frogs when exposed to atrazine in the laboratory undergo complete feminization.[e]

The story of wild leopard frogs in America and African clawed frogs in the laboratory dramatizes the importance of carefully evaluating the role of human-made chemicals in the environment. Populations of frogs and other amphibians are declining globally, and much research has been directed toward understanding why. Studies to evaluate past or impending extinctions of organisms often center on global processes such as climate change, but the story of leopard frogs leads us down another path, one associated with our use of the natural environment. It also raises a number of more disturbing questions: Are we participating in an unplanned experiment on how human-made chemicals, such as herbicides and pesticides, might transform the bodies of living beings, perhaps even people? Are these changes in organisms limited to only certain plants and animals, or are they a forerunner of what we might expect in the future on a much broader scale? Perhaps we will look back on this moment of understanding as a new beginning in meaningful studies that will answer some of these important questions.

The major sources of chronic heat pollution are electric power plants that produce electricity in steam generators and release large amounts of heated water into rivers. This changes the average water temperature and the concentration of dissolved oxygen (warm water holds less oxygen than cooler water), thereby changing a river's species composition (see the discussion of eutrophication in Chapter 19). Every species has a temperature range within which it can survive and an optimal temperature for living. For some species of fish, the range is small, and even a small change in water temperature is a problem. Lake fish move away when the water temperature rises more than about 1.5°C above normal; river fish can withstand a rise of about 3°C.

Heating river water can change its natural conditions and disturb the ecosystem in several ways. Fish spawning cycles may be disrupted, and the fish may have a heightened susceptibility to disease. Warmer water also causes physical stress in some fish, making them easier for predators to catch, and warmer water may change the type and abundance of food available for fish at various times of the year.

There are several solutions to chronic thermal discharge into bodies of water. The heat can be released into the air by cooling towers (Figure 8.8), or the heated water can be temporarily stored in artificial lagoons until it cools down to normal temperatures. Some attempts have been made to use the heated water to grow organisms of commercial value that require warmer water. Waste heat from a power plant can also be captured and used for a variety of purposes, such as warming buildings (see Chapter 14 for a discussion of cogeneration).

Particulates

Particulates here refer to small particles of dust (including soot and asbestos fibers) released into the atmosphere

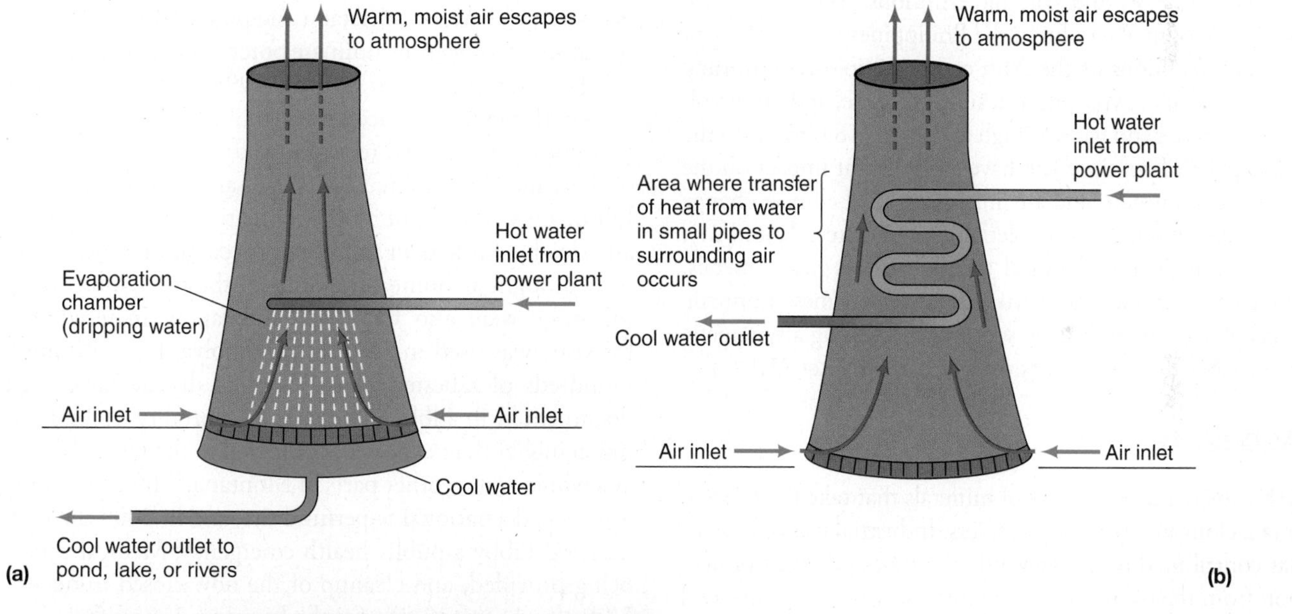

FIGURE 8.8 **Two types of cooling towers. (a) Wet cooling tower.** Air circulates through the tower; hot water drips down and evaporates, cooling the water. **(b) Dry cooling tower.** Heat from the water is transferred directly to the air, which rises and escapes the tower. **(c) Cooling towers emitting steam at Didcot power plant**, Oxfordshire, England. Red and white lines are vehicle lights resulting from long exposure time (photograph taken at dusk).

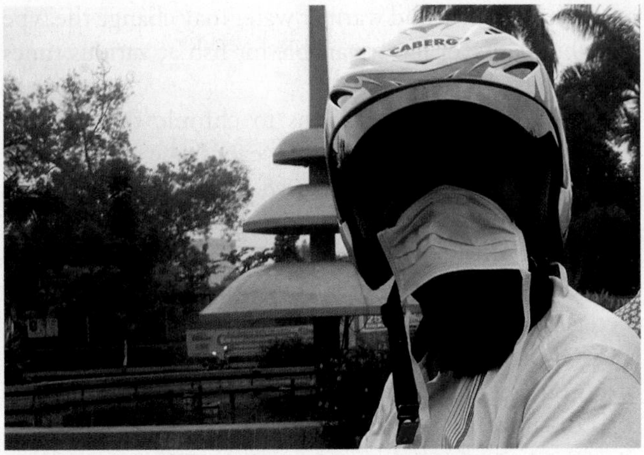

NANJLA TANJUNG/Reuters/Landov LLC

FIGURE 8.9 Fires in Indonesia in 1997 caused serious air pollution. The person shown here is wearing a surgical mask in an attempt to breathe cleaner air.

by many natural processes and human activities. Modern farming and the burning of oil and coal add considerable amounts of particulates to the atmosphere, as do dust storms, fires (Figure 8.9), and volcanic eruptions. The 1991 eruptions of Mount Pinatubo in the Philippines were the largest volcanic eruptions of the 20th century, explosively hurling huge amounts of volcanic ash, sulfur dioxide, and other volcanic material and gases as high as 30 km (18.6 mi) into the atmosphere. Eruptions can have a significant impact on the global environment and are linked to global climate change and stratospheric ozone depletion (see Chapters 20 and 21). In addition, many chemical toxins, such as heavy metals, enter the biosphere as particulates. Sometimes, nontoxic particulates link with toxic substances, creating a synergetic threat. (See discussion of particulates in Chapter 21.)

Asbestos

Asbestos is a term for several minerals that take the form of small, elongated particles, or fibers. Industrial use of asbestos has contributed to fire prevention and has provided protection from the overheating of materials. Asbestos is also used as insulation for a variety of other purposes. Unfortunately, however, excessive contact with asbestos has led to asbestosis (a lung disease caused by inhaling asbestos) and to cancer in some industrial workers. Experiments with animals have demonstrated that asbestos can cause tumors if the fibers are embedded in lung tissue.[26] The hazard related to certain types of asbestos under certain conditions is considered so serious that extraordinary steps have been taken to reduce the use of asbestos or ban it outright. The expensive process of asbestos removal from old buildings (particularly schools) in the United States is one of those steps.

There are several types of asbestos, and they are not equally hazardous. Most commonly used in the United States is white asbestos, which comes from the mineral chrysolite. It has been used to insulate pipes, floor and ceiling tiles, and brake linings of automobiles and other vehicles. Approximately 95% of the asbestos that is now in place in the United States is of the chrysolite type. Most of this asbestos was mined in Canada, and environmental health studies of Canadian miners show that exposure to chrysolite asbestos is not particularly harmful. However, studies involving another type of asbestos, known as crocidolite asbestos (blue asbestos), suggest that exposure to this mineral can be very hazardous and evidently does cause lung disease. Several other types of asbestos have also been shown to be harmful.[26]

A great deal of fear has been associated with nonoccupational exposure to chrysolite asbestos in the United States. Tremendous amounts of money have been spent to remove it from homes, schools, public buildings, and other sites, even though no asbestos-related disease has been recorded among those exposed to chrysolite in nonoccupational circumstances. It is now thought that much of the removal was unnecessary and that chrysolite asbestos doesn't pose a significant health hazard. Additional research into health risks from other varieties of asbestos is necessary to better understand the potential problem and to outline strategies to eliminate potential health problems.

For example, from 1979 to 1998 a strip mine near Libby, Montana, produced vermiculite (a natural mineral) that was contaminated (commingled) with a fibrous form of the mineral tremolite, classified as an asbestos. People in Libby, a town of about 3,000 population, were exposed to asbestos by workers in the mines (occupational exposure) who brought it home on clothes. Asbestos tailings from the mine were also found in landscaping material that for years was used in city parks, schoolyards, and homes. Hundreds of asbestos-related cases of disease have been documented in Libby and asbestos mortality in Libby is much higher than expected, compared to the United States as a whole and to other parts of Montana.[27] In 2002 Libby was named a national Superfund site, and in 2009 the EPA declared Libby a public health emergency. Medical care is being provided, and cleanup of the now closed mine and Libby has been ongoing for the last several years.[28]

Electromagnetic Fields

Electromagnetic fields (EMFs) are part of everyday urban life. Cell phones, electric motors, electric transmission lines for utilities, and our electrical appliances—toasters, electric blankets, computers, and so forth—all produce magnetic fields. There is currently a controversy over whether these fields produce a health risk.

Early on, investigators did not believe that magnetic fields were harmful because fields drop off quickly with distance from the source, and the strengths of the fields that most people come into contact with are relatively weak. For example, the magnetic fields generated by

- The effect of a chemical or toxic material on an individual depends on the dose. It is also important to determine tolerances of individuals, as well as acute and chronic effects of pollutants and toxins.
- Risk assessment involves identifying the hazard, assessing the exposure and the dose response, and characterizing the possible results.

REEXAMINING THEMES AND ISSUES

© Biletskiy_Evgeniy/iStockphoto

SUSTAINABILITY

Ensuring that future generations inherit a relatively unpolluted, healthy environment remains a challenging problem. Sustainable development requires that our use of chemicals and other materials not damage the environment.

© Anton Balazh 2011/iStockphoto

GLOBAL PERSPECTIVE

Releasing toxins into the environment can cause global patterns of contamination or pollution, particularly when a toxin or contaminant enters the atmosphere, surface water, or oceans and becomes widely dispersed. For example, pesticides, herbicides, and heavy metals emitted into the atmosphere in the midwestern United States may be transported by winds and deposited on glaciers in polar regions.

ssguy/ShutterStock

URBAN WORLD

Industrial processes in urban areas concentrate potentially toxic materials that may be inadvertently, accidentally, or deliberately released into the environment. Human exposure to a variety of pollutants—including lead, asbestos, particulates, organic chemicals, radiation, and noise—is often greater in urban areas.

B2M Productions/Getty Images, Inc.

PEOPLE AND NATURE

Feminization of frogs and other animals from exposure to human-produced, hormonally active agents (HAAs) is an early warning or red flag that we are disrupting some basic aspects of nature. We are performing unplanned experiments on nature, and the consequences to us and other living organisms with which we share the environment are poorly understood. Control of HAAs seems an obvious candidate for application of the Precautionary Principle, discussed in Chapter 1.

George Doyle/Getty Images, Inc.

SCIENCE AND VALUES

Because we value both human and nonhuman life, we are interested in learning all we can about the risks of exposing living things to chemicals, pollutants, and toxins. Unfortunately, our knowledge of risk assessment is often incomplete, and the dose response for many chemicals is poorly understood. What we decide to do about exposure to toxic chemicals reflects our values. Increased control of toxic materials in homes and the work environment is expensive. To reduce environmental hazards at worksites in other countries, are we willing to pay more for the goods those workers manufacture?

KEY TERMS

area sources 159
asbestos 170
biomagnification 161
carcinogen 159
contamination 159
disease 159
dose response 175
ecological gradient 177
ED-50 176

STUDY QUESTIONS

1. What kinds of life-forms would most likely survive in a highly polluted world? What would be their general ecological characteristics?
2. Some environmentalists argue that there is no such thing as a threshold for pollution effects. What do they mean? How would you determine whether it was true for a specific chemical and a specific species?
3. What is biomagnification, and why is it important in toxicology?
4. You are lost in Transylvania while trying to locate Dracula's castle. Your only clue is that the soil around the castle is known to have an unusually high concentration of the heavy metal arsenic. You wander in a dense fog, able to see only the ground a few meters in front of you. What changes in vegetation warn you that you are nearing the castle?
5. Distinguish between acute effects and chronic effects of pollutants.
6. Design an experiment to test whether tomatoes or cucumbers are more sensitive to lead pollution.
7. Why is it difficult to establish standards for acceptable levels of pollution? In giving your answer, consider physical, climatological, biological, social, and ethical reasons.
8. A new highway is built through a pine forest. Driving along the highway, you notice that the pines nearest the road have turned brown and are dying. You stop at a rest area and walk into the woods. One hundred meters away from the highway, the trees seem undamaged. How could you make a crude dose-response curve from direct observations of the pine forest? What else would be necessary to devise a dose-response curve from direct observation of the forest? What else would be necessary to devise a dose-response curve that could be used in planning the route of another highway?
9. Do you think as valid the hypothesis that some crime is caused in part by environmental pollution? Why? Why not? How might the hypothesis be further tested? What are the social ramifications of the tests?

FURTHER READING

Amdur, M., J. Doull, and C.D. Klaasen, eds., *Casarett & Doull's Toxicology: The Basic Science of Poisons,* 4th ed. (Tarrytown, NY: Pergamon, 1991). A comprehensive and advanced work on toxicology.

Carson, R., *Silent Spring* (Boston: Houghton Mifflin, 1962). A classic book on problems associated with toxins in the environment.

Schiefer, H.B., D.G. Irvine, and S.C. Buzik, *Understanding Toxicology: Chemicals, Their Benefits and Risks* (Boca Raton, FL: CRC Press, 1997). A concise introduction to toxicology as it pertains to everyday life, including information about pesticides, industrial chemicals, hazardous waste, and air pollution.

Selinus, O. (ed.), *Essentials of Medical Geology* (Burlington, MA: Elsevier Academic Press, 2005).

NOTES

1. Whitworth, K.W., E. Symanski, and A.L. Coker. 2008. Childhood lymphohematopoietic cancer incidence and hazardous air pollutants in southeast Texas, 1995–2004. *Environmental Health Perspectives* 116(11):1576–1580.
2. U.S. Department of Labor. 1,3-Butadiene. Health Effects. http://www.osha.gov/SLTC/butadiene/index.html. Accessed October 23, 2012.
3. Linder, S.H., Marko, D., and Sexton, K. 2008. Cumulative cancer risk from air pollution in Houston: Disparities in risk burden and social disadvantage. *Environmental Science & Technology* 42(12):4312–4322.
4. World Health Organization. Environmental Health. www.who.int/topics/environmental_health/en/. Accessed September 24, 2012.

5. Gunn, J., ed. 1995. *Restoration and Recovery of an Industrial Region: Progress in Restoring the Smelter-damaged Landscape near Sudbury, Canada.* New York: Springer-Verlag.

6. Gunn, J. Keller, W. Negusanti, J. Potvin, R., Beckett, P., and Winterhalder, K. 1995. Ecosystem recovery after emission reductions: Sudbury, Canada. *Water, Air, & Soil Pollution* 85(3):1783–1788.

7. National Center for Biotechnology Information http:/ www.ncbi.nlm.nih.gov/pubmedhealthPMH0002358/# adam_001382.disease.causes. Accessed October 6, 2012.

8. U.S. Geological Survey. 1995. *Mercury Contamination of Aquatic Ecosystems.* USGS FS pp. 216–295.

9. Greer, L., M. Bender, P. Maxson, and D. Lennett. 2006. Curtailing mercury's global reach. In L. Starke, ed., *State of the World 2006,* 96–114. New York: Norton.

10. Landers, H. Dixon, et al. 2008. *The Fate, Transport, and Ecological Impacts of Airborne Contaminants in Western National Parks (USA).* Western Airborne Contaminants Assessment Project. National Park Service. http://www.nature.nps.gov/air/Studies/air_toxics/wacap.cfm Accessed October 29, 2012.

11. McGinn, A.P. 2000 (April 1). POPs culture. *World Watch,* pp. 26–36.

12. Carlson, E.A. 1983. International symposium on herbicides in the Vietnam War: An appraisal. *BioScience* 33:507–512.

13. Cleverly, D., J. Schaum, D. Winters, and G. Schweer. 1999. Inventory of sources and releases of dioxin-like compounds in the United States. Paper presented at the 19th International Symposium on Halogenated Environmental Organic Pollutants and POPs, September 12–17, Venice, Italy. Short paper in *Organohalogen Compounds* 41:467–472.

14. Roberts, L. 1991. Dioxin risks revisited. *Science* 251:624–626.

15. Kaiser, J. 2000. Just how bad is dioxin? *Science* 5473: 1941–1944.

16. Johnson, J. 1995. SAB Advisory Panel rejects dioxin risk characterization. *Environmental Science & Technology* 29:302A.

17. National Research Council Committee on EPA's Exposure and Human Health Reassessment of TCDD and Related Compounds. 2006. *Health Risks from Dioxin and Related Compounds.* Washington, DC: National Academy Press.

18. Thomas, V.M., and T.G. Spiro. 1996. The U.S. dioxin inventory: Are there missing sources? *Environmental Science & Technology* 30:82A–85A.

19. U.S. Environmental Protection Agency. 1994 (June). Estimating exposure to dioxinlike compounds. Review draft. Office of Research and Development, EPA/600/6–88/005 Ca-c.

20. Krimsky, S. 2001. Hormone disrupters: A clue to understanding the environmental cause of disease. *Environment* 43(5): 22–31.

21. Committee on Hormonally Active Agents in the Environment, National Research Council, National Academy of Sciences. 1999. *Hormonally Active Agents in the Environment.* Washington, DC: National Academy Press.

22. Royte, E. 2003. Transsexual frogs. *Discover* 24(2):26–53.

23. Hayes, T.B., K. Haston, M. Tsui, A. Hong, C. Haeffele, and A. Vock. 2002. Hermaphrodites beyond the cornfield. Atrazine-induced testicular oogenesis in leopard frogs (*Rana pipiens*). *Nature* 419:895–896.

24. Hayes, T.B., et al. 2002. Hermaphroditic demasculinized frogs after exposure to the herbicide atrazine at low ecologically relevant doses. *Proceedings of the National Academy (PNAS)* 99(8):5476–5480.

25. Hayes, T.B. and ten others. 2010. Atrazine induces complete feminization and chemical castration in male African clawed frogs (*Xenopus laevis*). *Proceedings of the National Academy (PNAS)* 107(10):4612–4617.

26. Ross, M. 1990. Hazards associated with asbestos minerals. In B.R. Doe, ed., *Proceedings of a U.S. Geological Survey Workshop on Environmental Geochemistry,* pp. 175–176. U.S. Geological Survey Circular 1033.

27. Agency for Toxic Substances & Disease Registry. 2002. Mortality from asbestosis in Libby, Montana, 1979–1998. www.atsdr.cdc.gov.

28. Libby Asbestos. EPA website. http://www.epa.gov/region8/superfund/libby/index.html Accessed October 29, 2012.

29. Pool, R. 1990. Is there an EMF-cancer connection? *Science* 249:1096–1098.

30. Linet, M.S., E.E. Hatch, R.A. Kleinerman, L.L. Robison, W.T. Kaune, D.R. Friedman, R.K. Severson, C.M. Haines, C.T. Hartsock, S. Niwa, S. Wacholder, and R.E. Tarone. 1997. Residential exposure to magnetic fields and acute lymphoblastic leukemia in children. *New England Journal of Medicine* 337(1):1–7.

31. Kheifets, L.I., E.S. Gilbert, S.S. Sussman, P. Guaenel, S.D. Sahl, D.A. Savitz, and G. Thaeriault. 1999. Comparative analyses of the studies of magnetic fields and cancer in electric utility workers: Studies from France, Canada, and the United States. *Occupational and Environmental Medicine* 56(8):567–574.

32. Ahlbom, A., E. Cardis, A. Green, M. Linet, D. Savitz, and A. Swerdlow. 2001. Review of the epidemiologic literature on EMF and health. *Environmental Perspectives* 109(6): 911–933.

33. World Health Organization, International Agency for Research on Cancer. Volume 80: Non-ionizing radiation, Part 1, Static and extremely low-frequency (ELF) electric and magnetic fields. IARC Working Group on the Evaluation of Carcinogenic Risks to Humans. 2002: Lyon, France.

34. Kleinerman, R.A., W.T. Kaune, E.E. Hatch, et al. 2000. Are children living near high voltage power lines at increased risk of acute lymphocytic leukemia? *American Journal of Epidemiology* 15:512–515.

35. Greenland, S., A.R. Sheppard, W.T. Kaune, C. Poole, and M.A. Kelsh. 2000. A pooled analysis of magnetic fields, wire codes, and childhood leukemia. Childhood Leukemia-EMF Study Group. *Epidemiology* 11(6):624–634.

36. Schoenfeld, E.R., E.S. O'Leary, K. Henderson, et al. 2003. Electromagnetic fields and breast cancer on Long Island: A case-control study. *American Journal of Epidemiology* 158:47–58.

37. Kabat, G.C., E.S. O'Leary, E.R. Schoenfeld, et al. 2003. Electric blanket use and breast cancer on Long Island. *Epidemiology* 14(5):514–520.

38. Kliukiene, J., T. Tynes, and A. Andersen. 2004. Residential and occupational exposures to 50-Hz magnetic fields and breast cancer in women: A population-based study. *American Journal of Epidemiology* 159(9):852–861.

39. Zhu, K., S. Hunter, K. Payne-Wilks, et al. 2003. Use of electric bedding devices and risk of breast cancer in African-American women. *American Journal of Epidemiology* 158:798–806.

40. Orent, W. 2012 (March). The big, overlooked factor in the rise of pandemics: The human vector. *Discover* 23(2): 68–74.

41. Pew Charitable Trusts and Johns Hopkins Bloomberg School of Public Health. 2008. *Putting Meat on the Table: Industrial Farm Animal Production in America.* A report of the Pew Commission on Industrial Farm Animal Production. Final Report. www.ncifap.org/. Accessed October 23, 2012.

42. Gee, D. 2006. Late lessons from early warnings: Toward realism and precaution with endocrine-disrupting substances. *Environmental Health Perspective* 114:152–160.

43. Francis, B.M. 1994. *Toxic Substances in the Environment.* New York: John Wiley & Sons.

44. Poisons and poisoning. 1997. *Encyclopedia Britannica.* Vol. 25, p. 913. Chicago: Encyclopedia Britannica.

45. Endocrine Society Issues Position Statement on Endocrine-Disrupting Chemicals. June 29, 2009. The Endocrine Society. http://www.endo-society.org/media/press/2009/SocietyIssues-PositionStatementonEDC.cfm Accessed October 29, 2012.

46. Air Risk Information Support Center (Air RISC), U.S. Environmental Protection Agency. 1989. *Glossary of Terms Related to Health Exposure and Risk Assessment.* EPA/450/3–88/016. Research Triangle Park, NC.

47. Needleman, H.L., J.A. Riess, M.J. Tobin, G.E. Biesecker, and J.B. Greenhouse. 1996. Bone lead levels and delinquent behavior. *Journal of the American Medical Association* 275: 363–369.

48. Centers for Disease Control. 1991. *Preventing Lead Poisoning in Young Children.* Atlanta: Public Health Service, Centers for Disease Control.

49. Goyer, R.A. 1991. Toxic effects of metals. In M.O. Amdur, J. Doull, and C.D. Klaassen, eds., *Toxicology*, pp. 623–680. New York: Pergamon.

50. Bylinsky, G. 1972. Metallic nemesis. In B. Hafen, ed., *Man, Health and Environment,* pp. 174–185. Minneapolis: Burgess.

51. Hong, S., J. Candelone, C.C. Patterson, and C.F. Boutron. 1994. Greenland ice evidence of hemispheric lead pollution two millennia ago by Greek and Roman civilizations. *Science* 265:1841–1843.

52. Carson, R. 1962. *Silent Spring.* Boston: Houghton Mifflin Co.

53. Landrigan, P.J. and 10 others. 1999. Pesticides and inner-city children: Exposures, risks, and prevention. *Environmental Health Perspectives* 107:431–437.

54. Juruske, R., Mutel, C. I., Stossel, F., and Hellweg, S. 2009. Life cycle human toxicity assessment of pesticides: Comparing fruit and vegetable diets in Switzerland and the United States. *Chemosphere* 77:939–945.

55. U.S. Environmental Protection Agency. 2000. *America's Children and the Environment.* EPA 240-R-00-006.

56. Bouchard, M. F., Bellinger, D. C., Wright, R. O., and Weisskopf. 2010. Attention-deficit/hyperactivity disorder and urinary metabolites of organophosphate pesticides. *Pediatrics*: 125(6)1270–1277.

A CLOSER LOOK 8.1 NOTES

a. Ehrlich, P.R., A.H. Ehrlich, and J.P. Holdren. 1970. *Ecoscience: Population, Resources, Environment.* San Francisco: Freeman.

b. Waldbott, G.L. 1978. *Health Effects of Environmental Pollutants,* 2nd ed. Saint Louis: Mosby.

A CLOSER LOOK 8.2 NOTES

a. Committee on Hormonally Active Agents in the Environment, National Research Council, National Academy of Sciences. 1999. *Hormonally Active Agents in the Environment.* Washington, DC: National Academy Press.

b. Royte, E. 2003. Transsexual frogs. *Discover* 24(2):26–53.

c. Hayes, T.B., K. Haston, M. Tsui, A. Hong, C. Haeffele, and A. Vock. 2002. Hermaphrodites beyond the cornfield. Atrazine-induced testicular oogenesis in leopard frogs (*Rana pipiens*). *Nature* 419:895–896.

d. Hayes, T.B., et al. 2002. Hermaphroditic demasculinized frogs after exposure to the herbicide atrazine at low ecologically relevant doses. *Proceedings of the National Academy (PNAS)* 99(8):5476–5480.

e. Hayes, T.B. and ten others. 2010. Atrazine induces complete feminization and chemical castration in male African clawed frogs (*Xenopus laevis*). *Proceedings of the National Academy (PNAS)* 107(10): 4612–4617.

CHAPTER 9

Biological Diversity and Biological Invasions

LEARNING OBJECTIVES

Biological diversity has become one of the "hot-button" environmental topics—there is a lot of news about endangered species, loss of biodiversity, and its causes. This chapter provides a basic scientific introduction that will help you understand the background to this news, the causes of and solutions to species loss, and the problems that arise when we move species around the globe. Interest in the variety of life on Earth is not new; people have long wondered how the amazing diversity of living things on Earth came to be. This diversity has developed through biological evolution and is affected by interactions among species and by the environment. After reading this chapter, you should be able to . . .

- Identify the factors that are most responsible for the fact that islands tend to have fewer species than a continent
- List and explain the four major mechanisms of biological evolution
- Defend or present solid reasoning against the statement: The only reason to protect endangered species is that they provide products useful to people
- Describe how we can control invasive species that enter the United States carried by air travelers or in their checked baggage
- Explain the difference between a habitat and an ecological niche. Determine which is more important, if either, in the conservation of endangered species
- Determine whether symbiosis can be a significant force in biological evolution and how this could occur if evolution is based on the survival of the fittest
- Explain how species that appear to have identical needs can coexist
- Defend, or present solid reasoning against the likelihood, that dinosaurs can be reengineered using modern molecular biology and paleontology

Mamoudou Setamou, center, a citrus entomologist for the Texas A&M University Kingsville Citrus Center in Weslaco, walks through the grove where citrus greening disease has been found in San Juan, Texas. The California Department of Food and Agriculture announced Friday that citrus greening, also known as huanglongbing, has been discovered in lemon/pummelo tree in a residential neighborhood of Los Angeles County. The bacterial disease is carried by the Asian citrus psyllid and attacks the vascula.

The Monitor, Nathan Lambrecht, File/AP Photo

graphingactivity

CASE STUDY

Citrus Greening

In 2005 a tiny fruit fly that carries and disperses a bacterial disease of citrus plants arrived in the United States from China (Figure 9.1). This disease, known as "citrus greening" or Chinese *huanglongbing* (yellow dragon disease), had been extending its range and had reached India, many African countries, and Brazil (Figure 9.3a). Wherever the fly and the bacteria have gone, citrus crops have failed. The bacteria interfere with the flow of organic compounds in the phloem (the living part of the plant's bark). The larvae of the fruit fly sucks juices from the tree, inadvertently injecting the bacteria. Winds blow the adult flies from one tree to another, making control of the fly difficult. According to the U.S. Department of Agriculture (USDA), citrus greening is the most severe new threat to citrus plants in the United States and might end commercial orange production in Florida. Many Florida counties have tested positive for the disease and are under a quarantine that prevents citrus plants from being moved from one area to another (Figure 9.3b).[1]

Introductions of new species into new habitats have occurred as long as life has existed on Earth. And beginning with the earliest human travelers, our ancestors have moved species around the world, sometimes on purpose, sometimes unknowingly. Polynesians brought crops, pigs, and many other animals and plants from one Pacific island to another as they migrated and settled widely before A.D. 1000. The intentional spread of crop plants around the world has been one of the primary reasons that our species has been able to survive in so many habitats and has grown to such a huge number. But if invasion by species is as old as life, and often beneficial to people, why is it also the source of so many environmental problems? The answers lie in this chapter.

Courtesy USDA

FIGURE 9.1 Larvae of *Diaphorina citri*, the tiny fly that spreads citrus greening bacteria (*Candidatus Liberibacter asiaticus*).

© AP/Wide World Photos

FIGURE 9.2 The disease yellows leaves, turns fruit greenish brown, and eventually kills the tree.

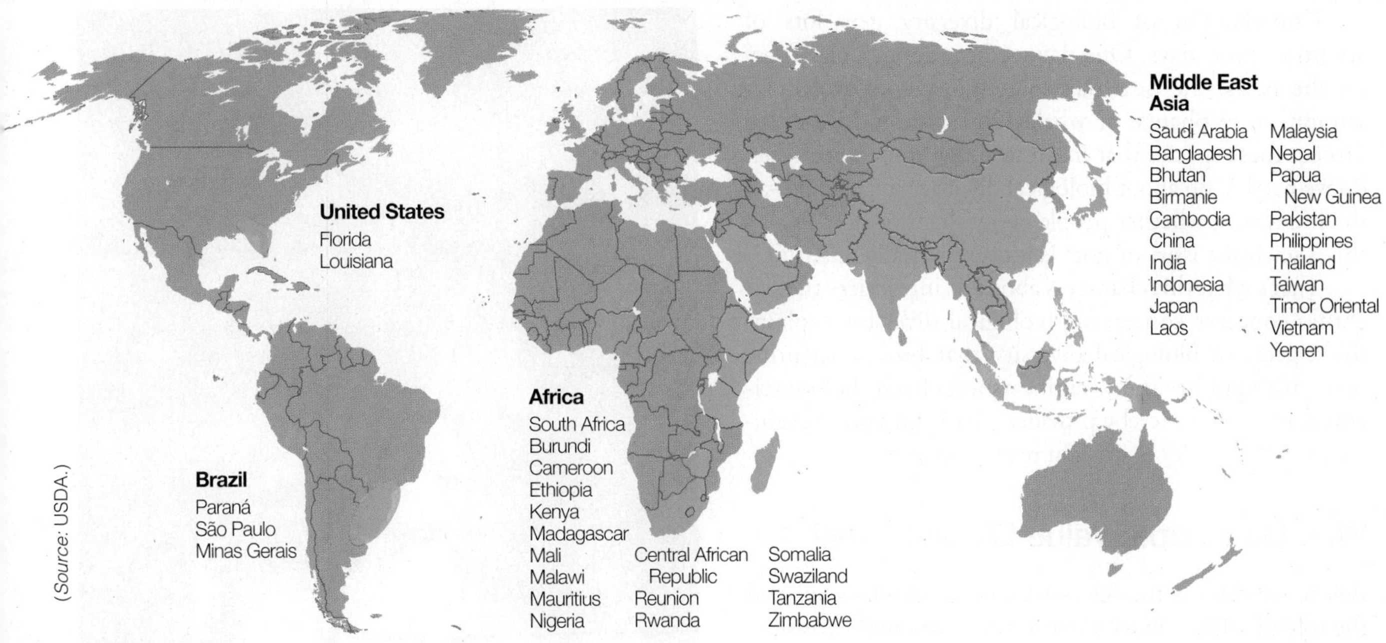

(Source: USDA.)

FIGURE 9.3 **(a) Where citrus greening has spread from its origin in China.** The yellow shows the disease locations.

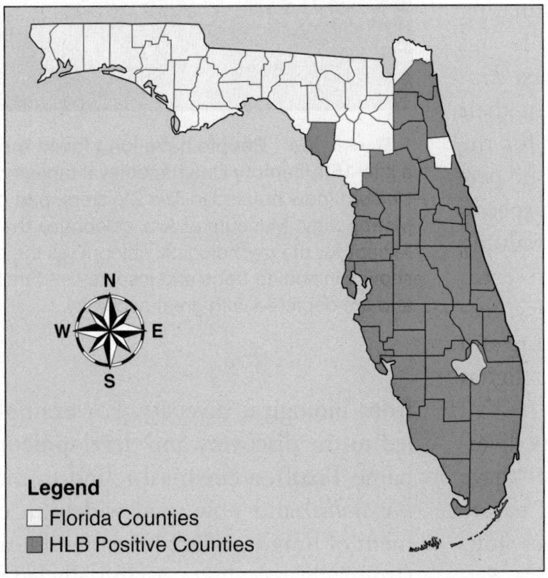

(Source: http://www.freshfromflorida.com/pi/chrp/greening/StatewidePositiveHLBSections.pdf)

(b) Counties in Florida that have tested positive for citrus greening.

9.1 What Is Biological Diversity?

Biological diversity refers to the variety of life-forms, commonly expressed as the number of species or the number of genetic types in an area. (We remind you of the definitions of *population* and *species* in Chapter 5: A **population** is a group of individuals of the same species living in the same area or interbreeding and sharing genetic information. A **species** is all individuals that are capable of interbreeding. A species is made up of populations.)

Conservation of biological diversity gets lots of attention these days. One day we hear about polar bears on the news, the next day something about wolves or salmon, or elephants, or whales. What should we do to protect these species that mean so much to people? What do we need to do about biological diversity in general—all the life-forms, whether people enjoy them or not? And is this a scientific issue or not? Is it even partially scientific?

That's what this chapter is about. It introduces the scientific concepts concerning biological diversity, explains the aspects of biological diversity that have a scientific base, distinguishes the scientific aspects from the nonscientific ones, and thereby provides a basis for you to evaluate the biodiversity issues you read about.

Why Do People Value Biodiversity?

Before we discuss the scientific basis of biodiversity and the role of science in its conservation, we should consider why people value it. There are nine primary reasons: utilitarian; public-service; ecological; moral; theological; aesthetic; recreational; spiritual; and creative.[2]

Utilitarian means that a species or group of species provides a product that is of direct value to people. *Public-service* means that nature and its diversity provide some service, such as taking up carbon dioxide or pollinating flowers, that is essential or valuable to human life and would be expensive or impossible to do ourselves. *Ecological* refers to the fact that species have roles in their ecosystems and that some of these are necessary for the persistence of their ecosystems, perhaps even for the persistence of all life. Scientific research tells us which species have such ecosystem roles. The *moral* reason for valuing biodiversity is the belief that species have a right to exist, independent of their value to people. The *theological* reason refers to the fact that some religions value nature and its diversity, and a person who subscribes to that religion supports this belief.

The last four reasons for valuing nature and its diversity—aesthetic, recreational, spiritual, and creative—have to do with the intangible (nonmaterial) ways that nature and its diversity benefit people (see Figure 9.4). These four are often lumped together, but we separate them here. *Aesthetic* refers to the beauty of nature, including the variety of life. *Recreational* is self-explanatory—people enjoy getting out into nature, not just because it is beautiful to look at but because it provides us with healthful activities that we enjoy. *Spiritual* describes the way contact with nature and its diversity often moves people, an uplifting often perceived as a religious experience. *Creative* refers to the fact that artists, writers, and musicians find stimulation for their creativity in nature and its diversity

Science helps us determine what are utilitarian, public-service, and ecosystem functions of biological diversity, and scientific research can lead to new utilitarian benefits from biological diversity. For example, medical research led to the discovery and development of paclitaxel (trade name Taxol), a chemical found in the Pacific yew (*Taxus brevifolia)* and now used widely in chemotherapy treatment of lung, ovarian, breast, and head and neck cancers. (Ironically, this discovery initially led to the harvest of this endangered tree species, creating an environmental controversy until the compound could be made artificially.)

The rise of the scientific and industrial age brought a great change in the way that people valued nature. Long ago, for example, when travel through mountains was arduous, people struggling to cross them were probably not particularly interested in the scenic vistas. But around the time of the Romantic poets, travel through the Alps became easier, and suddenly poets began to appreciate the "terrible joy" of mountain scenery. Thus scientific knowledge indirectly influences the nonmaterial ways that people value biological diversity.[3]

Rob Crandall/The Image Works

FIGURE 9.4 People have long loved the diversity of life. Here, a late-15th-century Dutch medieval tapestry, *The Hunting of the Unicorn* (now housed in The Cloisters, part of the New York City's Metropolitan Museum of Art), celebrates the great diversity of life. Except for the mythological unicorn, all the plants and animals shown, including frogs and insects, are familiar to naturalists today and are depicted with great accuracy.

9.2 Biological Diversity Basics

Biological diversity involves the following concepts:

- ***Genetic diversity***: the total number of genetic characteristics of a specific species, subspecies, or group of species. In terms of genetic engineering and our new understanding of DNA, this could mean the total base-pair sequences in DNA; the total number of genes, active or not; or the total number of active genes.
- ***Habitat diversity***: the different kinds of habitats in a given unit area.
- ***Species diversity***, which in turn has three qualities:

  species richness—the total number of species;

  species evenness—the relative abundance of species; and

  species dominance—the most abundant species.

To understand the differences between species richness, species evenness, and species dominance, imagine two ecological communities, each with 10 species and 100 individuals, as illustrated in Figure 9.5. In the first community (Figure 9.5a), 82 individuals belong to a single species, and the remaining nine species are represented by two individuals each. In the second community (Figure 9.5b), all the species are equally abundant; each therefore has 10 individuals. Which community is more diverse?

At first, one might think that the two communities have the same species diversity because they have the same number of species. However, if you walked through both communities, the second would appear more diverse. In the first community, most of the time you would see individuals only of the dominant species (elephants in Figure 9.5a); you probably wouldn't see many of the other species at all. The first community would appear to have relatively little diversity until it was subjected to careful study, whereas in the second community even a casual visitor would see many of the species in a short time. You can test the probability of encountering a new species in either community by laying a ruler down in any direction on Figures 9.5a and 9.5b and counting the number of species that it touches.

As this example suggests, merely counting the number of species is not enough to describe biological diversity. Species diversity has to do with the relative chance of seeing species as much as it has to do with the actual number present. Ecologists refer to the total number of species in an area as **species richness**, the relative abundance of species as **species evenness**, and the most abundant species as **dominant**.

The Number of Species on Earth

Many species have come and gone on Earth. But how many exist today? Some 1.5 million species have been named, but available estimates suggest there may be almost 3 million (Table 9.1), and some biologists believe the number will turn out to be much, much larger. No one knows the exact number because new species are discovered all the time, especially in little-explored areas such as tropical savannas and rain forests.[4]

For example, in the spring of 2008, an expedition sponsored by Conservation International and led by scientists from Brazilian universities discovered 14 new species in or near Serra Geraldo Tocantins Ecological Station, a 716,000-hectare (1.77-million-acre) protected area in the Cerrado, a remote tropical savanna region of Brazil, said to be one of the world's most biodiverse areas. They

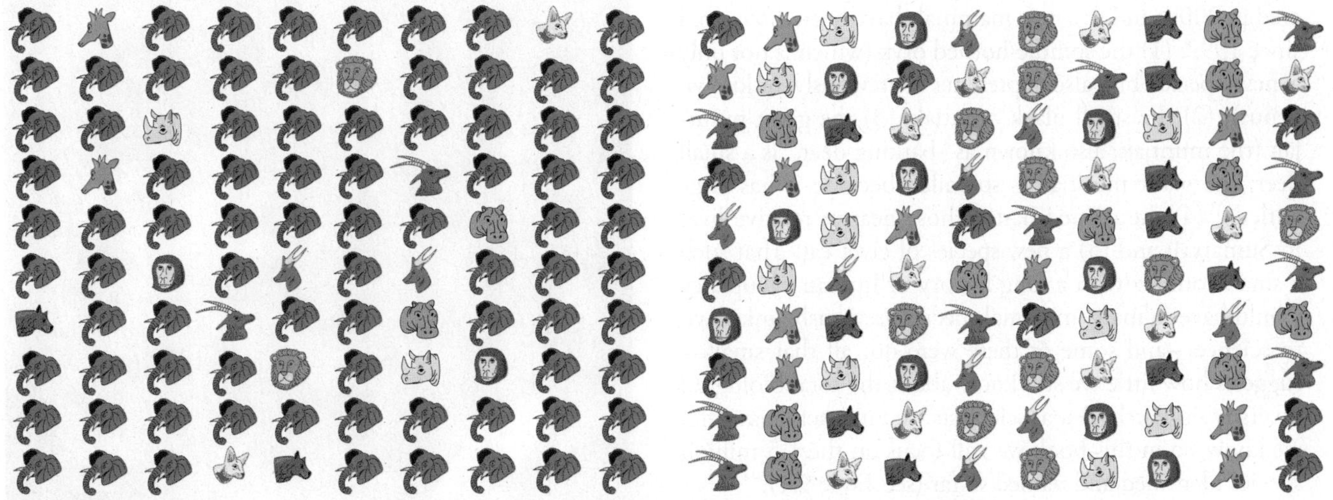

(a) **(b)**

FIGURE 9.5 **Diagram illustrating the difference between species evenness, which is the relative abundance of each species, and species richness, which is the total number of species.** Figures **(a)** and **(b)** have the same number of species but different relative abundances. Lay a ruler across each diagram and count the number of species the edge crosses. Do this several times, and determine how many species are in diagram **(a)** and diagram **(b)**. See text for explanation of results.

Table 9.1 NUMBER OF SPECIES BY MAJOR FORMS OF LIFE AND BY NUMBER OF ANIMAL SPECIES (FOR A DETAILED LIST OF SPECIES BY TAXONOMIC GROUP, SEE APPENDIX.)

A. NUMBER OF SPECIES BY MAJOR FORMS OF LIFE

		ESTIMATED NUMBER	
LIFE-FORM	EXAMPLE	MINIMUM	MAXIMUM
Monera/Bacteria	Bacteria	4,800	10,000
Fungi	Yeast	71,760	116,260
Lichens	Old man's beard	13,500	13,500
Prostista/Protoctista	Ameba	80,710	194,760
Plantae	Maple tree	478,365	529,705
Animalia	Honeybee	873,084	1,870,019
Total		1,522,219	2,734,244

B. NUMBER OF ANIMAL SPECIES

		ESTIMATED NUMBER	
ANIMALS	EXAMPLE	MINIMUM	MAXIMUM
Insecta	Honeybees	668,050	1,060,550
Chondrichthyes	Sharks, rays, etc.	750	850
Osteichthyes	Bony fish	20,000	30,000
Amphibia	Amphibians	200	4,800
Reptilia	Reptiles	5,000	7,000c
Aves	Birds	8,600	9,000
Mammalia	Mammals	4,000	5,000
Animal total	Total	873,084	1,870,019

found eight new fish, three new reptiles, one new amphibian, one new mammal, and one new bird.

In Laos, a new bird, the barefaced bulbul, was discovered in 2009, and five new mammals have been discovered since 1992: (1) the spindle-horned oryx (which is not only a new species but also represents a previously unknown genus); (2) the small black muntjak; (3) the giant muntjak (the muntjak, also known as "barking deer," is a small deer; the giant muntjak is so called because it has large antlers); (4) the striped hare (whose nearest relative lives in Sumatra); and (5) a new species of civet cat. That such a small country with a long history of human occupancy would have so many mammal species previously unknown to science—and some of these were not all that small—suggests how little we still know about the total biological diversity on Earth. But as scientists we must act from what we know, so in this book we will focus on the 1.5 million species identified and named so far (see Table 9.1).

Discovery of new species continues. In 2012, a previously unknown-to-science species of monkey, *Cercopithecus lomamiensis,* was discovered in the Democratic Republic of Congo (see Figure 9.6).[5]

HANDOUT/Reuters/Newscom

FIGURE 9.6 **New species of monkey discovered.** Here are two babies of *Cercopithecus lomamiensis*, a previously unknown-to-science species of monkey, discovered in the Democratic Republic of Congo in 2012.

All living organisms are classified into groups called *taxa,* usually on the basis of their evolutionary relationships or similarity of characteristics. (Carl Linnaeus, a Swedish physician and biologist, who lived from 1707 to 1778, was the originator of the classification system and played a crucial role in working all this out. He explained this system in his book *Systema Naturae.*)

The hierarchy of these groups (from largest and most inclusive to smallest and least inclusive) begins with a domain or kingdom. In the recent past, scientists classified life into five kingdoms: animals, plants, fungi, protists, and bacteria. Recent evidence from the fossil record and studies in molecular biology suggest that it may be more appropriate to describe life as existing in three major domains, one called Eukaryota or Eukarya, which includes animals, plants, fungi, and protists (mostly single-celled organisms); Bacteria; and Archaea.[4] As you learned in Chapter 6, Eukarya cells include a nucleus and other small, organized features called organelles; Bacteria and Archaea do not. (Archaea used to be classified among Bacteria, but they have substantial molecular differences that suggest ancient divergence in heritage—see Chapter 7, Figure 7.4.)

The plant kingdom is made up of divisions, whereas the animal kingdom is made up of phyla (singular: phylum). A phylum or division is, in turn, made up of classes, which are made up of orders, which are made up of families, which are made up of genera (singular: genus), which are made up of species.

Some argue that the most important thing about biological diversity is the total number of species and that the primary goal of biological conservation should be to maintain that number at its current known maximum. An interesting and important point to take away from Table 9.1 is that most of the species on Earth are insects (somewhere between 668,000 and more than 1 million) and plants (somewhere between 480,000 and 530,000), and also that there are many species of fungi (about 100,000) and protists (about 80,000 to almost 200,000). In contrast, our own kind, the kind of animals most celebrated on television and in movies—mammals—number a meager 4,000 to 5,000, about the same as reptiles. When it comes to numbers of species on Earth, our kind doesn't seem to matter much—we amount to about half a percent of all animals. If the total number in a species were the only gauge of a species' importance, we wouldn't matter.

9.3 Biological Evolution

The first big question about biological diversity is: How did it all come about? Before modern science, the diversity of life and the adaptations of living things to their environment seemed too amazing to have come about by chance. The great Roman philosopher and writer Cicero put it succinctly: "Who cannot wonder at this harmony of things, at this symphony of nature which seems to will the well-being of the world?" He concluded that "everything in the world is marvelously ordered by divine providence and wisdom for the safety and protection of us all."[6] The only possible explanation seemed to be that this diversity was created by God (or gods).

With the rise of modern science, however, other explanations became possible. In the 19th century, Charles Darwin found an explanation that became known as biological evolution. **Biological evolution** refers to the change in inherited characteristics of a population from generation to generation. It can result in new species—populations that can no longer reproduce with members of the original species but can (and at least occasionally do) reproduce with each other. Along with self-reproduction, biological evolution is one of the features that distinguish life from everything else in the universe. (The others are carbon-based, organic-compound-based, self-replicating systems.)

The word *evolution* in the term *biological evolution* has a special meaning. Outside biology, *evolution* is used broadly to mean the history and development of something. For example, book reviewers talk about the evolution of a novel's plot, meaning how the story unfolds. Geologists talk about the evolution of Earth, which simply means Earth's history and the geologic changes that have occurred over that history. Within biology, however, the term *biological evolution* means a one-way process: Once a species is extinct, it is gone forever. You can run a machine, such as a mechanical grandfather clock, forward and backward, but when a new species evolves, it cannot evolve backward into its parents.

Our understanding of evolution today owes a lot to the modern science of molecular biology and the practice of genetic engineering, which are creating a revolution in how we think about and deal with species. At present, scientists have at hand the complete DNA code for a number of species, and the list is growing rapidly. In 2012 the complete DNA genome had been determined for more than 20 species, including disease organisms: bacterium *Haemophilus influenzae*; the malaria parasite; its carrier the malaria mosquito; a fungi, baker's yeast (*Saccharomyces cerevisiae*); plants: thale cress (*Arabidopsis thaliana*); wild mustard (*Arabidopsis thalian*); rice (*Oryza sativa*); animals: the fruit fly (*Drosophila*); a nematode *C. elegans* (a very small worm that lives in water); the puffer fish (*Takifugu rubripes*); a tree, the black poplar (*Populus trichocarpa*), and ourselves—humans.[7,8] Scientists focused on these species either because they are of great interest to us or because they are relatively easy to study, having either few base pairs (the nematode worm) or having already well-known genetic characteristics (the fruit fly).

According to the theory of biological evolution, new species arise as a result of competition for resources and the differences among individuals in their adaptations to environmental conditions. Since the environment continually changes, which individuals are best adapted changes over time and space. As Darwin wrote, "Can it be doubted, from the struggle each individual has to obtain subsistence, that any minute variation in structure, habits, or instincts, adapting that individual better to the new [environmental] conditions, would tell upon its vigor and health? In the struggle it would have a better chance of surviving; and those of its offspring that inherited the variation, be it ever so slight, would also have a better chance."

Sounds plausible, but how does this evolution occur? Through four processes: mutation, natural selection, migration, and genetic drift.

The Four Key Processes of Biological Evolution

Mutation

Mutations are changes in genes. Contained in the chromosomes within cells, each **gene** carries a single piece of inherited information from one generation to the next, producing a **genotype**, the genetic makeup that is characteristic of an individual or a group.

Genes are made up of a complex chemical compound called deoxyribonucleic acid (DNA). DNA in turn is made up of chemical building blocks that form a code, a kind of alphabet of information. The DNA alphabet consists of four letters that stand for specific nitrogen-containing compounds, called bases, which are combined in pairs: (A) adenine, (C) cytosine, (G) guanine, and (T) thymine. Each gene has a set of the four base pairs, and how these letters are combined in long strands determines the genetic "message" interpreted by a cell to produce specific compounds.

The number of base pairs that make up a strand of DNA varies. To make matters more complex, some base pairs found in DNA seem to be nonfunctional—they are not active and do not determine any chemicals produced by the cell. Some of these have been discovered to have other functions, such as affecting the activity of others, turning those other genes on or off. And creatures such as ourselves have genes that limit the number of times a cell can divide, and thus determine the individual's maximum longevity.

When a cell divides, the DNA is reproduced and each new cell gets a copy. But sometimes an error in reproduction changes the DNA and thereby changes the inherited characteristics. Such errors can arise from various causes. Sometimes an external agent comes in contact with DNA and alters it. Radiation, such as X rays and gamma rays, can break the DNA apart or change its chemical structure. Certain chemicals also can change DNA. So can viruses. When DNA changes in any of these ways, it is said to have undergone **mutation**.

In some cases, a cell or an offspring with a mutation cannot survive (Figure 9.7a and b). In other cases, the mutation simply adds variability to the inherited characteristics (Figure 9.7c). But in still other cases, individuals with mutations are so different from their parents that they cannot reproduce with normal offspring of their species, so a new species has been created.

Natural Selection

When there is variation within a species, some individuals may be better suited to the environment than others (see A Closer Look 9.1). (Change is not always for the better. Mutation can result in a new species whether or not that species is better adapted than its parent species to the environment.) Organisms whose biological characteristics make them better able to survive and reproduce in their environment leave more offspring than others. Their descendants form a larger proportion of the next

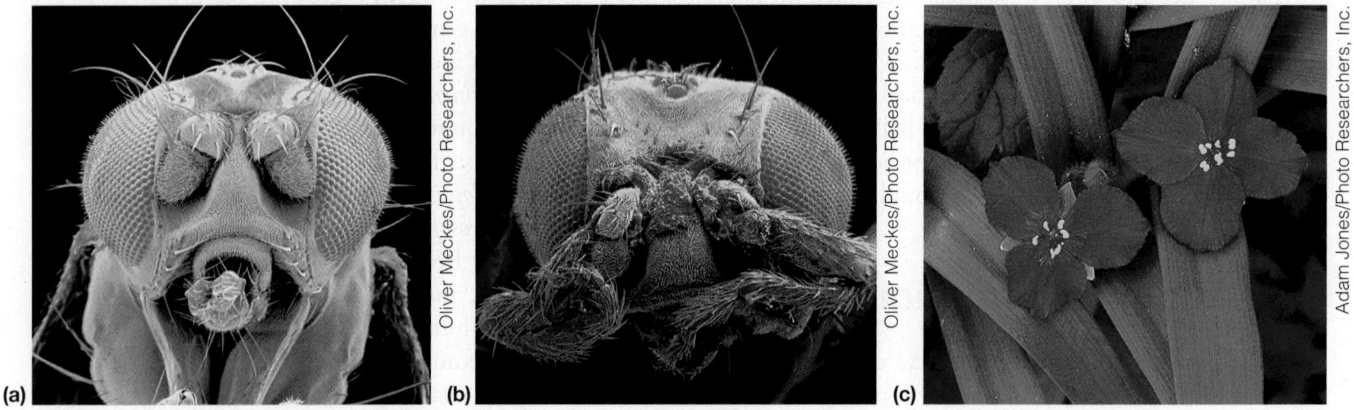

FIGURE 9.7 **(a) A normal fruit fly, (b) a fruit fly with an antennae mutation, and (c) Tradescantia, a small flowering plant used in the study of effects of mutagens.** The color of stamen hairs in the flower (pink versus clear) is the result of a single gene and changes when that gene is mutated by radiation or certain chemicals, such as ethylene chloride.

A CLOSER LOOK 9.1

Natural Selection: Mosquitoes and the Malaria Parasite

Malaria poses a great threat to 2.4 billion people—over one-third of the world's population—living in more than 90 countries, most of them in the tropics. In the United States, in 2003, Palm Beach County, Florida, experienced a small but serious malaria outbreak, and of particular concern is that the malaria was the result of bites from local mosquitoes, not those brought in by travelers from nations where malaria is a continual problem. Worldwide, an estimated 300–400 million people are infected each year, and 1.1 million of them die (Figure 9.8) [a] It is the fourth largest cause of death of children in developing nations—in Africa alone, more than 3,000 children die daily from this disease.[b] Once thought to be caused by filth or polluted air (hence the name *malaria*, from the Latin for "bad air"), malaria is actually caused by parasitic microbes (four species of the protozoon *Plasmodium*). These microbes affect and are carried by *Anopheles* mosquitoes, which then transfer the protozoa to people. One solution to the malaria problem, then, would be the eradication of *Anopheles* mosquitoes.

By the end of World War II, scientists had discovered that the pesticide DDT was extremely effective against *Anopheles* mosquitoes. They had also found chloroquine highly effective in killing *Plasmodium* parasites. (Chloroquine is an artificial derivative of quinine, a chemical from the bark of the quinine tree that was an early treatment for malaria.) In 1957 the World Health Organization (WHO) began a $6 billion campaign to rid the world of malaria using a combination of DDT and chloroquine.

At first, the strategy seemed successful. By the mid-1960s, malaria was nearly gone or had been eliminated from 80% of the target areas. However, success was short-lived. The mosquitoes began to develop a resistance to DDT, and the protozoa became resistant to chloroquine. In many tropical areas, the incidence of malaria worsened. For example, the WHO program had reduced the number of cases in Sri Lanka from 1 million to only 17 by 1963, but by 1975 600,000 cases had been reported, and the actual number is believed to be four times

Karel Prinsloo/AP Photo

(a)

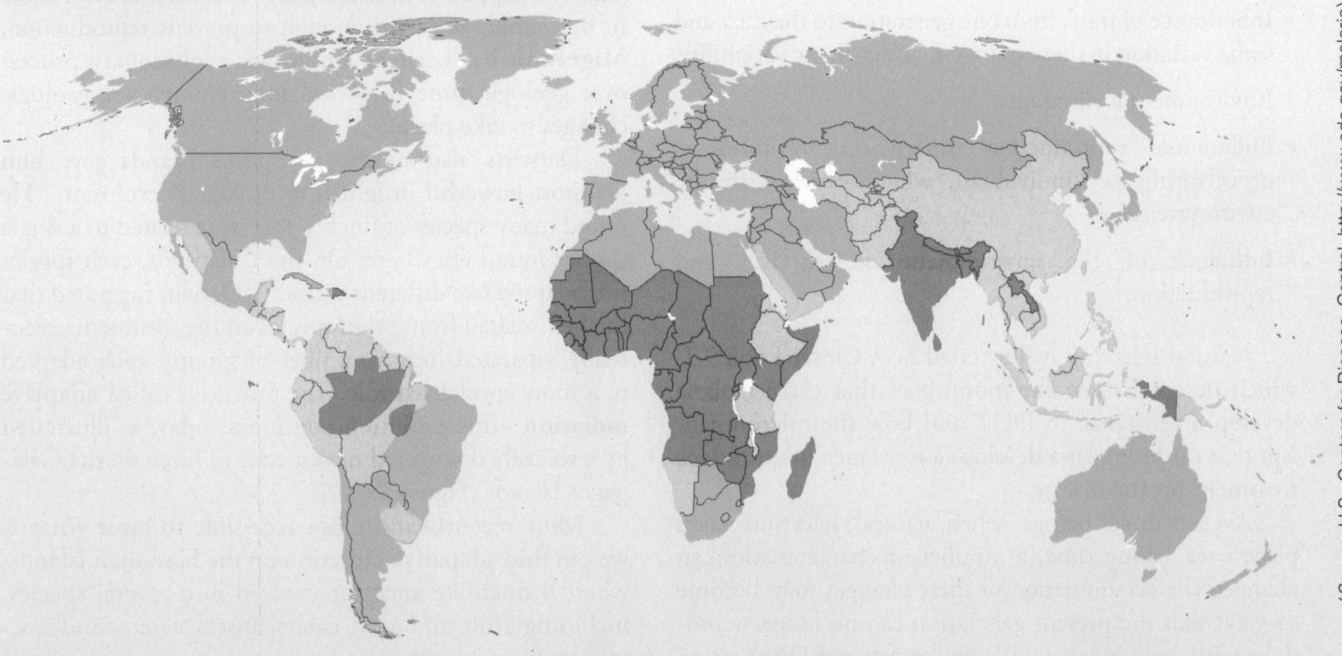

(b)

(*Source:* U.S. Centers for Disease Control. http://www.cdc.gov/malaria)

FIGURE 9.8 **(a)** A child with malaria and **(b)** where malaria primarily occurs today.

higher. Worldwide, in 2006 (most recent data available) there were 247 million cases of malaria, resulting in 881,000 deaths. The mosquitoes' resistance to DDT became widespread, and resistance of the protozoa to chloroquine was found in 80% of the 92 countries where malaria was a major killer.[c]

The mosquitoes and the protozoa developed this resistance through natural selection. When they were exposed to DDT and chloroquine, the susceptible individuals died; they left few or no offspring, and any offspring they left were susceptible. The most resistant survived and passed their resistant genes on to their offspring. Thus, a change in the environment—the human introduction of DDT and chloroquine—caused a particular genotype to become dominant in the populations.

A practical lesson from this experience is that if we set out to eliminate a disease-causing species, we must attack it completely at the outset and destroy all the individuals before natural selection leads to resistance. But sometimes this is impossible, in part because of the natural genetic variation in the target species. Since the drug chloroquine is generally ineffective now, new drugs have been developed to treat malaria. However, these second- and third-line drugs will eventually become unsuccessful, too, as a result of the same process of biological evolution by natural selection. This process is speeded up by the ability of the *Plasmodium* to rapidly mutate. In South Africa, for example, the protozoa became resistant to mefloquine immediately after the drug became available as a treatment.

An alternative is to develop a vaccine against the *Plasmodium* protozoa. Biotechnology has made it possible to map the genetic structure of these malaria-causing organisms. Scientists are currently mapping the genetic structure of *P. falciparum*, the most deadly of the malaria protozoa, and expect to finish within several years. With this information, they expect to create a vaccine containing a variety of the species that is benign in human beings but produces an immune reaction.[d] In addition, scientists are mapping the genetic structure of *Anopheles gambiae*, the carrier mosquito. This project could provide insight into genes, which could prevent development of the malaria parasite within the mosquito. In addition, it could identify genes associated with insecticide resistance and provide clues to developing a new pesticide.

generation and are more "fit" for the environment. This process of increasing the proportion of offspring is called **natural selection**. Which inherited characteristics lead to more offspring depends on the specific characteristics of an environment, and as the environment changes over time, the characteristics' "fit" will also change. In summary, natural selection involves four primary factors:

- Inheritance of traits from one generation to the next and some variation in these traits—that is, genetic variability.
- Environmental variability.
- Differential reproduction (differences in numbers of offspring per individual), which varies with the environment.
- Influence of the environment on survival and reproduction.

Natural selection is illustrated in A Closer Look 9.1, which describes how the mosquitoes that carry malaria develop a resistance to DDT and how the microorganism that causes malaria develops a resistance to quinine, a treatment for the disease.

As explained before, when natural selection takes place over a long time, a number of characteristics can change. The accumulation of these changes may become so great that the present generation can no longer reproduce with individuals that have the original DNA structure, resulting in a new species.

Migration and Geographic Isolation

Sometimes two populations of the same species become geographically isolated from each other for a long time. During that time, the two populations may change so much that they can no longer reproduce together even when they are brought back into contact. In this case, two new species have evolved from the original species. This can happen even if the genetic changes are not more fit but simply different enough to prevent reproduction. **Migration** has been an important evolutionary process over geologic time (a period long enough for geologic changes to take place).

Darwin's visit to the Galápagos Islands gave him his most powerful insight into biological evolution.[9] He found many species of finches that were related to a single species found elsewhere. On the Galápagos, each species was adapted to a different niche.[9,10] Darwin suggested that finches isolated from other species on the continents eventually separated into a number of groups, each adapted to a more specialized role. The process is called **adaptive radiation**. This evolution continues today, as illustrated by a recently discovered new species of finch on the Galápagos Islands (Figure 9.9).

More recently and more accessible to most visitors, we can find adapative radiation on the Hawaiian Islands, where a finchlike ancestor evolved into several species, including fruit and seed eaters, insect eaters, and nectar eaters, each with a beak adapted for its specific food (Figure 9.10).[10,11,]

All Canada Photos/Alamy

FIGURE 9.9 One of Darwin's finches. Charles Darwin's observations of the adaptive radiation of finches on the Galápagos Islands was a key to development of the theory of biological evolution. This evolution continues today. In 1982 a finch new to the islands, a large ground finch, migrated there. Over the years since then, the beak of a smaller, native ground species of finch has been evolving to become larger, apparently in response to competition from the new arrival. Today, the offspring of the native ground finch have on average longer beaks than their ancestors. This is said to be the first time such a response has been observed in progress on the islands.

Ironically, the *loss* of geographic isolation can also lead to a new species. This can happen when one population of a species migrates into a habitat already occupied by another population of that species, thereby changing gene frequency in that habitat. Such a change can result, for example, from the migration of seeds of flowering plants blown by wind or carried in the fur of mammals. If the seed lands in a new habitat, the environment may be different enough to favor genotypes that are not as favored by natural selection in the parents' habitat. Natural selection, in combination with geographic isolation and subsequent migration, can thus lead to new dominant genotypes and eventually to new species.

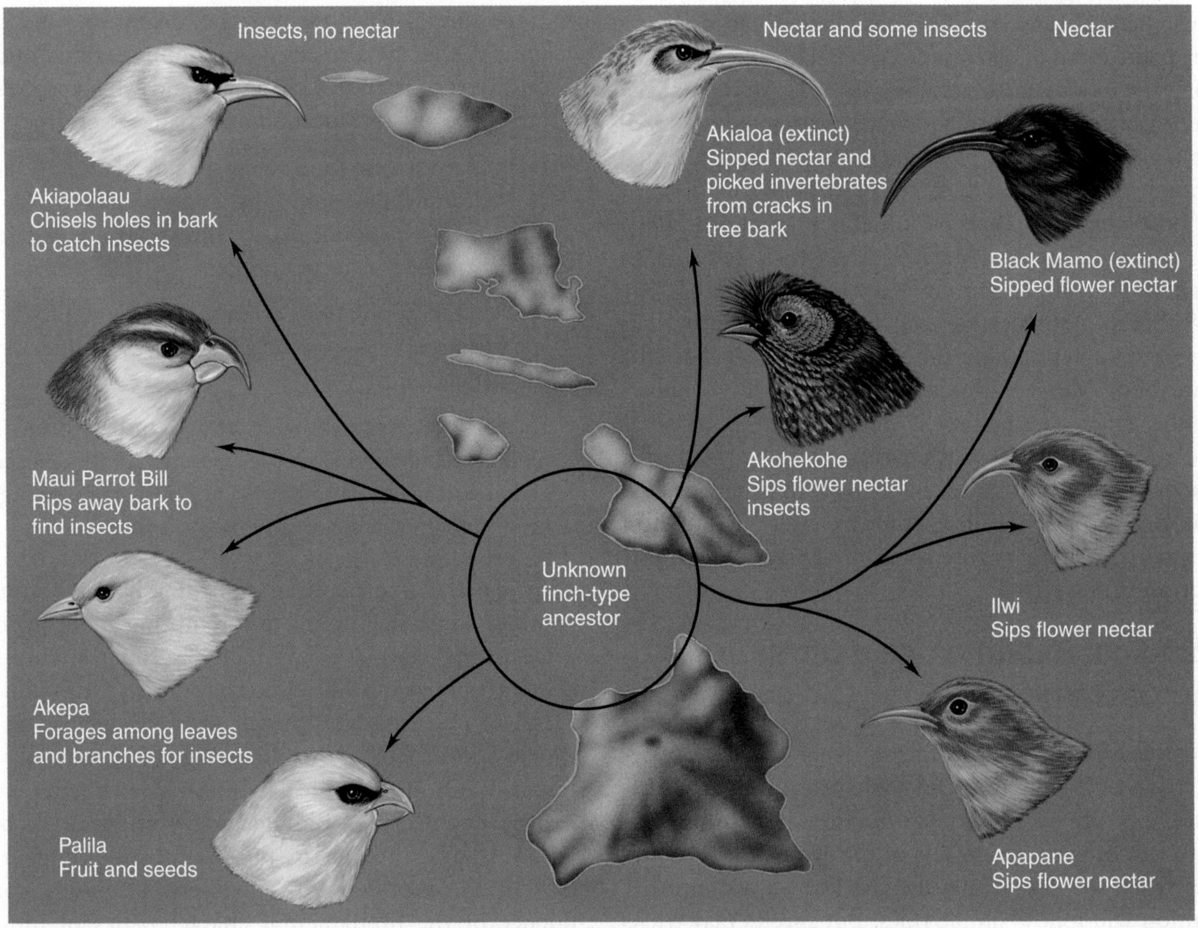

Source: From C.B. Cox, I.N. Healey, and P.D. Moore, *Biogeography.* New York: Halsted, 1973)

FIGURE 9.10 Evolutionary divergence among honeycreepers in Hawaii. Sixteen species of birds, each with a beak specialized for its food, evolved from a single ancestor. Nine of the species are shown here. The species evolved to fit ecological niches that, on the North American continent, had been filled by other species not closely related to the ancestor.

Genetic Drift

Genetic drift refers to changes in the frequency of a gene in a population due not to mutation, selection, or migration, but simply to chance. One way this happens is through the **founder effect**. The founder effect occurs when a small number of individuals are isolated from a larger population; they may have much less genetic variation than the original species (and usually do), and the characteristics that the isolated population has will be affected by chance. In the founder effect and genetic drift, individuals may not be better adapted to the environment—in fact, they may be more poorly adapted or neutrally adapted. Genetic drift can occur in any small population and may present conservation problems when it is by chance isolated from the main population.

For example, bighorn sheep live in the mountains of the southwestern deserts of the United States and Mexico. In the summer, these sheep feed high up in the mountains, where it is cooler, wetter, and greener. Before high-density European settlement of the region, the sheep could move freely and sometimes migrated from one mountain to another by descending into the valleys and crossing them in the winter. In this way, large numbers of sheep interbred. With the development of cattle ranches and other human activities, many populations of bighorn sheep could no longer migrate among the mountains by crossing the valleys. These sheep became isolated in very small groups—commonly, a dozen or so—and chance may play a large role in what inherited characteristics remain in the population.

This happened to a population of bighorn sheep on Tiburón Island in Mexico, which was reduced to 20 animals in 1975 but increased greatly to 650 by 1999. Because of the large recovery, this population has been used to repopulate other bighorn sheep habitats in northern Mexico. But a study of the DNA shows that the genetic variability is much less than in other populations in Arizona. Scientists who studied this population suggest that individuals from other isolated bighorn sheep populations should be added to any new transplants to help restore some of the greater genetic variation of the past.[12]

Biological Evolution as a Strange Kind of Game

Biological evolution is so different from other processes that it is worthwhile to spend some extra time exploring the topic. There are no simple rules that species must follow to win or even just to stay in the game of life. Sometimes when we try to manage species, we assume that evolution will follow simple rules. But species play tricks on us; they adapt or fail to adapt over time in ways that we did not anticipate. Such unexpected outcomes result from our failure to fully understand how species have evolved in relation to their ecological situations. Nevertheless, we continue to hope and plan as if life and its environment will follow simple rules. This is true even for the most recent work in genetic engineering.

Complexity is a feature of evolution. Species have evolved many intricate and amazing adaptations that have allowed them to persist. It is essential to realize that these adaptations have evolved not in isolation but in the context of relationships to other organisms and to the environment. The environment sets up a situation within which evolution, by natural selection, takes place. The great ecologist G.E. Hutchinson referred to this interaction in the title of one of his books, *The Ecological Theater and the Evolutionary Play*. Here, the ecological situation—the condition of the environment and other species—is the theater and the scenery within which natural selection occurs, and natural selection results in a story of evolution played out in that theater—over the history of life on Earth.[13] These features of evolution are another reason that life and ecosystems are not simple, linear, steady-state systems (see Chapter 4).

In summary, the theory of biological evolution tells us the following about biodiversity:

- Since species have evolved and do evolve, and since some species are also always becoming extinct, biological diversity is always changing, and which species are present in any one location can change over time.
- Adaptation has no rigid rules; species adapt in response to environmental conditions, and complexity is a part of nature. We cannot expect threats to one species to necessarily be threats to another.
- Species and populations do become geographically isolated from time to time, and undergo the founder effect and genetic drift.
- Species are always evolving and adapting to environmental change. One way they get into trouble—become endangered—is when they do not evolve fast enough to keep up with the environment.

Now that the manipulation of DNA is becoming routine, some scientists are attempting to bring back once extinct species. This includes some attempts to develop something like dinosaurs by starting with the DNA of chickens, very much as was imagined in the movie *Jurassic Park*.[14] But this process will be a very different one from biological evolution; it is better called reverse engineering. Biological evolution takes place without any conscious or purposeful intention on the part of the species or its members. It's just a consequence of the fundamental characteristics of life and its environment. Attempts to reconstruct a dinosaur are conscious and purposeful actions. This distinction between biological evolution and consciousness of reverse engineering is useful as a way to help understand both.

9.4 Competition and Ecological Niches

Why there are so many species on Earth has become a key question since the rise of modern ecological and evolutionary sciences. In the next sections we discuss the answers. They partly have to do with how species interact. Speaking most generally, they interact in three ways: competition, in which the outcome is negative for both; symbiosis, in which the interaction benefits both participants; and predation–parasitism, in which the outcome benefits one and is detrimental to the other.

The Competitive Exclusion Principle

The **competitive exclusion principle** supports those who argue that there should be only a few species. It states that *two species with exactly the same requirements cannot coexist in exactly the same habitat.* Garrett Hardin expressed the idea most succinctly: "Complete competitors cannot coexist."[15]

This principle is illustrated by the introduction of the American gray squirrel into Great Britain. It was introduced intentionally because some people thought it was attractive and would be a pleasant addition to the landscape. About a dozen attempts were made, the first perhaps as early as 1830 (Figure 9.11). By the 1920s, the American gray squirrel was well established in Great Britain, and in the 1940s and 1950s its numbers expanded greatly. It competes with the native red squirrel and is winning—there are now about 2.5 million gray squirrels in Great Britain, and only 140,000 red squirrels, most them in Scotland, where the gray squirrel is less abundant.[16] The two species have almost exactly the same habitat requirements.

One reason for the shift in the balance of these species may be that in the winter the main source of food for red squirrels is hazelnuts, while gray squirrels prefer acorns. Thus, red squirrels have a competitive advantage in areas with hazelnuts, and gray squirrels have the advantage in oak forests. When gray squirrels were introduced, oaks were the dominant mature trees in Great Britain; about 40% of the trees planted were oaks. But that is not the case today. This difference in food preference may allow the coexistence of the two, or perhaps not.

The competitive exclusion principle suggests that there should be very few species. We know from our discussions of ecosystems (Chapter 6) that food webs have at least four levels—producers, herbivores, carnivores, and decomposers. Suppose we allowed for several more levels of carnivores, so that the average food web had six levels. Since there are about 20 major kinds of ecosystems, one would guess that the total number of winners on Earth would be only 6 × 20, or 120 species.

Being a little more realistic, we could take into account adaptations to major differences in climate and other environmental aspects within kinds of ecosystems. Perhaps we could specify 100 environmental categories: cold and dry; cold and wet; warm and dry; warm and wet; and so forth. Even so, we would expect that within each environmental category, competitive exclusion would result in the survival of only a few species. Allowing six species per major environmental category would result in only 600 species.

That just isn't the case. How did so many different species survive, and how do so many coexist? Part of the answer lies in the different ways in which organisms interact, and part of the answer lies with the idea of the ecological niche.

David Tipling/Getty Images, Inc.

Photo by D. B. Botkin

(a) (b)

FIGURE 9.11 **(a) British red squirrel, which is being outcompeted by the (b) American gray squirrel introduced into Great Britain.**

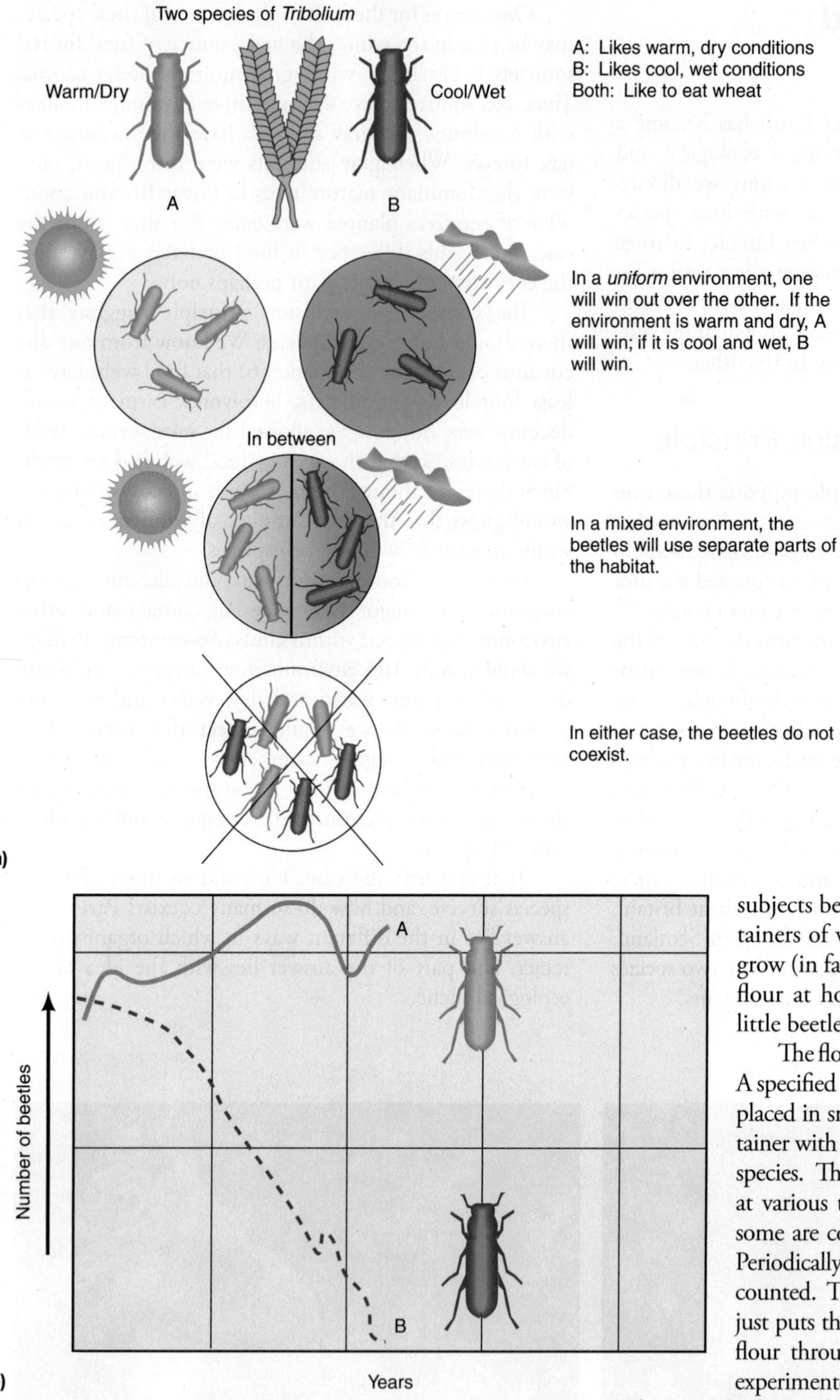
Two species of Tribolium
A: Likes warm, dry conditions
B: Likes cool, wet conditions
Both: Like to eat wheat
Warm/Dry
Cool/Wet
A
B
In a uniform environment, one will win out over the other. If the environment is warm and dry, A will win; if it is cool and wet, B will win.
In between
In a mixed environment, the beetles will use separate parts of the habitat.
In either case, the beetles do not coexist.
(a)
A
B
Number of beetles
Years
(b)

FIGURE 9.12 A classical experiment with flour beetles. Two species of flour beetles are placed in small containers of flour. Each container is kept at a specified temperature and humidity. Periodically, the flour is sifted, and the beetles are counted and then returned to their containers. Which species persists is observed and recorded. **(a)** The general process illustrating competitive exclusion in these species; **(b)** results of a specific, typical experiment under warm, dry conditions.

Niches: How Species Coexist

The **ecological niche** concept explains how so many species can coexist, and this concept is introduced most easily by experiments done with a small, common insect—the flour beetle (*Tribolium*), which, as its name suggests, lives on wheat flour. Flour beetles make good experimental subjects because they require only small containers of wheat flour to live and are easy to grow (in fact, too easy; if you don't store your flour at home properly, you will find these little beetles happily eating in it).

The flour beetle experiments work like this: A specified number of beetles of two species are placed in small containers of flour—each container with the same number of beetles of each species. The containers are then maintained at various temperature and moisture levels—some are cool and wet, others warm and dry. Periodically, the beetles in each container are counted. This is very easy. The experimenter just puts the flour through a sieve that lets the flour through but not the beetles. Then the experimenter counts the number of beetles of each species and puts the beetles back in their container to eat, grow, and reproduce for another interval. Eventually, one species always wins—some of its individuals continue to live in the container while the other species goes extinct. So far, it would seem that there should be only one species of *Tribolium*. But which species survives depends on temperature and moisture. One species does better when it is cold and wet, the other when it is warm and dry (Figure 9.12).

Curiously, when conditions are in between, sometimes one species wins and sometimes the other, seemingly randomly; but invariably one persists while the second becomes extinct. So the competitive exclusion principle holds for these beetles. Both species can survive in a complex environment—one that has cold and wet habitats as well as warm and dry habitats. In no location, however, do the species coexist.

The little beetles provide us with the key to the coexistence of many species. Species that require the same resources can coexist by using those resources under different environmental conditions. So it is habitat *complexity* that allows complete competitors—and not-so-complete competitors—to coexist because they avoid competing with each other.[17]

The flour beetles are said to have the same ecologically functional *niche,* which means they have the same *profession*—eating flour. But they have different *habitats.* Where a species lives is its habitat, but what it does for a living (its profession) is its ecological niche. Suppose you have a neighbor who drives a school bus. Where your neighbor lives and works—your town—is his habitat. What your neighbor does—drive a bus—is his niche. Similarly, if someone says, "Here comes a wolf," you think not only of a creature that inhabits the northern forests (its habitat) but also of a predator that feeds on large mammals (its niche).

Understanding the niche of a species is useful in assessing the impact of land development or changes in land use. Will the change remove an essential requirement for some species' niche? A new highway that makes car travel easier might eliminate your neighbor's bus route (an essential part of his habitat) and thereby eliminate his profession (or niche). Other things could also eliminate this niche. Suppose a new school were built and all the children could now walk to school. A school bus driver would not be needed; this niche would no longer exist in your town. In the same way, cutting a forest may drive away prey and eliminate the wolf's niche.

Measuring Niches

An ecological niche is often described and measured as the set of all environmental conditions under which a species can persist and carry out its life functions. It is illustrated by the distribution of two species of flatworm that live on the bottom of freshwater streams. A study of two species of these small worms in Great Britain found that some streams contained one species, some the other, and still others both.[18]

The stream waters are cold at their source in the mountains and become progressively warmer as they flow downstream. Each species of flatworm occurs within a specific range of water temperatures. In streams where species A occurs alone, it is found from 6° to 17°C (42.8°–62.6°F) (Figure 9.13a). Where species B occurs alone, it is found from 6° to 23°C (42.8°–73.4°F) (Figure 9.13b). When they occur in the same stream, their temperature ranges are much narrower. Species A lives in the upstream sections, where the temperature ranges from 6° to 14°C (42.8°–57.2°F), and species B lives in the warmer downstream areas, where temperatures range from 14° to 23°C (57.2°–73.4°F) (Figure 9.13c).

The temperature range in which species A occurs when it has no competition from B is called its *fundamental temperature niche.* The set of conditions under which it persists in the presence of B is called its *realized temperature niche.* The flatworms show that species divide up their habitat so that they use resources from different parts of it. Of course, temperature is only one aspect of the environment. Flatworms also have requirements relating to the acidity of the water and other factors. We could create graphs for each of these factors, showing the range within which A and B occurred. The collection of all those graphs would constitute the complete Hutchinsonian description of the niche of a species.

A Practical Implication

From the discussion of the competitive exclusion principle and the ecological niche, we learn something

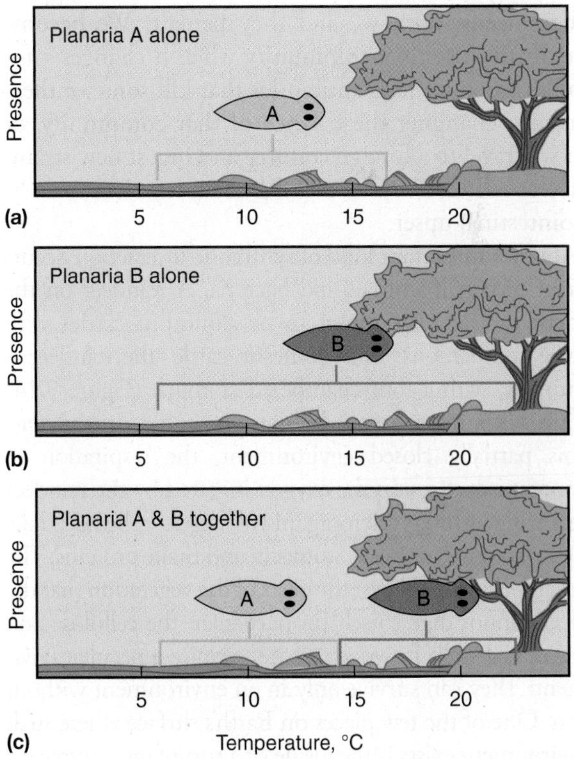

FIGURE 9.13 Fundamental and realized niches. The occurrence of freshwater flatworms in cold mountain streams in Great Britain. **(a)** The presence of species A in relation to temperature in streams where it occurs alone. **(b)** The presence of species B in relation to temperature in streams where it occurs alone. **(c)** The temperature range of both species in streams where they occur together. Inspect the three graphs: What is the effect of each species on the other?

important about the conservation of species: If we want to conserve a species in its native habitat, we must make sure that all the requirements of its niche are present. Conservation of endangered species is more than a matter of putting many individuals of that species into an area. All the life requirements for that species must also be present—we have to conserve not only a population but also its habitat and its niche.

9.5 Symbiosis

Our discussion up to this point might leave the impression that species interact mainly through competition—by interfering with one another. But symbiosis is also important. This term is derived from a Greek word meaning "living together." In ecology, **symbiosis** describes a relationship between two organisms that is beneficial to both and enhances each organism's chances of persisting. Each partner in symbiosis is called a **symbiont**.

Symbiosis is widespread and common; most animals and plants have symbiotic relationships with other species. We, too, have symbionts—microbiologists tell us that about 10% of our body weight is actually the weight of symbiotic microorganisms that live in our intestines. They help our digestion, and we provide a habitat that supplies all their needs; both we and they benefit. We become aware of this intestinal community when it changes—for example, when we take antibiotics that kill some of these organisms, changing the balance of that community, or when we travel to a foreign country and ingest new strains of bacteria. Then we suffer a well-known traveler's malady, gastrointestinal upset.

Another important kind of symbiotic interaction occurs between certain mammals and bacteria. A reindeer on the northern tundra may appear to be alone but carries with it many companions. Like domestic cattle, the reindeer is a ruminant, with a four-chambered stomach (Figure 9.14) teeming with microbes (a billion per cubic centimeter). In this partially closed environment, the respiration of microorganisms uses up the oxygen ingested by the reindeer while eating. Other microorganisms digest cellulose, take nitrogen from the air in the stomach, and make proteins. The bacterial species that digest the parts of the vegetation that the reindeer cannot digest itself (in particular, the cellulose and lignins of cell walls in woody tissue) require a peculiar environment: They can survive only in an environment without oxygen. One of the few places on Earth's surface where such an environment exists is the inside of a ruminant's stomach.[3] The bacteria and the reindeer are symbionts, each providing what the other needs, and neither could survive without the other. They are therefore called **obligate symbionts**.

Crop plants illustrate another kind of symbiosis. Plants depend on animals to spread their seeds and have evolved symbiotic relationships with them. That's why fruits are so edible; it's a way for plants to get their seeds spread, as Henry David Thoreau discussed in his book *Faith in a Seed*.

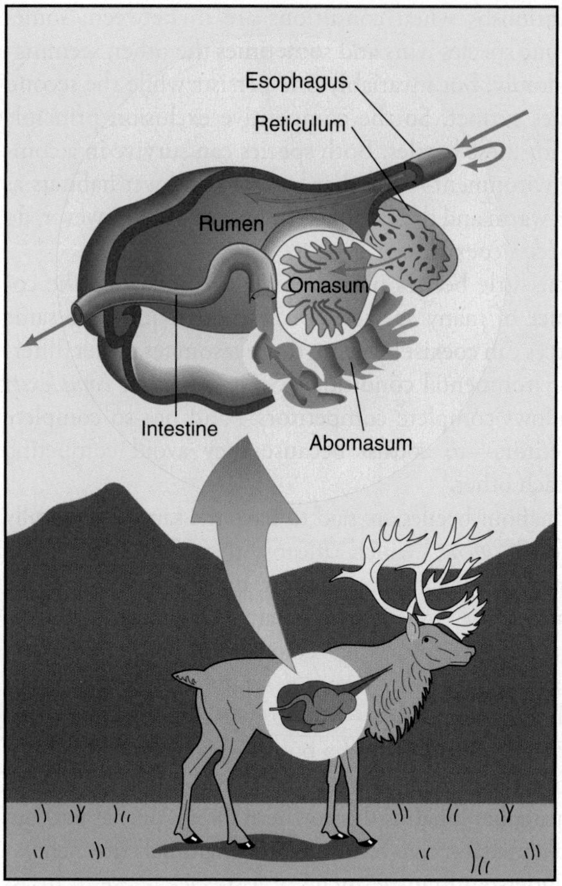

FIGURE 9.14 The stomach of a reindeer illustrates complex symbiotic relationships. For example, in the rumen, bacteria digest woody tissue the reindeer could not otherwise digest. The result is food for the reindeer and food and a home for the bacteria, which could not survive in the local environment outside.

A Broader View of Symbiosis

So far we have discussed symbiosis in terms of physiological relationships between organisms of different species. But symbiosis is much broader, and includes social and behavioral relationships that benefit both populations. Consider, for example, dogs and wolves. Wolves avoid human beings and have long been feared and disliked by many peoples, but dogs have done very well because of their behavioral connection with people. Being friendly, helpful, and companionable to people has made dogs very abundant. This is another kind of symbiosis.

A Practical Implication

We can see that symbiosis promotes biological diversity, and that if we want to save a species from extinction, we must save not only its habitat and niche but also its symbionts. This suggests another important point that will become more and more evident in later chapters: *The*

S. Greg Panosian/iStockphoto

(a)

Jim Parkin/iStockphoto

(b)

imagebroker/Alamy

(c)

FIGURE 9.18 **Joshua tree of North America and giant Euphorbia of Africa illustrate convergent evolution.** Given sufficient time and similar climates in different areas, species similar in shape will tend to occur. The Joshua tree **(a)** and saguaro cactus **(b)** of North America look similar to the giant Euphorbia **(c)** of East Africa. But these plants are not closely related. Their similar shapes result from evolution under similar climates, a process known as convergent evolution.

differences between these look-alike plants can be found in their flowers, fruits, and seeds, which change the least over time and thus provide the best clues to the genetic history of a species. Geographically isolated for 180 million years, these plants have been subjected to similar climates, which imposed similar stresses and opened up similar ecological opportunities. On both continents, desert plants evolved to adapt to these stresses and potentials, and have come to look alike and prevail in like habitats. Their similar shapes result from evolution in similar desert climates, a process known as **convergent evolution**.

Another important process that influences life's geography is **divergent evolution**. In this process, a population is divided, usually by geographic barriers. Once separated into two populations, each evolves separately, but the two groups retain some characteristics in common. It is now believed that the ostrich (native to Africa), the rhea (native to South America), and the emu (native to Australia) have a common ancestor but evolved separately (Figure 9.19). In open savannas and grasslands, a large bird that can run quickly but feed efficiently on small seeds and insects has certain advantages over other organisms seeking the same food. Thus, these species maintained the same characteristics in widely separated areas. Both convergent and divergent evolution increase biological diversity.

People make use of convergent evolution when they move decorative and useful plants around the world. Cities that lie in similar climates in different parts of the world now share many of the same decorative plants. Bougainvillea, a spectacularly bright flowering shrub originally native to Southeast Asia, decorates cities as distant from each other as Los Angeles and the capital of Zimbabwe. In New York City and its outlying suburbs, Norway maple from Europe and the tree of heaven and gingko tree from China grow with native species such as sweet gum, sugar maple, and pin oak. People intentionally introduced the Asian and European trees.

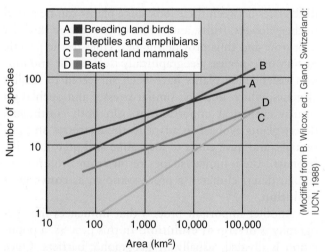

FIGURE 9.19 Divergent evolution. Three large, flightless birds evolved from a common ancestor but are now found in widely separated regions: **(a)** the ostrich in Africa, **(b)** the rhea in South America, and **(c)** the emu in Australia.

9.8 Invasions, Invasive Species, and Island Biogeography

Ever since Darwin's voyage on *The Beagle,* which took him to the Galápagos Islands, biologists have been curious about how biological diversity can develop on islands: Do any rules govern this process? How do such invasions happen? And how is biological diversity affected by the size of and distance to a new habitat? E.O. Wilson and R. MacArthur established a theory of island biogeography that sets forth major principles about biological invasion of new habitats,[24] and as it turns out, the many jokes and stories about castaways on isolated islands have a basis in fact.

- Islands have fewer species than continents. The two sources of new species on an island are migration from the mainland and evolution of new species in place.
- The smaller the island, the fewer the species, as can be seen in the number of reptiles and amphibians in various West Indian islands (Figure 9.20).
- The farther the island is from a mainland (continent), the fewer the species (Figure 9.21).[25]

Clearly, the farther an island is from the mainland, the harder it will be for an organism to travel the distance, and the smaller the island, the less likely that it will be found by individuals of any species. In addition, the smaller the island, the fewer individuals it can support. Small islands tend to have fewer habitat types, and some habitats on a small island may be too small to support a population large enough to have a good chance of surviving for a long time. Generally, the smaller the population, the greater its risk of extinction. It might be easily extinguished by a storm, flood, or other catastrophe or disturbance, and every species is subject to the risk of extinction by predation, disease (parasitism), competition, climatic change, or habitat alteration.

A final generalization about island biogeography is that over a long time, an island tends to maintain a rather constant number of species, which is the result of the rate at which species are added minus the rate at which they become extinct. These numbers follow the curves shown

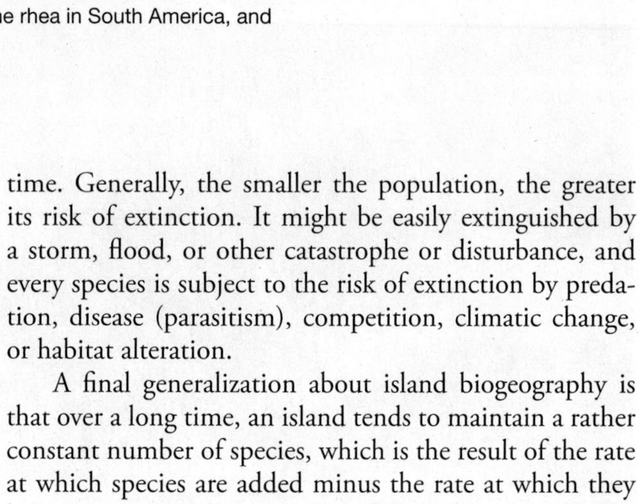

FIGURE 9.20 Islands have fewer species than do mainlands. The larger the island, the greater the number of species. This general rule is shown by a graph of the number of species of birds, reptiles, and amphibians, recent land mammals, and bats for islands in the Caribbean.

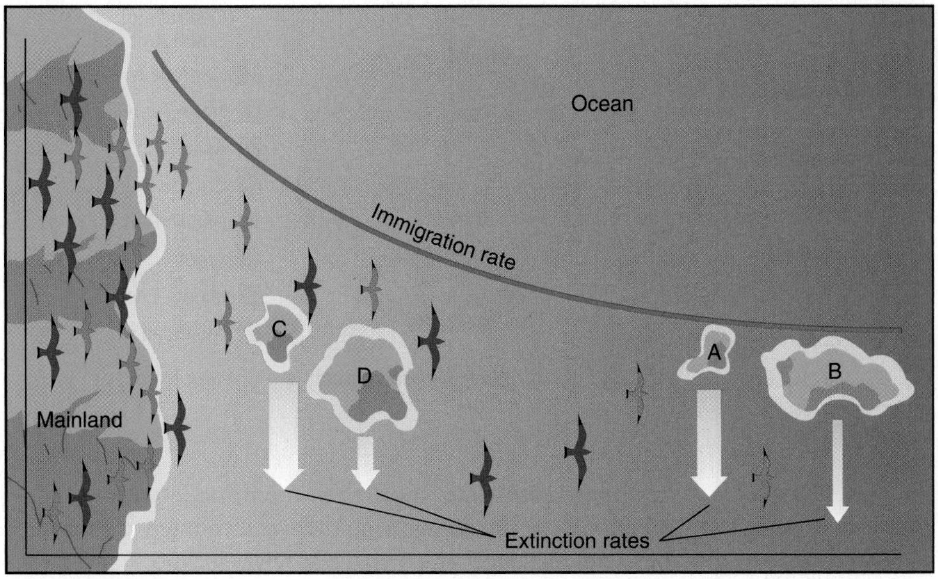

FIGURE 9.21 Idealized relation of an island's size, distance from the mainland, and number of species. The nearer an island is to the mainland, the more likely it is to be found by an individual, and thus the higher the rate of immigration. The larger the island, the larger the population it can support and the greater the chance of persistence of a species—small islands have a higher rate of extinction. The average number of species therefore depends on the rate of immigration and the rate of extinction. Thus, a small island near the mainland may have the same number of species as a large island far from the mainland. The thickness of the arrow represents the magnitude of the rate.

(*Source:* Modified from R.H. MacArthur and E.O. Wilson, *The Theory of Island Biogeography*. Princeton, NJ: Princeton University Press, 1967)

in Figure 9.21. For any island, the number of species of a particular life-form can be predicted from the island's size and distance from the mainland.

The concepts of island biogeography apply not just to real islands in an ocean but also to ecological islands. An **ecological island** is a comparatively small habitat separated from a major habitat of the same kind. For example, a pond in the Michigan woods is an ecological island relative to the Great Lakes that border Michigan. A small stand of trees within a prairie is a forest island. A city park is also an ecological island. Is a city park large enough to support a population of a particular species? To know whether it is, we can apply the concepts of island biogeography.

Biogeography and People

Benefits of Biological Invasions

We have seen that biogeography affects biological diversity. Changes in biological diversity in turn affect people and the living resources on which we depend. These effects extend from individuals to civilizations. For example, the last ice ages had dramatic effects on plants and animals and thus on human beings. Europe and Great Britain have fewer native species of trees than other temperate regions of the world. Only 30 tree species are native to Great Britain (that is, they were present prior to human settlement), although hundreds of species grow there today.

Why are there so few native species in Europe and Great Britain? Because of the combined effects of climate change and the geography of European mountain ranges. In Europe, major mountain ranges run east–west, whereas in North America and Asia the major ranges run north–south. During the past 2 million years, Earth has experienced several episodes of continental glaciation, when glaciers several kilometers thick expanded from the Arctic over the landscape. At the same time, glaciers formed in the mountains and expanded downward. Trees in Europe, caught between the ice from the north and the ice from the mountains, had few refuges, and many species became extinct. In contrast, in North America and Asia, as the ice advanced, tree seeds could spread southward, where they became established and produced new plants. Thus, the tree species "migrated" southward and survived each episode of glaciation.[16]

Since the rise of modern civilization, these ancient events have had many practical consequences. As we mentioned earlier, soon after Europeans discovered North America, they began to bring exotic North American species of trees and shrubs into Europe and Great Britain. These exotic imports were used to decorate gardens, homes, and parks and formed the basis of much of the commercial forestry in the region. For example, in the famous gardens of the Alhambra in Granada, Spain, Monterey cypress from North America are grown as hedges and cut in elaborate shapes. In Great Britain and Europe, Douglas fir and Monterey pine are important commercial timber trees today. These are only two examples of how knowledge of biogeography—enabling people to predict what will grow where based on climatic similarity—has been used for both aesthetic and economic benefits.

Table 9.3 REPORTABLE PLANT PEST INTERCEPTIONS, 2007

PLANT PEST INTERCEPTIONS	NUMBER	PERCENT
Airport	42,003	61%
Express carrier	6	0.01%
Inspection station	2,763	4.01%
Land border	14,394	20.91%
Maritime	4,518	6.56%
Other government programs	86	0.12%
Pre-departure	4,869	7.07%
Rail	16	0.02%
USPS Mail	184	0.27%
Total	**68,839**	**100%**

Source: McCullough, D. G., T. T. Works, J. F. Cavey, A. M. Liebold, and D. Marshall. 2006. Interceptions of nonindigenousplant pests at U.S. ports of entry and border crossings over a 17-year period. *Biological Invasions* 8: 611–630.

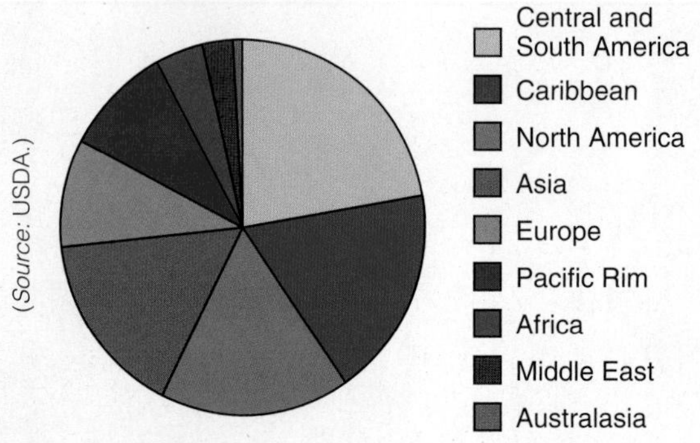

FIGURE 9.22 **Where invasive pests are coming from.**

Why Invasive Species Are a Serious Problem Today

The ease and speed of long-distance travel have led to a huge rate of introductions, with invasive pests (including disease-causing microbes) arriving from all around the world both intentionally and unintentionally (Table 9.3 and Figure 9.22). Table 9.3 shows the number of plant pests intercepted by various means in 2007 by the U.S. government. The majority of interceptions—42,003—were at airports, ten times more than maritime interceptions, which before the jet age would have accounted for most of them. Passenger ships arrive at fewer locations and far less frequently than do commercial aircraft today. Bear in mind that the 42,003 were just those intercepted—no doubt many passed undetected—and that these are only for pests of plants, not for such things as zebra mussels dumped into American waters from cargo ships. According to the USDA, the present situation is not completely controllable.

Another major avenue of species invasions has been the international trade in exotic pets, like the Burmese python. Many of these pets are released outdoors when they get to be too big and too much trouble for their owners.

The upshot of this is that we can expect the invasion of species to continue in large numbers, and some will cause problems not yet known in the United States.

CRITICAL THINKING ISSUE

Polar Bears and the Reasons People Value Biodiversity

In 2008, the U.S. Endangered Species Act listed polar bears as a threatened species. Worldwide, an estimated 20,000 to 25,000 polar bears roam the Arctic, hunting ringed and bearded seals, their primary food. About 5,000 of these polar bears live within the United States. There has been debate about whether the bears should be listed as threatened under the Endangered Species Act.

In 2012, the debate intensified when the state of Alaska argued that the polar bears are not threatened. The Alaskan newspaper *The Courier-Mail* ran the following article about this court case on October 20, 2012, written by Sean Cockerham:

ALASKA ARGUES POLAR BEAR NOT THREATENED

A lawyer arguing for the US State of Alaska that Polar Bears are not a threatened species has run into Sceptical Appeals Court Judges.

Alaska, along with hunting groups and others, is appealing the 2011 decision by a federal judge that the government correctly listed polar bears as threatened, under the federal Endangered Species Act. Polar bears are the first and only species listed solely on the basis of threats from global warming. The U.S. Fish and Wildlife Service says melting sea ice means two-thirds of the world's polar bears could be gone by 2050.

Murray Feldman, a lawyer from Boise, Idaho, who is representing Alaska and the other appellants, argued on Friday that polar bears are doing fine and don't need the protection of the threatened status. "Polar bears occupy the entirety of their historic range, with population at an all-time high," Feldman argued in front of the U.S. Court of Appeals in the District of Columbia. The three-judge panel did not immediately rule. But at least two of the judges were sceptical of some of Feldman's arguments. Judge Harry Edwards sounded unimpressed with Feldman's claims that the Fish and Wildlife Service used flawed population forecasting models.

The judge said those models simply confirmed other findings and were not a key part of the decision to list the bears. "It's beating up on something that appears not critical," Edwards said. He and Judge Merrick Garland also questioned Feldman's repeated use of U. S. Geological Survey statements to argue against the listing.

Edwards said that argument ignored the agency's conclusion that the bears should have been listed as not only threatened, but endangered.

Arctic sea ice melted to a record low this summer, according to the National Snow and Ice Data Center, which said the seasonal melting is more rapid than expected.

Polar bears spend much of their lives hunting seals from sea ice. Biologists say the melting makes it harder to find the seals and forces the bears to swim tremendous distances between ice, putting them at risk for drowning.

Feldman, arguing for Alaska, told the judges that Fish and Wildlife used uncertain predictions and failed to connect the dots between habitat loss and the huge predicted drop in bears. But Katherine Hazard,a lawyer for the Fish and Wildlife Service, said the listing was based on decades of research and an extensive look at sea ice melting.

"They identified no science the agency should have considered and didn't," Hazard told the judges of the appellants' argument.

Refer back to the reasons that people value biodiversity. Read up on polar bears (we've listed some sources below) and decide which of these reasons apply to this species. In particular, consider the following questions.

Critical Thinking Questions

1. As a top predator, is the polar bear a necessary part of its ecosystem? (*Hint:* Consider the polar bear's ecological niche.)
2. Do the Inuit who live among polar bears value them as part of the Arctic diversity of life? (This will take some additional study on your part.)
3. Of the nine reasons we discussed earlier for conserving biological diversity, which ones are the primary reasons for conservation of polar bears?
4. Based on the current status of polar bears, do you think that they should continue to be listed as threatened under the Endangered Species Act? Explain your reasons.

Some Additional Sources of Information about Polar Bears

U.S. Department of Interior Ruling on the Polar Bear: http://alaska.fws.gov/fisheries/mmm/polarbear/pdf/Polar_Bear_Final_Rule.pdf. Accessed November 12, 2012.

About Polar Bears as a Species and Their Habitat and Requirements: Polar Bears: Proceedings of the 14th Working Meeting of the IUCN/SSC Polar Bear Specialist Group, June 20–24, 2005, Seattle, Washington. Available at the International Union for the Conservation of Nature (IUCN) website: http://www.iucn.org/

Derocher, A.E., Nicholas J. Lunn, and Ian Stirling. 2004. Global Warming and Polar Bears: Polar bears in a warming climate. *Integrative and Comparative Biology*, 44:13–176.

SUMMARY

- Biological evolution—the change in inherited characteristics of a population from generation to generation—is responsible for the development of the many species of life on Earth. Four processes that lead to evolution are mutation, natural selection, migration, and genetic drift.
- Biological diversity involves three concepts: genetic diversity (the total number of genetic characteristics), habitat diversity (the diversity of habitats in a given unit area), and species diversity. Species diversity, in turn, involves three ideas: species richness (the total number of species), species evenness (the relative abundance of species), and species dominance (the most abundant species).
- About 1.5 million species have been identified and named. Insects and plants make up most of these species. With further explorations, especially in tropical areas, the number of identified species, especially of invertebrates and plants, will increase.
- Species engage in three basic kinds of interactions: competition, symbiosis, and predation–parasitism. Each type of interaction affects evolution, the persistence of species, and the overall diversity of life. It is important to understand that organisms have evolved together, so predator, parasite, prey, competitor, and symbiont have adjusted to one another. Human interventions frequently upset these adjustments.
- The competitive exclusion principle states that two species that have exactly the same requirements cannot coexist in exactly the same habitat; one must win. The reason more species do not die out from competition is that they have developed a particular niche and thus avoid competition.
- The number of species in a given habitat is determined by many factors, including latitude, elevation, topography, severity of the environment, and diversity of the habitat. Predation and moderate disturbances, such as fire, can actually increase the diversity of species. The number of species also varies over time. Of course, people affect diversity as well.

REEXAMINING THEMES AND ISSUES

Sean Randall/Getty Images, Inc.

HUMAN POPULATION

The growth of human populations has decreased biological diversity. If the human population continues to grow, pressures will continue on endangered species, and maintaining existing biological diversity will be an ever-greater challenge.

© Biletskiy_Evgeniy/iStockphoto

SUSTAINABILITY

Sustainability involves more than just having many individuals of a species. For a species to persist, its habitat must be in good condition and must provide that species' life requirements. A diversity of habitats enables more species to persist.

© Anton Balazh 2011/iStockphoto

GLOBAL PERSPECTIVE

For several billion years, life has affected the environment on a global scale. These global effects have in turn affected biological diversity. Life added oxygen to the atmosphere and removed carbon dioxide, thereby making animal life possible.

ssguy/ShutterStock

URBAN WORLD

People have rarely thought about cities as having any beneficial effects on biological diversity. However, in recent years there has been a growing realization that cities can contribute in important ways to the conservation of biological diversity. This topic will be discussed in Chapter 22.

B2M Productions/Getty Images, Inc.

PEOPLE AND NATURE

George Doyle/Getty Images, Inc.

SCIENCE AND VALUES

People have always treasured the diversity of life, but we have been one of the main causes of the loss in diversity.

Perhaps no environmental issue causes more debate, is more central to arguments over values, or has greater emotional importance to people than biological diversity. Concern about specific endangered species has been at the heart of many political controversies. Resolving these conflicts and debates will require a clear understanding of the values at issue, as well as knowledge about species and their habitat requirements and the role of biological diversity in life's history on Earth.

KEY TERMS

adaptive radiation 194
biogeography 202
biological diversity 187
biological evolution 191
biome 204
biotic province 203
competitive exclusion principle 197
convergent evolution 205
divergent evolution 205
dominant species 189
ecological island 207
ecological niche 198
founder effect 196
gene 192
genetic drift 196
genotype 192
migration 194
mutation 192
natural selection 194
obligate symbionts 200
parasitism 201
population 187
predation 201
species 187
species evenness 189
species richness 189
symbiont 200
symbiosis 200

STUDY QUESTIONS

1. Why do introduced species often become pests?
2. On which of the following planets would you expect a greater diversity of species? (a) a planet with intense tectonic activity; (b) a tectonically dead planet. (Remember that *tectonics* refers to the geologic processes involving the movement of tectonic plates and continents, processes that lead to mountain building and so forth.)
3. You are going to conduct a survey of national parks. What relationship would you expect to find between the number of species of trees and the size of each park?
4. A city park manager has run out of money to buy new plants. How can the park's labor force alone be used to increase the diversity of (a) trees and (b) birds in the park?
5. A plague of locusts visits a farm field. Soon after, many kinds of birds arrive to feed on the locusts. What changes occur in animal dominance and diversity? Begin with the time before the locusts arrive and end after the birds have been present for several days.
6. What will happen to total biodiversity if (a) the emperor penguin becomes extinct? (b) the grizzly bear becomes extinct?
7. What is the difference between a habitat and a niche?
8. More than 600 species of trees grow in Costa Rica, most of them in the tropical rain forests. What might account for the coexistence of so many species with similar resource needs?
9. Which of the following can lead to populations that are less adapted to the environment than were their ancestors?
   (a) Natural selection
   (b) Migration
   (c) Mutation
   (d) Genetic drift

FURTHER READING

Botkin, D.B., ***No Man's Garden: Thoreau and a New Vision for Civilization and Nature*** (Washington, DC: Island Press, 2001). Discusses why people have valued biological diversity from both a scientific and cultural point of view.

Charlesworth, B., and C. Charlesworth, ***Evolution: A Very Short Introduction*** (Oxford, UK: Oxford University Press, 2003).

Darwin, C.A., ***The Origin of Species by Means of Natural Selection, or the Preservation of Proved Races in the Struggle for Life*** (London: Murray, 1859, reprinted variously). A book that marked a revolution in the study and understanding of biotic existence.

Dawkins, R. ***The Selfish Gene*** (New York, NY: Oxford University Press; 3rd edition, 2008). Now considered a classic in the discussion of biological evolution for those who are not specialists in the field.

Flannery, M.A. ***Alfred Russel Wallace: A Rediscovered Life*** (Seattle, WA: Discovery Institute Press, 2011).

Leveque, C., and J. Mounolou, ***Biodiversity*** (New York, NY: John Wiley, 2003).

Margulis, L., K.V. Schwartz, M. Dolan, K. Delisle, and C. Lyons, ***Diversity of Life: The Illustrated Guide to the Five Kingdoms*** (Sudbury, MA: Jones & Bartlett, 1999).

Novacek, M.J. ed., ***The Biodiversity Crisis: Losing What Counts*** (An American Museum of Natural History Book. New York, NY: New Press, 2001).

Wacey, D. ***Early Life on Earth: A Practical Guide*** (New York, NY: Springer, 2009). A new and thoughtful book about the beginnings of life on our planet.

NOTES

1. Dowdy, Alan K. 2008. Current issues in invasive species management. PowerPoint Presentation USDA, APHIS Plant Protection and Quarantine, Emergency and Domestic Programs. Provided courtesy of Dowdy.
2. Botkin, D.B. 2001. *No Man's Garden: Thoreau and a New Vision for Civilization and Nature*. Washington, DC: Island Press.
3. Botkin, D. B. 2012. *The Moon in the Nautilus Shell: Discordant Harmonies Reconsidered*. New York, NY: Oxford University Press.
4. Woese, C.R., O. Kandler, and M.L. Wheelis. 1990. Towards a natural system of organisms: Proposals for the domains Archaea, Bacteria, and Eucharya. *Proceedings of the National Academy of Sciences* (USA) 87:4576–4579.
5. Hart, J.A., K.M. Detwiler, et al. 2012. Lesula: A new species of *Cercopithecus* monkey endemic to the Democratic Republic of Congo and implications for conservation of Congo's central basin. *PLoS ONE* 7(9):e44271.
6. Cicero. *The Nature of the Gods* (44 B.C.). Reprinted by Penguin Classics. New York, 1972.
7. International Fugu Genome Consortium http://www.fugu-sg.org/project/info.html Accessed November 12, 2012.
8. *Complete DNA coding opens new ways to beat malaria. 2008 Christian Science Monitor*. Published: 10/2/2002. © Guardian News & Media. Boston and various other sources
9. Darwin, C.R. 1859. *The Origin of Species by Means of Natural Selection or the Preservation of Favored Races in the Struggle for Life*. London: John Murray. (Originally published as: Darwin, C. 1859. *On the Origin of Species by Means of Natural Selection*. London, UK: John Murray.)
10. Grant, P.R. 1986. *Ecology and Evolution of Darwin's Finches*. Princeton, NJ: Princeton University Press.
11. Cox, C., I.N. Healey, and P.D. Moore. 1973. *Biogeography*. New York, NY: Halsted.
12. Hedrick, P.W., G.A. Gutierrez-Espeleta, and R.N. Lee. 2001. Founder effect in an island population of bighorn sheep. *Molecular Ecology* 10:851–857.
13. Hutchinson, G.E. 1965. *The Ecological Theater and the Evolutionary Play*. New Haven, CT: Yale University Press.
14. Hayden, T. (September 26, 2011). How to hatch a dinosaur. *Wired*. http://www.wired.com/magazine/2011/09/ff_chickensaurus/all/Accessed November 12, 2012. And paleontologist Jack Horner, personal communication.
15. Hardin, G. 1960. The competitive exclusion principle. *Science* 131:1292–1297.
16. British Forestry Commission. http://www.forestry.gov.uk/forestry/Redsquirrel. Accessed May 20, 2008.
17. Miller, R.S. 1967. Pattern and process in competition. *Advances in Ecological Research* 4:1–74.
18. Park, T., 1954, Experimental studies of interspecific competition. II. Temperature, humidity, and competition in two species of *Tribolium*, *Physiol. Zool.* 27:177–328.
19. Wallace, A.R. 1896. *The Geographical Distribution of Animals*. Vol. 1. New York, NY: Hafner. Reprinted 1962.

20. Pielou, E.C. 1979. *Biogeography.* New York, NY: John Wiley & Sons.

21. Takhtadzhian, A.L. 1986. *Floristic Regions of the World.* Berkeley: University of California Press; and Udvardy, M. 1975. *A Classification of the Biogeographical Provinces of the World.* IUCN Occasional Paper 18. Morges, Switzerland: IUCN.

22. Lentine, J.W. 1973. Plates and provinces, a theoretical history of environmental discontinuity. In N.F. Huges, ed., *Organisms and Continents through Time.* Special Papers in Paleontology 12, pp. 79–92.

23. Mather, J.R., and G.A. Yoshioka. 1968. The role of climate in the distribution of vegetation. *Annals of the Association of American Geographers,* 58:29–41; and Prentice, I.C., W. Cramer, S.P. Harrison, R. Leemans, R.A. Monserud, and A.M. Solomon. 1992. A global biome model based on plant physiology and dominance, soil properties and climate. *Journal of Biogeography* 19:117–134.

24. Wilson, E.O., and R.H. MacArthur. 2001. *The Theory of Island Biogeography.* Princeton, NJ: Princeton University Press

25. Tallis, J.H. 1991. *Plant Community History.* London: Chapman & Hall.

A CLOSER LOOK 9.1 NOTES

a. United Nations World Health Organization. http://www.who.int/features/factfiles/malaria/en/index.html

b. World Health Organization. 2008. World malaria 2008; and U.S. Centers for Disease Control. Malaria, April 11, 2007. http://www.cdc.gov/malaria/facts.htm; United Nations World Health Organization. 2003. http://www.who.int/mediacentre/releases/2003/pr33/en/; World Health Organization. 2000. *Overcoming antimicrobial resistance: World Health Report on Infectious Diseases 2000.*

c. World Health Organization. 2008. World malaria 2008; and U.S. Centers for Disease Control. Malaria, April 11, 2007. http://www.cdc.gov/malaria/facts.htm; United Nations World Health Organization. 2003. http://www.who.int/mediacentre/news/releases/2003/pr33/en/ World Health Organization. 2000. *Overcoming Antimicrobial Resistance: World Health Report on Infectious Diseases 2000.*

d. James, A.A. 1992. Mosquito molecular genetics: The hands that feed bite back. *Science* 257:37–38; Kolata, G., 1984. The search for a malaria vaccine. *Science* 226:679–682; Miller, L.H. 1992. The challenge of malaria. *Science* 257:36–37; World Health Organization. 1999 (June). Using malaria information. News Release No. 59. World Health Organization. 1999 (July). Sequencing the *Anopheles gambiae genome.* News Release No. 60. WHO.

CHAPTER 10

Ecological Restoration

LEARNING OBJECTIVES

Ecological restoration is the part of ecosystem management that deals with the recovery of ecosystems that have been damaged by human activities. It is a relatively new field. In this chapter, we explore the concepts of ecological restoration. After reading this chapter, you should be able to . . .

- **Develop a plan for Yellowstone National Park that increases the number of elk above the current level while sustaining streamside vegetation**
- **Using one of the examples described in this chapter, explain how people can maintain a traditional or modern way of life within a restored ecosystem**
- **Describe how bison might once again roam free over expanses of the American West; if you think it is not possible or the wrong thing to do, present solid reasoning against the idea**
- **Describe how we can manage an ecosystem type of your choosing to avoid the need for restoration**
- **Explain the difference between preservation and restoration**
- **Determine whether computers can play a significant role in ecological restoration**
- **Develop a plan for restoring Hutcheson Memorial Forest. You should be able to state the goal, the methods to achieve it, and essential measurements that will need to be made**
- **Explain how a constructed ecosystem could be part of ecological restoration; if you think this is not possible, present solid reasons against it**

A great white heron fishes in the complex mangrove woodlands of the Florida Everglades. The Everglades National Park involves one of the largest ecological restoration projects in the United States.

Photo by Daniel B. Botkin

CASE STUDY

The Florida Everglades

The Florida Everglades is one of the nation's most valuable ecological treasures, listed by the United Nations as a World Heritage Site. The Everglades is also interconnected with a large area of the rest of Florida, beginning with a series of small lakes near Orlando, Florida (near the center of the state), and extending southward to the Florida Bay. The area south of Lake Okeechobee—about 175 km (110 mi) south of Orlando—is a long, wide system of shallow wetlands with slow-moving water. You can imagine the Everglades as a very wide, grass-filled, slow-moving, shallow river. Everglades National Park is at the very southern end of the system and extends out into Florida Bay to about the Florida Keys. Much of the flow of the Everglades is funneled through a location known as Shark Slough, and the velocity of flow of the lower Everglades, while still slow, is greater than that to the north.

Tourists from all over the world come to the Everglades to see its unusual landscape and wildlife. It is home to more than 11,000 species of plants, several hundred species of birds, and numerous species of fish and mammals (Figure 10.1). It is the last remaining habitat for approximately 70 threatened or endangered species or subspecies, including the Florida manatee (a subspecies of the West Indian manatee), the Florida panther (a subspecies of the American mountain lion), and the American crocodile (which lives on the ocean side of the park, while alligators live on the fresh water, northern part).

Unfortunately, in the past century, much of the Everglades has been drained for agriculture and urban development; today only 50% of the original wetlands remain (Figure 10.2a and b).

Restoration of the Everglades is complicated by agriculture and urbanization—several million people live in south Florida. All—agriculture, people, and the national park—compete for the water resources. One of the major issues involving the Everglades today is to somehow arrive at a plan that will ensure long-term sustainability and quality of the water supply for the Everglades and also supply water for agriculture, towns, and cities.[1] This plan will involve the following:

- Reducing the area in agriculture and/or the water use per hectare in agriculture.
- Reducing the flow of agricultural fertilizers and pesticides from farmland into the Everglades.
- Managing land development that encroaches upon the Everglades.
- Managing access to the Everglades by people.
- Removing introduced exotic species that are dangerous to people, or threaten native species with extinction, or disrupt the presettlement ecosystems.
- Developing scientific methods and theory to better predict the possible consequences of changes to the geologic, hydrologic, and biologic parts of the Everglades as restoration goes forward.
- Restoring the original large-scale water flow in south Florida—that grass-filled, slow-moving, shallow river. (Today the flow of water in the northern part of the Everglades is controlled by a complex system of canals, locks, and levees for a variety of purposes, including agriculture, flood control, and water supply.)

The Everglades restoration is the largest wetland-restoration project in the world. Known as the

Photo by Daniel B. Botkin

(a)

Photo by Daniel B. Botkin

(b)

FIGURE 10.1 **Everglades wildlife** includes **(a)** alligators and **(b)** the magnificent frigiate bird.

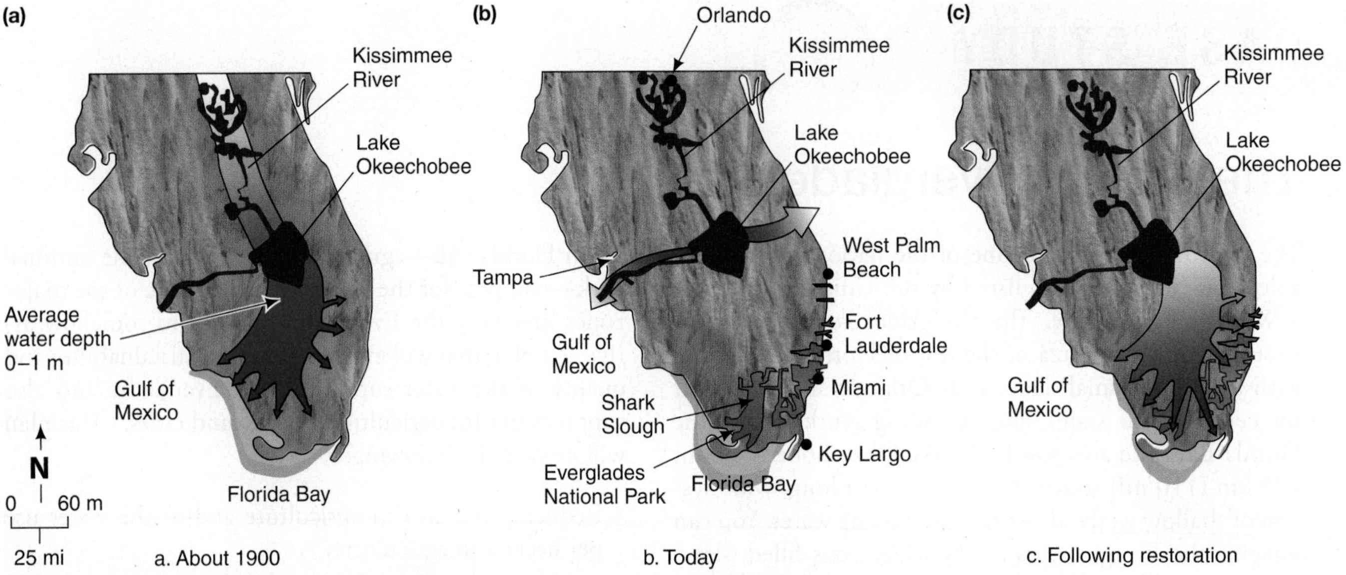

(*Sources:* Modified from the U.S. Geological Survey and Comprehensive Everglades Restoration Plan.)

FIGURE 10.2 **The Florida Everglades from about 1900 to the present (a and b) and projected into the future following restoration (c).** Water flow (both surface and groundwater) is shown in blue. Land development and water diversions changed the dominant flow from north–south to east–west and the new project will return the north-south flow.

Comprehensive Everglades Restoration Plan, the project was developed by a number of government agencies, both local and federal, and is slated to continue over a 30-year period at a total cost of about $10 billion. To date, about $2.4 billion has been spent. Land acquired in 2008 that will be removed from agriculture and returned as part of the Everglades will enlarge the park's area by approximately 728 km^2 (281 mi^2). The acquired land will allow restoration of the hydrology of the Everglades in a much more significant way, as it is in the northern part of the system, where much of the water was used for agricultural purposes. Goals for the Florida Everglades include the following:[1]

- Restoration of a more natural water flow (Figure 10.2c).
- Recovery of native and endangered species.
- Removal of invasive exotic species that are causing problems to ecosystems and people.
- Restoration of water quality, especially control of nutrients from agricultural and urban areas.
- Habitat restoration for wildlife that use the Everglades.

Progress to date has been significant. The amount of pollutants flowing into the Everglades from a variety of sources has been reduced by about 50%. Thousands of hectares have been treated to remove invasive, exotic species, including the Brazilian pepper tree and tilapia (a fish from Africa). In recent years, the Burmese python, which can grow to over 7 m (21 ft) and weigh over 100 kg (220 lb), has become established in the Everglades (Figure 10.3). The snakes are an exotic, invasive species introduced by people who purchased small pythons as pets and released them into the wild when they became too large and difficult to keep. They have established a rapidly growing population in the Everglades (and elsewhere in Florida) and are becoming a top predator there, with increasing capacity to deplete populations of native birds and mammals. Recently, a large python killed and attempted to swallow a seven-foot alligator! A program is now under way to attempt to control the growing population of pythons (estimated to exceed 100,000), but it's unlikely that they can be completely eradicated.

A purpose of this chapter is to provide an understanding of what is possible and what can be done to restore damaged ecosystems.

National Park Service

FIGURE 10.3 **Asian Burmese python and native alligator in battle.** The python, an exotic, invasive species, was released by people who purchased them as pets, and escaped from stores.

10.1 What Is Ecological Restoration?

Ecological restoration is defined as providing assistance to the recovery of an ecosystem that has been degraded, damaged, or destroyed.[2] Originally until near the end of the 20th century, restoration seemed simple: Just remove all human actions and let nature take care of itself. But this led to surprising and undesirable results. A classic example is the conservation of Hutcheson Memorial Forest, the last remaining known uncut, therefore primeval, forest in New Jersey. This forest has been owned since 1701 by the Mettler family, who farmed and kept this forest as a woodlot that they never harvested, as careful family records showed. In 1954, Rutgers University obtained the forest, and ecologist Murray Buell, who arranged for the purchase, planned that it would be left undisturbed and therefore would represent an old-growth oak-hickory forest, the kind that was supposed to be the final endpoint of forest succession (see Chapter 6 for a discussion of succession).

What was this forest supposed to be like? In 1749 to 1750, the Swedish botanist Peter Kalm traveled from Philadelphia to Montreal, collecting plants for Carl Linnaeus. Kalm traveled through this area of New Jersey and described the forests as being composed of large oaks, hickories, and chestnuts, so free of underbrush that one could drive a horse and carriage through the woods.[3]

An article in *Audubon* in 1954 described this wood as "a climax forest . . . a cross-section of nature in equilibrium in which the forest trees have developed over a long period of time. The present oaks and other hardwood trees have succeeded other types of trees that went before them. Now these trees, after reaching old age, die and return their substance to the soil and help their replacements to sturdy growth and ripe old age in turn."[4] But this was not how the forest looked in the 1950s nor how it looks today (Figure 10.4). There are some old trees, many of them in poor condition, and the forest is dense with young tree stems of many sizes. Few oaks have regenerated. In the 1960s, the majority of the seedlings in the forest were maples.

What went wrong? Reconstruction of the forest history showed that prior to 1701 when Europeans took over the land, the Indians had burned this forest on average every ten years. These frequent light fires keep the land relatively open and supported oaks and hickories, resistant to fire, and suppressed maples, easily killed by fire.

These findings created a dilemma. The nature preserve was set up to provide an example of the way the forests were before European alternation of the land, and therefore would never be subjected to cutting, planting, fires, or any other human action. But the forest wasn't like that at all. What should be done? Should it be left alone to become whatever it would, probably a forest nobody had ever seen and really didn't want? Or should it be manipulated to make it look like the forests of the 18th century? It is a puzzle. The forest management has basically ignored the question since Murray Buell's time, but the default is following the first option.

Photo by Daniel B. Botkin

FIGURE 10.4 Hutcheson Memorial Forest in 2007. The forest was supposed to look like it did in the 18th century, occupied only by huge old trees widely scattered in otherwise open country.

The story of Hutcheson Memorial Forest makes clear that restoration cannot mean a return to some imaginary past single original state, nor can it mean management with no actions. This is a hard lesson to learn, and it continues to be violated. But the best of modern ecological restoration takes this into account.

Today, a distinction is made between restoration ecology and ecological restoration. **Restoration ecology** is the science of restoration; **ecological restoration** is the application—the actual activity of restoring ecosystems. Some general principles for restoration are:

- Ecosystems are dynamic, not static (change and natural disturbance are expected).
- No simple set of rules will be applicable to a specific restoration project.

- Adaptive management, using the best science, is necessary for restoration to succeed.
- Careful consideration of ecosystems (life), geology (rocks, soils), and hydrology (water) plays an important role in all restoration projects.

Of particular importance is the principle that *ecosystems are dynamic*, not static—that is, they are always changing and subject to natural disturbance. Any restoration plan must consider disturbance and how resilient the restored system will be. Also important is **adaptive management**, the application of science to the management process. Hypotheses may be tested, and, as restoration goes forward, flexible plans may be developed to accommodate change. Probably the most common projects involve river restoration and the restoration of freshwater and coastal wetlands.[5]

There is also **ecological engineering**, which is defined as the design of ecosystems for the mutual benefit of humans and nature. It is a multidisciplinary activity involving habitat reconstruction; stream and river restoration; wetland restoration and construction; pollution control by ecosystems; and development of sustainable agro-ecosystems.

10.2 Goals of Restoration: What Is "Natural"?

If an ecosystem passes naturally through many different states and all of them are "natural," and if the change itself, caused by wildfire, flood, and windstorm, is natural, then what is its natural state? And how can restoration that involves such disturbance occur without damage to human life and property? Can we restore an ecological system to any one of its past states and claim that this is natural and successful restoration?

In Chapters 4 and 6, we discussed the ideas of steady-state and non–steady-state ecological systems. We argued that until the second half of the 20th century the predominant belief in Western civilization was that any natural area—a forest, a prairie, an intertidal zone—left undisturbed by people achieved a single condition that would persist indefinitely, just as was assumed for the Hutcheson Memorial Forest. This condition, as mentioned in Chapter 4, has been known as the *balance of nature*. The major tenets of a belief in the balance of nature are as follows:

- Left undisturbed, nature achieves a permanency of form and structure that persists indefinitely.
- If it is disturbed and the disturbance is removed, nature returns to exactly the same permanent state.
- In this permanent state of nature, there is a "great chain of being," with a place for each creature (a habitat and a niche) and each creature in its appropriate place.

Attempts to save nature following these beliefs are called preservation. These ideas have their roots in Greek and Roman philosophies about nature, but they have played an important role in modern environmentalism as well. In the early 20th century, ecologists formalized the belief in the balance of nature. At that time, people thought that wildfires were always detrimental to wildlife, vegetation, and natural ecosystems. *Bambi*, a 1942 Walt Disney movie, expressed this belief, depicting a fire that brought death to friendly animals. In the United States, Smokey Bear is a well-known symbol used for many decades by the U.S. Forest Service to warn visitors to be careful with fire and avoid setting wildfires. The message is that wildfires are always harmful to wildlife and ecosystems.

All of this suggests a belief that the balance of nature exists. But if that were true, the answer to the question "restore to what?" would be simple: restore to the original, natural, permanent condition. The way to do it would be simple, too: Get out of the way and let nature take its course. Since the second half of the 20th century, though, ecologists have learned that nature is not constant, and that forests, prairies—all ecosystems—undergo change. But sometimes, in subtle ways, there is a retreat to the balance of nature idea. For example, a scientific paper published in 2009 suggested that carbon storage by forests could be based on the amount stored in old growth.[6,4] The assumption is that forests, left to grow without human interference, will "restore" themselves to a single biomass and therefore carbon storage, and will remain there forever. This is a balance of nature assumption.

Change has been a part of natural ecological systems for millions of years, and many species have adapted to change. Indeed, many require specific kinds of change in order to survive. This means that we can restore ecosystem processes (flows of energy, cycling of chemical elements) and help populations of endangered and threatened species increase on average, but the abundances of species and conditions of ecosystems will change over time as they are subjected to internal and external changes, and following the process of succession discussed in Chapter 6.

Dealing with change—natural and human induced—poses questions of human values as well as science. This is illustrated by wildfires in forests, grasslands, and shrublands, which can be extremely destructive to human life and property. From 1990 to 2009, three wildfires that started in chaparral shrubland in Santa Barbara, California, burned about 1,000 homes. The wildfire hazard can be minimized but not eliminated. Scientific understanding tells us that fires are natural and that some species require them. But whether we choose to allow fires to burn, or even light fires ourselves, is a matter of values. Restoration ecology depends on science to discover what used to be, what is possible, what an ecosystem or species requires

Table 10.1 SOME POSSIBLE RESTORATION GOALS

GOAL	APPROACH
1. Preindustrial	Maintain ecosystems as they were in A.D. 1500
2. Presettlement (e.g., of North America)	Maintain ecosystems as they were about A.D. 1492
3. Preagriculture	Maintain ecosystems as they were about 5000 B.C.
4. Before any significant impact of human beings	Maintain ecosystems as they were about 10,000 B.C.
5. Maximum production	Independent of a specific time
6. Maximum diversity	Independent of a specific time
7. Maximum biomass	Independent of old growth
8. Preserve a specific endangered species	Whatever stage it is adapted to
9. Historical range of variation	Create the future like the known past

to persist, and how different goals can be achieved. But selecting goals for restoration is a matter of human values.

Some possible goals of restoration are listed in Table 10.1. Which state we attempt to restore a landscape to (preindustrial to modern) depends on more-specific goals and possibilities that, again, are linked to values. For example, restoring the Florida Everglades to a preindustrial state is not possible or desirable (a value), given the present land and water use that supports the people of Florida. The goal instead is to improve biodiversity, water flow through the Everglades, and water quality (see opening Case Study).

10.3 What Is Usually Restored?

Ecosystems of all types have undergone degradation and need restoration. However, certain kinds of ecosystems have undergone especially widespread loss and degradation and are therefore a focus of attention today. Table 10.2 gives examples of ecosystems that are commonly restored.

Attention has focused on forests, wetlands, and grasslands, especially the North American prairie; streams and rivers and the riparian zones alongside them; lakes; beaches; and habitats of threatened and endangered species. Also included are areas that people desire to restore for aesthetic and moral reasons, showing once again that restoration involves values. In this section, we briefly discuss the restoration of rivers and streams, wetlands, and prairies.

Rivers, Streams, and Wetlands Restoration: Some Examples

Rivers and streams and wetlands probably are restored more frequently than any other systems. Thousands of streams have been degraded by urbanization, agriculture, timber harvesting, and channelization (shortening, widening, and even paving over or confining the channel to culverts). In North America, large areas of both freshwater and coastal wetlands have been greatly altered during the past 200 years. It is estimated that California, for example, has lost more than 90% of its wetlands, both freshwater and coastal, and that the total wetland loss for the United States is about 50%. Not only the United States has suffered; wetlands around the world are affected.

Table 10.2 SELECTED EXAMPLES OF RESTORATION PROJECTS

SYSTEM	OBJECTIVE
Rivers/Streams	Improve biodiversity, water quality, bank stability. Very common practice across the United States.
Coastal Wetlands	Improve biodiversity and water quality, store water, provide a buffer to erosion from storms to inland areas. Very common practice along all U.S. coastal areas.
Freshwater Wetlands	Improve biodiversity and water quality, store water, and, for river systems, reduce the flood hazard.
Beaches	Sustain beaches and their ecosystems. Most often involve sand supply.
Sand Dunes	Improve biodiversity in both inland and coastal areas.
Landscape	Increase biodiversity and conserve endangered species. Often a very complex process.
Land Disturbed by Mining	Reestablish desired ecosystems, reduce erosion, and improve water quality.

Rivers and Streams

One of the largest and most expensive restoration projects in the United States is the restoration of the Kissimmee River in Florida. This river was channelized, or straightened, by the U.S. Army Corps of Engineers to provide ship passage through Florida. Although the river and its adjacent ecosystems were greatly altered, shipping never developed, and now several hundred million dollars must be spent to put the river back as it was before. The task includes restoring the meandering flow of the river and replacing the soil layers in the order in which they had lain on the bottom of the river prior to channelization.[7]

The Kissimmee, which originally meandered 103 miles in central Florida, had an unusual hydrology because it inundated its floodplain for prolonged periods (Figure 10.5a). The floodplain and river supported a biologically diverse ecosystem, including wetland plants; 38 species of wading birds and waterfowl; fish; and other wildlife. Few people lived in the Kissimmee basin before about 1940, and the land use was mostly agricultural. Due to rapid development and growth and a growing flood hazard as a result of inappropriate land use, people asked the federal government to design a flood-control plan for southern Florida. The channelization of the Kissimmee River occurred between 1962 and 1971 as part of the flood-control plan. About two-thirds of the floodplain was drained, and a straight canal was excavated (Figure 10.5b). Turning the meandering river into a straight canal degraded the river ecosystem and greatly reduced the wetlands and populations of birds, mammals, and fish. The Florida Audubon Society estimates that ducks using the river decreased by 93%.[8]

Criticism of the loss of the river ecosystem went on for years, finally leading to the current restoration efforts. The purpose of the restoration is to return part of the river to its historical meandering riverbed and wide floodplain. Specifics of the restoration plan (shown in Figure 10.6) include restoring as much as possible of the historical biodiversity and ecosystem function; recreating patterns of wetland plant communities as they existed before channelization; reestablishing prolonged flooding of the floodplain; and recreating a river floodplain environment and connection to the main river similar to the way it used to be.[7]

The restoration project was authorized by the U.S. Congress in 1992, in partnership with the South Florida Water Management District and the U.S. Army Corps of Engineers. It is an ongoing project, and by 2012 approximately 23 km

Cameron Davidson/Alamy

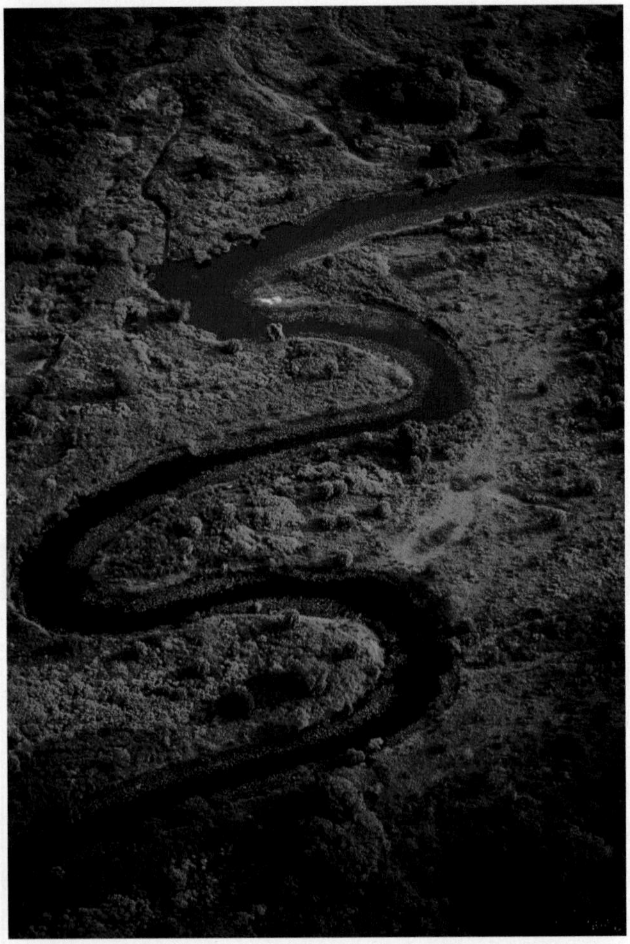

(a)

Cameron Davidson/Alamy

(b)

FIGURE 10.5 **(a)** The Kissimmee River before channelization. **(b)** The Kissimmee River after channelization that produced a wide, straight ditch.

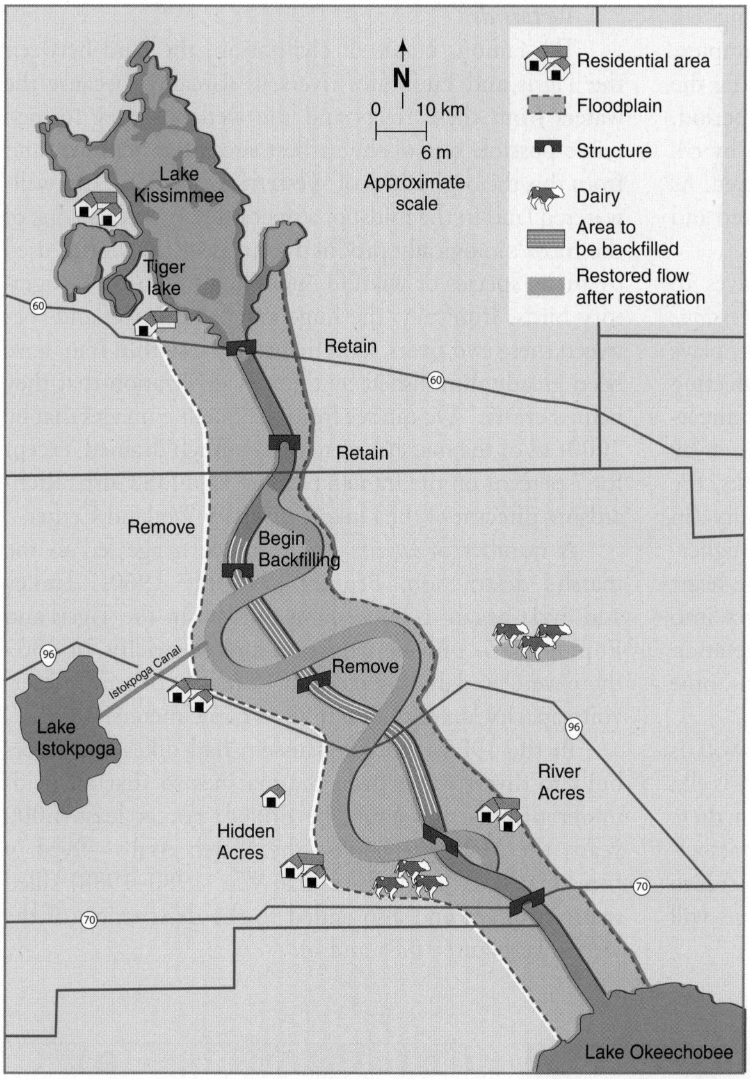

(*Source*: Modified from South Florida Water Management District.)

FIGURE 10.6 **Map showing part of the Kissimmee River restoration plan.** A major objective is to recreate the historical river floodplain environment that is wet much of the year, improving biodiversity and ecosystem function.

of the nearly straight channel had been restored to a meandering channel with floodplain wetlands about 24 km long. As a result, water again flowed through a meandering channel and onto the floodplain, wetland vegetation was reestablished, and birds and other wildlife returned. The potential flood hazard is being addressed. Some of the structures that control flooding will be removed, and others will be maintained. Flood protection is a main reason the entire river will not be returned to what it was before channelization.

The cost of the restoration of the Kissimmee River is several times greater than it was to channelize it. Thus, the decision to go forward with the restoration reflects the high value that people in south Florida place on conserving biological diversity and providing for recreational activities in a more natural environment.

Yellowstone National Park is another interesting case. Here, an unanticipated stream restoration has taken place, resulting from the reintroduction of wolves. Wolves were eliminated from Yellowstone by the 1920s and were introduced back into the park in 1995–1996. Initially, 66 wolves were introduced, and by 2007 the wolf population had grown to 1,774, with about 98 in Yellowstone itself (Figure 10.7), 736 in Idaho (outside the park), 328 in Wyoming, 653 in Montana, and the rest in other areas.[9]

Mountain streams in Yellowstone generally consist of stream channels, beds, and banks composed of silt, sand, gravel, and bedrock. Cool, clean water is supplied via the hydrologic and geologic system from snowmelt and rain that infiltrate the rocks and soil to seep into streams. The water supports life in the stream, including fish and other organisms, as well as streamside vegetation adjacent to stream channels. The riparian vegetation is very different from vegetation on adjacent uplands. Stream bank vegetation helps retard erosion of the banks and thus the amount of sediment that enters the stream.

Riparian vegetation, such as cottonwood and willow trees, is a popular food source for animals, including elk. Extensive browsing dramatically reduces the abundance of riparian plants, damaging the stream environment by reducing shade and increasing bank erosion, which introduces fine sediment into the water. Fine sediment, such as fine sand and silt, not only degrades water quality but also may fill the spaces between gravel particles or seal the bed with mud, damaging fish and aquatic insect habitat.[10] Before the wolves arrived in the mid-1990s, willows and other streamside plants were nearly denuded by browsing elk. Since the arrival of the wolves, the elk population has dropped from almost 17,000 to approximately 4,600 in 2010.[11,9]

age fotostock/SuperStock

FIGURE 10.7 **Wolves in Yellowstone National Park.**

Also, wolves were most successful in hunting elk along streams, where the elk had to negotiate the complex, changing topography. The elk responded by avoiding the dangerous stream environment. Over a four-year period, from 1998 to 2002, the number of willows eaten by elk declined greatly, and the riparian vegetation recovered. As it did so, the stream channel and banks also recovered and became more productive for fish and other animals.

In sum, although the reintroduction of wolves to Yellowstone is controversial, the wolves are a *keystone species*—a species that, even if not overly abundant, plays an important role in the ecological community, affecting the abundance of some species in the ecosystem indirectly. By hunting and killing elk and by scaring them away from the streams, wolves improve the stream banks, the water quality, and the broader ecologic community (in this case, the stream ecosystem). The result is a higher-quality stream environment.[12] This takes us to the heart of a controversy over the reintroduction of wolves into Yellowstone National Park: improved riparian vegetation in the park but fewer elk to hunt outside the park. Some hunters are quite unhappy about this result.

Still, the debate about introduction of the wolf is complex. In Yellowstone, just over 90% of wolf prey is elk, and today there are far fewer bison and deer than there were before the wolf introduction. Land-use issues associated with grazing for cattle and sheep are more difficult to assess. How we choose to manage wolf populations will reflect both science and values.[10, 12]

Wetlands

The famous cradle of civilization, the land between the Tigris and Euphrates rivers, is so called because the waters from these rivers and the wetlands they formed made possible one of the earliest sites of agriculture, and from this the beginnings of Western civilization. This well-watered land in the midst of a major desert was also one of the most biologically productive areas in the world, used by many species of wildlife, including millions of migratory birds. Ironically, the huge and famous wetlands between these two rivers, land that today is within Iraq, have been greatly diminished by the very civilization that they helped create. "We can see from the satellite images that by 2000, all of the marshes were pretty much drained, except for 7 percent on the Iranian border," said Dr. Curtis Richardson, director of the Duke University Wetland Center.[13]

A number of events of the modern age led to the marsh's destruction. Beginning in the 1960s, Turkey and Syria began to build dams upriver, in the Tigris and Euphrates, to provide irrigation and electricity. In 2009 there were 56 dams higher than 15 meters or with a reservoir capacity larger than 3 million cubic meters![14]

In the 1980s Saddam Hussein had dikes and levees built to divert water from the marshes so that oil fields under the marshes could be drilled. For at least 5,000 years, the Ma'adan people—the Marsh Arabs—lived in these marshes. But the Iran–Iraq War (1980–1988) killed many of them and also added to the destruction of the wetlands (Figure 10.8a and b).[15]

Nik Wheeler/© Corbis

(a)

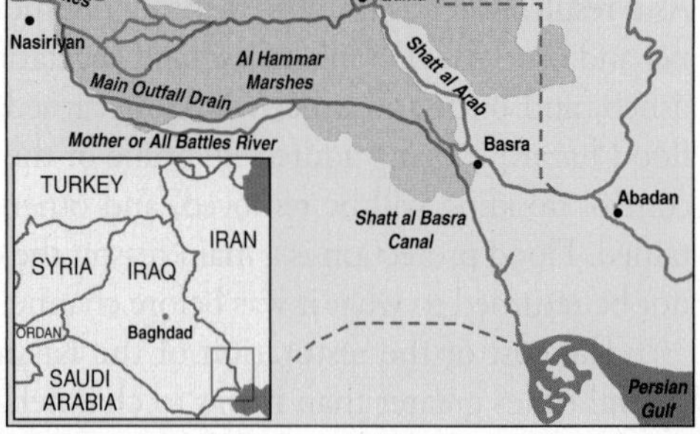

(b)

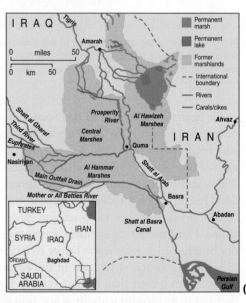

FIGURE 10.8 **(a)** A Marsh Arab village in the famous wetlands of Iraq, said to be one of the places where Western civilization originated. The people in this picture are among an estimated 100,000 Ma'adan people who now live in their traditional marsh villages, many having returned recently. These marshes are among the most biologically productive areas on Earth. **(b)** Map of the Fertile Crescent, where the Marsh Arabs live, called the cradle of civilization. It is the land between the Tigris (the eastern river) and Euphrates (the western river) in what is now Iraq. Famous cities of history, such as Nineveh, developed here, made possible by the abundant water and good soil. The gray area shows the known original extent of the marshes, the bright green their present area.

Today efforts are under way to restore the wetlands. According to the United Nations Environment Program, since the early 1970s the area of the wetlands has increased by 58%.[15] But some scientists believe that there has been little improvement, and the question remains: Can ecosystems be restored once people have seriously changed them?

Prairie Restoration

Tallgrass prairie is also being restored. Prairies once occupied more land in the United States than any other kind of ecosystem. Today, only a few small remnants of prairie exist. Prairie restoration is of two kinds. In a few places, one can still find original prairie that has never been plowed. Here, the soil structure is intact, and restoration is simpler. One of the best known of these areas is the Konza Prairie near Manhattan, Kansas. In other places, where the land has been plowed, restoration is more complicated. Nevertheless, the restoration of prairies has gained considerable attention in recent decades, and restoration of prairie on previously plowed and farmed land is occurring in many midwestern states. The Allwine Prairie, within the city limits of Omaha, Nebraska, has been undergoing restoration from farm to prairie for many years. Prairie restoration has also been taking place near Chicago.

About 10% of North American original tallgrass prairie remains in scattered patches from Nebraska east to Illinois and from southern Canada south to Texas (in the Great Plains physiographic province of the United States). A peculiarity of prairie history is that although most prairie land was converted to agriculture, this was not done in cemeteries, or along roads and railroads; thus, long, narrow strips of unplowed native prairie remain on these rights-of-way. In Iowa, for example, prairie once covered more than 80% of the state—11 million hectares (28 million acres). More than 99.9% of the prairie land has been converted to other uses, primarily agriculture, but along roadsides there are 242,000 hectares (600,000 acres) of prairie—more than in all of Iowa's county, state, and federal parks. These roadside and railway stretches of prairie provide some of the last habitats for native plants, and restoration of prairies elsewhere in Iowa is making use of these habitats as seed sources (Figure 10.9a).[16]

Studies suggest that the species diversity of tallgrass prairie has declined as a result of land-use changes that have led to the loss or fragmentation of habitat.[16] For example, human-induced changes that nearly eliminated bison (another keystone species) from the prairie greatly changed the community structure and biodiversity of the ecosystem. Scientists evaluating the effects of grazing suggest that, managed properly, grazing by bison can restore or improve the biodiversity of tallgrass prairie[16] (Figure 10.9b). The effect of grazing by cattle is not as clear. Range managers have for years maintained that cattle grazing is good for ecosystems, but such grazing must be carefully managed. Cattle are not bison. Bison tend to range over a wider area and in a different pattern. Cattle tend to stay longer in a particular area and establish grazing trails with denuded vegetation. However, both cattle and bison, if too many of them are left too long in too small an area, will cause extensive damage to grasses.

Fire is another important factor in tallgrass prairies. Spring fires enhance the growth of the dominant tall grasses that bison prefer. Tallgrass prairie is a mixture of taller grasses, which prefer warm, dry environments, and other grasses, forbs, and woody plants, which prefer cooler, wetter conditions. The tall grasses often dominate and, if not controlled by fire and grazing, form a thick cover (canopy) that makes it more difficult for shorter plants to survive. Grazing opens the canopy and allows more light to reach closer to the ground and support more species. This increases biodiversity. Long ago, fires set by lightning and/or people helped keep the bison's grazing lands from turning into forests. Today, ecological restoration has attempted to use controlled burns to remove exotic species and woody growth (trees). However, fire alone is

Photo by Daniel B. Botkin

Corbis/SuperStock

(a) (b)

FIGURE 10.9 American Prairie. (a) Restoration of prairie grassland. On the Dixon Ranch where bison are raised prairie grasses are also being restored. **(b)** American bison grazing in tallgrass prairie ecosystem, Custer State Park, South Dakota.

not sufficient in managing or restoring prairie ecosystem biodiversity. Moderate grazing is the hypothetical solution. Grazing of bison on degraded grassland will have negative impacts, but moderate grazing by bison or cattle on "healthy prairies" may work.

One of the newest threats to tallgrass prairie ecosystems is atmospheric nitrogen from automobile emissions. Nitrogen helps some species, but too much of it causes problems for tallgrass prairie ecosystems, whose diversity and productivity are significantly influenced by the availability of nitrogen. Fire and millions of grazing bison regulated nitrogen availability during prehistoric and pre-automobile times.

Another example of restoration is found in A Closer Look 10.1; in this case ecologists managed species to save the island fox.

A CLOSER LOOK 10.1

Island Fox on Santa Cruz Island

The island fox is found on the Channel Islands, eight islands off the coast of southern California (Figure 10.10a). The fox evolved over the past 20,000 years into a separate species of its recent ancestors, the California gray fox. Due to isolation on the islands, as the island fox evolved, it became smaller, and today it is about the size of a house cat (Figure 10.10b).[a]

Island fox most likely reached the islands off the coast of Santa Barbara about 20,000 years ago, when sea levels were more than 120 meters lower than they are today and the distance to the mainland was much shorter, increasing the likelihood of the animals' reaching the offshore environment. At that time, this consisted of one large island known as Santa Rosa. By the time Native Americans arrived, about 12,000 years ago, the island fox had become well established. Native Americans evidently kept the foxes as pets, and some burial sites suggest that foxes were, in fact, buried with their owners. The island fox in the Channel Islands lived to ages unheard of for mainland gray foxes. Many of them lived more than 10 years, and a few even about 12 years. A number of them became blind in their old age, from either cataracts or accident, and could be seen feeding on beaches and other areas despite their handicap.

A subspecies of the island fox evolved on the six islands on which they are found today, and until fairly recently they had no natural enemies. But in the 1990s, the populations of Island fox on several islands suddenly plummeted. On San Miguel Island, for example, a population of approximately 400 foxes in 1994 was reduced to about 15 in only five years. Similar declines occurred on Santa Rosa and Santa Cruz islands. At first it seemed that some disease must be spreading rapidly through the fox population. On Catalina Island, in fact, an occurrence of canine distemper did lead to a decline in the number of foxes on that island. On other islands, particularly in the Santa Barbara Channel, the cause was not so easily determined.

Ecologists eventually solved the mystery by discovering that foxes were being killed and eaten by golden eagles

Gary Crabbe/Alamy

(a)

© Kevin Schafer / Alamy

(b)

FIGURE 10.10 **(a) The Santa Barbara Channel and Channel Islands; (b) Island fox.**

Kaido Kärner/Alamy

(a)

Tom Brakefield/Stock Connection/Aurora

(b)

FIGURE 10.11 (a) Golden eagle eating an island fox on Santa Cruz Island and (b) a bald eagle hunting fish.

(Figure 10.11a), which had only recently arrived on the islands after the apparent demise of the islands' bald eagles (Figure 10.11b). The bald eagles primarily eat fish and hadn't bothered the foxes. Bald eagles are also territorial and kept golden eagles off the islands. It is believed that the bald eagles became endangered because the use of DDT in the 1970s later led to increasing concentrations of the pesticide in fish that bald eagles ate, causing their eggshells to become too soft to protect the embryos. The golden eagles moved in and colonized the islands in the 1990s, apparently attracted by the amount of food they could easily obtain from their daylight hunting, as well as by the absence of bald eagles. The golden eagles found young feral pigs much to their liking and evidently also found island foxes to be easy targets.

Remains of island foxes have been found in eagle nests, and it is now generally agreed that the golden eagles are responsible for the decline in the fox populations. In fact, of 21 fox carcasses studied on Santa Cruz Island in the 1990s, 19 were apparently victims of golden eagle predation.[b, c]

To conserve the island fox on the three Channel Islands in Santa Barbara Channel, which are part of Channel Islands National Park, a management program was developed. The plan has five steps:[c]

1. Capture remaining island foxes and place them in protected areas.
2. Begin a captive breeding program to rebuild the fox populations.
3. Capture golden eagles and transfer them to the mainland. (The idea is to put them in suitable habitat far from the islands so they will not return.)
4. Reintroduce bald eagles into the island ecosystem. (It is hoped that the birds will establish territories and that this will prevent the return of golden eagles.)
5. Remove populations of feral pigs, which attract golden eagles to the islands.

This five-step program has been put into effect; foxes are now being bred in captivity at several sites, and new kits have been born. By 2009, golden eagles on Santa Cruz Island had been mostly removed, the pigs had also been removed, bald eagles had been reintroduced, and foxes raised in captivity had been released. The historical average population of fox on the islands is 1,400. By 2004 fewer than 100 were present, but by 2009 there were about 700—a remarkable recovery. If all the steps necessary to save the island fox are successful, then the island fox will again take its place as one of the keystone species on the Channel Islands.

10.4 Applying Ecological Knowledge to Restore Heavily Damaged Lands and Ecosystems

An example of how ecological succession can aid in the restoration (termed **reclamation** for land degraded by mining) of heavily damaged lands is the ongoing effort to undo mining damage in Great Britain, where some mines have been used since medieval times and approximately 55,000 hectares (136,000 acres) have been damaged. Restoration of coal-mined areas, especially from strip mining, is an important need in the United States as well. The U.S. Surface Mining Law—the Abandoned Mine Land Reclamation Program—provides for the restoration of lands mined and abandoned or left inadequately restored before August 3, 1977. More than $3.5 billion has been spent through this law on restoration of coal mining land.[17]

Restoration of lands that are abandoned coal mines is now accomplished through a coal production tax, levied on operators of mine at \$0.12/ton of underground mined coal and \$0.135/ton of surface mined coal. By 2011, more than \$7.2 billion had been spent from these funds to reclaim more than 295,000 acres.[18] According to the U.S. Office of Surface Mining Reclamation and Enforcement, "Environmental problems associated with abandoned mine lands include surface and ground water pollution, entrances to open mines, water-filled pits, unreclaimed or inadequately reclaimed refuse piles and mine sites (including some with dangerous high walls), sediment-clogged streams, damage from landslides, and fumes and surface instability resulting from mine fires and burning coal refuse. Environmental restoration activities under the abandoned mine reclamation program correct or mitigate these problems.[18]

The success of coal mining restoration remains controversial. On one side are reports such as that from the Clean Air Task Force, which states that "hundreds of thousands of acres of surface mines have not been reclaimed, and reclamation of steep terrain, such as found in Appalachia, is difficult.[19] Finally, despite reclamation efforts, an ecosystem may be destroyed and replaced by a totally different habitat."[20]

On the other side, associations of coal mining corporations claim restoration has been done successfully. An example of a successfully restored surface mine is shown in Figure 10.12.[21]

Recently, programs have been initiated to remove toxic pollutants from the mines and mine tailings, to restore these damaged lands to useful biological production, and to reclaim the attractiveness of the landscape.[22]

One area damaged by a long history of mining lies within the British Peak District National Park, where lead has been mined since the Middle Ages and waste tailings are as much as 5 m (16.4 ft) deep. The first attempts to restore this area used a modern agricultural approach: heavy application of fertilizers and planting of fast-growing agricultural grasses to revegetate the site rapidly. These grasses quickly green on the good soil of a level farm field, and it was hoped that, with fertilizer, they would do the same in this situation. But after a short period of growth, the grasses died. The soil, leached of its nutrients and lacking organic matter, continued to erode, and the fertilizers that had been added were soon leached away by water runoff. As a result, the areas were shortly barren again.

When the agricultural approach failed, an ecological approach was tried, using knowledge about ecological

Coal News, Inc.

Coal News, Inc.

(a) (b)

FIGURE 10.12 Coal mine restoration in Logan County, West Virginia. (a) Valley fill created in 1982 to reclaim coal-mined land. **(b)** Same valley, August 2010.

Courtesy Ed Keller

FIGURE 10.13 **Restoration of a limestone quarry on Vancouver Island, Canada,** in 1921 resulted in a tourist attraction known as Butchart Gardens that draws about 1 million visitors each year.

succession. Instead of planting fast-growing but vulnerable agricultural grasses, ecologists planted slow-growing native grasses, known to be adapted to mineral-deficient soils and the harsh conditions that exist in cleared areas. In choosing these plants, the ecologists relied on their observations of what vegetation first appeared in areas of Great Britain that had undergone succession naturally.[22] The result of the ecological approach has been the successful restoration of damaged lands (Figure 10.13).

Heavily damaged landscapes can be found in many places. Restoration similar to that in Great Britain is done in the United States, Canada, and many other places to reclaim lands damaged by strip mining. Butchart Gardens, an open-pit limestone quarry on Vancouver Island, Canada, is an early-20th-century example of mine restoration. The quarry, a large, deep excavation where Portland cement was produced, was transformed into a garden that attracts a large number of visitors each year (Figure 10.13). The project was the vision of one person, the wife of the mine owner. One person can make all the difference!

10.5 Criteria Used to Judge the Success of Restoration

Criteria used to evaluate whether a specific restoration has been successful, and, if so, how successful, will vary, depending on details of the project and the ecosystem to which the restoration is compared. Criteria used to judge the success of the Everglades restoration (with issues of endangered species) will be much different from criteria used to evaluate the success of **naturalization** of an urban stream to produce a greenbelt. However, some general criteria apply to both.[2]

- The restored ecosystem has the general structure and processes of the target (reference) ecosystem.
- The physical environment (hydrology, soils, rocks) of the restored ecosystem is capable of sustainably supporting the system.
- The restored ecosystem is linked with and appropriately integrated into the larger landscape community of ecosystems.
- Potential threats to the restored ecosystem have been minimized to an acceptable level of risk.
- The restored ecosystem is sufficiently adapted to normally respond to expected disturbances that characterize the environment, such as windstorms or fire.
- The restored ecosystem is, as nearly as possible, as self-sustaining as the target (reference) ecosystem. It therefore undergoes the natural range of variation over time and space; otherwise it cannot be self-sustaining.

In the final analysis, restoration, broadly defined to include naturalization, is successful if it improves the environment and the well-being of the people who are linked to the environment. An example is development of city parks that allow people to better communicate with nature.

CRITICAL THINKING ISSUE

How Can We Evaluate Constructed Ecosystems?

What happens when restoring damaged ecosystems is not an option? In such cases, those responsible for the damage may be required to establish alternative ecosystems to replace the damaged ones. An example involves some saltwater wetlands on the coast of San Diego County, California.

In 1984, construction of a flood-control channel and two projects to improve interstate freeways damaged an area of saltwater marsh. The projects were of concern because California had lost 91% of its wetland area since 1943, and the few remaining coastal wetlands were badly fragmented.

Table 10.3 GOALS, PROGRESS, AND STATUS OF MITIGATION IN THE UNITED STATES AS OF 2006

SPECIES	MITIGATION GOALS	PROGRESS IN MEETING REQUIREMENTS	STATUS AS OF 2006
California	Tidal channels with 75% of the fish species and 75% of the number of fish found in natural channels	Met standards	FWS recommended change from endangered to threatened
Salt-marsh bird's beak	Through reintroduction, at least 5 patches (20 plants each) that remain stable or increase for 3 years	Did not succeed on constructed islands but an introduced population on natural Sweetwater Marsh thrived for 3 years (reached 140,000 plants); continue to monitor because plant is prone to dramatic fluctuations in population	Still listed as endangered
Light-footed clappeer rail	Seven home ranges (82 ha), each having tidal channels with:	Constructed	Still listed as endangered; in 2005, eight captive-raised birds were released
	a. Forage Species equal to 75% of the invertebrate species and 75% of the number of invertebrates in natural areas	Met standards	
	b. High marsh areas for rails to find refuge during high tides	Sufficient in 1996 but two home ranges fell short in 1997	
	c. Low marsh for nesting with 50% coverage by tall cordgrass	All home ranges met low marsh acreage requirement and all but one met cordgrass requirement and six lacked sufficient tall cordgrass Plant height can be increased with continual use of fertilizer but tall cordgrass is not self-sustaining	
	d. Population of tall cordgrass that is self-sustaining for 3 years		

Note: FWS stands for U.S. Fish and Wildlife Service.

In addition, the damaged area provided habitat for three endangered species: the California least tern, the light-footed clapper rail, and a plant called the salt-marsh bird's beak. The California Department of Transportation, with funding from the Army Corps of Engineers and the Federal Highway Administration, was required to compensate for the damage by constructing new areas of marsh in the Sweetwater Marsh National Wildlife Refuge. To meet these requirements, eight islands, known as the Connector Marsh, with a total area of 4.9 hectares, were constructed in 1984. An additional 7-hectare area, known as Marisma de Nación, was established in 1990. Goals for the constructed marsh, which were established by the U.S. Fish and Wildlife Service, included the following: [23]

1. Establishment of tide channels with sufficient fish to provide food for the California least tern.
2. Establishment of a stable or increasing population of salt-marsh bird's beak for three years.
3. Selection of the Pacific Estuarine Research Laboratory (PERL) at San Diego State University to monitor progress on the goals and conduct research on the constructed marsh.

In 1997, PERL reported that goals for the least tern and bird's beak had been met, but that attempts to establish a habitat suitable for the rail had been only partially successful (see Table 10.3).

During the past decade, PERL scientists have conducted extensive research on the constructed marsh to determine the reasons for its limited success. They found that rails live, forage, and nest in cordgrass more than 60 cm tall. Nests are built of dead cordgrass attached to stems of living cordgrass so that the nests can remain above the water as it rises and falls. If the cordgrass is too short, the nests are not high enough to avoid being washed out during high tides.[23]

Researchers suggested that the coarse soil used to construct the marsh did not retain the amount of nitrogen needed for cordgrass to grow tall. Adding nitrogen-rich fertilizer to the soil resulted in taller plants in the constructed marsh, but only if the fertilizer was added on a continuing basis.

Another problem is that the diversity and numbers of large invertebrates, which are the major food source of the rails, are lower in the constructed marsh than in natural marshes. PERL researchers suspect that this, too, is linked to low nitrogen levels. Because nitrogen stimulates the growth of algae and plants, which provide food for small invertebrates, and these in turn provide food for larger invertebrates, low nitrogen can affect the entire food chain.[24-26]

Critical Thinking Questions

1. Make a diagram of the food web in the marsh showing how the clapper rail, cordgrass, invertebrates, and nitrogen are related.
2. The headline of an article about the Sweetwater Marsh project in the April 17, 1998, issue of *Science* declared, "Restored Wetlands Flunk Real-World Test." Based on the information you have about the project, would you agree or disagree with this judgment? Explain your answer.
3. How do you think one can decide whether a constructed ecosystem is an adequate replacement for a natural ecosystem?
4. The term *adaptive management* refers to the use of scientific research in ecosystem management. In what ways has adaptive management been used in the Sweetwater Marsh project? What lessons from the project could be used to improve similar projects in the future?

SUMMARY

- Ecological restoration is the process of helping degraded ecosystems recover and become more self-sustaining, and therefore able to pass through their natural range of conditions.
- Overarching goals of ecological restoration are to help transform degraded ecosystems into sustainable ecosystems and to develop new relationships between the natural and human-modified environments.
- Adaptive management, which applies science to the restoration process, is necessary if restoration is to be successful.
- Restoration of damaged ecosystems is a relatively new emphasis in environmental sciences that is developing into a new field. Restoration involves a combination of human activities and natural processes. It is also a social activity.
- Disturbance, change, and variation in the environment are natural, and ecological systems and species have evolved in response to these changes. These natural variations must be part of the goals of restoration.

REEXAMINING THEMES AND ISSUES

Sean Randall/Getty Images, Inc.

HUMAN POPULATION

If we degrade ecosystems to the point where their recovery from disturbance is slowed or they cannot recover at all, then we have reduced the local carrying of those areas for human beings. For this reason, an understanding of the factors that determine ecosystem restoration is important to developing a sustainable human population.

© Biletskiy_Evgeniy/iStockphoto

SUSTAINABILITY

Heavily degraded land, such as land damaged by pollution or overgrazing, loses the capacity to recover. By helping degraded ecosystems to recover, we promote sustainability. Ecological principles are useful in restoring ecosystems and thereby achieving sustainability.

© Anton Balazh 2011/iStockphoto

GLOBAL PERSPECTIVE

Each degradation of the land takes place locally, but such degradation has been happening around the world since the beginnings of civilization. Ecosystem degradation is therefore a global issue.

ssguy/ShutterStock

URBAN WORLD

In cities, we generally eliminate or damage the processes of succession and the ability of ecosystems to recover. As our world becomes more and more urban, we must learn to maintain these processes within cities, as well as in the countryside. Ecological restoration is an important way to improve city life.

B2M Productions/Getty Images, Inc.

PEOPLE AND NATURE

Restoration is one of the most important ways that people can compensate for their undesirable effects on nature.

George Doyle/Getty Images, Inc.

SCIENCE AND VALUES

Because ecological systems naturally undergo changes and exist in a variety of conditions, there is no single "natural" state for an ecosystem. Rather, there is the process of succession, with all of its stages. In addition. there are major changes in the species composition of ecosystems over time. While science can tell us what conditions are possible and have existed in the past, which ones we choose to promote in any location is a question of values. Values and science are intimately integrated in ecological restoration.

KEY TERMS

adaptive management 218
ecological engineering 218
ecological restoration 217
reclamation 225
restoration ecology 217

STUDY QUESTIONS

1. Develop a plan to restore an abandoned field in your town to natural vegetation for use as a park. The following materials are available: bales of hay; artificial fertilizer; and seeds of annual flowers, grasses, shrubs, and trees.
2. Oil has leaked for many years from the gasoline tanks of a gas station. Some of the oil has oozed to the surface. As a result, the gas station has been abandoned and revegetation has begun to occur. What effects would you expect this oil to have on the process of succession?
3. Refer to the Everglades in the opening case study. Assume there is no hope of changing water diversion from the upstream area that feeds water to the Everglades. Develop a plan to restore the Everglades, assuming the area of wetlands will decrease by another 30% as more water is diverted for people and agriculture in the next 20 years.
4. How can adaptive management best be applied to restoration projects?

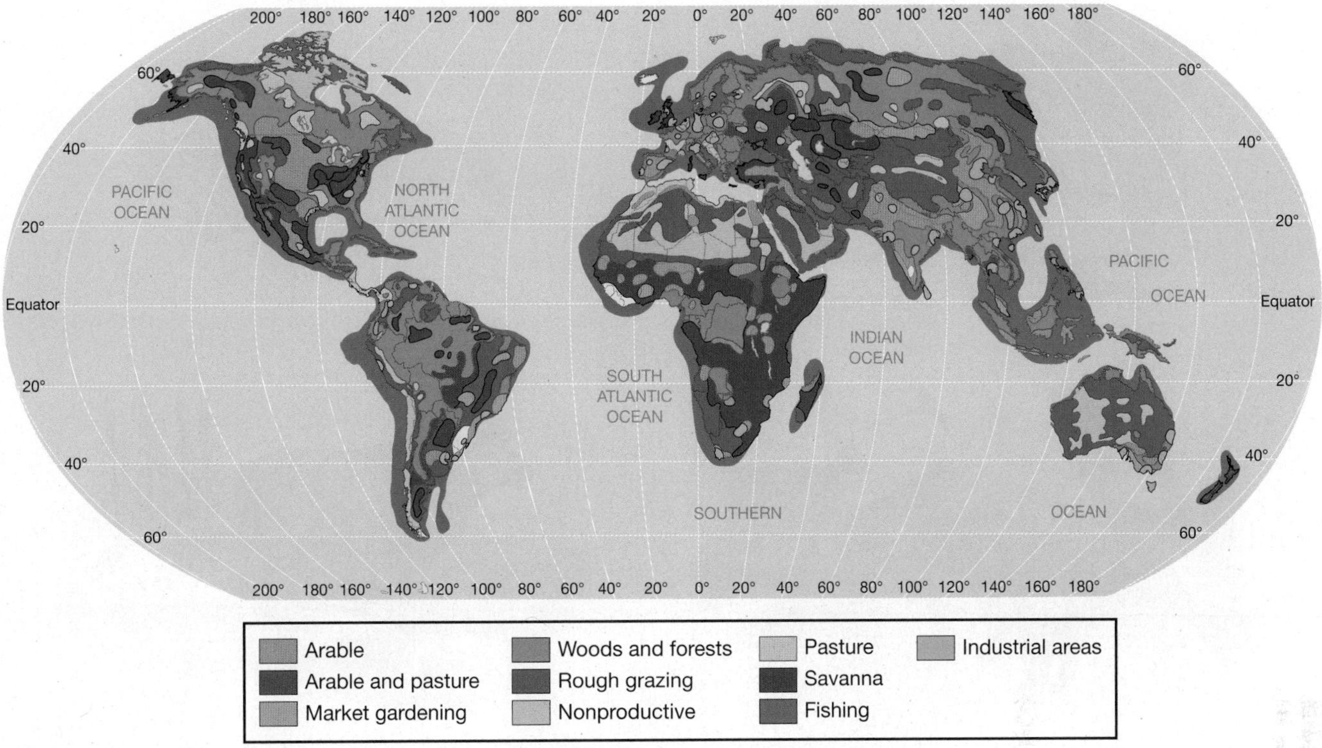

FIGURE 11.1 **World land use showing arable (farmable) land.** The percentage of land in agriculture varies considerably among continents, from 22% of the land in Europe to 57% in Australia. In the United States, cropland occupies 18% of the land, and an additional 26% is used for pasture and rangeland, so agriculture uses 44% of the land.

total annual value of agricultural production is on the order of $320 to $350 billion.[5]

Agriculture has many environmental effects. In the United States, 248 million farmland acres are treated with commercial fertilizers, 65 million acres with chemical insect pesticides, and 194 million acres with chemicals to control weeds.[5]

Farming creates novel ecological conditions, referred to as **agroecosystems**. Agroecosystems differ from natural ecosystems in six ways (Figure 11.2).

- **Ecological succession is halted to keep the agroecosystem in an early-successional state** (see Chapter 6). Most crops are early-successional species, which means that they grow fast, spread their seeds widely and rapidly, and do best when sunlight, water, and chemical nutrients in the soil are abundant. Under natural conditions, crop species would eventually be replaced by later-successional plants. Slowing or stopping natural ecological succession requires time and effort on our part.
- **Biological diversity and food chains are simplified. The focus is on monoculture, one plant species rather than many.** Large areas are planted with a single species or even a single strain or subspecies, such as a single hybrid of corn. The downside of **monoculture** is that it makes the entire crop vulnerable to attack by a single disease or a single change in environmental conditions. Repeated planting of a single species can reduce the soil content of certain essential elements, reducing overall soil fertility.
- **Crops are planted in neat rows and fields.** These simple geometric layouts make life easy for pests because the crop plants have no place to hide. In natural ecosystems, many different species of plants grow mixed together in complex patterns, so it is harder for pests to find their favorite victims.
- **Most agroecosystems require plowing, which is unlike any natural soil disturbance**—nothing in nature repeatedly and regularly turns over the soil to a specific depth. Plowing exposes the soil to erosion and damages its physical structure, leading to a decline in organic matter and a loss of chemical elements.
- **Genetically modified crops are becoming common,** especially in the United States, and are changing agriculture in ways not possible before. We will discuss this later.
- **Agricultural globalization:** As Charles Mann explains in his important book, *1493*, agriculture became globalized with the arrival of Columbus in the New World and with rapidly increasing trade and exchange of agricultural crops around the world. It is not possible to understand modern agriculture without recognizing that it is the result of this globalization, and that most of what we eat and what people in other nations eat comes from crops that were ecological transplants, invasive species that have become of great benefit.[6]

FIGURE 11.2 How farming changes an ecosystem. It converts complex ecosystems of high structural and species diversity to a monoculture of uniform structure and greatly modifies the soil. See text for additional information about the agricultural effects on ecosystems.

11.2 Can We Feed the World?

Can we produce enough food to feed Earth's growing human population? The answer has a lot to do with the environment and our treatment of it: Can we grow crops sustainably, so that both crop production and agricultural ecosystems remain viable? Can we produce this food without seriously damaging other ecosystems that receive the wastes of agriculture? And to these concerns we must now add: Can we produce all this food and also grow crops used only to produce fuels? To answer these questions, let us begin by considering how crops grow and how productive they can be.

The history of agriculture is a series of human attempts to overcome environmental limitations and problems. Each new solution has created new environmental problems, which in turn have required their own solutions. Thus, in seeking to improve agricultural systems, we should expect some undesirable side effects and be ready to cope with them.

The world's food supply is also greatly influenced by social disruptions and social attitudes, which affect the environment and in turn affect agriculture. In Africa, social disruptions since 1960 have included more than 20 major wars and more than 100 coups.[7] Such social instability makes sustained agricultural yields difficult—indeed, it makes any agriculture difficult, if not impossible.[8] So does variation in weather, the traditional bane of farmers.

How We Starve

People "starve" in two ways: undernourishment and malnourishment. World food production must provide adequate nutritional quality, not just total quantity. **Undernourishment** results from insufficient calories in available food, so that one has little or no ability to work or even move and eventually dies from the lack of energy. **Malnourishment** results from a lack of specific chemical components of food, such as proteins, vitamins, or other essential chemical elements. Widespread undernourishment manifests itself as famines that are obvious, dramatic, and fast-acting. Malnourishment is long term and insidious. Although people may not die outright, they are less productive than normal and can suffer permanent impairment and even brain damage.

Among the major problems of undernourishment are marasmus, which is progressive emaciation caused by a lack of protein and calories; kwashiorkor, which results from a lack of sufficient protein in the diet and in infants leads to a failure of neural development and thus to learning disabilities (Figure 11.3); and chronic hunger, when people have enough food to stay alive but not enough to lead satisfactory and productive lives (see Figures 11.4 and 11.5).

The supply of protein has been the major nutritional-quality problem. Animals are the easiest protein food source for people, but depending on animals for protein raises several questions of values. These include ecological ones (Is it better to eat lower on the food chain?), environmental ones (Do domestic animals erode soil faster than crops do?), and ethical ones (Is it morally right to eat animals?). How people answer these questions affects approaches to agriculture and thereby influences the environmental effects of agriculture. Once again, the theme of science and values arises.

FIGURE 11.3 **Photograph of a child suffering from kwashiorkor.**

Since the end of World War II, rarely has a year passed without a famine somewhere in the world.[9] Food emergencies affected 34 countries worldwide at the end of the 20th century. Varying weather patterns in Africa, Latin America, and Asia, as well as an inadequate international trade in food, contributed to these emergencies.

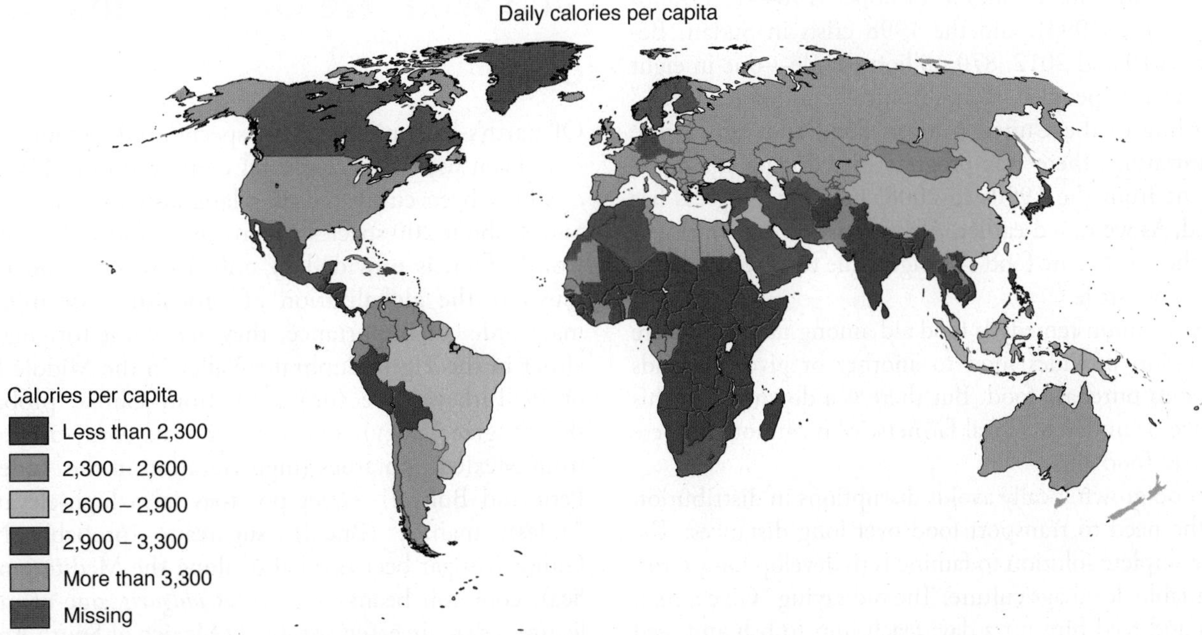

FIGURE 11.4 **Daily intake of calories worldwide.**

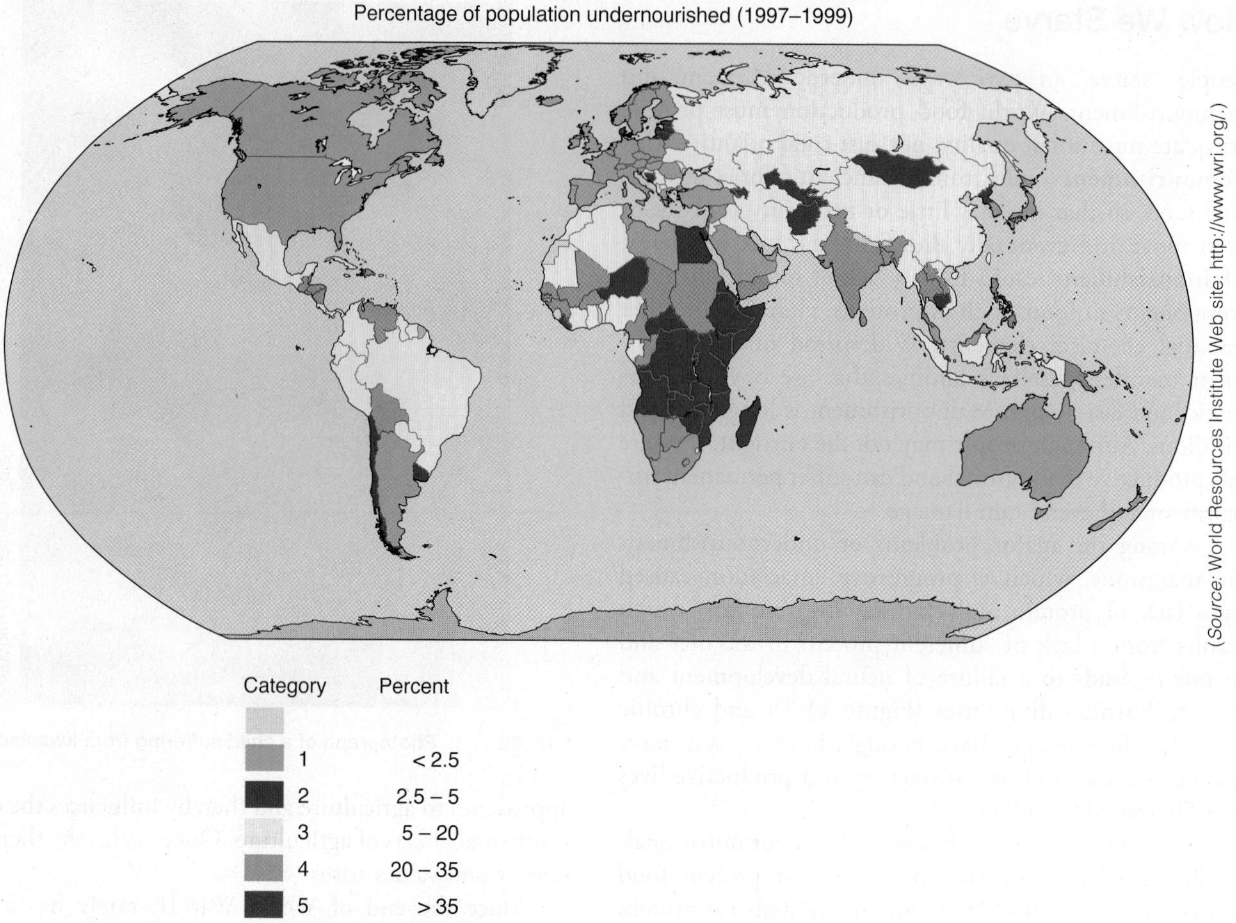

FIGURE 11.5 Where people are undernourished. The percentage is the portion of the country's total population that is undernourished.

Examples include famines in Ethiopia (1984–1985), Somalia (1991–1993), and the 1998 crisis in Sudan. Between 2010 and 2012, 870 million people—one in eight of the world's people—were chronically undernourished.[9] According to the United Nations Food and Agriculture Organization, there was progress in reducing malnourishment from the 1990s to 2008, but that progress has slowed. As we noted earlier, Africa remains the continent with the most acute food shortages, due to adverse weather and civil strife.[9]

A common remedy is food aid among nations, where one nation provides food to another or gives or lends money to purchase food. But there is a downside to this practice: It undercuts local farmers, who cannot compete with free food aid.

Food grown locally avoids disruptions in distribution and the need to transport food over long distances. The only complete solution to famine is to develop long-term, sustainable, local agriculture. The old saying "Give a man a fish and feed him for a day; teach him to fish and feed him for life" is true.

11.3 What We Grow on the Land

Crops

Of Earth's half-million plant species, only about 3,000 have been used as agricultural crops and only 150 species have been cultivated on a large scale. In the United States, about 200 species are grown as crops. Most of the world's food is provided by only 14 species, and these illustrate the globalization of agriculture. In approximate order of importance, they are wheat (originating either in the Tigris-Euphrates Valley in the Middle East or in Turkey), rice (originally from China, probably the Yangtze Valley), maize (American corn, originally from Mexico), potatoes (high elevation in the Andes of Peru and Bolivia), sweet potatoes (also high-elevation Andes), manioc (Brazil), sugarcane (probably New Guinea), sugar beet (probably along the Mediterranean Sea), common beans—*Phaseolus vulgaris*, one of many legumes that are eaten (probably Mexico or South America), soybeans (Southeast Asia, probably China), barley

© Anna Huchak/iStockphoto

picturegarden/Digital Vision/Getty Images

chungking/istockphoto

FIGURE 11.6 **Among the world's major crops are (a) wheat, (b) rice, and (c) soybeans.** See text for a discussion of the relative importance of these three crops.

(Tigris-Euphrates Valley), sorghum (North Africa), coconuts (probably the Western Pacific), and bananas (Southeast Asia, possibly Papua New Guinea) (Figure 11.6). Of these, six provide more than 80% of the total calories that people consume either directly or indirectly.[10]

There is a large world trade in small grains. The United States is the world's largest exporter of wheat and corn (maize), providing 41% of all the wheat traded (Figure 11.7). Other major wheat exporting nations are Canada, Argentina, China, India, and Australia. Major corn exporting nations are Argentina, Brazil and China.[11] (see Figures 11.8 and 11.9). The rest of the world's nations are net importers. World small-grain production increased greatly in the second half of the 20th century—from 0.8 billion metric tons in 1961 to 1 billion in 1966—and doubled to 2 billion in 1996, a remarkable increase in 30 years. In 2005, world small-grain production was 2.2 billion tons, a record crop.[11] But production has remained relatively flat since then. The question we must ask, and cannot answer at this time, is whether this means that the world's carrying

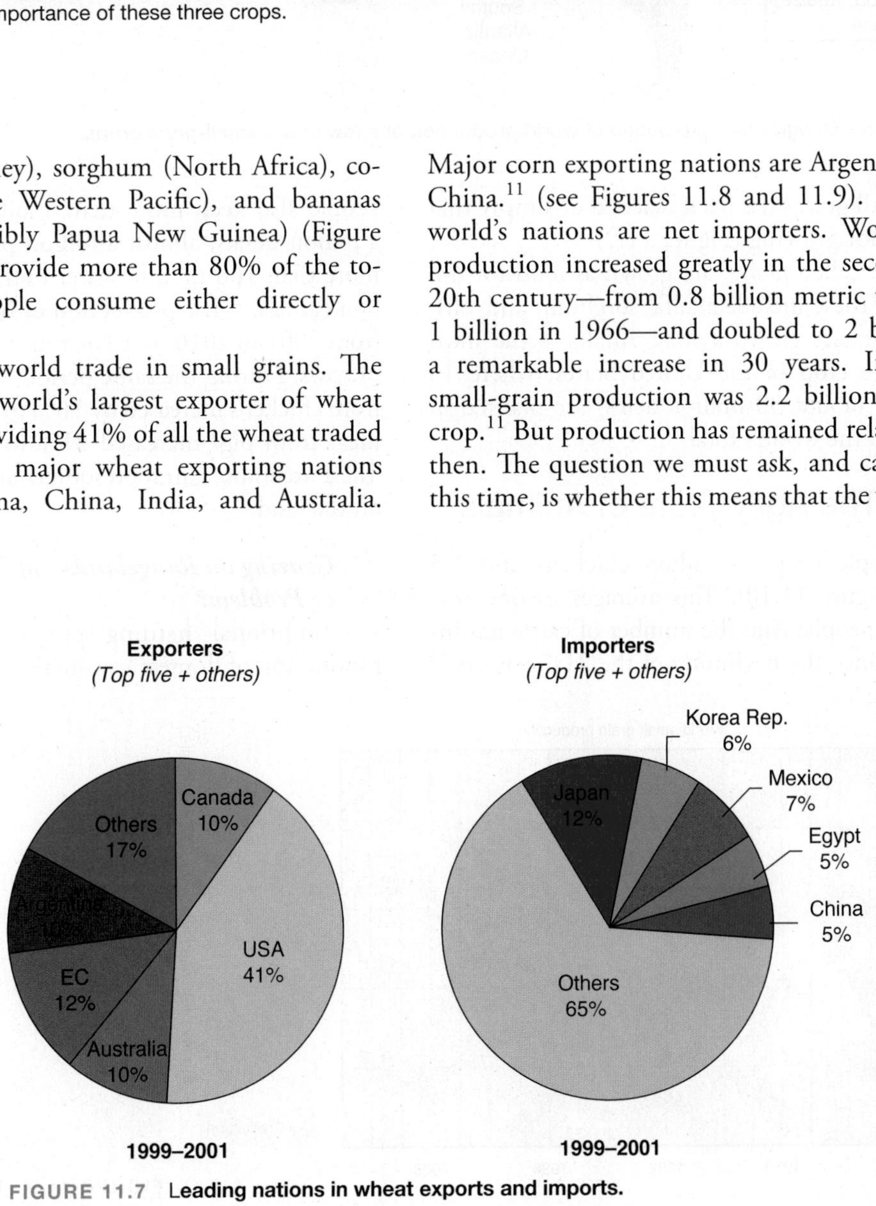

(*Source*: UN Food and Agriculture Organization, Agricultural Commodities: Profiles and Relevant to Negotiating Issues.)

FIGURE 11.7 **Leading nations in wheat exports and imports.**

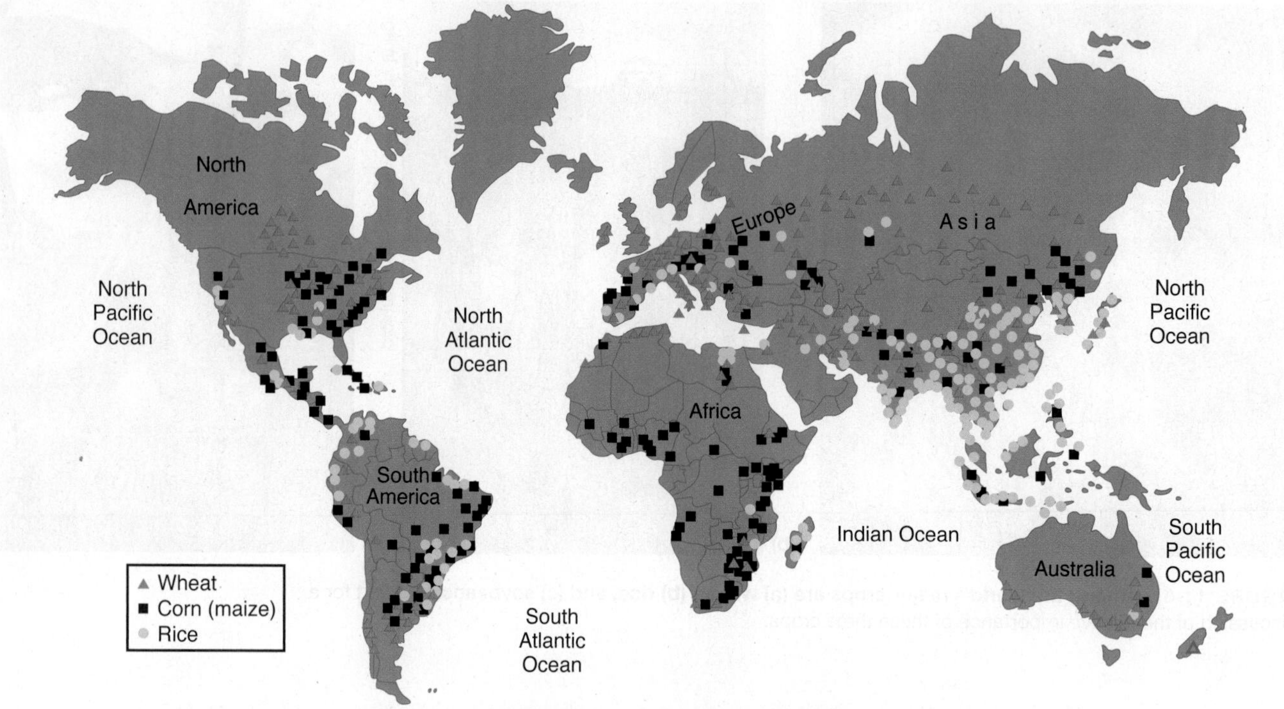

FIGURE 11.8 **Geographic distribution of world production of a few major small-grain crops.**

capacity for small grains has been reached or simply that the demand is not growing (Figure 11.9).

Some crops, called *forage*, are grown as food for domestic animals. These include alfalfa, sorghum, and various species of grasses grown as hay. Alfalfa is the most important forage crop in the United States, where 14 million hectares (about 30 million acres) are planted in alfalfa—one-half the world's total.

Livestock: The Agriculture of Animals

Worldwide, people keep 14 billion chickens and 1.5 billion cattle (Figure 11.10). This averages to one cow for every 4 to 5 people! And the number of cattle has increased by 9% since the beginning of the 21st century.[12]

People also keep more than 1 billion sheep, more than a billion ducks, almost a billion pigs, 700 million goats, more than 160 million water buffalo, and about 18 million camels.[12] The production of beef, however, increased from 2001 to 2010 by 19%, much faster than the number of cows. During the same period, the production of meat from chickens increased by 16% to 64 million tonnes, and meat from pigs increased 21% to 106 million tonnes.[12] These are important food sources and have a major impact on the land.

Grazing on Rangelands: An Environment Benefit or Problem?

Traditional herding practices and industrialized production of domestic animals have different effects on

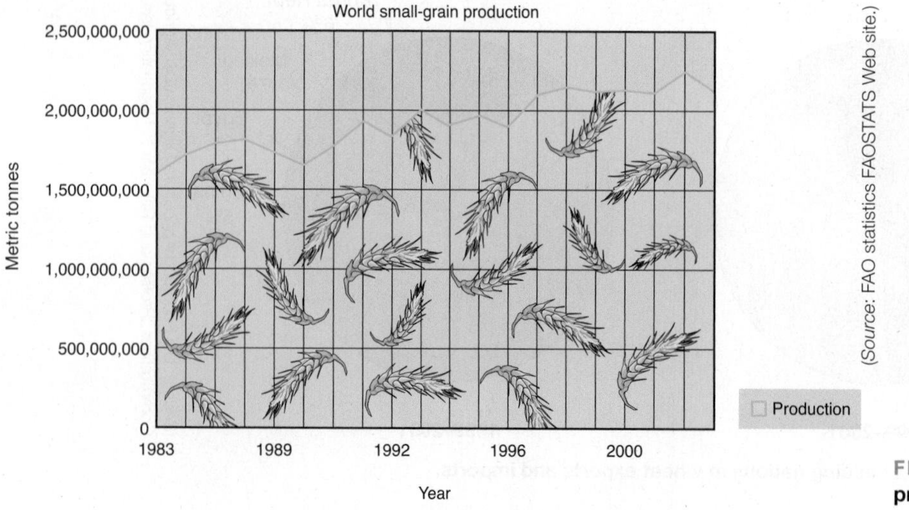

FIGURE 11.9 **World small-grain production since 1983.**

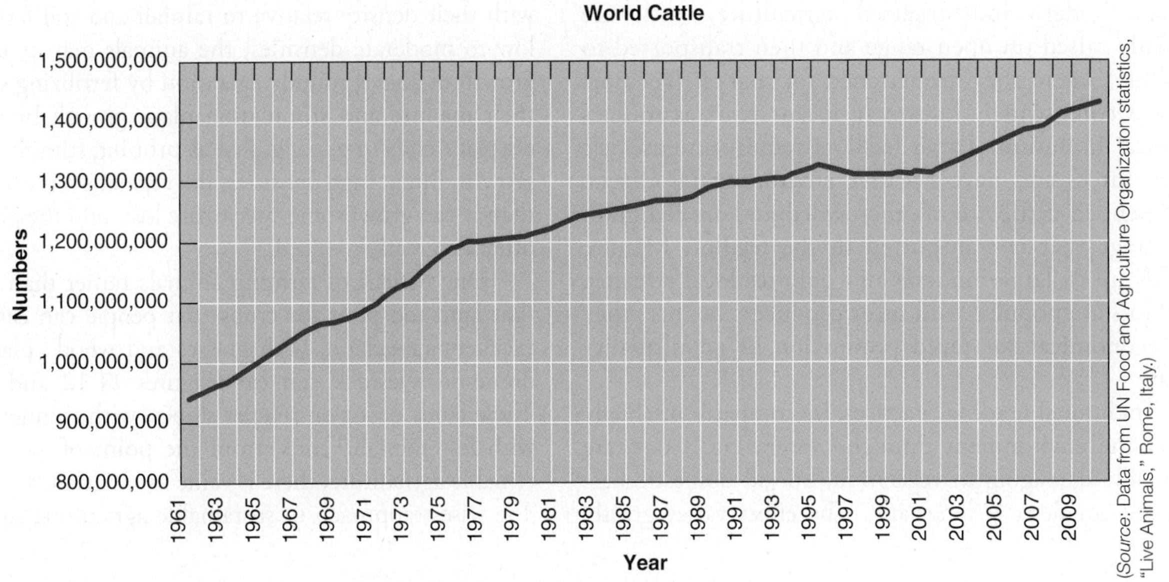

FIGURE 11.10 **Total number of cattle in the world, 1961 to 2012.**

the environment. Most cattle live on rangeland or pasture. **Rangeland** provides food for grazing and browsing animals without plowing and planting. **Pasture** is plowed, planted, and harvested to provide forage for animals. More than 34 million square kilometers (km^2) are in permanent pasture worldwide—an area larger than the combined sizes of Canada, the United States, Mexico, Brazil, Argentina, and Chile.[3]

About 30% of Earth's land is arid rangeland, land easily damaged by grazing, especially during drought. Much of the world's rangeland is in poor condition from overgrazing. In the United States, where more than 99% of rangeland is west of the Mississippi River, rangeland conditions have improved since the 1930s, especially in upland areas. However, land near streams and the streams themselves continue to be heavily affected by grazing.

Grazing cattle trample stream banks and release their waste into stream water. Therefore, maintaining a high-quality stream environment requires that cattle be fenced behind a buffer zone. The upper Missouri River is famous for its beautiful "white cliffs," but private lands along the river that are used to graze cattle take away from the scenic splendor. The large numbers of cattle that come down to the Missouri River to drink damage the land along the river, and the river sometimes runs visibly with manure (Figure 11.11). These effects extend to an area near a federally designated wild and scenic portion of the upper Missouri River, and tourists traveling on the Missouri have complained. In recent years, fencing along the upper Missouri River has increased, with small openings to allow cattle to drink, but otherwise restricting what they can do to the shoreline.

Associated Press

FIGURE 11.11 **Cattle graze along the upper Missouri River, polluting it with their manure and increasing erosion of trampled ground near the river.**

In modern industrialized agriculture, cattle are initially raised on open range and then transported to feedlots, where they are fattened for market. Feedlots have become widely known in recent years as sources of local pollution. Large feedlots require intense use of resources and have negative environmental effects. The penned cattle are often crowded and are fed grain or forage that is transported to the feedlot. Manure builds up in large mounds and pollutes local streams. Feedlots are popular with meat producers because they are economical for rapid production of good-quality meat.

Traditional herding practices, by comparison, chiefly affect the environment through overgrazing. Goats are especially damaging to vegetation, but all domestic herbivores can destroy rangeland. The effect varies greatly with their density relative to rainfall and soil fertility. At low to moderate densities, the animals may actually aid growth of aboveground vegetation by fertilizing soil with their manure and stimulating plant growth by clipping off plant ends in grazing, just as pruning stimulates plant growth. But at high densities, the vegetation is eaten faster than it can grow; some species are lost, and the growth of others is greatly reduced.

One benefit of farming animals rather than crops is that land too poor for crops that people can eat can be excellent rangeland, with grasses and woody plants that domestic livestock can eat (Figures 11.12 and 11.13). These lands occur on steeper slopes, with thinner soils or with less rainfall. Thus, from the point of view of sustainable agriculture, there is value in rangeland or pasture. The wisest approach to sustainable agriculture involves a

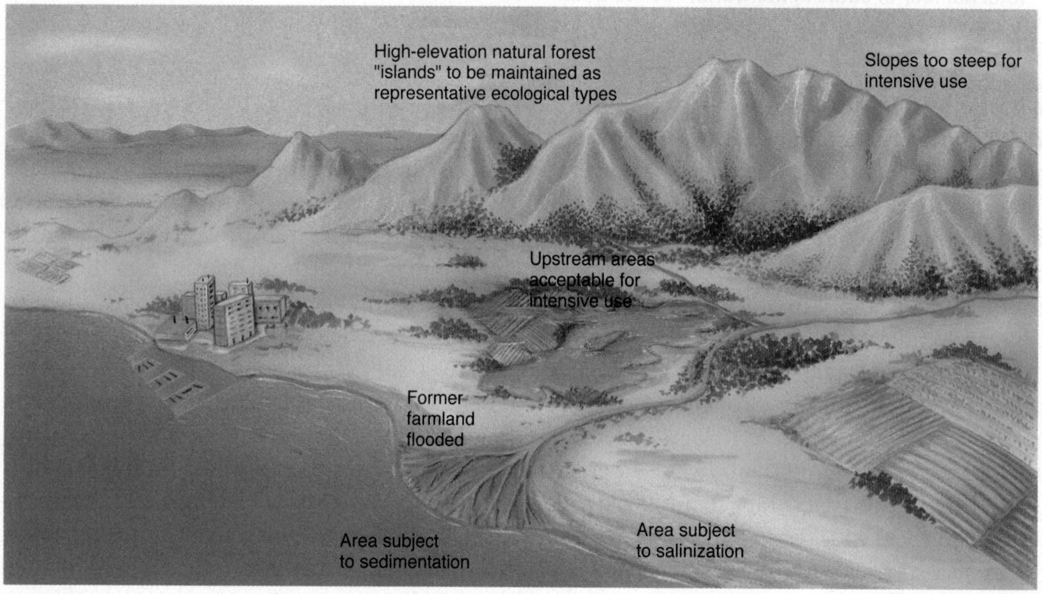

FIGURE 11.12 **Physical and ecological considerations in watershed development**—such as slope, elevation, floodplain, and river delta location—limit land available for agriculture.

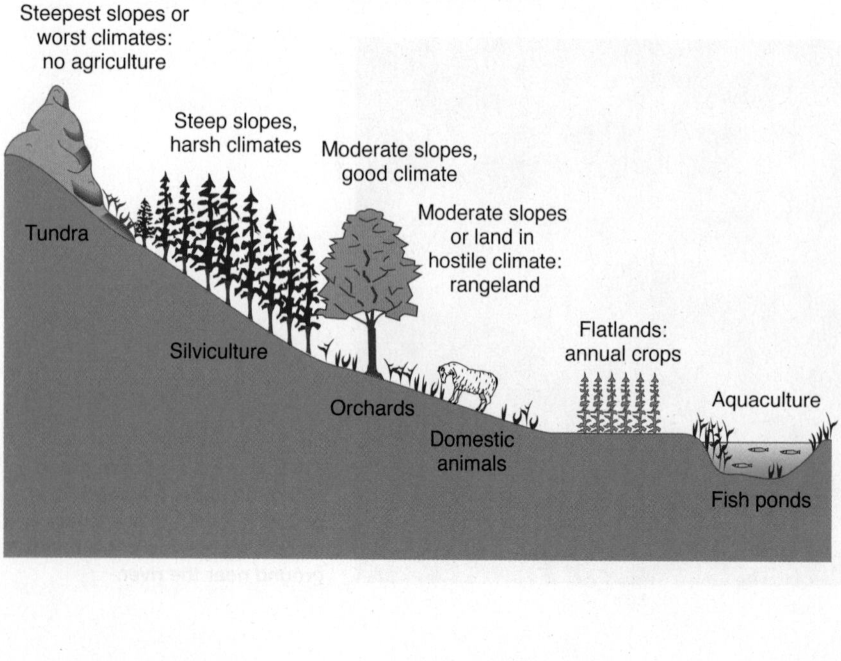

FIGURE 11.13 **Land unsuitable for crop production on a sustainable basis can be used for grazing and for other purposes.**

combination of different kinds of land use: using the best agricultural lands for crops, using poorer lands for pasture and rangeland.

11.4 Soils

To most of us, soils are just "dirt." But in fact soils are a key to life on the land, affecting life and affected by it. Soils develop over a very long time, perhaps thousands of years, and if you look at them closely, they are quite remarkable. You won't find anything like Earth soil on Mars or Venus or the moon. The reason is that water and life have greatly altered the land surface. (Recent findings on Mars by the NASA rovers show long-ago stream erosion, but no Earth-like soils.)

Geologically, soils are earth materials modified over time by physical, chemical, and biological processes into a series of layers. Each kind of soil has its own chemical composition. If you dig carefully into a soil so that you leave a nice, clean vertical cut, you will see the soil's layers. In a northern forest, a soil is dark at the top, then has a white powdery layer, pale as ash, then a brightly colored layer, usually much deeper than the white one and typically orangish. Below that is a soil whose color is close to that of the bedrock (which geologists call "the parent material," for obvious reasons). We call the layers *soil horizons.* The soil horizons shown in Figure 11.14 are not necessarily all present in any one soil. Very young soils may have only an upper *A* horizon over a *C* horizon, whereas mature soils may have nearly all the horizons shown.

(a)

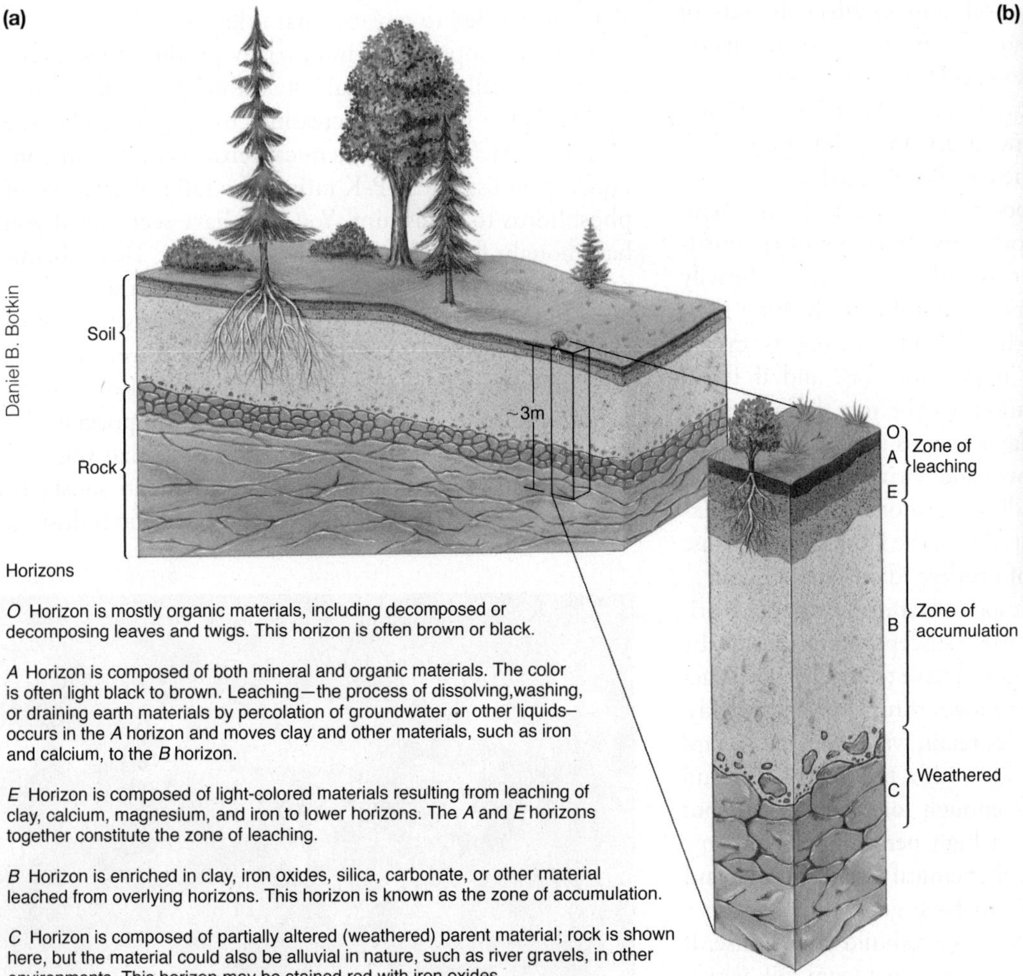

Horizons

O Horizon is mostly organic materials, including decomposed or decomposing leaves and twigs. This horizon is often brown or black.

A Horizon is composed of both mineral and organic materials. The color is often light black to brown. Leaching—the process of dissolving,washing, or draining earth materials by percolation of groundwater or other liquids—occurs in the *A* horizon and moves clay and other materials, such as iron and calcium, to the *B* horizon.

E Horizon is composed of light-colored materials resulting from leaching of clay, calcium, magnesium, and iron to lower horizons. The *A* and *E* horizons together constitute the zone of leaching.

B Horizon is enriched in clay, iron oxides, silica, carbonate, or other material leached from overlying horizons. This horizon is known as the zone of accumulation.

C Horizon is composed of partially altered (weathered) parent material; rock is shown here, but the material could also be alluvial in nature, such as river gravels, in other environments. This horizon may be stained red with iron oxides.

R Unweathered (unaltered) parent material. (Not shown)

Photo by Daniel B. Botkin

(b)

FIGURE 11.14 **(a) Idealized diagram of a soil, showing soil horizons. (b) Photograph of a soil in New Zealand temperature rain forests.** The many years of heavy rain, made slightly acid by carbon dioxide in the rainwater and acids leached from the vegetation, transport minerals from the upper or A (white) horizon, depositing them into the B horizon (the next lower yellowish horizon). The original parent material—what the soil would have looked like from the top before the vegetation began to grow—is the darker color below.

Rainwater is slightly acid (it has a pH of about 5.5) because it has some carbon dioxide from the air dissolved in it, and this forms carbonic acid, a mild acid. Also, rainwater can leach some organic acids from vegetation. As a result, when rainwater moves down into the soil, iron, calcium, magnesium, and other nutritionally important elements are leached from the upper horizons (*A* and *E*) and may be deposited in a lower horizon (*B*). The upper horizons are usually full of life and are viewed by ecologists as complex ecosystems, or ecosystem units (horizons *O* and *A*).

Within soil, decomposition is the name of the game as fungi, bacteria, and small animals live on what plants and animals on the surface produce and deposit. Bacteria and fungi, the great chemical factories of the biosphere, decompose organic compounds from the surface. Soil animals, such as earthworms, eat leaves, twigs, and other remains, breaking them into smaller pieces that are easier for the fungi and bacteria to process. In this way, the earthworms and other soil animals affect the rate of chemical reactions in the soil. There are also predators on soil animals, so there is a soil ecological food chain.

Soil *fertility* is the combination of the capacity of a soil to supply nutrients necessary for plant growth, to store water but not become too waterlogged, and to have a structure that allows good flow of air. Soils that have formed on geologically young materials are often nutrient-rich. Soils in humid areas and tropics may be heavily leached and relatively nutrient-poor due to the high rainfall (Figure 11.14b). In such soils, nutrients may be cycled through the organic-rich upper horizons; and if forest cover is removed, reforestation may be very difficult. Soils that accumulate certain clay minerals in semiarid regions may swell when they get wet and shrink as they dry out, cracking roads, walls, buildings, and other structures. Expansion and contraction of soils in the United States cause billions of dollars' worth of property damage each year.

Coarse-grained soils, especially those composed primarily of sand, are particularly susceptible to erosion by water and wind. Sand and gravel have relatively large spaces between grains, so water moves through them quickly. Soils with small clay particles retain water well and retard the movement of water. Soils with a mixture of clay and sand can retain water well enough for plant growth but also drain well. Soils with a high percentage of organic matter also retain water and chemical nutrients for plant growth. It is an advantage to have good drainage, so a coarse-grained soil is a good place to build your house. If you are going to farm, you'll do best in a loam soil, which has a mixture of particle sizes.

Restoring Our Soils

Soil erosion and its effects became famous in the 1930s with the Dust Bowl (A Closer Look 11.1). Because of improved farming practices, soil erosion is estimated to have slowed 43% on cropland in the United States.[13] However, these estimates are based heavily on models, and the integration of models with representative monitoring of soil is incomplete.[13] One outstanding example is the watershed of Coon Creek, Wisconsin, an area of 360 km^2 that has been heavily farmed for more than a century. This stream's watershed was the subject of a detailed study in the 1930s by the U.S. Soil Conservation Service and was restudied in the 1970s and 1990s. Soil erosion in the 1990s was only 6% of what it had been in the 1930s.[14] The bad news is that, even so, the soil is eroding faster than new soil is being generated.[15]

Fertilizers

Traditionally, farmers combated the decline in soil fertility by using organic fertilizers, such as animal manure, which improve both chemical and physical characteristics of soil. But organic fertilizers have drawbacks, especially under intense agriculture on poor soils. In such situations, they do not provide enough of the chemical elements needed to replace what is lost.

The development of industrially produced fertilizers, commonly called "chemical" or "artificial" fertilizers, was a major factor in greatly increasing crop production in the 20th century. Typically, chemical fertilizers come in containers that list the N-P-K rating, the ratio of nitrogen to phosphorus to potassium. You may have seen this if you have bought fertilizers for home gardening. These chemical elements are the big three nutrients for plants.

One of the most important advances was the invention of the method to convert molecular nitrogen gas in the atmosphere to nitrate, which can be used directly by plants. Phosphorus, another biologically important element, is mined, usually from a fossil source that was biological in origin, such as deposits of bird guano on islands used for nesting (Figure 11.15). The scientific-industrial

Daniel Valla/Alamy

FIGURE 11.15 Boobies on a guano island stand on centuries of bird droppings. The birds feed on ocean fish and nest on islands. In dry climates, their droppings accumulate, becoming a major source of phosphorus for agriculture for centuries.

A CLOSER LOOK 11.1

The Great American Dust Bowl

Soil erosion became a national issue in the United States in the 1930s, when intense plowing, combined with a major drought, loosened the soil over large areas. The soil blew away, creating dust storms that buried automobiles and houses, destroyed many farms, impoverished many people, and led to a large migration of farmers from Oklahoma and other western and midwestern states to California. The human tragedies of the Dust Bowl were made famous by John Steinbeck's novel *The Grapes of Wrath,* later a popular movie starring Henry Fonda (Figure 11.16).

The land that became the Dust Bowl had been part of America's great prairie, where grasses rooted deep, creating a heavily organic soil a meter or more down. The dense cover provided by grass stems and the anchoring power of roots protected the soil from the erosive forces of water and wind. When the plow turned over those roots, the soil was exposed directly to sun, rain, and wind, which further loosened the soil.

© Bettmann/CORBIS

FIGURE 11.16 The Dust Bowl. Poor agricultural practices and a major drought created the Dust Bowl, which lasted about ten years during the 1930s. Heavily plowed lands lacking vegetative cover blew away easily in the dry winds, creating dust storms and burying houses.

age brought with it mechanized mining of phosphates and their long-distance transport, which, at a cost, led to short-term increases in soil fertility. Nitrogen, phosphorus, and other elements are combined in proportions appropriate for specific crops in specific locations.

Today there is concern about the increasing demand for phosphate as a fertilizer, especially given the increasing world food demand. Some claim that there may be a phosphorus supply crisis in the near future as agricultural demand exceeds supply.[16] The Food and Agriculture Organization forecasts that the demand for phosphate as a fertilizer will grow 2.4% annually through 2015.[17,18] Phosphate demand and shortages need to be watched carefully; it may be one of the next major limitations to food production.

Limiting Factors

Crops require about 20 chemical elements. These must be available in the right amounts, at the right times, and in the right proportions to each other. It is customary to divide these life-important chemical elements into two groups, macronutrients and micronutrients. A **macronutrient** is a chemical element required by all living things in relatively large amounts. Macronutrients are sulfur, phosphorus, magnesium, calcium, potassium, nitrogen, oxygen, carbon, and hydrogen. A **micronutrient** is a chemical element required in small amounts—either in extremely small amounts by all forms of life or in moderate to small amounts for some forms of life. Micronutrients are often rarer metals, such as molybdenum, copper, zinc, manganese, and iron.

High-quality agricultural soil has all the chemical elements required for plant growth and also has a physical structure that lets both air and water move freely through the soil and yet retain water well. The best agricultural soils have a high organic content and a mixture of sediment particle sizes. But some crops require quite different soils. For example, lowland rice grows in flooded ponds and requires a heavy, water-saturated soil, while watermelons grow best in very sandy soil. Soils rarely have everything a crop needs. The question for a farmer is: What needs to be added or done to make a soil more productive for a crop? The traditional answer is that, at any time, just one factor is limiting. If that **limiting factor** can be improved, the soil will be more productive; if that single factor is not improved, nothing else will make a difference.

The idea that some single factor determines the growth and therefore the presence of a species is known as **Liebig's law of the minimum**, after Justus von Liebig, a 19th-century agriculturalist credited with first stating this idea. A general statement of Liebig's law is: The growth of a plant is affected by one limiting factor at a time—the one whose availability is the least in comparison to the needs of a plant.

Sometimes trace elements are required in extremely small amounts. Striking cases of soil-nutrient limitations have been found in Australia—which has some of the oldest soils in the world—on land that has been above sea level for many millions of years, during which time severe leaching has taken place. For example, it is estimated that in certain Australian soils adding an ounce of molybdenum to a field increases the yield of grass by 1 ton/year.

The idea of a limiting growth factor, originally used in reference to crop plants, has been extended by ecologists to include all life requirements for all species in all habitats. If Liebig were always right, then environmental factors would always act one by one to limit the distribution of living things. But there are exceptions. For example, nitrogen is a necessary part of every protein, and proteins are essential building blocks of cells. Enzymes, which make many cell reactions possible, contain nitrogen. A plant given little nitrogen and phosphorus might not make enough of the enzymes involved in taking up and using phosphorus. Increasing nitrogen to the plant might therefore increase the plant's uptake and use of phosphorus. If this were so, the two elements would have a **synergistic effect**, in which a change in the availability of one resource affects the response of an organism to some other resource.

So far we have discussed the effects of chemical elements when they are in short supply. But it is also possible to have too much of a good thing—most chemical elements become toxic in concentrations that are too high. As a simple example, plants die when they have too little water but also when they are flooded, unless they have specific adaptations to living in water. So it is with chemical elements and compounds required for life.

Agricultural Sustainability

At this point in our discussion, we have reached a partial answer to the question "How could farming be sustained for thousands of years, while the soil has been degraded?" In the first chapter of this book, we discussed sustainability. We wrote that, strictly speaking, harvesting a resource at a certain rate is sustainable if we can continue to harvest that resource at that same rate for some specified time well into the future. We also pointed out two warnings:

- Commonly, in discussions about environmental problems, the time period is not specified and is assumed to be very long—mathematically an infinite planning time, but in reality as long as it could possibly matter to us. For conservation of the environment and its resources to be based on quantitative science, both a rate of removal and a planning time horizon must be specified.
- Ecosystems and species are always undergoing change, and a completely operational definition of *sustainability* will have to include such variation over time.

We must recognize a distinction between the sustainability of a product (in this case crops) and the sustainability of the ecosystem. *In agriculture, crop production can be sustained, but the ecosystem may not be. And if the ecosystem is not sustained, then people must provide additional inputs of energy and chemical elements to replace what is lost.* This is what has happened in Italy and other places where farming has been sustained for very long periods.

Selecting an appropriate site for a crop is crucial to sustainability. There are some striking counterexamples. As Charles C. Mann explains in his book *1493: Uncovering the World Columbus Created*, sweet potatoes, native to Central and South America, were introduced into China, which is today the world's biggest producer of that crop.

In contrast to lowland rice, which grows only in lowland ponds and was previously the main crop of the country, sweet potatoes are upland crops. Cultivation of sweet potatoes made possible farming land that had not previously been farmed—hilly and mountainous areas. Unfamiliar with such upland crops, Chinese planted sweet potatoes (as well as American corn) by cutting down forests on steep slopes where soils were too fragile for intense growing of these crops. Erosion during heavy rainfalls increased. This degraded the soil where the crops were grown, and also the eroded soil destroyed farmlands in the valleys.[19] This was unsustainable, and over many years improvements were made in some areas so that sweet potatoes could be grown.

Why, in contrast have farms in Italy been sustained for several thousand years? The answer in part is that Italian farmers have provided additional inputs of energy and chemical elements to replace what has been lost—fertilizing the land. But there is more to it. Some of Rome's great writers and philosophers, whom today you would not think of as having written about farming, did exactly that in their time. Cicero wrote that farming was the best of all occupations. Marcus Cato the Elder, known primarily as a statesman and politician, wrote in the second century B.C. *On Farming*.[20] In that book he wrote that the best farms were vineyards, which are perennial vines well rooted in the soil and therefore resistant to erosion. The primary Roman grain was wheat. He wrote that in buying a farm, you should seek one "at the root of a mountain, south-facing, in a healthy position." In short, do not try to farm on steep slopes, where erosion will be a serious problem, and seek a south-facing location that gets the best sun, important given the Italian climate. He urged care in choosing crops: "If you ask me what would make a farm the first choice, I will say this: varied ground, . . . then, first the vineyard (or an abundance of wine), second an irrigated kitchen garden, third a willow wood, fourth an olive field, fifth a meadow, sixth a grain-field, seventh a plantation of trees, eighth an orchard, ninth an acorn wood.[21] Perhaps agriculture in Italy has been the subject of considerable thought, care, and planning about crops. And perhaps this is the key to sustainability.

11.5 Controlling Pests

From an ecological point of view, pests are undesirable competitors, parasites, or predators. The major agricultural pests are insects, which feed mainly on the live parts of plants, especially leaves and stems; nematodes (small worms), which live mainly in the soil and feed on roots and other plant tissues; bacterial and viral diseases; weeds (plants that compete with the crops); and vertebrates (mainly rodents and birds) that feed on grain or fruit.

Because a farm is maintained in a very early stage of ecological succession and is enriched by fertilizers and water, it is a good place not only for crops but also for other early-successional plants. These noncrop and therefore undesirable plants are what we call weeds. A weed is just a plant in a place we do not want it to be. There are about 30,000 species of weeds, and in any year a typical farm field is infested with between 10 and 50 species of them. Some weeds can have a devastating effect on crops. For example, the production of soybeans is reduced by 60% if a weed called cocklebur grows three individuals per meter (one individual per foot).[22]

Even today, with modern technology, the total losses from all pests are huge; in the United States, pests account for an estimated loss of one-third of the potential harvest and about one-tenth of the harvested crop. Preharvest losses are due to competition from weeds, diseases, and herbivores; postharvest losses are largely due to herbivores.[23]

Pesticides

Before the Industrial Revolution, farmers could do little to prevent pests except remove them or use farming methods that tended to decrease their density. Preindustrial farmers planted aromatic herbs and other vegetation that repelled insects. The scientific industrial revolution brought major changes in agriculture pest control, which we can divide into four stages:

Stage 1: Broad-Spectrum Inorganic Toxins

With the beginning of modern science-based agriculture, people began to search for chemicals that would reduce the abundance of pests. Their goal was a "magic bullet"—a chemical (referred to as a narrow-spectrum pesticide) that would have a single target, just one pest, and not affect anything else. But this proved elusive. The earliest pesticides were simple inorganic compounds that were widely toxic. One of the earliest was arsenic, a chemical element toxic to all life, including people. It was certainly effective in killing pests, but it killed beneficial organisms as well and was very dangerous to use.

Stage 2: Petroleum-Based Sprays and Natural Plant Chemicals (1930s on)

Many plants produce chemicals as a defense against disease and herbivores, and these chemicals are effective pesticides. Nicotine, from the tobacco plant, is the primary agent in some insecticides still widely used today. However, although natural plant pesticides are comparatively safe, they were not as effective as desired.

Stage 3: Artificial Organic Compounds

Artificial organic compounds have created a revolution in agriculture, but they have some major drawbacks (see Risk-Benefit Analysis in Chapter 3). One problem is secondary pest outbreaks, which occur after extended use (and possibly because of extended use) of a pesticide. Secondary pest outbreaks can come about in two ways: (1) Reducing one target species reduces competition with a second species, which then flourishes and becomes a pest, or (2) the pest develops resistance to the pesticides through evolution and natural selection, which favor those with a greater immunity to the chemical. Resistance has developed to many pesticides. For example, Dasanit (fensulfothion), an organophosphate first introduced in 1970 to control maggots that attack onions in Michigan, was originally successful but is now so ineffective that it is no longer used for that crop.

Some artificial organic compounds, such as DDT, are broad-spectrum, but more effective than natural plant chemicals. However, they also have had unexpected environmental effects. For example, aldrin and dieldrin have been widely used to control termites as well as pests on corn, potatoes, and fruits. Dieldrin is about 50 times as toxic to people as DDT. These chemicals are designed to remain in the soil and typically do so for years. Therefore, they have spread widely.

In 1999, world use of pesticides totaled 262,000 metric tonnes. By 2007, it had increased 21% to 316,000 metric tonnes[23] In contrast, use during that same time in the United States decreased by 17%, from 94,347 to 78,000 metric tonnes, suggesting that modern methods of pest control, discussed later in this chapter, are having some positive effects (Figure 11.17).

But the magic bullet has remained elusive. Once applied, these chemicals may decompose in place or may be blown by the wind or transported by surface and subsurface waters, meanwhile continuing to decompose. Sometimes the initial breakdown products (the first, still complex chemicals produced from the original pesticides) are toxic, as is the case with DDT. Eventually, the toxic compounds are decomposed to their original inorganic or simple, nontoxic organic compounds, but for some chemicals this can take a very long time.

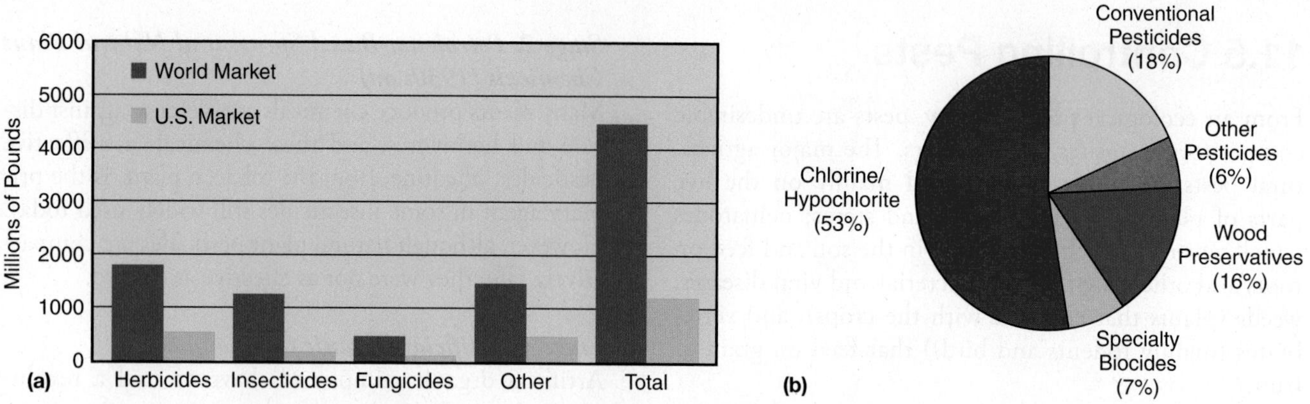

FIGURE 11.17 World use of pesticides. (a) Total amounts, (b) percentage by main type.

Public-health standards and environmental-effects standards have been established for some of these compounds. The United States Geological Survey has established a network for monitoring 60 sample watersheds throughout the nation. These are medium-size watersheds, not the entire flow from the nation's major rivers. One such watershed is that of the Platte River, a major tributary of the Missouri River.[24]

The most common herbicides used for growing corn, sorghum, and soybeans along the Platte River were alachlor, atrazine, cyanazine, and metolachlor, all organonitrogen herbicides. Monitoring of the Platte near Lincoln, Nebraska, suggested that during heavy spring runoff, concentrations of some herbicides might be reaching or exceeding established public-health standards. But this research is just beginning, and it is difficult to reach definitive conclusions as to whether present concentrations are causing harm in public water supplies or to wildlife, fish, algae in fresh water, or vegetation. Advances in knowledge give us much more information, on a more regular basis, about how much of many artificial compounds are in our waters, but we are still unclear about their environmental effects. A wider and better program to monitor pesticides in water and soil is important to provide a sound scientific basis for dealing with pesticides.

Stage 4: Integrated Pest Management and Biological Control

Integrated pest management (IPM) uses a combination of methods, including biological control, certain chemical pesticides, and some methods of planting crops (Figure 11.18). A key idea underlying IPM is that the goal can be control rather than complete elimination of a

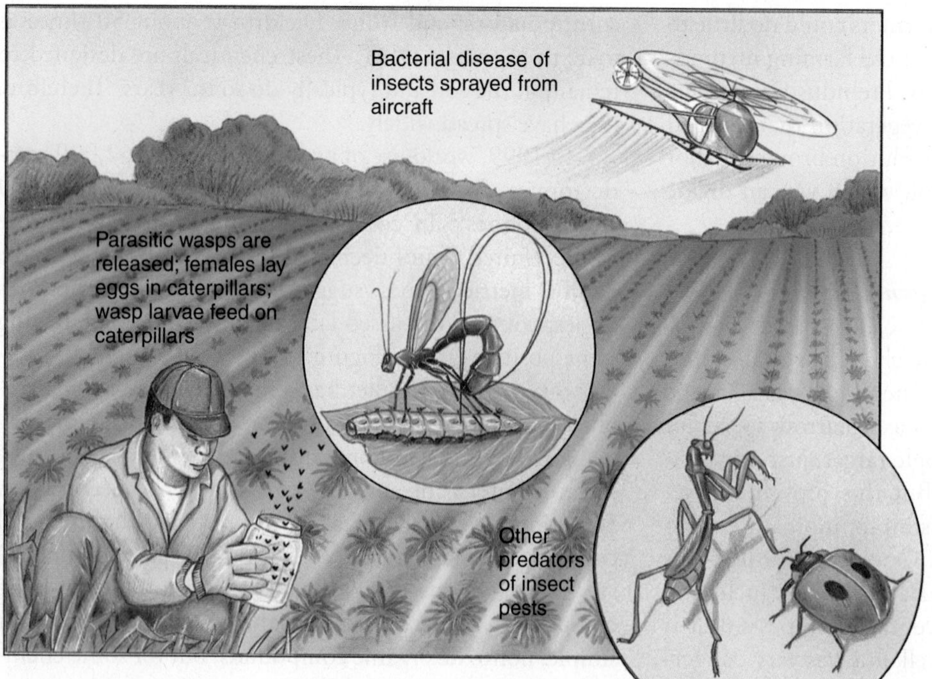

FIGURE 11.18 Integrated pest management. The goal is to reduce the use of artificial pesticides, lower costs, and efficiently control pests.

pest. This is justified for several reasons. Economically, it becomes more and more expensive to eliminate a greater and greater percentage of a pest, while the value of ever-greater elimination, in terms of crops to sell, becomes less and less. This suggests that it makes economic sense to eliminate only enough to provide benefit and leave the rest. In addition, allowing a small but controlled portion of a pest population to remain does less damage to ecosystems, soils, water, and air.

Integrated pest management also moves away from monoculture growing in perfectly regular rows. Studies have shown that just the physical complexity of a habitat can slow the spread of parasites. In effect, a pest, such as a caterpillar or mite, is trying to find its way through a maze. If the maze consists of regular rows of nothing but what the pest likes to eat, the maze problem is easily solved by the dumbest of animals. But if several species, even two or three, are arranged in a more complex pattern, pests have a hard time finding their prey.

No-till or low-till agriculture, which we discussed in the opening case study, is another feature of IPM. While it reduces erosion, it has a downside: It helps the natural enemies of some pests to build up in the soil, whereas plowing destroys the habitats of these enemies.

Biological control includes using one species that is a natural enemy of another. One of the most effective is the bacterium ***Bacillus thuringiensis***, known as BT, which causes a disease that affects caterpillars and the larvae of other insect pests. Spores of BT are sold commercially—you can buy them at your local garden store and use them in your home garden. BT has been one of the most important ways to control epidemics of gypsy moths, an introduced moth whose larvae periodically strip most of the leaves from large areas of forests in the eastern United States. BT has proved safe and effective—safe because it causes disease only in specific insects and is harmless to people and other mammals, and because, as a natural biological "product," its presence and its decay are nonpolluting.

Another group of effective biological-control agents are small wasps that are parasites of caterpillars. Control of the oriental fruit moth, which attacks a number of fruit crops, is an example of IPM biological control. The moth was found to be a prey of a species of wasp, *Macrocentrus ancylivorus*, and introducing the wasp into fields helped control the moth. Interestingly, in peach fields the wasp was more effective when strawberry fields were nearby. The strawberry fields provided an alternative habitat for the wasp, especially important for overwintering.[25] As this example shows, spatial complexity and biological diversity also become part of the IPM strategy.

In the list of biological-control species we must not forget ladybugs, which are predators of many pests. You can buy these, too, at many garden stores and release them in your garden.

Another biological control uses sex pheromones, chemicals released by most species of adult insects (usually the female) to attract members of the opposite sex. In some species, pheromones have been shown to be effective up to 4.3 km (2.7 mi) away. These chemicals have been identified, synthesized, and used as bait in insect traps, in insect surveys, or simply to confuse the mating patterns of the insects involved.

While biological control works well, it has not solved all problems with agricultural pests.

11.6 The Future of Agriculture

Today, there are three major technological approaches to agriculture. One is modern mechanized agriculture, where production is based on highly mechanized technology that has a high demand for resources—including land, water, and fuel—and makes little use of biologically based technologies. Another approach is resource-based—that is, agriculture based on biological technology and conservation of land, water, and energy. An offshoot of this second approach is organic food production—growing crops without artificial chemicals (including pesticides) or genetic engineering but instead using ecological control methods. The third approach is genetic engineering.

In mechanized agriculture, production is determined by economic demand and limited by that demand, not by resources. In resource-based agriculture, production is limited by environmental sustainability and the availability of resources, and economic demand usually exceeds production.

With these methods in mind, we can consider what can be done to help crop production keep pace with human population growth. Here are some possibilities.

Increased Production per Acre

Some agricultural scientists and agricultural corporations believe that production per unit area will continue to increase, partially through advances in genetically modified crops. This new methodology, however, raises some important potential environmental problems, which we will discuss later.

Increased Farmland Area

A United Nations Food and Agriculture Organization conference held in early 2008 considered the coming world food crisis. It was reported that 23 million hectares (more than 46 million acres) of farmland had been withdrawn from production in Eastern Europe and the Commonwealth of Independent States (CIS) region, especially in countries such as Kazakhstan, Russia, and Ukraine, and that at least half—13 million hectares

(15 million acres)—could be readily put back into production with little environmental effect.[26] This would be like adding all the farmland in Iowa, Illinois, and Indiana.[27]

The need for additional farmland brings up, once again, the problem that agrifuels bring to agriculture: taking land away from food production to produce fuels instead.

New Crops and Hybrids

Since there are so many plant species, perhaps some yet unused ones could provide new sources of food and grow in environments little used for agriculture. Those interested in conserving biological diversity urge a search for such new crops on the grounds that this is one utilitarian justification for the conservation of species. It is also suggested that some of these new crops may be easier on the environment and therefore more likely to allow sustainable agriculture. But it may be that over the long history of human existence those species that are edible have already been found, and the number is small. Research is under way to seek new crops or plants that have been eaten locally but whose potential for widespread, intense cultivation has not been tested.

Among the likely candidates for new crops are amaranth for seeds and leaves; *Leucaena*, a legume useful for animal feed; and triticale, a synthetic hybrid of wheat and rye.[28] A promising source of new crops is the desert; none of the 14 major crops are plants of arid or semiarid regions, yet there are vast areas of desert and semidesert. The United States has 200,000 million hectares (about 500,000 million acres) of arid and semiarid rangeland. In Africa, Australia, and South America, the areas are even greater. Several species of plants can be grown commercially under arid conditions, allowing us to use a biome for agriculture that has been little used in this way in the past. Examples of these species are guayule (a source of rubber), jojoba (for oil), bladderpod (for oil from seeds), and gumweed (for resin). Jojoba, a native shrub of the American Sonoran Desert, produces an extremely fine oil, remarkably resistant to bacterial degradation, which is useful in cosmetics and as a fine lubricant. Jojoba is now grown commercially in Australia, Egypt, Ghana, Iran, Israel, Jordan, Mexico, Saudi Arabia, and the United States.[29] Although these examples are not food crops, they release other lands, now used to produce similar products, to grow food.

Among the most successful developments of new hybrids have been those developed by the **green revolution**, the name attached to post–World War II programs that have led to the development of new strains of crops with higher yields, better resistance to disease, or better ability to grow under poor conditions. These crops include superstrains of rice (at the International Rice Research Institute in the Philippines) (see Figure 11.19) and strains of maize with improved disease resistance (at the International Maize and Wheat Improvement Center in Mexico).

Better Irrigation

Drip irrigation—from tubes that drip water slowly—greatly reduces the loss of water from evaporation and increases yield. However, it is expensive and thus most likely to be used in developed nations or nations with a large surplus of hard currency—in other words, in few of the countries where hunger is most severe.

Organic Farming

Organic farming is typically considered to have three qualities: It is more like natural ecosystems than monoculture; it minimizes negative environmental impacts; and the food that results from it does not contain artificial compounds. According to the U.S. Department of Agriculture (USDA), organic farming has been one of the fastest-growing sectors in U.S. agriculture, although it still occupies a small fraction of U.S. farmland and contributes only a small amount of agriculture income. By the end of the 20th century, it amounted to about $6 billion—much less than the agricultural production of California. By 2008, sales of organic products had increased to $21.1 billion, more than 3% of total food sales.[30] USDA certification of organic farming became mandatory in 2002. After that, organic cropland that was listed as certified more than doubled, and the number of farmers certifying their products rose 40%. In the United States today, more than 1.3 million acres are certified as organic. There are about 12,000 organic farmers in the United States, and the number is growing 12% per year.[31]

TED ALJIBE/AFP/Getty Images

FIGURE 11.19 Experimental rice plots at the International Rice Research Institute, Philippines, showing visual crop variation based on the use of fertilizers.

Eating Lower on the Food Chain

Some people believe it is ecologically unsound to use domestic animals as food, on the grounds that eating each step farther up a food chain leaves much less food to eat per acre. This argument is as follows: No organism is 100% efficient; only a fraction of the energy in food taken in is converted to new organic matter. Crop plants may convert 1–10% of sunlight to edible food, and cows may convert only 1–10% of hay and grain to meat. Thus, the same area could produce 10 to 100 times more vegetation than meat per year. This holds true for the best agricultural lands, which have deep, fertile soils on level ground. However, there are large areas of farmland that are too fragile for intensive plowed crop production but suitable for grazing; production of meat, dairy, and eggs is a sound use of such land.

11.7 Genetically Modified Food: Biotechnology, Farming, and Environment

The discovery that DNA is the universal carrier of genetic information has led to development and use of **genetically modified crops** (GMCs), which has given rise to new environmental controversies as well as a promise of increased agricultural production. In the United States, the land area planted with GMCs has grown rapidly from the first plantings in 1996 to now occupying 85% of the land planted in corn, 94% of cotton, and 93% of soybeans—three of America's major crops (Figure 11.20).[32]

At first, GMCs created a huge controversy, especially in Europe (Figure 11.21), but the controversy has died down in the United States, rarely making news. In contrast, the European Union has strict rules about genetically modified crops, requiring that if a food product is more than 0.9% genetically modified it must be labeled as such.[33] The European approach is based on freedom of choice: When consumers see that a food product contains GMCs, the choice is theirs whether to purchase the item. This is not the case in the United States; it is difficult for the average consumer to avoid food and nonfood agricultural products from GMCs. Today, foods listed as organic are not supposed to contain genetically modified crops, but otherwise it is not possible to separate GMC products from non-GMC products when they reach the retail market.

Genetic engineering in agriculture involves several different practices, which we can group as follows: (1) faster and more efficient ways to develop new hybrids; (2) introduction of the "terminator gene"; and (3) transfer of genetic properties from widely divergent kinds of life. These three practices have quite different potentials and problems.

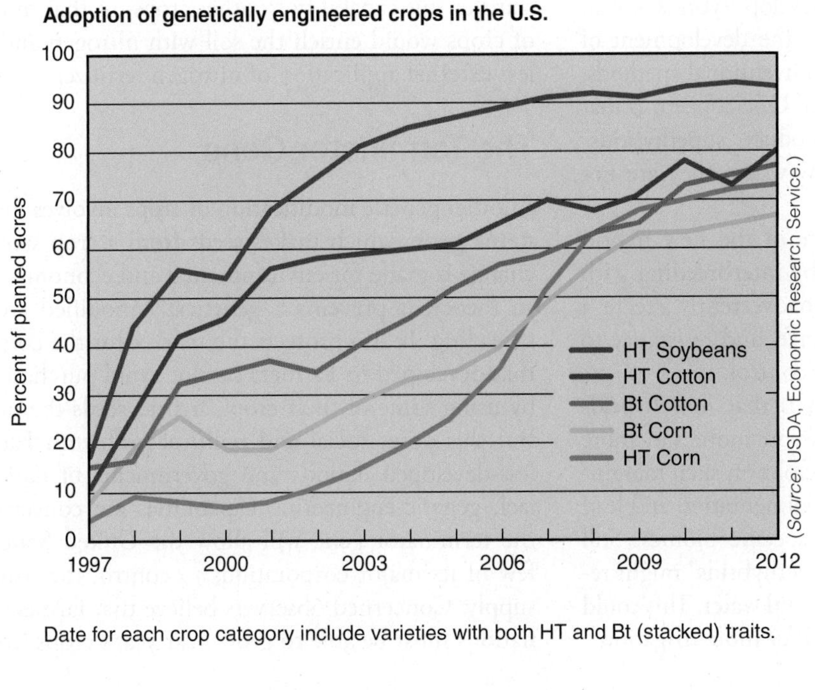

FIGURE 11.20 Percentage of area planted in genetically modified crops for corn, cotton, and soybeans.

EPA/Pedro Armestre/Corbis Images

FIGURE 11.21 Genetically modified crops have created some major controversies, as indicated by this demonstration in a cornfield in Villanueva de Gallego in northeastern Spain in 2003. The slogan on the left banner reads "Stop genetic pollution." Miguel Arias Canete, the Spanish Minister of Agriculture, Food, and Fishing at that time, is portrayed close to a huge corn ear.

The development of hybrids within a species is a natural phenomenon (see Chapter 9), and the development of hybrids of major crops, especially of small grains, has been a major factor in the great increase in productivity of 20th-century agriculture. So, strictly from an environmental perspective, genetic engineering to develop hybrids within a species is likely to be as benign as the development of agricultural hybrids has been with conventional methods. There is an important caveat, however. One concern is that genetic modification methods may produce "superhybrids" that are so productive they can grow where they are not wanted and become pests.

There is also concern that some of the new hybrid characteristics could be transferred by interbreeding with closely related weeds. This could inadvertently create a "superweed" whose growth, persistence, and resistance to pesticides would make it difficult to control.

Another environmental concern is that new hybrids might be developed that could grow on more and more marginal lands. The development of crops on such marginal lands might increase erosion and sedimentation and lead to decreased biological diversity in specific biomes. Still another potential problem is that "superhybrids" might require much more fertilizer, pesticides, and water. This could lead to greater pollution and the need for more irrigation.

On the other hand, genetic engineering could lead to hybrids that require less fertilizer, pesticide use, and water. For example, at present, only legumes (peas and their relatives) have symbiotic relationships with bacteria and fungi that allow them to fix nitrogen. Attempts are under way to transfer this capability to other crops, so that more kinds of crops would enrich the soil with nitrogen and require less external application of nitrogen fertilizer.

The Terminator Gene

Another genetic modification of crops involves the **terminator gene**, which makes seeds from a crop sterile. This change is made for environmental and economic reasons. In theory, it prevents a genetically modified crop from spreading. It also protects the market for the corporation that developed it: Farmers cannot avoid purchasing seeds by using some of their crops' hybrid seeds the next year. But this poses social and political problems. Farmers in less-developed nations and governments of nations that lack genetic-engineering capabilities are concerned that the terminator gene will allow the United States and a few of its major corporations to control the world food supply. Concerned observers believe that farmers in poor nations must be able to grow next year's crops from their own seeds because they cannot afford to buy new seeds every year. This is not directly an environmental problem, but it can become such a problem indirectly by affecting total world food production, which then affects the human population and how land is used in areas that have been in agriculture.

Transfer of Genes from One Major Form of Life to Another

Most environmental concerns have to do with the third kind of genetic modification of crops: the transfer of genes from one major kind of life to another. This is a novel effect and, as we have explained, is therefore more likely to have undesirable results. In several cases, in fact, this type of genetic modification has affected the environment in unforeseen and undesirable ways. Perhaps the best-known example involves potatoes and corn, caterpillars that eat these crops, a disease of caterpillars that controls these pests, and an endangered species, monarch butterflies. Here is what happened.

As discussed earlier, the bacterium *Bacillus thuringiensis* is a successful pesticide that causes a disease in many caterpillars. With the development of biotechnology, agricultural scientists studied the bacteria and discovered the toxic chemical and the gene that caused its production within the bacteria. This gene was then transferred to potatoes and corn so that the biologically engineered plants produced their own pesticide. At first, this was believed to be a constructive step in pest control because it was no longer necessary to spray a pesticide. However, the genetically engineered potatoes and corn produced the toxic BT substance in every cell—not just in the leaves that the caterpillars ate, but also in the potatoes and corn sold as food, in the flowers, and in the pollen. This has a potential—not yet demonstrated—to create problems for species that are not intended targets of the BT.

A strain of rice has been developed that produces beta-carotene, important in human nutrition. This rice has added nutritional benefits that are particularly valuable for the poor of the world who depend on rice as a primary food. The gene that enables rice to make beta-carotene comes from daffodils, but the modification actually required the introduction of four specific genes and would likely be impossible without genetic-engineering techniques. That is, genes were transferred between plants that would not exchange genes in nature. Once again, the rule of natural change suggests that we should monitor such actions carefully. Indeed, although the genetically modified rice appears to have beneficial effects, the government of India has in the past refused to allow it to be grown in that country and has proposed a ten-year moratorium, originally planned to go into effect in 2013, but has not yet been finalized at the time we are writing.

11.8 Aquaculture

In contrast to food obtained on land, we still get most of our marine and freshwater food by hunting. Hunting wild fish has not been sustainable (see Chapter 13), and thus **aquaculture**, the farming of this important source of protein in both marine and freshwater habitats, is growing rapidly and could become one of the major ways to provide food of high nutritional quality. Popular aquacultural animals include carp, tilapia, oysters, and shrimp, but in many nations other species are farm-raised and culturally important, such as yellowtail (important in Japan and perhaps just one of several species); crayfish (United States); eels and minnows (China); catfish (southern and midwestern United States); salmon (Canada, Chile, Norway, and the United States); trout (United States); plaice, sole, and the Southeast Asian milkfish (Great Britain); mussels (Canada, France, Spain, and Southeast Asian countries); and sturgeon (Ukraine). A few species—trout and carp—have been subject to genetic breeding programs.[34]

Although relatively new in the United States, aquaculture has a long history elsewhere, especially in China, where at least 50 species are grown, including finfish, shrimp, crab, other shellfish, sea turtles, and sea cucumbers (not a vegetable but a marine animal).[34] In the Szechuan area of China, fish are farmed on more than 100,000 hectares (about 250,000 acres) of flooded rice fields. This is an ancient practice that can be traced back to a treatise on fish culture written by Fan Li in 475 B.C.[34] In China and other Asian countries, farmers often grow several species of fish in the same pond, exploiting their different ecological niches. Ponds developed mainly for carp, a bottom-feeding fish, also contain minnows, which feed at the surface on leaves added to the pond.

Aquaculture can be extremely productive on a per-area basis, in part because flowing water brings food from outside into the pond or enclosure. Although the area of Earth that can support freshwater aquaculture is small, we can expect this kind of aquaculture to increase and become a more important source of protein.

Sometimes fishponds use otherwise wasted resources, such as fertilized water from treated sewage. Other fishponds exist in natural hot springs (Idaho) or use water that is still warm after it is used to cool electric power plants (Long Island, New York; Great Britain).[34]

Mariculture, the farming of ocean fish, though producing a small part of the total marine fish catch, has grown rapidly in the last decades and will likely continue to do so. Oysters and mussels are grown on rafts lowered into the ocean, a common practice in the Atlantic Ocean in Portugal and in the Mediterranean in such nations as France. These animals are filter feeders—they obtain food from water that moves past them. Because a small raft is exposed to a large volume of water, and thus a large volume of food, rafts can be extremely productive. Mussels grown on rafts in the bays of Galicia, Spain, produce 300 metric tonnes per hectare, whereas public harvesting grounds of wild shellfish in the United States yield only about 10 kg/ha (that's just a hundredth of a metric tonnes). Oysters and mussels are grown on artificial pilings in the intertidal zone in the state of Washington (Figure 11.22).

FIGURE 11.22 **An oyster farm in Poulsbo, Washington.** Oysters are grown on artificial pilings in the intertidal zone.

Some Negatives

Although aquaculture has many benefits and holds great promise for our food supply, it also causes environmental problems. Fishponds and marine fish kept in shallow enclosures connected to the ocean release wastes from the fish and chemicals such as pesticides, polluting local environments. In some situations, aquaculture can damage biological diversity. This is a concern with salmon aquaculture in the Pacific Northwest, where genetic strains not native to a stream are grown and some are able to mix with wild populations and breed. Problems with salmon aquaculture demonstrate the need for improved methods and greater care about environmental effects.

CRITICAL THINKING ISSUE

Should Rice Be Grown in a Dry Climate?

As this chapter's introductory case study explains, water is a precious resource, especially in California, where the average rainfall is low (38–51 cm/year, or 15–20 in./year), the population—37 million—is large and growing, and agricultural water use is high. (Farmers use 46% of the state's water to irrigate 3.5 million hectares [8.6 million acres], more than in any other state.)[35]

With cities and industries, not to mention fish and other wildlife, in need of water, growing crops that require a lot of water has come under heavy criticism, especially because much of the water that farmers receive is subsidized by the government. Some farmers have responded by reducing the acreage of water-intensive crops and switching to crops that require less water, such as fruits, vegetables, and nuts.

California produces 21% of the nation's rice, making it the second-largest rice-growing state in the United States, second only to Arkansas.[36] The rice produced has a market value of more than $300 million, and it uses enough water to supply one-fourth of the state's population. Although rice growers are not the biggest water users when evapotranspiration is taken into account, they have been particular targets of attack because the flooded fields required for rice are a visible reminder of the amount of water used by agriculture. In addition to high water use, rice growing has had other adverse environmental effects: Its high pesticide and herbicide use contaminates rivers and drinking supplies, and the burning of stubble left after harvesting contributes to air pollution in the valley.

Rice growers have responded to pressure to clean up their act in a number of ways. In the 1990s they decreased water use by 32% and pesticide use by 98%. According to the California Rice Commission, in 2012 "16 gallons of water are needed to produce one serving of brown rice. 25 gallons for white rice."[37] Rice was once a thirstier crop, but a series of innovations over the last three decades have improved our water-use efficiency.

Growers have switched to biodegradable pesticides. And they have decreased the burning of stubble by plowing it under, harvesting it, or flooding the fields in winter when evapotranspiration is much lower, allowing the organic matter to rot. Experts are also trying to find ways to protect young salmon, which run in the rivers of the Sacramento Valley, from being pumped into the channels leading to the rice fields. Although drawing water out of the rivers might have a negative impact on salmon, the release of water at the end of the winter, when the rivers are low, could help the spring run of salmon.

The fields also provide a wetland habitat for many migrating birds, including snow geese, and other species, so that winter flooding benefits an even wider diversity and greater number of species (Figure 11.23).

Critical Thinking Questions

1. Most of the areas where rice is grown have alkaline, hardpan soil, unsuited to other crops. If rice were not grown on the land, it probably would be developed for housing. Each acre of rice requires five acre-feet of water. Less than one acre-foot could supply water for a family of four for a year. If housing lots were one-eighth acre and all housed families of four, how many acre-feet of water per acre of housing would be used in a year? Which uses more, an acre of rice or an acre of people? How would real estate development affect wildlife habitat?
2. Farmers find that the presence of waterfowl in flooded fields speeds up the rotting of stubble. Can you think of at least two reasons for this?
3. Although birds can feed on rice grains in dry fields, they get a more balanced diet by feeding in flooded fields. Why would this be so?
4. Two of the unknowns in this system are the long-term effects of flooding on the ability of the soil to support rice growing and the long-term effects on dryland species, such as rattlesnakes and rats. What is a scientific way of investigating one of these questions?
5. The new rice-growing practices are referred to as "win–win." What is meant by the term in general, and how does this situation illustrate the term?

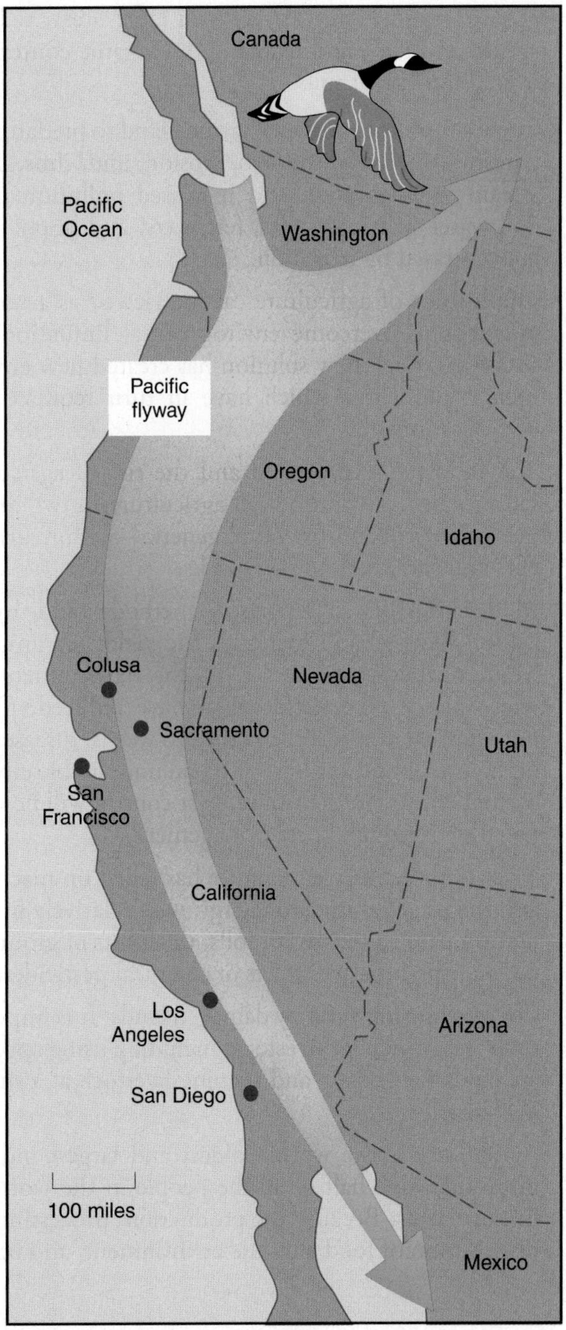

FIGURE 11.23 The Pacific flyway, used by many birds that stop at agricultural wetlands in California.

(*Source*: California Rice Commission 2003.)

SUMMARY

The basic environmental questions about agriculture are: Can we produce enough food to feed Earth's growing human population? Can we grow crops sustainably, so that both crop production and agricultural ecosystems maintain their viability? Can we produce this food without seriously damaging other ecosystems that receive the wastes of agriculture?

- Agriculture changes the environment; the more intense the agriculture, the greater the changes.
- From an ecological perspective, agriculture is an attempt to keep an ecosystem in an early-successional stage.
- Farming greatly simplifies ecosystems, creating short and simple food chains, growing a single species or

genetic strain in regular rows in large areas, reducing species diversity, and reducing the organic content and overall fertility of soils.

- These simplifications open farmed land to predators and parasites, increased soil loss, erosion, and, thus, downstream sedimentation and increased pollution of soil and water with pesticides, fertilizers, and heavy metals concentrated by irrigation.
- The history of agriculture can be viewed as a series of attempts to overcome environmental limitations and problems. Each new solution has created new environmental problems, which have in turn required their own solutions.
- The Industrial Revolution and the rise of agricultural sciences have revolutionized agriculture in two areas—one ecological and the other genetic—with many benefits and some serious drawbacks.
- Modern fertilizers, irrigation methods, and hybridization have greatly increased the yield per unit area. Modern chemistry has led to the development of a wide variety of pesticides that have reduced, though not eliminated, the loss of crops to weeds, diseases, and herbivores, but these have also had undesirable environmental effects. In the future, pest control will be dominated by integrated pest management.
- Most 20th-century agriculture has relied on machinery and the use of abundant energy, with relatively little attention paid to the loss of soils, the limits of groundwater, and the negative effects of chemical pesticides.
- Overgrazing has severely damaged lands. It is important to properly manage livestock, including using appropriate lands for grazing and keeping livestock at a sustainable density.
- Agriculture is the world's oldest and largest industry; more than one-half of all the people in the world still live on farms. Because the production, processing, and distribution of food alter the environment, and because of the size of the industry, large effects on the environment are unavoidable.
- Alternative agricultural methods appear to offer the greatest hope of sustaining agricultural ecosystems and habitats over the long term, but more tests and better methods are needed.
- Agriculture has numerous global effects. It changes land cover, affecting climate at regional and global levels, increasing carbon dioxide in the atmosphere, and adding to the buildup of greenhouse gases, which in turn affects climate. Fires to clear land for agriculture may significantly affect the climate by adding small particulates to the atmosphere. Genetic modification is a new global issue that has not only environmental but also political and social effects.
- The agricultural revolution makes it possible for fewer and fewer people to produce more and more food and leads to greater productivity per acre. Freed from dependence on farming, people flock to cities, which leads to increased urban effects on the land. Thus, agricultural effects on the environment indirectly extend to the cities.
- Farming is one of the most direct and large-scale ways that people affect nature. Our own sustainability, as well as the quality of our lives, depends heavily on how we farm.
- Human activities have seriously damaged one-fourth of the world's total land area, impacting one-sixth of the world's population (about 1 billion people). Overgrazing, deforestation, and destructive farming practices have caused so much damage that recovery in some areas will be difficult, and restoration of the rest will require serious actions. A major value judgment we must make in the future is whether our societies will allocate funds to restore these damaged lands. Restoration requires scientific knowledge, both about present conditions and about actions required for restoration. Will we seek this knowledge and pay for it?

REEXAMINING THEMES AND ISSUES

Sean Randall/Getty Images, Inc.

HUMAN POPULATION

Our modern food problem is the result of the great increase in human population growth, which outstrips local food production in many areas, and an inadequate food distribution system for this growing population. Human population growth will eventually be limited by total food production.

© Biletskiy_Evgeniy/iStockphoto

SUSTAINABILITY

A major goal of agriculture must be to achieve sustainable food production in any location. This requires the development of farming methods that do not damage soils, eliminate water supplies, cause extinction of wild relatives of crops or of potential new food species, or lead to permanent pollution downstream. The world as a whole is moving from demand-based to resource-based agriculture. The latter is more consistent with a sustainable approach to agriculture. From an ecological perspective, agriculture is an attempt to keep an ecosystem in a specific stage, usually an early, highly productive successional stage. In this way, agriculture works against natural mechanisms of sustainability, and we must compensate for this. Alternative agricultural methods appear to offer the greatest hope of sustaining agricultural ecosystems and habitats over the long term, but more tests and better methods are needed. As the experience with European agriculture shows, crops can be produced on the same lands for thousands of years as long as sufficient fertilizers and water are available; however, the soils and other aspects of the original ecosystem are greatly changed—these are not sustained. In agriculture, production can be sustained, but the ecosystem may not be.

© Anton Balazh 2011/iStockphoto

GLOBAL PERSPECTIVE

Most of the world's food is provided by only 14 crop species. A challenge for the future is to search the world's diversity of life for additional food species. Food is a global resource in that it is traded internationally, and the availability of food in any area is the result of both local production and global agricultural markets.

ssguy/ShutterStock

URBAN WORLD

The artificiality of urban environments leads many city dwellers to believe that they are independent of the environment. But when urban people in Brazil were starving, they broke into food-storage facilities and caused social disruption, thus illustrating the intimate connection between urban life and agricultural production.

B2M Productions/Getty Images, Inc.

PEOPLE AND NATURE

Agriculture is one of the major ways in which people have changed nature. Through agriculture, people have directly changed more than 10% of the Earth's land surface and indirectly affected a larger area. The development of agriculture allowed the human population to increase greatly, and modern agriculture is necessary to sustain the huge number of people on Earth.

George Doyle/Getty Images, Inc.

SCIENCE AND VALUES

The fundamental ethical questions we face concerning our food supply are: Shall we continue to attempt to feed more and more people? Should we attempt this at the risk of sacrificing the habitats of noncrop species, natural ecosystems, and landscapes; reducing water supply for other purposes; and increasing the erosion of landscape? Or shall we attempt to limit our numbers and also limit total food production?

KEY TERMS

macronutrient 245
malnourishment 237
mariculture 253
micronutrient 245
monoculture 235
organic farming 250
pasture 241
rangeland 241
synergistic effect 246
terminator gene 252
undernourishment 237

STUDY QUESTIONS

1. Design an integrated pest management scheme for a small vegetable garden in a city lot behind a house. How would this scheme differ from integrated pest management used on a large farm? What aspects of IPM could not be used? How might the artificial structures of a city be put to use to benefit IPM?
2. Under what conditions might grazing cattle be sustainable when growing wheat is not? Under what conditions might a herd of bison provide a sustainable supply of meat when cows might not?
3. Pick one of the nations in Africa that has a major food shortage. Design a program to increase its food production. Discuss how reliable that program might be given the uncertainties that nation faces.
4. Should genetically modified crops be considered acceptable for "organic" farming?
5. You are about to buy your mother a bouquet of 12 roses for Mother's Day, but you discover that the roses were genetically modified to give them a more brilliant color and to produce a natural pesticide through genetic energy. Do you buy the flowers? Explain and justify your answer based on the material presented in this chapter.
6. A city garbage dump is filled, and it is suggested that the area be turned into a farm. What factors in the dump might make it a good area to farm, and what might make it a poor area to farm?
7. You are sent into the Amazon rain forest to look for new crop species. In what kinds of habitats would you look? What kinds of plants would you look for?

FURTHER READING

Borgstrom, G., *The Hungry Planet: The Modern World at the Edge of Famine* (New York: Macmillan, 1965). A classic book by one of the leaders of agricultural change.

Cunfer, G., *On the Great Plains: Agriculture and Environment* (College Station, TX: Texas A&M University Press, 2005). Uses the history of European agriculture applied to the American Great Plains as a way to discuss the interaction between nature and farming.

Mann, C., *1493: Uncovering the New World Columbus Created* (New York, NY: Vintage, 2012 reprint). Charles C. Mann is the author of *1493*, a New York Times best-seller, and *1491*, which won the U.S. National Academy of Sciences' Keck Award for the best book of the year.

Manning, R., *Against the Grain: How Agriculture Has Hijacked Civilization* (New York, NY: North Point Press, 2004). An important iconoclastic book in which the author attributes many of civilization's ills—from war to the spread of disease—to the development and use of agriculture. In doing so, he discusses many of the major modern agricultural issues.

Mazoyer, M., and L. Roudar, *A History of World Agriculture: From the Neolithic Age to the Current Crisis* (New York, NY: Monthly Review Press, 2006). By two French professors of agriculture, this book argues that the world is about to reach a new farming crisis, which can be understood from the history of agriculture.

McNeely, J.A., and S.J. Scherr, *Ecoagriculture* (Washington, DC: Island Press, 2003).

Seymour, John, and Deirdre Headon, eds., *The Self-sufficient Life and How to Live It* (Cambridge: DK ADULT, 2003). Ever think about becoming a farmer and leading an independent life? This book tells you how to do it. It is an interesting, alternative way to learn about agriculture. The book is written for a British climate, but the messages can be applied generally.

Smil, V., *Feeding the World* (Cambridge, MA: MIT Press, 2000).

NOTES

1. Ruen, J. November 2012. No-till continues to grow. *AgProfessional Magazine,* pp. 1–4. Also available at: http://www.agprofessional.com/agprofessional-magazine/No-Till-Continues-to-Grow-179527291.html
2. Ostendorf Mintz, M. March 2011. No-till brings challenges, surprises and new options, *No-Till Farmer*, pp. 10–12.
3. The World Bank. 2012. Agricultural Land (% of land area). http://data.worldbank.org/indicator/AG.LND.AGRI.ZS

4. U.S. Department of Agriculture (USDA). 2007. National Resources Inventory 2003 Annual NRI Soil Erosion Report February 2007. Accessed December 23, 2008 from http:// www.nrcs.usda.gov/technical/NRI/2003/SoilErosion mrb.pdf

5. U.S. Department of Agriculture. 2012. Statistical Abstracts, Table 841. Value Added to Economy by Agricultural Sector: 1990 to 2009. http://www.census.gov/compendia/statab/2012/tables/12s0841.pdf

6. Mann, C. 2012 (reprint). *1493: Uncovering the New World Columbus Created.* New York, NY: Vintage.

7. Information on African wars is from Globalsecurity.org http://www.globalsecurity.org/military/world/war/index.html Accessed June 16, 2008. Information on African coups is from http://www.crisisgroup.org/home/index.cfm?id=4040.

8. UN Food and Agriculture Organization. 2003. Agricultural Commodities: Profiles and Relevant to Negotiating Issues. Rome, Italy.

9. Field, J.O., ed. 1993. *The Challenge of Famine: Recent Experience, Lessons Learned.* Hartford, CT: Kumarian Press.

10. World Food Programme. 1998. *Tackling Hunger in a World Full of Food: Tasks Ahead for Food Aid*; and UN FAO. FAOSTAT 2003 Web site.

11. Anonymous. 2012. *The State of Food Insecurity in the World.* UN Food and Agriculture Organization. Rome, Italy.

12. FAO Statistics. 2012. Live animals. http://faostat.fao.org/site/573/DesktopDefault.aspx?PageID=573#ancor

13. Anonymous. 2010. Soil Erosion on Cropland. 2007 National Resources Inventory. http://www.nrcs.usda.gov/Internet/FSE_DOCUMENTS/nrcs143_012269.pdf

14. Trirnble, S.W., and P. Crosson. 2000. U.S. soil erosion rates—myth and reality. *Science* 289:248–250.

15. USDA, National Resources Inventory. 2003. Annual NRI. http://www.nrcs.usda.gov/technical/NRI/2003/images/eros chart.

16. Vaccari, D.A. June 3, 2009. Phosphorus famine: The threat to our food supply. *Scientific America.*

17. Anonymous. 2011. Current World Fertilizer Trends and Outlook to 2015. UN Food and Agriculture Organization Report. Rome, Italy.

18. 2007 National Resources Inventory. http://www.nrcs.usda.gov/Internet/FSE_DOCUMENTS/nrcs143_012269.pdf

19. Mann, C. 2012 (reprint). *1493: Uncovering the New World Columbus Created.* New York, NY: Vintage. (Section title: The Mountains Reveal Their Stones).

20. Cato the Censor. (1933). *Columbia University Records of Civilization: On Farming*, translated by Ernest Brehaut. New York, NY: Columbia University Press.

21. Andrew Dalby (translator). CATO On Agriculture (*De Re Rustica).* http://www.soilandhealth.org/01aglibrary/010121cato/catofarmtext.htm

22. Barfield, C.S., and J.L. Stimac. 1980. Pest management: An entomological perspective. *BioScience* 30:683–688.

23. FAO Statistics. Pesticide Use Through 2010. UN Food and Agriculture Organization. Rome, Italy.

24. Botkin, D.B. 2012. *Beyond the Stony Mountains: Nature in the American West from Lewis and Clark to Today* (ebook, New York, NY: Croton River Publishers) (Originally published in 2004 and still available from the author in hardback by Oxford University Press, New York, NY.)

25. Lashof, J.C., ed. 1979. *Pest Management Strategies in Crop Protection.* Vol. 1. Washington, DC: Office of Technology Assessment, U.S. Congress; and Baldwin, F.L., and P.W. Santelmann. 1980. Weed science in integrated pest management. *BioScience* 30:675–678.

26. EBRD and FAO call for bold steps to contain soaring food prices, UN FAO http://www.fao.org/newsroom/en/news/2008/1000808/index.html, FAO newsroom, March 10, 2008.

27. USDA, National Agricultural Statistics Service Research and Development Division. 1997 Census of agriculture acreage by state for harvested cropland, corn, soybeans, wheat, hay, and cotton. http://www.nass.usda.gov/research/sumpant.htm. Accessed March 10, 2008.

28. USDA. U.S. organic farming emerges in the 1990s. Available at http://www.ers.usda.gov/publications/aib770/aib770.pdf.

29. PEW Biotechnology Initiative, Pew Charitable Trusts. Web site. http://pewagbiotech.org/resources/factsheets/display.

30. USDA. 2012. Organic Farming Economic Research Service. http://www.ers.usda.gov/topics/natural-resources-environment/organic-agriculture/organic-market-overview.aspx. Updated June 19, 2012.

31. USDA. July 2000. Magriet Caswell interview.

32. USDA. Anonymous. 2012. Genetically engineered varieties of corn, upland cotton, and soybeans, by state and for the United States, 2000-12. http://www.ers.usda.gov/data-products/adoption-of-genetically-engineered-crops-in-the-us.aspx. Last updated July 12, 2012.

33. Davison, J. 2010. GM plants: Science, politics and EC regulations. *Plant Science* 178(2):94–98.

34. USDA. 2003. Alternative Farming Systems Information Center. Available at http://www.nal.usda.gov/afsic/ofp/

35. USDA, National Resources Inventory. 2003. Annual NRI. http://www.nrcs.usda.gov/technical/NRI/2003/nri03eros-mrb.html

36. Martin, G. 1992 (June 29). Rice grower proud of his bird habitat. *San Francisco Chronicle,* pp. A1, A6.

37. California Rice Commission Statistical Report, 009. (Updated May 13, 2009). California Rice Statistics and Related National and International Data. http://www.calrice.org/pdf/2009-Statistical-Report-FINAL.pdf

CHAPTER 12

Landscapes: Forests, Parks, and Wilderness

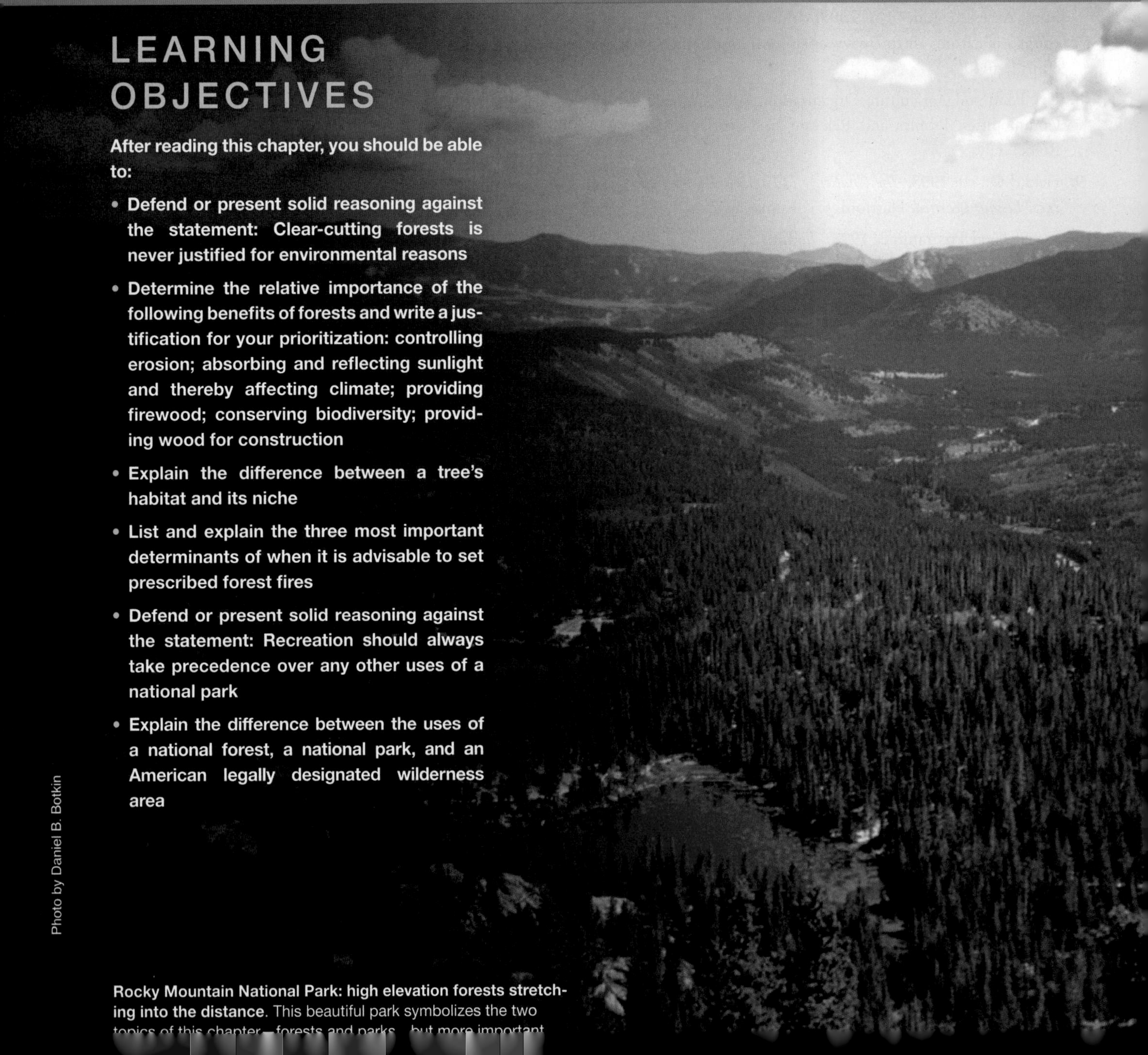

LEARNING OBJECTIVES

After reading this chapter, you should be able to:

- **Defend or present solid reasoning against the statement: Clear-cutting forests is never justified for environmental reasons**
- **Determine the relative importance of the following benefits of forests and write a justification for your prioritization: controlling erosion; absorbing and reflecting sunlight and thereby affecting climate; providing firewood; conserving biodiversity; providing wood for construction**
- **Explain the difference between a tree's habitat and its niche**
- **List and explain the three most important determinants of when it is advisable to set prescribed forest fires**
- **Defend or present solid reasoning against the statement: Recreation should always take precedence over any other uses of a national park**
- **Explain the difference between the uses of a national forest, a national park, and an American legally designated wilderness area**

Photo by Daniel B. Botkin

Rocky Mountain National Park: high elevation forests stretching into the distance. This beautiful park symbolizes the two topics of this chapter—forests and parks—but more important

CASE STUDY

Jamaica Bay National Wildlife Refuge: Nature and the Big City

The largest bird sanctuary in the northeastern United States is—surprise!—in New York City. It is the Jamaica Bay Wildlife Refuge, covering more than 9,000 acres—14 square miles of land, 20,000 acres in total, within view of Manhattan's Empire State Building (see Figure 12.1). Jamaica Bay is run by the National Park Service, and you can get there by city bus or subway.[1] More than 300 bird species have been seen there, including the glossy ibis (common farther south) and the curlew sandpiper, which breeds in northern Siberia. Clearly, this wildlife refuge, like the city itself, is a major transportation crossroads. In fact, it is one of the major stopovers on the Atlantic bird migration flyway.

We are not as likely to think of viewing nature near a big city as we are to think of taking a trip far away to wilderness, but as more and more of us become urban dwellers, parks and preserves within easy reach of cities are going to become more important. Also, cities like New York usually lie at important crossroads, not just for people but for wildlife, as illustrated by Jamaica Bay's many avian visitors.

In the 19th century, this bay was a rich source of shellfish, but these were fished out and their habitats destroyed by urban development of many kinds. And like so many other natural areas, parks, and preserves, Jamaica Bay Wildlife Refuge has troubles. The estuary that it is part of is today only half the size it was in colonial times, and the refuge's salt marshes are disappearing at a rate that alarms conservationists. Some of the wetlands have been filled, some shorelines bulkheaded to protect developments, and channels dredged. A lot of marshland disappeared with the building of Kennedy International Airport, just a few miles away. The salt marshes and brackish waters of the bay are also damaged by a large flow of fresh water from treated sewage. Contrary to what you may think, the only difficulty with this water is that it is fresh, which is a problem to the bay's ecosystems.

Help may be on the way. A watershed protection plan has been written, and there is growing interest in this amazing refuge. The good news is that plentiful wildlife viewing is within a commuter's trip for more than 10 million people. Still, natural areas like the wetlands and bay near New York City and the forests and prairies throughout North America present a conflict. On the one hand, they have been valued for the profits to be made from developing the land for other uses. On the other hand, people value and want to preserve

Photo by Daniel B. Botkin

FIGURE 12.1 Jamaica Bay National Wildlife Refuge, the largest wildlife refuge in the northeastern United States, is within the city limits of New York.

the wildlife and vegetation, the natural ecosystems, for all the reasons discussed in Chapter 7 on biological diversity.

In 2012, Hurricane Sandy had some effect on the refuge. Two freshwater ponds were breached by the bay's salt water, converting them into a channel of the bay. Some believe that the ponds must be fixed—returned to the way they were by direct action. Others suggest that natural processes will accomplish the restoration. An important point has been generally overlooked. There is a benefit to maintaining a nature preserve along a coast, where intense urban development is not wise; a nature preserve along the city's shore protects the urban development inland while providing all the benefits described here.

In the 17th century, when the first Europeans arrived in what is now New York City and Long Island, they found a landscape already occupied by the Lenape Indians, who farmed, hunted, fished, and made trails that ran from Manhattan to Jamaica Bay.[1] Much of the land, especially land extending north along the Hudson River, was forested, and the forests, too, were occupied and used for their resources by the Lenape and other Indians. The dual uses of landscapes were already established: They were both harvested for many resources and appreciated for their beauty and variety.

Since then, the entire landscape has been heavily altered, but those dual uses of the land are still with us and give rise to conflicts about which should dominate. For that reason, in this chapter we discuss both landscapes set aside for biological conservation and forests open to harvests.

12.1 Forests

How People Have Viewed Forests

Forests have always been important to people; indeed, forests and civilization have always been closely linked. Since the earliest civilizations—in fact, since some of the earliest human cultures—wood has been one of the major building materials and the most readily available and widely used fuel. Forests provided materials for the first boats and the first wagons. Even today, nearly half the people in the world depend on wood for cooking, and in many developing nations wood remains the primary heating fuel.[2]

At the same time, people have appreciated forests for spiritual and aesthetic reasons. There is a long history of sacred forest groves. When Julius Caesar was trying to conquer the Gauls in what is now southern France, he found the enemy difficult to defeat on the battlefield, so he burned the society's sacred groves to demoralize them—an early example of psychological warfare. In the Pacific Northwest, the great forests of Douglas fir provided the Indians with many practical necessities of life, from housing to boats, but they were also important to them spiritually.

Today, forests continue to benefit people and the environment indirectly through what we call *public-service functions*. Forests retard erosion and moderate the availability of water, improving the water supply from major watersheds to cities. Forests are habitats for endangered species and other wildlife. They are important for recreation, including hiking, hunting, and bird and wildlife viewing. At regional and global levels, forests may also be significant factors affecting the climate.

12.2 Forestry

Forestry has a long history as a profession. The professional growing of trees is called **silviculture** (from *silvus*, Latin for "forest," and *cultura*, for "cultivate"). People have long practiced silviculture, much as they have grown crops, but forestry developed into a science-based activity and into what we today consider a profession in the late 19th and early 20th centuries. The first modern U.S. professional forestry school was established at Yale University around the turn of the 20th century, spurred by growing concerns about the depletion of America's living resources. In the early days of the 20th century, the goal of silviculture was generally to maximize the yield in the harvest of a single resource. The ecosystem was a minor concern, as were nontarget, noncommercial species and associated wildlife.

In this chapter, we approach forestry as professionals who make careful use of science and whose goals are the conservation and preservation of forests and the sustainability of timber harvest and of forest ecosystems. Unfortunately, these goals sometimes conflict with the goals of others.

Modern Conflicts over Forestland and Forest Resources

What is the primary purpose of national forests? A national source of timber? The conservation of living resources? Recreation?

How should we handle forest fires? How can we and should we restore forests?

Who should own and manage our forests and their resources? The people? Corporations? Government agencies?

In the past quarter century a revolution has taken place as to who owns America's forests, and this has major implications for how, and how well, our forests will be managed, conserved, sustained, and used in the future. In 1981, there were 15 major forest products companies that both owned forestland and processed timber into commercial products. By 2010 all but one had sold off its forestland, focusing on processing timber grown elsewhere.[3]

The state of Maine illustrates the change. About 80% of forestland owned by industrial forest companies was sold in that state between 1994 and 2000. Most of it (60%) was purchased by timber investment management organizations (TIMOs). The rest was sold to nongovernment entities, primarily conservation and environmental organizations.

Industrial forest companies, such as International Paper and Weyerhaeuser, owned the forestland, harvested the timber and planned how to do it, and made products from it. They employed professional foresters, and the assumption within the forest industry was that the profession of forestry and the science on which it was based played an important role in improving harvests and maintaining the land. Although the practices of timber companies were often heavily criticized by environmental groups, both sides shared a belief in sound management of forests, and in the 1980s and 1990s the two sides made many attempts to work together to improve forest ecosystem sustainability.

In contrast, TIMOs are primarily financial investors who view forestland as an opportunity to profit by buying and selling timber. It is unclear how much sound forestry will be practiced on TIMO-owned land, but there is less emphasis on professional forestry and forest science,[4] and far fewer professional foresters have been employed. The danger is that forestland viewed only as a commercial commodity will be harvested and abandoned once the resource is used. If this happens, it will be the exact opposite of what most people involved in forestry, both in the industry and in conservation groups, hoped for and thought was possible throughout the 20th century.

Meanwhile, funding for forest research by the U.S. Forest Service has also been reduced. Our national forests, part of our national heritage, may also be less well managed and therefore less well conserved in the future.

How could this have come about? It is an ironic result of political and ideological activities. Ultimately, the conflict between industrial forestry and environmental conservation seems to have led timberland owners to decide it was less bothersome and less costly to just sell off forestland, buy wood from whomever owned it, and let them deal with the consequences of land use. Consistent with this rationale, much forest ownership by organizations in the United States has moved offshore, to places with fewer environmental constraints and fewer and less powerful environmental groups. This change should be all the more worrisome to those interested in environmental conservation because it has happened without much publicity and is relatively little known by the general public except where forestry is a major livelihood, as it is in the state of Maine.

In sum, then, modern conflicts about forests center on the following questions:

- Should a forest be used only as a resource to provide materials for people and civilization, or should a forest be used only to conserve natural ecosystems and biological diversity (see Figure 12.2), including specific endangered species?
- Can a forest serve some of both of these functions at the same time and in the same place?
- Can a forest be managed sustainably for either use? If so, how?
- What role do forests play in our global environment, such as climate?
- When are forests habitats for specific endangered species?
- When and where do we need to conserve forests for our water supply?

Radius/SuperStock

Photo by Daniel B. Botkin

(a) (b)

FIGURE 12.2 **(a) The dual human uses of forests.** This temperate rain forest on Vancouver Island illustrates the beauty of forests. Its tree species are also among those most desired for commercial timber production. **(b)** Scientists view a timber harvest in the Pacific Northwest.

12.3 World Forest Area and Global Production and Consumption of Forest Resources

At the beginning of the 21st century, approximately 26% of Earth's surface was forested—about 3.8 billion hectares (15 million square miles) (Figure 12.3).[2] This works out to about 0.6 hectare (about 1 acre) per person. The forest area is up from 3.45 billion hectares (13.1 million square miles, or 23% of the land area) estimated in 1990, but down from 4 billion hectares (15.2 million square miles, or 27%) in 1980.

Countries differ greatly in their forest resources, depending on the suitability of their land and climate for tree growth and on their history of land use and deforestation. Ten nations have two-thirds of the world's forests. In descending order, these are the Russian Federation, Brazil, Canada, the United States, China, Australia, the Democratic Republic of the Congo, Indonesia, Angola, and Peru (Figure 12.4).

Developed countries account for 70% of the world's total production and consumption of industrial wood products; developing countries produce and consume about 90% of wood used as firewood. Timber for construction, pulp, and paper makes up approximately 90% of the world timber trade; the rest consists of hardwoods used for furniture, such as teak, mahogany, oak, and maple. North America is the world's dominant supplier. Total global production/consumption is about 1.5 billion m^3 annually. To think of this in terms easier to relate to, a cubic meter of timber is a block of wood 1 meter thick on each side. A billion cubic meters would be a block of wood 1 meter (39 inches) thick in a square 1,000 km (621 miles) long on each side. This is a distance greater than that between Washington, DC, and Atlanta, Georgia, and longer than the distance between San Diego and Sacramento, California. The great pyramid of Giza, Egypt, has a volume of more than 2.5 million cubic meters, so the amount of timber consumed in a year would fill 600 great pyramids of Egypt.

The United States has approximately 304 million hectares (751 million acres) of forests, of which 86 million hectares (212 million acres) are considered commercial-grade forest, defined as forest capable of producing at least 1.4 m^3/ha (20 ft^3/acre) of wood per year.[4] Commercial timberland occurs in many parts of the United States. Nearly 75% is in the eastern half of the country (about equally divided between the North and South); the rest is in the West (Oregon, Washington, California, Montana, Idaho, Colorado, and other Rocky Mountain states) and in Alaska.[4]

In the last several decades, world trade in timber does not appear to have grown much, if at all, based on the information reported by nations to the United Nations Food and Agriculture Organization. Thus, the amount traded annually (about 1.5 billion m^3, as mentioned earlier) is a reasonable estimate of the total present world demand for

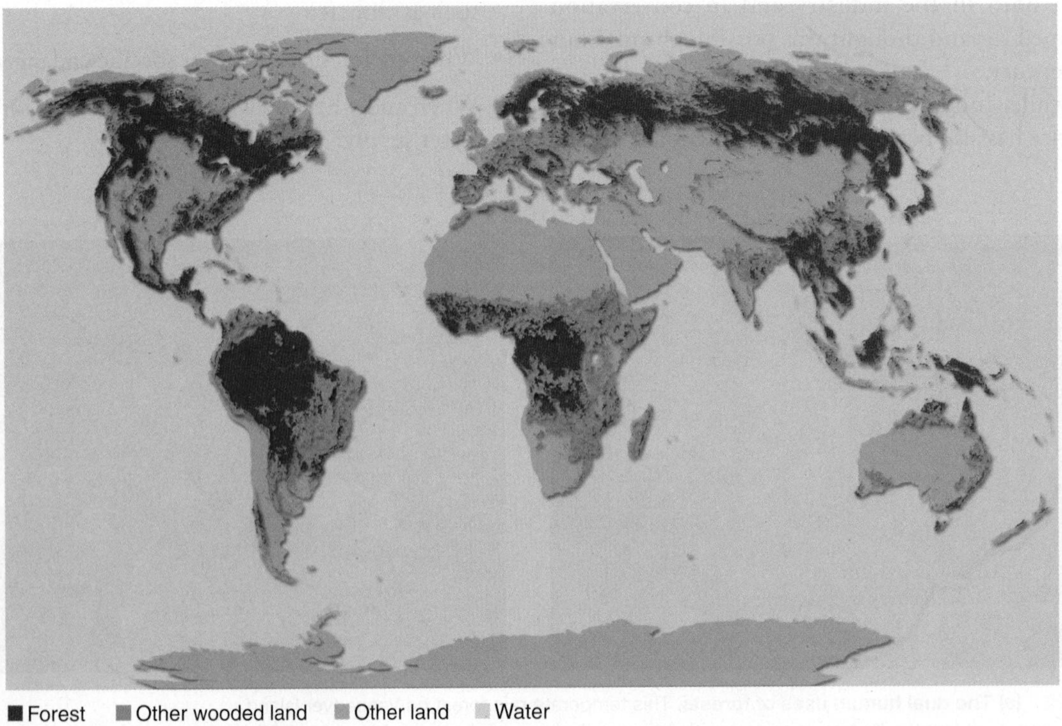

(*Source*: Food and Agriculture Organization of the United Nations, Viale delle Terme di Caracalla, 00153 Rome, Italy.)

FIGURE 12.3 Forests of the world.

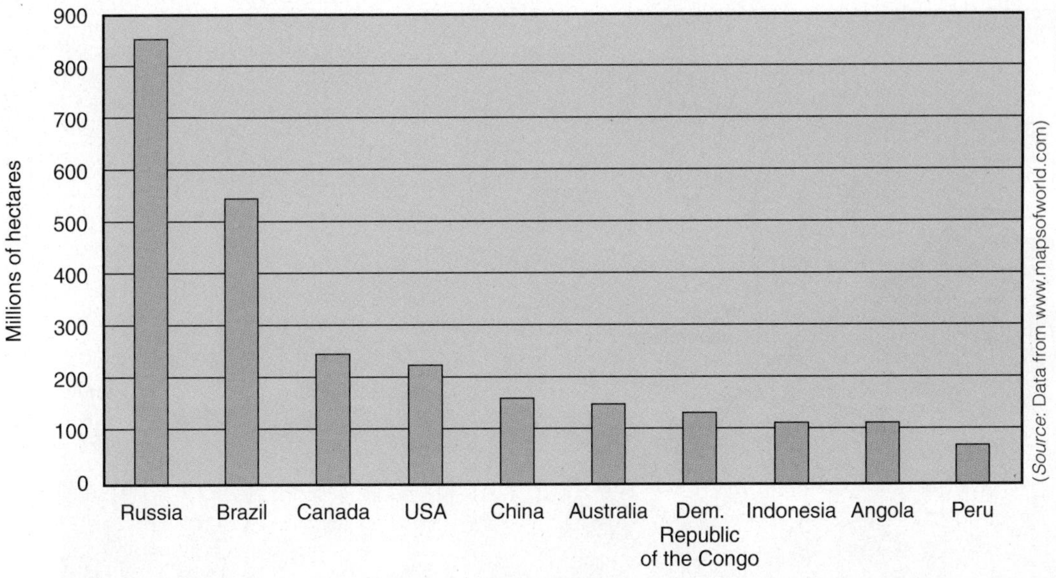

FIGURE 12.4 **Countries with the largest forest areas.**

the 7 billion people on Earth, at their present standards of living. The fundamental questions are whether and how Earth's forests can continue to produce at least this amount of timber for an indefinite period, and whether and how they can produce even more as the world's human population continues to grow and as standards of living rise worldwide. Keep in mind, all of this has to happen while forests continue to perform their other functions, which include public-service functions, biological conservation functions, and functions involving the aesthetic and spiritual needs of people.

In terms of the themes of this book, the question is: How can forest production be sustainable while meeting the needs of people *and* nature? The answer involves science and values.

As we mentioned, wood is a major energy source in many parts of the world. Some 63% of all wood produced in the world, or 2.1 million m^3, is used for firewood. Firewood provides 5% of the world's total energy use,[5] 2% of total commercial energy in developed countries, but 15% of the energy in developing countries, and is the major source of energy for most countries of sub-Saharan Africa, Central America, and continental Southeast Asia.[6]

As the human population grows, the use of firewood increases. In this situation, management is essential, including management of woodland stands (an informal term that foresters use to refer to groups of trees) to improve growth. However, well-planned management of firewood stands has been the exception rather than the rule.

12.4 How Forests Affect the Whole Earth

Trees affect the Earth by evaporating water, slowing erosion, and providing habitat for wildlife (see Figure 12.5). Trees can also affect climate. Indeed, vegetation of any kind can affect the atmosphere in four ways, and since forests cover so much of the land, they can play an especially important role in the biosphere:

1. By changing the color of the surface and thus the amount of sunlight reflected and absorbed.
2. By increasing the amount of water transpired and evaporated from the surface to the atmosphere.
3. By changing the rate at which greenhouse gases are taken up, stored, and released from Earth's surface into the atmosphere, especially carbon dioxide.
4. By changing "surface roughness," which affects wind speed at the surface.

In general, vegetation warms the Earth by making the surface darker, so it absorbs more sunlight and reflects less. The contrast is especially strong between the dark needles of conifers and winter snow in northern forests and between the dark green of shrublands and the yellowish soils of many semiarid climates. Vegetation in general and forests in particular tend to evaporate more water than bare surfaces. This is because the total surface area of the many leaves is many times larger than the area of the soil surface.

Is this increased evaporation good or bad? That depends on one's goals. Increasing evaporation means that less water runs off the surface. This reduces erosion. Although increased evaporation also means that less water is available for our own water supply and for streams, in most situations the ecological and environmental benefits of increased evaporation outweigh the disadvantages.

Today there is great interest in the potential for trees in forests to take up carbon dioxide. This potential, referred to as "carbon sequestering," is suggested as one of the major ways to counter the buildup of carbon dioxide in the atmosphere. (See Chapter 20.)

FIGURE 12.5 **A forested watershed**, showing the effects of trees in evaporating water, retarding erosion, and providing wildlife habitat.

Vegetation makes a much rougher surface than bare soil or ice, but the effect of forest on wind speed is considered to be relatively minor and is little discussed.

12.5 The Ecology of Forests

Each species of tree has its own niche (see Chapter 6) and is thus adapted to specific environmental conditions. For example, in boreal forests, one of the determinants of a tree niche is the water content of the soil. White birch grows well in dry soils; balsam fir in well-watered sites; and northern white cedar in bogs (Figure 12.6). (See A Closer Look 12.1 for information about how trees grow.)

Another determinant of a tree's niche is its tolerance of shade. Some trees, such as birch and cherry, can grow only in the bright sun of open areas and are therefore found in clearings and are called "shade intolerant." Other species, such as sugar maple and beech, can grow in deep shade and are called "shade tolerant."

Most of the big trees of the western United States require open, bright conditions and certain kinds of disturbances in order to germinate and survive the early stages of their lives. These trees include coastal redwood, which wins in competition with other species only if both fires and floods occasionally occur; Douglas fir, which begins its growth in openings; and the giant sequoia, whose seeds

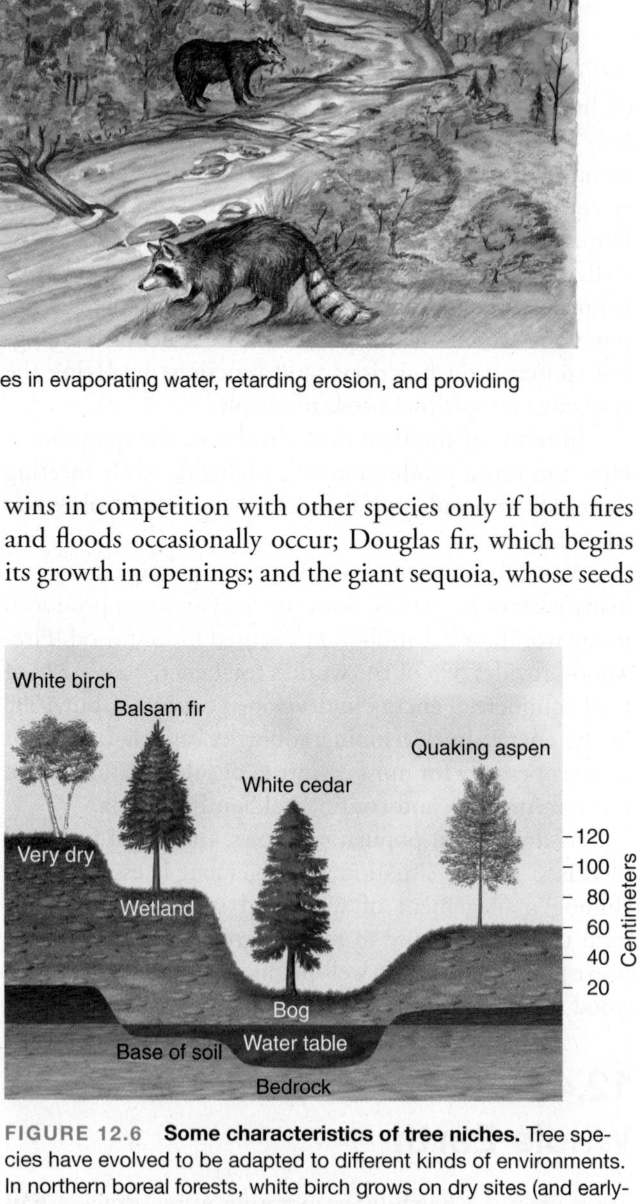

FIGURE 12.6 **Some characteristics of tree niches.** Tree species have evolved to be adapted to different kinds of environments. In northern boreal forests, white birch grows on dry sites (and early-successional sites); balsam fir grows in wetter soils, up to wetlands; and white cedar grows in even the wetter sites of northern bogs.

A CLOSER LOOK 12.1

The Life of a Tree

To solve the big issues about forestry, we need to understand how a tree grows, how an ecosystem works, and how foresters have managed forestland (Figure 12.7). Leaves of a tree take up carbon dioxide from the air and absorb sunlight. These, in combination with water transported up from the roots, provide the energy and chemical elements for leaves to carry out *photosynthesis*. Through photosynthesis, the leaves convert carbon dioxide and water into a simple sugar and molecular oxygen. This simple sugar is then combined with other chemical elements to provide all the compounds that the tree uses.

Tree roots take up water, along with chemical elements dissolved in the water and small inorganic compounds, such as the nitrate or ammonia necessary to make proteins. Often the process of extracting minerals and compounds from the soil is aided by symbiotic relationships between the tree roots and fungi. Tree roots release sugars and other compounds that are food for the fungi, and the fungi benefit the tree as well.

Leaves and roots are connected by two transportation systems. Phloem, on the inside of the living part of the bark, transports sugars and other organic compounds down to stems and roots. Xylem, farther inside (Figure 12.7), transports water and inorganic molecules upward to the leaves. Water is transported upward by a sun-powered pump—that is, sunlight provides energy to pump the water up the tree by heating leaves so that they evaporate water. Water from below is then pulled upward to replace water that evaporated.

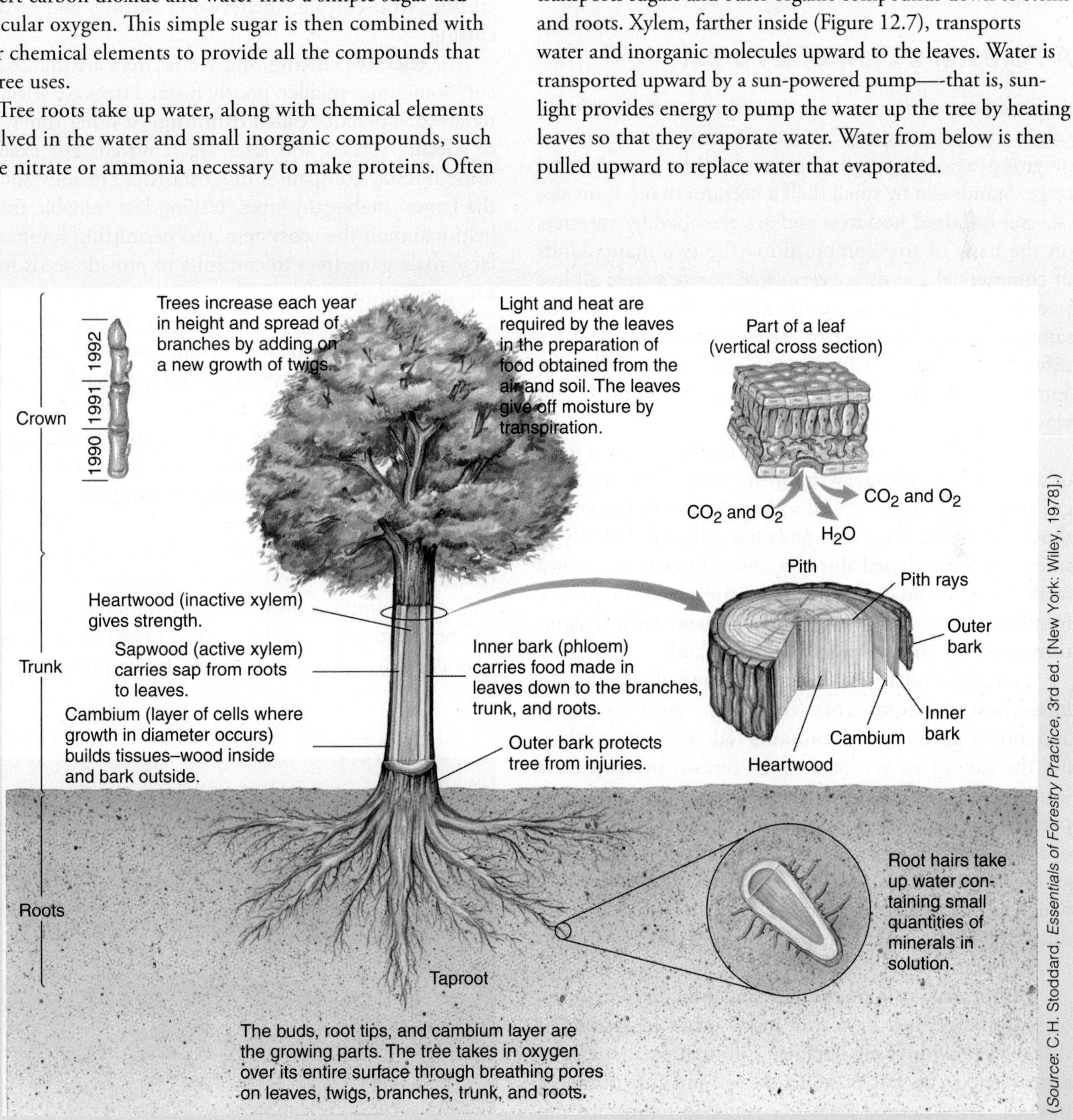

FIGURE 12.7 **How a tree grows.**

will germinate only on bare, mineral soil. Where there is a thick layer of organic mulch, the sequoia's seeds cannot reach the surface and will die before they can germinate. Some trees are adapted to early stages of succession, where sites are open and there is bright sunlight. Others are adapted to later stages of succession, where there is a high density of trees. (See the discussion of ecological succession in Chapter 6.)[7]

Understanding the niches of individual tree species helps us to determine where we might best plant them as a commercial crop—and where they might best contribute to biological conservation or to landscape beauty.

12.6 Forest Management

A Forester's View of a Forest

Traditionally, foresters have managed trees locally in stands. Trees in a **stand** are usually of the same species or group of species and often at the same successional stage. Stands can be small (half a hectare) to medium size (several hundred hectares) and are classified by foresters on the basis of tree composition. The two major kinds of commercial stands are *even-aged stands*, where all live trees began growth from seeds and roots germinating the same year, and *uneven-aged stands*, which have at least three distinct age classes. In even-aged stands, trees are approximately the same height but differ in girth and vigor.

A forest that has never been cut is called a *virgin forest* or sometimes an **old-growth forest**. A forest that has been cut and has regrown is called a **second-growth forest**. Although the term *old-growth forest* has gained popularity in several well-publicized disputes about forests, it is not a scientific term and does not yet have an agreed-on, precise meaning. Another important management term is **rotation time**, the time between cuts of a stand.

Foresters and forest ecologists group the trees in a forest into the **dominants** (the tallest, most common, and most vigorous), **codominants** (fairly common, sharing the canopy or top part of the forest), **intermediate** (forming a layer of growth below dominants), and **suppressed** (growing in the understory). The productivity of a forest varies according to soil fertility, water supply, and local climate. Foresters classify sites by **site quality**, which is the maximum timber crop the site can produce in a given time. Site quality can decline with poor management.

Although forests are complex and difficult to manage, one advantage they have over many other ecosystems is that trees provide easily obtained information that can be a great help to us. For example, the age and growth rate of trees can be measured from tree rings. In temperate and boreal forests, trees produce one growth ring per year.

Harvesting Trees

Managing forests that will be harvested can involve removing poorly formed and unproductive trees (or selected other trees) to permit larger trees to grow faster, planting genetically controlled seedlings, controlling pests and diseases, and fertilizing the soil. Forest geneticists breed new strains of trees just as agricultural geneticists breed new strains of crops. There has been relatively little success in controlling forest diseases, which are primarily fungal.

Harvesting can be done in several ways. **Clear-cutting** (Figure 12.8) is the cutting of all trees in a stand at the same time. Alternatives to clear-cutting are selective cutting, strip-cutting, shelterwood cutting, and seed-tree cutting.

In **selective cutting**, individual trees are marked and cut. Sometimes smaller, poorly formed trees are selectively removed, a practice called **thinning**. At other times, trees of specific species and sizes are removed. For example, some forestry companies in Costa Rica cut only some of the largest mahogany trees, leaving less valuable trees to help maintain the ecosystem and permitting some of the large mahogany trees to continue to provide seeds for future generations.

In **strip-cutting**, narrow rows of forest are cut, leaving wooded corridors whose trees provide seeds. Strip-cutting offers several advantages, such as protection against erosion.

Shelterwood cutting is the practice of cutting dead and less desirable trees first, and later cutting mature trees. As a result, there are always young trees left in the forest.

Seed-tree cutting removes all but a few seed trees (mature trees with good genetic characteristics and high seed production) to promote regeneration of the forest.

Scientists have tested the effects of clear-cutting, which is one of the most controversial forest practices.[7, 8] For

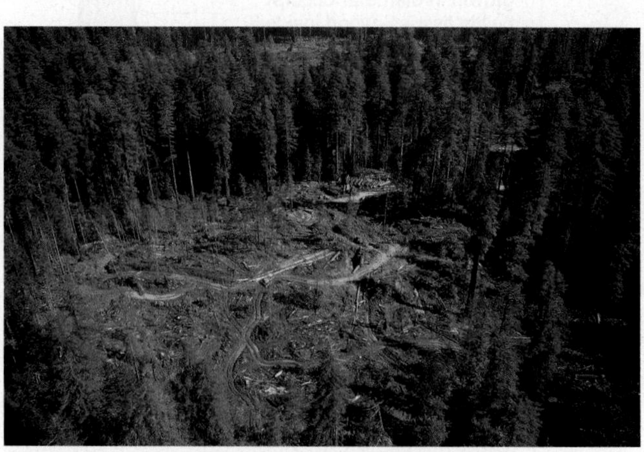

FIGURE 12.8 A clear-cut forest in western Washington.

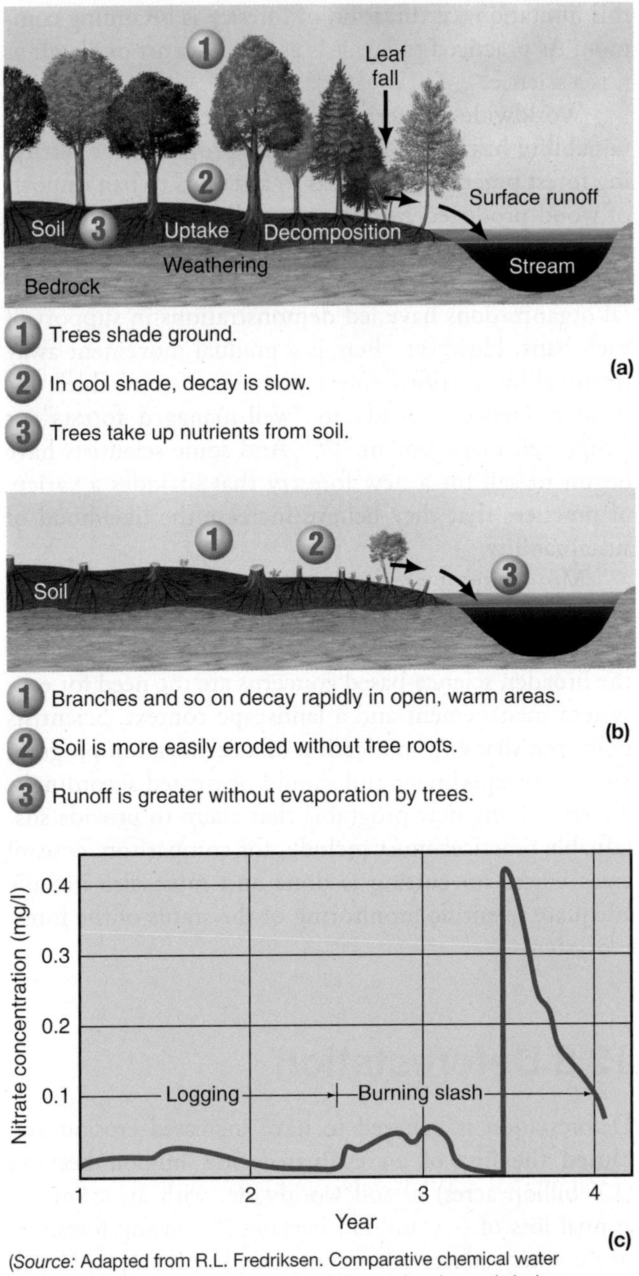

(*Source:* Adapted from R.L. Fredriksen. Comparative chemical water quality—natural and disturbed streams following logging and slash burning, *Forest Land Use and Stream Environment*. Corvallis: Oregon State University, 1971, pp. 125–137.)

FIGURE 12.9 **Effects of clear-cutting on forest chemical cycling.** Chemical cycling **(a)** in an old-growth forest and **(b)** after clear-cutting. **(c)** Increased nitrate concentration in streams after logging and the burning of slash (leaves, branches, and other tree debris).

example, in the U.S. Forest Service Hubbard Brook experimental forest in New Hampshire, an entire watershed was clear-cut, and herbicides were applied to prevent regrowth for two years.[9] The results were dramatic. Erosion increased, and the pattern of water runoff changed substantially. The exposed soil decayed more rapidly, and the concentrations of nitrates in the stream water exceeded public-health standards. In another experiment, at the U.S. Forest Service H.J. Andrews experimental forest in Oregon, a forest where rainfall is high (about 240 cm, or 94 in., annually), clear-cutting greatly increased the frequency of landslides, as did the construction of logging roads.[10]

Clear-cutting also changes chemical cycling in forests and can open the way for the soil to lose chemical elements necessary for life. Exposed to sun and rain, the ground becomes warmer. This accelerates the process of decay, with chemical elements, such as nitrogen, converted more rapidly to forms that are water-soluble and thus readily lost in runoff during rains (Figure 12.9).[11]

The Forest Service experiments show that clear-cutting can be a poor practice on steep slopes in areas of moderate to heavy rainfall. The worst effects of clear-cutting resulted from the logging of vast areas of North America during the 19th and early 20th centuries. Clear-cutting on such a large scale is neither necessary nor desirable for the best timber production. However, where the ground is level or slightly sloped, where rainfall is moderate, and where the desirable species require open areas for growth, clear-cutting on an appropriate spatial scale may be a useful way to regenerate desirable species. The key here is that clear-cutting is neither all good nor all bad for timber production or forest ecosystems. Its use must be evaluated on a case-by-case basis, taking into account the size of cuts, the environment, and the available species of trees.

Plantations

Sometimes foresters grow trees in a **plantation**, which is a stand of a single species, typically planted in straight rows (Figure 12.10). Usually plantations are fertilized, sometimes by helicopter, and modern machines harvest trees rapidly—some remove the entire tree, root and all. In short, plantation forestry is a lot like modern

Craig Hedman

FIGURE 12.10 **A modern forest plantation of longleaf pine stand supports a rich, herbaceous understory.** Note that the trees are evenly spaced and similar, if not identical, in size.

agriculture. Intensive management like this is common in Europe and parts of the northwestern United States and offers an important alternative solution to the pressure on natural forests. If plantations were used where forest production was high, then a comparatively small percentage of the world's forestland could provide all the world's timber. For example, high-yield forests produce 15–20 m^3/ha/yr. According to one estimate, if plantations were put on timberland that could produce at least 10 m^3/ha/yr, then 10% of the world's forestland could provide enough timber for the world's timber trade.[12] In the United States today about 8% of the forests are plantations, but these provide half of all the softwood (conifers) harvested.[13] This could reduce pressure on old-growth forests, on forests important for biological conservation, and on forestlands important for recreation.

12.7 Can We Achieve Sustainable Forestry?

There are two basic kinds of ecological sustainability: (1) sustainability of the harvest of a specific resource that grows within an ecosystem; and (2) sustainability of the entire ecosystem—and therefore of many species, habitats, and environmental conditions. For forests, this translates into sustainability of the harvest of timber and sustainability of the forest as an ecosystem. Although sustainability has long been discussed in forestry, we don't have enough scientific data to show that sustainability of either kind has been achieved in modern forests except in a few cases.

Certification of Forest Practices

If the data do not indicate whether a particular set of practices has led to sustainable forestry, what can be done? The general approach today is to compare the actual practices of specific corporations or government agencies with practices that are believed to be consistent with sustainability. This has become a formal process called **certification of forestry**, and there are organizations whose main function is to certify forest practices. For example, interest in sustainable foresty grew in the last decades of the 20th century, with the establishment of such organizations as the nonprofit Sustainable Forestry Initiative, which provides, based on its own analyses, certification that a forest is being managed according to sustainable forestry principles.[14]

The catch here is that nobody actually knows whether the beliefs are correct and therefore whether the practices will turn out to be sustainable. Since trees take a long time to grow, and a series of harvests is necessary to prove sustainability, the proof lies in the future. Despite this limitation, certification of forestry is becoming common. As practiced today, it is as much an art or a craft as it is a science.

Worldwide concern about the need for forest sustainability has led to international programs for certifying forest practices, as well as to attempts to ban imports of wood produced from purportedly unsustainable forest practices. Some European nations have banned the import of certain tropical woods, and some environmental organizations have led demonstrations in support of such bans. However, there is a gradual movement away from calling certified forest practices "sustainable"; instead reference is made to "well-managed forests" or "improved management."[14, 15] And some scientists have begun to call for a new forestry that includes a variety of practices that they believe increase the likelihood of sustainability.

Most basic is accepting the dynamic characteristics of forests—that to remain sustainable over the long term, a forest may have to change in the short term. Some of the broader, science-based concerns are the need for ecosystem management and a landscape context. Scientists point out that any application of a certification program creates an experiment and should be treated accordingly. Therefore, any new programs that claim to provide sustainable practices must include, for comparison, control areas where no cutting is done and must also include adequate scientific monitoring of the status of the forest ecosystem.

12.8 Deforestation

Deforestation is believed to have increased erosion and caused the loss of an estimated 562 million hectares (1.4 billion acres) of soil worldwide, with an estimated annual loss of 5–6 million hectares.[16] Cutting forests in one country affects other countries. For example, Nepal, one of the most mountainous countries in the world, is estimated to have lost more than half its forest cover between 1950 and 1980, and another 25% from 1990 to 2010.[17] This destabilized soil, increasing the frequency of landslides, amount of runoff, and sediment load in streams. Many Nepalese streams feed rivers that flow into India (Figure 12.11). Heavy flooding in India's Ganges Valley caused about a billion dollars' worth of property damage a year and is blamed on the loss of large forested watersheds in Nepal and other countries.[18] If present trends continue, little forestland will remain in Nepal, thus permanently exacerbating India's flood problems.[17, 18]

The good news is that the rate of deforestation has slowed over the last 10 years, according to the United Nations Environment Program. How bad is it today? Every year an area about as big as Costa Rica is deforested.

Steve McCurry/Magnum Photos, Inc

(a)

John Warburton-Lee/Photography/Alamy

(b)

FIGURE 12.11 **(a) Planting pine trees on the steep slopes in Nepal to replace entire forests that were cut.** The dark green in the background is yet-uncut forest, and the contrast between foreground and background suggests the intensity of clearing that is taking place. **(b)** The Indus River in northern India carries a heavy load of sediment, as shown by the sediments deposited within and along the flowing water and by the color of the water itself. This scene, near the headwaters, shows that erosion takes place at the higher reaches of the river.

History of Deforestation

People have cut down and burned forests for thousands of years. Fossil records suggest that prehistoric farmers in Denmark cleared forests so extensively that early-successional weeds occupied large areas. In the New World, Mayas were clearing tropical forests in what is now Guatemala as early as the first century B.C.[19] In the Near East, human-induced forest clearing appears to have occurred as early as 1800 B.C.[20] Removal of forests continued northward in Europe as civilization advanced. In medieval times, Great Britain's forests were cut, and many forested areas were eliminated. With colonization of the New World, much of North America was cleared.

The greatest losses in the present century have taken place in South America, where 4.3 million acres have been lost on average per year since 2000 (Figure 12.12). Many of these forests are in the tropics, mountain regions, or high latitudes, places difficult to exploit before the advent of modern transportation and machines. The problem is especially severe in the tropics because of rapid human population growth. Satellite images provide a new way to detect deforestation (Figure 12.12a).

Photo Researchers

(a)

Gregory G. Dimijian, M.D./Photo Researchers

(b)

FIGURE 12.12 **(a)** A satellite image showing clearings in the tropical rain forests in the Amazon in Brazil. The image is in false infrared. Rivers appear black, and the bright red is the leaves of the living rain forest. The straight lines of other colors, mostly light blue to gray, show deforestation by people extending from roads. Much of the clearing is for agriculture. The distance across the image is about 100 km (63 mi). **(b)** An intact South American rain forest with its lush vegetation of many species and a complex vertical structure. This one is in Peru.

Causes of Deforestation

As we have made clear in Chapter 6, forests, like all ecosystems, undergo change, and there are various natural causes of deforestation. During Ice Age maximum, forested areas at mid- to high latitudes are covered by ice and destroyed. Fires are persistent in most forests. Lowland and streamside forests are cleared by floods. Historically, the two most common reasons people remove forests are to clear land for agriculture and settlement and to use or sell timber for lumber, paper products, or fuel. Logging by large timber companies and local cutting by villagers are both major causes of deforestation. Agriculture is a principal cause of deforestation in such nations as Indonesia, Nepal, and Brazil, and was one of the major reasons for clearing forests in New England during the first settlement by Europeans. A more subtle cause of the loss of forests is indirect deforestation—the death of trees from pollution or disease.

(a)

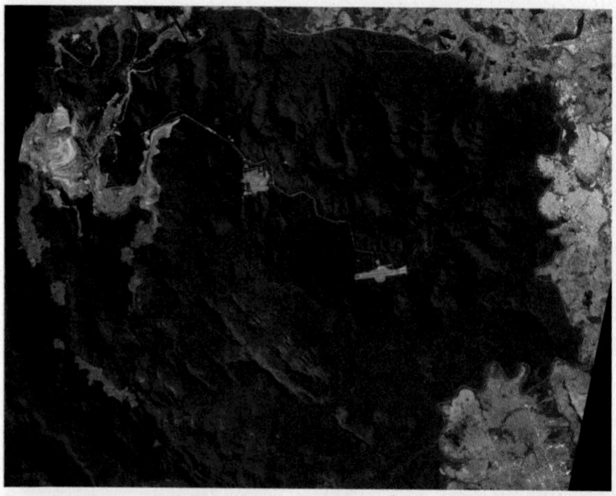

NASA image created by Jesse Allen, using EO-1 ALI data provided courtesy of the NASA EO-1 Team.

(b)

Jacques Jangoux/Alamy

FIGURE 12.13 **Carajás, Brazil: Strip mining for iron ore deforests some of Brazil's Amazon basin forests.** The Carajás Mine and Carajás mountain range (Serra dos Carajás) in northeastern Brazil are shown in these photos. The deforested red-tan exposed soil and bedrock contrast with the deep green of the rain forest. Companhia Vale do Rio Doce, the company that runs the mine, preserves the forest immediately around the mine. The large image, which encompasses a broader area, shows deforestation north and east of the mine. **(a)** Seen from satelite; **(b)** air photo.

Mining and fuel extraction clear forests. In the U.S. Appalachian Mountains, strip mining for coal destroyed mountain forests. The mining of iron ore and other minerals is deforesting parts of the Amazon basin tropical forests. For example, one of the world's largest copper reserves lies in the Carajás Mineral Province, Brazil, where iron ore, manganese, and gold are also mined (Figure 12.13). In 2007, 296 million metric tons of iron ore were pulled from the mine. The mine is estimated to contain about 18 billion tons of iron ore, plus gold, manganese, copper, and nickel. Thus only 2% of the mineral value has been removed. Mining has multiple effects: direct removal of trees at the mine site, cutting of wood from surrounding forests for charcoal used in iron and steel production, and damage to forests downstream from mine runoff.[21]

If global warming occurs as projected, indirect forest damage might take place over large regions, with major die-offs in many areas and major shifts in the areas of potential growth for each species of tree due to altered combinations of temperature and rainfall.[22] The extent of this effect is controversial. Some suggest that global warming would merely change the location of forests, not their total area or production.

12.9 Dealing with Forest Fires

Fire is a repeated occurrence in most forests of the world, to the extent that many tree species are adapted to fire and many require fires for regeneration and other functions, as we will discuss in Chapter 13. The recognition of the naturalness of forest fires grew in the 20th century among ecologists and forest scientists, but throughout most of the 19th and 20th centuries, fire suppression was the rule. Many in government and the public sphere saw fires as only damaging, thus the then familiar poster of Smokey Bear and his slogan, "Only you can prevent forest fires" (Figure 12.14).

The 1988 fire in Yellowstone National Park was the largest ever recorded there. It burned more than one-third of the park—3,213 km^2 (1,240 square miles), four-fifths the size of Rhode Island. The long suppression of fire in the park, part of the Smokey Bear philosophy, had created a superabundance of fuel that needed only a very dry summer for a large fire to become likely. And that's what happened.

Looking backward, the National Park Service issued the following explanation in 2008:

> *In the 1950s and 1960s, national parks and forests began to experiment with controlled burns, and by the 1970s Yellowstone and other parks had instituted a natural fire management plan to allow the process of lightning-caused fire to continue influencing*

UIG via Getty Images

FIGURE 12.14 Smokey Bear poster. Posters of Smokey Bear influenced public policy in America for decades, even though scientists and foresters knew that the message tied to the image was false. Many forests require fire, but the conditions must be appropriate to prevent serious damage. This poster needs to be revised to indicate that the human-caused fires can be destructive—not that only people can prevent them. The issue of how and when to allow and promote forest fires remains controversial and a subject of research.

wildland succession. . . . In the first sixteen years of Yellowstone's natural fire policy (1972–1987), 235 fires were allowed to burn 33,759 acres. Only 15 of those fires were larger than 100 acres, and all of the fires were extinguished naturally. Public response to the fires was good, and the program was considered a success. The summers of 1982–1987 were wetter than average, which may have contributed to the relatively low fire activity in those years.[23]

An especially fire-prone vegetation is found in the chaparral of southern California. It is dense and mostly shrubland, with scattered oaks and pines such as occur in the Las Padres National Forest and the hills behind Santa Barbara. This vegetation grows in the semiarid climate of this part of the state: hot dry summers and cool winters with some rain, averaging about 18 inches a year, sometimes occurring in a few major storms. The abundant vegetation has adapted to this climate by requiring fire to reproduce; it also increases the chances of fire. Wildfires occur about every half century. When this area was wilder country without permanent buildings, towns, and cities, these fires did not affect people seriously. But today in modern Santa Barbara, these fires can be terribly destructive. For example, in 1990 a wildfire name the "Painted Cave Fire" started in the hills and burned 5,000 acres in three hours, destroying 500 houses and doing an estimated $750 million damage.[24] Another fire near Los Angeles in September 1993 burned 40,500 acres. Attempts to stop and control the fire involved ground crews and aircraft, and fire-fighting costs tallied more than $2 million a day. The cost was borne by state and federal governments, primarily to try to save houses.

Although it is well acknowledged that many kinds of forests require fire and that comparatively frequent light fires are characteristic and least damaging to forests, it has been difficult to manage forests in the modern world with so much human development—houses, towns, cities, farms, and industrial complexes. A project in the Ponderosa pine forests near Flagstaff, Arizona, illustrates the difficulties. Ponderosa pine woodlands are characteristic of that area. These are beautiful, open woodlands with scattered pines surrounded by grasslands, the countryside often seen in cowboy and Indian movies. Fire had been suppressed in these forests with the usual result—high fuel loads. When a fire occurs in such conditions, it likely kills seed-bearing trees, destroys the soil, and delays forest recovery for a very long time, if it happens at all.

In this project, what the forests were like before European settlement was analyzed and the modern forest was returned to those conditions, including the density of trees and the average tree size. Ground cover was also cleared of its large build-up of leaves and dead wood. This is a labor-intensive activity. Once the forest area was reset to its smaller fuel load, fires were lit, and they burned the way fires burned before European settlement—swiftly and lightly, keeping the Ponderosa woodlands open, beautiful, and—with sufficiently frequent fires—capable of looking that way in the future (Figure 12.15). Note that this planning was not for a forest that was always the same, not for a perfect balance of nature, but for a forest that changed within a range of conditions, that people liked and considered natural.[25]

Ian Horvath

FIGURE 12.15 A controlled burn in Arizona's Coconino National Forest.

12.10 Parks, Nature Preserves, and Wilderness

As suggested by this chapter's opening case study about the Jamaica Bay Wildlife Refuge, governments often protect landscapes from harvest and other potentially destructive uses by establishing parks, nature preserves, and legally designated wilderness areas. So do private organizations, such as the Nature Conservancy, the Southwest Florida Nature Conservancy, and the Land Trust of California, which purchase lands and maintain them as nature preserves. Whether government or private conservation areas succeed better in reaching the goals listed in Table 12.1 is a matter of considerable controversy.

In most cases, public parks became established comparatively recently, most since the rise of modern industrial civilizations. (See A Closer Look 12.2.) Parks, natural areas, and wilderness provide benefits within their boundaries and can also serve as migratory corridors between other natural areas. Originally, parks were established for specific purposes related to the land within the park boundaries (discussed later in this chapter). In the future, the design of large landscapes to serve a combination of land uses—including parks, preserves, and wilderness—needs to become more important and a greater focus of discussion.

What's the Difference between a Park and a Nature Preserve?

A park is an area set aside for use by people. A nature preserve, although it may be used by people, has as its primary purpose the conservation of some resource, typically a biological one. Every park or preserve is an ecological island of one kind of landscape surrounded by a different kind of landscape, or several different kinds. Ecological and physical islands have special ecological qualities, and concepts of island biogeography are used in the design and management of parks. Specifically, the size of the park and the diversity of habitats determine the number of species that can be maintained there. Also, the farther the park is from other parks or sources of species, the fewer species are found. Even the shape of a park can determine what species can survive within it.

In the United States, marine sanctuaries are an unusual kind of park. The National Marine Sanctuary System was created by Title III of the Marine Protection, Research, and Sanctuaries Act of 1972, administered by the National Oceanic and Atmospheric Administration. It consists of 13 sanctuaries and one national monument and includes areas along the Pacific and Atlantic coasts of the continental United States, the Hawaiian Islands, American Samoa, and the Great Lakes. Similar to national parks on land, these underwater protected areas range in size from less than one square mile to more than 137,000 square miles. In total, the sanctuary system covers more than 150,000 square miles.

Sanctuaries protect thriving ecosystems like coral reefs and kelp forests, along with important breeding and feeding areas for marine life such as whales, seabirds, sharks, and sea turtles. Within sanctuary boundaries lie countless links to our heritage—from historic shipwrecks to traditional Native American fishing grounds. Each sanctuary site is unique, with its own distinct collection of resources and management challenges.

Table 12.1 GOALS OF PARKS, NATURE PRESERVES, AND WILDERNESS AREAS

Parks are as old as civilization. The goals of park and nature-preserve management can be summarized as follows:

1. Preservation of unique geological and scenic wonders of nature, such as Niagara Falls and the Grand Canyon
2. Preservation of nature without human interference (preserving wilderness for its own sake)
3. Preservation of nature in a condition thought to be representative of some prior time (e.g., the United States prior to European settlement)
4. Wildlife conservation, including conservation of the required habitat and ecosystem of the wildlife
5. Conservation of specific endangered species and habitats
6. Conservation of the total biological diversity of a region
7. Maintenance of wildlife for hunting
8. Maintenance of uniquely or unusually beautiful landscapes for aesthetic reasons
9. Maintenance of representative natural areas for an entire country
10. Maintenance for outdoor recreation, including a range of activities from viewing scenery, to wilderness recreation (hiking, cross-country skiing, rock climbing), and tourism (car and bus tours, swimming, downhill skiing, camping)
11. Maintenance of areas set aside for scientific research, both as a basis for park management and for the pursuit of answers to fundamental scientific questions
12. Provision of corridors and connections between separated natural areas

A CLOSER LOOK 12.2

A Brief History of Parks Explains Why Parks Have Been Established

The French word *parc* once referred to an enclosed area for keeping wildlife to be hunted. Such areas were set aside for the nobility and excluded the public. An example is Coto Doñana National Park on the southern coast of Spain (Figure 12.16). Originally a country home of nobles, today it is one of Europe's most important natural areas, used by 80% of birds migrating between Europe and Africa.

The first major *public* park of the modern era was Victoria Park in Great Britain, authorized in 1842. The concept of a *national* park, whose purposes would include protection of nature as well as public access, originated in North America in the 19th century.[a] The world's first national park was Yosemite National Park in California (Figure 12.17), made a park by an act signed by President Lincoln in 1864. The term *national park*, however, was not used until the establishment of Yellowstone in 1872.

The purpose of the earliest national parks in the United States was to preserve the nation's unique, awesome landscapes—a purpose that Alfred Runte, a historian of national parks, refers to as "monumentalism." In the 19th century, Americans considered their national parks a contribution to civilization equivalent to the architectural treasures of the Old World and sought to preserve them as a matter of national pride.[a]

In the second half of the 20th century, the emphasis of park management became more ecological, with parks established both to conduct scientific research and to maintain examples of representative natural areas. For instance, Zimbabwe established Sengwa National Park (now called Matusadona National Park) solely for scientific research. It has no tourist areas, and tourists are not generally allowed; its purpose is the study of natural ecosystems with as little human interference as possible so that the principles of wildlife and wilderness management can be better formulated and understood. Other national parks in the countries of eastern and southern Africa—including those of Kenya, Uganda, Tanzania, Zimbabwe, and South Africa—have been established primarily for viewing wildlife and for biological conservation.

In recent years, the number of national parks throughout the world has increased rapidly. The law establishing national parks in France was first enacted in 1960. Taiwan had no

(a)

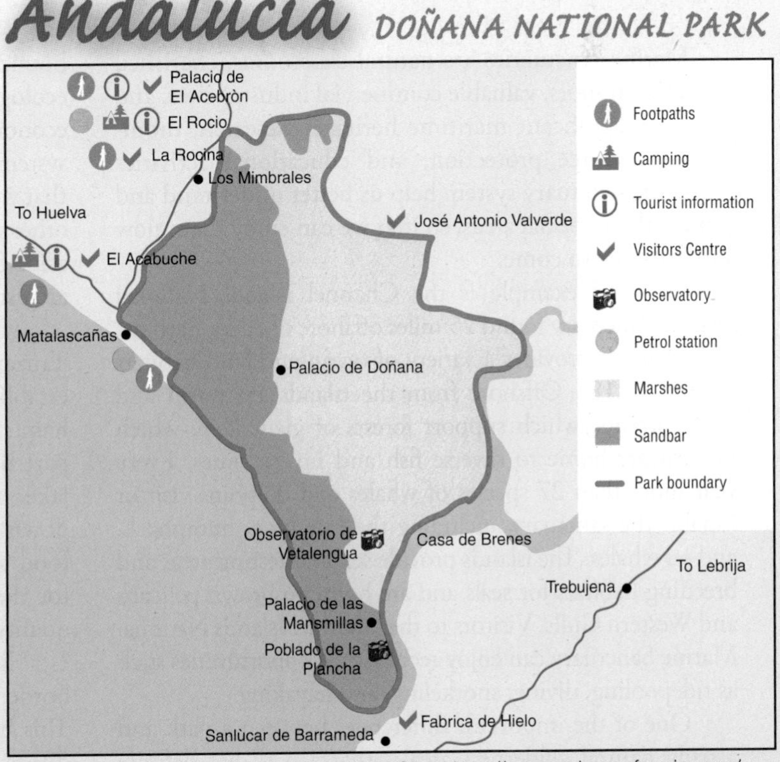

(b)

(*Source*: Colours of Spain. World Heritage Sites. http://www.coloursofspain.com/travelguidedetail/17/andalucia_andalusia/world_heritage_sites_donana_national_park.)

FIGURE 12.16 **(a) Flamingos are among the many birds that use Coto Doñana National Park,** a major stopover on the Europe-to-Africa flyway. **(b)** Map of Coto Doñana National Park, Spain.

Chee-Onn Leong/iStockphoto

FIGURE 12.17 The famous main valley of Yosemite National Park.

national parks prior to 1980 but now has six. In the United States, the area in national and state parks has expanded from less than 12 million hectares (30 million acres) in 1950 to nearly 83.6 million acres today, with much of the increase due to the establishment of parks in Alaska.[b]

Conserving representative natural areas of a country is an increasingly common goal of national parks. For example, the goal of New Zealand's national park planning is to include at least one area representative of each major ecosystem of the nation, from seacoast to mountain peak. In some cases, such as Spain's Coto Doñana National Park, national parks are among the primary resting grounds of major bird flyways (Figure 12.16) or play other crucial roles in the conservation of biodiversity.

Marine sanctuaries are natural classrooms, cherished recreational spots, valuable commercial industry sites, and places of significant maritime heritage. Research, monitoring, resource protection, and educational activities across the sanctuary system help us better understand and protect these special areas, so that we can enjoy them now and for years to come.

A good example is the Channel Islands National Marine Sanctuary found 25 miles offshore of Santa Barbara, California. It provides a variety of ocean and land habitats (Figure 12.18). Offshore from the islands are warm and cool currents, which support forests of giant kelp, which in turn are home to diverse fish and invertebrates. Every year more than 27 species of whales and dolphins visit or inhabit the sanctuary, including the rare blue, humpback, and sei whales. The islands provide seabird nesting areas and breeding habitats for seals and are home to brown pelicans and Western Gulls. Visitors to the Channel Islands National Marine Sanctuary can enjoy recreational opportunities such as tidepooling, diving, snorkeling, and kayaking.

One of the important differences between a park and a truly natural wilderness area is that a park has definite boundaries. These boundaries are usually arbitrary from an ecological viewpoint and have been established for political, economic, or historical reasons unrelated to the natural ecosystem. In fact, many parks have been developed on areas that would have been considered wastelands, useless for any other purpose. Even where parks or preserves have been set aside for the conservation of some species, the boundaries are usually arbitrary, and this has caused problems.

For example, Lake Manyara National Park in Tanzania, famous for its elephants, was originally established with boundaries that conflicted with elephant habits. Before this park was established, elephants spent part of the year feeding along a steep incline above the lake. At other times of the year, they would migrate down to the valley floor, depending on the availability of food and water. These annual migrations were necessary for the elephants to obtain food of sufficient nutritional quality throughout the year. However, when the park was established, farms that were laid out along its northern border crossed the traditional pathways of the elephants. This had two negative effects. First, elephants came into direct conflict with farmers. Elephants crashed through

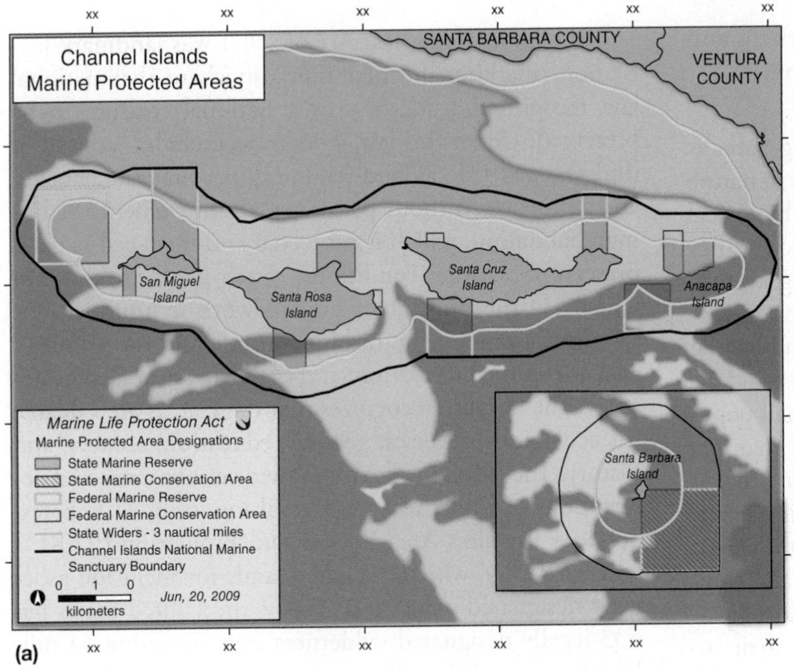

(a)

(b) Photo by Daniel B. Botkin

FIGURE 12.18 Channel Island Marine Sanctuary **(a)** map and **(b)** college students on a field trip see seals on San Miguel Island, part of the sanctuary.

farm fences, eating corn and other crops and causing general disruption. Second, whenever the farmers succeeded in keeping elephants out, the animals were cut off from reaching their feeding ground near the lake.

When it became clear that the park boundaries were arbitrary and inappropriate, the boundaries were adjusted to include the traditional migratory routes. This eased the conflicts between elephants and farmers.

Conflicts Relating to Parks

Size, Access, and Types of Activities

The idea of a national, state, county, or city park is well accepted in North America, but conflicts arise over what kinds of activities and what intensity of activities should be allowed in parks. These uses include recreation, conservation of scenery, conservation of biological diversity (especially of stages of the ecosystem that represent times past), scientific research, conservation of specific endangered species or species of special interest, and more recently, sites with kinds of DNA not found elsewhere that can be of considerable economic value. Any and all of these uses can conflict. In addition, there are uses outside of and under parks, such as mining for oil and natural gas and use of water aquifers.

The 1916 law that created the U.S. National Park Service, known as the Organic Act, stated that the "purpose is to conserve the scenery and the natural and historic objects and the wildlife therein and to provide for the enjoyment of the same in such manner and by such means as will leave them unimpaired for the enjoyment of future generations."

Often, biological conservation and the needs of individual species require limited human access, but people want to go to these areas because they are beautiful and desirable places for recreation.

As a recent example, travel into Yellowstone National Park by snowmobile in the winter has become popular, but this has led to noise and air pollution and has marred the experience of the park's beauty for many visitors. In 2003 a federal court determined that snowmobile use should be phased out in this park.

Alfred Runte explained the heart of the conflict. "This struggle was not against Americans who like their snowmobiles, but rather against the notion that anything goes in the national parks," he said. "The courts have reminded us that we have a different, higher standard for our national parks. Our history proves that no one loses when beauty wins. We will find room for snowmobiles, but just as important, room without them, which is the enduring greatness of the national parks."[26]

Many of the recent conflicts relating to national parks have concerned the use of motor vehicles. Voyageurs National Park in northern Minnesota, established in 1974—fairly recently compared with many other national parks—occupies land that was once used by a variety of recreational vehicles and provided livelihoods for hunting and fishing guides and other tourism businesses. These people felt that restricting motor-vehicle use would destroy their livelihoods. Voyageurs National Park has 100 miles of snowmobile trails and is open to a greater variety of motor-vehicle recreation than Yellowstone.[27]

Interactions Between People and Wildlife

While many people like to visit parks to see wildlife, some wildlife, such as grizzly bears and bison in Yellowstone National Park, can be dangerous. There has been conflict in the past between conserving the grizzly and making the park as open as possible for recreation.

How Much Land Should Be in Parks?

Another important controversy in managing parks is what percentage of a landscape should be in parks or nature preserves, especially with regard to the goals of biological diversity. Because parks isolate populations genetically, they may provide too small a habitat for maintaining a minimum safe population size. If parks are to function as biological preserves, they must be adequate in size and habitat diversity to maintain a population large enough to avoid the serious genetic difficulties that can develop in small populations. An alternative, if necessary, is for a park manager to move individuals of one species—say, lions in African preserves—from one park to another to maintain genetic diversity. But park size is a source of conflicts, with conservationists typically wanting to make parks bigger and commercial interests typically wanting to keep them smaller.

Nations differ widely in the percentage of their total area set aside as national parks. Costa Rica, a small country with high biological diversity, has more than 12% of its land in national parks.[28] Kenya, a larger nation that also has numerous biological resources, has 7.6% of its land in national parks.[29] In France, an industrialized country in which civilization has altered the landscape for several thousand years, only 0.7% of the land is in the nation's six national parks. However, France has 38 regional parks that encompass 11% (5.9 million hectares) of the nation's area.

The total amount of protected natural area in the United States is more than 104 million hectares (about 240 million acres), approximately 11.2% of the total U.S. land area.[30] However, the states differ greatly in the percentage of land set aside for parks, preserves, and other conservation areas. The western states have vast parks, whereas the six Great Lakes states (Michigan, Minnesota, Illinois, Indiana, Ohio, and Wisconsin), covering an area approaching that of France and Germany combined, allocate less than 0.5% of their land to parks and less than 1% to designated wilderness.[31]

12.11 Conserving Wilderness

What It Is, and Why It Is of Growing Importance

As a modern legal concept, **wilderness** is an area undisturbed by people. The only people in a wilderness are visitors, who do not remain. The conservation of wilderness is a new idea introduced in the second half of the 20th century. It is one that is likely to become more important as the human population increases and the effects of civilization become more pervasive throughout the world.

The U.S. Wilderness Act of 1964 was landmark legislation, marking the first time anywhere that national law recognized wilderness as a national treasure to be preserved. Under this law, wilderness includes "an area of undeveloped federal land retaining its primeval character and influence, without permanent improvements or human habitation, which is protected and managed so as to preserve its natural conditions." Such lands are those in which (1) the imprint of human work is unnoticeable, (2) there are opportunities for solitude and for primitive and unconfined recreation, and (3) there are at least 5,000 acres. The law also recognizes that these areas are valuable for ecological processes, geology, education, scenery, and history. The Wilderness Act required certain maps and descriptions of wilderness areas, resulting in the U.S. Forest Service's Roadless Area Review and Evaluation (RARE I and RARE II), which evaluated lands for inclusion as legally designated wilderness. Today, the United States has 633 legally designated wilderness areas, covering 44 million hectares (106 million acres)—more than 4% of the nation. Another 200 million acres meet the legal requirements and could be protected by the Wilderness Act. Half of this area is in Alaska, including the largest single area, Wrangell–St. Elias (Figure 12.19), covering 3.7 million hectares (9 million acres).[30, 32]

Where You'll Find It and Where You Won't

Countries with a significant amount of wilderness include New Zealand, Canada, Sweden, Norway, Finland, Russia, and Australia; some countries of eastern and southern Africa; many countries of South America, including parts of the Brazilian and Peruvian Amazon basin; the mountainous high-altitude areas of Chile and Argentina; some of the remaining interior tropical forests of Southeast Asia; and the Pacific Rim countries (parts of Borneo, the Philippines, Papua New Guinea, and Indonesia). In

Rich Reid/Getty Images, Inc.

FIGURE 12.19 Wrangell–St. Elias Wilderness Area, Alaska, Designated in 1980 and now covering 9,078,675 acres. As the photograph suggests, this vast area gives a visitor a sense of wilderness as a place where a person is only a visitor and human beings seem to have no impact.

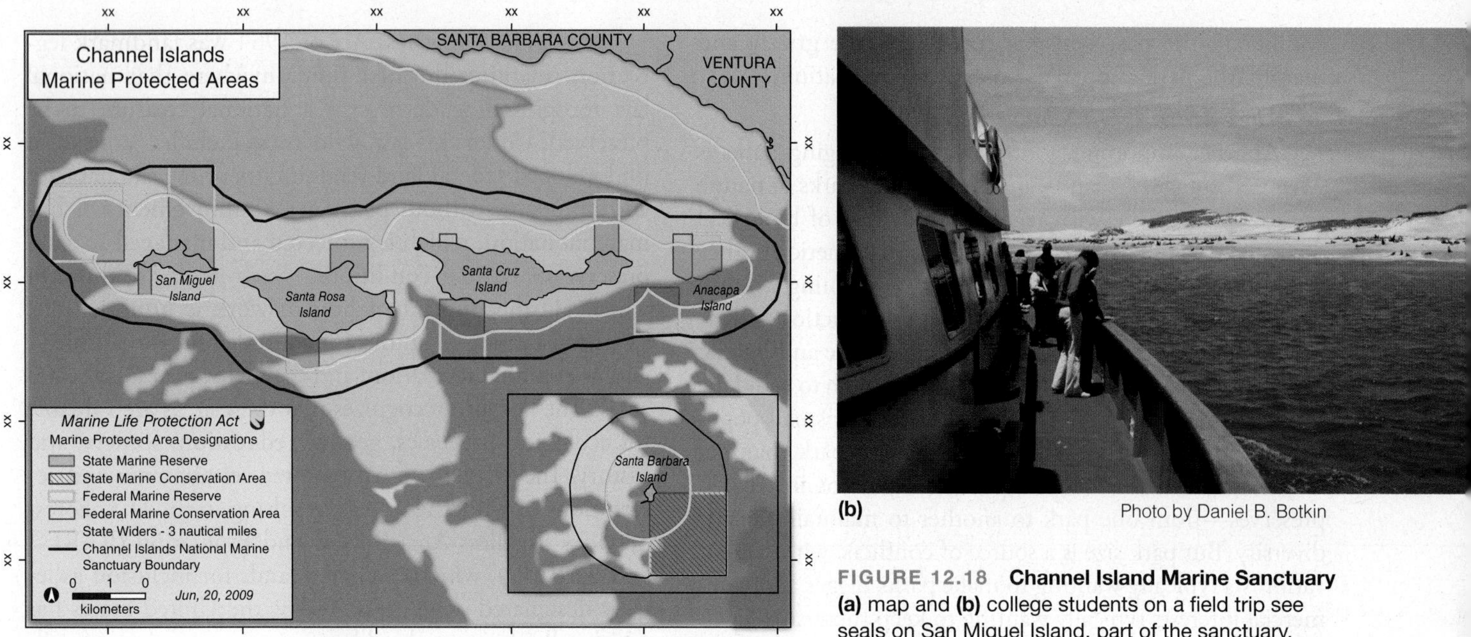

(a) **(b)** Photo by Daniel B. Botkin

FIGURE 12.18 Channel Island Marine Sanctuary **(a)** map and **(b)** college students on a field trip see seals on San Miguel Island, part of the sanctuary.

farm fences, eating corn and other crops and causing general disruption. Second, whenever the farmers succeeded in keeping elephants out, the animals were cut off from reaching their feeding ground near the lake.

When it became clear that the park boundaries were arbitrary and inappropriate, the boundaries were adjusted to include the traditional migratory routes. This eased the conflicts between elephants and farmers.

Conflicts Relating to Parks

Size, Access, and Types of Activities

The idea of a national, state, county, or city park is well accepted in North America, but conflicts arise over what kinds of activities and what intensity of activities should be allowed in parks. These uses include recreation, conservation of scenery, conservation of biological diversity (especially of stages of the ecosystem that represent times past), scientific research, conservation of specific endangered species or species of special interest, and more recently, sites with kinds of DNA not found elsewhere that can be of considerable economic value. Any and all of these uses can conflict. In addition, there are uses outside of and under parks, such as mining for oil and natural gas and use of water aquifers.

The 1916 law that created the U.S. National Park Service, known as the Organic Act, stated that the "purpose is to conserve the scenery and the natural and historic objects and the wildlife therein and to provide for the enjoyment of the same in such manner and by such means as will leave them unimpaired for the enjoyment of future generations."

Often, biological conservation and the needs of individual species require limited human access, but people want to go to these areas because they are beautiful and desirable places for recreation.

As a recent example, travel into Yellowstone National Park by snowmobile in the winter has become popular, but this has led to noise and air pollution and has marred the experience of the park's beauty for many visitors. In 2003 a federal court determined that snowmobile use should be phased out in this park.

Alfred Runte explained the heart of the conflict. "This struggle was not against Americans who like their snowmobiles, but rather against the notion that anything goes in the national parks," he said. "The courts have reminded us that we have a different, higher standard for our national parks. Our history proves that no one loses when beauty wins. We will find room for snowmobiles, but just as important, room without them, which is the enduring greatness of the national parks."[26]

Many of the recent conflicts relating to national parks have concerned the use of motor vehicles. Voyageurs National Park in northern Minnesota, established in 1974—fairly recently compared with many other national parks—occupies land that was once used by a variety of recreational vehicles and provided livelihoods for hunting and fishing guides and other tourism businesses. These people felt that restricting motor-vehicle use would destroy their livelihoods. Voyageurs National Park has 100 miles of snowmobile trails and is open to a greater variety of motor-vehicle recreation than Yellowstone.[27]

Interactions Between People and Wildlife

While many people like to visit parks to see wildlife, some wildlife, such as grizzly bears and bison in Yellowstone National Park, can be dangerous. There has been

conflict in the past between conserving the grizzly and making the park as open as possible for recreation.

How Much Land Should Be in Parks?

Another important controversy in managing parks is what percentage of a landscape should be in parks or nature preserves, especially with regard to the goals of biological diversity. Because parks isolate populations genetically, they may provide too small a habitat for maintaining a minimum safe population size. If parks are to function as biological preserves, they must be adequate in size and habitat diversity to maintain a population large enough to avoid the serious genetic difficulties that can develop in small populations. An alternative, if necessary, is for a park manager to move individuals of one species—say, lions in African preserves—from one park to another to maintain genetic diversity. But park size is a source of conflicts, with conservationists typically wanting to make parks bigger and commercial interests typically wanting to keep them smaller.

Nations differ widely in the percentage of their total area set aside as national parks. Costa Rica, a small country with high biological diversity, has more than 12% of its land in national parks.[28] Kenya, a larger nation that also has numerous biological resources, has 7.6% of its land in national parks.[29] In France, an industrialized country in which civilization has altered the landscape for several thousand years, only 0.7% of the land is in the nation's six national parks. However, France has 38 regional parks that encompass 11% (5.9 million hectares) of the nation's area.

The total amount of protected natural area in the United States is more than 104 million hectares (about 240 million acres), approximately 11.2% of the total U.S. land area.[30] However, the states differ greatly in the percentage of land set aside for parks, preserves, and other conservation areas. The western states have vast parks, whereas the six Great Lakes states (Michigan, Minnesota, Illinois, Indiana, Ohio, and Wisconsin), covering an area approaching that of France and Germany combined, allocate less than 0.5% of their land to parks and less than 1% to designated wilderness.[31]

12.11 Conserving Wilderness

What It Is, and Why It Is of Growing Importance

As a modern legal concept, **wilderness** is an area undisturbed by people. The only people in a wilderness are visitors, who do not remain. The conservation of wilderness is a new idea introduced in the second half of the 20th century. It is one that is likely to become more important as the human population increases and the effects of civilization become more pervasive throughout the world.

The U.S. Wilderness Act of 1964 was landmark legislation, marking the first time anywhere that national law recognized wilderness as a national treasure to be preserved. Under this law, wilderness includes "an area of undeveloped federal land retaining its primeval character and influence, without permanent improvements or human habitation, which is protected and managed so as to preserve its natural conditions." Such lands are those in which (1) the imprint of human work is unnoticeable, (2) there are opportunities for solitude and for primitive and unconfined recreation, and (3) there are at least 5,000 acres. The law also recognizes that these areas are valuable for ecological processes, geology, education, scenery, and history. The Wilderness Act required certain maps and descriptions of wilderness areas, resulting in the U.S. Forest Service's Roadless Area Review and Evaluation (RARE I and RARE II), which evaluated lands for inclusion as legally designated wilderness. Today, the United States has 633 legally designated wilderness areas, covering 44 million hectares (106 million acres)—more than 4% of the nation. Another 200 million acres meet the legal requirements and could be protected by the Wilderness Act. Half of this area is in Alaska, including the largest single area, Wrangell–St. Elias (Figure 12.19), covering 3.7 million hectares (9 million acres).[30, 32]

Where You'll Find It and Where You Won't

Countries with a significant amount of wilderness include New Zealand, Canada, Sweden, Norway, Finland, Russia, and Australia; some countries of eastern and southern Africa; many countries of South America, including parts of the Brazilian and Peruvian Amazon basin; the mountainous high-altitude areas of Chile and Argentina; some of the remaining interior tropical forests of Southeast Asia; and the Pacific Rim countries (parts of Borneo, the Philippines, Papua New Guinea, and Indonesia). In

Rich Reid/Getty Images, Inc.

FIGURE 12.19 **Wrangell–St. Elias Wilderness Area, Alaska,** Designated in 1980 and now covering 9,078,675 acres. As the photograph suggests, this vast area gives a visitor a sense of wilderness as a place where a person is only a visitor and human beings seem to have no impact.

addition, wilderness can be found in the polar regions, including Antarctica, Greenland, and Iceland.

Many countries have no wilderness left to preserve. In the Danish language, the word for wilderness has even disappeared, although that word was important in the ancestral languages of the Danes.[32] Switzerland is a country in which wilderness is not a part of preservation. For example, a national park in Switzerland lies in view of the Alps—scenery that inspired the English romantic poets of the early 19th century to praise what they saw as wilderness and to attach the adjective *awesome* to what they saw. But the park is in an area that has been heavily exploited for such activities as mining and foundries since the Middle Ages. All the forests are planted.[33]

The Wilderness Experience: Natural versus Naturalistic

In a perhaps deeper sense, wilderness is an idea and an ideal that can be experienced in many places, such as Japanese gardens, which might occupy no more than a few hundred square meters. Henry David Thoreau distinguished between "wilderness" and "wildness." He thought of wilderness as a physical place and wildness as a state of mind. During his travels through the Maine woods in the 1840s, he concluded that wilderness was an interesting place to visit but not to live in. He preferred long walks through the woods and near swamps around his home in Concord, Massachusetts, where he was able to experience a *feeling* of wildness. Thus, Thoreau raised a fundamental question: Can one experience true wildness only in a huge area set aside as a wilderness and untouched by human actions, or can wildness be experienced in small, heavily modified, and, though not entirely natural, *naturalistic* landscapes, such as those around Concord in the 19th century?[34]

As Thoreau suggests, small, local, naturalistic parks may have more value than some of the more traditional wilderness areas as places of solitude and beauty. In Japan, for instance, there are roadless recreation areas, but they are filled with people. One two-day hiking circuit leads to a high-altitude marsh where people can stay in small cabins. Trash is removed from the area by helicopter. People taking this hike experience a sense of wildness.

In some ways, the answer to the question raised by Thoreau is highly personal. We must discover for ourselves what kind of natural or naturalistic place meets our spiritual, aesthetic, and emotional needs. This is yet another area in which one of our key themes, science and values, is evident.

Conflicts in Managing Wilderness

The legal definition of *wilderness* has given rise to several controversies. The wilderness system in the United States began in 1964 with 3.7 million hectares (9.2 million acres) under U.S. Forest Service control. Today, the United States has 633 legally designated wilderness areas, covering 44 million hectares (106 million acres)—more than 4% of the nation. Another 200 million acres meet the legal requirements and could be protected by the Wilderness Act.

Those interested in developing the natural resources of an area, including mineral ores and timber, have argued that the rules are unnecessarily stringent, protecting too much land from exploitation when there is plenty of wilderness elsewhere. Those who wish to conserve additional wild areas have argued that the interpretation of the U.S. Wilderness Act is too lenient and that mining and logging are inconsistent with the wording of the act. These disagreements are illustrated by the argument over drilling in the Arctic National Wildlife Refuge, a dispute that reemerged with the rising price of petroleum.

The notion of managing wilderness may seem paradoxical—is it still wilderness if we meddle with it? In fact, though, with the great numbers of people in the world today, even wilderness must be defined, legally set aside, and controlled. We can view the goal of managing wilderness in two ways: in terms of the wilderness itself and in terms of people. In the first instance, the goal is to preserve nature undisturbed by people. In the second, the purpose is to provide people with a wilderness experience.

Legally designated wilderness can be seen as one extreme in a spectrum of environments to manage. The spectrum ranges from wilderness rarely disturbed by anyone to preserves in which some human activities are allowed to be visible—parks designed for outdoor recreation, forests for timber production and various kinds of recreation, hunting preserves, and urban parks—and finally, at the other extreme, open-pit mines. You can think of many stages in between on this spectrum.

Wilderness management should involve as little direct action as possible, so as to minimize human influence. This also means, ironically, that one of the necessities is to control human access so that a visitor has little, if any, sense that other people are present.

Consider, for example, the Desolation Wilderness Area in California, consisting of more than 24,200 hectares (60,000 acres), which in one year had more than 250,000 visitors. Could each visitor really have a wilderness experience there, or was the human carrying capacity of the wilderness exceeded? This is a subjective judgment. If, on one hand, all visitors saw only their own companions and believed they were alone, then the actual number of visitors did not matter for each visitor's wilderness experience. On the other hand, if every visitor found the solitude ruined by strangers, then the management failed, no matter how few people visited.

Wilderness designation and management must also take into account adjacent land uses. A wilderness next to a garbage dump or a power plant spewing smoke is a contradiction in terms. Whether a wilderness can be adjacent

to a high-intensity campground or near a city is a more subtle question that must be resolved by citizens.

Today, those involved in wilderness management recognize that wild areas change over time and that these changes should be allowed to occur as long as they are natural. This is different from earlier views that nature undisturbed was unchanging and should be managed so that it did not change. In addition, it is generally argued now that in choosing what activities can be allowed in a wilderness, we should emphasize activities that depend on wilderness (the experience of solitude or the observation of shy and elusive wildlife) rather than activities that can be enjoyed elsewhere (such as downhill skiing).

Another source of conflict is that wilderness areas frequently contain economically important resources, including timber, fossil fuels, and mineral ores. There has been heated debate about whether wilderness areas should be open to the extraction of these resources.

Still another controversy involves the need to study wilderness versus the desire to leave wilderness undisturbed. Those in favor of scientific research in the wilderness argue that it is necessary for the conservation of wilderness. Those opposed argue that scientific research contradicts the purpose of a designated wilderness as an area undisturbed by people. One solution is to establish separate research preserves.

CRITICAL THINKING ISSUE

Can Tropical Forests Survive in Bits and Pieces?

Although tropical rain forests occupy only about 7% of the world's land area, they provide habitat for at least half of the world's species of plants and animals. Approximately 100 million people live in rain forests or depend on them for their livelihood. Tropical plants provide products such as chocolate, nuts, fruits, gums, coffee, wood, rubber, pesticides, fibers, and dyes. Drugs for treating high blood pressure, Hodgkin's disease, leukemia, multiple sclerosis, and Parkinson's disease have been made from tropical plants, and medical scientists believe many more are yet to be discovered.

In the United States, most of the interest in tropical rain forests has focused on Brazil, whose forests are believed to have more species than any other geographic area. Estimates of destruction in the Brazilian rain forest range from 6 to 12%, but numerous studies have shown that deforested area alone does not adequately measure habitat destruction because surrounding habitats are also affected (refer back to Figure 12.12a). For example, the more fragmented a forest is, the more edges there are, and the greater the impact on the living organisms. Such edge effects vary depending on the species, the characteristics of the land surrounding the forest fragment, and the distance between fragments. For example, a forest surrounded by farmland is more deeply affected than one surrounded by abandoned land in which secondary growth presents a more gradual transition between forest and deforested areas. Some insects, small mammals, and many birds find only 80 m (262.5 ft) to be a barrier to movement from one fragment to another, whereas one small marsupial has been found to cross distances of 250 m (820.2 ft). Corridors between forested areas also help to offset the negative effects of deforestation on plants and animals of the forest.

Critical Thinking Questions

1. Look again at Figure 12.12a, the satellite image of part of the Brazilian rain forest. You are asked to make a plan that will allow 50% of the area to be cut and the rest established as a national park. Make a design for how you think this would best be done, taking into account conservation of biological diversity, the difficulty of travel in tropical rain forests, and the needs of local people to make a living. In your plan, the areas to be harvested will not change over time once the design is in place.
2. You are asked to create a park like the one in question 1, taking into account that the forested areas cut for timber will be allowed to regenerate and during that time, until actual harvest, could be used for recreation. Modify your design to take that into account.
3. The forest fragments left uncut in Figure 12.12 are sometimes compared with islands. What are some ways in which this is an appropriate comparison? Some ways in which it is not?

SUMMARY

- In the past, land management for harvesting resources and conserving nature was mostly local, with each parcel of land considered independently.
- Today, a landscape perspective has developed, and lands used for harvesting resources are seen as part of a matrix that includes lands set aside for the conservation of biological diversity and for landscape beauty.
- Forests are among civilization's most important renewable resources. Forest management seeks a sustainable harvest and sustainable ecosystems. Because examples of successful sustainable forestry are rare, "certification of sustainable forestry" has developed to determine which methods appear most consistent with sustainability and then compare the management of a specific forest with those standards.
- Given their rapid population growth, continued use of firewood as an important fuel in developing nations is a major threat to forests. It is doubtful that these nations can implement successful management programs in time to prevent serious damage to their forests and severe effects on their people.
- Clear-cutting is a major source of controversy in forestry. Some tree species require clearing to reproduce and grow, but the scope and method of cutting must be examined carefully in terms of the needs of the species and the type of forest ecosystem.
- Properly managed plantations can relieve pressure on forests.
- Managing parks for biological conservation is a relatively new idea that began in the 19th century. The manager of a park must be concerned with its shape and size. Parks that are too small or the wrong shape may have too small a population of the species for which the park was established and thus may not be able to sustain the species.
- A special extreme in conservation of natural areas is the management of wilderness. In the United States, the 1964 Wilderness Act provided a legal basis for such conservation. Managing wilderness seems a contradiction—trying to make sure it will be undisturbed by people requires interference to limit user access and to maintain the natural state, so an area that is not supposed to be influenced by people actually is.
- Parks, nature preserves, wilderness areas, and actively harvested forests affect one another. The geographic pattern of these areas on a landscape, including corridors and connections among different types, is part of the modern approach to biological conservation and the harvest of forest resources.

REEXAMINING THEMES AND ISSUES

Sean Randall/Getty Images, Inc.

HUMAN POPULATION

Forests provide essential resources for civilization. As the human population grows, there will be greater and greater demand for these resources. Because forest plantations can be highly productive, we are likely to place increasing emphasis on them as a source of timber. This would free more forestland for other uses.

© Biletskiy_Evgeniy/iStockphoto

SUSTAINABILITY

Sustainability is the key to conservation and management of wild living resources. However, sustainable harvests have rarely been achieved for timber production, and sustained ecosystems in harvested forests are even rarer. Sustainability must be the central focus for forest resources in the future.

© Anton Balazh 2011/iStockphoto

GLOBAL PERSPECTIVE

Forests are global resources. A decline in the availability of forest products in one region affects the rate of harvest and economic value of these products in other regions. Biological diversity is also a global resource. As the human population grows, the conservation of biological diversity is likely to depend more and more on legally established parks, nature preserves, and wilderness areas.

ssguy/ShutterStock

URBAN WORLD

We tend to think of cities as separated from living resources, but urban parks are important in making cities pleasant and livable; if properly designed, they can also help to conserve wild living resources.

B2M Productions/Getty Images, Inc.

PEOPLE AND NATURE

Forests have provided essential resources, and often people have viewed them as perhaps sacred but also dark and scary. Today, we value wilderness and forests, but we rarely harvest forests sustainably. Thus, the challenge for the future is to reconcile our dual and somewhat opposing views so that we can enjoy both the deep meaningfulness of forests and their important resources.

George Doyle/Getty Images, Inc.

SCIENCE AND VALUES

Many conflicts over parks, nature preserves, and legally designated wilderness areas also involve science and values. Science tells us what is possible and what is required in order to conserve both a specific species and total biological diversity. But what society desires for such areas is, in the end, a matter of values and experience, influenced by scientific knowledge.

KEY TERMS

STUDY QUESTIONS

1. What environmental conflicts might arise when a forest is managed for the multiple uses of (a) commercial timber, (b) wildlife conservation, and (c) a watershed for a reservoir? In what ways could management for one use benefit another?
2. What arguments could you offer for and against the statement "Clear-cutting is natural and necessary for forest management"?
3. Can a wilderness park be managed to supply water to a city? Explain your answer.
4. A park is being planned in rugged mountains with high rainfall. What are the environmental considerations if the purpose of the park is to preserve a rare species of deer? If the purpose is recreation, including hiking and hunting?
5. What are the environmental effects of decreasing the rotation time (accelerating the rate of cutting) in forests from an average of 60 years to 10 years? Compare these effects for (a) a woodland in a dry climate on a sandy soil and (b) a rain forest.
6. In a small but heavily forested nation, two plans are put forward for forest harvests. In Plan A, all the forests to be harvested are in the eastern part of the nation, while all the forests of the West are set aside as wilderness areas, parks, and nature preserves. In Plan B, small areas of forests to be harvested are distributed throughout the country, in many cases adjacent to parks, preserves, and wilderness areas. Which plan would you choose? Note that in Plan B, wilderness areas would be smaller than in Plan A.
7. The smallest legally designated wilderness in the United States is Pelican Island, Florida, covering 5 acres. Do you think this can meet the meaning of *wilderness* and the intent of the Wilderness Act?

FURTHER READING

Botkin, D.B., *The Moon in the Nautilus Shell: Discordant Harmonies Reconsidered* (New York, NY: Oxford University Press, 2012).

Botkin, D.B., *No Man's Garden: Thoreau and a New Vision for Civilization and Nature* (Washington, DC: Island Press, 2001).

Hendee, J.C., *Wilderness Management: Stewardship and Protection of Resources and Values* (Golden, CO: Fulcrum Publishing, 2002). Considered the classic work on this subject.

Kimmins, J.P., *Forest Ecology*, 3rd ed. (Upper Saddle River, NJ: Prentice Hall, 2003). A textbook that applies recent developments in ecology to the practical problems of managing forests.

Runte, A., *National Parks: The American Experience* (Lincoln, NE: Bison Books of the University of Nebraska, 1997). Considered the classic book about the history of national parks in America and the reasons for their development.

NOTES

1. Lloyd, E.C. 2007. Jamaica Bay Watershed Protection Plan Volume I-Regional Profile, New York City Department of Environmental Protection, Emily Lloyd, Commissioner, October 1, 2007.
2. United Nations Food and Agriculture Organization. 2001. Rome: UN FAO.ftp://ftp.fao.org/docrep/fao/003/y0900e/y0900e02.pdf.
3. Stein, P.R. (2011). Trends in forestland ownership and conservation. *Forest History Today Spring-Fall* 83–86.
4. U.S. Forest Service. 2008. Forest Ownership Patterns and Family Forest Highlights from the National Woodland Owner Survey. U.S. Department of Agriculture: Northern Research Station.
5. World Firewood Supply, World Energy Council website. www.worldenergy.org. April 24, 2006. Accessed 6/01/08.
6. World Resources Institute. 1993. *World Resources 1992–93.* New York, NY: Oxford University Press.
7. Botkin, D.B. 2012. *The Moon in the Nautilus Shell: Discordant Harmonies Reconsidered.* New York, NY: Oxford University Press.
8. Likens. G.E., F.H. Borman, R.S. Pierce, J.S. Eaton, and N.M. Johnson. 1977. *The Biogeochemistry of a Forested Ecosystem.* New York, NY: Springer-Verlag.
9. The Hubbard Brook ecosystem continues to be one of the most active and long-term ecosystem studies in North America. An example of a recent publication is: Bailey, S.W., D.C. Buso, and G.E. Likens. 2003. Implications of sodium mass balance for interpreting the calcium cycle of a forested ecosystem. *Ecology* 84(2):471–484.
10. Swanson, F.J., and C.T. Dyrness. 1975. Impact of clearcutting and road construction on soil erosion by landslides in the western Cascade Range, Oregon. *Geology* 3:393–396.
11. Fredriksen, R.L. 1971. Comparative chemical water quality—natural and disturbed streams following logging and slash burning. *Forest Land Use and Stream Environments,* pp. 125–137. Corvallis: Oregon State University.
12. Sedjo, R.A., and D.B. Botkin. 1997. Using forest plantations to spare the natural forest environment. *Environment* 39(10):14–20.
13. Sheffield, R. 2007. *Planted Trees and Plantations. Forest Resources of the United States—2007.* Washington, DC: USFS, pp. 67–69.
14. Sustainable Forestry Initiative Inc., 900 17th Street, NW, Suite 700, Washington, DC 20006.
15. Jenkins, M. B. 1999. *The Business of Sustainable Forestry.* Washington, DC: Island Press.
16. World Resources Institute. *Disappearing Land: Soil Degradation.* Washington, DC: WRI.
17. Kimmins, H. 1995. Proceedings of the conference on certification of sustainable forestry practices. Malaysia.
18. Jenkins, Michael B. 1999. *The Business of Sustainable Forestry.* Washington, DC: Island Press.
19. Rosenmeier, M. F., D. A. Hodell, et al. 2002. A 4000-year lacustrine record of environmental change in the southern Maya lowlands, Pet´en, Guatemala. *Quaternary Research* 57: 183–190.
20. Deckers, K., and H. Pessin. 2010. Vegetation development in the Middle Euphrates and Upper Jazirah (Syria/Turkey) during the Bronze Age. *Quaternary Research* 74:216–226.
21. International Conference on Engineering Education. 1997. Carajás mining in the Amazon region. Mining-technology.com. Accessed July 28, 2009. Carajás Iron Ore Mine, Brazil; NASA image created by Jesse Allen, using EO-1 ALI data provided courtesy of the NASA EO-1 Team. Caption by Holli Riebeek.
22. Botkin, D. B. 1992. Global warming and forests of the Great Lakes states. In J. Schrnandt, ed., *The Regions and Global Warming: Impacts and Response Strategies.* New York, NY: Oxford University Press.
23. National Park Service, 2008. Wildland Fire in Yellowstone. http://www.nps.gov/yell/naturescience/wildlandfire.htm
24. Moseley, R. J., History of the Santa Barbara County Fire Department. http://www.sbcfire.com/au/dphist/history.htm
25. Malakoff, D. 2002. Arizona ecologist puts stamp on forest restoration debate. *Science* 297:2194–2196.
26. Quotations from Alfred Runte cited by the Wilderness Society on its Web site, at http://www.wilderness.org/-NewsRoom/Statement/20031216.cm.

27. Witzig, Fred T. 2004, *Voyageurs National Park: The Battle to Create Minnesota's National Park*. Minneapolis, MN: University of Minnesota Press.

28. Costa Rica's TravelNet.National parks of Costa Rica. 1999. Costa Rica: Costa Rica's TravelNet.

29. Anonymous. 2013. National Parks and Reserves, Kenya. URL http://www.masterseek.com/national-parks-and-game-reserves-in-kenya/company2-23512587

30. National Park Service Web site. http://www.wilderness.net/NWPS/fastfacts.

31. The National Wilderness Preservation System. 2006. http://www.wilderness.net/NWPS/static

32. Hendee, J.C., G.H. Stankey, and R.C. Lucas. 1978. *Wilderness Management*. United States Forest Service Misc. Pub. No. 1365.

33. Perlin, J. 1989. *A Forest Journey: The Role of Wood in the Development of Civilization*. New York, NY: Norton.

34. Botkin, D.B. 2012. ebook. *No Man's Garden: Thoreau and a New Vision for Civilization and Nature*. New York, NY: Croton River Publishers. (Originally published in hardback in 2001 by Island Press, Washington, DC.)

A CLOSER LOOK 12.2 NOTES

a. Runte, A. 1997. *National Parks: The American Experience*. Lincoln, NE: Bison Books of the University of Nebraska.

b. National Park Service Web site. http://www.wilderness.net/NWPS/fastfacts.

CHAPTER 13

Wildlife, Fisheries, and Endangered Species

LEARNING OBJECTIVES

Wildlife, fish, and endangered species are among the most popular environmental issues today. People love to see wildlife; many people enjoy fishing, make a living from it, or rely on fish as an important part of their diet. Since the 19th century the fate of endangered species has drawn public attention. You would think that by now we would be doing a good job of conserving and managing these kinds of life, but often we are not. This chapter tells you how we are doing and how we can improve our conservation and management of wildlife, fisheries, and endangered species.

After studying this chapter, you should be able to ...

- Explain why it is impossible to manage a wildlife species so that it sustains its single carrying capacity
- Analyze how white-tailed deer could be managed so that they are neither overabundant nor threatened with extinction and are not a problem in suburbia
- Support or argue against the statement: The best way to determine if a commercial fishery is in decline is to obtain an estimate of the total population size
- Consider the conservation of a rare species where you live and explain how it should be managed
- Consider the reasons for the conservation of biological diversity and decide which two are the most important
- Determine whether it is possible to meet all the requirements of sustainability for fisheries: sustainable fish populations, harvests, and the population of fishermen and their industry

A white- tailed deer feeds calmly in front of a suburban home in Croton-on-Hudson, New York, just 25 miles north of New York City. Many of the town's residents commute to the city to work and probably did not expect to live where wildlife was a problem.

Photo by Daniel B. Botkin

CASE STUDY

How Can There Be Both Too Few and Too Many Wildlife?

A great irony of the modern conservation of nature is that on the one hand we do a poor job of saving endangered species, exemplified by a recent huge increase in the slaughter of African elephants. Their ivory has greatly increased in value, leading to large-scale illegal trade. But on the other hand our rural and suburban areas are being inundated with too many wild animals, including white-tailed deer (chapter opening photo), coyotes, bears, and even mountain lions (although more rarely).

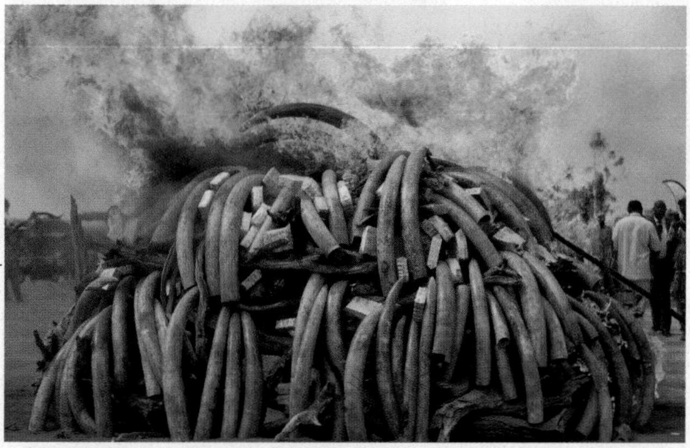
DAI KUROKAWA/epa/Corbis

FIGURE 13.1 **Pile of elephant ivory**

Elephant ivory, always valuable, especially in Asian cultures, has skyrocketed to as much as $1,000 a pound in illegal markets. The result has been massive killings of African elephants and the discovery of ivory smuggled into the Far East—as much as 24 tons in one uncovered shipment into Malaya. In 1963 the International Union for the Conservation of Nature (IUCN) set up the Convention on International Trade in Endangered Species (CITES) to halt such trade. By 2012, 176 nations had signed on to this agreement, but violations continue (Figure 13.1).

An obvious question arises: How can we stop the killing of animals of endangered species while reducing too abundant wildlife to an acceptable level? The concept that connects these two extremes is sustainability. The goal conservationists choose for both the rare and threatened wildlife and too abundant wildlife is sustainable population levels. In the past, it was believed that the path to wildlife sustainability was simple—just let nature alone, do nothing. Populations would, if undisturbed by people, come into a balance. But today two things are clear: (1) valuable wildlife must be protected, and (2) in the modern world wildlife unmanaged easily reaches unsustainably high abundances.

13.1 Traditional Single-Species Wildlife Management

Wildlife, fisheries, and endangered species are considered together in this chapter because they have a common history of exploitation, management, and conservation, and because modern attempts to manage and conserve them follow the same approaches. Although any form of life—from bacteria and fungi to flowering plants and animals—can become endangered, concern about endangered species has tended to focus on wildlife. We will maintain that focus, but we ask you to remember that the general principles apply to all forms of life.

Attempts to apply science to the conservation and management of wildlife and fisheries, and therefore to endangered species, began around the turn of the 20th century, when each species was viewed as a single population in isolation.

Carrying Capacity and Sustainable Yields

The classical, early-20th-century idea of wildlife and fisheries was formalized in the S-shaped logistic growth curve (Figure 13.2). The logistic growth curve assumes that changes in the size of a population are simply the result of the population's size in relation to a maximum, called the *carrying capacity*. The logistic characterizes the population only by its total size, nothing else—not the ratio of young to old, healthy to sick, males to females, and the

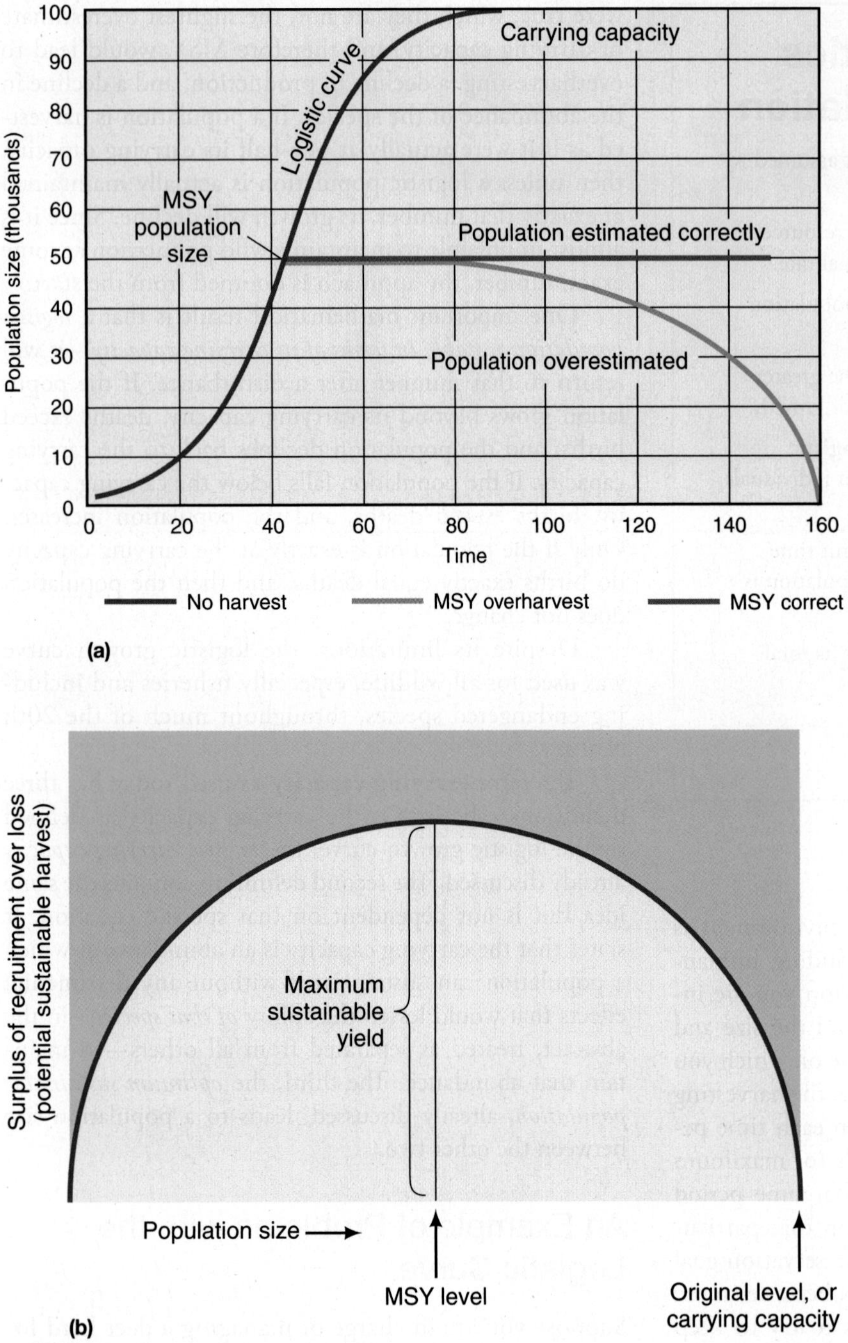

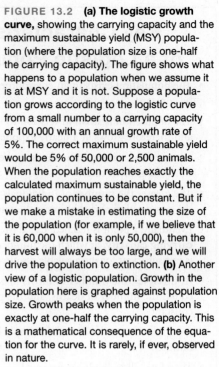

FIGURE 13.2 **(a) The logistic growth curve,** showing the carrying capacity and the maximum sustainable yield (MSY) population (where the population size is one-half the carrying capacity). The figure shows what happens to a population when we assume it is at MSY and it is not. Suppose a population grows according to the logistic curve from a small number to a carrying capacity of 100,000 with an annual growth rate of 5%. The correct maximum sustainable yield would be 5% of 50,000 or 2,500 animals. When the population reaches exactly the calculated maximum sustainable yield, the population continues to be constant. But if we make a mistake in estimating the size of the population (for example, if we believe that it is 60,000 when it is only 50,000), then the harvest will always be too large, and we will drive the population to extinction. **(b)** Another view of a logistic population. Growth in the population here is graphed against population size. Growth peaks when the population is exactly at one-half the carrying capacity. This is a mathematical consequence of the equation for the curve. It is rarely, if ever, observed in nature.

environment doesn't appear at all; it is just assumed to be constant. (See the accompanying box Key Characteristics of a Logistic Population.) The carrying capacity is defined simply as the maximum population that can be sustained indefinitely. Implicit in this definition is the idea that if the population exceeds the carrying capacity, it will damage its environment and/or its own ability to grow and reproduce, and therefore the population will decline.

Two management goals resulted from these ideas and the logistic equation: For a species that we intend to harvest, the goal was maximum sustainable yield (MSY); for a species that we wish to conserve, the goal was to have that species reach, and remain at, its carrying capacity. **Maximum sustainable yield** is defined as the maximum growth rate (measured either as a net increase in the number of individuals or in biomass over a specified time period) that the population could sustain indefinitely. The **maximum sustainable-yield population** is defined as the population size at which the maximum growth rate occurs. More simply, the population was viewed as a factory that could keep churning out exactly the same quantity of a product year after year.

Today a broader view is developing. It acknowledges that a population exists within an ecological community

Key Characteristics of a Logistic Population

- The population exists in an environment assumed to be constant.
- The population is small in relation to its resources and therefore grows at a nearly exponential rate.
- Competition among individuals in the population slows the growth rate.
- The greater the number of individuals, the greater the competition and the slower the rate of growth.
- Eventually, a point is reached, called the **logistic carrying capacity**, at which the number of individuals is just sufficient for the available resources.
- At this level, the number of births in a unit time equals the number of deaths, and the population is constant.
- A population can be described simply by its total number.
- Therefore, all individuals are equal.

and within an ecosystem, and that the environment is always, or almost always, changing (including human-induced changes). Therefore, the population you are interested in interacts with many others, and the size and condition of those others can affect the one on which you are focusing. With this new understanding, the harvesting goal is to harvest sustainably, removing in each time period the maximum number of individuals (or maximum biomass) that can be harvested within that time period without diminishing either the population that particularly interests you or its ecosystem. The preservation goal for a threatened or an endangered species becomes more open, with more choices. One option is to try to keep the population near its carrying capacity. Another is to sustain a **minimum viable population**, which is the estimated smallest population that can maintain itself and its genetic variability indefinitely. A third option that has been suggested, which leads to a population size somewhere between the other two, is the **optimum sustainable population**.

In the logistic curve, the greatest production occurs when the population is exactly one-half of the carrying capacity (see Figure 13.2). This is nifty because it makes everything seem simple—all you have to do is figure out the carrying capacity and keep the population at one-half of it. But what seems simple can easily become troublesome. Even if the basic assumptions of the logistic curve were true, which they are not, the slightest overestimate of carrying capacity, and therefore MSY, would lead to overharvesting, a decline in production, and a decline in the abundance of the species. If a population is harvested as if it were actually at one-half its carrying capacity, then unless a logistic population is actually maintained at exactly that number, its growth will decline. Since it is almost impossible to maintain a wild population at some exact number, the approach is doomed from the start.

One important mathematical result is that a *logistic population is stable in terms of its carrying capacity*—it will return to that number after a disturbance. If the population grows beyond its carrying capacity, deaths exceed births, and the population declines back to the carrying capacity. If the population falls below the carrying capacity, births exceed deaths, and the population increases. Only if the population is exactly at the carrying capacity do births exactly equal deaths, and then the population does not change.

Despite its limitations, the logistic growth curve was used for all wildlife, especially fisheries and including endangered species, throughout much of the 20th century.

The term **carrying capacity** as used today has three definitions. The first is the carrying capacity as defined by the logistic growth curve, the *logistic carrying capacity* already discussed. The second definition contains the same idea but is not dependent on that specific equation. It states that the carrying capacity is an abundance at which a population can sustain itself without any detrimental effects that would lessen the ability *of that species*—in the abstract, treated as separated from all others—to maintain that abundance. The third, the *optimum sustainable population*, already discussed, leads to a population size between the other two.

An Example of Problems with the Logistic Curve

Suppose you are in charge of managing a deer herd for recreational hunting in one of the 50 U.S. states. Your goal is to maintain the population at its MSY level, which, as you can see from Figure 13.2, occurs at exactly one-half of the carrying capacity. At this abundance, the population increases by the greatest number during any time period.

To accomplish this goal, first you have to determine the logistic carrying capacity. You are immediately in trouble because only in a few cases has the carrying capacity ever been determined by legitimate scientific methods (see Chapter 2) and, second, we now know that the carrying capacity varies with changes in the environment. The procedure in the past was to estimate the carrying capacity by *nonscientific* means and then attempt to maintain the

population at one-half that level. This method requires accurate counts each year. It also requires that the environment not vary, or, if it does, that it vary in a way that does not affect the population. Since these conditions cannot be met, the logistic curve has to fail as a basis for managing the deer herd.

An interesting example of the staying power of the logistic growth curve can be found in the United States Marine Mammal Protection Act of 1972. This act states that its primary goal is to conserve "the health and stability of marine ecosystems," which is part of the modern approach, and so the act seems to be off to a good start. But then the act states that the secondary goal is to maintain an "optimum sustainable population" of marine mammals. What is this? The wording of the act allows two interpretations. One is the logistic carrying capacity, and the other is the MSY population level of the logistic growth curve. So the act takes us back to square one, the logistic curve.

13.2 Improved Approaches to Wildlife Management

In the past, the U.S. Council on Environmental Quality (an office within the executive branch of the federal government), the World Wildlife Fund of the United States, the Ecological Society of America, the Smithsonian Institution, and the International Union for the Conservation of Nature (IUCN) proposed four principles of wildlife conservation:

- A safety factor in terms of population size, to allow for limitations of knowledge and imperfections of procedures. An interest in harvesting a population should not allow the population to be depleted to some theoretical minimum size.
- Concern with the entire community of organisms and all the renewable resources, so that policies developed for one species are not wasteful of other resources.
- Maintenance of the ecosystem of which the wildlife are a part, minimizing risk of irreversible change and long-term adverse effects as a result of use.
- Continual monitoring, analysis, and assessment. The application of science and the pursuit of knowledge about the wildlife of interest and its ecosystem should be maintained and the results made available to the public.

These principles continue to be useful. They broaden the scope of wildlife management from a narrow focus on a single species to inclusion of the ecological community and ecosystem. They call for a safety net in terms of population size, meaning that no population should be held at exactly the MSY level or reduced to some theoretical minimum abundance. These new principles provide a starting point for an improved approach to wildlife management.

New approaches to wildlife conservation and management include: (1) historical range of abundance; (2) estimation of the probability of extinction based on historical range of abundance; (3) use of age-structure information (see Chapter 5); and (4) better use of harvests as sources of information. These, along with an understanding of the ecosystem and landscape context for populations, are improving our ability to conserve wildlife. We discuss these approaches next.

Time Series and Historical Range of Variation

Successful scientific management and conservation of wildlife requires monitoring—an ongoing measurement of population size over a number of years. This set of estimates is called a **time series** and could provide us with a measure of the **historical range of variation**—the known range of abundances of a population or species over some past time interval. But shockingly, such monitoring exists for relatively few species of concern (see A Closer Look 13.1, Stories Told by the Grizzly Bear and the Bison). One of the few species with excellent monitoring is the American whooping crane (Figure 13.3), America's tallest bird, standing about 1.6 m (5 ft) tall. Because this species became so rare and because it migrated as a single flock, people began counting the total population in the late 1930s. At that time, they saw only 14 whooping cranes, a population that winters in Aransas National Wildlife Refuge in Texas and summers in Wood Buffalo National Park, in Alberta, Canada. They counted all the individuals in the population each year—the total number *and* the number born that year. The difference between these two numbers gives the number dying each year as well. And from this time series, we can estimate the probability of extinction.

The first estimate of the probability of extinction based on the historical range of variation, made in the early 1970s, was a surprise. Although the birds were few, the probability of extinction was less than one in a billion.[1] How could this number be so low? Use of the historical range of variation carries with it the assumption that causes of variation in the future will be only those that occurred during the historical period.

Not that the whooping crane is without threats—current changes in its environment may not be simply repeats of what happened in the past. For the whooping cranes, one catastrophe—such as a long, unprecedented

drought on the wintering grounds—could cause a population decline not observed in the past. Unfortunately, at present, mathematical estimates of the probability of extinction have been done for just a handful of species. The good news is that the wild whooping cranes on the main flyway have continued to do well.

Unfortunately, the original complete census method, counting every bird, was abandoned in 2009, on the grounds that the cranes had become too abundant and widespread for this method to be economical. Since then, a standard aerial sampling method has been used, from which the total population is estimated. Unfortunately, the two methods have not been compared, so the new estimation method is not calibrated against the complete census. This decreases the reliability of the new method. However, the good news is that the population appears to be doing well, increasing from 263 in 2010 to 279 in 2011, the most recent date for population estimates.[2]

Age Structure as Useful Information

An additional key to successful wildlife management is monitoring of the population's age structure (see Chapter 5), which can provide many different kinds of information. For example, the age structures of the catch of salmon from the Columbia River in Washington for two different periods, 1941–1943 and 1961–1963, were quite different. In the first period, most of the catch (60%) consisted of four-year-olds; the three-year-olds and five-year-olds each made up about 15% of the population. Twenty years later, in 1961 and 1962, half the catch consisted of three-year-olds, the number of five-year-olds had declined to about 8%, and the total catch had declined considerably. During the period 1941–1943, 1.9 million fish were caught. During the second period, 1961–1963, the total catch dropped to 849,000, just 49% of the total caught in the earlier period. The shift in catch toward younger ages, along with an overall decline in catch, suggests that the

jimkruger/iStockphoto

(a)

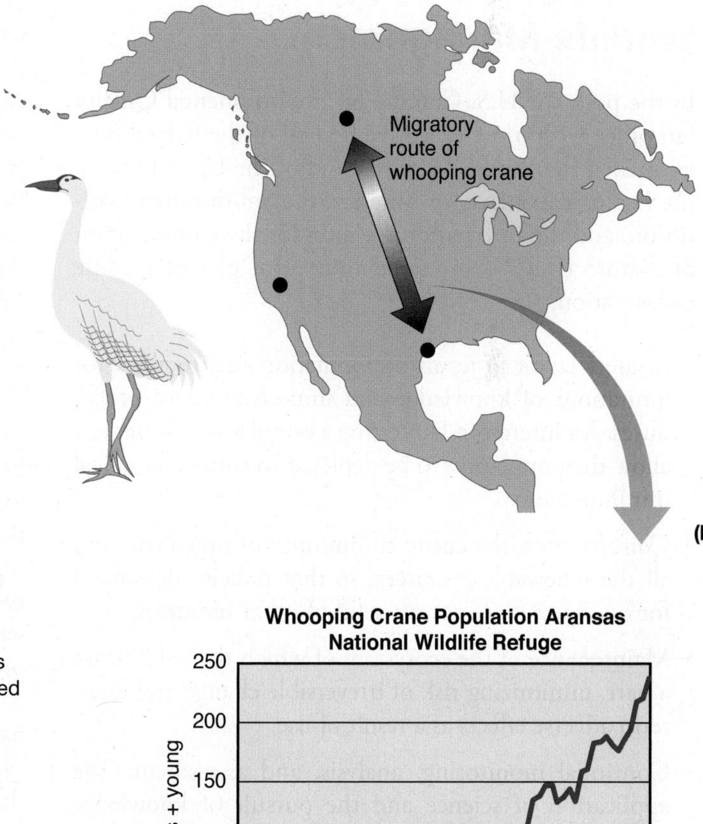

(b)

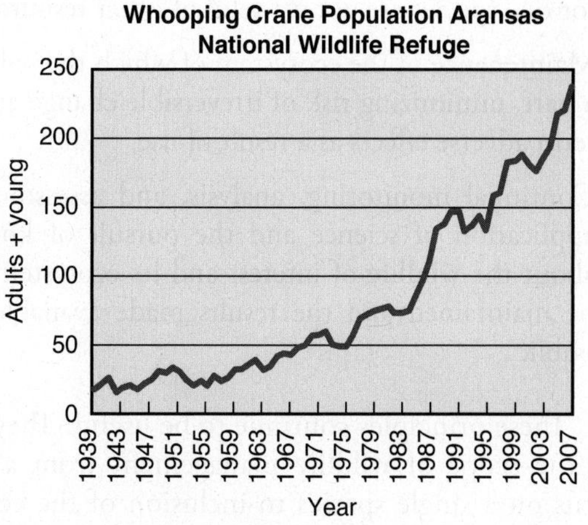

(c)

Data courtesy of Tom Stehn, Whooping Crane Coordinator, U.S. Fish and Wildlife Service, Aransas NWR, P.O. Box 100 Austwell, TX 77950.

FIGURE 13.3 The whooping crane (a) is one of many species that appear to have always been rare. Rarity does not necessarily lead to extinction, but a rare species, especially one that has undergone a rapid and large decrease in abundance, needs careful attention and assessment as to threatened or endangered status; **(b)** migration route; and **(c)** change in population from 1940 to 2007.

A CLOSER LOOK 13.1

Stories Told by the Grizzly Bear and the Bison

The grizzly bear and the American bison illustrate many of the general problems of conserving and managing wildlife and endangered species. In Chapter 2 we pointed out that the standard scientific method sometimes does not seem suited to studies in the environmental sciences. This is also true of some aspects of wildlife management and conservation. Several examples illustrate the needs and problems.

The Grizzly Bear

A classic example of wildlife management is the North American grizzly bear. An endangered species, the grizzly has been the subject of efforts by the U.S. Fish and Wildlife Service to meet the requirements of the U.S. Endangered Species Act, which include restoring the population of these bears.

The grizzly became endangered as a result of hunting and habitat destruction. It is arguably the most dangerous North American mammal, famous for unprovoked attacks on people, and has been eliminated from much of its range for that reason. Males weigh as much as 270 kg (600 lb), females as much as 160 kg (350 lb). When they rear up on their hind legs, they are almost 3 m (8 ft) tall. No wonder they are frightening. Despite this, or perhaps because of it, grizzlies intrigue people, and watching grizzlies from a safe distance has become a popular recreation (Figure 13.4).

At first glance, restoring the grizzly seems simple enough. But then that old question arises: Restore to what? One answer is to restore the species to its abundance at the time of the European discovery and settlement of North America. But it turns out that there is very little historical information about the abundance of the grizzly at that time, so it is not easy to determine how many (or what density per unit area) could be considered a "restored" population. We also lack a good estimate of the grizzly's present abundance, and thus we don't know how far we will have to take the species to "restore" it to some hypothetical past abundance. Moreover, the grizzly is difficult to study—it is large and dangerous and tends to be reclusive. The U.S. Fish and Wildlife Service attempted to count the grizzlies in Yellowstone National Park by installing automatic flash cameras that were set off when the grizzlies took a bait. This seemed a good idea, but the grizzlies didn't like the cameras and destroyed them,[a] so we still don't have a good scientific estimate of their present number. The National Wildlife Federation lists 1,200 in the contiguous states, 32,000 in Alaska, and about 25,000 in Canada, but these are crude estimates.[b]

FIGURE 13.4 **Photo of a grizzly bear.**

Bob Bennett/Oxford Scientific/Getty Images

How do we arrive at an estimate of a population that existed at a time when nobody thought of counting its members? Where possible, we use historical records, as discussed in Chapter 2. We can obtain a crude estimate of the grizzly's abundance at the beginning of the 19th century from the journals of the Lewis and Clark expedition. Lewis and Clark did not list the numbers of most wildlife they saw; they simply wrote that they saw "many" bison, elk, and so forth. But the grizzlies were especially dangerous and tended to travel alone, so Lewis and Clark noted each encounter, stating the exact number they met. On that expedition, they saw 37 grizzly bears over a distance of approximately 1,000 miles (their records were in miles).[a]

Lewis and Clark saw grizzlies in an area that stretched from near what is now Pierre, South Dakota, to what is today Missoula, Montana. A northern and southern geographic limit to the grizzly's range can be obtained from other explorers. Assuming Lewis and Clark could see a half-mile to each side of their line of travel on average, the density of the bears was approximately 3.7 per 100 square miles. If we estimate that the bears' geographic range was 320,000 square miles in the mountain and western plains states, we arrive at a total population of 320,000 x 0.37, or about 12,000 bears.

Suppose we phrase this as a hypothesis: "The number of grizzly bears in 1805 in what is now the United States was 12,000." Is this open to disproof? Not without time travel. Therefore, it is not a scientific statement; it can only be taken as an educated guess or, more formally, an assumption or premise. Still, it has some basis in historical documents, and it is better than no information, since we have few alternatives to determine what used to be. We can use this assumption to create a plan to restore the grizzly to that abundance. But is this the best approach?

Another approach is to ask what is the minimum viable population of grizzly bears—forget completely about what

might have been the situation in the past and make use of modern knowledge of population dynamics and genetics, along with food requirements and potential production of that food. Studies of existing populations of brown and grizzly bears suggest that only populations larger than 450 individuals respond to protection with rapid growth.[b] Using this approach, we could estimate how many bears appear to be a "safe" number—that is, a number that carries small risk of extinction and loss of genetic diversity. More precisely, we could phrase this statement as "How many bears are necessary so that the probability that the grizzly will become extinct in the next ten years (or some other period that we consider reasonable for planning) is less than 1% (or some other percentage that we would like)?"

With appropriate studies, this approach could have a scientific basis. Consider a statement of this kind phrased as a hypothesis: "A population of 450 bears (or some other number) results in a 99% chance that at least one mature male and one mature female will be alive ten years from today." We can disprove this statement, but only by waiting for ten years to go by. Although it is a scientific statement, it is a difficult one to deal with in planning for the present.

The American Bison

Another classic case of wildlife management, or mismanagement, is the demise of the American bison, or buffalo (Figure 13.5a). The bison was brought close to extinction in the 19th century for two reasons: They were hunted because coats made of bison hides had become fashionable in Europe, and they also were killed as part of warfare against the Plains peoples (Figure 13.5b). U.S. Army Colonel R.I. Dodge was quoted in 1867 as saying, "Kill every buffalo you can. Every buffalo dead is an Indian gone."[a]

Unlike the grizzly bear, the bison has recovered, in large part because ranchers have begun to find them profitable to raise and sell for meat and other products. Informal estimates, including herds on private and public ranges, suggest there are 200,000–300,000, and bison are said to occur in every state in the United States, including Hawaii, a habitat quite different from their original Great Plains home range.[c] About 20,000 roam wild on public lands in the United States and Canada.[d]

How many bison were there before European settlement of the American West? And how low did their numbers drop? Historical records provide insight. In 1865 the U.S. Army, in response to Indian attacks in the fall of 1864, set fires to drive away the Indians and the buffalo, killing vast numbers of animals.[e] The speed with which bison were almost eliminated was surprising—even to many of those involved in hunting them.

Many early writers tell of immense herds of bison, but few counted them. One exception was General Isaac I. Stevens, who, on July 10, 1853, was surveying for the transcontinental railway in North Dakota. He and his men climbed a high hill and saw "for a great distance ahead every square mile" having "a herd of buffalo upon it." He wrote

ASSOCIATED PRESS

(a)

(b)

Buffalo Hunt Chase (colour litho);Catlin, George (1794-1872); colour lithograph;600 X 414;©Butler Institute of American Art, Youngstown, OH, USA;PERMISSION REQUIRED FOR NON EDITORIAL USAGE;Out of copyright

FIGURE 13.5 **(a) A bison ranch in the United States.** In recent years, interest in bison ranches has increased greatly. In part, the goal is to restore bison to a reasonable percentage of its numbers before the Civil War. In part, too, bison are ranched because people like them. In addition, there is a growing market for bison meat and other products, including cloth made from bison hair. **(b)** Painting of a buffalo hunt by George Catlin in 1832–1833 at the mouth of the Yellowstone River.

that "their number was variously estimated by the members of the party—some as high as half a million. I do not think it any exaggeration to set it down at 200,000."[a] In short, his estimate of just one herd was about the same number of bison that exist in total today!

One of the better attempts to estimate the number of buffalo in a herd was made by Colonel R.I. Dodge, who took a wagon from Fort Zarah to Fort Larned on the Arkansas River in May 1871, a distance of 34 miles. For at least 25 of those miles, he found himself in a "dark blanket" of buffalo. He estimated that the mass of animals he saw in one day totaled 480,000. At one point, he and his men traveled to the top of a hill from which he estimated that he could see 6 to 10 miles, and from that high point there appeared to be a single solid mass of buffalo extending over 25 miles. At ten animals

per acre, not a particularly high density, the herd would have numbered 2.7–8.0 million animals.[a]

In the fall of 1868, "a train traveled 120 miles between Ellsworth and Sheridan, Wyoming, through a continuous, browsing herd, packed so thick that the engineer had to stop several times, mostly because the buffaloes would scarcely get off the tracks for the whistle and the belching smoke."[e] That spring, a train had been delayed for eight hours while a single herd passed "in one steady, unending stream." We can use accounts like this one to set bounds on the possible number of animals seen. At the highest extreme, we can assume that the train bisected a circular herd with a diameter of 120 miles. Such a herd would cover 11,310 square miles, or more than 7 million acres. If we suppose that people exaggerated the density of the buffalo, and there were only ten per acre, this single herd would still have numbered 70 million animals!

Some might say that this estimate is probably too high, because the herd would more likely have formed a broad, meandering, migrating line rather than a circle. The impression remains the same—there were huge numbers of buffalo in the American West even as late as 1868, numbering in the tens of millions and probably 50 million or more. Ominously, that same year, the Kansas Pacific Railroad advertised a "Grand Railway Excursion and Buffalo Hunt."[e] Some say that many hunters believed the buffalo could never be brought to extinction because there were so many. The same was commonly believed about all of America's living resources throughout the 19th century.

We tend to view environmentalism as a social and political movement of the 20th century, but it is said that after the Civil War there were angry protests in every legislature over the slaughter of buffalo. In 1871 the U.S. Biological Survey sent George Grinnell to survey the herds along the Platte River. He estimated that only 500,000 buffalo remained there and that at the then-current rate of killing, the animals would not last long. As late as the spring of 1883, a herd of an estimated 75,000 crossed the Yellowstone River near Miles City, Montana, but fewer than 5,000 reached the Canadian border.[e] By the end of that year—only 15 years after the Kansas Pacific train was delayed for eight hours by a huge herd of buffalo—only a thousand or so buffalo could be found, 256 in captivity and about 835 roaming the plains. A short time later, there were only 50 buffalo wild on the plains.

Today, more and more ranchers are finding ways to maintain bison, and the market for bison meat and other bison products is growing, along with an increasing interest in reestablishing bison herds for aesthetic, spiritual, and moral reasons. The history of the bison once again raises the question of what we mean by "restore" a population. Even with our crude estimates of original abundances, the numbers would have varied from year to year. So we would have to "restore" bison not to a single number independent of the ability of its habitat to support the population, but to some range of abundances. How do we approach that problem and estimate the range?

fish were being exploited to a point at which they were not reaching older ages. Such a shift in the age structure of a harvested population is an early sign of overexploitation and of a need to alter allowable catches.

A modification of measuring or estimating the entire age structure that is simpler in practice is to count the number or percentage of the females in a fish population that are at the minimum size that can reproduce. If most or all the female fish caught are at this size, the population is in trouble.[3]

Harvests as an Estimate of Numbers

Another way to estimate animal populations is to use the number harvested. Records of the number of buffalo killed were neither organized nor all that well kept, but they were sufficient to give us some idea of the number taken. In 1870, about 2 million buffalo were killed. In 1872, one company in Dodge City, Kansas, handled 200,000 hides. Estimates based on the sum of reports from such companies, together with guesses at how many animals were likely taken by small operators and not reported, suggest that about 1.5 million hides were shipped in 1872 and again in 1873.[4] In those years, buffalo hunting was the main economic activity in Kansas. The Indians were also killing large numbers of buffalo for their own use and for trade. According to estimates, as many as 3.5 million buffalo were killed per year, nationwide, during the 1870s.[5] The bison numbered at least in the low millions.

Still another way harvest counts are used to estimate previous animal abundance is the **catch per unit effort**. This method assumes that the same effort is exerted by all hunters/harvesters per unit of time, as long as they have the same technology. So if you know the total time spent in hunting/harvesting and you know the catch per unit of effort, you can estimate the total population. This method leads to rather crude estimates with a large observational error; but where there is no other source of information, it can offer unique insights.

An interesting application of this method is the reconstruction of the harvest of the bowhead whale and, from that, an estimate of the total bowhead population. Taken traditionally by Eskimos, the bowhead was the

object of "Yankee," or American, whaling from 1820 until the beginning of World War I. (See A Closer Look 13.4 later in this chapter for a general discussion of marine mammals.) Every ship's voyage was recorded, so we know essentially 100% of all ships that went out to catch bowheads. In addition, on each ship a daily log was kept, with records including sea conditions, ice conditions, visibility, number of whales caught, and their size in terms of barrels of oil. Some 20% of these logbooks still exist, and their entries have been computerized. Using some crude statistical techniques, it was possible to estimate the abundance of the bowhead in 1820 at 20,000, plus or minus 10,000.[6] Indeed, it was possible to estimate the total catch of whales and the catch for each year—and therefore the entire history of the hunting of this species.

13.3 Fisheries

Fish are important to our diets—they provide about 16% of the world's protein and are especially important protein sources in developing countries. Fish supply 6.6% of food in North America (where people are less interested in fish than are people in most other areas), 8% in Latin America, 9.7% in western Europe, 21% in Africa, 22% in central Asia, and 28% in the Far East.

Scientists estimate that there are 27,000 species of fish and shellfish in the oceans. People catch many of these species for food, but only a few kinds provide most of the food—anchovies, herrings, and sardines account for almost 20% (Table 13.1). Fishing is an international trade, but a few countries dominate. Japan, China, Russia, Chile, and the United States are among the major fisheries nations. And commercial fisheries are concentrated in relatively few areas of the world's oceans (Figure 13.6). Continental shelves, which make up only 10% of the oceans, provide more than 90% of the fishery harvest. Fish are abundant where their food is abundant and, ultimately, where there is high production of algae at the base of the food chain. Algae are most abundant in areas with relatively high concentrations of the chemical elements necessary for life, particularly nitrogen and phosphorus. These areas occur most commonly along the continental shelf, particularly in regions of wind-induced upwellings and sometimes quite close to shore.

The world's total fish harvest has increased greatly since the middle of the 20th century. The total harvest was 35 million metric tons (MT) in 1960. It more than doubled in just 20 years (an annual increase of about 3.6%)

Table 13.1 WORLD FISHERIES CATCHES

KIND	HARVEST (MILLIONS OF METRIC TONS)	PERCENT	ACCUMULATED PERCENTAGE
Herring, sardines, and anchovies	25	19.23%	19.23%
Carp and relatives	15	11.54%	30.77%
Cod, hake, and haddock	8.6	6.62%	37.38%
Tuna and their relatives	6	4.62%	42.00%
Oysters	4.2	3.23%	45.23%
Shrimp	4	3.08%	48.31%
Squid and octopus	3.7	2.85%	51.15%
Other mollusks	3.7	2.85%	54.00%
Clams and relatives	3	2.31%	56.31%
Tilapia	2.3	1.77%	58.08%
Scallops	1.8	1.38%	59.46%
Mussels and relatives	1.6	1.23%	60.69%
Subtotal	78.9	60.69%	
Total All Species	130	100%	

Source: National Oceanic & Atmospheric Administration World.

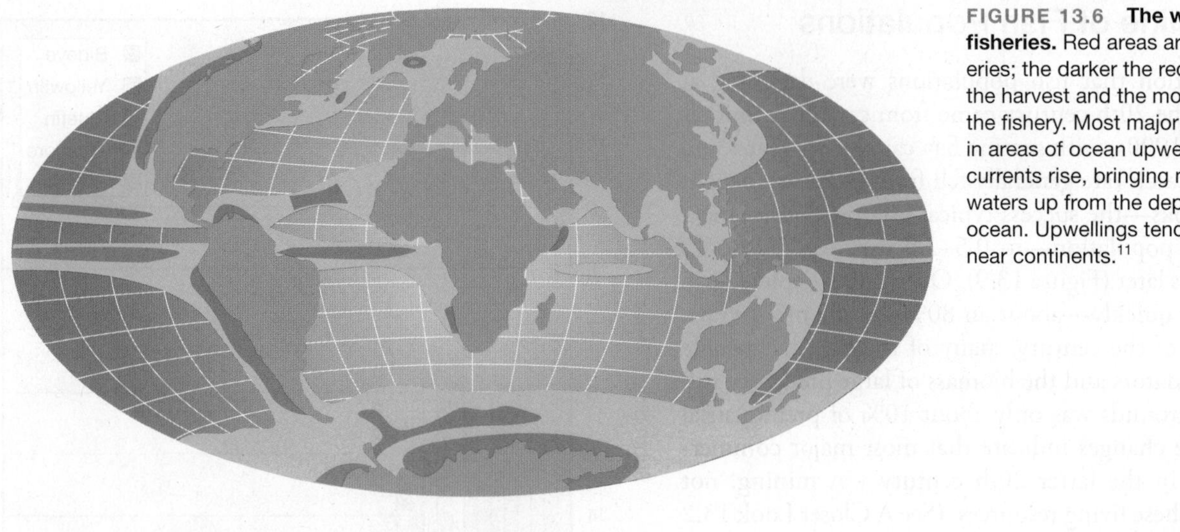

FIGURE 13.6 The world's major fisheries. Red areas are major fisheries; the darker the red, the greater the harvest and the more important the fishery. Most major fisheries are in areas of ocean upwellings, where currents rise, bringing nutrient-rich waters up from the depths of the ocean. Upwellings tend to occur near continents.[11]

to 76.6 MT in 1980 (Figure 13.7), then almost doubled again by 1990. The rate of increase slowed slightly in the next ten years—to 42% in 2000, then slowed to a 29% increase by 2009 and almost leveled off in 2010. The total global fish harvest doubled in 20 years because of increases in the number of boats, improvements in technology, and especially increases in aquaculture production. The recent leveling off in fish catch suggests that the harvest may be becoming sustainable. To understand whether this is the case, we need to step back and think about the species that make up the catch and what has been happening to them.

Although the total marine fisheries catch has increased during the past half-century, the effort required to catch a fish has increased as well. More fishing boats with better and better gear sail the oceans (Figure 13.8). That is why the total catch can increase while the total population of a fish species declines.

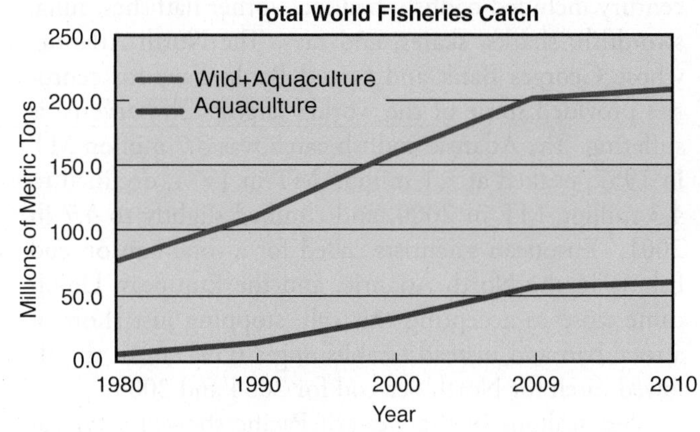

(*Source:* United Nations Food and Agriculture Organization, Fishery and Aquaculture Statistics 2010, Rome: FAO.)

FIGURE 13.7 Total World Fish Catch The blue line is the sum of wild and farmed (aquaculture) fish; the red line is aquaculture only.

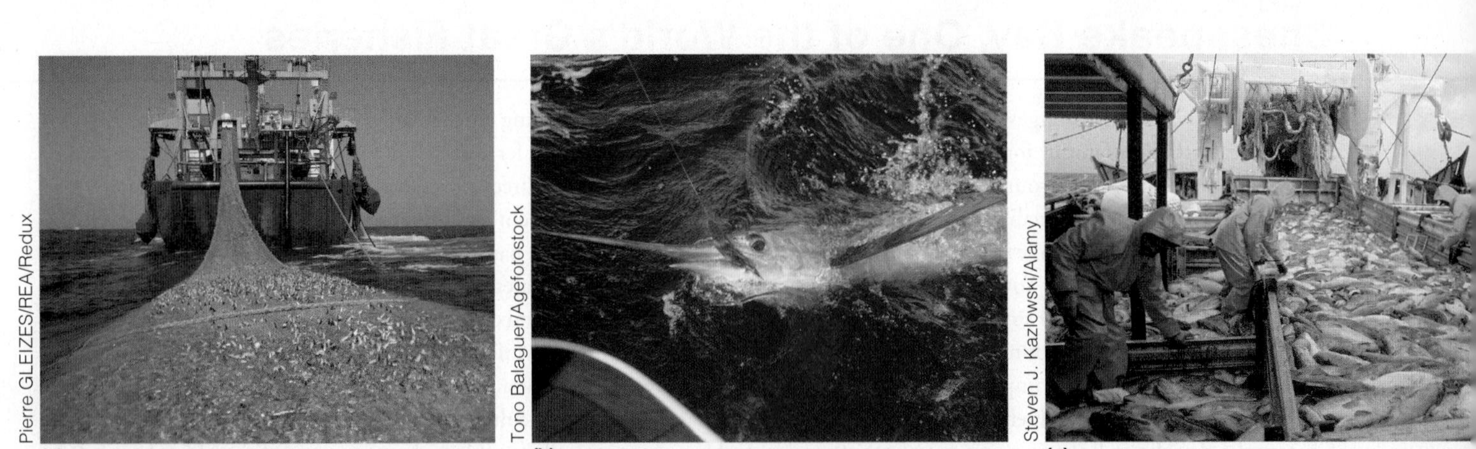

Pierre GLEIZES/REA/Redux
(a)
Tono Balaguer/Agefotostock
(b)
Steven J. Kazlowski/Alamy
(c)

FIGURE 13.8 Some modern commercial fishing methods. (a) Trawling using large nets; **(b)** longlines have caught a swordfish; **(c)** workers on a factory ship.

The Decline of Fish Populations

One indication that fish populations were declining at the end of the 20th century came from catch per unit effort. In the 1990s, with marine fish caught with lines and hooks, the catch rate generally fell from 6–12 fish caught per 100 hooks—the success typical of a previously unexploited fish population—to 0.5–2.0 fish per 100 hooks just ten years later (Figure 13.9). Overfishing depleted fish populations quickly—about an 80% decline in 15 years. By the end of the century, many of the fish that people ate were predators and the biomass of large predatory fish on fishing grounds was only about 10% of preindustrial levels. These changes indicate that most major commercial fishing in the latter 20th century was mining, not sustaining, these living resources. (See A Closer Look 13.2 about similar declines in Chesapeake Bay.)

Species suffering these declines at the end of the 20th century included codfish, haddock, other flatfishes, tuna, swordfish, sharks, skates, and rays. The North Atlantic, whose Georges Bank and Grand Banks have for centuries provided some of the world's largest fish harvests, is suffering. The Atlantic codfish catch was 3.7 million MT in 1957, peaked at 7.1 million MT in 1974, declined to 4.3 million MT in 2000, and climbed slightly to 4.7 in 2001.[7] European scientists called for a total ban on cod fishing in the North Atlantic, and the European Union came close to accepting this call, stopping just short of a total ban and instead establishing a 65% cut in the allowed catch for North Sea cod for 2004 and 2005.[7]

Sea scallops in the western Pacific showed a typical harvest pattern, starting with a very low harvest in 1964 at

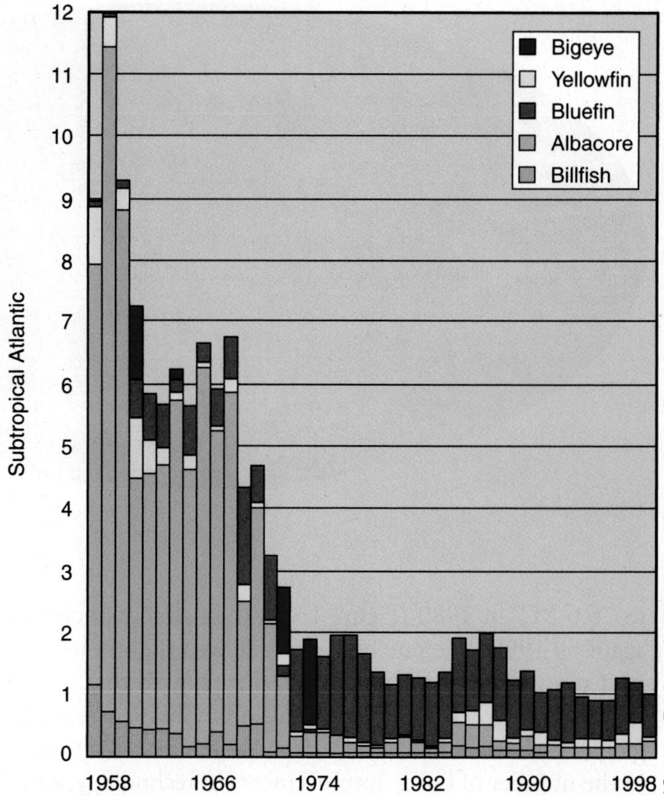

(*Source:* Ransom A. Meyers and Boris Worm, "Rapid Worldwide Depletion of Predatory Fish Communities," Nature [May 15, 2003].)

FIGURE 13.9 **Tuna catch decline.** The catch per unit of effort is represented here as the number of fish caught per 100 hooks for tuna and their relatives in the subtropical Atlantic Ocean. The vertical axis shows the number of fish caught per 100 hooks. The catch per unit of effort was 12 in 1958, when heavy modern industrial fishing for tuna began, and declined rapidly to about 2 by 1974. This pattern occurred worldwide in all the major fishing grounds for these species.

A CLOSER LOOK 13.2

Chesapeake Bay, One of the World's Great Fisheries

Chesapeake Bay, America's largest estuary, was one of the world's great fisheries, famous for oysters and crabs and as the breeding and spawning ground for bluefish, sea bass, and many other commercially valuable species (Figure 13.10). The bay, 200 miles long and 30 miles wide, drains an area of more than 165,000 km^2 from New York State to Maryland and is fed by 48 large rivers and 100 small ones.

Adding to the difficulty of managing the Chesapeake Bay fisheries, food webs are complex. Typical of marine food webs, the food chain of the bluefish that spawns and breeds in Chesapeake Bay shows links to a number of other species, each requiring its own habitat within the Bay and depending on processes that have a variety of scales of space and time.

Furthermore, Chesapeake Bay is influenced by many factors on the surrounding lands in its watershed—runoff from farms, including chicken and turkey farms, that are highly polluted with fertilizers and pesticides; introductions of exotic species; and direct alteration of habitats from fishing and the development of shoreline homes. There is also the varied salinity of the bay's waters—freshwater inlets from rivers and streams, seawater from the Atlantic, and brackish water resulting from the mixture of these.

Just determining which of these factors, if any, are responsible for a major change in the abundance of any fish species is difficult enough, let alone finding a solution that is economically feasible and maintains traditional levels of fisheries employment in the bay. The Chesapeake Bay's fisheries resources are at the limit of what environmental sciences can deal with at this time. Scientific theory remains inadequate, as do observations, especially of fish abundance.

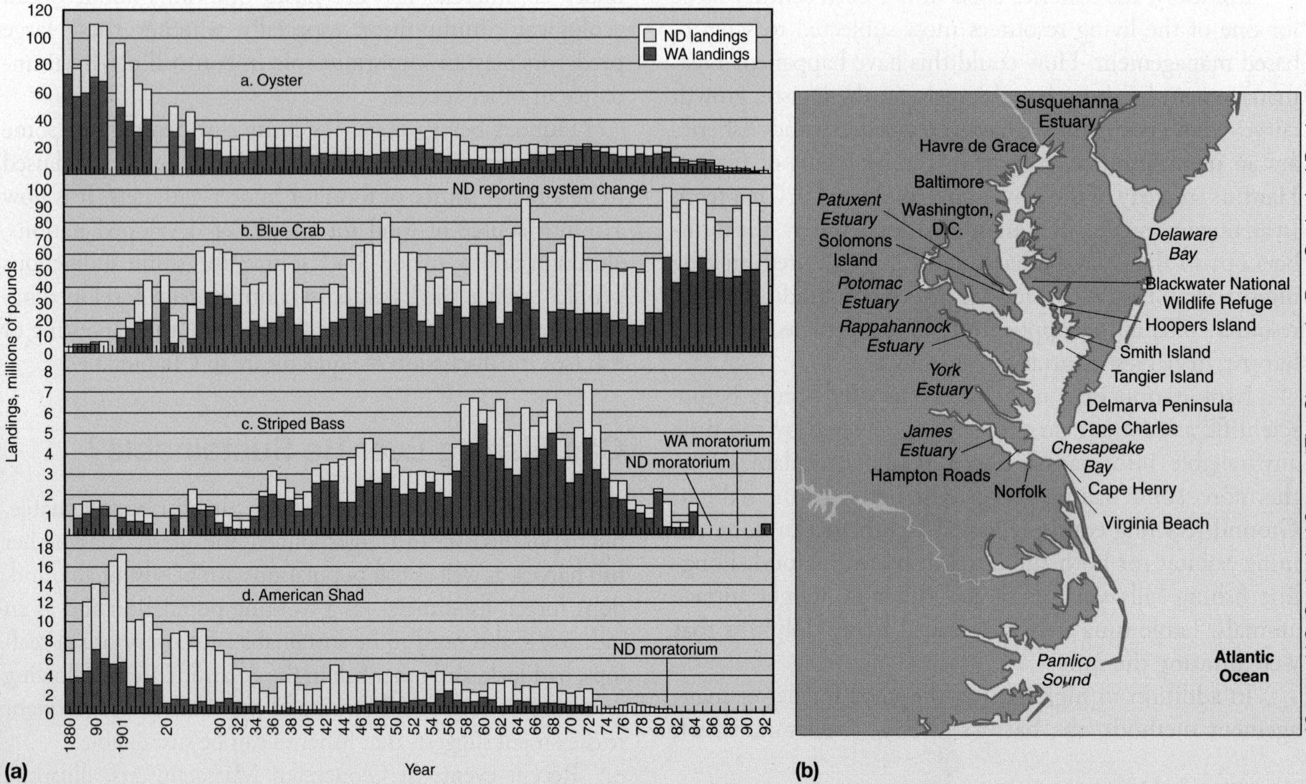

(*Source:* The Chesapeake Bay Foundation.)

(*Source:* U.S. Geological Survey, "The Chesapeake Bay: Geological Product of Rising Sea Level," 1998.)

FIGURE 13.10 **(a) Fish catch in the Chesapeake Bay.** (ND is the Virigina Catch; WA the Maryland catch) Oysters have declined dramatically. **(b)** Map of the Chesapeake Bay Estuary.

200 MT, increasing rapidly to 5,887 MT in 1975, declining to 1,489 in 1974, rising to about 7,670 MT in 1993, and then declining to 2,964 in 2002.[8] Catch of tuna and their relatives peaked in the early 1990s at about 730,000 MT and fell to 680,000 MT in 2000, a decline of 14% (Figure 13.9).

But during that time, changes in management practices led to remarkable recoveries for some species. For example, Atlantic sea scallops on Georges Bank increased from a few thousand in 1982 to more than 20,000 by 2002. Management changes including closing some areas to fishing for the scallops, reducing the number of days a vessel was allowed to fish for scallops each year, restricting the kinds of fishing gear (which limited the number of small scallops and finfish escaping), and limiting crew size. Apparently, closing off part of the sea scallop habitat led to increases in population not only within the closed area, but also in areas open to fishing.[9]

There is also good news for Georges Bank fisheries, following the severe restrictions that started at the end of the 20th century (Figure 13.11). Haddock have been fished commercially on Georges Bank since the 19th century and reliable catch statistics began early, in 1904. Annual catch reached 73,200 metric tons (MT) in the 1920s and again in the 1960s, then crashed to 5,600 MT from 1985 to 2000. Once again, severe management restrictions similar to those for scallops have led to the start of a recovery, with the catch exceeding 20,000 MT by 2005 (Figure 13.11). But Georges Bank haddock are still considered overfished, and restrictions continue.[10]

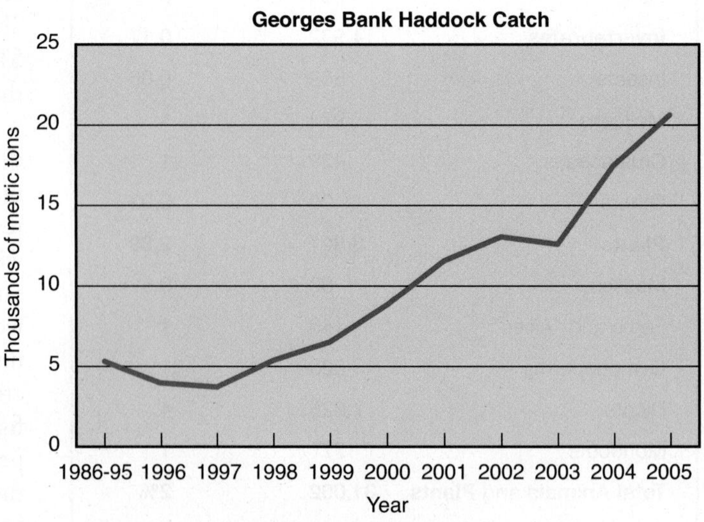

(*Source:* J. Brodziak and M. Traver, "Status of Fishery Resources off the Northeastern U.S." New England Fisheries Management Council., Washington, DC: NOAA, 2006.)

FIGURE 13.11 **Haddock caught in Georges Bank are recovering.**

Ironically, the fisheries crisis in the 20th century arose for one of the living resources most subjected to science-based management. How could this have happened? First, management has been based largely on the logistic growth curve, whose problems we have discussed. Second, fisheries are an open resource, subject to the problems of Garrett Hardin's "tragedy of the commons," discussed in Chapter 3. In an open resource, often in international waters, the numbers of fish that may be harvested can be limited only by international treaties, which are not tightly binding. Open resources offer ample opportunity for unregulated or illegal harvest, or harvest contrary to agreements.

Exploitation of a new fishery usually occurs before scientific assessment, so the fish are depleted by the time any reliable information about them is available. Furthermore, some fishing gear is destructive to the habitat. Ground-trawling equipment destroys the ocean floor, ruining habitat for both the target fish and its food. Long-line fishing kills sea turtles and other nontarget surface animals. Large tuna nets have killed many dolphins that were hunting the tuna.

In addition to highlighting the need for better management methods, the harvest of large predators, which many commercial fish are raises questions about ocean ecological communities, especially whether these large predators play an important role in controlling the abundance of other species.

Human beings began as hunter-gatherers, and some hunter-gatherer cultures still exist. Wildlife on the land used to be a major source of food for hunter-gatherers. It is now a minor source of food for people of developed nations, although still a major food source for some indigenous peoples, such as the Eskimo. In contrast, even developed nations are still primarily hunter-gatherers in the harvesting of fish (see the discussion of aquaculture in Chapter 11).

Table 13.2 NUMBER OF THREATENED SPECIES

LIFE-FORM	NUMBER THREATENED	PERCENT OF SPECIES KNOWN
Vertebrates	**5,188**	**9**
Mammals	1,101	20
Birds	1,213	12
Reptiles	304	4
Amphibians	1,770	31
Fish	800	3
Invertebrates	**1,992**	**0.17**
Insects	559	0.06
Mollusks	974	1
Crustaceans	429	1
Others	30	0.02
Plants	**8,321**	**2.89**
Mosses	80	0.5
Ferns and "Allies"	140	1
Gymnosperms	305	31
Dicots	7,025	4
Monocots	771	1
Total Animals and Plants	**31,002**	**2%**

Source: IUCN Red List www.iuenredlist.org/info/table/table1 (2004).

Can Fishing Ever Be Sustainable?

We need to consider whether fishing can ever be sustainable, but to put this issue in a larger context, the question is whether fish harvest as well as fish populations can be sustainable and, don't forget, fishermen—as a working population and as an industry—also need to be sustainable. Recovery of sea scallops, haddock, and other fish in the Atlantic Ocean following severe restrictions based on careful monitoring and frequent reassessment suggests that fisheries can be sustainable.

Recent events in Gloucester, Massachusetts, illustrate the larger question. Although, as a result of severe management restrictions, some of the Atlantic ground fish improved through 2006, the fishery has declined again. In January 2013, the New England Fishery Management Council recommended a 77% reduction from last year's catch for the next three years for cod in the Gulf of Maine and a 66% reduction from last year's catch for one year for Georges Bank. These measures appear necessary to save the cod population, as well as the long-term harvest of cod, but the short-term effect may be to put many fishermen out of business.

The discussion in Gloucester, long a port for cod fishermen, is whether the port should support other forms of commerce, biotech companies, and warehouse apartments. To help the fishermen, the federal government allocated $150 million, but the city's mayor suggests that some of that money should be spent developing other industries.[11]

An even larger question arises, namely: Will America decide to give up any support for the last remaining major commercial harvest of wildlife—its fisheries? As a result, will we then seek much less fish in our diet? Or will we attempt to expand aquaculture to provide the same amount of fish the country now consumes? What would be the effect on agriculture? Would it put an undesirable pressure on land to produce high-protein crops or domestic animals to replace the protein from fish? Should we just eat lower on the food chain? And people concerned about cultural heritage may ask: Should the nation abandon an entire way of life that has been rich in story and song and loved by those engaged in fishing? None of these are simple questions, and the entire set is rarely discussed together.

Suppose you went into fishing as a business and expected reasonable growth in that business in the first 20 years. The world's ocean fish catch rose from 39 million MT in 1964 to 68 million MT in 2003, an average of 3.8% per year for a total increase of 77%.[12] From a business point of view, even assuming all fish caught are sold, that is not considered rapid sales growth. But it is a heavy burden on a living resource.

There is a general lesson to be learned here: Few wild biological resources can sustain a harvest large enough to meet the requirements of a growing business. Although these growth rates begin to look pretty good in times when the overall economy is poor, as was the case during the economic downturn of 2008–2012, most wild biological resources do not prove to be a good business over the long run. We also learned this lesson from the near demise of the bison, discussed earlier, and it is true for whales as well (see A Closer Look 13.3). There have been a few exceptions, such as the several hundred years of fur trading by the Hudson's Bay Company in northern Canada. However, past experience suggests that economically beneficial sustainability is unlikely for most wild populations.

With that in mind, we note that farming fish—aquaculture (discussed in Chapter 11)—has been an important source of food in China for centuries and is an increasingly important food source worldwide. But aquaculture can create its own environmental problems.

In sum, fish are an important food and world harvests of fish are large, but the fish populations on which the harvests depend are generally declining, easily exploited, and difficult to restore. We desperately need new approaches to forecasting acceptable harvests and establishing workable international agreements to limit catch. This is a major environmental challenge, needing solutions within the next decade.

A CLOSER LOOK 13.3

Reasons for Conserving Endangered Species—and All Life on Earth

Important reasons for conserving endangered species are of two types: those having to do with tangible qualities and those dealing with intangible ones (see Chapter 3 for an explanation of tangible and intangible qualities). The tangible ones are utilitarian and ecological; the intangible are aesthetic, moral, recreational, spiritual, inspirational, and cultural.[a]

Utilitarian Justification

Many of the arguments for conserving endangered species, and for conserving biological diversity in general, have focused on the utilitarian justification: that many wild species have proved useful to us and many more may yet prove useful now or in the future, and therefore we should protect every species from extinction.

One example is the need to conserve wild strains of grains and other crops because disease organisms that attack crops evolve continually, and as new disease strains develop, crops become vulnerable. Crops such as wheat and corn depend on the continued introduction of fresh genetic characteristics from wild strains to create new, disease-resistant genetic hybrids. Related to this justification is the possibility of finding new crops among the many species of plants (see Chapter 11).

Another utilitarian justification is that many important chemical compounds come from wild organisms. Medicinal use of plants has an ancient history, going back into human prehistory. For example, a book titled *Materia Medica*, about the medicinal use of plants, was written in the 6th century A.D. in Constantinople by a man named Dioscorides (Figure 13.12).[b] To avoid scurvy, Native Americans advised early

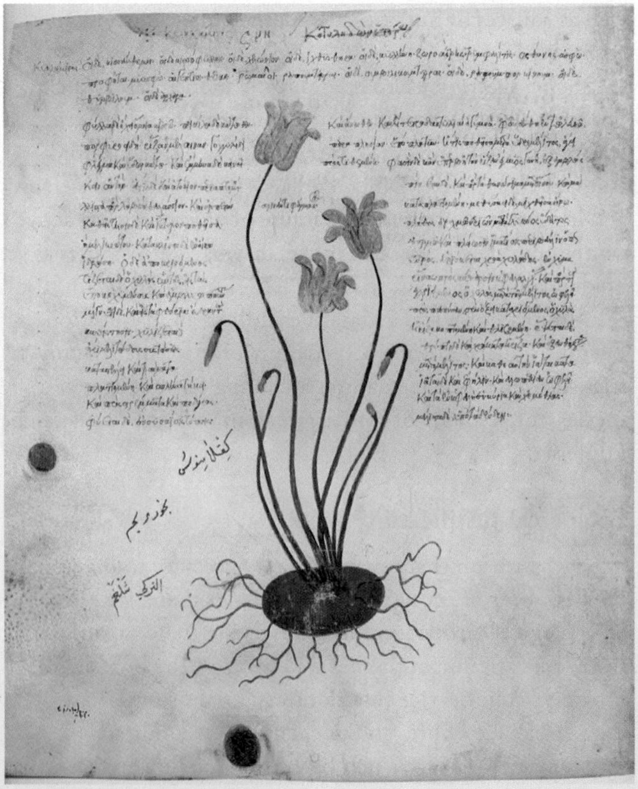

FIGURE 13.12 Sowbread *(Sow cyclamen)*, a small flowering plant, was believed useful medically at least 1,500 years ago, when this drawing of it appeared in a book published in Constantinople. Whether or not it is medically useful, the plant illustrates the ancient history of interest in medicinal plants.

Courtesy of Historical Collections & Services, Claude Moore Health Sciences Library, University of Virginia.

European explorers to chew on the bark of eastern hemlock trees (*Tsuga canadensis*); we know today that this was a way to get a little vitamin C.

Digitalis, an important drug for treating certain heart ailments, comes from purple foxglove, and aspirin is a derivative of willow bark. A more recent example was the discovery of a cancer-fighting chemical, paclitaxel, in the Pacific yew tree (genus name *Taxus*; hence the trade name Taxol). Well-known medicines derived from tropical forests include anticancer drugs from rosy periwinkles, steroids from Mexican yams, antihypertensive drugs from serpentwood, and antibiotics from tropical fungi.[c] Some 25% of prescriptions dispensed in the United States today contain ingredients extracted from vascular plants,[c] and these represent only a small fraction of the estimated 500,000 existing plant species. Other plants and organisms may produce useful medical compounds that are as yet unknown.

Scientists are testing marine organisms for use in pharmaceutical drugs. Coral reefs offer a promising area of study for such compounds because many coral-reef species produce toxins to defend themselves. According to the National Oceanic and Atmospheric Administration (NOAA), "Creatures found in coral ecosystems are important sources of new medicines being developed to induce and ease labor; treat cancer, arthritis, asthma, ulcers, human bacterial infections, heart disease, viruses, and other diseases; as well as sources of nutritional supplements, enzymes, and cosmetics."[d]

Some species are also used directly in medical research. For example, the armadillo, one of only two animal species (the other is us) known to contract leprosy, is used to study cures for that disease. Other animals, such as horseshoe crabs and barnacles, are important because of physiologically active compounds they make. Still others may have similar uses as yet unknown to us.

Tourism provides yet another utilitarian justification. Ecotourism is a growing source of income for many countries. Ecotourists value nature, including its endangered species, for aesthetic or spiritual reasons, but the result can be utilitarian.

Ecological Justification

When we reason that organisms are necessary to maintain the functions of ecosystems and the biosphere, we are using an ecological justification for conserving these organisms. Individual species, entire ecosystems, and the biosphere provide public-service functions essential or important to the persistence of life, and as such they are indirectly necessary for our survival. When bees pollinate flowers, for example, they provide a benefit to us that would be costly to replace with human labor. Trees remove certain pollutants from the air; and some soil bacteria fix nitrogen, converting it from molecular nitrogen in the atmosphere to nitrate and ammonia that can be taken up by other living things. That some such functions involve the entire biosphere reminds us of the global perspective on conserving nature and specific species.

Aesthetic Justification

An aesthetic justification asserts that biological diversity enhances the quality of our lives by providing some of the most beautiful and appealing aspects of our existence. Biological diversity is an important quality of landscape beauty. Many organisms—birds, large land mammals, and flowering plants, as well as many insects and ocean animals—are appreciated for their beauty. This appreciation of nature is ancient. Whatever other reasons Pleistocene people had for creating paintings in caves in France and Spain, their paintings of wildlife, done about 14,000 years ago, are beautiful. The paintings include species that have since become extinct, such as mastodons. Poetry, novels, plays, paintings, and sculpture often celebrate the beauty of nature. It is a very human quality to appreciate nature's beauty and is a strong reason for the conservation of endangered species.

Moral Justification

Moral justification is based on the belief that species have a right to exist, independent of our need for them; consequently, in our role as global stewards, we are obligated to promote the continued existence of species and to conserve biological diversity. This right to exist was stated in the U.N. General Assembly World Charter for Nature, 1982. The U.S. Endangered Species Act also includes statements concerning the rights of organisms to exist. Thus, a moral justification for the conservation of endangered species is part of the intent of the law.

Moral justification has deep roots within human culture, religion, and society. Those who focus on cost-benefit analyses tend to downplay moral justification, but although it may not seem to have economic ramifications, in fact it does. As more and more citizens of the world assert the validity of moral justification, more actions that have economic effects are taken to defend a moral position.

The moral justification has grown in popularity in recent decades, as indicated by the increasing interest in the deep-ecology movement. Arne Næss, one of its principal philosophers, explains: "The right of all the forms [of life] to live is a universal right which cannot be quantified. No single species of living being has more of this particular right to live and unfold than any other species."[e]

Cultural Justification

Certain species, some threatened or endangered, are of great importance to many indigenous peoples, who rely on these species of vegetation and wildlife for food, shelter, tools, fuel, materials for clothing, and medicine. Reduced biological diversity can severely increase the poverty of these people. For example, for poor indigenous people who depend on forests, there may be no reasonable replacement except continual outside assistance, which development projects are supposed to eliminate. Urban residents, too, share in the cultural benefits of biological diversity. (See Chapter 22)

Other Intangible Justifications: Recreational, Spiritual, Inspirational

As any mountain biker, scuba diver, or surfer will tell you, the outdoors is great for recreation, and the more natural, the better. Beyond improving muscle tone and cardiovascular strength, many people find a spiritual uplifting and a connectedness to nature from the outdoors, especially where there is a lot of diversity of living things. It has inspired poets, novelists, painters, and even scientists.

13.4 Endangered Species: Current Status

When we say that we want to save a species, what is it that we really want to save? There are four possible answers:

- A wild creature in a wild habitat, as a symbol to us of wilderness.
- A wild creature in a managed habitat, so that the species can feed and reproduce with little interference and so that we can see it in a naturalistic habitat.
- A population in a zoo, so that the genetic characteristics are maintained in live individuals.
- Genetic material only—frozen cells containing DNA from a species for future scientific research.

The goals we choose involve not only science but also values. As discussed, people have different reasons for wishing to save endangered species—utilitarian, ecological, cultural, recreational, spiritual, inspirational, aesthetic, and moral (see A Closer Look 13.3). Policies and actions differ widely, depending on which goal is chosen.

We have managed to turn some once-large populations of wildlife, including fish, into endangered species. With the expanding public interest in rare and endangered species, especially large mammals and birds, it is time for us to turn our attention to these species.

First some definitions: What does it mean to call a species "threatened" or "endangered"? The terms can have strictly biological meanings, or they can have legal meanings. The U.S. Endangered Species Act of 1973 defines *endangered species* as "any species which is in danger of extinction throughout all or a significant portion of its range other than a species of the Class Insecta determined by the Secretary to constitute a pest whose protection under the provisions of this Act would present an overwhelming and overriding risk to man." In other words, if certain insect species are pests, we want to be rid of them. It is interesting that insect pests can be excluded from protection by this legal definition, but there is no mention of disease-causing bacteria or other microorganisms.

Threatened species, according to the act, "means any species which is likely to become an endangered species within the foreseeable future throughout all or a significant portion of its range." It has been the practice since the 1960s to consider an animal species threatened if its population dropped below 500 individuals for a short time and endangered if it dropped below 50 individuals for a considerable length of time. But these have always been taken as informal, not precise, scientific definitions. Since species differ in so many ways, the determination has been on an individual basis, having to do with the abundance and habitat conditions, including predators, parasites, and human alternations.

Next, the basic facts: It is difficult to estimate accurately the total number of species that are threatened and endangered, because we are not sure how many species there are on Earth, as we learned in Chapter 9 (see Table 9.1). Estimates vary from about 1.5 million to 2.7 million, and some conservationists believe there are many more. Another consideration that makes the estimate difficult is that being rare is different from being on the verge of extinction. Some species, like the California condor and the whooping crane, seem to always have been rare. In comparison, some species that have been very abundant have gone extinct rapidly.

One of the most extreme examples is the passenger pigeon, which was so abundant its flocks were said to darken the sky as they flew over but were brought to extinction by hunting. There may have been as many as 3 to 5 billion of these birds in Columbus's time, but by 1990 none were known to exist in the wild. A few were saved at first, but this species was highly social and the few did not survive; the last,

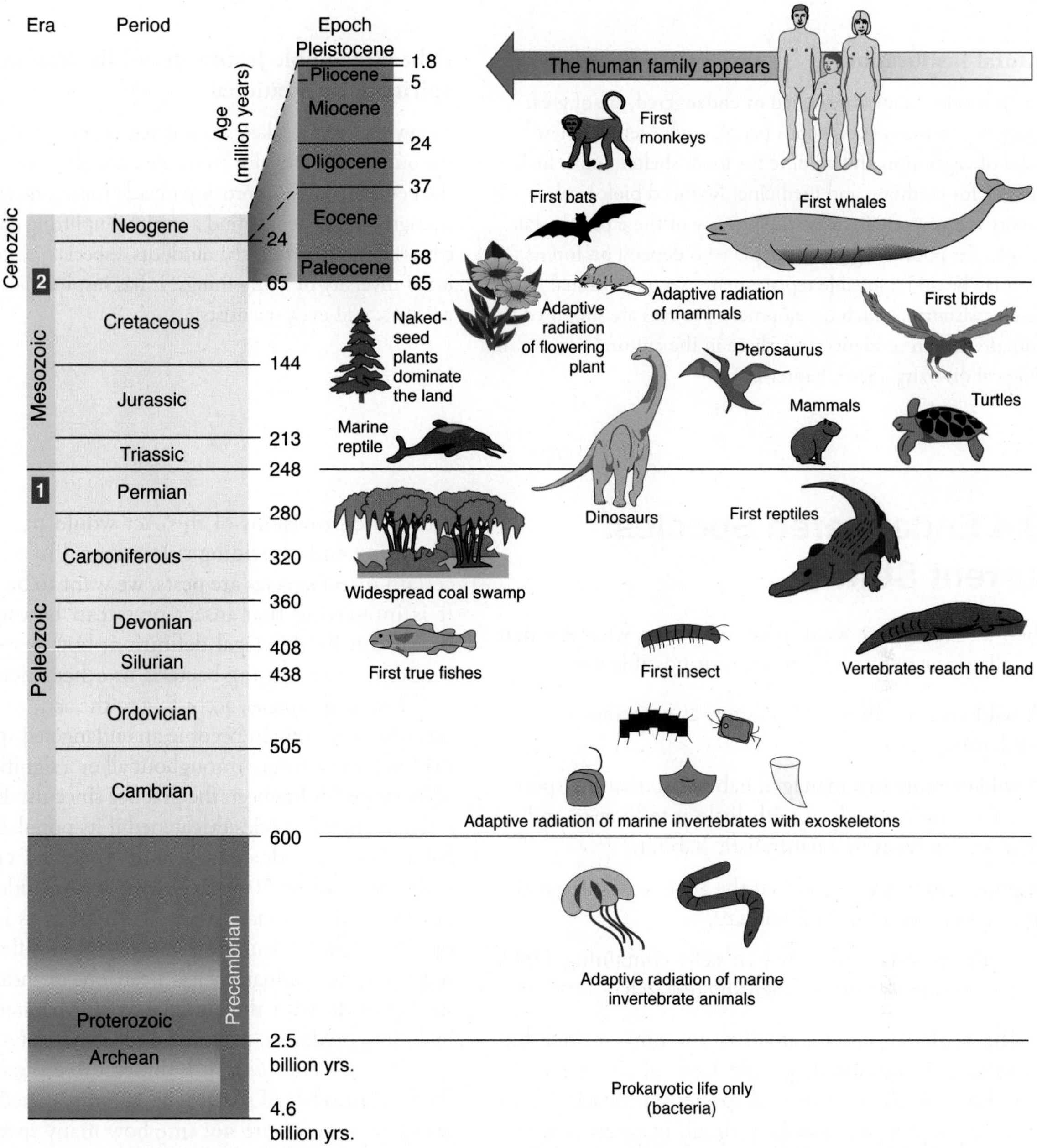

(*Source:* D.M. Raup, "Diversity Crisis in the Geological Past," in E.O. Wilson, ed., *Biodiversity* [Washington, DC: National Academy Press, 1988], p. 53; derived from S.M. Stanley, *Earth and Life through Time* [New York: W.H. Freeman, 1986]. Reprinted with permission.)

FIGURE 13.13 **(a) A brief diagrammatic history of evolution and extinction of life on Earth.** There have been periods of rapid evolution of new species and episodes of catastrophic losses of species. Two major catastrophes were the Permian loss, which included 52% of marine animals, as well as land plants and animals, and the Cretaceous loss of dinosaurs. **(b)** Graph of the number of families of marine animals in the fossil records, showing long periods of overall increase in the number of families punctuated by brief periods of major declines. **(c)** Extinct vertebrate species and subspecies, 1760–1979. The number of species becoming extinct increases rapidly after 1860. Note that most of the increase is due to the extinction of birds.

named "Martha," died in the Cincinnati Zoo in 1914.[13] But some species are known to be in danger of extinction, based, as noted before, not only on their abundances, but also on habitat requirements, pests, predators, and other threats.

One of the best scientific lists of threatened and endangered species is the 2012 International Union for the Conservation of Nature (IUCN) *Red List of Threatened Species*. IUCN has assessed 63,837 species and concludes that "19,817 are threatened with extinction, including 41% of amphibians, 33% of reef-building corals, 25% of mammals, 13% of birds, and 30% of conifers."[14] The U. S. Fish and Wildlife Service provides another of the best lists of threatened and endangered species, listing more than 1,200 animals worldwide.

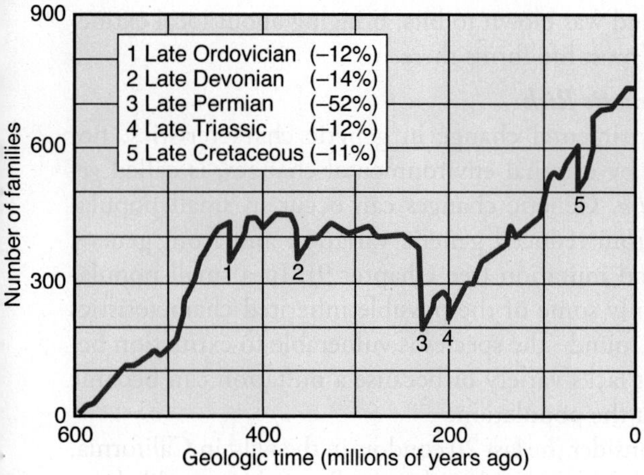

(*Source:* D.M. Raup and J.J. Sepkoski Jr., "Mass Extinctions in the Marine Fossil Record," Science 215 [1982]:1501–1502).

(b)

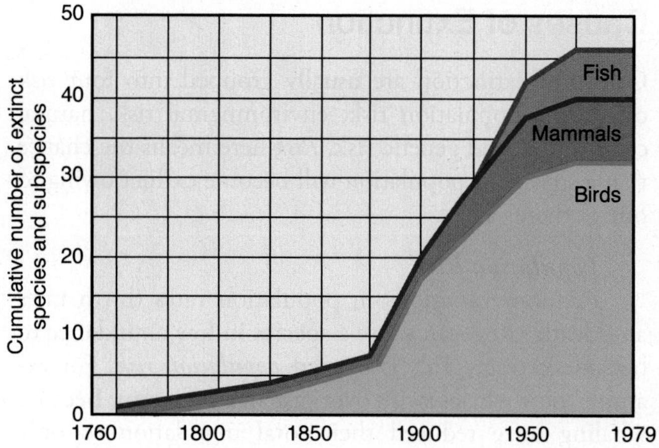

(*Source:* Council on Environmental Quality; additional data from B. Groombridge England: IUCN, 1993].)

(c)

Past *Red Lists* have estimated that 33,798 species of vascular plants (the familiar kind of plants—trees, grasses, shrubs, flowering herbs), or 12.5% of those known, have recently become extinct or endangered[13] and more than 8,000 (approximately 3%) of plant species are currently threatened.

13.5 How a Species Becomes Endangered and Extinct

Extinction is the rule of nature (see the discussion of biological evolution in Chapter 9). **Local extinction** means that a species disappears from a part of its range but persists elsewhere. **Global extinction** means a species can no longer be found anywhere. Although extinction is the ultimate fate of all species, the rate of extinctions has varied greatly over geologic time and has accelerated since the Industrial Revolution.

From 580 million years ago until the beginning of the Industrial Revolution, about one species per year, on average, became extinct. Over much of the history of life on Earth, the rate of evolution of new species equaled or slightly exceeded the rate of extinction. The average longevity of a species has been about 10 million years.[15] However, as discussed in Chapter 7, the fossil record suggests that there have been periods of catastrophic losses of species and other periods of rapid evolution of new species (Figures 13.13 and 13.14), which some refer to as "punctuated extinctions." About 250 million years ago, a mass extinction occurred in which approximately 53% of marine animal species disappeared; about 65 million years ago, most of the dinosaurs became extinct. Interspersed with the episodes of mass extinctions, there seem to have been periods of hundreds of thousands of years with comparatively low rates of extinction.

During the past 2.5 million years, which includes our species and its near humanoid relatives, relatively few species are known to have gone extinct.[16] But an intriguing example of large animal extinctions occurred about 10,000 years ago, at the end of the last great continental glaciation. At that time, 33 genera of large mammals—those weighing 50 kg (110 lb) or more—became extinct in North America and northern Europe, whereas only 13 genera had become extinct in the preceding 1 or 2 million years (Figure 13.13a). Smaller mammals were not as affected, nor were marine mammals and plants. As early as 1876, Alfred Wallace, an English biological geographer, noted that "we live in a zoologically impoverished world, from which all of the hugest, and fiercest, and strangest forms have recently disappeared." It has been suggested that these sudden extinctions coincided with the arrival, on different continents at different times, of Stone Age people and therefore may have been caused by hunting.[17]

Gerard Lacz Images / SuperStock

FIGURE 13.14 Artist's restoration of an extinct saber-toothed cat with prey. The cat is an example of one of the many large mammals that became extinct about 10,000 years ago.

Causes of Extinction

Causes of extinction are usually grouped into four risk categories: population risk, environmental risk, natural catastrophe, and genetic risk. *Risk* here means the chance that a species or population will become extinct owing to one of these causes.

Population Risk

Random variations in population rates (birth rates and death rates) can cause a species in low abundance to become extinct. This is termed *population risk.* For example, blue whales swim over vast areas of ocean. Because whaling once reduced their total population to only several hundred individuals, the success of individual blue whales in finding mates probably varied from year to year. If in one year most whales were unsuccessful in finding mates, then births could be dangerously low. Such random variation in populations, typical among many species, can occur without any change in the environment. It is a risk especially to species that consist of only a single population in one habitat. Mathematical models of population growth can help calculate the population risk and determine the minimum viable population size.

Environmental Risk

Population size can be affected by changes in the environment that occur from day to day, month to month, and year to year, even though the changes are not severe enough to be considered environmental catastrophes. Environmental risk involves variation in the physical or biological environment, including variations in predator, prey, symbiotic species, or competitor species. Some species are so rare and isolated that such normal variations can lead to their extinction.

For example, Paul and Anne Ehrlich described the local extinction of a population of butterflies in the Colorado mountains.[18] These butterflies lay their eggs in the unopened buds of a single species of lupine (a member of the legume family), and the hatched caterpillars feed on the flowers. One year, however, a very late snow and freeze killed all the lupine buds, leaving the caterpillars without food and causing local extinction of the butterflies. Had this been the only population of that butterfly, the entire species would have become extinct.

Natural Catastrophe

A sudden change in the environment that is not caused by human action is a natural catastrophe. Fires, major storms, earthquakes, and floods are natural catastrophes on land; changes in currents and upwellings are ocean catastrophes. For example, the explosion of a volcano on the island of Krakatoa in Indonesia in 1883 caused one of recent history's worst natural catastrophes. Most of the island was blown to bits, bringing about local extinction of most life-forms there.

Genetic Risk

Detrimental change in genetic characteristics, not caused by external environmental changes, is called *genetic risk.* Genetic changes can occur in small populations from reduced genetic variation and from genetic drift and mutation (see Chapter 9). In a small population, only some of the possible inherited characteristics will be found. The species is vulnerable to extinction because it lacks variety or because a mutation can become fixed in the population.

Consider the last 20 condors in the wild in California. It stands to reason that this small number was likely to have less genetic variability than the much larger population that existed several centuries ago. This increased the condor's vulnerability. Suppose that the last 20 condors, by chance, had inherited characteristics that made them less able to withstand lack of water. If left in the wild, these condors would have been more vulnerable to extinction than a larger, more genetically varied population.

13.6 The Good News: We Have Improved the Status of Some Species

Thanks to the efforts of many people, a number of previously endangered species, such as the Aleutian goose, have recovered. Other recovered species include the following:

- The elephant seal, which had dwindled to about a dozen animals around 1900 and now numbers in the hundreds of thousands.
- The sea otter, reduced in the 19th century to several hundred and now numbering approximately 10,000.

Many species of birds became endangered because the insecticide DDT caused thinning of eggshells and failure of reproduction. With the elimination of DDT in the United States, many bird species recovered, including some that are no longer on the U.S. Endangered Species list: the bald eagle, brown pelican, white pelican, and osprey.[19] Other examples include the following:

- The blue whale, thought to have been reduced to about 400 when whaling was still actively pursued by a number of nations. Today, 400 blue whales are sighted annually in the Santa Barbara Channel along the California coast, a sizable fraction of the total population.

Table 13.3 RECOVERED SPECIES IN THE UNITED STATES

	DATE SPECIES FIRST LISTED	DATE DELISTED	SPECIES NAME
1	7/27/1979	6/4//1987	Alligator, American (*Alligator mississippiensis*)
2	9/17/1980	8/27/2002	Cinquefoil, Robbins' (*Potentilla robbinsiana*)
3	7/5/1979	9/2/2011	Coneflower, Tennessee purple (*Echinacea tennesseensis*)
4	6/2/1970	5/23/2012	Crocodile, Morelet's (*Crocodylus moreletti*)
5	9/5/1985	2/18/2011	Daisy, Maguire (*Erigeron maguirei*)
6	7/24/2003	7/24/2003	Deer, Columbian white-tailed Douglas County DPS (*Odocoileus Virginianus leucurus*)
7	6/2/1970	9/12/1985	Dove, Palau ground (*Gallicolumba canifrons*)
8	3/11/1967	7/9/2007	Eagle, bald lower 48 states (*Haliaeetus leucocephalus*)
9	6/2/1970	8/25/1999	Falcon, American peregrine (*Falco peregrinus anatum*)
10	6/2/1970	10/5/1994	Falcon, Arctic peregrine (*Falco peregrinus tundrius*)
11	6/2/1970	9/12/1985	Flycatcher, Palau fantail (*Rhipidura lepida*)
12	3/11/1967	3/20/2001	Goose, Aleutian Canada (*Branta Canadensis leucopareia*)
13	12/30/1974	3/9/1995	Kangaroo, eastern gray (*Macropus giganteus*)
14	12/30/1974	3/9/1995	Kangaroo, western red (*Macropus rufus*)
15	12/30/1974	3/9/1995	Kangaroo, western gray (*Macropus fuliginosus*)
16	6/2/1970	9/21/2004	Monarch, Tinia (old world flycatcher) (*Monarcha takatsukasae*)
17	6/2/1970	9/12/1985	Owl, Palau (*Pyrroglaux podargina*)
18	6/2/1970	2/4/1985	Pelican brown U.S. Atlantic coast, FL, AL (*Pelecanus occidentails*)
19	9/3/1986	11/28/2011	Snake, Concho water (*Nerodia paucimaculata*)
20	8/30/1999	9/15/2011	Snake, Lake Erie water (*Nerodia sipedon insularum*)
21	7/1/1985	8/26/2008	Squirrel, Virginia northern flying (*Glaucomys sabrinus fuscus*)
22	5/22/1997	8/18/2005	Sunflower, Eggert's (*Helianthus eggertii*)
23	6/16/1994	6/16/1994	Whale, gray except where listed (*Eschrichtius robustus*)
24	4/10/1979	1/27/2012	Wolf, gray Minnesota (*Canis lupus*)
25	5/5/2011	5/5/2011	Wolf, gray Northern Rocky Mountains DPS, except WY (*Canis lupus*)
26	5/5/2011	9/30/2012	Wolf, gray Wyoming, EXPN population (*Canis lupus*) Northern Rocky Mountain DPS (*Canis lupus*)
27	7/19/1990	10/7/2003	Woolly-star, Hoover's (*Eriastrum hooveri*)

Source: U.S. Fish and Wildlife Service, Delisting Report, http://ecos.fws.gov/tess_public/DelistingReport.do.

- The gray whale, which was hunted to near-extinction but is now abundant along the California coast and in its annual migration to Alaska. (See A Closer Look 13.4.)

Since the U.S. Endangered Species Act became law in 1973, 13 species within the United States have officially recovered (Table 13.3), according to the U.S. Fish and Wildlife Service. The agency has also "delisted" from protection of the act 9 species because they have gone extinct, and 17 because they were listed in error or because it was decided that they were not a unique species or a genetically significant unit within a species. (The act allows listing of subspecies so genetically different from the rest of the species that they deserve protection.)[20]

A CLOSER LOOK 13.4

Conservation of Whales and Other Marine Mammals

Fossil records show that all marine mammals were originally inhabitants of the land. During the last 80 million years, several separate groups of mammals returned to the oceans and underwent adaptations to marine life. Each group of marine mammals shows a different degree of transition to ocean life. Understandably, the adaptation is greatest for those that began the transition longest ago. Some marine mammals—such as dolphins, porpoises, and great whales—complete their entire life cycle in the oceans and have organs and limbs that are highly adapted to life in the water; they cannot move on the land. Others, such as seals and sea lions, spend part of their time on shore.

Cetaceans

Whales

Whales fit into two major categories: baleen and toothed (Figure 13.15a and b). The sperm whale is the only great whale that is toothed; the rest of the toothed group is comprised of smaller whales, dolphins, and porpoises. The other great whales, in the baleen group, have highly modified teeth that look like giant combs and act as water filters. Baleen whales feed by filtering ocean plankton.

Drawings of whales have been dated as early as 2200 B.C.[a] Eskimos used whales for food and clothing as long ago as 1500 B.C. In the 9th century, whaling by Norwegians was reported by travelers whose accounts were written down in the court of the English king Alfred. The earliest whale hunters killed these huge mammals from the shore or from small boats near shore, but gradually whale hunters ventured farther out. In the 11th and 12th centuries, Basques hunted the Atlantic right whale from open boats in the Bay of Biscay, off the western coast of France. Whales were brought ashore for processing, and the boats returned to land once the search for whales was finished.

The Industrial Revolution eventually made whaling pelagic—almost entirely conducted on the open sea. Whalers searched for whales from ships that remained at sea for long periods, and the whales were brought on board and processed there by newly invented furnaces and boilers for extracting whale oil at sea. With these inventions, whaling grew as an industry. American fleets developed in the 18th century in New England, and by the 19th century the United States dominated the industry, providing most of the whaling ships and even more of the crews.[b, c]

Whales provided many 19th-century products. Whale oil was used for cooking, lubrication, and lamps. Whales provided the main ingredients for the base of perfumes. The elongated teeth (whale-bone or baleen) that enable baleen whales to strain the ocean waters for food are flexible and springy and were used for corset stays and other products before the invention of inexpensive steel springs.

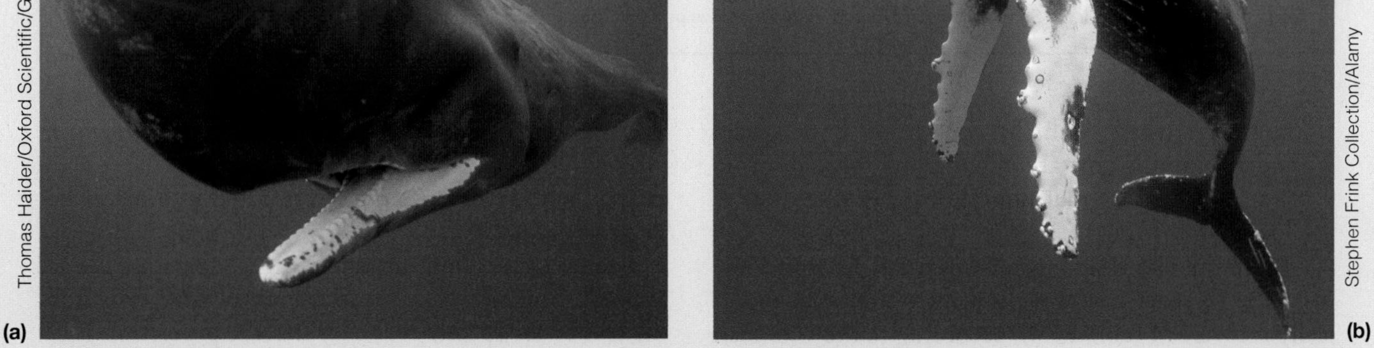

FIGURE 13.15 (a) A sperm whale, one of the toothed whales; (b) a blue whale, a baleen whale.

Although the 19th-century whaling ships were made famous by novels such as *Moby Dick*, more whales were killed in the 20th century than in the 19th. The resulting worldwide decline in most species of whales made this a global environmental issue.

Conservationists have been concerned about whales for many years. Attempts to control whaling began with the League of Nations in 1924. The first agreement, the Convention for the Regulation of Whaling, was signed by 21 countries in 1931. In 1946 a conference in Washington, DC, initiated the International Whaling Commission (IWC), and in 1982 the IWC established a moratorium on commercial whaling. Currently, 12 of approximately 80 species of whales are protected.[b]

The IWC has played a major role in reducing (almost eliminating) the commercial harvesting of whales. Since its formation, no whale species has become extinct, the total take of whales has decreased, and harvesting of whale species considered endangered has ceased. Endangered species protected from hunting have had a mixed history (see Table 13.4). Blue whales appear to have recovered somewhat but remain rare and endangered. Gray whales are now relatively abundant, numbering between 21,000 and 32,400 (Table 13.4).[d] However, global climate change, pollution, and ozone depletion now pose greater risks to whale populations than does whaling.

Table 13.4 ESTIMATES OF THE NUMBER OF WHALES

Whales are difficult to count, and the wide range of estimates indicates this difficulty. Of the baleen whales, the most numerous are the smallest—the minke and the pilot. The only toothed great whale, the sperm whale, is thought to be relatively numerous, but actual counts of this species have been made only in comparatively small areas of the ocean.

	RANGE OF ESTIMATES	
SPECIES	MINIMUM	MAXIMUM
Blue	400	1,400
Bowhead	6,900	9,200
Fin	27,700	82,000
Gray	21,900	32,400
Humpback	5,900	16,800
Minke	510,000	1,140,000
Pilot	440,000	1,370,000
Sperm	200,000	1,500,000

Source: International Whaling Commission, August 29, 2006, http://www.iwcoffice.org/conservation/estimate.htm. The estimate for sperm whale is from U.S. NOAA http://www.nmfs.noaa.gov/pr/species/mammals/cetaceans/spermwhale.htm.

The establishment of the IWC was not only vitally important to whales but also a major landmark in wildlife conservation. It was one of the first major attempts by a group of nations to agree on a reasonable harvest of a biological resource. The annual meeting of the IWC has become a forum for discussing international conservation, working out basic concepts of maximum and optimum sustainable yields, and formulating a scientific basis for commercial harvesting. The IWC demonstrates that even an informal commission whose decisions are accepted voluntarily by nations can function as a powerful force for conservation.

In the past, each marine mammal population was treated as if it was isolated, had a constant supply of food, and was subject only to the effects of human harvesting. That is, its growth was assumed to follow the logistic curve. We now realize that management policies for marine mammals must be expanded to include ecosystem concepts and the understanding that populations interact in complex ways.

The goal of marine mammal management is to prevent extinction and maintain large population sizes rather than to maximize production. As mentioned before, the Marine Mammal Protection Act, enacted by the United States in 1972, has as its goal an optimum sustainable population (OSP) rather than a maximum or an optimum sustainable yield. As we explained before, an OSP is the largest population that can be sustained indefinitely without deleterious effects on the ability of the population or its ecosystem to continue to support that same level.

Dolphins and Other Small Cetaceans

Among the many species of small "whales" are dolphins and porpoises, more than 40 species of which have been hunted commercially or have been killed inadvertently by other fishing efforts. A classic case is the inadvertent catch of the spinner, spotted, and common dolphins of the eastern Pacific. Because these carnivorous, fish-eating mammals often feed with yellowfin tuna, a major commercial fish, more than 7 million dolphins have been netted and killed inadvertently in the past 40 years.

The U.S. Marine Mammal Commission and commercial fishermen have cooperated in seeking ways to reduce dolphin mortality. Research into dolphin behavior helped in the design of new netting procedures that trapped far fewer dolphins. The attempt to reduce dolphin mortality illustrates cooperation among fishermen, conservationists, and government agencies and indicates the role of scientific research in managing renewable resources.

13.7 Can a Species Be Too Abundant? If So, What to Do?

All marine mammals are protected in the United States by the Federal Marine Mammal Protection Act of 1972, which has improved the status of many marine mammals. Sometimes, however, we succeed too well in increasing the populations of a species. Case in point: Sea lions now number more than 190,000 and have become so abundant as to be local problems. (Recall also the problems with white-tailed deer discussed in the opening case study.)[21,22] In San Francisco Harbor and in Santa Barbara Harbor, for example, sea lions haul out and sun themselves on boats and pollute the water with their excrement near shore. In one case, so many hauled out on a sailboat in Santa Barbara Harbor that they sank the boat, and some of the animals were trapped and drowned.

Mountain lions, too, have become locally overabundant. In the 1990s, California voters passed an initiative that protected the endangered mountain lion but contained no provisions for managing the lion if it became overabundant, unless it threatened human life and property. Few people thought the mountain lion could ever recover enough to become a problem, but in several cases in recent years mountain lions have attacked and even killed people. Current estimates suggest there may be as many as 4,000 to 6,000 in California.[23] These attacks become more frequent as the mountain lion population grows and as the human population grows and people build houses in what was mountain lion habitat.

13.8 How People Cause Extinctions and Affect Biological Diversity

People have become an important factor in causing species to become threatened, endangered, and finally extinct. We do this in several ways:

- By intentional hunting or harvesting (for commercial purposes, for sport, or to control a species that is considered a pest).
- By disrupting or eliminating habitats.
- By introducing exotic species, including new parasites, predators, or competitors of a native species.
- By generating pollution.

People have caused extinctions over a long time, not just in recent years. The earliest people probably caused extinctions through hunting. This practice continues, especially for specific animal products considered valuable, such as elephant ivory and rhinoceros horns. When people learned to use fire, they began to change habitats over large areas. The development of agriculture and the rise of civilization led to rapid deforestation and other habitat changes. Later, as people explored new areas, the introduction of exotic species became a greater cause of extinction (see Chapter 9), especially after Columbus's voyage to the New World, Magellan's circumnavigation of the globe, and the resulting spread of European civilization and technology. The introduction of thousands of novel chemicals into the environment made pollution an increasing threat to species in the 20th century, and pollution control has proved a successful way to help species.

The IUCN estimates that 75% of the extinctions of birds and mammals since 1600 have been caused by human beings. Hunting is estimated to have caused 42% of the extinctions of birds and 33% of the extinctions of mammals. The current extinction rate among most groups of mammals is estimated to be 1,000 times greater than the extinction rate at the end of the Pleistocene epoch.[14]

13.9 Ecological Islands and Endangered Species

Recall from our discussion in Chapter 9 that an ecological island is an area that is biologically isolated, so that a species living there cannot mix (or only rarely mixes) with any other population of the same species (Figure 13.16). Mountaintops and isolated ponds are ecological islands. Real geographic islands may also be ecological islands. Insights gained from studies of the biogeography of islands have important implications for the conservation of endangered species and for the design of parks and preserves for biological conservation.

Almost every park is a biological island for some species. A small city park between buildings may be an island for trees and squirrels. At the other extreme, even a large national park is an ecological island. For example, the Masaai Mara Game Reserve in the Serengeti Plain, which stretches from Tanzania to Kenya in East Africa, and other great wildlife parks of eastern and southern Africa are becoming islands of natural landscape surrounded by human settlements. Lions and other great cats exist in these parks as isolated populations, no longer able to roam with complete freedom and to mix over large areas. Other examples are islands of uncut forests left by logging operations, and oceanic islands, where intense fishing has isolated some fish populations.

(a) Richard Elliott/Getty Images, Inc.

(b) ©T. ULRICH/ClassicStock/The Image Works

(c) Tim Bewer/Lonely Planet Images/Getty Images

FIGURE 13.16 **Ecological islands: (a)** Central Park in New York City; **(b)** a mountaintop in Arizona where there are bighorn sheep; **(c)** an African wildlife park.

How large must an ecological island be to ensure the survival of a species? The size varies with the species but can be estimated. Some islands that seem large to us are too small for species we wish to preserve. For example, a preserve was set aside in India in an attempt to reintroduce the Indian lion into an area where it had been eliminated by hunting and by changing patterns of land use. In 1957, a male and two females were introduced into a 95-km^2 (36-mi^2) preserve in the Chakia forest, known as the Chandraprabha Sanctuary. The introduction was carried out carefully, and the population was counted annually. There were four lions in 1958, five in 1960, seven in 1962, and eleven in 1965, after which they disappeared and were never seen again.

Why did they go? Although 95 km^2 seems large to us, male Indian lions have territories of 130 km^2 (50 mi^2), within which females and young also live. A population that could persist for a long time would need a number of such territories, so an adequate preserve would require 640–1,300 km^2 (247–500 mi^2). Various other reasons were also suggested for the disappearance of the lions, including poisoning and shooting by villagers. But regardless of the immediate cause, a much larger area was required for long-term persistence of the lions.

13.10 Using Spatial Relationships to Conserve Endangered Species

The red-cockaded woodpecker (Figure 13.17a) is an endangered species of the American Southeast, numbering approximately 15,000.[24] The woodpecker makes its nests in old dead or dying pines, and one of its foods is the pine bark beetle (Figure 13.17b). To conserve this species of woodpecker, these pines must be preserved. But the old pines are home to the beetles, which are pests to the trees and damage them for commercial logging. This presents an intriguing problem: How can we maintain the woodpecker and its food (which includes the pine bark beetle) and also maintain productive forests?

The classic 20th-century way to view the relationship among the pines, the bark beetle, and the woodpecker would be to show a food chain (see Chapter 6). But this alone does not solve the problem for us. A newer approach is to consider the habitat requirements of the pine bark beetle and the woodpecker. Their requirements are somewhat different, but if we overlay a map of one's habitat requirements over a map of the other's, we can compare the co-occurrence of habitats. Beginning with such maps, it becomes possible to design a landscape that would allow the maintenance of all three—pines, beetles, and birds.

FIGURE 13.17 (a) Endangered red-cockaded woodpecker, and (b) the pine bark beetle, food for the woodpecker.

CRITICAL THINKING ISSUE

Should Wolves Be Reestablished in the Adirondack Park?

With an area slightly over 24,000 km^2, the Adirondack Park in northern New York is the largest park in the lower 48 states. Unlike most parks, however, it is a mixture of private (60%) and public (40%) land and is home to 130,000 people. When European colonists first came to the area, it was, like much of the rest of North America, inhabited by gray wolves. By 1960, wolves had been exterminated in all of the lower 48 states except for northern Minnesota. The last official sighting of a wolf in the Adirondacks was in the 1890s.

Although the gray wolf was not endangered globally—there were more than 60,000 in Canada and Alaska—it was one of the first animals listed as endangered under the 1973 Endangered Species Act. As required, the U.S. Fish and Wildlife Service developed a plan for recovery that included protection of the existing population and reintroduction of wolves to wilderness areas. The recovery plan would be considered a success if survival of the Minnesota wolf population was assured and at least one other population of more than 200 wolves had been established at least 320 km from the Minnesota population.

Under the plan, Minnesota's wolf population increased, and some wolves from that population, as well as others from southern Canada, dispersed into northern Michigan and Wisconsin, each of which had populations of approximately 100 wolves in 1998. Also, 31 wolves from Canada were introduced into Yellowstone National Park in 1995, and that population grew to over 100.

In 1992, when the results of the recovery plan were still uncertain, the Fish and Wildlife Service proposed to investigate the possibility of reintroducing wolves to northern Maine and the Adirondack Park. A survey of New York State residents in 1996 funded by Defenders of Wildlife found that 76% of people living in the park supported reintroduction. However, many residents and organizations within the park vigorously opposed reintroduction and questioned the validity of the survey. Concerns focused primarily on the potential dangers to people, livestock, and pets and the possible impact on the deer population. In response to the public outcry, Defenders of Wildlife established a citizens' advisory committee that initiated two studies by outside experts, one on the social and economic aspects of reintroduction and another on whether there were sufficient prey and suitable habitat for wolves.

Wolves prey primarily on moose, deer, and beaver. Moose have been returning to the Adirondacks in recent years and now number about 40, but this is far less than the moose population in areas where wolves are successfully reestablishing. Beaver are abundant in the Adirondacks, with an estimated population of over 50,000. Because wolves feed on beaver primarily in the spring and the moose population is small, the main food source for Adirondack wolves would be deer.

Deer thrive in areas of early-successional forest and edge habitats, both of which have declined in the Adirondacks as logging has decreased on private forestland and has been eliminated altogether on public lands. Furthermore, the Adirondacks are at the northern limit of the range for white-tailed deer, where

harsh winters can result in significant mortality. Deer density in the Adirondacks is estimated at 3.25 per square kilometer, fewer than in the wolf habitat in Minnesota, which also has 8,500 moose. If deer were the only prey available, wolves would kill between 2.5% and 6.5% of the deer population, while hunters take approximately 13% each year. Determining whether there is a sufficient prey base to support a population of wolves is complicated by the fact that coyotes have moved into the Adirondacks and occupy the niche once filled by wolves. Whether wolves would add to the deer kill or replace coyotes, with no net impact on the deer population, is difficult to predict.

An area of 14,000 km in various parts of the Adirondack Park meets criteria established for suitable wolf habitat, but this is about half of the area required to maintain a wolf population for the long term. Based on the average deer density and weight, as well as the food requirements of wolves, biologists estimate that this habitat could support about 155 wolves. However, human communities are scattered throughout much of the park, and many residents are concerned that wolves would not remain on public land and would threaten local residents as well as back-country hikers and hunters. Also, private lands around the edges of the park, with their greater density of deer, dairy cows, and people, could attract wolves.

Critical Thinking Questions

1. Who should make decisions about wildlife management, such as returning wolves to the Adirondacks—scientists, government officials, or the public?
2. Some people advocate leaving the decision to the wolves—that is, waiting for them to disperse from southern Canada and Maine into the Adirondacks. Study a map of the northeastern United States and southeastern Canada. What do you think is the likelihood of natural recolonization of the Adirondacks by wolves?
3. Do you think wolves should be reintroduced to the Adirondack Park? If you lived in the park, would that affect your opinion? How would removal of the wolf from the Endangered Species list affect your opinion?
4. Some biologists recently concluded that wolves in Yellowstone and the Great Lakes region belong to a different subspecies, the Rocky Mountain timber wolf, from those that formerly lived in the northeastern United States, the eastern timber wolf. This means that the eastern timber wolf is still extinct in the lower 48 states. Would this affect your opinion about reintroducing wolves into the Adirondacks?

SUMMARY

- Modern approaches to management and conservation of wildlife use a broad perspective that considers interactions among species as well as the ecosystem and landscape contexts.
- To successfully manage wildlife for harvest, conservation, and protection from extinction, we need certain quantitative information about the wildlife population, including measures of total abundance and of births and deaths, preferably recorded over a long period. The age structure of a population can also help. In addition, we have to characterize the habitat in quantitative terms. However, it is often difficult to obtain these data.
- A common goal of wildlife conservation today is to "restore" the abundance of a species to some previous number, usually a number thought to have existed prior to the influence of modern technological civilization. Information about long-ago abundances is rarely available, but sometimes we can estimate numbers indirectly—for example, by using the Lewis and Clark journals to reconstruct the 1805 population of grizzly bears, or using logbooks from whaling ships. Adding to the complexity, wildlife abundances vary over time even in natural systems uninfluenced by modern civilization. Also, historical information often cannot be subjected to formal tests of disproof and therefore does not by itself qualify as scientific. Adequate information exists for relatively few species.
- Another approach is to seek a minimum viable population, a carrying capacity, or an optimal sustainable population or harvest based on data that can be obtained and tested today. This approach abandons the goal of restoring a species to some hypothetical past abundance.
- The good news is that many formerly endangered species have been restored to an abundance that suggests they are unlikely to become extinct. Success is achieved when the habitat is restored to conditions required by a species. The conservation and management of wildlife present great challenges but also offer great rewards of long standing and deep meaning to people.

REEXAMINING THEMES AND ISSUES

Sean Randall/Getty Images, Inc.

HUMAN POPULATION

Human beings are a primary cause of species extinctions today and also contributed to extinctions in the past. Nonindustrial societies have caused extinction by such activities as hunting and the introduction of exotic species into new habitats. With the age of exploration in the Renaissance and with the Industrial Revolution, the rate of extinctions accelerated. People altered habitats more rapidly and over greater areas. Hunting efficiency increased, as did the introduction of exotic species into new habitats. As the human population grows, conflicts over habitat between people and wildlife increase. Once again, we find that the human population problem underlies this environmental issue.

© Biletskiy_Evgeniy/iStockphoto

SUSTAINABILITY

At the heart of issues concerning wild living resources is the question of sustainability of species and their ecosystems. One of the key questions is whether we can sustain these resources at a constant abundance. In general, it has been assumed that fish and other wildlife that are hunted for recreation, such as deer, could be maintained at some constant, highly productive level. Constant production is desirable economically because it would provide a reliable, easily forecast income each year. But despite attempts at direct management, few wild living resources have remained at constant levels. New ideas about the intrinsic variability of ecosystems and populations lead us to question the assumption that such resources can or should be maintained at constant levels.

© Anton Balazh 2011/iStockphoto

GLOBAL PERSPECTIVE

Although the final extinction of a species takes place in one locale, the problem of biological diversity and the extinction of species is global because of the worldwide increase in the rate of extinction and because of the growth of the human population and its effects on wild living resources.

ssguy/ShutterStock

URBAN WORLD

We have tended to think of wild living resources as existing outside of cities, but there is a growing recognition that urban environments will be more and more important in conserving biological diversity. This is partly because cities now occupy many sensitive habitats around the world, such as coastal and inland wetlands. It is also because appropriately designed parks and backyard plantings can provide habitats for some endangered species. As the world becomes increasingly urbanized, this function of cities will become more important.

B2M Productions/Getty Images, Inc.

PEOPLE AND NATURE

Wildlife, fish, and endangered species are popular issues. We seem to have a deep feeling of connectedness to many wild animals—we like to watch them, and we like to know that they still exist even when we cannot see them. Wild animals have always been important symbols to people, sometimes sacred ones. The conservation of wildlife of all kinds is therefore valuable to our sense of ourselves, both as individuals and as members of a civilization.

George Doyle/Getty Images, Inc.

SCIENCE AND VALUES

The reasons that people want to save endangered species begin with human values, including values placed on the continuation of life and on the public-service functions of ecosystems. Among the most controversial environmental issues in terms of values are the conservation of biological diversity and the protection of endangered species. Science tells us what is possible with respect to conserving species, which species are likely to persist, and which are not. Ultimately, therefore, our decisions about where to focus our efforts in sustaining wild living resources depend on our values.

KEY TERMS

carrying capacity 288
catch per unit effort 293
global extinction 303
historical range of variation 289
local extinction 303
logistic carrying capacity 288
maximum sustainable yield 287
maximum sustainable-yield population 287
minimum viable population 288
optimum sustainable population 288
time series 289

STUDY QUESTIONS

1. Why are we so unsuccessful in making rats an endangered species?
2. What have been the major causes of extinction (a) in recent times and (b) before people existed on Earth?
3. Refer back to Closer Look 13.1 about the American buffalo and grizzly bear. The U.S. Fish and Wildlife Service suggested three key indicators of the status of the grizzly bear: (1) sufficient reproduction to offset the existing levels of human-caused mortality, (2) adequate distribution of breeding animals throughout the area, and (3) a limit on total human-caused mortality. Are these indicators sufficient to assure the recovery of this species? What would you suggest instead?
4. This chapter discussed eight justifications for preserving endangered species. Which of them apply to the following? (You can decide that none apply.)
   (a) The black rhinoceros of Africa
   (b) The Furbish lousewort, a rare small flowering plant of New England, seen by few people
   (c) An unnamed and newly discovered beetle from the Amazon rain forest
   (d) Smallpox
   (e) Wild strains of potatoes in Peru
   (f) The North American bald eagle
5. Locate an ecological island close to where you live and visit it. Which species are most vulnerable to local extinction?

6. Oysters were once plentiful in the waters around New York City. Create a plan to restore them to numbers that could be the basis for commercial harvest.
7. Using information available in libraries, determine the minimum area required for a minimum viable population of the following:
   (a) Domestic cats
   (b) Cheetahs
   (c) The American alligator
   (d) Swallowtail butterflies
8. Both a ranch and a preserve will be established for the North American bison. The goal of the ranch owner is to show that bison can be a better source of meat than introduced cattle and at the same time have a less detrimental effect on the land. The goal of the preserve is to maximize the abundance of the bison. How will the plans for the ranch and preserve differ, and how will they be similar?

FURTHER READING

Botkin, D.B., *No Man's Garden: Thoreau and a New Vision for Civilization and Nature* (Washington, DC: Island Press, 2001). A work that discusses deep ecology and its implications for biological conservation, as well as reasons for the conservation of nature, both scientific and beyond science.

Caughley, G., and A.R.E. Sinclair, *Wildlife Ecology and Management* (London: Blackwell Scientific, 1994). A valuable textbook based on new ideas of wildlife management.

Decker, D. J., S. J. Riley, and W. F. Siemer, *Human Dimensions of Wildlife Management* (Baltimore, MD: Johns Hopkins University Press, 2012) Discusses interfaces between people and wildlife, including some that are mutually beneficial and those that have led to conflicts

Mackay, R., *The Atlas of Endangered Species: Revised and Updated* (Berkeley: University of California Press, 2008). A surprisingly short and easily readable book.

Pauly, D., J. Maclean, and J.L. Maclean, *The State of Fisheries and Ecosystems in the North Atlantic Ocean* (Washington, DC: Island Press, 2002).

NOTES

1. Miller, R.S., D.B. Botkin, and R. Mendelssohn. 1974. The whooping crane (*Grusamericana*) population of North America, Biol. *Conservation* 6:106–111.
2. McConnell, C. 2012. Aransas refuge estimates 245 whooping cranes. Report.Whooping Crane Conservation Association, Washington, D.C.
3. Cummins, K. Personal communication. February 2013.
4. Haines, F. 1970. *The Buffalo*. New York, NY: Thomas Y. Crowell.
5. The National Zoo website. April 10, 2006.
6. Bockstoce, J.R., D.B. Botkin, A. Philp, B.W. Collins, and J.C. George. 2007. The geographic distribution of bowhead whales in the Bering, Chukchi, and Beaufort seas: Evidence from whaleship records, 1849–1914, *Marine Fisheries Review* 67(3):1–43.
7. United Nations Food and Agriculture Organization. 2010. Fishery and Aquaculture Statistics. Rome: UN FAO.
8. Specific data obtained from National Marine Fisheries Service. http://swr.nmfs.noaa.gov/biologic.htm
9. Anonymous. 2002. Sea Scallop Fishery Management Plan. New England Fisheries Management Council. http://www.nefmc.org/scallops/summary/scallop.pdf
10. Brodziak, J., and M. Traver. 2006. Status of Fishery Resources off the Northeastern US. New_England_Fishery_Management_Council, Washington, DC: NOAA.
11. Shelley, P. 2013. Peter Shelley to NEFMC: Shut down New England cod fishery. http://www.Talkingfish.org
12. FAO Statistics. 2006. http://www.fao.org/waicent/portalstatistics_en.asp.

13. Anonymous. 2001. The Passenger Pigeon, Encyclopedia Smithsonian.http://www.si.edu/encyclopedia_Si/nmnh/passpig.htm

14. International Union for Conservation of Nature. *Red List of Threatened Species*, 2012. IUCN, Geneva. http://www.iucnredlist.org.

15. Regan, H.M., R. Lupia, A.N. Drinnan, and M.A. Burgman. 2001. The currency and tempo of extinction. *American Naturalist* 157(1):1–10.

16. Botkin, D.B., et al., 2007. Forecasting effects of global warming on biodiversity. *BioScience* 57(3): 227–236.

17. Martin, P.S. 1963. *The Last 10,000 Years*. Tucson, AZ: University of Arizona Press.

18. Ehrlich, P.R., et al. 1980. Extinction, reduction, stability and increase: The responses of checkerspot butterfly (*Euphydryas*) populations to the California drought. *Oecologia* 46(1): 101–105, and Ehrlich, P.R., and P.H. Raven. 1964. Butterflies and plants: A study in coevolution. *Evolution* 18(4):586–608.

19. U.S. Fishand WildlifeService. 2013. Endangered and threatened wildlife and plants. U.S. Department of the Interior.

20. U.S. Fish and Wildlife Service Delisting Report. http://ecos.fws.gov/tess_public/DelistingReport.do

21. "Impacts of California sea lion and Pacific harbor seal on Salmonids and west coast ecosystems". February 10, 1999. U.S. Department of Commerce, National Oceanic and Atmospheric Administration, National Marine Fisheries Service. p. Appendix-7.

22. NOAA delisted species (http://www.nmfs.noaa.gov/pr/species/esa.htm#delisted) and U.S. Fish and Wildlife Service Threatened and Endangered Species System (TESS).

23. California Department of Fish and Game. http://www.dfg.ca.gov/news/issues/lion/lion_faq.html, mountain lion abundances

24. U.S. Department of Defense and U.S. Fish and Wildlife Service. 2006. Red-cockaded woodpecker (*Picoides borealis*). http://www.fws. gov/endangered/pdfs/DoD/RCW_fact_sheet-Aug06.pdf

A CLOSER LOOK 13.1 NOTES

a. Botkin, D.B. 2004. *Beyond the Stony Mountains: Nature in the American West from Lewis and Clark to Today*. New York, NY: Oxford University Press.

b. Mattson, D.J., and M.W. Reid. 1991. Conservation of the Yellowstone grizzly bear. *Conservation Biology* 5:364–372.

c. The National Zoo. April 10, 2006. http://nationalzoo.si.edu/support/adoptspecies/Animalinfo/biosn/default.cfm.

d. www.bisoncentral.com. Accessed April 10, 2006.

e. Haines, F. 1970. *The Buffalo*. New York, NY: Thomas Y. Crowell.

A CLOSER LOOK 13.3 NOTES

a. Botkin, D. B. 2001. *No Man's Garden: Thoreau and a New Vision for Civilization and Nature*. Washington, DC: Island Press.

b. O'Donnell, James J. 2008. *The Ruin of the Roman Empire*. New York, NY: ECCO Press (HarperCollins), p. 190.

c. Botkin, D. B. 2012. *The Moon in the Nautilus Shell: Discordant Harmonies Reconsidered*. New York, NY: Oxford University Press, hardback and ebook.

d. Bruckner, A. 2002. Life-saving products from coral reefs. *Issues in Science and Technology*. Spring edition.

e. Naess, A. 1989. *Ecology, Community, and Lifestyle*. Cambridge, UK: Cambridge University Press.

A CLOSER LOOK 13.4 NOTES

a. Friends of the Earth. 1979. *The Whaling Question: The Inquiry by Sir Sidney Frost of Australia*. San Francisco, CA: Friends of the Earth.

b. United Nations Food and Agriculture Organization. 1978. *Mammals in the Seas*. Report of the FAO Advisory Committee on Marine Resources Research, Working Party on Marine Mammals. FAO Fisheries Series 5(1). Rome: UNFAO.

c. National Oceanic and Atmospheric Association. 2003. World fisheries. Available at http://www.st.nmfs. gov/st1/fus/current/04_world2002.pdf.

d. Reynolds, J., et. al. 2010. Marine Mammal Commission Annual Report to Congress 2009. M. M. Commission. Bethesda, Maryland, and National Oceanic and Atmospheric Association. 2012. Gray Whale (*Eschrichtius robustus*), NOAA Fisheries, Office of Protected Resources, http://www.nmfs.noaa.gov/pr/species/mammals/cetaceans/graywhale.htm NOAA says the range over recent years has been somewhere between 18,000 and 30,000.

CHAPTER 14

Energy: Some Basics

LEARNING OBJECTIVES

Understanding the basics of energy, as well as the sources and uses of energy, is essential for effective energy planning. After reading this chapter, you should be able to . . .

- **Argue that energy independence in the United States is closer to becoming a reality**
- **Compare and contrast the first and second laws of thermodynamics**
- **Compare and contrast first-law energy efficiency with second-law efficiency**
- **Defend or criticize this statement: "To be happy and productive, people in industrialized countries need to consume a disproportionately large share of the world's energy"**
- **Define the difference between efficient use and conservation of energy**
- **Compare and contrast the business-as-usual approach to energy with the approach suggested by Amory Lovins**
- **Rank the three major components of sustainable energy development in terms of importance to society. Justify your ranking**

A New England common illustrates the tragedy of the Commons, one of the key ideas of Environmental Economics.

Nuon Solar Team/Hans-Peter van Velthoven

CASE STUDY

Racing with the Sun (Power)

Every other October since 1987, solar-powered cars have raced from Darwin to Adelaide, Australia, in the World Solar Challenge, an 2,900 km (1,800 mi) route that puts the latest alternative-energy technology to the test. The cars can run only on sunlight that their solar cells capture and convert to electricity. Electric motors that are at least 90% efficient are necessary. Racing teams are usually comprised of college students, and teams are backed by major aerospace and high-tech corporations. The eleventh race, held in 2011, was won by a Japanese team; a Netherlands team finished second, and the fastest U.S. team, from the University of Michigan, finished third. Drivers had to avoid a bushfire, wallabies, cattle, sheep, lizards, and strong winds.[1] Top speeds ranged from 143–154 km/hr (89–95 mph), and the average speed of the winning car was about 70 mph.

Suppose you decided to organize a team from your university, design and build a solar-powered car, and enter the race. Here's the challenge: The roof of an automobile is just barely large enough to hold a solar panel that can gather enough energy to drive a car. It can't power a regular sedan or SUV, and it can just barely power any car at all. How would you win? Should you build a car that, under the race rules, has the largest solar-powered area and tries to gather as much sunlight as possible, making the car as heavy as you can? Or would you opt for energy efficiency and build the lightest car, trading off a larger energy input for greater energy efficiency? Would you spend money and add weight to make the car's shape as aerodynamic as possible, so that it would have the least resistance from the wind? And how about reliability? Would you build a stronger, therefore heavier car, or would you place your bet on the sleekest, lightest, car?

The car built by the Netherlands Nuon team, the Luna 6, had three wheels, a body made of carbon fiber, and solar panels covering nearly every inch of its top surface. The students who designed it tested a model in a wind tunnel by covering it with oil; the oil activated an ultraviolet light used to highlight any impediments to its aerodynamic design. The Luna 6 was 140 kg (308 lb), 20 kg (44 lb) lighter than the Luna 5 designed two years before.

The Tokai Challenger 2, also a three-wheeler, featured carbon monocoque construction on its top surface that incorporated a 6-m square array of silicon solar cells.

The Netherlands team lost out to the Tokai Challenger 2 largely because of the gains the Japanese team made on Day 3 of the four-and-a-half-day race, wrote British Diederik Kinds in *The Register*.[2] He analyzed the race and found that very small measures of efficiency combined to put the Japanese car over the top. The Tokai team was able to draw about 7% more from the car's battery; the Japanese car weighed 10 kg (22 lb) less than the Luna 6 and had 5% better aerodynamics; its solar panel was able to power the equivalent of two small light bulbs more than the Luna 6. The end result, according to Diederik, was that the Japanese car had available about 10% more energy than its closest competition, which it used to increase its lead by 30 minutes on Day 3 and ultimately win the race.

A few months after the race, the University of Michigan team merged with the Nuon team, forming the Nuum team, and agreed to work together to create a $3 million solar car for the next World Solar Challenge in 2013.[3] Part of the impetus for all teams is to be part of an effort to develop zero emission vehicles that can be mass-produced, a goal of international importance as we look to decrease dependence on fossil fuels for transportation.

In deciding how to design and build a winning solar-powered car, the teams needed to understand the fundamental laws of physics as well as the basic concepts of energy and how to apply them. That understanding is also important to our ability to decide what kinds of energy sources are best for people and for the environment. In this first chapter on energy we introduce those basic concepts.

14.1 Outlook for Energy

Energy Today and Tomorrow

The decisions we make today will affect energy use for generations. Should we choose complex, centralized energy production methods, or simpler and widely dispersed methods, or a combination of the two? Which energy sources should be emphasized? Which uses of energy should be explored for increased efficiency? How can we develop a sustainable energy policy? There are no easy answers.

The use of fossil fuels, especially oil, has improved sanitation, medicine, and agriculture and is helping to make possible the global human population increase that we have discussed in other chapters. Many of us are living longer, with a higher standard of living, than people before us. However, burning fossil fuels imposes growing environmental costs, ranging from urban pollution to a change in the global climate.

One thing certain about the energy picture for tomorrow is that it will involve living with uncertainty when it comes to energy availability and cost. The sources of energy and the patterns of energy use will undoubtedly change. It is clear that we need to examine our entire energy policy in terms of sources, supply, consumption, and environmental concerns. Meanwhile, let's take a moment to consider a key energy question of the 21st century.

Can the United States Achieve Energy Independence? When? How?

Until recently, the idea of the United States becoming independent with respect to energy seemed a far-fetched idea. A prevailing idea, based on rates of discovery and production of oil, was that oil would become increasingly scarce. After all, until just a few years ago, we were importing about two-thirds of the oil we used annually, and, in many years, some of the countries that supplied the oil were not particularly friendly to the United States. During the last several years, we have begun to import less oil and from countries that are friendlier with us. But that is not why we may begin to ask the question: not if, but when will we reach energy independence?

One idea advanced by many in the early part of this century was that the world would soon reach the time when approximately one-half of the available, conventional oil resource, defined as relatively light oil, will have been pumped and burned to fuel society. This is known as the concept of **peak oil**. With respect to conventional oil from most oil fields around the world, this concept is essentially correct; as such, it would indicate a major supply crisis looming.[4]

The concept of peak oil was labeled a myth by some and a red flag to others in the oil industry.[5] Neither position was entirely correct, but production of conventional oil, according to the International Energy Agency, did likely peak in 2006. Other studies suggest that the peak has not arrived and might not occur for decades.[6-8] Total production from all sources today is about 85 million barrels per day (mbd).

So, if oil production has peaked, why is there not a serious gap between nearly constant production and increasing world demand for oil? The answer is that the gap (the difference between production and demand) in oil is being filled by new sources of oil (and natural gas) now being developed and produced in ever growing amounts. One example is oil in tight shale rocks in the Bakken formation in North Dakota.[7,8] The shale is dense, and the oil and gas are tightly held in the fine-grained rock. Oil and natural gas in shale is abundant in the contiguous United States (Figure 14.1) as well as in other locations such as eastern Europe. Significantly, these sources of the shale oil and natural gas are the fastest growing energy resources in

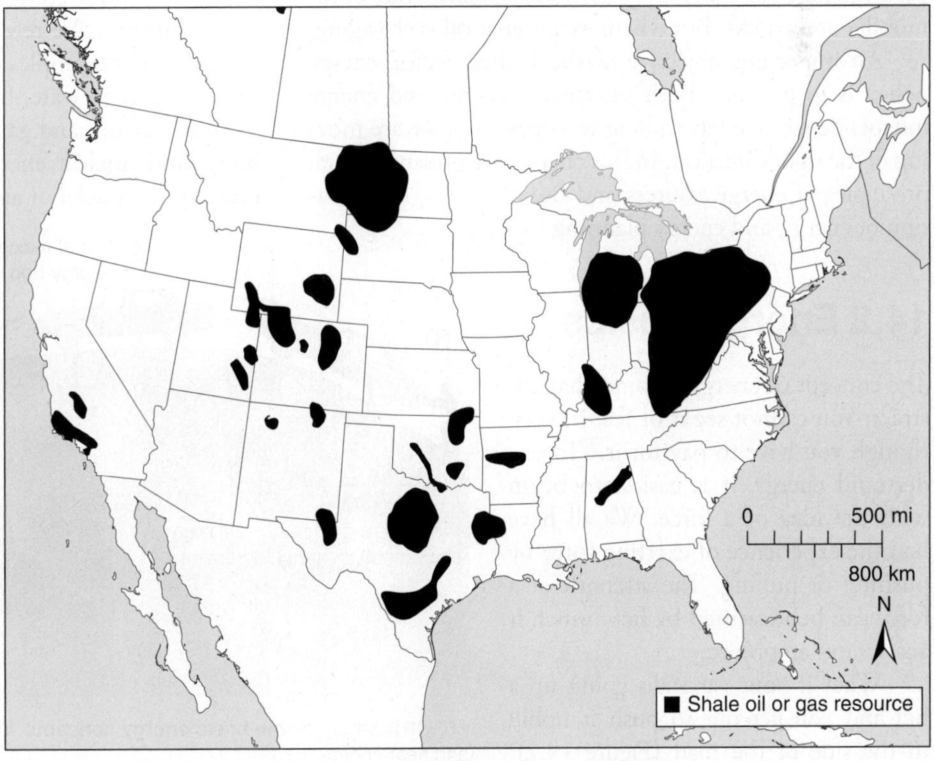

FIGURE 14.1 Map of the United States showing current locations of shale oil and/or natural gas resources.

the United States today. Many in the industry are touting the projected increase in domestic oil and gas as the major path to U.S. energy independence. In the short term, this is a valid point.

However, there are particular problems associated with these sources: They are more difficult to obtain and more expensive, and likely have higher environmental costs than do conventional methods from standard wells because they require special drilling techniques (these problems are discussed in Chapter 15).

Most environmentalists do not consider reaching energy independence through drilling and burning more and more oil or natural gas as an acceptable long-term energy strategy, especially when environmental costs are factored in, which include the possibility of water, soil, and air pollution. Environmentalists prefer pursuing a sustainable energy policy that can endure for generations and not harm the environment. The emerging use of more alternative energy, particularly wind and solar energy, is the preferred path if sustainable energy is the objective. Our continued dependence on fossil fuels (even our own) will mean we will be destined to endure: air pollution with adverse health effects; damage to ecosystems through disturbance of the land and the oceans caused by occasional serious accidents (oil spills); and climate change due to increasing carbon dioxide emissions in the atmosphere as fossil fuels are burned.

Even with increased production in the United States, achieving energy independence will not happen in the next year or two and probably not for several decades or longer. We will continue to be dependent on oil, which is used for most of our transportation, primarily our automobiles and trucks. But where we get that oil is changing.

An important objective of the United States' energy policy is to promote both sustainable energy and energy independence. We have a long way to go, but we are moving in the right direction. In this chapter we present general principles of energy, sources and use of energy, conservation of energy, and energy planning.

14.2 Energy Basics

The concept of energy is somewhat abstract: You cannot see it or feel it, even though you have to pay for it.[9] To understand energy, it is easiest to begin with the idea of a force. We all have had the experience of exerting force by pushing or pulling. The strength of a force can be measured by how much it accelerates an object.

What if your car stalls going up a hill and you get out to push it uphill to the side of the road (Figure 14.2)? You apply a force against gravity, which would otherwise cause the car to roll downhill. If the brake is on, the brakes, tires, and bearings may heat up from friction. The longer the distance over which you exert force in pushing the car, the greater the change in the car's position and the greater the amount of heat from friction in the brakes, tires, and bearings. In physicists' terms, exerting the force over the distance moved is work. That is, **work** is the product of a force times a distance. Conversely, energy is the ability to do work. Thus, if you push hard, but the car doesn't move, you have exerted a force but have not done any work (according to the definition), even if you feel very tired and sweaty.[9]

In pushing your stalled car, you have moved it against gravity and caused some of its parts (brakes, tires, bearings) to become heated. These effects have something in common: They are forms of energy. You have converted chemical energy in your body to the energy of motion of the car (kinetic energy). When the car is higher on the hill, the potential energy of the car has been increased, and friction produces heat energy.

Energy is often converted or transformed from one kind to another, but the total energy is always conserved. The principle that energy cannot be created or destroyed but is always conserved is known as the **first law of thermodynamics**. Thermodynamics is the science that keeps track of energy as it undergoes various transformations from one type to another. We use the first law to keep track of the quantity of energy.[10]

To illustrate the conservation and conversion of energy, think about a tire swing over a creek (Figure 14.3). When the tire swing is held in its highest position, it is not moving. It does contain stored energy, however, owing to its position. We refer to the stored energy as *potential energy*. Other examples of potential energy are the gravitational energy in water behind a dam; the chemical energy in coal, fuel oil, and gasoline, as well as in the fat in your body; and nuclear energy, which is related to the forces binding the nuclei of atoms.[9]

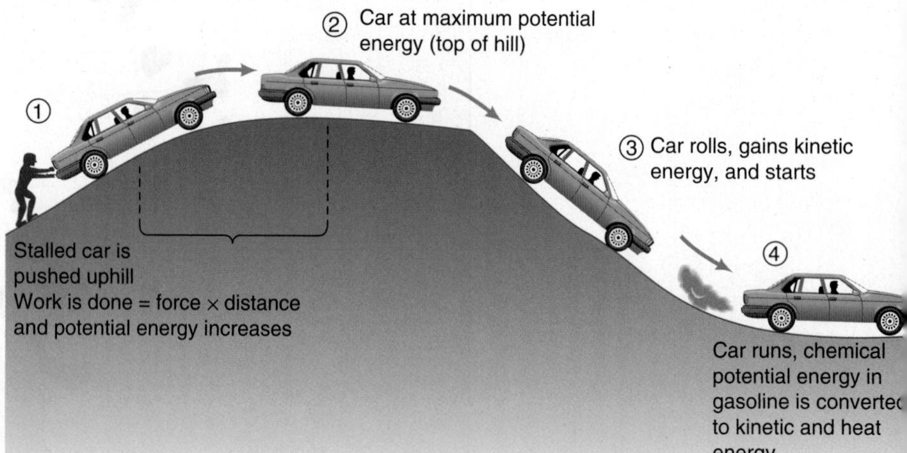
② Car at maximum potential energy (top of hill)
①
③ Car rolls, gains kinetic energy, and starts
④
Stalled car is pushed uphill
Work is done = force × distance and potential energy increases
Car runs, chemical potential energy in gasoline is converted to kinetic and heat energy

FIGURE 14.2 **Some basic energy concepts**, including potential energy, kinetic energy, and heat energy.

1
3
2
3
1
① Energy is all potential.
② Energy is all kinetic.
③ Energy is potential and kinetic.

FIGURE 14.3 **Diagram of a tire swing**, illustrating the relation between potential and kinetic energy.

The tire swing, when released from its highest position, moves downward. At the bottom (straight down), the speed of the tire swing is greatest, and no potential energy remains. At this point, all the swing's energy is the energy of motion, called *kinetic energy*. As the tire swings back and forth, the energy continuously changes between the two forms, potential and kinetic. However, with each swing, the tire slows down a little more and goes a little less high because of friction created by the movement of the tire and rope through air and friction at the pivot where the rope is tied to the tree. The friction slows the swing, generating *heat energy*, which is energy from random motion of atoms and molecules. Eventually, all the energy is converted to heat and emitted to the environment, and the swing stops.[9]

The example of the swing illustrates the tendency of energy to dissipate and end up as heat. Indeed, physicists have found that it is possible to change all the gravitational energy in a tire swing (a type of pendulum) to heat. However, it is impossible to change all the heat energy thus generated back into potential energy. Energy is conserved in the tire swing. When the tire swing finally stops, all the initial gravitational potential energy has been transformed by way of friction to heat energy. If the same amount of energy, in the form of heat, were returned to the tire swing, would you expect the swing to start again? The answer is no! What, then, is used up? It is not energy because energy is always conserved. What is used up is the energy *quality*—the availability of the energy to perform work. The higher the quality of the energy, the more easily it can be converted to work; the lower the energy quality, the more difficult it is to convert it to work.

This example illustrates another fundamental property of energy: Energy always tends to go from a more usable (higher-quality) form to a less usable (lower-quality) form. This is the **second law of thermodynamics**, and it means that when you use energy, you lower its quality.

Let's return to the example of the stalled car, which you have now pushed to the side of the road. Having pushed the car a little way uphill, you have increased its potential energy. You can convert this to kinetic energy by letting it roll back downhill. You engage the gears to restart the car. As the car idles, the potential chemical energy (from the gasoline) is converted to waste heat energy and other energy forms, including electricity to charge the battery and play the radio.

Why can't we collect the wasted heat and use it to run the engine? Again, as the second law of thermodynamics tells us, once energy is degraded to low-quality heat, it can never regain its original availability or energy grade. When we refer to low-grade heat energy, we mean that relatively little of it is available to do useful work. High-grade energy, such as that of gasoline, coal, or natural gas, has high potential to do useful work. The biosphere continuously receives high-grade energy from the sun and radiates low-grade heat to the depths of space.[9,10]

14.3 Energy Efficiency

Two fundamental types of energy efficiencies are derived from the first and second laws of thermodynamics: first-law efficiency and second-law efficiency. **First-law efficiency** deals with the amount of energy without any consideration of the quality or availability of the energy. It is calculated as the ratio of the actual amount of energy delivered where it is needed to the amount of energy supplied to meet that need. Expressions for efficiencies are given as fractions; multiplying the fraction by 100 converts it to a percentage. As an example, consider a furnace system that keeps a home at a desired temperature of 18°C (65°F) when the outside temperature is 0°C (32°F). The furnace, which burns natural gas, delivers 1 unit of heat energy to the house for every 1.5 units of energy extracted from burning the fuel. That means it has a first-law efficiency of 1 divided by 1.5, or 67% (see Table 14.1 for other examples).[10] The "unit" of energy for our furnace is arbitrary for the purpose of discussion; we also could use the British thermal unit (Btu) or some other units (see A Closer Look 14.1).

Table 14.1 EXAMPLES OF FIRST-AND SECOND-LAW EFFICIENCIES

ENERGY (END USE)	FIRST-LAW EFFICIENCY (%)	WASTE HEAT (%)	SECOND-LAW EFFICIENCY (%)	POTENTIAL FOR SAVINGS
Incandescent light bulb	5	95		
Fluorescent light	20	80		
Automobile	20-25	75-80	10	Moderate
Power plants (electric); fossil fuel and nuclear	30-40	60-70	30	Low to moderate
Burning fossil fuels (used directly for heat)	65	35		
Water heating			2	Very high
Space heating and cooling			6	Very high
All energy (U.S.)	50	50	10-15	High

Source:

A CLOSER LOOK 14.1

Energy Units

When we buy electricity by the kilowatt-hour, what are we buying? We say we are buying energy, but what does that mean? Before we go deeper into the concepts of energy and its uses, we need to define some basic units.

The fundamental energy unit in the metric system is the *joule;* 1 joule is defined as a force of 1 newton* applied over a distance of 1 meter. To work with large quantities, such as the amount of energy used in the United States in a given year, we use the unit *exajoule,* which is equivalent to 10^{18} (a billion billion) joules, roughly equivalent to 1 quadrillion, or 10^{15}, Btu, referred to as a *quad.* To put these big numbers in perspective, the United States today consumes approximately 100 exajoules (or quads) of energy per year, and world consumption is about 500 exajoules (quads) annually.

In many instances, we are particularly interested in the rate of energy use, or *power,* which is energy divided by time. In the metric system, power may be expressed as joules per second, or *watts* (W); 1 joule per second is equal to 1 watt. When larger power units are required, we can use multipliers, such as *kilo*-(thousand), *mega*-(million), and *giga*-(billion). For example, a modern nuclear power plant's electricity production rate is 1,000 megawatts (MW) or 1 gigawatt (GW).

Sometimes, it is useful to use a hybrid energy unit, such as the watt-hour (Wh); remember, energy is power multiplied by time. Electrical energy is usually expressed and sold in *kilowatt-hours* (kWh, or 1,000 Wh). This unit of energy is 1,000 W applied for 1 hour (3,600 seconds), the equivalent energy of 3,600,000 J (3.6 MJ).

The average estimated electrical energy in kilowatt-hours used by various household appliances over a period of a year is shown in Table 14.2. The total energy used annually is the power rating of the appliance multiplied by the time the appliance was actually used. The appliances that use most of the electrical energy are water heaters, refrigerators, clothes driers, and washing machines. A list of common household appliances and the amounts of energy they consume is useful in identifying the ones that might help save energy through conservation or improved efficiency.

*A newton (N) is the force necessary to produce an acceleration of 1 m per sec (m/s^2) to a mass of 1 kg.

Table 14.2 POWER USE OF TYPICAL HOUSEHOLD APPLIANCES IN WATTS (W) AND ANNUAL COST

Here are some examples of the range of nameplate wattages for various household appliances: units are Watts.

Aquarium = 50–1,210

Clock radio = 10

Coffee maker = 900–1,200

Clothes washer = 350–500

Clothes dryer = 1,800–5,000

Dishwasher = 1,200–2,400 (using the drying feature greatly increases energy consumption)

Dehumidifier = 785

Electric blanket- *Single/Double* = 60/100

Fans

- Ceiling = 65–175
- Window = 55–250
- Furnace = 750
- Whole house = 240–750

Hair dryer = 1,200–1,875

Heater *(portable)* = 750–1,500

Clothes iron = 1,000–1,800

Microwave oven = 750–1,100

Personal computer

- CPU—awake/asleep = 120/30 or less
- Monitor—awake/asleep = 150/30 or less
- Laptop = 50

Radio *(stereo)* = 70–400

Refrigerator *(frost-free, 16 cu ft)* = 725

Televisions (color)

- 19" = 65–110
- 27" = 113
- 36" = 133
- 53"-61" Projection = 170
- Flat screen = 120

Toaster = 800–1,400

Toaster oven = 1,225

VCR/DVD = 17–21/20–25

Vacuum cleaner = 1,000–1,440

Water heater *(40 gallon)* = 4,500–5,500

Water pump *(deep well)* = 250–1,100

Water bed *(with heater, no cover)* = 120–380

Estimating Appliance and Home Electronic Energy Use

If you're trying to decide whether to invest in a more energy-efficient appliance or if you'd like to determine your electricity loads, you may want to estimate appliance energy consumption.

Formula for Estimating Energy Consumption

You can use this formula to estimate an appliance's energy use:

(wattage × hours used per day) ÷ 1,000 = daily kilowatt-hour (kWh) consumption

1 kilowatt (kW) = 1,000 watts

Multiply the kW value by the number of days you use the appliance during the year for the annual consumption. You can then calculate the annual cost to run an appliance by multiplying the kWh per year by your local utility's rate per kWh consumed.

Note: To estimate the number of hours that a refrigerator actually operates at its maximum wattage, divide the total time the refrigerator is plugged in by three. Refrigerators, though turned "on" all the time, actually cycle on and off as needed to maintain interior temperatures.

Examples:

Window fan:

(200 watts × 4 hours/day × 120 days/year) ÷ 1,000
= 96 kWh × 8.5 cents/kWh
= $8.16/year

Personal Computer and Monitor:

[(120 watts + 150 watts) × 4 hours/day × 365 days/year] ÷ 1,000
= 394 kWh × 8.5 cents/kWh
= $33.51/year

Wattage

You can usually find the wattage of most appliances stamped on the bottom or back of the appliance or on its nameplate. The wattage listed is the maximum power drawn by the appliance. Since many appliances have a range of settings (for example, the volume on a radio), the actual amount of power consumed depends on the setting used at any one time.

Source: Modified slightly after: http://www.energysavers.gov/your_home/appliances/index.cfm/mytopic = 10040, accessed May 12, 2012.

First-law efficiencies are misleading because a high value suggests (often incorrectly) that little can be done to save energy through additional improvements in efficiency. This problem is addressed by the use of second-law efficiency. **Second-law efficiency** refers to how well matched the energy end use is with the quality of the energy source. For our home-heating example, the second-law efficiency would compare the minimum energy necessary to heat the home to the energy actually used by the gas furnace. If we calculated the second-law efficiency (which is beyond the scope of this discussion), the result might be 6%—much lower than the first-law efficiency of 65%.[10] (We will see why later.) Table 14.1 also lists some second-law efficiencies for common uses of energy.

Values of second-law efficiency are important because low values indicate where improvements in energy technology and planning may save significant amounts of high-quality energy. Second-law efficiency tells us whether the energy quality is appropriate to the task. For example, you could use a welder's acetylene blowtorch to light a candle, but a match is much more efficient (and safer as well).

We are now in a position to understand why the second-law efficiency is so low (6%) for the house-heating example discussed earlier. This low efficiency implies that the furnace is consuming too much high-quality energy in carrying out the task of heating the house. In other words, the task of heating the house requires heat at a relatively low temperature, near 18°C (65°F), not heat with temperatures in excess of 1,000°C (1,832°F), such as is generated inside the gas furnace. Lower-quality energy, such as solar energy, could do the task and yield higher second-law efficiency because there is a better match between the required energy quality and the house-heating end use. Through better energy planning, such as matching the quality of energy supplies to the end use, higher second-law efficiencies can be achieved, resulting in substantial savings of high-quality energy.

Examination of Table 14.1 indicates that electricity-generating plants have nearly the same first-law and second-law efficiencies. These generating plants are examples of heat engines. A heat engine produces work from heat. Most of the electricity generated in the world today comes from *heat engines* that use nuclear fuel, coal, gas, or other fuels. Our own bodies are examples of heat engines, operating with a capacity (power) of about 100 watts and fueled indirectly by solar energy. (See A Closer Look 14.1 for an explanation of watts and other units of energy.) The internal combustion engine (used in automobiles) and the steam engine are additional examples of heat engines. A great deal of the world's energy is used in heat engines, with profound environmental effects, such as thermal pollution, urban smog, acid rain, and global warming.

The maximum possible efficiency of a heat engine, known as *thermal efficiency*, was discovered by the French engineer Sadi Carnot in 1824, before the first law of thermodynamics was formulated.[11] Modern heat engines have thermal efficiencies that range between 60 and 80% of their ideal Carnot efficiencies. Modern 1,000-megawatt (MW) electrical generating plants have thermal efficiencies ranging between 30 and 40%; that means at least 60 to 70% of the energy input to the plant is rejected as waste heat. For example, assume that the electric power output from a large generating plant is 1 unit of power (typically 1,000 MW). Producing that 1 unit of power requires 3 units of input (such as burning coal) at the power plant, and the entire process produces 2 units of waste heat, for a thermal efficiency of 33%. The significant number here is the waste heat, 2 units, which amounts to twice the actual electric power produced.

Electricity may be produced by large power plants that burn coal or natural gas, by plants that use nuclear fuel, or by smaller producers, such as geothermal, solar, or wind sources (see Chapters 15, 16, and 17). Once produced, the electricity is fed into the grid, which is the network of power lines or the distribution system. Eventually, it reaches homes, shops, farms, and factories, where it provides light and heat and also drives motors and other machinery used by society. As electricity moves through the grid, losses take place. The wires that transport electricity (power lines) have a natural resistance to electrical flow. Known as *electrical resistivity*, this resistance converts some of the electric energy in the transmission lines to heat energy, which is radiated into the environment surrounding the lines.

14.4 Energy Sources and Consumption

People living in industrialized countries make up a relatively small percentage of the world's population but consume a disproportionate share of the total energy consumed in the world. There is a direct relationship between a country's standard of living (as measured by gross national product) and energy consumption per capita.

When petroleum production eventually declines near the end of the 21st century, oil and gasoline will be in shorter supply and more expensive. Before then, use of these fuels may be curtailed in an effort to lessen global climate change. As a result, within the next 30 years, both developed and developing countries will need to find innovative ways to obtain energy. In the future, affluence may be related as closely to more efficient use of a wider variety of energy sources as it is now to total energy consumption.

Fossil Fuels and Alternative Energy Sources

Today, approximately 85% of the energy consumed in the United States is derived from petroleum, natural gas, and coal. Because they originated from plant and animal material that existed millions of years ago, they are called fossil fuels. They are forms of stored solar energy that are part of our geologic resource base, and they are essentially nonrenewable. Other sources of energy—geothermal, nuclear, hydropower, and solar, among others—are referred to as *alternative* energy sources because they may serve as alternatives to fossil fuels in the future. Some of them, such as solar and wind, are not depleted by consumption and are known as *renewable energy* sources.

The shift to alternative energy sources may be gradual as fossil fuels continue to be used, or it could be accelerated by concern about potential environmental effects of burning fossil fuels. Regardless of which path we take, one thing is certain: Fossil fuels are finite. It took millions of years to form them, but they will be depleted in only a few hundred years of human history. Using even the most optimistic predictions, the fossil fuel epoch that started with the Industrial Revolution will represent only about 500 years of human history. Therefore, although fossil fuels have been extremely significant in the development of modern civilization, their use will be a brief event in the span of human history.

Energy Consumption in the United States

Energy consumption in the United States from 1980 and projected to 2035 is shown in Figure 14.4a. World energy consumption (for comparison) is shown in Figure 14.5. The United States, with about 5% of the world's population, uses about 20% of the world's energy. These figures dramatically illustrate the ongoing dependence on the three major fossil fuels (coal, natural gas, and petroleum) in the United States and the world. From approximately 1950 through the late 1970s, energy consumption in the United States soared from about 30 exajoules to 75 exajoules annually. (Energy units are defined in A Closer Look 14.1.) Since about 1980, energy consumption in the United States has risen by only about 25 exajoules (Figure 14.4b). This is encouraging because it suggests that policies promoting energy-efficiency improvements (such as requiring new automobiles to be more fuel efficient and buildings to be better insulated) have been at least partially successful.

What is not shown in Figure 14.4a, however, is the huge energy loss. For example, energy consumption in the United States in 1965 was approximately 50 exajoules, of which only about half was used effectively. Energy losses were about 50% (the number shown earlier in Table 14.1 for all energy). In 2011, energy consumption in the United States was about 100 exajoules, and, again, about 50% was lost in conversion processes. Energy losses in 2011 were about equal to total U.S. energy consumption in 1965! The largest energy losses are associated with the production of electricity and with transportation, mostly through the use of heat engines, which produce waste heat that is lost to the environment.

Another way to examine energy use is to look at the generalized energy flow of the United States by end use for a particular year. In 2008, we imported considerably more oil than we produced (we imported about 60% of the oil we used), and our energy consumption was fairly evenly distributed in three sectors: residential/commercial, industrial, and transportation. In 2010, our imports

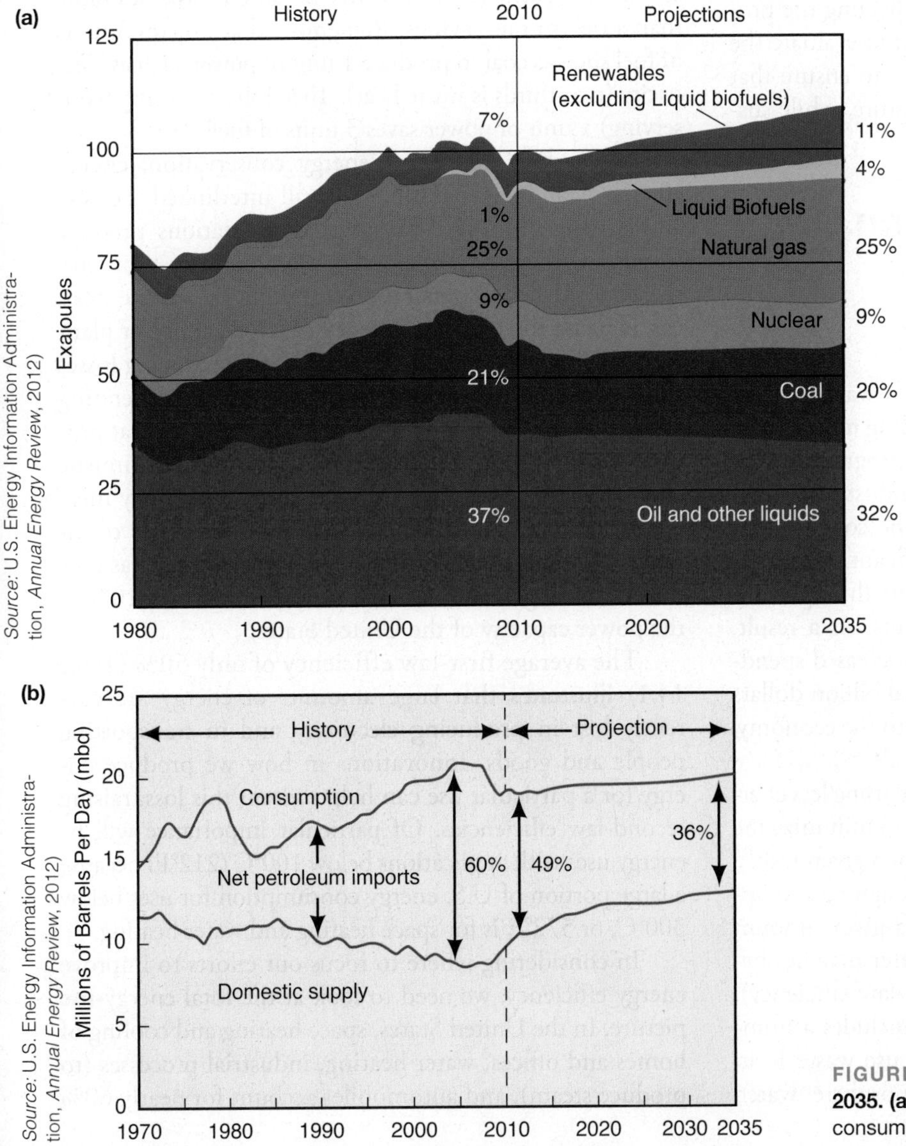

FIGURE 14.4 U.S. energy from 1970 projected to 2035. (a) total consumption by source; **(b)** production, consumption,and imports of oil, 1970-2035.

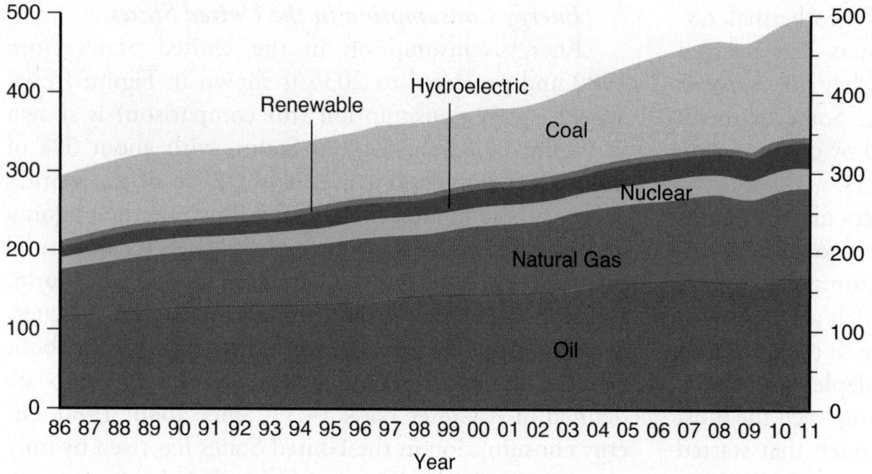

(*Source:* 2012, BP Statistical Review of World Energy—no permission necessary)

FIGURE 14.5 World energy consumption 1986-2011.

had been reduced to about 49% and are expected to drop to 36% by 2035 as we produce more domestic oil (Figure 14.4b). However, it is clear that we remain dangerously vulnerable to changing world conditions affecting the production and delivery of crude oil. We need to evaluate the entire spectrum of potential energy sources to ensure that sufficient energy will be available in the future, while sustaining environmental quality.

14.5 Energy Conservation, Increased Efficiency, and Cogeneration

We have come to realize that there are two ways to meet our growing demand for electricity—building more power plants or implementing energy efficiency programs at the customer level (homes, businesses, and industries). Utility companies have evidence to support the concept that energy conservation programs are a significant energy resource. It is far cheaper to reduce demand through efficiency programs than to build power plants. As a result, utility companies in the past decade have increased spending on energy efficiency programs by several billion dollars per year. Efficiency saves money, adds jobs to the economy, and reduces emissions of greenhouse gasses.[12]

Conservation of energy refers simply to using less energy and adjusting our energy needs and uses to minimize the amount of high-quality energy necessary for a given task.[13] Increased **energy efficiency** involves designing equipment to yield more energy output from a given amount of energy input (first-law efficiency) or better matches between energy source and end use (second-law efficiency). Another concept is **cogeneration**, which includes a number of processes designed to capture and use waste heat, rather than simply releasing it into the atmosphere, water, or other parts of the environment as a thermal pollutant. In other words, we design energy systems and power plants to provide energy more than once[14]—that is, to use it a second time, at a lower temperature, and, possibly, to use it in more than one way as well.

An example of cogeneration is the *natural gas combined cycle power plant* that produces electricity in two ways: gas cycle and steam cycle. In the gas cycle, the natural gas fuel is burned in a gas turbine to produce electricity. In the steam cycle, hot exhaust from the gas turbine is used to create steam that is fed into a steam generator to produce additional electricity. The combined cycles capture waste heat from the gas cycle, nearly doubling the efficiency of the power plant from about 30% to 50–60%. Energy conservation is particularly attractive because it provides more than a one-to-one savings. Remember that it takes 3 units of fuel such as coal to produce 1 unit of power such as electricity (two-thirds is waste heat). Therefore, not using (conserving) 1 unit of power saves 3 units of fuel!

These three concepts—energy conservation, energy efficiency, and cogeneration—are all interlinked. For example, when big, coal-burning power stations produce electricity, they may release large amounts of heat into the atmosphere. Cogeneration, by using that waste heat, can increase the overall efficiency of a typical power plant from 33% to as much as 75%, effectively reducing losses from 67 to 25%. Cogeneration also involves generating electricity as a by-product of industrial processes that produce steam as part of their regular operations. Optimistic energy forecasters estimate that, eventually, we may meet approximately one-half the electrical power needs of industry through cogeneration.[13,14] Another source has estimated that cogeneration could provide more than 10% of the power capacity of the United States.

The average first-law efficiency of only 50% (Table 14.1) illustrates that large amounts of energy are currently lost in producing electricity and in transporting people and goods. Innovations in how we produce energy for a particular use can help prevent this loss, raising second-law efficiencies. Of particular importance will be energy uses with applications below 100°C (212°F) because a large portion of U.S. energy consumption for uses below 300°C, or 572°F, is for space heating and water heating.

In considering where to focus our efforts to improve energy efficiency, we need to look at the total energy-use picture. In the United States, space heating and cooling of homes and offices, water heating, industrial processes (to produce steam), and automobiles account for nearly 60%

of the total energy use, whereas transportation by train, bus, and airplane accounts for only about 5%. Therefore, the areas we should target for improvement are building design, industrial energy use, and automobile design. We note, however, that debate continues as to how much efficiency improvements and conservation can reduce future energy demands and the need for increased energy production from traditional sources, such as fossil fuel.

Building Design

A spectrum of possibilities exists for increasing energy efficiency and conservation in residential buildings. For new homes, the answer is to design and construct homes that require less energy for comfortable living.[15] For example, we can design buildings to take advantage of passive solar potential, as did the early Greeks and Romans and the Native American cliff dwellers. (Passive solar energy systems collect solar heat without using moving parts.) Windows and overhanging structures can be positioned so that the overhangs shade the windows from solar energy in summer, thereby keeping the house cool, while allowing winter sun to penetrate the windows and warm the house.

The potential for energy savings through architectural design for older buildings is extremely limited. The position of the building on the site is already established, and reconstruction and modifications are often not cost-effective. The best approach to energy conservation for these buildings is insulation, caulking, weather stripping, installation of window coverings and storm windows, and regular maintenance.

Ironically, buildings constructed to conserve energy are more likely to develop indoor air pollution due to reduced ventilation. In fact, air pollution is emerging as one of our most serious environmental problems. Potential difficulties can be reduced by better designs for air circulation systems that purify indoor air and bring in fresh, clean air. Construction that incorporates environmental principles is more expensive, owing to higher fees for architects and engineers, as well as higher initial construction costs. Nevertheless, moving toward improved design of homes and residential buildings to conserve energy remains an important endeavor.

Industrial Energy

The rate of increase in energy use (consumption) leveled off in the early 1970s. Nevertheless, industrial production of goods (automobiles, appliances, etc.) continued to grow significantly. Today, U.S. industry consumes about one-third of the energy produced. The reason we have had higher productivity with lower growth of energy use is that more industries are using cogeneration and more energy-efficient machinery, such as motors and pumps designed to use less energy.[13,15]

Automobile Design

The development of fuel-efficient automobiles has steadily improved during the last 30 years. In the early 1970s, the average U.S. automobile burned approximately 1 gallon of gas for every 14 miles traveled. By 1996, the miles per gallon (mpg) had risen to an average of 28 for highway driving and as high as 49 for some automobiles.[16] Fuel consumption rates did not improve much from 1996 to 1999. In 2004, many vehicles sold were SUVs and light trucks with fuel consumption of 10–20 mpg. A loophole in regulations permits these vehicles to have poorer fuel consumption than conventional automobiles.[16] As a result of higher gasoline prices, sales of larger SUVs declined in 2006, but smaller SUVs remain popular as consumers are apparently sacrificing size for economy (up to a point). In 2012, the government (through executive action, without congressional approval) agreed on new Corporate Average Fuel Economy (CAFE) regulations that are visionary in scope. Starting in 2017 and reached by 2025, the new proposed CAFE standard will be 54.5 mpg, a significant increase from the previous standard of 35.5 mpg by 2016. This increased fuel efficiency will be reached with new innovative technology, including cars with lighter diesel engines and transmissions; use of lighter materials and improved tires; hybrid cars, which combine a fuel-burning engine and an electric motor; and all-electric vehicles and vehicles that use natural gas as a fuel. Demand for hybrid vehicles is growing rapidly and will benefit from the development of more advanced rechargeable batteries (plug-in hybrids; see Figure 14.6).

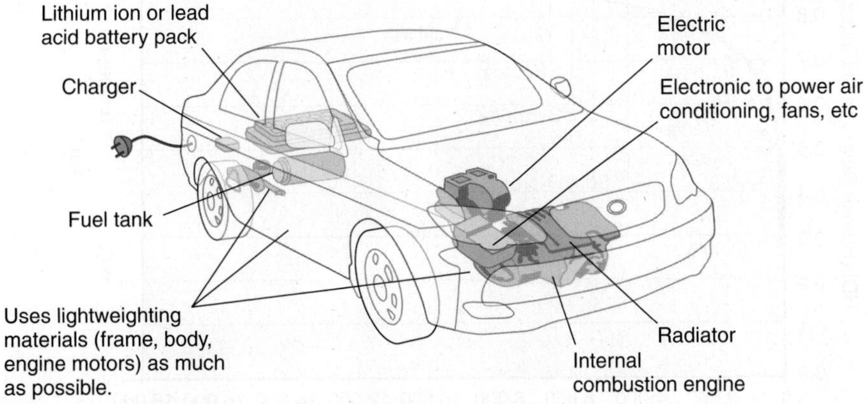

FIGURE 14.6 **Idealized diagram of a plug-in hybrid car**

A real change in cars is coming. What it will be and when are not entirely known, but it may be a transformation to all-electric cars. Where and how we produce the electricity to power those cars will be an issue.

Values, Choices, and Energy Conservation

What is the relationship between human development (a measure of life expectancy, education, and wealth) and use of energy per person? The results may surprise you. The United Nations has developed the Index of Human Development (HDI) for many countries. HDI varies from about 0.3 (low) to 0.5 (medium) to 0.9 (high). The HDI and annual electricity use per person can be compared for various countries (see Figure 14.7). The graph produced from this comparison shows that the HDI for western European countries such as France, Germany, the U.K., Spain, and Italy is 9—about the same as that for the United States—but per capita use of electricity in these countries (shown on the bottom line of the graph) is one-third to one-half of U.S. consumption. The graph shows that human development, in terms of life expectancy, education, and income per person, peaks at an electricity use of about 4,000 kWh (remember that 1 kWh is equivalent to 3.6 Mj; see a Closer Look 14.1). However, U.S. and Canadian usage to achieve this HDI is about 13,000 kWh and 16,000 kWh, respectively. Does this mean that the United States uses too much electricity? The answer is likely yes. With energy conservation, we should be able to reduce our per capita use of electricity to be more in line with other industrial countries such as Germany, the U.K., France, Japan, and South Korea.[17]

Changing behavior to conserve energy involves our values and the choices we make to act at a local level; these choices, in turn, address global environmental problems, such as human-induced warming caused by burning fossil fuels. For example, we make choices as to how far we commute to school or work and what method of transport we use to get there. Some people commute more than an hour by car to get to work, while others ride a bike, walk, or take a bus or train. Other ways of modifying behavior to conserve energy include the following:

- Using carpools to travel to and from work or school
- Purchasing a hybrid or all-electric car
- Turning off lights when leaving rooms
- Taking shorter showers (conserves hot water)
- Putting on a sweater and turning down the thermostat in winter
- Using energy-efficient compact fluorescent light bulbs or light-emitting diode (LED) lights
- Purchasing energy-efficient appliances
- Sealing drafts in buildings with weather stripping and caulk
- Better insulating your home
- Washing clothes in cold water and hanging them to dry whenever possible
- Purchasing local foods rather than foods that must be brought to market from afar
- Reducing standby power for electronic devices and appliances by using power strips and turning them off when not in use
- Installing solar water heaters or collectors

What other ways of modifying your behavior would help conserve energy?

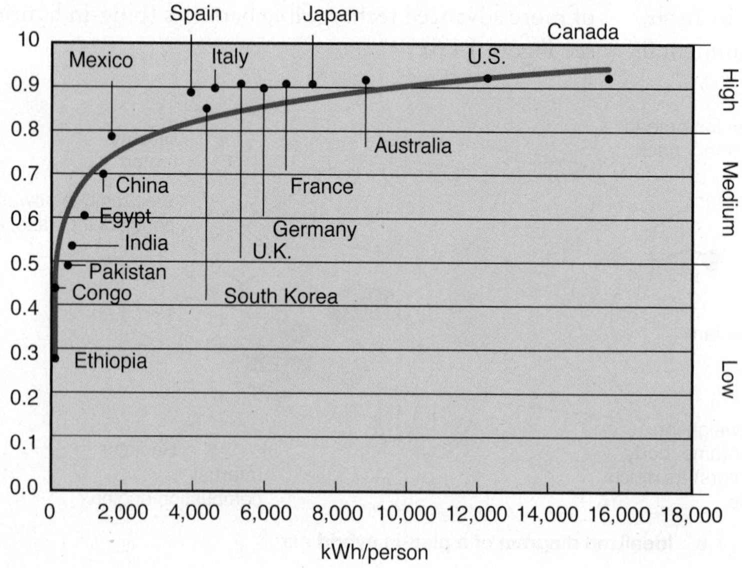

(*Source:* Modified (with updates) after Pasternak, D. 2000. *Global Energy Futures and Human Development: A Framework for Analysis.* U.S. Department of Energy. UCRL-ID-140773.)

FIGURE 14.7 Relationship between HDI and annual electricity use per capita for selected countries.

14.6 Sustainable-Energy Policy

Presidents of the United States since the mid-1970s have attempted to address energy problems and the question of how to become independent of foreign energy sources. The Energy Policy Act of 2005, passed by Congress and signed into law by then President George W. Bush in the summer of 2005, has been followed by heated debate about energy policy in the 21st century. A number of topics related to energy are being discussed, including the American Clean Energy and Security Act of 2009, which took a serious step toward energy self-sufficiency in the United States.

The 2009 act (not yet passed) has four parts: (1) *clean energy*, which involves renewable energy, sequestration of carbon, development of clean fuels and vehicles, and a better electricity transmission grid; (2) *energy efficiency*, for buildings, homes, transportation, and utilities; (3) *reduction of carbon dioxide and other greenhouse gases associated with global warming*, including programs to reduce global warming by reducing emissions of carbon dioxide in coming years; and (4) *making the transition to a clean energy economy*, including economic incentives for development of green energy jobs, exporting clean technology, increasing domestic competitiveness, and finding ways to adapt to global warming.

The United States faces serious energy problems. Energy policy, from a local to global scale, has emerged as a central economic concern, a national security issue, and an environmental question. How we respond to energy issues will largely define who we are and what sort of world we will live in during this century.

Today energy policy is at a crossroads. One path leads to the "business-as-usual" approach, which consists of finding greater amounts of fossil fuels, building larger power plants, and continuing to use energy as freely as we always have. The business-as-usual path is more comfortable—it requires no new thinking; no realignment of political, economic, or social conditions; and little anticipation of coming reductions in oil production.

People heavily invested in the continued use of fossil fuels and nuclear energy often favor the traditional path. They argue that much environmental degradation around the world has been caused by people who have been forced to use local resources, such as wood, for energy, leading to the loss of plant and animal life and increasing soil erosion. They argue that the way to solve these environmental problems is to provide cheap, high-quality energy, such as fossil fuels or nuclear energy.

In countries like the United States, with sizable resources of oil, natural gas, and coal, people supporting the business-as-usual path argue that we should exploit those resources while finding ways to reduce their environmental impact. According to these proponents, we should (1) allow the energy industry to develop the available energy resources and (2) let industry, free from government regulations, provide a steady supply of energy with less total environmental damage.

The energy plan put forth by President Bush was largely a business-as-usual approach: Find and use more coal, oil, and natural gas; use more nuclear power; and build more than 1,000 new fossil fuel plants in the next 20 years. Energy conservation and development of alternative energy sources, while encouraged, were not considered of primary importance.

A visionary path for energy policy was suggested more than 30 years ago by Amory Lovins.[18] That path focuses on energy alternatives that emphasize energy quality and are renewable, flexible, and environmentally more benign than those of the business-as-usual path. As defined by Lovins, these alternatives have the following characteristics:

- They rely heavily on renewable-energy resources, such as sunlight, wind, and biomass (wood and other plant material).
- They are diverse and are tailored for maximum effectiveness under specific circumstances.
- They are flexible, accessible, and understandable to many people.
- They are matched in energy quality and geographic distribution and are scaled to end-use needs, increasing second-law efficiency.

Lovins points out that people are not particularly interested in having a certain amount of oil, gas, or electricity delivered to their homes; they are more interested in comfortable homes, adequate lighting, food on the table, and energy for transportation.[18] According to Lovins, only about 5% of end uses require high-grade energy, such as electricity. Nevertheless, a lot of electricity is used to heat homes and water. Lovins shows that there is an imbalance in using nuclear reactions at extremely high temperatures and in burning fossil fuels at high temperatures simply to meet needs where the necessary temperature increase may be only a few 10s of degrees. He considers such large discrepancies wasteful and a misallocation of high-quality energy.

Energy for Tomorrow

The availability of energy supplies and the future demand for energy are difficult to predict because the technical, economic, political, and social assumptions underlying predictions are constantly changing. In addition, seasonal and regional variations in energy consumption must also be considered. For example, in areas with cold winters and hot, humid summers, energy consumption peaks during the winter months (from heating) and again in the summer (from air-conditioning). Regional variations in energy consumption are significant. For example, in the United States as a whole, the transportation sector uses about one-fourth of the energy consumed. However, in California, where people often commute long distances to work, about one-half of the energy is used for transportation, more than double the national average. Energy sources, too, vary by region. For example, in the eastern and southwestern United States, the fuel of choice for power plants is often coal, but power plants on the West Coast are more likely to burn oil or natural gas or use hydropower from dams to produce electricity.

Future changes in population densities, as well as intensive conservation measures, will probably alter existing patterns of energy use. This might involve a shift to greater reliance on alternative (particularly renewable) energy sources.[19,20] The United States' energy consumption in the year 2050 may be about 110 exajoules (Figure 14.4a). What will be the energy sources for the anticipated growth in energy consumption? Will we follow our past policy of business as usual (coal, oil, nuclear), or will we turn more to alternative energy sources (wind, solar, geothermal)? What is clear is that the mix of energy sources in 2035 will be different from today's and more diversified.[19-21]

All projections of specific sources and uses of energy in the future must be considered speculative. Perhaps most speculative of all is the idea that we really can meet most of our energy needs with alternative, renewable energy sources in the next several decades.

The energy decisions we make in the very near future will greatly affect both our standard of living and our quality of life. From an optimistic point of view, we have the necessary information and technology to ensure a bright, warm, lighted, and mobile future. But time may be running out, and we need action now. We can continue to take things as they come and live with the results of our present dependence on fossil fuels, or we can build a more sustainable energy future based on careful planning, innovative thinking, and a willingness to move from our dependence on petroleum.

U.S. energy policy for the 21st century is being discussed seriously, and significant change in policy is likely. Some of the recommendations are as follows:

- Promote conventional energy sources: Use more natural gas to reduce our reliance on energy from foreign countries.
- Encourage alternative energy: Support and subsidize wind, solar, geothermal, hydrogen, and biofuels (ethanol and biodiesel).
- Provide for energy infrastructure: Ensure that electricity is transmitted over dependable, modern infrastructure.
- Promote conservation measures: Set higher efficiency standards for buildings and for household products. Require that waste heat from power generation and industrial processes be used to produce electricity or other products. Support stronger fuel-efficiency standards for cars, trucks, and SUVs. Provide tax credits for installing energy-efficient windows and appliances in homes and for purchasing fuel-efficient hybrids, all-electric cars, and clean diesel vehicles.
- Carefully evaluate the pros and cons of nuclear power, which can generate large amounts of electricity without emitting greenhouse gases, but which has serious negatives as well.
- Promote research: Develop new alternative energy sources; find new, innovative ways to improve existing coal plants and to help construct cleaner coal plants; determine whether it is possible to extract vast amounts of oil trapped in oil shale and tar sands without harming the environment; and develop pollution-free, electric automobiles.

Which of the above points will become policy in future years is not known, but parts of the key ideas will move us toward sustainable energy.

Integrated, Sustainable Energy Management

The concept of **integrated energy management** recognizes that no single energy source can provide all the energy required by the various countries of the world.[22] A range of options that vary from region to region will have to be employed. Furthermore, the mix of technologies and sources of energy will involve both fossil fuels and alternative, renewable sources.

A basic goal of integrated energy management is to move toward **sustainable energy development** that is implemented at the local level. Sustainable energy development would have the following characteristics:

- It would provide reliable sources of energy.
- It would not destroy or seriously harm our global, regional, or local environments.
- It would help ensure that future generations inherit a quality environment with a fair share of the Earth's resources.

To implement sustainable energy development, leaders in various regions of the world will need energy plans based on local and regional conditions. The plans will integrate the desired end uses for energy with the energy sources that are most appropriate for a particular region and that hold potential for conservation and efficiency. Such plans will recognize that preserving resources can be profitable and that degradation of the environment and poor economic conditions go hand in hand.[22] In other words, degradation of air, water, and land resources depletes assets and ultimately will lower both the standard of living and the quality of life. A good energy plan recognizes that energy demands can be met in environmentally preferred ways and is part of an aggressive environmental policy whose goal is a quality environment for future generations. The plan should do the following:[22]

- Provide for sustainable energy development.
- Provide for aggressive energy efficiency and conservation.
- Provide for diversity and integration of energy sources.
- Develop and use the "smart grid" to optimally manage energy flow on the scale of buildings to regions.
- Provide for a balance between economic health and environmental quality.
- Use second-law efficiencies as an energy policy tool—that is, strive to achieve a good balance between the quality of an energy source and end uses for that energy.

An important element of the plan involves the energy used for automobiles. This builds on policies of the past 30 years to develop hybrid vehicles that use both an electric motor and an internal combustion engine and to improve fuel technology to reduce both fuel consumption and emission of air pollutants. Finally, the plan should factor in the marketplace through pricing that reflects the economic cost of using the fuel, as well as its cost to the environment. In sum, the plan should be an integrated energy-management statement that moves toward sustainable development. Those who develop such plans recognize that a diversity of energy supplies will be necessary and that the key components are (1) improvements in energy efficiency and conservation and (2) matching energy quality to end uses.[23]

The global pattern of ever-increasing energy consumption led by the United States and other nations cannot be sustained without a new energy paradigm that includes changes in human values, not just a breakthrough in technology. Choosing to own lighter, more fuel-efficient automobiles and living in more energy-efficient homes is consistent with a sustainable energy system that focuses on providing and using energy to improve human welfare. A sustainable energy paradigm establishes and maintains multiple linkages among energy production, energy consumption, human well-being, and environmental quality.[23]

CRITICAL THINKING ISSUE

Use of Energy Today and in 2030

Note: Before proceeding with this exercise, refer back to A Closer Look 4.1 to be sure you are comfortable with the units and big numbers.

The Organization for Economic Cooperation and Development (OECD) is a group of 30 countries, 27 of which are classified by the World Bank as having high-income economies. Non-OECD members are not all low-income countries, but many are. The developing countries (all of which are non-OECD) have most of the world's 7 billion people and are growing in population faster than the more affluent countries. The average rate of energy use in 2010 for an individual in non-OECD countries is 46 billion joules per person per year (1.5 kW per person), whereas for the OECD countries it is 210 billion joules per person per year (6.7 kW per person). In other words, people in OECD countries use about 4.5 times more energy per person than those in non-OECD countries. In 2010, each group—OECD and non-OECD—used about 250 EJ (1 EJ is 10^{18} J). The world average is 74 billion joules per person per year (2.3 kW per person).[24]

If the current annual population growth rate of 1.1% continues, the world's population will double in 64 years. However, as we learned in Chapters 1 and 4, the human population may not double again. It is expected to be about 8.5 billion by 2030. More people will likely mean more energy use. People in non-OECD countries will need to consume more energy per capita if the less-developed countries are to achieve a higher standard of living; thus, energy consumption in non-OECD countries as a group is projected to increase by 2030 to about 55 billion joules per person per year (1.7 kW per person). On the other hand, energy use in OECD countries is projected to decline to about 203 billion joules per person per year (6.4 kW per person). This would bring the global average in 2030 to about 80 billion joules per person per year (2.5 kW per person), up from 74 billion joules in 2010. If these projections are correct, 58% of the energy will be consumed in the non-OECD countries, compared with 50% today.

With worldwide average energy use of 2.3 kW per person in 2010, the 6.8 billion people on Earth use about 16 trillion watts annually. A projected population of 8.5 billion in 2030 with an estimated average per capita energy use rate of 2.5 kW would use about 21 trillion watts annually, an increase of about 33% from today.[24]

A realistic goal is for annual per capita energy use to remain about 2.5 kW, with the world population peaking at 8.5 billion people by the year 2030. If this goal is to be achieved, non-OECD countries will be able to increase their populations by no more than about 50% and their energy use by about 70%; OECD nations can increase their population by only a few percent and will have to reduce their energy use slightly.

Critical Thinking Questions

1. Using only the data presented in this exercise, how much energy, in exajoules, did the world use in 2010, and what would you project global energy use to be in 2030?
2. The average person emits as heat 100 watts of power (the same as a 100 W bulb). If we assume that 25% of it is emitted by the brain, how much energy does your brain emit as heat in a year? Calculate this in joules and kWh. What is the corresponding value for all people today, and how does that value compare with world energy use per year? Can this help explain why a large, crowded lecture hall (independent of the professor pontificating) might get warm over an hour?
3. Can the world supply one-third more energy by 2030 without unacceptable environmental damage? How?
4. What would the rate of energy use be if all people on Earth had a standard of living supported by energy use of 10 kW per person, as in the United States today? How do these totals compare with the present energy-use rate worldwide?
5. In what specific ways could energy be used more efficiently in the United States? Make a list of the ways and compare your list with those of your classmates. Then compile a class list.
6. In addition to increasing efficiency, what other changes in energy consumption might be required to provide an average energy-use rate in 2030 of 6.4 kW per person in OECD countries?
7. Would you view the energy future in 2030 as a continuation of the business-as-usual approach with larger, centralized energy production based on fossil fuels, or a softer path, with more use of alternative, distributed energy sources? Justify your view.

SUMMARY

- The first law of thermodynamics states that energy is neither created nor destroyed but is always conserved and is transformed from one kind to another. We use the first law to keep track of the quantity of energy.
- The second law of thermodynamics tells us that as energy is used, it always goes from a more usable (higher-quality) form to a less usable (lower-quality) form.
- Two fundamental types of energy efficiency are derived from the first and second laws of thermodynamics. In the United States today, first-law efficiencies average about 50%, which means that about 50% of the energy produced is returned to the environment as waste heat. Second-law efficiencies average 10–15%, so there is a high potential for saving energy through better matching of the quality of energy sources with their end uses.
- Energy conservation and improvements in energy efficiency can have significant effects on energy consumption. It takes three units of a fuel such as oil to produce one unit of electricity. As a result, each unit of electricity conserved or saved through improved efficiency saves three units of fuel.
- Arguments can be made for both the business-as-usual path and changing to a new path. The first path has a long history of success and has produced the highest standard of living ever experienced. However, present sources of energy (based on fossil fuels) are causing serious environmental degradation and are not sustainable (especially with respect to conventional oil). A second path, based on alternative energy sources that are renewable, decentralized, diverse, and flexible, provides a better match between energy quality and end use and emphasizes second-law efficiencies.
- The transition from fossil fuels to other energy sources requires sustainable, integrated energy management. The goal is to provide reliable sources of energy that do not cause serious harm to the environment and ensure that future generations will inherit a quality environment.
- Due to discovery of large amounts of oil and natural gas in the United States (mostly from tight, shale continuous deposits), we are approaching a time, when in a few decades, we will likely achieve energy independence. However, using more fossil fuels will come with the increased possibility of additional environmental consequences—from air and water pollution to climate change.

REEXAMINING THEMES AND ISSUES

Sean Randall/Getty Images, Inc.

HUMAN POPULATION

The industrialized and urbanized countries produce and use most of the world's energy. As societies change from rural to urban, energy demands generally increase. Controlling the increase of human population is an important factor in reducing total demand for energy (total demand is the product of average demand per person and number of people).

© Biletskiy_Evgeniy/iStockphoto

SUSTAINABILITY

It will be impossible to achieve sustainability in the United States if we continue with our present energy policies. The present use of fossil fuels is not sustainable. We need to rethink the sources, uses, and management of energy. Sustainability is the central issue in our decision to continue on the hard path or change to the soft path.

© Anton Balazh 2011/iStockphoto

GLOBAL PERSPECTIVE

Understanding global trends in energy production and consumption is important if we are to directly address the global impact of burning fossil fuels with respect to air pollution and global warming. Furthermore, the use of energy resources greatly influences global economics as these resources are transported and utilized around the world.

ssguy/ShutterStock

URBAN WORLD

A great deal of the total energy demand is in urban regions, such as Tokyo, Beijing, London, New York, and Los Angeles. How we choose to manage energy in our urban regions greatly affects the quality of urban environments. Burning cleaner fuels results in far less air pollution. This has been observed in several urban regions, such as London. Burning of coal in London once caused deadly air pollution; today, natural gas and electricity heat homes, and the air is cleaner. Burning coal in Beijing continues to cause significant air pollution and health problems for millions of people living there.

B2M Productions/Getty Images, Inc.

PEOPLE AND NATURE

Our development and use of energy are changing nature in significant ways. For example, burning fossil fuels is changing the composition of the atmosphere, particularly through the addition of carbon dioxide. The carbon dioxide is contributing to the warming of the atmosphere, water, and land (see Chapter 20 for details). A warmer Earth is, in turn, changing the climates of some regions and affecting weather patterns and the intensity of storms.

George Doyle/Getty Images, Inc.

SCIENCE AND VALUES

Public opinion polls consistently show that people value a quality environment. In response, energy planners are evaluating how to use our present energy resources more efficiently, practice energy conservation, and reduce adverse environmental effects of energy consumption. Science is providing options in terms of energy sources and uses; our choices will reflect our values.

KEY TERMS

cogeneration 326
conservation 326
energy efficiency 326
first-law efficiency 321
first law of thermodynamics 320
integrated energy management 330
peak oil 329
second-law efficiency 324
second law of thermodynamics 321
sustainable energy development 330
work 320

STUDY QUESTIONS

1. Can we now contemplate a time when the United States will achieve energy independence?
2. What evidence supports the notion that, although present energy problems are not the first in human history, they are unique in other ways?
3. How do the terms *energy*, *work*, and *power* differ in meaning?
4. Compare and contrast the potential advantages and disadvantages of a major shift from hard-path to soft-path energy development.
5. You have just purchased a 100-hectare wooded island in Puget Sound. Your house is built of raw timber and is not insulated. Although the island receives some wind, trees over 40 m tall block most of it. You have a diesel generator for electric power, and hot water is produced by an electric heater run by the generator. Oil and gas can be brought in by ship. What steps would you take in the next five years to reduce the cost of the energy you use with the least damage to the island's natural environment?
6. How might better matching of end uses with potential sources yield improvements in energy efficiency?
7. Complete an energy audit of the building you live in and then develop recommendations that might lead to lower utility bills.
8. How might plans using the concept of integrated energy management differ for the Los Angeles area and the New York City area? How might both of these plans differ from an energy plan for Mexico City, which is quickly becoming one of the largest urban areas in the world?
9. A recent energy scenario for the United States (see Figure 14.4a) suggests that, by 2035, energy sources might be oil (32%), natural gas (25%), coal (20%), nuclear (9%), and the sum of solar power, hydropower, wind power, biomass, and geothermal energy (15%). Do you think this is a likely scenario? What would be the major difficulties and points of resistance or controversy? How could renewable sources increase beyond 15%?

FURTHER READING

Botkin, D.B. 2010. *Powering the Future.* Upper Saddle River, NJ: FT Press. An up-to-date summary of energy sources and planning for the future.

Lindley, D. The energy should always work twice. *Nature* 458, no. 7235 (2009):138–141. A good paper on cogeneration.

Lovins, A.B. *Soft Energy Paths: Towards a Durable Peace.* 1979. New York, NY: Harper & Row. A classic energy book.

McKibben, B. Energizing America. *Sierra* 92, no. 1(2007): 30–38, 112–113. A good recent summary of energy for the future.

Miller, P. Saving energy. *National Geographic* 251, no. 3 (2009):60–81. Many ways to conserve energy.

Wald, M.L. 2009. The power of renewables. *Scientific American* 300, no. 3 (2009):57–61. A good summary of renewable energy.

NOTES

1. Based on material from Botkin, D.B. 2010. *Powering the Future: A Scientist's Guide to Energy Independence.* Upper Saddle River, NJ: FT Press.
2. Kinds, D. World solar challenge: Why the winners were so good. *The Register*, October 22, 2011. http://www.theregister.co.uk/2011/10/22/solar_car_tokai_advantages/
3. UMsolar and Nuon agree to merge, creating the first ever international solar car team, April 1, 2011 University of Michigan news release. http://solarcar.engin.umich.edu/tag/nuon-solar-team/
4. Roberts, P. 2008. Tapped out. *National Geographic* 213(6): 86–91.

5. Cavanay, R. 2006. Global oil about to peak? A recurring myth. *World Watch* 19(1):13–15.
6. U.S. Energy Information Administration. 2012. *Annual Energy Outlook 2012* Early Release. http://www.eia.gov.
7. Whitney, G., Behrens, C. E., and Glover, C. 2010. *U.S. fossil fuel resources: termonology, reporting, and summary*. CRS Report to Congress. Congressional Research Service. 7-5700. R40872. Available at: budget.house.gov/UploadedFiles/CRS_NOVEMBER2010 (pdf).
8. Maugeri, L. 2012. *Oil: The next revolution, the unprecedented upsurge of oil production capacity and what it means for the world.* Discussion Paper #2012-10. Harvard University: John F. Kennedy School of Government.
9. Morowitz, H.J. 1979. *Energy Flow in Biology.* New Haven, CT: Oxbow Press.
10. Ehrlich, P.R., A.H. Ehrlich, and J.P. Holdren. 1970. *Ecoscience: Population, Resources, Environment.* San Francisco, CA: W.H. Freeman.
11. Feynman, R.P., R.B. Leighton, and M. Sands. 1964. *The Feynman Lectures on Physics.* Reading, MA: Addison-Wesley.
12. York, D., P. Witte, S. Norwalk, and M. Kushler. 2012. *Three Decades of Counting: A Historical Review and Current Assessment of Electric Utility Energy Activity in the States.* American Council for an Energy-Efficient Economy. Research Report U123.
13. Miller, P. 2009. Saving energy. *National Geographic* 251(3): 60–81.
14. Lindley, D. 2009. The energy should always work twice. *Nature* 458(7235):138–141.
15. Flavin, C. 1984. *Electricity's Future: The Shift to Efficiency and Small-scale Power.* Worldwatch Paper 61. Washington, DC: Worldwatch Institute.
16. Consumers' Research. 1995. Fuel economy rating: 1996 mileage estimates. 78:22–26.
17. Pasternak, D. 2000. *Global Energy Futures and Human Development: A Framework for Analysis.* U.S. Department of Energy. UCRL-ID-140773
18. Lovins, A.B. 1979. *Soft Energy Paths: Towards a Durable Peace.* New York, NY: Harper & Row.
19. Duval, J. 2007. The fix. *Sierra* 92(1):40–41.
20. Wald, M.L. 2009. The power of renewables. *Scientific American* 300(3):57–61.
21. Flavin, C. 2008. *Low-Carbon Energy: A Roadmap*. Worldwatch Report 178. Washington, DC: Worldwatch Institute.
22. California Energy Commission. 1991. *California's Energy Plan: Biennial Report.* Sacramento, CA.
23. Flavin, C., and S. Dunn. 1999. Reinventing the energy system. In L.R. Brown et al., eds., *State of the World 1999: A Worldwatch Institute Report on Progress toward a Sustainable Society.* New York, NY: W.W. Norton.
24. U.S. Energy Information Administration. International Energy Outlook, 2010. Accessed May 13, 2012.

CHAPTER 15

Fossil Fuels and the Environment

LEARNING OBJECTIVES

We rely almost completely on fossil fuels—oil, natural gas, and coal—for our energy needs. However, these are nonrenewable resources and their production and use have a variety of serious environmental impacts. After reading this chapter, you should be able to . . .

- **Discuss why the future supply of oil and gas will be uncertain**
- **Compare and contrast how conventional and continuous oil and natural gas form with how coal forms**
- **Summarize why the adverse environmental effects of producing and using oil, natural gas, and coal are so different**

An oil platform offshore in the Gulf of Mexico

Eric Kulin/ageFotostock

graphingactivity

CASE STUDY

Oil Boom in North Dakota

Until recently, we thought that the oil resources of the world, and particularly those of North America, were becoming harder to find and that we were searching more and more for less and less oil. All that has changed in the past decade with the realization that oil in deep shale rock is becoming available and those resources are many times more abundant than what we previously thought.[1] For example, compared to an evaluation in 1995, the U.S. Geological Survey now estimates that there is at least a 25-fold increase in the amount of oil that can be recovered in North Dakota in what is called the Bakken Play (oil people label a discovery and recovery event a play). Many wells are being drilled in North Dakota, and this is having a significant impact on the environment of the state, in particular on small agricultural towns (Figure 15.1).

The economic recession that began in the United States in 2008 has taken a tremendous toll on most of the states; budgets are strapped, and many people are unable to find employment. That is not the case in North Dakota, where the unemployment rate in late 2012 was below 4% and the state budget had a significant surplus. This is due to the discovery and exploitation of oil that is spread deep in the ground over a very wide area in the tight shale rock of the Bakken formation. North Dakota's population did not grow significantly in 80 years. Now the western part of the state is experiencing boom conditions, and more jobs (most of them high-paying) are available than can be filled.

Jeff Wheeler/ZUMA Press/Newscom

FIGURE 15.1 A busy North Dakota highway. People are moving to North Dakota to take employment in the oil fields. The Bakken formation in western North Dakota is one of the biggest oil finds in recent history.

Counties in western North Dakota with populations of about 20,000 have grown by that amount in a very short time, and as a result local environments are being stressed at all levels. Often, roads are inadequate for the increased traffic, wastewater treatment becomes a problem, and sewage treatment facilities are overwhelmed. Schools and other institutions and support facilities are also experiencing stress. At any one time about 200 new wells are being drilled. and each of these is reported to produce more than 100 jobs. This amounts to employment for about 25,000 people and several thousand more support jobs related to the oil industry. Recovery of the oil requires hydrologic fracturing, which is more expensive than the process used to extract oil in other regions where more traditional oil fields are found.

People in North Dakota are seeing their farming and idyllic, pre-boom lives changing quickly. Some small towns are growing so rapidly that they can't even keep track of the increase in population and the amount of drilling in the Bakken oil fields. This has occurred because the size of the area underlain by oil is vast, covering about one-third of North Dakota and into Montana and Canada. The implications of all the newfound oil deposits in North Dakota are profound. At a time when environmentalists are striving to move the United States to energy self-sufficiency through renewable energy sources, with natural gas as a transition fuel, there is concern that exploitation of vast amounts of new oil will exacerbate the carbon loading in the environment, as well as contribute to well-known environmental problems associated with fossil fuels. In this chapter, we will examine some of the consequences of additional exploitation, recovery, and use of fossil fuels. In subsequent chapters, we will discuss alternative energy and nuclear power.

15.1 Fossil Fuels

Fossil fuels are forms of stored solar energy. Plants are solar energy collectors because they can convert solar energy to chemical energy through photosynthesis (see Chapter 7). The fossil fuels used today were created from incomplete biological decomposition of dead organic matter (mostly land and marine plants). Buried organic matter that was not completely oxidized was converted by chemical reactions over hundreds of millions of years to oil, natural gas, and coal. Biological and geologic processes in various parts of the geologic cycle produce the sedimentary rocks in which we find these fossil fuels.[2,3]

The fossil fuels—crude oil, natural gas, and coal—are our primary energy sources; they provide approximately 87% of the energy consumed worldwide. World energy consumption grew about 2.5% in 2011. The largest increase was in China due to burning more coal. In 2011 China mined and burned about 50% of the total coal used in the world. In the United States, consumption of fossil fuels has remained nearly constant since 2010, even as the U.S. population increased by about 3 million people.[1] There is significant variability in the use of fossil fuels. China uses a lot of coal, and the Middle East relies on oil and gas to provide nearly all of its energy. The energy use per person also varies significantly. The United States, with 314 million people, uses about the same amount of energy as China with its population of 1.3 billion. Thus the U.S. energy use per person is more than four times that of China.

15.2 Oil

Most geologists accept the hypothesis that **crude oil** and **natural gas** are derived from organic materials (mostly plants) that were buried with marine or lake sediments in what are known as *depositional basins.* Oil is found primarily along geologically young tectonic belts at plate boundaries, where large depositional basins are likely to occur (see Chapter 7). However, there are exceptions, such as in Texas, the Gulf of Mexico, North Dakota, and the North Sea, where oil has been discovered in depositional basins (some geologically very old) far from active plate boundaries.

The source material, or *source rock*, for oil and gas is fine-grained (less than 1/16 mm, or 0.0025 in., in diameter), organic-rich sediment buried to a depth of at least 500 m (1,640 ft), where it is subjected to increased heat and pressure. The elevated temperature and pressure initiate the chemical transformation of the sediment's organic material into oil and gas.

Conventional and Continuous Oil and Gas Resources

When we speak of the oil and gas resources of a particular field, area, or region, we are talking about the total amount of oil and gas present in the rocks. **Resources** now or can in the future be extracted at a profit. Oil and gas **reserves** are that portion of the resource that is identified and is currently available to be legally extracted at a profit. The distinction between resources and reserves is based on geologic, legal, and economic factors. As the price of oil increases, more of the resource may be considered a reserve. New technology to extract oil and gas that previously was not available under the old technology may increase both the resource and reserves. A personal analogy may help illustrate this idea: Your reserves are the money you have in your bank account while your resource is the potential money you can earn in the future. Your resources cannot pay your bills today.

Conventional oil resources are located in discrete oil fields where the oil and gas are trapped geologically. The pressure of deep burial compresses the sediment; this, along with the elevated temperature in the source rock, initiates the upward migration of the oil, which is relatively light, to a lower-pressure environment (known as the *reservoir rock*). The reservoir rock is coarser-grained and relatively porous (it has more and larger spaces between the grains). Sandstone and porous limestone, which have a relatively high proportion (about 30%) of empty space in which to store oil, are common reservoir rocks.

As already mentioned, oil is light; if the upward mobility of oil and gas is not blocked, they will escape to the atmosphere. This explains why oil in conventional resources is not generally found in geologically old rocks. Oil and gas in rocks older than about 0.5 billion years have had ample time to migrate to the surface, where they have either vaporized or eroded away.[3]

The oil fields from which we extract conventional oil resources are places where the natural upward migration of the oil to the surface is interrupted or blocked by what is known as a *trap* (Figure 15.2). The rock that helps form the trap, known as the *cap rock* (seal) is usually a very fine-grained sedimentary rock, such as shale, composed of silt and clay-sized particles. A favorable rock structure, such as an anticline (arch-shaped fold) or a fault (fracture in the rock along which displacement has occurred), is necessary to form traps, as shown in Figure 15.2. The important concept is that the combination of favorable rock structure and the presence of a cap rock allow deposits of oil to accumulate in the geologic environment, where they are then discovered and extracted.[3]

Continuous Oil Resources

Continuous oil resources, as compared to conventional resources, are regional in extent, occurring in broad geologic basins—the Bakken formation in North Dakota, for example (see the Case Study at the beginning of this chapter and Figure 15.3). These have no obvious seals or traps, have a large resource volume, and a low recovery factor. When favorable geologic conditions converge, "sweet spots" are identified in the continuous resource where economic extraction of oil is favorable. The rocks in which continuous oil resources are found are usually fine-grained shales

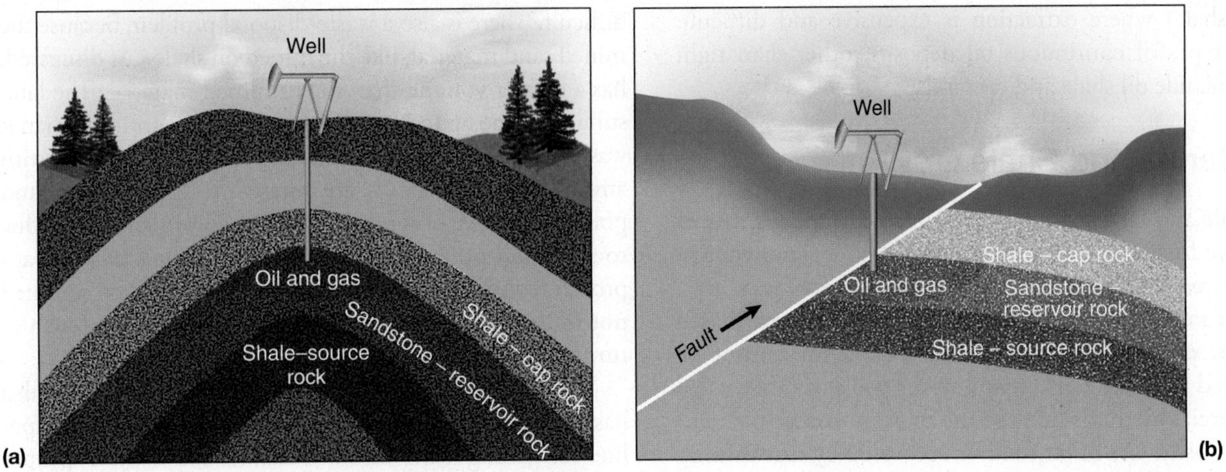

FIGURE 15.2 Two types of oil and gas traps: (a) anticline and **(b)** fault.

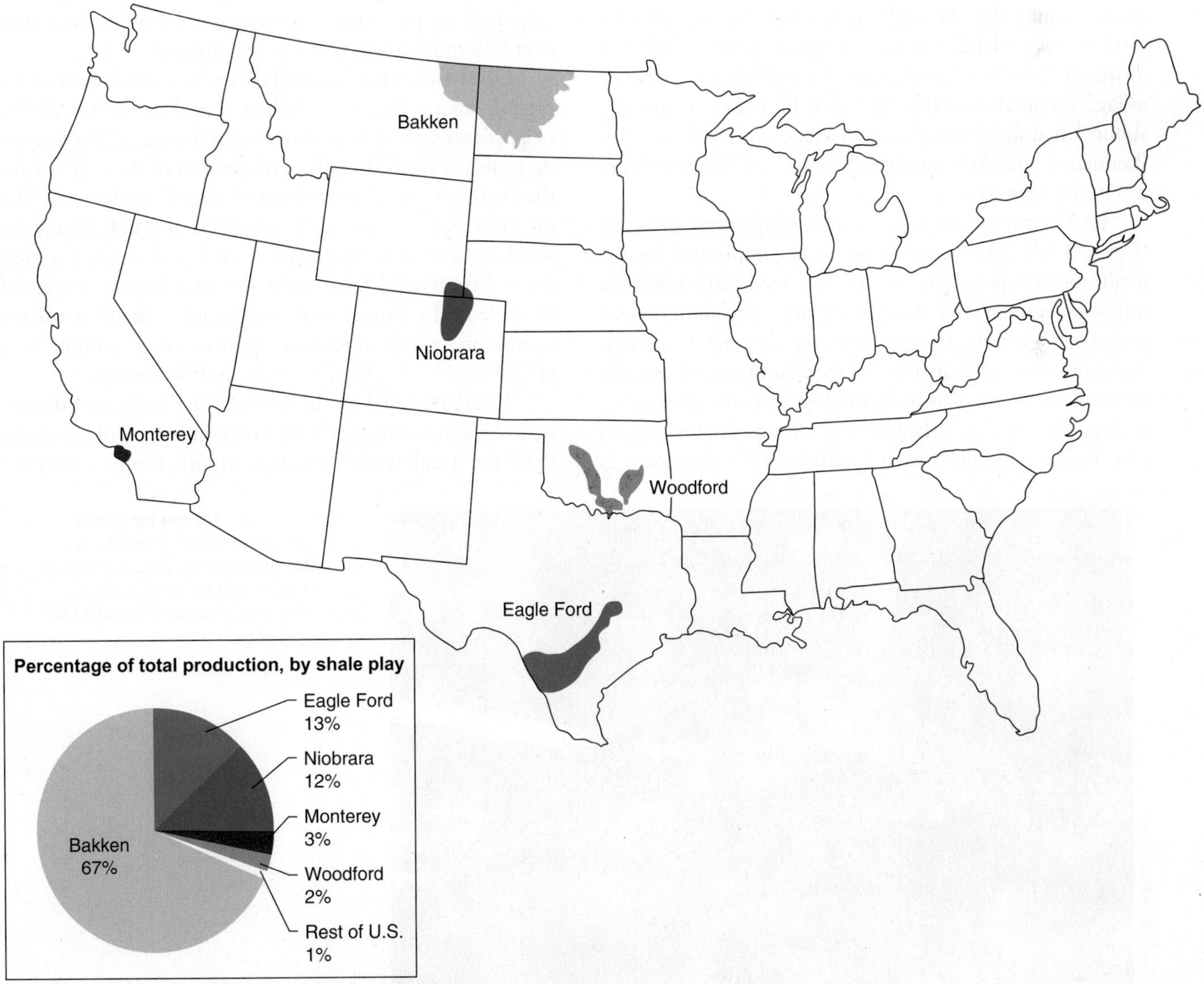

FIGURE 15.3 Location and percentage of U.S. shale production 2007–2011. The Bakken play in North Dakota represents two-thirds of total production to date.

(tight shale) where extraction is expensive and difficult. Other types of continuous oil deposits, other than tight shale, include oil shale and tar sands.

Tar Sands and Shale Oil

Oil shale and tar sands play a minor role in today's mix of available fossil fuels, but they may be more significant in the future, when traditional oil from wells becomes scarce.

Tar sands are sedimentary rocks or sands impregnated with tar oil, asphalt, or bitumen. Petroleum cannot be recovered from tar sands by pumping wells or other usual commercial methods because the oil is too viscous (thick) to flow easily. Oil in tar sands is recovered by first mining the sands—which are very difficult to remove—and then washing the oil out with hot water. It takes about two tons of tar sand to produce one barrel of oil.

Some 75% of the world's known tar sand deposits are in the Athabasca Tar Sands in the province of Alberta, Canada. About 19% of U.S. oil imports come from Canada, and about one-half of this is from tar sands.[4] The total Canadian resource that lies beneath approximately 78,000 km^2 (30,116 mi^2) of land is about 300 billion barrels. About half of this (173 billion barrels) can be economically recovered today.[4] Production of the Athabasca Tar Sands is currently about 1.6 million barrels of synthetic crude oil per day. Production will likely increase to about 3 million barrels per day in the next decade.

In Alberta, tar sand is mined in a large open-pit mine (Figure 15.4). The mining process is complicated by the fragile native vegetation—a water-saturated mat known as a muskeg swamp, a kind of wetland that is difficult to remove except when frozen. The mining of the tar sands has serious environmental consequences, ranging from a rapid increase in the human population in mining areas to the need to reclaim the land disturbed by the mining. Restoration of this fragile, naturally frozen (permafrost) environment is difficult. There is also a waste-disposal problem because the mined sand material, like the mined oil shale just discussed, has a greater volume than the unmined material. The land surface can be up to 20 m (66 ft) higher after mining than it was originally. Finally, recovery of oil from tar sand, as with any oil development, is an inherently dirty process, and pollution of air, water, and soil is a continual hazard due to spills and other industrial accidents. The Canadian approach requires that following mining the land be returned not to its original use but to some equivalent use, such as turning what was forestland into grazing land.[5]

Transporting oil from tar sands through pipelines also has potential environmental problems. In 2010, a pipeline containing heavy oil from Canada's tar sands ruptured and spilled about 4,000 cu m (1 mil gal) of oil into the Kalamazoo River in Michigan. The heavy oil sank to the riverbed, making cleanup difficult and expensive. The river was closed to the public for nearly two years. About 58 km (36 mi) of the river was heavily impacted as people decided to abandon their homes during the worst of the spill due to toxic odors. The spill is expected to have long-term health effects on people. Some oil remains in the river, even after over $75 million was spent in the cleanup.

Oil shale is a fine-grained sedimentary rock containing organic matter (kerogen). When heated to 500°C (900°F) in a process known as *destructive distillation*, oil shale yields up to nearly 60 L (14 gal) of oil per ton of shale. If not for the heating process, the oil would remain in the rock. The oil from shale is one of the so-called **synfuels** (from the words *synthetic* and *fuel*), which are liquid or gaseous fuels derived from solid fossil fuels. The best-known sources of oil shale in the United States are found in the Green River formation, which underlies approximately 44,000 km^2 (17,000 mi^2) of Colorado, Utah, and Wyoming.

Total identified world oil shale resources are estimated to be equivalent to about 3 trillion barrels of oil (more than the total world resource of other oil). However,

Wally Bauman/Alamy

FIGURE 15.4 Mining tar sands north of Fort McMurray in Alberta, Canada. The large shovel-bucket holds about 100 tons of tar sand. It takes about two tons of tar sand to produce one barrel of oil.

evaluation of the oil grade and the feasibility of economic recovery with today's technology is not complete. Oil shale resources in the United States amount to about 2 trillion bbl of oil, or two-thirds of the world total. Of this total, 90%, or 1.8 trillion bbl, is located in the Green River oil shales. The total oil that could be removed from U.S. oil shale deposits is about 100 billion barrels. This exceeds the oil reserves of the Middle East! But extraction is not easy, and environmental impacts would be serious.[6,7]

The environmental impact of developing oil shale varies with the recovery technique used. Both surface and subsurface mining techniques have been considered. Surface mining is attractive to developers because nearly 90% of the shale oil can be recovered, compared to less than 60% by underground mining. However, direct land and ecosystem destruction, which is difficult to restore, and waste disposal are major problems with either surface mining. Subsurface mining generally has smaller direct land and ecosystem destruction. But both surface and subsurface require that oil shale be processed, or *retorted* (crushed and heated), at the surface. The volume of waste will exceed the original volume of shale mined by 20–30% because crushed rock has pore spaces and, thus, more volume than the solid rock had. (If you doubt this, pour some concrete into a milk carton, remove it when it hardens, and break it into small pieces with a hammer. Then try to put the pieces back into the carton.) Thus, the mines from which the shale is removed will not be able to accommodate all the waste, and its disposal will become a problem.[8]

Although it is much more expensive and energy consuming to extract a barrel of oil from shale than it is to pump it from a well, interest in oil shale was heightened by an oil embargo in 1973 and by fear of continued shortages of crude oil. In the 1980s through the mid-1990s, however, plenty of cheap conventional oil was available, so oil-shale development was put on the back burner. Today, when it is clear that we will face oil shortages in the future, we are seeing renewed interest in oil shale,[7] and it is clear that any steep increases in oil prices will likely heighten this interest. This would result in significant environmental, social, and economic impacts in the oil-shale areas, including rapid urbanization to house a large workforce, construction of industrial facilities, and increased demand on water resources.

Production of Oil

Production wells in a conventional oil field recover oil through both primary and enhanced methods. *Primary production* involves simply pumping the oil from wells, but this method can recover only about 25% of the petroleum in the reservoir. To increase the amount of oil recovered to about 60%, enhanced methods are used. In *enhanced* recovery, steam, water, or chemicals, such as carbon dioxide or nitrogen gas, are injected into the oil reservoir to push the oil toward the wells, where it can be more easily recovered by pumping.

Production of oil from continuous oil deposits is more complicated and expensive than production of conventional oil. The potential resource that may become reserves in the future is much greater than 2 trillion barrels. For example, the Bakken formation in North Dakota, extending into Montana and Canada, has a total resource of about 500 billion barrels, but what can be obtained now is thought to be as low as 4 billion barrels (the amount the world uses in about 45 days). Recent estimates suggest that as much as 50%, or about 250 billion barrels, may eventually be produced from the Bakken shale. If that proves to be the case, then oil in the Bakken will be very significant in terms of global oil productivity.[1]

Next to water, oil is the most abundant fluid in the upper part of the Earth's crust. Most of the known, proven oil reserves, however, are in a few fields. *Proven* oil reserves are the part of the total resource that has been identified and can be extracted now at a profit. Of the total reserves, 48% are in the Middle East, 20% in South and Central America, 13% in North America, 8.5% in Europe and Eurasia, 8% in Africa, and 2.5% in Asia-Pacific.[9]

The total resource of conventional oil always exceeds known reserves; it includes oil that cannot be extracted at a profit and oil that is suspected but not proved to be present. Several decades ago, the amount of oil that ultimately could be recovered (the total resource) was estimated to be about 1.6 trillion barrels. Today, that estimate is just over 2 trillion barrels.[9] The increases in proven reserves of oil in the last few decades have primarily been due to discoveries of conventional oil in the Middle East, Venezuela, Brazil, Kazakhstan, and discoveries of continuous oil in the United States and other areas.

Oil in the 21st Century

A decade ago estimates of proven oil reserves suggested that at present production rates oil and natural gas would last only a few decades.[10,11] The important question, however, is not how long oil is likely to last at present and future production rates, but when we will reach peak production. This consideration is important because, following peak production, less oil will be available, leading to shortages and price shocks. Until recently, it was thought that a peak in production would likely occur between 2020 and 2050, within the lifetime of many people living today.[12] Even those who think peak oil production in the near future is a myth acknowledge that the peak is coming and that we need to be prepared.[10] Whichever projections are correct, there is a finite amount of time left in which to adjust to potential changes in lifestyle and economies in a post-petroleum era. We will never entirely run out of crude oil in the sense that at least smaller amounts will remain, though they may not economically be worth obtaining for large-scale use as fuels. Petroleum in the future will remain an important commodity because it is a precursor to many useful chemicals, including many plastics.

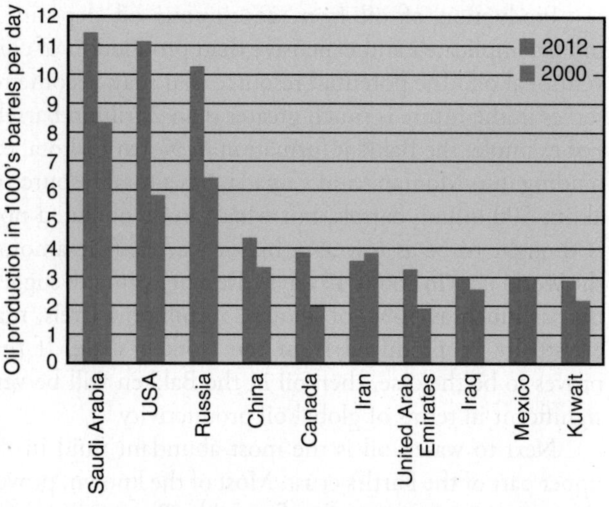

Data from U.S. Energy Information Administration for 2000 and 2012

FIGURE 15.5 The World's top 10 producers of oil in 2000 and 2012. The U.S. is now one of the two top producers.

There is an argument that oil should be saved for other purposes than burning it in our automobiles. Nevertheless, people of the world today depend on oil for nearly 33% of their energy, and significant shortages will cause major problems.[1,9]

The world production of oil in 2000 and 2012 and is shown in Figure 15.5. In 2000 U.S. production was about 25% less that the world leader Saudi Arabia. Notice that U.S. production in 2012 is just below Saudi Arabia, with Russia not far behind.[1] The U.S. increase in oil production is due in large part to development of continuous oil deposits in places such as North Dakota. Forecasts of energy production are risky because prices may change due to political and economic factors that are difficult to predict, and potential reserves may not materialize.[1] The U.S. may become the worlds top producer of oil in the next decade, but the Middle East Region total production will continue to significantly exceed that of the U.S.and the rest of north America.

What is an appropriate response to the likelihood that oil production will likely decline by the late 21st century? First, we need an improved educational program to inform or remind people and governments of the potential depletion of crude oil and the consequences of shortages. Presently, many people seem to be in denial. Planning and appropriate action are necessary to avoid military confrontation (we have already had one oil war), food shortages (oil is used to make the fertilizers modern agriculture depends on), and social disruption. Before significant oil shortages occur, we need to develop alternative energy sources, such as solar energy and wind power, and perhaps rely more on nuclear energy. This is a proactive response to a potentially serious situation.

15.3 Natural Gas

Natural gas is a mixture of hydrocarbon gases. The most common gas is methane (CH_4), but natural gas also includes propane (C_3H_8) and butane (C_4H_{10}). We have only begun to seriously search for natural gas and to utilize the natural gas resource to its full potential. One reason for the slow start is that natural gas is transported primarily by pipelines, and only in the last few decades have these pipelines been constructed in large numbers. In fact, until recently, natural gas found with petroleum was often simply burned off as waste; in some cases this practice continues.[8]

New supplies of natural gas are being found in large amounts in the United States and other locations. Optimistic estimates of the total resource suggest that, at current rates of consumption, the global supply may last more than 100 years. Today, over half of all natural gas produced in the United States is from wells drilled since about 2005.

This possibility has important implications. Natural gas is considered a clean fuel; burning it produces fewer pollutants than does burning oil or coal, so it causes fewer environmental problems than do the other fossil fuels. As a result, it is being considered as a possible transition fuel from other fossil fuels (oil and coal) to alternative energy sources, such as solar power, wind power, and ocean power.

There are several types of natural gas resources, including conventional gas fields often associated with drilling for oil. Another group of natural gas resources are unconventional or what is known as continuous gas resources. These include, in order of abundance and probable future use: shale gas, tight gas, and coalbed methane. The final natural gas resource we will briefly discuss is methane hydrates. Figure 15.6 is a schematic diagram illustrating conventional and continuous natural gas deposits that are being exploited today. U.S. natural gas production of some of these sources is shown in Figure 15.7. As of 2010, tight gas was the dominant supplier of natural gas, with other sources and shale gas not far behind. If we look forward to 2035, it is projected that nearly 50% of the total gas resources that are utilized will come from shale gas. Notice that natural gas from Alaska is a very small percentage of the total gas production, but that there is significant production from offshore and inshore sources that are not associated with oil. By 2035, it is expected that coalbed methane, as well as gas associated with oil fields and other sources that are not associated with oil fields, will be relatively small players, compared to shale gas and tight gas. Whether this expectation materializes will depend on economic factors and how readily shale gas may be obtained from the many potential sources in the United States. The Marcellus shale and the Devonian shale below the Marcellus is only one of 19 other basins being explored for shale gas from Michigan to Texas, and the resource is vast. Should all the different sources become available, the United States will have sufficient natural gas for at least a century. However, resources are not reserves, and projections may change in

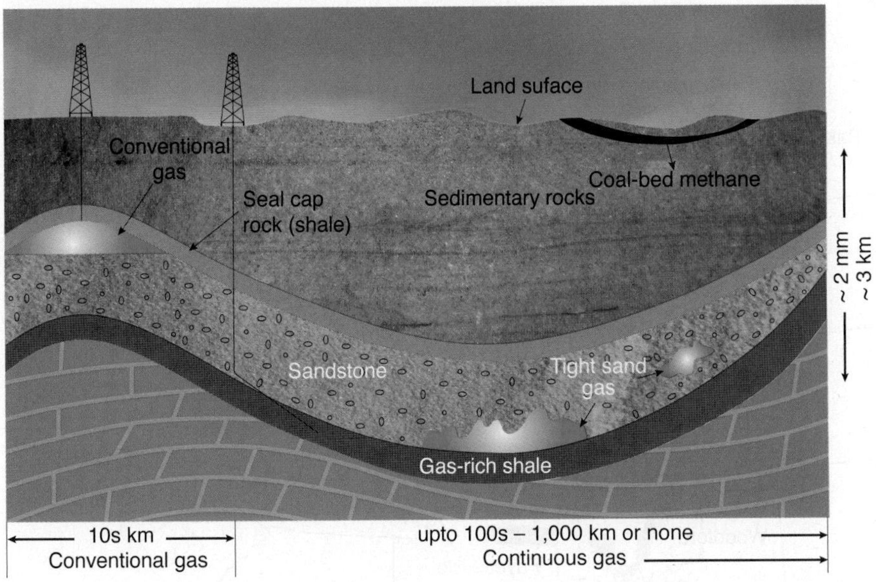

FIGURE 15.6 Conventional and continuous natural gas deposits (reservoirs).

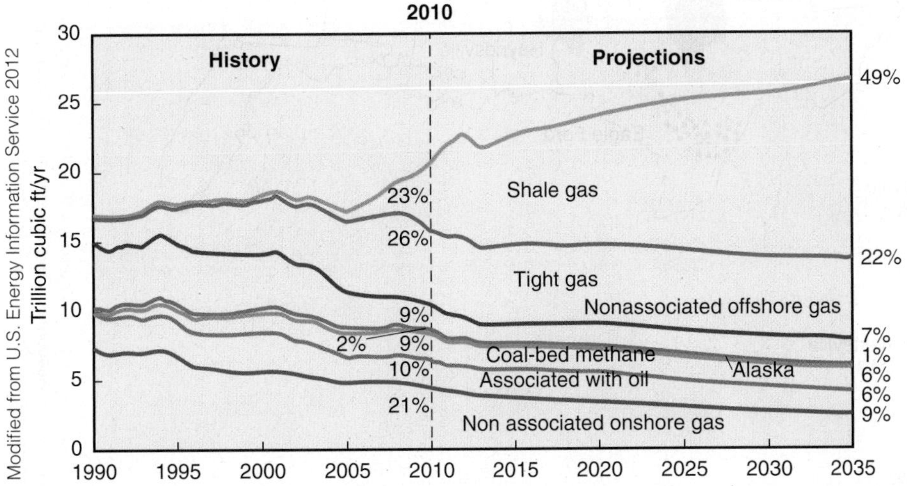

FIGURE 15.7 U.S. natural gas production, 1990–2035. Nonassociated gas is not associated with oil. One trillion cubic ft (tcf) per year is about 28.3 million cu m per year.

the future, depending on a variety of circumstances related to the pricing of natural gas and to the cost of obtaining it, including environmental costs.

Shale Gas

Shale gas is natural gas within tiny openings of shale rock. Shale has a lot of open space between grains, but fluids are held tightly.[13] According to the U.S. Geological Survey, Black Devonian shale (Marcellus shale) is over 350 million years old and buried a kilometer or so beneath northern Appalachia. The Marcellus shale resource that may be ultimately recovered is about 13.5 trillion cu m (482 million cu ft) of natural gas (mostly methane). At present consumption, this resource could supply all U.S. demand for natural gas for about 20 years. Of course, there are other sources of shale gas in other parts of the United States. The truth is that we are in the early stage of evaluating the shale gas resource and reserves, and numbers will change as more is known about the geology.

Methane is distributed throughout the black shale as a continuous gas resource found in geologic basins in parts of Ohio, New York, Pennsylvania, Virginia, and Kentucky as well as other areas in the United States (Figure 15.8). Recovery of the methane is costly, because deep wells that turn at depth to a horizontal position are necessary to extract the gas. Water and other chemicals are used to fracture the rocks **(hydrologic fracturing,** sometimes called **fracking)** to recover the gas. An energy rush is underway to develop the recovery of natural gas. Hundreds of gas wells have already been permitted in Pennsylvania alone. There is concern that drilling and hydrologic fracturing could result in water pollution because the fluids used to fracture the rock must be recovered from wells and disposed of before gas production starts (see A Closer Look 15.1). Yet another concern is that contaminated water could migrate upward and leak from wells to pollute water supplies. The city of New York is very concerned that drilling might contaminate the water supply in upstate New York.[14]

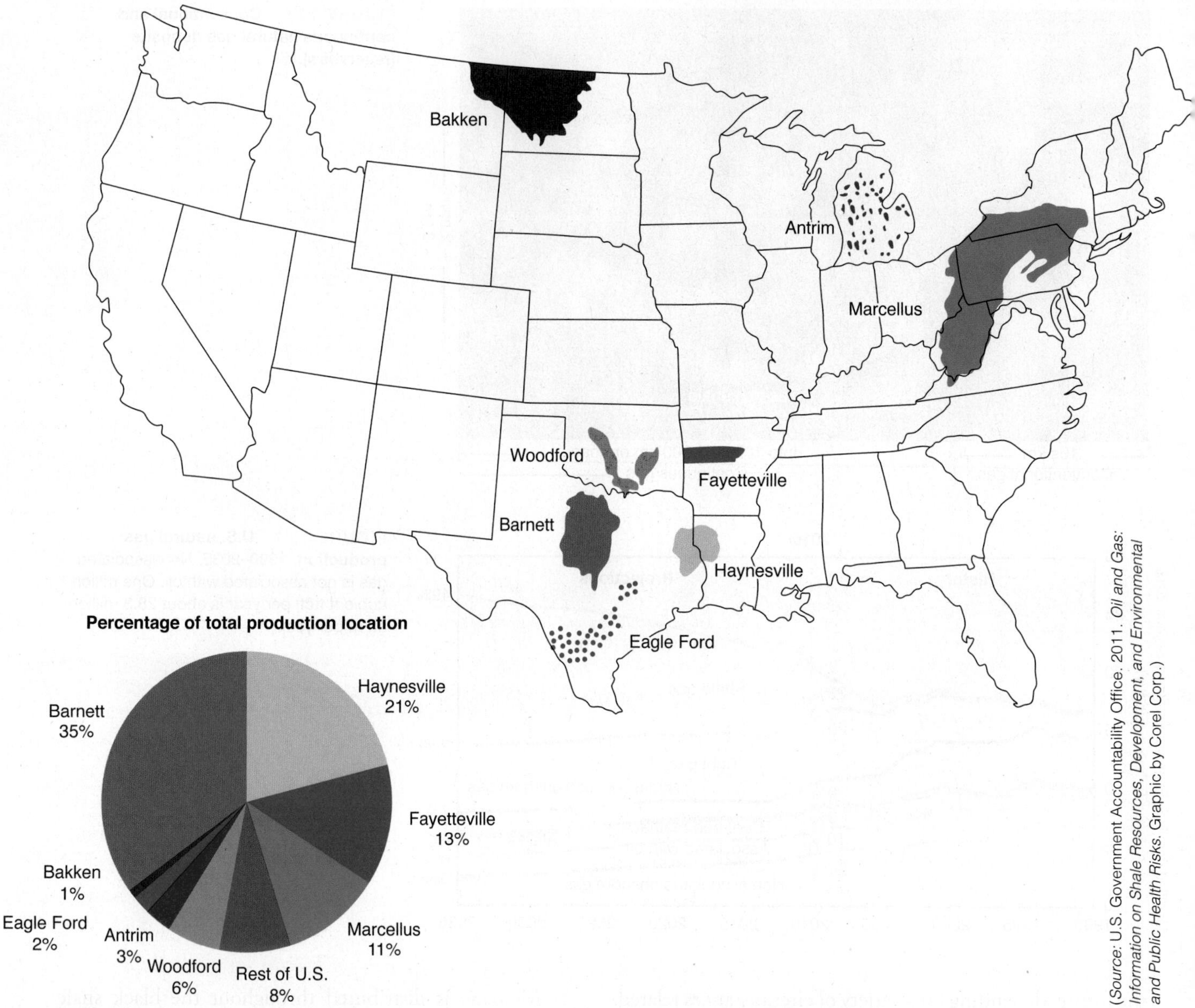

(Source: U.S. Government Accountability Office. 2011. *Oil and Gas: Information on Shale Resources, Development, and Environmental and Public Health Risks*. Graphic by Corel Corp.)

FIGURE 15.8 Shale gas plays in the United States. Production of shale gas has increased at a rapid pace since 2007.

A CLOSER LOOK 15.1

Hydrologic Fracturing To Recover Oil And Gas

Vast regions of the United States are underlain by very tight, fine-grained sedimentary rocks, such as shale and sandstone, in geologic basins at depths often of several kilometers. These rocks often contain large amounts of fossil fuels—particularly gas and oil—that is tightly held by the fine grains of the shale and is difficult to extract. In order to obtain these resources, the drilling extends deep and then turns horizontally in the target rocks that contain the oil and gas (Figure 15.9). Large amounts of fluids (sometimes millions of gallons, several thousand cubic meters) are injected under high pressure into a single well as the rock is fractured (sometimes multiple times). The fluids are mostly water (90%) but contain a wide variety of other chemicals to assist in the hydrologic fracturing and to keep the fractures open and so forth (see Table 15.1). A single well that is fractured several times may use several hundred thousand gallons (several hundred cubic meters) of fluid chemicals other than water. Many of these chemicals are known to create potential, if not outright, health risks (some are known to cause cancer) if they are released at the surface and people are exposed to the chemicals. There has also been concern that opening the fractures allows natural

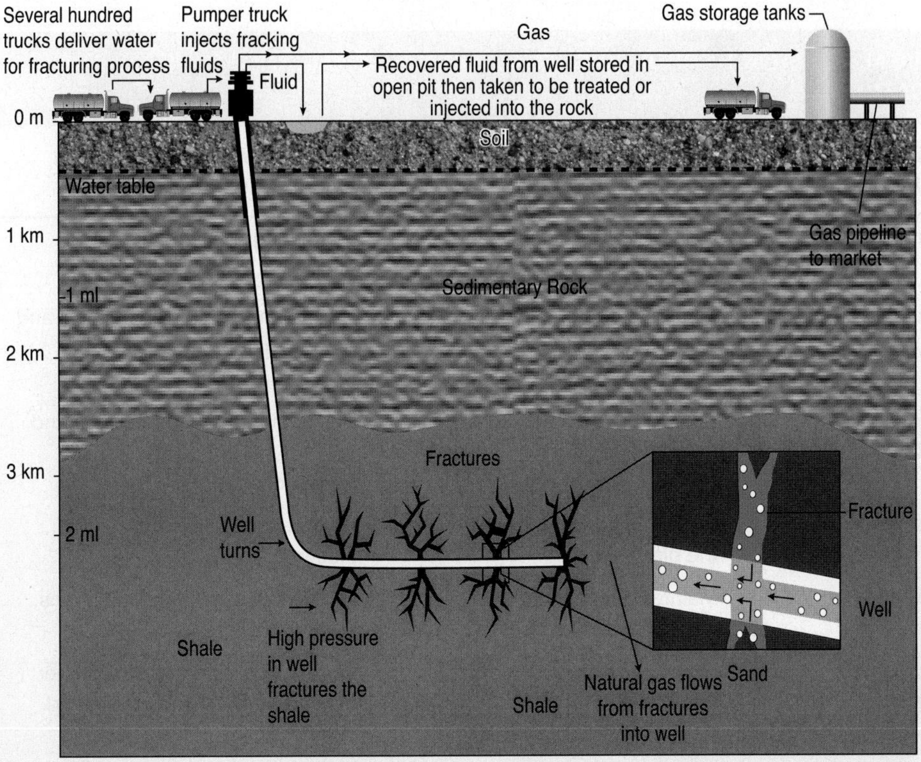

FIGURE 15.9 Idealized diagram of the process of hydraulic fracturing (fracking).

gas to seep up above the shale rock formations, perhaps to enter groundwater resources near the surface.

Natural gas migrating up through natural fractures or fractures from fracking deep below is probably not happening. However, natural gas sometimes is migrating up, perhaps through old, abandoned wells or failure of the wells that are drilled. Furthermore, the fluids that are brought up are sometimes stored at the surface in lined, open pits where they may leak into the environment or be flooded during high precipitation events, polluting rivers and the land. Figure 15.10 shows a drilling site and the associated pit that stores drilling fluids. In several instances, pollution from hydrologic fracturing has been recorded. The Environmental Protection Agency has recently studied this problem and is attempting to minimize potential problems through tighter regulation of open pits at the surface and methods of handling drilling and hydrologic fracturing fluids from the time they are pumped into the ground, returned to the surface, stored in tanks to be reused, or eventually eliminated.[a,b] There has also been concern that hydrologic fracturing may reactivate old faults and cause small earthquakes. High fluid pressure is known to favor earthquake occurrence. One small earthquake in Oklahoma is cited as an example of a hydrologic fracturing-induced earthquake. Many people who study earthquakes report that the pressures with hydrologic fracturing are not high enough to cause much of a change in regional seismicity or earthquakes in areas that have a low initial hazard. However, when fluids used in hydrologic fracturing are collected and disposed of in deep injection wells near faults, small earthquakes have occurred. This points to the need to understand the subsurface environment well before injecting fluids down specially designed disposal wells.[c]

At some locations in Pennsylvania, West Virginia, and other states, people are reporting methane seeping into their groundwater supply, with sufficient amounts of methane to cause fear of potential fires. Of course, methane has always been found in some areas, and there are many early records of people mentioning natural gas seeping from deep rock formations. However, there is concern that with the many wells being drilled, some of the casings near the surface of the wells may fail, allowing methane to seep into higher levels and contaminate the groundwater.

A tremendous amount of hydrologic fracturing is taking place in the United States in deep rock formations where oil and gas exist, and that trend is not likely to decrease in the future. Therefore, we need to develop adequate regulations to control recovery of the gas and oil as we move from the boom period into standard operating procedures of obtaining the resource. In the summer of 2012, the U.S. Environmental Protection Agency announced some new regulations aimed at reducing the adverse environmental effects of hydrologic fracturing of gas and oil wells. If the new regulations can be followed responsibly, oil and gas reserves in the United States will likely continue to increase.

MLADEN ANTONOV/AFP/Getty Images/Newscom

FIGURE 15.10 Natural gas well site in Pennsylvania.

Table 15.1 TYPICAL CHEMICALS USED IN HYDROLOGIC FRACTURING—"FRACKING" (NOT ALL ARE USED IN A SINGLE APPLICATION)

ADDITIVE	MAIN COMPOUND(S)	PURPOSE	COMMON USE OF MAIN COMPOUND
Diluted acid	Hydrochloric acid or muriatic acid	Help dissolve minerals and initiate cracks in the rock	Swimming pool chemical and cleaner
Biocide	Glutaraldehyde	Eliminate bacteria in the water that produce corrosive by-products	Disinfectant; sterilizes medical and dental equipment
Breaker	Ammonium persulfate	Allow a delayed breakdown of the gel polymer chains	Bleaching agent in detergent and hair cosmetics, production of household plastics
Corrosion inhibitor	N,n-dimethyl formamide	Prevent the corrosion of the pipe	Pharmaceuticals, acrylic fibers, plastics
Crosslinker	Borate salts	Maintain fluid viscosity as temperature increases	Laundry detergents, hand soaps, and cosmetics
Friction reducer	Polyacrylamide Mineral oil	Minimize friction between the fluid and the pipe	Water treatment, soil conditioner makeup remover, laxatives, and candy
Gel	Guar gum or hydroxyethyl cellulose	Thicken the water in order to suspend the sand	Cosmetics, toothpaste, sauces, baked goods, ice cream
Iron control	Citric acid	Prevent precipitation of metal oxides	Food additive, flavoring in food and beverages; lemon juice ~7% citric acid
KCl	Potassium chloride	Create a brine carrier fluid	Low-sodium table salt substitute
Oxygen scavenger	Ammonium bisulfite	Remove oxygen from the water to protect the pipe from corrosion	Cosmetics, food and beverage processing, water treatment
pH Adjusting agent	Sodium or potassium carbonate	Help maintain the effectiveness of other components	Washing soda, detergents, soap, water softener, glass and ceramics
Proppant	Silica, quartz sand	Allow the fractures to remain open so the gas can escape	Drinking water filtration, play sand, and concrete
Scale inhibitor	Ethylene glycol	Prevent scale deposits in the pipe	Automotive antifreeze, household cleansers, and deicing agent
Surfactant	Isopropanol	Increase the viscosity of the fracture fluid	Glass cleaner, antiperspirant, and hair color

Note: The specific compounds used in a given fracturing operation vary depending on source water quality and site-specific characteristics of the rock being fractured. The compounds shown above are representative of the major compounds used in hydraulic fracturing of tight shale.

Source: U.S. Department of Energy. 2009. *Modern Shale Gas: A Primer*. Washington, DC.

Tight Gas

Tight gas is natural gas that is produced from continuous deposits (reservoirs) of dense sandstone or limestone. The gas is originally produced in organic rich sediments and over geologic time has migrated to the reservoir rock. The gas is held tightly (hence the name tight gas). Many gas recovery wells in tight gas rock reservoirs are drilled horizontally at depth and are hydrologically fractured to enhance production (see A Closer Look 15.1). The tight gas resource in the United States is very large and supplies about 25% of all the natural gas produced in the country today.

Coal-bed Methane

The processes responsible for the formation of coal include partial decomposition of plants buried by sediments that slowly convert the organic material to coal. This process also releases a lot of methane (natural gas) that is stored within the coal. The methane is actually stored on the surfaces of the organic matter in the coal, and because coal has many large internal surfaces, the amount of methane for a given volume of rock is about seven times more than could be stored in gas reservoirs associated with petroleum. The estimated amount of coal-bed methane in the United States is more than 20 trillion cu m, of which about 3 trillion cu m could be recovered economically today with existing technology. At current rates of consumption in the United States, this represents about a five-year supply of methane.[15] Coal-bed methane constitutes about 9% of the U.S. natural gas production (see Figure 15.7).

Two areas within the nation's coalfields that are producing methane are the Wasatch Plateau in Utah and the Powder River Basin in Wyoming. The Powder River Basin is one of the world's largest coal basins and is experiencing an energy boom. The technology to recover coal-bed methane is a young one, but it is developing quickly. An advantage of coal-bed methane wells is that they only need to be drilled to shallow depths (about 100 m, or a few hundred feet). Drilling can be done with conventional water-well technology, and the cost is about $100,000 per well, compared to several million dollars for an oil well.[16]

The coal industry promotes the idea that coal-bed methane is a promising energy source that comes at a time when the United States is importing vast amounts of energy and attempting to evaluate a transition from fossil fuels to alternative fuels. However, coal-bed methane presents several serious environmental concerns, including: (1) disposal of large volumes of water produced when the methane is recovered and (2) migration of methane, which may contaminate groundwater or migrate into residential areas. With water becoming a limiting resource, this source of coal may not be as practical as some proponents suggest.

Of particular environmental concern in Wyoming is the safe disposal of salty water that is produced with the methane (the wells bring up a mixture of methane and water that contains dissolved salts from contact with subsurface rocks). Often, the water is reinjected into the subsurface, but in some instances the water flows into surface drainages or is placed in evaporation ponds.[15]

Some of the environmental conflicts that have arisen are between those producing methane from wells and ranchers trying to raise cattle on the same land. Frequently, the ranchers do not own the mineral rights, and although energy companies may pay fees for the well, the funds are not sufficient to cover damage resulting from producing the gas. The problem results when the salty water produced is disposed of in nearby streams. When ranchers use the surface water to irrigate crops for cattle, the salt damages the soil, reducing crop productivity. Although it has been argued that ranching is often a precarious economic venture and that ranchers have in fact been saved by new money from coal-bed methane, many ranchers oppose coal-bed methane production without an assurance that salty waters will be safely disposed of.

People are also concerned about the sustainability of water resources as vast amounts of water are removed from the groundwater aquifers. In some instances, springs have been reported to have dried up after coal-bed methane extraction in the area.[16] In other words, the "mining" of groundwater for coal-bed methane extraction will remove water that has perhaps taken hundreds of years to accumulate in the subsurface environment.

Another concern is the migration of methane away from the well sites, possibly to nearby urban areas. The problem is that, unlike the foul-smelling variety in homes, methane in its natural state is odorless as well as explosive. For example, in the 1970s an urban area near Gallette, Wyoming, had to be evacuated because methane was migrating into homes from nearby coal mines.

Finally, coal-bed methane wells, with their compressors and other equipment, have caused people living a few hundred meters away to report serious and distressing noise pollution.[16] In sum, coal-bed methane is a tremendous source of energy and relatively clean burning, but its extraction must be closely evaluated and studied to minimize environmental degradation.

Methane Hydrates

Beneath the seafloor, at depths of about 1,000 m, there exist deposits of **methane hydrate**, a white, ice-like compound made up of molecules of methane gas (CH_4), molecular "cages" of frozen water. The methane has formed as a result of microbial digestion of organic matter in the sediments of the seafloor and has become trapped in these ice cages. Methane hydrates in the oceans were discovered over 30 years ago and are widespread in both the Pacific and Atlantic oceans. Methane hydrates are also found on land; the first ones discovered were in permafrost areas of Siberia and North America, where they are known as marsh gas.[17]

Methane hydrates in the ocean occur where deep, cold seawater provides high pressure and low temperatures. They are not stable at lower pressure and warmer temperatures. At a water depth of less than about 500 m, methane hydrates decompose rapidly, freeing methane gas from the ice cages to move up as a flow of methane bubbles (like rising helium balloons) to the surface and the atmosphere.

In 1998, researchers from Russia discovered the release of methane hydrates off the coast of Norway. During the release, scientists documented plumes of methane gas as tall as 500 m being emitted from methane hydrate deposits on the seafloor. It appears that there have been many large emissions of methane from the sea. The physical evidence includes fields of depressions, looking something like bomb craters that pockmark the seafloor near methane hydrate deposits. Some of the craters are as large as 30 m deep and 700 m in diameter, suggesting that they were produced by rapid, if not explosive, eruptions of methane.

Methane hydrates in the marine environment are a potential energy resource with approximately twice as much energy as all the known natural gas, oil, and coal deposits on Earth.[16] Methane hydrates are particularly attractive to countries such as Japan that rely exclusively on foreign oil and coal for their fossil fuel needs. Unfortunately, mining methane hydrates will be a difficult task, at least for the near future. The hydrates tend to be found along the lower parts of the continental slopes, where water is often deeper than 1 km. The deposits themselves extend into the seafloor sediments another few hundred meters. Drilling rigs have more problems operating safely at these depths, and developing a way to produce the gas and transport it to land will be challenging.

15.4 The Environmental Effects of Oil and Natural Gas

Recovering, refining, and using oil—and to a lesser extent natural gas—cause well-known, documented environmental problems, such as air and water pollution, acid rain, and global warming. People have benefited in many ways from abundant, inexpensive energy, but at a price to the global environment and human health. The following discussion applies to conventional oil and gas, oil and gas in tight shale, and tight gas. Environmental concerns of oil shale, tar sand, and coal-bed methane were discussed with those resources where they were defined in Sections 15.2 and 15.3.

Gamma-Keystone via Getty Images

(a) (b)

FIGURE 15.11 **Drilling for oil in (a)** the Sahara Desert of Algeria; **(b)** the Cook Inlet of southern Alaska.

Recovery

Development of oil and gas fields involves drilling wells on land or beneath the seafloor (Figure 15.11).

Possible environmental impacts on land include the following:

- Use of land to construct pads for wells, pipelines, and storage tanks and to build a network of roads and other production facilities. Forestlands are of particular concern. The state of Pennsylvania has determined that of the 61,000 ha (1,500,000 acres) of state forestland underlain by shale gas deposits, nearly half is currently leased or sold for gas development and the other half is not leased because it is in ecologically sensitive areas.
- Pollution of surface waters and groundwater from: (1) leaks from broken pipes or tanks containing oil or other oil-field chemicals; (2) salty water (brine) brought to the surface in large volumes with the oil; and (3) hydraulic fracturing concerns (see A Closer Look 15.1). (The brine in salty water brought to the surface is toxic and may be disposed of by evaporation in lined pits, which could leak. Pumping it into the ground is another disposal method, using deep wells outside the oil fields. However, disposal wells may pollute groundwater,)

- Accidental release of air pollutants, such as hydrocarbons and hydrogen sulfide (a toxic gas).
- Land subsidence (sinking) as oil and gas are withdrawn from shallow conventional oil and gas fields.
- Loss or disruption of and damage to fragile ecosystems, such as wetlands or the marine environment. This is the center of the controversy over the development of petroleum resources in pristine environments such as the U.S. Arctic Ocean offshore of Alaska, where vast conventional oil and gas reserves are thought to exist.[18] However, the onshore and offshore Arctic region is a remote and fragile environment with unique fish and wildlife reserves. There also are potential social concerns regarding indigenous people who rely on the environment to maintain their way of life. As a result, Arctic oil and gas production remains controversial.

Environmental impacts associated with oil production in the marine environment include the following:

- Oil seepage into the sea from normal operations or large spills from accidents, such as blowouts or pipe ruptures (Figure 15.12). The very serious oil spill in the Gulf of Mexico (April 20, 2010) began from a blowout when equipment designed to prevent a blowout for this well drilled in very deep water (over 1.5 km, 1 mi) failed to operate properly. The platform was destroyed by a large explosion, and 11 oil workers were killed. By the middle of May, oil started to make landfall in Louisiana and other areas, and fishing was shut down over a large area. The oil spread despite the many and continuing efforts (chemical dispersants, skimming, and burning, among others) to deter the oil from spreading. The spill was the largest and potentially most damaging in U.S. history (see A Closer Look 15.2).
- Release of drilling muds (heavy liquids injected into the borehole during drilling to keep the hole open). These contain heavy metals, such as barium, which may be toxic to marine life.
- Aesthetic degradation from the presence of offshore oil-drilling platforms, which some people consider unsightly.

FIGURE 15.12 A Louisiana wetland after the *Deepwater Horizon* oil spill in the Gulf of Mexico.

Refining

Refining crude oil and converting it to products also have environmental impacts. At refineries, crude oil is heated so that its components can be separated and collected (this process is called *fractional distillation*). Other industrial processes then make products such as gasoline and heating oil.

Refineries may have accidental spills and slow leaks of gasoline and other products from storage tanks and pipes. Over years of operation, large amounts of liquid hydrocarbons may be released, polluting soil and groundwater below the site. Massive groundwater-cleaning projects have been required at several West Coast refineries.

Crude oil and its distilled products are used to make fine oil, a wide variety of plastics, and organic chemicals used by society in huge amounts. The industrial processes involved in producing these chemicals have the potential to release a variety of pollutants into the environment.

Delivery and Use

Some of the most extensive and significant environmental problems associated with oil and gas occur when the fuel is delivered and consumed. Crude oil, as with natural gas, is mostly transported on land in pipelines or across the ocean by tankers; both methods present the danger of oil spills. For example, a bullet from a high-powered rifle punctured the Trans-Alaska Pipeline in 2001, causing a small but damaging oil spill. Strong earthquakes may pose a problem for pipelines in the future, but proper engineering can minimize earthquake hazard. The large 2002 Alaskan earthquake ruptured the ground by several meters where it crossed the Trans-Alaska Pipeline. The pipeline's design prevented damage to the pipeline and to the environment. Although most effects of oil spills are relatively short-lived (days to years), marine spills have killed thousands of seabirds, spoiled beaches for decades (especially beneath the surface of gravel beaches), and caused loss of tourist and fishing revenues.

A CLOSER LOOK 15.2

The Oil Spill in the Gulf of Mexico, 2010

America's biggest oil spill began on April 20, 2010, about 66 km (41 mi) south of the Louisiana coast in the Gulf of Mexico. Everything about the spill was big, very big (Figure 15.13). It happened on the *Deepwater Horizon,* a floating, semi-submerged drilling platform with a surface area larger than a football field—121 m (396 ft) long and 78 m (256 ft) wide. Built in 2001 at a cost of $600 million and owned by Transocean, the *Deepwater Horizon* had previously dug the deepest offshore gas and oil well ever—down 10,685 m (35,055 ft). In February 2010, Transocean began a new job under lease by British Petroleum (BP), drilling in waters 1,500 m (5,000 ft) deep. BP's plan was to use this platform to drill an exploratory well into the bedrock below to a depth of 5,600 m (18,360 ft)—almost 3½ mi into the rock! Pipes descended from the platform through the seawater to the bedrock below, and then drilling began. The wellhead, which sits atop the seafloor, contained devices to control the drilling, to insert drilling fluids (called muds) into the hole, and to control the upward flow of oil and gas once those deposits were reached.

On April 20, things went wrong. Methane (natural gas) from the oil and gas deposits that were being drilled into from the platform broke through the wellhead at the surface far below. It rose rapidly, reaching the platform in a short time, starting a fire there at 9:56 P.M. local time and then causing a major explosion. Eleven of the 126 crew members were killed, and many others were injured; some saved themselves by diving off the collapsing rig into the ocean. The fire was big—so big and bright that people in boats who came to help said it was hard to look at and that it melted the paint off the boats (see Figure 15.14).

The *Deepwater Horizon* burned for 36 hours and then, on April 22, it sank, after which the oil spill began in earnest. At first, the U.S. Coast Guard reported that 8,000 barrels a day were leaking, but it was difficult to determine just how much oil was pouring out thousands of feet below. By July, the best estimate was about 60,000 barrels per day. The oil spread widely; by mid-June 2010, medium to heavy amounts of oil had reached more than 160 km (100 mi) east of the platform's position (Figure 15.15).[a]

The total amount of oil spilled by mid-July when the leak was stopped was about 5 million barrels (210 mil gal). At this rate of release, the BP spill equaled the *Exxon Valdez* oil spill (until then the largest spill in U.S. history) every four

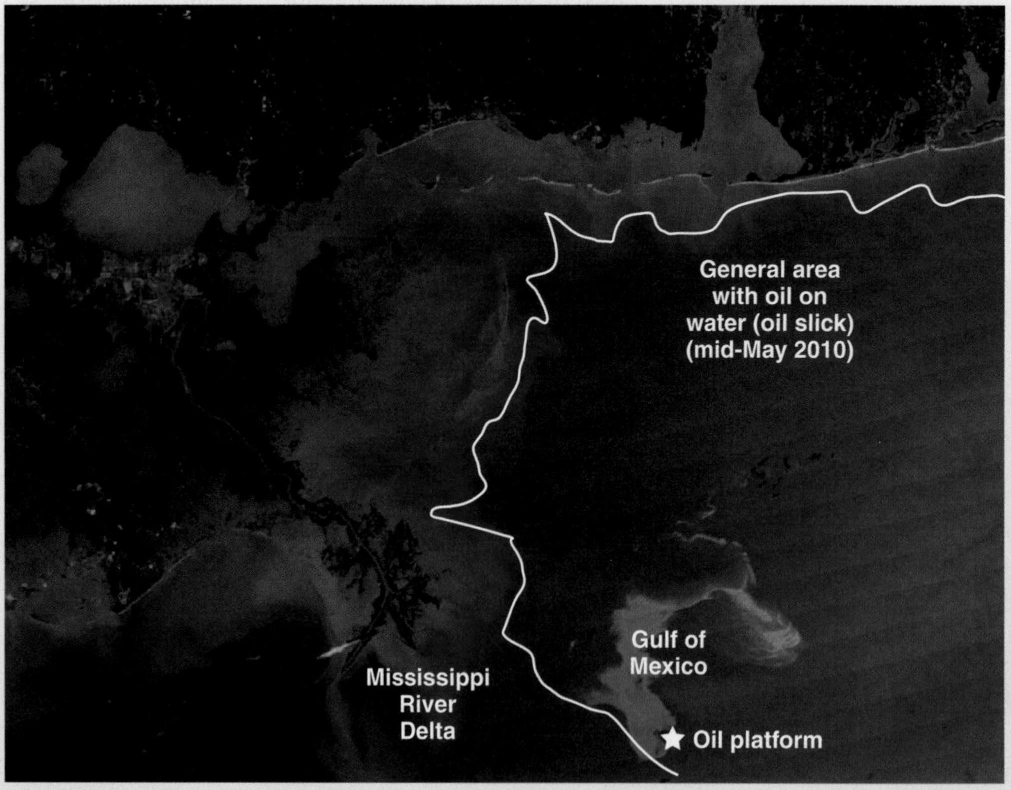

FIGURE 15.13 General area with oil on the water after the *Deepwater Horizon* oil spill in the Gulf of Mexico (mid-May 2010).

FIGURE 15.14 The *Deepwater Horizon* oil drilling platform on fire, April 22, 2010. Eleven workers were killed.

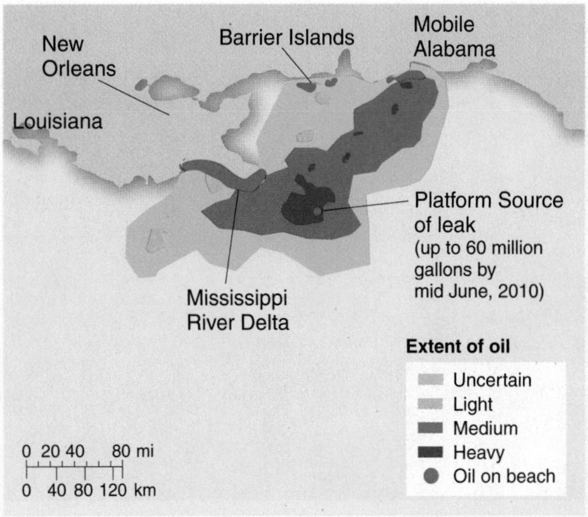

FIGURE 15.15 Extent of Gulf oil spill as of mid-June 2010.

and a half days. To put these large numbers in perspective, the average school gymnasium would hold about 1.3 million gallons of oil. Thus, the oil spilled in just the first two months of the BP spill would fill more than 100 school gymnasiums.

How does this compare to other blowouts and oil spills? The largest known blowout on land, which happened in Iran in 1956, involved about 120,000 barrels per day and lasted 3 months before being capped, releasing a total of almost 11 million barrels. A number of other "gushers" in the history of oil drilling released about 100,000 barrels per day.

Any way you look at it, the BP spill was a lot of oil released into the Gulf's fragile marine and coastal environments. Spilled oil that remains near the water surface moves with currents and winds, some of it ending up on shorelines, partly covering plants and animals, infiltrating the sediment, and doing other kinds of ecological damage (Figure 15.16). Although most effects of oil spills are relatively short-lived (days to a few years), previous marine spills have killed thousands of seabirds, temporarily spoiled beaches, and caused loss of tourist and fishing revenues. Complicating matters, four species of sea turtles (loggerheads, Kemp's ridley, leatherback, and green) that lay their eggs along Gulf State coasts were impacted by the oil. By June 25, 2010, 555 turtles had been found within the spill, and 417 of them were dead.[a]

A large volume of natural gas (methane) had been released with the oil, some dissolved in the oil and some from gas pockets. Eventually, much of the methane in the water was degraded by bacteria, whose increased respiration decreases oxygen levels in the water. Scientists studying the oxygen content of the deep water near the oil rig found oxygen depletion of 2% to 30% at depths of about 304m (1,000 ft.)

The spill was an economic disaster for BP, the oil industry in general, states along the Gulf Coast, and the people living there. If the oil that was spilled in the first 60 days had been obtained and sold, it would have provided BP with about $500 million (commercial value: 5 million barrels at $100 per barrel). By the end of the spill, BP had admitted that the spill had cost the company $1 billion, and it had agreed to provide the U.S. government with $20 billion to repay those who suffered damage from the spill. Oil drilling from platforms was curtailed after the spill, and oil production dropped (production rebounded in 2012 with improved safety). Commercial fishermen were put out of work because a large area of the Gulf's waters—from Morgan City west of New Orleans, Louisiana, to well east of Panama City, Florida, and south along the Florida Keys—were closed to fishing: an area approximately 480 km (300 mi) east–west and 480 km (300 mi) north–south. Some closed areas were opened by August 2010. People in tourist areas lost money because vacationers were choosing other locations where they wouldn't have to contend with oil on beaches or perception of an environmental problem. Supporting enterprises for commercial fishing and tourism also lost business. In late 2012, the U.S. Department of Justice brought criminal charges against BP and fined the corporation $4.5 billion. The two highest-ranking supervisors onboard *Deepwater Horizon* when the explosion occurred were charged with 23 criminal counts, including seaman's manslaughter. BP admitted guilt of 11counts of manslaughter and agreed to pay the $4.5 fine. Civil cases are now pending against BP that involve many billions of dollars in potential liability .

Will the Gulf recover from the 2010 oil spill? Certainly it will. Scientific studies of previous oil spills show that there is always an immediate (scientists call it an "acute") effect, killing fish,

NOAA

(a)

Patrick Semansky/AP/Wide World Photos

(b)

NOAA

(c)

NOAA

(d)

Win McNamee/Getty Images, Inc.

(e)

FIGURE 15.16 Oil on land. **(a)** Chandeleur Beach, Louisiana, 2010; **(b)** oil invades a Louisiana coastal wetland marsh in 2010; **(c)** dolphins swimming through some of the BP oil spill; **(d)** Kemp's ridley sea turtle at a rehabilitation center; **(e)** oil-covered seabird in Louisiana.

birds, and marine mammals and damaging vegetation and near-shore algae. Over the long run, the oil decomposes, and much of it becomes food for bacteria or nutrients for algae and plants. However, studies of previous oil spills also show that some oil remains even decades after spills, and, therefore, the effects on people, economics, and the environment that people value and enjoy are, from a human perspective, damaged for a long time.

By late August 2010, it was hard to find much floating oil in the Gulf, and some saltmarsh plants were showing signs of recovery. Recovery may be quicker in the Gulf than in Alaska because the warm water favors biologic decomposition of the oil, the oil is light, and the Gulf is a very large, deep, body of water subject to active surface processes from storm-generated wind and waves.

How did the BP spill happen? As with many major environmental disasters, a series of poor decisions were involved, including a failure to take advantage of the safest and best technology. Before the blowout, workers and others were aware of problems with the well, and they expressed concern about being able to prevent an incident in which oil or natural gas would escape. Indeed, the *Deepwater Horizon* had problems prior to the blowout and received 18 government citations for pollution. The BP wellhead had been fitted with a blowout preventer, but not with remote control or acoustically activated triggers for use in an emergency. The blowout preventer malfunctioned shortly after the heavy drilling mud had been withdrawn from the wellhead (the function of the drilling mud is to help keep oil from moving up the well to the surface). This is considered one of the major mistakes because without the mud and with the blowout-preventer malfunction, there was only water pressure to keep the oil and gas from escaping. And that was a recipe for disaster.

In addition, the response to the oil spill was inadequate. Rather than being proactive, the response was reactive: Each time something went wrong, there was spur-of-the-moment action. Also lacking was a clear line of authority and responsibility. The drilling was being done offshore by a private corporation, but it had large-scale effects, many of which were on government lands and waters and thus came under government control. Some available technologies that could have been applied were not; others were applied in too limited and tentative a way (Figure 15.17). News reports were rife with speculation by poorly informed people about all sorts of things that might be done, from gathering the oil with hay to blowing up the well with an atomic bomb.

About 6,000 boats were deployed with about 25,000 workers to try to minimize the spread of the spill by collecting it in the sea and on land. Some oil was burned, and chemical dispersants were applied from aircraft as well as at the bottom of the sea where the leak was occurring. These dispersants are chemicals and have environmental impacts themselves. It is known that these chemicals can damage marine ecosystems, but in this situation, some scientists considered dispersants the lesser of two evils. Dispersants were used, but their long-term impact, particularly on the deep-sea bed and in the seawater, is largely unknown. This brings up an important point: The science of the deep-ocean basin has not progressed enough to be able to adequately predict the processes there and how they will interact with the oil and dispersants.[a]

The *Deepwater Horizon* was just one of nearly 4,000 other platforms in the Gulf off the coast of the United States. From the perspective of environmental science, what lessons can we take home from the BP oil spill?

First of all, it did not have to happen. Best practices—those that take advantage of the best and safest modern technology, developed from modern science—were not followed. Second, modern industrialized nations use huge amounts of petroleum, and even with widespread movements away from petroleum, the need will not cease quickly. Therefore, it is essential that oil exploration and development make use of the best available technology and science, including the sciences that inform us about the environmental and ecological effects of an oil spill. And third, after decades of concern about offshore oil spills, the technologies to deal with their cleanup remain insufficient. What is needed is an oversight program that includes advance planning, early warning, and rapid and sufficient response. Given the huge amount of money spent on energy within the United States and the importance of energy to our nation's standard of living, creativity, and productivity, we can no longer deal with such things as oil spills in a haphazard way.

Win McNamee/Getty Images, Inc.

Spencer Platt/Getty Images, Inc.

(a) (b)

FIGURE 15.17 **Cleaning up 2010 Gulf Oil Spill: (a)** boats use red booms and skimmers to collect oil; **(b)** cleaning a Louisiana beach by hand.

15.5 Coal

Partially decomposed vegetation, when buried in a sedimentary environment, may be slowly transformed into the solid, brittle, carbonaceous rock we call **coal**. This process is shown in Figure 15.18. Coal is by far the world's most abundant fossil fuel, with a total recoverable resource of about 860 billion metric tons (Figure 15.19). The annual world consumption of coal is about 4 billion metric tons. Therefore there is sufficient coal for about 200 years at the current rate of use.[9] There are about 18,500 coal mines in the United States with combined reserves of 237 billion tons; 2011 production was 0.57 billion tons. At present rates of mining, U.S. reserves will last about 400 years. By comparison, China produces over three times as much coal as the United States, which is about one-half of total world production.[9,19] If, however, consumption of coal increases in the coming decades, the resource will not last nearly as long.

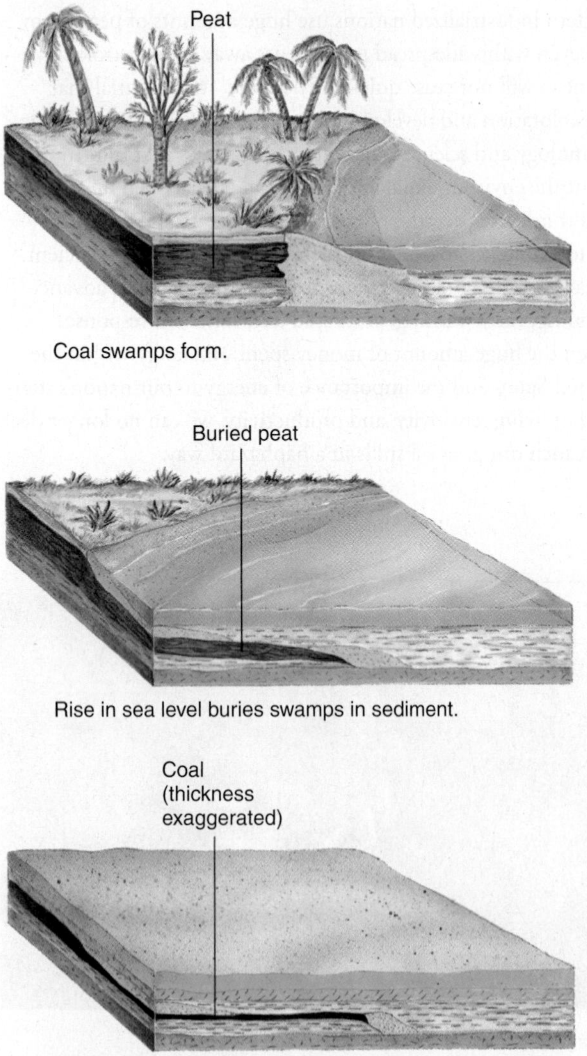
Peat
Coal swamps form.
Buried peat
Rise in sea level buries swamps in sediment.
Coal
(thickness
exaggerated)
Compression of peat forms coal.

FIGURE 15.18 Processes by which buried plant debris (peat) is transformed into coal.

Coal is classified, depending on its energy and sulfur content, as anthracite, bituminous, subbituminous, or lignite (see Table 15.2). The energy content is greatest in anthracite coal and lowest in lignite coal. The distribution of coal in the contiguous United States is shown in Figure 15.20.

The sulfur content of coal is important because low-sulfur coal emits less sulfur dioxide (SO_2) and is therefore more desirable as a fuel for power plants. Most low-sulfur coal in the United States is the relatively low-grade, low-energy lignite and subbituminous coal found west of the Mississippi River. Power plants on the East Coast treat the high-sulfur coal mined in their own region to lower its sulfur content before, during, or after combustion and, thus, avoid excessive air pollution. Although it is expensive, treating coal to reduce pollution may be more economical than transporting low-sulfur coal from the western states.

Coal Mining and the Environment

In the United States, coal mining has disturbed thousands of square kilometers of land, and only about half this land has been reclaimed. Reclamation is the process of restoring and improving disturbed land, often by reforming the surface and replanting vegetation (see Chapter 9). Unreclaimed coal dumps from open-pit mines are numerous and continue to cause environmental problems. Because little reclamation occurred before about 1960 and mining started much earlier, abandoned mines are common in the United States. One surface mine in Wyoming, abandoned more than 40 years ago, caused a disturbance so intense that vegetation has still not been reestablished on the waste dumps. Such barren, ruined landscapes emphasize the need for reclamation.

Strip Mining

Over half of the coal mining in the United States is done by *strip mining*, a surface mining process in which the overlying layer of soil and rock is stripped off to reach the coal (Figure 15.21). The practice of strip mining started in the late 19th century and has steadily increased because it tends to be cheaper and easier than underground mining. More than 40 billion metric tons of coal reserves are now accessible to surface mining techniques. In addition, approximately 90 billion metric tons of coal within 50 m (165 ft) of the surface are potentially available for strip mining. More and larger strip mines will likely be developed as the demand for coal increases.

The impact of large strip mines varies from region to region, depending on topography, climate, and reclamation practices. One serious problem in the eastern United States that gets abundant rainfall is *acid mine drainage*—the drainage of acidic water from mine sites (see Chapter 19). Acid mine drainage occurs when surface water (H_2O) infiltrates the spoil banks (rock debris left after the coal is

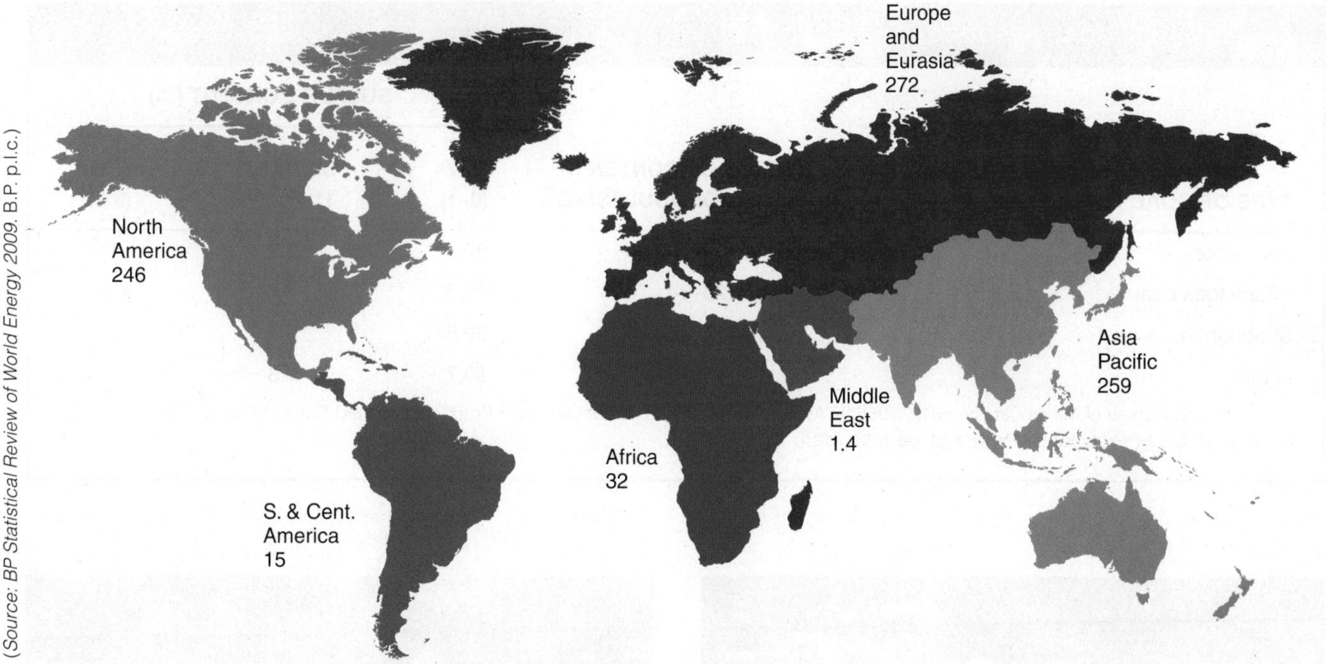

FIGURE 15.19 World coal reserves (billions of tons) in 2008.

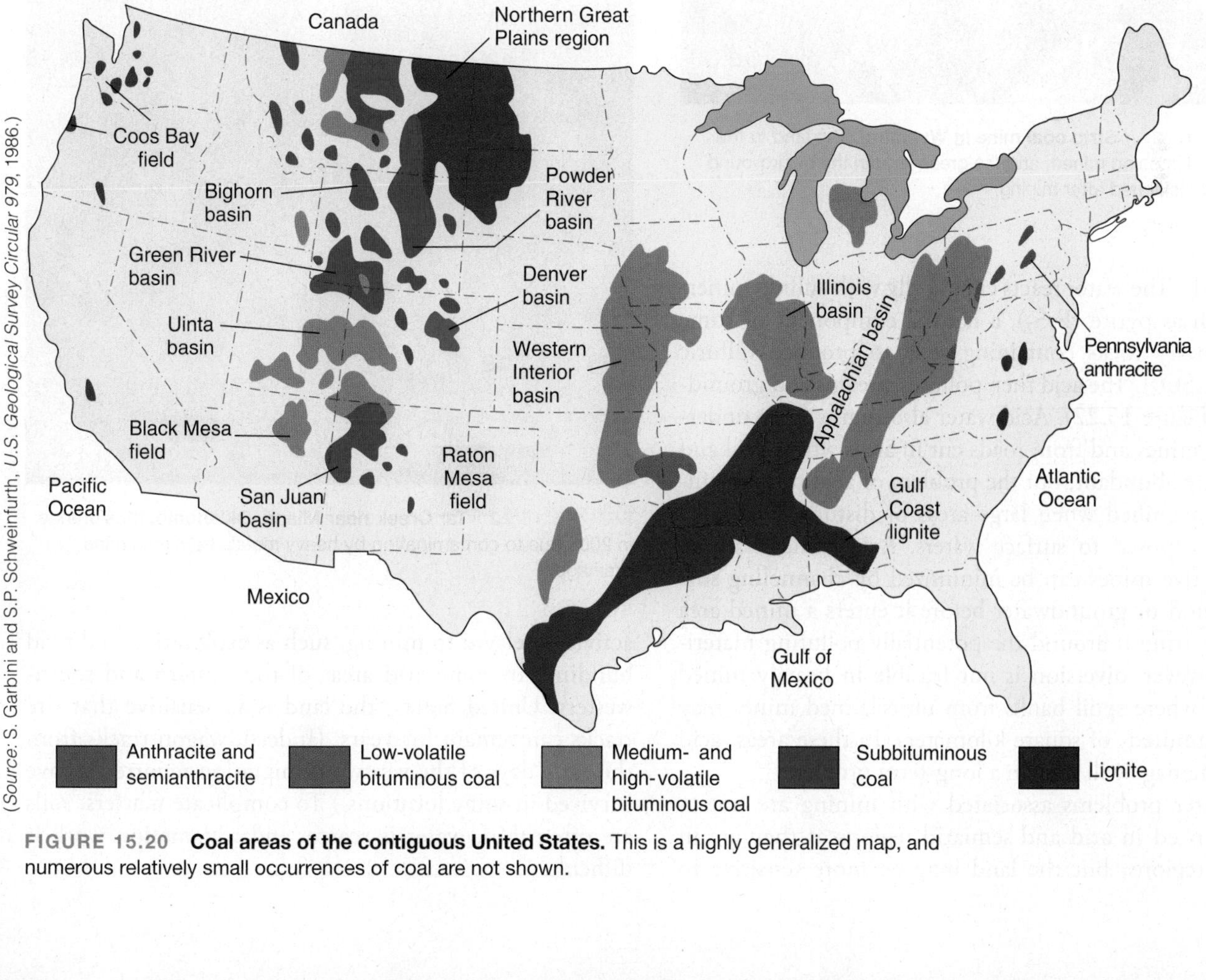

FIGURE 15.20 Coal areas of the contiguous United States. This is a highly generalized map, and numerous relatively small occurrences of coal are not shown.

Table 15.2 U.S. COAL RESOURCES

			SULFUR CONTENT (%)		
TYPE OF COAL	RELATIVE RANK	ENERGY OF CONTENT (MILLIONS OF JOULES/KG)	LOW (0–1)	MEDIUM (1.1–3.0)	HIGH (3 +)
Anthracite	1	30–34	97.1	2.9	–
Bituminous coal	2	23–34	29.8	26.8	43.4
Subbituminous coal	3	16–23	99.6	0.4	–
Lignite	4	13–16	90.7	9.3	–

Sources: U.S. Bureau of Mines Circular 8312, 1966: P. Averitt, "Coal" in D.A. Brobst and W.P Pratt, eds., United States Mineral Resources, *U.S. Geological Survey, Professional Paper* 820, pp. 133–142.

IndexStock/SuperStock

FIGURE 15.21 **Strip coal mine in Wyoming.** The land in the foreground is being mined, and the green land in the background has been reclaimed after mining.

©AP/Wide World Photos

FIGURE 15.22 **Tar Creek near Miami, Oklahoma,** runs orange in 2003 due to contamination by heavy metals from acid mine drainage.

removed). The water reacts chemically with sulfide minerals, such as pyrite (FeS_2), a natural component of some sedimentary rocks containing coal, to produce sulfuric acid (H_2SO_4). The acid then pollutes streams and groundwater (Figure 15.22). Acid water also drains from underground mines and from roads cut in areas where coal and pyrite are abundant, but the problem of acid mine drainage is magnified when large areas of disturbed material remain exposed to surface waters. Acid mine drainage from active mines can be minimized by channeling surface runoff or groundwater before it enters a mined area and diverting it around the potentially polluting materials. However, diversion is not feasible in heavily mined regions where spoil banks from unreclaimed mines may cover hundreds of square kilometers. In these areas, acid mine drainage will remain a long-term problem.

Water problems associated with mining are not as pronounced in arid and semiarid regions as they are in wetter regions, but the land may be more sensitive to activities related to mining, such as exploration and road building. In some arid areas of the western and southwestern United States, the land is so sensitive that tire tracks can remain for years. (Indeed, wagon tracks from the early days of the westward migration reportedly have survived in some locations.) To complicate matters, soils are often thin, water is scarce, and reclamation work is difficult.

Strip mining has the potential to pollute or damage water, land, and biological resources. However, good reclamation practices can minimize the damage. Reclamation practices required by law necessarily vary by site. For example, mining and reclamation of the Trapper Mine in Colorado (Figure 15.23) has been successful in part because there is sufficient precipitation (mostly snow) to reestablish vegetation without artificially applying water.

Large surface coal mining is almost always controversial. One of the most controversial has been the Black Mesa Mine in Arizona. The mine is in the Black Mesa area of the Hopi Reservation and was the only supplier of coal to the very large 1.5 MW Mohave Generating Station, a power plant at Laughlin, Nevada (144 km, or 90 mi, southeast of Las Vegas). The coal was delivered to the plant by a 440-km (275-mi) pipeline that transported slurry (crushed coal and water). The pipeline used over 1 billion gallons of water pumped from the ground per year—water that nourishes sacred springs and water for irrigation. Both the mine and the power plant suspended operation on December 31, 2005.

Mountaintop Removal

Coal mining in the Appalachian Mountains of West Virginia is a major component of the state's economy. However, there is growing environmental concern about a strip-mining technique known as mountaintop removal (Figure 15.24). This technique is very effective in obtaining coal as it levels the tops of mountains. But as mountaintops are destroyed, valleys are filled with waste rock and other mine waste, and the flood hazard increases as toxic wastewater is stored behind coal-waste sludge dams. Several hundred mountains have been destroyed, and by 2103 over 3,840 km (2,400 mi) of stream channels will likely have been damaged or destroyed.[20]

In October 2000, one of the worst environmental disasters in the history of mining in the Appalachian Mountains occurred in southeastern Kentucky. About 1 million cu m (250 mil gal) of toxic, thick black coal sludge, produced when coal is processed, was released into the environment. Part of the bottom of the impoundment (reservoir) where the sludge was being stored collapsed, allowing the sludge to enter an abandoned mine beneath the impoundment. The abandoned mine had openings to the surface, and sludge emerging from the mine flowed across people's yards and roads into a stream of the Big Sandy River drainage. About 100 km (65 mi) of stream was severely contaminated, killing several hundred thousand fish and other life in the stream.

Mountaintop removal also produces voluminous amounts of coal dust that settles on towns and fields, polluting the land and causing or exacerbating lung diseases, including asthma. Protests and complaints by communities in the path of mining were formerly ignored but are now getting more attention from state mining boards. As people become better educated about mining laws, they are more effective in confronting mining companies to get them to reduce potentially adverse consequences of mining. However, much more needs to be done.

Prof. Ed Keller

(a)

Prof. Ed Keller

(b)

FIGURE 15.23 **(a)** Mining an exposed coal bed at the Trapper Mine in Colorado; and **(b)** the land during restoration following mining. Topsoil (lower right) is spread prior to planting vegetation.

Mandel Ngan/Getty Images, Inc.

FIGURE 15.24 **Mountaintop mining in West Virginia** has been criticized as damaging to the environment as vegetation is removed, stream channels are filled with rock and sediment, and the land is changed forever.

Those in favor of mountaintop mining emphasize its value to the local and regional economy. They further argue that only the mountaintops are removed, leaving most of the mountain, with only the small headwater streams filled with mining debris. They go on to say that the mining, following reclamation, produces flat land for a variety of uses, such as urban development, in a region where flat land is mostly on floodplains with fewer potential uses.

Since the adoption of the Surface Mining Control and Reclamation Act of 1977, the U.S. government has required that mined land be restored to support its pre-mining use. The regulations also prohibit mining on prime agricultural land and give farmers and ranchers the opportunity to restrict or prohibit mining on their land, even if they do not own the mineral rights. Reclamation includes disposing of wastes, contouring the land, and re-planting vegetation.

Reclamation is often difficult and unlikely to be completely successful. In fact, some environmentalists argue that reclamation success stories are the exception and that strip mining should not be allowed in the semiarid south-western states because reclamation is uncertain in that fragile environment.

Underground Mining

Underground mining accounts for approximately 40% of the coal mined in the United States and poses special risks both for miners and for the environment. The dangers to miners have been well documented over the years in news stories, books, and films. Hazards include mine shaft collapses (cave-ins), explosions, fires, and respiratory illnesses, especially the well-known black lung disease, which is related to exposure to coal dust, which has killed or disabled many miners over the years.

Some of the environmental problems associated with underground mining include the following:

- Acid mine drainage and waste piles have polluted thousands of kilometers of streams (see Chapter 19).
- Land subsidence can occur over mines. Vertical subsidence occurs when the ground above a coal-mine tunnel collapses, often leaving a crater-shaped pit at the surface. Coal-mining areas in Pennsylvania and West Virginia, for example, are well known for serious subsidence problems. In recent years, a parking lot and crane collapsed into a hole over a coal mine in Scranton, Pennsylvania; and damage from subsidence caused condemnation of many buildings in Fairmont, West Virginia.
- Coal fires in underground mines, either naturally caused or deliberately set, may belch smoke and hazardous fumes, causing people in the vicinity to suffer from a variety of respiratory diseases. For example, in Centralia, Pennsylvania, a trash fire set in 1961 lit nearby underground coal seams on fire. They are still burning today and have turned Centralia into a ghost town.

Transporting Coal

Transporting coal from mining areas to large population centers where energy is needed is a significant environmental issue. Although coal can be converted at the production site to electricity, synthetic oil, or synthetic gas, these alternatives have their own problems. Power plants for converting coal to electricity require water for cooling, and in semiarid coal regions of the western United States, there may not be sufficient water. Furthermore, transmitting electricity over long distances is inefficient and expensive. Converting coal to synthetic oil or gas also requires a huge amount of water, and the process is expensive.[21,22]

Freight trains and coal-slurry pipelines (designed to transport pulverized coal mixed with water) are options to transport the coal itself over long distances. Trains are typically used, and will continue to be used, because they provide relatively low-cost transportation compared with the cost of constructing pipelines. The economic advantages of slurry pipelines are tenuous, especially in the western United States where large volumes of water needed to transport the slurry are not easily available. In summary, the mining and use of about 900 million tons of coal per year in the United States is perhaps the biggest single source of pollution in the country today. Figure 15.25 summarizes some of the impacts discussed above. Promoting increased use of coal without providing much tighter environmental control of pollutants will continue to degrade our air, water, and land resources.[23]

The Future of Coal

The burning of coal produces about 40% of the electricity used in the United States today. This is down from 50% in recent years. More U.S. power plants are switching from coal to natural gas as the fuel of choice for economic and environmental reasons.

Coal, compared to natural gas, is much more polluting. Emissions of nitrogen oxides are about five times as high, sulfur dioxide is 4,000 times as high, and particulates are 390 times as high for a unit energy output.

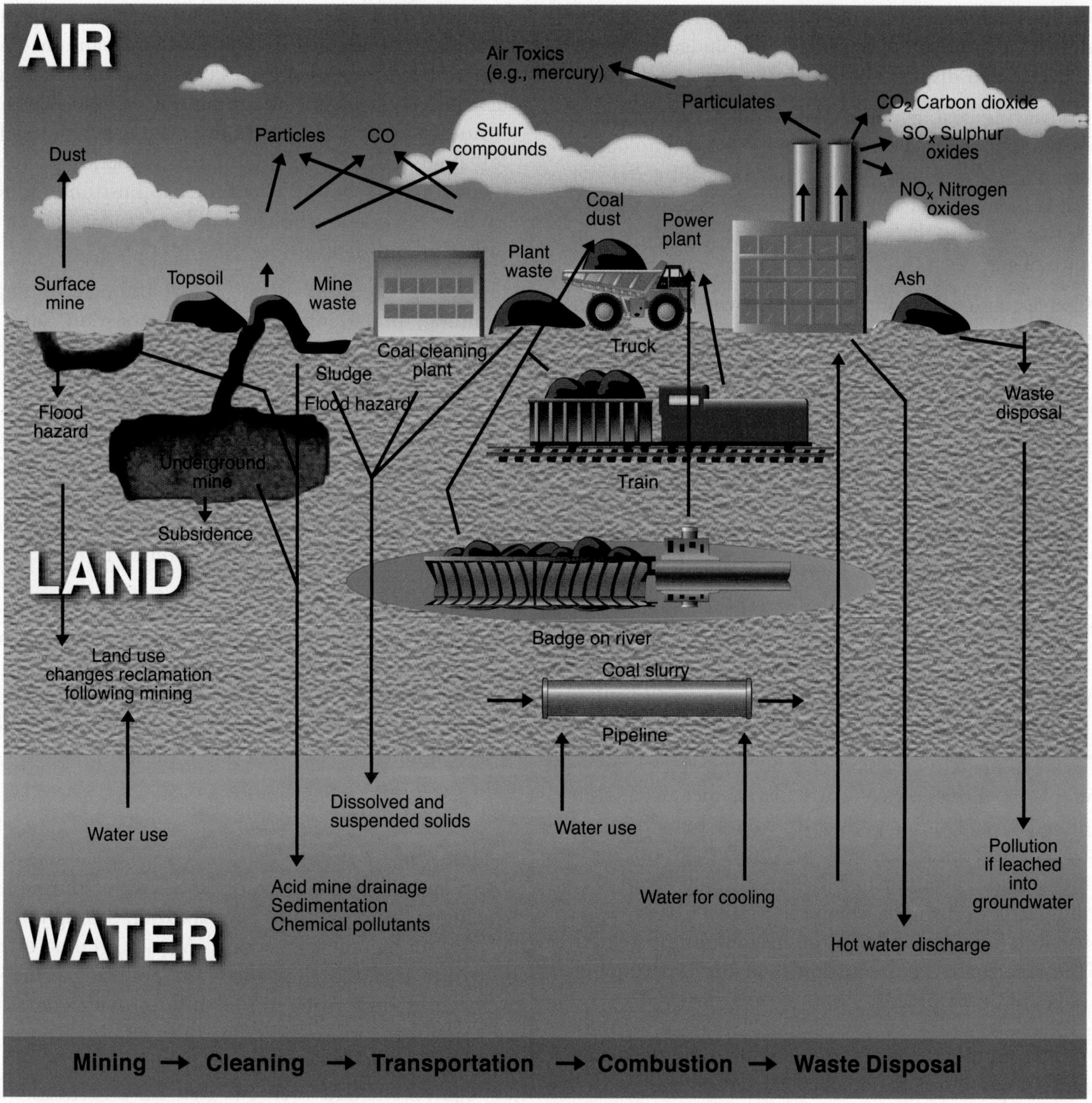

FIGURE 15.25 **Impacts from coal.** (*Source:* Modified after Clean Air Task Force 2011.)

Legislation as part of the Clean Air Amendments of 1990 mandated that sulfur dioxide emissions from coal-burning power plants be eventually cut by 70–90%, depending on the sulfur content of the coal, and that nitrogen oxide emissions be reduced by about 2 million metric tons per year. As a result of this legislation, utility companies that burn coal are struggling with various new technologies designed to reduce emissions of sulfur dioxide and nitrogen oxides from burning coal. Options being used or developed include the following:[22,24]

- Chemical and/or physical cleaning of coal prior to combustion.
- Producing new boiler designs that permit a lower temperature of combustion, reducing emissions of nitrogen oxides.
- Injecting material rich in calcium carbonate (such as pulverized limestone or lime) into the gases produced by the burning of coal. This practice, known as **scrubbing**, removes sulfur dioxides. In the scrubber—a large, expensive component of a power plant—the

carbonate reacts with sulfur dioxide, producing hydrated calcium sulfite as sludge. The sludge has to be collected and disposed of, which is a major problem.

- Converting coal at power plants into a gas (syngas, a methane-like gas) before burning. This technology is being tested and may become commercial by 2013 at the Polk Power Station in Florida. The syngas, though cleaner burning than coal, is still more polluting than natural gas.
- Converting coal to oil: We have known how to make oil (gasoline) from coal for decades. Until now, it has been thought to be too expensive. South Africa is performing this conversion now, producing over 150,000 barrels of oil per day from coal, and in 2009 China finished construction of a conversion plant in Mongolia. In the United States, we could produce 2.5 million barrels per day, which would require about 500 million tons of coal per year by 2020. There are environmental consequences, however, as superheating coal to produce oil generates a lot of carbon dioxide (CO_2), the major greenhouse gas.
- Educating consumers about energy conservation and efficiency to reduce the demand for energy and, thus, the amount of coal burned and emissions released.
- Developing zero-emission coal-burning electric power plants. Emissions of particulates, mercury, sulfur dioxides, and other pollutants would be eliminated by physical and chemical processes. Carbon dioxide would be eliminated by injecting it deep into the earth or using a chemical process to sequester it (tie it up) with calcium or magnesium as a solid. The concept of zero emission is in the experimental stages of development.

Coal mining and burning coal will continue to have significant environmental impacts for several reasons:

- More and more land will be stripmined and will therefore require careful and expensive restoration.
- Unlike oil and gas, burning coal produces large amounts of air pollutants. It also creates ash, which can be as much as 20% of the coal burned; boiler slag, a rock-like cinder produced in the furnace; and calcium sulfite sludge, formed when removing sulfur through scrubbing. Coal-burning power plants in the United States today produce about 90 million tons of these materials per year. Calcium sulfite from scrubbing can be used to make wallboard (by converting calcium sulfite to calcium sulfate, which is gypsum) and other products. Gypsum is being produced for wallboard in this way in Japan and Germany, but the United States can make wallboard less expensively from abundant natural gypsum deposits. Another waste product, boiler slag, can be used for fill along railroad tracks and at construction projects. Nevertheless, about 75% of the combustion products of burning coal in the United States today end up in waste piles or landfills.
- Handling large quantities of coal through all stages (mining, processing, shipping, combustion, and final disposal of ash) has adverse environmental effects. These include aesthetic degradation, noise, dust, and—most significant from a health standpoint—release of toxic or otherwise harmful trace elements into the water, soil, and air. For example, in late December 2008, the retaining structure of an ash pond at the Kingston Fossil Plant in Tennessee failed, releasing a flood of ash and water that destroyed several homes, ruptured a gas line, and polluted a river.[25]

All of these negative effects notwithstanding, it seems unlikely that the United States will abandon coal in the near future because we have so much of it and have spent so much time and money developing coal resources. Regardless, it remains a fact that coal is the most polluting of all the fossil fuels.

Allowance Trading

An approach to managing U.S. coal resources and reducing pollution is **allowance trading**, through which the Environmental Protection Agency grants utility companies tradable allowances for polluting: One allowance is good for one ton of sulfur dioxide emissions per year. In theory, some companies wouldn't need all their allowances because they use low-sulfur coal or new equipment and methods that have reduced their emissions. Their extra allowances could then be traded and sold by brokers to utility companies that are unable to stay within their allocated emission levels. The idea is to encourage competition in the utility industry and reduce overall pollution through economic market forces.[24]

Some environmentalists do not accept the concept of allowance trading. They argue that although buying and selling may be profitable to both parties in the transaction, it is less acceptable from an environmental viewpoint. They believe that companies should not be able to buy their way out of taking responsibility for pollution problems.

CRITICAL THINKING ISSUE

U.S. Oil Supply for the Future

In this exercise, we will critically think about the oil supply for the United States. You may start this exercise by revisiting Figure 15.5, which estimates U.S. production by the year. Figure 15.26 provides a broader view of production of energy (electricity) from fossil fuels from 1980 and projected to 2035. Notice that oil production in the United States may level off at about 2020, and that natural gas, which is about equal to coal production in 2020, may exceed the use of coal by 2035. In fact, in 2012 the use of coal decreased, and natural gas is increasing rapidly as the fuel of choice to produce electricity. The main reason for the tremendous increase in the use of natural gas is economic. The price of natural gas has dropped significantly, and it is perceived as being a much cleaner fuel. Coal companies and the industry in general continue to advocate for continued and greater use of coal and argue that it is becoming a cleaner fuel.

Critical Thinking Questions

1. Synthesize geological, political, and economic factors that will largely determine whether the projections in Figure 15.26 will, in fact, materialize.
2. Defend or criticize the statement that coal may become a much cleaner fuel and, therefore, will be used much more in the future for generating electricity, compared to other fossil fuels.
3. Defend the statement that the more and longer we use fossil fuels as the dominant fuel in the world, the greater the adverse impacts of that use will be in terms of air and water pollution and the greater the potential disruption for society as the increasing worldwide population competes for limited fossil fuels.

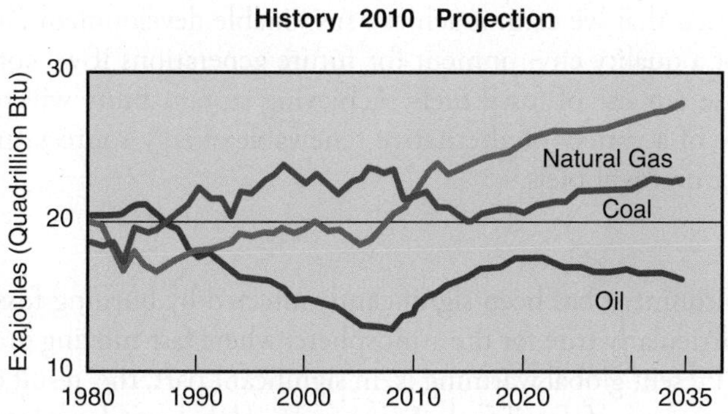

(*Source:* Data from U.S. Energy Information Administration—Annual Energy Outlook 2012.)

FIGURE 15.26 U.S. electrical energy production from fossil fuels, 1980 to 2035.

SUMMARY

- Fossil fuels are forms of stored solar energy. Most of these fuels are created from the incomplete biological decomposition of dead and buried organic material that is converted by complex chemical reactions in the geologic cycle.
- Because fossil fuels are nonrenewable, we will eventually have to develop other sources to meet our energy demands. We must decide when the transition to alternative fuels will occur and what the impacts of the transition will be.
- Environmental impacts related to oil and natural gas include those associated with exploration and development (damage to fragile ecosystems, water pollution, air pollution, and waste disposal); those associated with refining and processing (pollution of soil, water, and air); and those associated with burning oil and gas for energy to power automobiles, produce electricity, run industrial machinery, heat homes, and so on (air pollution).
- Today, the United States is undergoing a fossil fuel revolution as abundant new supplies of conventional and continuous oil and gas are being developed. The revolution will have profound economic and environmental consequences.

- Coal remains an important resource for producing electric power, but the new abundance of cleaner burning natural gas is likely to surpass coal as the fuel of choice in the coming decades.
- Coal is an energy source that is particularly damaging to the environment. The environmental impacts of mining, processing, transporting, and using coal are many. Mining coal can cause fires, subsidence, acid mine drainage, and difficulties related to land reclamation. Burning coal can release air pollutants, including sulfur dioxide and carbon dioxide, and it produces a large volume of combustion products and by-products, such as ash, slag, and calcium sulfite (from scrubbing).
- It has been argued that we cannot achieve sustainable development and maintenance of a quality environment for future generations if we continue to increase our use of fossil fuels. Achieving sustainability will require wider use of a variety of alternative renewable energy sources and less dependence on fossil fuels.

REEXAMINING THEMES AND ISSUES

Sean Randall/Getty Images, Inc.

HUMAN POPULATION

As the human population has increased, so has the total impact from the use of fossil fuels. Total impact is the impact per person times the total number of people. Reducing the impact will require continuing the present transition to less-polluting fossil fuels such as natural gas and eventual transition to alternative energy sources.

© Biletskiy_Evgeniy/iStockphoto

SUSTAINABILITY

It has been argued that we cannot achieve sustainable development and maintenance of a quality environment for future generations if we continue to increase our use of fossil fuels. Achieving sustainability will require wider use of a variety of alternative renewable energy sources and less dependence on fossil fuels.

© Anton Balazh 2011/iStockphoto

GLOBAL PERSPECTIVE

The global environment has been significantly affected by burning fossil fuels. This is particularly true for the atmosphere, where fast-moving processes operate. Present global warming is, in significant part, the result of burning huge amounts of fossil fuels. Solutions to global problems from burning fossil fuels are best implemented at the local and regional levels, where the fuels are consumed.

ssguy/ShutterStock

URBAN WORLD

The burning of fossil fuels in urban areas has a long history of problems. Not too many years ago, black soot from burning coal covered the buildings of most major cities of the world, and historical pollution events killed thousands of people. Today, we are striving to improve our urban environments and reduce urban environmental degradation from burning fossil fuels.

B2M Productions/Getty Images, Inc.

PEOPLE AND NATURE

Our exploration, extraction, and use of fossil fuels have changed nature in fundamental ways—from the composition of the atmosphere and the disturbance of coal mines to the pollution of ground and surface waters. Some people still buy giant SUVs supposedly to connect with nature, but using them causes more air pollution than do automobiles and, if used off-road, often degrades nature.

George Doyle/Getty Images, Inc.

SCIENCE AND VALUES

Scientific evidence of the adverse effects of burning fossil fuels is well documented. The controversy over their use is linked to our values. Do we value burning huge amounts of fossil fuels to increase economic growth more highly than we value living in a quality environment? Economic growth is possible without damaging the environment; developing a sustainable energy policy that doesn't harm the environment is possible with present technology. What is required are changes in values and lifestyle that are linked to energy production and use, human well-being, and environmental quality.

KEY TERMS

allowance trading 360
coal 354
continuous oil resources 338
conventional oil resources 338
crude oil 338
fossil fuels 338
hydrologic fracturing (fracking) 343
methane hydrate 348
natural gas 348
oil shale 340
peak oil 340
reserves 338
resources 338
scrubbing 359
shale gas 343
synfuels 340
tar sands 340
tight gas 347

STUDY QUESTIONS

1. What are the potential social, economic, and environmental consequences of the oil and gas revolution that is starting?
2. Compare the potential environmental consequences of burning oil, burning natural gas, and burning coal.
3. What actions can you personally take to reduce consumption of fossil fuels?
4. What environmental and economic problems could result from a rapid transition from coal to natural gas?
5. The transition from wood to fossil fuels took about 100 years. How long do you think the transition from fossil fuels to alternative energy sources will take? What will determine the time of transition?
6. What are some of the technical solutions to reducing air-pollutant emissions from burning coal? Which are best? Why?
7. Defend or criticize the statement that "coal is dead" or at least dying in terms of a fuel in the United States.
8. What are some of the ethical issues or questions associated with use of fossil fuels?

FURTHER READING

Boyl, G., B. Everett, and J. Ramage. 2003. ***Energy Systems and Sustainability.*** Oxford, UK: Oxford University Press. An excellent discussion of fossil fuel.

British Petroleum Company. 2012. ***B.P. Statistical Review of World Energy.*** London, UK: British Petroleum Company. Good, up-to-date statistics on fossil fuels.

Keating, M. 2011. ***Cradle to Grave: The Environmental Impacts from Coal. Clean Air Task Force.*** Boston, MA. Excellent summary of environmental impacts from coal.

Maugeri, L. 2012. ***Oil: The Next Revolution.*** Discussion Paper 2012-10. John F. Kennedy School of Government, Harvard University. Excellent discussion of the future of fossil fuels.

NOTES

1. Maugeri, L. 2012. *Oil: The Next Revolution.* Discussion Paper 2012-10. John F. Kennedy School of Government, Harvard University.
2. Van Koevering, T.E., and N.J. Sell. 1986. *Energy: A Conceptual Approach.* Englewood Cliffs, NJ: Prentice-Hall.
3. McCulloh, T.H. 1973. In D.A. Brobst and W.P. Pratt, eds., *Oil and Gas in United States Mineral Resources.* U.S. Geological Survey Professional Paper 820:477–496.
4. Kunzig, R. 2009. The Canadian oil boom. *National Geographic* 215(3):34–59.
5. Environmental Impact Statement. 2008. About the oil shale and tar sands programmatic EIS. http://ostseis.anl.gov/eis/index.cfm Accessed November 7, 2012.
6. Knapp, D.H. 1995. Non-OPEC oil supply continues to grow. *Oil & Gas Journal* 93:35–45.
7. Dyni, J.R. 2006. Geology and resources of some world oil-shale deposits. *U.S. Geological Survey Scientific Investigations Report 2005–5294.* Reston, VA .
8. Darmstadter, J., H.H. Landsberg, H.C. Morton, and M.J. Coda. 1983. *Energy Today and Tomorrow: Living with Uncertainty.* Englewood Cliffs, NJ: Prentice-Hall.
9. British Petroleum Company. 2012. *B.P. Statistical Review of World Energy 2008.* London, UK: British Petroleum Company.
10. Maugeri, L. 2004. Oil: Never cry wolf—when the petroleum age is far from over. *Science* 304:1114–1115.
11. Youngquist, W. 1998. Spending our great inheritance. Then what? *Geotimes* 43(7):24–27.
12. Edwards, J.D. 1997. Crude oil and alternative energy production forecast for the twenty-first century: The end of the hydrocarbon era. *American Association of Petroleum Geologists Bulletin* 81(8):1292–1305.
13. U.S. Department of Energy. 2009. *Modern Shale Gas: A Primer.* Washington, DC.
14. Milici, R.C., and C.S. Swezey. 2006. *Assessment of Appalachian Basin Oil and Gas Resources: Devonian Shale-Middle and Upper Paleozoic Total Petroleum System. U.S. Geological Survey Open File Report* 2006-1237. Reston, VA.
15. Nuccio, V. 2000. *Coal-bed Methane: Potential Environmental Concerns.* U.S. Geological Survey. USGS Fact Sheet. FS-123-00.
16. Wood, T. 2003 (February 2). Prosperity's brutal price. *Los Angeles Time Magazine.*
17. Suess, E., G. Bohrmann, J. Greinert, and E. Lauch. 1999. Flammable ice. *Scientific American* 28(5):76–83.
18. U.S. Geological Survey. 2008. *Circum-Arctic Resource Appraisal: Estimates of Undiscovered Oil and Gas North of the Arctic Circle.* USGS Fact Sheet 2008–3049.
19. U.S. Energy Information Administration 2012. *Annual Energy Outlook 2012.* http://www.eia.gov/forecasts/aeo/ Accessed November 7, 2012.
20. Stewart Burns, S. 2009 (September 30). Mountain top removal in Central Appalachia. *Southern Spaces.*
21. Webber, M.E. 2009. Coal-to-liquids: The good, the bad and the ugly. *Earth* 54(4):44–47.
22. Berlin Snell, M. 2007. Can coal be clean? *Sierra* 92(1):32–33.
23. Keating, M. 2011. *Cradle to Grave: The Environmental Impacts from Coal.* Boston, MA: Clean Air Task Force.
24. Corcoran, E. 1991. Cleaning up coal. *Scientific American* 264:106–116.
25. Environmental Protection Agency. 2009. EPAs response to the TVA Kingston Fossil Plant Fly ash release. http://www.epa.gov/region4/kingston Accessed November 7, 2012.

A CLOSER LOOK 15.1 NOTES

a. U.S. Department of Energy. 2009. *Modern Shale Gas: A Primer.* Washington, DC.

b. Maugeri, L. 2012. *Oil: The Next Revolution.* Discussion Paper 2012–10. John F. Kennedy School of Government, Harvard University.

c. Frohlich, C. 2012. Two-year survey comparing earthquake activity and injection-well locations in the Barnett Shale, Texas. Published online. *Proceedings of the National Academy of Sciences (PNAS).* Doi:10.1073/pnas.

A CLOSER LOOK 15.2 NOTES

a. NOAA Deepwater Horizon/BP Oil Spill Archive. http://response.restoration.noaa.gov Accessed November 16, 2012.

CHAPTER 16

Alternative Energy and the Environment

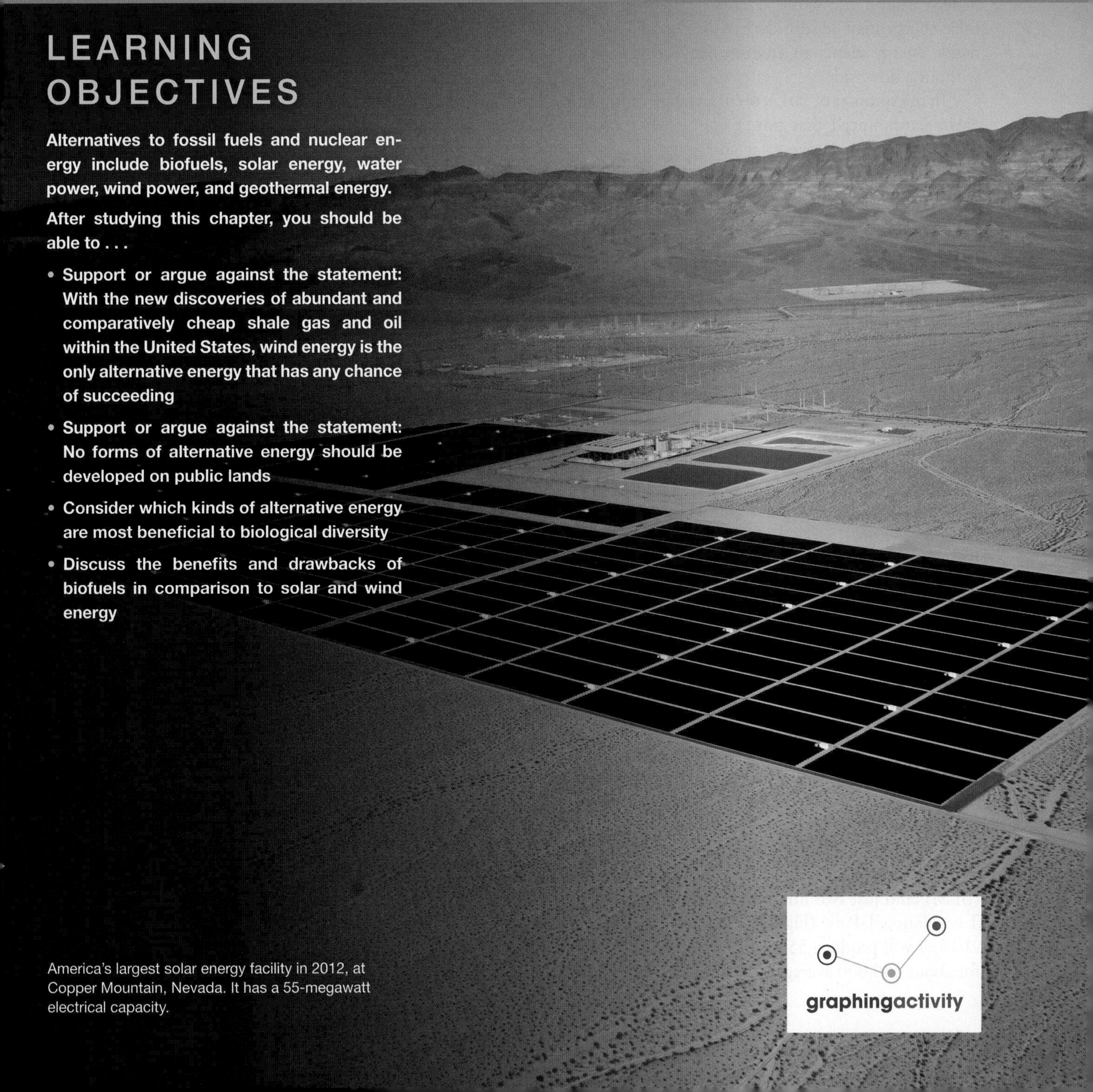

LEARNING OBJECTIVES

Alternatives to fossil fuels and nuclear energy include biofuels, solar energy, water power, wind power, and geothermal energy.

After studying this chapter, you should be able to . . .

- Support or argue against the statement: With the new discoveries of abundant and comparatively cheap shale gas and oil within the United States, wind energy is the only alternative energy that has any chance of succeeding
- Support or argue against the statement: No forms of alternative energy should be developed on public lands
- Consider which kinds of alternative energy are most beneficial to biological diversity
- Discuss the benefits and drawbacks of biofuels in comparison to solar and wind energy

America's largest solar energy facility in 2012, at Copper Mountain, Nevada. It has a 55-megawatt electrical capacity.

graphingactivity

CASE STUDY

Alternative Energy on Public Lands in The American Southwest: Good Idea or a Conflict Among Environmentalists?

Solar and wind energy are now proven sources of electricity, with costs dropping rapidly and installations being built widely around the world. Deutsche Bank projects that solar energy may be subsidy free as soon as 2014.[1] Alternative energy seems an environmental dream, a benefit without a downside.

In the summer of 2012, then Secretary of Interior Ken Salazar announced that power companies could submit plans for building large solar energy facilities on public lands administered by the Bureau of Land Management in six southwestern states: Arizona, California, Colorado, Nevada, New Mexico, and Utah. The idea was to promote alternative energy where it is readily available—those states are among the sunniest—and where the land was not under other industrial or residential use and power lines were available for widespread distribution. It seemed a perfect solution, benefiting the environment without having other negative environmental or economic effects. Since the facilities would be on public lands, there would be no issues associated with private land ownership.

But, ironically, some of the most active opponents of these installations are environmentalists devoted to national parks and America's open space. They argue that some of the locations under consideration are either so close to national parks and national wildlife refuges to harm them, or that the public lands where they would be built provide unique and irreplaceable value as open space, wildlife habitat, and recreational resources for those people who want to get in close touch with nature. In a report titled *Solar Energy, National Parks, and Landscape Protection in the Desert Southwest,* the National Parks Conservation Association (NPCA), writes, "The Mojave and Sonoran Deserts of the American Southwest are at the center of a rush to develop critical renewable energy resources . . . the sites selected for development are located near protected areas that could be compromised by such developments."

For example, Desert Sunlight Solar Farm, California, is under construction on Bureau of Land Management (BLM) land just two miles from the south side of Joshua Tree National Park (Figure 16.1). When completed in 2015, it will produce 550 megawatts (MW) of electricity for about 160,000 average California homes. Its projected benefits include displacing approximately 300,000 metric tons of carbon dioxide (CO_2) per year. That's the amount of CO_2 produced by about 60,000 cars a year.

But NPCA writes that "this project will also potentially interfere with wildlife migration routes, especially for desert bighorn sheep, and it will impair the view from wilderness lands within the national park . . . . These conflicts are unnecessary. Developing renewable energy resources, such as solar energy, can be accomplished while protecting our shared investments in our Southwestern national parks and protected landscapes."[2]

The question is not whether to go with solar and wind, but how to determine the appropriate sites for these installations. Although it would seem that very sunny places like America's desert Southwest would be the best place for solar, in fact successful solar energy installations are functioning in much more northern regions that are not usually thought of as very sunny. These include a 10-MW facility in Bavaria, Germany, which was the world's largest when it was built in 2004; and installations in the city of San Francisco, which is cloudy many days of the year. The key here is to take a multifactor approach, to make decisions about all sources of energy based on a range of environmental, social, and economic factors, as we will discuss in this chapter.

FIGURE 16.1 **A Joshua tree in Joshua Tree National Park, California.** Is it a good idea to have a major solar energy installation near this park? Some environmentalists raise the question.

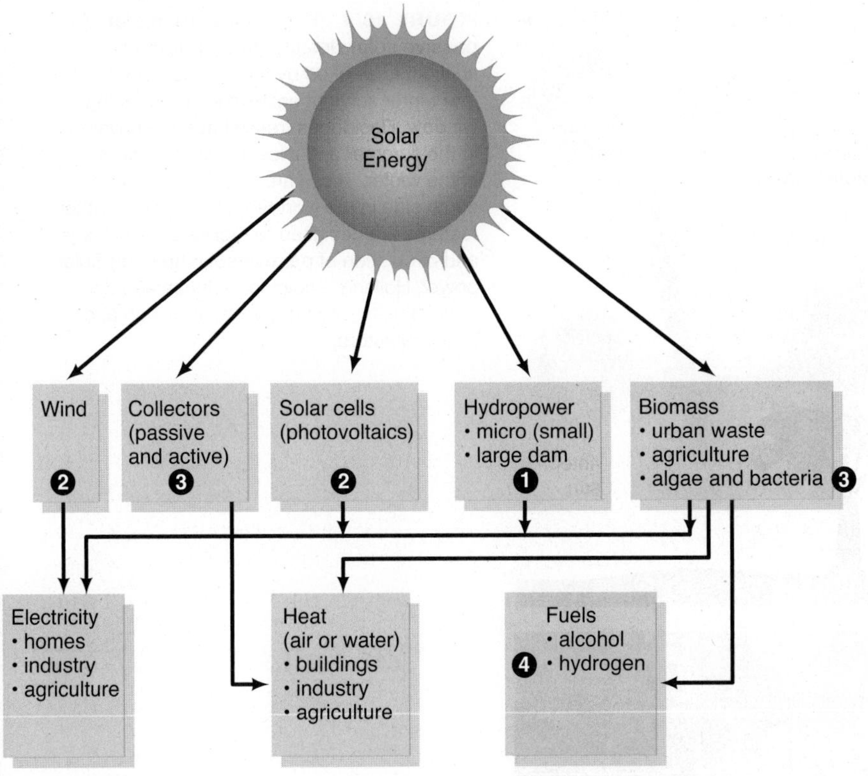

1 Produces in 2013 the most electricity from renewable solar energy

2 Rapidly growing, strong potential (wind and solar are growing at 30% per year)

3 Used today; important energy source

4 Some kinds can be a useful fuel to transition from fossil fuels

FIGURE 16.2 **Routes of various types of renewable solar energy.**

16.1 Introduction to Alternative Energy Sources

Today fossil fuels supply approximately 90% of the energy consumed by people. Fossil fuels and nuclear power are considered conventional energy sources. All other sources are considered **alternative energy** and are divided into renewable energy and nonrenewable energy. **Nonrenewable alternative energy** sources include deep-earth geothermal energy (the energy from the Earth's geological processes). This kind of geothermal energy is considered nonrenewable for the most part because heat can be extracted from Earth faster than it is naturally replenished—that is, output exceeds input (see Chapter 4).

The **renewable energy** sources are: direct solar (both active and passive); freshwater (hydro); wind; ocean; low-density, near-surface geothermal; and biofuels. Low-density, near-surface geothermal is simply solar energy stored by soil and rock near the surface. It is widespread and easily obtained and is renewed by the sun. The energy of biofuels comes from photosynthesis—biological carbon fixation. Sources include trees, grasses, peat; agrifuels; organic waste, including landfill gases; and ethanol—alcohol produced directly by some algae or as a by-product of bacteria's decomposition of organic wastes. These are renewable because they are regenerated by the energy from the sun (Figure 16.2).

16.2 Solar Energy

The total amount of solar energy reaching Earth's surface is tremendous. For example, on a global scale, ten weeks of solar energy is roughly equal to the energy stored in all known reserves of coal, oil, and natural gas on Earth. Solar energy is absorbed at Earth's surface at an average rate of 90,000 terawatts (1 TW equals 10^{12} W), which is about 7,000 times the total global demand for energy.[3] In the United States, on average, 13% of the sun's original energy entering the atmosphere arrives at the surface (equivalent to approximately 177 W/m^2, or about 16 W/ft^2, on a continuous basis). The estimated year-round availability of solar energy in the United States is shown in Figure 16.3. However, solar energy is site-specific, and detailed observation of a potential site is necessary to evaluate the daily and seasonal variability of its solar energy potential. It often surprises people that some major solar energy installations are in areas not considered especially sunny, such as

(*Source:* Modified from Solar Energy Research Institute, 1978)

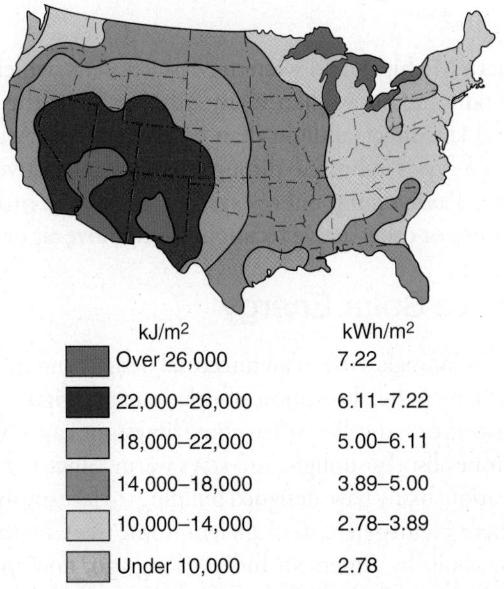

FIGURE 16.3 **Estimated solar energy for the contiguous United States.**

(a)

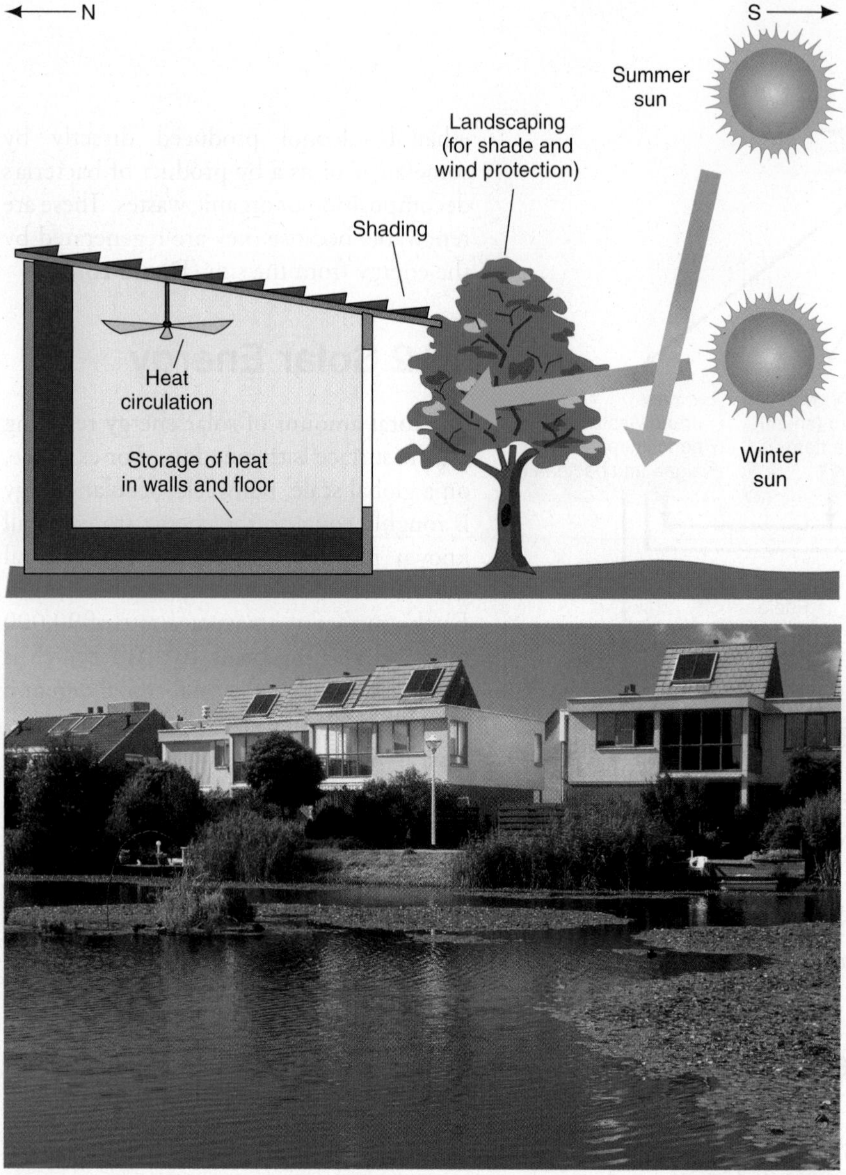

(b)

MARTIN BOND/stillpictures/Aurora Photos

FIGURE 16.4 **Essential elements of passive solar design. (a)** High summer sunlight is blocked by the overhang, but low winter sunlight enters the south-facing window. Deciduous trees shade the building in the summer and allow sunlight to warm it in the winter. These are best located on the south side in the Northern Hemisphere. Other features are designed to facilitate the storage and circulation of passive solar heat. **(b)** Solar power, Holland. Ecolonia, Aalphen aan den Rijn. Houses with passive solar design and solar collectors.

northern Michigan and western Massachusetts, which have more rain and snow than many other parts of the country, and less direct sunlight than desert areas like Arizona.[4] Solar energy is captured through passive and active solar systems. **Passive solar energy systems** do not use mechanical pumps or other active technologies to move air or water.

Passive Solar Energy

Passive solar makes use of architectural designs and materials that enhance the absorption of solar energy (Figure 16.4). Prehistoric cave dwellers often chose limestone caves because limestone absorbs sunlight and stays warm. Since the rise of civilization, many have designed buildings to face south in the Northern Hemisphere, as in ancient Rome (see Chapter 22). Houses built by American Indians long ago, now in Mesa Verde National Park, lie below the surface of the mesa and therefore are protected from much of the area winds; they face south, to gather sunlight. The mesa overhangs the

Photo by Daniel B. Botkin

FIGURE 16.5 **Prehistoric American Indian dwellings, now in Mesa Verde National Park, were positioned to benefit from passive solar energy, facing south.**

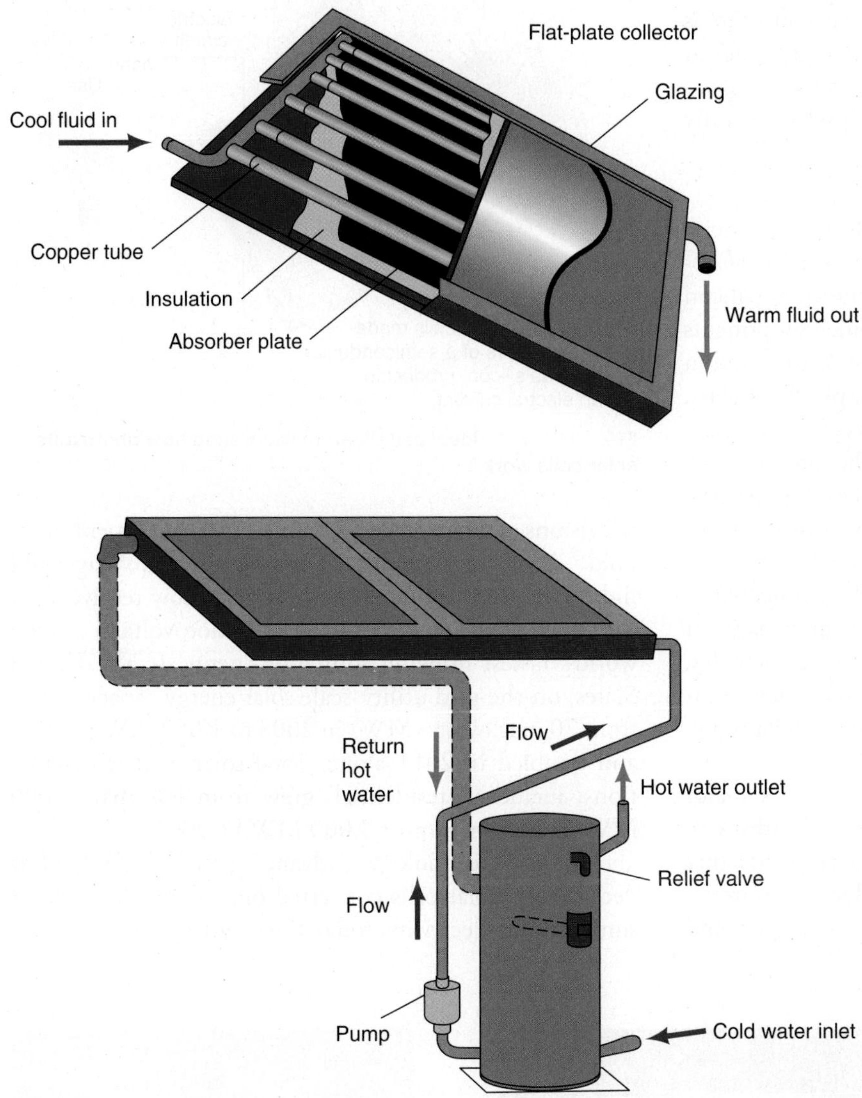

FIGURE 16.6 Detail of a flat-plate solar collector and pumped solar water heater which is one kind of Active Solar energy.

houses, providing some shade during the summer when the sun is high overhead (Figure 16.5) The early settlers of the American West sometimes lived in dugouts, which were homes cut into the soil, because soil absorbs and stores solar energy and the structures are out of the wind, so that they do not readily lose heat by convection.

Today, thousands of buildings in the United States—not just in the sunny Southwest but in other parts of the country, such as New England—use passive solar systems. Passive solar energy promotes cooling in hot weather and retaining heat in cold weather. Methods include (1) adding overhangs on buildings to block summer (high-angle) sunlight but allow winter (low-angle) sunlight to penetrate and warm rooms; (2) building a wall that absorbs sunlight during the day and radiates heat that warms the room at night; and (3) planting deciduous trees on the sunny side of a building. In summer, these shade and cool the building; in the winter, with the leaves gone, they let the sunlight in.

Passive solar energy also provides natural lighting to buildings through windows and skylights. Modern window glass can have a special glazing that transmits visible light, blocks infrared, and provides insulation.

Active Solar Energy

Active solar energy systems require mechanical power, such as electric pumps, to circulate air, water, or other fluids from solar collectors to a location where the heat is stored and then pumped to where the energy is used.

Solar Collectors

Solar collectors that provide space heating or hot water are usually flat, glass-covered plates over a black background where a heat-absorbing fluid (water or some other liquid) is circulated through tubes (Figure 16.6). Solar radiation enters the glass and is absorbed by the black background. Heat is emitted from the black material, heating the fluid in the tubes.

A second type is the evacuated tube collector, which is similar to the flat-plate collector, except that each tube, along with its absorbing fluid, passes through a larger tube that helps reduce heat loss. The use of solar collectors is expanding very rapidly; the global market grew about 50% from 2001 to 2004. In the United States, solar water-heating systems generally pay for themselves in only four to eight years.[4]

Photovoltaics

Photovoltaics convert sunlight directly into electricity (Figure 16.7). The systems use solar cells, also called *photovoltaic cells,* made of thin layers of semiconductors (silicon or other materials) and solid-state electronic components with few or no moving parts. The cells are constructed in standardized modules and encapsulated in plastic or glass, which can be combined to produce systems of various sizes so that power output can be matched to the intended use. Electricity is produced when sunlight strikes the cell. The different electronic properties of the layers cause electrons to flow out of the cell through electrical wires.

Large photovoltaic installations may be connected to an electrical grid. Off-the-grid applications can be large or small and include powering satellites and space vehicles, and powering electric equipment, such as water-level sensors, meteorological stations, and emergency telephones in remote areas (Figure 16.8).

Off-the-grid photovoltaics are emerging as a major contributor to developing countries that can't afford to build electrical grids or large central power plants that burn fossil fuels. One company in the United States is manufacturing photovoltaic systems that power lights and televisions at an installed cost of less than $400 per household.[5] About half a million homes, mostly in villages not linked to a countrywide electrical grid, now receive their electricity from photovoltaic cells.[6] Photovoltaics are the world's fastest growing source of energy. In the United States, on-the-grid utility-scale solar energy capacity grew from 70 megawatts (MW) in 2008 to 1,052 MW in 2011 and doubled in 2011 alone. Total solar electric installations, including residential, grew from less than 2,000 MW in 2008 to almost 3,000 MW by 2011.[7]

Solar cell technology is advancing rapidly. While a few decades ago solar cells converted only about 1 or 2% of sunlight into electricity, today they convert 20% or more.

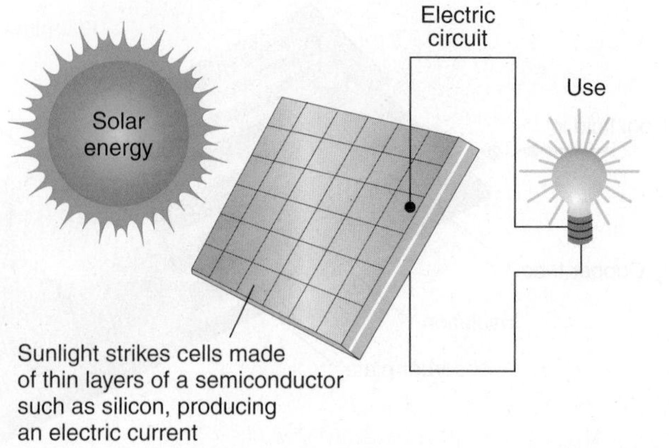
Electric circuit
Use
Solar energy
Sunlight strikes cells made of thin layers of a semiconductor such as silicon, producing an electric current

FIGURE 16.7 **Idealized diagram illustrating how photovoltaic solar cells work.**

(a)
(b)
H. Gruyaeart/Magnum Photos, Inc.
Prof. Ed Keller

FIGURE 16.8 **(a) Panels of photovoltaic cells** are used here to power a small refrigerator to keep vaccines cool. The unit is designed to be carried by camels to remote areas in Chad. **(b) Photovoltaics** are used to power emergency telephones along a highway on the island of Tenerife in the Canary Islands.

Photo by Daniel B. Botkin

FIGURE 16.9 Solar power tower at Barstow, California. Sunlight is reflected and concentrated at the central collector, where the heat is used to produce steam to drive turbines and generate electric power.

And the costs are dropping fast as well. As recently as 2008, installation of solar cells cost an average of \$6.71/watt, but by 2013 had dropped to about \$4.00/watt, a decrease of 40% and in some situations even more, depending on the complexity of the site.[8,4] And some forecast that the manufacturing costs of solar cells might drop to less than \$0.50/watt by 2015.[1] (This cost is for the cells only and doesn't include the cost of installation, inverters, etc.)

Solar Thermal Generators

Solar thermal generators focus sunlight onto water-holding containers. The water boils and is used to run such machines as conventional steam-driven electrical generators. These include solar power towers, shown in Figure 16.9. The first large-scale test of using sunlight to boil water and using the steam to run an electric generator was "Solar One," funded by the U.S. Department of Energy. It was built in 1981 by Southern California Edison and operated by that company along with the Los Angeles Department of Water & Power and the California Energy Commission. Sunlight was concentrated onto the top of the tower by 1,818 large mirrors (each about 20 ft in diameter) that were mechanically linked to each other and tracked the sun. At the end of 1999, this power tower was shut down, in part because the plant was not economically competitive with other sources of electricity. New solar thermal generators are being built with very large output (Figure 16.10).

More recently, solar devices that heat a liquid and produce electricity from steam have used many mirrors without a tower, each mirror concentrating sunlight onto

(a)

Paul Harris/NewsCom

(b)

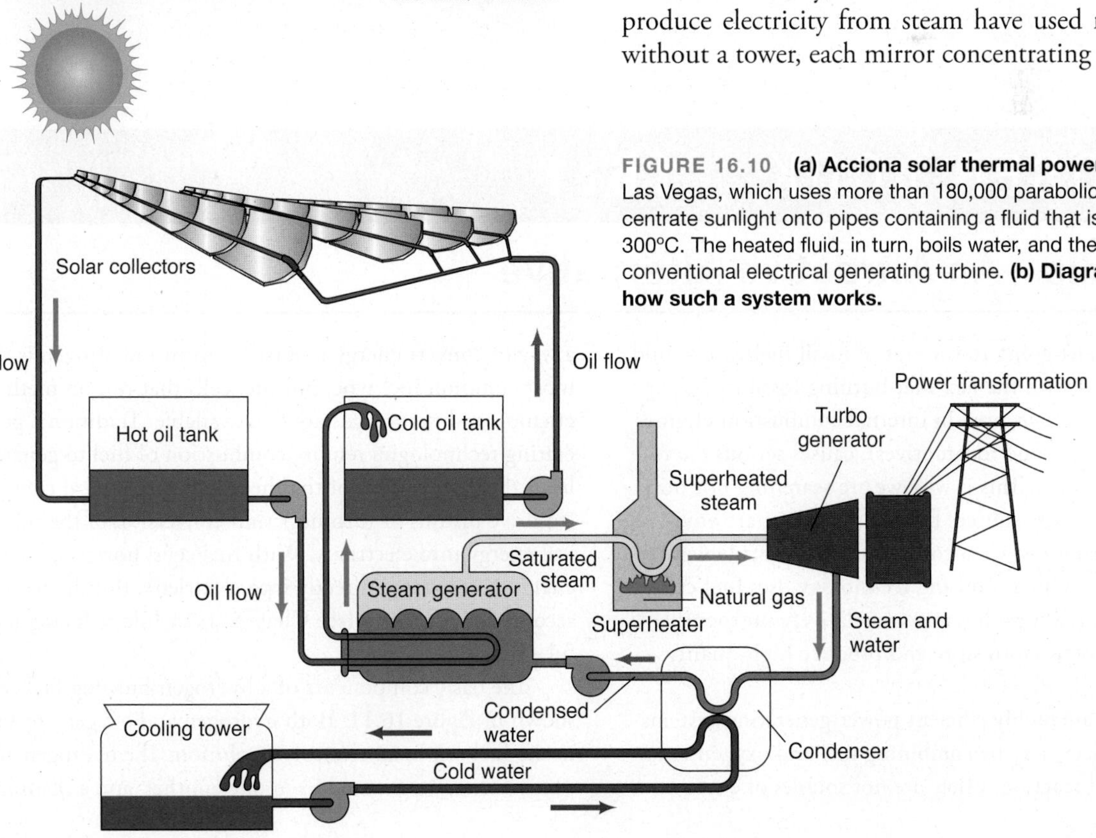

(*Source:* Courtesy of LUZ International.)

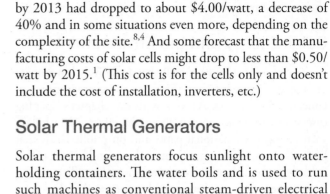

FIGURE 16.10 (a) Acciona solar thermal power plant, south of Las Vegas, which uses more than 180,000 parabolic mirrors to concentrate sunlight onto pipes containing a fluid that is heated above 300°C. The heated fluid, in turn, boils water, and the steam runs a conventional electrical generating turbine. **(b) Diagram illustrating how such a system works.**

a pipe containing the liquid (as shown in Figures 16.9 and 16.10). This is a simpler system and has been considered cheaper and more reliable.

Solar Energy and the Environment

The use of solar energy generally has a relatively low impact on the environment, but there are some environmental concerns nonetheless. As we saw in the chapter's opening case study, there is concern about the effects of installing solar energy near national parks and on public lands that have other important environmental uses and benefits. This is likely to be the primary environmental conflict among environmentalists concerning solar energy. Another concern is the large variety of metals, glass, plastics, and fluids used in the manufacture and use of solar equipment. Some of these substances may cause environmental problems through production and by accidental release of toxic materials.

16.3 Converting Electricity from Renewable Energy into a Fuel for Vehicles

An obvious question about solar energy, as well as energy from wind, oceans, and freshwater, is how to convert this energy into a form that we can easily transport and can use to power our motor vehicles. Basically, there are three choices: Use the electricity directly; convert the energy to mechanical motion, such as a spinning cylinder; or convert the energy to another form, a liquid or gaseous fuel. Direct use includes storing the electricity in batteries.

Producing Liquid and Gaseous Fuels from Electricity

An electrical current can be used to separate water into hydrogen and oxygen. The hydrogen can be used directly—burned within an engine—or it can power fuel cells (see A Closer Look 16.1). In a fuel cell the hydrogen is combined again with oxygen, and the energy is released as electricity

In theory, another option is to combine the hydrogen with carbon to form methane (CH_4), the major constituent of natural gas. Or the methane can be combined with oxygen to form ethanol (C_2H_6O)—the alcohol already used as part of standard gasoline in the United States. This would be a kind of reverse refinery, which is not yet in use.

Hydrogen, like natural gas, can be transported in pipelines and stored in tanks. Furthermore, it is a clean fuel; the combustion of hydrogen produces water, so it does not contribute to global warming, air pollution, or acid rain.

A CLOSER LOOK 16.1

Fuel Cells—An Attractive Alternative

Even if we weren't going to run out of fossil fuels, we would still have to deal with the fact that burning fossil fuels, particularly coal and fuels used in internal combustion engines (cars, trucks, ships, and locomotives), causes serious environmental problems. This is why we are searching not only for alternative energy sources but for those that are environmentally benign ways to convert a fuel to usable ways to generate power.[a] One promising technology uses fuel cells, which produce fewer pollutants, are relatively inexpensive, and have the potential to store and produce high-quality energy.

Fuel cells are highly efficient power-generating systems that produce electricity by combining fuel and oxygen in an electrochemical reaction. (They are not sources of energy but a way to convert energy to a useful form.) Hydrogen is the most common fuel type, but fuel cells that run on methanol, ethanol, and natural gas are also available. Traditional generating technologies require combustion of fuel to generate heat, then conversion of that heat into mechanical energy (to drive pistons or turbines), and conversion of the mechanical energy into electricity. With fuel cells, however, chemical energy is converted directly into electricity, thus increasing second-law efficiency (see Chapter 17) while reducing harmful emissions.

The basic components of a hydrogen-burning fuel cell are shown in Figure 16.11. Both hydrogen and oxygen are added to the fuel cell in an electrolyte solution. The hydrogen and oxygen remain separated from one another, and a platinum

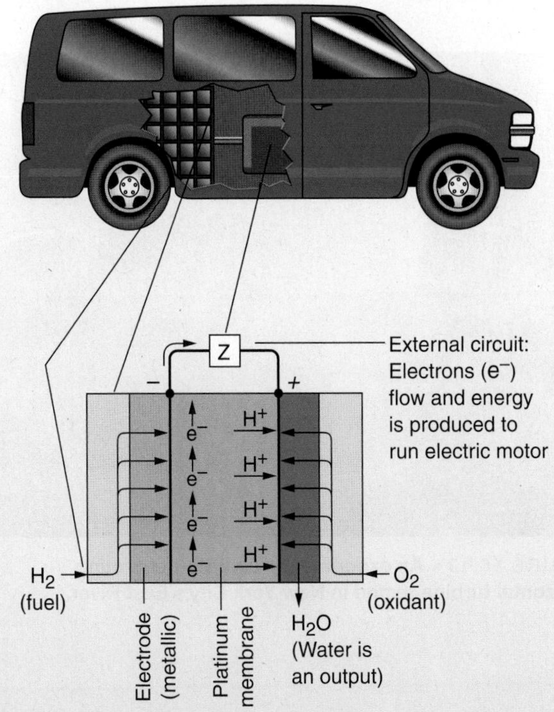

FIGURE 16.11 Idealized diagram showing how a fuel cell works and its application to power a vehicle.

membrane prevents electrons from flowing directly to the positive side of the fuel cell. Instead, they are routed through an external circuit, and along the way from the negative to the positive electrode, they are diverted into an electrical motor, supplying current to keep the motor running.[a, b] To maintain this reaction, hydrogen and oxygen are added as needed. When hydrogen is used in a fuel cell, the only waste product is water.

Fuel cells are efficient and clean, and they can be arranged in a series to produce the appropriate amount of energy for a particular task. In addition, the efficiency of a fuel cell is largely independent of its size and energy output. They can also be used to store energy to be used as needed. Fuel cells are used in many locations. For example, they power buses at Los Angeles International Airport and in Vancouver. They also power a few buses in east San Francisco Bay[c] and provide heat and power at Vandenberg Air Force Base in California.[d] But this experimental technology is expensive, with buses estimated to cost more than $1 million each.[e]

Technological improvements in producing hydrogen are certain, and the fuel price of hydrogen may be substantially lower in the future.

16.4 Water Power

Water power is a form of stored solar energy that has been successfully harnessed since at least the time of the Roman Empire. Waterwheels that convert water power to mechanical energy were turning in western Europe in the Middle Ages. During the 18th and 19th centuries, large waterwheels provided energy to power grain mills, sawmills, and other machinery in the United States.

Today, hydroelectric power plants use the water stored behind dams. In the United States, hydroelectric plants generate about 80,000 MW of electricity—about 10% of the total electricity produced in the nation. In some countries, such as Norway and Canada, hydroelectric power plants produce most of the electricity used. Figure 16.12a shows the major components of a hydroelectric power station.

Hydropower can also be used to store energy produced by other means, through the process of pump storage (Figure 16.12b and c). During times when demand for power is low, excess electricity produced from oil, coal, or nuclear plants is used to pump water uphill to a higher reservoir (high pool). When demand for electricity is high (on hot summer days, for instance), the stored water flows back down to a low pool through generators to help provide energy. The advantage of pump storage lies in the timing of energy production and use. However, pump storage facilities are generally considered ugly, especially at low water times.

Small-Scale Hydropower Systems

In the coming years, the total amount of electrical power produced by running water from large dams will probably not increase in the United States, where most of the acceptable dam sites are already in use. It may decrease because some dams are being dismantled. They are being removed either because they have reached their useful lifetime or because of concerns that a dam interferes with the migration of fish, such as salmon and shad, that lay their eggs in freshwater but spend most of their adult life in the ocean and migrate up and down rivers (see Chapter 18). Because small dams can damage stream environments by blocking fish passage and changing downstream flow, careful consideration must be given to their construction. A few small dams cause little environmental degradation beyond the specific sites, but a large number of dams can have an appreciable impact on a region. (This principle

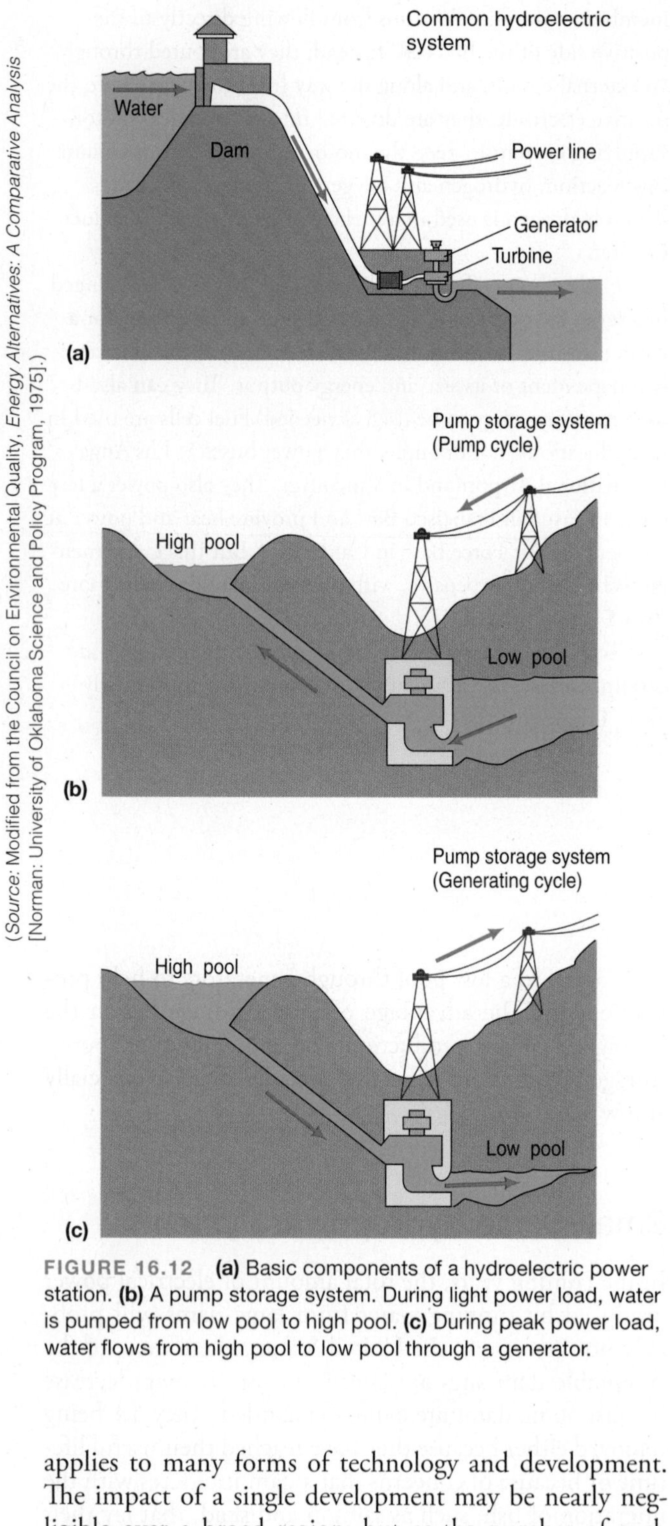

FIGURE 16.12 **(a)** Basic components of a hydroelectric power station. **(b)** A pump storage system. During light power load, water is pumped from low pool to high pool. **(c)** During peak power load, water flows from high pool to low pool through a generator.

Kris Unger/Verdant Power, Inc.

FIGURE 16.13 **An experimental electric-generating horizontal turbine tested in New York City's East River.**

applies to many forms of technology and development. The impact of a single development may be nearly negligible over a broad region; but as the number of such developments increases, the total impact may become significant.)

Hydroturbines are an interesting development and are in use in early tests. These work like the large turbines that operate in hydroelectric dams, but are smaller and lie horizontally within a river. They are fixed in place but are not part of a dam, so they allow the normal river flow and migration of fish (Figure 16.13).

Water Power and the Environment

Water power is clean power in that it requires no burning of fuel, does not pollute the atmosphere, produces no radioactive or other waste, and is efficient. However, there are environmental prices to pay:

- Large dams and reservoirs flood large tracts of land that could have had other uses. For example, towns and agricultural lands may be lost.
- As we noted earlier, dams block the migration of some fish, such as salmon, and the dams and reservoirs greatly alter habitat for many kinds of fish.
- Dams trap sediment that would otherwise reach the sea and eventually replenish the sand on beaches.
- Reservoirs with large surface areas increase evaporation of water compared to pre-dam conditions. In arid regions, evaporative loss of water from reservoirs is more significant than in more humid regions.
- For a variety of reasons, many people do not want to turn wild rivers into a series of lakes.

For all these reasons, and because many good sites already have a dam, the growth of large-scale water power in the future (with the exception of a few areas, including Africa, South America, and China) appears limited. Indeed, in the United States there is an emerging social movement to remove dams. Hundreds of dams, especially those with few useful functions, are being considered for removal, and a few have already been dismantled (see Chapters 18 and 19). The U.S. Department of Energy forecasts that electrical generation from large hydropower dams will decrease significantly.

16.5 Ocean Energy

The motion of waves, currents, and tides in oceans involves a lot of energy. Many have dreamed of harnessing this energy, but it's not easy, for the obvious reasons that ocean storms are destructive and ocean waters are corrosive. The most successful development of energy from the ocean has been **tidal power**. Use of the power of ocean tides can be traced back to the Roman occupation of Great Britain around Julius Caesar's time, when the Romans built a dam that captured tidal water and let it flow out through a waterwheel. By the 10th century, tides were used once again to power coastal mills in Britain.[8] However, only in a few places with favorable topography—such as the north coast of France, the Bay of Fundy in Canada, and the northeastern United States—are the tides strong enough to produce commercial electricity. The tides in the Bay of Fundy have a maximum range of about 15 m (49 ft). A minimum range of about 8 m (26 ft) appears necessary with present technology for development of tidal power.

Traditionally, to harness tidal power, a dam is built across the entrance to a bay or an estuary, creating a reservoir. As the tide rises (flood tide), water is initially prevented from entering the bay landward of the dam. Then, when there is sufficient water (from the oceanside high tide) to run the turbines, the dam is opened, and water flows through it into the reservoir (the bay), turning the blades of the turbines and generating electricity. When the bay is filled, the dam is closed, stopping the flow and holding the water in that reservoir. When the tide falls (ebb tide), the water level in the reservoir is higher than that in the ocean. The dam is then opened to run the turbines (which are reversible), and electric power is produced as the water is let out of the reservoir. Figure 16.14 shows the Rance tidal power plant on the north coast of France.

Yann Arthus-Bertrand/©Corbis

FIGURE 16.14 **Tidal power station on the river Rance near Saint-Malo, France.**

Constructed in the 1960s, it is the first and largest modern tidal power plant and has remained in operation since that time. The plant at capacity produces about 240,000 kW from 24 power units spread out across the dam. At the Rance power plant, most electricity is produced from the ebb tide, which is easier to control.

Tidal power, too, has environmental impacts. The dam changes the hydrology of a bay or an estuary, which can adversely affect the vegetation and wildlife. The dam restricts upstream and downstream passage of fish, and the periodic rapid filling and emptying of the bay as the dam opens and closes with the tides rapidly changes habitats for birds and other organisms.

New Ocean Energy Technologies

In September, 2012, an ocean turbine similar in principle to the horizontal turbine installed in New York City's East River began delivering electricity to the Bangor Hydro Electric Company of Maine. This horizontal turbine, designed and built by the Ocean Renewable Energy Project, generated electricity from the movement of tidal waters in Cobscook Bay, an estuary in Washington County, Maine, which opens into the Bay of Fundy. The Bay of Fundy has one of the largest tidal flows in the world. This marked the first power from any ocean energy project to provide electricity to a U.S. electrical grid (Figure 16.15).[9]

Another new approach is the use of ocean waves to power ocean-going vessels. One of the first is the Suntory Mermaid I. In 2008 its inventor, Yutaka Terao, of the Department of Naval Architecture and Ocean Engineering at the Tokai University School of Marine Science and Technology, sailed the Suntory Mermaid I from Honolulu, Hawaii, 7,000 km (3,700 nautical miles) to the western shore of Japan (Figure 16.16). The trip had been done before, of course, but this boat's propulsion system was new—the *Suntory Mermaid* has two horizontal fins that move up and down with the waves and generate the power to push the boat forward. Solar energy provides electricity.[10]

Another example of imaginative ocean engineering is a robotic boat called the Wave Glider. Using wave and solar energy, it sailed 10,000 miles from San Francisco to Australia in the fall of 2012. This boat makes use of wave motion in a somewhat different way than does the *Suntory Mermaid*. It is made of two units—a floater and a sub. The waves raise and lower the floater, and this motion is transferred to the sub in a way that moves the two forward. Using a small computer and GPS, the Wave Glider was able to make its way past sharks and the Great Barrier Reef of Australia. These two vessels demonstrate that there are opportunities for imaginative engineering designs to change the way boats are powered, independent of fossil fuels.

(a)

Photo courtesy of Ocean Renewable Power Company

FIGURE 16.15 The first ocean turbine providing electricity to a U.S. grid began operation in September 2012. **(a)** the turbine being readied for installation; **(b)** a diagram of the turbine.

(b)

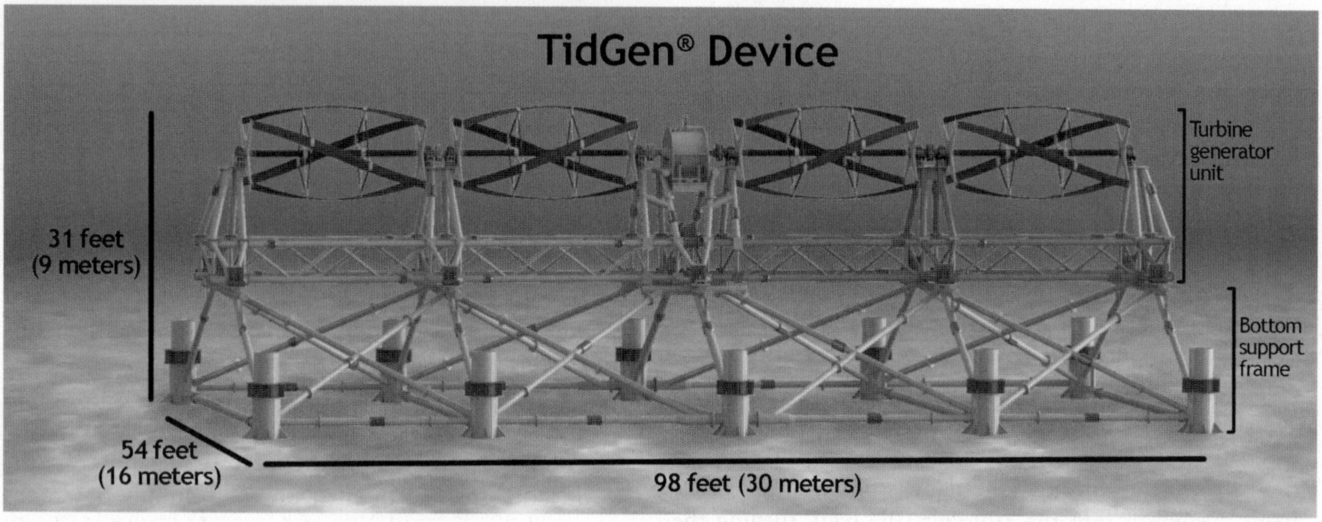

(*Source:* Ocean Renewable Power Company's TidGen® Device)

(a)

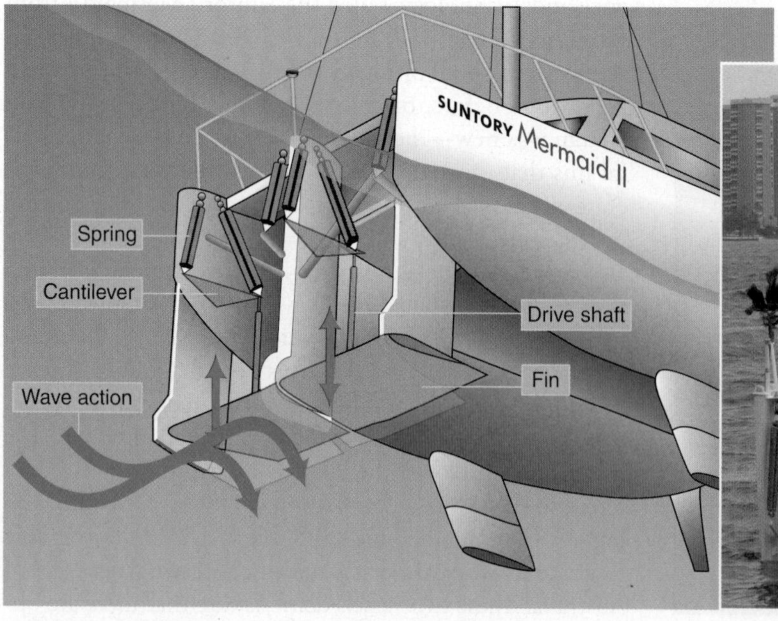

(*Source:* KEVIN HAND/POPULAR SCIENCE)

REUTERS/Shigeo Yamada/Landov Photographers

(b)

FIGURE 16.16 Wave energy propels boats. (a) On the *Suntory Mermaid*, the up-and-down wave motion is transferred mechanically to two fins mounted side by side beneath the bow, which use dolphin-like kicks to propel the boat; **(b)** the *Suntory Mermaid* sailing.

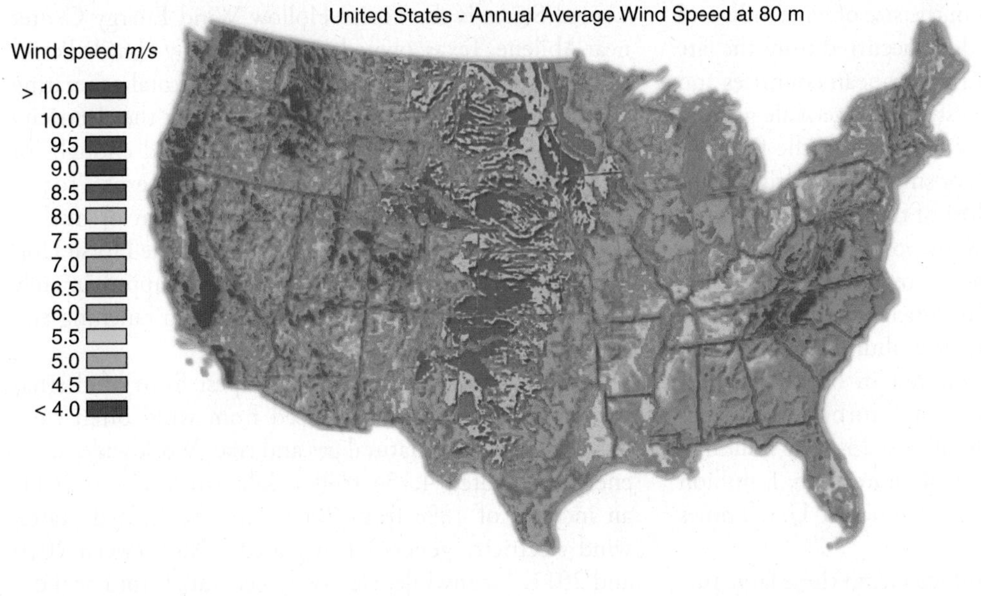

FIGURE 16.17 Wind energy potential in the United States.

16.6 Wind Power

Wind power, like solar power, has evolved over a long time. From early Chinese and Persian civilizations to the present, wind has propelled ships and has driven windmills to grind grain and pump water. In the past, thousands of windmills in the western United States were used to pump water for ranches. More recently, wind has been used to generate electricity. The trouble is that wind tends to be highly variable in time, place, and intensity.

Basics of Wind Power

Winds are produced when differential heating of Earth's surface creates air masses with differing heat contents and densities. The potential for energy from the wind is large, and thus wind "prospecting" has become an important endeavor. On a national scale, regions with the greatest potential are the Pacific Northwest coastal area, the coastal region of the northeastern United States, and a belt within the Great Plains extending from northern Texas through the Rocky Mountain states and the Dakotas (Figure 16.17). Other windy sites include mountain areas in North Carolina and the northern Coachella Valley in Southern California. A site with average wind velocity of about 18 km/h (11 mph) or greater is considered a good prospect for development of wind energy, although starting speeds for modern wind turbines can be considerably lower.

In any location, the wind's direction, velocity, and duration may be quite variable, depending on local topography and temperature differences in the atmosphere. For example, wind velocity often increases over hilltops and when wind is funneled through a mountain pass (Figure 16.18). The increase in wind velocity over a mountain is due to a vertical convergence of wind, whereas in a pass the increase is partly due to a horizontal convergence. Because the shape of a mountain or a pass is often related to the local or regional geology, prospecting for wind energy is a geologic as well as a geographic and meteorological task. The wind energy potential of a region or site is determined by instruments that measure and monitor over time the strength, direction, and duration of the wind.

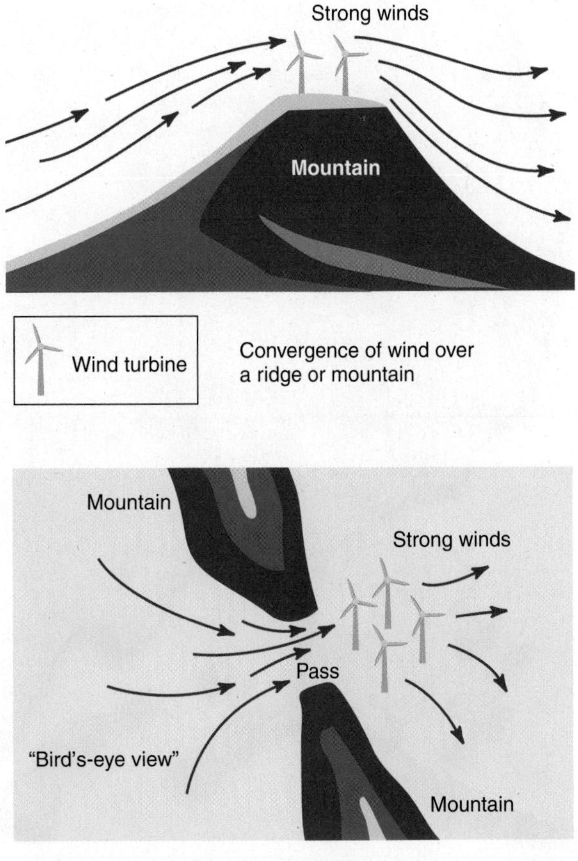

FIGURE 16.18 Idealized diagram showing how wind energy is concentrated by topography.

Significant improvements in the size of windmills and the amount of power they produce occurred from the late 1800s to the present, when many European countries and the United States became interested in large-scale generators driven by wind. In the United States in the first half of the 20th century, thousands of small, wind-driven generators were used on farms. Most of these small windmills generated approximately 1 kW of power, which is much too little to be considered for central power-generation needs. Interest in wind power declined for several decades prior to the 1970s because of the abundance of cheap fossil fuels. Since the 1980s, interest in building windmills has revived greatly. Modern wind turbines are big, as much as 70 m (230 ft) high, as tall as a 23-story building, and have a generating capacity of more than 1 million watts—enough electricity for 500 modern U.S. homes (Figure 16.19).[11]

Much of the electricity produced from these large turbines is connected to the electrical grid, and many of them are installed in so-called wind farms, which are providing large amounts of electricity. One of the biggest in the United States is the Horse Hollow Wind Energy Center near Abilene, Texas, owned and operated by Florida Power & Light. It has 421 wind turbines with a total generating capacity of 735 megawatts, enough to meet the electricity needs of approximately 220,000 homes and enough for all domestic use for a city the size of Austin, Texas. (To put this in perspective, one large fossil fuel plant or nuclear power plant produces about 1,000 MW.) The Horse Hollow wind turbines are spread widely across approximately 47,000 acres, and the land is used for both ranching and energy production.[8]

Today, wind energy is the cheapest form of alternative energy. Electricity produced from wind often costs less than that from natural gas and coal. Worldwide, wind energy generated 4,154 trillion kilowatt-hours in 2011, an increase of 18% from 2007.[11] In the United States, wind electricity generated increased 27% between 2010 and 2011. Meanwhile, electricity generated from coal declined 6.2%. However, wind still provided only a small percentage of U.S. electricity—1.4% in 2011 compared to 21% provided by coal.[11]

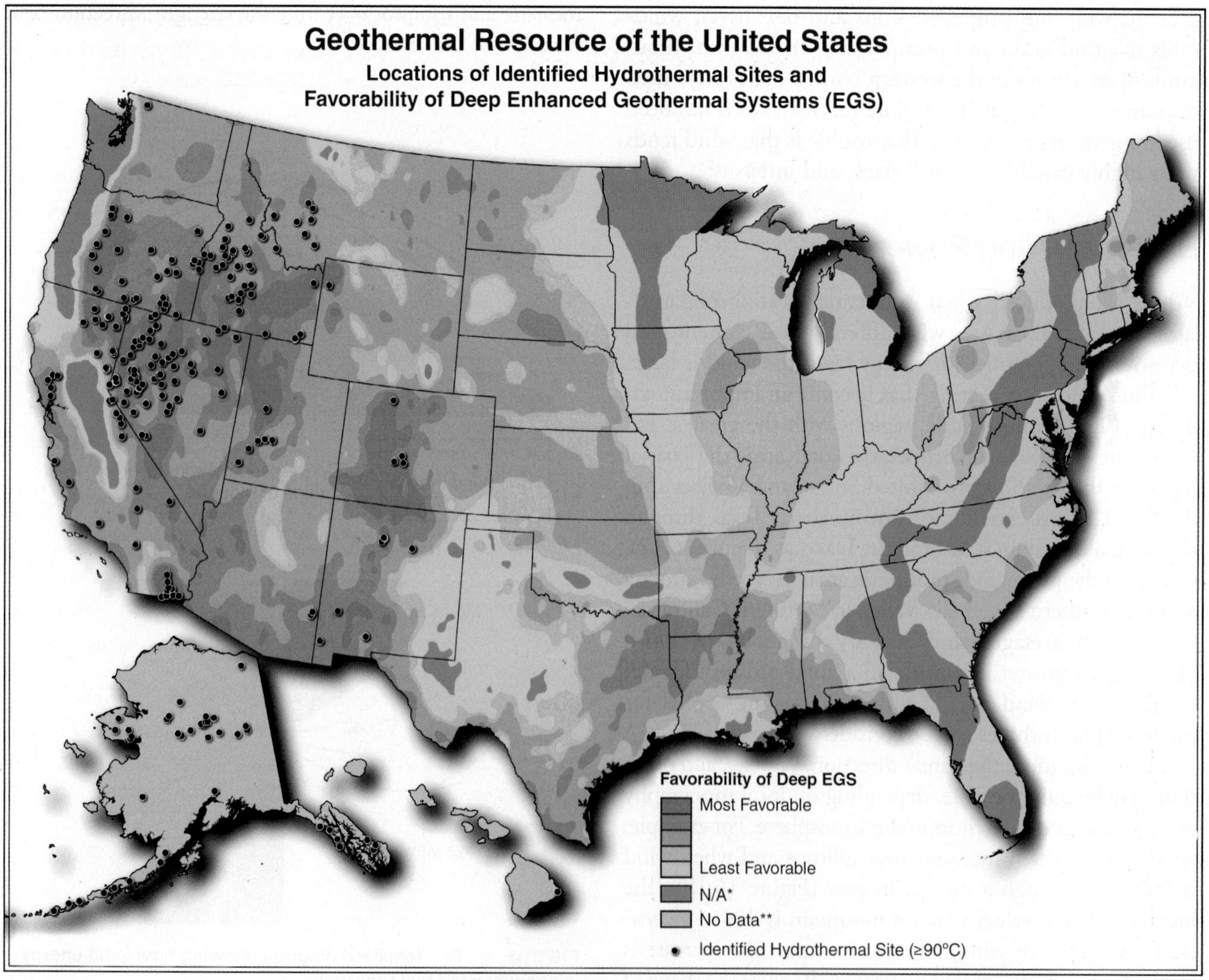

FIGURE 16.19 Geothermal resources of the United States.

Wind Power and the Environment

Wind energy does have a few disadvantages:

- Wind turbines can kill birds. (Birds of prey, such as hawks and falcons, are particularly vulnerable.)
- Wind turbines and wind farms may degrade an area's scenery, as we saw in the chapter's opening case study.

However, although wind farms must often compete with other land uses, in many cases wind turbines can share land used for farms, military bases, and other facilities. Everything considered, wind energy has a relatively low environmental impact.

The Future of Wind Power

Wind power's continued use is certain. As noted earlier, the use of wind energy has been growing fast. It is believed that there is sufficient wind energy in Texas, South Dakota, and North Dakota to satisfy the electricity needs of the entire United States. Consider the implications for nations such as China. China burns tremendous amounts of coal at a heavy environmental cost that includes exposing millions of people to hazardous air pollution. In rural China, exposure to the smoke from burning coal in homes has increased the rate of lung cancer by a factor of nine or more. Estimates are that China could probably double its current capacity to generate electricity with wind alone!

One scenario suggests that wind power could supply 10% of the world's electricity in the coming decades and, in the long run, could provide more energy than hydropower, which today supplies approximately 20% of the electricity in the world. The wind energy industry has created thousands of jobs in recent years. Worldwide, more than half a million people are employed in wind energy, and according to the World Wind Energy Association, wind energy "creates many more jobs than centralized, non-renewable energy sources."[11] Technology, meanwhile, is producing more efficient wind turbines, thereby reducing the price of wind power. All told, wind power is becoming a major investment opportunity.

16.7 Biofuels

Biofuel is energy recovered from biomass (organic matter). We can divide biofuels into five groups:

1. unmanaged growth harvested by people, including firewood, grasses, and peat;
2. organic wastes used directly, including cooking oil, which can fuel diesel engines, and methane, emitted from bacterial decomposition of waste;
3. agrifuels, which are crops grown to be converted into liquid fuels;
4. ethanol produced by some algae as a by-product of photosynthesis;
5. ethanol produced by bacteria as a by-product of bacteria's decomposition of organic wastes.

Biofuels and Human History

Biomass is the oldest fuel used by people. Our Pleistocene ancestors burned wood in caves to keep warm and cook food. Biofuels remained a major source of energy throughout most of the history of civilization. When North America was first settled, there was more wood fuel than could be used. Forests often were cleared for agriculture by girdling trees (cutting through the bark all the way around the base of a tree) to kill them and then burning the forests.

Until the end of the 19th century, wood was the major fuel in the United States. During the mid-20th century, when coal, oil, and gas were plentiful, burning wood became old-fashioned and quaint, done just for pleasure in an open fireplace, even though most of the heat went up the chimney. Now, with other fuels reaching a limit in abundance and production, there is renewed interest in using natural organic materials for fuel.

More than 1 billion people in the world today still use wood as their primary source of energy for heat and cooking.[12] Although firewood is the most familiar and most widely used biomass fuel, there are many others. In India and other countries, cattle dung is burned for cooking.

Peat, a form of compressed dead vegetation, provides heating and cooking fuel in northern countries, such as Scotland, where it is abundant. Renewed interest in biofuels grew during the last decades of the 20th century and continues today. For example, on December 10, 2012, a ship carrying nearly 20,000 metric tons of wood pellets produced in eastern Canada arrived at the Wismar seaport in Germany. The pellets will be used in steam-generating electric power plants.[13] Why are wood pellets becoming a new energy source? Because electricity produced from burning them releases less carbon dioxide into the atmosphere than coal and oil. Nations that want to show that they are reducing carbon emissions are therefore turning to wood pellets. The problem is that the total carbon release involved in growing and harvesting trees and energy used to produce the pellets and shipping the pellets are not accounted for by the nations burning the pellets. It may be that in the end more carbon dioxide is released in total using wood pellets to fuel power plants than is produced by fossil fuels. This carbon accounting has been the subject of only a few studies. One, issued by the Massachusetts Institute of Technology, stated that burning wood pellets releases a large amount of CO_2, creating a carbon debt.[13]

In recent years, biofuels have become controversial. Do biofuels offer a net benefit or disbenefit? (See Table 16.1.) In brief:

- Using wastes as a fuel is a good way to dispose of them. Making them takes more energy than they yield; on the other hand, they reduce the amount of energy we must obtain from other sources. Firewood that regenerates naturally or in plantations that require little energy input will remain an important energy source in developing nations and locally in industrialized nations. Despite pressure from some agricultural corporations and some governments to promote crops grown solely for conversion into liquid fuels (called agrifuels), at present these are poor sources of energy. Most scientific research shows that producing agrifuels takes more energy than they yield. In some cases, there appears to be a net benefit, but the energy produced per unit of land area is low, much lower than can be obtained from solar and wind.

What it boils down to is that photosynthesis, though a remarkable natural process, is less efficient than modern photovoltaic cells in converting sunlight to electricity. Some algae and bacteria appear to provide a net energy benefit and can yield ethanol directly, but production of ethanol from these sources is just beginning and is experimental.[8]

Biofuels and the Environment

The conversion of farmland from food crops to biofuels appears to be one of the main reasons that food prices have risen rapidly worldwide and that worldwide food production no longer exceeds demand. It also has environmental effects.

Biofuel agriculture competes for water with all other uses, and the main biofuel crops require heavy use of artificial fertilizers and pesticides.

Biofuels are supposed to reduce the production of greenhouse gases, but when natural vegetation is removed to grow biofuel crops, the opposite may be the case. It is estimated that 16 million hectares (40 million acres) have been deforested in Indonesia alone since 1970, a considerable proportion of which has been converted to oil palm plantation. This is an area equal to more than 10% of all the cropland in the United States, as large as Oklahoma and larger than Florida.[8]

The use of biofuels can pollute the air and degrade the land. For most of us, the smell of smoke from a single campfire is part of a pleasant outdoor experience. Under certain weather conditions, however, wood smoke from many campfires or chimneys in narrow valleys can lead to air pollution. The use of biomass as fuel places pressure on an already heavily used resource. A worldwide shortage of firewood is adversely affecting natural areas and endangered species. For example, the need for firewood has threatened the Gir Forest in India, the last remaining habitat of the Indian lion (not to be confused with the Indian tiger). The world's forests will also shrink and in some cases vanish if our need for their products exceeds their productivity.

Table 16.1 SELECTED EXAMPLES OF BIOMASS ENERGY SOURCES, USES, AND PRODUCTS

SOURCES	EXAMPLES	USES/PRODUCTS	COMMENT
Forest products	Wood, chips	Direct burning,[a] charcoal[b]	Major source in undeveloped nations; growing source in developed nations.
Agriculture residues	Coconut husks, sugarcane waste, corncobs, peanut shells	Direct burning	Minor source
Energy crops	Sugarcane, corn, sorghum	Ethanol (alcohol)[c], gasification[d]	Ethanol is major source of fuel in Brazil for automobiles
Algae and bacteria	Special farms	Experimental	
Trees	Palm oil	Biodiesel	Fuel for vehicles
Animal residues	Manure	Methane[e]	Used to run farm machinery
Urban waste	Waste paper, organic household waste	Direct burning of methane from wastewater treatment or from landfills[f]	Minor source

[a]Principal biomass conversion.

[b] Secondary product from burning wood.

[c] Ethanol is an alcohol produced by fermentation, which uses yeast to convert carbohydrates into alcohol in fermentation chambers (distillery).

[d] Biogas from gasification is a mixture of methane and carbon dioxide produced by pyroclytic technology, Which involves a thermochemical process that breaks down solid biomass into an oil-like liquid and almost pure carbon char.

[e] Methane is produced by anaerobic fermentation in a digester.

[f] Naturally produced in landfills by anaerobic fermentation.

Biofuels do have some potential benefits. One is that certain kinds of crops, such as nuts produced by trees, may provide a net energy benefit in environments that are otherwise not suited to the growth of food crops. Another environmental plus is that combustion of biofuels generally releases fewer pollutants, such as sulfur dioxide and nitrogen oxides, than does combustion of coal and gasoline. This is not always the case for burning urban waste, however. Although plastics and hazardous materials are removed before burning, some inevitably slip through the sorting process and are burned, releasing air pollutants, including heavy metals. There is a conflict in our society as to whether it is better, in terms of the environment, to burn urban waste to recover energy and risk some increase in air pollution, or dump these wastes in landfills, which can then pollute soil and groundwater.

16.8 Geothermal Energy

There are two kinds of **geothermal energy**: deep-earth, high-density; and shallow-earth, low-density. The first makes use of energy within the Earth. The second is a form of solar energy: When the sun warms the surface soils, water, and rocks, some of this heat energy is gradually transmitted down into the ground.

The first kind of geothermal energy—deep-earth, high-density—is natural heat from the interior of the Earth. It is mined and then used to heat buildings and generate electricity. The idea of harnessing Earth's internal heat goes back more than a century. As early as 1904, geothermal power was used in Italy. Today, Earth's natural internal heat is being used to generate electricity in 21 countries, including Russia, Japan, New Zealand, Iceland, Mexico, Ethiopia, Guatemala, El Salvador, the Philippines, and the United States. Total worldwide production is approaching 9,000 MW (equivalent to nine large, modern coal-burning or nuclear power plants)—double the amount in 1980. Some 40 million people today receive their electricity from geothermal energy at a cost competitive with that of other energy sources. In El Salvador, geothermal energy is supplying 25% of the total electric energy used. However, at the global level, geothermal energy accounts for less than 0.15% of the total energy supply.[14] In 2011 geothermal energy provided only 0.4% of the electricity in the United States.

This kind of geothermal energy may be considered a nonrenewable energy source when rates of extraction are greater than rates of natural replenishment. However, geothermal energy has its origin in the natural heat production within Earth, and only a small fraction of the vast total resource base is being used today. Although most geothermal energy production involves tapping high-heat sources, people are also using the low-temperature geothermal energy of groundwater in some applications.

Geothermal Systems

The average heat flow from the interior of the Earth is very low, about 0.06 watts per square meter (W/m^2). This amount is trivial compared with the 177 W/m^2 from sunlight at the surface, but in some areas the heat flow is high enough to be useful.[15] For the most part, high heat flow occurs mostly at plate tectonic boundaries (see Chapter 4, including oceanic ridge systems—divergent plate boundaries) and areas where mountains are being uplifted and volcanic islands are forming (convergent plate boundaries). You can see the effects of such natural heat flow at Yellowstone National Park. On the basis of geologic criteria, several types of hot geothermal systems (with temperatures greater than about 80°C, or 176°F) have been defined (Figure 16.19).

Some communities in the United States, including Boise, Idaho, and Klamath Falls, Oregon, are using deep-earth, high-density geothermal heating systems. A common type of geothermal system uses hydrothermal convection, where the circulation of steam and/or hot water transfers heat from the depths to the surface. An example is the Geysers Geothermal Field, 145 km (90 mi) north of San Francisco. It is the largest geothermal power operation in the world (Figure 16.20), producing about 1,000 MW of electrical energy. At the Geysers, the hot water is maintained in part by injecting treated wastewater from urban areas into hot rocks.

Kim Steele/Getty Images, Inc.

FIGURE 16.20 Geysers Geothermal Field, north of San Francisco, California, is the largest geothermal power operation in the world and produces energy directly from steam.

The second kind of geothermal energy—shallow-earth, low-density—is at much lower temperatures than geothermal sources and is used not to produce electricity but for heating buildings and swimming pools, and for heating soil to boost crop production in greenhouses. The potential for this kind of energy is huge, and it is cheap to obtain.[8] Such systems are extensively used in Iceland.

It may come as a surprise to learn that most groundwater can be considered a source of shallow-earth, low-density geothermal energy. It is geothermal because the normal heat flow from the sun to the Earth keeps the temperature of groundwater, at a depth of 100 m (320 ft), at about 13°C (55°F). This is cold for a shower but warmer than winter temperatures in much of the United States, where it can help heat a house. In warmer regions, with summer temperatures of 30–35°C (86–95°F), groundwater at 13°C (55°F) is cool and thus can be used to provide air conditioning, as it does today in coastal Florida and elsewhere. In summer, heat can be transferred from the warm air in a building to the cool groundwater. In winter, when the outdoor temperature is below about 4°C (40°F), heat can be transferred from the groundwater to the air in the building, reducing the need for heating from other sources (Figure 16.21). "Heat pumps" for this kind of heat transfer are used in warm locations such as Florida and as far north as Juneau, Alaska, but are limited by extreme temperatures and can't function in the cold winters of northern New Hampshire or interior Alaska, such as in Fairbanks. (Juneau, on the coast, has a much more moderate climate because of the influence of the ocean waters.)

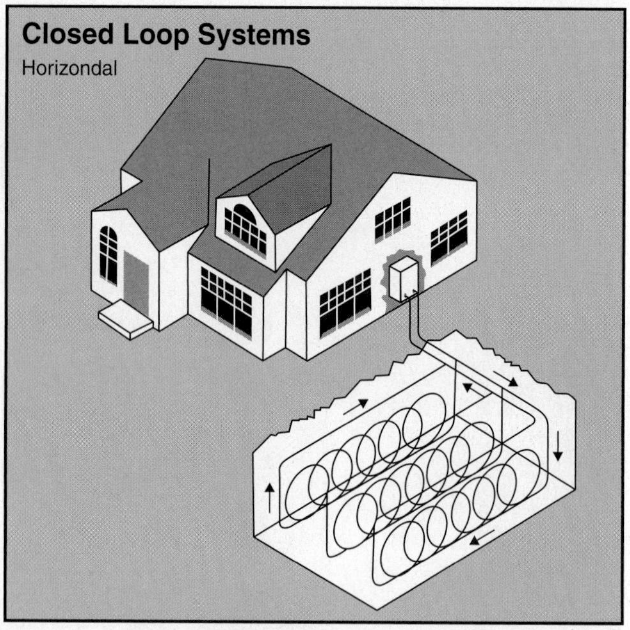

FIGURE 16.21 Heat and cooling right under your feet. How a geoexchange system works.

Geothermal Energy and the Environment

Deep-earth, high-density, geothermal energy development produces considerable thermal pollution from its hot wastewaters, which may be saline and highly corrosive. Other environmental problems associated with this kind of geothermal energy use include on-site noise, emissions of gas, and disturbance of the land at drilling sites, disposal sites, roads and pipelines, and power plants. The good news is that the use of deep-earth, high-density geothermal energy releases almost 90% less carbon dioxide and sulfur dioxide than burning coal releases to produce the same amount of electricity.[16] Furthermore, development of geothermal energy does not require large-scale transportation of raw materials or refining of chemicals, as development of fossil fuels does. Nor does geothermal energy produce the atmospheric pollutants associated with burning fossil fuels or the radioactive waste associated with nuclear energy.

Even so, deep-earth, high-density geothermal power is not always popular. For instance, on the island of Hawaii, where active volcanoes provide abundant near-surface heat, some argue that the exploration and development of geothermal energy degrade the tropical forest as developers construct roads, build facilities, and drill wells. In Hawaii, geothermal energy also raises religious and cultural issues. Some people, for instance, are offended by the use of the "breath and water of Pele," the volcano goddess, to make electricity. This points out the importance of being sensitive to the values and cultures of people where development is planned.

The Future of Geothermal Energy

By 2011, the United States produced only 10,898 MW of geothermal energy.[8] Globally, the likelihood is that high-density, deep-earth geothermal can be only a minor contributor to world energy demand, but that low-density, shallow-earth geothermal can be a major source of alternative and renewable energy.[17]

CRITICAL THINKING ISSUE

Should Governments Support Renewable Energy?

In 1991 the German Legislature passed the Erneuerbare Energien Gesetz (renewable energy law), guaranteeing small hydroelectric power generators a market for their electricity. It required utility companies to plug all renewable power producers into the national grid and buy their power at a fixed, slightly above-market rate. The rate was designed to provide a guaranteed return over the long term. The concept was that the feed-in tariff, which it is commonly called, would balance out subsidies given to coal power and the hidden health costs of pollution. Solar and wind developers, seeing there was a guaranteed, relatively lucrative market for their power, got onboard.

Soon homeowners were embracing the idea of generating their own power, and 65% of Germany's renewable energy is now generated by individuals, cooperatives, and communities by solar photovoltaic (PVC) arrays on roofs; Germany's utilities own only 6.5% of the nation's renewable power.[18]

Small installations, such as residential and small commercial PV arrays, are considered distributed (or decentralized) renewable energy. Concentrated (centralized) solar—what is generated in large installations that produce a megawatt or more of power—often get the attention in the United States. However, there is also great potential in distributed solar power, as demonstrated in Germany.

Germany has set a goal of 80% renewable power by 2050. Today, about 22% of Germany's power is generated from renewables, and about 25% of that is from solar. On one day in the spring of 2012, Germany produced 22 GW of power from the sun—half the world's total and the equivalent of 20 nuclear power stations.[19]

Solar has also grown in the United States, but not nearly at the same rate. Following a surge in growth, about 3,000 MW of solar were online in 2011. The United States brought more new solar capacity online in 2012 than in the three prior years combined.[20]

Still, solar only provided 1% of the nation's power in 2012. There have been some federal subsidies for solar, including incentives to homeowners, most of them centering on tax credits. Many U.S. utilities also have policies that allow the meter to "run backward" when a residential PV installation provides more power than is needed. The owner is given credit for the power produced and provided to the utility. And in several cities optional financing methods have been developed that allow homeowners to finance a solar system over a period of time, often like a mortgage payment. However, the incentives are not as attractive as Germany's feed-in tariff.

In the United States, solar and wind energy are heavily criticized by many people and organizations who claim they only work because of heavy subsidies, and in a free market. Many feel this energy should not be subsidized. But all forms of energy production in the United States are subsidized one way or another.[8] A study by the Environmental Law Institute, "Estimating U.S. Government Subsidies to Energy Sources: 2002—2008," shows that during that six-year period traditional fossil fuels received about $70.2 billion in federal subsidies, whereas corn ethanol received $16.8 billion and "traditional renewables," which include solar and wind received, $12.2 billion, 17% of the amount received by fossil fuels.

A report by the Department of Energy's own Energy Information Agency, "Direct Federal Financial Interventions and Subsidies in Energy in Fiscal Year 2010," showed that in 2010 solar energy received about $1.1 billion in federal subsidies; coal $1.4 billion; natural gas and petroleum $2.8 billion; nuclear $2.5 billion; wind $5 billion; and biofuels (mainly corn ethanol) $6.6 billion. In short, fossil fuels and nuclear together received about $6.7 billion in direct subsidies, while solar and wind together received $6.17 billion.

In addition, there is a major difference in the way the direct federal subsidies help nonrenewables and renewables. According to the Environmental Law Institute, "Most of the largest subsidies to fossil fuels were written into the U.S. Tax Code as permanent provisions. By comparison, many subsidies for renewables are time-limited initiatives implemented through energy bills, with expiration dates that limit their usefulness to the renewables industry."[8]

Critical Thinking Questions

In answering these questions, you might find it helpful to refer back to Chapter 3.

1. Should the United States subsidize home solar and wind energy? If so, by how much? If not, what will be our environmentally best future?
2. Is the United States missing an opportunity in becoming a leader and innovator in the energy transition to countries like Germany and China by not providing large subsidies, or should the United States just let the matters depend on the free market?
3. What would you do, other than provide government subsidies, to encourage development of solar and other renewables?
4. What are the pros and cons of centralized and distributed solar power?

SUMMARY

- The use of renewable alternative energy sources, such as wind and solar energy, is growing rapidly. These energy sources do not cause air pollution, health problems, or climate changes. They offer our best chance to replace fossil fuels and develop a sustainable energy policy.
- Passive solar energy systems often involve architectural designs that enhance absorption of solar energy without requiring mechanical power or moving parts.
- Some active solar energy systems use solar collectors to heat water for homes.
- Systems to produce heat or electricity include power towers and solar farms.
- Photovoltaics convert sunlight directly into electricity.
- Hydrogen gas may be an important fuel of the future, especially when used in fuel cells.
- Water power today provides about 10% of the total electricity produced in the United States. Except in some developing nations, good sites for large dams are already in use. Water power is clean, but there is an environmental price to pay in terms of disturbance of ecosystems, sediment trapped in reservoirs, loss of wild rivers, and loss of productive land.
- Wind power has tremendous potential as a source of electrical energy in many parts of the world. Many utility companies are using wind power as part of energy production or as part of long-term energy planning. Environmental impacts of wind installations include loss of land, killing of birds, and degradation of scenery.
- Biofuels are of three kinds: firewood, wastes, and crops grown to produce fuels. Firewood has been important historically and will remain so in many developing nations and many rural parts of developed nations. Burning wastes is a good way to dispose of them, and their energy is a useful by-product. Crops grown solely as biofuels appear to be a net energy sink or of only marginal benefit, at considerable environmental costs, including competition for land, water, and fertilizers, and the use of artificial pesticides. At present they are not a good option.
- Geothermal energy, natural heat from Earth's interior, can be used as an energy source. The environmental effects of developing geothermal energy relate to specific site conditions and the type of heat—steam, hot water, or warm water. Environmental impacts may involve on-site noise, industrial scars, emission of gas, and disposal of saline or corrosive waters.

REEXAMINING THEMES AND ISSUES

Sean Randall/Getty Images, Inc.

HUMAN POPULATION

As the human population continues to increase, so does global demand for energy. Environmental problems from increased use of fossil fuels could be minimized by controlling population growth, increasing conservation efforts, and using alternative, renewable energy sources that do not harm the environment.

© Biletskiy_Evgeniy/iStockphoto

SUSTAINABILITY

Use of fossil fuels is not sustainable. To plan for energy sustainability, we need to rely more on alternative energy sources that are naturally renewable and do not damage the environment. To do otherwise is antithetical to the concept of sustainability.

© Anton Balazh 2011/iStockphoto

GLOBAL PERSPECTIVE

To evaluate the potential of alternative energy sources, we need to understand global Earth systems and identify regions likely to produce high-quality alternative energy that could be used in urban regions of the world.

ssguy/ShutterStock

URBAN WORLD

Alternative renewable energy sources have a future in our urban environments. For example, the roofs of buildings can be used for passive solar collectors or photovoltaic systems. Patterns of energy consumption can be regulated through use of innovative systems such as pump storage to augment production of electrical energy when demand is high in urban areas.

B2M Productions/Getty Images, Inc.

PEOPLE AND NATURE

Many environmentalists perceive alternative energy sources, such as solar and wind, as being linked more closely with nature than are fossil fuels, nuclear energy, or even water power. This is because solar and wind energy development requires less human modification of the environment. Solar and wind energy allow us to live more in harmony with the environment, and thus we feel more connected to the natural world.

George Doyle/Getty Images, Inc.

SCIENCE AND VALUES

We are seriously considering alternative energy today because we value environmental quality and energy independence and want to plan for a future when we run out of fossil fuels. Recognizing that burning fossil fuels creates many serious environmental problems and that petroleum will soon become less available, we are trying to increase our scientific knowledge and improve our technology to meet our energy needs for the future while minimizing environmental damage. Our present science and technology can lead to a sustainable energy future, but we will need to change our values and our behavior to achieve it.

KEY TERMS

active solar energy systems 369
alternative energy 367
biofuel 379
fuel cells 372
geothermal energy 381
nonrenewable alternative energy 367
passive solar energy systems 368
photovoltaics 370
renewable energy 367
solar collectors 369
tidal power 375
water power 373
wind power 377

STUDY QUESTIONS

1. What types of government incentives might encourage use of alternative energy sources? Would their widespread use affect our economic and social environment?
2. Your town is near a large river that has a nearly constant water temperature of about 15°C (60°F). Could the water be used to cool buildings in the hot summers? How? What would be the environmental effects?
3. Which has greater future potential for energy production, wind or water power? Which causes more environmental problems? Why?
4. What are some of the problems associated with producing energy from biomass?
5. It is the year 2500, and natural oil and gas are rare curiosities that people see in museums. Given the technologies available today, what would be the most sensible fuel for airplanes? How would this fuel be produced to minimize adverse environmental effects?
6. When do you think the transition from fossil fuels to other energy sources will (or should) occur? Defend your answer.

FURTHER READING

Botkin, D.B. 2010. ***Powering the Future: A Scientist's Guide to Energy Independence*** (Upper Saddle River, NJ: FT Press).

Boyle, G. 2004. ***Renewable Energy*** (New York: Oxford University Press).

Scudder, T. 2006. ***The Future of Large Dams: Dealing with Social, Environmental, Institutional and Political Costs*** (London, UK: Earthscan). The author has been a member of a World Bank team that tried to make a new dam and reservoir in Laos environmentally and culturally sound. The book is one of the best summaries of the present limitations of water power.

NOTES

1. Beetz, B. 2013. Deutsche Bank: Sustainable solar market expected in 2014, *PV Magazine.*
2. National Parks and Conservation Association. 2013. *Solar Energy, National Parks, and Landscape Protection in the Desert Southwest.* Fort Collins, CO: NPCA, Center for Park Research.
3. Eaton, W.W. 1978. Solar energy. In L.C. Ruedisili and M.W. Firebaugh, eds., *Perspectives on Energy,* 2nd ed., pp. 418–436. New York, NY: Oxford University Press.
4. U.S. Department of Energy, Energy Information Agency. October 13, 2012. *Utility-scale Installations Lead Solar Photovoltaic Growth.* http://www.eia.gov/todayinenergy/detail.cfm?id=8570
5. Berger, J.J. 2000. *Beating the Heat.* Berkeley, CA: Berkeley Hills Books.
6. Demeo, E.M., and P. Steitz. 1990. The U.S. electric utility industry's activities in solar and wind energy. In K.W. Bšer, ed., *Advances in Solar Energy,* Vol. 6, pp. 1–218. New York, NY: American Solar Energy Society.
7. Energy Information Agency. 2013. *Total Renewable Electricity Net Generation* (table). Washington, DC: U.S. Department of Energy. http://www.eia.gov/cfapps/ipdbproject/IEDIndex3.cfm?tid=6 & pid=29 & aid=12
8. Botkin, D.B. 2010. *Powering the Future: A Scientist's Guide to Energy Independence* Indianapolis, IN: Pearson FT Press Science.
9. Ocean Power Research Corporation, http://www.orpc.co/newsevents_orpcnews.aspx. Accessed March 20, 2013
10. AFP. 2008 (July 5). Japanese sailor first to cross Pacific in wave-powered boat. http://afp.google.com/article/ALeqM5g-0joaEAMTHqfOtl2MAFNrhmW646A.
11. Anonymous. 2009. *World Wind Energy Report 2008.* Bonn, Germany: World Wind Energy Association. http://www.wwindea.org/home/index.php.
12. World Energy Council. 2006 (April 24). World Firewood Supply. www.worldenergy.org.
13. Simet, A. 2012. German port receives 20,000-ton wood pellet trans-shipment. *Biomass Magazine* and Manomet Center for Conservation Science. 2010. Biomass Sustainability and Carbon Policy Study: Report to the Commonwealth of Massachusetts Department of Energy Resources.
14. Energy Information Agency. 2013. Electric Power Annual Report (with data for 2011).Washington, DC: Energy Information Agency.
15. Wright, P. 2000. Geothermal energy. *Geotimes* 45(7):16–18.
16. Duffield, W.A., J.H. Sass, and M.L. Sorey. 1994. *Tapping the Earth's Natural Heat.* U.S. Geological Survey Circular 1125.
17. Johnson, J.T. 1990 (May). The hot path to solar electricity. *Popular Science,* pp. 82–85.
18. Davidson, O. 2012 (November 13). Germany has built clean energy economy that U.S. rejected 30 years ago, *Inside Climate News.* http://insideclimatenews.org/news/20121113/germany-energiewende-clean-energy-economy-renewables-solar-wind-biomass-nuclear-renewable-energy-transformation.
19. Curry, A. 2013 (March 29). Can you have too much solar energy? *Slate.* (http://www.slate.com/articles/health_and_science/alternative_energy/2013/03/solar_power_in_germany_how_a_cloudy_country_became_the_world_leader_in_solar.2.html.
20. Handley, M. 2013 (April 12). Report: Solar scores big gains in electricity generation, *U.S. News and World Report.* http://www.usnews.com/news/articles/2013/04/12/report-solar-scores-big-gains-in-electricity-generation.

A CLOSER LOOK 16.1 NOTES

a. Kartha, S., and P. Grimes. 1994. Fuel cells: Energy conversion for the next century. *Physics Today* 47:54–61.

b. Piore, A. 2002 (April 15). Hot springs eternal: Hydrogen power. *Newsweek,* p.32H.

c. Schatz Solar Hydrogen Project. N. D. Pamphlet. Arcata, CA: Humboldt State University.

d. Fuel cell buses. http://www.fuelcells.org/info/charts/buses.pdf, created by Fuel Cells 2000.

e. Fuel Cell Bus Programs Worldwide. http://www.cleanairnet.org/infopool/1411/propertyvalue-19516.html.

CHAPTER 17

Nuclear Energy and the Environment

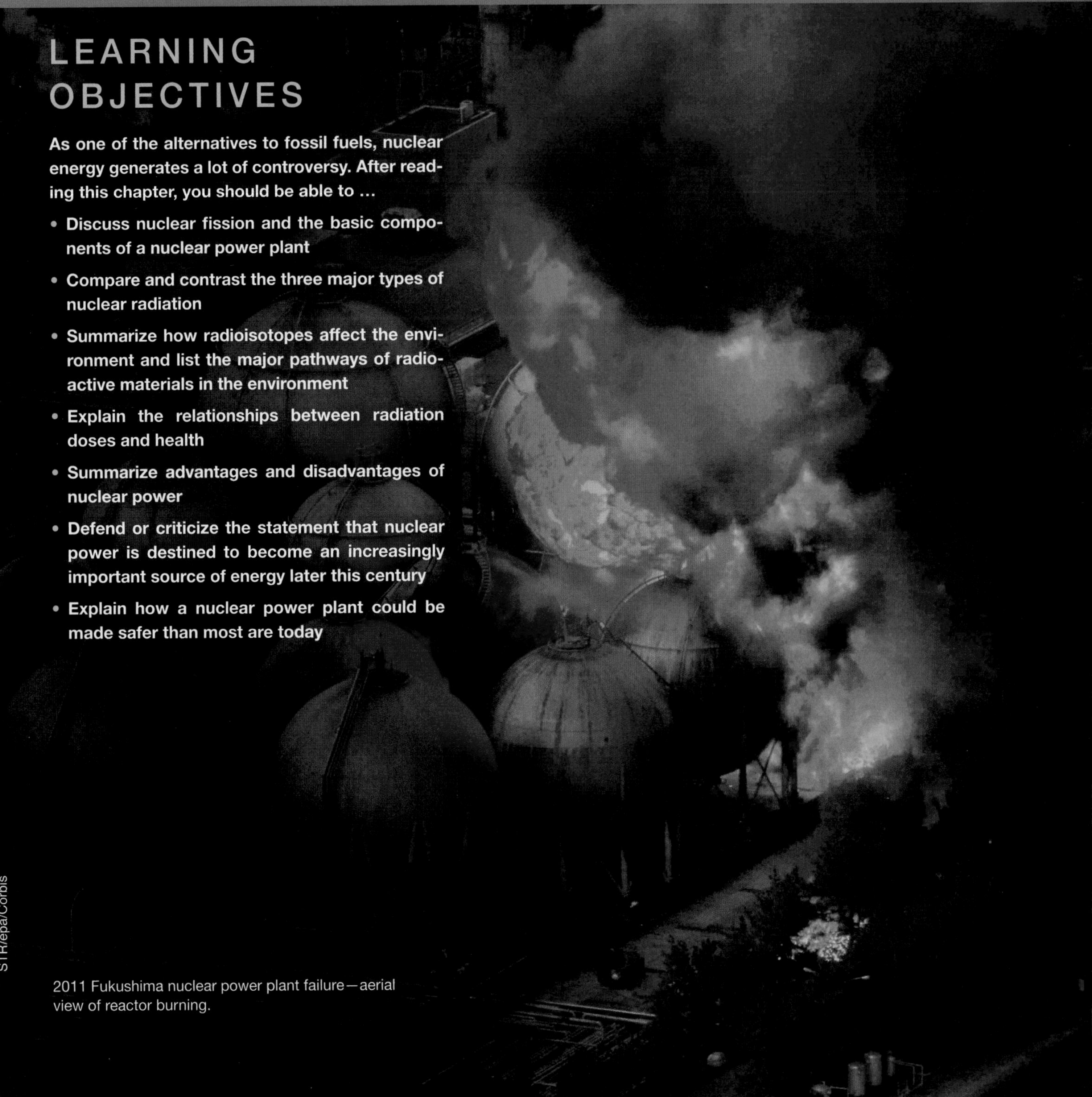

LEARNING OBJECTIVES

As one of the alternatives to fossil fuels, nuclear energy generates a lot of controversy. After reading this chapter, you should be able to ...

- **Discuss nuclear fission and the basic components of a nuclear power plant**
- **Compare and contrast the three major types of nuclear radiation**
- **Summarize how radioisotopes affect the environment and list the major pathways of radioactive materials in the environment**
- **Explain the relationships between radiation doses and health**
- **Summarize advantages and disadvantages of nuclear power**
- **Defend or criticize the statement that nuclear power is destined to become an increasingly important source of energy later this century**
- **Explain how a nuclear power plant could be made safer than most are today**

STR/epa/Corbis

2011 Fukushima nuclear power plant failure—aerial view of reactor burning.

CASE STUDY

Japan 2011: Aftermath of a Nuclear Disaster

Japan has several large nuclear power plants near Sendai, a few hundred miles northeast of Tokyo on the country's eastern coast (Figure 17.1). Major earthquake faults lie about 100 km offshore from the reactors—where the Pacific plate is forced beneath the Eurasian plate (see Chapter 7 for a discussion of plate tectonics). The faults have been extensively studied for decades and were known to be capable of producing magnitude 7–8 earthquakes. Tsunamis (large sea waves generated by earthquake activity) are also likely in this area. Hence, the nuclear reactors built there were designed to shut down when an earthquake occurred, and walls were constructed to protect against coastal flooding from a tsunami. On March 11, 2011, an earthquake of magnitude 9.1 occurred offshore. The earthquake displaced the ocean bottom (vertically by up to 9 m—30 ft) and generated a large set of tsunami waves (up to 30 m high—100 ft) that quickly reached the Japanese shore where the Fukushima-Daiichi Nuclear Power Plants were located (Figure 17.1).

The three operating nuclear power plants were immediately shut down as a result of the earthquake. However, inundation of seawater from tsunami waves overtopped defenses, flooding inland areas and the nuclear plants. Loss of cooling water in the plants from the flooding caused a nuclear disaster that released harmful nuclear radiation into the environment. (Specifics of the accident are discussed later in this chapter when we consider past nuclear power plant accidents.)

About 140,000 people were living within an area 30 km or less from the tsunami-damaged Fukushima nuclear power plants. At first, people were advised to stay in their homes and lock their doors and windows in an attempt to block out the radiation, but evacuation orders came quickly (Figure 17.2). Leaks of radiation occurred for several months following the accident; further precautions included a ban on consumption of food and drinks from the region, To ward off potential effects of radioactive iodine, the government issued pills with nonradioactive potassium iodine, which lessen the potential

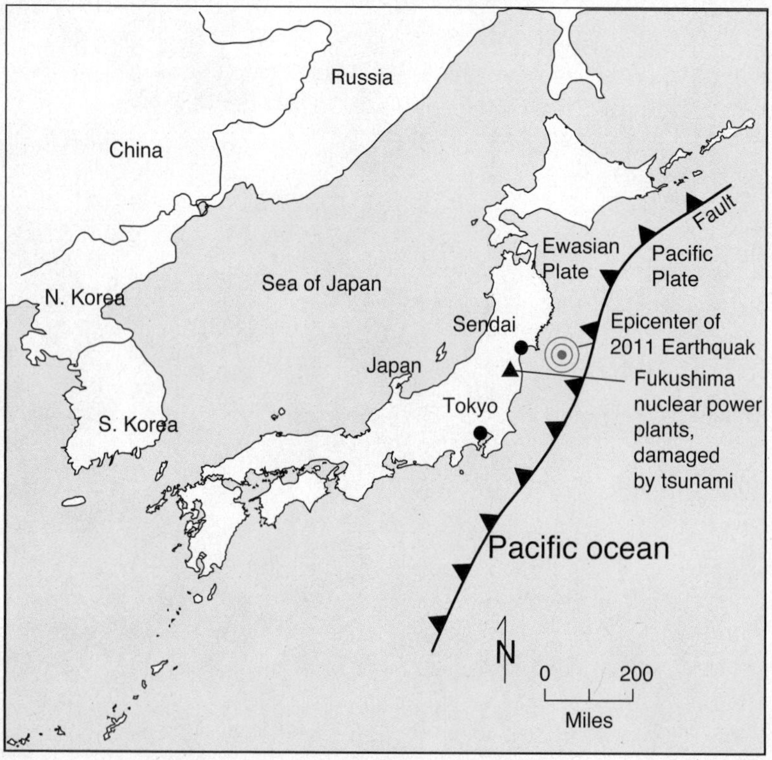

FIGURE 17.1 Map of Japan showing the locations of the epicenter of the 2011 earthquake, offshore fault, and location of damaged nuclear power plants.

Gamma-Rapho via Getty Images

FIGURE 17.2 Evacuees of 2011 Japan nuclear accident.

for radiation-induced thyroid disease. Thousands of people were evacuated from near the damaged nuclear power plants, out to a distance of approximately 30 km. However, some hotspots extended beyond the 30-km zone to as much as 45 km as the result of radioactive rain, wind gusts, and falling leaves from trees. It may be 100 years or longer before some locations in the areas affected by the greatest contamination of radioactive materials are safe for people to return to and live in. In other locations, people may return in a few decades.[1–3] We will discuss the more technical aspects of the accident later in this chapter.

Following the nuclear accidents, the Japanese people became very angry with the government and its response. They demanded action and a comprehensive evaluation of nuclear power in Japan. In May of 2012, all 50 reactors in Japan were temporally taken offline; only two were restarted by September 2012. Before the accident, Japan produced approximately 30% of its electric power from nuclear power plants and was planning to increase nuclear power to provide 50% of its electricity by 2030.

The 2011 accident in Japan is the most serious nuclear accident in the past three decades. Tokyo Electric Power Company (TEPCO), the utility company that managed the damaged reactors, was hit with billions of dollars of compensation and cleanup costs. When it couldn't pay the expenses, the company was nationalized and received a multibillion dollar government bailout. Before the accident, TEPCO had started to provide a broader energy mix, including solar and wind energy. The accident left the company with little hope of financing alternative energy sources. Nevertheless, TEPCO has stated that, while it prefers to continue a nuclear energy program, it will follow whatever future energy policy Japan adopts.

Three options to manage nuclear power in Japan were considered in 2012: (1) eliminate all nuclear power in Japan as soon as possible; (2) plan for the nuclear portion of electricity generation to be reduced to 15% by 2030; and (3) reduce electrical power to 25–30% of total power production by 2030. It is clear that the majority of people in Japan prefer eliminating nuclear power completely. However, a complete elimination of nuclear power quickly might cause significant economic problems, including higher rates for electricity that would force companies to relocate to other countries, taking jobs with them. Nevertheless, in the fall of 2012 the Japanese government stated a policy to phase out all nuclear power by 2040. With three nuclear power plants now under construction and strong opposition from business and industry to ending nuclear power, it is likely that the phase-out will extend past 2040.

The Japanese catastrophe of 2011 is a lesson to us all. We would like to be able to ensure that nuclear reactors are safer, but is this possible? Should nuclear power be part of America's future? Germany, partly in response to the Fukushima accident, decided to shut down all nuclear power by 2022. In this chapter, we will explore the concepts of nuclear energy in terms of what it is, our past experiences, and future prospects.

17.1 Current Role of Nuclear Power Plants in World Energy Production

Today, nuclear power provides about 14% of the world's electricity and 4.8% of the world's total energy. In the United States, 104 nuclear power plants produce about 20% of the country's electricity and about 8% of the total energy used. Worldwide, there are 435 operating nuclear power plants.[4,5] Nations differ greatly in the amount of energy they obtain from these plants. France ranks first, with about 78% of its electricity produced by nuclear energy. The United States ranks seventeenth in the percentage of electricity it obtains from nuclear power plants, but is the world leader in total energy produced from nuclear power (Figure 17.3). The United States produces more than 30% of the nuclear power produced in the world today.[4,5]

Most of the world's nuclear power plants are in North America, western Europe, Russia, China, and India. Most of the U.S. nuclear power plants are in the eastern half of the nation. The very few west of the Mississippi River are in Washington, California, Arizona, Nebraska, Kansas, and Texas. The last nuclear plant to be completed in the United States went online in 1996. In 2012 two new reactors in Georgia at an existing site (Vogtle) were approved to begin construction. Worldwide, 65 new reactors were being constructed in 2012. China is the leader with 26, followed by Russia with 10, and India with 7.[6]

17.2 What Is Nuclear Energy?

Hard as it may be to believe, **nuclear energy** is the energy contained in an atom's nucleus. Two nuclear processes can be used to release that energy to do work: **fission** and **fusion**. Nuclear fission is the splitting of atomic nuclei, and nuclear fusion is the fusing, or combining, of atomic nuclei. A by-product of both fission and fusion is the release of enormous amounts of energy. (Radiation and related terms are explained in A Closer Look 17.1. You may also wish to review the discussion of matter and energy in Chapter 14's A Closer Look 14.1.) (Note that at this time no fusion power reactor exists; research to develop such a power plant, which started in the middle of the twentieth century, continues—so far without success.)

Nuclear energy for commercial use is produced by splitting atoms in **nuclear reactors**, which are devices that produce controlled nuclear fission. In the United States, almost all of these reactors use a form of uranium oxide as fuel. Despite decades of research to try to develop it, nuclear fusion remains only a theoretical possibility.

Conventional Nuclear Reactors

The first human-controlled nuclear fission, demonstrated in 1942 by Italian physicist Enrico Fermi at the University of Chicago, led to the development of power plants that

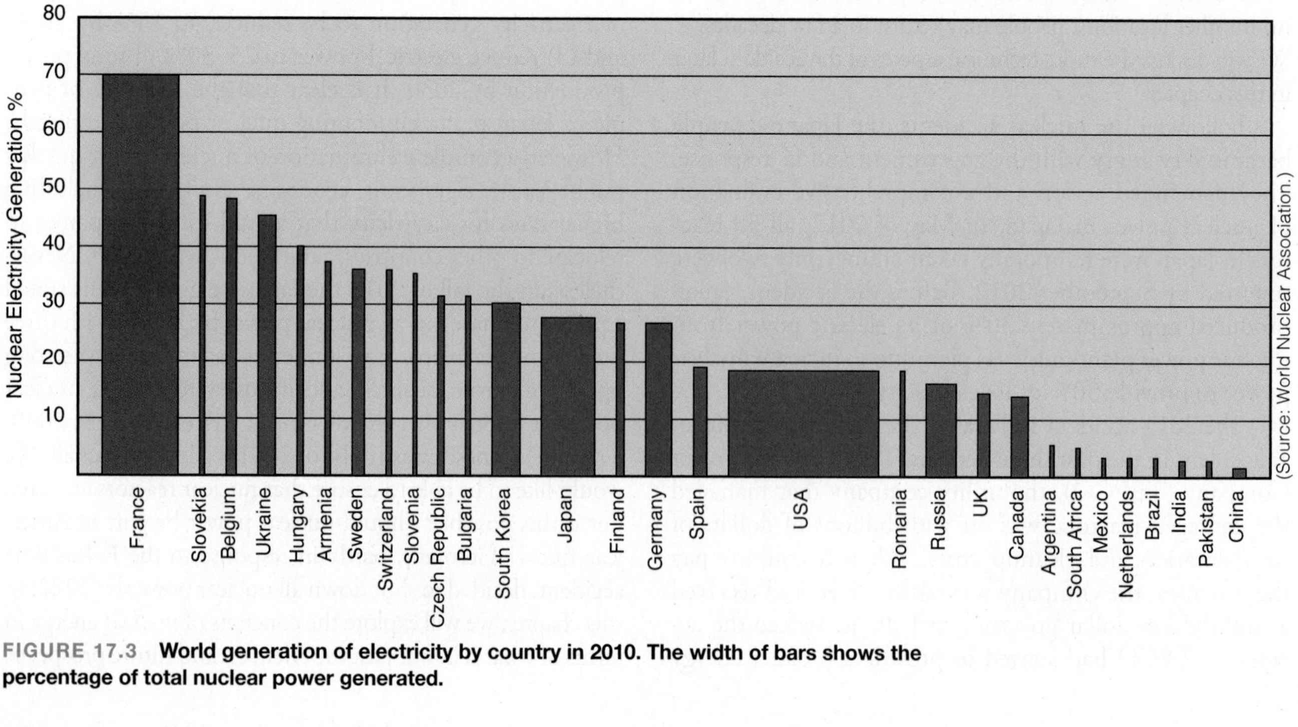

FIGURE 17.3 World generation of electricity by country in 2010. The width of bars shows the percentage of total nuclear power generated.

could use nuclear energy to produce electricity. Today, in addition to power plants to supply electricity for homes and industry, nuclear reactors power submarines, aircraft carriers, and icebreaker ships.

Nuclear fission produces much more energy per kilogram of fuel than other fuel-requiring sources, such as biomass and fossil fuels. For example, 1 kg (2.2 lb) of uranium oxide produces about the same amount of heat as 16 metric tons of coal.

Three types—isotopes—of uranium occur in nature: uranium-238, which accounts for approximately 99.3% of all natural uranium; uranium-235, which makes up about 0.7%; and uranium-234, about 0.005%. However, uranium-235 is the only naturally occurring fissionable (or fissile) material and is therefore essential to the production of nuclear energy. A process called enrichment increases the concentration of uranium-235 from 0.7% to about 3%. This enriched uranium is used as fuel.

The spontaneous decay of uranium atoms emits neutrons. Fission reactors split uranium-235 by neutron bombardment. This releases more neutrons than it took to create the first splitting (Figure 17.4). These released neutrons strike other uranium-235 atoms, releasing still more neutrons, other kinds of radiation, fission products, and heat. This is the "chain reaction" that is so famous, both for nuclear power plants and nuclear bombs—as the process continues, more and more uranium is split, releasing more neutrons and more heat. The neutrons released are fast moving and must be slowed down slightly (moderated) to increase the probability of fission.

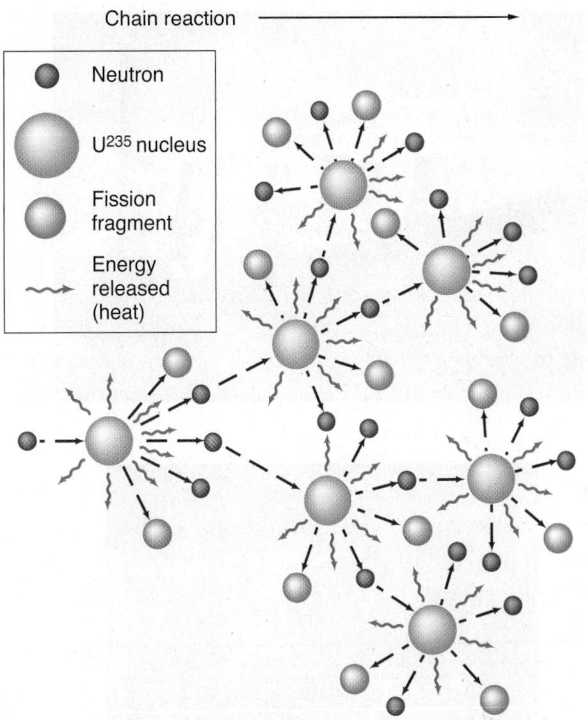

FIGURE 17.4 Fission of uranium-235. A neutron strikes the U-235 nucleus, producing fission fragments and free neutrons and releasing heat. The released neutrons may then strike other U-235 atoms, releasing more neutrons, fission fragments, and energy. As the process continues, a chain reaction develops.

All nuclear power plants use coolants to remove excess heat produced by the fission reaction. The rate of generation of heat in the fuel must match the rate at which heat is carried away by the coolant. Major nuclear accidents have occurred when something went wrong with the balance and heat built up in the reactor core.[5] The well-known term **meltdown** refers to a nuclear accident in which the coolant system fails, allowing the nuclear fuel to become so hot that it forms a molten mass that breaches the containment of the reactor and contaminates the outside environment with radioactivity.

The nuclear steam supply system includes heat exchangers (which extract heat produced by fission) and primary coolant loops and pumps (which circulate the coolant through the reactor). The heat is used to boil water, releasing steam that runs conventional steam-turbine electrical generators (Figure 17.5). In most common reactors, ordinary water is used as the coolant as well as the moderator. Reactors that use ordinary water are called "light water reactors" because there is also "heavy water," which combines deuterium with oxygen.[6]

Most reactors now in use consume more fissionable material than they produce and are known as **burner reactors**. Figure 17.6 shows the main components of a reactor: the core (consisting of fuel and moderator), control rods, coolant, and reactor vessel. The core is enclosed in the heavy, stainless steel reactor vessel; then, for safety and security, the entire reactor is contained in a reinforced concrete building.

In the reactor core, fuel pins—enriched uranium (uranium dioxide)—are fabricated into strong ceramic pellets capable of withstanding operating temperatures and intense radiation in the reactor. Pellets are loaded into hollow tubes (3–4 m long and less than 1 cm, or 0.4 in., in diameter). These tubes are packed together (40,000 or more in a reactor) in fuel subassemblies. A minimum fuel concentration is necessary to keep the reactor critical—that is, to achieve a self-sustaining chain reaction. A stable fission chain reaction in the core is maintained by controlling the number of neutrons that cause fission. Control rods, which contain materials that capture neutrons, are used to regulate the chain reaction. As the control rods are moved out of the core, the chain reaction increases; as they are moved into the core, the reaction slows. Full insertion of the control rods into the core stops the fission reaction.[7]

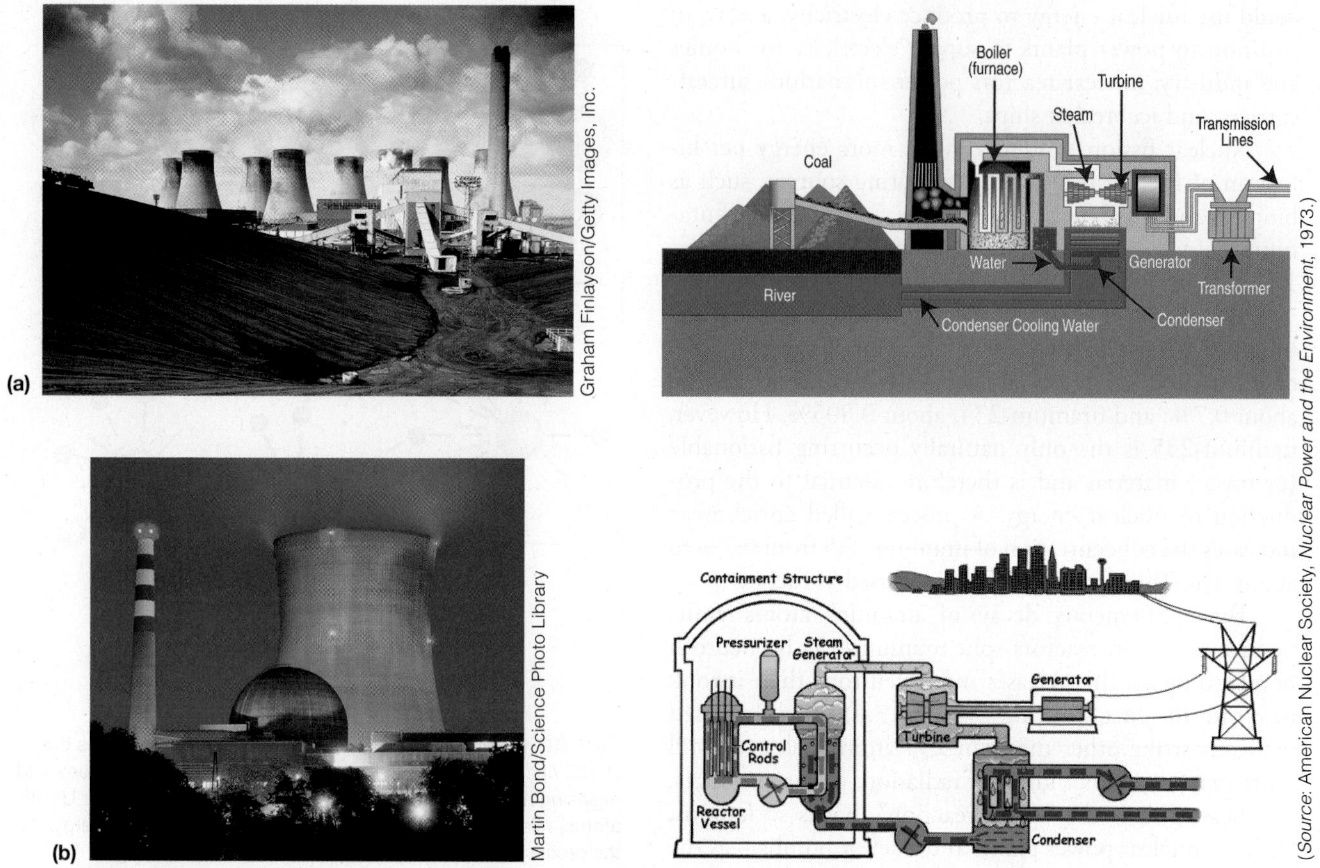

FIGURE 17.5 **Comparison of (a) a fossil fuel power plant and (b) a nuclear power plant with a boiling-water reactor.** Notice that the nuclear reactor has exactly the same function as the boiler in the fossil fuel power plant. The coal-burning plant **(a)** is in Ratcliffe-on-Saw, in Nottinghamshire, England, and the nuclear power station **(b)** is in Leibstadt, Switzerland.

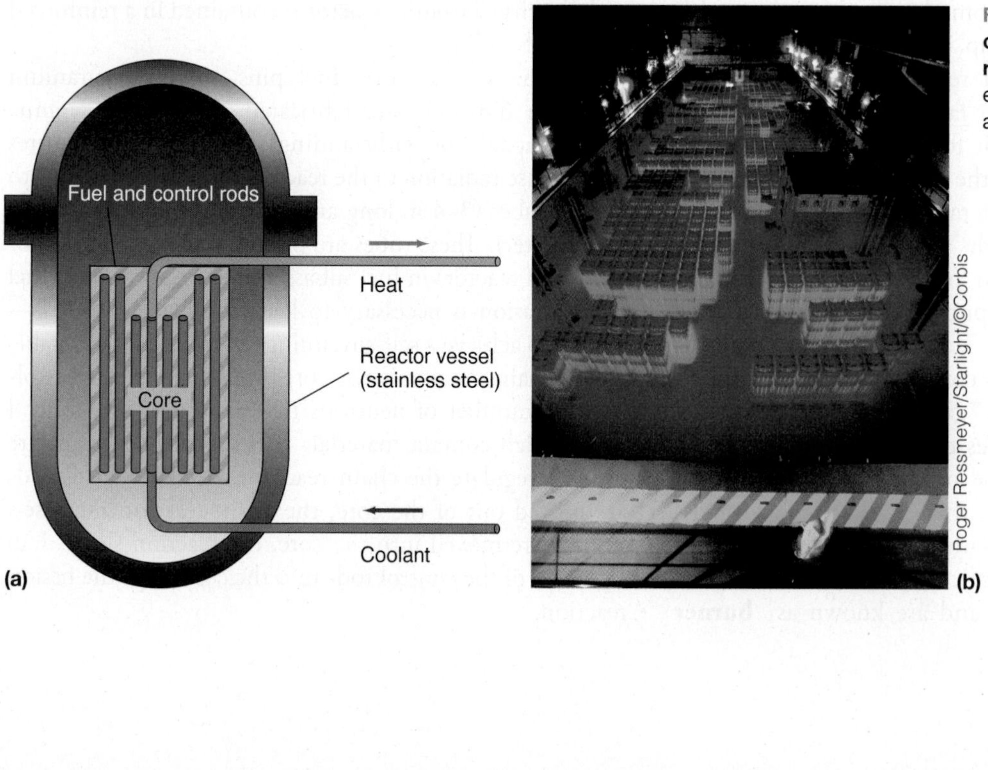

FIGURE 17.6 **(a) Main components of a nuclear reactor. (b)** Glowing spent fuel elements being stored in water at a nuclear power plant.

A CLOSER LOOK 17.1

Radioactive Decay

To many people, radiation is a subject shrouded in mystery. They feel uncomfortable with it, learning from an early age that nuclear energy may be dangerous because of radiation and that nuclear fallout from detonation of atomic bombs can cause widespread human suffering. One thing that makes radiation scary is that we cannot see it, taste it, smell it, or feel it. In this closer look, we try to demystify some aspects of radiation by discussing the process of radiation, or radioactivity.

First, we need to understand that radiation is a natural process, as old as the universe. Understanding the process of radiation involves understanding the **radioisotope**, a form of a chemical element that spontaneously undergoes **radioactive decay**. During the decay process, the radioisotope changes from one isotope to another and emits one or more kinds of radiation (Figure 17.7).

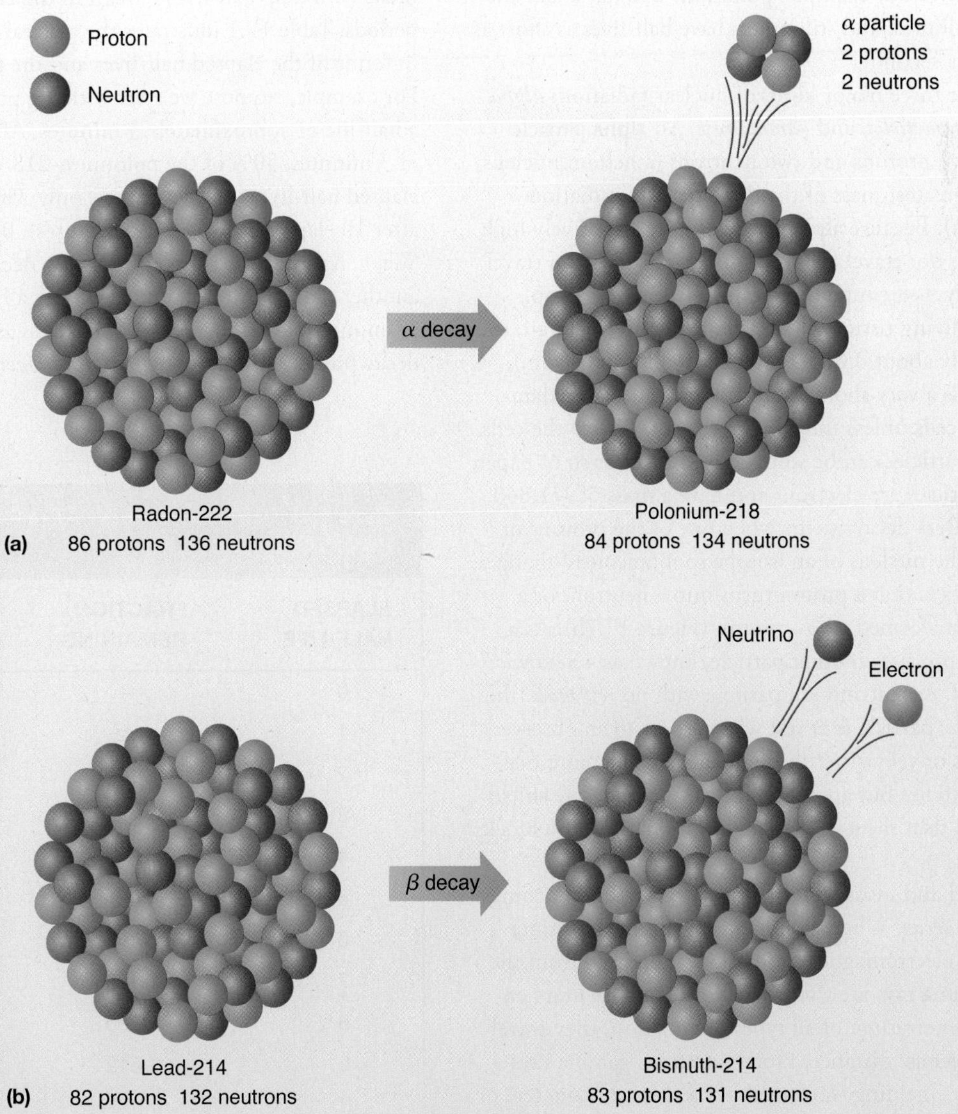

(*Source:* D.J. Brenner, *Radon: Risk and Remedy* [New York: Freeman, 1989]. Copyright 1989 by W.H. Freeman & Company. Reprinted with permission.)

FIGURE 17.7 Idealized diagrams showing (a) alpha and (b) beta decay processes.

You may recall from Chapter 7 that isotopes are atoms of an element that have the same atomic number (the number of protons in the nucleus) but that vary in atomic mass number (the number of protons plus neutrons in the nucleus). For example, two isotopes of uranium are $^{235}U_{92}$ and $^{238}U_{92}$. The atomic number for both isotopes of uranium is 92 (revisit A Closer Look 7.1); however, the atomic mass numbers are 235 and 238. The two different uranium isotopes may be written as uranium-235 and uranium-238 or ^{235}U and ^{238}U, respectively.

An important characteristic of a radioisotope is its *half-life*, the time required for one-half of a given amount of the isotope to decay to another form. Uranium-235 has a half-life of 700 million years, a very long time indeed! Radioactive carbon-14 has a half-life of 5,570 years, which is in the intermediate range, and radon-222 has a relatively short half-life of 3.8 days. Other radioactive isotopes have even shorter half-lives; for example, polonium-218 has a half-life of about 3 minutes, and still others have half-lives as short as a fraction of a second.

There are three major kinds of nuclear radiation: *alpha particles, beta particles,* and *gamma rays*. An alpha particle consists of two protons and two neutrons (a helium nucleus) and has the greatest mass of the three types of radiation (Figure 17.7a). Because alpha particles have a relatively high mass, they do not travel far. In air, alpha particles can travel approximately 5–8 cm (about 2–3 in.) before they stop. However, in living tissue, which is much denser than air, they can travel only about 0.005–0.008 cm (0.002–0.003 in.). Because this is a very short distance, they can't cause damage to living cells unless they originate very close to the cells. Also, alpha particles can be stopped by a sheet or so of paper.

Beta particles are electrons and have a mass of 1/1,840 of a proton. Beta decay occurs when one of the protons or neutrons in the nucleus of an isotope spontaneously changes. What happens is that a proton turns into a neutron, or a neutron is transformed into a proton (Figure 17.7b). As a result of this process, another particle, known as a *neutrino*, is also ejected. A neutrino is a particle with no *rest mass* (the mass when the particle is at rest with respect to an observer).[a] Beta particles travel farther through air than the more massive alpha particles but are blocked by even moderate shielding, such as a thin sheet of metal (aluminum foil) or a block of wood.

The third and most penetrating type of radiation comes from *gamma decay*. When gamma decay occurs, a gamma ray, a type of electromagnetic radiation, is emitted from the isotope. Gamma rays are similar to X rays but are more energetic and penetrating; of all types of radiation, they travel the longest average distance. Protection from gamma rays requires thick shielding, such as about a meter of concrete or several centimeters of lead.

Each radioisotope has its own characteristic emissions: Some emit only one type of radiation, whereas others emit a mixture. In addition, the different types of radiation have different toxicities (potential to harm or poison). In terms of human health and the health of other organisms, alpha radiation is most toxic when inhaled or ingested. Because alpha radiation is stopped within a very short distance by living tissue, much of the damaging radiation is absorbed by the tissue. When alpha-emitting isotopes are stored in a container, however, they are relatively harmless. Beta radiation has intermediate toxicity, although most beta radiation is absorbed by the body when a beta emitter is ingested. Gamma emitters are dangerous either inside or outside the body; but when they are ingested, some of the radiation passes out of the body.

Each radioactive isotope has its own half-life. Isotopes with very short half-lives are present only briefly, whereas those with long half-lives remain in the environment for long periods. Table 17.1 illustrates the general pattern for decay in terms of the elapsed half-lives and the fraction remaining. For example, suppose we start with 1 g polonium-218 with a half-life of approximately 3 minutes. After an elapsed time of 3 minutes, 50% of the polonium-218 remains. After 5 elapsed half-lives, or 15 minutes, only 3% is still present; and after 10 elapsed half-lives (30 minutes), 0.1% is still present. Where has the polonium gone? It has decayed to lead-214, another radioactive isotope, which has a half-life of about 27 minutes. The progression of changes associated with the decay process is often known as a *radioactive decay chain*.

Table 17.1 GENERALIZED PATTERN OF RADIOACTIVE DECAY

ELAPSED HALF-LIFE	FRACTION REMAINING	PERCENT REMAINING
0	--	100
1	1/2	50
2	1/4	25
3	1/8	13
4	1/16	6
5	1/32	3
6	1/64	1.5
7	1/128	0.8
8	1/256	0.4
9	1/512	0.2
10	1/1024	0.1

Now suppose we had started with 1 g uranium-235, with a half-life of 700 million years. Following 10 elapsed half-lives, 0.1% of the uranium would be left—but this process would take 7 billion years.

Some radioisotopes, particularly those of very heavy elements such as uranium, undergo a series of radioactive decay steps (a decay chain) before finally becoming stable, nonradioactive isotopes. For example, uranium decays through a series of steps to the stable nonradioactive isotope of lead. A decay chain for uranium-238 (with a half-life of 4.5 billion years) to stable lead-206 is shown in Figure 17.8. Also listed are the half-lives and types of radiation that occur during the transformations. Note that the simplified radioactive decay chain shown in Figure 17.8 involves 14 separate transformations and includes several environmentally important radioisotopes, such as radon-222, polonium-218, and lead-210. The decay from one radioisotope to another is often stated in terms of parent and daughter products. For example, uranium-238 is the parent of daughter product thorium-234.

Radioisotopes with short half-lives initially have a more rapid rate of change (nuclear transformation) than do radioisotopes with long half-lives. Conversely, radioisotopes with long half-lives have a less intense and slower initial rate of nuclear transformation but may be hazardous much longer.[b]

To sum up, when considering radioactive decay, two important facts to remember are (1) the half-life and (2) the type of radiation emitted.

Radioactive Elements	Radiation Emitted			Half-life		
	Alpha	Beta	Gamma	Minutes	Days	Years
Uranium-238 ↓	☢		☢			4.5 billion
Thorium-234 ↓		☢	☢		24.1	
Protactinium-234 ↓		☢	☢	1.2		
Uranium-234 ↓	☢		☢			247,000
Thorium-230 ↓	☢		☢			80,000
Radium-226 ↓	☢		☢			1,622
Radon-222 ↓	☢				3.8	
Polonium-218 ↓	☢	☢		3.0		
Lead-214 ↓		☢	☢	26.8		
Bismuth-214 ↓		☢	☢	19.7		
Polonium-214 ↓	☢			0.00016 (sec)		
Lead-210 ↓		☢	☢			22
Bismuth-210 ↓		☢			5.0	
Polonium-210 ↓	☢		☢		138.3	
Lead-206	None			Stable		

(*Source:* F. Schroyer, ed., *Radioactive Waste*, 2nd printing [American Institute of Professional Geologists, 1985].)

FIGURE 17.8 Uranium-238 decay chain.

17.3 Nuclear Energy and the Environment

The **nuclear fuel cycle** begins with the mining and processing of uranium, its transportation to a power plant, its use in controlled fission, and the disposal of radioactive waste. Ideally, the cycle should also include the reprocessing of spent nuclear fuel, and it must include the decommissioning of power plants. Since much of a nuclear power plant becomes radioactive over time from exposure to radioisotopes, disposal of radioactive wastes eventually involves much more than the original fuel.

Throughout this cycle, radiation can enter and affect the environment (Figure 17.9).

Problems with the Nuclear Fuel Cycle

- Uranium mines and mills produce radioactive waste that can expose mining workers and the local environment to radiation. Radioactive dust produced at mines and mills can be transported considerable distances by wind and water, so pollution can be widespread. Tailings—materials removed by mining but not processed—are generally left at the site, but in some instances, radioactive mine tailings were used in foundations and other building materials, contaminating dwellings.
- Uranium-235 enrichment and the fabrication of fuel assemblies also produce radioactive waste that must be carefully handled and disposed of.
- Site selection and construction of nuclear power plants in the United States are highly controversial. The environmental review process is extensive and expensive, often centering on hazards related to such events as earthquakes.
- The power plant or reactor is the site most people are concerned about because it is the most visible part of the cycle. It is also the site of past accidents, including partial meltdowns that have released harmful radiation into the environment.
- The United States at this time does not reprocess spent fuel from reactors to recover uranium and plutonium. However, many problems are associated with the handling and disposal of nuclear waste, as discussed later in this chapter.
- Waste disposal is controversial because no one wants a nuclear-waste-disposal facility nearby. The problem is that no one has yet figured out how to isolate nuclear waste for the millions of years that it remains hazardous.
- Nuclear power plants have a limited lifetime, usually estimated at only several decades, but decommissioning a plant (removing it from service) or modernizing it is a controversial part of the cycle and one with which we have little experience. For one thing, like nuclear waste, contaminated machinery must be safely disposed of or securely stored indefinitely.

Decommissioning or refitting a nuclear plant is very expensive (costing perhaps several hundred million

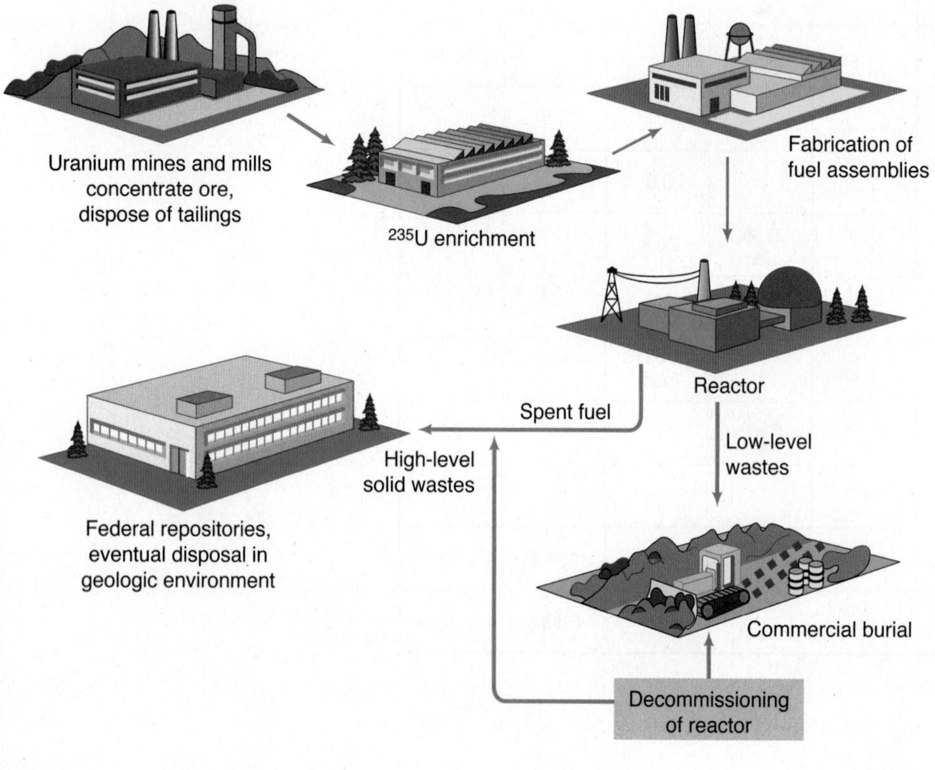

(*Source:* Office of Industry Relations, The Nuclear Industry, 1974.)

FIGURE 17.9 Idealized diagram showing the nuclear fuel cycle for the U.S. nuclear energy industry. Disposal of tailings, which because of their large volume may be more toxic than high-level waste, was treated casually in the past.

dollars) and is an important aspect of planning for the use of nuclear power. It costs more to dismantle a nuclear reactor than to build it. At present, power companies are filing to extend the licenses of several nuclear power plants originally slated to be decommissioned. Recently, the Nuclear Regulatory Commission (NRC), in response to a U.S. Court of Appeals decision, has started a study of the environmental effects of storing used nuclear fuel at reactors around the country. The NRC will not make a final decision on relicensing any nuclear plant until the study is completed (about 2014). Other potential problems that may impact relicensing of reactors include the discovery of active faults. Active faults have been found near the Diablo Canyon Nuclear Power Station in southern California, and earthquake studies are underway there. Farther south in California, the San Onofre Nuclear Power Plant was shut down on January 31, 2011, following a small release of radiation that resulted from excessive wear of steam-generator tubes. Cost of repairs is extensive, and the future of the San Onofre Power Plant is uncertain. (Licenses for the San Onofre and Diablo plants expire in 2022 and 2024, respectively.)

In addition to the above list of hazards in transporting and disposing of radioactive material, there are potential hazards in supplying other nations with reactors. Terrorist activity and the possibility of irresponsible people in governments add risks that are not present in any other form of energy production. For example, Kazakhstan inherited a large nuclear weapons-testing facility, covering hundreds of square kilometers, from the former Soviet Union. The soil in several sites contains "hotspots" of plutonium that pose a serious problem of toxic contamination. The facility also poses a security problem. There is international concern that this plutonium could be collected and used by terrorists to produce "dirty" bombs (conventional explosives that disperse radioactive materials). There may even be enough plutonium to produce small nuclear bombs.

Nuclear energy may indeed be one answer to some of our energy needs, but with nuclear power comes a level of responsibility not required by any other energy source.

17.4 Nuclear Radiation in the Environment and Its Effects on Human Health

Ecosystem Effects of Radioisotopes

As explained in A Closer Look 17.1, a radioisotope is an isotope of a chemical element that spontaneously undergoes radioactive decay. Radioisotopes affect the environment in two ways: by emitting radiation that affects other materials and by entering the normal pathways of mineral cycling and ecological food chains.

The explosion of a nuclear weapon does damage in many ways. At the time of the explosion, intense radiation of many kinds and energies is sent out, killing organisms directly. The explosion generates large amounts of radioactive isotopes, which are dispersed into the environment. Nuclear bombs exploding in the atmosphere produce a huge cloud that sends radioisotopes directly into the stratosphere, where the radioactive particles are widely dispersed by winds. Atomic fallout—the deposit of these radioactive materials around the world—was an environmental problem in the 1950s and 1960s, when the United States, the former Soviet Union, China, France, and Great Britain were testing and exploding nuclear weapons in the atmosphere.

The pathways of some of these isotopes illustrate the second way in which radioactive materials can be dangerous in the environment: They can enter ecological food chains (Figure 17.10). Let's consider an example. One of the radioisotopes emitted and sent into the stratosphere by atomic explosions was cesium-137. This radioisotope was deposited in relatively small concentrations but was widely dispersed in the Arctic region of North America. It fell on reindeer moss, a lichen that is a primary winter food of the caribou. A strong seasonal trend in the levels of cesium-137 in caribou was discovered; the level was highest in winter, when reindeer moss was the principal food, and lowest in summer. Eskimos who obtained a high percentage of their protein from caribou ingested the radioisotope by eating the meat, and their bodies concentrated the cesium. The more that members of a group depended on caribou as their primary source of food, the higher the level of the isotope in their bodies.

We are exposed to a variety of radiation sources from the sky, the air, and the food we eat (Figure 17.11). We receive natural background radiation from cosmic rays entering Earth's atmosphere from space and from naturally occurring radioisotopes in soil and rock. The average American receives about 2 to 4 mSv/yr. Of this, about 1 to 3 mSv/yr, or 50–75%, is natural. The differences are primarily due to elevation and geology. More cosmic radiation from outer space (which delivers about 0.3–1.3 mSv/yr) is received at higher elevations.

Radiation from rocks and soils (such as granite and organic shales) containing radioactive minerals delivers about 0.3 to 1.2 mSv/yr. The amount of radiation delivered from rocks, soils, and water may be much larger in areas where radon gas (a naturally occurring radioactive gas) seeps into homes. As a result, mountain states that also have an abundance of granitic rocks, such as Colorado,

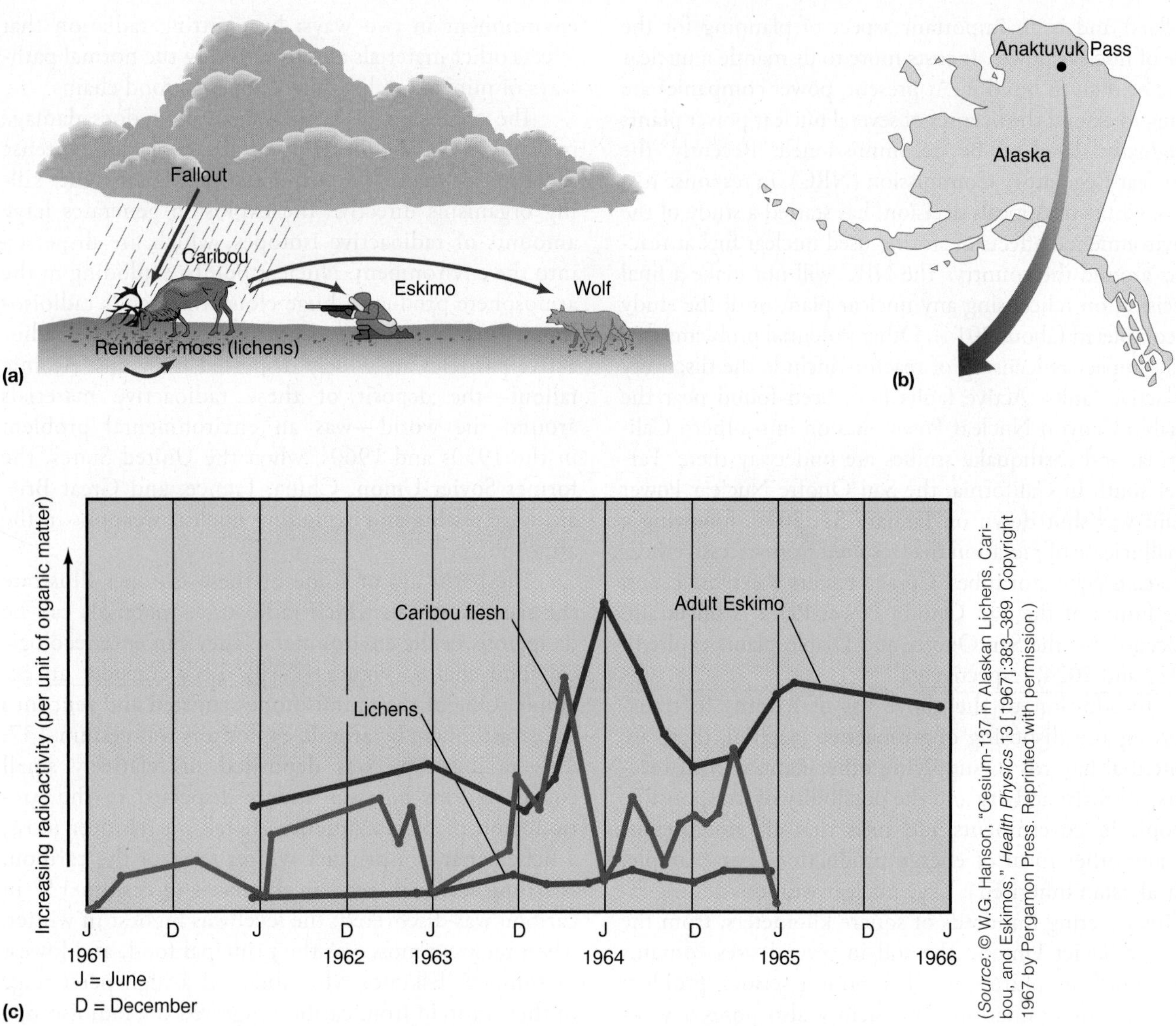

(*Source:* © W.G. Hanson, "Cesium-137 in Alaskan Lichens, Caribou, and Eskimos," *Health Physics* 13 [1967]:383–389. Copyright 1967 by Pergamon Press. Reprinted with permission.)

FIGURE 17.10 **Cesium-137 released into the atmosphere by atomic bomb tests was part of the fallout deposited on soil and plants. (a)** The cesium fell on lichens, which were eaten by caribou. The caribou were in turn eaten by Eskimos. **(b)** Measurements of cesium were taken in the lichens, caribou, and Eskimo in the Anaktuvuk Pass of Alaska. **(c)** The cesium was concentrated by the food chain. Peaks in concentrations occurred first in the lichens, then in the caribou, and last in the Eskimos.

have greater background radiation than do states that have a lot of limestone bedrock and are low in elevation, such as Florida. Despite this general pattern, locations in Florida where phosphate deposits occur have above-average background radiation because of a relatively high uranium concentration in the phosphate rocks.[8]

The amount of radiation we receive from our own bodies and other people is about 1.35 mSv/yr. Two sources are naturally occurring radioactive potassium-40 and carbon-14, which are present in our bodies and produce about 0.35 mSv/yr. Potassium is an important electrolyte in our blood, and one isotope of potassium (potassium-40) has a very long half-life. Although potassium-40 makes up only a very small percentage of the total potassium in our bodies, it is present in all of us. In short, we are all slightly radioactive, and if you choose to share your life with another person, you are also exposing yourself to a little bit more radiation.

To understand the effects of radiation, you need to be acquainted with the units used to measure radiation and the amount or dose of radiation that may cause a health problem. These are explained in A Closer Look 17.2.

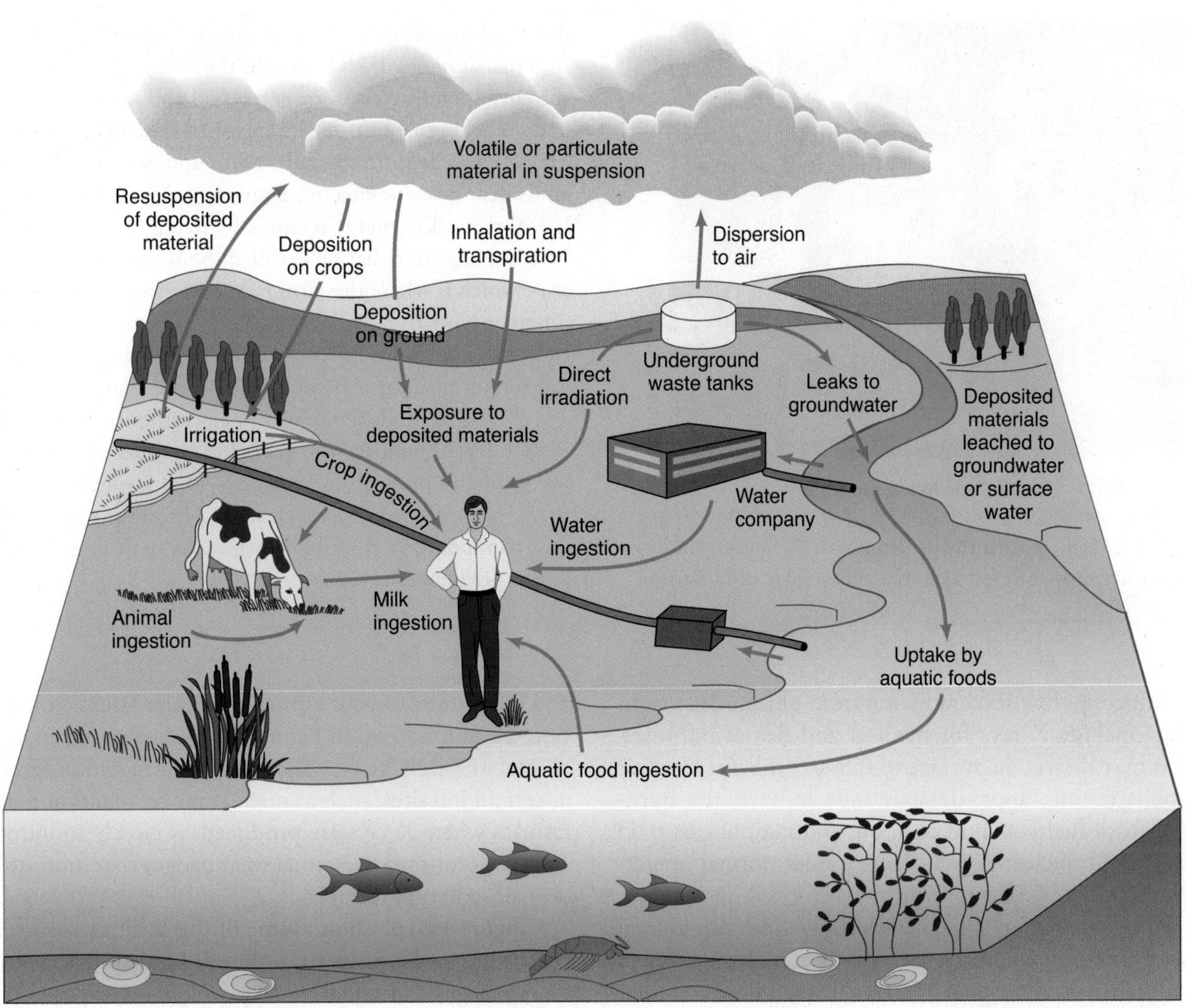

(*Source:* F. Schroyer, ed., *Radioactive Waste,* 2nd printing [American Institute of Professional Geologists, 1985].)

FIGURE 17.11 How radioactive substances reach people.

A CLOSER LOOK 17.2

Radiation Units and Doses

The units used to measure radioactivity are complex and somewhat confusing. Nevertheless, a modest acquaintance with them is useful in understanding and talking about radiation's effects on the environment.

A commonly used unit for radioactive decay is the curie (Ci), a unit of radioactivity defined as 37 billion nuclear transformations per second. The curie is named for Marie Curie and her husband, Pierre, who discovered radium in the 1890s. They also discovered polonium, which they named after Marie's homeland, Poland. The harmful effects of radiation were not known at that time, and both Marie Curie and her daughter died of radiation-induced cancer.[a] Her laboratory (Figure 17.12) is still contaminated today.

In the International System (SI) of measurement, the unit commonly used for radioactive decay is the *becquerel* (Bq), which is one radioactive decay per second. Units of measurement often used in discussions of radioactive isotopes, such as radon 222, are becquerels per cubic meter and *picocuries* per liter (pC/l). A picocurie is one-trillionth (10^{-12}) of a curie. Becquerels per cubic meter or picocuries per liter are therefore measures of the number of radioactive decays that occur each second in a cubic meter or liter of air.

FIGURE 17.12 **Marie Curie in her laboratory.**

When dealing with the environmental effects of radiation, we are most interested in the actual dose of radiation delivered by radioactivity. That dose is commonly measured in terms of *rads* (rd) and *rems*. In the International System, the corresponding units are *grays* (Gy) and *sieverts* (Sv). Rads and grays are the units of the absorbed dose of radiation; 1 gray is equivalent to 100 rads. Rems and sieverts are units of equivalent dose, or effective equivalent dose, where 1 sievert is 100 rems. The energy retained by living tissue that has been exposed to radiation is called the *radiation absorbed dose*, which is where the term rad comes from. Because different types of radiation have different penetrations and thus cause different degrees of damage to living tissue, the rad is multiplied by a factor known as the *relative biological effectiveness* to produce the rem or sievert units. When very small doses of radioactivity are being considered, the millirem (mrem) or millisievert (mSv)—that is, one-thousandth (0.001) of a rem or sievert—is used. For gamma rays, the unit commonly used is the roentgen, or, in SI units, coulombs per kilogram (C/kg).

Sources of low-level radiation from our modern technology include X rays for medical and dental purposes, which may deliver an average of 0.8–0.9 mSv/yr; nuclear weapons testing, approximately 0.04 mSv/yr; the burning of fossil fuels, such as coal, oil, and natural gas, 0.03 mSv/yr; and nuclear power plants (under normal operating conditions), 0.002 mSv/yr.[9]

Your occupation and lifestyle can affect the annual dose of radiation you receive. If you fly at high altitudes in jet aircraft, you receive an additional small dose of radiation—about 0.05 mSv for each flight across the United States. If you work at a nuclear power plant, you can receive up to about 3 mSv/yr. Living next door to a nuclear power plant adds 0.01 mSv/year, and sitting on a bench watching a truck carrying nuclear waste pass by would add 0.001 mSv to your annual exposure. Sources of radiation are summarized in Figure 17.13a, assuming an annual total of 3 mSv/yr.[10,11] The amount of radiation received at certain job sites, such as nuclear power plants and laboratories where X rays are produced, is closely monitored. At such locations, personnel wear badges that indicate the dose of radiation received.

Figure 17.13b shows some of the common sources of radiation to which we are exposed. Notice that exposure to radon gas can equal what people were exposed to as a result of the Chernobyl nuclear power accident, which occurred in the Soviet Union in 1986. In other words, in some homes, people are exposed to about the same radiation as that experienced by the people evacuated from the Chernobyl area.

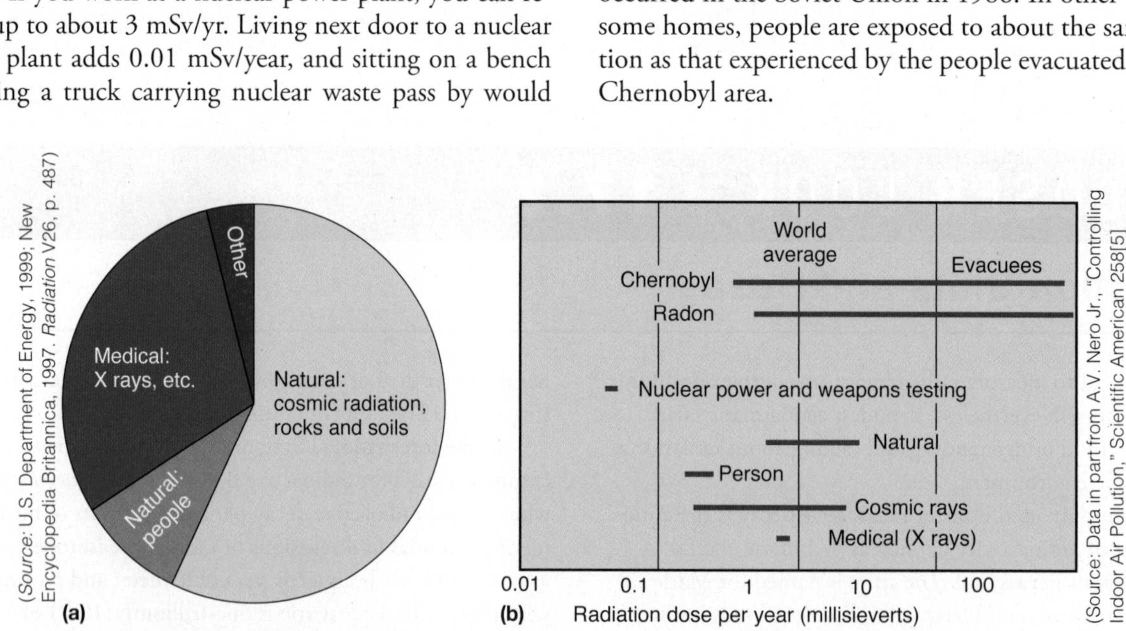

FIGURE 17.13 **(a)** Sources of radiation received by people; assumes annual dose of 3.0 mSv/yr, with 66% natural and 33% medical and other (occupational, nuclear weapons testing, television, air travel, smoke detectors, etc.). **(b)** Range in annual radiation dose to people from major sources.

Radiation Doses and Health

The most important question in studying radiation exposure in people is: At what point does the exposure or dose becomes a hazard to health? (See again A Closer Look 17.2.) Unfortunately, there are no simple answers to this seemingly simple question. We do know that a dose of about 5,000 mSv (5 sieverts) is considered lethal to 50% of people exposed to it. Exposure to 1,000–2,000 mSv is sufficient to cause health problems, including vomiting, fatigue, potential abortion of pregnancies of less than two months' duration, and temporary sterility in males. At 500 mSv, physiological damage is recorded. The maximum allowed dose of radiation per year for workers in industry is 50 mSv, approximately 30 times the average natural background radiation we all receive.[12] For the general public, the maximum permissible annual dose (for infrequent exposure) is set in the United States at 5 mSv, about three times the annual natural background radiation.[13] For continuous or frequent exposure, the limit for the general public is 1 mSv.

Most information about the effects of high doses of radiation comes from studies of people who survived the atomic bomb detonations in Japan at the end of World War II. We also have information about people exposed to high levels of radiation in uranium mines, workers who painted watch dials with luminous paint containing radium, and people treated with radiation therapy for disease.[14] Starting around 1917 in New Jersey, approximately 2,000 young women were employed painting watch dials with luminous paint. To maintain a sharp point on their brushes, they licked them and as a result were swallowing radium, which was in the paint. Many of the women died of anemia or bone cancer.[15] By 1924, dentists in New Jersey were reporting cases of jaw rot; within five years radium was known to be the cause.

Workers in uranium mines who were exposed to high levels of radiation have been shown to suffer a significantly higher rate of lung cancer than the general population. Studies show that there is a delay of 10 to 25 years between the time of exposure and the onset of disease.

Although there is vigorous, ongoing debate about the nature and extent of the relationship between radiation exposure and cancer mortality, most scientists agree that radiation can cause cancer. Some scientists believe that there is a linear relationship, such that any increase in radiation beyond the background level will produce an additional hazard. Others believe that the body can handle and recover from low levels of radiation exposure but that health effects (toxicity) become apparent beyond some threshold. The verdict is still out on this subject, but it seems prudent to take a conservative viewpoint and accept that there may be a linear relationship. Unfortunately, chronic health problems related to low-level exposure to radiation are neither well known nor well understood.

Radiation has a long history in the field of medicine. Drinking waters that contain radioactive materials goes back to Roman times. By 1899, the adverse effects of radiation had been studied and were well known; and in that year, the first lawsuit for malpractice in using X rays was filed. Because science had shown that radiation could destroy human cells, however, it was a logical step to conclude that drinking water containing radioactive material such as radon might help fight diseases such as stomach cancer. In the early 1900s it became popular to drink water containing radon, and the practice was supported by doctors, who stated that there were no known toxic effects. Although we now know that was incorrect, radiotherapy, which uses radiation to kill cancer cells in humans, has been widely and successfully used for a number of years.[15]

17.5 Nuclear Power Plant Accidents

Although the chance of a disastrous nuclear accident is estimated to be very low, the probability that an accident will occur increases with every reactor put into operation. According to the U.S. Nuclear Regulatory Commission's performance goal for a single reactor, the probability of a large-scale core meltdown in any given year should be no greater than 0.01%—one chance in 10,000. However, if there were 1,500 nuclear reactors (about three and a half times the present world total), a meltdown could be expected (at the low annual probability of 0.01%) every seven years. This is clearly an unacceptable risk.[16] Increasing safety by about 10 times would result in lower, more manageable risk, but the risk would still be appreciable because the potential consequences remain large.

Three Mile Island

One of the most dramatic events in the history of U.S. radiation pollution occurred on March 28, 1979, at the Three Mile Island nuclear power plant near Harrisburg, Pennsylvania. The malfunction of a valve, along with human errors (thought to be the major problem), resulted in a partial core meltdown. Intense radiation was released to the interior of the containment structure. Fortunately, the containment structure functioned as designed, and only a relatively small amount of radiation was released into the environment. Average exposure from the radiation emitted into the atmosphere has been estimated at 1 mSv, which is low in terms of the amount required to cause acute toxic effects. Average exposure to radiation in the surrounding area is estimated to have been approximately 0.012 mSv, which is only about 1% of the natural background radiation that people receive. However, radiation levels were much higher near the site. On the third day after the accident, 12 mSv/hour were measured at ground level near the site. By comparison, the average American receives about 2 mSv/year from natural radiation.

Next, we discuss the two most well-known and serious nuclear accidents: the Chernobyl Nuclear Power Plant in 1986 and the Fukushima Nuclear Power Plant in 2011. It is important to understand that these serious accidents resulted in part from human error.

Chernobyl

Lack of preparedness to deal with a serious nuclear power plant accident was dramatically illustrated by events that began unfolding on Monday morning, April 28, 1986. Workers at a nuclear power plant in Sweden, frantically searching for the source of elevated levels of radiation near their plant, concluded that it was not their installation that was leaking radiation; rather, the radioactivity was coming from the Soviet Union by way of prevailing winds. When confronted, the Soviets announced that an accident had occurred at a nuclear power plant at Chernobyl two days earlier, on April 26. This was the first notice to the world of the worst accident in the history of nuclear power generation.

It is speculated that the system that supplied cooling waters for the Chernobyl reactor failed as a result of human error, causing the temperature of the reactor core to rise to over 3,000°C (about 5,400°F), melting the uranium fuel, setting fire to the graphite surrounding the fuel rods that were supposed to moderate the nuclear reactions, and causing explosions that blew off the top of the building over the reactor. The fires produced a cloud of radioactive particles that rose high into the atmosphere. There were 237 confirmed cases of acute radiation sickness, and 31 people died of radiation sickness.[17]

In the days following the accident, nearly 3 billion people in the Northern Hemisphere received varying amounts of radiation from Chernobyl. With the exception of the 30-km (19-mi) zone surrounding Chernobyl, the world human exposure was relatively small. Even in Europe, where exposure was highest, it was considerably less than the natural radiation received during one year.[18]

In that 30-km zone, approximately 115,000 people were evacuated, and as many as 24,000 people were estimated to have received an average radiation dose of 0.43 Sv (430 mSv). Studies have found that since the accident the number of childhood thyroid cancer cases per year has risen steadily in Belarus, Ukraine, and the Russian Federation, the three countries most affected by Chernobyl. A total of 1,036 thyroid cancer cases have been diagnosed in children under 15 in the region. These cancer cases are believed to be linked to the released radiation from the accident, but other factors, such as environmental pollution, may also have played a role. It is predicted that a small percent of the roughly 1 million children exposed to the radiation eventually will contract thyroid cancer. Outside the 30-km zone, the increased risk of contracting cancer is very small and not likely to be detected from an ecological evaluation.[18,19]

To date, 4,000 deaths can be directly attributed to the Chernobyl accident, and according to one estimate, Chernobyl will ultimately be responsible for approximately 16,000 to 39,000 deaths. Proponents of nuclear power point out that this is fewer than the number of deaths caused each year by burning coal. The World Health Organization estimates there are several hundred thousand premature deaths worldwide annually as the result of air pollution from burning coal

Vegetation within 7 km of the power plant was either killed or severely damaged by the accident. Pine trees examined in 1990 around Chernobyl showed extensive tissue damage and still contained radioactivity. The distance between annual rings (a measure of tree growth) had decreased since 1986.

Scientists returning to the evacuated zone in the mid-1990s found, to their surprise, thriving and expanding animal populations. Species such as wild boar, moose, otters, waterfowl, and rodents seemed to be having a population boom in the absence of people. The wild boar population had increased tenfold since the evacuation. However, these animals may be paying a genetic price for living within the contaminated zone.

In areas surrounding Chernobyl, radioactive materials continue to contaminate soils, vegetation, surface water, and groundwater, presenting a hazard to plants and animals. The evacuation zone may be uninhabitable for a very long time (Figure 17.14) unless a way is found to remove the radioactivity. For example, the city of Prypyat, 5 km from Chernobyl, which had a population of 48,000 prior to the accident, is a "ghost city." It is abandoned,

Igor Kostin/©Corbis

FIGURE 17.14 Guard halting entry of people into the forbidden zone evacuated in 1986 as a result of the Chernobyl nuclear accident.

with blocks of vacant apartment buildings and rusting vehicles. Roads are cracking, and trees are growing as new vegetation transforms the urban land back to green fields.

The final story of the world's most serious nuclear accident is yet to completely unfold.[20] Estimates of the total cost of the Chernobyl accident vary widely, but it will probably exceed $200 billion. Chernobyl is the most serious nuclear accident to date; it certainly was not the first and is unlikely to be the last. Although the probability of a serious accident is very small at a particular site, the consequences may be great, perhaps posing an unacceptable risk to society. This is really not so much a scientific issue as a political one, involving values.

Japan's Nuclear Accident, 2011

In 2011, a very large earthquake of magnitude 9 occurred offshore of Japan and displaced the ocean bottom, which generated a large set of tsunami waves that quickly reached the Japanese shore (Figure 17.15) where several nuclear power plants were located. (See this chapter's opening case study.) When the earthquake occurred, the nuclear power plants immediately went into shutdown mode as a standard practice. What occurred shortly thereafter and produced a nuclear disaster more serious than anything that has occurred in the last several decades is a story with several lessons.

The Japanese have studied earthquakes and tsunamis for decades and were aware that large earthquakes in the region were possible. However, a magnitude 9.1 event was a surprise in that region, as was the extent of the tsunami, which overwhelmed sea walls protecting the plant and flooded the power stations. When the pumps that sent cooling water into the reactors (shut down but still very hot) failed, large amounts of heat and steam built up from water not circulating, and partial meltdowns at several reactors took place. Steam explosions and subsequent fires damaged buildings and reactor confinement structures, and radioactive materials were released into the environment. A number of other systems also failed, but the net result was that the people residing in the vicinity of the nuclear power plants were exposed to radiation sufficient to endanger human health.

The catastrophe could probably have been avoided. At the center of the controversy is the fact that some Japanese scientists had suggested that large tsunamis in the region were possible, and the tsunami protection barriers may not have been large enough. The possibility of very large tsunamis in the region with long return periods of about a thousand years was suggested as early as 2001. The problem was that these scientific findings were not believed or not taken seriously. The Japanese nuclear industry simply did not believe that the power plants were vulnerable. Even in the face of scientific evidence of the possibility of much larger tsunamis, the nuclear industry and government did not want to do the very expensive upgrades at the plants. In other words, warnings were ignored, perhaps to save money. It also has been suggested that the relationship between the nuclear industry and government is too closely aligned for proper regulation.[21,22]

If the nuclear power industry and government had taken heed of the warnings of their scientists, they might also have better protected the pumps and backup systems so that the plants would not flood should a tsunami overtop their defenses. In other words, steps could have been taken to locate the backup generators on higher ground.[22]

The Japanese catastrophe of 2011 is a lesson to us all that we are vulnerable to natural processes, and that we need to heed potential warnings of events that happen very infrequently. In the United States, we should learn from Japan's catastrophe and ensure that our reactors are safer—in this case, by closely evaluating each reactor site for the possibility of flooding and damage to pumps and backup pumps that circulate cooling water.

REUTERS/YOMIURI /LANDOV

FIGURE 17.15 Tsunami waves inundate Japan in 2011, flooding towns and cities and nuclear power plants.

17.6 Radioactive-Waste Management

Examination of the nuclear fuel cycle (refer back to Figure 17.9) illustrates some of the sources of waste that must be disposed of as a result of using nuclear energy to produce electricity. Radioactive wastes are by-products of nuclear reactors. The U.S. Federal Energy Regulatory Commission (FERC) defines two main categories of radioactive waste: low-level and high-level.

Low-Level Radioactive Waste

Low-level radioactive waste contains radioactivity in such low concentrations or quantities that it does not present a significant environmental hazard if properly handled. Low-level waste includes a wide variety of items, such as residuals or solutions from chemical processing; solid or liquid plant waste, sludges, and acids; and slightly contaminated equipment, tools, plastic, glass, wood, and other materials.

Low-level waste has been buried in near-surface burial areas in which the hydrologic and geologic conditions were thought to severely limit the migration of radioactivity. However, monitoring has shown that several U.S. disposal sites for low-level radioactive waste have not adequately protected the environment, and leaks of liquid waste have polluted groundwater. Of the original six burial sites, three were closed prematurely by 1979 due to unexpected leaks, financial problems, or loss of license, and as of 1995 only two remaining government low-level nuclear-waste repositories were still operating in the United States, one in Washington and the other in South Carolina. In addition, a private facility in Utah, run by Envirocare, accepts low-level waste. Construction of new burial sites, such as the Ward Valley site in southeastern California, has been met with strong public opposition, and controversy continues as to whether low-level radioactive waste can be disposed of safely.[23]

High-Level Radioactive Waste

High-level radioactive waste consists of commercial and military spent nuclear fuel; uranium and plutonium derived from military reprocessing; and other radioactive nuclear weapons materials. It is extremely toxic, and a sense of urgency surrounds its disposal as the total volume of spent fuel accumulates. At present, in the United States, tens of thousands of metric tons of high-level waste are being stored at more than a hundred sites in 40 states. Seventy-two of the sites are commercial nuclear reactors.[24–26]

These storage arrangements are at best a temporary solution, and serious problems with radioactive waste have occurred where it is being stored. Although improvements in storage tanks and other facilities will help, eventually some sort of disposal program must be initiated. Some scientists believe the geologic environment can best provide safe containment of high-level radioactive waste. Others disagree and have criticized proposals for long-term underground disposal of high-level radioactive waste. A comprehensive geologic disposal development program should have the following objectives:[26]

- Identification of sites that meet broad geologic criteria, including ground stability and slow movement of groundwater with long flow paths to the surface.
- Intensive subsurface exploration of possible sites to positively determine geologic and hydrologic characteristics.
- Predictions of the behavior of potential sites based on present geologic and hydrologic situations and assumptions about future changes in climate, groundwater flow, erosion, ground movements, and other variables.
- Evaluation of risk associated with various predictions.
- Political decision making based on risks acceptable to society.

What Should the United States Do with Its Nuclear Wastes?

For decades in the United States, the focal point for debates over nuclear wastes has been the plan to bury them deep in the earth at Yucca Mountain, Nevada. Extensive scientific evaluations were carried out, but the plan generated considerable resistance from the state and people of Nevada as well as from scientists not confident with the plan.[26] Some of the scientific questions at Yucca Mountain have concerned natural processes and hazards that might allow radioactive materials to escape, such as surface erosion, groundwater movement, earthquakes, and volcanic eruptions.

In 2002, Congress voted to submit a license of application for Yucca Mountain to the Nuclear Regulatory Commission, but in 2010 the Obama administration rejected that plan, and then Secretary of Energy Steven Chu set up a blue ribbon panel to consider the alternatives. At present, there are 70,000 tons of radioactive waste from nuclear power plants, and federally authorized temporary storage facilities for these are said to be full. That is to say, there is no government-sanctioned and locally approved place to put any more nuclear wastes. Yet they continue to build up.

Why was the Yucca Mountain repository so controversial, and why has it finally been canceled, or at least put on hold? The Nuclear Waste Policy Act of 1982 initiated a high-level nuclear waste-disposal program. The Department of Energy was given the responsibility to investigate several potential sites and make a recommendation. The 1982 act was amended in 1987; the amendment, along with the Energy Power Act of 1992, specified that high-level waste was to be disposed of underground in a deep, geologic waste repository. It also specified that the Yucca Mountain site in Nevada was to be the only site evaluated. Costs to build the facility reached $77 billion, but no nuclear wastes have ever been sent there.[26]

Evaluation of the safety and utility of a new waste repository would have to consider factors such as the following:[25–27]

- Probability and consequences of volcanic eruptions.
- Earthquake hazard.
- Flood hazard from storms.
- Estimation of how long the waste may be contained and the types and rates of radiation that may escape from deteriorated waste containers.
- How heat generated by the waste may affect moisture in and around the repository and the design of the repository.
- Characterization of groundwater flow near the repository.
- Identification and understanding of major geochemical processes that control the transport of radioactive materials.
- Possible rise of radioactive contaminated water to the surface after cracking in the bedrock from heat-generated radioactive wastes

One of the problems is just transporting the present amount of nuclear waste from power plants to any repository. According to previous U.S. government plans, beginning in 2010 some 70,000 tons of highly radioactive nuclear waste were going to be moved across the country to Yucca Mountain, Nevada by truck and train, one to six trainloads or truck convoys every day for 24 years. These train and truck convoys would have to be heavily guarded against terrorism and protected as much as possible against accidents.

A major question about the disposal of high-level radioactive waste is this: How credible are extremely long-range geologic predictions—those covering several thousand to a few million years?[27] Unfortunately, there is no easy answer to this question because geologic processes vary over both time and space. Climates change over long periods, as do areas of erosion, deposition, and groundwater activity. For example, large earthquakes even thousands of kilometers from a site may permanently change groundwater levels. The earthquake record for most of the United States extends back only a few hundred years; therefore, estimates of future earthquake activity are tenuous at best.

The bottom line is that geologists can suggest sites that have been relatively stable in the geologic past, but they cannot absolutely guarantee future stability. This means that policymakers (not geologists) need to evaluate the uncertainty of predictions in light of pressing political, economic, and social concerns.[26] In the end, the geologic environment may be deemed suitable for safe containment of high-level radioactive waste, but care must be taken to ensure that the best possible decisions are made on this important and controversial issue.

17.7 The Future of Nuclear Energy

The United States would need 1,000 new nuclear power plants of the same design and efficiency as existing nuclear plants to completely replace fossil fuels. The International Atomic Energy Agency, which promotes nuclear energy, says a total of just 4.7 million tons of "identified" conventional uranium stock can be mined economically. If we switched from fossil fuels to nuclear today, that uranium would run out in four years. Even the most optimistic estimate of the quantity of uranium ore is not reassuring: It would last only 29 years.[28]

Nevertheless, nuclear energy as a power source for electricity is now being seriously evaluated. Its advocates argue that nuclear power is good for the environment because:

- It does not contribute to potential global warming through release of carbon dioxide (see Chapter 20)
- It does not cause the kinds of air pollution or emit precursors (sulfates and nitrates) that cause acid rain (see Chapter 21).

The argument against nuclear power is based on political and economic considerations as well as scientific uncertainty about safety issues:

- Opponents emphasize that more than half the U.S. population lives within 75 miles of one of the nation's 104 nuclear power plants.
- Converting from coal-burning plants to nuclear power plants for the purpose of reducing carbon dioxide emissions would require an enormous investment in nuclear power to make a real impact.
- Given that safer nuclear reactors are only just being developed, there will be a time lag, so nuclear power is unlikely to have a real impact on environmental problems—such as air pollution, acid rain, and potential global warming—before at least the year 2050.
- Uranium ore to fuel conventional nuclear reactors is limited. The International Nuclear Energy Association estimates that at the 2004 rate of use, there would be 85 years of uranium fuel from known reserves, but if nations attempt to build many new power plants in the next decade, known reserves of uranium ore would be used up much more quickly.[29]

- Some nations may use nuclear reactors as a path to nuclear weapons. Reprocessing used nuclear fuel from a power plant produces plutonium that can be used to make nuclear bombs. There is concern that rogue nations with nuclear power could divert plutonium to make weapons or might sell plutonium to others, even terrorists, who would make nuclear weapons.[30]

Until 2001, proponents of nuclear energy were losing ground. Nearly all energy scenarios were based on the expectation that nuclear power would grow slowly or perhaps even decline in coming years. Since the Chernobyl and Japan accidents, many European countries have been reevaluating the use of nuclear power, and in most instances the number of nuclear power plants being built has significantly declined.

There is also a problem with present nuclear technology: Today's light-water reactors use uranium very inefficiently; only about 1% of it generates electricity, and the other 99% ends up as waste heat and radiation. Therefore, our present reactors are part of the nuclear-waste problem and not a long-term solution to the energy problem.

One way for nuclear power to be sustainable for at least hundreds of years would be to use a process known as breeding. **Breeder reactors** are designed to produce new nuclear fuel by transforming waste or lower-grade uranium into fissionable material. Although proponents of nuclear energy suggest that breeder reactors are the future of nuclear power, only a few are known to be operating anywhere in the world. Bringing breeder reactors online to produce safe nuclear power will take planning, research, and advanced reactor development. Also, fuel for the breeder reactors will have to be recycled because reactor fuel must be replaced every few years. What is needed is a new type of breeder reactor comprising an entire system that includes reactor, fuel cycle (especially fuel recycling and reprocessing), and less production of waste. Such a reactor appears possible but will require redefining our national energy policy and turning energy production in new directions. It remains to be seen whether this will happen.

New Kinds of Fission Reactors

One design philosophy that has emerged in recent decades in the nuclear industry is to build less complex, smaller reactors that are safer. Large nuclear power plants, which produce about 1,000 MW of electricity (sufficient power for a medium-size city), require an extensive set of pumps and backup equipment to ensure that adequate cooling is available to the reactor. Smaller reactors producing 10 to 50 MW are much less expensive than large reactors and can be designed to be safer. Small reactors have an economy of scale and potentially could be mass-produced. A 10 MW reactor can provide the electricity needs of 8,000 homes and could hold sufficient fuel to run 30 years before needing refueling. Thus a single small nuclear (10 MW) reactor could provide power for a village, while several linked together could power a small city. Small reactors can be designed with cooling systems that work by gravity and thus are less vulnerable to pump failure caused by power loss. Such cooling systems are said to have passive stability, and the reactors are said to be passively safe. Another approach is the use of helium gas to cool reactors that have specially designed fuel capsules capable of withstanding temperatures as high as 1,800°C (about 3,300°F). The idea is to design the fuel assembly so that it can't hold enough fuel to reach this temperature and thus the reactor is self-limiting and will not experience a core meltdown.[31]

Several new designs for conventional nonbreeder fission nuclear power plants are in development and the object of widespread discussion. Among these are the Advanced Boiling Water Reactor and the High Temperature Gas Reactor.[32,33] None are yet installed or operating anywhere in the world. The general goals of these designs are to increase safety, energy efficiency, and ease of operation. Some are designed to shut down automatically if there is any failure in the cooling system, rather than require the action of an operator. Although proponents of nuclear power believe these will offer major advances, it will be years, perhaps decades, until even one of each kind achieves commercial operation, so planning for the future cannot depend on them.

Fusion Reactors

In contrast to fission, which involves splitting heavy nuclei (such as uranium), fusion involves combining the nuclei of light elements (such as hydrogen) to form heavier ones (such as helium). As fusion occurs, heat energy is released. Nuclear fusion is the source of energy in our sun and other stars.

In a hypothetical fusion reactor, two isotopes of hydrogen—deuterium and tritium—are injected into the reactor chamber, where the necessary conditions for fusion are maintained. Products of the deuterium–tritium (DT) fusion include helium, producing 20% of the energy released, and neutrons, producing 80%.

Several conditions are necessary for fusion to take place. First, the temperature must be extremely high (approximately 100 million degrees Celsius for DT fusion). Second, the density of the fuel elements must be sufficiently high. At the temperature necessary for fusion, nearly all atoms are stripped of their electrons, forming a plasma—an electrically neutral material consisting of positively charged nuclei, ions, and negatively charged electrons. Third, the plasma must be confined long enough to ensure that the energy released by the fusion reactions exceeds the energy supplied to maintain the plasma.

The potential energy available when and if fusion reactor power plants are developed is nearly inexhaustible. One gram of DT fuel (from a water and lithium fuel supply) has the energy equivalent of 45 barrels of oil. Deuterium can be extracted economically from ocean water, and tritium can be produced in a reaction with lithium in a fusion reactor. Lithium can be extracted economically from abundant mineral supplies.

Many problems remain to be solved before nuclear fusion can be used on a large scale. Research is still in the first stage, which involves basic physics, testing of possible fuels (mostly DT), and magnetic confinement of plasma.

CRITICAL THINKING ISSUE

Indian Point: Should a Nuclear Power Installation Operate Near One of America's Major Cities?

There are two contradictory political movements regarding nuclear power plants in the United States. The federal government has supported an increase in the number of plants. The George W. Bush administration did, and the Obama administration has allocated $ 25.5 million for new "next-generation" nuclear power plants. But in February 2010, the Vermont Senate voted to prevent relicensing of the Yankee Power Plant, the state's only nuclear plant, after its current license expires in 2012. In 2009 the power plant leaked radioactive tritium into groundwater, and the plant's owners have been accused of misleading state regulators about underground pipes that carry cooling water at the plant.[34] Today in New York State there are major political pressures at the state, county, and local level to prevent the relicensing of Indian Point Power Plant on the Hudson River near New York City (Figure 17.16).

In 1962, after a series of contentious public hearings, Consolidated Edison began operating the first of three nuclear reactors at Indian Point, on the eastern shore of the Hudson River in Buchanan, New York, 38 km (24 mi) north of New York City. Indian Point's second and third reactors began operating in 1974 and 1976, respectively (Figure 17.13). The first unit had major problems and was finally shut down in 1974. The second and third have been operating since then, but their licenses expire in 2013 and 2015, respectively, and under U.S. law nuclear power plants must be relicensed. All three units are owned by Entergy Nuclear Northeast, a subsidiary of Entergy Corporation.

Twenty million people live within 80 km (50 mi) of this power plant, a proximity that causes considerable concern. Joan Leary Matthews, a lawyer for the New York State Department of Environmental Conservation, warned that "whatever the chances of a failure at Indian Point, the consequences could be catastrophic in ways that are almost too horrific to contemplate."[35]

The federal Nuclear Regulatory Commission (NRC) announced the beginning of the relicensing process on May 2, 2007. By 2008, the relicensing of the plant had become a regional controversy, opposed by the New York State government, Westchester County (where the plant is located), and a number of nongovernmental environmental organizations. The plant has operated for almost 50 years, so what's the problem?

Photo by Daniel B. Botkin

Photo by Daniel B. Botkin

FIGURE 17.16 Indian Point Energy Center (upper photograph) is a two-unit nuclear power plant on the eastern bank of the Hudson River within 40 km (25 mi) of New York City. The power plant faces relicensing that is generating controversy (lower photograph). One point of the debate is if a nuclear power plant should be so close to tens of millions of people.

There have been some: In 1980, one of the plant's two units filled with water (an operator's mistake). In 1982, the piping of the same unit's steam generator leaked and released radioactive water. In 1999, it shut down unexpectedly, but operators didn't realize it until the next day, when the batteries that automatically took over ran down.

In April 2007, a transformer burned in the second unit, radioactive water leaked into groundwater, and the source of the leak was difficult to find. Most recently, in 2009, a leak in the cooling system allowed 100,000 gallons of water to escape from the main system. Uneasiness about the plant's location increased after the terror attack on September 11, 2001. One of the hijacked jets flew close to the plant, and diagrams of unspecified nuclear plants in the United States have since been found in al Qaeda hideouts in Afghanistan.[36]

Proponents of nuclear power say these are minor problems, and there has been no major one. As far as they can tell, the plant is safe. The Energy Policy Act of 2005 promoted nuclear energy, and the Obama administration is moving ahead with federal funding of nuclear power plants. Others, however, such as New York State's attorney general Andrew Cuomo, believe the location is just too dangerous, and he has asked the Nuclear Regulatory Commission to deny Indian Point's relicensing, saying that it has "a long and troubling history of problems."

The conflict at Indian Point illustrates the worldwide debate about nuclear energy. Growing concern about fossil fuels has led to calls for increased use of nuclear power despite unanswered questions and unsolved problems regarding its use.

Critical Thinking Questions

1. Defend or support the position of the U.S. Nuclear Regulatory Commission that no nuclear power plant should be relicensed until a new study is completed on the environmental impacts of temporarily storing used reactor fuel on-site at reactors around the country.
2. Why or why shouldn't the Indian Point Nuclear Power Plant be permanently shut down?
3. Defend or criticize the statement that the potential dangers from nuclear power plants exceed the potential benefits of providing a sustainable source of electricity.
4. Summarize your opinions concerning the future of nuclear power in the United States.

SUMMARY

- Nuclear fission is the process of splitting an atomic nucleus into smaller fragments. As fission occurs, energy is released. The major components of a fission reactor are the core, control rods, coolant, and reactor vessel.
- Nuclear radiation occurs when a radioisotope spontaneously undergoes radioactive decay and changes into another isotope.
- The three major types of nuclear radiation are alpha, beta, and gamma.
- Each radioisotope has its own characteristic emissions. Different types of radiation have different toxicities; and in terms of the health of humans and other organisms, it is important to know the type of radiation emitted and the half-life.
- The nuclear fuel cycle consists of mining and processing uranium, generating nuclear power through controlled fission, reprocessing spent fuel, disposing of nuclear waste, and decommissioning power plants. Each part of the cycle is associated with characteristic processes, all with different potential environmental problems.
- The present burner reactors (mostly light-water reactors) use uranium-235 as a fuel. Uranium is a nonrenewable resource mined from the Earth. If many more burner reactors were constructed, we would face fuel shortages. Nuclear energy based on burning uranium-235 in light-water reactors is thus not sustainable. For nuclear energy to be sustainable, safe, and economical, we will need to develop breeder reactors.
- Radioisotopes affect the environment in two major ways: by emitting radiation that affects other materials and by entering ecological food chains.
- Major environmental pathways by which radiation reaches people include uptake by fish ingested by people, uptake by crops ingested by people, inhalation from air, and exposure to nuclear waste and the natural environment.
- The dose–response for radiation is fairly well established. We know the dose–response for higher exposures, when illness or death occurs. However, there are vigorous debates about the health effects of low-level exposure to radiation and what relationships exist between exposure and cancer. Most scientists believe that radiation can cause cancer. But, ironically, radiation can be used to kill cancer cells, as in radiotherapy treatments.
- We have learned from accidents at nuclear power plants that it is difficult to plan for the human factor. People make mistakes. We have also learned that we are not as prepared for accidents as we would like to think. Some believe that people are not ready for the responsibility of nuclear power. Others believe that we can design much safer power plants where serious accidents are impossible.

- There is a consensus that high-level nuclear waste may be safely disposed of in the geologic environment. The problem has been to locate a site that is safe and not objectionable to the people who make the decisions and to those who live in the region.
- Nuclear power is again being seriously evaluated as an alternative to fossil fuels. On the one hand, it has advantages: It emits no carbon dioxide, will not contribute to global warming or cause acid rain, and can be used to produce alternative fuels such as hydrogen. On the other hand, people are uncomfortable with nuclear power because of waste-disposal problems and possible accidents.

REEXAMINING THEMES AND ISSUES

Sean Randall/Getty Images, Inc.

HUMAN POPULATION

As the human population has increased, so has demand for electrical power. In response, a number of countries have turned to nuclear energy. Though relatively rare, accidents at nuclear power plants such as Chernobyl and Fukushima have exposed people to increased radiation, and there is considerable debate over potential adverse effects of that radiation. The fact remains that as the world population increases, and if the number of nuclear power plants increases, the total number of people exposed to a potential release of toxic radiation will increase as well.

© Biletskiy_Evgeniy/iStockphoto

SUSTAINABILITY

Some argue that sustainable energy will require a return to nuclear energy because it doesn't contribute to a variety of environmental problems related to burning fossil fuels. However, if nuclear energy is to significantly contribute to sustainable energy development, we cannot depend on burner reactors that will quickly use Earth's uranium resources; rather, development of safer breeder reactors will be necessary.

© Anton Balazh 2011/iStockphoto

GLOBAL PERSPECTIVE

Use of nuclear energy fits into our global management of the entire spectrum of energy sources. In addition, testing of nuclear weapons has spread radioactive isotopes around the entire planet, as have nuclear accidents. Radioactive isotopes that enter rivers and other waterways may eventually enter the oceans of the world, where oceanic circulation may further disperse and spread them.

ssguy/ShutterStock

URBAN WORLD

Development of nuclear energy is a product of our technology and our urban world. In some respects, it is near the pinnacle of our accomplishments in terms of technology.

B2M Productions/Getty Images, Inc.

PEOPLE AND NATURE

Nuclear reactions are the source of heat for our sun and are fundamental processes of the universe. Nuclear fusion has produced the heavier elements of the universe. Our use of nuclear reactions in reactors to produce useful energy is a connection to a basic form of energy in nature. However, abuse of nuclear reactions in weapons could damage or even destroy nature on Earth.

George Doyle/Getty Images, Inc.

SCIENCE AND VALUES

We have a good deal of knowledge about nuclear energy and nuclear processes. Still, people remain suspicious and in some cases frightened by nuclear power—in part because of the value we place on a quality environment and our perception that nuclear radiation is toxic to that environment. As a result, the future of nuclear energy will depend in part on how much risk is acceptable to society. It will also depend on research and development to produce much safer nuclear reactors.

KEY TERMS

STUDY QUESTIONS

1. If exposure to radiation is a natural phenomenon, why are we worried about it?
2. What is a radioisotope, and why is it important to know its half-life?
3. What is the normal background radiation that people receive? Why is it variable?
4. What are the possible relationships between exposure to radiation and adverse health effects?
5. What processes in our environment may result in radioactive substances reaching people?
6. Suppose it is recommended that high-level nuclear waste be disposed of in the geologic environment of the region in which you live. How would you go about evaluating potential sites?
7. Are there good environmental reasons to develop and build new nuclear power plants? Discuss both sides of the issue.

FURTHER READING

Botkin, D.B., ***Powering the Future: A Scientist's Guide to Energy Independence*** (Indianapolis, IN: Pearson FT Press, 2010).

Hore Lacy, I., ***Nuclear Energy in the 21st Century*** (New York: Academic Press, 2006). A pro-nuclear power plant book.

World Nuclear Association, 2012. Radioactive waste management. http://www.world-nuclear.org/info/Nuclear-Fuel-Cycle/Nuclear-Wastes/Radioactive-Waste-Management/#.UktZLCSRI5I

NOTES

1. Gupta, T. 2011. Exploring the nuclear accidents of Japan: A tremor, a tsunami, and the dark cloud it left behind. *USCience Review*. http://www-scf.usc.edu/~uscience/radiation_nuclear_accidents.html. Accessed 5/21/12.
2. Meyers, C. 2012. Japan earthquake anniversary: *Nuclear evacuees scarred by disaster one year later.* http://www.huffingtonpost.com/2012/02/13/japan-earthquake-anniversary_n_1272447.html#254326. Accessed 5/21/12.
3. Author unknown. 2012. Hot spots and blind spots. *The Economist*, October 8, 2011.
4. World Nuclear Association. 2012. *Nuclear Power in the USA.* http://www.world-nuclear.org.
5. World Nuclear Association. 2012. *Nuclear Power in the world today.* http://www.world-nuclear.org.
6. World Nuclear Association. 2012. *Plans for nuclear reactors worldwide.* http://www.world-nuclear.org.

7. Duderstadt, J.J. 1978. Nuclear power generation. In L.C. Ruedisili and M.W. Firebaugh, eds., *Perspectives on Energy*, 2d ed., pp. 249–273. New York, NY: Oxford University Press.

8. New Encyclopedia Britannica. 1997. Radiation, 26: 487.

9. Waldbott, G.L. 1978. *Health Effects of Environmental Pollutants*, 2d ed. St. Louis, MO: C.V. Mosby.

10. U.S. Department of Energy. 1999. Radiation (in) waste isolation pilot plant. Carlsbad, New Mexico. Accessed at http://www.wipp.energy.gov/.

11. New Encyclopedia Britannica. 1997. Radiation, 26: 487.

12. Waldbott, G.L. 1978. *Health Effects of Environmental Pollutants*, 2d ed. St. Louis: C.V. Mosby.

13. Ehrlich, P.R., A.H. Ehrlich, and J.P. Holdren. 1970. *Ecoscience: Population, Resources, Environment.* San Francisco, CA: W.H. Freeman.

14. University of Maine and Maine Department of Human Services. 1983 (February). *Radon in water and air.* Resource Highlights.

15. Brenner, D.J. 1989. *Radon: Risk and Remedy.* New York, NY: W.H. Freeman.

16. Till, O.E. 1989. Advanced reactor development. *Annals of Nuclear Energy* 16(6):301–305.

17. Anspaugh, L.R. , R.J. Catlin, and M. Goldman. 1988. The global impact of the Chernobyl reactor accident. *Science* 242:1513–1518.

18. Nuclear Energy Agency. 2002. *Chernobyl Assessment of Radiological and Health Impacts: 2002 Update of Chernobyl: Ten Years On.*

19. Anspaugh, L.R., R.J. Catlin, and M. Goldman. 1988. The global impact of the Chernobyl reactor accident. *Science* 242:1513–1518

20. Fletcher, M. 2000 (November 14). The last days of Chernobyl. *Times 2* London, pp. 3–5.

21. Normile, D. 2011. Scientific consensus on great quake came too late. *Science* 322:22–23.

22. Fackler, M. 2012. Nuclear disaster in Japan was avoidable, critics contend. *The New York Times,* March 9.

23. Weisman, J. 1996. Study inflames Ward Valley controversy. *Science* 271:1488.

24. Roush, W. 1995. Can nuclear waste keep Yucca Mountain dry—and safe? *Science* 270:1761.

25. Hanks, T.C., I.J. Winograd, R.E. Anderson, T.E. Reilly, and E.P. Weeks. 1999. *Yucca Mountain as a Radioactive Waste Repository.* U.S. Geological Survey Circular 1184.

26. Hanks et al., 1999. Nevada: A shift at Yucca Mountain. *The New York Times.* Nuclear Regulatory Commission. 2000 NRC's high-level waste program. Accessed July 18, 2000 at http://www.nrc.gov/NMSS/ DWM/hlw.htm.

27. Nuclear Regulatory Commission, 2000 NRC's high-level waste program.

28. Botkin, D.B. 2010. *Powering the Future: A Scientist's Guide to Energy Independence.* Indianapolis, IN: Pearson FT Press.

29. Botkin, D.B. October 20, 2008. The limits of nuclear power. *International Herald Tribune.*

30. Starke, L., ed. 2005. *Vital Signs 2005.* New York, NY: W.W. Norton, p. 139.

31. Carroll, C. 2010. Small town nukes. *National Geographic* 217(3):31–33.

32. U.S. Nuclear Regulatory Commission. 2008. *Backgrounder on New Nuclear Plant Designs.* http://www.nrc.gov/reading-rm/doc-collections/fact-sheets/new-nuc-plant-des-bg.html. Accessed October 16, 2012.

33. Kadak, A.C., R.G. Ballinger, T. Alvey, C.W. Kang, P. Owen, A. Smith, M. Wright and X. Yao. 1998. *Nuclear Power Plant Design Project: A Response to the Environmental and Economic Challenge of Global Warming Phase 1 Review of Options & Selection of Technology of Choice.* http://web.mit.edu/pebble-bed/background.html. Accessed October 24, 2012.

34. Anonymous, 2010. Vermont senate votes to close nuke plant in 2012. *The New York Times.*

35. Rosa, E.A., and R.E. Dunlap. 1994. Nuclear power: Three decades of public opinion. *Public Opinion Quarterly* 58(2):295–324.

36. Author unknown. May 6, 2009. Indian Point nuclear power plant. *The New York Times.* Updated: October 12, 2012. http://topics.nytimes.com/top/reference/timestopics/subjects/i/indian_point_nuclear_power_plant_ny/index.html?=8qa&scp=1-spot&sq=indian+point&st=nyt

A CLOSER LOOK 17.1 NOTES

a. Brenner, D.J. 1989. *Radon: Risk and Remedy.* New York, NY: W.H. Freeman.

b. Ehrlich, P.R., A.H. Ehrlich, and J.P. Holdren. 1970. *Ecoscience: Population, Resources, Environment.* San Francisco, CA: W.H. Freeman.

A CLOSER LOOK 17.2 NOTES

a. Brenner, D.J. 1989. *Radon: Risk and Remedy.* New York, NY: W.H. Freeman.

CHAPTER 18

Water Supply, Use, and Management

Associated Press

LEARNING OBJECTIVES

Although water is one of the most abundant resources on Earth, water management involves many important issues and problems. After reading this chapter, you should be able to . . .

- **Discuss why water is one of the major resource issues of the 21st century**
- **Define what a water budget is and state why it is useful in analyzing water supply problems and potential solutions**
- **Summarize how water can be conserved at home, in industry, and in agriculture**
- **Support the argument that sustainable water management will become more difficult as the demand for water increases**
- **Summarize the concept of virtual water and the link to water management and conservation**
- **Summarize the argument that natural service functions of wetlands are driving their conservation**
- **Support the argument that we are facing a growing global water shortage linked to our food supply**

Gavin Hellier/Robert Harding

Singapore River (1960 and today). A river transformed.

CASE STUDY

Singapore Water Supply: Emergence of a Water Ethic

Singapore (Figure 18.1) is one of the world's leading economies. The story of its emergence from a small, polluted 650-sq-km (250-sq-mi) tropical island in the 1960s to an environmentally aware economic force today is a remarkable one. With about 190 km (120 mi) of coastline, Singapore, located at the southern end of the Malaysian Peninsula, is the second smallest country in Southeast Asia—about one-fourth the size of Rhode Island. Since the1960s, Singapore has expanded its land area for housing and industry by reclaiming land from the ocean and may increase the land area by a total of 100 sq km (40 sq mi) by 2030. Land area in 2012 was about 700 sq km (270 sq mi). Reclaiming land in deeper water and infringing on shipping lanes will limit how much new land may ultimately be added.

When the country came into being in 1965, it was poverty-stricken and densely populated with not much going for it other than a great location for the shipping industry. The island, being tropical, has abundant rainfall of approximately 400 mm (100 in.) per year, but it suffered from a water shortage that was greatly restricting the country's ability to develop. The problem was that the substrata, largely granite with a mixture of marine sediments amid clay and sandy soil, did not hold water very well, and much of the water was lost. The highest elevation on the island is about 165 m (550 ft) above sea level (on granite rock); most of the island is less than about 20 m above sea level (66 ft). In 1965, Singapore had a few small rivers and some reservoirs. Most water was imported from Malaysia.

Water pollution was a major problem in 1965. When Singapore became independent, a high density of people lived near the harbor area and the local river near the sea. There were duck farms and pig farms intermixed with a variety of small businesses in substandard housing where both animal and human waste was simply dumped into the rivers. Waste from buildings and boats was thrown directly into the water. All sorts of garbage floated in the water, and the boats ferrying supplies back and forth between waiting ships leaked oil and dumped other waste into the water. Reportedly, the smell from the water was almost overwhelming.[1]

The transformation of Singapore was linked to establishing a quality environment and sustainable water supply. The Republic of Singapore has a strong elected government, and it has a lot to say about how its citizens live their lives. As a result, the government was able to make changes

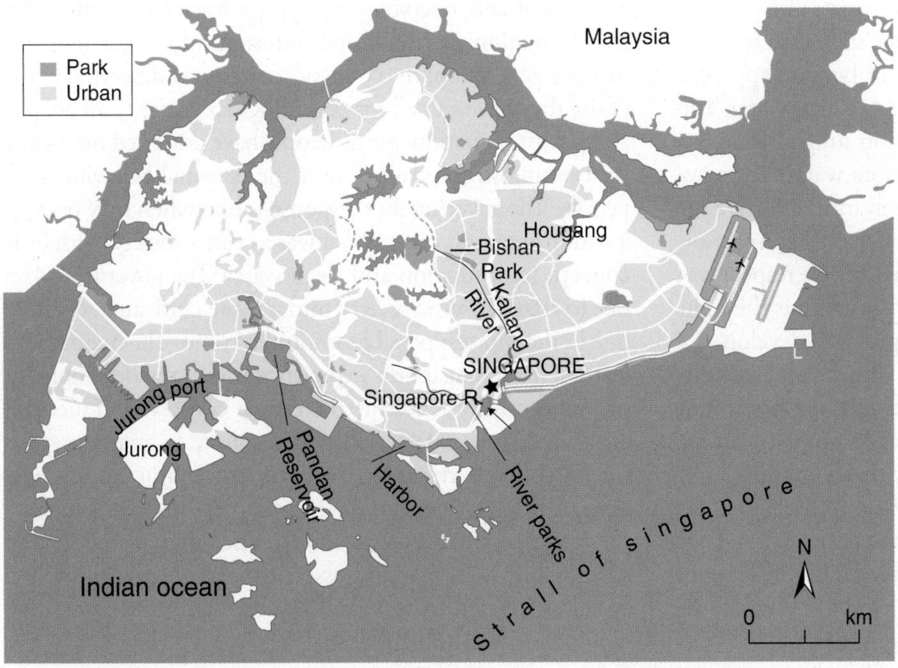

FIGURE 18.1 Map of Singapore showing urban areas, parks, and reservoirs.

quickly to the environment. As the first step in cleaning up the water, farmers and vendors living along the river were moved into apartment houses and low-income housing.

Next the government implemented a dredging program to remove all the garbage and replace the bed of the harbor and river with sand. Everything done was designed to create a more pleasant environment and, more importantly, provide a sustainable water supply. The basic plan was to ensure that not one drop of water would escape the water system. Wastewater is collected, treated, and reused as drinking water.

Singapore still imports some water from Malaysia, but that may eventually end. Storm water runoff is collected, treated, and used. The country has a number of reservoirs and has constructed a desalination plant (see Figure 18.1). There is a large reservoir near the ocean in the capital city that now stores fresh water (rather than the previous brackish saltwater) and provides recreational opportunities for the people. Upstream, river parks have been created adjacent to urban lands, and Singapore has made a complete economic transformation to one of the richest countries in the world.[1]

Singapore has a number of parks, and one of the most popular is Bishan Park (Figure 18.2), located along the Kallang River at an inland urban site. A restoration project transformed a 2.7-km-long straight concrete drainage channel into the meandering, more natural looking 3-m-long river. The adjacent 62-hectare (150 acres) Bishan Park is designed to create open spaces for the community to enjoy and to enable people to get close to the water. After completion of the park, there was no new deliberate introduction of wildlife, but the naturalized river produced habitats that increased the park's biodiversity significantly and wildlife arrived naturally—especially species of wildflowers, birds, and dragonflies, as well as migratory birds from Indonesia and Africa.[2]

Water usage by Singapore's populace started to increase as water became more abundant. It soon became apparent that a strategy was needed to conserve and use less water. The government instigated water conservation programs, and a tax was charged to people who used more water. The government strategy that seems to have been most effective for water conservation has been to build a personal connection between the water and the people, so that people respect water and want to protect and conserve it because it is the right thing to do—in other words, the people have developed a water ethic. In 1977, Luna Leopold (a famous U.S. hydrologist) introduced the concept of a reverence for rivers and thus a water ethic. He pointed out that history was on his side: Among other things he reported that Herodotus (an ancient Greek historian living in the year 430 BCE—2,400 years ago) wrote about how the Persians respected and revered rivers. Persians, Herodotus said, never degraded a river with waste from their bodies, not even washing their hands in it. In other words, the Persians had a river ethic.[3]

FIGURE 18.2 Bishan Park on the Kallang River in Singapore.

A program to establish a water ethic in Singapore started in the mid-1990s. Since then, water consumption per capita has declined by about 20%. The education program starts with children, and the goal is to bring the people closer to their water resources through using water, having waterfront areas for public use and recreation, and teaching a reverence for water resources. Children are encouraged to visit their city river parks, get their feet wet, take boat rides, and enjoy the water environment. As a result of this program, the small island nation has successfully managed to build a water ethic among its citizens, who now care much more about their water resources.[1,3]

Singapore's water program has had some ecological consequences in that some species of fish have been eliminated, and environments have changed. At the same time, new urban environments for wildlife are emerging along urban river parks and reservoirs. Singapore was able to effect the transformation of its society, industry, and water usage in just one generation; the country's citizens are deserving of the praise they have gained for providing safe water to all people, along with sanitation. Some people have criticized Singapore for exerting strong control over the personal behavior of its people. The government's response is that when they became a nation about 50 years ago, it was a cruder society, with little concern for sanitation and clean water. The government set out to transform the society into a more cultivated, civilized one in the shortest possible amount of time. The main lesson learned from Singapore's experience is that the development of a water ethic was a key to their water conservation success.[1]

Water is a critical, limited resource in many regions on Earth. As a result, it is one of the major resource issues of the 21st century.[4] This chapter discusses our water resources in terms of supply, use, management, and sustainability. It also addresses important environmental concerns related to water: wetlands, dams and reservoirs, channelization, and flooding.

18.1 Water

To understand water as a necessity, as a resource, and as a factor in the pollution problem, we must first understand its characteristics, its role in the biosphere, and its part in sustaining life. Water is a unique liquid; without it, life as we know it is impossible. Consider the following:

- Compared with most other common liquids, water has a high capacity to absorb and store heat. Its capacity to hold heat has important climatic significance. Solar energy warms the oceans, storing huge amounts of heat. The heat can be transferred to the atmosphere, developing hurricanes and other storms. The heat in warm oceanic currents, such as the Gulf Stream, warms Great Britain and western Europe, making these areas much more hospitable for humans than would otherwise be possible at such high latitudes.
- Water is the universal solvent. Because many natural waters are slightly acidic, they can dissolve a great variety of compounds, ranging from simple salts to minerals, including sodium chloride (common table salt) and calcium carbonate (calcite) in limestone rock. Water also reacts with complex organic compounds, including many amino acids found in the human body.
- Compared with other common liquids, water has a high surface tension, a property that is extremely important in many physical and biological processes that involve moving water through, or storing water in, small openings or pore spaces.
- Water is the only common compound whose solid form is lighter than its liquid form. (It expands by about 8% when it freezes, becoming less dense.) That is why ice floats. If ice were heavier than liquid water, it would sink to the bottom of the oceans, lakes, and rivers. If water froze from the bottom up, shallow seas, lakes, and rivers would freeze solid. All life in the water would die because cells of living organisms are mostly water, and as water freezes and expands, cell membranes and walls rupture. If ice were heavier than water, the biosphere would be vastly different from what it is, and life, if it existed at all, would be greatly altered.[5]
- Sunlight penetrates water to variable depths, permitting photosynthetic organisms to live below the surface.

A Brief Global Perspective

In brief, the water supply problem we are facing is a growing global water shortage that is linked to our food supply. Increasing water demand due to population growth is projected to reduce the per capita available water resource by more than 30% in the next 50 years. More than 2 billion people today live in countries with moderate to severe water stress (potential shortage of supply), and this number may grow to exceed 4 billion people by 2025. Poorer countries with large populations and harsh climates along with unreliable water sources, combined with lack of appropriate technology and water experience, are most vulnerable to future water problems.[4,6,7]

A review of the global hydrologic cycle, introduced in Chapter 7, is important here. The main process in the cycle is the global transfer of water from the atmosphere to the land and oceans and back to the atmosphere (Figure 18.3). Table 18.1 lists the relative amounts of water in the major storage compartments of the cycle. Notice that more than 97% of Earth's water is in the oceans; the next largest storage compartment, the ice caps and glaciers, accounts for another 2%. Together, these sources account for more than 99% of the total water, and both are generally unsuitable for human use because of salinity (seawater) and location (ice caps and glaciers). Only about 0.001% of the total water on Earth is in the atmosphere at any one time. However, this relatively small amount of water in the global water cycle, with an average atmosphere

Table 18.1 THE WORLD'S WATER SUPPLY (SELECTED EXAMPLES)

LOCATION	SURFACE AREA (KM^2)	WATER VOLUME (KM^3)	PERCENTAGE OF TOTAL WATER	ESTIMATED AVERAGE RESIDENCE TIME OF WATER
Oceans	361,000,000	1,230,000,000	97.2	Thousands of years
Atmosphere	510,000,000	12,700	0.001	9 days
Rivers and streams	–	1,200	0.0001	2 weeks
Groundwater (shallow to depth of 0.8 km)	130,000,000	4,000,000	0.31	Hundreds to many thousands of years
Lakes (fresh water)	855,000	123,000	0.01	Tens of years
Ice caps and glaciers	28,200,000	28,600,000	2.15	Tens of thousands of years and longer

Source: U.S. Geological Survey

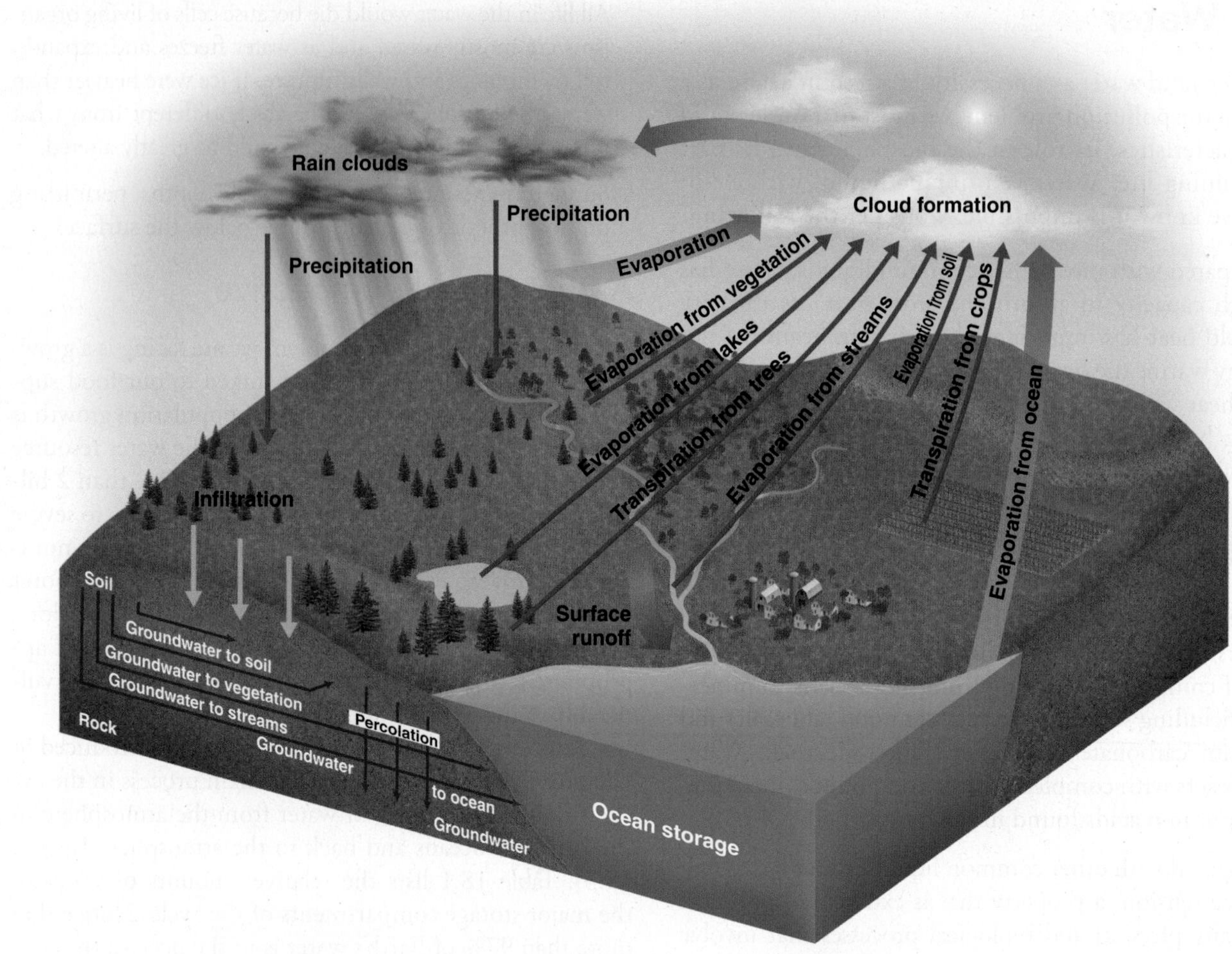

(*Source:* Modified form Council on Environment Quality and Department of State, The Global 2000 Report to the President, vol. 2 (Washington,DC).)

FIGURE 18.3 The hydrologic cycle, showing important processes and transfer of water.

residence time of only about nine days, produces all our freshwater resources through the process of precipitation.

Water can be found in liquid, solid, or gaseous form at a number of locations at or near Earth's surface. Depending on the specific location, the water's residence time may vary from a few days to many thousands of years (see Table 18.1). However, as already mentioned, more than 99% of Earth's water in its natural state is unavailable or unsuitable for beneficial human use. Thus, the amount of water for which all the people, plants, and animals on Earth compete is much less than 1% of the total.

As the world's population and industrial production of goods increase, the use of water will also accelerate. The average global per capita use of water (water footprint from 1996 to 2005) is shown in Figure 18.4. The lowest use of water per person is mostly in Central Africa and Asia, while high per capita use is mostly in North America, central South America, northern Africa, and some European countries.[8]

Compared with other resources, water is used in very large quantities. In recent years, the total mass (or weight) of water used on Earth per year has been approximately 1,000 times the world's total production of minerals, including petroleum, coal, metal ores, and nonmetals. Where it is abundant and readily available, water is generally a very inexpensive resource. In places where it is not abundant, such as the southwestern United States, the cost of water has been kept artificially low by government subsidies and programs.

Because the quantity and quality of water available at any particular time are highly variable, water shortages have occurred, and they will probably occur with increasing frequency, sometimes causing serious economic disruption and human suffering.[6,7] In the Middle East and northern Africa, scarce water has led to harsh exchanges and threats between countries and could even lead to war. The U.S. Water Resources Council estimates that water use in the United States by the year 2020 may exceed surface-water resources by 13%.[6,7] Therefore, an important question is: How can we best manage our water resources, use, and treatment to maintain adequate supplies?

Groundwater and River/Stream Flow

Before moving on to issues of water supply and management, we introduce groundwater and surface water and the terms used in discussing them. You will need to be familiar

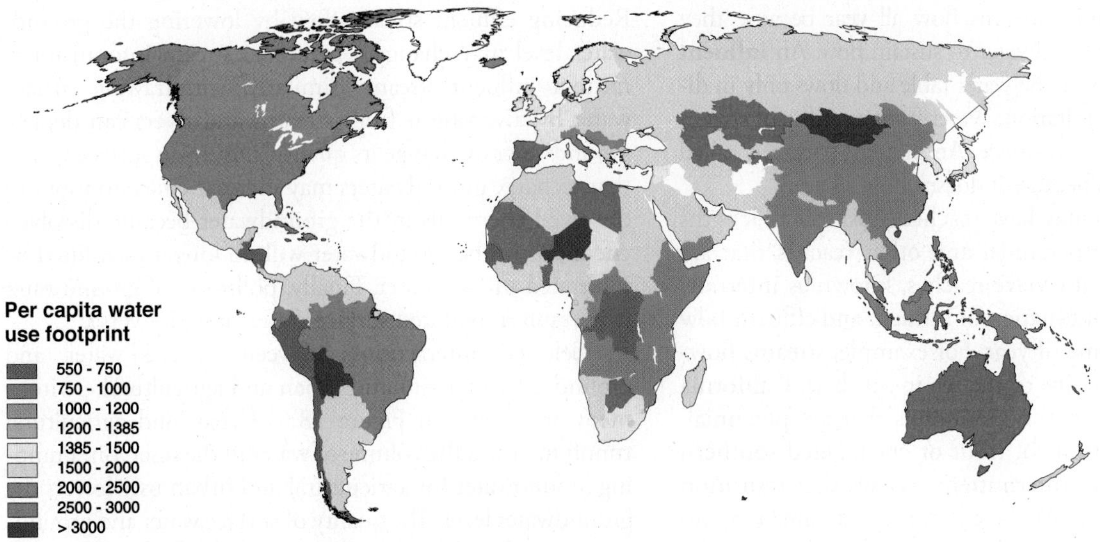

(Source: modified from Henderson, L.J. 1913. *The Fitness of the Environment: An Inquiry into the Biological Significance of the Properties of Matter*. New York: Macmillan.)

FIGURE 18.4 **Per capita use of water. Units are (m^3/year) per person**

with this terminology to understand many environmental issues, problems, and solutions.

The term **groundwater** usually refers to the water below the water table, where saturated conditions exist. The upper surface of the groundwater is called the *water table*.

Rain that falls on the land evaporates, runs off the surface, or moves below the surface and is transported underground. Locations where surface waters move into (infiltrate) the ground are known as *recharge zones*. Places where groundwater flows or seeps out at the surface, such as springs, are known as *discharge zones* or *discharge points*.

Water that moves into the ground from the surface first seeps through pore spaces (empty spaces between soil particles or rock fractures) in the soil and rock known as the *vadose zone*. This area is seldom saturated (not all pore spaces are filled with water). The water then enters the groundwater system, which is saturated (all of its pore spaces are filled with water).

An *aquifer* is an underground zone or body of earth material from which groundwater can be obtained (from a well) at a useful rate. Loose gravel and sand with lots of pore space between grains and rocks or many open fractures generally make good aquifers. Groundwater in aquifers usually moves slowly at rates of centimeters or meters per day. When water is pumped from an aquifer, the water table is depressed around the well, forming a cone of *depression*. Figure 18.5 shows the major features of a groundwater and surface-water system.

Streams may be classified as effluent or influent. In an **effluent stream**, the flow is maintained during the dry season by groundwater seepage into the stream channel from the subsurface. A stream that flows all year is called a *perennial*

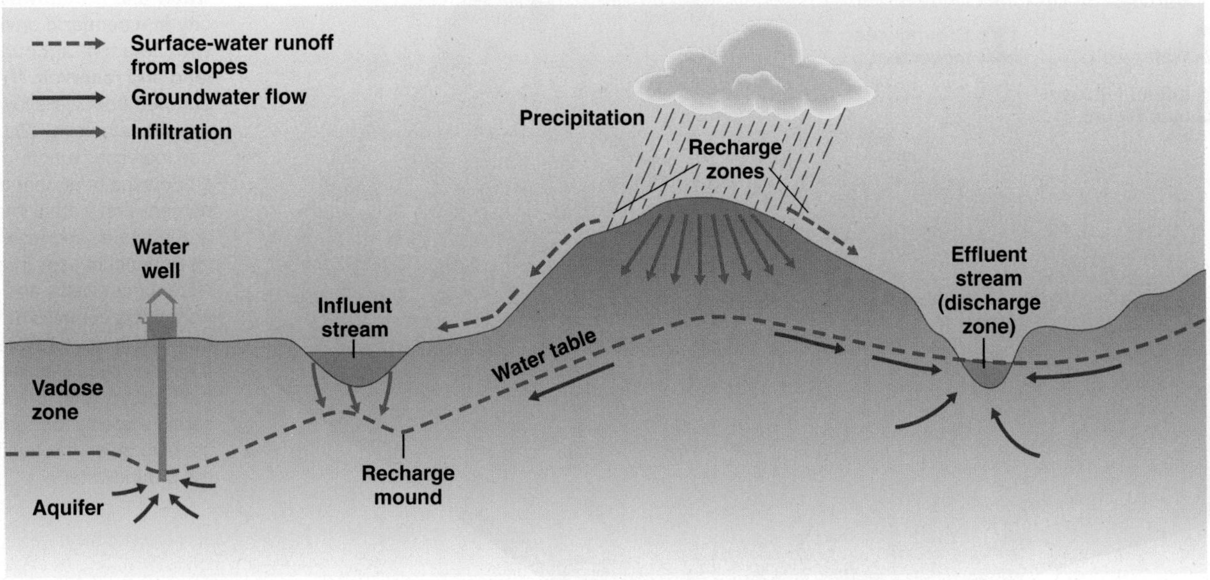

FIGURE 18.5 **Groundwater and surface-water flow system.**

stream. Most perennial streams flow all year because they constantly receive groundwater to sustain flow. An **influent stream** is entirely above the water table and flows only in direct response to precipitation. Water from an influent stream seeps down into the subsurface. An influent stream is called an *ephemeral* stream because it doesn't flow all year.

A given stream may have reaches (unspecified lengths of stream) that are perennial and other reaches that are ephemeral. It may also have reaches, known as intermittent, that have a combination of influent and effluent flow varying with the time of year. For example, streams flowing from the mountains to the sea in southern California often have reaches in the mountains that are perennial, supporting populations of trout or endangered southern steelhead, and lower intermittent reaches that transition to ephemeral reaches. At the coast, these streams may receive fresh or salty groundwater and tidal flow from the ocean to become a perennial lagoon.

Interactions between Surface Water and Groundwater

Surface water and groundwater interact in many ways and should be considered part of the same resource. Nearly all natural surface-water environments, such as rivers and lakes, as well as human-made water environments, such as reservoirs, have strong linkages with groundwater. For example, pumping groundwater from wells may reduce stream flow, lower lake levels, or change the quality of surface water. Reducing effluent stream flow by lowering the groundwater level may change a perennial stream into an intermittent influent stream. Similarly, withdrawing surface water by diverting it from streams and rivers can deplete groundwater or change its quality. Diverting surface waters that recharge groundwaters may increase concentrations of dissolved chemicals in the groundwater because dissolved chemicals in the groundwater will no longer be diluted by infiltrated surface water. Finally, pollution of groundwater may result in polluted surface water, and vice versa.[9]

Selected interactions between surface water and groundwater in a semiarid urban and agricultural environment are shown in Figure 18.6. Urban and agricultural runoff increases the volume of water in the reservoir. Pumping groundwater for agricultural and urban uses lowers the groundwater level. The quality of surface water and groundwater is reduced by urban and agricultural runoff, which adds nutrients from fertilizers, oil from roads, and nutrients from treated wastewaters to streams and groundwater.

18.2 Water Resources

The common sources of water are surface water and groundwater. Additional sources are desalination and treated wastewater. Conservation of water is also a source of water. Which sources are used at a particular place depends on factors such as location, water quality, water availability, technology available, water demand, and in some cases preferences.

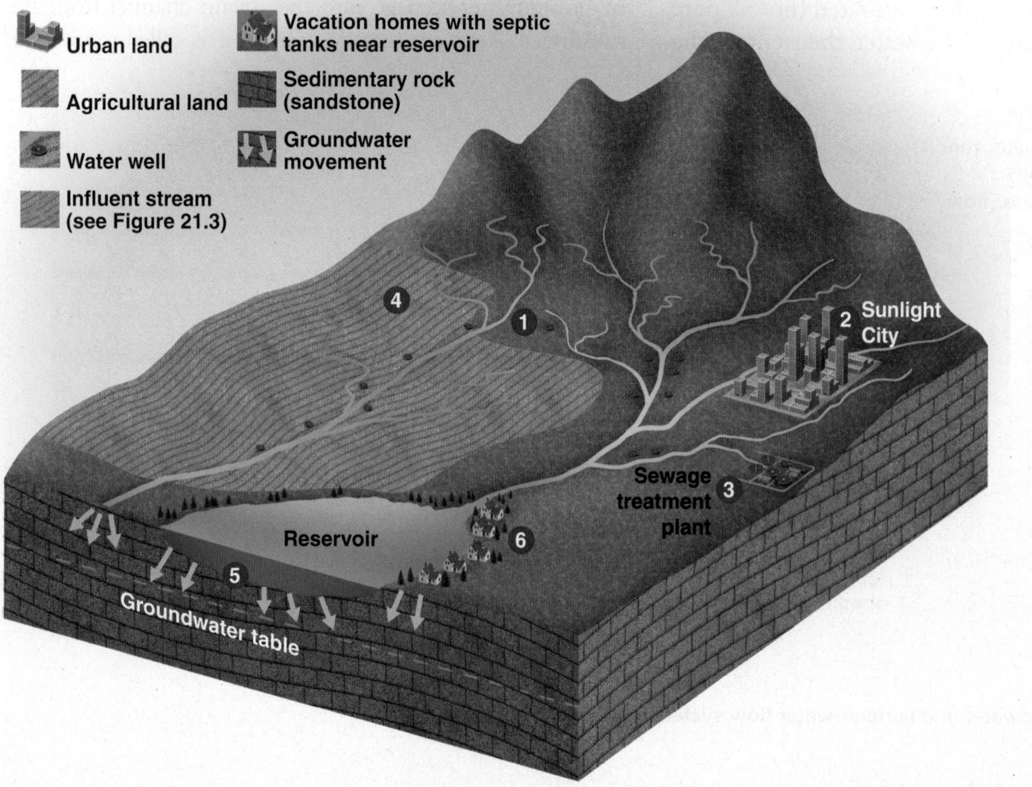

FIGURE 18.6 Idealized diagram illustrating some interactions between surface water and groundwater for a city in a semiarid environment with adjacent agricultural land and reservoir. (1) Water pumped from wells lowers the groundwater level. (2) Urbanization increases runoff to streams. (3) Sewage treatment discharges nutrient-rich waters into streams, groundwater, and reservoirs. (4) Agriculture uses irrigation waters from wells, and runoff from fields contains nutrients from fertilizers. (5) Water from the reservoir is seeping down to the groundwater. (6) Water from septic systems for homes is seeping down through the soil to the groundwater.

Surface Runoff

The water supply at any particular point on the land surface depends on several factors in the hydrologic cycle, including the rates of precipitation, evaporation, transpiration (water in vapor form that directly enters the atmosphere from plants through pores in leaves and stems), stream flow, and subsurface flow. A concept useful in understanding water supply is the **water budget** (Figure 18.7). A water budget balances the inputs, outputs, and storage of water in a system such that water inflow minus outflow equals the change in water storage.[6] Water systems include lakes, rivers, or aquifers. Simple annual water budgets (precipitation – evaporation = runoff) for North America and other continents are shown in Table 18.2. The total average annual water yield (runoff) from Earth's rivers is approximately 47,000 km^3 (1.2×10^{16} gal), but its distribution is far from uniform (see Table 18.2). Some runoff occurs in relatively uninhabited regions, such as Antarctica, which produces about 5% of Earth's total runoff. South America, which includes the relatively uninhabited Amazon basin, provides about 25% of Earth's total runoff. Total runoff in North America is about two-thirds that of South America. Unfortunately, much of the North American runoff occurs in sparsely settled or uninhabited regions, particularly in the northern parts of Canada and Alaska.

In developing water budgets for water resources management, it is useful to consider annual precipitation and runoff patterns. Potential problems with water supply can be predicted in areas where average precipitation and runoff are relatively low, such as the arid and semi-arid parts of the southwestern and Great Plains regions of the United States. Surface-water supply can never be as high as the average annual runoff because not all runoff can be successfully stored, due to evaporative losses from river channels, ponds, lakes, and reservoirs. Water shortages are common in areas that have naturally low precipitation and runoff, coupled with strong evaporation. In such areas, rigorous conservation practices are necessary to help ensure an adequate supply of water.[6,7]

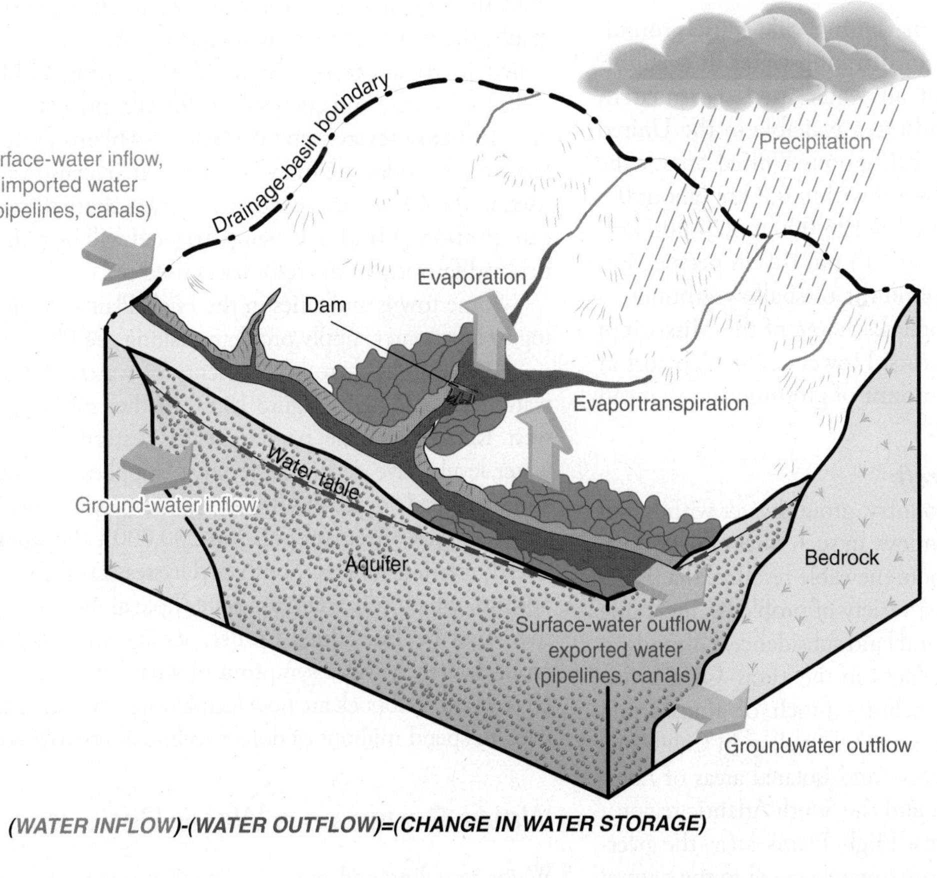

Typical water budget components

WATER INFLOW

- Precipitation
- Surface–water flow into basin
- Imported water
- Groundwater inflow

WATER OUTFLOW

- Evaporation
- Transpiration by vegetation (evapotranspiration)
- Surface–water outflow
- Exported water
- Groundwater outflow

CHANGE IN WATER STORAGE, increased/decreased water in:

- Snowpack
- Streams,rivers,reservoirs
- Aquifers

FIGURE 18.7 **Concept of the water budget.**

(*Source:* Modified from U.S. Geological Survey. 2002. *Concepts for National Assessment for Water Availability and Use*. Circular 1223: Reston,Virginia.)

Table 18.2 ANNUAL WATER BUDGETS FOR THE CONTINENTS[a]

	PRECIPITATION		EVAPORATION		RUNOFF
CONTINENTAL	mm/yr	km^3	mm/yr	km^3	km^3/yr
North America	756	18,300	418	10,000	8,180
South America	1,600	28,400	910	16,200	12,200
Europe	790	8,290	507	5,320	2,970
Asia	740	32,200	416	18,100	14,100
Africa	740	22,300	587	17,700	4,600
Australia and Oceania	791	7,080	511	4,570	2,510
Antarctica	165	2,310	0	0	2,310
Earth (entire land area)	800	119,000	485	72,000	47,000[b]

[a] Precipitation 2 evaporation 5 runoff.

[b] Surface runoff is 44,800; groundwater runoff is 2,200.

Source: I. A. Shiklomanov, "World Fresh Water Resources," in P. H. Gleick, ed., Water in Crisis (New York: Oxford University Press, 1993), pp. 3–12.

Groundwater Resources

Nearly half the people in the United States use groundwater as a primary source of drinking water. It accounts for approximately 20% of all water used. Fortunately, the total amount of groundwater available in the United States is enormous. In the contiguous United States, the amount of shallow groundwater within 0.8 km (about 0.5 mi) of the surface is estimated at 125,000 to 224,000 km^3 (3.3×10^{16} to 5.9×10^{16} gal). To put this in perspective, the lower estimate of the amount of shallow groundwater is about equal to the total discharge of the Mississippi River during the last 200 years. However, the high cost of pumping limits the total amount of groundwater that can be economically recovered.[6,7]

Groundwater Overdraft:

In many parts of the country, groundwater withdrawal from wells exceeds natural inflow. In such cases of **overdraft**, we can think of water as a nonrenewable resource that is being *mined*. This can lead to a variety of problems, including damage to river ecosystems and land subsidence. Groundwater overdraft is a serious problem in the Texas–Oklahoma–High Plains area (which includes much of Kansas and Nebraska and parts of other states), as well as in California, Arizona, Nevada, New Mexico, and isolated areas of Louisiana, Mississippi, Arkansas, and the South Atlantic region.

In the Texas–Oklahoma–High Plains area, the overdraft amount per year is approximately equal to the natural flow of the Colorado River for the same period.[6] The Ogallala Aquifer (also called the High Plains Aquifer), which is composed of water-bearing sands and gravels that underlie an area of about 400,000 km^2 from South Dakota into Texas, is the main groundwater resource in this area. Although the aquifer holds a tremendous amount of groundwater, it is being used in some areas at a rate up to 20 times higher than the rate at which it is being naturally replaced. As a result, the water table in many parts of the aquifer has declined in recent years (Figure 18.8), causing yields from wells to decrease and energy costs for pumping the water to rise. The most severe water depletion problems in the Ogallala Aquifer today are in locations where irrigation was first used in the 1940s. There is concern that eventually a significant portion of land now being irrigated will be returned to dryland farming as the resource is used up.

Some towns and cities in the High Plains are also starting to have water supply problems. Along the Platte River in northern Kansas there is still plenty of water, and groundwater levels are high (Figure 18.8). Farther south, in southwest Kansas and the panhandle in western Texas, where water levels have declined the most, supplies may last only another decade or so. In Ulysses, Kansas (population 6,000), and Lubbock, Texas (population 200,000), the situation is already getting serious. South of Ulysses, Lower Cimarron Springs, which was a famous water hole along a dry part of the Santa Fe Trail, dried up decades ago due to pumping groundwater. It was a symptom of what was coming. Both Ulysses and Lubbock are now facing water shortages and will need to spend millions of dollars to find alternative sources.

Water Reuse as a Water Resource

Water recycling and reuse, in locations where water is becoming scarce, is a growing source of water for both urban areas and agriculture. From Florida to California, numerous communities reuse treated wastewater for a variety of noncontact uses. In some more limited cases, treated wastewater is highly treated by natural flow through the ground and high-technology filtration to reclaim water

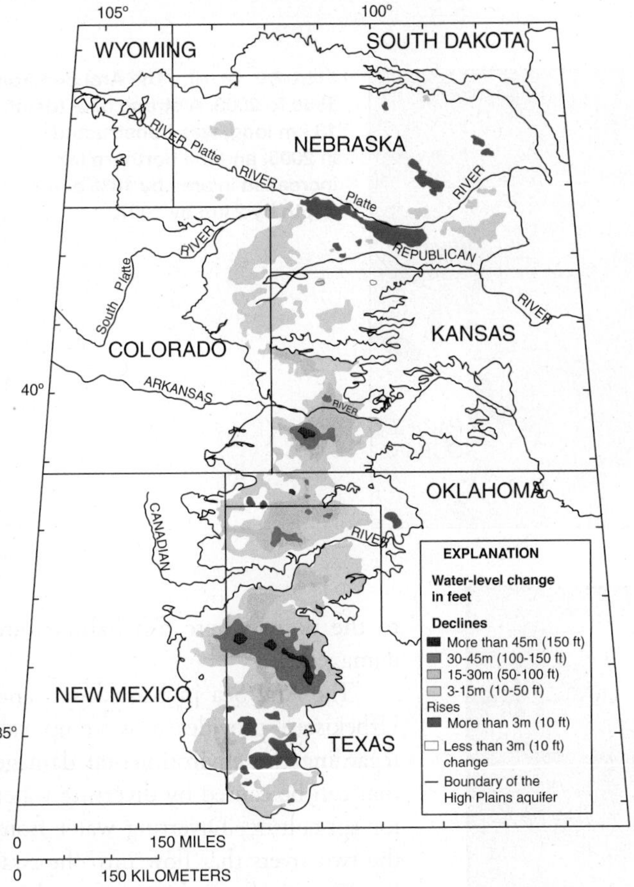

(Source: Modified from U.S. Geological Survey. 2002. *Concepts for National Assessment for Water Availability and Use*. Circular 1223: Reston, Virginia.)

FIGURE 18.8 Groundwater level changes as a result of pumping in the Texas–Oklahoma–High Plains region.

for all desired uses, including human consumption. For example, Singapore (see the opening case study) treats all wastewater, and it is used again. We will return to wastewater treatment and water reuse in Chapter 19.

Water Conservation as a Water Source

When we conserve water, we save water for other uses. Thus, conservation may be considered a water resource. For example, when an agricultural operation saves water through improved irrigation, the saved water may be made available for urban uses or use by ecosystems. We will discuss water conservation as a major subject in Section 18.4.

Desalination as a Water Resource

Seawater is about 3.5% salt; that means each cubic meter of seawater contains about 40 kg (88 lb) of salt. **Desalination**, a technology that removes salt from water, is being used at several hundred plants around the world to produce water with reduced salt. To be used as a freshwater resource, the salt content must be reduced to about 0.05%. Large desalination plants produce 20,000–30,000m^3 (about 5–8 million gal) of water per day. Today, more than 15,000 desalination plants in over 100 countries are in operation; improving technology is significantly lowering the cost of desalination.

Even so, desalinated water costs several times as much as traditional water supplies in the United States. Desalinated water has a *place value*, which means that the price rises quickly with the transport distance and the cost of moving water from the plant. Because the various processes that remove the salt require large amounts of energy, the cost of the water is also tied to ever-increasing energy costs. For these reasons, desalination will remain an expensive process, used only when alternative water sources are not available.

Desalination also has environmental impacts. Discharge of very salty water from a desalination plant into another body of water, such as a bay, may locally increase salinity and kill some plants and animals. The discharge from desalination plants may also cause wide fluctuations in the salt content of local environments, which may damage ecosystems.

18.3 Water Use

In discussing water use, it is important to distinguish between off-stream and in-stream uses. **Off-stream use** refers to water removed from its source (such as a river or reservoir) for use. Much of this water is returned to the source after use; for example, the water used to cool industrial processes may go to cooling ponds and then be discharged to a river, lake, or reservoir. **Consumptive use** is an off-stream use in which water is consumed by plants and animals or used in industrial processes. The water enters human tissue or products or evaporates during use and is not returned to its source.

In-stream use includes the use of rivers for navigation, hydroelectric power generation, fish and wildlife habitats, and recreation. These multiple uses usually create controversy because each requires different conditions. For example, fish and wildlife require certain water levels and flow rates for maximum biological productivity. These levels and rates will differ from those needed for hydroelectric power generation, which requires large fluctuations in discharges to match power needs. Similarly, in-stream uses of water for fish and wildlife will likely conflict with requirements for shipping and boating. Figure 18.9

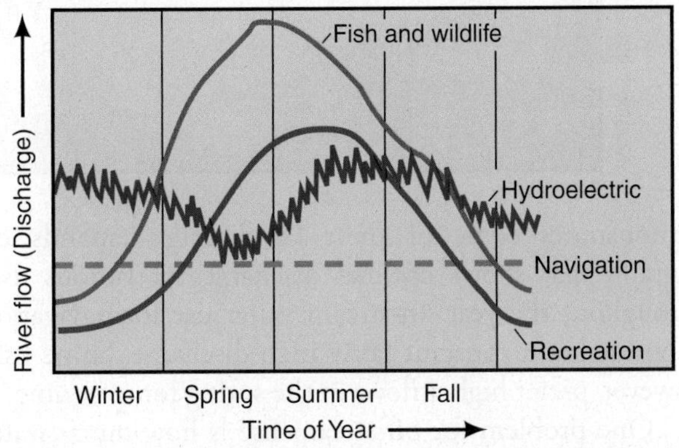

FIGURE 18.9 In-stream water uses and optimal discharge (volume of water flowing per second) for each use. Discharge is the amount of water passing by a particular location and is measured in cubic meters per second. Obviously, all these needs cannot be met simultaneously.

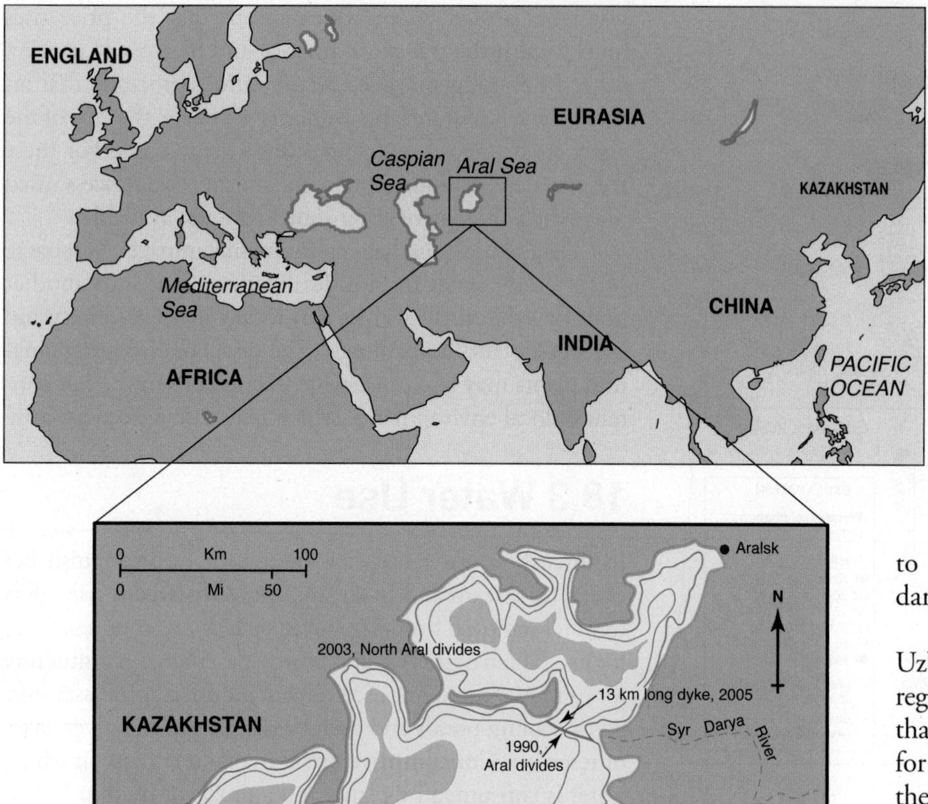

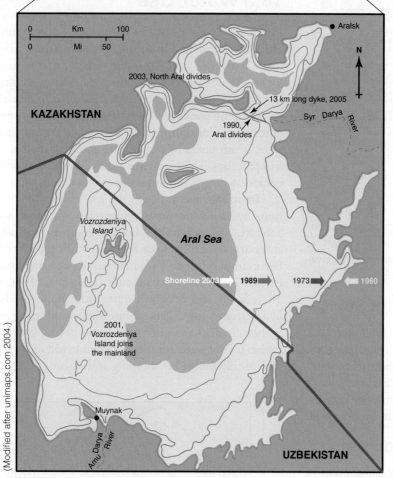

FIGURE 18.10 The Aral Sea from 1960 to 2003. A strong dike (dam), 13 km long, was constructed in 2005, and the northern lake increased in area by 18% and in depth by 2 km by 2007.

demonstrates some of these conflicting demands on a graph that shows optimal discharge for various uses throughout the year. In-stream water use for navigation is optimal at a constant fairly high discharge. Some fish, however, prefer higher flows in the spring for spawning.[6,7]

One problem for off-stream use is how much water can be removed from a stream or river without damaging the stream's ecosystem. This is an issue in the Pacific Northwest, where fish, such as steelhead trout and salmon, are on the decline partly because diversions for agricultural, urban, and other uses have reduced stream flow to the point where fish habitats are damaged.

The Aral Sea in Kazakhstan and Uzbekistan provides a wake-up call regarding the environmental damage that can be caused by diverting water for agriculture. Diverting water from the two rivers that flow into the Aral Sea has transformed one of the largest bodies of inland water in the world from a vibrant ecosystem into a dying sea. The present shoreline is surrounded by thousands of square kilometers of salt flats that formed as the sea's surface area shrank about 90% in volume in the past 50 years (Figure 18.10). The volume of the sea was reduced by about 70%, and the salt content increased to more than twice that of seawater, causing fish kills, including sturgeon, an important component of the economy. Dust raised by winds from the dry salt flats is producing a regional air-pollution problem, and the climate in the region has changed as the moderating effect of the sea has been reduced. Winters have grown colder and summers warmer. Fishing centers, such as Muynak in the south and Aralsk to the north that were once on the shore of the sea, are now many kilometers inland (Figure 18.11). Loss of fishing, along with a decline in tourism, has damaged the local economy.[4,10]

Restoration of the small northern port of the Aral Sea is ongoing. In 2005 a low, long dam was constructed across the lakebed just south of where the Syr Darya River enters the lake, and another dam further north is scheduled for completion in 2013. Conservation of water and the construction of the dam are producing and will continue to produce dramatic improvements to the northern port of

AFP/EPA/NewsCom

FIGURE 18.11 **Ships grounded in the dry seabed as the fishing industry collapsed.**

the lake, and some fishing is returning there. The future of the lake has improved, but great concern remains.[4,10]

Transport of Water

In many parts of the world, demands are being made on rivers to supply water to agricultural and urban areas. This is not a new trend—ancient civilizations, including the Romans and Native Americans, constructed canals and aqueducts to transport water from distant rivers to where it was needed. In our modern civilization, as in the past, water is often moved long distances from areas with abundant rainfall or snow to areas of high use (usually agricultural areas). For instance, in California, two-thirds of the state's runoff occurs north of San Francisco, where there is a surplus of water. However, two-thirds of the water use in California occurs south of San Francisco, where three-fourths of the people in the state live and there is a deficit of water. In recent years, canals of the California Water Project have moved great quantities of water from the northern to the southern part of the state, mostly for agricultural uses, but increasingly for urban uses as well.

On the opposite coast, New York City has imported water from nearby areas for more than 100 years. Water use and supply in New York City show a repeating pattern. Originally, local groundwater, streams, and the Hudson River itself were used. However, as the population increased and the land was paved over, surface waters were diverted to the sea rather than percolating into the soil to replenish groundwater. Furthermore, what water did infiltrate the soil was polluted by urban runoff. Water needs in New York exceeded local supply, and in 1842 the first large dam was built.

As the city rapidly expanded from Manhattan to Long Island, water needs increased. The shallow aquifers of Long Island were at first a source of drinking water, but this water was used faster than the infiltration of rainfall could replenish it. At the same time, the groundwater became contaminated with urban and agricultural pollutants and from saltwater seeping in underground from the ocean. (The pollution of Long Island groundwater is explored in more depth in the next chapter.) Further expansion of the population created the same pattern: initial use of groundwater, pollution, salinization, and overuse of the resource. A larger dam was built in 1900 about 50 km (30 mi) north of New York City at Croton-on-Hudson (Figure 18.12), and later new, larger dams were built farther and farther upstate in forested areas.

From a broader perspective, the cost of obtaining water for large urban centers from far-off sources, along with competition for available water from other sources and users, will eventually place an upper limit on the water supply of New York City. As with other resources, as the water supply shrinks and demand for water rises, so does the price. As shortages develop, stronger conservation measures are implemented, and the cost of water increases. If the price goes high enough, costlier sources may be developed—for example, pumping from deeper wells or desalinating.

Transfer of Water: Real and Virtual

Here we consider transfer of water to be different from transport of water in canals and pipes discussed above. Water has been transferred long distances in containers for hundreds of years—for example, water bags on camels crossed the deserts in ancient caravans. Ice has also been harvested from frozen lakes, rivers, and glaciers (and still is) to be sold. Water stored in crops has also been used

USGS

FIGURE 18.12 **Croton-on-Huson Dam** was constructed in 1900 to supply water to New York City.

throughout human history, and those crops may be used in a place far away from where they were grown.

Real Water

The transfer of water through its export in plastic and glass bottles grew very rapidly beginning in 1960, increasing from about 5 billion liters in 1960 to 27 billion liters in 2000. Bottled water costs about 1,000 to 5,000 times more than tap water. A major complaint is that plastic water bottles are discarded routinely by users, a practice that pollutes the environment. When recycled, plastic bottles are a resource for other plastic production. (Typically, only about one-third are recycled.) Some universities and other places are discouraging bottled water by providing free dispensing units of purified water that people can use to fill their water bottles.[4]

Some people believe tap water is less pure than bottled water or take exception to the common practice of adding fluoride to municipal water as a means of reducing dental cavities. They believe that bottled water is safer (even providing a health benefit) than tap water and better tasting. The taste of chlorine in some tap water is a major reason people turned to bottled water. Depending on your location and local water supply, bottled water may be better tasting. However, it is not always safer.[4]

Bottled water is very useful following emergencies such as floods and earthquakes when local water supplies may be damaged or polluted. It may be stored with other emergency supplies to be quickly taken to where it is needed.[4]

Virtual Water

Virtual water is water embodied in goods and crops. It is measured as the volume of virtual water imported or exported to other regions or countries.

It is useful to consider the amount of water necessary to produce a product, such as an automobile, or a crop, such as rice.[11–14] This content is measured at the place where the product is produced or the crop is grown.[12]

The amount of water necessary to produce crops and animals may be surprisingly large and variable. A few years ago, the question of how much water is required to produce a cup of coffee was asked. The answer is not trivial. Coffee is an important crop for many countries and the major social drink in much of the world. Many a romance has been initiated with the question, "Would you like a cup of coffee?"

How much water is necessary to produce a cup of coffee requires knowing how much water is necessary to produce the coffee berries (that contain the bean) and the roasted coffee. The question is complicated by the fact that water used to raise coffee varies from location to location, as does the yield of berries. Much of the water in coffee-growing areas is free; it comes from rain. However, that doesn't mean the water has no value. People are usually surprised to learn that it takes about 140 L (40 gal) of water to produce one cup of coffee. The amount of water that is needed to produce a ton of a crop varies from a low of about 175 m^3 for sugarcane to 1,300 m^3 for wheat, 3,400 m^3 for white rice, and 21,000 m^3 for roasted coffee. For the meat we eat, the amount per ton is 3,900 m^3 for chicken, 4,800 m^3 for pork, and 15,500 m^3 for beef.[12]

The United States produces food that is exported around the world (as discussed in Chapter 11). As a result, people consuming imported U.S. crops in western Europe directly affect the regional water resources of the United States. Similarly, our consumption of imported foods—such as cantaloupes grown in Mexico or blueberries in Chile—affects the regional water supply and groundwater resources of the countries that grew and exported them.

The amount of water required to produce a crop of domestic animals is useful in water resource planning from the local to global scale. A country with an arid climate and restricted water resources can choose between developing those resources for agriculture or for other water uses—for example, to support wetland ecosystems or a growing human population. Since the average global amount of water necessary to produce a ton of white rice is about 3,400 m^3 (nearly 900,000 gal), growing rice in countries with abundant water resources makes sense. For countries with a more arid environment, it might be prudent to import rice and save local and regional water resources for other purposes. Jordan, for example, imports about 7 billion m^3 of water per year by importing foods that require a lot of water to produce. As a result, Jordan withdraws only about 1 billion m^3 of water per year from its own water resources. Egypt, on the other hand, has the Nile River and imports only about one-third as much water stored in food and goods it imports as it withdraws from its own domestic supply. Egypt has a goal of water independence and is much less dependent on water imported in food and goods (often called virtual water with units in volume of water per year).[12]

Another way to think of this is that Jordon derives a higher percentage of its water needs (water to produce food and goods) from other countries than does Egypt. In the United States, California (with a need for more water resources in the future as its population increases) grows rice, and that uses a lot of water. If the United States imported more rice from countries with abundant water resources, the California water used to grow rice could be conserved for other uses.

Examination of global water resources and potential global water conservation is an important part of sustaining our water supply. For example, by trading virtual water, the international trade markets reduce agriculture's global water use by about 5%.[12] Figure 18.13 shows net virtual water budgets (balances) for major trades. The

fish and other aquatic life may die. Acidic water can also seep into and pollute groundwater.

Acid mine drainage is produced by complex geochemical and microbial reactions. The general equation is as follows:

$$4\,FeS_2 + 15\,O_2 + 14\,H_2O \rightarrow 4\,Fe(OH)_3\;\;8\,H_2SO_4$$

Pyrite + Oxygen + Water → Ferric Hydroxide + Sulfuric Acid

Acid mine drainage is a significant water pollution problem in Wyoming, Indiana, Illinois, Kentucky, Tennessee, Missouri, Kansas, and Oklahoma and is probably the most significant water pollution problem in West Virginia, Maryland, Pennsylvania, Ohio, and Colorado. The total impact is significant because thousands of kilometers of streams have been damaged.

Even abandoned mines can cause serious problems. Subsurface mining for sulfide deposits containing lead and zinc began in the tristate area of Kansas, Oklahoma, and Missouri in the late 19th century and ended in some areas in the 1960s. When the mines were operating, they were kept dry by pumping out the groundwater that seeped in. However, since the mining ended, some of them have flooded and overflowed into nearby creeks, polluting the creeks with acidic water. The problem was so severe in the Tar Creek area of Oklahoma that it was at one time designated by the U.S. Environmental Protection Agency as the nation's worst hazardous-waste site.

One solution being used in Tar Creek and other areas is a passive treatment method that uses naturally occurring chemical and/or biological reactions in controlled environments to treat acid mine drainage. The simplest and least expensive method is to divert acidic water to an open limestone channel, where it reacts with crushed limestone and the acid is neutralized. A general reaction that neutralizes the acid is

$$H_2SO_4 + CaCO_3 \rightarrow CaSO_4 + H_2O + CO_2$$

Sulfuric Acid + Calcium Carbonate (crushed limestone) → Calcium Sulfate + Water + Carbon Dioxide

Another solution is to divert the acidic water to a bioreactor (an elongated trough) containing sulfate-reducing bacteria and a bacteria nutrient to encourage bacterial growth. The sulfate-reducing bacteria are held in cells that have a honeycomb structure, forcing the acidic water to follow a tortuous path through the bacteria-laden cells of the reactor. Complex biochemical reactions between the acidic water and bacteria in the reactor produce metal sulfides and in the process reduce the sulfuric acid content of the water. Both methods result in cleaner water with a lower concentration of acid being released into the environment.

19.8 Surface-Water Pollution

Pollution of surface water occurs when too much of an undesirable or harmful substance flows into a body of water, exceeding that body of water's natural ability to remove it, dilute it to a harmless concentration, or convert it to a safe form (see A Closer Look 19.3).

Water pollutants, like other pollutants, are categorized as being emitted from point or nonpoint sources (see Chapter 8). **Point sources** are distinct and confined, such as pipes from industrial and municipal sites that empty into streams or rivers (Figure 19.13). In general, point source pollutants from industries are controlled through on-site treatment or disposal and are regulated by permit. Municipal point sources are also regulated by permit. In older cities in the northeastern and Great Lakes areas of the United States, most point sources are outflows from combined sewer systems. As mentioned earlier, such systems combine storm water flow with municipal wastewater. During heavy rains, urban storm runoff may exceed the capacity of the sewer system, causing it to overflow and deliver pollutants to nearby surface waters.

Nonpoint sources, such as runoff, are diffused and intermittent and are influenced by factors such as land use, climate, hydrology, topography, native vegetation, and geology. Common urban nonpoint sources include runoff from streets or fields; such runoff contains all sorts of pollutants, from heavy metals to chemicals and sediment. Rural sources of nonpoint pollution are generally associated with agriculture, mining, or forestry. Nonpoint sources are difficult to monitor and control.

Photoshot Holdings Ltd/Alamy

FIGURE 19.13 This pipe is a point source of chemical pollution from an industrial site entering a river in England.

A CLOSER LOOK 19.3

Water Pollution from Industrial Livestock Farms

Industrial (giant) livestock farms that raise hogs, chickens, turkeys, or dairy cows generate huge amounts of waste. The most costly by-product comes from manure that often is placed in very large (several acres) waste lagoons. Some liquid waste is sprayed on fields. The lagoons can leak or wash out during storms, polluting ground and surface water with nitrogen, ammonia, antibiotics (fed to animals), and disease-causing organisms (*E. coli* and salmonella, for example).[a] Consider the example of hog farms in North Carolina. Hurricane Floyd struck the Piedmont area of North Carolina in September 1999. The killer storm took a number of lives while flooding many homes and forcing some 48,000 people into emergency shelters. The storm had another, more unusual effect. Floodwaters containing thousands of dead pigs, along with their feces and urine, flowed through schools, churches, homes, and businesses. The stench was reportedly overwhelming, and the count of pig carcasses may have been as high as 30,000. The storm waters had overlapped and washed out more than 38 pig lagoons with as much as 950 million L (250 million gal) of liquid pig waste, which ended up in flooded creeks, rivers, and wetlands. In all, nearly 250 commercial pig farms flooded out, drowning hogs whose floating carcasses had to be collected and disposed of (Figure 19.14).

Prior to Hurricane Floyd, the pig farm industry in North Carolina had been involved in a scandal reported by newspapers and television—and even by the television news show *60 Minutes*. North Carolina has a long history of hog production, and the population of pigs swelled from about 2 million in 1990 to nearly 10 million in 1997. At that time, North Carolina became the second-largest pig-farming state in the nation.[b] As the number of large commercial pig farms grew, the state allowed the hog farmers to build automated and very confining farms housing hundreds or thousands of pigs. There were no restrictions on farm location, and many farms were constructed on floodplains.

Each pig produces approximately 2 tons of waste per year. The North Carolina herd was producing approximately

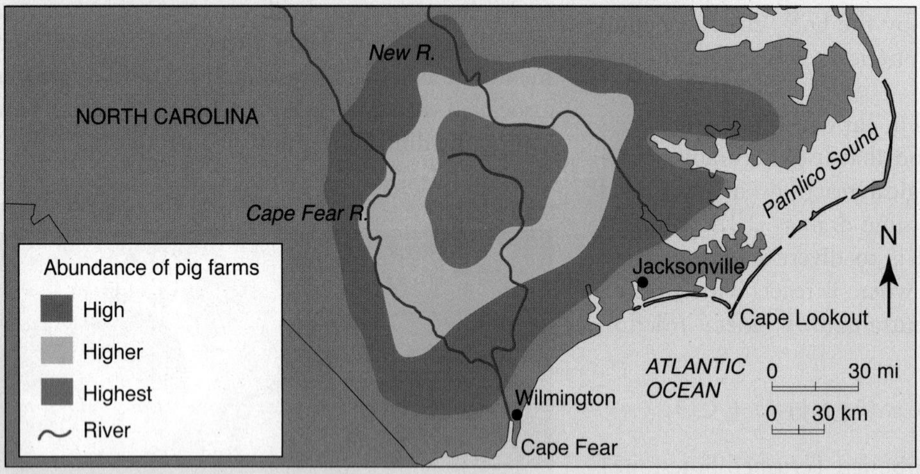

(a)

Associated Press

(b)

FIGURE 19.14 North Carolina's "Bay of Pigs." (a) Map of areas flooded by Hurricane Floyd in 1999 with relative abundance of pig farms. **(b)** Collecting dead pigs near Boulaville, North Carolina. The animals were drowned when floodwaters from the Cape Fear River inundated commercial pig farms.

20 million tons of waste a year, mostly manure and urine, which was flushed out of the pig barns and into open, unlined lagoons about the size of football fields. Favorable regulations, along with the availability of inexpensive waste-disposal systems (the lagoons), were responsible for the tremendous growth of the pig population in North Carolina in the 1990s.

After the hurricane, mobile incinerators were moved into the hog region to burn the carcasses, but there were so many that hog farmers had to bury some animals in shallow pits. The pits were supposed to be at least 1 m (3 ft.) deep and dry, but there wasn't always time to find dry ground, and for the most part the pits were dug and filled on floodplains. As these pig carcasses rot, bacteria will leak into the groundwater and surface water for some appreciable time.

An early warning occurred in 1995, when a pig-waste lagoon failed and sent approximately 950 million L (250 million gal) of concentrated pig feces down the New River past the city of Jacksonville, South Carolina, and into the New River estuary. The spill's adverse effects on marine life lasted approximately three months.

The lesson to be learned from North Carolina's so-called Bay of Pigs is that we are vulnerable to environmental catastrophes caused by large-scale industrial agriculture. Economic growth and production of livestock must be carefully planned to anticipate problems, and waste-management facilities must be designed so as not to pollute local streams, rivers, and estuaries.

Was the lesson learned in North Carolina? The pig farmers had powerful friends in government and big money. Incredible as it may seem, following the hurricane, the farmers asked for $1 billion in grants to help repair and replace the pig facilities, including waste lagoons, destroyed by the hurricane. Furthermore, they asked for exemptions from the Clean Water Act for a period of six months so that waste from the pig lagoons could be discharged directly into streams.[c] This exemption was not allowed.

With regard to future management, considering that North Carolina is frequently struck by hurricanes, barring pig operations from floodplains seems obvious. However, this is only the initial step. The whole concept of waste lagoons needs to be rethought and alternative waste-management practices put into effect if pollution of surface water and groundwater is to be avoided. To this end, North Carolina in 2007 passed legislation to ban construction or expansion of new waste lagoons and encouraged pig farms to treat pig waste to extract methane (gas) as an energy source. Other methods of on-site treatment to reduce organic matter and nutrients are ongoing.

North Carolina's pig problem led to the formation of what is called the "Hog Roundtable," a coalition of civic, health, and environmental groups with the objective of controlling industrial-scale pig farming. Its efforts, along with others, resulted in a mandate to phase out pig-waste lagoons and expand regulations to require buffers between pig farms and surface waters and water wells. The coalition also halted construction of a proposed slaughterhouse that would have allowed more pig farms to be established.

Reducing Surface-Water Pollution

From an environmental view, two approaches to dealing with surface-water pollution are: (1) reduce the sources of pollution; and (2) treat the water to remove pollutants or convert them to forms that can be disposed of safely. Which option is used depends on the specific circumstances of the pollution problem. Reduction at the source is the environmentally preferred way of dealing with pollutants. For example, air-cooling towers, rather than water-cooling towers, may be used to dispose of waste heat from power plants, thereby avoiding thermal pollution of water. The second method—water treatment—is used for a variety of pollution problems. Water treatments include chlorination to kill microorganisms such as harmful bacteria and filtering to remove heavy metals.

There is a growing list of success stories in the treatment of water pollution. One of the most notable is the cleanup of the Thames River in Great Britain. For centuries, London's sewage had been dumped into that river, and there were few fish to be found downstream in the estuary. In recent decades, however, improved water treatment has led to the return of a number of species of fish, some of which have not been seen in the river for centuries.

Many large cities in the United States—such as Boston, Miami, Cleveland, Detroit, Chicago, Portland, and Los Angeles—grew on the banks of rivers, but the rivers were often nearly destroyed by pollution and concrete. Today, there are grassroots movements all around the country dedicated to restoring urban rivers and adjacent lands as greenbelts, parks, and other environmentally sensitive developments. For example, the Cuyahoga River in Cleveland, Ohio, was so polluted by 1969 that sparks from a train ignited oil-soaked wood in the river, setting the surface of the river on fire! The burning of an American river became a symbol for a growing environmental consciousness. The Cuyahoga River today is cleaner and no longer flammable—from Cleveland to Akron, it is a beautiful greenbelt (Figure 19.15). The greenbelt changed part of the river from a sewer into a valuable public resource and focal point for economic and environmental renewal.[24] However, in downtown Cleveland and Akron, the river remains an industrial stream and parts remain polluted.

J. BLANK/ClassicStock/The Image Works

FIGURE 19.15 The Cuyahoga River (lower left) flows toward Cleveland, Ohio, and the Erie Canal (lower right) is in the Cuyahoga National Park. The skyline is that of industrial Cleveland.

Two of the newer surface-water cleanup techniques involve nanotechnology and bioengineering. **Nanotechnology** uses extremely small material particles (10^{-9}m size, about 100,000 times thinner than human hair) designed for a number of purposes. Some nano particles can capture heavy metals such as lead, mercury, and arsenic from water. The nano particles have a tremendous surface area to volume. One cubic centimeter of particles has a surface area exceeding a football field and can take up over 50% of its weight in heavy metals.[25]

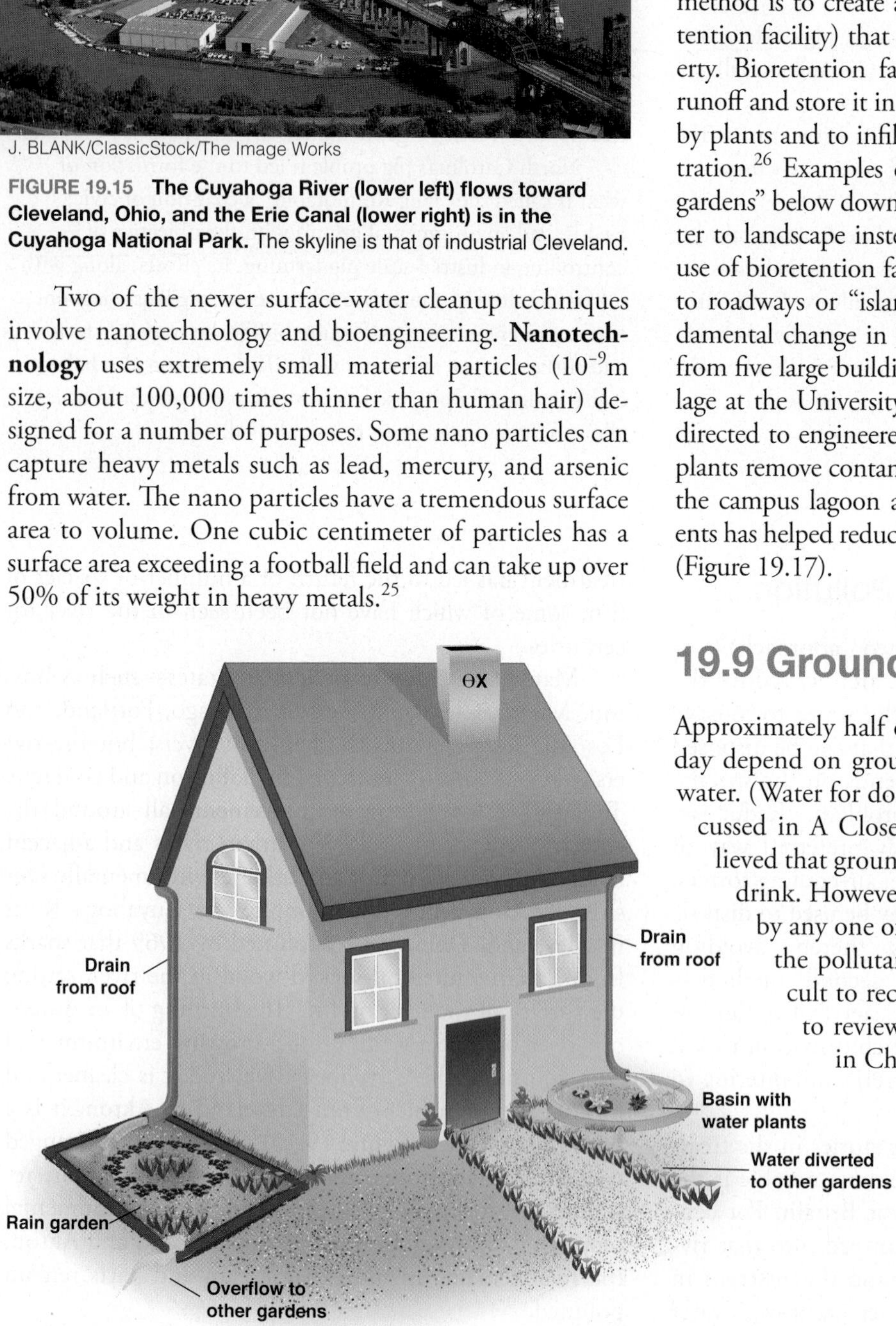

FIGURE 19.16 Water from roof runoff is part of a closed loop where water remains on the site and is used in rain gardens. Runoff from parking areas is diverted to other gardens.

Bioengineering uses plants and soil in an engineered landscape (especially urban lands) to minimize water pollution. Urbanization significantly changes the hydrology of a landscape. Open land (forest or grasslands) with native vegetation is replaced by surfaces such as rooftops, parking lots, streets, and sidewalks that are impervious (water can't infiltrate them). Lawns and gardens are planted, and pesticides and fertilizers are applied. Urban runoff from streets and parking lots contains trash, oil, gasoline, and other pollutants that are directed to storm water drains that may directly enter local streams, lakes, or the ocean.

Engineering technology is being used to treat urban runoff before it reaches streams, lakes, or the ocean. One method is to create a "closed-loop" local landscape (bioretention facility) that does not allow runoff to leave a property. Bioretention facilities are designed to collect urban runoff and store it in landscapes that allow water to be used by plants and to infiltrate the soil for additional natural filtration.[26] Examples of bioretention facilities include "rain gardens" below downspouts and parking lots that direct water to landscape instead of the street (Figure 19.16).[27] The use of bioretention facilities such as landscape areas parallel to roadways or "islands" in parking lots represents a fundamental change in how urban land is developed. Runoff from five large building complexes, such as Manzaneta Village at the University of California, Santa Barbara, can be directed to engineered wetlands (bioswales) where wetland plants remove contaminants before water is discharged into the campus lagoon and then the ocean. Removing nutrients has helped reduce cultural eutrophication of the lagoon (Figure 19.17).

19.9 Groundwater Pollution

Approximately half of all people in the United States today depend on groundwater as their source of drinking water. (Water for domestic use in the United States is discussed in A Closer Look 19.4.) People have long believed that groundwater is, in general, pure and safe to drink. However, groundwater can be easily polluted by any one of several sources (see Table 19.1), and the pollutants, though very toxic, may be difficult to recognize. (At this point, you may wish to review groundwater processes as discussed in Chapter 18, Section 18.1.)

Photo courtesy Ed Keller

Photo courtesy Ed Keller

FIGURE 19.17 Bioswales collect runoff from the Manzaneta Village Dormitory Complex at the University of California, Santa Barbara. (a) Plants in bioswales **(b)** help filter water and remove nutrients, reducing cultural eutrophication.

In the United States today, only a small portion of the groundwater is known to be seriously contaminated. However, as mentioned earlier, the problem may become worse as human population pressure on water resources increases. Our realization of the extent of the problem is growing as the testing of groundwater becomes more common. For example, Atlantic City and Miami are two eastern cities threatened by polluted groundwater that is slowly migrating toward their wells.

It is estimated that 75% of the 175,000 known waste-disposal sites in the United States may be spewing plumes of hazardous chemicals that are migrating into

A CLOSER LOOK 19.4

Water for Domestic Use: How Safe Is It?

Water for domestic use in the United States is drawn from surface waters and groundwater. Although some groundwater sources are high quality and need little or no treatment, most are treated to conform to national drinking water standards (revisit Table 19.2).

Before treatment, water is usually stored in reservoirs or special ponds. Storage allows for solids, such as fine sediment and organic matter, to settle out, improving the clarity of water. The water is then run through a water plant, where it is filtered and chlorinated before it is distributed to individual homes. Once in people's homes, it may be further treated. For example, many people run their tap water through readily available charcoal filters before using it for drinking and cooking.

A growing number of people prefer not to drink tap water and instead drink bottled water. As a result, bottled water has become a multibillion-dollar industry. A lot of bottled water is filtered tap water delivered in plastic containers, and health questions have arisen regarding toxins leaching from the plastic, especially if bottles are left in the sun. Hot plastics can leach many more chemicals into the water than cool plastic. In any case, the plastic bottles should be used only once, then recycled.[a]

Some people prefer not to drink water that contains chlorine or that runs through metal pipes. Furthermore, water supplies vary in clarity, hardness (concentration of calcium and magnesium), and taste; and the water available locally may not be to some people's liking. A common complaint about tap water is a chlorine taste, which may be detectable at chlorine concentrations as low as 0.2–0.4 mg/l. People may also fear contamination by minute concentrations of pollutants.

The drinking water in the United States is among the safest in the world. There is no doubt that treating water with chlorine has nearly eliminated waterborne diseases, such as typhoid and cholera, which previously caused widespread suffering and death in the developed world and still do in many parts of the world. However, we need to know much more about the long-term effects of exposure to low concentrations of toxins in our drinking water. How safe is the water in the United States? It's much safer than it was 100 years ago, but low-level contamination (below what is thought dangerous) of organic chemicals and heavy metals is a concern that requires continued research and evaluation.

groundwater resources. Because many of the chemicals are toxic or are suspected carcinogens, it appears that we have inadvertently been conducting a large-scale experiment on how people are affected by chronic low-level exposure to potentially harmful chemicals. The final results of the experiment will not be known for many years.[28]

The hazard presented by a particular groundwater pollutant depends on several factors, including the concentration or toxicity of the pollutant in the environment and the degree of exposure of people or other organisms to the pollutants.[29] (See Section 8.6 on risk assessment in Chapter 8.)

Principles of Groundwater Pollution: An Example

Some general principles of groundwater pollution are illustrated by an example. Pollution from leaking underground gasoline tanks belonging to automobile service stations is a widespread environmental problem that no one thought very much about until only a few years ago. Underground tanks are now strictly regulated. Many thousands of old, leaking tanks have been removed, and the surrounding soil and groundwater have been treated to remove the gasoline. Cleanup can be a very expensive process, involving removal and disposal of soil (as a hazardous waste) and treatment of the water using a process known as vapor extraction (Figure 19.18). Treatment may also be accomplished underground by microorganisms that consume the gasoline. This is known as **bioremediation** and is much less expensive than removal, disposal, and vapor extraction.

Pollution from leaking buried gasoline tanks emphasizes some important points about groundwater pollutants:

- Some pollutants, such as gasoline, are lighter than water and thus float on the groundwater.
- Some pollutants have multiple phases: liquid, vapor, and dissolved. Dissolved phases chemically combine with the groundwater (e.g., salt dissolves into water).
- Some pollutants are heavier than water and sink or move downward through groundwater. Examples of sinkers include some particulates and cleaning solvents. Pollutants that sink may become concentrated deep in groundwater aquifers.
- The method used to treat or eliminate a water pollutant must take into account the physical and chemical properties of the pollutant and how these interact with surface water or groundwater. For example, the extraction well for removing gasoline from a groundwater resource (Figure 19.18) takes advantage of the fact that gasoline floats on water.
- Because cleanup or treatment of water pollutants in groundwater is very expensive, and because undetected or untreated pollutants may cause environmental damage, the emphasis should be on preventing pollutants from entering groundwater in the first place.

Groundwater pollution differs in several ways from surface-water pollution. Groundwater often lacks

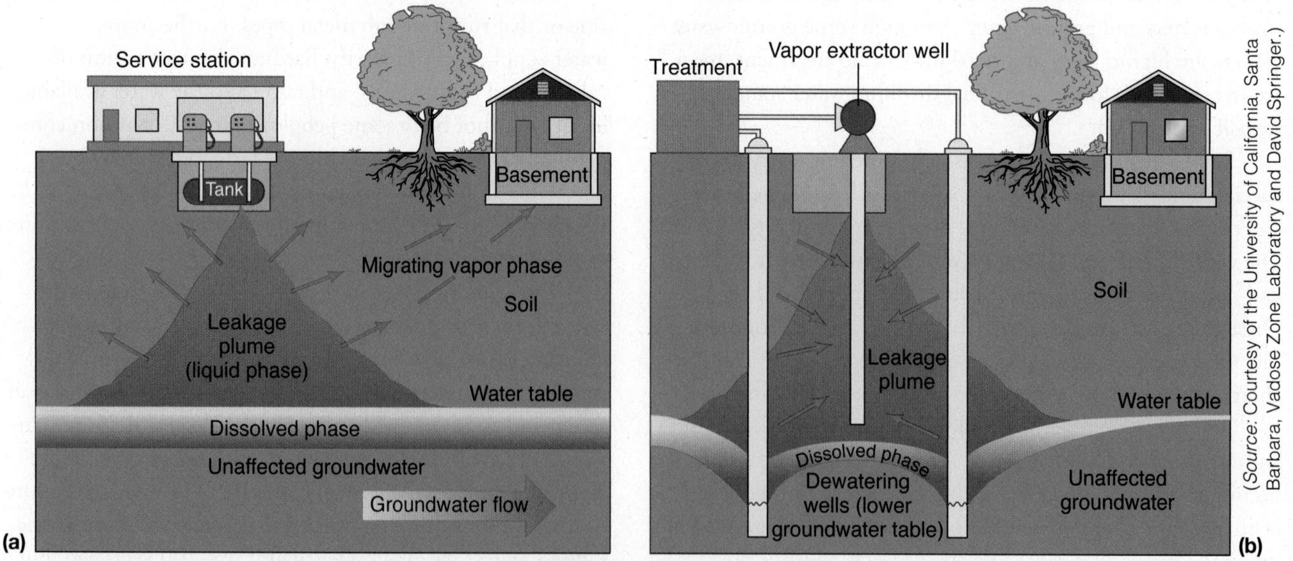

FIGURE 19.18 Diagram illustrating (a) a leak from a buried gasoline tank and (b) possible remediation using a vapor extractor system. Notice that the liquid gasoline and the vapor from the gasoline are above the water table; a small amount dissolves into the water. All three phases of the pollutant (liquid, vapor, and dissolved) float on the denser groundwater. The extraction well takes advantage of this situation. The function of the dewatering wells is to pull the pollutants in where the extraction is most effective.

oxygen, a situation that kills aerobic types of microorganisms (which require oxygen-rich environments) but may provide a happy home for anaerobic varieties (which live in oxygen-deficient environments). The breakdown of pollutants that occurs in the soil and in material a meter or so below the surface does not occur readily in groundwater. Furthermore, the channels through which groundwater moves are often very small and variable. Thus, the rate of movement is low in most cases, and the opportunity for dispersion and dilution of pollutants is limited.

Long Island, New York

Another example—that of Long Island, New York—illustrates several groundwater pollution problems and how they affect people's water supply. Two counties on Long Island, New York (Nassau and Suffolk), with a population of several million people, depend entirely on groundwater. Two major problems with the groundwater in Nassau County are intrusion of saltwater and shallow aquifer contamination.[30] Saltwater intrusion, where subsurface salty water migrates to wells being pumped, is a problem in many coastal areas of the world. The general movement of groundwater under natural conditions for Nassau County is illustrated in Figure 19.19. Salty groundwater is restricted from migrating inland by the large wedge of fresh water moving beneath the island. Notice also that the aquifers are layered, with those closest to the surface being the most salty.

In spite of the huge quantities of water in Nassau County's groundwater system, intensive pumping has lowered water levels by as much as 15 m (50 ft) in some areas. As groundwater is removed near coastal areas, the subsurface outflow to the ocean decreases, allowing saltwater to migrate inland. Saltwater intrusion has become a problem for south shore communities, which now must pump groundwater from a deeper aquifer, below and isolated from the shallow aquifers that have saltwater intrusion problems.

The most serious groundwater problem on Long Island is shallow aquifer pollution associated with urbanization. Sources of pollution in Nassau County include urban runoff, household sewage from cesspools and septic tanks, salt used to deice highways, and industrial and solid waste. These pollutants enter surface waters and then migrate downward, especially in areas of intensive pumping and declining groundwater

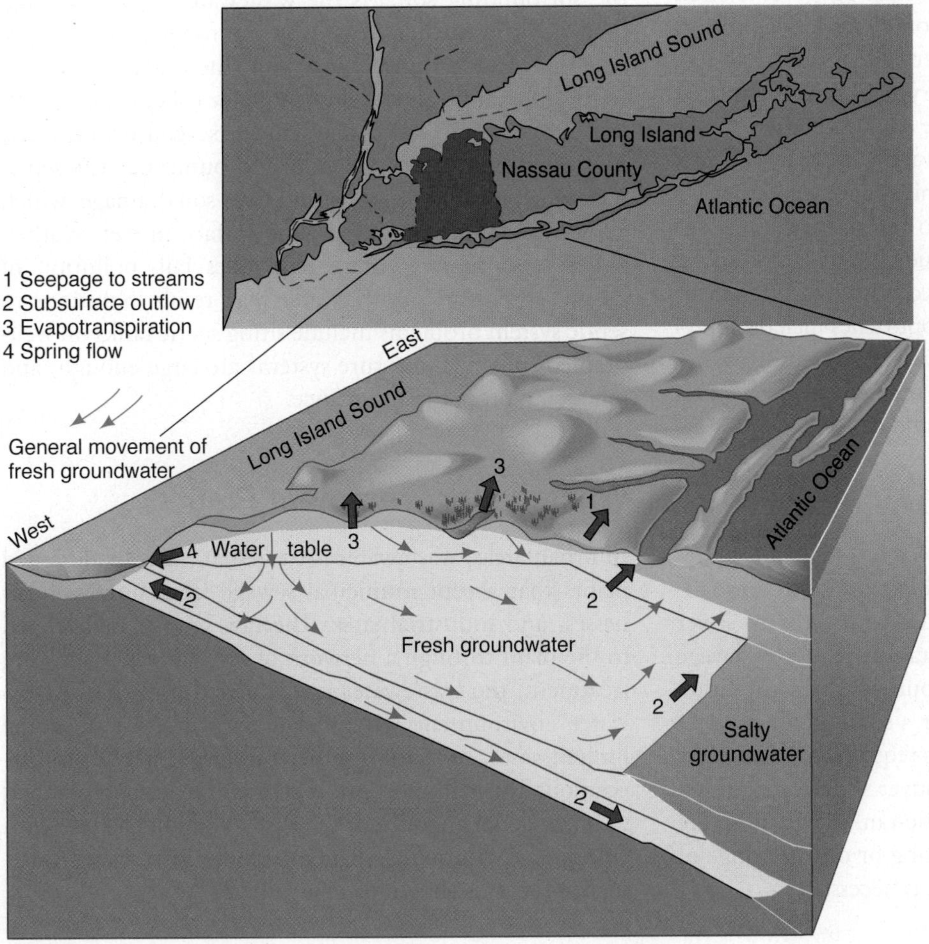

FIGURE 19.19 The general movement of fresh groundwater for Nassau County, Long Island.

(*Source:* G.L. Foxworth, Nassau County, Long Island, New York. Water problems in humid country, in G.D. Robinson and A.M. Spieke, eds., *Nature to Be Commanded*, U.S. Geological Survey Professional Paper 950, 1978, pp.55–68.)

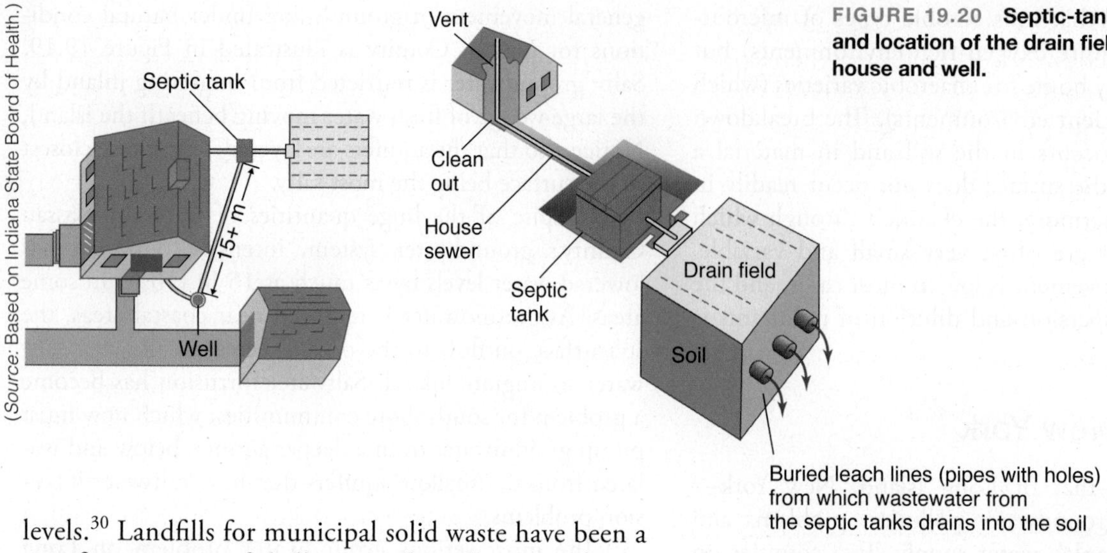

FIGURE 19.20 Septic-tank sewage system and location of the drain field with respect to the house and well.

levels.[30] Landfills for municipal solid waste have been a significant source of shallow-aquifer pollution on Long Island because pollutants (garbage) placed on sandy soil can quickly enter shallow groundwater. For this reason, most Long Island landfills were closed in the past few decades.

19.10 Wastewater Treatment

Water used for industrial and municipal purposes is often degraded during use by the addition of suspended solids, salts, nutrients, bacteria, and oxygen-demanding material. In the United States, by law, these waters must be treated before being released back into the environment.

Wastewater treatment—sewage treatment—costs about $40 billion per year in the United States, and the cost keeps rising, but it will continue to be big business. Conventional wastewater treatment includes septic-tank disposal systems in rural areas and centralized wastewater treatment plants in cities. Recent innovative approaches include applying wastewater to the land and renovating and reusing wastewater. We discuss the conventional methods in this section and some newer methods in later sections.

Septic-Tank Disposal Systems

In many rural areas, no central sewage systems or wastewater treatment facilities are available. As a result, individual septic tank disposal systems, not connected to sewer systems, continue to be an important method of sewage disposal in rural areas as well as outlying areas of cities. Because not all land is suitable for a septic-tank disposal system, an evaluation of each site is required by law before a permit can be issued. An alert buyer should make sure that the site is satisfactory for septic-tank disposal before purchasing property in a rural setting or on the fringe of an urban area where such a system is necessary.

The basic parts of a septic-tank disposal system are shown in Figure 19.20. The sewer line from the house leads to an underground septic tank in the yard. The tank is designed to separate solids from liquid, digest (biochemically change) and store organic matter through a period of detention, and allow the clarified liquid to discharge into the drain field (absorption field) from a piping system through which the treated sewage seeps into the surrounding soil. As the wastewater moves through the soil, it is further treated by the natural processes of oxidation and filtering. By the time the water reaches any freshwater supply, it should be safe for other uses.

Sewage drain fields may fail for several reasons. The most common causes are failure to pump out the septic tank when it is full of solids, and poor soil drainage, which allows the effluent to rise to the surface in wet weather. When a septic-tank drain field does fail, pollution of groundwater and surface water may result. Solutions to septic system problems include siting septic tanks on well-drained soils, making sure systems are large enough, and practicing proper maintenance.

Wastewater Treatment Plants

In urban areas, wastewater is treated at specially designed plants that accept municipal sewage from homes, businesses, and industrial sites. The raw sewage is delivered to the plant through a network of sewer pipes. Following treatment, the wastewater is discharged into the surface-water environment (river, lake, or ocean) or, in some limited cases, used for another purpose, such as crop irrigation. The main purpose of standard treatment plants is to break down and reduce the BOD and kill bacteria with chlorine. A simplified diagram of the wastewater treatment process is shown in Figure 19.21.

Wastewater treatment methods are usually divided into three categories: **primary treatment**, **secondary treatment**, and **advanced wastewater treatment**. Primary and secondary treatments are required by federal law for all municipal plants in the United States. However, treatment plants may qualify for a waiver exempting them from secondary treatment if installing secondary treatment facilities poses an excessive financial burden. Where secondary treatment is not sufficient to protect the quality of the surface water into which the treated water is discharged—for example, a river with endangered fish species that must be protected—advanced treatment may be required.[31]

Primary Treatment

Incoming raw sewage enters the plant from the municipal sewer line and first passes through a series of screens to remove large floating organic material. The sewage next enters the "grit chamber," where sand, small stones, and grit are removed and disposed of. From there, it goes to the primary sedimentation tank, where particulate matter settles out to form sludge. Sometimes, chemicals are used to help the settling process. The sludge is removed and transported to the "digester" for further processing. Primary treatment removes approximately 30 to 40% of BOD by volume from the wastewater, mainly in the form of suspended solids and organic matter.[31]

Secondary Treatment

There are several methods of secondary treatment. The most common treatment is known as *activated sludge*, because it uses living organisms—mostly bacteria. In this procedure, the wastewater from the primary sedimentation tank enters the aeration tank (Figure 19.21), where it is mixed with air (pumped in) and with some of the sludge from the final sedimentation tank. The sludge contains aerobic bacteria that consume organic material (BOD) in the waste. The wastewater then enters the final sedimentation tank, where sludge settles out. Some of this "activated sludge," rich in bacteria, is recycled and mixed again in the aeration tank with air and new, incoming wastewater acting as a starter. The bacteria are used again and again. Most of the sludge from the final sedimentation tank, however, is transported to the sludge digester. There, along with sludge from the primary sedimentation

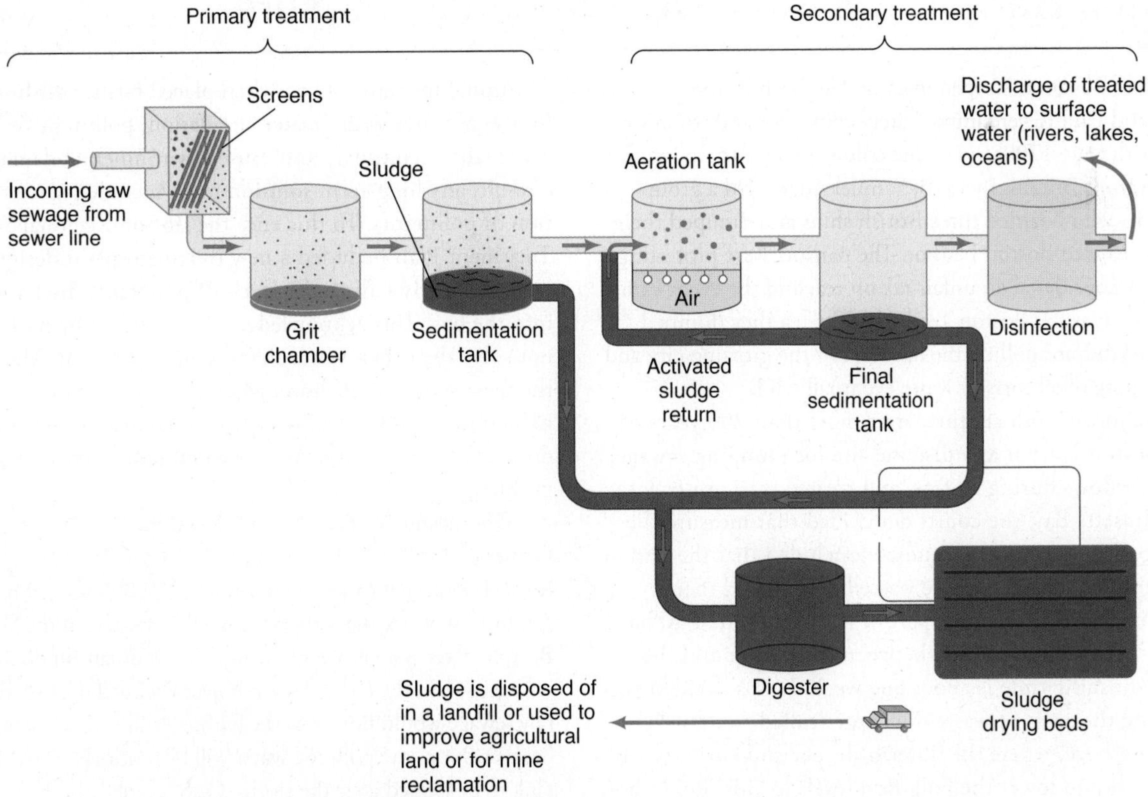

FIGURE 19.21 **Diagram of sewage treatment processes.** The use of digesters is relatively new, and many older treatment plants do not have them.

tank, it is treated by anaerobic bacteria (bacteria that can live and grow without oxygen), which further degrade the sludge by microbial digestion.

Methane gas (CH_4) is a product of the anaerobic digestion and may be used at the plant as a fuel to run equipment or to heat and cool buildings. In some cases, it is burned off. Wastewater from the final sedimentation tank is next disinfected, usually by chlorination, to eliminate disease-causing organisms. The treated wastewater is then discharged into a river, lake, or ocean (see A Closer Look 19.5), or in some limited cases used to irrigate farmland. Secondary treatment removes about 90% of BOD that enters the treatment plant in the sewage.[31]

The sludge from the digester is dried and disposed of in a landfill or applied to improve soil. In some instances, treatment plants in urban and industrial areas contain many pollutants, such as heavy metals, that are not removed in the treatment process. Sludge from these plants is too polluted to use in the soil and must be disposed of. Some communities, however, require industries to pretreat sewage to remove heavy metals before the sewage is sent to the treatment plant; in these instances, the sludge can be more safely used for soil improvement.

Advanced Wastewater Treatment

As noted above, primary and secondary treatments do not remove all pollutants from incoming sewage. Some additional pollutants, however, can be removed by taking more treatment steps. For example, phosphates and nitrates, organic chemicals, and heavy metals can be removed by specifically designed treatments, such as sand filters, carbon filters, and chemicals applied to assist in the removal process.[31] Treated water is then discharged into surface water or may be used for irrigating agricultural lands or municipal properties, such as golf courses, city parks, and grounds surrounding wastewater treatment plants.

Advanced wastewater treatment is used when it is particularly important to maintain good water quality. For example, if a treatment plant discharges treated wastewater into a river and there is concern that nutrients remaining after secondary treatment may cause damage to

A CLOSER LOOK 19.5

Boston Harbor: Cleaning Up a National Treasure

The city of Boston is steeped in early American history. Samuel Adams and Paul Revere immediately come to mind when we consider the late 1700s, when the colonies were struggling for freedom from Britain. In 1773, Samuel Adams led a group of patriots who boarded three British ships and dumped their cargo of tea into Boston Harbor. The patriots were protesting what they considered an unfair tax on tea, and the event came to be known as the Boston Tea Party. The tea they dumped overboard did not pollute the harbor, but the growing city and the dumping of all sorts of waste eventually did.

Late in the 20th century, after more than 200 years of using Boston Harbor as a disposal site for dumping sewage, sewer overflows during storms, and treated wastewater into Massachusetts Bay, the courts demanded that measures be taken to clean up the bay. Studies concluded that the harbor had become polluted because waste being placed there moved into a small, shallow part of Massachusetts Bay, and despite vigorous tidal action between the harbor and the bay, the flushing time is about one week. It was decided that relocating the areas of waste discharge (called "outfalls") farther offshore, where the water is deeper and currents are stronger, would lower the pollution levels in Boston Harbor.

Moving the wastewater outfalls offshore was definitely a step in the right direction, but the long-term solution to protecting the marine ecosystem from pollutants will require additional measures. Even when placed farther offshore, in deeper water with greater circulation, pollutants will eventually accumulate and cause environmental damage. As a result, any long-term solution must include source reduction of pollutants. To this end, the Boston Regional Sewage Treatment Plan included a new treatment plant designed to significantly reduce the levels of pollutants discharged into the bay. This acknowledges that dilution by itself cannot solve the urban waste-management problem. Moving the sewage outfall offshore, when combined with source reduction of pollutants, is a positive example of what can be done to better manage our waste and reduce environmental problems.[a]

The cleanup of Boston Harbor and the beaches went a step further in 2011 with the completion of a 3.4 km (1.9 mi) long (tunnel) tank that can temporarily store 72,000 m^3 (19 million gal) of wastewater that otherwise would overwhelm the South Boston sewer system, which includes both urban runoff and sewage. When the old system was overwhelmed during storms, raw sewage would flow into the harbor, polluting the water and beaches. Now, the polluted water will be routinely stored in the tank to be treated after the storm. Only a rare, large runoff event can now cause problems. Boston beaches that previously closed due to pollution several times each summer will now close much less frequently.

the river ecosystem (eutrophication), advanced treatment may be used to reduce the nutrients.

Chlorine Treatment

As mentioned earlier, chlorine is frequently used to disinfect water as part of wastewater treatment. Chlorine is very effective in killing the pathogens responsible for outbreaks of serious waterborne diseases that over time have killed many thousands of people. However, chlorine has been discovered to have another, less benign potential: Chlorine treatment also produces minute quantities of chemical by-products, some of which are potentially hazardous to people and other animals. For example, a recent study in Britain revealed that in some rivers, male fish sampled downstream from wastewater treatment plants had testes containing both eggs and sperm. This is likely related to the concentration of sewage effluent and the treatment method used.[32] Evidence also suggests that these by-products in the water may pose a risk of cancer and other human health effects. The degree of risk is controversial and is currently being debated.[33]

19.11 Land Application of Wastewater

The practice of applying wastewater to the land arose from the fundamental belief that waste is simply a resource out of place. Land application of untreated human waste was practiced for hundreds if not thousands of years before the development of wastewater treatment plants, which have sanitized the process by reducing BOD and using chlorination.

Many sites around the United States are now recycling wastewater, and the technology for wastewater treatment is rapidly evolving. An important question is: Can we develop environmentally preferred, economically viable wastewater treatment plants that are fundamentally different from those in use today? An idea for such a plant, called a resource-recovery wastewater treatment plant, is shown in Figure 19.22. The term *resource recovery* here refers to the production of resources, including methane gas (which can be burned as a fuel), as well as ornamental plants and flowers that have commercial value.[34]

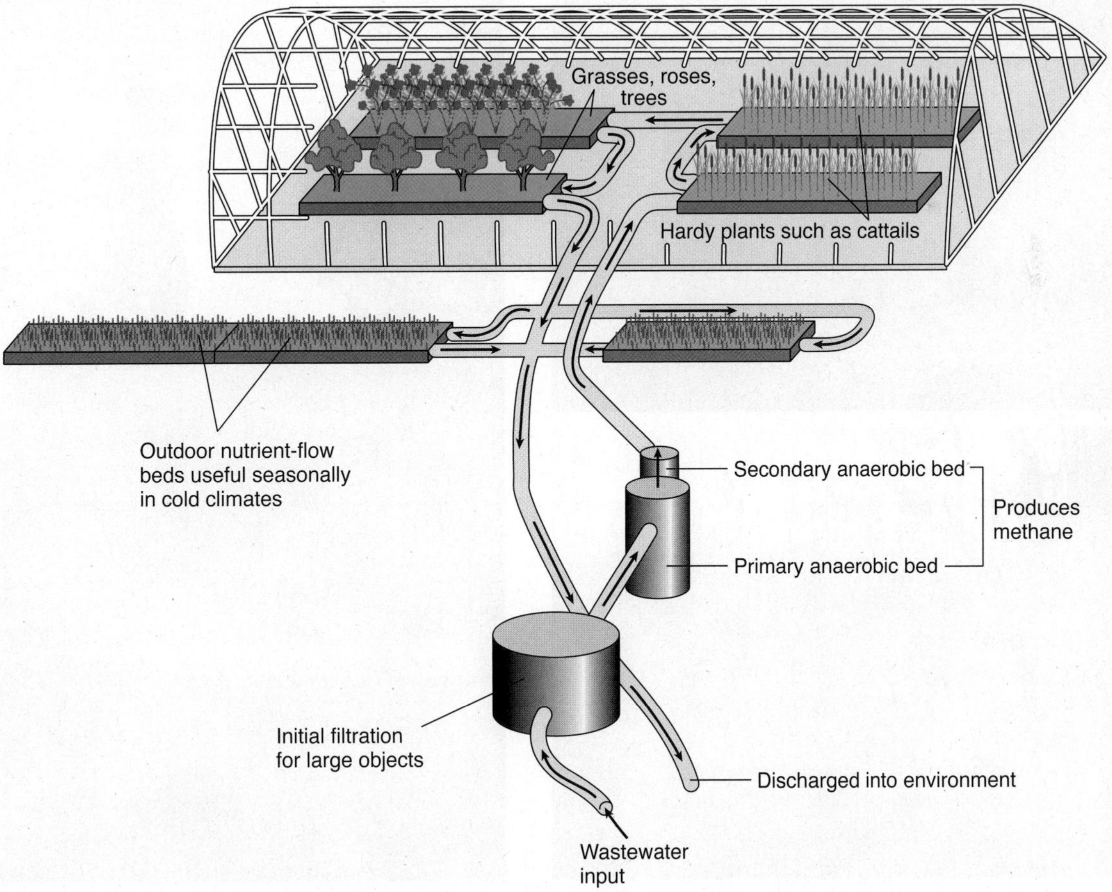

(*Source:* Based on W.J. Jewell, "Resource-Recovery Wastewater Treatment," *American Scientist* [1994] 82:366–375.)

FIGURE 19.22 Components of a resource-recovery wastewater treatment plant. For this model, two resources are recovered: methane, which can be burned to produce energy from the anaerobic beds; and ornamental plants, which can be sold.

Wastewater and Wetlands

Wastewater is being applied successfully to natural and constructed wetlands at a variety of locations.[35–37] Natural or human-made wetlands can be effective in treating the following water-quality problems:

- municipal wastewater from primary or secondary treatment plants (BOD, pathogens, phosphorus, nitrate, suspended solids, metals)
- storm water runoff (metals, nitrate, BOD, pesticides, oils)
- industrial wastewater (metals, acids, oils, solvents)
- agricultural wastewater and runoff (BOD, nitrate, pesticides, suspended solids)
- mining waters (metals, acidic water, sulfates)
- groundwater seeping from landfills (BOD, metals, oils, pesticides)

Using wetlands to treat wastewater is particularly attractive to communities that find it difficult to purchase traditional wastewater treatment plants. For example, the city of Arcata, in northern California, makes use of a wetland as part of its wastewater treatment system. The wastewater comes mostly from homes, with minor inputs from the numerous lumber and plywood plants in Arcata. It is treated by standard primary and secondary methods, then chlorinated and dechlorinated before being discharged into Humboldt Bay.[35]

Louisiana Coastal Wetlands

The state of Louisiana, with its abundant coastal wetlands, is a leader in the development of advanced treatment using wetlands after secondary treatment (Figure 19.23). Wastewater rich in nitrogen and phosphorus, applied to coastal wetlands, increases the production of wetland plants, thereby improving water quality as these nutrients are used by the plants. When the plants die, their organic

(a) Courtesy John Day, Louisiana State University

(b) Courtesy John Day, Louisiana State University

(c) Courtesy John Day, Louisiana State University

FIGURE 19.23 **(a) Wetland Pointe au Chene Swamp, three miles south of Thibodaux, Louisiana, receives wastewater; (b)** one of the outfall pipes delivering wastewater; and **(c)** ecologists doing field work at the Pointe au Chene Swamp to evaluate the wetland.

(a)

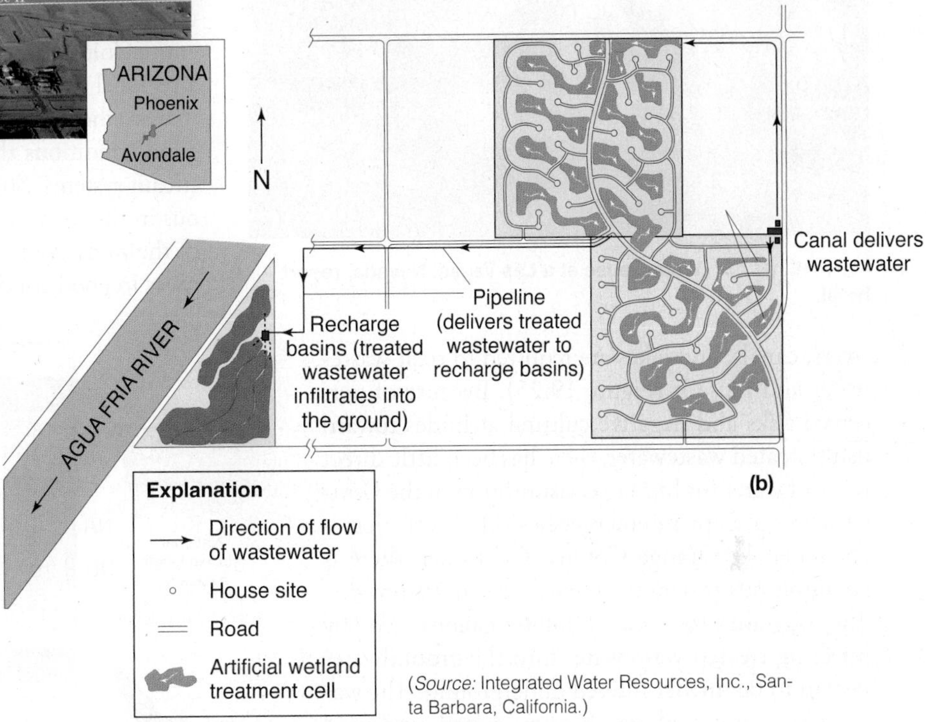

(b)

(*Source:* Integrated Water Resources, Inc., Santa Barbara, California.)

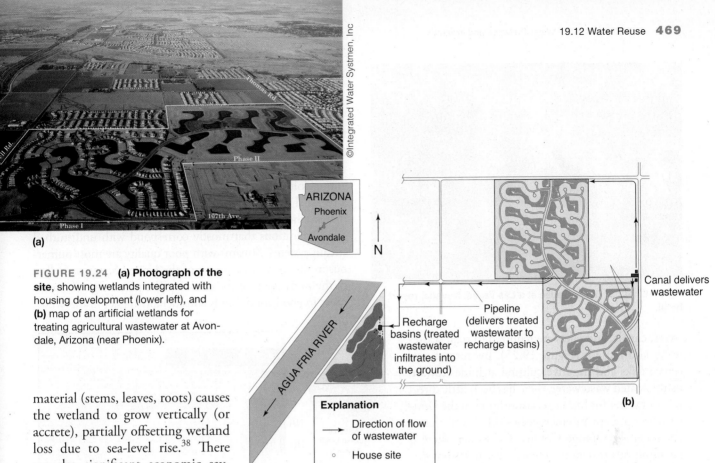

FIGURE 19.24 **(a) Photograph of the site**, showing wetlands integrated with housing development (lower left), and **(b)** map of an artificial wetlands for treating agricultural wastewater at Avondale, Arizona (near Phoenix).

material (stems, leaves, roots) causes the wetland to grow vertically (or accrete), partially offsetting wetland loss due to sea-level rise.[38] There are also significant economic savings in applying treated wastewater to wetlands, because the financial investment is small compared with the cost of advanced treatment at conventional treatment plants. Over a 25-year period, a savings of about $40,000 per year is likely.[37]

In sum, the use of isolated wetlands, such as those in coastal Louisiana, is a practical way to improve water quality in small, widely dispersed communities in the coastal zone. As water-quality standards are tightened, wetland wastewater treatment will become a viable, effective alternative that is less costly than traditional treatment.[38,39]

Phoenix, Arizona: Constructed Wetlands

Wetlands can be constructed in arid regions to treat poor-quality water. For example, at Avondale, Arizona, near Phoenix, a wetland treatment facility for agricultural wastewater is sited in a residential community (Figure 19.24). The facility is designed to eventually treat about 17,000 m^3/day (4.5 million gal/day) of water. Water entering the facility has nitrate (NO_3) concentrations as high as 20 mg/l. The artificial wetlands contain naturally occurring bacteria that reduce the nitrate to below the maximum contaminant level of 10 mg/l. Following treatment, the water flows by pipe to a recharge basin on the nearby Agua Fria River, where it seeps into the ground to become a groundwater resource. The cost of the wetland treatment facility was about $11 million, about half the cost of a more traditional treatment facility.

19.12 Water Reuse

Water reuse results when water is withdrawn, treated, used, treated, and returned to the environment, followed by further withdrawals and use. Water reuse is very common and a fact of life for millions of people who live along large rivers. Many sewage treatment plants are located along rivers and discharge treated water into the rivers. Downstream, other communities withdraw, treat, and consume the water.

Water reuse is a planned endeavor. In fact, reused water is a water source (see Chapter 18). For example, in the United States, treated wastewater is applied to numerous sites to recharge groundwater and then reused for agricultural and municipal purposes.

Direct water reuse refers to use of treated wastewater that is piped directly from a treatment plant to the next user. In most cases, the water is used in industry, in agricultural activity, or in irrigation of golf courses, institutional grounds (such as university campuses), and parks. Direct water reuse is growing rapidly and is the norm for industrial processes in factories. In Las Vegas, Nevada, new resort hotels that use a great deal of water for fountains,

Photo.courtesy Ed Keller

FIGURE 19.25 Water reuse at a Las Vegas, Nevada, resort hotel.

rivers, canals, and lakes are required to treat wastewater and reuse it (Figure 19.25). Because of perceived risks and negative cultural attitudes toward using treated wastewater, there has been little direct reuse of water for human consumption in the United States, except in emergencies. However, that is changing. In Orange County, California, there is an ambitious program to reuse treated wastewater. The program processes 70 million gallons a day by injecting treated wastewater into the groundwater system to be further filtered underground. The water is then pumped out, further treated, and used in homes and businesses.[40]

19.13 Conditions of Stream Ecosystems in the United States

Assessment of stream ecosystem conditions in the United States has been an important research goal since passage of the Clean Water Act of 1977. Until recently, no straightforward way to perform this evaluation had been seriously attempted, so the condition of small streams that can be waded was all but unknown. That void has now been partly filled by recent studies aimed at providing a credible, broad-scale assessment of small streams in the United States. A standardized field collection of data was an important step, and the data at each site include the following:

- measurement of stream channel morphology and habitat characteristics
- measurement of the streamside and near-stream vegetation, known as the *riparian vegetation*
- measurement of water chemistry
- measurement of the assemblage and composition of the stream environment (biotic environment)

The evaluation includes two key biological indicators: (1) an index of how pristine a stream ecosystem is and (2) an index that represents a loss of biodiversity. Results of the study are shown in Figure 19.26. The top graph is for the entire United States, while the three lower graphs are done on a regional basis. The ratings range from poor conditions—that is, those most disturbed by environmental stress—to good conditions that mostly correspond with undisturbed stream systems. Streams with poor quality are most numerous in the northeastern part of the United States, as well as in the midsection of the country. The percentage of stream miles in good condition is considerably higher in the West.

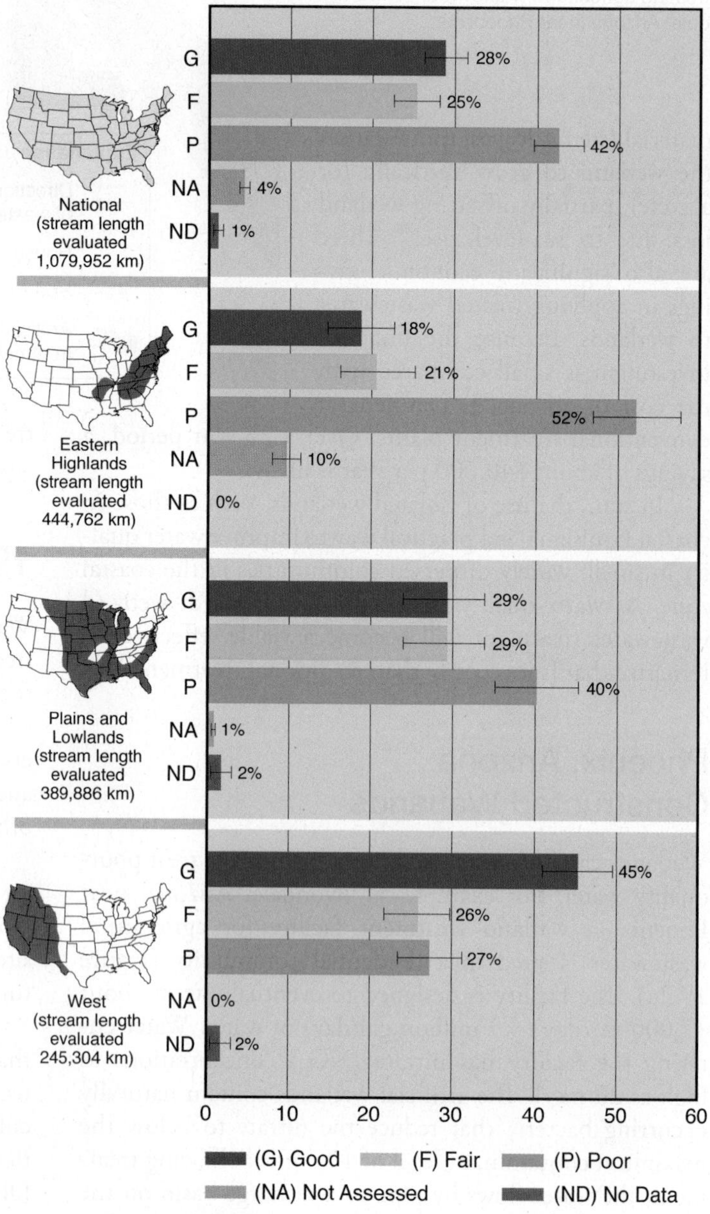

(Source: Modified after J.M. Faustini et al., 2009, "Assessing Stream Ecosystem Condition in the United States. *EOS, Transactions, American Geophysical Union* 36:309–310).

FIGURE 19.26 Condition of ecosystems of small streams that can be waded in the United States.

This finding is not surprising, given the extent of stream channel modifications and changes in land use in the eastern half of the country, compared to the western half. Western states tend to have more mountains and more areas of natural landscape that have not been modified by agriculture and other human activities. However, streams in the West were deemed to have a higher risk of future degradation than those in other areas because the West has more pristine streams to measure changes against and because it also has more high-quality stream ecosystem conditions that can potentially be degraded than those in other parts of the country.[41]

19.14 Water Pollution and Environmental Law

Environmental law, the branch of law dealing with conservation and use of natural resources and control of pollution, is very important as we debate environmental issues and make decisions about how best to protect our environment. In the United States, laws at the federal, state, and local levels address these issues.

Federal laws to protect water resources go back to the Refuse Act of 1899, which was enacted to protect navigable streams, rivers, and lakes from pollution. Table 19.5 lists major federal laws that have a strong water resource/pollution component. Each of these major pieces of legislation has had a significant impact on water-quality issues. Many federal laws have been passed with the purpose of cleaning up or treating pollution problems or treating wastewater. However, there has also been a focus on preventing pollutants from entering water. Prevention has the advantage of avoiding environmental damage and costly cleanup and treatment.

From the standpoint of water pollution, the mid-1990s in the United States was a time of debate and controversy. In 1994, Congress attempted to rewrite major environmental laws, including the Clean Water Act (1972, amended in 1987). The purpose was to give

Table 19.5 FEDERAL WATER LEGISLATION

DATE	LAW	OVERVIEW
1899	Refuse Act	Protects navigable water from pollution
1956	Federal Water and Pollution Control Act	Enhances the quality of water resources and prevents, controls, and abates water pollution.
1958	Fish and Wildlife Coordination Act	Mandates the coordination of water resources projects such as dams, power plants, and flood control must coordinate with U.S. Fish and Wildlife Service to enact wildlife conservation measures
1969	National Environmental Policy Act	Requires environmental impact statement prior to federal actions (development) that significantly affect the quality of the environment. Included are dams and reservoirs, channelization, power plants, bridges, and so on.
1970	Water Quality Improvement Act	Expands power of 1956 act through control of oil pollution and hazardous pollutants and provides for research and development to eliminate pollution in Great Lakes and acid mine drainage.
1972 (amended in 1977)	Federal Water Pollution Control Act (Clean Water Act)	Seeks to clean up nation's water. Provides billions of dollars in federal grants for sewage treatment plants. Encourages innovative technology, including alternative water treatment methods and aquifer recharge of wastewater.
1974	Federal Safe Drinking Water Act	Aims to provide all Americans with safe drinking water. Sets contaminant levels for dangerous substances and pathogens
1980	Comprehensive Environmental Response, Compensation, and Liability Act	Established revolving fund (Superfund) to clean up hazardous waste disposal sites, reducing ground water pollution.
1984	Hazardous and Solid Waste Amendments to the Resource Conservation and Recovery Act	Regulates underground gasoline storage tanks. Reduces potential for gasoline storage tanks. Reduces potential for gasoline to pollute groundwater
1987	Water Quality Act	Established national policy to control nonpoint sources of water pollution. Important in development of state management plants to control nonpoint water pollution sources.
2009	American Recovery And Reinvestment Act	Provides funds for water and wastewater projects.

industry greater flexibility in choosing how to comply with environmental regulations concerning water pollution. On the one hand, industry interests favored the proposed new regulations that, in their estimation, would be more cost effective without causing increased environmental degradation. Environmentalists, on the other hand, viewed attempts to rewrite the Clean Water Act as a giant step backward in the nation's fight to clean up our water resources. Apparently, Congress had incorrectly assumed it knew the public's values on this issue. Survey after survey has established that there is strong support for a clean environment in the United States and that people are willing to pay to have clean air and clean water. There is a consensus that the Clean Water Act was a huge success in cleaning U.S. waters. However, achieving water sustainability will require modification of the Clean Water Act in order to:[42]

- better target pollution abatement, based on watersheds
- better integrate the Clean Water Act with drinking water standards
- better integrate water pollution to global change
- better control nutrients causing widespread problems in the Gulf of Mexico and other areas
- better manage storm water runoff and pollution.

Congress has continued to debate changes in environmental laws, but little has been resolved.[43]

CRITICAL THINKING ISSUE

Can Safe Drinking Water and Sanitation Be Provided for All People on Earth?

Depending on what sources you read, about 1 billion people on Earth today do not have access to safe drinking water. In many places, obtaining safe water costs many more times than it would in cities like New York, with its municipal water system. In addition, about 2 billion people do not have basic sanitation consisting of a toilet. It's been pointed out that diseases caused by polluted water and lack of sanitation are, for the most part, avoidable and treatable (and have been for many decades), and yet millions of people are still dying in some of the poorest countries on Earth.

The cost of providing safe drinking water and a toilet per person in places were these do not exist varies from about $10 to $20 per person. Thus, if water and toilets for 2 billion people are needed, then the problem is how to raise $20–$40 billion to finance the program worldwide. Agencies and charities have been involved in this issue for decades, but the problem is nearly overwhelming in terms of the expense and the fact that it is growing, due to increasing populations that may double in some poor countries in the next 30 years. Some of the options available are:

- continuing work with charities
- investment opportunity for entrepreneurs
- microfinancing
- pressure from social media
- foreign aid programs from more wealthy countries

Should the problem be attacked from the top down—say, in building treatment plants and facilities to distribute water and sewers—or should it be from the bottom up from individual people living in particular areas of concern? One interesting way to look at this issue is not as a $20–$40 billion expense, but as an opportunity to invest in a capital improvement project of immense size with potential for profits. One method that has been discussed is microfinancing, where grant money is used as seed money to provide small loans to people interested in obtaining clean water and sanitation facilities. Interestingly, in many countries with water and sanitation problems, more people have cell phones than sanitary facilities. Thus, it has been suggested that people in these countries could use the social media to help gather support for programs to provide clean water and sanitation.

Critical Thinking Questions

1. Summarize possible solutions to the problem of providing clean drinking water and sanitation for approximately 2 billion people who lack such amenities in the world today.
2. How might the social media be used today in countries where clean drinking water and sanitation are lacking to help build support for financing water programs?
3. Considering all the elements and potential problems that might arise from trying to provide clean drinking water and sanitation for people of the Earth, can you come up with a plan and outline its basic components?

SUMMARY

- The primary water pollution problem in the world today is the lack of disease-free drinking water.
- Water pollution is degradation of quality that renders water unusable for its intended purpose.
- Major categories of water pollutants include disease-causing organisms, dead organic material, heavy metals, organic chemicals, acids, sediment, heat, and radioactivity.
- Sources of pollutants may be point sources, such as pipes that discharge into a body of water, or nonpoint sources, such as runoff, which are diffused and intermittent.
- Eutrophication of water is a natural or human-induced increase in the concentration of nutrients, such as phosphorus and nitrogen, required for living things. A high concentration of such nutrients may cause a population explosion of photosynthetic bacteria. As the bacteria die and decay, the concentration of dissolved oxygen in the water is lowered, leading to the death of fish.
- Sediment is a twofold problem: Soil is lost through erosion, and water quality suffers when sediment enters a body of water.
- Acid mine drainage is a serious water pollution problem because when water and oxygen react with sulfide minerals that are often associated with coal or metal sulfide deposits, they form sulfuric acid. Acidic water draining from mines or tailings pollutes streams and other bodies of water, damaging aquatic ecosystems and degrading water quality.
- Urban processes—for example, waste disposal in landfills, application of fertilizers, and dumping of chemicals such as motor oil and paint—can contribute to shallow aquifer contamination. Overpumping of aquifers near the ocean may cause saltwater, found below the fresh water, to rise closer to the surface, contaminating the water resource by a process called saltwater intrusion.
- Wastewater treatment at conventional treatment plants includes primary, secondary, and, occasionally, advanced treatment. In some locations, natural ecosystems, such as wetlands and soils, are being used as part of the treatment process.
- Water reuse is the norm for millions of people living along rivers where sewage treatment plants discharge treated wastewater back into the river. People who withdraw river water downstream are reusing some of the treated wastewater.
- Industrial reuse of water is the norm for many factories.
- Deliberate use of treated wastewater for irrigating agricultural lands, parks, golf courses, and the like is growing rapidly as demand for water increases.
- Use of treated wastewater that is applied to the land and later withdrawn as a municipal water source is becoming more common.
- Cleanup and treatment of both surface-water and groundwater pollution are expensive and may not be completely successful. Furthermore, environmental damage may result before a pollution problem is identified and treated. Therefore, we should continue to focus on preventing pollutants from entering water, which is a goal of much water-quality legislation.

REEXAMINING THEMES AND ISSUES

Sean Randall/Getty Images, Inc.

HUMAN POPULATION

We state in this chapter that the number one water pollution problem in the world today is the lack of disease-free drinking water. This problem is likely to get worse in the future as the number of people, particularly in developing countries, continues to increase. As population increases, so does the possibility of continued water pollution from a variety of sources relating to agricultural, industrial, and urban activities.

© Biletskiy_Evgeniy/iStockphoto

SUSTAINABILITY

Any human activity that leads to water pollution—such as the building of pig farms and their waste facilities on floodplains—is antithetical to sustainability. Groundwater is fairly easy to pollute and, once degraded, may remain polluted for a long time. Therefore, if we wish to leave a fair share of groundwater resources to future generations, we must ensure that these resources are not polluted, degraded, or made unacceptable for use by people and other living organisms.

© Anton Balazh 2011/iStockphoto

GLOBAL PERSPECTIVE

Several aspects of water pollution have global implications. For example, some pollutants may enter the atmosphere and be transported long distances around the globe, where they may be deposited and degrade water quality. Examples include radioactive fallout from nuclear reactor accidents or experimental detonation of nuclear devices. Waterborne pollutants from rivers and streams may enter the ocean and circulate with marine waters around the ocean basins of the world.

ssguy/ShutterStock

URBAN WORLD

Urban areas are centers of activities that may result in serious water pollution. A broad range of chemicals and disease-causing organisms are present in large urban areas and may enter surface waters and groundwaters. An example is bacterial contamination of coastal waters, resulting in beach closures. Many large cities have grown along the banks of streams and rivers, and the water quality of those streams and rivers is often degraded as a result. There are positive signs that some U.S. cities are viewing their rivers as valuable resources, with a focus on environmental and economic renewal. Thus, rivers flowing through some cities are designated as greenbelts, with parks and trail systems along river corridors. Examples include New York City; Cleveland, Ohio; San Antonio, Texas; Corvallis, Oregon; and Sacramento and Los Angeles, California.

B2M Productions/Getty Images, Inc.

PEOPLE AND NATURE

Polluting our water resources endangers people and ecosystems. When we dump our waste in rivers, lakes, and oceans, we are doing what other animals have done for millions of years—it is natural. For example, a herd of hippopotamuses in a small pool may pollute the water with their waste, causing problems for other living things in the pond. The difference is that we understand that dumping our waste damages the environment, and we know how to reduce our impact.

George Doyle/Getty Images, Inc.

SCIENCE AND VALUES

It is clear that the people of the United States place a high value on the environment and, in particular, on critical resources such as water. Attempts to weaken water-quality standards are viewed negatively by the public. There is also a desire to protect water resources necessary for the variety of ecosystems found on Earth. This has led to research and development aimed at finding new technologies to reduce, control, and treat water pollution. Examples include development of new wastewater treatments and support of laws and regulations that protect water resources.

KEY TERMS

acid mine drainage 456
advanced wastewater treatment 465
bioengineering 460
biological or biochemical oxygen demand (BOD) 450
bioremediation 462
cultural eutrophication 453
ecosystem effect 452
environmental law 471
eutrophication 451
fecal coliform bacteria 454
nanotechnology 460
nonpoint sources 457
outbreaks 454
point sources 457
primary treatment 465
secondary treatment 465
wastewater treatment 464
water reuse 469

STUDY QUESTIONS

1. Do you think outbreaks of waterborne diseases will be more common or less common in the future? Why? Where are outbreaks most likely to occur?
2. What was learned from the *Exxon Valdez* oil spill that might help reduce the number of future spills and their environmental impact?
3. What is meant by the term *water pollution*, and what are several major processes that contribute to water pollution?
4. Compare and contrast point and nonpoint sources of water pollution. Which is easier to treat, and why?
5. What is the twofold effect of sediment pollution?
6. In the summer, you buy a house with a septic system that appears to function properly. In the winter, effluent discharges at the surface. What could be the environmental cause of the problem? How could the problem be alleviated?
7. Describe the major steps in wastewater treatment (primary, secondary, advanced). Can natural ecosystems perform any of these functions? Which ones?
8. In a city along an ocean coast, rare waterbirds inhabit a pond that is part of a sewage treatment plant. How could this have happened? Is the water in the sewage pond polluted? Consider this question from the birds' and your own point of view.
9. How does water that drains from coal mines become contaminated with sulfuric acid? Why is this an important environmental problem?
10. What is eutrophication, and why is it an ecosystem effect?
11. How safe do you believe the drinking water is in your home? How did you reach your conclusion? Are you worried about low-level contamination by toxins in your water? What could be the sources of contamination?
12. Do you think our water supply is vulnerable to terrorist attacks? Why? Why not? How could potential threats be minimized?
13. Would you be willing to use treated wastewater in your home for personal consumption, as they are doing in Orange County, California? Why? Why not?
14. How would you design a system to capture runoff where you live before it enters a storm drain?

FURTHER READING

Dunne, T., and L.B. Leopold, *Water and Environmental Planning* (San Francisco: W.H. Freeman, 1978). A great summary and detailed examination of water resources and problems.

Hester, R.E., and R.M. Harrison, eds., *Agricultural Chemicals and the Environment* (Cambridge: Royal Society of Chemistry, Information Services, 1996). A good source of information about the impact of agriculture on the environment, including eutrophication and the impact of chemicals on water quality.

Jones, J.A.A., *Water Sustainability* (London: Hudder Education, 2010). See Chapter 5 for a good discussion of water pollution.

NOTES

1. Environmental Protection *Agency. American Heritage Rivers, Hudson River*. Accessed February 19, 2005 www.epa.gov/rivers/98rivers/hudson.html.
2. Bopp, R.F., and H.J. Simpson, 1989. *Contamination of the Hudson River: The Sediment Record in Contaminated Marine Sediments, an Assessment and Remediation*. Washington, DC: National Academy Press.
3. Wall, G.R., K. Riva-Murray, and P.J. Phillips. 1998. 2009. *Water Quality in the Hudson River Basin. New York and Adjacent States, 1992–95*. U.S. Department of the Interior, U.S. Geological Survey. Circular 1165.
4. Environmental Protection Agency Region 2. 2010. EPA IV number: NYN000206222.
5. Environmental Protection Agency Region 2. 2012. *Hurricane Sandy Sampling Results. 3. Gowanus Canal. 2012*. www.epa.gov/region2/superfund/npl/**gowanus/sandysampling**.pdf
6. Jones, J.A.A. 2010. *Water Sustainability*. London, UK: Hudder Education. p. 452.
7. Morrison, J. 2005. How much is clean water worth? *National Wildlife* 43(2):22–28.
8. Gleick, P.H. 1993. An introduction to global fresh water issues. In P. H. Gleick, ed., *Water in Crisis*, pp. 3–120 New York, NY: Oxford University Press.
9. Hileman, B. 1995. Pollution tracked in surface and groundwater. *Chemical & Engineering News* 73:5.

10. Maugh, T.H. 1979. Restoring damaged lakes. *Science* 203:425–127.

11. Day, Jr., J.W. Gilliam, P.M. Groffman, D.L. Hey, G.W. Randall, and N. Wang. 1999. *Reducing Nutrient Loads, Especially to Surface Water, Ground Water, and the Gulf of Mexico Nitrate-Nitrogen.* Topic 5 Report for the Integrated Assessment on Hypoxia in the Gulf of Mexico. www.cop.noaa.gov/pubs/das/das19.pdf

12. Hinga, K.R. 1989. Alteration of phosphorus dynamics during experimental eutrophication of enclosed-marine ecosystems. *Marine Pollution Bulletin* 20:624–628.

13. Richmond, R.H. 1993. Coral reefs: Present problems and future concerns resulting from anthropogenic disturbance. *American Zoologist* 33:524–536.

14. Bell, P.R. 1991. Status of eutrophication in the Great Barrier Reef Lagoon. *Marine Pollution Bulletin* 23:89–93.

15. Hunter, C.L., and C.W. Evans. 1995. Coral reefs in Kaneohe Bay, Hawaii: Two centuries of Western influence and two decades of data. *Bulletin of Marine Science* 57:499.

16. Lewis, S.A. 1995. Trouble on tap. *Sierra* 80:54–58.

17. Smith, R.A. 1994. Water quality and health. *Geotimes* 39:19–21.

18. MacKenzie, W.R., et al. 1994. A massive outbreak in Milwaukee of Cryptosporidium infection transmitted through the public water supply. *New England Journal of Medicine* 331:161–167.

19. Kluger, J. 1998. Anatomy of an outbreak. *Time* 152 (5):56–62.

20. Department of Alaska Fish and Game 1989. *Alaska Fish and Game* 21(4), Special Issue.

21. Holway, M. 1991. Soiled shores. *Scientific American* 265: 102–106.

22. Robinson, A.R. 1973. Sediment, our greatest pollutant? In R.W. Tank, ed., *Focus on Environmental Geology*, pp. 186–192. New York, NY: Oxford University Press.

23. Yorke, T.H. 1975. Effects of sediment control on sediment transport in the northwest branch, Anacostia River basin, Montgomery Country, Maryland. *Journal of Research* 3: 481–494.

24. Poole, W. 1996. Rivers run through them. *Land and People* 8:16–21.

25. Haikin, M. 2007. Nanotechnology takes on water pollution. *Business 2.0 Magazine*. Accessed May 30, 2008 @ money.cnn.com.

26. Davis, A.P. August 2005. Green engineering principles promote low impact development. *Environmental Science and Technology*, pp.339A–344A.

27. Natural Resources Defense Council. 2008. Mimicking nature to solve a water pollution problem. Accessed May 30, 2008 @ www.nrdc.org.

28. Carey, J. 1984 (February/March). Is it safe to drink? *National Wildlife*, Special Report, pp. 19–2l.

29. Pye, U.I., and R. Patrick. 1983. Groundwater contamination in the United States. *Science* 221:713–718.

30. Foxworthy, G. L. 1978. Nassau County, Long Island, New York—Water problems in humid country. In G.D. Robinson and A.M. Spieker, eds., *Nature to Be Commanded*, pp. 55–68. U.S. Geological Survey Professional Paper 950. Washington, DC: U.S. Government Printing Office.

31. Van der Leeden, F., F.L. Troise, and D.K. Tood. 1990. *The Water Encyclopedia*, 2d ed. Chelsea, MI: Lewis Publishers.

32. Jobling, S., M. Nolan, C.R. Tyler, G. Brighty, and J.P. Sumpter. 1998. Widespread sexual disturbance in wild fish. *Environmental Science and Technology* 32(17):2498–2506.

33. Environmental Protection Agency, Drinking Water Committee of the Science Advisory Board. 1995. An SAB Report: *Safe Drinking Water. Future Trends and Challenges.* Washington, DC: Environmental Protection Agency.

34. Jewell, W.J. 1994. Resource-recovery wastewater treatment. *American Scientist* 82:366–375.

35. Task Force on Water Reuse. 1989. *Water Reuse: Manual of Practice* SM-3. Alexandria, VA: Water Pollution Control Federation.

36. Kadlec, R.H., and R.L. Knight. 1996. *Treatment Wetlands.* New York, NY: Lewis Publishers.

37. Breaux, A.M., and J.W. Day Jr. 1994. Policy considerations for wetland wastewater treatment in the coastal zone: A case study for Louisiana. *Coastal Management* 22:285–307.

38. Day, J.W., Jr., J.M. Rybczyk, L. Carboch, W.H. Conner, P. Delgado-Sanchez, R.T. Pratt, and W. Westphal. 1998. A review of recent studies of the ecology and economic aspects of the application of secondary treated municipal effluent wetlands in southern Louisiana. In L.P. Rozas et al., eds., *Symposium on Recent Research in Coastal Louisiana*, February 1998, Louisiana Sea Great College Program, pp. 1–12.

39. Breaux, A., S. Fuber, and J. Day. 1995. Using natural coastal wetland systems: An economic benefit analysis. *Journal of Environmental Management* 44:285–291.

40. Barone, J. May 2008. Better water. *Discover.* Better Planet Special Issue, pp. 31–32.

41. Faustini, J.M., et al. 2009. Assessing stream ecosystem condition in the United States. *EOS, Transactions, American Geophysical Union* 36:309–310.

42. Schopenhauer, A. October 1, 2012. The Clean Water Act at 40. *Water Environment and Technology.* pp. 29–38.

43. Hileman, B. 1995. Rewrite of Clean Water Act draws praise, fire. *Chemical & Engineering News* 73:8.

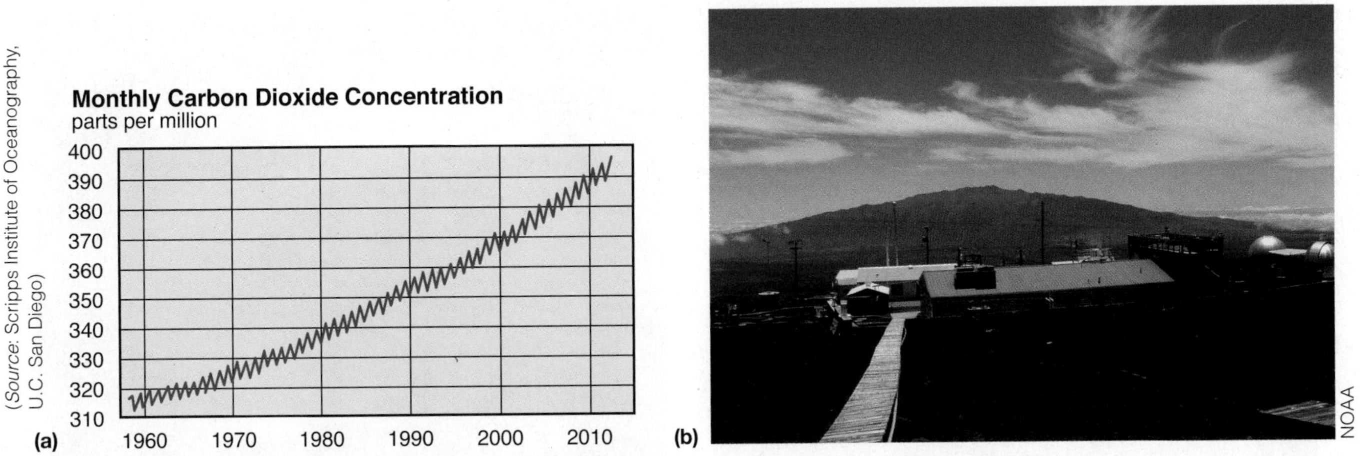

FIGURE 20.2 **(a) Carbon dioxide concentrations in the air above Mauna Loa, Hawaii; (b) the NOAA Observatory on Mauna Loa, where these measurements were made.**

The great breakthrough came during the International Geophysical Year (actually 18 months) of scientific observations of the global environment. As part of that project, an atmospheric observatory was established near the top of Mauna Loa volcano, at 11,500 ft above sea level. It was put there because scientists believed that this location represented the background chemistry of the atmosphere, unaffected by regional or local changes caused by the photosynthesis of green plants and oceanic algae, the respiration of animals, and the industrial and settlement activities of people. Charles Keeling began measuring the concentration of carbon dioxide and discovered that it was increasing annually. This increase was first reported in 1973 at a conference titled "Carbon and the Biosphere," one of the first times that a group of scientists from many disciplines met to discuss the possibility of a global warming (Figure 20.2).[7]

As far as modern records at the time indicated, the climate had been warming since the early 1960s (a cooling trend had lasted from approximately 1940 to 1960). At this stage, it seemed that the increase in atmospheric carbon dioxide was indeed warming the Earth.

20.2 Twentieth-Century Methods to Reconstruct Past Temperatures

Keeling's observations on Mauna Loa raised the fundamental question of how the climate in the 20th century compared to climate in past times. In the latter decades of the 20th century, attempts to answer this question began in earnest. Fortunately, there has always been a lot of interest in past climates, because climates affect so much of human life and the living things around us. One method is to evaluate temperature records. However, the trouble with using these records to accurately estimate the global average is that few places have a complete record since 1850, and the places that do—such as London and Philadelphia, where people were especially interested in the weather and where scientists could readily make continual measurements—are not very representative of the global average. Until the advent of satellite remote sensing, air temperature over the oceans was measured only where ships traveled (which is not the kind of sampling that makes a statistician happy), and many of Earth's regions have never had good long-term, ground-based temperature measurements. As a result, when we want or need to know what the temperatures were like in the 19th century, before CO_2 concentrations began to rise from the burning of fossil fuels, experts seek ways to extrapolate, interpolate, and estimate.

Several groups have tried to reconstruct Earth's average surface temperature by using available observations. For example, the Hadley Meteorological Center in Great Britain created a data set that divides Earth's surface into areas 5° longitude wide and 5° latitude high for every month of each year, starting with 1850. Where historical records exist, these data are placed within the appropriate geographic rectangle. Where they do not exist, either the rectangle is left empty or various attempts are made to estimate what might be reasonable for that location and time. Recently, the extrapolation methods used to make these reconstructions have come under criticism, and today controversy has arisen over the reliability and usefulness of such attempts.[8]

Temperature measurement has improved greatly in recent years, thanks to such devices as ocean platforms with automatic weather-monitoring equipment, coordinated by the World Meteorological Organization. Thus, we have more accurate records since about 1960 (Figure 20.3).

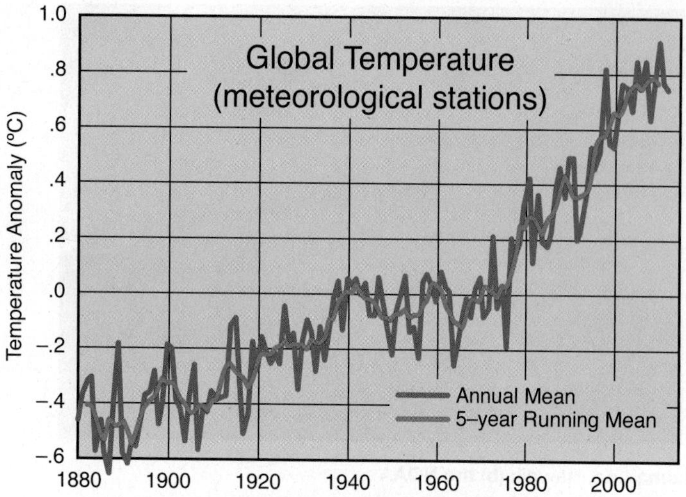

(*Source:* Global Temperature. NASA, 2012, GISS Surface temperature analysis. Http://data.giss.nasa.gov)

FIGURE 20.3 **The temperature difference between the average at the end of the 19th century and the years between 1880 and today.** This graph shows the difference between calculated annual mean world surface temperatures for each year (are green) and the 5-year running mean (red). Temperature departure refers to changes in mean global temperature from some standard, such as 1951–1980. Climatologists studying climate change prefer, in general, to look at the difference between temperatures at one time compared to another, rather than the actual temperature, for a variety of technical reasons.

Historical Documents

A better way of reconstructing Earth's average surface temperature than relying solely on existing temperature measurement (prior to computers and satellite remote sensing) was to use many kinds of historical records. These records included paintings of the Alps throughout the centuries, which showed the extent and range of mountain glaciers; in some paintings, glaciers came down into valley towns, indicating cold times; in others, the same glacier ended high up in the mountain, indicating warm times. Some historians used the dates of crop harvests, including wine harvests, because the time for plants to mature varies with temperature and rainfall. A variety of documents are available from the historical records, which in some cases go back several centuries. Included here would be people's written recollections in books, newspapers, journal articles, personal journals, ships' logs, travelers' diaries, and farmers' logs, along with dates of wine harvests and small-grain harvests.[3,8] Although these sources are mostly recorded as qualitative data, we can sometimes get quantitative information from them. For example, a painting of a mountain glacier in Switzerland can be used to determine the elevation to which the glacier had descended by the year it was painted. Then, as the climate cooled further, someone may have written that the same glacier reached farther down the mountain, eventually blocking a river in the valley, which flooded and destroyed a town, whose elevation is also known (Figure 20.4).[3]

Archive VAW, ETH Zurich

FIGURE 20.4 **Argentiere glacier in the French Alps.** (Top): Etching made between 1850 and 1860 when the glacier was near its greatest extent, and (Bottom): a modern photograph of the glacier taken from a similar vantage point in 1966.

Discovery of Continental Glaciation and Ice Ages: What the Rocks Told Swiss Farmers

That there had been vast changes in climate, including long-past ice ages, had been discovered by 1837. The scientist Jean Louis (Rodolphe) Agassiz, who was born in Switzerland, discovered continental glaciation when he roamed the Swiss hillsides and spoke with the farmers there. The farmers were familiar with mountain glaciers and the kinds of debris the glaciers deposited at their farther extents, along their sides, and underneath. At first, Agassiz was skeptical that there could have been such major changes in the land, but the farmers showed him rocks and strangely shaped hills of soil debris that looked just like deposits left by mountain glaciers plowing their way over the land but were far from any mountain glaciers. Puzzling over what could have moved those deposits long distances, Agassiz realized that these landforms created by mountain glaciers also occurred widely on flatter lands and could only have been produced by giant glaciers. It was soon recognized that glaciers had indeed covered vast areas in Great Britain, mainland Europe, and North America.[9] This was quite a shocking realization, but it made clear that Earth's climate had varied greatly over the eons.

Throughout the 20th century, other direct geological records greatly extended this knowledge. Places now cold were discovered to have had vegetation typical of the tropics in the far distant past. The developing theory of plate tectonics contributed to understanding the great changes that had occurred both in climate and in the location of continents.

Scientific data that are not strictly climatic in nature but can be correlated with climate data, such as temperature of the land or sea, are useful in studying the past climate (these are called proxy data). Such data include sediments, tree rings, and stable isotopes of oxygen and hydrogen from ice cores, corals, and carbon 14 (^{14}C).[9]

Sediments

Biological material, including pollen from plants, is deposited on the land and stored for very long periods in lake, bog, and pond sediments and, once transported downstream to the coast, in the oceans. Samples of these very small fossils and of chemicals in the sediments can be interpreted to study past climates and extend our knowledge back hundreds of thousands of years (Figure 20.5). Pollen is useful because (1) the quantity of pollen is an indicator of the relative abundance of each plant species; (2) the pollen can be dated, and since the grains are preserved in sedimentary layers that also might be dated, we can develop a chronology; and (3) based on the types of plants found at different times, we can construct a climatic history.

Science Source

FIGURE 20.5 Scientist examining a sediment core taken by drilling into the seafloor.

Tree Rings

The growth of trees is influenced by climate—both temperature and precipitation. Many trees put on one growth ring per year, and patterns in the tree rings—their width, density, and isotopic composition—tell us something about the variability of the climate. When conditions are good for growth, a ring is wide; when conditions are poor, the ring is narrow. Interest in the use of tree rings to reconstruct past climates preceded the modern concern with global warming. It was stimulated in part by archaeologists studying ancient cultures and wanting to know what the climate was like during those times and in part by ecologists interested in forest histories. A Tree Ring Research Laboratory established at the University of Arizona in 1937 specializes in these measurements. Tree ring chronology, known as dendrochronology, has produced a proxy record of climate that extends back over 10,000 years (Figure 20.6).

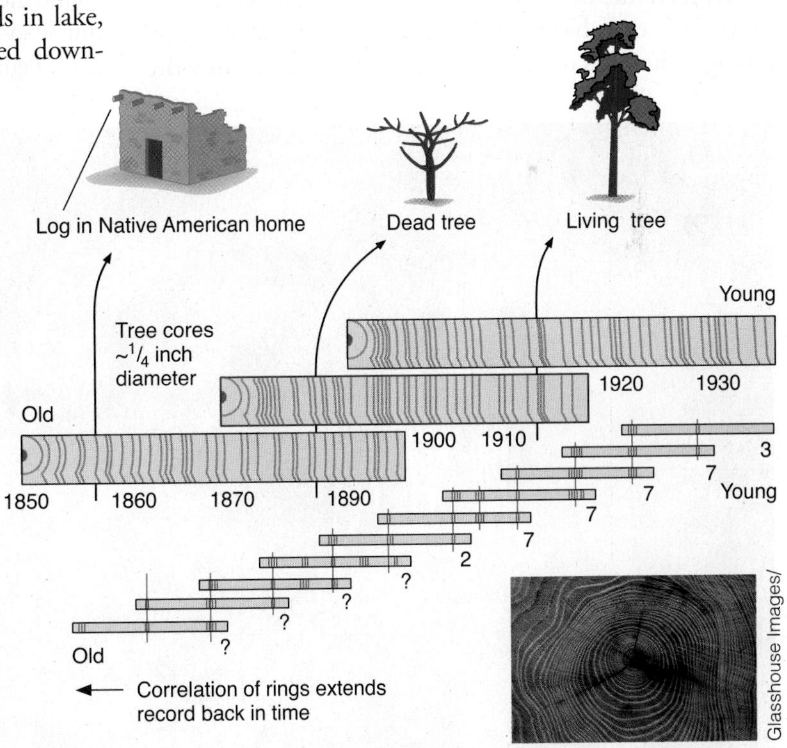

Glasshouse Images/ SuperStock

FIGURE 20.6 Dendrochronology, the study of tree growth rings, can be used as an indicator of past climate. The spacing (relative volume of wood) of rings and the isotopic content (^{14}C, for example) of wood can provide information about past rainfall and solar activity.

It is a good technique and it is accurate, but, as with all scientific methods, it has limitations. Tree rings are useful for local and regional climate reconstructions where trees are primarily affected by temperature and rainfall, often in very sunny climates.

Ice Cores

Polar ice caps and mountain glaciers have an accumulation record of snow that has been transformed into glacial ice over hundreds to thousands of years. Ice cores (Figure 20.7) often contain small bubbles of air deposited at the time of the snow, through which we can measure the atmospheric gases. Two important gases being measured in ice cores are (CO_2) and methane (CH_4). The temperature of the atmosphere in various glacial layers can be estimated by using analysis of stable isotopes of oxygen and hydrogen that correlate well with temperature. We will return to ice cores when we discuss changes in glaciers later in this chapter.

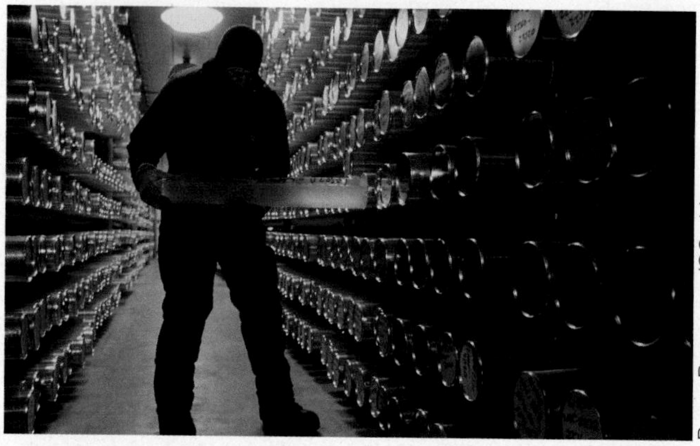

Roger Ressmeyer/©Corbis

FIGURE 20.7 Scientist examining an ice core from a glacier. The core was stored in a freezer so that ice bubbles could be extracted from it to provide data about the atmosphere in the past (CO_2, dust, lead, etc.).

Corals

Corals have hard skeletons composed of calcium carbonate ($CaCO_3$), a mineral extracted by the corals from seawater. The carbonate contains isotopes of oxygen, as well as a variety of trace metals, which have been used to determine the temperature of the water in which the coral grew. The growth of corals has been dated directly with a variety of dating techniques over short time periods of coral growth, thereby revealing the chronology of climate change over variable time periods.

Carbon 14

Radioactive carbon 14 (^{14}C) is produced in the upper atmosphere by the collision of cosmic rays and nitrogen 14 (^{14}N). Cosmic rays come from outer space; those the Earth receives are predominantly from the sun. The abundance of cosmic rays varies with the number of sunspots, so-called because they appear as dark areas on the sun (Figure 20.8). The frequency of sunspots has been

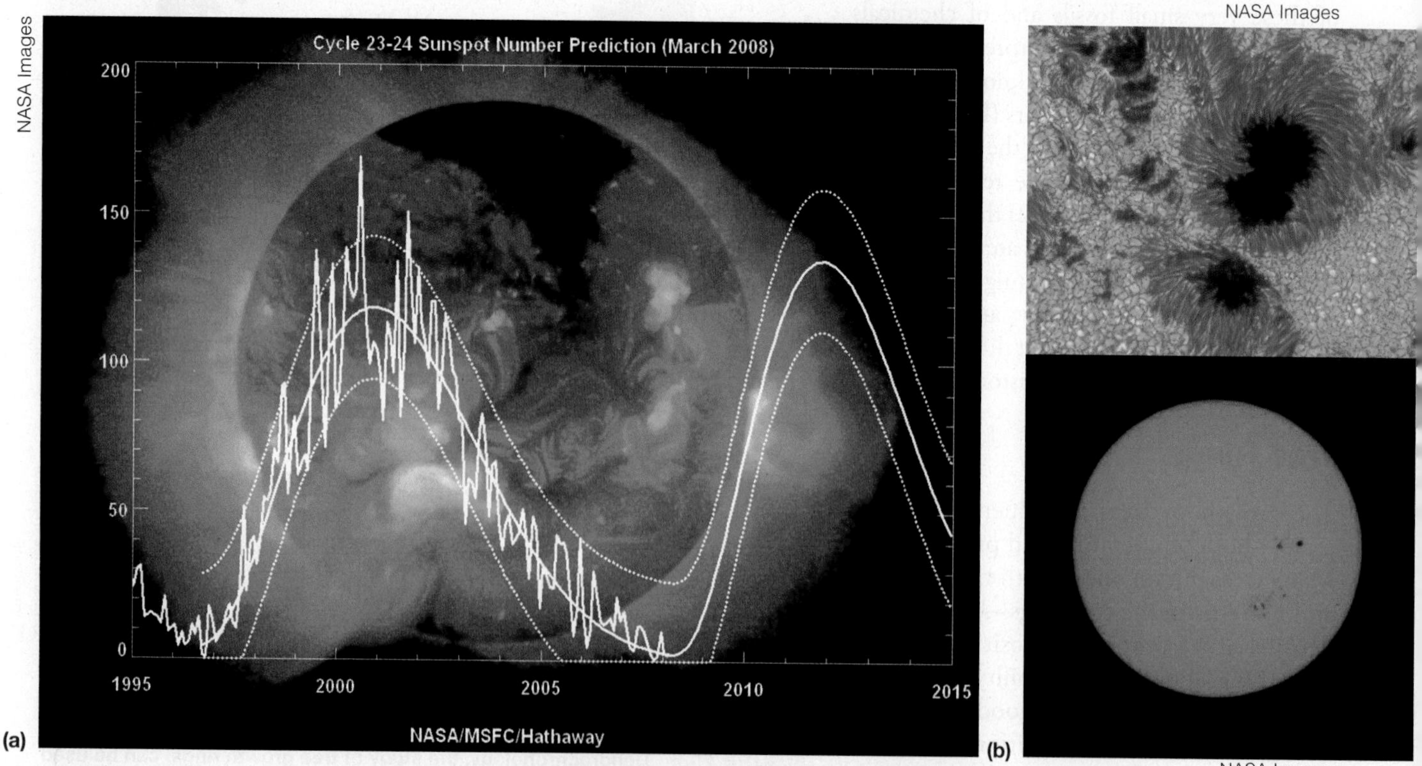

NASA Images

NASA Images

NASA Images

FIGURE 20.8 (a) Sunspot cycle with number of sunspots per year, from 1995 to 2009 and (b) photos of sunspots in 2003. The large sunspots are about 150,000 km across (for comparison, Earth's diameter is about 12,750 km). The sunspots are the dark areas; each is surrounded by an orange ring and then a hot gold color. Notice the great variety in the number of sunspots from year to year in the 11-year cycle (over 150 in 2001 and very few in 2008 and 2009).

accurately measured for decades and observed by people for nearly 1,000 years. As sunspot activity increases, more energy from the sun reaches Earth. There is an associated solar wind, which produces ionized particles consisting mostly of protons and electrons, emanating from the sun. The radioactive ^{14}C is taken up by photosynthetic organisms—green plants, algae, and some bacteria—and stored in them. If these materials become part of sediments (see above), the year in which they were deposited can be estimated from the decay rate of the ^{14}C.

The record of ^{14}C in the atmosphere has been correlated with tree ring chronology. Each ring of wood of known age contains carbon, and the amount of ^{14}C can be measured. Then, given the climatic record, it may be correlated with ^{14}C, and that correlation has been shown to be very strong.[9]

Thus, we can examine the output of the sun, going back thousands of years, by studying tree rings and the ^{14}C they contain. This connects to our opening case study about the Medieval Warm Period. Based on these records, it appears that the production of solar energy was slightly higher around A.D. 1000 during the Medieval Warm Period and slightly lower during the Little Ice Age that followed several hundred years later and lasted from A.D. 1300 to 1850.

By the end of the 20th century, direct and indirect empirical evidence made it clear that climate had changed greatly over many millions of years and that climate change was the norm. Yet unanswered from these kinds of evidence were questions about how great the climatic effects would be from the amount of CO_2 that was likely to be added in the atmosphere from the future burning of fossil fuels and whether those changes might be unique in Earth's history, or at least unique since the rise of mammals and flowering plants or during the evolution of our species, both in the amount of change and the rate of change.

To understand how greenhouse gases work and affect climate, we have to first understand how the atmosphere, as a huge system, works and how it is affected by the rest of the environment—oceans, land, life, and the inputs and outputs from the cosmos.

20.3 How the Atmosphere Works

To forecast what effects the burning of fossil fuels could have on the future climate requires an understanding of how the atmosphere works as a system, as well as an understanding that climate and weather are not the same thing. **Weather** is what is happening now or over some short time period—this hour, today, this week—in the atmosphere near the ground: its temperature, pressure, cloudiness, precipitation, and winds. **Climate** is the average weather and usually refers to average weather conditions over long periods, at least seasons, but more often, years or decades. When we say it's hot and humid in New York today or raining in Seattle, we are speaking of weather. When we say Los Angeles has cool, wet winters and warm, dry summers, we are referring to the Los Angeles climate.

Here is how the atmosphere works in terms of climate. The major determinant of climate is how much radiant energy Earth receives from the sun and how much it loses by reradiating energy back into space. Since climates are characteristic of certain latitudes (and other factors that we will discuss later), they are classified mainly by latitude—tropical, subtropical, midlatitudinal (continental), sub-Arctic (continental), and Arctic—but also by wetness/dryness, such as humid continental, Mediterranean, monsoon, desert, and tropical wet–dry (Figure 20.9). Recall from the discussion of biogeography in Chapter 9 that similar climates produce similar kinds of ecosystems. Therefore, knowing the climate, we can make pretty good predictions about what kinds of life we will find there and what kinds could survive there if introduced.

Earth's atmosphere is the thin layer of gases that envelop the planet. These gases are almost always in motion, sometimes rising, sometimes falling, and most of the time moving across Earth's surface. The atmosphere's gas molecules are held near to Earth's surface by gravity and pushed upward by thermal energy that heats the molecules. Approximately 90% of the weight of the atmosphere is in the first 12 km above Earth's surface. Major gases in the atmosphere include nitrogen (78%), oxygen (21%), argon (0.9%), carbon dioxide (0.03%), and water vapor in varying concentrations in the lower few kilometers. The atmosphere also contains trace amounts of methane, ozone, hydrogen sulfide, carbon monoxide, oxides of nitrogen and sulfur, and a number of small hydrocarbons, as well as synthetic chemicals, such as chlorofluorocarbons (CFCs). Methane, at about 0.00017% of the atmosphere, is emerging as an important gas that tracks closely with climate change (more so than CO_2).

Thus, the atmosphere is a dynamic system, changing continuously. It is a vast, chemically active system, fueled by sunlight and affected by high-energy compounds emitted by living things (for example, oxygen, methane, and carbon dioxide) and by our industrial and agricultural activities. Many complex chemical reactions take place in the atmosphere, changing from day to night and with the chemical elements available.

Structure of the Atmosphere

You might think that the atmosphere is homogeneous, since it is a collection of gases that mix and move continuously. Actually, however, it has a surprisingly complicated structure. The **atmosphere** is made up of several vertical layers, beginning at the bottom with the **troposphere**, most familiar to us because we spend most of our lives in it. Above the troposphere is the **stratosphere**, which we

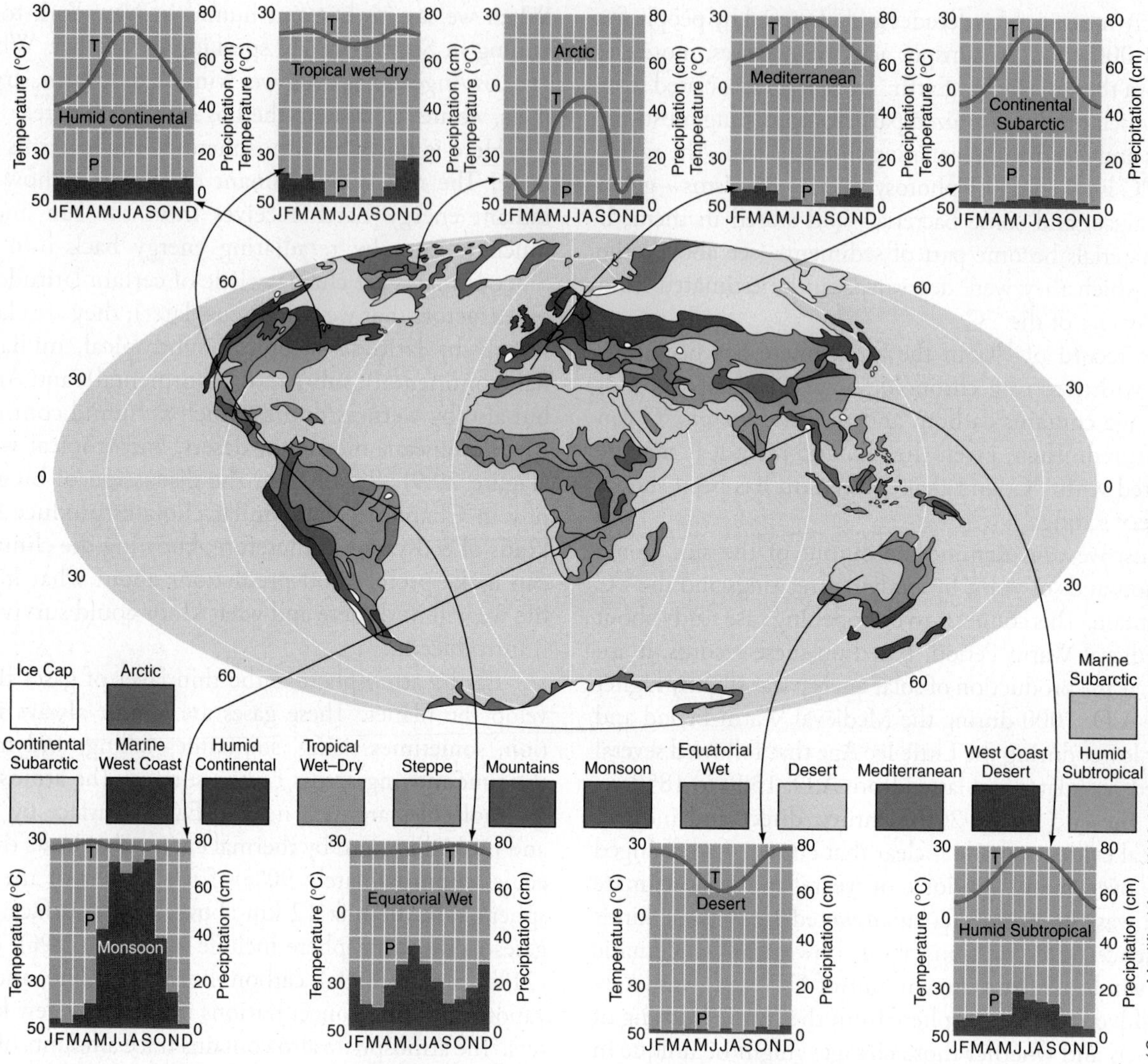

(*Source:* Modified from W.M. Marsh and J. Dozier, *Landscape* [New York: John Wiley & Sons, 1981. Reprinted with permission of John Wiley & Sons, Inc.].)

FIGURE 20.9 The climates of the world and some of the major climate types in terms of characteristic precipitation and temperature conditions.

visit occasionally when we travel by jet airplane, and then several other layers at higher altitudes, less familiar to us, each characterized by a range of temperatures and pressures (Figure 20.10).

The troposphere, which extends from the ground up to 10–20 km, is where weather occurs. Within the troposphere, the temperature decreases with elevation, from an average of about 15°C (57°F) at the surface to −60°C (−76°F) at 12 km elevation. At the top of the troposphere is a boundary layer called the *tropopause*, which has a constant temperature of about −60°C and acts as a lid, or cold trap, on the troposphere because it is where almost all remaining water vapor condenses.

Another important layer for life is the *stratospheric ozone layer*, which extends from the tropopause to an elevation of approximately 40 km (25 mi), with a maximum concentration of ozone above the equator at about 25–30 km (16–19 mi) (Figure 20.10). Stratospheric ozone (O_3) protects life in the lower atmosphere from receiving harmful doses of ultraviolet radiation (see Chapter 21).

Atmospheric Processes: Temperature, Pressure, and Global Zones of High and Low Pressure

Two important qualities of the atmosphere are *pressure* and *temperature*. Pressure is force per unit area. Atmospheric pressure is caused by the weight of overlying atmospheric gases on those below and therefore decreases with altitude. At sea level, atmospheric pressure is 10^5 N/m^2 (Newtons per square meter) (14.7 lb/in.). We are familiar with this as **barometric pressure**, which the weatherperson gives to us in units that are the height to which a column of mercury is raised by that pressure. We are also familiar with low- and high-pressure systems in the atmosphere. When the air pressure is low, air tends to rise, cooling as it rises and condensing its water vapor; it is therefore characterized by clouds and precipitation. When air pressure is high, it is moving downward, which warms the air, changing the condensed water drops in clouds to vapor; therefore high-pressure systems are clear and sunny.

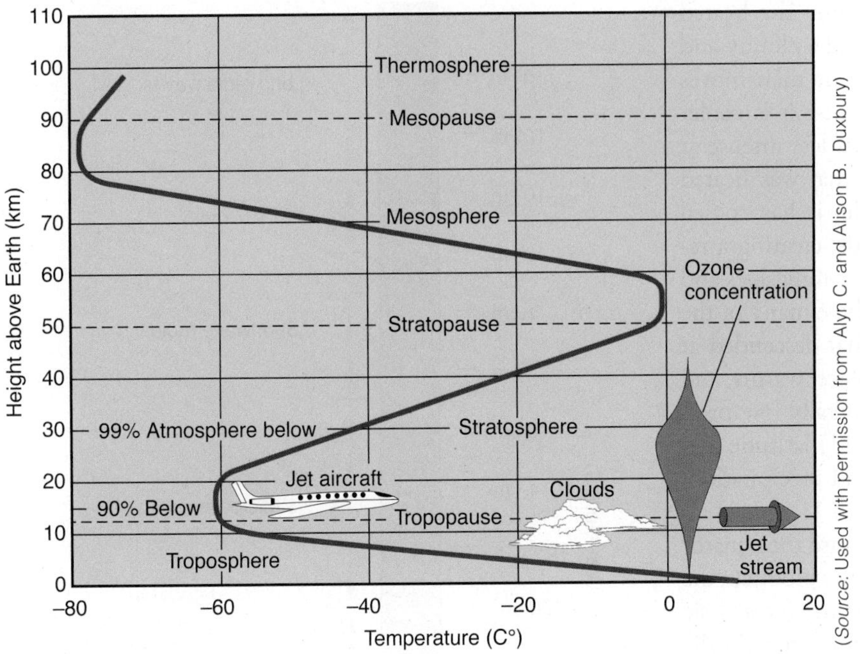

FIGURE 20.10 An idealized diagram of the structure of the atmosphere showing temperature profile and the ozone layer of the atmosphere to an altitude of 110 km. Note that 99% of the atmosphere (by weight) is below 30 km, the ozone layer is thickest at about 25–30 km, and the weather occurs below about 11 km—about the elevation of the jet stream.

Temperature, familiar to us as the relative warmth or coldness of materials, is a measure of thermal energy, which is the *kinetic* energy—the motion of atoms and molecules in a substance.

Water vapor content is another important characteristic of the lower atmosphere. It varies from less than 1% to about 4% by volume, depending on air temperature, air pressure, and availability of water vapor from the surface.

The atmosphere moves because of Earth's rotation and the differential heating of Earth's surface and atmosphere. These produce global patterns that include prevailing winds and latitudinal belts of low and high air pressure from the equator to the poles. Three cells of atmospheric circulation (Hadley cells) are present in each hemisphere (see Figure 20.11 for more details). In general, belts of low air pressure develop at the equator, where the air is warmed

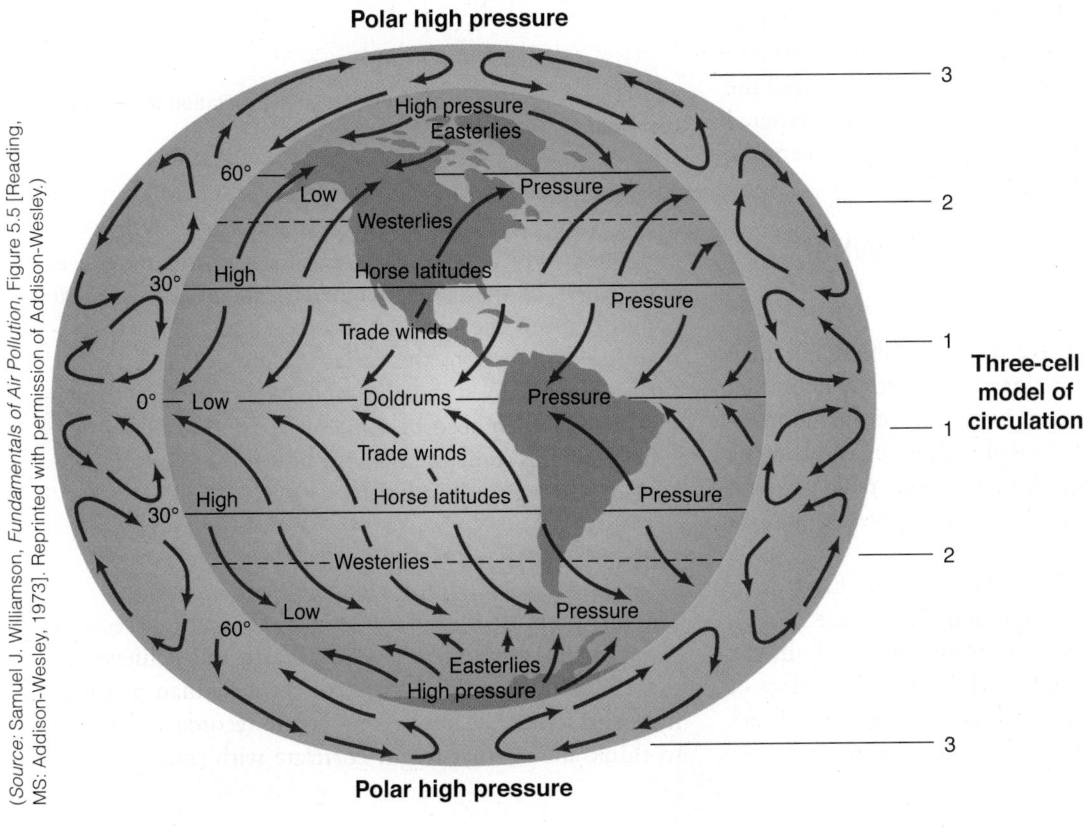

FIGURE 20.11 Generalized circulation of the atmosphere. The heating of the surface of the Earth is uneven, producing pressure differences (warm air is less dense than cooler air). There is rising warm air at the equator and sinking cool air at the poles. With rotation of Earth, three cells of circulating air are formed in each hemisphere (called Hadley cells after George Hadley who first proposed a model of atmospheric circulation in 1735).

most of the time during the day by the sun. The heated air rises, creating an area of low pressure and a cloudy and rainy climate (cell 1, Figure 20.11). This air then moves to higher latitudes (toward the poles) because it is cooler at higher elevations and because sunlight is less intense at the higher latitudes. By the time the air that was heated at the equator reaches about 30° latitude, it has cooled enough to become heavier, and it descends, creating a region of high pressure with its characteristic sunny skies and low rainfall and forming a latitude belt where many of the world's deserts are found. Then, the air that descended at 30° latitude moves poleward along the surface, warms, and rises again, creating another region of generally low pressure around 50° to 60° (cell 2, Figure 20.11) latitude and once again becoming a region of clouds and precipitation. Atmospheric orientation in cell 3 (Figure 20.11) moves air toward the poles at higher elevation and toward the equator along the surface. Sinking cool air at the poles produces the polar high-pressure zones at both poles. At the most basic level, warm air rises at the equator and moves toward the poles, where it sinks after going through (cell 2); its return flow is along the surface of Earth toward the equator.

Of course, the exact locations of the areas of rising (low-pressure) and falling (high-pressure) air vary with the season, as the sun's position moves north and south relative to Earth's surface. You can begin to understand that what would seem at first glance to be just a simple container of gases has complicated patterns of movement, and that these change all the time for a variety of reasons.

The latitudinal belts (cells) just described have names, most of which came about during the days of sailing ships. Such names include the "doldrums," regions at the equator with little air movement; "trade winds," northeast and southeast winds important when clipper ships moved the world's goods; and "horse latitudes," two belts centered about 30° north and south of the equator with descending air and high pressure.

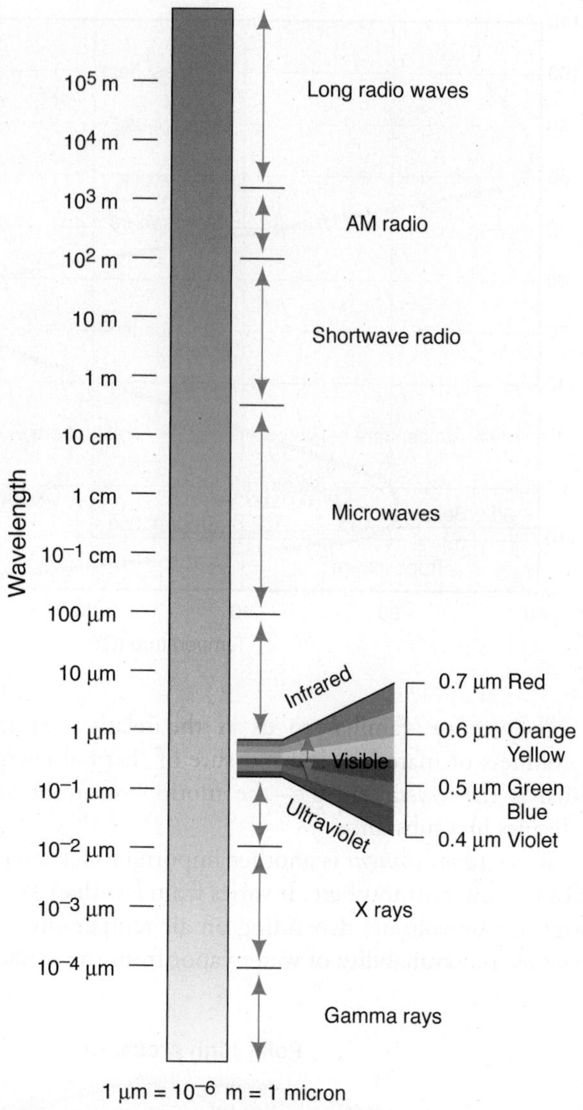

FIGURE 20.12 Kinds of electromagnetic radiation the Earth receives.

Energy and the Atmosphere: What Makes the Earth Warm?

Almost all the energy Earth receives is from the sun (a small amount comes from the interior of the Earth and an even smaller amount from frictional forces due to the moon revolving around the Earth). Sunlight comes in a wide range of electromagnetic radiation, from very long radio waves to short infrared waves, then shorter visible light, and on to X rays and gamma rays (Figure 20.12).

Most of the sun's radiation that reaches Earth is in the visible and near-infrared wavelengths (Figure 20.13), while Earth, much cooler, radiates energy mostly in the longer far-infrared wavelengths. (The hotter the surface of any object, the shorter the dominant wavelengths. That's why a hot flame is blue and a cooler flame is red.)

Under typical conditions, Earth's atmosphere reflects about 30% of the electromagnetic (radiant) energy that comes in from the sun and absorbs about 25%. The remaining 45% gets to the surface (Figure 20.14). As the surface warms up, it radiates more energy back to the atmosphere, which absorbs some of it. The warmed atmosphere radiates some of its energy upward into outer space and some downward to Earth's surface.

Milankovitch Cycles

One of the most important insights about the effects of variations in sunlight received by Earth was achieved in the 1920s by Milutin Milankovitch, a Serbian astronomer who looked at long-term climate records and began to think about what might correlate with these records.

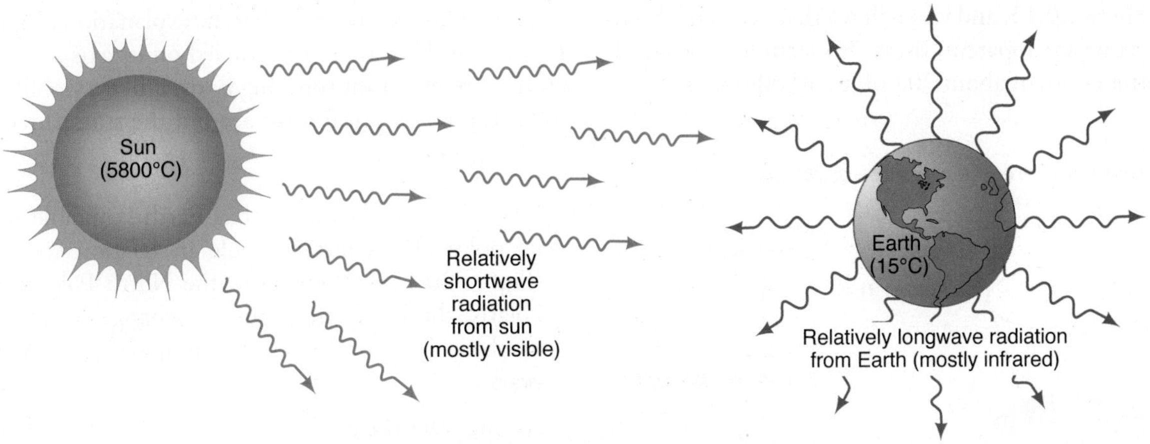

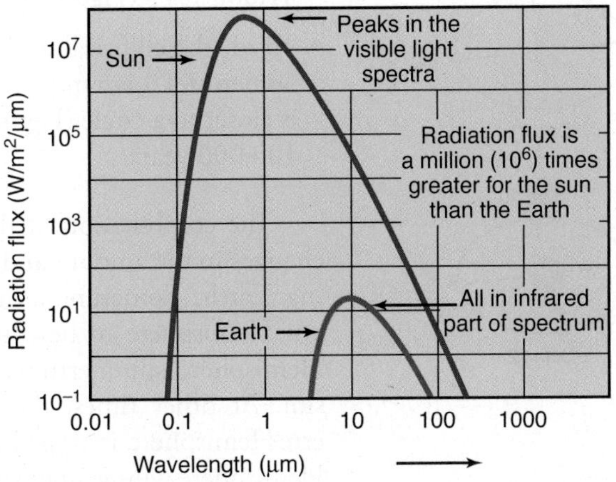

FIGURE 20.13 **The sun, much hotter than the Earth, mostly emits energy in the visible and near infrared. The cooler Earth emits energy, mostly in the far (longer wavelength) infrared.**

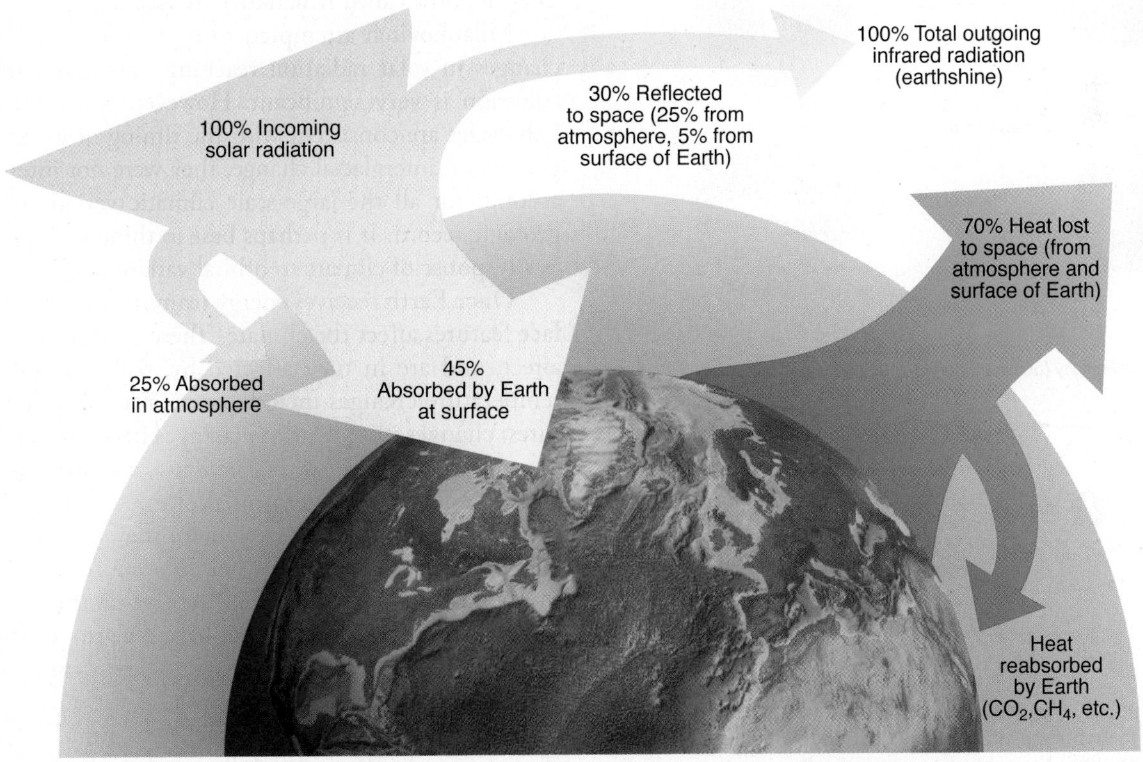

FIGURE 20.14 **Earth's energy budget.**

Look at Figure 20.15, and you will see that cycles of about 100,000 years are apparent; these also seem to be divided into shorter cycles of about 20,000 to 40,000 years.

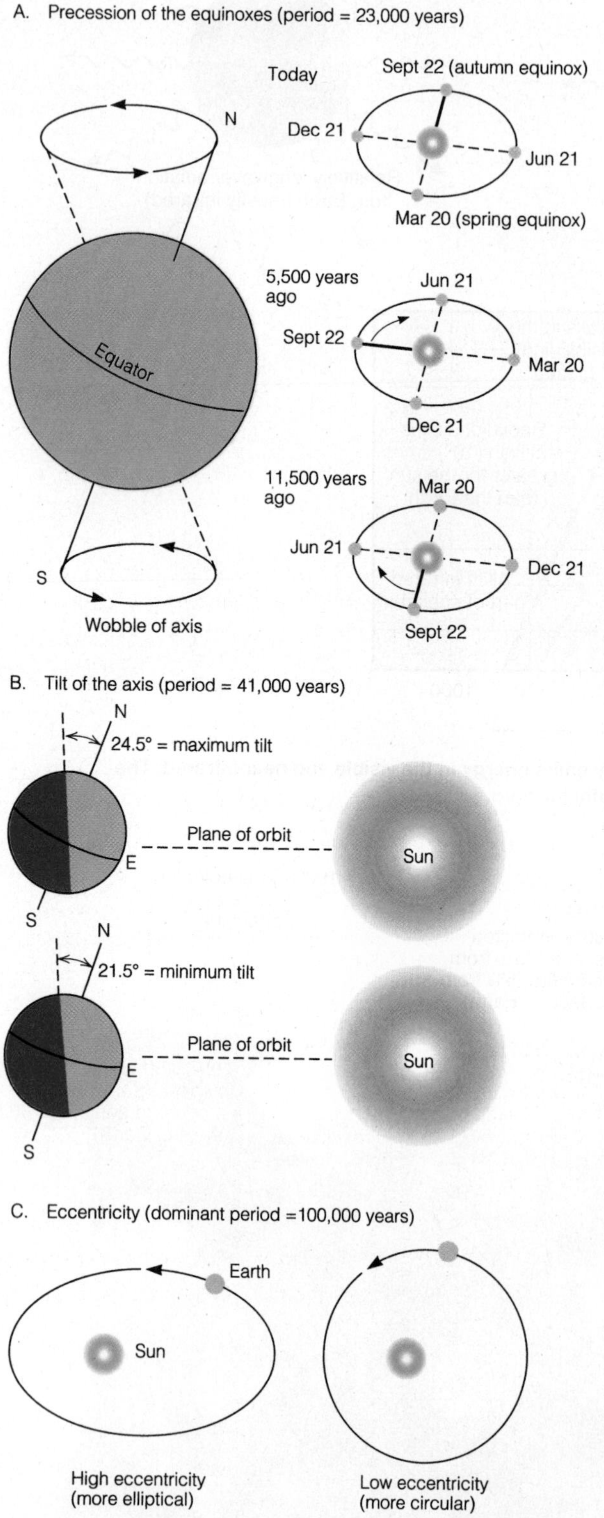

FIGURE 20.15 The Melankovitch cycles, which are the result of the Earth wobbling, changing its tilt, and following an elliptical orbit that changes the orbit shape periodically.

(*Source:* Skinner, B., Porter, S. and Botkin, D., Blue Planet: An Introduction to Earth System Science, Second Edition. New York: John Wiley & Sons, 2000, p. 335. This material is reproduced with permission of John Wiley & Sons, Inc.)

Milankovitch realized that the explanation might have to do with the way Earth revolved on its axis and rotated around the sun. Our spinning Earth is like a wobbling top following an elliptical orbit around the sun. Three kinds of changes occur:

- First, the wobble means that Earth is unable to keep its poles at a constant angle in relation to the sun (Figure 20.15a). Right now, the North Pole points to Polaris, the North Star, but this changes as the planet wobbles. The wobble makes a complete cycle in 26,000 years.
- Second, the tilt of Earth's axis varies over a period of 41,000 years (Figure 20.15b).
- Third, the elliptical orbit around the sun also changes. Sometimes it is a more extreme ellipse; at other times, it is closer to a circle (Figure 20.15c), and this occurs over 100,000 years.

The combination of these changes leads to periodic changes in the amount and distribution of sunlight reaching Earth. Sometimes, the wobble causes the Northern Hemisphere to be tilted toward the sun (Northern Hemisphere summertime) when Earth is closest to the sun. At other times, the opposite occurs—the Northern Hemisphere is tipped away from the sun (Northern Hemisphere wintertime) when Earth is closest to the sun. Milankovitch showed that these variations correlated with the major glacial and interglacial periods (Figure 20.15). They are now called Milankovitch cycles.[10]

Milankovitch attempted to explain ice ages through changes in solar radiation reaching Earth, and his contribution is very significant. However, while Milankovitch cycles are consistent with the timing of variations in glacial and interglacial change, they were not intended to account for all the large-scale climatic variations in the geologic record. It is perhaps best to think of these cycles as a response of climate to orbital variations.

Once Earth receives energy from the sun, Earth's surface features affect the climate. These earthly factors that affect, and are in turn affected by, regional and global temperature changes include warmer ice sheet temperatures; changes in vegetation; changes in atmospheric gases, such as carbon dioxide, methane, and nitrous oxide; and particulates and aerosols. Volcanoes inject aerosols into the upper atmosphere, where they reflect sunlight and cool Earth's surface. A helpful way to think about the Milankovitch cycles is that they set up the basic situation where an ice age is possible or a warm, interglacial age is possible. Then the other factors that influence climate come into play. The question, as yet unresolved, is how much do Milankovitch cycles account for climate change compared to other influences.

Solar Cycles

The sun goes through cycles too, sometimes growing hotter, sometimes colder. Today, solar intensity is observed directly with telescopes and other instruments. Past variations in the sun's intensity can be determined because hotter and cooler sun periods emit different amounts of **radionuclides**—atoms with unstable nuclei that undergo radioactive decay (such as beryllium 10 and carbon 14), which are trapped in glacial ice and can then be measured. Evaluation of these radionuclides in ice cores from glaciers reveals that during the Medieval Warm Period, from approximately A.D. 950 to 1250, the amount of solar energy reaching Earth was relatively high, and that minimum solar activity occurred during the 14th century, coincident with the beginning of the Little Ice Age. Thus, it appears that the variability of solar energy input explains a small part of Earth's climatic variability.[11,12] Since about 1880, solar input has increased about 0.5%.

20.4 How Earth's Atmosphere, Oceans, and Land Affect Climate

Atmospheric Transparency Affects Climate and Weather

How transparent the atmosphere is to the radiation coming to it, from both the sun and Earth's surface, affects the temperature of the Earth. Dust and aerosols absorb light, cooling Earth's surfaces. Volcanoes and large forest fires put dust into the atmosphere, as do various human activities, such as plowing large areas.

How much a volcanic eruption can affect climate was demonstrated in 1991 with the eruption of Mount Pinatubo in the Philippines, one of the largest eruptions of the 20th century (Figure 20.16). The volcano put 20 million tons of sulfur dioxide into the stratosphere and an estimated 250 million tons of CO_2. Sulfur dioxide particles decrease the transparency of the atmosphere, as any dust will, and the result was an average decrease in Earth's surface temperature in 1991 of about 0.1°C to 0.2°C.[13]

Each gas compound has its own absorption spectrum (the electromagnetic radiation that is absorbed by a gas as it passes through the atmosphere). Thus, the chemical and physical composition of the atmosphere can make things warmer or cooler.

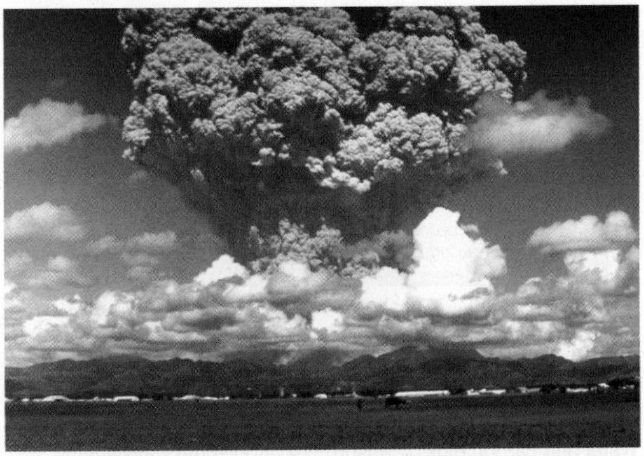

(*Source:* U.S. Geological Survey Fact Sheet 113-97, http://pubs.usgs.gov/fs/1997/fs113-97.)

FIGURE 20.16 Mount Pinatubo, Philippines, erupting in 1991. The dust that was put into the atmosphere cooled the Earth in 1992 by 0.1° to 0.2°C.

The Surface of Earth and Albedo Effects

Albedo is the reflectivity of an object, measured as the percentage of incoming radiation that is reflected. For example, consider the following approximate albedos: that of Earth (as a whole) is 30%; clouds, depending on type and thickness, 40–90%; fresh snow, 85%; glacial ice, 20–40%; a pine forest, 10%; dark rock, 5–15%; dry sand, 40%; and a grass-covered meadow, 15%.

An exposed dark rock surface near the North Pole absorbs more of the sunlight it receives than it reflects in the summer, warming the surface and the air passing over it. When a glacier spreads out and covers that rock, the glacier reflects more of the incoming sunlight than the darker rock, cooling both the surface and the air that comes in contact with it.

Vegetation also affects the climate and weather in the same way. If vegetation is a darker color than the soil, it warms the surface. If it is a lighter color than the soil, it cools the surface. Now you know why if you walk barefoot on dark asphalt on a hot day, you feel the heat radiating from the surface (and you may even burn the bottom of your feet).

Roughness of Earth's Surface Affects the Atmosphere

Above a completely smooth surface, air flows smoothly—a flow called laminar. A rough surface causes air to become turbulent—to spin, rotate, reverse, and so forth. Turbulent air gives up some of the energy in its motion (its kinetic energy), and that energy is turned into heat. This affects the weather above. Forests are a much rougher surface than smooth rock or glaciers, so in this way, too, vegetation affects weather and climate.

The Chemistry of Life Affects the Atmosphere

The emission and uptake of chemicals by living things affect the weather and climate, as we will discuss in detail in the next section. Thus, a planet with water vapor, liquid

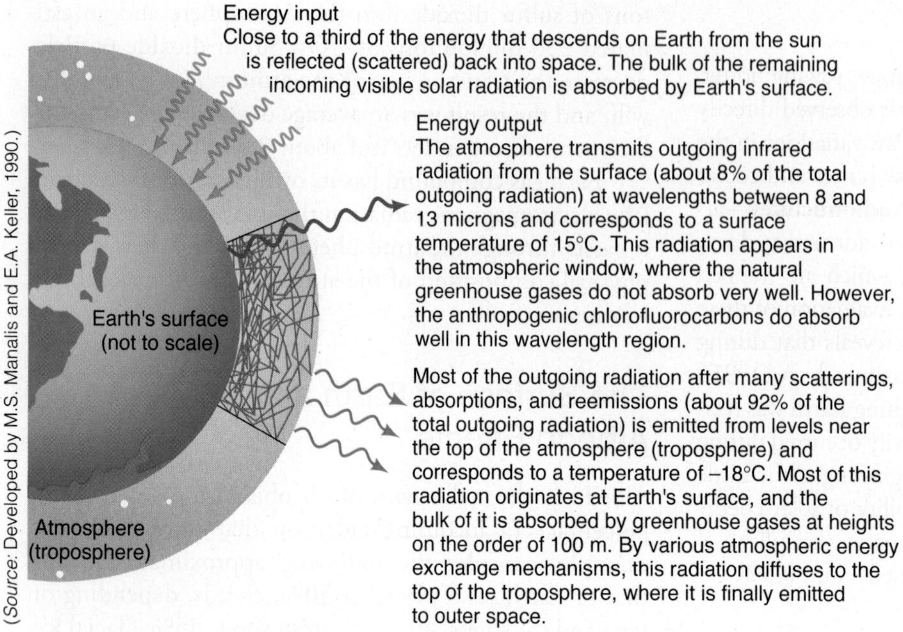

(Source: Developed by M.S. Manalis and E.A. Keller, 1990.)

FIGURE 20.17 Idealized diagram showing the greenhouse effect. Incoming visible solar radiation is absorbed by Earth's surface, to be reemitted in the infrared region of the electromagnetic spectrum. Most of this reemitted infrared radiation is absorbed by the atmosphere, maintaining the greenhouse effect.

water, frozen water, and living things has a much more complex energy exchange system than a lifeless, waterless planet. This is one reason (of many) why it is difficult to forecast climate change.

20.5 The Greenhouse Effect

Each gas in the atmosphere has its own absorption spectrum—which wavelengths it absorbs and which ones it transmits. Certain gases in Earth's atmosphere are especially strong absorbers in the infrared and therefore absorb radiation emitted by the warmed surfaces of the Earth. Warmed by this, the gases reemit this radiation. Some of it reaches back to the surface, making Earth warmer than it otherwise would be. The process by which the heat is trapped is not the same as it is in a greenhouse (air in a closed greenhouse has restricted circulation and will heat up). Still, in trapping heat this way, the gases act a little like a greenhouse's glass panes, which is why it is called the **greenhouse effect**. The major **greenhouse gases** are water vapor, carbon dioxide, methane, some oxides of nitrogen, and chlorofluorocarbons (CFCs). The greenhouse effect is a natural phenomenon that occurs on Earth and on other planets in our solar system. Most natural greenhouse warming is due to water in the atmosphere—water vapor and small particles of water in the atmosphere produce about 85% and 12%, respectively, of the total greenhouse warming.

How the Greenhouse Effect Works

Figure 20.17 is a highly idealized diagram of some important aspects of the greenhouse effect. The arrows labeled "energy input" represent energy from the sun absorbed at or near Earth's surface. The arrows labeled "energy output" represent energy emitted from the upper atmosphere and Earth's surface, which balances input, consistent with Earth's energy balance. The highly contorted lines near the surface of the Earth represent the absorption of infrared radiation (IR) occurring there and producing the 15°C (59°F) near-surface temperature. Following many scatterings and absorptions and reemissions, the infrared radiation emitted from levels near the top of the atmosphere (troposphere) corresponds to a temperature of approximately −18°C (0°F). The one output arrow that goes directly through Earth's atmosphere represents radiation emitted through what is called the atmospheric window. The atmospheric window (Figure 20.18), centered

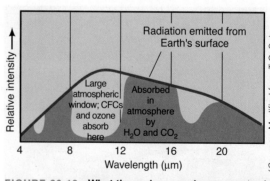

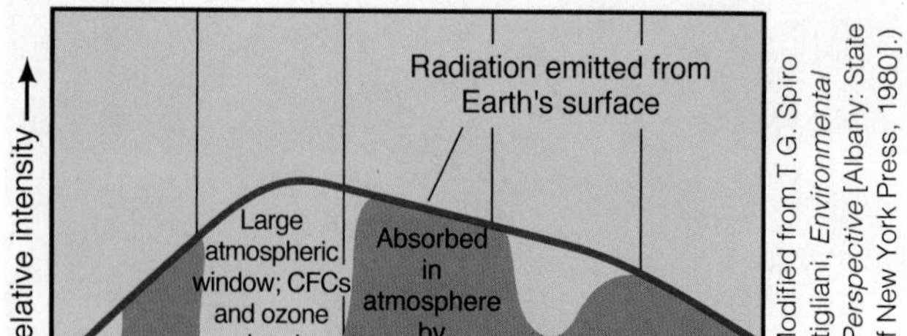

(Source: a Modified from T.G. Spiro and W.M. Stigliani, Environmental Science in Perspective [Albany: State University of New York Press, 1980].)

FIGURE 20.18 What the major greenhouse gases absorb in the Earth's atmosphere. Earth's surface radiates mostly in the infrared, which is the range of electromagnetic energy shown here. Water and carbon dioxide absorb heavily in some wavelengths within this range, making them major greenhouse gases. The other greenhouse gases, including methane, some oxides of nitrogen, CFCs, and ozone, absorb smaller amounts but in wavelengths not absorbed by water and carbon dioxide.

on a wavelength of 10 μm, is a region of wavelengths (8–12 μm) where outgoing radiation from Earth is not absorbed well by natural greenhouse gases (water vapor and carbon dioxide). Anthropogenic CFCs do absorb in this region, however, and CFCs significantly contribute to the greenhouse effect in this way.

20.6 The Major Greenhouse Gases

The major anthropogenic greenhouse gases are listed in Table 20.1. The table also lists the recent rate of increase for each gas and its currently estimated relative contribution to human influences on the greenhouse effect.

Carbon Dioxide

Current estimates suggest that approximately 200 billion metric tons of carbon in the form of carbon dioxide (CO_2) enter and leave Earth's atmosphere each year as a result of a number of biological and physical processes: of greenhouse gases emitted by human activities, an estimated 50% to 60% is attributed to this gas. Measurements of carbon dioxide trapped in air bubbles in the Antarctic ice sheet suggest that 160,000 years before the Industrial Revolution, the atmospheric concentration of carbon dioxide varied from approximately 200 to 300 ppm.[14] The highest concentration of carbon dioxide in the atmosphere, before the present, occurred during the major interglacial period about 125,000 years ago.

About 140 years ago, just before the major use of fossil fuels began as part of the Industrial Revolution, some of the first chemical measurements of the atmosphere showed that the carbon dioxide concentration was approximately 280 ppm.[15] Since then, and especially in the past few decades, the concentration of CO_2 in the atmosphere has grown rapidly. Today, the CO_2 concentration is about 396 ppm, and its current rate of increase is about 0.5% per year. At current rates, the level could rise to as much as 450 ppm by the year 2050—more than 1.5 times the preindustrial level.[15]

Methane

The concentration of methane (CH_4) in the atmosphere has more than doubled over the past 200 years and is estimated to contribute approximately 12% to 20% of the anthropogenic greenhouse effect.[16,17] Certain bacteria that can live only in oxygenless atmospheres produce methane and release it. These bacteria live in the guts of termites and the intestines of ruminant mammals, such as cows, which produce methane as they digest woody plants. These bacteria also live in oxygenless parts of freshwater wetlands where they decompose vegetation, releasing methane as a decay product. Methane is also released with seepage from oil fields and seepage from methane hydrates (see Chapter 15).

Our activities also release methane. These include landfills (the major methane source in the United States), the burning of biofuels, production of coal and natural gas, and agriculture, such as raising cattle and cultivating rice. (Methane is also released by anaerobic activity in

Table 20.1 MAJOR GREENHOUSE GASES

TRACE GASES	RELATIVE CONTRIBUTION (%)	GROWTH RATE (%/YR)
CFC	15^a-25^b	5
CH_4	12^a-20^b	0.4^c
O_3(troposphere)	8^d	0.5
N_2O	5^d	0.2
Total	40–50	
Contribution of CO_2	50–60	0.3^e–$0.5^{d,f}$

[a] W. A. Nierenberg, "Atmospheric CO_2: Causes, Effects, and Options," *Chemical Engineering Progress* 85, no.8 (August 1989): 27

[b] J. Hansen, A. Lacis, and M. Prather, "Greenhouse Effect of Chlorofluorocarbons and Other Trace Gases," *Journal of Geophysical Research* 94 (November 20, 1989): 16, 417.

[c] Over the past 200 yrs.

[d] H. Rodha, "A Comparison of the Contribution of Various Gases to the Greenhouse Effect," *Science* 248 (1990): 1218, Table 2.

[e] W. W. Kellogg, "Economic and Political Implications of Climate Change," paper presented at Conference on Technology-based Confidence Building: Energy and Environment, University of California, Los Alamos National Laboratory, July 9–14, 1989.

[f] H. Abelson, "Uncertainties about Global Warming," *Science* 247 (March 30,1990): 1529.

flooded lands where rice is grown.) As with carbon dioxide, there are important uncertainties in our understanding of the sources and sinks of methane in the atmosphere.

Chlorofluorocarbons

Chlorofluorocarbons (CFCs) are inert, stable compounds that have been used in spray cans as aerosol propellants and in refrigerators. The rate of increase of CFCs in the atmosphere in the recent past was about 5% per year, and it has been estimated that approximately 15% to 25% of the anthropogenic greenhouse effect may be related to CFCs. Because they affect the stratospheric ozone layer and also play a role in the greenhouse effect, the United States banned their use as propellants in 1978. In 1987, 24 countries signed the Montreal Protocol to reduce and eventually eliminate production of CFCs and accelerate the development of alternative chemicals. As a result of the treaty, production of CFCs was nearly phased out by 2000.

Potential global warming from CFCs is considerable because each CFC molecule may absorb hundreds or even thousands of times more infrared radiation emitted from the surface than is absorbed by a molecule of carbon dioxide. Furthermore, because CFCs are highly stable, their residence time in the atmosphere is long. Even though their production was drastically reduced, their concentrations in the atmosphere will remain significant (although lower than today's) for many years, perhaps for as long as a century.[18,19] (CFCs are discussed in Chapter 21, which examines stratospheric ozone depletion.)

Nitrous Oxide

Nitrous oxide (N_2O) is increasing in the atmosphere and probably contributes as much as 5% of the anthropogenic greenhouse effect.[18] Anthropogenic sources of nitrous oxide include agricultural application of fertilizers and the burning of fossil fuels. This gas, too, has a long residence time; even if emissions were stabilized or reduced, elevated concentrations of nitrous oxide would persist for at least several decades.

20.7 Greenhouse Gases and Climate

With this background, we can now consider the extent to which greenhouse gases might affect climate. Look again at Figure 20.14, which presents Earth's energy budget. It shows that approximately 30% of the solar radiation received by our planet is reflected back to space from the atmosphere as shortwave solar radiation, while 70% is absorbed by Earth's surface and atmosphere. The 70% that is absorbed is eventually reemitted into space as infrared radiation (IR). The sum of the reflected solar radiation and the outgoing infrared radiation balances with the energy arriving from the sun.

Within Earth's biosphere—the planetary system that includes living things—infrared radiation can be transmitted back and forth between the atmosphere, oceans, ice, and solid surfaces (Figure 20.19). Infrared radiation emitted by the land can be absorbed by gases in the atmosphere. The temperature of these gases rises and also radiate in the infrared, some of which goes back down to the surface. The amount of IR absorbed at Earth's surface from the greenhouse effect is estimated to be greater than twice the amount of shortwave solar radiation absorbed by Earth's surface (Figure 20.19).

The greenhouse effect keeps Earth's lower atmosphere approximately 33°C warmer than it would otherwise be and performs other important service functions as well. For example, without the strong downward emission of IR from the greenhouse effect, the land surface would cool much faster at night and warm much more quickly during the day. In sum, the greenhouse effect helps to limit temperature swings from day to night and maintain relatively comfortable surface temperatures. It is, then, not the greenhouse effect itself but the changes in greenhouse gases that have become a concern.

20.8 The Oceans and Climate Change

The oceans play an important role in climate because two-thirds of the Earth is covered by water. Moreover, water has the highest heat-storage capacity of any compound, so a very large amount of heat energy can be stored in the world's oceans. There is a complex, dynamic, and ongoing relationship between the oceans and the atmosphere. If carbon dioxide increases in the atmosphere, it will also increase in the oceans, and over time the oceans can absorb a very large quantity of CO_2. This can cause seawater to become more acidic ($H_2O + CO_2 \longrightarrow H_2CO_3$) as carbonic acid increases.

20.9 Forecasting Climate Change Using Computer Simulation

Scientists have been trying to use mathematics to predict the weather since the beginning of the 20th century. They began by trying to forecast the weather a day in advance, using the formal theory of how the atmosphere

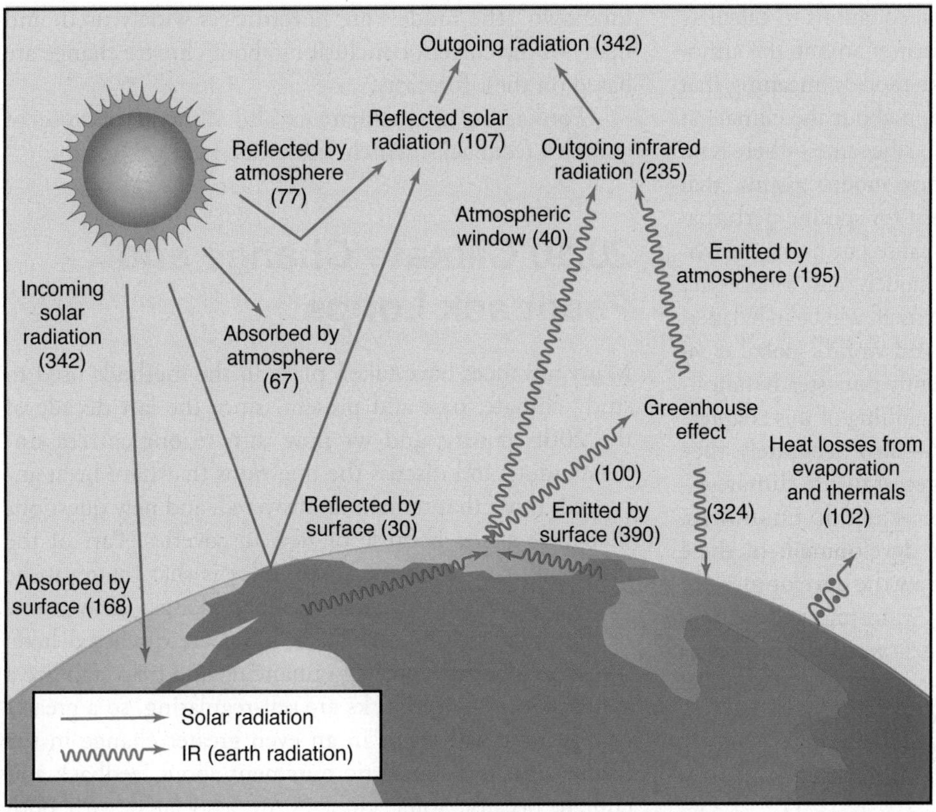

(Source: IPCC, *Climate Change 2007: The Physical Science Basis. Working Group I. Contribution to the Forth Assessment Report.* New York: Cambridge University Press, 2007. Reprinted with permission.)

FIGURE 20.19 **Idealized diagram showing some details of Earth's greenhouse effect.** Incoming solar radiation in units of w/m^2 is balanced by outgoing radiation. Notice that some of the fluxes (rates of transfer) of infrared radiation (IR) are greater than incoming solar radiation, reflecting the role of the greenhouse effect. Some of these fluxes are explained in the diagram.

functioned. The first person to try this approach eventually went off to fight in World War I and never completed the forecast. By the early 1970s, computers had gotten fast enough and models sophisticated enough to forecast the next day's weather in two days—not much help in practice but a start. At least they knew whether their forecast of yesterday's weather had been right!

Computers are much faster today, and the major theoretical method used today to forecast climate change is a group of computer models called **general circulation models (GCMs)**. Mathematically, these are deterministic differential equation models. The dominant computer models of Earth's climate are all based on the general idea shown in Figure 20.20. The atmosphere is divided into three-dimensional rectangular units, each a few kilometers high and several kilometers north and south. For each of these, the flux of energy and matter is calculated for each of the adjacent cells. Since there are many cells, and each cell has six sides, a huge number of calculations have to be made. A number of these models were developed in the second half of the 20th century. In design, they were similar. Each attempted to take into account as many details of the exchange of energy and matter that occur in the atmosphere and between the atmosphere, solid surfaces, and oceans as possible.

Today, many such computer simulations are in use around the world, but they are all very similar mathematically.

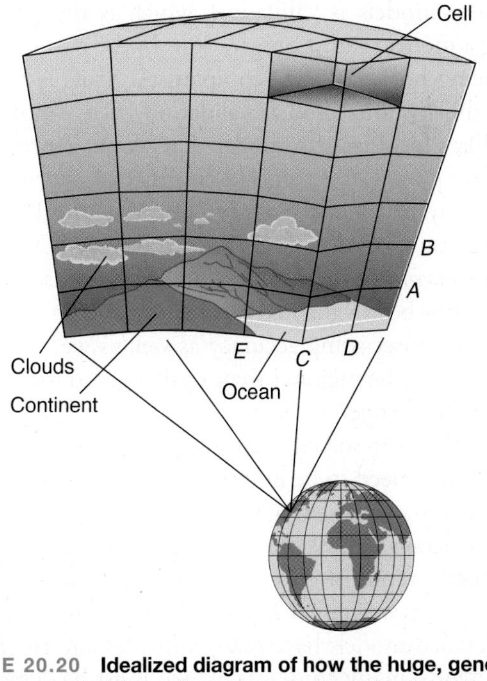

FIGURE 20.20 **Idealized diagram of how the huge, general circulation models (models of the entire Earth's climate) are viewed by the computers that run the programs.** The atmosphere is divided into hundreds of rectangular cells, and the flow of energy and material is calculated for the transfer between each boundary of every cell to its adjacent cells.

They all use deterministic differential equations to calculate the rate of exchange of energy and matter among the atmospheric cells. They are all steady-state models, meaning that for any given set of input information about the climate at the beginning, the result will always be the same—there is no chance or randomness involved. These models assume that the climate is in a steady state, except for specific perturbations, especially those believed to be caused by human activities. Thus, an assumption of these models, and a necessary outcome, is that the climate, if left to itself, will be in balance, in a steady state. This is unlike the real world's global environmental systems, which are inherently non–steady state—always changing, as we saw at the beginning of this chapter.

How well these models work—how accurately they reproduce past climates and can forecast future climates—continues to be one of the major scientific unknowns. One of the original leaders in the development of these models, Steve Schneider, who oversaw the development of a GCM at the National Center for Atmospheric Research, said that on the one hand the models were as complex as current computers could handle—a single run of a model could takes weeks to months of computer time even in the fastest computers. But at the same time, they were still rather crude representations of reality. He would say that scientists basically had two choices in looking to the future: Either trust the output of these admittedly incomplete models or trust the judgment of experienced individual meteorologists.[20]

A key to deciding the value of forecasts from these computer models is validation, which is the process of testing a model against observations in the past or present and seeing how well they compare. A. Scott Armstrong, an authority on model validation, lists more than 100 characteristics of a model that would represent the most complete validation. He says that of course no one expects any real model to meet all these criteria, but his examination of GCMs leads him to believe that none of them meets *any* of the criteria.[20] In other words, the models have not been validated adequately, and so we do not know their forecasting accuracy as well as we would like. This is one of the major current challenges to understanding climate change.

Another unresolved scientific question is how detailed these models need to be in terms of giving the most accurate forecasts. The existing GCMs are designed to include as many details as possible, but there is evidence that simpler models tend, in general, to give more accurate forecasts than very detailed models. In part, this is because very detailed models have many variables and their equations involve many constants, which must be determined empirically. However, many of these constants have not been adequately measured; this has led to the introduction of approximations whose effects on the forecasts are unknown. The models are nevertheless widely used, and many of the current conclusions about climate change are based on their forecasts.

Forecasting the climate is also difficult because of complex feedbacks, which we discuss next.

20.10 Climate Change and Feedback Loops

Many advances have taken place in the methods used to study climate, past and present, since the last decade of the 20th century, and we now turn to our current understanding and discuss the questions that have been answered, those that remain unanswered, and new questions that have arisen because of new discoveries. Part of the reason climate change is so complex is that there can be many positive and negative feedback loops. Only a few of the many possible feedback loops are discussed here. Negative feedbacks are self-enhancing and help stabilize a system. Positive feedbacks are self-regulating, so a greater change now will result in an even greater change in the future. This is a simplistic statement about feedback and climate because some changes are associated with both positive and negative feedback. With this caveat stated, we will discuss positive and negative feedbacks with respect to climate change. You may also wish to review the concepts presented in our discussion of feedback in Chapter 3.

Here are some feedback loops that have been suggested for climate change.[21]

Possible Negative Feedback Loops for Climate Change

- As global warming occurs, the warmth and additional carbon dioxide could stimulate the growth of algae. The algae, in turn, could absorb carbon dioxide, reducing the concentration of CO_2 in the atmosphere and cooling Earth's climate.
- Increased CO_2 concentration with warming might similarly stimulate growth of land plants, leading to increased CO_2 absorption and reducing the greenhouse effect.
- If polar regions receive more precipitation from warmer air carrying more moisture, the increasing snowpack and ice buildup could reflect solar energy away from Earth's surface, causing cooling.
- Warming could increase water evaporation from the ocean and the land, leading to cloudier conditions (the water vapor condenses), and the clouds would reflect more sunlight and cool the surface.

Possible Positive Feedback Loops for Climate Change

- The warming Earth increases water evaporation from the oceans, adding water vapor to the atmosphere. Water vapor is a major greenhouse gas that, as it increases, causes additional warming. If more clouds form from the increased water vapor, and more solar radiation is reflected, this would cause cooling, as our discussion of negative feedback shows. Thus, water vapor is associated with both positive and negative feedback. This makes study of clouds and global climate change complex.
- The warming Earth could melt a large amount of permafrost at high latitudes, which would in turn release the greenhouse gas methane, a by-product of decomposition of organic material in the melted permafrost layer. This would cause additional warming.
- Replacing some of the summer snowpack or glacial ice with darker vegetation and soil surfaces decreases the albedo (reflectivity), increasing the absorption of solar energy and further warming the surface. This is a powerful positive feedback, explaining, in part, why the Arctic is warming faster than lower latitudes.
- In warming climates, people use more air conditioning and thus more fossil fuels. The resulting increase in carbon dioxide could lead to additional global warming.
- Since negative and positive feedback can occur simultaneously in the atmosphere, the dynamics of climate change are all the more complex. Research is ongoing to better understand the negative feedback processes associated with clouds and their water vapor.

Climate Forcing

It can be helpful to view climate change in terms of **climate forcing**—defined as an imposed perturbation of Earth's energy balance.[15] As a way to visualize forcing, consider a checking account that you are free to use but earns no interest. Assume you initially deposit $1,000, and each year you deposit $500 and write checks for $500. You do this for many years, and at the end of that time you still have $1,000 in your account. The point is that for any system, when input equals output for some material (in this case, dollars), the amount in the system remains constant. In our bank account example, if we increased the total amount in the account by only $3 per year (a 0.3% increase per year), it would double to $2,000 over a period of about 233 years. In short, a small imbalance over many years can cause significant change.

Forcings change the states of the atmosphere and the surface, which feeds back to affect climate. Positive forcings cause warming, and negative forcings cause cooling. For example, as ice sheets grow, they reflect more incoming solar radiation, which enhances cooling (Figure 20.21a). Other forcings occur when the areas covered by vegetation change reflectivity and absorption of solar energy, as a result of the uptake and release of atmospheric gases. Climate forcings affect Earth's energy balance and provide an important quantitative tool with which to evaluate global climate change in the past.[15]

Climate forcing during the industrial age is shown in Figure 20.21b. In recent decades, human-caused forcings are estimated to have dominated over natural forcings. The total climate forcing from greenhouse gases is about 1.6 W/m^2. The aerosol black carbon (soot) is about 0.8 W/m^2. Recent revised estimates are higher, at about 1.1 W/m^2.[22] Thus, total climate forcing in very recent times may have increased from 1.6 to about 1.9 W/m^2.

20.11 New Understanding of the Interplay Between the Oceans and the Atmosphere

Natural oscillations of the ocean are linked to the atmosphere and can produce warmer or cooler periods of a few years to a decade or so. Ocean currents of the world have oscillations related to changes in water temperature, air pressure, storms, and weather over periods of a year or so to decades. Some of the insights into these oscillations are new and have changed some of the ways scientists think about climate change. They occur in the North Pacific, South Pacific, Indian, and North Atlantic oceans and can influence the climate. The Pacific Decadal Oscillation (PDO) for the North Pacific from 1900 to about 2010 is shown in Figure 20.22.

The effect of the oscillations can be ten times as strong (in a given year) as long-term warming that we have observed over the past century—larger, over a period of a few decades, than human-induced climate change. Some scientists attribute the cool winters of 2009–2010 to natural ocean–atmosphere oscillations and also suggest that these caused a cool year in 1911 that froze Niagara Falls. By comparison, the annual increase in warming estimated to be due to human activity is about two hundredths of a degree Celsius per year.[23]

Two other natural climate phenomena that can have an influence on storm tracks and, thus, weather are the Arctic Oscillation (AO) and the North Atlantic Oscillation (NAO). These oscillations have a positive and a negative state, and usually are present or absent at the same time. The positive AO and NAO are characterized by a strong polar vortex that produces a strong subtropic high-pressure

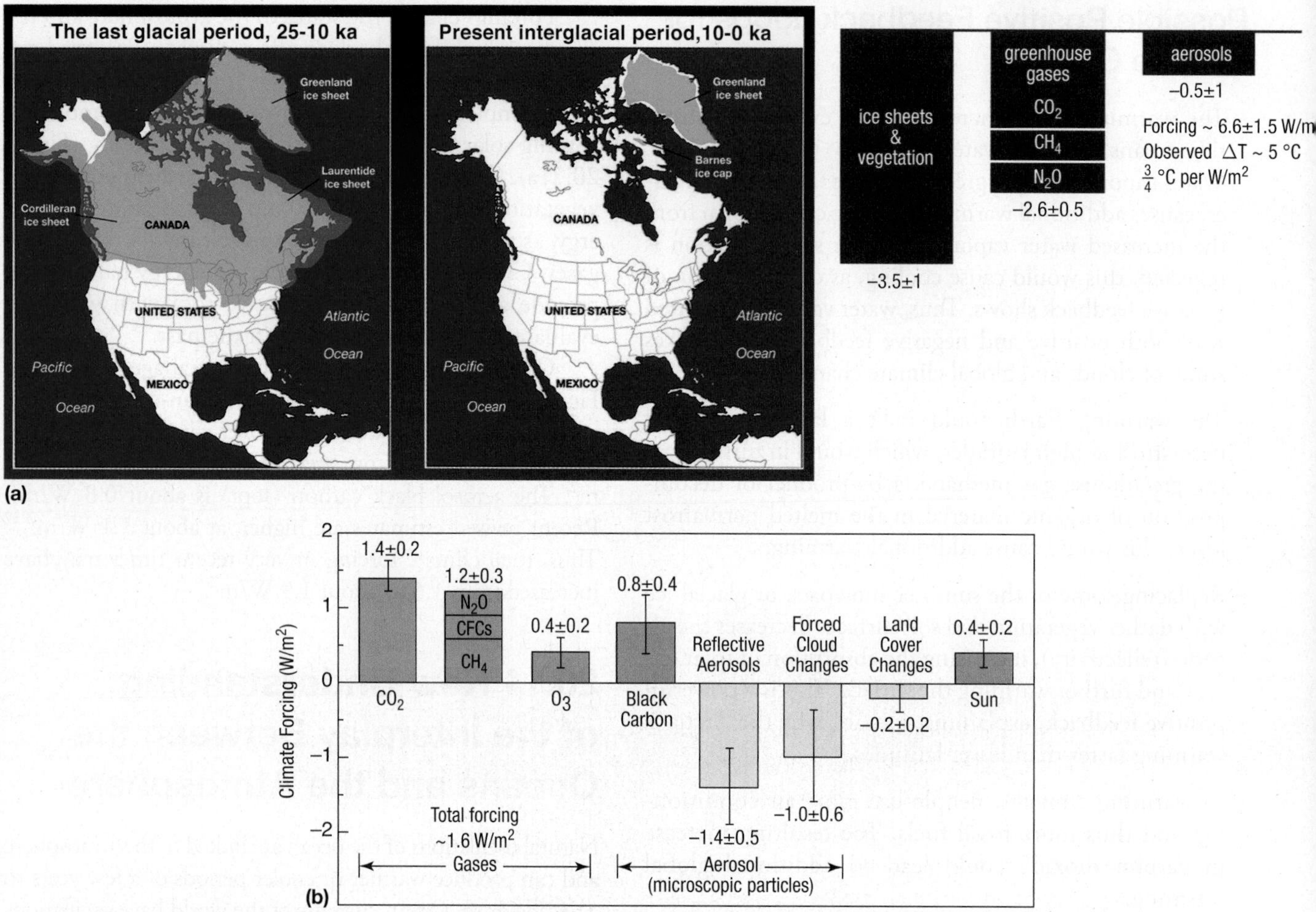

- Increases of greenhouse gases (except O_3) are known from observations and bubbles of air trapped in ice sheets. The increase of CO_2 from 285 parts per million (ppm) in 1850 to 392 in 2010 is accurate to about 5 ppm. The conversion of this gas change to a climate forcing (1.4 W/m^2), from calculation of the infrared opacity, adds about 10% to the uncertainty.
- Increase of CH_4 since 1850, including its effect on stratospheric H_2O and tropospheric O_3, causes a climate forcing about half as large as that by CO_2. Main sources of CH_4 include landfills, coal mining, leaky natural gas lines, increasing ruminant (cow) population, rice cultivation, and waste management. Growth rate of CH_4 has slowed in recent years.
- Tropospheric O_3 is increasing. The United States and Europe have reduced O_3 precursor emissions (hydrocarbons) in recent years, but increased emissions are occurring in the developing world.
- Black carbon ("soot"), a product of incomplete combustion, is visible in the exhaust of diesel trucks. It is also produced by biofuels and outdoor biomass burning. Black carbon aerosols are not well measured, and their climate forcing is estimated from measurements of total aerosol absorption. The forcing includes the effect of soot in reducing the reflectance of snow and ice.
- Human-made reflective aerosols include sulfates, nitrates, organic carbon, and soil dust. Sources include burning fossil fuel and agricultural activities. Uncertainty in the forcing by reflective aerosols is at least 35%.
- Indirect effects of aerosols on cloud properties are difficult to compute, but satellite measurements of the correlation of aerosol and cloud properties are consistent with the estimated net forcing of −1 W/m^2, with uncertainty of at least 50%.

FIGURE 20.21 **(a) Climate forcing during the last major glaciations about 22,000 years ago was minus 6.6 ± 1.5 W/m^2, which produced a drop in global lower atmospheric temperature; (b) climate forcing during the industrial age.**

(*Source:* (a) U.S. Geological Survey , Department of the Interior/USGS. (b) Modified from J. Hansen, Can We Defuse the Global Warming Time Bomb? Edited version of the presentation to the Council on Environmental Quality. June 12, 2003. Washington, DC, also Natural Science [http://www.naturalscience.com.])

center and strong low-pressure center in the sub-Arctic. These help keep cold arctic air in the north. The result of the positive AO and NAO is that warmer air from the southern latitudes reaches farther north in the United States and Europe.[24]

Negative AO and NAO conditions have resulted in severe cold weather striking the eastern coast of the United States. What has been perplexing to scientists is how this happens in a warming world. The decade from 2002 to 2012 appears to be the warmest since climate data have been collected and yet, during recent winters, there have been pronounced cold spells. The answer may be due to the loss of sea ice in the Arctic in recent years, favoring a negative AO.[24]

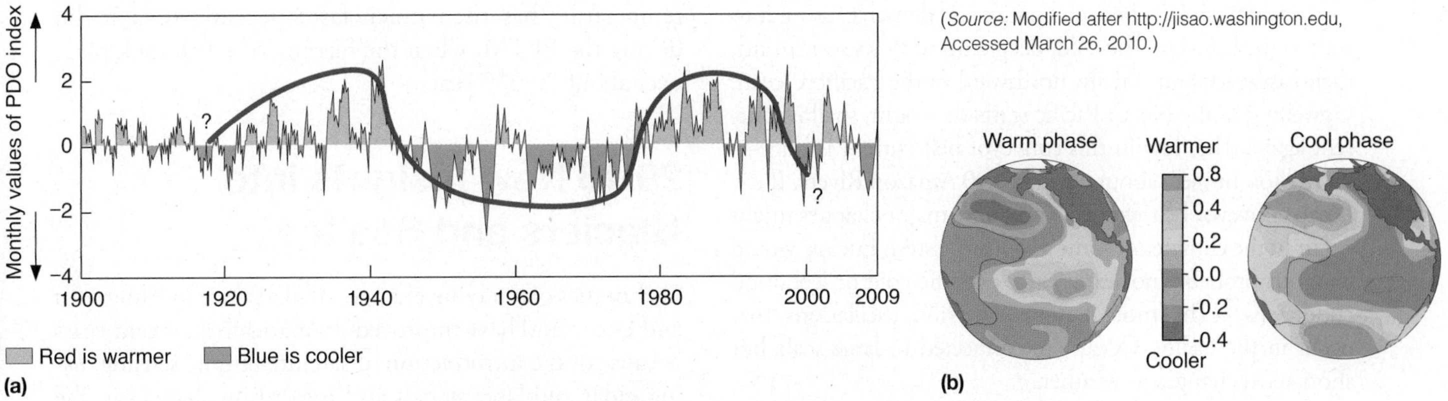

FIGURE 20.22 **(a) The Pacific Decadal Oscillation (PDO) from 1900 to 2009.** Three oscillations are clear (heavy line); (b) PDO warm and cool phases with characteristic changes in water temperature (°C) in the North Pacific and along the coast of the Pacific Northwest.

El Niño and Climate

A curious and historically important climate change linked to variations in ocean currents is the Southern Oscillation, known informally as El Niño. From the time of the early Spanish settlement of the west coast of South America, people observed a strange event that occurred about every seven years. Usually starting around Christmas (hence the Spanish name El Niño, referring to the little Christ Child), the ocean waters would warm up, fishing would become poor, and seabirds would disappear.

Under normal conditions, there are strong vertical, rising currents, called upwellings, off the shore of Peru. These are caused by prevailing winds coming westward off the South American continent, which move the surface water away from the shore and allow cold water to rise from the depths, along with important nutrients that promote the growth of algae (the base of the food chain) and thus produce lots of fish. Seabirds feed on those fish and nest in great numbers on small islands just offshore.

El Niño occurs when those cold upwellings weaken or cease altogether. As a result, nutrients decline, algae grow poorly, and so do the fish, which either die, fail to reproduce, or move away. The seabirds, too, either leave or die. Rainfall follows warm water eastward during El Niño years, so there are high rates of precipitation and flooding in Peru, while droughts and fires are common in Australia and Indonesia. Because warm ocean water provides an atmospheric heat source, El Niño changes global atmospheric circulation, which causes changes in weather in regions far removed from the tropical Pacific.[25]

The "Ocean Conveyor Belt"

The ocean conveyor belt—a global circulation of ocean waters characterized by strong northward movement of upper warm waters of the Gulf Stream in the Atlantic Ocean—was understood in the 20th century to play a role in climate. The temperature of these waters is approximately 12°–13°C when they arrive near Greenland, and they are cooled in the North Atlantic to a temperature of 2°–4°C (Figure 20.23).[26]

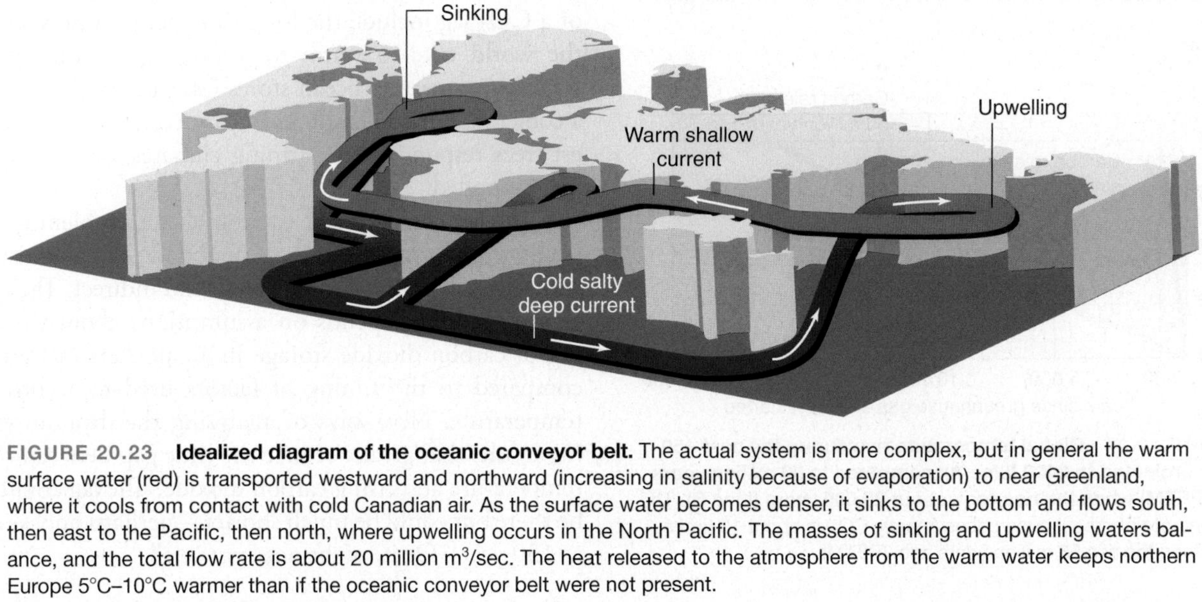

(*Source:* Modified from W. Broker, "Will Our Ride into the Greenhouse Future Be a Smooth One?" *Geology Today* 7, No. 5 [1997]: 2–6.)

FIGURE 20.23 **Idealized diagram of the oceanic conveyor belt.** The actual system is more complex, but in general the warm surface water (red) is transported westward and northward (increasing in salinity because of evaporation) to near Greenland, where it cools from contact with cold Canadian air. As the surface water becomes denser, it sinks to the bottom and flows south, then east to the Pacific, then north, where upwelling occurs in the North Pacific. The masses of sinking and upwelling waters balance, and the total flow rate is about 20 million m^3/sec. The heat released to the atmosphere from the warm water keeps northern Europe 5°C–10°C warmer than if the oceanic conveyor belt were not present.

As the water cools, it becomes saltier and denser, causing it to sink to the bottom. The cold, deep current flows southward, then eastward, and, finally, northward in the Pacific Ocean. Upwelling in the North Pacific starts the warm, shallow current again. The flow in this conveyor belt current is huge—20 million m^3/sec, about equal to 100 Amazon Rivers. If the ocean conveyor belt were to stop, some major changes might occur in the climates of some regions. Western Europe would cool but probably not experience extreme cold or icebound conditions.[27] The more famous El Niño oscillations that occur in the Pacific Ocean are connected to large-scale but short-term changes in weather.

20.12 New Insights from the Fossil Record

The Paleocene-Eocene Thermal Maximum

The **Paleocene-Eocene Thermal Maximum (PETM)** occurred about 56 million years ago. Geologists have studied the PETM by evaluating sedimentary rocks and sediments from the bottom of the ocean. These suggest a massive increase in carbon dioxide, probably stemming from the final stages of the breakup of the supercontinent Pangaea: As the tectonic plates spread apart, huge volumes of volcanic rock and heat rose through the land and oceans, "cooking" carbonate-rich sediments and rocks, which released the two strong greenhouse gases: carbon dioxide and methane. Average global temperature is estimated to have risen as much as 5-8 degrees Celsius, and glacial ice was all but absent. The PETM may provide some insights into the present warming that is occurring.[28–30] Major differences between the present warming and the warming that occurred 56 million years ago are summarized in Figure 20.24. One difference is that global temperature has risen much faster recently than it did during the PETM, when the heating of Earth took place over about 20,000 years.

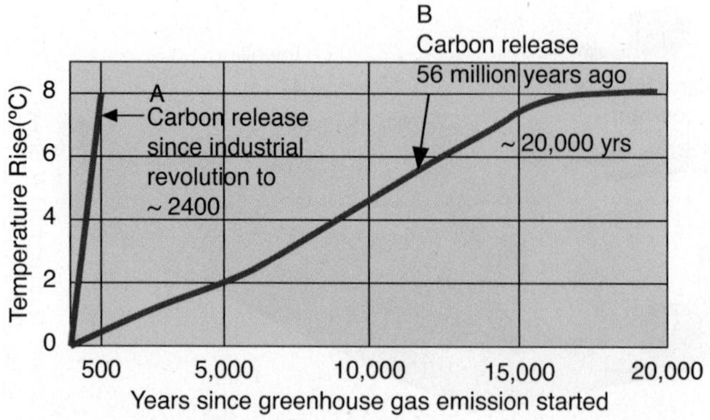

FIGURE 20.24 **Global temperature rise during the past 150 years projected to 2400 (Line A); compared to 56 million years ago (PETM) when the rise took about 20,000 years (Line B).** The present rate of greenhouse gases (mostly CO_2) is about 15 times more than the rise 56 million years ago. (PETM)

20.13 New Insights into Glaciers and Sea Ice

Techniques of studying changes of glacial ice in Antarctica and Greenland have improved dramatically in recent years because of the introduction of satellite remote sensing, using radar and laser signals and measuring changes in the polar gravity.[31]

Polar ice caps and mountain glaciers accumulate snow that is transformed into glacial ice over hundreds to thousands of years. The ice contains small air bubbles deposited with the snow, and we can measure the atmospheric gases in these bubbles, including carbon dioxide (CO_2) and methane (CH_4). The record of CO_2 and temperature from Antarctic ice cores is shown in Figure 20.25. To estimate what the atmospheric temperature was 400,000 years ago requires evaluation of stable isotopes of oxygen and hydrogen that have been shown to correlate well with temperature. The age of glacial ice is estimated by correlating ice accumulation rates linked to the geologic record of climate change from other proxy sources.

Information from the famous high-resolution Vostok Antarctic ice cores, along with two other ice cores, has raised some new questions. The study of these ice cores began in the 20th century, but in recent years new results suggest that changes in the concentration of carbon dioxide lagged changes in temperature between a statistically valid range of 600 to 1,000 years.[32] There are two possible explanations: Either the lag is the result of natural physical and chemical processes, or there is a measurement or scientific error in interpretation of the data. Natural causes of a CO_2 lag include the long time ocean waters around the world take to warm up following an atmospheric warming; a warmer ocean stores less carbon dioxide than a cooler one. It could also result from a lag in the way forest trees responded to warming climates, or the way sea ice responded.

The lag could also be the result of a problem in analyzing the ice cores from the Antarctic.[31] The chemistry is complicated, highly sophisticated, and indirect. The question of the lag depends on assumptions about the timing of carbon dioxide storage in air pockets in ice cores compared to the timing of factors used to reconstruct temperature. New ways of analyzing the data are yielding results that greatly reduce the time gap between rising temperature and rising carbon dioxide. The lag could still be there but could be much shorter—perhaps only as long as 130 years. Or it might not exist at all.[31]

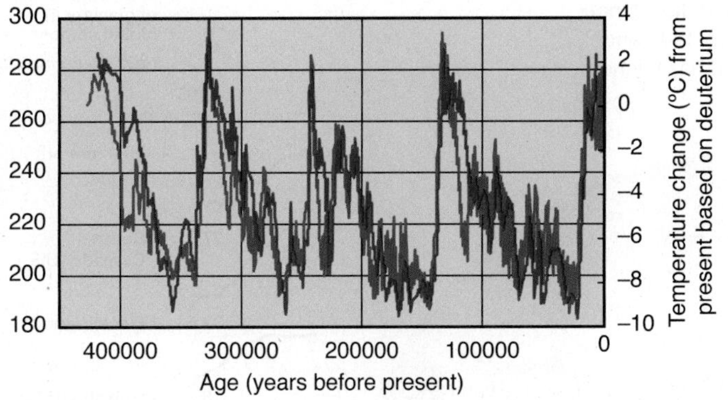

(Source: IPCC. *Climate Change 2007: The Physical Science Basis: Working Group I. Contribution to the Fourth Assessment Report*. New York: Cambridge University Press, 2007. Reprinted with permission.)

FIGURE 20.25 **Carbon dioxide (red line) and temperature (blue line) over the past several hundred thousand years.**

Ice cores also contain a variety of chemicals and materials, such as volcanic ash and dust, which may provide additional insights into possible causes of climate change. Ice cores are obtained by drilling into the ice.

The amount of ice on Earth's surface changes in complicated ways. A major concern is that global warming may lead to a great decline in the volume of water stored as ice; this is an especially worrisome possibility because the melting of glacial ice raises the mean sea level and because mountain glaciers are often significant sources of water for lower-elevation ecosystems. At present, many more glaciers in North America, Europe, and other areas are retreating than are advancing (Figure 20.26). In the Cascades of the Pacific Northwest and the Alps in Switzerland and Italy retreats are accelerating. For example, all eight glaciers on Mount Baker in the Northern Cascades of Washington were advancing in 1976. Today, all eight are retreating.[33] As another example, if present trends continue, all glaciers in Glacier National Park in Montana could be gone by 2030, and most glaciers in the European Alps could be gone by the end of the century.[34]

Not all reduction of glacial ice is due to global warming. For example, since first observed in 1912, the glaciers of Mount Kilimanjaro in Africa have decreased in area by about 80%. The glaciers formed during an African Humid Period about 4,000 to 11,000 years ago. Although there have been wet periods since then—notably in the 19th century, which appears to have led to a secondary increase in ice—conditions have generally been drier.[35]

The ice is disappearing not because of warmer temperatures at the top of the mountain, which are almost always below freezing, but because less snowfall is occurring, and ice is being depleted by solar radiation and sublimation (ice is transformed from solid state to water vapor without melting). Much of the ice depletion had occurred by the mid-1950s.[35]

Northern Hemisphere sea ice coverage in September, the time of the ice minimum, based on satellite images, has declined an average of 10% per decade (13.0 ± 2.9%) since 1980 (Figure 20.27). If present trends were to continue, the Arctic Ocean might be seasonally ice-free by 2030.[36] Since 1980, the smallest area of Arctic ice occurred in 2012, when sea ice covered about 3.4 million km^2.

Changes in sea ice involve more than total area; also involved are the depth and age of the ice. Newer ice is thinner and therefore contains less water than ice that has persisted for more than one year.

Some of the largest ice masses on land occur on Greenland and Antarctica. Glacial ice loss from Greenland and Antarctica between 2005 and 2010 is estimated to be at 296–293 Gt/yr. Glacial ice loss on Antarctica was greatest at West Antarctica, which has been losing ice since 1992. Meanwhile, the East Antarctic ice sheet has actually been growing since 2000, increasing from 2005 to 2010 between 27 and 89 Gt/yr. The increase in the volume of the East Antarctica ice sheet appears due in part to increased snowfall in that part of Antarctica, which in turn may be a result of increased moisture in the atmosphere due to warming temperatures.[31]

Greenland's ice sheets have been one of the major focuses of the debate on whether Earth's glaciers are melting, with some scientific papers claiming these glaciers are melting and others that they are increasing. Standard estimates indicate that between 1992 and 2011, Greenland's ice sheets lost 1,770 to 2,630 billion metric tons.[31] One study concluded that the current Greenland warming is not unprecedented in the history of the ice sheet. Although there is a general agreement that close to its margins the ice is thinning, it is thickening in the interior. There is a documented increase in glacial ice melting during recent years. At the same time, there is evidence of glacial acceleration in some parts of Greenland.[20]

Glacial ice loss from Antarctica and Greenland in 1992–2011 caused an estimated global sea-level rise of about 11.2 ± 3.8 mm (0.59 ± 0.20 mm/yr).[31]

One of the major concerns in the global-warming debate is whether the satellite observations show novel changes—meaning changes that have not likely happened

FIGURE 20.26 (a) The thinning of selected glaciers (m^2) since 1977 (National Geographic Maps); (b) Muir Glacier in 1941 and 2004. The glacier retreated more than 12 km (7 mi) and has thinned by over 800 m (2625 ft).

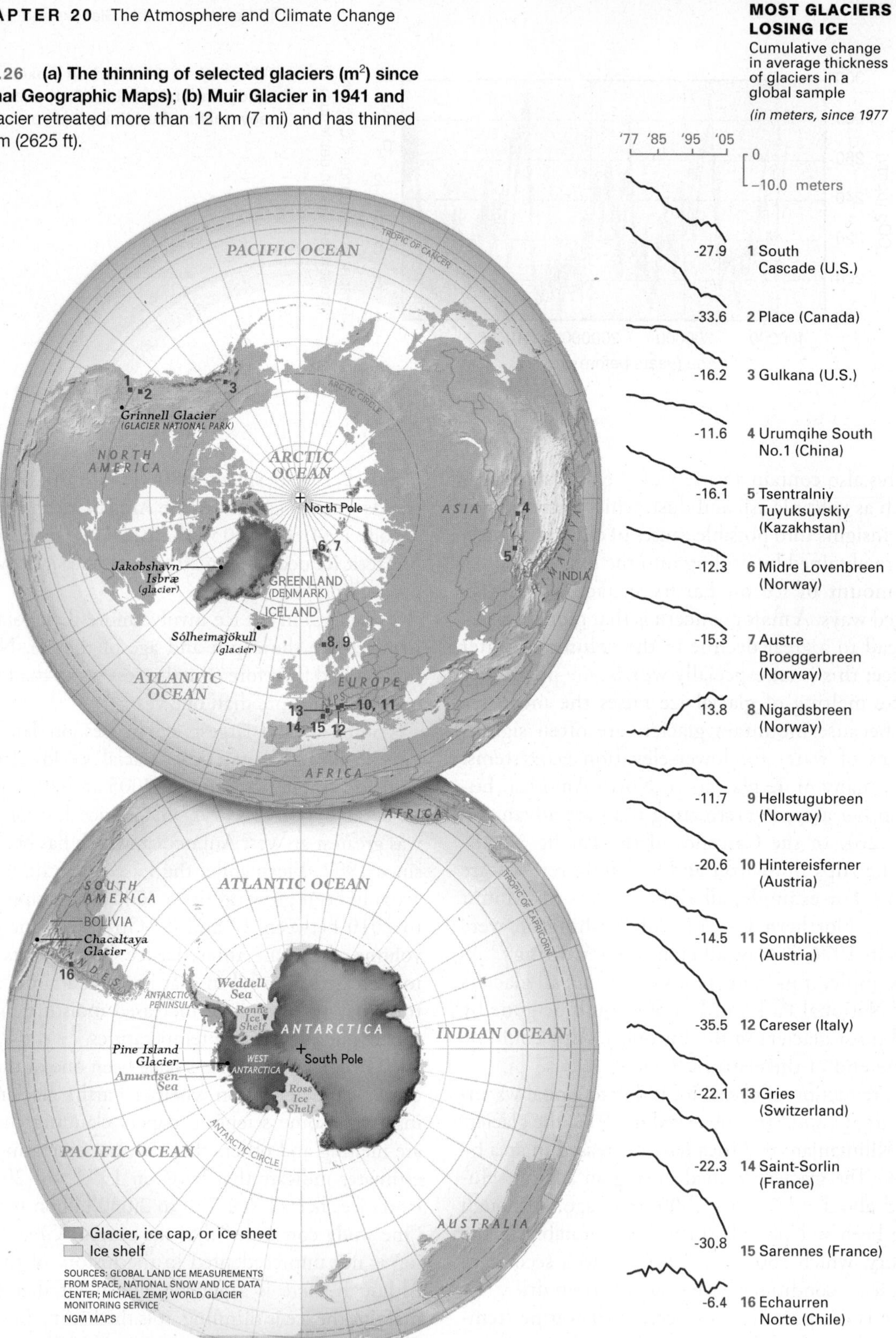

(a)

(b) 1941 2004

(*Source*: National Snow and Ice Data Center W.O. Field [1941] and Molina [2004].)

(a)

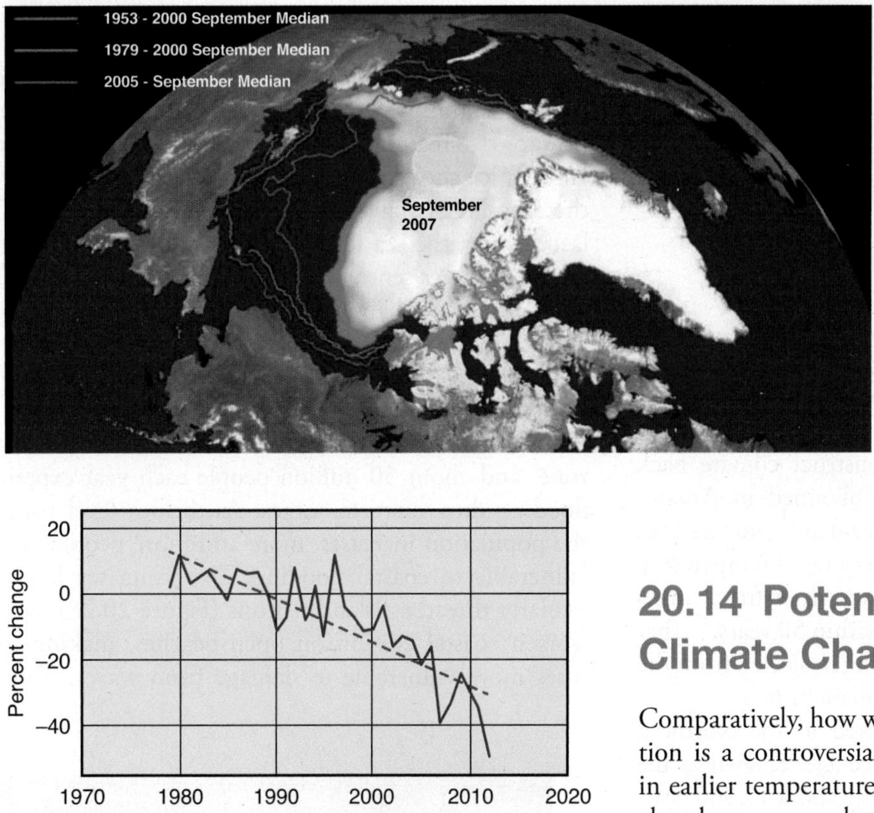

(b)

(*Source:* Modified after Stroever et al., 2008. EOS 89 [2] 13–14.) (b) Change in sea ice extent in the Arctic by percent from 1980 to 2012 (a).

FIGURE 20.27 **(a) Satellite observations, which began in 1977, show that Arctic sea ice reached a minimum in September 2007 and has increased since then.** The sea ice coverage varies greatly between summer and winter, with July marking the summer minimum. The rapid decline in 2007 was partly due to atmospheric circulation that favored melting.

in tens of thousands of years. Historical records shed some light on this issue: Notably, whalers kept records in their sailing ships' logbooks each day as they hunted bowhead whales between Alaska and Siberia from 1850 to 1910. The comparison of historical and modern ship observations shows that at the end of winter, in May, the southern extent of sea ice in the middle of the 19th century was similar to that in the 1970s. In contrast, Arctic sea ice between Alaska and Siberia extended much farther south in the middle of the 19th century than in the 1970s.[20]

Changes in the total area are not the only concern about sea ice. Because Arctic sea ice is melting back and then recovering each winter, the ice is new and much thinner and more fragile than ice that has remained in place for a number of years. The thinner ice is more easily broken up by storms, as happened in the winter of 2013.[37]

In conclusion, although there is considerable variability in the rate of loss and gain of glacial ice in West Antarctica and Greenland, and in some years there is an increase, available observations suggest that, on balance, there has been a large net loss of glacial ice in Antarctica and Greenland.

20.14 Potential Rates of Global Climate Change

Comparatively, how warm has it been recently? This question is a controversial one because gaps and limitations in earlier temperature records do not make it completely clear how accurately past measurements reflect the average Earth surface temperature. Some of the most important advances in climate science have resulted from ever wider use of remote sensing of Earth's atmosphere, oceans, and land surfaces by satellites. For example, beginning in 1978, satellites have measured the distribution of sea ice around the globe.

The global average temperature since 1900 has risen by about 0.8°C (1.5°F), and the global surface temperature has risen about 0.2°C (0.36°F) per decade in the past 30 years.[21] The warmest year since direct surface-air temperature has been measured was 2005. Virtually tied for second were 2002, 2003, 2006, 2007, and 2009. The period 2001–2010 is the warmest on record. Temperature increase paused from 1998-2013, but sea level and ocean temperatures continued to rise, did not change much. Similar apparent pauses in global temperature increases have been observed in the past (see Figure 20.3) due to variations in the amount of light emitted by the sun and received by Earth. The global temperature will continue to vary, no matter what the long-term trend.

According to current global climate models, if the concentrations of all greenhouse gases and aerosols had been kept constant at year 2000 levels, warming of about 0.1°C per decade would be expected.[21] Based on current and expected rates of CO_2 release by human activities, it is estimated that by 2030 the concentration of carbon dioxide in the atmosphere will be double the pre-Industrial Revolution level. If so, the GCMs forecast that the average global temperature will rise approximately 1°−2°C (2°−4°F), with greater increases toward the poles.[38]

In recent decades, the surface-air temperature has risen more in some polar regions, in part because of positive feedback. As snow and ice melt, solar energy that used to be reflected outward by ice and snow is now absorbed by vegetation and water, resulting in enhanced warming. This phenomenon is termed **polar amplification**.

How Fast Has Climate Changed in the Past?

Until recently, it was thought that the climate changed very slowly before the Industrial Revolution—temperatures were believed to change no faster than 2°F (1°C) in 1,000 years. But studies that reconstruct climate back 400,000 years from the ice cores obtained in Antarctica and deep-ocean waters off Greenland indicate that there have been many intervals of very rapid temperature change in Greenland. Some of the most dramatic changes were those of 7°–12°C (14°–24°F) within 50 years.[39] That is a lot faster and greater than anything projected from the global climate models for the immediate future.

One rapid cooling event occurred in the Northern Hemisphere at the end of the last ice age. Known as the Younger Dryas Event, it began abruptly about 12,900 years ago and lasted several centuries. The cause of the abrupt cooling is controversial, but it is known to have involved large-scale changes in atmospheric and oceanic circulation. One hypothesis is that the Younger Dryas Event was triggered by a large explosion in the atmosphere from an extraterrestrial object, such as an asteroid, that injected a huge amount of terrestrial ash and particles into the atmosphere, reflecting some solar radiation and cooling the climate.[40]

20.15 Potential Changes in River Flow and Sea Levels

With continued global warming, the melting of glacial ice and reductions in snow cover are anticipated to accelerate throughout the 21st century. This is also expected to reduce water availability and hydropower potential and to change the seasonality of flows in regions supplied by meltwater from major mountain ranges (e.g., the Hindu Kush, Himalaya, and Andes), where more than one-sixth of the world population currently lives.

California, which depends on snowmelt from the Sierra Nevada to irrigate one of the richest agricultural regions in the world, will have problems storing water in reservoirs if these forecasts come true. Rainfall will likely increase, but there will be less snowpack with warming. Runoff will be faster than it is when snow slowly melts, so reservoirs will fill sooner, and more water will escape to the Pacific Ocean. Lower runoff is projected for much of Mexico, South America, southern Europe, India, southern Africa, and Australia.

Rise in Sea Level

The sea level reached a minimum during the most recent glacial maximum. Since then, it has risen slowly. Sea level rises for two reasons: (1) liquid water expands as it warms, and (2) ice sheets and glaciers on land that melt increase the amount of water in the oceans. Since the end of the last ice age, the sea level has risen approximately 23 cm (about 1 ft) per century. Climatologists forecast that global warming could about double that rate. Various models predict that the sea level may rise anywhere from 20 cm to approximately 2 m (8–80 in.) in the next century; the most likely rise is probably 20–40 cm (8–16 in.).[38]

About half the people on Earth live in a coastal zone, and about 50 million people each year experience flooding due to storm surges. As the sea level rises and the population increases, more and more people become vulnerable to coastal flooding. The rising sea level particularly threatens island nations (Figure 20.28) and will worsen coastal erosion on open beaches, making structures more vulnerable to damage from waves. This will

(a) TORSTEN BLACKWOOD/AFP/Getty Images

(b) Ashley Cooper/Alamy

FIGURE 20.28 One of the world's smallest nations, Tuvalu, may succumb to sea-level rise. Tuvalu consists of nine coral islands in the South Pacific, with a total area smaller than Manhattan, and its highest elevation above sea level is 4.5 m. Sea levels have been rising since the end of the last ice age, a natural response. But global warming could accelerate this rise, possibly making the 12,000 citizens of Tuvalu the world's first sea-level rise refugees.

lead to further investments to protect cities in the coastal zone by constructing seawalls, dikes, and other structures to control erosion. Groundwater supplies for coastal communities could also be threatened by saltwater intrusion (see Chapter 19). In short, coastal erosion is a difficult problem that is very expensive to deal with. In many cases, it is best to allow erosion to take place naturally where feasible and defend against coastal erosion only where absolutely necessary.

20.16 Changes in Biological Diversity

Some of the greatest uncertainties about the consequences of global warming have to do with changes in biodiversity. Because organisms are complex, their responses to change can be complex (A Closer Look 20.1). Warming is one change, but others—such as availability of nutrients, relations with other organisms (predator and prey), and competition for habitat and niches in ecosystems—also affect biodiversity. Because we lack adequate theoretical models to link specific climate changes to specific changes in overall biodiversity, our best insights come from empirical evidence. Surprisingly few species went extinct as a result of climate change during the past 2.5 million years, even though the amount of change was about the same as that forecast for today and the next few decades.[41,42] Warming will certainly change some areas, and plants and animals will experience stress. Many will adapt, as many apparently did during the Medieval Warm Period (see the opening case study). Polar bears were undoubtedly stressed during this period but did not become extinct. Salmon in the Pacific Northwest have adapted to changing climate for tens of thousands of years. They have evolved, adapted, and continue to persist despite a complex system of changes.[8]

On the other hand, black guillemots (*Cepphus grylle*), birds that nest on Cooper Island, Alaska, illustrate the concerns some scientists have about global warming and certain species (Figure 20.29). Black guillemots in their entire range are not a threatened or endangered species. However, the abundance of this species on Cooper Island has declined since temperature increases in the 1990s caused the sea ice to recede farther from the island each spring. The parent birds feed on Arctic cod found under the sea ice and must then return to the nest to feed their chicks, who are not yet mature enough to survive on their own. For the parents to do this, the distance from feeding grounds to nest must be less than about 30 km, but in recent years, the ice in the spring has been receding as much as 500–800 km (300–500 mi) from the island. As a result, the black guillemots on the island have lost an important source of food. The birds have sometimes targeted sculpin, but it is not as abundant as cod. Polar bears have also

David Tipling/Getty Images

Modified after Stroeve, J., et al., (2008), Arctic Sea Ice Extent Plummets in 2007, Eos Trans. Copyright by AGU. This material was reprinted by permission of John Wiley & Sons, Inc.

FIGURE 20.29 **Black guillemots, medium-size birds also called sea pigeons, nest in the Far North (blue in the map above), including Cooper Island, Alaska.**

been a problem; with less sea ice, bears have gone ashore and eaten young birds. In 2009, of the 180 guillemots that hatched, only one on the island fledged (flew away). In 2010, bear-proof nesting boxes resulted in about 100 birds that fledged.[43] The future of black guillemots on Cooper Island depends on future springtime weather: Too warm, and the birds may disappear; too cold, and there may be too few snow-free days for breeding, in which case they also will disappear. In 2012, ten birds were fitted with geolocators to learn more about their movements after breeding.

Corals

A concern with the warming of near-surface water of the tropical oceans is loss or damage of coral reefs, possibly with a dramatic decline in their number and size. The concern is that corals may not be able to adapt to the warmer water associated with projected climate change and will bleach white from this stress. However, the genetics of corals suggests that some are more resilient to warmer water and can adapt to expected temperature change.[44]

As emission of carbon dioxide continues to increase, the ocean is taking up more and more of the chemical, which is reacting with water to produce carbonic acid, a weak acid that is increasing the acidity of the ocean water. The oceans of the world are thought to be able to take up as much as one-third of the total emissions of carbon dioxide. The acidity increase takes place mainly in the upper zone of the oceans that are reached by sunlight and affected by winds and tides. The average pH of the oceans is about 8.2, which is basic (the opposite of acidic), but it has declined to about 7.5 in some areas. This increase in acidity may be contributing to the bleaching of corals and lower productivity of shellfish. Some scientists fear that coral reef ecology may eventually decline in some areas. Coral reefs are tremendously productive, providing food fish for marine ecosystems and people. All sorts of shellfish and other organisms, such as small, snail-like plankton, live in coral reefs and other areas, and they all require calcium carbonate for their shells in order to live and thrive. Coral reefs were already under threat from many human activities, including intense fishing and scuba diving; some of the corals are killed or seriously injured by even the slightest weight placed on them by a diver's foot or gear.

Agricultural Productivity

Globally, agricultural production will likely increase in some regions and decline in others.[21] In the Northern Hemisphere, some of the more northern areas, such as Canada and Russia, may become more productive. Although global warming might move North America's prime farming climate north from the Midwestern United States to the region of Saskatchewan, Canada, the U.S. loss would not simply be translated into a gain for Canada. Saskatchewan would have the optimum climate for growing, but Canadian soils are thinner and less fertile than the prairie-formed soils of the U.S. Midwest. Therefore, a climate shift could have serious negative effects on midlatitude food production. Meanwhile, lands in the southern part of the Northern Hemisphere may become more arid. Prolonged drought as a result of future warming, as evidently occurred in some areas during the

A CLOSER LOOK 20.1

Some Animals and Plants in Great Britain Are Adjusting to Global Warming

Two of the longest studies of animals and plants in Great Britain show that at least some species are adjusting to recent and rapid climate change. The first is a 47-year study of the bird *Parus major*. This study, one of the longest ever conducted for any bird species, shows that these birds are responding behaviorally to rapid climate change. It's a case of the early bird gets the worm. A species of caterpillar that is one of the main foods of this bird during egg-laying has been emerging earlier as the climate has warmed. In response, females of this bird species are laying their eggs an average of two weeks earlier (Figure 20.30). Both birds and caterpillars are doing okay so far.[a]

The second study, one of the longest experiments about how vegetation responds to temperature and rainfall, shows that long-lived small grasses and sedges are highly resistant to climate change. The authors of the study report that changes in temperature and rainfall during the past 13 years "have had little effect on vegetation structure and physiognomy."[b]

These studies demonstrate something that ecologists have known for a long time and that has been one focus of their studies, as described in Chapters 5, 6, and 7: Individuals, populations, and species have evolved with, are adapted to, and adjust to environmental change, including climate change. However, as we learned from the niche concept, each species persists within a certain range of each environmental condition, so there are limits to the adjustment any one species can make over a short time. Larger changes require biological evolution, which for long-lived animals and plants can take a long time. Whether most species will be able to adjust fast enough to global warming is a hotly debated topic.

age fotostock/SUPERSTOCK

FIGURE 20.30 This pretty bird, *Parus major*, a native of Great Britain, is apparently adjusting to rapid climate change.

Medieval Warm Period (see the opening case study), and an accompanying loss of agricultural productivity could be one of the serious impacts of global warming.

Human Health Effects

Like other biological and ecological responses, the effects of global warming on human health are difficult to forecast. The IPCC *Climate Change 2007: Synthesis Report* is cautious about these possible effects, stating only that one needs to be thinking about "some aspects of human health, such as excess heat-related mortality in Europe, changes in infectious disease vectors in parts of Europe, and earlier onset of and increases in seasonal production of allergenic pollen in Northern Hemisphere high and mid-latitudes."[38] Some observers have suggested that global warming might increase the incidence of malaria. However, this has been shown not to be the case in past and present circumstances because temperature alone is not a good correlate for malaria.[45] The same finding has been shown for tick-borne encephalitis, another disease that some thought might increase from global warming.[46]

20.17 Adjusting to Potential Global Warming

People can adjust to the threat of global warming in two ways:

Adapt: Learn to live with global climate change over the next 50 years because there is warming in the pipeline from greenhouse gases already emitted.

Mitigate: Work to reduce the emissions of greenhouse gases and take actions to reduce the undesirable effects of a global warming.

How can carbon dioxide emissions be reduced? By increasing energy conservation and efficiency, along with using alternative energy sources. Rebalancing our use of fossil fuels so that we burn more natural gas would also be helpful because natural gas releases 28% less carbon per unit of energy than does oil and 50% less than coal.[47,48] The United States has reduced carbon emission in the production of energy by 9% in the past five years (4% in 2012 alone). The reduction occurred as some coal-burning power plants switched to natural gas. This trend will continue if natural-gas prices remain favorable.

Strategies to reduce CO_2 emissions include making greater use of mass transit and less use of automobiles; providing larger economic incentives to energy-efficient technology; setting higher fuel economy standards for cars, trucks, and buses; and establishing higher standards of energy efficiency.

Because clearing forests for agriculture reduces storage of carbon dioxide, protecting the world's forests would help reduce the threat of global warming, as would reforestation.[49]

Geologic (rock) sequestration is another way to reduce the amount of carbon dioxide that would otherwise enter the atmosphere. The idea is to capture carbon dioxide from power plants and industrial smokestacks and inject it into deep subsurface geologic reservoirs. Two geologic environments suitable for carbon sequestration are sedimentary rocks that contain saltwater and sedimentary rocks at the sites of depleted oil- and gasfields. To significantly mitigate the adverse effects of CO_2 emissions that result in global warming, we need to sequester approximately 2 gigatons of CO_2 per year.[50]

The process of carbon sequestration involves compressing carbon dioxide and changing it to a mixture of both liquid and gas, then injecting it deep underground. Individual injection projects can sequester approximately 1 million tons of CO_2 per year. A carbon sequestration project is under way in Norway beneath the North Sea. The carbon dioxide from a large natural-gas production facility is injected approximately 1 km into sedimentary rocks below a natural-gas field. The project, begun in 1996, injects about 1 million tons of CO_2 every year, and it is estimated that the entire reservoir can hold up to about 600 billion tons of CO_2—about as much as is likely to be produced from all of Europe's fossil fuel plants in the next several hundred years. Sequestering carbon dioxide beneath the North Sea is expensive, but it saves the company from paying carbon dioxide taxes for emissions into the atmosphere.

Pilot projects to demonstrate the potential of sequestering CO_2 in sedimentary rocks have been initiated in Texas beneath depleted oilfields. The storage potential at sites in Texas and Louisiana is immense: One estimate is that 200–250 billion tons of CO_2 could be sequestered in this region.[51]

International Agreements to Mitigate Global Warming

There are several approaches to seeking international agreements to limit greenhouse gas emissions. One major approach is for each nation to agree to some specific limit on emissions. However, it is unclear whether the major fossil fuel users will agree to such a plan, and even if they do, whether the plan will actually be carried out. There are some contrary examples. Right now, some European nations are importing wood pellets produced from U.S. forests. When burned, these pellets produce less CO_2 than an equivalent amount of coal. The European governments use this option to claim they have reduced carbon emissions, but their claims do not take into account the energy required to log the trees, produce the pellets, which require heating, and ship them across the ocean. In short, some ways to do carbon accounting can look good on paper but will not yield a true reduction in carbon emissions.

Another major approach is carbon trading, in which a nation agrees to cap its carbon emissions at a certain total amount and then issues emission permits to its corporations and other entities, allowing each to emit a certain quantity. These permits can be traded. For example, a power company that wants to build a new fossil fuel power plant might trade permits with a company that does reforestation, based on estimates of the amount of CO_2 the power plant would release and of an area of forest that could take up that amount. One of the most important programs of this kind is the European Climate Exchange. Carbon trading in the United States has come under criticism from both sides of the debate over what to do about potential global warming. That is, should we use "cap-and-trade" or not? Those in favor argue that we need to control CO_2 emissions to be proactive and reduce potential adverse inputs from global warming. Those opposed to cap-and-trade say that reducing emissions and changing energy policy is too expensive and will result in economic disaster, and in any case is unlikely to be followed by enough nations no matter what the international agreements.

Attempts to establish international treaties limiting greenhouse gas emissions began in 1988 at a major scientific conference on global warming held in Toronto, Canada, at which scientists recommended a 20% reduction in carbon dioxide emissions by 2005. The meeting was a catalyst for scientists to work with politicians to initiate international agreements for reducing emissions of greenhouse gases.

In 1992, at the Earth Summit in Rio de Janeiro, Brazil, a general blueprint for reducing global emissions was suggested. Some in the United States, however, objected that the reductions in CO_2 emissions would be too costly. Agreements from the Earth Summit did not include legally binding limits. After the meetings in Rio de Janeiro, governments worked to strengthen a climate-control treaty that included specific limits on the amounts of greenhouse gases that each industrialized country could emit into the atmosphere.

Legally binding emission limits were discussed in Kyoto, Japan, in December 1997, but specific aspects of the agreement divided the delegates. The United States eventually agreed to cut emissions to about 7% below 1990 levels, a goal reached by substituting natural gas in many power plants that formerly burned coal. However, that was far short of the reductions suggested by leading global-warming scientists, who recommended reductions of 60–80% below 1990 levels. A Kyoto Protocol resulted from this meeting, was signed by 166 nations, and became a formal international treaty in February 2006.

In July 2008, the leaders of the G8 nations, meeting in Japan, agreed to "consider and adopt" reductions of at least 50% in greenhouse gas emissions as part of a new UN treaty to be discussed in Copenhagen in 2009. This was the first time the United States agreed in principle to such a reduction. (In practice, the United States has not gone along with it.)

The United States, with 5% of the world's population, emits about 20% of the world's atmospheric carbon dioxide. The fast-growing economies of China and India are rapidly increasing their CO_2 emissions and are not bound by the Kyoto Protocol. China now emits more carbon than any other country. In China, carbon emission grew about 9% in 2011. California, which by itself is twelfth in the world in CO_2 emissions, passed legislation in 2006 to reduce emissions 25% by 2020. Some have labeled the action a "job killer," but environmentalists point out that the legislation will bring opportunities and new jobs to the state. California is often an environmental leader, and other states are considering how to control greenhouse gases. The United States, as of 2013, has not agreed to any international agreements to address climate change. New energy bills to reduce greenhouse gas emissions and turn to alternative energy to reduce our dependency on fossil fuels have been stopped in Congress. Failure to address global change will compromise our ability to be proactive and will instead require a reactive response as change occurs. This is not effective environmental planning.

CRITICAL THINKING ISSUE

Global Change: Critical Thinking Element

Editorials by science journalists concerning global change frequently ask the question, "What's going on with the weather today?" Many of the stories come to the conclusion that recent climate change (warming) is most likely responsible for changes in everything from droughts and increased precipitation to intense storms and cold spells in the midlatitudes. The argument that droughts and storms are due to climate change is based primarily on what has happened in the last four decades. There remains confusion about the difference between weather and climate. The Internet, newspapers, television news, and even movies are information sources for many people.[52]

Earth's average temperature has risen about 0.4 degrees Celsius since 1970; moisture in the atmosphere has risen a few percentages; the number of heat waves has increased by nearly one-third; and extreme rainfall has increased by about 7%.[52]

Some journalists have suggested that Hurricane Sandy in 2012 and other recent hurricanes are the result of human-induced climate change. Journalists have pointed out that billion dollar plus weather catastrophes were much more common between 1996 and 2011 than between 1980 and 1995.[52] But can these events be attributed to human-induced global warming or even to a climatic warming trend of any kind? No. We do not have sufficient evidence to suggest that particular events, such as Hurricane Sandy, are a result of global warming. The city of New York has experienced larger hurricanes in the past, and there have been warnings for decades that the city is vulnerable to just such an event as Sandy.

Sometimes over long time periods, past disasters become more like myth than reality. The more we learn about very large hazardous events of the distant past (sometimes, from the geologic record in sediments left from storms), the more we realize that very large storms that reoccur infrequently (perhaps about 200 years or more) are rare but not unexpected.

The population of the United States since 1980 has increased significantly—from about 227 million in 1980 to 315 million in 2012. As population increased, there was also more infrastructure and buildings. Many more people are living in harm's way today than in 1980. The most expensive disasters continue to be hurricanes, while the most frequent are tornadoes with hailstorms and thunderstorms. However, no agreement has been reached that, for example, the number of hurricanes has increased as a result of global warming, though some evidence points to the greater intensity of some hurricanes.

These statistics and others make up what is known to have happened over the past 40 years. Many people who learn about science from popular media sources believe that these and even last year's weather can be attributed completely to human causes.[52] But climatologists have long pointed out that climate change is a long-term process, and it is a mistake to try to interpret recent weather patterns as proof of any specific causalities.

After reading this chapter on global change and weather, you might think about the following questions in more detail: Is a dangerous change in climate occurring, or have we simply experienced a recent decade or so of bad luck in the United States with respect to weather disasters?

Critical Thinking Questions

1. Is the time period over the last 25 years long enough to conclude that the weather is changing as a result of global warming and that more weather disasters are the result?
2. Do you think the increase in population in the United States, along with where people have chosen to live and how we build our homes, is partly responsible for the increase in the weather disasters? If so, how much?
3. Can you outline what sort of data over what time duration would help resolve the question of whether or not the recent weather disasters are setting a new trend or are just part of the natural background rate of processes associated with a changing world?
4. If a friend of yours or your parents asks you if global warming is causing the recent weather disasters, how would you answer with scientific statements?

SUMMARY

- The atmosphere, a layer of gases that envelops Earth, is a dynamic system that is constantly changing. A great number of complex chemical reactions take place in the atmosphere, and atmospheric circulation takes place on a variety of scales, producing the world's weather and climates.
- Nearly all the compounds found in the atmosphere either are produced primarily by biological activity or are greatly affected by life.
- Major climate changes have occurred throughout Earth's history. Of special interest to us is that periodic glacial and interglacial episodes have characterized Earth since the evolution of our species.
- During the past 1,000 years, several warming and cooling trends have affected civilizations.
- During the past 100 years, the mean global surface air temperature has risen by about 0.8°C. About 0.5°C of this increase has occurred since around 1960.
- Water vapor, carbon dioxide, methane, some oxides of nitrogen, and CFCs are the major greenhouse gases. The vast majority of the greenhouse effect is produced by water vapor, a natural constituent of the atmosphere. Carbon dioxide and other greenhouse gases also occur naturally in the atmosphere. However, especially since the Industrial Revolution, human activity has added substantial amounts of carbon dioxide to the atmosphere, along with such greenhouse gases as methane and CFCs.

- Climate models suggest that a doubling of carbon dioxide concentration in the atmosphere could raise the mean global temperature 1°–2°C in the next few decades and 1.5°–4.5°C by the end of this century.
- Many complex positive feedback and negative feedback cycles affect the atmosphere. Natural cycles, solar forcing, aerosol forcing, particulate forcing from volcanic eruptions, and El Niño events also affect the temperature of Earth.
- There are concerns, based on scientific evidence, that global warming is leading to changes in climate patterns, a rise in sea level, melting of glaciers, and changes in the biosphere. A potential threat from future warming, as in the Medieval Warm Period, is the occurrence of prolonged drought that would compromise our food supply.
- Adjusting to global warming includes learning to live with the changes and attempting to mitigate warming by reducing emissions of greenhouse gases.

REEXAMINING THEMES AND ISSUES

Sean Randall/Getty Images, Inc.

HUMAN POPULATION

The burning of fossil fuels and trees has increased emissions of carbon dioxide into the atmosphere. As the human population increases and standards of living rise, the demand for energy increases; as long as fossil fuels are used, greenhouse gases will also increase.

© Biletskiy_Evgeniy/iStockphoto

SUSTAINABILITY

Through our emissions of greenhouse gases, we are conducting global experiments, the final results of which are difficult to predict. As a result, achieving sustainability in the future will be more difficult. If we do not know in detail what the consequences or magnitude of human-induced climate change will be, then it is difficult to predict how we might achieve sustainable development for future generations.

© Anton Balazh 2011/iStockphoto

GLOBAL PERSPECTIVE

Climate change and global warming is a global problem.

ssguy/ShutterStock

URBAN WORLD

Many urban regions are impacted by climate change that is changing water resources, food production, heat wave frequency, and diseases. If sea levels rise as climate models forecast, coastal cities will be affected by higher storm surges.

B2M Productions/Getty Images, Inc.

PEOPLE AND NATURE

The study of warming 55 million years ago during the PETM suggests that, while warming was about ten times slower than it is today, many species adapted, others evolved, and, at the bottom of the deep oceans, some went extinct. Our ancestors adapted to natural climate change over the past million years. During that period, Earth experienced glacial and interglacial periods that were colder and warmer than today. Burning fossil fuels has led to human-induced climate changes different in scope from past human experience.

George Doyle/Getty Images, Inc.

SCIENCE AND VALUES

Responding to global warming requires choices based on value judgments. Scientific information, especially geologic data and the instrumental (historical) record, along with modern computer simulations, is providing a solid foundation for the belief that global warming is happening. The extent to which scientific information of this kind is accepted involves value decisions.

KEY TERMS

atmosphere 485
barometric pressure 486
climate 485
climate forcing 497
general circulation models (GCMs) 495
greenhouse effect 492
greenhouse gases 492
Medieval Warm Period (MWP) 479
Paleocene-Eocene Thermal Maximum (PETM) 500
polar amplification 504
radionuclides 491
stratosphere 485
troposphere 485
weather 485

STUDY QUESTIONS

1. Summarize the scientific data that indicate global warming is occurring as a result of human activity.
2. What is the composition of Earth's atmosphere, and how has life affected the atmosphere during the past several billion years?
3. What is the greenhouse effect? What is its importance to global climate?
4. What is an anthropogenic greenhouse gas? Discuss the various anthropogenic greenhouse gases in terms of their potential to cause global warming.
5. What are some of the major negative feedback cycles and positive feedback cycles that might increase or decrease global warming?
6. In terms of the effects of global warming, do you think that a potential change in climate patterns and storm frequency and intensity is likely to be more serious than a global rise in sea level? Illustrate your answer with specific problems and areas where the problems are likely to occur.
7. How would you summarize the potential biological effects of global warming?
8. What lessons can we take away from the Paleocene-Eocene Thermal Maximum (PETM)?
9. How would you refute or defend the statement that the best adjustment to global warming is to do little or nothing and learn to live with change?

FURTHER READING

Botkin, D.B. ***The Moon in the Nautilus Shell*** (New York, NY: Oxford University Press, 2012). Chapters 12 and 13 discuss climate change and life.

Fay, J.A., and D. Golumb. ***Energy and the Environment*** (New York, NY: Oxford University Press, 2002). See Chapter 10 on global warming.

IPCC. ***Climate Change 2007. The Physical Science Basis*** (New York, NY: Cambridge University Press, 2007). A report by the international panel that was awarded the Nobel Prize for its work on global warming.

Lovejoy, T.E., and L. Hannah. ***Climate Change and Biodiversity*** (New Haven, CT: Yale University Press, 2005). A discussion, continent by continent, of what has happened to biodiversity in the past when climate has changed.

Rohli, R.V., and A.J. Vega. ***Climatology*** (Sudbury, MA: Jones & Bartlett, 2008). An introduction to the basic science of how the atmosphere works.

Weart, S.R. ***The Discovery of Global Warming*** (Cambridge, MA: Harvard University Press, 2008). A discussion of how the possibility of global warming was discovered and the history of the controversies about it.

NOTES

1. Mann, M.E., and eight others. 2009. Global signatures and dynamical origins of the Little Ice Age and Medieval Climate Anomaly. *Science* 326:1256–1260.
2. Fagan, B.M. 2008. *The Great Warming: Climate Change and the Rise and Fall of Civilizations.* New York, NY: Bloomsbury Press.
3. Le Roy Ladurie, E. 1971. *Times of Feast, Times of Famine: A History of Climate Since the Year 1000.* Garden City, NY: Doubleday & Co.
4. National Research Council. 2006. Surface temperature reconstructions for the last 2,000 years. Washington DC: National Academy Press.
5. Fagan, B. 2004. *The Long Summer: How Climate Changed Civilization.* New York, NY: Basic Books.
6. Callendar, C.S. 1938. The artificial production of carbon dioxide and its influence on temperature. *Quarterly Journal of the Royal Meteorological Society* 84:223–237; Can carbon dioxide influence climate? *Weather* 4(1949):310–314; On the amount of carbon dioxide in the atmosphere, *Talus* 10(1958):243.
7. Scripps Institute of Oceanography. 2013. Scripps CO_2 Program. Scripsco2.ucsd.edu.
8. Botkin, D.B. 1990. *Discordant Harmonies: A New Ecology for the 21st Century.* New York, NY: Oxford University Press.
9. NOAA. 2009. Paleo proxy data. In *Introduction to Paleoclimatology.* Accessed March 24, 2010. www.ncdc.noaa.gov.
10. Bennett, K. 1990. Milankovitch cycles and their effects on species in ecological and evolutionary time. *Paleobiology* 16(1):11–12.
11. Foukal, P., C. Frohlich, H. Sprint, and T.M.L. Wigley. 2006. Variations in solar luminosity and their effect on the Earth's climate. *Nature* 443:151–166.
12. Soon, W. 2007. Implications of the secondary role of carbon dioxide and methane forcing in climate change: Past, present and future. *Physical Geography* 28(2):97–125.
13. U.S. Geological Survey. 1997. Fact Sheet 113-97. http://pubs.usgs.gov/fs/1997/fs113-97.
14. Encyclopedia Britannica online. Accessed January 2, 2009. http://www.britannica.com/EBchecked/topic-art/174962/69345/Carbon-dioxide-concentrations-in-Earths-atmosphere-plotted-over-the-past.
15. Hansen, J., et al. 2005. Efficiency of climate forcing. *Journal of Geophysical Research* 110 (D18104):45P.
16. Brook, E. 2008. Palaeoclimate—windows on the greenhouse. *Nature* 453(7193):291–292, Doi 10.1038/453291a.
17. Dlugokencky, E.J., L.P. Steele, P.M. Lang, and K.A. Masarie. 1994. The growth rate and distribution of atmospheric methane. *Journal of Geophysical Research* 99(D8):17021–17043.
18. Hansen, J., A. Lacis, and M. Prather. 1989. Greenhouse effect of chlorofluorocarbons and other trace gases. *Journal of Geophysical Research* 94(D13):16417–16421.
19. Rodhe, H. 1990. A comparison of the contribution of various gases to the greenhouse effect. *Science* 248:1217–1219.
20. Botkin, D.B. 2012. *The Moon in the Nautilus Shell: Discordant Harmonies Reconsidered.* New York, NY: Oxford University Press.
21. IPCC. 2007. *Climate Change 2007: The Physical Science Basis: Working Group I Contribution to the Fourth Assessment Report, IPCC.* New York, NY: Cambridge University Press.
22. Bond, T.C., and twenty-four others. 2013. Bounding the role of black carbon in the climate system: A scientific assessment. *Journal of Geographical Research – Atmosphere* (in press).
23. Hidore, J., J.E. Oliver, M. Snow, and R. Snow. 2010. *Climatology.* Upper Saddle River, NJ: Prentice-Hall.
24. Greene, C.H. The winters of our discontent. *Scientific American* 307(6):50–55.
25. Jet Propulsion Laboratory. *El Niño–when the Pacific Ocean speaks, Earth listens.* wmv.jpl.nasa.gov, accessed September 25, 2008.
26. Broker, W. 1997. Will our ride into the greenhouse future be a smooth one? *Geology Today* 7 No.5:2–6.
27. Steager, R. 2006. The source of Europe's mild climate. *American Scientist* 94:334–341.
28. McInery, F.A. and Wing, S.L., 2011. The Paleocene-Eocene thermal max: A perturbation of carbon, cycle, climate, and biosphere with implications for the future. *Annual Review of Earth Planet Science* 39:489–516.
29. Ying, C., and eight others. 2011. Slow release of fossil carbon during the Paleocene-Eocene Thermal Maximum. *Nature Geoscience* 4:481–485.
30. Kump, L.R. 2011. The last great global warming. *Scientific American* 305:57.
31. Shepherd, A., et al. 2012. A reconciled estimate of ice-sheet mass balance. *Science* 338:1183.
32. Caillon, N., and five others. 2003. Timing of atmospheric CO_2 and Antarctic temperature changes across Termination III. *Science* 299(5613):1728–1731.
33. Pelto, M.S. 1996. Recent changes in glacier and alpine runoff in the North Cascades, Washington. *Hydrological Processes* 10:1173–1180.
34. Appenzeller, T. 2007. The big thaw. *National Geographic* 211(6):56–71.

35. Mote, P.W., and G. Kasen, 2007. The shrinking glaciers of Kilimanjaro: Can global warming be blamed? *American Scientist* 95(4):218–225.

36. Stroeve, J., M. Serreze, and S. Drobot. 2008. Arctic Sea ice extent plummets in 2007. *EOS, Transactions*. American Geophysical Union 89(2):13–14.

37. Freedman, A. 2013 (March 6). Large fractures spotted in vulnerable Arctic Sea ice. *Climate Central.* http://www.climatecentral.org/news/large-fractures-spotted-in-arctic-sea-ice-15728. Accessed March 28, 2013. Data from National Snow and Ice Data Center.

38. IPCC. 2007. *Climate Change 2007: Synthesis Report.* Valencia, Spain: IPCC.

39. MacDougall, J.D. 2004. *Frozen Earth: The Once and Future Story of Ice Ages.* Berkeley, CA: University of California Press.

40. Firestone, R.B., and twenty-five others. 2007. Evidence for an extraterrestrial impact 12,900 years ago that contributed to the magafauna extinctions and the Younger Dryas cooling. *Proceedings of the National Academy of Sciences* 104:(41):16016–16021.

41. Botkin, D.B. 2007. Forecasting effects of global warming on biodiversity. *Bio Science* 57(3):227–236.

42. Botkin, D.B. 2012. *The Moon in the Nautilus Shell.* New York, NY: Oxford University Press.

43. Divokt, G. 2011. Black guillemots in a melting Arctic: Responding to shifts in prey, competitors, and predators. *Transcriptions* pp. 125–130 in R.T. Watson, T. J. Cade, M. Fuller, G. Hunt, and E. Potapov (Eds.), *Gyrfalcons and Ptarmigan in a Changing World.* Volume I. Boise, Idaho: The Peregrine Fund.

44. Brashis, D.J., and six others. 2013. Genomic basis for coral resilience to climate change. *Proceedings of the National Academy of Sciences* 110(4):1387–1392.

45. Rogers, D.J., and S.E. Randolph. 2000. The global spread of malaria in a future, warmer world. *Science* 289:1763–1766.

46. Sumilo, D., L. Asokliene, A. Bormane, V. Vasilenko, and I. Golovijova. 2007. Climate change cannot explain the upsurge of tick-borne encephalitis in the Baltics. *PLoS ONE* 2(6):e500 doi:10.1371/journa.pone.0000500.

47. Botkin, D.B. 2010. *Powering the Future: A Scientist's Guide to Energy Independence.* Upper Saddle River, NJ: Pearson FT Press.

48. Dunn, S. 2001. Decarbonizing the energy economy. In *World Watch Institute State of the World 2001.* New York: W.W. Norton.

49. Rice, C.W. 2002. Storing carbon in soil: Why and how. *Geotimes* 47(1):14–17.

50. Friedman, S.J. 2003. Storing carbon in earth. *Geotimes* 48(3):16–20.

51. Bartlett, K. 2003. Demonstrating carbon sequestration. *Geotimes* 48 (3):22–23. 48 on P66 http://www. juliantrubin.com/bigten/photosynthesisexperiments.html; and a variety of other sources.

52. Theissen, K.M. 2011. What do U.S. students know about climate change? *EOS, Transactions.* American Geophysical Union 92(51):477–478.

A CLOSER LOOK 20.1 NOTES

a. Charmantier, A., Robin H. McCleery, Lionel R. Cole, Chris Perrins, Loeske E. B. Kruuk, and Ben C. Sheldon. 2008. Adaptive phenotypic plasticity in response to climate change in a wild bird population. *Science* 320(5877):800–803.

b. Grime, J.P., J.D. Fridley, A.P. Askew, K. Thompson, J.G. Hodgson, and C.R. Bennett. 2008. Long-term resistance to simulated climate change in an infertile grassland. *PNAS* 105(29):10028–10032. Earth System Science Committee (1988). *Earth System Science: A Preview.* Boulder, CO: University Corporation for Atmospheric Research.

CHAPTER 21

Air Pollution

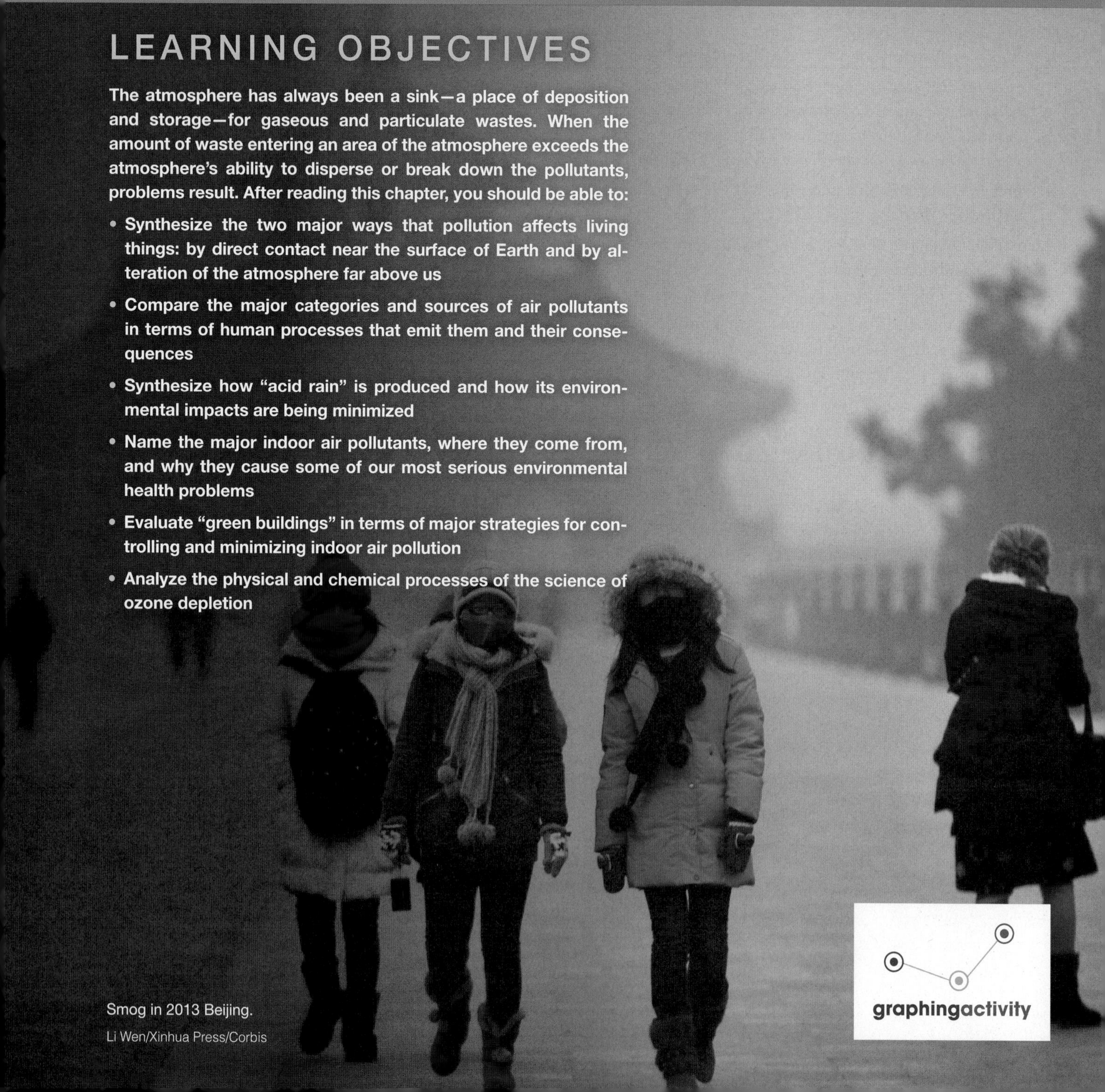

LEARNING OBJECTIVES

The atmosphere has always been a sink—a place of deposition and storage—for gaseous and particulate wastes. When the amount of waste entering an area of the atmosphere exceeds the atmosphere's ability to disperse or break down the pollutants, problems result. After reading this chapter, you should be able to:

- Synthesize the two major ways that pollution affects living things: by direct contact near the surface of Earth and by alteration of the atmosphere far above us
- Compare the major categories and sources of air pollutants in terms of human processes that emit them and their consequences
- Synthesize how "acid rain" is produced and how its environmental impacts are being minimized
- Name the major indoor air pollutants, where they come from, and why they cause some of our most serious environmental health problems
- Evaluate "green buildings" in terms of major strategies for controlling and minimizing indoor air pollution
- Analyze the physical and chemical processes of the science of ozone depletion

Smog in 2013 Beijing.
Li Wen/Xinhua Press/Corbis

graphingactivity

CASE STUDY

China Air Pollution 2013

For several decades, Chinese cities, such as Beijing, Harbin and Shanghai, have suffered from air-pollution events that are hazardous to the health of people living there. Some health advisers have stated that thousands of people in these cities will suffer premature death each year as a result of air pollution, as well as increased risk from heart and lung disease. In recent decades, the economy of China has grown dramatically, often at rates of about 10% per year (doubling in about 7 years). Recently, the economy has slowed to about 7% (doubling in about 10 years), but the rate of growth has been difficult to sustain without damage to the environment. China burns more coal than any other country in the world, and coal is a dirty fuel, releasing harmful pollutants, including small particles that can be particularly harmful to lungs. China also has many fleets of aging trucks that, as they burn fossil fuels, release large amounts of small-particle pollutants. This situation in Beijing came to a head in early January 2013 when air pollution spiked to an unknown record high, well above what is considered hazardous (see the opening photo). Then in October of 2013, the city of Harbin with 10 million people suffered a hazardous air pollution event. Schools and the airport were shut down as visibility was reduced in some places to less than 10 m (32 ft.). The Chinese people responded by buying and wearing face masks, and the general level of concern for health grew dramatically.

As a result of its economic expansion, China has a growing middle class (perhaps totaling 300 million or more people), and they desire to live in a cleaner environment. In other words, the people of China are becoming more health conscious and, in spite of the perceived necessity of increased economic growth, are striving for more environmental control. For example, in January 2013, the city of Hong Kong announced that it was ready to spend \$1.3 billion to retire about 80,000 older trucks that burn diesel fuels and emit harmful particles.

21.1 Air Pollution in the Lower Atmosphere

A Brief Overview

As the fastest moving fluid medium in the environment, the atmosphere has always been one of the most convenient places to dispose of unwanted materials. The atmosphere has been a sink for waste disposal ever since we first used fire, and people have long recognized the existence of atmospheric pollutants, both natural and human-induced. Leonardo da Vinci wrote in 1550 that a blue haze formed from materials emitted into the atmosphere by trees. What he had observed is a natural photochemical smog from hydrocarbons given off by living trees. This haze, whose cause is still not fully understood, gave rise to the name Smoky Mountains for the range in the southeastern United States.

The phenomenon of acid rain was first described in the 17th century, and, by the 18th century, it was known that plants in London were damaged by air pollution. Beginning with the Industrial Revolution in the 18th century, air pollution became more noticeable. The word *smog* was introduced by a physician at a public health conference in 1905 to denote poor air quality resulting from a mixture of smoke and fog. In the past few decades air quality in the United States has improved significantly, but there remains room for more improvements.[1]

Stationary and Mobile Sources of Air Pollution

The two major categories of air-pollution sources are stationary sources and mobile sources. **Stationary sources** have a relatively fixed location and include point sources, fugitive sources, and area sources.

Point sources, discussed in Chapter 8, emit pollutants from one or more controllable sites, such as power plant smokestacks (Figure 21.1).

F. Hoffman/The Image Works

FIGURE 21.1 **This steel mill in Beijing, China, is a major source of air pollution.**

FIGURE 21.2 Burning sugarcane fields, Maui, Hawaii—an example of a fugitive source of air pollution.

- *Fugitive sources* generate air pollutants from open areas exposed to wind. Examples include burning for agricultural purposes (Figure 21.2), as well as dirt roads, construction sites, farmlands, storage piles, surface mines, and other exposed areas.
- *Area sources*, also discussed in Chapter 8, are well-defined areas within which are several sources of air pollutants—for example, small urban communities, areas of intense industrialization within urban complexes, and agricultural areas sprayed with herbicides and pesticides.

Mobile sources of air pollutants include automobiles, trucks, buses, aircraft, ships, trains, and anything else that pollutes as it moves from place to place.[2]

General Effects of Air Pollution

Air pollution affects many aspects of our environment, including its visual qualities, vegetation, animals, soils, water quality, natural and artificial structures, and human health. Air pollutants affect visual resources by discoloring the atmosphere and by reducing visual range and atmospheric clarity. We cannot see as far in polluted air, and what we do see has less color contrast. These effects were once limited to cities but now extend to some wide-open spaces of the United States. For example, near the Four Corners, where New Mexico, Arizona, Colorado, and Utah meet, emissions from two large fossil-fuel-burning power plants have altered visibility in a region where visibility used to be 80 km (50 mi) from a mountaintop on a clear day.[2] The power plants are two of the largest sources of pollution in the United States.

Air pollution's numerous effects on vegetation include damage to leaves, needles, and fruit; reduced or suppressed growth; increased susceptibility to diseases, pests, and adverse weather; and disruption of reproductive processes.[1,2]

Air pollution is a significant factor in the human death rate in many large cities. For example, it has been estimated that in Athens, Greece, the number of deaths is several times higher on days when the air is heavily polluted; and in Hungary, where air pollution has been a serious problem in recent years, it may contribute to as many as 1 in 17 deaths. The United States is certainly not immune to health problems related to air pollution. The most polluted air in the nation is in the Los Angeles urban area, where millions of people are exposed to it. An estimated 175 million people live in areas of the United States where exposure to air pollution contributes to lung disease, which causes more than 300,000 deaths per year. Air pollution in the United States is directly responsible for annual health costs of over $50 billion. In China, whose large cities have serious air-pollution problems, mostly from burning coal, the health cost is now about $50 billion per year and may rise to about $100 billion per year by 2020.

Air pollutants can affect our health in several ways, depending on the dose or concentration and other factors, including individual susceptibility (see the discussion of dose–response in Chapter 8). Some of the primary effects are cancer, birth defects, eye and respiratory system irritation, greater susceptibility to heart disease, and aggravation of chronic diseases, such as asthma and emphysema. People suffering from respiratory diseases are the most likely to be affected. Healthy people tend to acclimate to pollutants, but this is a physiological tolerance; as explained in Chapter 8, it doesn't mean that the pollutants are doing no harm (Figure 21.3).

Many air pollutants have *synergistic effects*—that is, the combined effects are greater than the sum of the separate effects. For example, sulfate and nitrate may attach to small particles in the air, facilitating their inhalation deep into lung tissue. There, they may do greater damage than a combination of the two pollutants would be expected to, based on their separate effects. This phenomenon has obvious health consequences; consider joggers breathing deeply and inhaling particulates as they run along the streets of a city. The effects of air pollutants on vertebrate animals in general include impairment of the respiratory system; damage to eyes, teeth, and bones; increased susceptibility to disease, parasites, and other stress-related environmental hazards; decreased availability of food sources (such as vegetation affected by air pollutants); and reduced ability for successful reproduction.[2]

Air-pollution deposits can also make soil and water toxic. In addition, soils may be leached of nutrients by pollutants that form acids. Air pollution's effects on human-made structures include discoloration, erosion, and decomposition of building materials (see the discussion of acid rain later in this chapter).

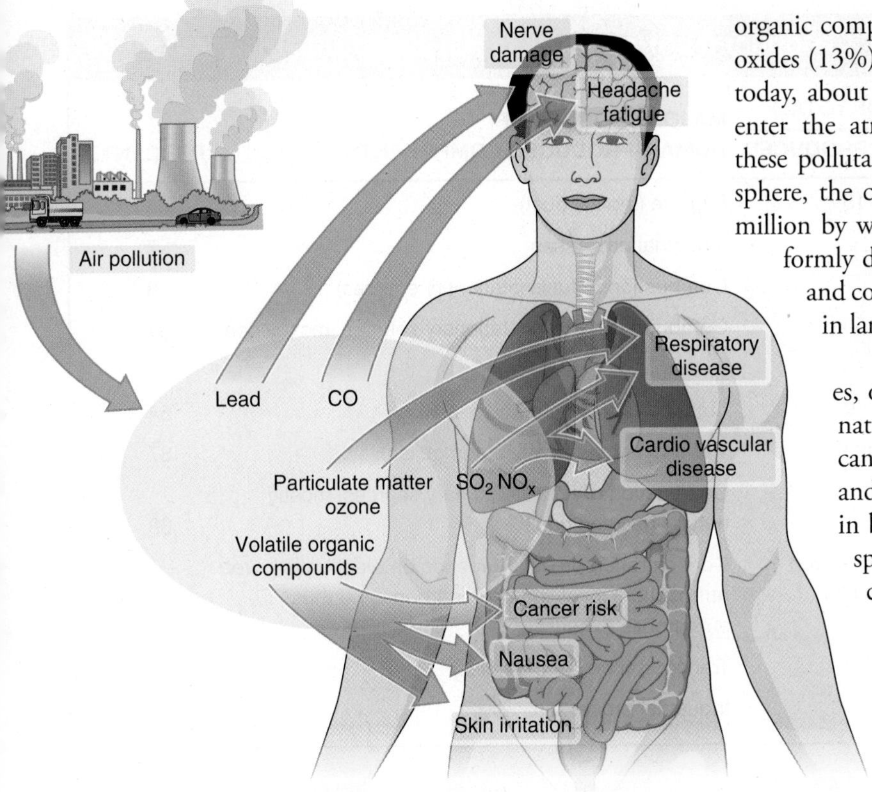

FIGURE 21.3 Idealized diagram showing some of the parts of the human body (brain, cardiovascular system, and pulmonary system) that can be damaged by common air pollutants. The most severe health risks from normal exposures are related to particulates. Other substances of concern include carbon monoxide, photochemical oxidants, sulfur dioxide, and nitrogen oxides. Toxic chemicals and tobacco smoke also can cause chronic or acute health problems.

The Major Air Pollutants

Nearly 200 air pollutants are recognized and assessed by the EPA and listed in the Clean Air Act. They can be classified as primary or secondary. **Primary pollutants** are emitted directly into the air. They include particulates, sulfur dioxide, carbon monoxide, nitrogen oxides, and hydrocarbons. **Secondary pollutants** are produced by reactions between primary pollutants and normal atmospheric compounds. For example, ozone forms over urban areas through reactions of primary pollutants, sunlight, and natural atmospheric gases. Thus, ozone is a secondary pollutant.

The major air pollutants occur either as particulate matter (PM) or in gaseous forms. Particulates are very small particles of solid or liquid substances and may be organic or inorganic. Gaseous pollutants include sulfur dioxide (SO_2), nitrogen oxides (NO_x), carbon monoxide (CO), ozone (O_3), and volatile organic compounds (VOCs), such as hydrocarbons (compounds containing only carbon and hydrogen that include petroleum products), hydrogen sulfide (H_2S), and hydrogen fluoride (HF).

The primary pollutants that account for nearly all air-pollution problems are carbon monoxide (58%), volatile organic compounds (11%), nitrogen oxides (15%), sulfur oxides (13%), and particulates (3%). In the United States today, about 140 million metric tons of these substances enter the atmosphere from human-related processes. If these pollutants were uniformly distributed in the atmosphere, the concentration would be only a few parts per million by weight. Unfortunately, pollutants are not uniformly distributed but tend to be produced, released, and concentrated locally or regionally—for example, in large cities.

In addition to pollutants from human sources, our atmosphere contains many pollutants of natural origin, such as sulfur dioxide from volcanic eruptions; hydrogen sulfide from geysers and hot springs, as well as from biological decay in bogs and marshes; ozone in the lower atmosphere as a result of unstable meteorological conditions, such as violent thunderstorms; a variety of particles from wildfires and windstorms;[2] and natural hydrocarbon seeps, such as La Brea Tar Pits in Los Angeles.

The data in Table 21.1 suggest that, except for sulfur and nitrogen oxides, natural emissions of air pollutants exceed human-produced emissions. Nevertheless, it is the human component that is most abundant in urban areas and leads to the most severe problems for human health.

Criteria Pollutants

The six most common pollutants are called **criteria pollutants** because the EPA has set specific limits on the levels of these six and they are responsible for most of our air-pollution problems. The six are sulfur dioxide, nitrogen oxides, carbon monoxide, ozone, particulates, and lead.

Sulfur Dioxide

Sulfur dioxide (SO_2) is a colorless and odorless gas normally present at Earth's surface in low concentrations. A significant feature of SO_2 is that once emitted into the atmosphere, it can be converted into fine particulate sulfate (SO_4) and removed from the atmosphere by wet or dry deposition. The major anthropogenic (human) source of sulfur dioxide is the burning of fossil fuels, mostly coal in power plants (see Table 21.1). Another major source comprises a variety of industrial processes, ranging from petroleum refining to the production of paper, cement, and aluminum.[1–4]

Adverse effects of sulfur dioxide depend on the dose or the concentrations (see Chapter 8) and include injury or death to animals and plants, as well as corrosion of paint and metals. Crops, such as alfalfa, cotton, and barley, are especially susceptible. Sulfur dioxide can severely damage the lungs of people and other animals, especially in the sulfate form. It is also an important precursor to acid rain (see A Closer Look 21.1).[1–4]

Table 21.1 MAJOR NATURAL AND HUMAN-PRODUCED COMPONENTS OF SELECTED AIR POLLUTANTS

AIR POLLUTANTS	EMISSIONS (% OF TOTAL) NATURAL	HUMAN PRODUCED	MAJOR SOURCES OF HUMAN-PRODUCED COMPONENTS	PERCENT
Particulates	85	15	Fugitive (mostly dust)	85
			Industrial processes	7
			Combustion of fuels (stationary sources)	8
Sulfur oxides (SO_x)	50	50	Combustion of fuels (stationary sources, mostly coal)	84
			Industrial processes	9
Carbon monoxide (CO)	91	9	Transportation (automobiles)	54
Nitrogen dioxide (NO_2)		Nearly all	Transportation (mostly automobiles)	37
			Combustion of fuels (stationary sources, mostly natural gas and coal)	38
Ozone (O_3)	A secondary pollutant derived from reaction with sunlight NO_2, and oxygen (O_2)		Concentration present depends on reaction in lower atmosphere involving hydrocarbons and thus automobile exhaust	
Hydrocarbons (HC)	84	16	Transportation (automobiles)	27
			Industrial processes	7

U.S. emission of SO_2 peaked at about 32 million tons in the early 1970s. Since then, emissions have fallen about 70% as a result of effective emission controls.[1]

Nitrogen Oxides

Although nitrogen oxides (NO_x) occur in many forms in the atmosphere, they are emitted largely as nitric oxide (NO) and nitrogen dioxide (NO_2), and only these two forms are subject to emission regulations. The more important of the two is NO_2, a yellow-brown to reddish-brown gas. A major concern with NO_2 is that it may be converted by complex reactions in the atmosphere to an ion, NO_3^{2-}, within small water particulates, impairing visibility. As mentioned earlier, both NO and NO_2 are major contributors to smog, and NO_2 is also a major contributor to acid rain (see A Closer Look 21.1). Nitrogen oxides contribute to nutrient enrichment and eutrophication of water in ponds, lakes, rivers, and the ocean (see Chapter 19). Nearly all NO_2 is emitted from anthropogenic sources. The two main sources are automobiles and power plants that burn fossil fuels.[1,2]

Nitrogen oxides have various effects on people, including irritation of eyes, nose, throat, and lungs and increased susceptibility to viral infections, including influenza (which can cause bronchitis and pneumonia).[1,2] Dissolved in water, nitrogen oxides form acids that can harm vegetation. But when the oxides are converted to nitrates, they can promote plant growth.

U.S. emission rates of NO_x (Table 21.1) are primarily from combustion of fuels in power plants and vehicles. They have been reduced by about 52% since 1980.

Carbon Monoxide

Carbon monoxide (CO) is a colorless, odorless gas that, even at very low concentrations, is extremely toxic to humans and other animals. The high toxicity results from a physiological effect: Carbon monoxide and hemoglobin have a strong natural attraction for one another; if there is any carbon monoxide in the vicinity, the hemoglobin in our blood will take it up nearly 250 times faster than it will oxygen and carry mostly carbon monoxide, rather than oxygen, from the atmosphere to the internal organs. Effects range from dizziness and headaches to death. Many people have been accidentally asphyxiated by carbon monoxide from incomplete combustion of fuels in campers, tents, and houses. Carbon monoxide is particularly hazardous to people with heart disease, anemia, or respiratory disease. It may also cause birth defects, including mental retardation and impaired fetal growth. Its effects tend to be worse at higher altitudes, where oxygen levels are lower. Detectors (similar to smoke detectors) are now commonly used to warn people if CO in a building reaches a dangerous level.

Approximately 90% of the carbon monoxide in the atmosphere comes from natural sources. The other 10% comes mainly from fires, automobiles, and other sources of incomplete burning of organic compounds, but these are easily concentrated locally, especially by enclosures, so this 10% causes most of the health problems. Emissions of CO peaked in the early 1970s at about 200 million metric tons and declined about 70% by 2010. This significant reduction stemmed largely from cleaner burning engines despite an increased number of vehicles.

A CLOSER LOOK 21.1

Acid Rain

Acid rain is precipitation in which the pH is below 5.6. The pH, a measure of acidity and alkalinity, is the negative logarithm of the concentration of the hydrogen ion (H^+). Because the pH scale is logarithmic, a pH value of 3 is 10 times more acidic than a pH value of 4 and 100 times more acidic than a pH value of 5. Automobile battery acid has a pH value of 1. Many people are surprised to learn that all rainfall is slightly acidic; water reacts with atmospheric carbon dioxide to produce weak carbonic acid. Thus, pure rainfall has a pH of about 5.6, where 2 is highly acidic and 7 is neutral (see Figure 21.4). (Natural rainfall in tropical rain forests has been observed in some instances to have a pH of less than 5.6; this is probably related to acid precursors emitted by the trees.)

Acid rain includes both wet (rain, snow, fog) and dry (particulate) acidic depositions. The depositions occur near and downwind of areas where the burning of fossil fuels produces major emissions of sulfur dioxide (SO_2) and nitrogen oxides (NO_x). These oxides are the primary contributors to acid rain.

Acid rain has likely been a problem at least since the beginning of the Industrial Revolution. In recent decades, however, it has gained more and more attention, and today it is a major, global environmental problem affecting all industrial countries. In the United States, nearly all of the eastern states are affected, as well as West Coast urban centers, such as Seattle, San Francisco, and Los Angeles. The problem is also of great concern in Canada, Germany, Scandinavia, and Great Britain. Developing countries that rely heavily on coal, such as China, are facing serious acid rain problems as well.

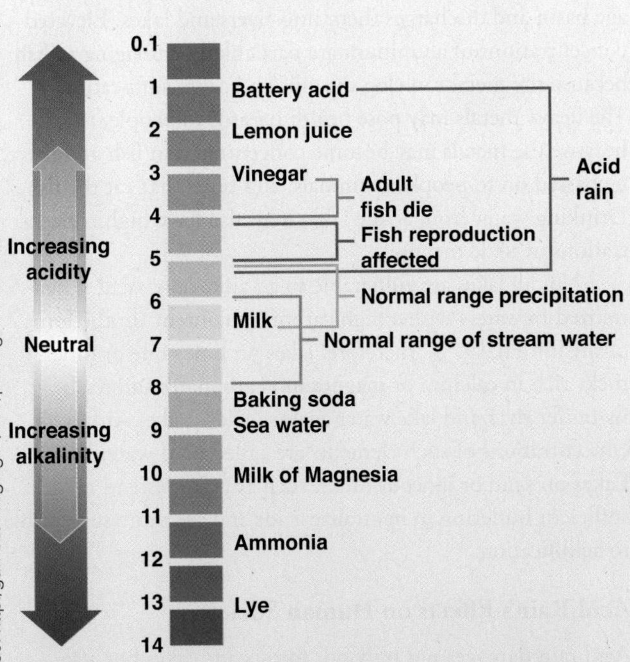

(*Source:* http://ga.water.usgs.gov/edu/phdiagram.html. Accessed March 25, 2013.)

FIGURE 21.4 The pH scale shows the levels of acidity in various fluids. The scale ranges from less than 1 to 14, with 7 being neutral: pHs lower than 7 are acidic; pHs greater than 7 are alkaline (basic). Acid rain can be very acidic and harmful to the environment.

Causes of Acid Rain

As we have said, sulfur dioxide and nitrogen oxides are the major contributors to acid rain. Emissions of SO_2 peaked in the 1970s and has declined since then by about 70% due to pollution control at power plants. Nitrogen oxides leveled off at about 25 million metric tons per year in the mid-1980s and had dropped by about 50% since 1980.

In the atmosphere, reactions with oxygen and water vapor transform SO_2 NO_x into sulfuric and nitric acids, which may travel long distances with prevailing winds and be deposited as acid precipitation—rainfall, snow, or fog (Figure 21.5). Sulfate and nitrate particles may also be deposited directly on the surface of the land as dry deposition and later be activated by moisture to become sulfuric and nitric acids.

Again, sulfur dioxide is emitted primarily by stationary sources, such as power plants that burn fossil fuels, whereas nitrogen oxides are emitted by both stationary and mobile sources, such as automobiles. Approximately 80% of sulfur dioxide and 65% of nitrogen oxides in the United States come from states east of the Mississippi River.

Sensitivity to Acid Rain

Geology, climate, vegetation, and soil help determine the effects of acid rain because these differ widely in their "buffers"—chemicals that can neutralize acids. Sensitive areas are those in which the type of bedrock (such as granite) or soils (such as those consisting largely of sand) cannot buffer acid input. Limestone bedrock provides the best buffering because it is made up mainly of calcium carbonate ($CaCO_3$), the mineral known as calcite. Calcium carbonate reacts with the hydrogen in the water and neutralizes the acid.

Soils may lose their fertility when exposed to acid rain, either because nutrients are leached out by acid water or because the acid in the soil releases elements that are toxic to plants.

Acid Rain's Effects on Forest Ecosystems

It has long been suspected that acid precipitation damages trees. Studies in Germany led scientists to cite acid rain and other air pollution as the cause of death for thousands of acres of evergreen trees in Bavaria. Similar studies in the

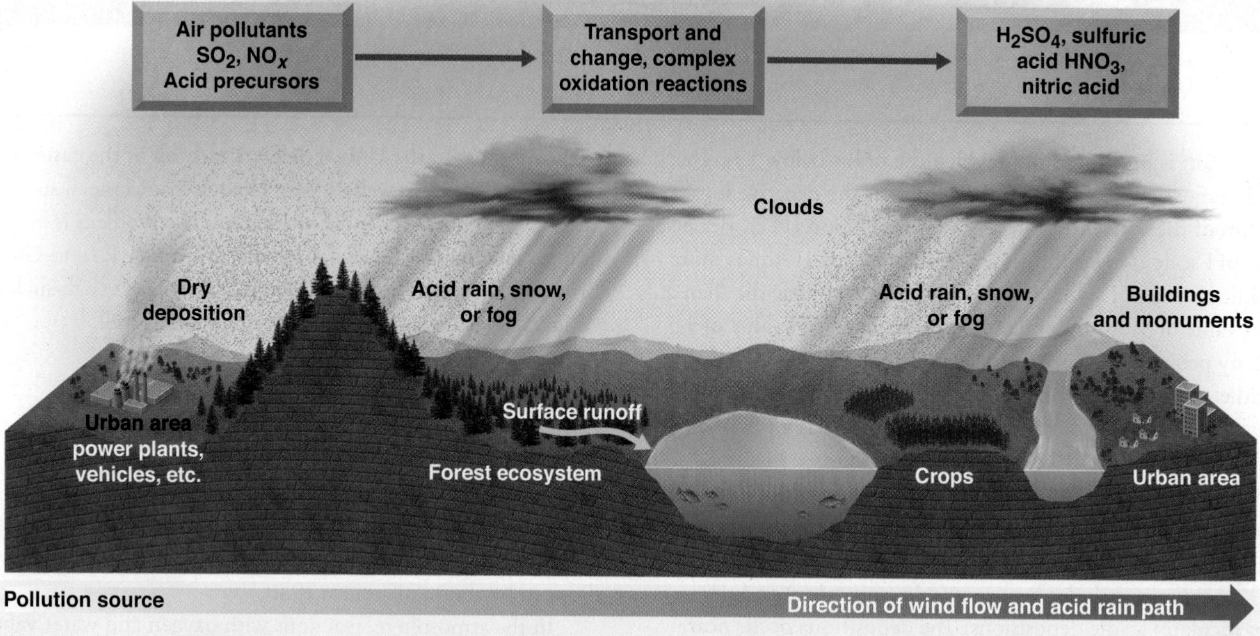

FIGURE 21.5 Idealized diagram showing selected aspects of acid rain formation and paths.

Appalachian Mountains of Vermont (where many soils are naturally acidic) suggest that in some locations half the red spruce trees have died in past years. Some high-elevation forests of the Appalachian Mountains, including the Great Smoky Mountains and Shenandoah National Park, have been impacted by acid rain, acid fog, and dry deposition of acid. Symptoms include slowed tree growth, leaves and needles that turn brown and fall off, and in extreme cases the death of trees. The acid rain does not directly kill trees; rather, it weakens them as essential nutrients are leached from soils or stripped from leaves by acid fog. Acidic rainfall also may release toxic chemicals, such as aluminum, that damage trees.[a]

Acid Rain's Effects on Lake Ecosystems

Records from Scandinavian lakes show an increase in acidity accompanied by a decrease in fish. The increased acidity has been traced to acid rain caused by industrial processes in other countries, particularly Germany and Great Britain. Thousands of lakes, ponds, and streams in the eastern United States are sensitive to acidification, including the Adirondacks and Catskill Mountains of New York State and others in the Midwest and in the mountains of the western United States. Little Echo Pond in Franklin, New York is one of the most acidic lakes, with a measured pH of 4.2.[a]

Acid rain affects lake ecosystems in three ways. First, it damages aquatic species (fish, amphibians, and crayfish) directly by disrupting their life processes in ways that limit growth or cause death. For example, crayfish produce fewer eggs in acidic water, and the eggs produced often grow into malformed larvae.

Second, acid rain dissolves chemical elements necessary for life in the lake. Once in solution, the necessary elements leave the lake with water outflow. Thus, elements that once cycled in the lake are lost. Without these nutrients, algae do not grow, animals that feed on the algae have little to eat, and animals that feed on these animals also have less food.[a,b]

Third, acid rain leaches metals, such as aluminum, lead, mercury, and calcium, from the soils and rocks in a drainage basin and discharges them into rivers and lakes. Elevated concentrations of aluminum are particularly damaging to fish because the metal can clog the gills and cause suffocation. The heavy metals may pose health hazards to people, too, because the metals may become concentrated in fish and then be passed on to people, mammals, and birds that eat the fish. Drinking water from acidic lakes may also have high concentrations of toxic metals.

Not all lakes are vulnerable to acidification. Acid is neutralized in waters with a high carbonate content (in the form of the ion HCO_3^-). Therefore, lakes on limestone or other rocks rich in calcium or magnesium carbonates can readily buffer river and lake water against acids. Lakes with high concentrations of such elements are called hard water lakes. Lakes on sand or igneous rocks, such as granite, tend to lack sufficient buffering to neutralize acids and are more susceptible to acidification.[c]

Acid Rain's Effects on Human Society

Acid rain damages not only our forests and lakes but also many building materials, including steel, galvanized steel, paint, plastics, cement, masonry, and several types of rock, especially limestone, sandstone, and marble. Classical buildings on the Acropolis in Athens and in other cities show considerable decay (chemical weathering) that accelerated in the 20th century as a result of air pollution. The problem has grown to such an extent that buildings require restoration, and the protective coatings on statues and other monuments must be replaced quite frequently, at a cost of billions of dollars a year. Particularly important statues in Greece and other areas have been removed and placed in protective glass containers, leaving replicas standing in their former outdoor locations for tourists to view.[b]

Stone decays about twice as rapidly in cities as it does in less urban areas. The damage comes mainly from acid rain and humidity in the atmosphere, as well as from corrosive groundwater.[d] This implies that measuring rates of stone decay will tell us something about changes in the acidity of rain and groundwater in different regions and ages. It is now possible, where the ages of stone buildings and other structures are known, to determine whether the acid rain problem has changed over time.

Control of Acid Rain

Acid rain is becoming an environmental success story. We know what causes acid precipitation—the solution is what we are struggling with. One solution to lake acidification is rehabilitation by the periodic addition of lime, as has been done in New York State, Sweden, and Ontario. This solution is not satisfactory over a long period, however, because the continuing effort is expensive. A better approach is to target the components of acid rain, the emissions of sulfur dioxide and nitrogen oxides. As noted, sulfur dioxide emissions in the United States are down about 70% since 1970—a big improvement that is significantly reducing acid rain. Emissions were lowered by a market-based SO_2 cap and trade program of the U.S. Environmental Protection Agency's Acid Rain Program, by which utilities receive pollution allowances that they can trade or sell if they lower emissions from their power plants (see Chapters 3 and 20 for discussions of cap-and-trade).[e]

Ozone and Other Photochemical Oxidants

Photochemical oxidants are secondary pollutants arising from atmospheric interactions of nitrogen dioxide and sunlight. Ozone, of primary concern here, is a form of oxygen in which three atoms of oxygen occur together rather than the normal two. A number of other photochemical oxidants, known as PANs (peroxyacyl nitrates), occur with photochemical smog.

Ozone is relatively unstable and releases its third oxygen atom readily, so it oxidizes or burns things more readily and at lower concentrations than does normal oxygen. Released into the air or produced in the air, ozone may injure living things. However, since these include bacteria and other organisms, it is sometimes used for sterilizing purposes—for example, bubbling ozone gas through water is one way to purify water.

Ozone in the lower atmosphere is a secondary pollutant produced on bright, sunny days in areas where there is significant primary pollution. The major sources of ozone, as well as other oxidants, are automobiles and industrial processes that release nitrogen dioxide by burning fossil fuels. Because of the nature of its formation, ozone is difficult to regulate and thus is the pollutant whose health standard is most frequently exceeded in U.S. urban areas.[5,6]

The adverse environmental effects of ozone and other oxidants, like those of other pollutants, depend in part on the dose or concentration of exposure and include damage to plants and animals, as well as to materials, such as rubber, paint, and textiles. Ozone's effects on plants can be subtle. At very low concentrations, it can slow growth without visible injury. At higher concentrations, it kills leaf tissue and, if pollutant levels remain high, whole plants. The death of white pine trees along highways in New England is believed to be due in part to ozone pollution.

Ozone is damaging trees in what at first glance would seem to be an unlikely place—California's Sequoia National Park. This park is in the Sierra Nevada, high above the Central Valley and the city of Fresno (population 500,000+). Fresno County is one of the fastest growing regions in the United States. Air pollutants (including ozone) collect in the natural depression of the Central Valley and, as in Los Angeles, the smog works its way up the mountains to elevations approaching 2 km (6,600 ft).

Sources of primary pollutants include freight trains, thousands of tractors and trucks, and the numerous vehicles in Fresno and and the valley that burn gasoline and/or diesel fuel. There are also agricultural chemical plants and processing plants.

Visibility from the park used to be more than160 km (100 mi)—across the agricultural and urbanizing valley to the coast ranges and even the Pacific Ocean. Today there are far fewer clear days. Giant Sequoia Redwood (*Sequoia gigantean*), some of the oldest (3,000+ years) and largest trees on the planet, form the groves of Sequoia National Park. The smog (particularly ozone) is harming redwood seedlings by inhibiting photosynthesis, making regeneration of the trees more difficult. In addition, the needles of Ponderosa and Jeffrey pines in the park are turning yellow because of ozone toxicity, which also inhibits photosynthesis.

Sequoia National Park has the worst air quality of all our national parks—and it is not in an urban area. The summer season is ozone season in the park, and in 2011 ozone levels exceeded health standards 87 days.

The ozone pollution problem in Sequoia is a difficult one and will not be reduced significantly until the pollution in the Central Valley is reduced—a difficult effort that will take a long time because there are so many sources.

Direct contact with ozone also damages animals, including people, especially from exposure to the eyes and respiratory system. Many millions of Americans are exposed to ozone levels that damage cell walls in lungs and airways. Tissue reddens and swells, and cellular fluids seep into the lungs. Eventually, the lungs lose elasticity and are more susceptible to bacterial infection; scars and lesions also may form in the airways. Even young, healthy people may be unable to breathe normally, and on especially polluted days breathing may be shallow and painful. In contrast to what is happening in Sequoia National Park, overall ground-level ozone in the United States has decreased by 9% since 1990.[1,2,6]

While too much ozone causes problems down here, too little of it has become a problem in the stratosphere. Because of the effect of sunlight on normal oxygen, ozone forms a natural layer high in the stratosphere that protects us from harmful ultraviolet radiation from the sun. However, the emission of certain chemicals in the lower atmosphere has led to serious ozone depletion in the stratosphere over the past few decades. The story of ozone (see A Closer Look 21.2) is becoming an environmental success story at the global level because:

- The problem was identified from ground and satellite measurements.
- Advances in atmospheric, physical, and chemical science have allowed us to understand ozone in the stratosphere.
- Management was put in place to eventually restore nature's ozone shield.

A CLOSER LOOK 21.2

High-Altitude (Stratospheric) Ozone Depletion

The problem of ozone depletion in the stratosphere starts down here in the lower atmosphere. About 21% of the air we breathe at sea level is *diatomic* oxygen (O_2), which is two oxygen atoms bonded together. **Ozone (O_3)** is a *triatomic* form of oxygen in which three atoms of oxygen are bonded. Ozone is a strong oxidant and reacts chemically with many materials in the atmosphere. In the lower atmosphere, as we have discussed, ozone is a pollutant produced by photochemical reactions involving sunlight, nitrogen oxides, hydrocarbons, and diatomic oxygen. In the stratosphere, however, ozone plays an entirely different role, protecting us from ultraviolet radiation. To reiterate, the stratosphere is a band of atmosphere about 9 to 25 km (5 to 15 mi) from the surface of the Earth.

Ultraviolet Radiation and Ozone

The ozone layer in the stratosphere is often called the **ozone shield** because it absorbs most of the potentially hazardous ultraviolet radiation that enters Earth's atmosphere from the sun. Ultraviolet radiation has wavelengths between 0.1 and 0.4 μm and is subdivided into ultraviolet A (UVA), ultraviolet B (UVB), and ultraviolet C (UVC). Ultraviolet radiation with a wavelength of less than about 0.3 μm can be very hazardous to life. If much of this radiation reached Earth's surface, it would injure or kill most living things.[a,b]

Ultraviolet C (UVC) has the shortest wavelength and is the most energetic of the three types. It has enough energy to break down diatomic oxygen (O_2) in the stratosphere into two oxygen atoms, each of which may combine with an O_2 molecule to create ozone. Ultraviolet C is strongly absorbed in the stratosphere, and negligible amounts reach Earth's surface.[a,b]

Ultraviolet A (UVA) radiation has the longest wavelength and the least energy of the three types. UVA can cause some damage to living cells, is not affected by stratospheric ozone, and is transmitted to the surface of Earth.[a]

Ultraviolet B (UVB) radiation is energetic and strongly absorbed by stratospheric ozone. In fact, ozone is the only known gas that absorbs UVB. Thus, depletion of ozone in the stratosphere allows more UVB to reach the Earth. Because UVB radiation is known to be hazardous to living things,[a,b,c] this increase in UVB is the hazard we are talking about when we discuss the problem of ozone depletion in the stratosphere.

The structure of the atmosphere and concentrations of ozone are shown in Figure 21.6. Approximately 90% of the ozone in the atmosphere is in the stratosphere, ranging from about 15 km to 40 km (9 to 25 mi) in altitude, with peak concentrations of about 400 ppb. The altitude of peak concentration varies from about 30 km (19 mi) near the equator to about 15 km (9 mi) in polar regions.[a]

Processes that produce ozone in the stratosphere are

- *Photodissociation*—intense ultraviolet radiation (UVC) breaks an oxygen molecule (O_2) into two oxygen atoms.
- These atoms then react with another oxygen molecule to form two ozone molecules. Ozone, once produced, may absorb UVC radiation, which breaks the ozone molecule into an oxygen molecule and an oxygen atom.

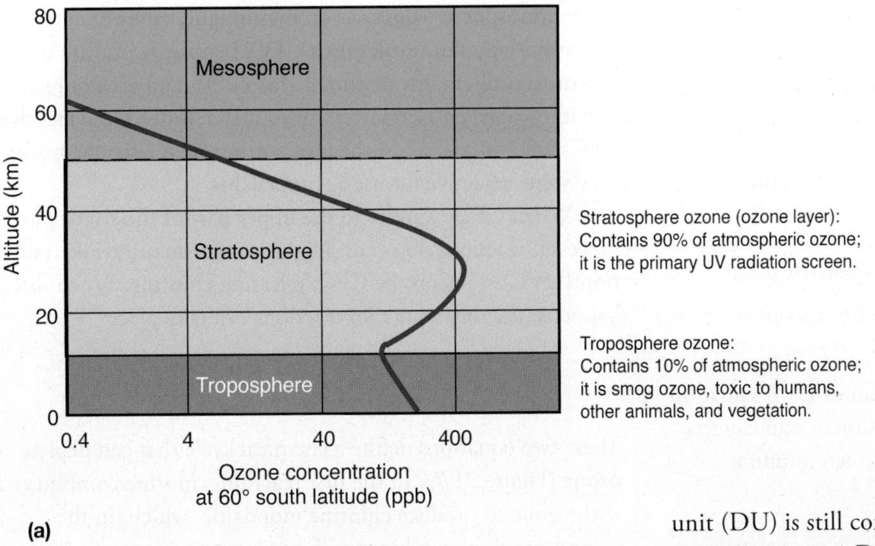

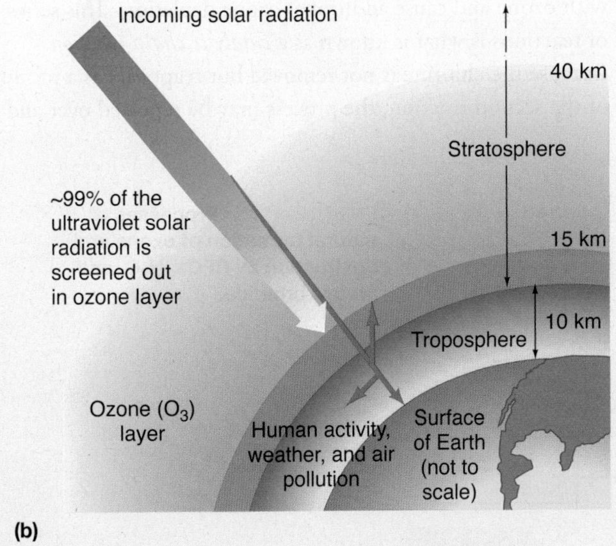

(*Source:* Ozone concentrations modified from R.T. Watson, "Atmospheric Ozone," in J.G. Titus, ed., *Effects of Change in Stratospheric Ozone and Global Climate*, vol. 1, *Overview*, p. 70 [U.S. Environmental Protection Agency].)

FIGURE 21.6 **(a) Structure of the atmosphere and ozone concentration.** (b) Reduction of the potentially most biologically damaging ultraviolet radiation by ozone in the stratosphere.

- This is followed by the recombination of the oxygen atom with another oxygen molecule to re-form into ozone. As part of this process, UVC radiation is converted to heat energy in the stratosphere. Natural conditions that prevail in the stratosphere result in a dynamic balance between the creation and destruction of ozone.

In sum, approximately 99% of all ultraviolet solar radiation (all UVC and most UVB) is absorbed or screened out in the ozone layer. The absorption of ultraviolet radiation by ozone is a natural service function of the ozone shield and protects us from the potentially harmful effects of ultraviolet radiation.

Measuring Stratospheric Ozone

Scientists first measured the concentration of atmospheric ozone in the 1920s from the ground, using an instrument known as a Dobson ultraviolet spectrometer. The Dobson unit (DU) is still commonly used to measure the ozone concentrations; 1 DU equals a concentration of 1 ppb O_3. Today, we have a record of ozone concentrations spanning about 50 years. Most of the measurement stations are in the midlatitudes, and the accuracy of the data varies with the levels of quality control.[a] Satellite measurements of atmospheric ozone concentrations began in 1970 and continue today.

Ground-based measurements first identified ozone depletion over the Antarctic. Members of the British Antarctic Survey began to measure ozone in 1957, and in 1985 they published the first data that suggested significant ozone depletion over Antarctica. The data are taken during October of each year—the Antarctic spring—and show that the concentration of ozone hovered around 300 DU from 1957 to about 1970, and then dropped sharply, to approximately 140 DU by 1986. Despite the variations, the direction of change, with minor exceptions, is clear: Ozone concentrations in the stratosphere during the Antarctic spring have been decreasing since the mid-1970s.[d–g] The depletion in ozone was dubbed the *ozone hole*. There is no actual hole in the ozone shield where all the ozone is depleted; rather, the term describes a relative depletion in the concentration of ozone that occurs during the Antarctic spring.

Ozone Depletion and CFCs

The hypothesis that ozone in the stratosphere is being depleted by **chlorofluorocarbons (CFCs)** was first suggested in 1974 by Mario Molina and F. Sherwood Rowland.[g] This hypothesis, based mostly on physical and chemical properties of CFCs and knowledge about atmospheric conditions, was immediately controversial and vigorously debated by scientists, companies producing CFCs, and other interested parties.[h,i] The major features of the Molina and Rowland hypothesis are as follows:[a,b]

- CFCs emitted in the lower atmosphere by human activity are very stable and nonreactive in the lower atmosphere and therefore have a very long residence time (about 100 years). No significant sinks for CFCs are known, with the possible exception of soils, which evidently do remove an unknown amount of CFCs from the atmosphere at Earth's surface. [i,j]

- Because of their long residence time in the lower atmosphere, and because the lower atmosphere is very fluid, the CFCs eventually disperse, wander upward, and enter the stratosphere. Once they reach altitudes above most of the stratospheric ozone, they may be destroyed by the highly energetic solar ultraviolet radiation. This releases chlorine, a highly reactive atom.
- The reactive chlorine may then enter into reactions that deplete ozone in the stratosphere.

Ozone depletion allows an increased amount of UVB radiation to reach Earth. Ultraviolet B is a cause of human skin cancers and is also believed to be harmful to the human immune system.

Simplified Stratospheric Chlorine Chemistry

CFCs are considered responsible for most of the ozone depletion. Let us look more closely at how it occurs.

Earlier, we noted that there are no tropospheric sinks for CFCs. That is, the processes that remove most chemicals in the lower atmosphere—destruction by sunlight, rain-out, and oxidation—do not break down CFCs because CFCs are transparent to sunlight, are essentially insoluble, and are nonreactive in the oxygen-rich lower atmosphere.[j] Indeed, the fact that CFCs are nonreactive in the lower atmosphere was one reason they were attractive for use as propellants.

When CFCs wander to the upper part of the stratosphere, however, reactions do occur. Highly energetic ultraviolet radiation (UVC) splits up the CFC, releasing chlorine. When this happens, the following two reactions can take place: [j]

$$(1)\ Cl + O_3 \longrightarrow ClO + O_2$$
$$(2)\ ClO + O \longrightarrow Cl + O_2$$

These two equations define a chemical cycle that can deplete ozone (Figure 21.7). In the first reaction, chlorine combines with ozone to produce chlorine monoxide, which, in the second reaction, combines with monatomic oxygen to produce chlorine again. The chlorine can then enter another reaction with ozone and cause additional ozone depletion. This series of reactions is what is known as a *catalytic chain reaction.* Because the chlorine is not removed but reappears as a product of the second reaction, the process may be repeated over and

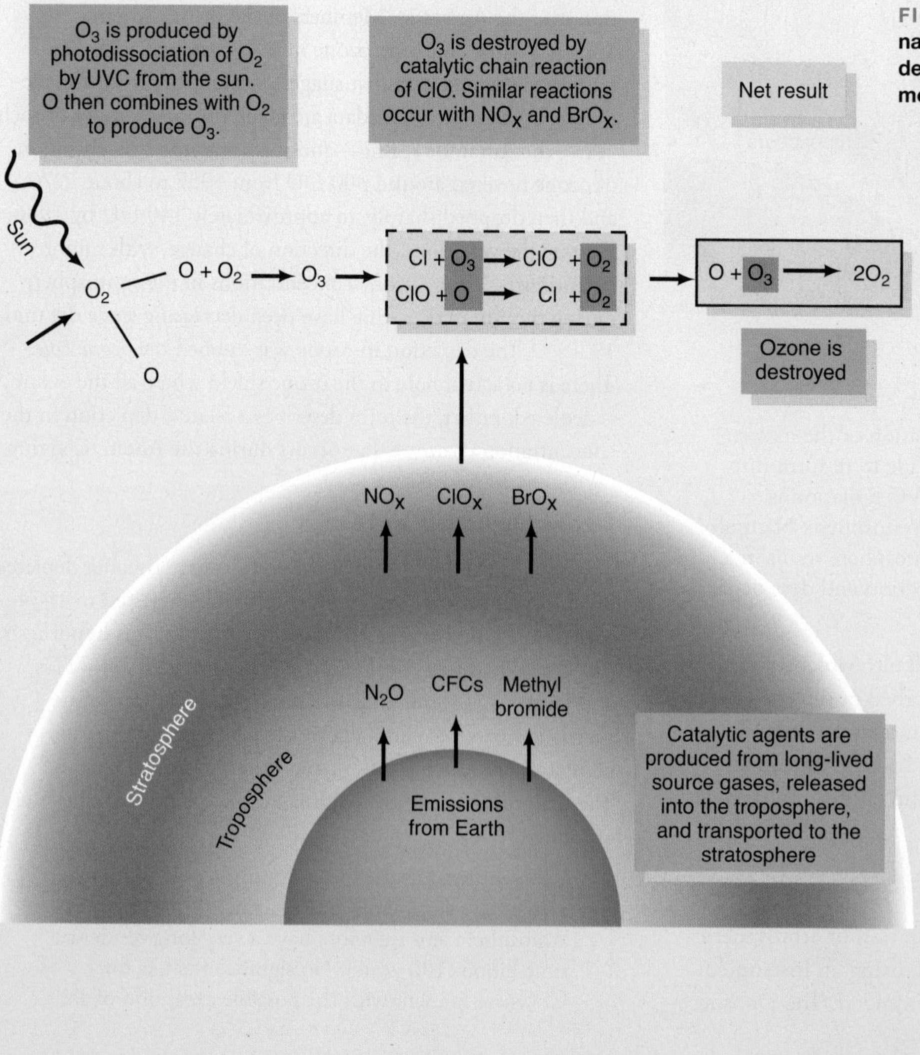
O3 is produced by photodissociation of O2 by UVC from the sun. O then combines with O2 to produce O3.
O3 is destroyed by catalytic chain reaction of ClO. Similar reactions occur with NOx and BrOx.
Net result
Sun
O2
O + O2 → O3 →
O
Cl + O3 → ClO + O2
ClO + O → Cl + O2
O + O3 → 2O2
Ozone is destroyed
NOx ClOx BrOx
N2O CFCs Methyl bromide
Stratosphere
Troposphere
Emissions from Earth
Catalytic agents are produced from long-lived source gases, released into the troposphere, and transported to the stratosphere

FIGURE 21.7 Processes of natural formation of ozone and destruction by CFCs, N_2O, and methyl bromide.

(*Source:* Modified from NASA-GSFC, "*Stratospheric Ozone*," accessed August 22, 2000, at http://see.gsfc.nasa.gov.)

over again. It has been estimated that each chlorine atom may destroy approximately 100,000 molecules of ozone in one or two years before the chlorine is finally removed from the stratosphere through other chemical reactions and rain-out.[j] The significance of these reactions is apparent when we realize how many metric tons of CFCs have been emitted into the atmosphere.

It should be noted that what actually happens chemically in the stratosphere is considerably more complex than the two equations shown here. The atmosphere is essentially a chemical soup in which a variety of processes related to aerosols and clouds take place (some of these processes are addressed in the discussion of the ozone hole). Nevertheless, these equations show us the basic chemical chain reaction that occurs in the stratosphere to deplete ozone.

The catalytic chain reaction just described can be interrupted through storage of chlorine in other compounds in the stratosphere. Two possibilities are as follows:

- Ultraviolet light breaks down CFCs to release chlorine, which combines with ozone to form chlorine monoxide (ClO), as already described. This is the first reaction discussed. The chlorine monoxide may then react with nitrogen dioxide (NO_2) to form a chlorine nitrate ($ClONO_2$). If this reaction occurs, ozone depletion is minimal. The chlorine nitrate, however, is only a temporary reservoir for chlorine. The compound may be destroyed, and the chlorine released again.
- Chlorine released from CFCs combine with methane (CH_4) to form hydrochloric acid (HCl). The hydrochloric acid may then diffuse downward. If it enters the troposphere, rain may remove it, thus removing the chlorine from the ozone-destroying chain reaction. This is the ultimate end for most chlorine atoms in the stratosphere. However, while the hydrochloric acid molecule is in the stratosphere, it may be destroyed by incoming solar radiation, releasing the chlorine for additional ozone depletion.

It has been estimated that the chlorine chain reaction that destroys ozone may be interrupted by the processes just described as many as 200 times while a chlorine atom is in the stratosphere.[a,k]

It is important to remember that for the Southern Hemisphere and under natural conditions, the highest concentration of ozone is in the polar regions (about 60° south latitude) and the lowest near the equator. At first this may seem strange because ozone is produced in the stratosphere by solar energy, and there is more solar energy near the equator. But although much of the world's ozone is produced near the equator, the ozone in the stratosphere moves from the equator toward the poles with global air-circulation patterns.[d]

In part as a result of ozone depletion, concentrations of ozone have declined in both northern and southern temperate latitudes. While remaining relatively constant at the equator, ozone has been significantly reduced in the Antarctic since the 1970s. Massive destruction of ozone in the Antarctic constitutes the "ozone hole."[a]

The Antarctic Ozone Hole

Since the Antarctic ozone hole was first reported in 1985, it has captured the interest of many people around the world. Every year since then, ozone depletion has been observed in the Antarctic in October, the spring season there. Because the thickness of the ozone layer above the Antarctic in springtime has been declining since the mid-1970s, the geographic area covered by the ozone hole has grown from a million or so square kilometers in the late 1970s and early 1980s to about 29 million km^2 by 1995—about the size of North America in 2000. It has since stabilized as the ozone concentration has ceased its steep decline.[d,l] In 2012 the size of the ozone hole was about 18 million km^2, the second smallest in 20 years.

Polar Stratospheric Clouds

The minimum concentration of ozone in the Antarctic since 1980 has varied from about 50% to 70% of that in the 1970s. Polar stratospheric clouds over the Antarctic appear to be one of the causes of this variation. Observed for at least the past hundred years about 20 km (12 mi) above the polar regions, the clouds have an eerie beauty and an iridescent glow, reminiscent of mother-of-pearl.[l] They form during the polar winter (called the polar night because the tilt of Earth's axis limits sunlight). During the polar winter, the Antarctic air mass is isolated from the rest of the atmosphere and circulates about the pole in what is known as the Antarctic *polar vortex*. The vortex forms as the isolated air mass cools, condenses, and descends.[d,e]

Clouds form in the vortex when the air mass reaches a temperature between 195 K and 190 K (-78° to -83°C; -108° to -117°F). At these very low temperatures, small sulfuric acid particles (approximately 01. μm) freeze and serve as seed particles for nitric acid (HNO_3). These clouds are called Type I polar stratospheric clouds. If temperatures drop below 190 K (-83°C; -117°F), water vapor condenses around some of the earlier-formed Type I cloud particles, forming Type II polar stratospheric clouds, which contain larger particles. Type II polar stratospheric clouds are the ones with the mother-of-pearl color.

During the formation of polar stratospheric clouds, nearly all the nitrogen oxides in the air mass are converted to the clouds as nitric acid particles, which grow heavy and descend below the stratosphere, leaving very little nitrogen oxide in the vicinity of the clouds.[a,k,l] This facilitates ozone-depleting reactions that may ultimately reduce stratospheric ozone in the polar vortex by as much as 1% to 2% per day in the early spring, when sunlight returns to the polar region (Figure 21.8).

An idealized diagram of the polar vortex that forms over Antarctica is shown in Figure 21.8a. The ozone-depleting reactions within the vortex are illustrated in Figure 21.8b. As shown, in the dark Antarctic winter almost all available nitrogen oxides are tied up on the edges of particles in the polar stratospheric clouds or have settled out. Hydrochloric acid and chlorine nitrate (the two important sinks of chlorine) act on particles of polar stratospheric clouds to form dimolecular chlorine (Cl_2) and nitric acid through the following reaction:[m]

$$HCl + ClONO_2 \longrightarrow Cl_2 + HNO_3$$

In the spring, when sunlight returns and breaks apart chlorine (Cl_2), the ozone-depleting reactions discussed earlier occur. Nitrogen oxides are absent from the Antarctic stratosphere in the spring, so the chlorine cannot be sequestered to form chlorine nitrate, one of its major sinks, and remains free to destroy ozone. In the early Antarctic spring, these ozone-depleting reactions can be rapid, producing the 50% reduction in ozone observed in recent years. Ozone depletion in the Antarctic vortex ceases later in spring as the environment warms and the polar stratospheric clouds disappear, releasing nitrogen back into the atmosphere, where it can combine with chlorine and thus be removed from ozone-depleting reactions. Stratospheric ozone concentrations then increase as ozone-rich air masses again migrate to the polar region.

A weaker, shorter polar vortex forms over the North Pole area and can lead to ozone depletion of as much as 30–40%. When the vortex breaks up, it can send ozone-deficient air masses southward to drift over areas of Europe and North America.[n]

Environmental Effects of Ozone Depletion

Ozone depletion damages some food chains on land and in the oceans and is dangerous to people, increasing the incidence of

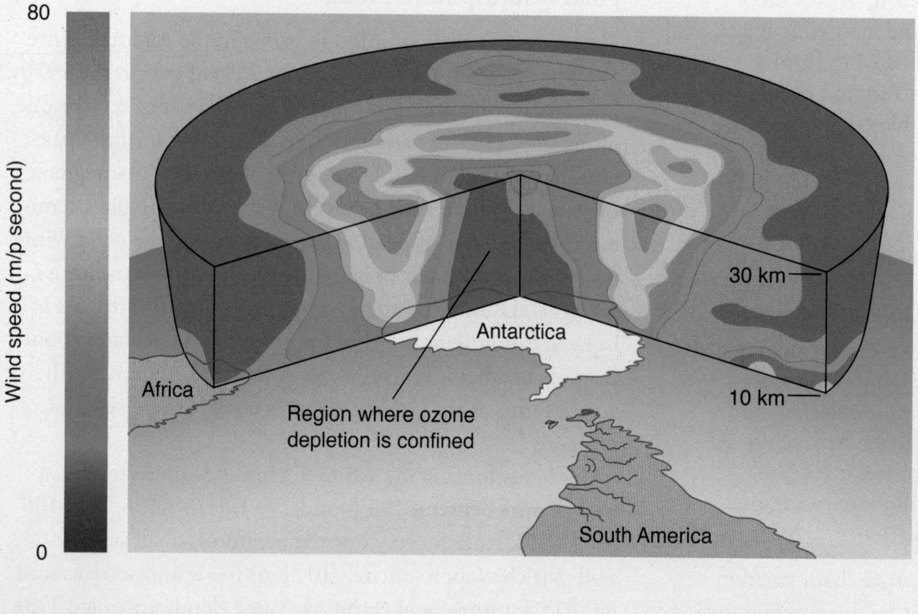

(a)

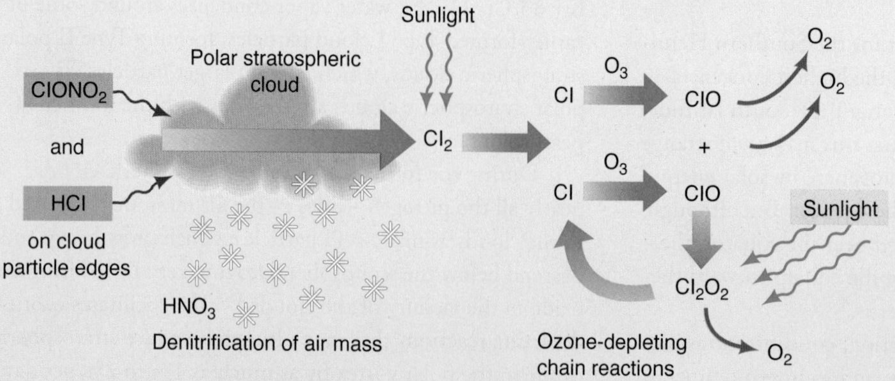

(b)

FIGURE 21.8 (a) Idealized diagram of the Antarctic polar vortex and (b) the role of polar stratospheric clouds in the ozone-depletion chain reaction.

(*Source:* Based on O.B. Toon and R.P. Turco, "Polar Stratospheric Clouds and Ozone Depletion," *Scientific American*, 264, no. 6 [1991]:68–74.)

skin cancers and cataracts and suppressing immune systems.[o,p] A 1% decrease in ozone can cause a 1–2% increase in UVB radiation and a 2% increase in skin cancer.[p] Because skin cancers have increased globally, health-conscious people today are replacing tanning oils with sunblocks and hats, and newspapers in the United States now provide the **Ultraviolet (UV) Index** (Table 21.2). Developed by the National Weather Service and EPA, the index predicts UV intensity on a scale from 1 to 11+ Some news agencies also use the index to recommend the level of sunblock. It is speculated that the incidence of skin cancer due to ozone depletion will rise until about 2060 and then decline as the ozone shield recovers as a result of controls on CFC emissions.[q,r]

You can lower your risk of skin cancer and other skin damage from UV exposure by taking a few simple precautions:

- Limit exposure to the sun between 10 A.M. and 4 P.M., the hours of intense solar radiation, and stay in the shade when possible.
- Use a sunscreen with an SPF of at least 30 (but remember that protection diminishes with increased exposure), or use clothing to cover up.
- Wear UV-protective sunglasses.
- Avoid tanning salons and sun lamps.
- Consult the UV Index before going out.

A simple guideline: If your shadow is longer than you are, such as in the evening or early morning, UV exposure is relatively low. If your shadow is shorter than you are, you are in the part of the day with highest UV exposure.

The Future of Ozone Depletion

The signing of the Montreal Protocol in September 1987 was an important diplomatic achievement: 27 nations signed the agreement originally, and an additional 119 signed later. The protocol outlined a plan to eventually reduce global emissions of CFCs to 50% of 1986 emissions. It originally called for eliminating production of CFCs by 1999, but the period was shortened because of scientific evidence that stratospheric ozone was being depleted faster than predicted. An eventual phase-out of all CFC consumption is part of the Montreal Protocol. Stratospheric concentrations of CFCs are expected to return to pre-1980 levels by about 2050, and the rate of increase of CFC emissions has already been reduced.[d,s,t] Of primary importance is developing substitutes for CFCs that are both safe and effective. Hydrofluorocarbons (HFCs) are the long-term substitute for CFCs because they do not contain chlorine.

However, a troubling aspect of ozone depletion is that if the manufacture, use, and emission of all ozone-depleting chemicals were to stop today, the problem would not go away—because millions of metric tons of those chemicals are now in the lower atmosphere, working their way up to the stratosphere. Several CFCs have atmospheric lifetimes of 75–140 years. Thus, an estimated 35% of the CFC-12 molecules in the atmosphere will likely still be there in 2100, and approximately 15% in 2200.[a] In addition, some 10–15% of the CFC molecules manufactured in recent years have not yet been admitted to the atmosphere because they remain in foam insulation, air-conditioning units, and refrigerators.[a] Nevertheless, indicators suggest that growth in the concentrations of CFCs has been slowed and in some cases reversed, and recovery of ozone should be noticeable by 2020 or later.[d]

Today by necessity, we are adapting to ozone depletion by learning to live with higher levels of exposure to ultraviolet radiation (for example, by using sunblock, wearing hats, and avoiding direct midday solar radiation). In the long term, achieving a sustainable level of stratospheric ozone will require management of human-made ozone-depleting chemicals.

Table 21.2 ULTRAVIOLET (UV) INDEX FOR HUMAN EXPOSURE

EXPOSURE CATEGORY	UV INDEX	COMMENT
Low	< 2	Sunblock recommended for all exposure
Moderate	3 to 5	Sunburn can occur quickly
High	6 to 7	Potentially hazardous
Very high	8 to 10	Potentially very hazardous
Extreme	11	Potentially very hazardous

Note: At moderate exposure to UV, sunburn can occur quickly, at high exposure, fair-skinned people may burn in 10 minutes or less of exposure.

Source: Modified after U.S. Environmental Protection Agency 2004 (with the National Weather Service). Accessed June 16, 2004 at www.epa.gov.

Particulate Matter: PM_{10}, $PM_{2.5}$, and Ultrafine Particles

Particulate matter (PM) is made up of tiny particles. The term *particulate matter* is used for varying mixtures of particles suspended in the air we breathe. In regulations these are divided into three categories: PM_{10}, particles up to 10 micrometers (μm) in diameter; $PM_{2.5}$, particles between 2.5 and 0.18 microns; and UP, **ultrafine particles** smaller than 0.18 micrometers in diameter, released into the air by vehicles on streets and freeways. For comparison, the diameter of a human hair is about 50 to 120 μm (Figure 21.9).

Nearly all industrial processes, as well as the burning of fossil fuels, release particulates into the atmosphere. Farming, too, adds considerable particulate matter to the atmosphere, as do windstorms in areas with little vegetation and volcanic eruptions. Particles are everywhere, and high concentrations and/or specific types of particles pose a serious danger to human health, including aggravation of cardiovascular and respiratory diseases. Major particulates include asbestos (especially dangerous, discussed in detail in Chapter 8)[1] and small particles of heavy metals, such as arsenic, copper, lead, and zinc, which are usually emitted from smelters and other industrial facilities. Particulates can reduce visibility and affect climate (see Chapter 20).[1] Much particulate matter is easily visible as smoke, soot, or dust; other particulate matter is not easily visible.

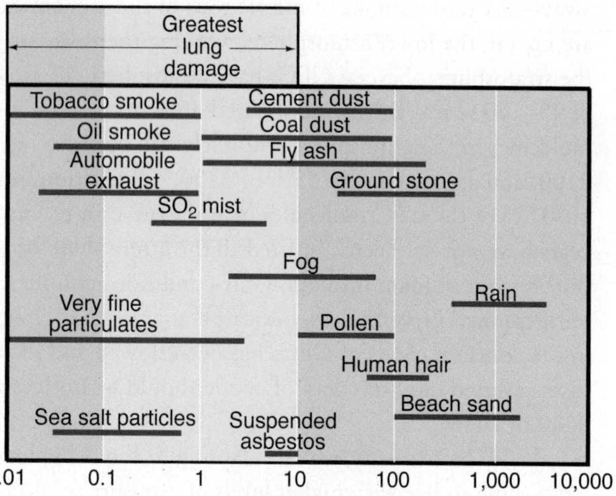

(a)

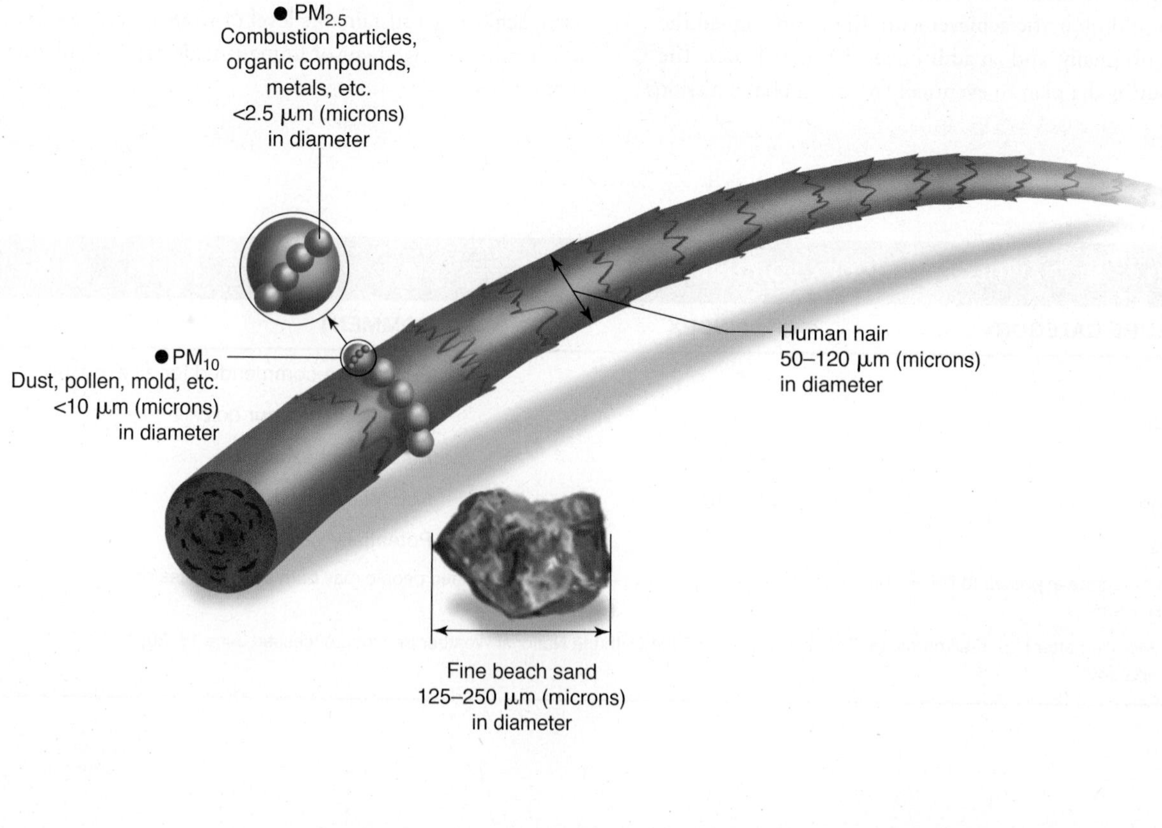

(b)

(*Source:* EPA, 2013.)

FIGURE 21.9 **(a) Sizes of selected particulates.** The shaded area shows the size range that produces the greatest lung damage.; (b) Idealized diagram showing relative sites of $PM_{2.5}$ and PM_{10}.

Fine particles—$PM_{2.5}$ and smaller—are easily inhaled into the lungs, where they can be absorbed into the bloodstream or remain embedded for a long time. Among the most significant of these particles are sulfates, nitrates, and soot. As already explained, with the exception of soot, these are mostly secondary pollutants produced in the atmosphere by chemical reactions between normal atmospheric constituents and sulfur dioxide and nitrogen oxides. These reactions are important in the formation of sulfuric and nitric acids in the atmosphere (and were discussed in A Closer Look 21.1 about acid rain).[1,2]

Ultrafine particles (UPs), released into the air by motor vehicles, are so small that they cannot be easily filtered and can enter the bloodstream. Rich in organic compounds and other reactive chemicals, they may be the most hazardous components of air pollution, especially with respect to heart disease. They evidently can contribute to inflammation (cell and tissue damage by oxidation), reducing the protective quality of "good" cholesterol and leading to plaque buildup in the arteries that can result in heart attack and stroke. Those most at risk are the young, the elderly, and individuals living near a freeway, or exercising near heavy traffic, or spending a lot of time in traffic (sitting in slow-moving traffic can roughly triple your short-term risk of a heart attack). The risk to an individual is very small, but when millions of people are exposed to a small risk, large numbers are affected. The prudent approach is to limit your exposure. For example, avoid jogging or bike riding near heavy traffic for extended periods.[3]

Particulate matter is measured as *total suspended particulates* (TSPs). Values for TSPs tend to be much higher in large cities in developing countries, such as Mexico, China, and India, than in developed countries, such as Japan and the United States.

Particulates, especially $PM_{2.5}$, affect human health, ecosystems, and the biosphere. $PM_{2.5}$ air pollution is one of the most serious air-pollution problems in the world today (see the opening case study). About 37 million people on Earth live in cities where $PM_{2.5}$ emissions exceed those recommended. In the United States, $PM_{2.5}$ emissions have declined about 55% since 1990. Nevertheless, particulate air pollution is estimated to contribute to the premature death of thousands annually in the United States.[7] Studies indicate that human mortality in cities is associated with particulate pollution, and that the mortality risk is about 15% to 25% higher in some cities with the highest levels of fine particulate pollution.[8] Particulates are linked to both lung cancer and bronchitis (see Figure 21.9) and are especially hazardous to the elderly and to people who have respiratory problems, such as asthma. There is a direct relationship between particulate pollution and increased hospital admissions for respiratory distress.

Dust raised by road building and plowing not only makes breathing more difficult for animals (including humans) but also can be deposited on green plants, interfering with absorption of carbon dioxide and oxygen and release of water (transpiration). On a larger scale, particulates associated with large construction projects—such as housing developments, shopping centers, and industrial parks—may injure or kill plants and animals and damage surrounding areas, changing species composition, altering food chains, and thereby affecting ecosystems. The terrorist attacks that destroyed the Twin Towers in New York City on September 11, 2001, sent huge amounts of particles of all sizes into the air, causing serious health problems that continue even today for people who were exposed to the particulates.

Modern industrial processes have greatly increased the total amount of suspended particulates in Earth's atmosphere. These particulates block sunlight and can cause **global dimming**, a gradual reduction in the solar energy that reaches Earth's surface. Global dimming cools the atmosphere and has lessened the global warming that has been predicted. Its effects are most apparent in the midlatitudes of the Northern Hemisphere, especially over urban regions or where jet air traffic is more common. Jet plane exhaust emits particulates high in the atmosphere. That this could affect the climate was suggested in 2001, when civil air traffic was shut down for two days after the September 11 attacks in New York. During those two days, the daily temperature range over the United States was about 1°C higher than usual.[9] Of course, this may have been just a coincidence.

Lead

Lead (a heavy metal) is an important constituent of automobile batteries and many other industrial products. Leaded gasoline (still used in some countries) helps protect engines and promotes more effective fuel consumption. However, the lead is emitted into the air with the exhaust and has thereby been spread widely around the world, reaching high levels in soils and waters along roadways. Once released, lead can be transported through the air as particulates to be taken up by plants through the soil or deposited directly on their leaves. Thus, it enters terrestrial food chains. When lead is carried by streams and rivers, deposited in quiet waters, or transported to oceans or lakes, it is taken up by aquatic organisms and enters aquatic food chains. Lead is toxic to wildlife and people. It can damage the nervous system, impair learning, and reduce IQ and memory. In children it can also contribute to behavioral problems. (Recall that this is the subject of the Critical Thinking section in Chapter 8.) In adults it can contribute to cardiovascular and kidney disease, as well as anemia.[1,2]

Lead reaches Greenland as airborne particulates and in seawater and is stored in glacial ice. The concentration of lead in Greenland glaciers was essentially zero in A.D. 800 and reached measurable levels with the beginning of the Industrial Revolution in the mid-18th century. The lead content of the glacial ice increased steadily from 1750 until about 1950, when there was a sudden upsurge in the rate of lead accumulation, reflecting rapid growth in the use of leaded gasoline. The accumulation of lead in Greenland's ice illustrates that our use of heavy metals in the 20th century reached the point of affecting the entire biosphere.

Lead has now been removed from nearly all gasoline in the United States, Canada, and much of Europe. In the United States, lead emissions have declined about 98% since the early 1980s.[1] The reduction and eventual elimination of lead in gasoline are a good start in reducing levels of anthropogenic lead in the biosphere.

Air Toxics

Toxic air pollutants, or **air toxics**, are among those pollutants known or suspected to cause cancer and other serious health problems after either long-term or short-term exposure. The most serious exposure to air toxics occurs in California and New York, with Oregon, Washington, DC, and New Jersey making up the rest of the top five. States with the cleanest air include Montana, Wyoming, and South Dakota.

Air toxics include gases, metals, and organic chemicals that are emitted in relatively small volumes. They cause respiratory, neurological, reproductive, or immune diseases, and some may be carcinogenic. The EPA estimates that the average risk of cancer from exposure to air toxics is about 1 in 21,000. The assessment concluded that benzene poses the most significant risk for cancer, accounting for 25% of the average individual cancer risk from all air toxics. Again, the effect on an individual's health depends on a number of factors, including duration and frequency of exposure, toxicity of the chemical, concentration of the pollutant the individual is exposed to, and method of exposure, as well as an individual's general health.[10]

Among the more than 150 known toxic air pollutants are hydrogen sulfide, hydrogen fluoride, various chlorine gases, benzene, methanol, and ammonia. In 2006 the EPA released an assessment of the national health risk from air toxics. It focused on exposure from breathing the pollutants; it did not address other ways people are exposed to them.

Standards and regulations established for more than 150 air toxics are expected to reduce annual emissions from 1990 levels. Even though vehicle miles will likely increase significantly by 2020, emissions of gaseous air toxics (such as benzene) from vehicles on highways are projected to decline about 80% from 1990 levels. Following are several examples of air toxics.

Hydrogen Sulfide

Hydrogen sulfide (H_2S) is a highly toxic corrosive gas, easily identified by its rotten egg odor. It is produced from natural sources, such as geysers, swamps, and bogs, and from human sources, such as petroleum refineries and metal smelters. The potential effects of hydrogen sulfide include functional damage to plants and health problems ranging from toxicity to death for humans and other animals.[4]

Hydrogen Fluoride

Hydrogen fluoride (HF) is a gas released by some industrial activities, such as aluminum production, coal gasification, and burning of coal in power plants. Hydrogen fluoride is extremely toxic; even a small concentration (as low as 1 ppb) may cause problems for plants and animals. It is potentially dangerous to grazing animals because some forage plants can become toxic when exposed to this gas.[2]

Mercury

Mercury is a heavy metal released into the atmosphere by coal-burning power plants, other industrial processes, and mining. Natural processes—such as volcanic eruptions and evaporation from soil, wetlands, and oceans—also release mercury into the air. Its toxicity to people is well-documented and includes neurological and development damage, as well as damage to the brain, liver, and kidneys. Mercury from the atmosphere may be deposited in rivers, ponds, lakes, and the ocean, where it accumulates through biomagnification and both wildlife and people are exposed to it.[1]

Volatile Organic Compounds

Volatile organic compounds (VOCs) include a variety of organic compounds. Some of these compounds are used as solvents in industrial processes, such as dry cleaning, degreasing, and graphic arts. Hydrocarbons (compounds of hydrogen and carbon) comprise one group of VOCs. Thousands of hydrocarbons exist, including natural gas, or methane (CH_4); butane (C_4H_{10}); and propane (C_3H_8). Analysis of urban air has identified many hydrocarbons, and their potential adverse effects are numerous. Some are toxic to plants and animals, and others may be converted to harmful compounds through complex chemical changes that occur in the atmosphere. Some react with sunlight to produce photochemical smog.

Globally, our activities produce only about 15% of hydrocarbon emissions. In the United States, however, nearly half the hydrocarbons entering the atmosphere are emitted from anthropogenic sources. The largest of these sources in the United States is automobiles. Anthropogenic sources are particularly abundant in urban regions. However, in some southeastern U.S. cities, such as Atlanta, Georgia, natural emissions (in Atlanta's case,

air pollution) has declined, even though the population nearly tripled and the number of motor vehicles quadrupled during this period. Nevertheless, exposure to ozone in southern California remains the nation's worst. Even if known and new controls in urban areas are implemented, air quality will continue to be a significant problem in coming decades, particularly if the urban population continues to increase.

We have focused on air pollution in southern California because its air quality is especially poor. However, most large and not-so-large U.S. cities have poor air quality for a significant part of the year. With the exception of the Pacific Northwest, no U.S. region is free from air pollution and its health effects.[6]

Developing Countries

The pessimistic view is that population pressures and environmentally unsound policies and practices will dictate what happens in many developing parts of the world, and the result will be poorer air quality. Developing countries often don't have the financial base necessary to fight air pollution and are more concerned about finding ways to house and feed their growing populations than about preserving their air quality.

Consider Mexico City. With a population of about 25 million, Mexico City is one of the four largest urban areas in the world. Cars, buses, industry, and power plants in the city emit hundreds of thousands of metric tons of pollutants into the atmosphere each year. The city is at an elevation of about 2,255 m (7,400 ft) in a natural basin surrounded by mountains, a perfect situation for a severe air-pollution problem. It is becoming a rare day in Mexico City when the mountains can be seen. Headaches, irritated eyes, and sore throats are common when the pollution settles in, and physicians report a steady increase in respiratory diseases. They advise parents to take their children out of the city permanently. The people in Mexico City do not need to be told they have an air-pollution problem; it is all too apparent. However, developing a successful strategy to improve the quality of the air is difficult.[14]

21.2 Controlling Common Pollutants of the Lower Atmosphere

The most reasonable ways to control the most common air pollutants in our cities include reducing emissions, capturing them before they reach the atmosphere, and removing them from the atmosphere. From an environmental viewpoint, reducing emissions through energy efficiency and conservation (such as burning less fuel) is the preferred strategy, with clear advantages over all other approaches (see Chapter 14). Here, we discuss control of selected air pollutants.

Particulates

Particulates emitted from fugitive, point, or area stationary sources are much easier to control than the very small particulates of primary or secondary origin released from mobile sources, such as automobiles. As we learn more about these very small particles, we will have to devise new methods to control them.

A variety of "settling chambers" or collectors are used to control emissions of coarse particulates from power plants and industrial sites (point or area sources) by providing a

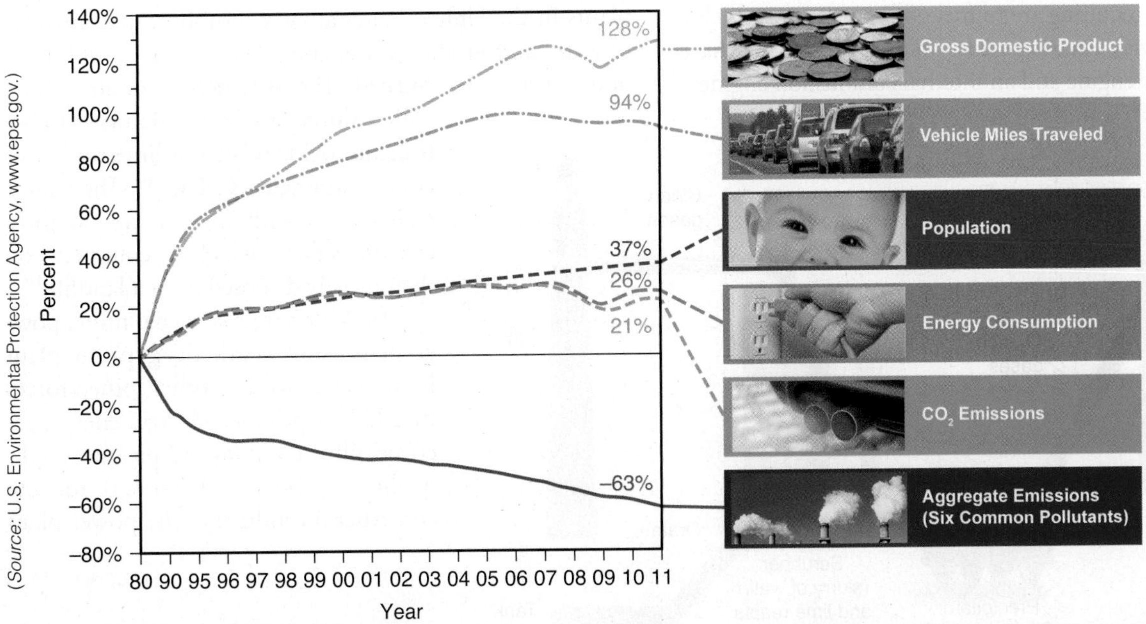

FIGURE 21.15 Change in U.S. population, gross domestic product, energy consumption, and aggregate emission of the six common air pollutants (ground-level ozone, particulates, lead, nitrogen dioxide, sulfur dioxide, and carbon monoxide).

mechanism that causes particles in gases to settle out in a location where they can be collected for disposal in landfills. In recent decades, we have made great strides in controlling particulates, such as ash, from power plants and industry.

Automobiles

Controlling such pollutants as carbon monoxide, nitrogen oxides, and hydrocarbons in urban areas is best achieved by pollution-control measures for automobiles. Control of these materials will also limit ozone formation in the lower atmosphere, since, as you have learned, ozone forms through reactions with nitrogen oxides and hydrocarbons in the presence of sunlight.

Nitrogen oxides from automobile exhausts are controlled by recirculating exhaust gas and diluting the air-to-fuel mixture burned in the engine. Dilution lowers the temperature of combustion and decreases the oxygen concentration in the burning mixture so that it produces fewer nitrogen oxides. Unfortunately, the same process increases hydrocarbon emissions. Nevertheless, exhaust recirculation to reduce nitrogen oxide emissions has been common practice in the United States for more than 20 years.[15]

The exhaust system's catalytic converter is the device most commonly used to reduce carbon monoxide and hydrocarbon emissions from automobiles. In the converter, oxygen from outside air is introduced, and exhaust gases from the engine are passed over a catalyst, typically platinum or palladium. Two important chemical reactions occur: Carbon monoxide is converted to carbon dioxide; and hydrocarbons are converted to carbon dioxide and water.

Other approaches to reducing air pollution from vehicles include reducing the number and types of cars on roads; developing cleaner fuels through use of fuel additives and reformulation; and requiring more fuel-efficient motor vehicles, such as those with electric engines and hybrid cars that have both an electric engine and an internal combustion engine.

Sulfur Dioxide

Sulfur dioxide emissions have been reduced by using abatement measures before, during, or after combustion. Technology to clean up coal so that it will burn more cleanly is already available. Although removing the sulfur makes fuel more expensive, the expense must be balanced against the long-term consequences of burning high sulfur coal. Switching from high-sulfur coal to low-sulfur coal seems an obvious way to reduce emissions of sulfur dioxide, and in some regions this will work. Unfortunately, however, most of the naturally low-sulfur coal in the United States is in the western part of the country, whereas most coal is burned in the East, so transportation is an issue, and using low-sulfur coal is a solution only in cases where it is economically feasible.

Sulfur emissions can also be reduced by washing coal. When finely ground coal is washed with water, iron sulfide (mineral pyrite) settles out because of its relatively high density. But this is ineffective for removing organic sulfur bound up with carbonaceous material, and it is expensive.

Another option is *coal gasification*, which converts relatively high-sulfur coal to a gas in order to remove the sulfur. The gas is quite clean and can be transported relatively easily, augmenting supplies of natural gas. True, it is still fairly expensive compared with gas from other sources, but its price may become more competitive in the future.

Desulfurization, or **scrubbing** (Figure 21.16), removes sulfur from stationary sources such as power plants. This technology was developed in the 1970s in the United States in response to passage of the Clean Air Act. However, the technology was not initially implemented in the United States; instead, regulators chose to allow plants to disperse pollutants through very tall smokestacks. This worsened the regional acid rain problem.

Nearly all scrubbers (90%) used at coal-burning power plants in the United States are wet scrubbers that use a lot of water and produce a wet end product. Wet scrubbing is done after coal is burned. The SO_2 rich gases are treated with a slurry (a watery mixture) of lime (calcium oxide, CaO) or limestone (calcium carbonate, $CaCO_3$). The sulfur oxides react with the calcium to form calcium sulfite, which is collected and then usually disposed of in a landfill.[16]

In West Virginia, a coal mine, power plant, and synthetic gypsum plant located close to each other joined forces in 2008 to produce electric energy, recover sulfur dioxide, and produce high-quality wallboard (sheetrock) for the construction industry. The power plant

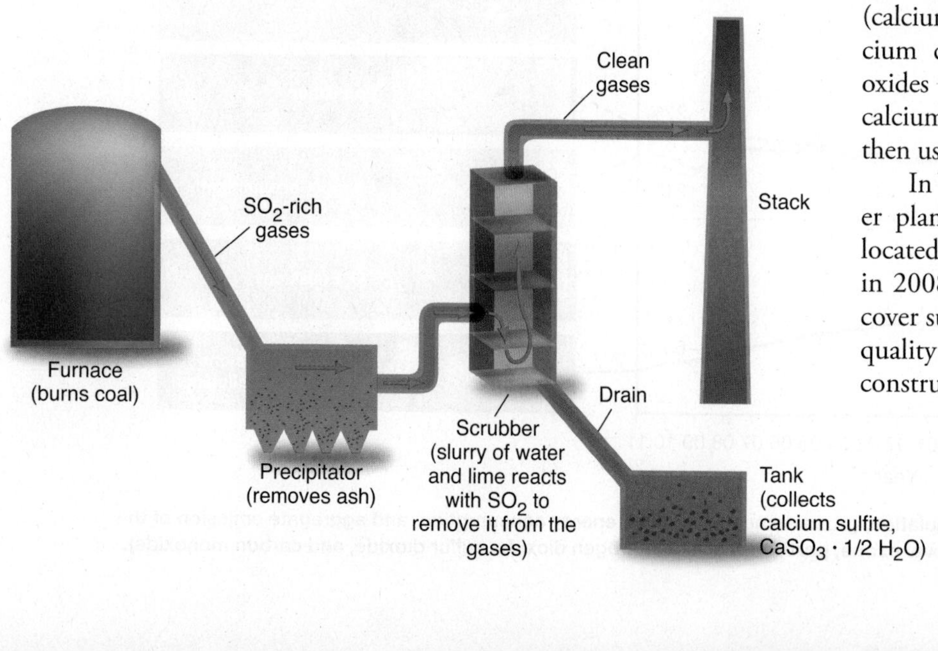

FIGURE 21.16 Scrubber used to remove sulfur oxides from the gases emitted by tall stacks.

benefits by selling the raw gypsum (from scrubbers) rather than paying to dispose of it in a landfill. The wallboard plant is right next to the power plant and uses gypsum that does not have to be mined from earth.[17]

Air-Pollution Legislation and Standards

Clean Air Act Amendments of 1990

The Clean Air Act Amendments of 1990 are comprehensive regulations enacted by the U.S. Congress that address the problems of acid rain, toxic emissions, ozone depletion, and automobile exhaust. In dealing with acid rain, the amendments establish limits on the maximum permissible emissions of sulfur dioxide from utility companies burning coal. The goal of the legislation—to reduce such emissions by about 50%, to 10 million tons a year, by 2000—was more than achieved.[1]

An innovative aspect of the legislation is the incentives it offers to utility companies to reduce emissions of sulfur dioxide. As explained earlier here and discussed in detail in Chapter 3, the incentives are marketable permits (allowances) that allow companies to buy and sell the right to pollute[18] (see A Closer Look 21.1). The 1990 amendments also call for reducing emissions of nitrogen dioxides by approximately 2 million tons from the 1980 level. The actual reduction has been 10 million tons—an air-pollution success story!

Ambient Air Quality Standards

Air quality standards are important because they are tied to emission standards that attempt to control the concentrations of various pollutants in the atmosphere. The many countries that have developed air quality standards include France, Japan, Israel, Italy, Canada, Germany, Norway, and the United States. National Ambient Air Quality Standards (NAAQS) for the United States, defined to comply with the Clean Air Act, are shown in Table 21.3. Tougher standards were set for ozone and $PM_{2.5}$ in recent years to reduce adverse health effects on children and elderly people, who are most susceptible to air pollution. The new standards are saving the lives of thousands and improving the health of hundreds of thousands of children. The ozone standard was significantly

Table 21.3 U.S. NATIONAL AMBIENT AIR QUALITY STANDARDS (NAAQS)

POLLUTANT	STANDARD VALUE[a]		STANDARD TYPE
Carbon monoxide (CO)			
8-hour average	9 ppm	(10 mg/m^3)	Primary [c]
1-hour average	35 ppm	(40 mg/m^3)	Primary
Nitrogen dioxide (NO_2)			
Annual arithmetic mean	0.053 ppm	(100 μg/m^3)	Primary and secondary [d]
Ozone (O_3)			
8-hour average	0.075 ppm	(147 μg/m^3)	Primary and secondary
Lead (Pb)			
Quarterly average	1.5 μg/m^3		Primary and secondary
Particulate (PM 10) *Particles with diameters of 10 micrometers or less*			
Annual arithmetic mean	50 μg/m^3		Primary and secondary
24-hour average	150 μg/m^3		Primary and secondary
Particulate (PM 2.5)[b] *Particles with diameters of 2.5 micrometers or less*			
Annual arithmetic mean	15 μg/m^3		Primary and secondary
24-hour average	65 μg/m^3		Primary and secondary
Sulfur dioxide (SO_2)			
Annual arithmetic mean	0.03 ppm	(80 μg/m^3)	Primary
24-hour average	0.14 ppm	(365 μg/m^3)	Primary
3-hour average	0.50 ppm	(1300 μg/m^3)	Secondary

[a] Parenthetical value is an approximately equivalent concentration.

[b] The ozone 8-hour standard and the PM 2.5 standards are included for information only. A 1999 federal court ruling blocked implementation of these standards, which the EPA proposed in 1997. EPA has asked the U.S. Supreme Court to reconsider that decision. (Note: In March 2001, the Court ruled in favor of the EPA, and the new standards are expected to take effect within a few years.)

[c] Primary standards set limits to protect public health, including the health of sensitive populations such as asthmatics, children, and the elderly.

[d] Secondary standards set limits to protect public welfare, including protection against decreased visibility and damage to animals, crops, vegetation, and buildings.

Source: U.S. Environmental Protection Agency.

Table 21.4 AIR QUALITY INDEX (AQI) AND HEALTH CONDITIONS

INDEX VALUES	DESCRIPTOR	CAUTIONARY STATEMENT	GENERAL ADVERSE HEALTH EFFECTS	ACTION LEVEL (AQI)[a]
0–50	Good	None	None	None
51–100	Moderate	Unusually sensitive people should consider limiting prolonged outdoor exertion.	Very few symptoms[b] for the most susceptible people[c]	None
101–150	Unhealthy for sensitive groups	Active children and adults, and people with respiratory disease, such as asthma, should limit prolonged outdoor exertion.	Mild aggravation of symptoms in susceptible people, few symptoms for healthy people	None
151–199	Unhealthy	Active children and adults, and people with respiratory disease, such as asthma, should avoid prolonged outdoor exertion; everyone else, especially children, should limit prolonged outdoor exertion.	Mild aggravation of symptoms in susceptible people, irritation symptoms for healthy people	None
200–300	Very unhealthy	Active children and adults, and people with respiratory disease, such as asthma, should avoid outdoor exertion; everyone else, especially children, should limit outdoor exertion.	Significant aggravation of symptoms in susceptible people, widespread symptoms in healthy people	Alert (200+)
Over 300	Hazardous	*Everyone* should avoid outdoor exertion.	300–400: Widespread symptoms in healthy people 400–500: Premature onset of some diseases Over 500: Premature death of ill and elderly people; healthy people experience symptoms that affect normal activity	Warning (300+) Emergency (400+)

[a] Triggers preventative action by state or local officials.

[b] Symptoms include eye, nose, and throat irritation; chest pain; breathing difficulty.

[c] Susceptible people are young, old, and ill people, and people with lung or heart disease.

AQI 51–100	Health advisories for susceptible individuals.
AQI 101–150	Health advisories for all.
AQI 151–200	Health advisories for all.
AQI 200+	Health advisories for all; triggers an alert; activities that cause pollution might be restricted.
AQI 300	Health advisories to all; triggers a warning; probably would require power plant operations to be reduced and carpooling to be used.
AQI 400+	Health advisories for all; triggers an emergency; cessation of most industrial and commercial activities, including power plants; nearly all private use of vehicles prohibited.

Source: U.S. Environmental Protection Agency.

strengthened in 2008. The change is expected to result in health benefits of more than $15 billion per year.

Air Quality Index

In the United States, the Air Quality Index (AQI) (Table 21.4) is used to describe air pollution on a given day. For example, air quality in urban areas is often reported as good, moderate, unhealthy for sensitive groups, unhealthy, very unhealthy, or hazardous, corresponding to a color code of the Air Quality Index. The AQI is determined by measuring the concentration of five major pollutants: particulate matter, sulfur dioxide, carbon monoxide, ozone, and nitrogen dioxide. An AQI value greater than 100 is unhealthy. In most U.S. cities, AQI values range between

0 and 100. Values above 100 are generally recorded for a particular city only a few times a year, but some cities with serious air pollution problems may exceed an AQI of 100 many times a year. In a typical year, AQI values above 200 (for all U.S. sites) are rare, and those above 300 are very rare. In large cities outside the United States with dense human populations and numerous uncontrolled sources of pollution, AQIs greater than 200 are frequent.

The Cost of Controlling Outdoor Air Pollution

The cost of outdoor air-pollution control varies widely from one industry to another. For example, the cost for incremental control in a fossil-fuel-burning utility is a few hundred dollars per additional ton of particulates removed. For an aluminum refinery, the cost to remove an additional ton of particulates may be as much as several thousand dollars. Some economists would argue that it is wise to raise the standards for utilities and relax them, or at least not raise them, for aluminum plants. This would lead to more cost-efficient pollution control while maintaining good air quality. However, the geographic distribution of various facilities will determine the possible trade-offs.[18,19]

Economic analysis of air pollution is not simple. There are many variables, some of which are hard to quantify. We do know the following:

- With increasing air-pollution controls, the capital cost for technology to control air pollution increases.
- As the controls for air pollution increase, the loss from pollution damages decreases.
- The total cost of air pollution is the cost of pollution-control plus the environmental damages of the pollution.

Although the cost of pollution abatement technology is fairly well known, it is difficult to accurately determine the loss from pollution damages, particularly when considering health problems and damage to vegetation, including food crops. For example, exposure to air pollution may cause or aggravate chronic respiratory diseases in people, at a very high cost. A recent study of the health benefits of cleaning up the air quality in the Los Angeles basin estimated that the annual cost of air pollution in the basin is 1,600 lives and about $10 billion.[20] Air pollution also leads to loss of revenue from people who choose not to visit some areas, such as Los Angeles and Mexico City, because of known air-pollution problems.[21,22]

Sustainable Air Quality

The U.S. Environmental Protection Agency (EPA) initiated a voluntary program, known as Sustainable Skylines whose objective is to achieve sustainable air quality by reducing the six major air pollutants, as well as other toxic air pollutants and greenhouse gases. Cities that participate in the program are encouraged to integrate energy, land use, transportation, and air quality planning in order to achieve measurable improvements within a three-year period. As of 2011, three cities have participated—Dallas, Texas, Kansas City, Kansas and Missouri (the greater Kansas City metropolitan area), and Philadelphia, Pennsylvania. Among the projects included in a particular Sustainable Skyline venture are the following:

- Reducing emissions from landscape equipment by improved irrigation of lawns and turf management, as well as retrofitting small off-road equipment to achieve reduced emissions of air pollutants.
- Reducing vehicle emissions by increasing public transportation and reducing the distances traveled in vehicles.
- Replacing existing taxis with "green taxis" that emit far less pollution.
- Encouraging "**green buildings**" with healthier interior environments and landscaping that benefit the local external environment.
- Reducing emissions from idling vehicles and retrofitting diesel engines to reduce emissions.
- Creating programs to encourage planting trees in the city to develop a tree canopy in as many areas as possible.

Each city that participates in the Sustainable Skylines Program will have its own local programs and policies, developed in collaboration with the city's inhabitants and city leaders, along with public and private partners. For example, in Dallas the description of activities has the goal of helping to reduce the urban "heat island" effect. Urban areas are often warmer than surrounding areas due to the abundance of equipment and lights, as well as surfaces that absorb heat. Cities with little vegetation also have less evaporative cooling. This is a particular problem in Dallas, which has a naturally warm climate much of the year. As a result, the goal of the Sustainable Skylines Program for Dallas is to increase the number of shaded surfaces and green vegetated surfaces of roofs and surrounding buildings in order to reduce the heat island effect and cool the city.

In the greater Kansas City area, the objectives are to encourage a variety of sustainable environmental projects with social benefits. The plan is to address such issues as transportation, energy, land use, resource efficiency, green buildings, and air quality, with a focus on projects that will result in cleaner, healthier air for this large urban area.

In Philadelphia, the first objective centers on climate and energy measures aimed at improving air quality. The city is merging the EPA initiative with the Philadelphia Greenworks Plan to transform Philadelphia into one of the "greenest" cities in the United States. Green, here, refers to sustainability of energy, environment, equity, and economy.

21.3 Indoor Air Pollution

We have discussed air pollution in the lower atmosphere and the depletion of stratospheric ozone by chemical emissions that rise from the lower atmosphere to cause depletion of O_3 in the stratosphere that produces a hazard from exposure to ultravioleted radiation from the sun. We turn next to air pollution in our homes, schools, and other buildings that we spend time in.

Indoor air pollution from fires for cooking and heating has affected human health for thousands of years. A detailed autopsy of a 4th-century Native American woman, frozen shortly after death, revealed that she suffered from **black lung disease** from breathing very polluted air over many years. The pollutants included hazardous particles from lamps that burned seal and whale blubber.[23] This same disease has long been recognized as a major health hazard for underground coal miners and has been called coal miners' disease. As recently as the mid-1970s, black lung disease was estimated to be responsible for about 4,000 deaths each year in the United States.[24]

People today spend between 70% and 90% of their time in enclosed places—homes, workplaces, automobiles, restaurants, and so forth—but only recently have we begun to fully study the indoor environment and how pollution of that environment affects our health. The World Health Organization has estimated that as many as one in three people may be working in a building that causes them to become sick, and as many as 20% of public schools in the United States have problems related to indoor air quality. The EPA considers indoor air pollution one of the most significant environmental health hazards people face in the modern workplace.[25]

Hurricane Katrina in 2005 (see Chapter 22) left a great number of people homeless. In response, the Federal Emergency Management Agency (FEMA) provided thousands of trailers for people to live in. That sounded like a great idea until complaints started to come in about health problems of people living in the trailers. A study by the Centers for Disease Control and Prevention (CDC) confirmed that the mobile homes suffered from indoor air pollution by formaldehyde in their construction materials. Formaldehyde is a chemical widely used in the manufacture of building materials, as well as a number of other products. It is considered a probable human carcinogen (a substance that causes or promotes cancer). Common symptoms of exposure to formaldehyde include irritation of the skin, nose, throat, and eyes. People with asthma may be more sensitive to the chemical, and their symptoms may be worse. Since discovery of the high levels of formaldehyde in mobile homes in late 2007, plans have gone forward to remove the remaining people, particularly those experiencing symptoms of formaldehyde toxicity.[26,27]

1. Heating, ventilation, and air-conditioning systems may be sources of indoor air pollutants, including molds and bacteria, if filters and equipment are not maintained properly. Gas and oil furnaces release carbon monoxide, nitrogen dioxide, and particles.
2. Restrooms may have a variety of indoor air pollutants, including secondhand smoke, and also molds and fungi due to humid conditions.
3. Furniture and carpets often contain toxic chemicals (formaldehyde, organic solvents, asbestos) that may be released over time in buildings.
4. Coffee machines, fax machines, computers, and printers can release particles and chemicals, including ozone (O_3), which is highly oxidizing.
5. Pesticides can contaminate buildings with cancer-causing chemicals.
6. Fresh-air intake that is poorly located—for example, above a loading dock or first-floor restaurant exhaust fan—can bring in air pollutants.
7. People who smoke indoors, perhaps in restaurants or offices, pollute the indoor environment, and even people who smoke outside buildings, particularly near open or revolving doors, may cause pollution as the smoke (secondhand smoke) is drawn into and up through the building by the chimney effect.
8. Remodeling, painting, and other such activities often bring a variety of chemicals and materials into a building. Fumes from such activities may enter the building's heating, ventilation, and air-conditioning system, causing widespread pollution.
9. A variety of cleaning products and solvents used in offices and other parts of buildings contain harmful chemicals whose fumes may circulate throughout a building.
10. People can increase carbon dioxide levels; they can emit bioeffluents and spread bacterial and viral contaminants.
11. Loading docks can be sources of organics from garbage containers, of particulates, and of carbon monoxide from vehicles.
12. Radon gas can seep into a building from soil; rising damp (water), which facilitates the growth of molds, can enter foundations and rise up walls.
13. Dust mites and molds can live in carpets and other indoor places.
14. Pollen can come from inside and outside sources.

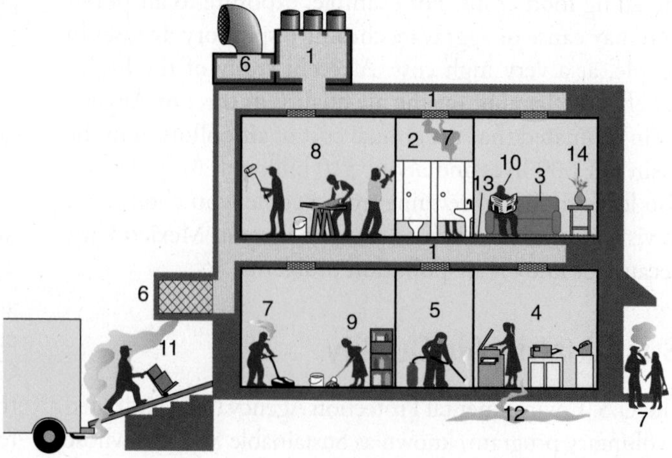

FIGURE 21.17 Some potential sources of indoor air pollution.

The history of formaldehyde in the mobile homes provided to Katrina victims is a sad legacy of the entire way our federal government responded to Hurricane Katrina and its aftermath. It is also important because it brings to the public consciousness the potential problems of indoor air pollution, which is often more significant than outdoor air pollution.

Sources of Indoor Air Pollution

The sources of indoor air pollution are incredibly varied (Figure 21.17) and can arise from both human activities and natural processes. Two common pollutants are shown in Figure 21.18. Other common indoor air pollutants, together with guidelines for allowable exposure, are listed in Table 21.5.

Table 21.5 SOURCES, CONCENTRATIONS, OCCURRENCES, AND POSSIBLE HEALTH EFFECTS OF INDOOR AIR POLLUTANTS

POLLUTANT	SOURCE	GUIDELINES (DOSE OR CONCENTRATIONS)	POSSIBLE HEALTH EFFECTS
Asbestos	Fireproofing; insulation, vinyl floor, and cement products; vehicle brake linings	0.2 fibers/mL for fibers larger than 5 μm	Skin irritation, lung cancer
Biological aerosols/ microorganisms	Infectious agents, bacteria in heating, ventilation, and air-conditioning systems: allergens	None available	Diseases, weakened immunity
Carbon dioxide	Motor vehicles, gas appliances, smoking	1,0000 ppm	Dizziness, headaches, nausea
Carbon monoxide	Motor vehicles, kerosene and gas space heaters, gas and wood stoves, fireplaces; smoking	10,000 $\mu g/m^2$ for 8 hours; 40,000 $\mu g/m^3$ for 1 hour	Dizziness, headaches, nausea, death
Formaldehyde	Foam insulation; plywood, particleboard, ceiling tile, paneling, and other construction materials	120 $\mu g/m^3$	Skin irritant, carcinogen
Inhalable particulates	Smoking, fireplaces, dust, combustion sources (wildfires, burning trash, etc.)	55-110 $\mu g/m^3$ annual; 350 $\mu g/m^2$ for 1 hour	Respiratory and mucous irritant, carcinogen
Inorganic particulates			
Nitrates	Outdoor air	None available	
Sulfates	Outdoor air	4 $\mu g/m^3$ annual; 12 $\mu g/m^3$ for 24 hours	
Metal particulates			
Arsenic	Smoking, pesticides, rodent poisons	None available	Toxic, carcinogen
Cadmium	Smoking, fungicides	2 $\mu g/m^3$ for 24 hours	
Lead	Automobile exhaust	1.5 $\mu g/m^3$ for 3 months	
Mercury	Old fungicides; fossil fuel combustion	2 $\mu g/m^3$ for 24 hours	
Nitrogen dioxide	Gas and kerosene space heaters, gas stoves, vehicular exhaust	100 $\mu g/m^3$ annual	Respiratory and mucous irritant
Ozone	Photocopying machines, electrostatic air cleaners, outdoor air	235 $\mu g/m^3$ for 1 hour	Respiratory irritant causes fatigue
Pesticides and other semivolatile organics	Sprays and strips, outdoor air	5 $\mu g/m^3$ for chlordane	Possible carcinogens
Radon	Soil gas that enters buildings, construction materials, groundwater	4pCi/L	Lung cancer
Sulfur dioxide	Coal and oil combustion, kerosene space heaters, outside air	80 $\mu g/m^3$ annual; 365 $\mu g/m^3$ for 24 hours	Respiratory and mucous irritant
Volatile organics	Smoking, cooking, solvents, paints, varnishes, cleaning sprays, carpets, furniture, draperies, clothing	None available	Possibe carcinogens

Source: N. L. Nagda, H. E. Rector, and M. D. Koontz, 1987; M. C. Baechler et al., 1991; E. J. Bardana Jr. and A. Montaro (eds.), 1997; M. Meeker,1996; D. W. Moffatt, 1997.

Andrew Syred/Photo Researchers Oliver Meckes/Photo Researchers

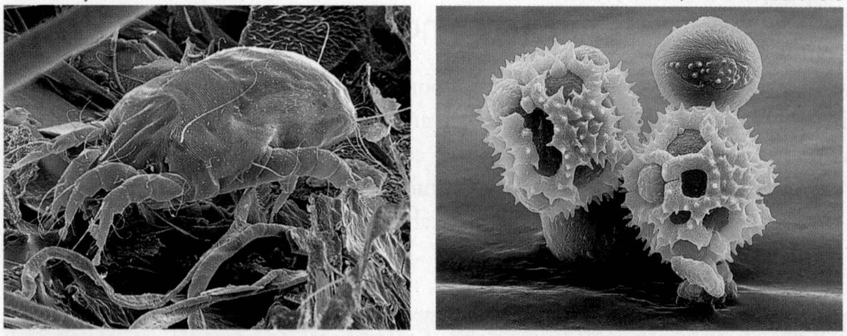

FIGURE 21.18 **(a) This dust mite (magnified about 140 times) is an eight-legged relative of spiders.** It feeds on human skin in household dust and lives in materials such as fabrics on furniture. Dead dust mites and their excrement can cause allergic reactions and asthma attacks in some people. **(b)** Microscopic pollen grains that in large amounts may be visible as a brown or yellow powder. The pollen shown here is from dandelions and horse chestnuts.

Many products and processes used in our homes and workplaces are sources of pollution. Other air pollutants—such as carbon monoxide, particulates, nitrogen dioxide, radon, and carbon dioxide—may enter a building by infiltration, either through cracks and other openings in the foundations and walls or by way of ventilation systems, and are generally found in much higher concentrations indoors than outdoors (see Figure 21.19). The reason is somewhat ironic: The steps we have taken to make our homes and offices energy efficient often trap pollutants inside. Two of the best ways to conserve energy in homes and other buildings are to increase insulation and decrease infiltration of outside air. But windows that don't open and extensive caulking and weather stripping, while reducing energy consumption, also reduce natural ventilation. With less natural ventilation, we must depend more on the ventilation systems that are part of heating and air-conditioning systems.

Pathways, Processes, and Driving Forces

Both natural and human processes create differential pressures that move air and contaminants from one area of a building to another. Areas of high pressure may develop on the windward side of a building, whereas pressure is lower on the leeward, or protected, side. As a result, air is drawn into a building from the windward side. Opening and closing doors produces pressure differentials that cause air to move within buildings. Wind, too, can affect the movement of air in a building, particularly if the structure is leaky.[28]

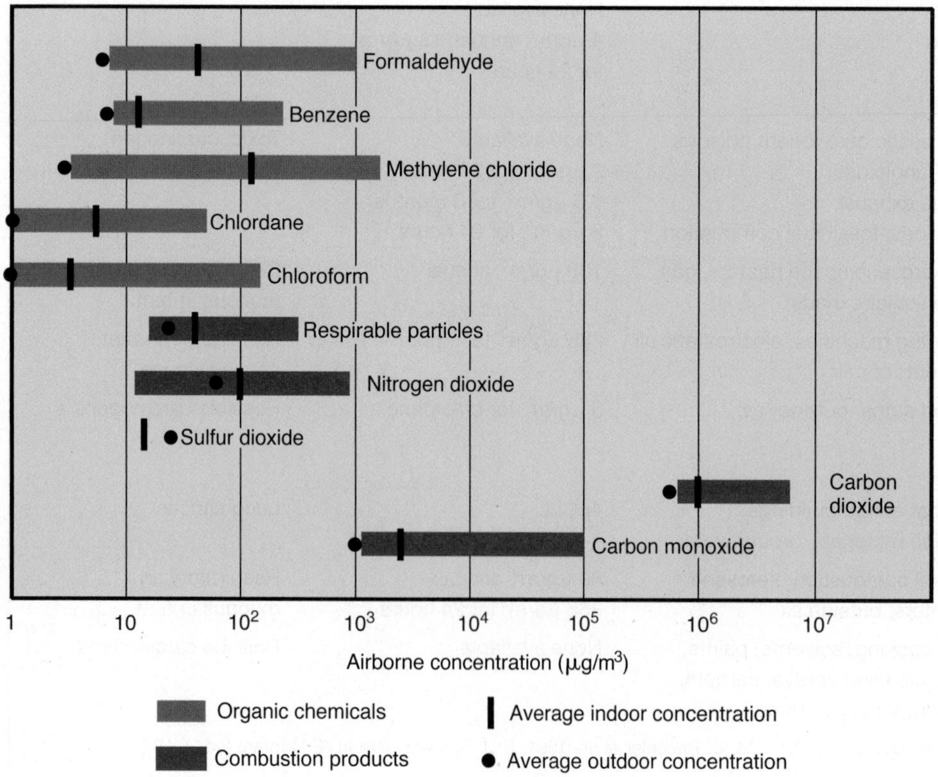

(*Source:* A.V. Nero Jr., "Controlling Indoor Air Pollution," *Scientific American*, 258, no. 5 [1998]:42–48.)

FIGURE 21.19 **Concentrations of common indoor air pollutants compared with outdoor concentrations plotted on a log scale, that is, $10^2 = 100$; $10^3 = 1{,}000$, $10^4 = 10{,}000$, etc.**

A **chimney effect (or stack effect)** occurs when the indoor and outdoor temperatures differ. Warm air rises within a building. If the indoor air is warmer than the outdoor air, then as the warmer air rises to the building's upper levels, it is replaced in the lower levels by outdoor air drawn in through various openings—windows, doors, and cracks in the foundations and walls. Because air is so fluid, the possible interactions between the driving forces and the building are complex, and the distribution of potential air contaminants and pollutants can be extensive. One outcome is that people in various parts of a building may complain about the air quality even if they are widely separated from each other and from potential sources of pollution.[28]

Heating, Ventilation, and Air-Conditioning Systems

Heating, ventilation, and air-conditioning systems are designed to provide a comfortable indoor environment. Their design depends on a number of variables, including the activity of people in the building, air temperature and humidity, and air quality. If the heating, ventilation, and air-conditioning system is designed correctly and functions properly, it will maintain a comfortable temperature and adequate ventilation (using outdoor air) and also remove common air pollutants via exhaust fans and filters.[28]

Regardless of the type of system used in a home or other building, its effectiveness depends on the proper design of the equipment for that building, proper installation, and correct maintenance and operating procedures. Indoor air pollution may result if any one of these factors concentrates pollutants from the many possible sources. Filters plugged or contaminated with fungi, bacteria, or other potentially infectious agents can cause serious problems. In addition, as we see later in this chapter, ventilation systems are not generally designed to reduce some types of indoor pollution.[28,29]

Environmental Tobacco Smoke

Environmental tobacco smoke (ETS), also known as *secondhand smoke*, comes from two sources: smoke exhaled by smokers and smoke emitted from burning tobacco in cigarettes, cigars, or pipes. People exposed to ETS are referred to as *passive smokers*.[30]

ETS is the most widely known hazardous indoor air pollutant. It is hazardous for the following reasons.[30,31]

- Tobacco smoke contains several thousand chemicals, many of which are irritants. Examples include NOx, CO, hydrogen cyanide, and about 40 carcinogenic chemicals.
- Studies of nonsmoking workers exposed to ETS found that they have impaired airway functions comparable to that caused by smoking up to ten cigarettes a day. They suffer more illnesses, such as coughs, eye irritation, and colds, and lose more work time than those not exposed to ETS.
- In the United States, about 3,000 deaths from lung cancer and 40,000 deaths from heart disease a year are thought to be associated with ETS.

The number of smokers in the United States has declined, but there about 60 million Americans still smoke. The rate is higher in the developing world, where health warnings are few or nonexistent. Smoking is extremely addictive because tobacco contains nicotine, a highly addictive substance. Nevertheless, education and social pressure have persuaded some thoughtful people to quit smoking and to encourage others to keep trying to quit.

Radon Gas

It has become apparent over the past few decades that radon gas—colorless, odorless, and tasteless—may be a significant environmental health problem in the United States.[32,33] **Radon** comes from natural processes, not from human activities. It is a naturally occurring radioactive gas that is a product of the radioactive decay chain from uranium to stable lead. Radon 222, which has a half-life of 3.8 days, is emitted during the radioactive decay of radium 226. Radon decays with emission of an alpha particle to polonium 218, which has a half-life of approximately 3 minutes. (The discussion of radiation, radiation units, radiation doses, and health problems related to radiation in Chapter 17 will help you understand the following discussion.)

Geology and radon gas. The concentration of radon gas that reaches the surface of the Earth and thus can enter our dwellings is related to the concentration of radon in the rocks and soil, as well as to the efficiency of the transfer processes from the rocks or soil to the surface. Some regions in the United States contain bedrock with an above-average natural concentration of uranium. A large area that includes parts of Pennsylvania, New Jersey, and New York—an area known as the Reading Prong—has many homes with elevated radon concentrations.[32] Such areas have also been identified in a number of other states, including Florida, Illinois, New Mexico, South Dakota, North Dakota, Washington, and California.

How dangerous is radon gas? Many people today are worried about radon in their homes because studies indicate that exposure to elevated concentrations increases the risk of lung cancer, and that the risk increases with the level and duration of exposure and also with certain habits, such as smoking.[33] Radon, combined with smoking, is thought to produce a synergistic effect that is particularly hazardous. One estimate is that the combination of exposure to radon gas and tobacco smoke is 10 to 20 times more hazardous than exposure to either pollutant by itself.[33]

The Environmental Protection Agency (EPA) estimates that 14,000 lung cancer deaths per year in the United States are related to exposure to radon and its daughter products (products that result from its radioactive decay), primarily polonium 218. (The estimate actually ranges from 7,000 to 30,000.) By comparison, approximately 140,000 people die of lung cancer in the United States each year. If these estimates are correct—and they are controversial—approximately 10% of the lung cancer deaths in the United States can be attributed to radon. Exposure to radon has also been linked to other forms of cancer, such as melanoma (a deadly skin cancer) and leukemia, but, again, such linkages are highly controversial.[34,35] The link between radon and cancer is mostly based on studies of uranium miners, a group of people exposed to high concentrations of radon in mines.

If the estimated risks from radon are anywhere close to the actual risk, then the hazard is a large one. The U.S. Surgeon General has stated that "indoor radon gas is a national health problem." The risks posed by radon are thought to be hundreds of times higher than risks from outdoor pollutants in air and water. Such pollutants are generally regulated to reduce the risk of premature death and disease to less than 0.001%. Risks from some indoor pollutants, such as organic chemicals, may be as high as 0.1%.[36] These risks still are very small compared with the risk for radon. For example, people who live in homes for about 20 years with an average concentration of radon of about 25 pCi/L are estimated to have a 1 to 2% chance of contracting lung cancer.[33,36]

How does radon enter homes and other buildings? Radon enters homes and other buildings in three main ways: (1) It migrates up from soil and rock into basements and lower floors; (2) dissolved in groundwater, it is pumped into wells and then into homes; and (3) radon-contaminated materials, such as building blocks, are used in construction. It is difficult to estimate how many homes in the United States may have elevated concentrations of radon. The EPA estimates that about 7% have elevated radon levels and recommends that all homes and schools be tested. The test is simple and inexpensive.

Symptoms of Indoor Air Pollution

People living or working in particular indoor environments may react to pollutants in different ways. Some are particularly susceptible to indoor air pollution; others report different symptoms from the same pollutant; and still others report symptoms that turn out not to stem from air pollution.

A wide variety of symptoms can result from exposure to indoor air pollutants (see Table 21.6). Some chemical pollutants can cause nosebleeds, chronic sinus infections, headaches, and irritation of the skin or eyes, nose, and throat. More serious problems include loss of balance and memory, chronic fatigue, difficulty in speaking, and allergic reactions, including asthma.

Sick Buildings

An entire building can be considered "sick" because of environmental problems. There are two types of "sick" buildings:

- Buildings with identifiable problems, such as toxic molds or bacteria known to cause disease. The diseases are known as *building-related illnesses* (BRIs).
- Buildings with **sick building syndrome (SBS)**, where the symptoms people report cannot be traced to any known cause.

A sick building's indoor environment appears to be unhealthy in that a number of people in the building report adverse health effects that they believe are related to the amount of time they spend in the building. Their complaints may range from funny odors to more serious symptoms, such as headaches, dizziness, nausea, and so forth. In addition, an unusual number of people in the building may feel sick, or may have contracted a serious disease, such as cancer.[37–39]

In many cases, it is difficult to establish what may be causing the sick building syndrome. It has sometimes been found to be related to poor management and low worker morale, rather than to toxins in the building. When the occupants of a building report adverse health effects and a study does not detect the cause, a number of other things may be happening.[28]

21.4 Controlling Indoor Air Pollution

As much as $250 billion per year might be saved by decreasing illnesses and increasing productivity through improving the work environment.[25] A good starting point would be environmental legislation requiring certain indoor air quality standards. At a minimum, these standards should include increasing the inflow of fresh air through ventilation. In Europe, systems of filters and pumps in many office buildings circulate air three times as frequently as is typical in the United States. Many building codes in Europe require that workers have access to fresh air (windows) and natural light. Unfortunately, no similar codes exist for U.S. workers, and many buildings use central air-conditioning with windows permanently sealed.[25]

You might think that heating, ventilating, and air-conditioning systems, operating properly and well maintained, will ensure good indoor air quality, but in fact these systems are not designed to maintain all aspects of air

Table 21.6 SOME SYMPTOMS OF INDOOR AIR POLLUTION

SYMPTOMS	ETS[a]	COMBUSTION PRODUCTS[b]	BIOLOGIC POLLUTANTS[c]	VOCS[d]	HEAVY METALS[e]	SBS[f]
Respiratory						
Inflammation of mucous membranes of the nose, nasal congestion	Yes	Yes	Yes	Yes	No	Yes
Nosebleed	No	No	No	Yes	No	Yes
Cough	Yes	Yes	Yes	Yes	No	Yes
Wheezing, worsening asthma	Yes	Yes	No	Yes	No	Yes
Labored breathing	Yes	No	Yes	No	No	Yes
Severe lung disease	Yes	Yes	Yes	No	No	Yes
Other						
Irritation of mucous membranes of eyes	Yes	Yes	Yes	Yes	No	Yes
Headache or dizziness	Yes	Yes	Yes	Yes	Yes	Yes
Lethargy, fatigue, malaise	No	Yes	Yes	Yes	Yes	Yes
Nausea, vomiting, anorexia	No	Yes	Yes	Yes	Yes	No
Cognitive impairment, personality change	No	Yes	No	Yes	Yes	Yes
Rashes	No	No	Yes	Yes	Yes	No
Fever, chills	No	No	Yes	No	Yes	No
Abnormal heartbeat	Yes	Yes	No	No	Yes	No
Retinal hemorrhage	No	Yes	No	No	No	No
Muscle pain, cramps	No	No	No	Yes	No	Yes
Hearing loss	No	No	No	Yes	No	No

[a] Environmental tobacco smoke.

[b] Combustion products include particles, NO_x, CO, and CO_2.

[c] Biologic pollutants include molds, dust mites, pollen, bacteria, and viruses.

[d] Volatile organic compounds, including formaldehyde and solvents.

[e] Heavy metals include lead and mercury.

[f] Sick building syndrome.

Source: Modified from American Lung Association, Environmental Protection Agency, and American Medical Association, "Indoor Air Pollution—An introduction for Health Professionals," 523-217/81322 (Washington, D.C.: GPO, 1994).

quality. For example, commonly used ventilation systems do not generally reduce radon gas. Other strategies include source removal, source modification, and air cleaning.[37]

Education also plays an important role in developing strategies to reduce indoor air pollution; it enables people to make informed decisions about exposure to chemicals, such as paints and solvents, and about strategies to avoid potentially hazardous conditions in the home and workplace.[37] At one level, this may involve deciding not to install unvented or poorly vented appliances. A surprising (and tragic) number of people are killed each year by carbon monoxide poisoning due to poor ventilation in homes, campers, and tents. Educated people are also more aware of their legal rights with respect to product liability and safety.

Making Homes and Other Buildings Radon Resistant

Protecting new homes from potential radon problems is straightforward and relatively inexpensive. It is also easy to upgrade an older home to reduce radon. The techniques vary according to the type of foundation the structure has. The basic strategy is to prevent radon from entering a home (usually sealing entry points) and ensure that radon is removed from the site (this generally involves designing a ventilation system).[40,41]

Designing Buildings to Minimize Indoor Air Pollution

There is a movement under way in the United States and the world to create buildings specifically designed to provide a healthful indoor environment for their occupants. The basic objectives of the design are to minimize indoor air pollutants; ensure that fresh air is supplied and circulated; manage moisture to avoid problems such as mold; reduce energy use; use materials whose origin is environmentally benign and can be recycled, as much as possible; create as pleasing a working environment as possible; use vegetation planted on roofs and wherever else possible to take up carbon dioxide, release oxygen, and add to the general pleasantness of the working environment.

CRITICAL THINKING ISSUE

Should Carbon Dioxide Be Regulated Along with Other Major Air Pollutants?

The six common pollutants, sometimes called the *criteria pollutants*, are ozone, particulate matter, lead, nitrogen dioxide, carbon monoxide, and sulfur dioxide. These pollutants have a long history with the EPA, and major efforts have been made to reduce them in the lower atmosphere over the United States. This effort has been largely successful—all of them have been significantly reduced since 1990.

In 2009, the EPA suggested that we add carbon dioxide to this list. Two years earlier, the U.S. Supreme Court had ordered the EPA to make a scientific review of carbon dioxide as an air pollutant that could possibly endanger public health and welfare. Following that review, the EPA announced that greenhouse gases pose a threat to public health and welfare. This proclamation makes it possible that greenhouse gases, especially carbon dioxide, will be regulated by the Clean Air Act, which regulates most other serious air pollutants.

The EPA's conclusion that greenhouse gases harm or endanger public health and welfare is based primarily on the role these gases play in climate change. The analysis states that the impacts include, but are not limited to, increased drought that will impact agricultural productivity; more intense rainfall, leading to a greater flood hazard; and increased frequency of heat waves that affect human health. The EPA's proposal program to regulate carbon dioxide as an air pollutant has been upheld by court decisions

The next step in adding carbon dioxide and other greenhouse gasses, such as methane, to the list of pollutants regulated by the EPA was a series of public hearings and feedback from a variety of people and agencies. Some people oppose listing carbon dioxide as an air pollutant because, first of all, it is a nutrient and stimulates plant growth; and, second, it does not directly affect human health in most cases (the exception being carbon dioxide emitted by volcanic eruption and other volcanic activity, which can be extremely toxic).

The EPA in late September of 2013 announced the initial steps to reduce carbon pollution under President Obama's Climate Action Plan. The objective will be standards for new coal burning power plants. Conversations are starting to develop standards for existing power plants.

Critical Thinking Questions

After going over the information concerning global climate change and the role of carbon dioxide in causing change, consider the following questions:

1. Do you think carbon dioxide, along with other greenhouse gases, should be controlled under the Clean Air Act? Why? Why not?
2. Assuming carbon dioxide and other greenhouse gases are to be controlled under the Clean Air Act, what sorts of programs might be used for such control? For example, the control of sulfur dioxide was primarily through a cap-and-trade program where the total amount of emissions were set, and companies bought and sold shares of allowed pollution up to the cap.
3. If the United States can curtail emissions of carbon dioxide under the Clean Air Act, how effective will this be in, say, reducing the global concentration of carbon dioxide to about 350 parts per million given what other countries are likely to do in the future with respect to emissions and given that the concentration today is about 390 parts per million?

SUMMARY

- There are two main kinds of air pollutants: primary and secondary. Primary pollutants are emitted directly into the air: particulates, sulfur dioxide, carbon monoxide, nitrogen oxides, and hydrocarbons. Secondary pollutants are produced through reactions between primary pollutants and other atmospheric compounds. Ozone is a secondary pollutant that forms over urban areas through photochemical reactions between primary pollutants and natural atmospheric gases.
- There are also two kinds of sources: stationary and mobile. Stationary sources have a relatively fixed position and include point sources, area sources, and fugitive sources.
- Meteorological conditions—in particular, restricted circulation in the lower atmosphere due to temperature inversion—greatly determine whether or not polluted air is a problem in an urban area.
- Pollution-control methods are tailored to specific pollution sources and types and vary from settling chambers for particulates to scrubbers that remove sulfur before it enters the atmosphere.
- Emissions of air pollutants in the United States are decreasing, but in large urban areas of developing countries it remains a serious problem.
- The concentration of atmospheric ozone has been measured for more than 70 years. Concentrations in the stratosphere have declined since the mid-1970s, allowing more ultraviolet radiation to reach the lower atmosphere, where it can damage living things.
- In 1974, Mario Molina and F. Sherwood Rowland hypothesized that stratospheric ozone might be depleted by emissions of chlorofluorocarbons (CFCs) into the lower atmosphere. Major features of the hypothesis are that CFCs are very stable and have a long residence time in the atmosphere. Eventually they reach the stratosphere, where they may be destroyed by solar ultraviolet radiation, releasing chlorine. The chlorine may then enter into a catalytic chain reaction that depletes ozone in the stratosphere.
- Banning chemicals that deplete stratospheric ozone is a step in the right direction. However, millions of tons of these are now in the lower atmosphere and working their way up, so even if all production, use, and emission of these chemicals stopped today, the problem would continue for a long time. The good news is that concentrations of CFCs in the atmosphere have apparently peaked and are now static or in slow decline.
- Possible sources of indoor air pollution are construction materials, furnishings, types of equipment used for heating and cooling, as well as natural processes that allow gases to seep into buildings.
- Indoor concentrations of air pollutants are generally greater than outdoor concentrations of the same pollutants.
- Ventilation is commonly used to control indoor air pollution, but tighter construction impedes ventilation, and many popular ventilation systems do not reduce certain types of indoor air pollutants.
- The most common natural process that affects interior air quality is the "chimney" or "stack effect" that occurs when the indoor and outdoor environments differ in temperature.
- People react to indoor air pollution in different ways, and so reported symptoms may vary.
- In some cases, reported symptoms have nothing to do with air pollution.
- Controlling indoor air pollution involves several strategies, including ventilation, source removal, source modification, and air cleaning equipment, as well as education.

REEXAMINING THEMES AND ISSUES

Sean Randall/Getty Images, Inc.

HUMAN POPULATION

Population growth will exacerbate air-pollution problems. As the number of people increases, so does the use of resources, many of which are related to emissions of air pollutants. This may be partially offset in developed countries, where the per capita emissions of air pollutants have been reduced in recent years.

© Biletskiy_Evgeniy/iStockphoto

SUSTAINABILITY

Ensuring that future generations inherit a quality environment is an important objective of sustainability. Thus, it is vital that we develop technology that minimizes air pollution.

© Anton Balazh 2011/iStockphoto

GLOBAL PERSPECTIVE

Atmospheric processes and atmospheric pollution occur on regional and global scales. Pollutants emitted into the atmosphere at a particular site may join the global circulation pattern and spread throughout the world, and pollutants emitted from urban or agricultural areas may be dispersed to pristine areas far removed from human activities. Therefore, an understanding of global atmospheric processes is critical to finding solutions to many air-pollution problems, including acid deposition.

ssguy/ShutterStock

URBAN WORLD

Cities and urban corridors are sites of intense human activity, and many of these activities contribute to air pollution. Some large cities have such severe air-pollution problems that the health and lives of people are being affected.

B2M Productions/Getty Images, Inc.

PEOPLE AND NATURE

Although we think of nature as unspoiled, in reality nature can be toxic. This is especially true with regard to air pollution—for example, the vast majority of particulates and carbon monoxide are generated by volcanic eruption and wildfire. Even hydrocarbons have local sources, such as seeps, which, in areas such as offshore Goleta, California, emit a significant amount of hydrocarbons that contribute to smog.

George Doyle/Getty Images, Inc.

SCIENCE AND VALUES

The science and technology necessary to reduce air pollution are well known; what we do with these tools involves a value judgment. It is clear that people value a high-quality environment, and clean air is at the top of the list. The developed countries have an obligation to take a leadership role in finding ways to use resources while minimizing air pollution. Of particular importance is finding methods and technologies that will allow for reducing air pollution while stimulating economies. What is considered waste in one part of the urban-industrial complex may be used as resources in another part. This idea is at the heart of what is sometimes called industrial ecology. The discovery, understanding, and management of ozone-depleting chemicals represent an environmental success, reflecting the value of the environment.

KEY TERMS

acid rain 519
air toxics 530
atmospheric inversion 532
black lung disease 540
chimney effect (or stack effect) 543
chlorofluorocarbons (CFCs) 523
criteria pollutants 517
environmental tobacco smoke (ETS) 543
global dimming 529
green building 539
mobile sources 516
ozone (O_3) 522
ozone shield 522
photochemical smog 532
primary pollutants 517
radon 543
scrubbing 536
secondary pollutants 517
sick building syndrome (SBS) 544
smog 532
stationary sources 515
sulfurous smog 532
Ultraviolet (UV) Index 527
ultrafine particles 528

STUDY QUESTIONS

1. Since the amount of pollution emitted into the air is a very small fraction of the total material in the atmosphere, why do we have air-pollution problems?
2. What are the differences between primary and secondary pollutants?
3. Carefully examine Figure 21.14, which shows a column of air moving through an urban area, and Figure 21.10, which shows relative concentrations of pollutants that develop on a typical warm day in Los Angeles. What linkages between the information in these two figures might be important in trying to identify and learn more about potential air pollution in an area?
4. Why is acid deposition a major environmental problem, and how can it be minimized?
5. Why will air-pollution abatement strategies in developed countries probably be much different in terms of methods, process, and results than air-pollution abatement strategies in developing countries?
6. In a highly technological society, is it possible to have 100% clean air? Is it likely?
7. Study Figure 21.6 carefully and discuss how the information in parts (a) and (b) are linked and related to the chapter in general. Apply Critical Thinking skills.
8. Discuss the processes responsible for stratospheric ozone depletion. Which are most significant? Where? Why?
9. What are some of the common sources of indoor air pollutants where you live, work, or attend classes?
10. Develop a research plan to complete an audit of the indoor air quality in your local library. How might that research plan differ from a similar audit for the science buildings on your campus?
11. What do you think about the concept of sick building syndrome? If you were working for a large corporation and a number of employees said they were getting sick and listed a series of symptoms and problems, how would you react? What could you do? Play the role of the administrator and develop a plan to look at the potential problem.
12. Suppose that next year our understanding of ozone depletion is changed by the discovery that concentrations of stratospheric ozone have natural cycles and that lower concentrations in recent years have resulted not from our activities but from natural processes. How would you put all the information in this chapter into perspective? Would you think science had let you down?

FURTHER READING

Boubel, R.W., D.L. Fox, D.B. Turner, and A.C. Stern. ***Fundamentals of Air Pollution,*** **4th ed.** (New York, NY: Academic, 2008). A thorough book covering the sources, mechanisms, effects, and control of air pollution.

Brenner, D.J. ***Radon: Risk and Remedy*** (New York, NY: W.H. Freeman, 1989). A wonderful book about the hazards of radon gas. It covers everything from the history of the problem to what was happening in 1989, as well as solutions.

Christie, M. ***The Ozone Layer: A Philosophy of Science Perspective*** (Cambridge, UK: Cambridge University Press, 2000). A complete look at the history of the ozone hole, from the first discovery of its existence to more recent studies of the hole over Antarctica.

Hamill, P., and O.B. Toon. ***"Polar Stratospheric Clouds and the Ozone Hole,"*** *Physics Today* 44, no. 12 (1991):34–42. A good review of the ozone problem and important chemical and physical processes related to ozone depletion.

Reid, S. ***Ozone and Climate Change: A Beginner's Guide*** (Amsterdam: Gordon & Breach Science Publishers, 2000). A look at the science behind the ozone hole and future predictions, written for a general audience.

Rowland, F.S. ***"Stratospheric Ozone Depletion by Chlorofluoro carbons."*** *wAMBIO*19, no. 6–7 (1990):281–292. An excellent summary of stratospheric ozone depletion, it discusses some of the major issues.

Wang, L."Paving out Pollution," ***Scientific American,*** **February 2002, p. 20.** Discussion of an innovative approach to reducing air pollution.

NOTES

1. U.S. Environmental Protection Agency. 2012. *National air quality trends*. www.epa.gov.
2. National Park Service. 1984. *Air Resources Management Manual.*
3. Araujo, J.A., and ten others. 2008. Ambient particulate pollutants in the ultrafine range promote early atherosclerosis and systematic oxidative stress. *Circulation Research* 102:589–596.
4. Godish, T. 1991. *Air Quality*, 2d ed. Chelsea, MI: Lewis Publishers.
5. Seitz, F., and C. Plepys. 1995. Monitoring air quality in healthy people 2000. *Healthy People 2000: Statistical Notes no. 9.* Atlanta: Centers for Disease Control and Prevention, National Center for Health Statistics.
6. American Lung Association. 2011. *State of the Air 2011*. www.stateoftheair.org.
7. Moore, C. 1995. Poisons in the air. *International Wildlife* 25:38–45.
8. Pope, C. A., III, D.V. Bates, and M.E. Raizenne. 1995. Health effects of particulate air pollution: Time for reassessment? *Environmental Health Perspectives* 103:472–480.
9. Travis, D.J., A.M. Carleton, and R.G. Lauritsen 2002. Contrails reduce daily temperature range. *Nature* 419:601.
10. U.S. Environmental Protection Agency 2006. *National scale air toxics assessment for 1999. Estimated emissions. Concentrations and risks*. Technical Fact Sheet. Accessed April 10, 2006 at www.epa.gov.
11. Tyson, P. 1990. Hazing the Arctic. *Earthwatch* 10:23–29.
12. NASA. 2008. NASA satellite measures pollution from East Asia to North America. www.nasa.gov.
13. Pittock, A.B., L.A. Frakes, D. Jenssen, J.A. Peterson, and J.W. Zillman, eds. 1978. *Climatic Change and Variability: A Southern Perspective*. New York, NY: Cambridge University Press.
14. Blake, D.R., and F.S. Rowland. 1995. Urban leakage of liquefied petroleum gas and its impact on Mexico City air quality. *Science* 269:953.
15. Pountain, D. May 1993. Complexity on wheels. *Byte*, pp. 213–220.
16. Moore, C. 1995. Green revolution in the making. *Sierra* 80:50.
17. U.S. Environmental Protection Agency. 2008. Case Study 22. AEP and Centain Teed put environmental process byproducts to beneficial use in wallboard. www.epa.gov.
18. Kolstad, C.D. 2000. *Environmental Economics*. New York, NY: Oxford University Press.
19. Crandall, R.W. 1983. *Controlling Industrial Pollution: The Economics and Politics of Clean Air*. Washington, DC: Brookings Institution.
20. Hall, J.V., A.M. Winer, M.T. Kleinman, F.W. Lurmann, V. Brajer, and S.D. Colome. 1992. Valuing the health benefits of clean air. *Science* 255:812–816.
21. Krupnick, A.J., and P.R. Portney. 1991. Controlling urban air pollution: A benefits-cost assessment. *Science* 252:522–528.
22. Lipfert, F.W., S.C. Morris, R.M. Friedman, and J.M. Lents. 1991. Air pollution benefit-cost assessment. *Science* 253:606.
23. Zimmerman, M.R. 1985. Pathology in Alaskan mummies. *American Scientist* 73:20–25.
24. Ehrlich, P.R., A.H. Ehrlich, and J.P. Holdren. 1970. *Ecoscience: Population, Resources, Environment.* San Francisco, CA: W.H. Freeman.
25. Conlin, M. June 5, 2000. Is your office killing you? *Business Week*, pp. 114–124.
26. U.S. Environmental Protection Agency. *Basic information: Formaldehyde*. Accessed April 28, 2008 at www.epa.gov.
27. Anonymous. February 15, 2008. *FEMA hurries hurricane survivors out of toxic trailers*. Environment News Service (ENS).
28. U.S. Environmental Protection Agency. 1991. *Building Air Quality: A Guide for Building Owners and Facility Managers.* EPA/400/ 1–91/033, DHHS (NIOSH) Pub. No. 91–114. Washington, DC: Environmental Protection Agency.
29. Zummo, S.M., and M.H. Karol. 1996. Indoor air pollution: Acute adverse health effects and host susceptibility. *Environmental Health* 58:25–29.
30. Godish, T. 1997. *Air Quality*, 3d ed. Boca Raton, FL: Lewis Publishers.
31. O'Reilly, J.T., P. Hagan, R. Gots, and A. Hedge. 1998. *Keeping Buildings Healthy*. New York, NY: John Wiley & Sons.
32. Brenner, D.J. 1989. *Radon: Risk and Remedy*. New York, NY: W.H. Freeman.
33. U.S. Environmental Protection Agency. 1992. *A Citizen's Guide to Radon: The Guide to Protecting Yourself and Your Family from Radon*, 2d ed. ANR-464. Washington, DC: Environmental Protection Agency.
34. Henshaw, D.L., J.P. Eatough, and R.B. Richardson. 1990. Radon as a causative factor in induction of myeloid leukaemia and other cancers. *The Lancet* 335:1008–1012.
35. Pershagen, G., G. Akerblom, O. Axelson, B. Clavensjo, L. Damber, G. Desai, A. Enflo, F. Lagarde, H. Mellander, M. Svartengren, and G.A. Swedjemark. 1994. Residential radon exposure and lung cancer in Sweden. *New England Journal of Medicine* 330:159–164.
36. Nero, A.V., Jr. 1988. Controlling indoor air pollution. *Scientific American* 258:42–48.
37. Committee on Indoor Air Pollution. 1981. *Indoor Pollutants.* Washington, DC: National Academy Press.
38. Massachusetts Department of Public Health, Bureau of Environmental Health Assessments. 1995. *Symptom Prevalence Survey Related to Indoor Air Concerns at the Registry of Motor Vehicles Building, Ruggles Station.*

CASE STUDY

Rivers, Seacoasts, And Cities

Floods, droughts, and shifting seacoasts and rivers can determine the success or failure of cities. A good harbor on a seacoast or on a river can make a great place for a city. But because the environment is always changing, coastlines and rivers can be, from a human point of view, fickle. In the Middle Ages, Venice, Italy, was one of the most powerful cities in the world, a center for trade. But after the discovery of America and alternative routes to the Far East, Venice's situation relative to other cities was no longer important for trade. It then became a famous center for tourism, crafts, industries such as glassmaking, and the arts. But in recent decades those roles have been threatened by the city's site, subject, as sea level rises, to winter high waters. During these storms, visitors have to travel above the flooded city on elevated walkways, as we see in this chapter's opening photograph. And the lagoon's waters are heavily polluted, making for unpleasant pedestrian walks.

The opposite has also happened—a good waterside location may cease to be functional for a city. A historic example begins with the Lewis and Clark expedition as they traveled up the Missouri River in 1804. Not far north of the present location of Kansas City, Missouri, on July 2, 1804, the expedition camped opposite an old Kansas Indian village where there was a large island in the river and an "extensive" prairie beyond it (Figure 22.1). The island, Clark wrote, appeared to have, "thrown the Current of the river against the place the Village formerly Stood," so that the current washed away

(a)
(b)
(c)

Photo by Daniel B. Botkin

Photo by Daniel B. Botkin

Photo by Daniel B. Botkin

FIGURE 22.1 **(a) The Missouri River passes an island near Weston, Missouri.** Changes in the path of the river led to a decline in the town, as explained in the text. **(b)** Abandoned by the river, Weston was soon abandoned by its residents. As a result, its buildings were not replaced by newer structures, leaving the town a historic place, as shown here. **(c)** Bruges, Belgium, was also once an active seaport, but the ocean waters gradually filled in the land, leaving Bruges high and dry. Like Weston and Venice, it became a historic city and a tourist attraction.

the bank, forming an arc or natural harbor. He realized that, as a river flowed around a curve of an obstacle, the water on the outside of the curve had to flow faster to keep up with the water on the inside of that curve. Faster-moving water is more erosive, and this is why, as Clark understood, the river's island had forced the river to create a harbor. "The Situation appears to be a verry elligable one for a Town, the valley rich & extensive, with a Small Brook Meanding through it and one part of the bank affording yet a good Landing for Boats," Clark wrote in his journal that day.

Just as Clark had predicted, a town named Weston developed at this location in 1837. Soldiers from nearby Fort Leavenworth saw the potential of the location, bought the land, and began to develop it. Weston's natural bay made a good port for boats to tie up, and there was good land nearby for farming as well as good upland locations—dry and well-drained for basements but with good water supply for houses. The soldiers established a dock and a main street that led from the dock away from the river. Settlers moved in quickly and set up a variety of shops and activities. The countryside was rich for farming. Tobacco farms were established, and their products shipped downstream on boats that tied up at Weston's harbor.

But in 1881 a bad flood occurred on the Missouri River, and the river cut a new main channel two and one-half miles southwest of Weston. In this one event, the river meandered away from the town, leaving Weston high and dry, with a harbor no longer at the foot of Main Street. The town was quickly abandoned, and many of its buildings were left intact. Today, Weston's old and picturesque houses have made it a tourist attraction.[1,2]

A similar thing happened to Bruges, Belgium, which first developed as an important center for commerce in the 13th century, because its harbor on the English Channel permitted trade with England and other European nations. By the 15th century, however, the harbor had seriously silted in, and the limited technology of the time did not make dredging possible. This problem, combined with political events, led to a decline in the importance of Bruges—a decline from which it never recovered. Nevertheless, today, Bruges still lives, a beautiful city with many fine examples of medieval architecture. Because these buildings were never replaced with modern ones, Bruges has become a modern tourist destination.

22.1 City Life

Changes in the environment that affect cities—especially rising sea levels that threaten cities—are much in the news these days. As the opening case study makes clear, cities have always been strongly affected by the environment. It is important in the modern world that we realize that cities are an important part of our environment in both positive and negative ways. Moreover, we must understand cities within an environmental context.

In the past, the emphasis of environmental action was most often on wilderness, wildlife, endangered species, and the impact of pollution on natural landscapes outside cities. Now it is time to turn more of our attention to city environments. In the development of the modern environmental movement in the 1960s and 1970s, it was fashionable to consider everything about cities bad and everything about wilderness good. Cities were viewed as polluted, dirty, lacking in wildlife and native plants, and artificial—therefore bad. Wilderness was viewed as unpolluted, clean, teeming with wildlife and native plants, and natural—therefore good.

Although it was fashionable to disdain cities, many people live in urban environments and have suffered directly from their decline. According to the United Nations Environment Program, in 1950 fewer than a third of the people of the world lived in a town or city; in contrast, today 45% almost half of the world's population is urban, and the forecasts are that in less than 20 years—by 2030—62 almost two-thirds of world's population will live in cities and towns.[3] (See Chapter 5.) In the United States, about 249 million—80% of the population—live in urban areas, and a surprisingly 24% live in the ten largest urban areas (Table 22.1). Perhaps even more striking is the fact that these ten urban areas occupy just 1% of the land area of the United States' lower 48 states (Figure 22.2).[4,5]

Economic development leads to urbanization; 75% of people in developed countries live in cities, but only 38% of people in the poorest of the developing countries are city dwellers.[6] Megacities—huge metropolitan areas with more than 8 million residents—are emerging more and more. In 1950 the world had only two: the New York City and nearby urban New Jersey metropolitan area (12.2 million residents altogether) and greater London (12.4 million). By 1975, Mexico City, Los Angeles, Tokyo, Shanghai, and São Paulo, Brazil, had joined this list. By 2013, the most recent date for which reliable data are available, 30 urban areas—24 cities—had more than 10 million people.[6]

Yet, comparatively little public concern has focused on urban ecology. Many urban people see environmental

Table 22.1 THE UNITED STATES' TEN MOST POPULOUS URBANIZED AREAS. 24% OF THE U.S. POPULATION LIVE IN THESE TEN URBANIZED AREAS.

URBANIZED AREA	POPULATION	LAND AREA (SQUARE MILES)	DENSITY (PEOPLE PER SQ. MI.)
New York–Newark, NY–NJ–CT	18,351,295	3,450.20	5,318.90
Los Angeles–Long Beach–Anaheim, CA	12,150,996	1,736.00	6,999.30
Chicago, IL–IN	8,608,208	2,442.70	3,524.00
Miami, FL	5,502,379	1,238.60	4,442.40
Philadelphia, PA–NJ–DE–MD	5,441,567	1,981.40	2,746.40
Dallas-Fort Worth-Arligton, TX	5,121,892	1,779.10	2,878.90
Houston, TX	4,944,332	1,660.00	2,978.50
Washington, DC-VA-MD	4,586,770	1,321.70	3,470.30
Atlanta, GA	4,515,419	2,645.40	1,706.90
Boston, MA–NH–RI	4,181,019	1,873.50	2,231.70
Average			3,442.04
Total	73,403,877	20,129	
Percent of U.S. Population	24%	1%	
Total U.S. Urban Population	249,253,271		
Grand total	311,566,589	2,959,064	

(*Source:* U.S. Census Bureau, http://www.census.gov/geo/www/ua/uafacts.html.)

issues as outside their realm, but the reality is just the opposite: City dwellers are at the center of some of the most important environmental issues. People are realizing that city and wilderness are inextricably connected. We cannot fiddle in the wilderness while our Romes burn from sulfur dioxide and nitrogen oxide pollution. Fortunately, we are experiencing a rebirth of interest in urban environments and urban ecology. The National Science Foundation has

NASA Earth Observatory/NOAA NGDC

FIGURE 22.2 **The World At Night** shows major urban areas. This is a composite made from the NASA-NOAA Suomi National Polar-orbiting Partnership (NPP) satellite in April and October 2012.

added two urban areas, Baltimore and Phoenix, to its Long-Term Ecological Research Program, a program that supports research on, and long-term monitoring of, specific ecosystems and regions.

In summary, today and in the future, most people will live in cities. For most people, living in an environment of good quality will mean living in a city that is managed carefully to maintain that environmental quality.

22.2 The City as a System

We need to analyze a city as the ecological system that it is—but of a special kind. Like any other life-supporting system, a city must maintain a flow of energy, provide necessary material resources, and have ways of removing wastes. These ecosystem functions are maintained in a city by transportation and communication with outlying areas. A city is not a self-contained ecosystem; it depends on other cities and rural areas. A city takes in raw materials from the surrounding countryside: food, water, wood, energy, mineral ores—everything that a human society uses. In turn, the city produces and exports material goods and, if it is a truly great city, also exports ideas, innovations, inventions, arts, and the spirit of civilization. A city cannot exist without a countryside to support it. As was said half a century ago, city and country, urban and rural, are one thing—one connected system of energy and material flows—not two things (see Figure 22.3).

As a consequence, if the environment of a city declines, almost certainly the environment of its surroundings will also decline. The reverse is also true as we saw in the chapter's opening case study: If the environment around a city declines, the city itself will be threatened.

Cities also export waste products to the countryside, including polluted water, air, and solids. The average city resident in an industrial nation annually uses (directly or indirectly) about 208,000 kg (229 tons) of water, 660 kg (0.8 ton) of food, and 3,146 kg (3.5 tons) of fossil fuels and produces 1,660,000 kg (1,826 tons) of sewage, 660 kg (0.8 ton) of solid wastes, and 200 kg (440 lb) of air pollutants. If these are exported without care, they pollute the countryside, reducing its ability to provide necessary resources for the city and making life in the surroundings less healthy and less pleasant.

Given such dependencies and interactions between city and surroundings, it's no wonder that relationships between people in cities and in the countryside have often been strained. Why, country dwellers want to know, should they have to deal with the wastes of those in the city? The answer is that many of our serious environmental problems occur at the interface between urban and rural areas. People who live outside but near a city have a vested interest in maintaining both a good environment for that city and a good system for managing the city's resources. The more concentrated the human population, the more land is available for other uses, including wilderness, recreation, conservation of biological diversity, and production of renewable resources. So cities benefit wilderness, rural areas, and so forth.

With the growing human population, we can imagine two futures. In one, cities are pleasing and livable, use resources from outside the city in a sustainable way, minimize pollution of the surrounding countryside, and allow room for wilderness, agriculture, and forestry. In the other future, cities continue to be seen as environmental

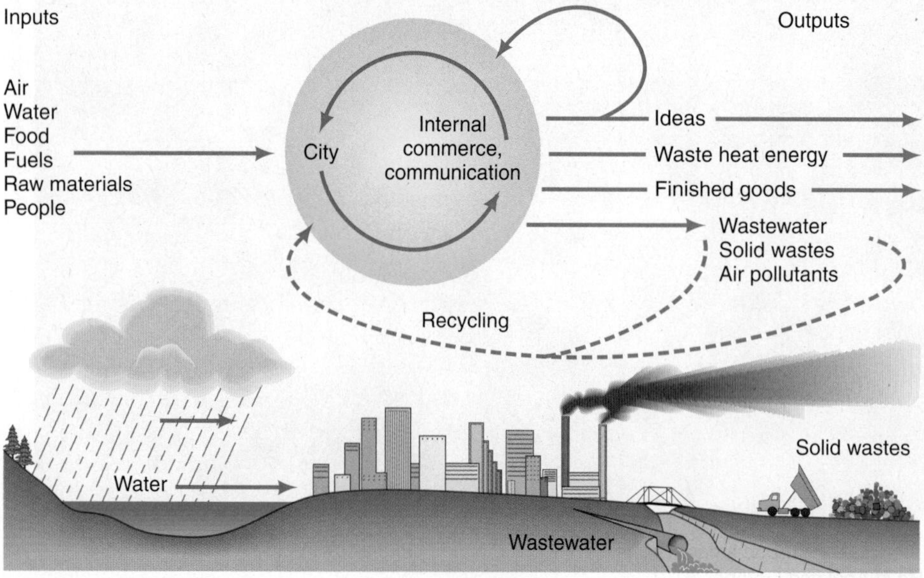

FIGURE 22.3 The city as a system. with flows of energy and materials. A city must function as part of a city–countryside ecosystem, with an input of energy and materials, internal cycling, and an output of waste heat energy and material wastes. As in any natural ecosystem, recycling materials can reduce the need for input and the net output of wastes.

negatives and are allowed to decay from the inside. People flee to grander and more expansive suburbs that occupy much land, and the poor who remain in the city live in an unhealthy and unpleasant environment. Without care for the city, its technological structure declines and it pollutes even more than in the past. Trends in both directions appear to be occurring.

In light of all these concerns, this chapter describes how a city can fit within, use, and avoid destroying the ecological systems on which it depends, and considers how the city itself can serve human needs and desires as well as environmental functions. With this information, you will have the foundation for making decisions, based on science and on what you value, about what kind of urban-rural landscape you believe will provide the most benefits for people and nature.

22.3 The Location of Cities: Site and Situation

Here is an idea that our modern life, with its rapid transportation and its many electronic tools, obscures but the opening case study illustrates: Cities are not located at random but develop mainly because of local conditions and regional benefits. In most cases, they grow up at crucial transportation locations—an aspect of what is called the city's situation—and at a good site, one that can be readily defended, with good building locations, water supplies, and access to resources. The primary exceptions are cities that have been located primarily for political reasons. Washington, DC, for example, was located to be near the geographic center of the area of the original 13 states; but the site was primarily swampland, and nearby Baltimore provided the major harbor of the region. Moscow, Russia, was similarly situated primarily for political rather than economic reasons.

The Importance of Site and Situation

The location of a city is influenced primarily by the **site**, which is the summation of all the environmental features of that location; and the **situation**, which is the placement of the city with respect to other areas. A good site includes a geologic substrate suitable for buildings, such as a firm rock base and well-drained soils that are above the water table; nearby supplies of drinkable water; nearby lands suitable for agriculture; and forests. Sometimes, however, other factors—such as the importance of creating a port city—can compensate for a poor geological site, as long as people are able to build an artificial foundation for the city and maintain that foundation despite nature's attempts to overwhelm it.

Cities influence and are influenced by their environment. The environment of a city affects its growth, success, and importance—and can also provide the seeds of its destruction. All cities are so influenced, and those who plan, manage, and live in cities must be aware of all aspects of the urban environment.

The environmental situation is especially important with respect to transportation and defense. Waterways, for example, are important for transportation. Before the era of railroads, automobiles, and airplanes, cities depended on water for transportation, so most early cities—including all the important cities of the Roman Empire—were on or near waterways. Waterways continue to influence the locations of cities; most major cities of the eastern United States are situated either at major ocean harbors, like New Orleans (see A Closer Look 22.1), or at the fall line on major rivers.

A **fall line** on a river occurs where there is an abrupt drop in elevation of the land, creating waterfalls (Figure 22.4), typically where streams pass from harder, more erosion-resistant rocks to softer rocks. In eastern North America, the major fall line occurs at the transition from the granitic and metamorphic bedrock that forms the Appalachian Mountains to the softer, more easily eroded, and more recent sedimentary rocks. In general, the transition from major mountain range bedrock to another bedrock forms the primary fall line on continents.

Cities have frequently been established at fall lines, especially the major continental fall lines, for a number of reasons. Fall lines provide water power, which was an important source of energy in the 18th and 19th centuries, when the major eastern cities of the United States were established or rose to importance. At that time, the fall line was the farthest inland that larger ships could navigate; and just above the fall line was the farthest downstream that the river could be easily bridged. Not until the development of steel bridges in the late 19th century did it become practical to span the wider regions of a river below the fall line. The proximity of a city to a river has another advantage: River valleys have rich, water-deposited soils that are good for agriculture. In early times, rivers also provided an important means of waste disposal, which today has become a serious problem.

Cities also are often founded at other kinds of crucial locations, growing up around a market, a river crossing, or a fort. Newcastle, England, and Budapest, Hungary, are located at the lowest bridging points on their rivers. Other cities, such as Geneva, are located where a river enters or leaves a major lake. Some well-known cities are at the confluence of major rivers: Saint Louis lies at the confluence of the Missouri and Mississippi rivers; Manaus (Brazil), Pittsburgh (Pennsylvania), Koblenz (Germany), and Khartoum (Sudan) are all at the confluence of several rivers.

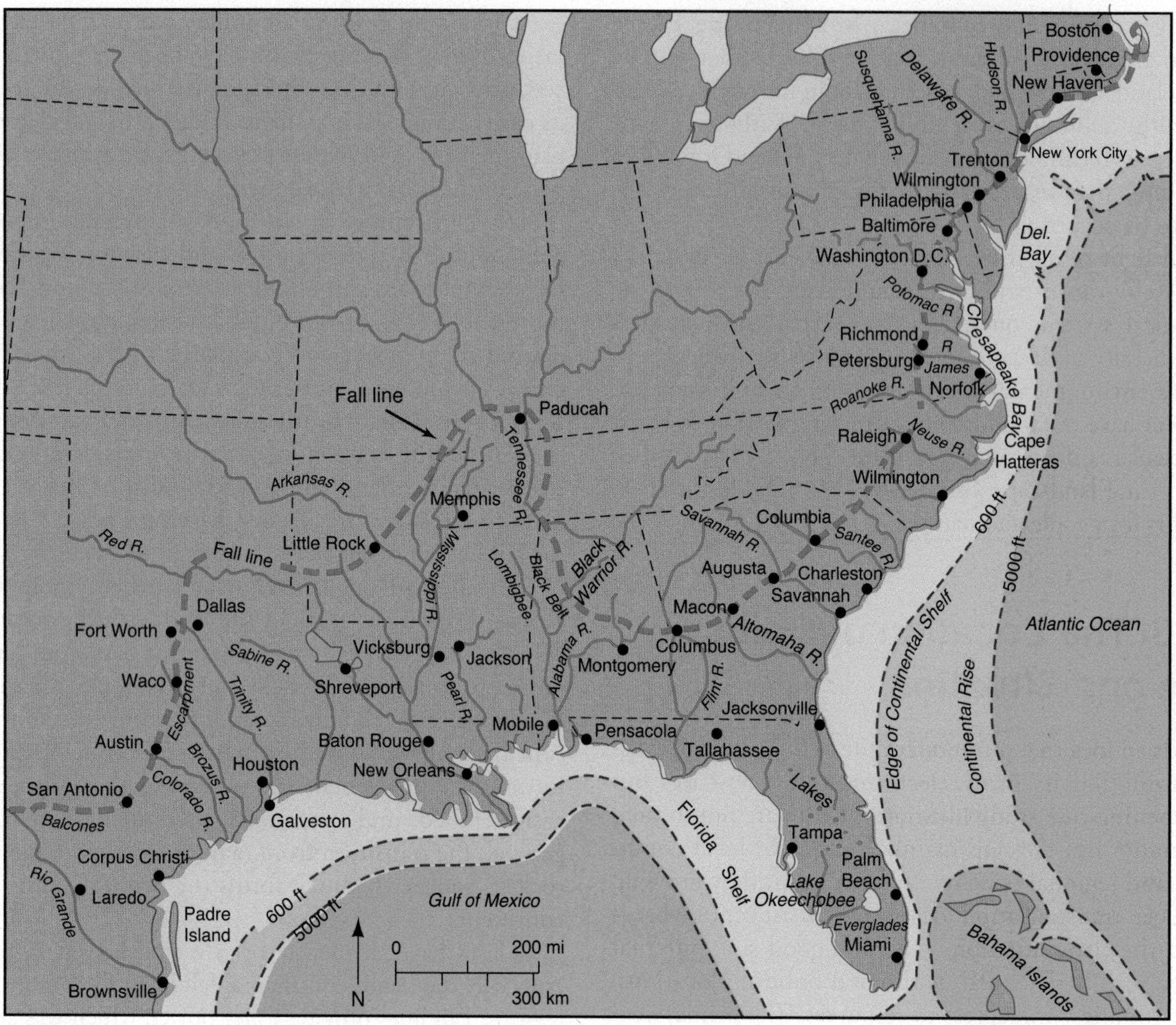

C.B. Hunt, *Natural Regions of the United States and Canada* [San Francisco: W.H. Freeman, 1974]. Copyright 1974 by W.H. Freeman & Co.

FIGURE 22.4 The fall line. Most major cities of the eastern and southern United States lie either at the sites of harbors or along a fall line (shown by the dashed line in the figure), which marks locations of waterfalls and rapids on major rivers. This is one way the location of cities is influenced by the characteristics of the environment.

A CLOSER LOOK 22.1

Should We Try to Restore New Orleans?

On August 29, 2005, Hurricane Katrina roared, slammed, and battered its way into New Orleans with 192 km/hr (120 mph) winds. Its massive storm surges breached the levees that had protected many of the city's residents from the Gulf Coast's waters, flooding 80% of the city and an estimated 40% of the houses. With so many people suddenly homeless and such major damage (Figure 22.5), the New Orleans mayor, Ray Nagin, ordered a first-time-ever complete evacuation of the city, an evacuation that became its own disaster. Some estimates claimed that 80% of the 1.3 million residents of the greater New Orleans metropolitan area evacuated.

By the time it was over, Katrina was the most costly hurricane in the history of the United States—between $75 billion and $100 billion, in addition to an estimated $200 billion in lost business revenue. A year after the hurricane, much of the damage remained, and even today much of New Orleans is not yet restored. Citizens remain frustrated by the lack of progress on many fronts.[a] An estimated 50,000 homes will have to be demolished. Many former residents are still living elsewhere, scattered across the nation. The storm affected the casino and entertainment industry, as many of the Gulf Coast's casinos were destroyed or sustained considerable damage. New Orleans also was home to roughly 115,000 small businesses, many of which were permanently damaged.

The problem with New Orleans is that it is built in the wetlands at the mouth of the Mississippi River, and much of it is below sea level (Figure 22.6). Although a port at the mouth of the Mississippi River has always been an important location

© David Frazier/Corbis

(a) (b)

© Smiley N. Pool/Dallas Morning News/Corbis

FIGURE 22.5 New Orleans. (a) air photo before Hurricane Katrina (b) air photo right after Hurricane Katrina.

for a city, there just wasn't a great place to build that city. The original development, the French Quarter, was just barely above sea level, about the best level that could be found.

Hurricane Katrina was rated a Category 3 hurricane, but discussions about protecting the city from future storms focus on an even worse scenario: a Category 5 hurricane with winds up to 249 km/hr (155 mph). As of 2009, 68,000 homes remained abandoned.[b] Clearly, New Orleans requires an expensive improvement in its site if it is to survive at all. And if it does survive, will it be restored to its former glory and importance? Will it continue in any fashion, even as a mere shadow of its former self? Fortunately, the city has many residents who love it and are working hard to restore it.

To know how to rebuild the city, to decide whether it is worth doing, and to forecast whether such a restoration is likely, we have to understand the ecology of cities, how cities fit into the environment, the complex interplay between a city and its surroundings, and the way a city acts as an environment for its residents.

Tim Vasquez/Weather Graphics.

FIGURE 22.6 Map of New Orleans showing how much of the city is below sea level. The city began with the French Quarter, which is above sea level. As the population grew and expanded, levees were built to keep the water out, and the city became an accident waiting to happen.

Many famous cities are at crucial defensive locations, such as on or adjacent to easily defended rock outcrops. Examples include Edinburgh, Athens, and Salzburg. Other cities and municipalities are situated on peninsulas—for example, Istanbul and Monaco. Cities also frequently arise close to a mineral resource, such as salt (Salzburg, Austria), metals (Kalgoorlie, Australia), or medicated waters and thermal springs (Spa, Belgium; Bath, Great Britain; Vichy, France; and Saratoga Springs, New York).

When a successful city grows and spreads over surrounding terrain, its original purpose may be obscured. Its original market or fort may have evolved into a square or a historical curiosity. In most cases, though, cities originated where the situation provided a natural meeting point for people.

An ideal location for a city has both a good site and a good situation, but such a place is difficult to find. Paris is perhaps one of the best examples of a perfect location for a city—one with both a good site and a good situation. Paris began on an island more than 2,000 years ago, the situation providing a natural moat for defense and waterways for transportation. Surrounding countryside, a fertile lowland called the Paris basin, affords good local agricultural land and other natural resources. New Orleans on the other hand, is an example of a city with an important situation but, as Hurricane Katrina made abundantly clear, a poor site (Figure 22.7).

Site Modification

Site is provided by the environment, but technology and environmental change can alter a site for better or worse. People can improve the site of a city and have done so when the situation of the city made it important and when its citizens could afford large projects. An excellent situation can sometimes compensate for a poor site. However, improvements are almost always required to the site so the city can persist.

Changes in a site over time can have adverse effects on a city, as we saw in the opening discussion about Weston, Missouri, and Bruges, Belgium.

Photo Daniel Botkin

(a)

age fotostock / SuperStock

(b)

FIGURE 22.7 **(a) Geologic, topographic, and hydrologic conditions greatly influence how successful the city can be.** If these conditions, known collectively as the city's site, are poor, much time and effort are necessary to create a livable environment. New Orleans has a poor site but an important situation. **(b)** In contrast, New York City's Manhattan is a bedrock island rising above the surrounding waters, providing a strong base for buildings and a soil that is sufficiently above the water table so that flooding and mosquitoes are much less of a problem. It has a good site and a good situation.

The City Park

Parks have become more and more important in cities. A significant advance for U.S. cities was the 19th-century planning and construction of Central Park in New York City, the first large public park in the United States. The park's designer, Frederick Law Olmsted, was one of the most important modern experts on city planning. He took site and situation into account and attempted to blend improvements to a site with the aesthetic qualities of the city.[9]

Central Park is an example of "design with nature," a term coined much later than Olmsted's time, and its design influenced other U.S. city parks. For Olmsted, the goal of a city park was to provide psychological and physiological relief from city life through access to nature and beauty (Figure 22.9). Vegetation was one of the keys to creating beauty in the park, and Olmsted carefully considered the opportunities and limitations of topography, geology, hydrology, and vegetation.

In contrast to the approach of a preservationist, who might simply have focused on returning the area to its natural, wild state, Olmsted created a naturalistic environment, keeping the rugged, rocky terrain but creating ponds where he thought they were desirable. To add variety, he constructed "rambles," walkways that were densely planted and followed circuitous patterns. He created a "sheep meadow" by using explosives to flatten the terrain. In the southern part of the park, where there were flat meadows, he created recreational areas. To meet the needs of the city, he built transverse roads through the park and also created depressed roadways that allowed traffic to cross the park without detracting from the vistas seen by park visitors.

Photo by Daniel B. Botkin

FIGURE 22.9 New York City's Central Park. was designed by Frederick Law Olmsted to provide psychological and physiological relief from city life, which you can see it does here.

Olmsted has remained a major figure in American city planning, and the firm he founded continued to be important in city planning into the 20th century. His skill in creating designs that addressed both the physical and aesthetic needs of a city is further illustrated by his work in Boston. Boston's original site had certain advantages: a narrow peninsula with several hills that could be easily defended, a good harbor, and a good water supply. But as Boston grew, demand increased for more land for buildings, a larger area for docking ships, and a better water supply. The need to control ocean floods and to dispose of solid and liquid wastes grew as well. Much of the original tidal flats area, which had been too wet to build on and too shallow to navigate, had been filled in (Figure 22.10). Hills had been leveled and the marshes filled with soil. The

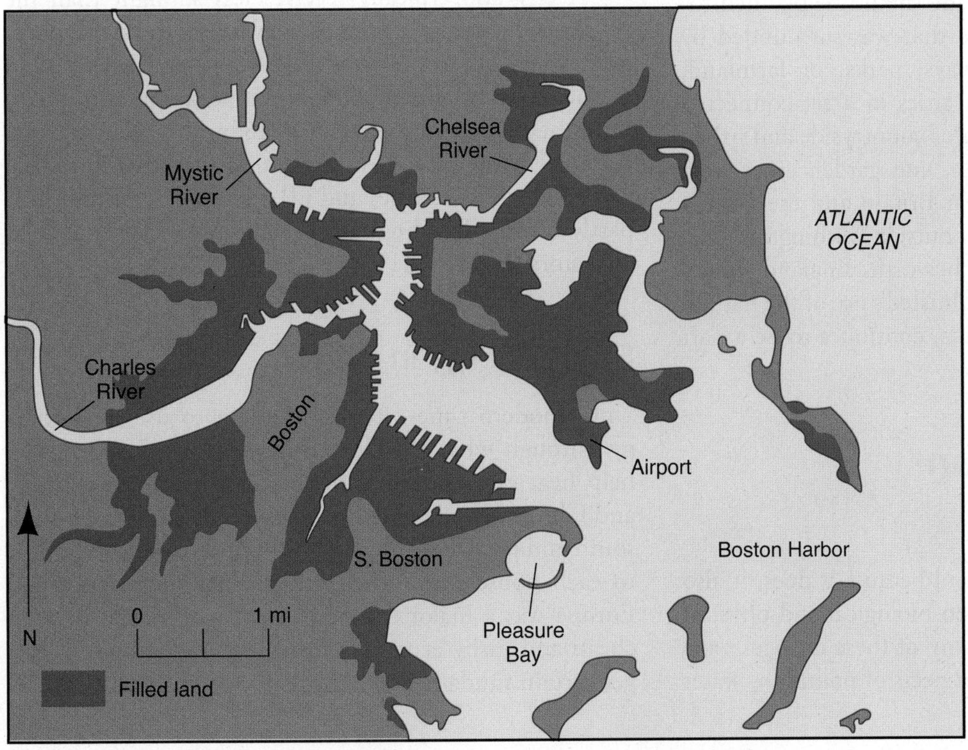

A.W. Spirn, *The Granite Garden: Urban Nature and Human Design* [New York: Basic Books, 1984].

FIGURE 22.10 Nature integrated into a city plan. Boston has been modified over time to improve the environment and provide more building locations. This map of Boston shows land filled in to provide new building sites as of 1982. Although such landfill allows for expansion of the city, it can also create environmental problems, which then must be solved.

largest project had been the filling of Back Bay, which began in 1858 and continued for decades. Once filled, however, the area suffered from flooding and water pollution.

Olmsted's solution to these problems was a water-control project called the "fens." His goal was to "abate existing nuisances" by keeping sewage out of the streams and ponds and building artificial banks for the streams to prevent flooding—and to do this in a natural-looking way. His solution included creating artificial watercourses by digging shallow depressions in the tidal flats, following meandering patterns like natural streams; setting aside other artificial depressions as holding ponds for tidal flooding; restoring a natural salt marsh planted with vegetation tolerant of brackish water; and planting the entire area to serve as a recreational park when not in flood. He put a tidal gate on the Charles River—Boston's major river—and had two major streams diverted directly through culverts into the Charles so that they flooded the fens only during flood periods. He reconstructed the Muddy River primarily to create new, accessible landscape. The result of Olmsted's vision was that control of water became an aesthetic addition to the city. The blending of several goals made the development of the fens a landmark in city planning. Although to the casual stroller it appears to be simply a park for recreation, the area serves an important environmental function in flood control.

Parks near rivers and the ocean are receiving more and more attention. For example, New York City is spending several hundred million dollars to build the Hudson River Park along the Hudson River, where previously abandoned docks and warehouses littered the shoreline and barred public access to the river.

An extension of the park idea was the "garden city," a term coined in 1902 by Ebenezer Howard. Howard believed that city and countryside should be planned together. A **garden city** was one that was surrounded by a **greenbelt**—a belt of parkways, parks, or farmland. The idea was to locate garden cities in a set connected by greenbelts, forming a system of countryside and urban landscapes. The idea caught on, and garden cities were planned and developed in Great Britain and the United States. Greenbelt, Maryland, just outside Washington, DC, is one of these cities, as is Lecheworth, England. Howard's garden city concept, like Olmsted's use of the natural landscape in designing city parks, continues to be a part of city planning today.

22.6 The City as an Environment

A city changes the landscape, and because it does, it also changes the relationship between biological and physical aspects of the environment. Many of these changes were discussed in earlier chapters as aspects of pollution, water management, or climate. You may find some mentioned again in the following sections, generally with a focus on how effective city planning can reduce the problems.

The Energy Budget of a City

Like any ecological and environmental system, a city has an "energy budget." The city exchanges energy with its environment in the following ways: (1) absorption and reflection of solar energy, (2) evaporation of water, (3) conduction of air, (4) winds (air convection), (5) transport of fuels into the city and burning of fuels by people in the city, and (6) convection of water (subsurface and surface stream flow). These in turn affect the climate in the city, and the city may affect the climate in the nearby surroundings, a possible landscape effect.

The Urban Atmosphere and Climate

Cities affect the local climate; as the city changes, so does its climate (see Chapter 20). Cities are generally less windy than nonurban areas because buildings and other structures obstruct the flow of air. But city buildings also channel the wind, sometimes creating local wind tunnels with high wind speeds. The flow of wind around one building is influenced by nearby buildings, and the total wind flow through a city is the result of the relationships among all the buildings. Thus, plans for a new building must take into account its location among other buildings as well as its shape. In some cases, when this has not been done, dangerous winds around tall buildings have blown out windows, as happened to the John Hancock Building in Boston on January 20, 1973, a famous example of the problem.

A city also typically receives less sunlight than the countryside because of the particulates in the atmosphere over cities—often over ten times more particulates than in surrounding areas.[10] Despite reduced sunlight, a city is a heat island, warmer than surrounding areas, for two reasons: (1) the burning of fossil fuels and other industrial and residential activities and (2) a lower rate of heat loss, partly because buildings and paving materials act as solar collectors (Figure 22.11).[11]

Solar Energy in Cities

Until modern times, it was common to use solar energy, through what is called today *passive solar energy*, to help heat city houses. Cities in ancient Greece, Rome, and China were designed so that houses and patios faced south and passive solar energy applications were accessible to each household.[11] The 20th century in America and Europe was a major exception to this approach because cheap and easily accessible fossil fuels led people to forget certain fundamental lessons. Today, the industrialized

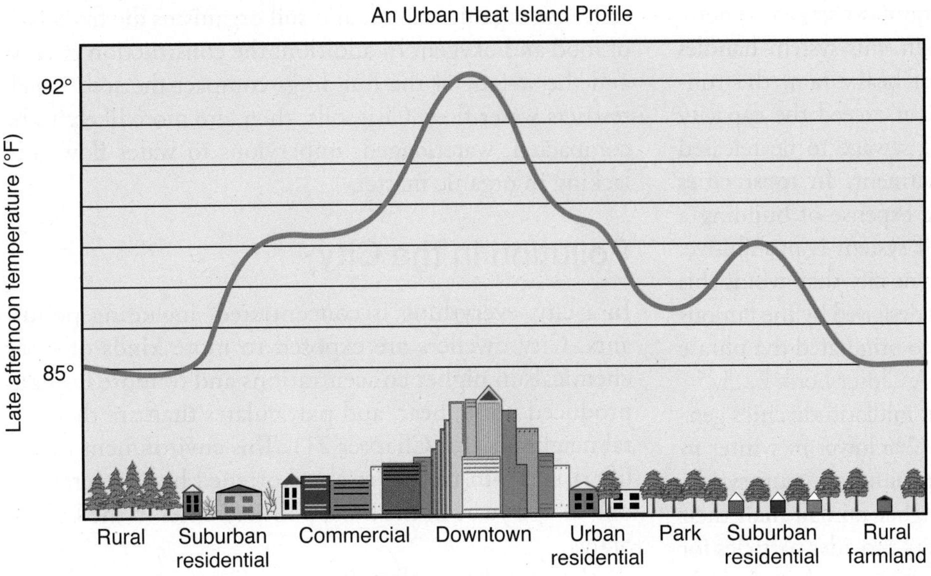

FIGURE 22.11 A typical urban heat island profile. The graph shows temperature changes correlated with the density of development and trees.

Andrasko and Huang in H. Akbari et al., *Cooling Our Communities: A Guidebook on Tree Planting and Light-Colored Surfacing* [Washington, DC: U.S. EPA Office of Policy Analysis, 1992].

nations are beginning to appreciate the importance of solar energy once again. Solar photovoltaic devices that convert sunlight to electricity are becoming a common sight in many cities, and some cities have enacted solar energy ordinances that make it illegal to shade another property owner's building in such a way that it loses solar heating capability. (See Chapter 16 for a discussion of solar energy.)

Water in the Urban Environment

Modern cities affect the water cycle, in turn affecting soils and consequently plants and animals in the city. Because city streets and buildings prevent water infiltration, most rain runs off into storm sewers. The streets and sidewalks also add to the heat island effect by preventing water in the soil from evaporating to the atmosphere, a process that cools natural ecosystems. Chances of flooding increase both within the city and downstream outside the city. New, ecological methods of managing storm water can alleviate these problems by controlling the speed and quality of water running off pavements and into streams. For example, a plan for the central library's parking lot in Alexandria, Virginia, includes wetland vegetation and soils that temporarily absorb runoff from the parking lot, remove some of the pollutants, and slow down the water flow (Figure 22.12).

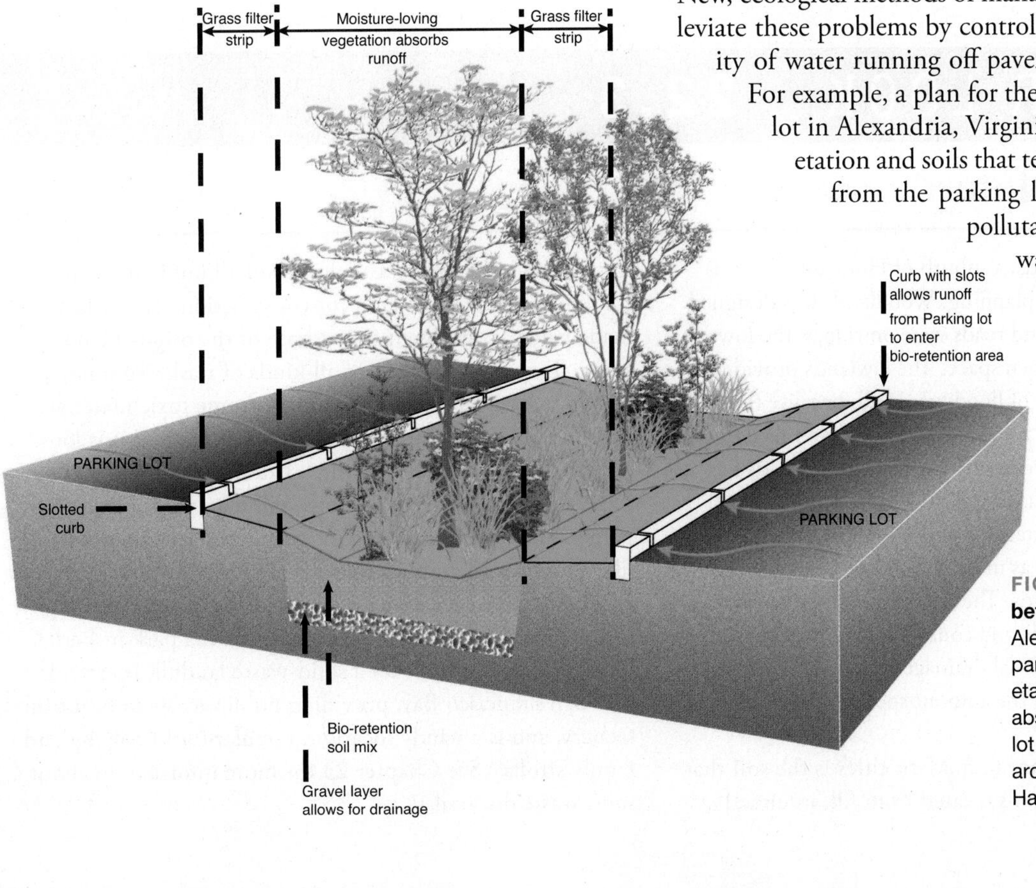

Modified after Rhodeside & Harwell Landscape Architects.

FIGURE 22.12 Planned for better drainage. A plan for the Alexandria, Virginia, central library parking lot includes wetland vegetation and soils that temporarily absorb runoff from the parking lot (see arrows). The landscape architecture firm of Rhodeside & Harwell planned the project.

Most cities have a single underground sewage system. During times of no rain or light rain, this system handles only sewage. But during periods of heavy rain, the runoff is mixed with the sewage and can exceed the capacity of sewage-treatment plants, causing sewage to be released downstream without sufficient treatment. In most cities that already have such systems, the expense of building a completely new and separate runoff system is prohibitive, so other solutions must be found. One city that avoids this problem is Woodlands, Texas. It was designed by the famous landscape architect Ian McHarg, who originated the phrase "design with nature," the subject of A Closer Look 22.3.[10]

Because of reduced evaporation, midlatitude cities generally have lower relative humidity (2% lower in winter to 8% lower in summer) than the surrounding countryside. At the same time, cities can have higher rainfall than their surroundings because dust above a city provides particles for condensation of water vapor. Some urban areas have 5–10% more precipitation and considerably more cloud cover and fog than their surrounding areas. Fog is particularly troublesome in the winter and may impede ground and air traffic.

Soils in the City

A modern city has a great impact on soils. Since most of a city's soil is covered by cement, asphalt, or stone, the soil no longer has its natural cover of vegetation, and the natural exchange of gases between the soil and air is greatly reduced. No longer replenished by vegetation growth, these soils lose organic matter, and soil organisms die from lack of food and oxygen. In addition, the construction process and the weight of the buildings compact the soil, which restricts water flow. City soils, then, are more likely to be compacted, waterlogged, impervious to water flow, and lacking in organic matter.

Pollution in the City

In a city, everything is concentrated, including pollutants. City dwellers are exposed to more kinds of toxic chemicals in higher concentrations and to more human-produced noise, heat, and particulates than are their rural neighbors (see Chapter 21). This environment makes life riskier—in fact, lives are shortened by an average of one to two years in the most polluted cities in the United States.

Some urban pollution comes from motor vehicles, which emit nitrogen oxides, ozone, carbon monoxide, and other air pollutants from exhaust. Electric power plants also produce air pollutants. Home heating is a third source, contributing particulates, sulfur oxides, nitrogen oxides, and other toxic gases. Industries are a fourth source, contributing a wide variety of chemicals. The primary sources of particulate air pollution—which consists of smoke, soot, and tiny particles formed from emissions of sulfur dioxide and volatile organic compounds—are older, coal-burning power plants, industrial boilers, and gas- and diesel-powered vehicles.[12]

A CLOSER LOOK 22.3

Design with Nature

The new town of Woodlands, a suburb of Houston, Texas, is an example of professional planning. Woodlands was designed so that most of its houses and roads were on ridges; the lowlands were left as natural open space. The lowlands provide areas for temporary storage of floodwater and, because the land is unpaved, allow rain to penetrate the soil and recharge the aquifer for Houston. Preserving the natural lowlands has other environmental benefits as well. In this region of Texas, low-lying wetlands are habitats for native wildlife, such as deer. Large, attractive trees, such as magnolias, grow here, providing food and habitat for birds. The innovative city plan has economic as well as aesthetic and conservational benefits. It is estimated that a conventional drainage system would have cost $14 million more than the amount spent to develop and maintain the wetlands.[a]

A kind of soil important in modern cities is the soil that occurs on **made lands**—lands created from fill, intended sometimes to serve as waste dumps of all kinds and sometimes to provide more land for construction. The soils of made lands are different from those of the original landscape. They may be made of all kinds of trash, from newspapers to bathtubs, and may contain some toxic materials. The fill material is unconsolidated, meaning that it is loose material without rock structure. Thus, it is not well suited to be a foundation for buildings. Fill material is particularly vulnerable to earthquake tremors and can act somewhat like a liquid and amplify the effects of the earthquake on buildings. However, some made lands have been turned into well-used parks. For example, a marina park in Berkeley, California, is built on a solid-waste landfill. It extends into San Francisco Bay, providing public access to beautiful scenery, and is a windy location, popular for kite flying and family strolls. (See Chapter 23 for more information about solid-waste disposal.)

Although it is impossible to eliminate exposure to pollutants in a city, exposure can be reduced through careful design, planning, and development. For example, when lead was widely used in gasoline, exposure to lead was greater near roads. Exposure could be reduced by placing houses and recreational areas away from roads and by developing a buffer zone using trees that are resistant to the pollutant and that absorb pollutants.

22.7 Cities and Their Rivers

Traditionally, rivers have been valued for their usefulness in transportation and as places to dump wastes and therefore not as places of beauty or recreation. The old story was that a river renewed and cleaned itself every mile or every three miles (depending on who said it). That may have been relatively correct when there was one person or one family per linear river mile, but it is not for today's cities, with their high population densities and widespread use and dumping of modern chemicals.

Kansas City, Missouri, at the confluence of the Kansas and Missouri rivers, illustrates the traditional disconnect between a city and its river. The Missouri River's floodplain provides a convenient transportation corridor, so the south shore is dominated by railroads, while downtown the north shore forms the southern boundary of the city's airport. Except for a small riverfront park, the river has little place in this city as a source of recreation and relief for its citizens or in the conservation of nature.

The same used to be true of the Hudson River in New York, but that river has undergone a major cleanup since the beginning of the project *Clearwater,* led in part by folksinger Pete Seeger and also by activities of the city's Hudson River Foundation and Metropolitan Waterfront Alliance. Not only is the river cleaner, but an extensive Hudson River Park is being completed, transforming Manhattan's previously industrial and uninviting riverside into a beautifully landscaped and inviting park (Figure 22.13) extending for miles from the southern end of Manhattan to near the George Washington Bridge.

The throngs of sunbathers, picnickers, older people, young couples, and parents with children relaxing on the grass and enjoying the river views are proof of city dwellers' need for contact with nature. A lesson we are learning is that for cities on rivers, one way to bring nature to the city is to connect the city to its river.

Vegetation in Cities

Trees, shrubs, and flowers add to the beauty of a city. Plants fill different needs in different locations. Trees provide shade, which reduces the need for air-conditioning and makes travel much more pleasant in hot weather. In parks, vegetation provides places for quiet contemplation; trees and shrubs can block some of the city sounds, and their complex shapes and structures create a sense of solitude. Plants also provide habitats for wildlife, such as birds and squirrels, which many urban residents consider pleasant additions to a city.

The use of trees in cities has expanded since the Renaissance. In earlier times, trees and shrubs were set apart in gardens, where they were viewed as scenery but not experienced as part of ordinary activities. Street trees were first used in Europe in the 18th century; among the first cities to line streets with trees were London and Paris (Figure 22.14). In many cities, trees are now considered an essential element of the urban visual scene, and major cities have large tree-planting programs. In New York City, for example, 11,000 trees are planted each year, and in Vancouver, Canada, 4,000

Photo by Daniel B. Botkin

FIGURE 22.13 The newly built Hudson River Park on Manhattan's West Side illustrates the changing view of rivers and the improved use of riverfronts for recreation and urban landscape beauty.

Robert Holmes/©Corbis

FIGURE 22.14 Paris was one of the first modern cities to use trees along streets to provide beauty and shade, as shown in this picture of the famous Champs-Elysées.

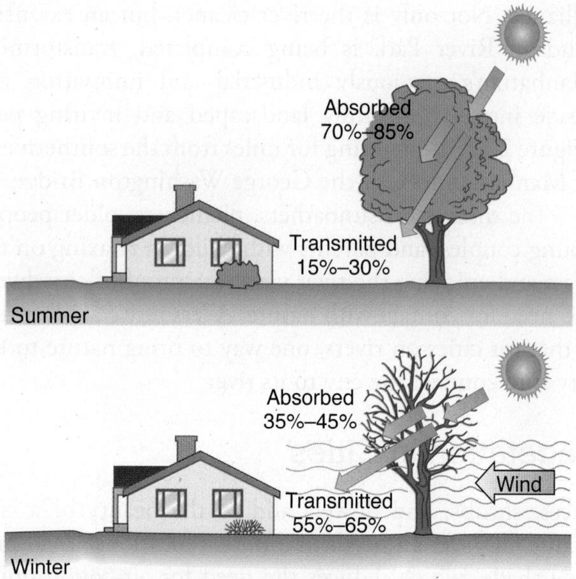

J. Huang and S. Winnett, in H. Akbari et al., *Cooling Our Communities: A Guidebook on Tree Planting and Light-Colored Surfacing* [Washington, DC: U.S. EPA Office of Policy Analysis, 1992].)

FIGURE 22.15 Trees cool homes. Trees can improve the microclimate near a house, protecting the house from winter winds and providing shade in the summer while allowing sunlight through in the winter.

are planted each year.[13] Trees are also increasingly used to soften the effects of climate near houses. In colder climates, rows of conifers planted to the north of a house can protect it from winter winds. Deciduous trees to the south can provide shade in the summer, reducing requirements for air-conditioning, yet allowing sunlight to warm the house in the winter (Figure 22.15).

Cities can even provide habitat for endangered plants. For example, Lakeland, Florida, uses endangered plants in local landscaping with considerable success. However, it is necessary to select species carefully because vegetation in cities must be able to withstand special kinds of stress, such as compacted soils, poor drainage, and air pollution. Because trees along city streets are often surrounded by cement, and because the soils tend to be compacted and drain poorly, the root systems are likely to suffer from extremes of drought on the one hand and soil saturation (immediately following or during a rainstorm) on the other. The solution to this particular problem is to specially prepare streets and sidewalks for tree growth. A tree-planting project was completed for the World Bank Building in Washington, DC, in 1996. Special care was taken to provide good growing conditions for trees, including aeration, irrigation, and adequate drainage so that the soils did not become waterlogged. The trees continue to grow and remain healthy.

Many species of trees and plants are very sensitive to air pollution and will not thrive in cities. The eastern white pine of North America, for example, is extremely sensitive to ozone pollution and does not do well in cities with heavy motor-vehicle traffic or along highways. Dust, too, can interfere with the exchange of oxygen and carbon dioxide necessary for photosynthesis and respiration of the trees. City trees also suffer direct damage from pets, from the physical impact of bicycles, cars, and trucks, and from vandalism. Trees subject to such stresses are more susceptible to attacks by fungus diseases and insects. The lifetime of trees in a city is generally shorter than in their natural woodland habitats unless they are given considerable care.

Some species of trees are more useful and successful in cities than are others. An ideal urban tree would be resistant to all forms of urban stress, have a beautiful form and foliage, and produce no messy fruit, flowers, or leaf litter that required removal. In most cities, in part because of these requirements, only a few tree species are used for street planting. However, reliance on one or a few species results in ecologically fragile urban planting, as we learned when Dutch elm disease spread throughout the eastern United States, destroying urban elms and leaving large stretches of a city treeless. It is prudent to use a greater diversity of trees to avoid the effects of insect infestations and tree diseases.[14]

Cities, of course, have many recently disturbed areas, including abandoned lots and the medians in boulevards and highways. Disturbed areas provide habitat for early-successional plants, including many that we call "weeds," which are often introduced (exotic) plants, such as European mustard. Therefore, wild plants that do particularly well in cities are those characteristic of disturbed areas and of early stages in ecological succession (see Chapter 6). City roadsides in Europe and North America have wild mustards, asters, and other early-successional plants.

Urban "Wilds": The City as Habitat for Wildlife and Endangered Species

We don't associate wildlife with cities—indeed, with the exception of some birds and small, docile mammals such as squirrels, most wildlife in cities are considered pests. But there is much more wildlife in cities, a great deal of it unnoticed. In addition, there is growing recognition that urban areas can be modified to provide habitats for wildlife that people can enjoy. This can be an important method of biological conservation.[15,16]

We can divide city wildlife into the following categories: (1) species that cannot persist in an urban environment and disappear; (2) those that tolerate an urban environment but do better elsewhere; (3) those that have adapted to urban environments, are abundant there, and are either neutral or beneficial to human beings; and (4) those that are so successful they become pests.

Cooper's hawks probably belong in the third category. They are doing pretty well in Tucson, Arizona, a city of 900,000 people. Although this hawk is a native of the surrounding Sonoran Desert, some of them are nesting in groves of trees within the city. Nest success in 2005 was 84%, between two-thirds and three-quarters of the juvenile hawks that left the nest were still alive six months later, and the population is increasing. Scientists studying the hawk in Tucson concluded that the "urbanized landscape can provide high-quality habitat."[17]

Cities can even be home to rare or endangered species. Peregrine falcons once hunted pigeons above the streets of Manhattan. Unknown to most New Yorkers, the falcons nested on the ledges of skyscrapers and dived on their prey in an impressive display of predation. The falcons disappeared when DDT and other organic pollutants caused a thinning of their eggshells and a failure in reproduction, but they have been reintroduced into the city. The first reintroduction took place in 1982. In 2013, 20 breeding pairs have been found to live in the city, some nesting on major bridges; the population in New York City is one of the largest concentrations of this species now in the United States. The reintroduction of peregrine falcons illustrates an important recent trend: the growing understanding that city environments can assist in the conservation of nature, including the conservation of endangered species.

In sum, cities are a habitat, albeit artificial. They can provide all the needs—physical structures and necessary resources such as food, minerals, and water—for many plants and animals. We can identify ecological food chains in cities, as shown in Figure 22.16 for insect-eating birds and for a fox. These can occur when areas cleared of buildings and abandoned begin to recover and are in an early stage of ecological succession. For some species, cities' artificial structures are sufficiently like their original habitat to be home. Chimney swifts, for example, which once lived in hollow trees, are now common in chimneys and other vertical shafts, where they glue their nests to the walls with saliva. A city can easily have more chimneys per square kilometer than a forest has hollow trees.

Cities also have natural habitats in parks and preserves. In fact, modern parks provide some of the world's best wildlife habitats. In New York City's Central Park, approximately 260 species of birds have been observed—100 in a single day. Urban zoos, too, play an important role in conserving endangered species, and the importance of parks and zoos will increase as truly wild areas shrink.

Finally, cities that are seaports often have many species of marine wildlife at their doorsteps. New York City's waters include sharks, bluefish, mackerel, tuna, striped bass, and nearly 250 other species of fish.

(a)

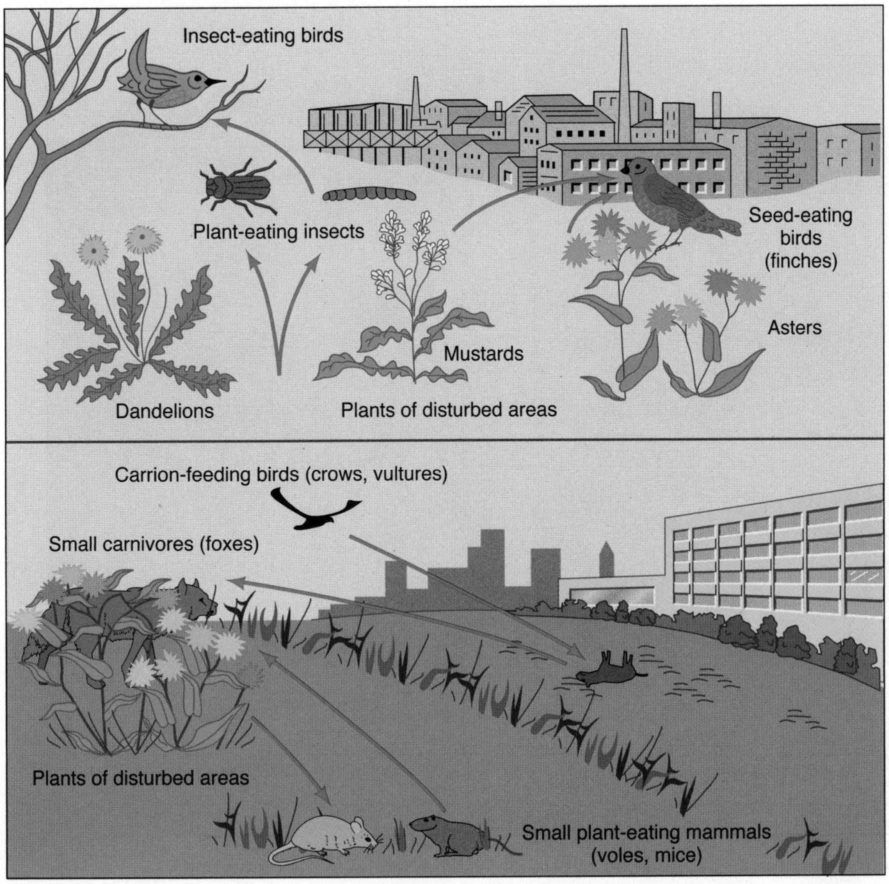

(b)

FIGURE 22.16 **(a) An urban food chain based on plants of disturbed places and insect herbivores. (b)** An urban food chain based on roadkill.

City environments can contribute to wildlife conservation in a number of ways. Urban kitchen gardens—backyard gardens that provide table vegetables and decorative plants—can be designed to provide habitats. For instance, these gardens can include flowers that provide nectar for threatened or endangered hummingbirds. Rivers and their riparian zones, ocean shorelines, and wooded parks can provide habitats for endangered species and ecosystems. For example, prairie vegetation, which once occupied more land area than any other vegetation type in the United States, is rare today, but one restored prairie exists within the city limits of Omaha, Nebraska. (Some urban nature preserves are not accessible to the public or they offer only limited access, as is the case with the prairie preserve in Omaha.)

Urban drainage structures can also be designed as wildlife habitats. A typical urban runoff design depends on concrete-lined ditches that speed the flow of water from city streets to lakes, rivers, or the ocean. However, as with Boston's Back Bay design, discussed earlier, these features can be planned to maintain or create stream and marsh habitats, with meandering waterways and storage areas that do not interfere with city processes. Such areas can become habitats for fish and mammals (Figure 22.17). Modified to promote wildlife, cities can provide urban corridors that allow wildlife to migrate along their natural routes.[18] Urban corridors also help to prevent some of the effects of ecological islands (see Chapter 9) and are increasingly important to biological conservation.

Animal Pests

Pests are familiar to urban dwellers. The most common city pests are cockroaches, fleas, termites, rats, pigeons, and (since banning DDT) bedbugs, but there are many more, especially species of insects. In gardens and parks, pests include insects, birds, and mammals that feed on fruit and vegetables and destroy foliage of shade trees and plants. Pests compete with people for food and spread diseases. Indeed, before modern sanitation and medicine, such diseases played a major role in limiting human population density in cities. Bubonic plague is spread by fleas found on rodents; mice and rats in cities promoted the spread of this disease. Bubonic plague continues to be

FIGURE 22.17 **Water drainage systems in a city can be modified to provide wildlife habitat in the community.**

a health threat in cities. Poor sanitation and high population densities of people and rodents set up a situation where the disease can strike.

An animal is a pest to people when it is in an undesired place at an undesirable time doing an unwanted thing. A termite in a woodland helps the natural regeneration of wood by hastening decay and speeding the return of chemical elements to the soil, where they are available to living plants. But termites in a house are pests because they threaten the house's physical structure.

Animals that do well enough in cities to become pests have certain characteristics in common. They are generalists in their food choice, so they can eat what we eat (including the leftovers we throw in the trash), and they have a high reproductive rate and a short average lifetime.

Controlling Pests

We can best control pests by recognizing how they fit their natural ecosystem and identifying the things that control them in nature. People often assume that the only way to control animal pests is with poisons, but there are limitations to this approach. Early poisons used in pest control were generally also toxic to people and pets (see Chapter 11). Another problem is that reliance on one toxic compound can cause a species to develop a resistance to it, which can lead to rebound—a renewed increase in that pest's population. A pesticide used once and spread widely will greatly reduce the population of the pest. However, when the pesticide loses its effectiveness, the pest population can increase rapidly as long as habitat is suitable and food plentiful. This is what happened when an attempt was made to control Norway rats in Baltimore.

One of the keys to controlling pests is to eliminate their habitats. For example, the best way to control rats is to reduce the amount of open garbage and eliminate areas to hide and nest. Common access areas used by rats are the spaces within and between walls and the openings between buildings where pipes and cables enter. Houses can be constructed to restrict access by rats. In older buildings, we can seal areas of access.

CRITICAL THINKING ISSUE

How Can Urban Sprawl Be Controlled?

As the world becomes increasingly urbanized, individual cities are growing in area as well as population. Residential areas and shopping centers move into undeveloped land near cities, impinging on natural areas and creating a chaotic, unplanned human environment. "Urban sprawl" has become a serious concern in communities all across the United States. According to the EPA, in a recent six-month period approximately 5,000 people left Baltimore City to live in suburbs, with the result that nearly 10,000 acres of forests and farmlands were converted to housing. At this rate, the state of Maryland could use as much land for development in the next 25 years as it has used in the entire history of the state.[19] In the past ten years, 22 states have enacted new laws to try to control urban sprawl.

The city of Boulder, Colorado, has been in the forefront of this effort since 1959, when it created the "blue line"—a line at an elevation of 1,761 m (5,778 ft) (the city itself is at 1,606 m (5269 ft)) above which it would not extend city water or sewer services. Boulder's citizens felt, however, that the blue line was insufficient to control development and maintain the city's scenic beauty in the face of rapid population growth. (Boulder's population had grown in the decade before 1959 from 29,000 to 66,000 and reached 96,000 by 1998.) To prevent uncontrolled development in the area between the city and the blue line, in 1967 Boulder began to use a portion of the city sales tax to purchase land, creating a 10,800-hectare greenbelt around the city proper.

In 1976 Boulder went one step further and limited increases in new residences to 2% a year. Two years later, recognizing that planned development requires a regional approach, the city and surrounding Boulder County adopted a coordinated development plan. By the early 1990s, it had become apparent that further growth control was needed for nonresidential building. The plan that the city finally adopted reduced the allowable density of many commercial and industrial properties, in effect limiting jobs rather than limiting building space.

Boulder's methods to limit the size of its population have worked. The 2010 census showed that the population had increased by a mere 3,000 people and totaled just a little more than 97,000.

The benefits of Boulder's controlled-growth initiatives have been a defined urban–rural boundary; rational, planned development; protection of sensitive environmental areas and scenic vistas; and large areas of open space within and around the city for recreation. And in spite of its growth-control measures, Boulder's economy has remained strong. However, restraints on residential growth forced many people who found jobs in Boulder to seek affordable housing in adjoining communities, where populations ballooned. The population of Superior, Colorado, for example, grew from 225 in 1990 to 9,000 in 2000. Further, as commuting workers—40,000 a day—tried to get to and from their jobs in Boulder, traffic congestion and air pollution increased. In addition, because developers had built housing but not stores in the outlying areas, shoppers flocked into Boulder's downtown mall. When plans for a competing mall in the suburbs were finally announced, however, Boulder officials worried about the loss of revenue if the new mall drew shoppers away from the city. At the same time, sprawl from Denver (only 48 km from Boulder), as well as its infamous "brown cloud" of polluted air, began to spill out along the highway connecting the two communities.

Critical Thinking Questions

1. Is a city an open or a closed system (see Chapter 4)? Use examples from the case of Boulder to support your answer.
2. As Boulder takes steps to limit growth, it becomes an even more desirable place to live, which subjects it to even greater growth pressures. What ways can you suggest to avoid such a positive-feedback loop?
3. Some people in Boulder think the next step is to increase residential density within the city. Do you think people living there will accept this plan? What are the advantages and disadvantages of increasing density?
4. To some, the story of Boulder is the saga of a heroic battle against commercial interests that would destroy environmental resources and a unique quality of life. To others, it is the story of an elite group building an island of prosperity and the good life for themselves. Which do you think it is?

SUMMARY

- As an urban society, we must recognize the city's relation to the environment. A city influences and is influenced by its environment and is an environment itself.
- Like any other life-supporting system, a city must maintain a flow of energy, provide necessary material resources, and have ways of removing wastes. These functions are accomplished through transportation and communication with outlying areas.
- Because cities depend on outside resources, they developed only when human ingenuity resulted in modern agriculture and thus excess food production. The history of cities divides into four stages: (1) the rise of towns; (2) the era of classic urban centers; (3) the period of industrial metropolises; and (4) the age of mass telecommunication, computers, and new forms of travel.
- Locations of cities are strongly influenced by environment. It is clear that cities are not located at random but in places of particular importance and environmental advantage. A city's site and situation are both important
- A city creates an environment that is different from surrounding areas. Cities change local climate; they are commonly cloudier, warmer, and rainier than surrounding areas.
- In general, life in a city is riskier because of higher concentrations of pollutants and pollutant-related diseases.
- Cities favor certain animals and plants. Natural habitats in city parks and preserves will become more important as wilderness shrinks.
- Trees are an important part of urban environments, but cities place stresses on trees. Especially important are the condition of urban soils and the supply of water for trees.
- Cities can help to conserve biological diversity, providing habitat for some rare and endangered species.
- As the human population continues to increase, we can envision two futures: one in which people are dispersed widely throughout the countryside and cities are abandoned except by the poor; and another in which cities attract most of the human population, freeing much landscape for conservation of nature, production of natural resources, and public-service functions of ecosystems.

REEXAMINING THEMES AND ISSUES

Sean Randall/Getty Images, Inc.

HUMAN POPULATION

As the world's human population increases, we are becoming an increasingly urbanized species. Present trends indicate that in the future, most citizens of most nations will live in their country's single largest city. Thus, concern about urban environments will be increasingly important.

Cities contain the seeds of their own destruction: The very artificiality of a city gives its inhabitants the sense that they are independent of their surrounding environment. But the opposite is the case: The more artificial a city, the more it depends on its surrounding environment for resources and the more susceptible it becomes to major disasters unless it recognizes and plans for this susceptibility. The keys to sustainable cities are an ecosystem approach to urban planning and a concern with the aesthetics of urban environments.

© Biletskiy_Evgeniy/iStockphoto

SUSTAINABILITY

Cities depend on the sustainability of all renewable resources and must therefore recognize that they greatly affect their surrounding environments. Urban pollution of rivers that flow into an ocean can affect the sustainability of fish and fisheries. Urban sprawl can have destructive effects on endangered habitats and ecosystems, including wetlands. At the same time, cities designed to support vegetation and some wildlife can contribute to the sustainability of nature.

© Anton Balazh 2011/iStockphoto

GLOBAL PERSPECTIVE

The great urban centers of the world produce global effects. As an example, because people are concentrated in cities and because many cities are located at the mouths of rivers, most major river estuaries of the world are severely polluted.

ssguy/ShutterStock

URBAN WORLD

The primary message of this chapter is that Earth is becoming urbanized and that environmental science must deal more and more with urban issues.

The modern tendency has been to focus environmental conservation efforts on wilderness, large parks, and preserves outside of cities. Meanwhile, city environments have been allowed to decay. As the world becomes increasingly urbanized, however, a change in values is necessary. If we are serious about conserving biological diversity, we must assign greater value to urban environments. The more pleasant city environments are, and the more recreation people can find in them, the less pressure there will be on the countryside.

B2M Productions/Getty Images, Inc.

PEOPLE AND NATURE

George Doyle/Getty Images, Inc.

SCIENCE AND VALUES

Modern environmental sciences tell us much that we can do to improve the environments of cities and the effects of cities on their environments. What we choose to do with this knowledge depends on our values. Scientific information can suggest new options, and we can select among them for the future of our cities, depending on our values.

KEY TERMS

city planning 562
fall line 567
garden city 564
greenbelt 564
made lands 566
site 557
situation 557

STUDY QUESTIONS

1. Should we try to save New Orleans, or should we just give up and move the port at the mouth of the Mississippi River elsewhere? Explain your answer in terms of environment and economics.
2. Which of the following cities are most likely to become ghost towns in the next 100 years? In answering this question, use your knowledge of changes in resources, transportation, and communications.
   (a) Honolulu, Hawaii
   (b) Fairbanks, Alaska
   (c) Juneau, Alaska
   (d) Savannah, Georgia
   (e) Phoenix, Arizona
3. Some futurists picture a world that is one giant biospheric city. Is this possible? If so, under what conditions?
4. The ancient Greeks said that a city should have only as many people as can hear the sound of a single voice. Would you apply this rule today? If not, how would you plan the size of a city?
5. You are the manager of Central Park in New York City and receive the following two offers. Which would you approve? Explain your reasons.
   (a) A gift of $1 billion to plant trees from all the eastern states.
   (b) A gift of $1 billion to set aside half the park to be forever untouched, thus producing an urban wilderness.
6. Your state asks you to locate and plan a new town. The purpose of the town is to house people who will work at a wind farm—a large area of many windmills, all linked to produce electricity. You must first locate the site for the wind farm and then plan the town. How would you proceed? What factors would you take into account?
7. Visit your town center. What changes, if any, would make better use of the environmental location? How could the area be made more livable?
8. In what ways does air travel alter the location of cities? The value of land within a city?
9. You are put in charge of ridding your city's parks of slugs, which eat up the vegetable gardens rented to residents. How would you approach controlling this pest?
10. It is popular to suggest that in the Information Age people can work at home and live in the suburbs and the countryside, so cities are no longer necessary. List five arguments for and five arguments against this point of view.
11. Determine whether it is possible for New Orleans to continue as a major port, or whether, like Venice, Italy, it will fade as a major city, only to be a tourist attraction.
12. Decide which, if any, is the most geographic feature that determines whether a city will be a success. In writing your analysis, consider if, with modern technology, geography no longer matters where a city is placed.

FURTHER READING

Beck, T. ***Principles of Ecological Landscape Design*** (Washington, DC: Island Press, 2013).

Beveridge, C.E., and P. Rocheleau. ***Frederick Law Olmsted: Designing the American Landscape*** (New York, NY: Rizzoli International, 1995). The most important analysis of the work of the father of landscape architecture.

Howard, E. ***Garden Cities of Tomorrow*** (Cambridge, MA: MIT Press, 1965, reprint). A classic work of the 19th century that has influenced modern city design, as in Garden City, New Jersey, and Greenbelt, Maryland. It presents a methodology for designing cities with the inclusion of parks, parkways, and private gardens.

McHarg, I.L. ***Design with Nature*** (New York, NY: John Wiley & Sons, 1995). A classic book about cities and environment.

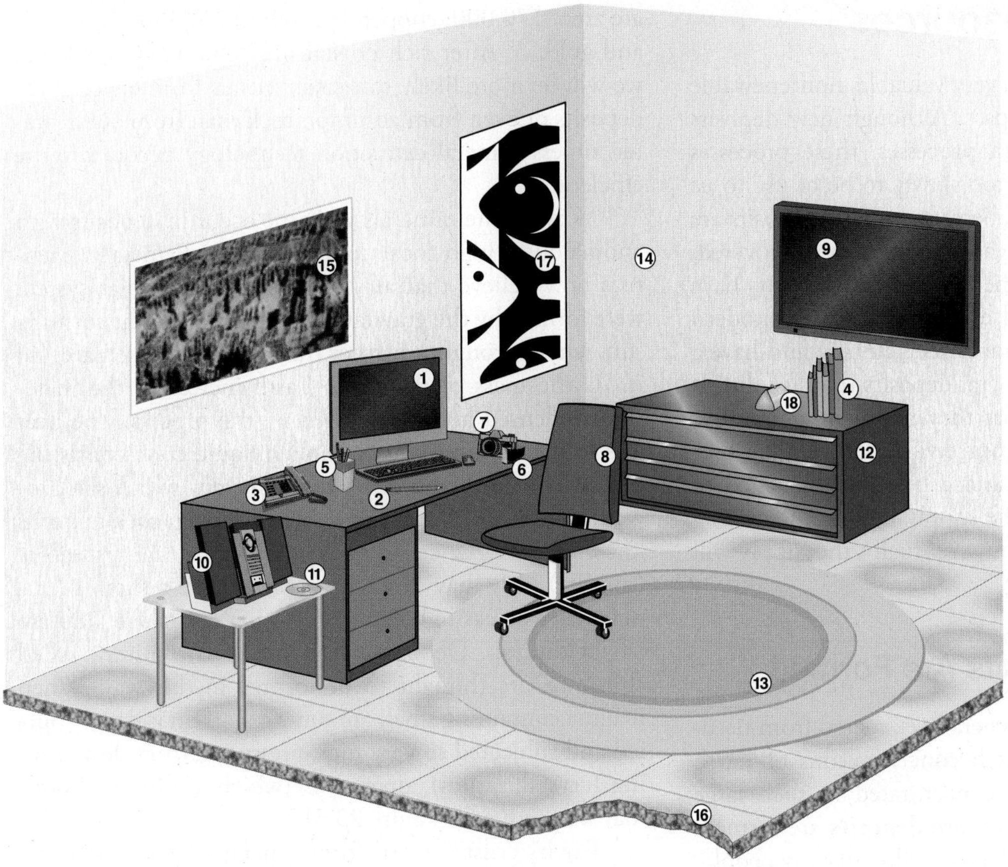

1. **Computer**—Includes gold, silica, nickel, aluminum, zinc, iron, petroleum products and about 30 other minerals.
2. **Pencil**—Includes graphite and clays.
3. **Telephone**—Includes copper, gold and petroleum products.
4. **Books**—Includes limestone and clays.
5. **Pens**—Includes limestone, mica, petroleum products, clays, silica, and talc.
6. **Film**—Includes petroleum products and silver.
7. **Camera**—Includes silica, zinc, copper, aluminum, and petroleum products
8. **Chair**—Includes aluminum and petroleum products.
9. **Television**—Includes aluminum, copper, iron, nickel, silica, rare earth, and strontium.
10. **Stereo**—Includes gold, iron, nickel, beryllium, and petroleum products.
11. **Compact Disc**—Includes aluminum and petroleum products.
12. **Metal Chest**—Includes iron and nickel. The brass trim is made of copper and zinc.
13. **Carpet**—Includes limestone, petroleum products, and selenium.
14. **Drywall**—Includes gypsum clay, vermiculite, calcium carbonate, and micas.
15. **Geologic Map**—Includes clays, petroleum products, and mineral pigments.
16. **Concrete Foundation**—Includes limestone, clays, sand, and gravel
17. **Paint-mineral Pigments**—Includes pigments (such as iron, zinc, and titanium).
18. **Cosmetics**—Includes mineral chemicals.

Modified from S.J. Kropschot and K.M. Johnson, 2006. U.S. 65. Mineral Resources Program. USGS Circular 1289. Menlo Park, CA.

FIGURE 23.2 **Mineral products used in a home office.**

- Eliminate subsidies for extracting virgin materials such as minerals, oil, and timber.
- Establish "green building" incentives that encourage the use of recycled content materials and products in new construction.
- Assess financial penalties for production that uses poor materials management practices.
- Provide financial incentives for industrial practices and products that benefit the environment by enhancing sustainability (for example, by reducing waste production and using recycled materials).
- Provide more incentives for people, industry, and agriculture to develop materials management programs that eliminate or reduce waste by using it as raw material for other products.

Materials management in the United States today is beginning to influence where industries are located. For example, because approximately 50% of the steel produced in the nation now comes from scrap, new steel mills are no longer located near resources such as coal and iron ore. New steel mills are now found in a variety of places, from California to North Carolina and Nebraska; their resource is the local supply of scrap steel. Because they are starting with scrap metal, the new industrial facilities use far less energy and cause much less pollution than older steel mills that must start with virgin iron ore.[7]

Similarly, the recycling of paper is changing where new paper mills are constructed. In the past, mills were built near forested areas where the timber for paper production was being logged. Today, they are being built near cities that have large supplies of recycled paper. New Jersey, for example, has 13 paper mills using recycled paper and 8 steel "mini mills" producing steel from scrap metal. Remarkably, New Jersey has little forested land and no iron mines; resources for the paper and steel mills come from materials already in use, exemplifying the power of materials management.[7]

We have focused on renewable resources in previous parts of this book (Chapter 11, agriculture; Chapter 12, forests; Chapter 13, wildlife; Chapter 18, water; and Chapter 21, air). We discussed nonrenewable resources with respect to fossil fuels in Chapter 15. The remainder of this chapter will discuss other nonrenewable mineral resources and how to sustain them as long as possible by intelligent waste management.

23.3 Mineral Resources

Minerals can be considered a very valuable, nonrenewable heritage from the geologic past. Although new deposits are still forming from Earth processes, these processes are producing new deposits too slowly to be of use to us today or anytime soon. Also, because mineral deposits are generally in small, hidden areas, they must be discovered, and unfortunately most of the easily found deposits have already been discovered and exploited. Thus, if modern civilization were to vanish, our descendants would have a harder time finding rich mineral deposits than we did. It is interesting to speculate that they might mine landfills for metals thrown away by our civilization. Unlike biological resources, minerals cannot be easily managed to produce a sustained yield; the supply is finite. Recycling and conservation will help, but, eventually, the supply will be exhausted.

How Mineral Deposits Are Formed

Metals in mineral form are generally extracted from naturally occurring, unusually high concentrations of Earth materials. When metals are concentrated in such large amounts by geologic processes, **ore deposits** are formed. The discovery of natural ore deposits allowed early peoples to exploit copper, tin, gold, silver, and other metals while slowly developing skills in working with metals.

The origin and distribution of mineral resources is intimately related to the history of the biosphere and to the entire geologic cycle (see Chapter 7). Nearly all aspects and processes of the geologic cycle are involved to some extent in producing local concentrations of useful materials. Earth's outer layer, or crust, is silica-rich, made up mostly of rock-forming minerals containing silica, oxygen, and a few other elements. The elements are not evenly distributed in the crust: Nine elements account for about 99% of the crust by weight (oxygen, 45.2%; silicon, 27.2%; aluminum, 8.0%; iron, 5.8%; calcium, 5.1%; magnesium, 2.8%; sodium, 2.3%; potassium, 1.7%; and titanium, 0.9%). In general, the remaining elements are found in trace concentrations.

The ocean, covering nearly 71% of Earth, is another reservoir for many chemicals other than water. Most elements in the ocean have been weathered from crustal rocks on the land and transported to the oceans by rivers. Others are transported to the ocean by wind or glaciers. Ocean water contains about 3.5% dissolved solids, mostly chlorine (55.1% of the dissolved solids by weight). Each cubic kilometer of ocean water contains about 2.0 metric tons of zinc, 2.0 metric tons of copper, 0.8 metric ton of tin, 0.3 metric ton of silver, and 0.01 metric ton of gold. These concentrations are low compared with those in the crust, where corresponding values (in metric tons/km^3) are zinc, 170,000; copper, 86,000; tin, 5,700; silver, 160; and gold, 5. After rich crustal ore deposits are depleted, we will be more likely to extract metals from lower grade deposits or even from common rock than from ocean water, unless mineral extraction technology becomes more efficient.

Why do the minerals we mine occur in deposits with anomalously high local concentrations? Planetary scientists now believe that all the planets in our solar system were formed by the gravitational attraction of the forming sun, which brought together the matter dispersed around it. As the mass of the proto-Earth increased, the material condensed and was heated by the process. The heat was sufficient to produce a molten liquid core, consisting primarily of iron and other heavy metals, which sank toward the center of the planet. When molten rock material known as *magma* cools, heavier minerals that crystallize (solidify) early may slowly sink toward the bottom of the magma, whereas lighter minerals that crystallize later are left at the top. Deposits of an ore of chromium, called chromite, are thought to be formed in this way. When magma containing small amounts of carbon is deeply buried and subjected to very high pressure during slow cooling (crystallization), diamonds (which are pure carbon) may be produced (Figure 23.3).[8,9]

Earth's crust formed from generally lighter elements and is a mixture of many different kinds. The elements in the crust are not uniformly distributed because geologic processes (such as volcanic activity, plate tectonics, and sedimentary processes), as well as some biological processes, selectively dissolve, transport, and deposit elements and minerals.

Sedimentary processes related to the transport of sediments by wind, water, and glaciers often concentrate materials in amounts sufficient for extraction. As sediments are transported, running water and wind help segregate them by size, shape, and density. This sorting is useful to

Helen Thompson/Animals AnimalsEarth Scenes

FIGURE 23.3 Diamond mine near Kimberley, South Africa. This is the largest hand-dug excavation in the world.

people. The best sand or sand and gravel deposits for construction, for example, are those in which the finer materials have been removed by water or wind. Sand dunes, beach deposits, and deposits in stream channels are good examples. The sand and gravel industry amounts to several billion dollars annually and, in terms of the total volume of materials mined, is one of the largest nonfuel mineral industries in the United States.[5]

Rivers and streams that empty into oceans and lakes carry tremendous quantities of dissolved material from the weathering of rocks. Over geologic time, a shallow marine basin may be isolated by tectonic activity that uplifts its boundaries; or climate variations, such as the ice ages, may produce large inland lakes with no outlets. As these basins and lakes eventually dry up, the dissolved materials drop out of solution and form a wide variety of compounds, minerals, and rocks that have important commercial value.[9]

Biological processes form some mineral deposits, such as phosphates and iron ore deposits. The major iron ore deposits exist in sedimentary rocks that were formed more than 2 billion years ago.[10] Although the processes are not fully understood, it appears that major deposits of iron stopped forming when the atmospheric concentration of oxygen reached its present level.[11]

Organisms, too, form many kinds of minerals, such as the calcium minerals in shells and bones. Some of these minerals cannot be formed inorganically in the biosphere. Thirty-one biologically produced minerals have been identified.[12]

Weathering, the chemical and mechanical decomposition of rock, concentrates some minerals in the soil, such as native gold and oxides of aluminum and iron. (The more soluble elements, such as silica, calcium, and sodium, are selectively removed by soil and biological processes.) If sufficiently concentrated, residual aluminum oxide forms an ore of aluminum known as bauxite. Important nickel and cobalt deposits are also found in soils developed from iron and magnesium-rich igneous rocks.

23.4 Figuring Out How Much Is Left

Estimating how much is left of our valuable and nonrenewable mineral resources will help us estimate how long they are likely to last at our present rate of use and motivate us to do everything we can to sustain them as long as possible for future generations. We can begin by looking at the classification of minerals as *resources* and *reserves.*

Mineral Resources and Reserves

Mineral **resources** are broadly defined as known concentrations of elements, chemical compounds, minerals, or rocks. Mineral **reserves** are concentrations that at the time of evaluation can be legally and economically extracted as a commodity that can be sold at a profit (Figure 23.4).

The main point here is that *resources are not reserves.* An analogy from a student's personal finances may help clarify this point. A student's reserves are liquid assets, such as money in the bank, whereas the student's resources include the total income the student can expect to earn during his or her lifetime. This distinction is often critical to the student in school because resources that may become available in the future cannot be used to pay this month's bills.[13] For planning purposes, it is important to continually reassess all components of a total resource, considering new technology, the probability of geologic discovery, and shifts in economic and political conditions.[4]

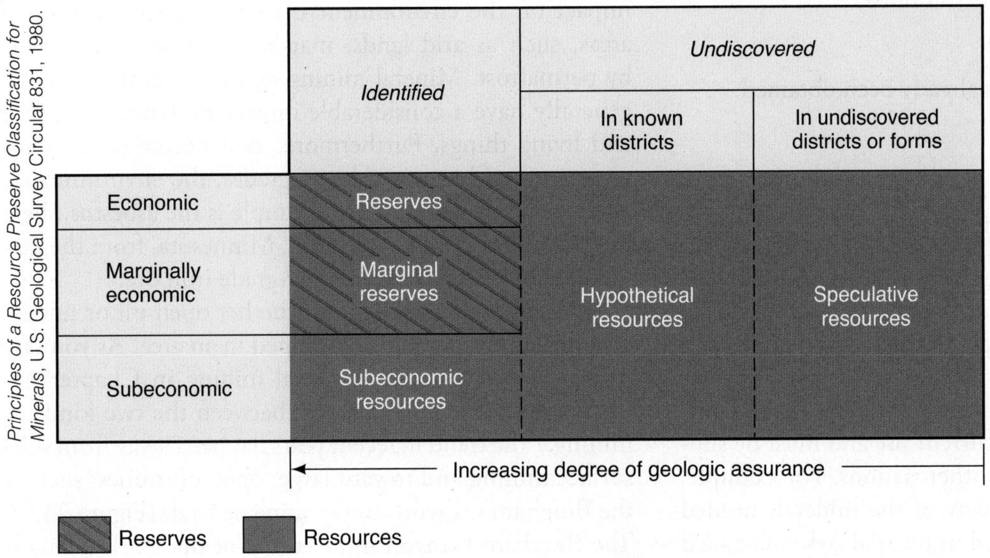

FIGURE 23.4 Classification of mineral resources used by the U.S. Geological Survey and the U.S. Bureau of Mines.

Availability and Use of Our Mineral Resources

Earth's mineral resources can be divided into broad categories according to their use: elements for metal production and technology, building materials, minerals for the chemical industry, and minerals for agriculture. Metallic minerals can be further classified by their abundance. Abundant metals include iron, aluminum, chromium, manganese, titanium, and magnesium. Scarce metals include copper, lead, zinc, tin, gold, silver, platinum, uranium, mercury, and molybdenum.

Some minerals, such as salt (sodium chloride), are necessary for life. Primitive peoples traveled long distances to obtain salt when it was not locally available. Other minerals are desired or considered necessary to maintain a particular level of technology.

When we think about minerals, we usually think of metals; but with the exception of iron, the predominant minerals are not metallic. Consider the annual world consumption of a few selected elements. Sodium and iron are used at a rate of approximately 100–1,000 million metric tons per year; and nitrogen, sulfur, potassium, and calcium at a rate of approximately 10–100 million metric tons per year, primarily as soil conditioners or fertilizers. Elements such as zinc, copper, aluminum, and lead have annual world consumption rates of about 3–10 million metric tons, and gold and silver are consumed at annual rates of 10,000 metric tons or less. Of the metallic minerals, iron makes up 95% of all the metals consumed; and nickel, chromium, cobalt, and manganese are used mainly in alloys of iron (as in stainless steel).

The basic issue associated with mineral resources is not actual exhaustion or extinction but the cost of maintaining an adequate stock by mining and recycling. At some point, the costs of mining exceed the worth of material. When the availability of a particular mineral becomes limited, there are four possible solutions:

- Find more sources.
- Recycle and reuse what has already been obtained.
- Reduce consumption.
- Find a substitute.

Which choice or combination of choices is made depends on social, economic, and environmental factors.

U.S. Supply of Mineral Resources

Domestic supplies of many mineral resources in the United States are insufficient for current use and must be supplemented by imports from other nations. For example, the United States imports many of the minerals needed for its complex military and industrial system, called strategic minerals (such as bauxite, manganese, graphite, cobalt, strontium, and asbestos). Of particular concern is the possibility that the supply of a much-desired or much-needed mineral will be interrupted by political, economic, or military instability in the supplying nation.

That the United States—along with many other countries—depends on a steady supply of imports to meet its domestic demand does not necessarily mean that sufficient kinds and amounts can't be mined domestically. Rather, it suggests economic, political, or environmental reasons that make it easier, more practical, or more desirable to import the material. This has resulted in political alliances that otherwise would be unlikely. Industrial countries often need minerals from countries whose policies they don't necessarily agree with; as a result, they make political concessions on human rights and other issues that they would not otherwise make.[3]

Moreover, the fact remains that mineral resources are limited, and this raises important questions. How long will a particular resource last? How much short-term or long-term environmental deterioration are we willing to accept to ensure that resources are developed in a particular area? How can we make the best use of available resources?

23.5 Impacts of Mineral Development

The impact of mineral exploitation depends on ore quality, mining procedures, local hydrologic conditions, climate, rock types, size of operation, topography, and many more interrelated factors. In addition, our use of mineral resources has a significant social impact.

Environmental Impacts

Exploration for mineral deposits generally has a minimal impact on the environment if care is taken in sensitive areas, such as arid lands, marshes, and areas underlain by permafrost. Mineral mining and processing, however, generally have a considerable impact on land, water, air, and living things. Furthermore, as it becomes necessary to use ores of lower and lower grades, the environmental effects tend to worsen. One example is the asbestos fibers in the drinking water of Duluth, Minnesota, from the disposal of waste from mining low-grade iron ore.

A major practical issue is whether open pit or underground mines should be developed in an area. As you saw in our earlier discussion of coal mining in Chapter 15, there are important differences between the two kinds of mining.[2] The trend in recent years has been away from subsurface mining and toward large, open pit mines, such as the Bingham Canyon copper mine in Utah (Figure 23.5). The Bingham Canyon mine is one of the world's largest

Bettmann/©Corbis

FIGURE 23.5 Aerial photograph of Bingham Canyon Copper Pit, Utah. It is one of the largest artificial excavations in the world.

human-made excavations, covering nearly 8 km^2 (3 mi^2) to a maximum depth of nearly 800 m (2,600 ft).

Surface mines and quarries today cover less than 0.5% of the total area of the United States, but even though their impacts are local, numerous local occurrences will eventually constitute a larger problem. Environmental degradation tends to extend beyond the immediate vicinity of a mine. Large mining operations remove material in some areas and dump waste in others, changing topography. At the very least, severe aesthetic degradation is the result. In addition, dust may affect the air quality, even though care is taken to reduce it by sprinkling water on roads and on other sites that generate dust.

A potential problem with mineral resource development is the possible release of harmful trace elements into the environment. Water resources are particularly vulnerable even if drainage is controlled and sediment pollution is reduced (see Chapter 15 for more about this issue, including a discussion of acid mine drainage). The white streaks in Figure 23.6 are mineral deposits apparently leached from tailings from a zinc mine in Colorado. Similar-looking deposits may cover rocks in rivers for many kilometers downstream from some mining areas.

Mining-related physical changes in the land, soil, water, and air indirectly affect the biological environment. Plants and animals killed by mining activity or by contact with toxic soil or water are some of the direct impacts. Indirect impacts include changes in nutrient cycling, total biomass, species diversity, and ecosystem stability. Periodic or accidental discharge of low-grade pollutants through failure of barriers, ponds, or water diversions, or through the breaching of barriers during floods, earthquakes, or volcanic eruptions, also may damage local ecological systems to some extent.

Social Impacts

The social impacts of large-scale mining result from the rapid influx of workers into areas unprepared for growth. This population influx places stress on local services, such as water supplies, sewage and solid-waste disposal systems, as well as on schools, housing, and nearby recreation and wilderness areas. Land use shifts from open range, forest, and agriculture to urban patterns. Construction and urbanization affect local streams through sediment pollution, reduced water quality, and increased runoff. Air quality suffers as a result of more vehicles, construction dust, and power generation.

Perversely, closing down mines also has negative social impacts. Nearby towns that have come to depend on the income of employed miners can come to resemble the well-known "ghost towns" of the old American West. The price of coal and other minerals also directly affects the livelihood of many small towns. This is especially evident in the Appalachian Mountain region of the United States,

Ed Keller

FIGURE 23.6 Tailings from a lead, zinc, and silver mine in Colorado. White streaks on the slope are mineral deposits apparently leached from the tailings.

where coal mines have closed partly because of lower prices for coal and partly because of rising mining costs. One of the reasons mining costs are rising is the increased level of environmental regulation of the mining industry. Of course, regulations have also helped make mining safer and have facilitated land reclamation. Some miners, however, believe the regulations are not flexible enough, and there is some truth to their arguments. For example, some mined areas might be reclaimed for use as farmland now that the original hills have been leveled. Regulations, however, may require that the land be restored to its original hilly state, even though hills make inferior farmland.

Minimizing the Environmental Impact of Mineral Development

Minimizing the environmental impacts of mineral development requires consideration of the entire cycle of mineral resources shown in Figure 23.7. This diagram reveals that waste is produced by many components of the cycle.

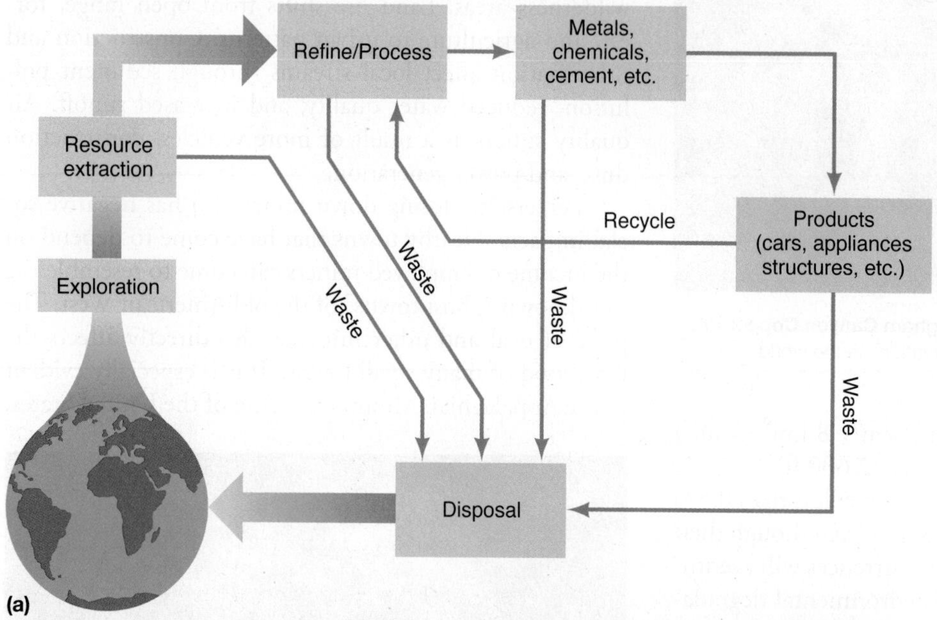

(a)

(b) (c) (d) (e) (f)

FIGURE 23.7 (a) Simplified flowchart of the resource cycle: (b) mining gold in South Africa; (c) copper smelter, Montana; (d) sheets of copper for industrial use; (e) appliances made in part from metals; and (f) disposal of mining waste from a Montana gold mine into a tailings pond.

In fact, the major environmental impacts of mineral use are related to waste products. Waste produces pollution that may be toxic to people, may harm natural ecosystems and the biosphere, and may be aesthetically displeasing. Waste may attack and degrade air, water, soil, and living things. Waste also depletes nonrenewable mineral resources and, when simply disposed of, provides no offsetting benefits for human society.

Environmental regulations at the federal, state, and local levels address pollution of air and water by all aspects of the mineral cycle, and may also address reclamation of land used for mining minerals. Today, in the United States, approximately 50% of the land used by the mining industry has been reclaimed.

Minimizing the environmental effects of mining takes several interrelated paths:[3]

- *Reclaiming* areas disturbed by mining (see A Closer Look 23.1).
- *Stabilizing soils* that contain metals to minimize their release into the environment. Often this requires placing contaminated soils in a waste facility.
- *Controlling air emissions* of metals and other materials from mining areas.
- *Treating contaminated water before it can leave a mining site or treating contaminated water that has left a mining site.*
- *Treating waste onsite and offsite.* Minimizing onsite and offsite problems by controlling sediment, water, and air pollution through good engineering and conservation practices is an important goal. Of particular interest is the development of biotechnological processes such as biooxidation, bioleaching, and biosorption, the bonding of waste to microbes, as well as genetic engineering of microbes. These practices have enormous potential for both extracting metals and minimizing environmental degradation. At several sites, for example, constructed wetlands use acid-tolerant plants to remove metals from mine wastewaters and neutralize acids by biological activity.[14] The Homestake Gold Mine in South Dakota uses biooxidation to convert contaminated water from the mining operation into substances that are environmentally safe; the process uses bacteria that have a natural ability to oxidize cyanide to harmless nitrates.[15]
- *Practicing the three R's of waste management.* That is, **R**educe the amount of waste produced, **R**euse waste as much as possible, and maximize **R**ecycling opportunities. Wastes from some parts of the mineral cycle, for example, may themselves be considered ores because they contain materials that might be recycled to provide energy or other products.[16–18]

We will look at the three R's in greater detail in Section 23.7, Integrated Waste Management.

A CLOSER LOOK 23.1

Golden, Colorado: Open Pit Mine Becomes a Golf Course

The city of Golden, Colorado, has an award-winning golf course on land that, for about 100 years, was an open pit mine (quarry) excavated in limestone rock (Figure 23.8). The mine produced clay for making bricks from clay layers between limestone beds. Over the life of the mine, the clay was used as a building material at many sites, including prominent buildings in the Denver area, such as the Colorado Governor's Mansion. The mine site included unsightly pits with vertical limestone walls as well as a landfill for waste disposal. However, it had spectacular views of the Rocky Mountain foothills. Today the limestone cliffs with their exposed plant and dinosaur fossils have been transformed into golf greens, fairways, and a driving range. The name Fossil Trace Golf Club reflects its geologic heritage. The course includes trails to fossil locations and also has channels, constructed wetlands, and three lakes that store floodwater runoff, helping to protect Golden from flash floods. The reclamation project started with a grassroots movement by the people of Golden to have a public golf course. The reclamation is now a moneymaker for the city and demonstrates that mining sites can not only be reclaimed, but also be transformed into valuable property.

Courtesy of Fossil Trace Golf Course

FIGURE 23.8 This award-winning golf course in Golden, Colorado, was for a century an open pit mine (quarry) for clay to produce bricks.

23.6 Materials Management and Our Waste

History of Waste Disposal

During the first century of the Industrial Revolution, the volume of waste produced in the United States was relatively small and could be managed using the concept of "dilute and disperse." Factories were located near rivers because the water provided a number of benefits, including easy transport of materials by boat, enough water for processing and cooling, and easy disposal of waste into the river. With few factories and a sparse population, dilute and disperse was sufficient to remove the waste from the immediate environment.[19]

As industrial and urban areas expanded, the concept of dilute and disperse became inadequate, and a new concept, "concentrate and contain," came into use. It has become apparent, however, that containment was, and is, not always achieved. Containers, whether simple trenches excavated in the ground or metal drums and tanks, may leak or break and allow waste to escape. Health hazards resulting from past waste-disposal practices have led to the present situation, in which many people have little confidence in government or industry to preserve and protect public health.[20]

In the United States and many other parts of the world, people are facing a serious solid-waste disposal problem. Basically, we are producing a great deal of waste and don't have enough acceptable space for disposing of it. It has been estimated that within the next few years approximately half the cities in the United States may run out of landfill space. Philadelphia, for example, is essentially out of landfill space now and is bargaining with other states on a monthly or yearly basis to dispose of its trash. The Los Angeles area has landfill space for only about ten more years.

To say we are actually running out of space for landfills isn't altogether accurate—land used for landfills is minute compared to the land area of the United States. Rather, existing sites are being filled, and it is difficult to site new landfills. After all, no one wants to live near a waste-disposal site, be it a sanitary landfill for municipal waste, an incinerator that burns urban waste, or a hazardous waste-disposal operation for chemical materials. This attitude is widely known as NIMBY ("not in my backyard").

The environmentally correct concept with respect to waste management is to consider wastes as resources out of place. Although we may not soon be able to reuse and recycle all waste, it seems apparent that the increasing cost of raw materials, energy, transportation, and land will make it financially feasible to reuse and recycle more resources and products. Moving toward this objective is moving toward an environmental view that there is no such thing as waste. Under this concept, waste would not exist because it would not be produced—or, if produced, it would be a resource to be used again. This is referred to as the "zero waste" movement.

Zero waste is the essence of what is known as **industrial ecology**, the study of relationships among industrial systems and their links to natural systems. Under the principles of industrial ecology, our industrial society would function much as a natural ecosystem functions. Waste from one part of the system would be a resource for another part.[21]

Until recently, zero waste production was considered unreasonable in the waste-management arena. However, it is catching on. The city of Canberra, Australia, may be the first community to propose a zero waste plan. Thousands of kilometers away, in the Netherlands, a national waste reduction goal of 70 to 90% has been set. How this goal is to be met is not entirely clear, but a large part of the planning involves taxing waste in all its various forms, from smokestack emissions to solids delivered to landfills. Already, in the Netherlands, pollution taxes have nearly eliminated discharges of heavy metals into waterways. At the household level, the government is considering programs—known as "pay as you throw"—that would charge people by the volume of waste they produce. Taxing waste, including household waste, motivates people to produce less of it.[22]

Of particular importance to waste management is the growing awareness that many of our waste-management programs involve moving waste from one site to another, and not really managing it. For example, waste from urban areas may be placed in landfills; but eventually these landfills may cause new problems by producing methane gas or noxious liquids that leak from the site and contaminate the surrounding areas. Managed properly, however, methane produced from landfills is a resource that can be burned as a fuel (an example of industrial ecology).

In sum, previous notions of waste disposal are no longer acceptable, and we are rethinking how we deal with materials, with the objective of eliminating the concept of waste entirely. In this way, we can reduce the consumption of minerals and other virgin materials, which depletes our environment, and live within our environment more sustainably.[21]

23.7 Integrated Waste Management

The dominant concept today in managing waste is known as **integrated waste management (IWM)**, which is best defined as a set of management alternatives that includes *reuse, source reduction, recycling, composting, landfill,* and *incineration.*[20]

Reduce, Reuse, Recycle

The ultimate objective of the three R's of IWM is to reduce the amount of urban (municipal solid waste) and other waste that must be disposed of in landfills, incinerators, and other waste-management facilities. Study of the *waste stream* (the waste produced) in areas that use IWM technology suggests that the amount (by weight) of urban refuse disposed of in landfills or incinerated can be reduced by at least 50% and perhaps as much as 70%. A 50% reduction by weight could be achieved by (1) source reduction, such as packaging better designed to reduce waste (10%reduction); (2) large-scale composting programs (10% reduction); and (3) recycling programs (30% reduction).[20]

As this list indicates, recycling is a major player in reducing the urban waste stream. Metals such as iron, aluminum, copper, and lead have been recycled for many years and are still being recycled today. The metal from almost all of the millions of automobiles discarded annually in the United States is recycled.[16,17] The total value of recycled metals is about $50 billion. Iron and steel account for approximately 90% by weight and 40% by total value of recycled metals. Iron and steel are recycled in such large volumes for two reasons. First, the market for iron and steel is huge, and as a result there is a large scrap collection and scrap processing industry. Second, an enormous economic and environmental burden would result from failure to recycle because over 50 million tons of scrap iron and steel would have to be disposed of annually.[17,18]

Today in the United States we recycle over 30% of our total municipal solid waste, up more than 10% from 25 years ago. This amounts to 99% of automobile batteries, 63% of steel cans, 71% office type papers, 63% of aluminum cans, 35% of tires, 28% of glass containers, and about 30% of various plastic containers (Table 23.2).[23] This is encouraging news. Can recycling actually reduce the waste stream by 50%? Recent work suggests that the 50% goal is reasonable. In fact, it has been reached in some parts of the United States, and the potential upper limit for recycling is considerably higher. It is estimated that as much as 80 to 90% of the U.S. waste stream might be recovered through what is known as intensive recycling.[24] A pilot study involving 100 families in

Table 23.2 RECYCLING REGENERATION AND RECOVERY BY WEIGHT (MILLIONS OF TONS) AND PERCENT OF MUNICIPAL SOLID WASTE IN 2010

MATERIAL	WEIGHT GENERATED	WEIGHT RECOVERED	RECOVERY AS PERCENT OF GENERATION
Paper and paperboard	71.31	44.57	62.5%
Glass	11.53	3.13	27.1%
Metals			
Steel	16.90	5.71	33.8%
Aluminum	3.41	0.68	19.9%
Other nonferrous metalst	2.10	1.48	70.5%
Total metals	**22.41**	**7.87**	**35.1%**
Plastics	31.04	2.55	8.2%
Rubber and leather	7.78	1.17	15.0%
Textiles	13.12	1.97	15.0%
Wood	15.88	2.30	14.5%
Other materials	4.79	1.41	29.4%
Total materials in products	**177.86**	**64.97**	**36.5%**
Other wastes			
Food other	34.76	0.97	2.8%
Yard trimmings	33.40	19.20	57.5%
Miscellaneous inorganic wastes	3.84	Negligile	Negligible
Total other wastes	**72.00**	**20.17**	**28.0%**
Total municipal solid waste	**249.86**	**85.14**	**34.1%**

Data from U.S. Environmental Protection Agency

East Hampton, New York, achieved a level of 84%. More realistic for many communities is partial recycling, which targets specific materials such as glass, aluminum cans, plastic, organic material, and newsprint. Partial recycling can provide a significant reduction, and in many places it is approaching or even exceeding 50%.[25,26]

Recycling is simplified with **single-stream recycling**, in which paper, plastic, glass, and metals are not separated before collection; the waste is commingled in one container and separated later at recycling centers. This is more convenient for homeowners, reduces the cost of collection, and increases the rate of recycling. Thus, single-stream recycling is growing rapidly.

Public Support for Recycling

There is enthusiastic public support for recycling in the United States today. Many people understand that managing our waste has many advantages to society as a whole and the environment in particular. People like the notion of recycling because they correctly assume they are helping to conserve resources, such as forests, that make up much of the nonurban environment of the planet.

An encouraging sign of public support for the environment is the increased willingness of industry and business to support recycling on a variety of scales. For example, fast-food restaurants are using less packaging and providing onsite bins for recycling paper and plastic. Groceries and supermarkets are encouraging the recycling of plastic and paper bags by providing bins for their collection, and some offer inexpensive reusable canvas shopping bags instead of disposables. Companies are redesigning products so that they can be more easily disassembled after use and the various parts recycled. As this idea catches on, small appliances such as electric frying pans and toasters may be recycled rather than ending up in landfills. The automobile industry is also responding by designing automobiles with coded parts so that they can be more easily disassembled (by professional recyclers) and recycled, rather than left to become rusting eyesores in junkyards.

On the consumer front, people are more likely to purchase products that can be recycled or that come in containers that are more easily recycled or composted. Many consumers have purchased small home appliances that crush bottles and aluminum cans, reducing their volume and facilitating recycling. The entire arena is rapidly changing, and innovations and opportunities will undoubtedly continue.

Large cities from New York to Los Angeles have initiated recycling programs, but many people are concerned that recycling is not yet "cost-effective." As with many other environmental solutions, implementing the IWM concept successfully can be a complex undertaking. In some communities where recycling has been successful, it has resulted in glutted markets for recycled products, which has sometimes required temporarily stockpiling or suspending the recycling of some items. It is apparent that if recycling is to be successful, markets and processing facilities will also have to be developed to ensure that recycling is a sound financial venture as well as an important part of IWM.

To be sure, there are success stories, such as a large urban paper mill on New York's Staten Island that recycles more than 1,000 tons of paper per day. It is claimed that this paper mill saves more than 10,000 trees a day and uses only about 10% of the electricity required to make paper from virgin wood processing. On the West Coast, San Francisco has an innovative and ambitious recycling program that diverts nearly 50% of the urban waste from landfills to recycling programs. City officials are even talking about the concept of zero waste, hoping to achieve total recycling of waste by 2020. In part, this goal is being achieved by instigating a "pay-as-you-throw" approach; businesses and individuals are charged for disposal of garbage but not for materials that are recycled. Materials from the waste of the San Francisco urban area are shipped as far away as China and the Philippines to be recycled into usable products; organic waste is sent to agricultural areas; and metals, such as aluminum, are sent around California and to other states where they are recycled.

To understand some of the issues concerning recycling and its cost, consider the following points:

- The average cost of disposal at a landfill is about $40/ton in the United States, and even at a higher price of about $80/ton may be cheaper than recycling.
- Landfill fees in Europe range from $200 to $300/ton.
- Europe has been more successful in recycling, in part because countries such as Germany hold manufacturers responsible for disposing of the industrial goods they produce, as well as the packaging.
- In the United States, packaging accounts for approximately one-third of all waste generated by manufacturing.
- The cost to cities such as New York, which must export its waste out of state, is steadily rising and is expected to exceed the cost of recycling within about ten years.
- Placing a 10-cent refundable deposit on all beverage containers except milk would greatly increase the number recycled. For example, states with a deposit system have an average recycling rate of about 70–95% of bottles and cans, whereas states that do not have a refundable-deposit system average less than 30%.
- When people have to pay for trash disposal at a landfill, but are not charged for materials that are recycled—such as paper, plastic, glass, and metals—the success of recycling is greatly enhanced.

- Beverage companies do not particularly favor requiring a refundable deposit for containers. They claim that the additional costs to do this would be several billion dollars, but they do agree that recovery rates would be higher, providing a steadier supply of recycled metal, such as aluminum, as well as plastic.
- Education is a big issue with recycling. Many people still don't know which items are recyclable and which are not.
- Global markets for recyclable materials, such as paper and metals, have potential for expansion, particularly for large urban areas on the seacoast, where shipping materials is economically viable. Recycling in the United States today is a $14 billion industry; if done right, it generates new jobs and revenue for participating communities.

There are a number of good reasons to recycle municipal solid waste:[27]

- **Saving money by recycling**: For many years we have known that it is less expensive to make products using materials that have been recycled. Using recycled aluminum has been mentioned more than many other materials, probably because about 90% more energy is expended to extract and use new aluminum resources from the planet than to recycle and reuse that aluminum. Products made from recycled metals and other materials are generally less expensive than those manufactured from materials mined for the first time from Earth.
- **Recycling increases employment**: As we recycle more, new businesses are generated in transporting, processing, and selling recycled materials to be used in manufacturing new goods. On the one hand, for every 100,000 tons of waste that enters the waste stream and ends up in a landfill, six new jobs are created. On the other hand, if that same 100,000 tons of waste is recycled, six times as many jobs may be created. The number of jobs required at landfills is less than the number created through recycling because, as materials are reused, more employment results, including jobs for people who sort, transport, and sell the material, as well as for engineers and chemists who work to create better uses for those recycled materials. Furthermore, these jobs are often created in urban environments where more people live and work. Today, more than one million people are employed in the recycling industry, and over 55,000 recycling and reuse facilities have been built.
- **Recycling politics**: Becoming energy self-sufficient has been a longtime U.S. objective. Recent oil and gas discoveries are heading us in that direction, but recycling of materials can also play a role. Today, the United States recycles just over 30% of municipal solid waste, and this recycling results in a savings of about 5 billion gallons of gasoline. Thus, it is argued that through greater use of recycling, the United States and other countries that recycle may be more self-reliant and independent.
- **Recycling and values**: One of the fundamental principles of this book centers on values and science. It is argued that recycling is the right thing to do because it provides for better management of the Earth's resources. Value judgments include the fact that it is far better to reuse and recycle than to simply discard materials and products we have used. Extraction of raw materials from the planet poses serious environmental concerns, and reusing those products reduces our impact. An example often cited is the use of recycled paper so that fewer trees need to be harvested for paper production, thereby reducing the impact of deforestation. Another example is the use of plastics that, when recycled and reprocessed, can produce a number of products we routinely use (see A Closer Look 23.2). The ethical and moral argument is that by recycling and reusing more we can help minimize potential damages resulting from resource exploitation.

Recycling of Human Waste

The use of human waste, or "night soil," on croplands is an ancient practice. In Asia, recycling of human waste has a long history. Chinese agriculture was sustained for thousands of years through collection of human waste, which was spread over agricultural fields. The practice grew, and by the early 20th century the land application of sewage was a primary disposal method in many metropolitan areas in countries, including Mexico, Australia, and the United States.[28] Early uses of human waste for agriculture occasionally spread infectious diseases through bacteria, viruses, and parasites in waste applied to crops. Today, with the globalization of agriculture, we still see occasional warnings and outbreaks of disease from contaminated vegetables (see Chapter 19).

A major problem associated with recycling human waste is that, along with human waste, thousands of chemicals and metals flow through our modern waste stream. Even garden waste that is composted may contain harmful chemicals, such as pesticides.[28]

23.8 Municipal Solid-Waste Management

Municipal solid-waste management continues to be a problem in the United States and other parts of the world. In many areas, particularly in developing countries, waste-management practices are inadequate. These practices, which include poorly controlled open dumps and illegal

A CLOSER LOOK 23.2

Life of A Plastic Water Bottle

Perhaps nothing is more common on beaches and in parks and other places where trash ends up than portable plastic water bottles. You can't drive along a highway or visit a park or a beach without seeing many discarded plastic water bottles. The consumption of these small water bottles is increasing on a global basis at about 10% per year. The growth of use is greatest in developing countries in Asia and South American, but the United States leads in consumption, with approximately 9 billion bottles used per year (Figure 23.9).

Of the billions of small plastic water bottles used, only about 13% are recycled. As a result, about 2 billion tons end up in landfills. When plastic water bottles are recycled, they may be flaked or go through other processes and used in a variety of commercial products, ranging from clothing to decking to producing new water bottles.

It has been argued that use of bottled water has increased so greatly because bottled water is perceived to be safer than any municipal supply of water. Studies have shown, however, that while bottled water is generally safe and reliable, so is municipal water. As a result, many people have moved to filtering municipal water in lieu of buying bottled water. For example, at the University of California, Santa Barbara, a number of water filtration facilities are being made available where students may refill their water bottles. Water bottles are generally safe for refilling, as long as they are periodically cleaned and kept in sanitary condition.

Some of the environmental problems related to plastic water bottles and pollution are as follows: [a,b]

- It takes hundreds of years for plastic water bottles to decompose. In other words, they are nearly a permanent part of the environment once they are placed in landfills or dumped along roadways or in other areas.
- Bottled water requires a high use of energy—from pumping or collecting the water (often at a remote location) to processing, transporting, and refrigerating it before sale.
- Plastic water bottles that end up on beaches, in rivers, and in the ocean cause serious water problems that have impacts for wildlife (see A Closer Look 23.4).
- Discarded plastic water bottles create aesthetic degradation of our environment. Many are retrieved by volunteers, highway personnel, and people working in parks, but proper disposal, reuse, or recycling is much preferred.

In the final analysis, it is discouraging that such a small percentage of plastic water bottles are being reused and recycled to make other useful products. Furthermore, the cost of the water in these bottles is much more expensive than the water from our municipal supplies. Decades ago, a person in the water industry, on giving this author a quarter, said that he had just purchased all the water I would drink for the rest of my life. Today, that cost is probably closer to $40 over the 80 years a person might live, but it is much, much less than the cost of bottled water. Of course, in some instances it does pay to purchase bottled water, especially in some coastal areas and islands where municipal sources of water, though safe, are not particularly attractive to drink. In those instances, purchasing water in larger-sized containers—at least one gallon—makes sense. This is exactly what recently happened in Concord, Maine. The city banned the sale of plastic water bottles that were less than one liter in size.[a,b]

In summary, it makes both financial sense and environmental sense to reduce our consumption of water in plastic bottles as much as possible. We also need to ensure that a much higher percentage are recycled and reused if we are to move beyond a plastic water bottle crisis of growing proportions.

Sandeep Pradhananga

FIGURE 23.9 Collecting and recycling plastic water bottles in Patten Dunbar Square, Nepal, a UNESCO Natural Heritage Site.

roadside dumping, can spoil scenic resources, pollute soil and water, and pose health hazards.

Illegal dumping is as much a social as a physical problem because many people are simply disposing of waste as inexpensively and as quickly as possible, perhaps not seeing their garbage as an environmental problem. If nothing else, this is a tremendous waste of resources, since much of what is dumped could be recycled or reused. In areas where illegal dumping has been reduced, the keys have been awareness, education, and alternatives. Education programs teach people about the environmental problems of unsafe, unsanitary dumping of waste, and funds are provided for cleanup and for inexpensive collection and recycling of trash at sites of origin.

Next, we look at the composition of solid waste in the United States, and then we describe specific disposal methods: onsite disposal, composting, incineration, open dumps, and sanitary landfills.

Composition of Solid Waste

The average content of solid waste that is not recycled and is likely to end up at a disposal site in the United States is shown in Figure 23.10. It is no surprise that paper is by far the most abundant component. However, considerable variation can be expected, based on factors such as land use, economic base, industrial activity, climate, and time of year.

People have many misconceptions about our waste stream.[29] With all the negative publicity about fast-food packaging, polystyrene foam, and disposable diapers, many people assume that these make up a large percentage of the waste stream and are responsible for the rapid filling of landfills. However, excavations into modern landfills using archaeological tools have cleared up some misconceptions. We now know that fast-food packaging accounts for only about 0.25% of the average landfill; disposable diapers, approximately 0.8%; and polystyrene products about 0.9%.[29] Paper is a major constituent in landfills, perhaps as much as 50% by volume and 40% by weight. The largest single item is newsprint, which accounts for as much as 18% by volume.[30] Newsprint is one of the major items targeted for recycling because big environmental dividends can be expected. However (and this is a value judgment), the need to deal with the major waste products doesn't mean that we need not cut down on our use of disposable diapers, polystyrene, and other paper products. In addition to creating a need for disposal, these products are made from resources that might be better managed.

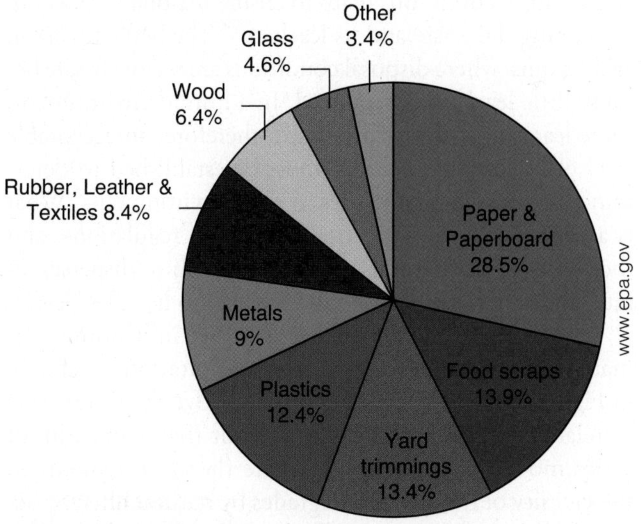

FIGURE 23.10 U.S. municipal solid-waste generation before recycling in 2010 was about 250 million tons, or about 4.6 lb (2 kg) per person.

Onsite Disposal

A common onsite disposal method in urban areas is the garbage disposal device installed in the wastewater pipe under the kitchen sink to grind garbage and flush it into the sewer system. This effectively reduces the amount of handling and quickly removes food waste. What's left of it is transferred to sewage treatment plants, where solids remaining as sewage sludge still must be disposed of.[31,32]

Composting

Composting is a biochemical process in which organic materials, such as lawn clippings and kitchen scraps, decompose to a rich, soil-like material. The process involves rapid partial decomposition of moist solid organic waste by aerobic organisms. Although simple backyard compost piles may come to mind, large-scale composting as a waste-management option is generally carried out in the controlled environment of mechanical digesters. This technique is popular in Europe and Asia, where intense farming creates a demand for compost. However, a major drawback of composting is the necessity of separating organic material from other waste. Therefore, it is probably economically advantageous only where organic material is collected separately from other waste. Another negative is that composting plant debris previously treated with herbicides may produce a compost toxic to some plants. Nevertheless, composting is an important component of IWM, and its contribution continues to grow.[31,32]

Incineration

Incineration burns combustible waste at temperatures high enough (900°–1,000°C, or 1,650°–1,830°F) to consume all combustible material, leaving only ash and noncombustibles to dispose of in a landfill. Under ideal conditions, incineration may reduce the volume of waste by 75–95%.[32] In practice, however, the actual decrease in volume is closer to 50% because of maintenance problems, as well as waste supply problems. Besides reducing a large volume of combustible waste to a much smaller volume of ash, incineration has another advantage: It can be used to supplement other fuels and generate electrical power.

Incineration of urban waste is not necessarily a clean process; it may produce air pollution and toxic ash. In the United States, for example, incineration is apparently a significant source of environmental dioxin, a carcinogenic toxin (see Chapter 8).[33] Smokestacks from incinerators also may emit oxides of nitrogen and sulfur, which lead to acid rain; heavy metals, such as lead, cadmium, and mercury; and carbon dioxide, which is related to global warming.

In modern incineration facilities, smokestacks fitted with special devices trap pollutants, but the process of pollutant abatement is expensive. The plants themselves are expensive, and government subsidization may be needed to aid in their establishment. Evaluation of the urban waste stream suggests that an investment of $8 billion could build enough incinerators in the United States to burn approximately 25% of the solid waste that is generated. However, a similar investment in source reduction, recycling, and composting could divert as much as 75% of the nation's urban waste stream away from landfills.[24]

The economic viability of incinerators depends on revenue from the sale of the energy produced by burning the waste. As recycling and composting increase, they will compete with incineration for their portion of the waste stream, and sufficient waste (fuel) to generate a profit from incineration may not be available. The main conclusion that can be drawn based on IWM principles is that a combination of reusing, recycling, and composting could reduce the volume of waste requiring disposal at a landfill by at least as much as incineration.[24]

Open Dumps (Poorly Controlled Landfills)

In the past, solid waste was often disposed of in open dumps (now called landfills), where refuse was piled up and left uncovered. Thousands of open dumps have been closed in recent years, and new open dumps are banned in the United States and many other countries. Nevertheless, many are still being used worldwide (Figure 23.11)

Sanitary Landfills

A **sanitary landfill** (also called a municipal solid-waste landfill) is designed to concentrate and contain refuse without creating a nuisance or hazard to public health or safety. The idea is to confine the waste to the smallest practical area, reduce it to the smallest practical volume, and cover it with a layer of compacted soil at the end of each day of operation, or more frequently if necessary. Covering the waste is what makes the landfill sanitary. The compacted layer restricts (but does not eliminate) continued access to the waste by insects, rodents, and other animals, such as seagulls. It also isolates the refuse, minimizing the amount of surface water seeping into it and the amount of gas escaping from it.[34]

Leachate

The most significant hazard from a sanitary landfill is pollution of groundwater or surface water. If waste buried

Ricardo Moraes/©AP/Wide World

FIGURE 23.11 **Urban garbage dump in Rio de Janeiro, Brazil.** At this site, people are going through the waste and recycling materials that can be reused or resold. This activity is all too common in dumps for large cities in the developing world. In some cases several thousand scavengers, including children, sift through tons of burning garbage to collect cans and bottles.

in a landfill comes into contact with water percolating down from the surface or with groundwater moving laterally through the refuse, **leachate**—a noxious, mineralized liquid capable of transporting bacterial pollutants—is produced.[35] For example, two landfills dating from the 1930s and 1940s on Long Island, New York, have produced subsurface leachate trails (plumes) several hundred meters wide that have migrated kilometers from the disposal site. The nature and strength of the leachate produced at a disposal site depend on the composition of the waste, the amount of water that infiltrates or moves through the waste, and the length of time that infiltrated water is in contact with the refuse.[32]

Site Selection

The siting of a sanitary landfill is very important and must take into consideration a number of factors, including topography, location of the groundwater table, amount of precipitation, type of soil and rock, and location of the disposal zone in the surface water and groundwater flow system. A favorable combination of climatic, hydrologic, and geologic conditions helps to ensure reasonable safety in containing the waste and its leachate.[36] The best sites are in arid regions, where disposal conditions are relatively safe because little leachate is produced. In a humid environment, some leachate is always produced; therefore, an acceptable level of leachate production must be established to determine the most favorable sites in such environments. What is acceptable varies with local water use, regulations, and the ability of the natural hydrologic system to disperse, dilute, and otherwise degrade the leachate to harmless levels.

Elements of the most desirable site in a humid climate with moderate to abundant precipitation are shown in Figure 23.12. The waste is buried above the water table in relatively impermeable clay and silt that water cannot easily move through. Any leachate therefore remains in the vicinity of the site and degrades by natural filtering action and chemical reactions between clay and leachate.[37,38]

Siting waste-disposal facilities also involves important social considerations. Often, planners choose sites where they expect minimal local resistance or where they perceive

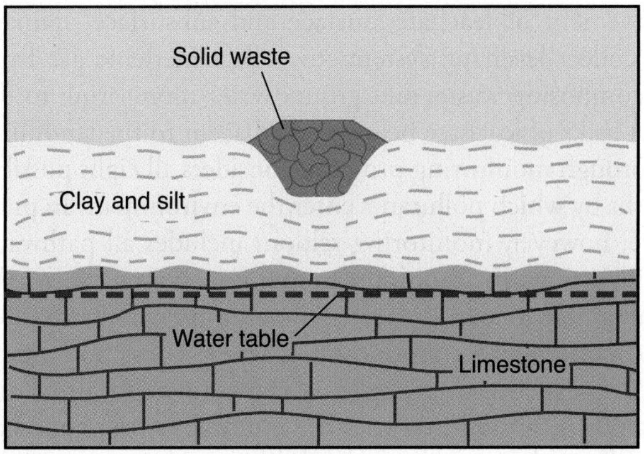

W.J. Schneider, *Hydraulic Implications of Solid Waste Disposal*, U.S. Geological Survey Circular 601F, 1970.

FIGURE 23.12 **The most desirable landfill site in a humid environment.** Waste is buried above the water table in a relatively impermeable environment.

land to have little value. Waste-disposal facilities are frequently located in areas where residents tend to have low socioeconomic status or belong to a particular racial or ethnic group. The study of social issues in siting waste facilities, chemical plants, and other such facilities is an emerging field known as **environmental justice**.[39,40]

Monitoring Pollution in Sanitary Landfills

Once a site is chosen for a sanitary landfill and before filling starts, monitoring the movement of groundwater should begin. Monitoring involves periodically taking samples of water and gas from specially designed monitoring wells. Monitoring the movement of leachate and gases should continue as long as there is any possibility of pollution, and it is particularly important after the site is completely filled and permanently covered. Continued monitoring is necessary because a certain amount of settling always occurs after a landfill is completed; if small depressions form, surface water may collect, infiltrate, and produce leachate. Monitoring and proper maintenance of an abandoned landfill reduce its pollution potential.[34]

How Pollutants Can Enter the Environment from Sanitary Landfills

Pollutants from a solid-waste disposal site can enter the environment through as many aseight paths (Figure 23.13):[41]

1. Methane, ammonia, hydrogen sulfide, and nitrogen gases can be produced from compounds in the waste and the soil and can enter the atmosphere.

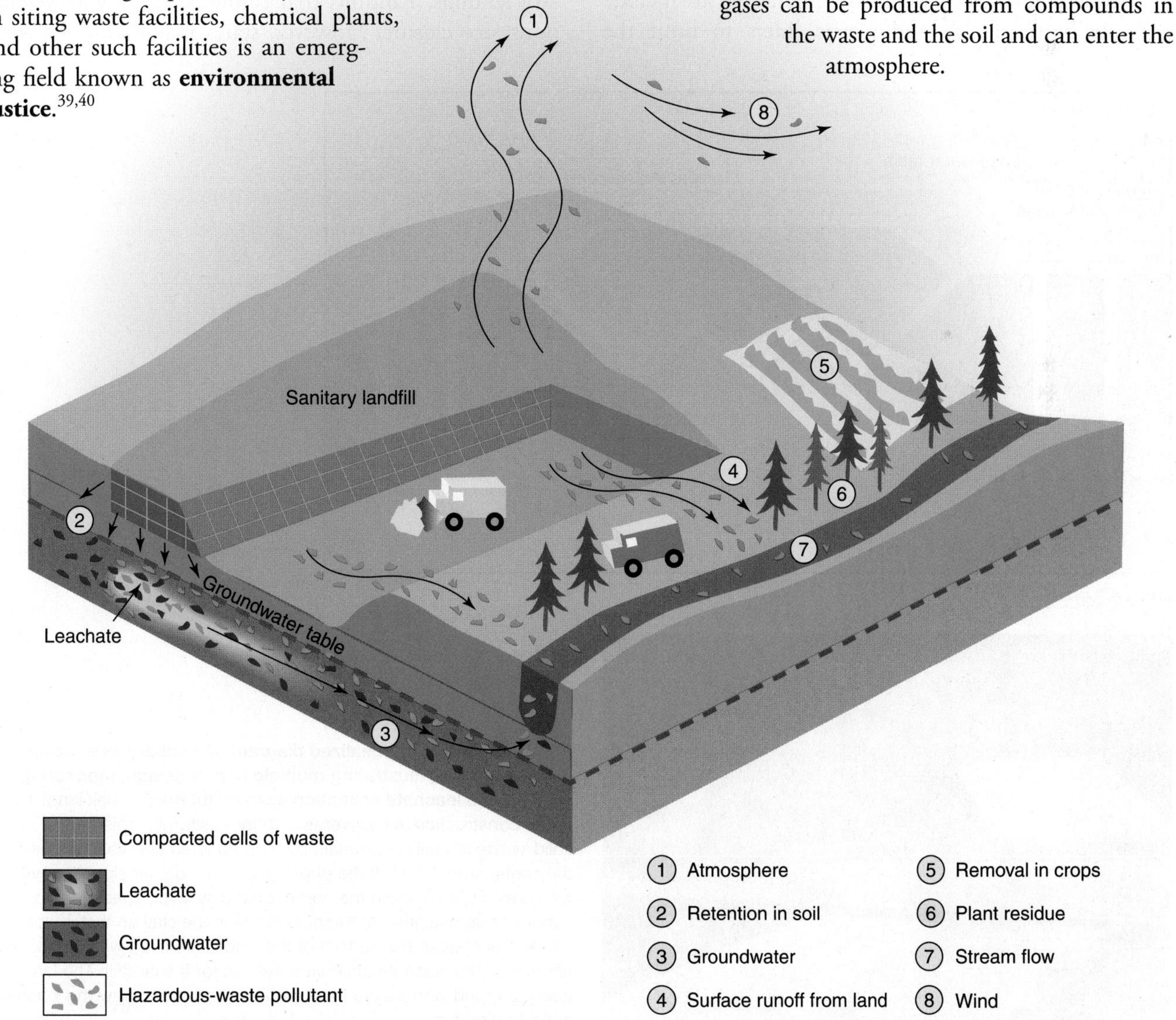

FIGURE 23.13 **Idealized diagram showing eight paths that pollutants from a sanitary landfill site may follow to enter the environment.**

2. Heavy metals, such as lead, chromium, and iron, can be retained in the soil.
3. Soluble materials, such as chloride, nitrate, and sulfate, can readily pass through the waste and soil to the groundwater system.
4. Overland runoff can pick up leachate and transport it into streams and rivers.
5. Some plants (including crops) growing in the disposal area can selectively take up heavy metals and other toxic materials. These materials are then passed up the food chain as people and animals eat the plants.
6. If plant residue from crops left in fields contains toxic substances, these substances return to the soil.
7. Streams and rivers may become contaminated by waste from groundwater seeping into the channel (3) or by surface runoff (4).
8. Wind can transport toxic materials to other areas.

Modern sanitary landfills are engineered to include multiple barriers: clay and plastic liners to limit the movement of leachate; surface and subsurface drainage to collect leachate; systems to collect methane gas from decomposing waste; and groundwater monitoring to detect leaks of leachate below and adjacent to the landfill. A thorough monitoring program considers all eight possible paths by which pollutants enter the environment. In practice, however, monitoring seldom includes all pathways. It is particularly important to monitor the zone above the water table to identify potential pollution before it reaches and contaminates groundwater, where correction would be very expensive. Figure 23.14 shows **(a)** an idealized diagram of a landfill that uses the multiple barrier approach and **(b)** a photograph of a landfill site under construction.

Federal Legislation for Sanitary Landfills

New landfills that opened in the United States after 1993 must comply with stricter requirements under the Resource Conservation and Recovery Act of 1980. The legislation, as its title states, is intended to strengthen and standardize the design, operation, and monitoring of sanitary landfills. Landfills that cannot comply with regulations face closure. However, states may choose between

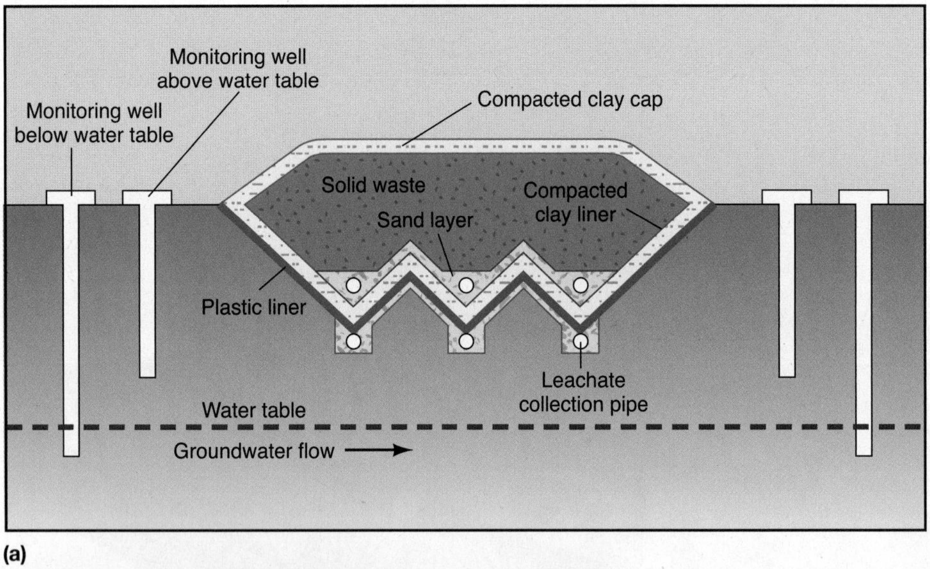

(a)

Courtesy EPA

(b)

FIGURE 23.14 (a) Idealized diagram of a solid-waste facility (sanitary landfill) illustrating multiple barrier design, monitoring system, and leachate collection system. (b) Rock Creek landfill under construction in Calaveras County, California. This municipal solid-waste landfill is underlain by a compacted clay liner, exposed in the center left portion of the photograph. The darker slopes, covered with gravel piles, overlie the compacted clay layer. These form a vapor barrier designed to keep moisture in the clay so that it won't crack. Trenches at the bottom of the landfill are lined with plastic and are part of the leachate collection system for the landfill. The landfill is also equipped with a system to monitor the water below the leachate collection system.

two options: (1) comply with federal standards or (2) seek EPA approval of solid-waste management plans.

The federal standards include the following:

- Landfills may not be sited on floodplains, wetlands, earthquake zones, unstable land, or near airports (birds drawn to landfill sites are a hazard to aircraft).
- Landfills must have liners.
- Landfills must have a leachate collection system.
- Landfill operators must monitor groundwater for many specified toxic chemicals.
- Landfill operators must meet financial assurance criteria to ensure that monitoring continues for 30 years after the landfill is closed.

EPA approval of a state's landfill program allows greater flexibility:

- Groundwater monitoring may be suspended if the landfill operator can demonstrate that hazardous constituents are not migrating from the landfill.
- Alternative types of daily cover over the waste may be used.
- Alternative groundwater protection standards are allowed.
- Alternative schedules for documentation of groundwater monitoring are allowed.
- Under certain circumstances, landfills in wetlands and fault zones are allowed.
- Alternative financial assurance mechanisms are allowed.

Given the added flexibility, it appears advantageous for states to develop EPA-approved waste-management plans.

Reducing the Waste that Ends Up in a Landfill

Most of the municipal solid waste we generate is from our homes, and over 50% of it could be diverted from the landfill by the 3 R's of waste management: reduce, reuse, and recycle. Diversion may eventually be increased to as much as 85% through improved waste management. In other words, the life of the landfill can be extended by keeping more waste out of the landfill through conservation and recycling, or through turning waste, even waste that is presently buried, into a source of clean energy. The latter involves first removing materials that can be recycled, then linking noncombustion thermal or biochemical processes with the remaining solid waste to produce electricity and alternative fuels (for example, biodiesel). The less waste in the landfill, the less potential for pollution of ground and surface water, along with the important fringe benefit of green energy.

The average waste per person in the United States increased from about 1 kg (2.2 lb) per day in 1960 to 2 kg (4.5 lb) per day in 2008. This is an annual growth rate of about 1.5% per year and is not sustainable because the doubling time for waste production is only a few decades. The 236 million tons we produced in 2003 would be close to 500 million tons by 2050, and we are already having big waste-management problems today. Table 23.3 lists some of the many ways you could reduce the waste you generate. What other ways can you think of?

Table 23.3 ACTIONS YOU CAN TAKE TO REDUCE THE WASTE YOU GENERATE

Keep track of the waste you personally generate: Know how much waste you produce. This will make you conscious of how to reduce it.

Recycle as much as is possible and practical: Take your cans, glass, and paper to a recycling center or use curbside pickup. Take your hazardous materials such as batteries, cell phones, computers, paint, used oil, and solvents to a hazardous waste collection site.

Reduce packaging: Whenever possible buy your food items in bulk or concentrated form.

Use durable products: Choose automobiles, light bulbs, furniture, sports equipment, and tools that will last a longer time.

Reuse products: Some things may be used several times. For example, you can reuse boxes and shipping "bubble wrap" to ship packages.

Purchase products made from recycled material: Many bottles, cans, boxes, containers, cartons, carpets, clothing, floor tiles, and other products are made from recycled material. Select these whenever you can.

Purchase products designed for ease in recycling: Products as large as automobiles along with many other items are being designed with recycling in mind. Apply pressure to manufacturers to produce items that can be easily recycled.

Source: Modified from U.S. Environmental Protection Agency. Accessed April 21, 2006 at www.epa.gov.

23.9 Hazardous Waste

Creation of new chemical compounds has proliferated in recent years. In the United States, approximately 1,000 new chemicals are marketed each year, and about 70,000 chemicals are currently on the market. Although many of these chemicals have been beneficial to people, approximately 35,000 chemicals used in the United States are classified as definitely or potentially hazardous to people or ecosystems if they are released into the environment as waste—and unfortunately, a lot of it is.

The United States currently produces about 700 million metric tons of hazardous chemical waste per year, referred to more commonly as **hazardous waste**. About 70% of it is generated east of the Mississippi River, and about half of the total by weight is generated by chemical products industries. The electronics industry (see A Closer Look 23.3 for a discussion of e-waste) and petroleum and coal products industries each contribute about 10%.[42–44] Hazardous waste may also enter the environment when buildings are destroyed by events such as fires and hurricanes, releasing paints, solvents, pesticides, and other chemicals that were stored in them, or when debris from damaged buildings is later burned or buried. As a result, collection of such chemicals after natural disasters is an important goal in managing hazardous materials.

In the mid-20th century, as much as half the total volume of hazardous waste produced in the United States was indiscriminately dumped.[43] Some was illegally dumped on public or private lands, a practice called midnight dumping. Buried drums of illegally dumped hazardous waste have been discovered at hundreds of sites by contractors while in the process of constructing buildings and roads. Cleanup has been costly and has delayed projects.[42]

The case of Love Canal is a well-known hazardous waste horror story. In 1976, in a residential area near Niagara Falls, New York, trees and gardens began to die. Rubber on tennis shoes and bicycle tires disintegrated. Puddles of toxic substances began to ooze through the soil. A swimming pool popped from its foundation and floated in a bath of chemicals.

The story of Love Canal started in 1892 when William Love excavated a canal 8 km (5 mi) long as part of the development of an industrial park. The development didn't need the canal when inexpensive electricity arrived, so the uncompleted canal remained unused for decades and became a dump for wastes. From 1920 to 1952, some 20,000 tons of more than 80 chemicals were dumped into the canal. In 1953 the Hooker Chemical Company—which produced the insecticide DDT as well as an herbicide and chlorinated solvents, and had dumped chemicals into the canal—was pressured to donate the land to the city of Niagara Falls for $1.00. The city knew that chemical wastes were buried there, but no one expected any problems. Eventually, several hundred homes and an elementary school were built on and near the site, and for years everything seemed fine. Then, in 1976–1977, heavy rains and snows triggered a number of events, making Love Canal a household word.[42]

A study of the site identified many substances suspected of being carcinogens, including benzene, dioxin, dichlorethylene, and chloroform. Although officials admitted that little was known about the impact of these chemicals, grave concern was voiced for people living in the area. Eventually, concern centered on alleged high rates of miscarriages, blood and liver abnormalities, birth defects, and chromosome damage. The government had to destroy about 200 homes and a school, and about 800 families were relocated and reimbursed. After about $400 million was spent on cleaning up the site, the EPA eventually declared the area clean, and about 280 remaining homes were sold.[45] Today, the community around the canal is known as Black Creek Village, and many people live there.

Uncontrolled or poorly controlled dumping of chemical waste has polluted soil and groundwater in several ways:

- In some places, chemical waste is still stored in barrels, either stacked on the ground or buried. The barrels may eventually corrode and leak, polluting surface water, soil, and groundwater.
- When liquid chemical waste is dumped into an unlined lagoon, contaminated water may percolate through soil and rock to the groundwater table.
- Liquid chemical waste may be illegally dumped in deserted fields or even along roads.

Some sites pose particular dangers. The floodplain of a river, for example, is not an acceptable site for storing hazardous waste. Yet, that is exactly what occurred at a site on the floodplain of the River Severn near a village in one of the most scenic areas of England. Several fires at the site in 1999 were followed by a large fire of unknown origin on October 30, 2000. Approximately 200 tons of chemicals, including industrial solvents (xylene and toluene), cleaning solvents (methylene chloride), and various insecticides and pesticides, produced a fireball that rose into the night sky. Wind gusts of hurricane strength spread toxic smoke and ash to nearby farmlands and villages, which had to be evacuated. People exposed to the smoke complained of a variety of symptoms, including headaches, stomachaches and vomiting, sore throats, coughs, and difficulty breathing.

A few days later, on November 3, the site flooded. The floodwaters interfered with cleanup after the fire

A CLOSER LOOK 23.3

"e-waste": A Growing Environmental Problem

The use of electronic products in the past few decades has changed our lives and the way we communicate with each other. One result is that hundreds of millions of computers and other electronic devices—such as cell phones, iPods, televisions, and computers games—are discarded every year. The average life of a computer is about three years, and it is not manufactured with recycling in mind. That is changing in the United States as the cost to recycle TV and computer screens is being charged to their manufacturers.

There are other ways to reduce the impact of electronic waste, called **e-waste**. The lives of these products can be extended by donation or resale to other people or by recycling. About one-half of the states in the United States have laws about disposal and recycling of electronic products. However, not all recycling providers use environmentally preferred methods.

When we take our e-waste to a location where computers are turned in, we assume that it will be handled properly, but this is not always what happens. In the United States, which helped start the technology revolution and produces most of the e-waste, its eventual disposal may cause serious environmental problems. The plastic housing for computers, for example, may produce toxins when burned. Computer parts also have small amounts of heavy metals—including gold, tin, copper, cadmium, and mercury—that are harmful and may cause cancer if inhaled, ingested, or absorbed through the skin. At present, many millions of computers are disposed of by what is billed as recycling, but there are not always adequate processes to ensure that this e-waste won't cause future problems. Many of these computers are being exported to countries such as Nigeria and China where they are being recycled in less than ideal conditions.

China's largest e-waste facility is in Guiyu, near Hong Kong. People in the Guiyu area process more than 1.5 million tons of e-waste each year with little thought to the potential toxicity of the material the workers are handling (Figure 23.15). About half of the e-waste at Guiya comes from China with the rest imported. Today about 25% of e-waste is now generated in poor countries.[a,b] In the United States, computers cannot be recycled profitably without charging the people who dump them a fee. Even with that, many U.S. firms ship their e-waste out of the country, where there is less regulation and greater profits are possible. The recycling revenue for the Guiyu area is several 10's of millions of dollars per year, so the central government is reluctant to regulate the activity. Workers earning about $8 per day at locations where computers are disassembled may be unaware that some of the materials they are handling are toxic and that they thus have a hazardous occupation. Altogether, in the Guiyu area, more than 5,000 family-run facilities specialize in scavenging e-waste for raw materials. While doing this, they are exposing themselves to a variety of toxins and potential health problems.

To date, the United States has not made an effective attempt to regulate the computer industry so that less waste is produced. In fact, the United States is the only major nation that did not ratify an international agreement that restricts and bans exports of hazardous e-waste.[a]

Our current ways of handling e-waste are not sustainable, and the value we place on a quality environment should include the safe handling and recycling of such waste. Hopefully, that is the path we will take in the future. There are positive signs. Some companies are now processing e-waste to reclaim metals such as gold and silver. Others are designing computers that use less toxic materials and are easier to recycle. The European Union is taking a leadership role in requiring more responsible management of e-waste.

Bob Sacha/©Corbis

FIGURE 23.15 e-waste being processed in China—a hazardous occupation.

and increased the risk of downstream contamination by waterborne hazardous wastes. In one small village, contaminated floodwaters apparently inundated farm fields, gardens, and even homes.[46] Of course, the solution to this problem is to clean up the site and move waste storage to a safer location.

23.10 Hazardous Waste Legislation

Recognition in the 1970s that hazardous waste was a danger to people and the environment and that the waste was not being properly managed led to important federal legislation in the United States.

Resource Conservation and Recovery Act

Management of hazardous waste in the United States began in 1976 with passage of the Resource Conservation and Recovery Act (RCRA). At the heart of the act is identification of hazardous wastes and their life cycles. The idea was to issue guidelines and assign responsibilities to those who manufacture, transport, and dispose of hazardous waste. This is known as "cradle-to-grave" management. Regulations require stringent record keeping and reporting to verify that the wastes are not a public nuisance or a health problem.

RCRA applies to solid, semisolid, liquid, and gaseous hazardous wastes. It considers a waste hazardous if its concentration, volume, or infectious nature may contribute to serious disease or death or if it poses a significant hazard to people and the environment as a result of improper management (storage, transport, or disposal).[42] The act classifies hazardous wastes in several categories: materials highly toxic to people and other living things; wastes that may ignite when exposed to air; extremely corrosive wastes; and reactive unstable wastes that are explosive or generate toxic gases or fumes when mixed with water.

Comprehensive Environmental Response, Compensation, and Liability Act

In 1980, Congress passed the Comprehensive Environmental Response, Compensation, and Liability Act (CERCLA). It defined policies and procedures for release of hazardous substances into the environment (for example, landfill regulations). It also mandated development of a list of sites where hazardous substances were likely to produce or already had produced the most serious environmental problems and established a revolving fund (Superfund) to clean up the worst abandoned hazardous waste sites. In 1984 and 1986, CERCLA was strengthened by amendments that made the following changes:

- Improved and tightened standards for disposal and cleanup of hazardous waste (for example, requiring double liners, monitoring landfills).
- Banned land disposal of certain hazardous chemicals, including dioxins, polychlorinated biphenyls (PCBs), and most solvents.
- Initiated a timetable for phasing out disposal of all untreated liquid hazardous waste in landfills or surface impoundments.
- Increased the Superfund. The fund was allocated about $8.5 billion in 1986; Congress approved $5.1 billion for fiscal year 1998, which almost doubled the Superfund budget.[47] Today (2010 to 2014) the EPA cost for Superfund sites is about $500 million per year and rising. The EPA does not include cost of cleanup at sites that are early in the process.

The Superfund has had management problems, and cleanup efforts are far behind schedule. Unfortunately, the funds available are not sufficient to pay for decontaminating all targeted sites. Furthermore, present technology may not be sufficient to treat all abandoned waste-disposal sites; it may be necessary to simply try to confine waste at those sites until better disposal methods are developed. It seems apparent that abandoned disposal sites are likely to remain problems for some time to come.

Federal legislation has also changed the ways in which real estate business is conducted. For example, there are provisions by which property owners may be held liable for costly cleanup of hazardous waste on their property, even if they did not directly cause the problem. As a result, banks and other lending institutions might be held liable for release of hazardous materials by their tenants.

The Superfund Amendment and Reauthorization Act (SARA) of 1986 permits a possible defense against such liability if the property owner completed an **environmental audit** before purchasing the property. Such an audit involves studying past land use at the site, usually determined by analyzing old maps, aerial photographs, and reports. It may also involve drilling and sampling groundwater and soil to determine whether hazardous materials are present. Environmental audits are now completed routinely before purchasing property for development.[47]

In 1990 the U.S. Congress reauthorized hazardous waste control legislation. Priorities include:

- Establishing who is responsible (liable) for existing hazardous waste problems.

- When necessary, assisting in or providing funding for cleanup at sites identified as having a hazardous waste problem.
- Providing measures whereby people who suffer damages from the release of hazardous materials are compensated.
- Improving the required standards for disposal and cleanup of hazardous waste.

23.11 Hazardous Waste Management: Land Disposal

Management of hazardous chemical waste involves several options, including recycling; onsite processing to recover by-products that have commercial value; microbial breakdown; chemical stabilization; high temperature decomposition; incineration; and disposal by **secure landfill** (Figure 23.16) or deep well injection. A number of technological advances have been made in toxic waste management; as land disposal becomes more expensive, the recent trend toward onsite treatment is likely to continue. However, onsite treatment will not eliminate all hazardous chemical waste; disposal of some waste will remain necessary.

Table 23.4 compares hazardous "waste" reduction technologies for treatment and disposal. Notice that all available technologies cause some environmental disruption. There is no simple solution for all waste-management issues.

Direct land disposal of hazardous waste is often not the best initial alternative. The consensus is that even with extensive safeguards and state-of-the-art designs, land disposal alternatives cannot guarantee that the waste will be contained and that it will not cause environmental disruption in the future. This concern holds true for all land disposal facilities, including landfills, surface impoundments, land application, and injection wells. Pollution of air, land, surface water, and groundwater may result if a land disposal site fails to contain hazardous waste. Pollution of groundwater is perhaps the most significant risk because groundwater provides a convenient route for pollutants to reach people and other living things.

Some of the paths that pollutants may take from land disposal sites to contaminate the environment include leakage and runoff to surface water or groundwater from improperly designed or maintained landfills; seepage, runoff, or air emissions from unlined lagoons; percolation and seepage from failure of surface land application of

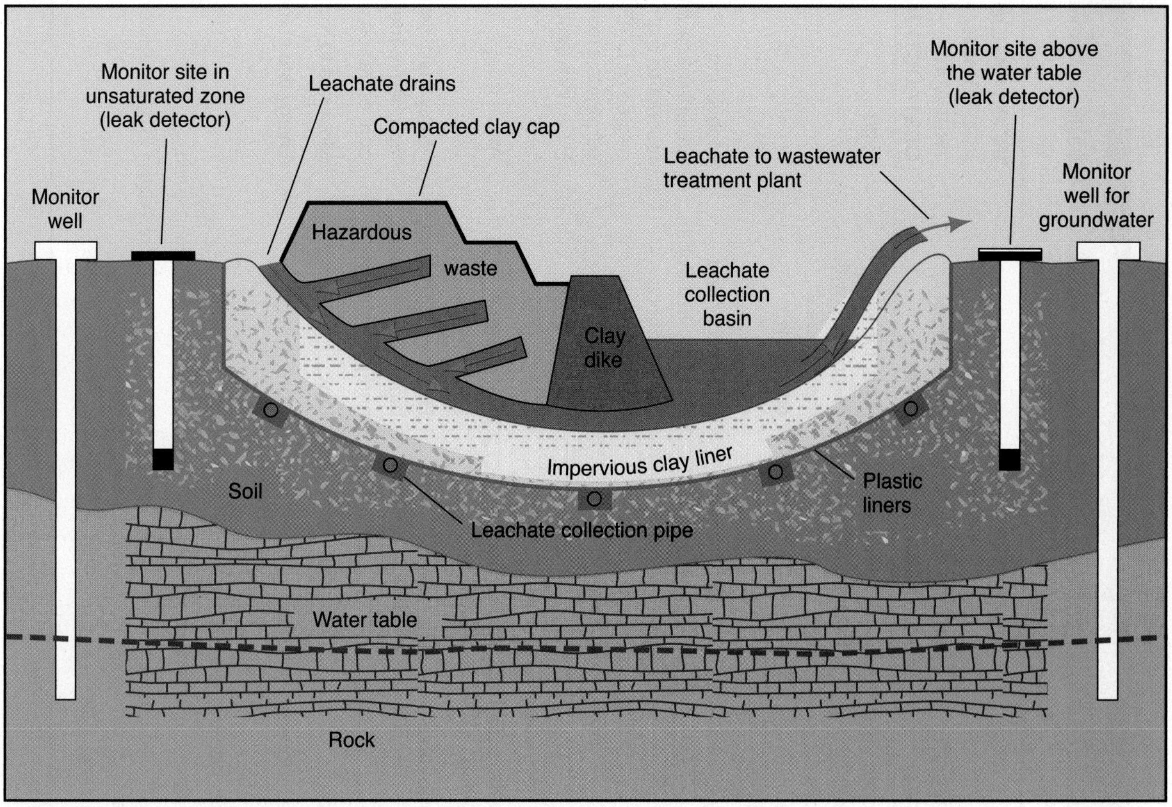

FIGURE 23.16 **A secure landfill for hazardous chemical waste.** The impervious liners, systems of drains, and leak detectors are integral parts of the system to ensure that leachate does not escape from the disposal site. Monitoring in the unsaturated zone is important and involves periodic collection of soil water.

Table 23.4 COMPARISON OF HAZARD REDUCTION TECHNOLOGIES

	DISPOSAL		TREATMENT			
PARAMETER COMPARED	LANDFILLS AND IMPOUNDMENTS	INJECTION WELLS	INCINERATION AND OTHER THERMAL DESTRUCTION	HIGH-TEMPERATURE DECOMPOSITION[a]	CHEMICAL STABILIZATION	MICROBIAL BREAKDOWN
Effectiveness: how well it contains or destroys hazardous characteristics	Low for volatiles, high for unsoluble solids	High, for waste compatible with the disposal environment	High	High for many chemicals	High for many metals	High for many metals and some organic waste such as oil
Reliability issues	Siting, construction, and operation Uncertainties: long-term integrity and cover	Site history and geology, well depth, construction, and operation	Monitoring uncertainties with respect to high degree of DRE: surrogate measures, PICs, incinerability[b]	Mobile units; on-site treament avoids hauling risks Operational simplicity	Some inorganics still soluble Uncertain leachate production	Monitoring uncertainties during construction and operation
Environment media most affected	Surface water and groundwater	Surface water and groundwater	Air	Air	Groundwater	Soil, groundwater
Least compatible wastes[c]	Highly toxic, persistent chemicals	Reactive; corrosive; highly toxic, mobile, and persistent	Highly toxic organics, high heavy-metal concentration	Some inorganics	Organics	Highly toxic persistent chemicals
Relative costs	Low to moderate	Low	Moderate to high	Moderate to high	Moderate	Moderate
Resource recovery potential	None	None	Energy and some acids	Energy and some metals	Possible building materials	Some metals

[a] Molten salt, high-temperature fluid well, and plasma arc treatments.

[b] DRE=destruction and removal efficiency; PIC = product of incomplete combustion.

[c] Wastes for which this method may be less effective for reducing exposure, relative to other technologies. Wastes listed do not necessarily denote common usage.

Source: Modified after Council on Environmental Quality, 1983.

waste to soils; leaks in pipes or other equipment associated with deep well injection; and leaks from buried drums, tanks, or other containers.[48–51]

23.12 Alternatives to Land Disposal of Hazardous Waste

Our handling of hazardous chemical waste should be multifaceted. In addition to the disposal methods just discussed, chemical waste management should include such processes as source reduction, recycling and resource recovery, treatment, and incineration. Recently, it has been argued that these alternatives to land disposal are not being used to their full potential—that is, the volume of waste could be reduced, and the remaining waste could be recycled or treated in some form prior to land disposal of the treatment residues.[52] The advantages of source reduction, recycling, treatment, and incineration include the following

- Useful chemicals can be reclaimed and reused.
- Treatment may make waste less toxic and therefore less likely to cause problems in landfills.
- The volume of waste that must eventually be disposed of is reduced.
- Because a reduced volume of waste is finally disposed of, there is less stress on the dwindling capacity of waste disposal sites.

Although some of the following techniques have been discussed as part of integrated waste management, they have special implications and complications in regard to hazardous wastes.

Source Reduction

The object of source reduction in hazardous waste management is to reduce the amount of hazardous waste generated by manufacturing or other processes. For example, changes in the chemical processes involved, equipment and raw materials used, or maintenance measures may successfully reduce the amount or toxicity of hazardous waste produced.[52]

Recycling and Resource Recovery

Hazardous chemical waste may contain materials that can be recovered for future use. For example, acids and solvents collect contaminants when they are used in manufacturing processes. These acids and solvents can be processed to remove the contaminants and then be reused in the same or different manufacturing processes.[52]

Hazardous waste resulting from industrial processes can be recycled effectively and safely. The Environmental Protection Agency (EPA) reported that in 2011, approximately 39 million tons of hazardous waste was managed, and of that, approximately 4% was recycled and another 4% was recovered. The EPA distinguishes between those wastes that are used or reused without reclamation and those that require reclamation before they may be used again. A particular material is said to be reclaimed if it is processed to recover usable products.[53] For example, hazardous waste reclamation includes the recovery of solvents such as acetone from industrial processes or recovery of metals such as lead.

The EPA keeps track of how much hazardous waste is generated, how it is managed, and the final disposition of those wastes as mandated by the Resource and Conservation and Recovery Act of 1976 discussed earlier. While the total percentage of hazardous waste that is recycled is low at 4%, it is encouraging that all hazardous waste is now managed. Hopefully, with continued improvements in technology and industrial processes, more hazardous waste will be reused and recycled.[53]

Treatment

Hazardous chemical waste can be treated by a variety of processes to change its physical or chemical composition and reduce its toxicity or other hazardous characteristics. For example, acids can be neutralized, heavy metals can be separated from liquid waste, and hazardous chemical compounds can be broken up through oxidation.[52]

Incineration

High-temperature incineration can destroy hazardous chemical waste. However, incineration is considered a waste treatment, not a disposal method, because the process produces an ash residue that must itself be disposed of in a landfill. Hazardous waste has also been incinerated offshore on ships, creating potential air pollution and ash—disposal problems in the marine environment—an environment we consider next.

23.13 Ocean Dumping

Oceans cover more than 70% of Earth. They play a part in maintaining our global environment and are of major importance in the cycling of carbon dioxide, which helps regulate the global climate. Oceans are also important in cycling many chemical elements important to life, such as nitrogen and phosphorus, and are a valuable resource because they provide us with such necessities as food and minerals.

It seems reasonable that such an important resource would receive preferential treatment, and yet oceans have long been dumping grounds for many types of waste, including industrial waste, construction debris, urban sewage, and plastics (see A Closer Look 23.4). Ocean dumping contributes to the larger problem of ocean pollution, which has seriously damaged the marine environment and caused a health hazard. Figure 23.17 shows locations in the oceans of the world that are accumulating pollution continuously; have intermittent pollution problems; or have potential for pollution from ships in the major shipping lanes. Notice that the areas with continual or intermittent pollution are near the shore.

Unfortunately, these are also areas of high productivity and valuable fisheries. Shellfish today often contain organisms that cause diseases such as polio and hepatitis. In the United States, at least 20% of the nation's commercial shellfish beds have been closed (mostly temporarily) because of pollution. Beaches and bays have been closed (again, mostly temporarily) to recreational uses. Lifeless zones in the marine environment have been created. Heavy kills of fish and other organisms have occurred, and profound changes in marine ecosystems have taken place (see Chapter 19).[54,55]

Marine pollution has a variety of specific effects on oceanic life, including the following:

- Death or retarded growth, vitality, and reproductivity of marine organisms.
- Reduction of dissolved oxygen necessary for marine life, due to increased biochemical oxygen demand.
- Eutrophication caused by nutrient-rich waste in shallow estuaries, bays, and parts of the continental shelf, resulting in oxygen depletion and subsequent killing of algae, which may wash up and pollute coastal areas. (See Chapter 19 for a discussion of eutrophication in the Gulf of Mexico.)
- Habitat change caused by waste-disposal practices that subtly or drastically change entire marine ecosystems.[54]

Marine waters of Europe are in particular trouble, in part because urban and agricultural pollutants have raised concentrations of nutrients in seawater. Blooms (heavy, sudden growth) of toxic algae are becoming more common. For example, in 1988 a bloom was responsible for killing nearly all marine life to a depth of about 15 m (50 ft), in the waterway connecting the North Sea to the Baltic Sea. It is believed that urban waste and agricultural runoff contributed to the toxic bloom.

Although oceans are vast, they are basically giant sinks for materials from continents, and parts of the marine environment are extremely fragile.[55] One area of concern is

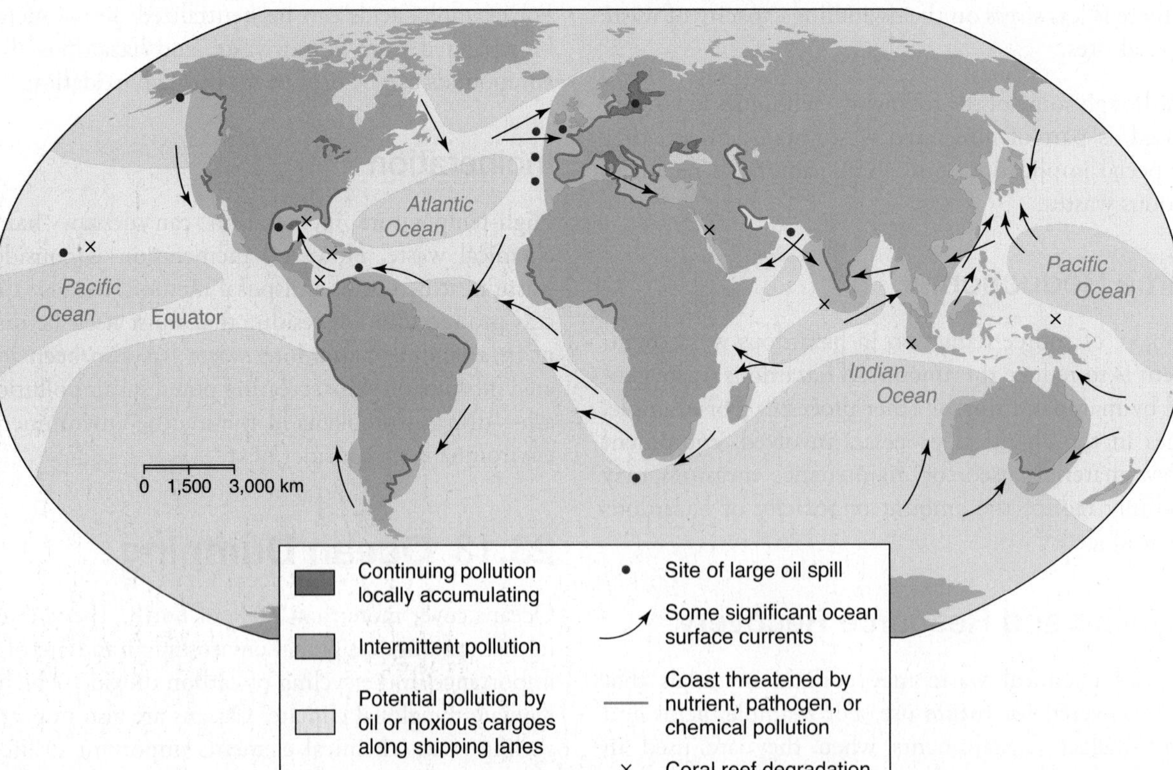

Modified from the Council on Environmental Quality, *Environmental Trends*, 1981, with additional data from A.P. McGinn, "Safeguarding the Health of the Oceans," Worldwatch Paper 145 [Washington, DC: Worldwatch Institute, 1999], pp. 22–23.

FIGURE 23.17 **Ocean pollution of the world.** Notice that the areas of continuing and locally accumulating pollution, as well as the areas with intermittent pollution, are in near-shore environments.

A CLOSER LOOK 23.4

Plastics in the Ocean

Vast quantities of plastic are used for a variety of products, ranging from beverage containers to cigarette lighters. For decades, people have been dumping plastics into the oceans. Some are dumped by passengers from ships; others are dropped as litter along beaches and swept into the water by the tides. Once in the ocean, plastics that float move with the currents and tend to accumulate where currents converge, concentrating the debris. Convergent currents of the Pacific (Figure 23.18) have a whirlpool-like action that concentrates debris near the center of these zones. One such zone is north of the equator, near the northwestern Hawaiian Islands. These islands are so remote that most people would expect them to be unspoiled, even pristine. Actually, there are literally hundreds of tons of plastics and other types of human debris on these islands. Recently, the National Oceanographic and Atmospheric Administration collected more than 80 tons of marine debris on Pearl and Hermes Atolls. Plastic debris is also widespread throughout the western North Atlantic Ocean. In the large North Atlantic subtropical a gyre about 1,200 km (750 mi) in diameter is centered about 1,000 km (625 mi) east of Florida.[a] Most plastic consists of small fragments of a few mm up to about half the size of a penny and apparently is being digested by microbes.

The island ecosystems include sea turtles, monk seals, and a variety of birds, including albatross. Marine scientist Jean-Michel Cousteau and his colleagues have been studying the problem of plastics on the northwestern Hawaiian Islands, including Midway Island and Kure Atoll. They have reported that the beaches of some of the islands and atolls look like a "recycling bin" of plastics. They found numerous cigarette lighters, some with fuel still in them, as well as caps from plastic bottles and all kinds of plastic toys and other debris. Birds on the islands pick up the plastic, attracted to it but not knowing what it is, and eat it. Figure 23.19 shows a dead albatross with debris in its stomach that caused its death. Plastic rings from a variety of products are also ingested by sea turtles and have been found around the snouts of seals, causing them to starve to death. In some areas, the carcasses of albatrosses litter the shorelines.

The solution to the problem is to be more careful about recycling plastic products to ensure they do not enter the marine environment. Collecting plastic items that accumulate on beaches is a step in the right direction, but it is a reactive response. Better to be proactive and reduce the source of the pollution.

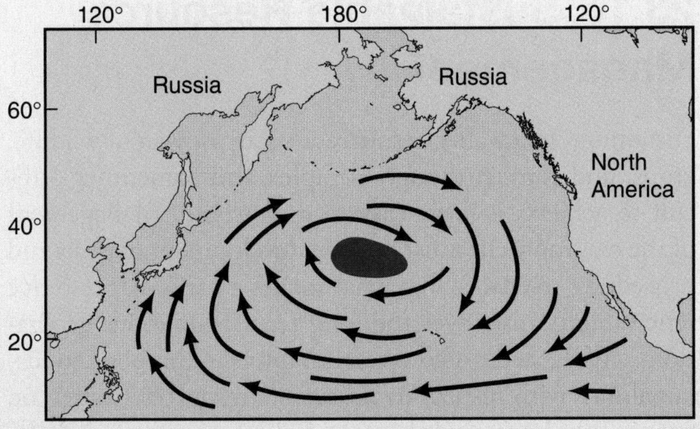

→ Direction of surface current

● Zone of accumulation of human-made plastic and other floating debris to remote islands

FIGURE 23.18 General circulation of the North Pacific Ocean. Arrows show the direction of the currents. Notice the tightening clockwise spiral pattern that carries floating debris to remote islands.

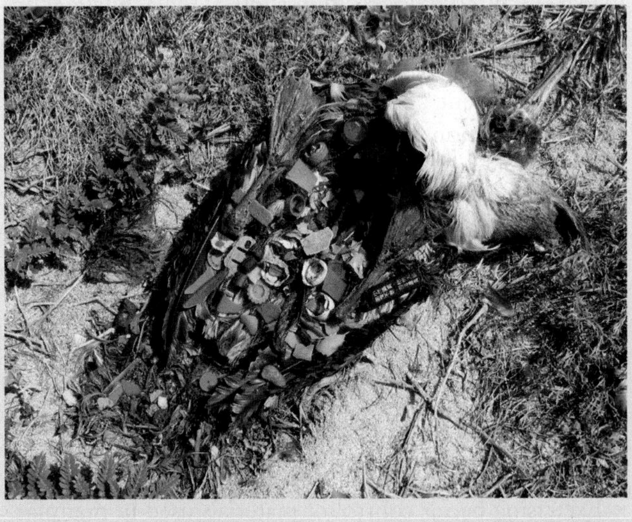

Courtesy of Cynthia Vanderlip/Algalita Marine Research Foundation

FIGURE 23.19 Albatross killed on a remote Pacific island by ingesting a large volume of plastic and other debris delivered by ocean currents. The photograph is not staged—the bird actually ingested all the plastic shown!

the *microlayer*, the upper 3 mm of ocean water. The base of the marine food chain consists of planktonic life abundant in the microlayer, and the young of certain fish and shellfish also reside there in the early stages of their life. Unfortunately, these upper few millimeters of the ocean also tend to concentrate pollutants, such as toxic chemicals and heavy metals. One study reported that concentrations of heavy metals—including zinc, lead, and copper—in the microlayer are from 10 to 1,000 times higher than in the deeper waters. It is feared that disproportionate pollution of the microlayer will have especially serious effects on marine organisms.[55] There is also concern that ocean pollution is a threat to some marine ecosystems, such as coral reefs, estuaries, salt marshes, and mangrove swamps.

Marine pollution can also have major impacts on people and society. Contaminated marine organisms, as we mentioned, may transmit toxic elements or diseases to people who eat them. In addition, beaches and harbors polluted by solid waste, oil, and other materials may not only damage marine life but also lose their visual appeal and other amenities. Economic loss is also considerable. Loss of shellfish from pollution in the United States, for example, amounts to many millions of dollars per year. In addition, a great deal of money is spent cleaning up solid waste, liquid waste, and other pollutants in coastal areas.[52]

23.14 Pollution Prevention

Approaches to waste management are changing. During the first several decades of environmental concern and management (the 1970s and 1980s), the United States approached the problem through government regulations and waste-control measures: chemical, physical, or biological treatment and collection (for eventual disposal), or transformation or destruction of pollutants after they had been generated. This was considered the most cost-effective approach to waste management.

With the 1990s came a growing emphasis on **pollution prevention**—ways to stop generating so much waste, rather than ways to dispose of it or manage it. This approach, which is part of materials management, includes the following:[56]

- Purchasing the proper amount of raw materials so that no excess remains to be disposed of.
- Exercising better control of materials used in manufacturing processes so that less waste is produced.
- Substituting nontoxic chemicals for hazardous or toxic materials currently used.
- Improving engineering and design of manufacturing processes so that less waste is produced.

These approaches are often called P-2 approaches, for "pollution prevention." Probably the best way to illustrate the P-2 process is through a case history.[56]

A Wisconsin firm that produced cheese was faced with the disposal of about 2,000 gallons a day of a salty solution generated during the cheese-making process. Initially, the firm spread the salty solution on nearby agricultural lands—then a common practice for firms that could not discharge wastewater into publicly owned treatment plants. This method of waste disposal, when done incorrectly, caused the level of salts in the soil to rise so much that it damaged crops. As a result, the Department of Natural Resources in Wisconsin placed limitations on this practice.

The cheese firm decided to modify its cheese-making processes to recover salt from the solution and reuse it in production. This involved developing a recovery process that used an evaporator. The recovery process reduced the salty waste by about 75% and at the same time reduced the amount of the salt the company had to purchase by 50%. The operating and maintenance costs for recovery were approximately 3 cents per pound of salt recovered, and the extra cost of the new equipment was recovered in only two months. The firm saved thousands of dollars a year by recycling its salt.

The case history of the cheese firm suggests that rather minor changes can often result in large reductions of waste produced. And this case history is not an isolated example. Thousands of similar cases exist today as we move from the era of recognizing environmental problems, and regulating them at a national level, to providing economic incentives and new technology to better manage materials.[56]

23.15 Sustainable Resource Management

Sustaining renewable resources, such as water, wildlife, crops, and forests, though complex and sometimes difficult to achieve, is fairly easy to understand. Management of the environment must include development of goals and procedures to ensure that what makes a particular resource renewable persists over the long term (numerous generations). We have devoted several chapters in this book to sustainability with respect to renewable resources (water, air, energy, crops, forests, fish, and wildlife). However, simultaneously considering sustainable development and mineral exploitation and use is problematic. This is because, even with the most careful use, nonrenewable mineral resources will eventually be used up, and sustainability is a long-term concept that requires finding ways to assure future generations a fair share of Earth's resources. Recently, it has been argued that, given human ingenuity and sufficient lead time, we can find solutions for sustainable development that incorporate nonrenewable mineral resources.

Human ingenuity is important because often it is not the mineral we need so much as what we use the mineral for. For example, we mine copper and use it to transmit electricity in wires or electronic pulses in telephone wires. It is not the copper itself we desire but the properties of copper that allow these transmissions. We can use fiberglass cables in telephone wires, eliminating the need for copper. Digital cameras have eliminated the need for film development that uses silver. The message is that it is possible to compensate for a nonrenewable mineral by finding new ways to do things. We are also learning that we can use raw mineral materials more efficiently. For example, in the late 1800s when the Eiffel Tower was constructed, 8,000 metric tons of steel were used. Today the tower could be built with only a quarter of that amount.[57]

Finding substitutes or more efficient ways to use nonrenewable resources generally requires several decades of research and development. A measure of how much time we have for finding solutions to the depletion of nonrenewable reserves is the ***R*-to-*C* ratio**, where *R* is the known reserves (for example, hundreds of thousands of tons of a metal) and *C* is the rate of consumption (for example, thousands of tons per year used by people). The *R*-to-*C* ratio is often misinterpreted as the time a reserve will last at the present rate of consumption. During the past 50 years, the *R*-to-*C* ratios for metals, such as zinc and copper, have fluctuated around 30 years. During that time, consumption of the metals roughly tripled, but we discovered new deposits. Although the *R*-to-*C* ratio is a *present* analysis of a dynamic system in which both the amount of reserves and consumption may change over time, it does provide a view of how scarce a particular mineral resource may be. Metals with relatively small ratios can be viewed as being in short supply, and it is those resources for which we should find substitutes through technological innovation.[57]

In sum, we may approach sustainable development and use of nonrenewable mineral resources by developing more efficient ways of mining resources and finding ways to more efficiently use available resources, recycling more and applying human ingenuity to find substitutes for a nonrenewable mineral.

CRITICAL THINKING ISSUE

Should We Buy a New Car or Keep the Old One?

People in the United States generally change cars about every four years. On the one hand, it has been argued that if you keep your car longer, you will save money and be kinder to the environment. On the other hand, some argue that, depending on your driving habits and how much you drive, it is more economical and better for the environment to have a lighter automobile that has very good fuel economy, such as a hybrid car or an all-electric car.

So which path is "greener"? Should you replace an older car with a younger one? To help approach this problem and think critically about it, here are a few facts:

- The Environmental Transport Association (ETA) is on record as saying it makes sense from an environmental and economic perspective for you to trade your older car for one that is lighter and has improved fuel efficiency if you drive a lot (say, 50 to 100 miles or more a day). If, however, you drive relatively few miles per day, they argue that it makes economic and even environmental sense to keep your older car, as long as it is in good operating condition. In other words, the more you drive, the greater the potential environmental benefits from changing to a more fuel-efficient automobile.
- For the average vehicle, about 20% of its lifetime energy use is for manufacturing; the remaining 80% is energy used for fuel.
- About 1.5 times more energy is needed to manufacture a hybrid automobile that uses electricity and fuel. On the other hand, these cars use about one-half the energy to operate as does a standard car.
- A typical hybrid car that may carry four adults is likely to have about a 125,000-mile lifetime. Assuming that a standard automobile might last the same number of miles as a hybrid, which required 1.5 more energy in manufacture and when new costs about $5,000 more, will be more economical (due to less fuel consumption) than a standard car after a few years of driving.
- If you are a person who seldom uses an automobile or only drives a few miles per day and your automobile is kept for, say, 15 years, then who pollutes more—the hybrid that is driven many miles a day or the standard car that is only driven a few miles? At that point, there may be a different balance point between efficiency, economy, and so forth, where the cost of producing a new car exceeds keeping an older vehicle for a longer period of time.

If we take some of our comparisons to a more extreme level, using a very large car such as a Hummer, compared to a typical smaller hybrid or standard car, and we estimate the cost of producing the car and fuel over, say, 150,000 miles, we come to the conclusion that the Hummer has an energy expenditure for production that is about twice that of the smaller cars. In terms of fuel efficiency, the ratios are even larger, with smaller cars using one-half to one-fourth of the fuel used by a large car (in this case, our example of a Hummer). Therefore, something like 90% of the energy consumed in the lifetime of the larger car is for burning gasoline or diesel, whereas for the smaller car, 75% of the total energy is used for fuel consumption.

Of course, another part of the argument about whether you should keep an older car or buy a newer one has to do with how you maintain and drive your automobile or if you do indeed choose to drive much less. You can improve your fuel efficiency by about one-fourth by proper maintenance of your automobile. You can also save a lot of gasoline by choosing to drive at times other than during peak-traffic hours and by being careful about how you drive, in terms of acceleration and your use of the air conditioner. You can decide to rearrange your life in such a way that you are nearly car-free and drive part of the way on your journey and use a carpool or public transportation or even a bicycle for the rest of the way. You may decide to work at home and thus drive a lot less. All these things will affect your decision. So, now let's think about it a bit.

Critical Thinking Questions

1. Summarize the arguments for and against keeping an automobile for, say, 15 years, versus replacing it with a more efficient automobile.
2. Would you recommend to your parents that they keep their current automobile for a much longer period of time or trade it in on a newer, more efficient model?
3. What value judgments have you made concerning your decision about whether to trade in your automobile or purchase a newer, more efficient, lighter model?

SUMMARY

- Mineral resources are usually extracted from naturally occurring, anomalously high concentrations of Earth materials. Such natural deposits allowed early peoples to exploit minerals while slowly developing technological skills.
- Mineral resources are not mineral reserves. Unless discovered and developed, resources cannot be used to ease present shortages.
- The availability of mineral resources is one measure of the wealth of a society. Modern technological civilization would not be possible without the exploitation of mineral resources. However, it is important to recognize that mineral deposits are not infinite and that we cannot maintain exponential population growth on a finite resource base.
- The United States and many other affluent nations rely on imports for their supplies of many minerals. As other nations industrialize and develop, such imports may be more difficult to obtain, and affluent countries may have to find substitutes for some minerals or use a smaller portion of the world's annual production.
- The mining and processing of minerals greatly affect the land, water, air, and biological resources and have social impacts as well, including increased demand for housing and services in mining areas.
- Sustainable development and use of nonrenewable resources are not necessarily incompatible. Reducing consumption, reusing, recycling, and finding substitutes are environmentally preferable ways to delay or alleviate possible crises caused by the convergence of a rapidly rising population and a limited resource base.
- The history of waste-disposal practices since the Industrial Revolution has progressed from dilution and dispersion to the concept of integrated waste management (IWM), which emphasizes the three R's: reducing waste, reusing materials, and recycling.
- One goal of the emerging concept of industrial ecology is to develop a system in which the concept of waste doesn't exist because waste from one part of the system would be a resource for another part.
- The most common way to dispose of solid waste is the sanitary landfill. However, around many large cities, space for landfills is hard to find, partly because few people wish to live near a waste-disposal site.
- Hazardous chemical waste is one of the most serious environmental problems in the United States. Hundreds or even thousands of abandoned, uncontrolled disposal sites could be time bombs that will eventually cause serious public health problems. We know that we will continue to produce some hazardous chemical waste. Therefore, it is imperative that we develop and use safe ways to dispose of it.
- Ocean dumping is a significant source of marine pollution. The most seriously affected areas are near shore, where valuable fisheries often exist.
- Pollution prevention (P-2)—identifying and using ways to prevent the generation of waste—is an important emerging area of materials management.

REEXAMINING THEMES AND ISSUES

Sean Randall/Getty Images, Inc.

HUMAN POPULATION

Materials management strategies are inextricably linked to the human population. As the population increases, so does the waste generated. In developing countries where population increase is the most dramatic, increases in industrial output, when linked to poor environmental control, produce, or aggravate waste-management problems.

© Biletskiy_Evgeniy/iStockphoto

SUSTAINABILITY

Assuring a quality environment for future generations is closely linked to materials management. Of particular importance here are the concepts of integrated waste management, materials management, and industrial ecology. Carried to their natural conclusion, the ideas behind these concepts would lead to a system in which the issue would no longer be waste management but instead resource management. Pollution prevention (P-2) is a step in this direction.

© Anton Balazh 2011/iStockphoto

GLOBAL PERSPECTIVE

Materials management is becoming a global problem. Improper management of materials contributes to air and water pollution and can cause environmental disruption on a regional or global scale. For example, waste generated by large inland cities and disposed of in river systems may eventually enter the oceans and be dispersed by the global circulation patterns of ocean currents. Similarly, soils polluted by hazardous materials may erode, and the particles may enter the atmosphere or water system, to be dispersed widely.

ssguy/ShutterStock

URBAN WORLD

Because so much of our waste is generated in the urban environment, cities are a focus of special attention for materials management. Where population densities are high, it is easier to implement the principles behind "reduce, reuse, and recycle." There are greater financial incentives for materials management where waste is more concentrated.

B2M Productions/Getty Images, Inc.

PEOPLE AND NATURE

Production of waste is a basic process of life. In nature, waste from one organism is a resource for another. Waste is recycled in ecosystems as energy flows and chemicals cycle. As a result, the concept of waste in nature is much different than that in the human waste stream. In the human system, waste may be stored in facilities such as landfills, where it may remain for long periods, far from natural cycling. Our activities to recycle waste or burn it for energy move us closer to transforming waste into resources. Converting waste into resources brings us closer to nature by causing urban systems to operate in parallel with natural ecosystems.

George Doyle/Getty Images, Inc.

SCIENCE AND VALUES

People today value a quality, pollution-free environment. The way materials have been managed continues to affect health and other environmental problems. An understanding of these problems has resulted in a considerable amount of work and research aimed at reducing or eliminating the impact of resource use. How a society manages its waste is a sign of its maturity and its ethical framework. Accordingly, we have become more conscious of environmental justice issues related to materials management.

KEY TERMS

composting 591
environmental audit 599
environmental justice 593
e-waste 597
hazardous waste 596
incineration 591
industrial ecology 586
integrated waste management (IWM) 586
leachate 592
materials management 578
nonrenewable resources 578
ore deposits 580
pollution prevention 604
R-to-*C* ratio 605
renewable resources 578
reserves 581
resources 581
sanitary landfill 592
secure landfill 599
single-stream recycling 588

STUDY QUESTIONS

1. What is the difference between a resource and a reserve?
2. Under what circumstances might sewage sludge be considered a mineral resource?
3. If surface mines and quarries cover less than 0.5% of the land surface of the United States, why is there so much environmental concern about them?
4. A deep-sea diver claims that the oceans can provide all our mineral resources with no negative environmental effects. Do you agree or disagree?
5. What factors determine the availability of a mineral resource?
6. Using a mineral resource involves four general phases: (a) exploration, (b) recovery, (c) consumption, and (d) disposal of waste. Which phase do you think has the greatest environmental effect?
7. Have you ever contributed to the hazardous waste problem through disposal methods used in your home, school laboratory, or other location? How big a problem do you think such events are? For example, how bad is it to dump paint thinner down a drain?
8. Why is it so difficult to ensure safe land disposal of hazardous waste?
9. Would you approve the siting of a waste-disposal facility in your part of town? If not, why, and where do you think such facilities should be?
10. Why might there be a trend toward onsite disposal rather than land disposal of hazardous waste? Consider the physical, biological, social, legal, and economic aspects of the question.
11. Considering how much waste has been dumped in the near-shore marine environment, how safe is it to swim in bays and estuaries near large cities?
12. Do you think we should collect household waste and burn it in special incinerators to make electrical energy? What problems and what advantages do you see for this method, compared with other waste-management options?
13. Should companies that dumped hazardous waste years ago, when the problem was not understood or recognized, be held liable today for health problems to which their dumping may have contributed?
14. Suppose you found that the home you had been living in for 15 years was atop a buried waste-disposal site. What would you do? What kinds of studies should be done to evaluate potential problems?

FURTHER READING

Allenby, B.R. ***Industrial Ecology: Policy Framework and Implementation*** (Upper Saddle River, NJ: Prentice Hall, 1999). A primer on industrial ecology.

Ashley, S. It's not easy being green. *Scientific American*, April 2002, pp. 32–34. A look at the economics of developing biodegradable products and a little of the chemistry involved.

Brookins, D.G. ***Mineral and Energy Resources*** (Columbus, Ohio: Charles E. Merrill, 1990). A good summary of mineral resources.

Kesler, S.F. ***Mineral Resources, Economics and the Environment*** (Upper Saddle River, NJ: Prentice Hall, 1994). A good book about mineral resources.

Kreith, F. ed., ***Handbook of Solid Waste Management*** (New York, NY: McGraw-Hill, 1994). Thorough coverage of municipal waste management, including waste characteristics, federal and state legislation, source reduction, recycling, and landfilling.

Watts, R.J. ***Hazardous Wastes*** (New York, NY: John Wiley & Sons, 1998). A to Z of hazardous wastes.

NOTES

1. Sullivan, D.E. 2006. *Recycled Cell Phones—A Treasure Trove of Valuable Metals.* U.S. Geological Survey Fact Sheet 2006–3097.
2. Kropschot, S.J., and K.M. Johnson. 2006. *U.S. Geological Survey Circular 1289.* Menlo Park, CA.
3. Hudson, T.L., F.D. Fox, and G.S. Plumlee. 1999. *Metal Mining and the Environment.* Alexandria, VA: American Geological Institute.
4. McKelvey, V.E. 1973. Mineral resource estimates and public policy. In D.A. Brobst and W.P. Pratt, eds., *United States Mineral Resources,* pp. 9–19. U.S. Geological Survey Professional Paper 820.
5. U.S. Department of the Interior, Bureau of Mines. 1993. *Mineral Commodity Summaries, 1993.* I 28.149:993. Washington, DC: U.S. Department of the Interior.
6. McGreery, P. 1995. Going for the goals: Will states hit the wall? *Waste Age* 26:68–76.
7. Brown, L.R. 1999 (March/April). Crossing the threshold. *Worldwatch,* pp. 12–22.
8. Meyer, H.O.A. 1985. Genesis of diamond: A mantle saga. *American Mineralogist* 70:344–355.
9. Kesler, S.F. 1994. *Mineral Resources, Economics, and the Environment.* New York, NY: Macmillan.
10. Awramik, S.A. 1981. The pre-Phanerozoic biosphere—three billion years of crises and opportunities. In M.H. Nitecki, ed., *Biotic Crises in Ecological and Evolutionary Time,* pp. 83–102. Spring Systematics Symposium. New York, NY: Academic Press.
11. Margulis, L., and J.E. Lovelock. 1974. Biological modulation of the Earth's atmosphere. *Icarus* 21:471–489.
12. Lowenstam, H.A. 1981. Minerals formed by organisms. *Science* 211:1126–1130.
13. Brobst, D.A., W.P. Pratt, and V.E. McKelvey, 1973. *Summary of United States Mineral Resources.* U.S. Geological Survey Circular 682.
14. Jeffers, T.H. 1991 (June). Using microorganisms to recover metals. *Minerals Today.* Washington, DC: U.S. Department of Interior, Bureau of Mines, pp. 14–18.
15. Haynes, B.W. 1990 (May). Environmental technology research. *Minerals Today.* Washington, DC: U.S. Bureau of Mines, pp. 13–17.
16. Sullivan, P.M., M.H. Stanczyk, and M.J. Spendbue. 1973. *Resource Recovery from Raw Urban Refuse.* U.S. Bureau of Mines Report of Investigations 7760.
17. Davis, F.F. 1972 (May). Urban ore. *California Geology,* pp. 99–112.
18. U.S. Geological Survey. 2010. *Minerals Yearbook 2007—Recycling Metals.* http://minerals.usgs.gov. Accessed April 3, 2010.
19. Galley, J.E. 1968. Economic and industrial potential of geologic basins and reservoir strata. In J.E. Galley, ed., *Subsurface Disposal in Geologic Basins: A Study of Reservoir Strata,* pp. 1–19. American Association of Petroleum Geologists Memoir 10. Tulsa, OK: American Association of Petroleum Geologists.
20. Relis, P., and A. Dominski. 1987. *Beyond the Crisis: Integrated Waste Management.* Santa Barbara, CA: Community Environmental Council.
21. Allenby, B.R. 1999. *Industrial Ecology: Policy Framework and Implementation.* Upper Saddle River, NJ: Prentice-Hall.
22. Garner, G., and P. Sampat. 1999 (May). Making things last: Reinventing of material culture. *The Futurist,* pp. 24–28.
23. U.S. Environmental Protection Agency. *Municipal Solid Waste.* www.epa.gov. Accessed April 21, 2006.
24. Relis, P., and H. Levenson. 1998. *Discarding Solid Waste as We Know It: Managing Materials in the 21st Century.* Santa Barbara, CA: Community Environmental Council.
25. Young, J.E. 1991. Reducing waste saving materials. In L.R. Brown, ed., *State of the World,* 1991, pp. 39–55. New York, NY: W.W. Norton.
26. Steuteville, R. 1995. The state of garbage in America: Part I. *BioCycle* 36:54.
27. Author unknown. 2013. *Recycling Benefits to the Economy.* www.all-recycling-facts.com. Accessed March 13, 2013.
28. Gardner, G. 1998 (January/February). Fertile ground or toxic legacy? *Worldwatch,* pp. 28–34.
29. Rathje, W.L., and C. Murphy. 1992. Five major myths about garbage, and why they're wrong. *Smithsonian* Magazine 23:113–122.
30. Rathje, W.L. 1991. Once and future landfills. *National Geographic* 179(5):116–134.
31. U.S. Environmental Protection Agency. 2002. *Solid Waste Management: A Local Challenge with Global Impacts.* www.epa.gov. Accessed May 26, 2010.
32. Schneider, W.J. 1970. *Hydrologic Implications of Solid Waste Disposal.* 135(22). U.S. Geological Survey Circular 601F. Washington, DC: U.S. Geological Survey.
33. Thomas, V.M., and T.G. Spiro. 1996. The U.S. dioxin inventory: Are there missing sources? *Environmental Science and Technology* 30:82A-85A.
34. Turk, L.J. 1970. Disposal of solid wastes—acceptable practice or geological nightmare? In *Environmental Geology,* pp. 1–42. Washington, DC: American Geological Institute Short Course, American Geological Institute.
35. Hughes, G.M. 1972. Hydrologic considerations in the siting and design of landfills. *Environmental Geology Notes,* no. 51. Urbana: Illinois State Geological Survey.
36. Bergstrom, R.E. 1968. Disposal of wastes: Scientific and administrative considerations. *Environmental Geology Notes,* no 20. Urbana, IL: Illinois State Geological Survey.
37. Cartwright, K., and Sherman, F.B. 1969. Evaluating sanitary landfill sites in Illinois. *Environmental Geology Notes,* no. 27. Urbana: Illinois State Geological Survey.
38. Rahn, P.H. 1996. *Engineering Geology,* 2nd ed. Upper Saddle River, NJ: Prentice-Hall.

39. Bullard, R.D. 1990. *Dumping in Dixie: Race, Class and Environmental Quality.* Boulder, CO: Westview Press.
40. Sadd, J.L., J.T. Boer, M Foster, Jr., and L.D. Snyder 1997. Addressing environmental justice: Demographics of hazardous waste in Los Angeles County. *Geology Today* 7(8):18–19.
41. Walker, W.H. 1974 Monitoring toxic chemical pollution from land disposal sites in humid regions. *Ground Water* 12:213–218.
42. Watts, R.J. 1998. *Hazardous Wastes.* New York, NY: John Wiley & Sons.
43. Wilkes, A.S. 1980. *Everybody's Problem: Hazardous Waste.* SW-826. Washington, DC: U.S. Environmental Protection Agency, Office of Water and Waste Management.
44. Harder, B. 2005 (November 8). Toxic e-waste is cashed in poor nations. *National Geographic News.*
45. Elliot, J. 1980. Lessons from Love Canal. *Journal of the American Medical Association* 240:2033–2034, 2040.
46. Whittell, G. 2000 (November 29). Poison in paradise. *(London) Times* 2, p. 4.
47. U.S. Environmental Protection Agency. 2010. *Summary of the Comprehensive Environmental Response, Compensation, and Liability (Superfund).* www.epa.gov. Accessed March 19, 2010.
48. Bedient, P.B., H.S. Rifai, and C J. Newell. 1994. *Ground Water Contamination.* Englewood Cliffs, NJ: Prentice-Hall.
49. Huddleston, R.L. 1979. Solid waste disposal: Land farming. *Chemical Engineering* 86:119–124.
50. McKenzie, G.D., and W.A. Pettyjohn. 1975. Subsurface waste management. In G.D. McKenzie and R.O. Utgard, eds., *Man and His Physical Environment: Readings in Environmental Geology,* 2nd ed., pp. 150–156. Minneapolis, MN: Burgess Publishing.
51. National Research Council, Committee on Geological Sciences 1972. *The Earth and Human Affairs.* San Francisco, CA: Canfield Press.
52. Cox, C. 1985. *The Buried Threat: Getting Away from Land Disposal of Hazardous Waste.* No. 115–5. California Senate Office of Research.
53. Environmental Protection Agency. 2013. *Hazardous Waste Recycling.* www.epa.gov. Accessed March 30, 2013.
54. Council on Environmental Quality. 1970. *Ocean Dumping: A National Policy: A Report to the President.* Washington, DC: U.S. Government Printing Office.
55. Lenssen, N. 1989 (July–August). The ocean blues. *Worldwatch,* pp. 26–35.
56. U.S. Environmental Protection Agency. 2000. *Forward Pollution Protection: The Future Look of Environmental Protection.* http://www.epa.gov/p2/p2case.htm#num4. Accessed August 12, 2000.
57. Wellmar, F.W., and M. Kosinowoski. 2003. Sustainable development and the use of nonrenewable sources. *Geotimes* 48(12):14–17.

A CLOSER LOOK 23.2 NOTES

a. Zhihua, H., L.W. Morton, and R.L. Mahler. 2011. Bottled water: United States consumers and their perception of water quality. *International Journal of Environmental Research, Public Health.* 8(2):868–578.
b. Didier, S. *Water Bottle Pollution Facts.* www.greenliving.nationalgeographic.com. Accessed March 30, 2013.

A CLOSER LOOK 23.3 NOTES

a. Harder, B. 2005 (November 8). Toxic e-waste is cashed in poor nations. *National Geographic News.*

A CLOSER LOOK 23.4 NOTES

a. Woods Hole Oceanographic Institute. 2010. Plastic particles permeate the Atlantic. *Oceanus.* www.whoi.edu. Accessed September 2, 2010.

CHAPTER 24

Our Environmental Future

LEARNING OBJECTIVES

Throughout this book we have emphasized:

- **How we think about the environment and how what we do will reflect our core values**
- **Change is inevitable, and a static Earth is no longer a tenable hypothesis**
- **Science can help solve environmental problems, but our values will determine the solutions we will choose**

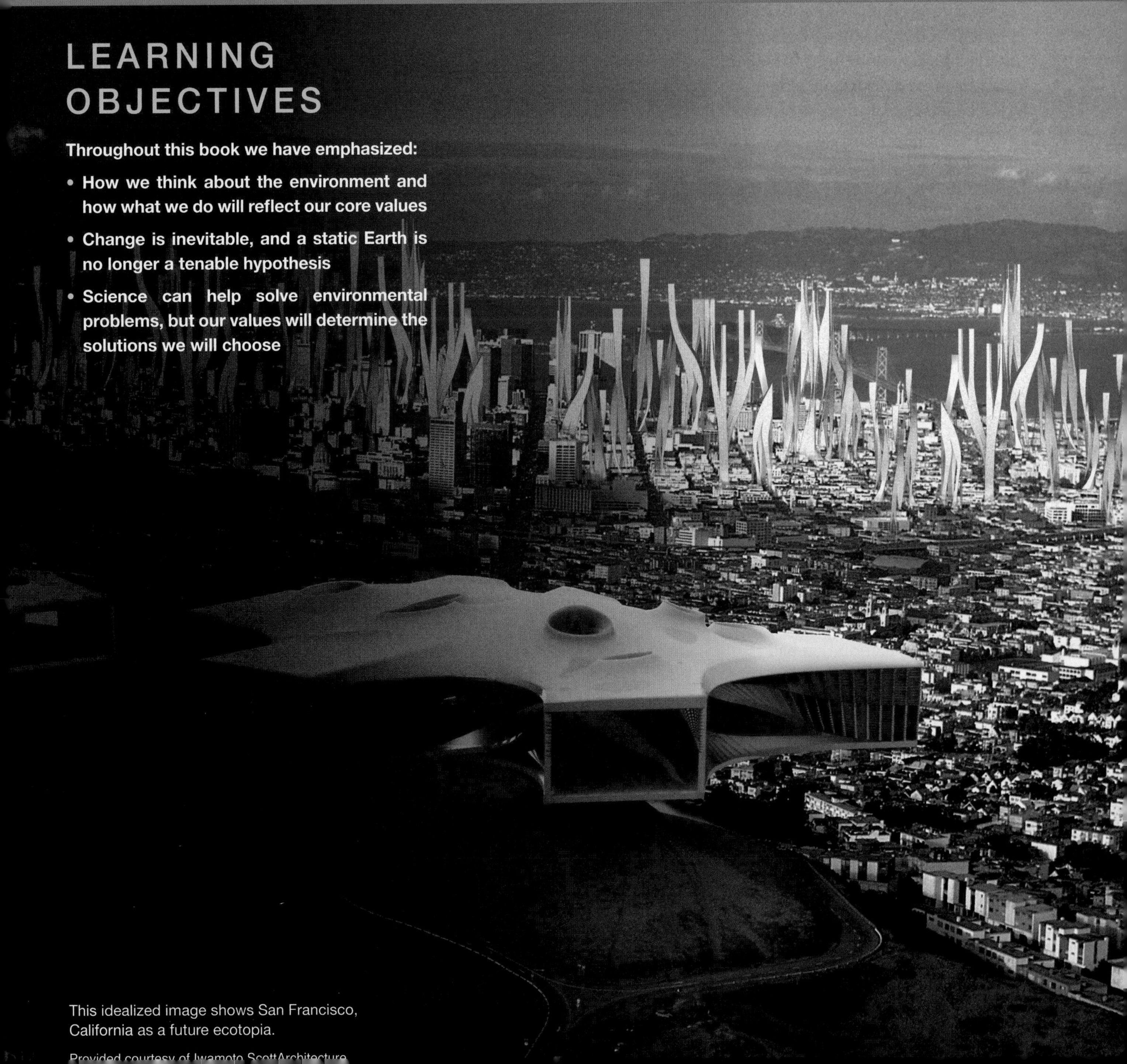

This idealized image shows San Francisco, California as a future ecotopia.

Provided courtesy of Iwamoto ScottArchitecture

CASE STUDY

Imagine an Ecotopia

Imagine a future in which we use our environment wisely—helping to create an "ecotopia." Some people have done so, as is seen in the opening photo of this chapter.

A learning objective of this chapter is to work out, to the best that present information allows, what you think is possible and desirable for this future world, focusing primarily on the United States, and then to describe how this might be accomplished. Having read this book, you may imagine a future in which, for example, we move away from fossil fuels and shift to renewable energy, no longer needing to damage the environment by mining and burning fossil fuels, nor forced to import them from uncertain and unfriendly sources. But which alternative energy sources would you favor?

This may seem an empty academic exercise, but unless we have an idea of what we want, we won't know in which direction to seek our future. Ideas are powerful, as history has proved. Wars have been fought over ideas. Ideas led Europeans to the New World and forged the American democracy. So what seems to be simply an academic exercise could be a powerful force for the future. It is not difficult today to imagine an "ecotopia"—a world in which the environment, human societies, and individuals are treated well in the present and helped to persist long into the future. But it would be extremely difficult to help it come about. What would that ecotopia be like? Here are qualities you would probably want to include:

- Since human population growth is the underlying environmental problem, an ecotopia would have to include a human population that had stabilized or even perhaps declined.
- All living resources would be sustainable, as would harvests of those resources.
- There would be enough wilderness and other kinds of natural or naturalistic areas for everyone to have opportunities for recreation and the enjoyment of nature.
- Pollution would be minimized.
- The risk of extinction of many species would be minimized.
- There would be enough functioning ecosystems to handle the public-service functions of ecosystems.
- Representatives of all natural ecosystems would be sustained in their dynamic ecological states.
- Poverty would be alleviated, benefiting both people and environment, because when you are poor it is hard to devote your resources to anything beyond immediate necessities.
- Energy would be abundant but, as much as possible, it would not cause pollution or otherwise damage land, water, and ecosystems.
- Water would be available to meet the needs of people and natural ecosystems,
- Natural resources, both finite and renewable, would also be available and recycled where possible.
- Violence between people and countries would be eliminated as a way to resolve differences.
- Societies would have ample resources to be creative and innovative.

Admittedly, developing all of these qualities—and/or whatever else you've thought of—will be far from easy.

24.1 The Process of Planning a Future

Both human societies and natural ecosystems are complex systems. One of the questions asked by modern science is the degree to which such systems are *self-organizing*. A seed of a plant, for example, is a self-organizing system: It can develop into a mature plant without any outside planning or rational effort. But plants grown in agriculture are not simply left to their self-organizing abilities. Farmers plan for them and carry out those plans, and in these ways a plant is no longer completely self-organizing.

In our discussion of ecosystems in Chapter 6, we said that an ecosystem is the basic unit that can sustain life, and, in that sense, is necessary for life to persist. To some extent ecosystems show self-organizing characteristics, as in ecological succession, but that process of succession isn't as fixed, neat, and perfect a pattern as the growth of a seed into a mature plant.

One of the major themes of this book is the connection between people and nature. We understand

Photo by Daniel B. Botkin

FIGURE 24.1 **New York City—a self-organizing system.**

Transocean/ZUMApress/Newscom

FIGURE 24.2 ***Deepwater Horizon* oil platform before the disaster of 2010—an example of a designed system.**

today that human societies are linked to natural ecosystems. To what degree can these linked, complex systems self-organize? In various chapters, we have reviewed some examples that appear as self-organization. For example, as we saw in Chapter 22 (urban environments), cities (Figure 24.1) developed at important transportation centers and where local resources could support a high density of people. In medieval Europe, bridges and other transportation aids developed in response to local needs. People arriving at a river would pay the farmer whose land lay along the river to row them across. Sometimes this service would become more profitable than farming, or at least an important addition to the farmer's income. Eventually, he might build a toll bridge. People would congregate naturally at such a crossing and begin to trade. A town would develop.[1] The combination of environment and society led in a self-organizing way to cities.[1]

In contrast, the oil-drilling platform accident that caused the 2010 *Deepwater Horizon* disaster in the Gulf of Mexico (see Chapter 15) was not self-organizing at all. The *Deepwater Horizon* was imagined, designed, and built by a large manufacturing corporation with a planned purpose: to serve as a floating platform for drilling into difficult oil and gas reserves (Figure 24.2). It functioned within the laws of the United States and international treaties that affected activities in the Gulf of Mexico. These are external plans and agreements to regulate and control how the complex structure of the *Deepwater Horizon* could and would be used. The failure of this platform was not the result of self-organization but of external (human) decisions.

In a democracy, planning with the environment in mind leads to a tug-of-war between individual freedom and the welfare of society as a whole. On one hand, citizens of a democracy want freedom to do what they want, wherever they want, especially on land that, in Western civilizations, is "owned" by the citizens or where citizens have legal rights to water or other resources. On the other hand, land and resource development and use affect society at large, and in either direct or indirect ways everyone benefits or suffers from a specific development. Society's concerns lead to laws, regulations, bureaucracies, forms to fill out, and limitations on land use.

Our society has formal planning processes for land use. These processes have two qualities: a set of rules (laws, regulations, etc.) requiring forms to be filled out and certain procedures to be followed; and an imaginative attempt to use land and resources in ways that are beautiful, economically beneficial, and sustainable. All human civilizations plan the development and use of land and resources in one way or another—through custom or by fiat of a king or emperor, if not by democratic processes. For thousands of years, experts have created formal plans for cities (see Chapter 22) and for important buildings and other architectural structures, such as bridges.

How can we balance freedom of individual action with effects on society? How can we achieve a sustainable use of Earth's natural resources, making sure that they will still be available for future generations to use and enjoy? In short, the questions are: Who speaks for nature? Who legally represents the environment? The landowner? Society at large? At this time, we have no definitive answers. Planning is a social experiment in which we all participate. Planning occurs at every level of activity, from a garden to a house, a neighborhood, a city park and its surroundings, a village, town, or city, a county, state, or nation. However, the history of our laws provides insight into our modern dilemma.

Issues of environmental planning and review are closely related to how land is used. Land use in the United States is dominated by agriculture and forestry; only a small portion of land (about 3%) is urban. However, rural lands are being converted to nonagricultural uses at about 9,000 km^2 (about 3,500 mi^2) per year. About half the conversion is for wilderness areas, parks, recreational areas, and wildlife refuges; the other half is for urban development, transportation networks, and other facilities. On a national scale, there is relatively little conversion of rural lands to urban uses. But in rapidly growing urban areas, increasing urbanization may be viewed as destroying agricultural land and exacerbating urban environmental problems, and urbanization in remote areas with high scenic and recreational value may be viewed as potentially damaging to important ecosystems.

24.2 Environment and Law: A Horse, a Gun, and a Plan

The legal system of the United States has historical origins in the British common law system—that is, laws derived from custom, judgment, and decrees of the courts rather than from legislation. The U.S. legal system preserved and strengthened British law to protect the individual from society—expressed best perhaps in the frontier spirit of "Just give me a little land, a horse, and a gun and leave me alone." Individual freedom—nearly unlimited discretion to use one's own property as one pleases—was given high priority, and the powers of the federal government were strictly limited.

But there was a caveat: When individual behavior infringed on the property or well-being of others, the common law provided protection through doctrines prohibiting trespass and nuisance. For example, if your land is damaged by erosion or flooding caused by your neighbor's improper management of his land, then you have recourse under common law. If the harm is more widespread through the community, creating a public nuisance, then only the government has the authority to take action—for instance, to limit certain air and water pollution.

The common law provides another doctrine, that of public trust, which both grants and limits the authority of government over certain natural areas of special character. Beginning with Roman law, navigable and tidal waters were entrusted to the government to hold for public use. More generally, "The public trust doctrine makes the government the public guardian of those valuable natural resources which are not capable of self-regeneration and for which substitutes cannot be made by man."[2] For such resources, the government has the strict responsibility of a trustee to provide protection and is not permitted to transfer such properties into private ownership. This doctrine was considerably weakened by the exaltation of private-property rights and by strong development pressures in the United States, but in more recent times it has shown increased vitality, especially concerning the preservation of coastal areas. This is the basis for much modern environmental law, policy, regulation, and planning: common law with respect to you and your neighbors and the public trust doctrine.

The Three Stages in the History of U.S. Environmental Law

The history of federal legislation affecting land and natural resources occurred in three stages. In the first stage, the goal for public lands was to convert them to private uses. During this phase, Congress passed laws that were not intended to address environmental issues but did affect land, water, minerals, and living resources—and thereby had large effects on the environment. In 1812, Congress established the General Land Office, whose original purpose was to dispose of federal lands. The government disposed of federal lands through the Homestead Act of 1862 and other laws. As an example of Stage 1, in the 19th century the U.S. government granted rights-of-way to railroad companies to promote the development of rapid transportation. In addition to rights-of-way, the federal government granted the railroads every other square mile along each side of the railway line, creating a checkerboard pattern. The square miles in between were kept as federal land and are administered today by the Bureau of Land Management. These lands are difficult to manage for wildlife or vegetation because their artificial boundaries rarely fit the habitat needs of species, especially those of large mammals.

The second stage began in the second half of the 19th century, when Congress began to pass laws that conserved public lands for recreation, scenic beauty, and historic preservation. Late in the 19th century, Americans came to believe that the nation's grand scenery should be protected and that public lands provided benefits, some directly economic, such as rangelands for private ranching.

Federal laws created the National Park Service in the second half of the 19th century in response to Americans' growing interest in their scenic resources. Congress made Yosemite Valley a California state park in 1864 and created Yellowstone National Park in 1872 "as a public park or pleasuring-ground for the benefit and enjoyment of the people."[3] Interest in Native American ruins led soon after to the establishment in 1906 of Mesa Verde National Park, putting into public lands the prehistoric cliff dwellings of early North Americans and at the same time creating national monuments. The National Park System was created by Congress in 1916. Today it consists of several hundred areas.

Also in the second stage, the United States Forest Service began in 1898, and President Grover Cleveland appointed Gifford Pinchot to be head of the Division of Forestry, soon renamed the U.S. Forest Service. Pinchot believed that the purpose of national forests was "the art of producing from the forest whatever it can yield for the service of man." The focus was on production of useful products.

Although the term *sustainability* had not yet become popular, in 1937 the federal government passed the Oregon and California Act, which required that timberland in western Oregon be managed to give sustained yields.[4]

In the third stage, Congress enacted laws whose primary purpose was environmental. This stage has antecedents in the 1930s but didn't get going in force until the 1960s and it continues today. The acknowledged need to regulate the use of land and resources has been filled by legislation enacted at all levels of government. In the late 1960s, public awareness and concern in the United States that our environment was deteriorating reached a high level. Congress responded by passing the National Environmental Protection Act (NEPA) in 1969 and a series of other laws in the 1970s (Figure 24.3). Federal laws relating to land management proliferated to the point where they became confusing. By the end of World War II, there were 2,000 laws about managing public lands, often contradicting one another. In 1946 Congress set up the Bureau of Land Management (BLM) to help correct this confusion.

Government regulation of land and resources has also given rise to controversy: How far should the government be allowed to go to protect what appears to be the public good against what have traditionally been private rights and interests? Today, the BLM attempts to balance the traditional uses of public lands—grazing and mining—with the environmental era's interest in outdoor recreation, scenic beauty, and biological conservation. Part of achieving a sustainable future in the United States will be finding a balance among these uses, as well as a balance between the amount of land that should be public and the amount of land that need not be public.

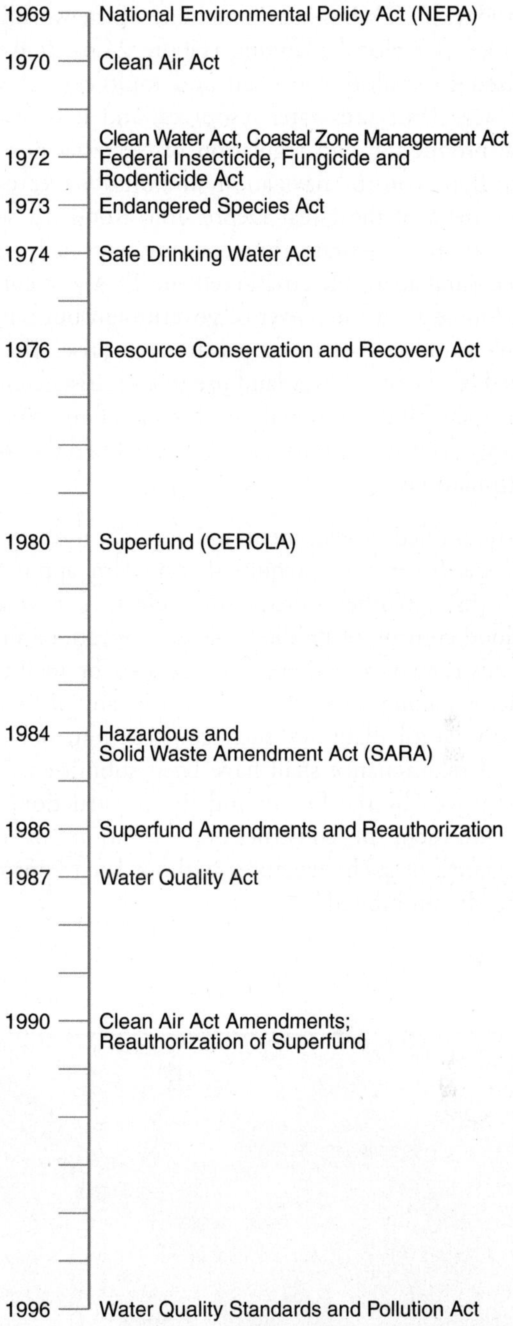

FIGURE 24.3 **Major federal environmental legislation and the year enacted.** Most of the important environmental legislation was adopted from 1969 to 1996. Some laws were enacted earlier in a much less comprehensive form (e.g., the Clean Air Act in 1963), and most were amended subsequently.

24.3 Planning to Provide Environmental Goods and Services

One important experiment of the 20th century was regional planning. In the United States, this means planning across state boundaries. One of the best-known regional plans in the United States began in 1933, when President Franklin D. Roosevelt proposed the establishment of the Tennessee Valley Authority (TVA), a semi-independent agency responsible for promoting economic growth and social well-being for the people throughout parts of seven states, which were economically depressed at the time. There had been rampant exploitation of timber and fossil-fuel resources in the region, and the people living there were among the poorest in the country.[5]

Today, the TVA is considered one of the world's best examples of regional planning (Figure 24.4). It is characterized by multidimensional and multilevel planning to manage land and water resources, and it is involved in the production and regulation of electrical power, as well as flood control, navigation, and outdoor recreation. In the midst of the Great Depression, Roosevelt sought new ways to invigorate the economy, especially in depressed rural areas. He envisioned the TVA as a corporation clothed with the power of government but with the flexibility and initiative of a private enterprise. The TVA granted legal control over land use to a multistate authority of a new kind and posed novel issues of governmental authority. The act creating the TVA contained the following stipulations:

> The unified development and regulation of the Tennessee River system require that no dam, appurtenant works, or other obstruction, affecting navigation, flood control, or public lands or reservations shall be constructed, and thereafter operated or maintained across, along, or in the said river or any of its tributaries until plans for such construction, operation, and maintenance shall have been submitted to and approved by the Board; and the construction, commencement of construction, operation, or maintenance of such structures without such approval is hereby prohibited.[6]

24.4 Planning for Recreation on Public Lands

Today, management of public lands for recreational activities requires planning at a variety of levels, with considerable public input. For example, when a national forest is developing management plans, public meetings are often held to inform people about the planning process and to ask for ideas and suggestions. Maximizing public input promotes better communication between those responsible for managing resources and those using them for recreational purposes.

Government officials and scientists involved in developing plans for public lands are often faced with land-use problems so complex that no easy answers can be found. Nonetheless, because action or inaction today can have serious consequences tomorrow, it is best to have at least some plans to protect and preserve a quality environment for future generations. Plans for many of the national forests and national parks in the United States have been or are being developed, generally taking into account a spectrum of recreational activities and attempting to balance the desires of several user groups.

Severe winter floods in Yosemite National Park during 1996–1997 damaged roads, campgrounds, bridges, and other structures. The flood led to a rethinking of the goals and objectives of park management, and one result was that some land claimed by the floods was returned to

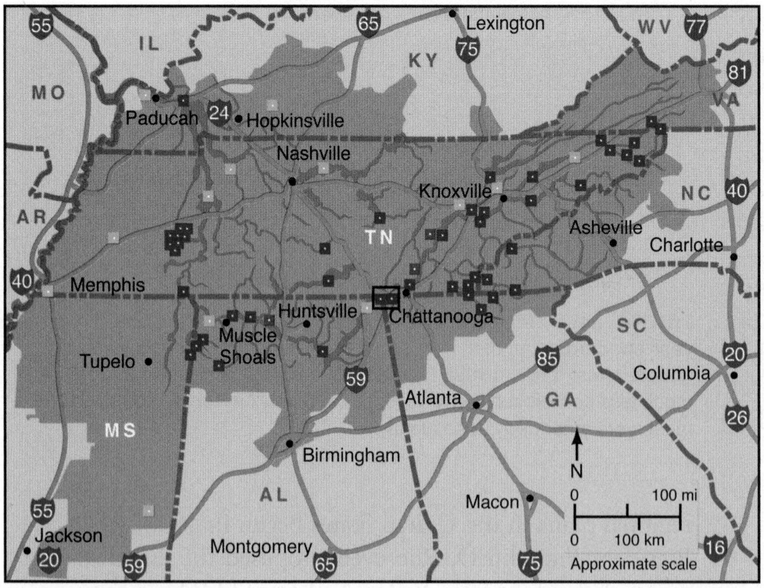

■ Reservoirs

Raccoon Mountain Dam

(a)

Tennessee Valley Authority

(b)

FIGURE 24.4 **(a)** A map showing the region encompassed by the TVA (darker area) and one of the major impoundments, the Raccoon Mountain Dam; **(b)** a large reservoir created by one of the TVA dams.

natural ecosystems. Another result was the elimination of private vehicles in parts of the park. Many other important policies have also been implemented in U.S. forests and parks. For example:

- In wilderness areas, only a limited number of people are admitted.
- In coastal areas, regulations may limit such activities as jet skiing and surfing in swimming areas.
- Regions that are home to endangered species, or to spcies that may pose a danger to people, may have more stringent regulations governing the activities of visitors. In Yellowstone National Park in Wyoming and Montana, for example, special consideration is given to grizzly bear habitats through controls on where people may venture.

Other recreational activities that are, or may become, subject to increased regulation include hiking, camping, fishing, boating, skiing, snowmobiling, and such recently popularized activities as treasure hunting, which includes panning for gold. At the extremes, certain areas have been set aside for intensive off-road-vehicle use, while other areas have been closed entirely. Activities on government lands can be more easily regulated than those occurring elsewhere. However, park management may be difficult if goals are not clear and natural processes are not understood.

Who Stands for Nature? Skiing at Mineral King

Planning for recreational activities on U.S. government lands (including national forests and national parks) is controversial. At the heart of the controversy are two different moral positions, both of which have wide support in the United States. On one side, some argue that public land must be open to public use, and therefore the resources within those lands should be available to citizens and corporations for economic benefit. On the other side are those who argue that public lands should serve the needs of society first and individuals second, and that public lands can and must provide for land uses not possible on private lands.

A classic example of this controversy concerned a plan by the Disney Corporation in the 1960s and 1970s to develop a ski resort with a multimillion-dollar complex of recreational facilities on federal land in a part of California's Sierra Nevada called Mineral King Valley (Figure 24.5), which had been considered a wilderness area. The Sierra Club, arguing that such a development would adversely affect the aesthetics of this wilderness, as well as its ecological balance, brought a suit against the government.

The case raised a curious question: If a wrong was being done, who was wronged? Christopher D. Stone, a lawyer, discussed this idea in an article entitled "Should Trees Have Standing? Toward Legal Rights for Natural Objects." The California courts decided that the Sierra Club itself could not claim direct harm from the development, and because the government owned the land but also represented the people, it was difficult to argue that the people in general were wronged. Stone said that the Sierra Club's case might be based, by common-law analogy, on the idea that in some cases inanimate objects have been treated as having legal standing—as, for example, in lawsuits involving ships, where ships have legal standing. Stone suggested that trees should have that legal standing, that although the Sierra Club was not able to claim direct damage to itself, it could argue on behalf of the nonhuman wilderness.

B. Anthony Stewart/National Geographic/Getty Images, Inc.

FIGURE 24.5 Mineral King Valley is now part of Sequoia National Park after nearly 20 years of controversy about the development of a ski resort in the valley.

The case was taken to the U.S. Supreme Court, which concluded that the Sierra Club itself did not have a sufficient "personal stake in the outcome of the controversy" to bring the case to court. But in a famous dissenting statement, Justice William O. Douglas addressed the question of legal standing (*standing* is a legal term relating here to the right to bring suit). He proposed establishing a new federal rule that would allow, "environmental issues to be litigated before federal agencies or federal courts in the name of the inanimate object about to be despoiled, defaced, or invaded by roads and bulldozers and where injury is the subject of public outrage." In other words, trees would have legal standing.

While trees did not achieve legal standing in that case, it was a landmark in that legal rights and ethical values were explicitly discussed for wilderness and natural systems. This subject in ethics still evokes lively controversy. Should our ethical values be extended to nonhuman, biological communities and even to Earth's life-support system? What position you take will depend in part on your understanding of the characteristics of wilderness, natural systems, and other environmental factors and features, and in part on your values.

Mineral King Valley and surrounding peaks of Mineral King, about 6,000 ha (12,600 acres), were transferred from the national forest to Sequoia National Park in September 1978. The transfer ended nearly 20 years of controversy over proposed development of a ski resort.

How Big Should Wildlands Be? Planning a Nation's Landscapes

Recent thinking about the environment has focused on the big picture: What is necessary at a national scale, or at some landscape scale, to achieve our goals? We are not the first to ask this question. John Wesley Powell, the famous one-armed American explorer who was the first to lead men down the Colorado River through the Grand Canyon, observed the dry American West and suggested that the land should be organized around major watersheds rather than laid out for political and social reasons, as the states ultimately were (Figure 24.6). His utopian vision was of a landscape where farmers spent their own money on dams and canals, doing so because the land was organized politically around watersheds. They could use, but not sell, their water. This plan seemed to impose too much control from the top and never happened.[7] Instead, in 1902 Congress passed an act that began the 20th-century construction of large dams and canals funded with federal dollars. Water rights could be sold, and cities like Los Angeles could assert the right to water hundreds of miles away.

While we cannot go back to Powell's vision completely, our society is gradually thinking more and more in terms of planning around large watersheds. This regional approach may help us move closer to the dream of our ecotopia. Modern scientific studies of ecosystems and landscapes also lead to speculation about the best way to conserve biological resources. Some argue that nature can be saved only in the large. A group called the Wildlands Project maintains that big predators, referred to as "umbrella species," are keys to ecosystems and that these predators require large home ranges. The assumption is that big, wide-ranging carnivores offer a wide umbrella of land protection under which many species that are more abundant but smaller and less charismatic find safety and resources.[8] Leaders of the Wildlands Project argue that even the biggest national parks, such as Yellowstone, are not big enough, and that America needs "rewilding." They propose that large areas of the United States be managed around the needs of big predators and that we replan our landscapes to provide a combination of core areas, corridors, and inner and outer buffers (Figure 24.7). No human activities would take place in the core areas, and even in the corridors and buffers human activity would be restricted.

The Wildlands Project has created a major controversy, with some groups seeing the project as a fundamental threat to American democracy. Another criticism of the Wildlands Project is directed at its scientific foundation. These critics say that although some ecological research suggests that large predators may be important, what controls populations in all ecosystems is far from understood. Similarly, the idea of keystone species, central to the rationale of the Wildlands Project, lacks an adequate scientific base.

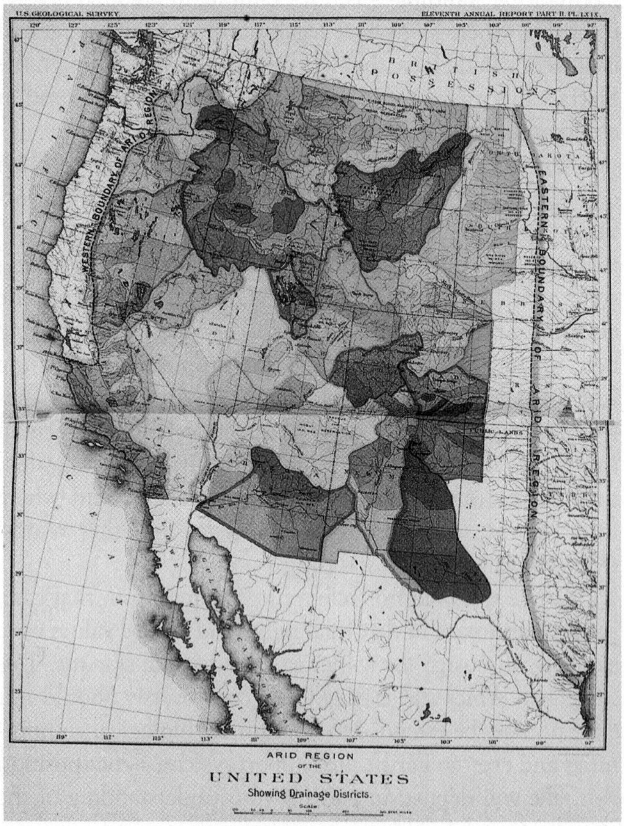

Library of Congress Prints and Photographs Division

FIGURE 24.6 **Powell's map of water in the West.**

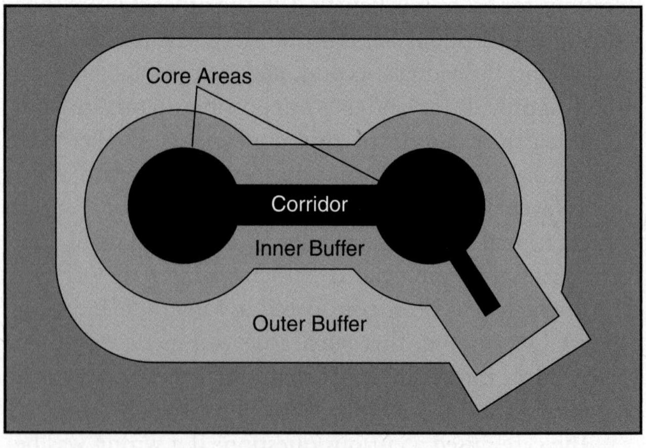

Process

FIGURE 24.7 **Wildlands Project diagram of land divisions.**

A related idea that developed in the last two decades was *rewilding*—that is, returning the land that was once American prairie to land without towns and cities, where bison are once again allowed to roam free. As Reed Noss, one of the founders of the Wildlands Project, has written:

> A cynic might describe rewilding as an atavistic obsession with the resurrection of Eden. A more sympathetic critic might label it romantic. We contend, however, that rewilding is simply scientific realism, assuming that our goal is to insure the long-term integrity of the land community. Rewilding with extirpated carnivores and other keystone species is a means as well as an end. The "end" is the moral obligation to protect wilderness and to sustain the remnants of the Pleistocene—animals and plants—not only for our human enjoyment but because of their intrinsic value.[8]

Proposals for the environment of the future thus involve science and values, and people and nature. So what do you want? A vast area of the United States returned to what might be self-functioning ecosystems? Or some open system of conservation that integrates people and allows for more freedom of action? The choices lie with your generation and the next, and tests of those choices' validity are also yours. The implications for the environment and for people are huge.[8]

24.5 Planning For Disasters

Disasters resulting from natural hazards, including flooding, tsunamis, earthquakes, tornadoes, and hurricanes, occur frequently over the surface of Earth. Many of these disasters are local to regional in extent and may cause billions of dollars of property loss, as well as loss of human life. Examples include Hurricane Katrina in 1995 (Figure 24.8), Hurricane Sandy in 2012 (Figure 24.9), and tsunamis in 2004 (Indonesia) and 2011 (Japan). Disasters may occur so infrequently in a particular area that people may relegate them to past history—characterizing them as more myth than reality. For example, the 2011 tsunami in Japan was thought to be a big and rare event that would likely never happen again, so it was not considered necessary to include the possibility of it occurring in emergency planning. However, prior to the 2011 event previous studies had shown that large tsunamis (similar in size to the 2011 tsunami) had occurred at least twice in the previous 2,000 years. Planning included no preparation for such an event; the 2011 tsunami flooded critical machinery, and partial meltdowns of several nuclear power plants resulted (see Chapter 17).

A second example comes from California. During the winter of 1860 to 1861, large floods occurred from San Diego to San Francisco. The damages to Southern California were considerable, even though its population was comparatively low. In northern California, a real catastrophe occurred. Over a period of weeks, storms dumped incredible amounts of precipitation in the Central Valley and on the western slopes of the Sierra Nevada. At that time, Sacramento was the capital of the state, and intense flooding from rivers to the east and north caused catastrophic flooding. Sacramento was literally underwater. The capital had to be moved for several months to San Francisco. A giant lake developed in the Central Valley of California that was several hundred miles long and several tens of miles wide. Overall, about one-fourth of the total housing in the state was damaged or destroyed.[9]

That event is not generally taken into consideration in planning for future flood disasters. Until recently it was

Mario Tama/Getty Images

FIGURE 24.8 In 2005 Hurricane Katrina severely damaged much of New Orleans, killing many people and causing thousands to evacuate.

RAMIN TALAIE/EPA/Newscom

FIGURE 24.9 Hurricane Sandy caused massive damage to New York City and the New Jersey coast in 2012.

more myth than a real event to planners. Such a huge flood is rare, but recent evaluation of sediments in the San Francisco and Santa Barbara areas suggest that, while it is rare, something similar is likely to occur again in the future.

These large floods in California occur when large atmospheric rivers—thousands of miles long, several hundred miles wide, and several thousand feet deep—move from the equatorial area across the Pacific to reach the shores of California. Large atmospheric rivers can carry as much water in vapor form as is in the Amazon River—about 10 times that of the Mississippi River. They also may have high (lower hurricane strength) winds and long duration (days) rainfall.[9]

In 1860, Sacramento was not populated by many people, but today it has well over one million. Similarly, cities of Southern California have grown very rapidly, and millions of people are at risk. It is estimated that if a similar event were to occur today, damages would likely be in the vicinity of $500 billion.

Planning for disasters is difficult but certainly possible. First, we have to recognize that such events may well occur in the future and plan for them. We can plan for large infrequent tsunamis in Japan and other vulnerable areas, including the Pacific Northwest of the United States. Many coastal cities have tsunami-warning signs, and there are sensors in the seabed that would warn us of a pending tsunami, unless it is generated close to the shore, in which case there would be little warning. Proactive steps would include locating critical facilities such as power plants and vulnerable equipment above sea level and ensuring that coastal defenses are sufficiently high. The same strategies would be useful for cities in coastal zones subject to large tsunamis (defined as tsunamis that are likely to move inland several miles or kilometers). With sufficient warning, evacuation is an option—but requires careful planning long before a tsunami arrives.

Hurricanes are familiar along the Gulf Coast, and it has been known for decades that hurricanes will occasionally strike the eastern seaboard. New York City had been warned for a long time that a storm the size of Hurricane Sandy was likely. Hopefully we will take steps to ensure that damage from these storms will be minimized in the future through better disaster planning.

Floods of the intensity of the 1860–1861 event in California can also be planned for, and steps can be taken to minimize potential damages. In particular, low-lying areas likely to be flooded can be protected by appropriate flood proofing, which might be as simple as constructing walls around critical facilities above predicted flood levels. In the Central Valley of California, rivers such as the American and Sacramento are lined with levees, which now provided a false sense of security. Many of these levees are aging and unable to withstand prolonged flooding from future large floods (or perhaps even lesser floods if combined with an earthquake that shakes the soil of the levees). Steps should be taken to strengthen the levee system and provide ways to bleed off excess water to areas where flooding will not cause property damage to urban areas and threaten lives. In particular areas, flood proofing would provide additional protection.

In summary, planning for disasters in the future requires us to be proactive rather than reactive. Our urban areas are getting larger and more populated. As a result, more people are in harm's way. Recognizing this reality and taking proper planning steps to minimize future disasters are paramount. For example, in California, scenarios have been developed for extensive flooding resulting from the hypothetical arrival of a large atmospheric river. Results from these scenarios will help plan for future disasters.

24.6 How You Can Be an Actor in the Environmental Law Processes

The case of Mineral King raises the question: What is the role of our legal system—laws, courts, judges, lawyers—in achieving environmental goals? The current answer is that environmental groups working through the courts have been a powerful force in shaping the direction of environmental quality control since the early 1970s. Their influence arose in part because the courts, appearing to respond to the national sense of environmental crisis of that time, took a more activist stance and were less willing to defer to the judgment of government agencies. At the same time, citizens were granted unprecedented access to the courts and, through them, to environmental policy.

Citizen Actions

Even without specific legislative authorization for citizens' suits, courts have allowed citizen actions in environmental cases as part of a trend to liberalize standing requirements.[10]

In the 1980s, a new type of environmentalism (which some people would label radical) arose, based in part on the premise that when it comes to the defense of wilderness, there can be no compromise. Methods used by these new environmentalists have included sit-ins to block roads into forest areas where mining or timber harvesting is scheduled; sitting in trees to block timber harvesting; implanting large steel spikes in trees to discourage timber harvesting; and sabotaging equipment, such as bulldozers (a practice known as "ecotage").

Ecotage and other forms of civil disobedience have undoubtedly been responsible for millions of dollars' worth of damage to a variety of industrial activities related to the use of natural resources in wilderness areas. One result of civil disobedience by some environmental groups is that other environmental groups, such as the Sierra Club, are now considered moderate in their approach to protecting the environment. There is no doubt, however, that civil disobedience has been successful in defending the environment in some instances. For example, members of the group Earth First succeeded in halting construction of a road being built to allow timber harvesting in an area of southwestern Oregon. Earth First's tactics included blockading the road by sitting or standing in front of the bulldozers, which slowed the pace of road work considerably. In conjunction with this action, the group filed a lawsuit against the U.S. Forest Service.

Environmentalists are now relying more on the law when arguing for ecosystem protection. The Endangered Species Act has been used as a tool in attempts to halt activities such as timber harvesting and development. Although the presence of an endangered species is rarely responsible for stopping a proposed development, those species are increasingly being used as weapons in attempts to save remaining portions of relatively undisturbed ecosystems.

Mediation

The expense and delay of litigation have led people to seek other ways to resolve disputes. In environmental conflicts, an alternative that has recently received considerable attention is mediation, a negotiation process between the adversaries guided by a neutral facilitator. The task of the mediator is to clarify the issues, help each party understand the position and needs of the other parties, and attempt to arrive at a compromise whereby each party gains enough to prefer a settlement to the risks and costs of litigation. Often, a citizens' suit, or the possibility that a suit might be filed, gives an environmental group a place at the table in mediation. Litigation, which may delay a project for years, becomes something that can be bargained away in return for concessions from a developer. Some states require mediation as an alternative or prior to litigation in the highly contentious siting of waste-treatment facilities. In Rhode Island, for example, a developer who wishes to construct a hazardous-waste treatment facility must negotiate with representatives of the host community and submit to arbitration of any issues not resolved by negotiation. The costs of the negotiation process are borne by the developer.

A classic example of a situation in which mediation could have saved millions of dollars in legal costs and years of litigation is the Storm King Mountain case, which represented a conflict between a utility company and conservationists. In 1962, the Consolidated Edison Company of New York announced plans for a new hydroelectric project in the Hudson River Highlands, an area that has thriving fisheries and is also considered to have unique aesthetic value (Figure 24.10). The utility company argued that it needed the new facility, and the environmentalists fought to preserve the landscape and the fisheries. Litigation began with a suit filed in 1965 and ended in 1981 after 16 years of intense courtroom battles that left a paper trail exceeding 20,000 pages. After spending millions of dollars and untold hours, the various parties finally managed to forge an agreement with the assistance of an outside mediator. If they had been able to sit down and talk at an early stage, mediation might have settled the issue much sooner and at much less cost to the parties and to society.[11] The Storm King Mountain case is often cited as a major victory for environmentalists, but the cost was great to both sides.

Stone/Getty Images

FIGURE 24.10 Storm King Mountain and the Hudson River Highlands in New York State were the focus of environmental conflict between a utility company and conservationists for nearly 20 years before a dispute about building a power plant was finally resolved by mediation.

24.7 International Environmental Law and Diplomacy

Legal issues involving the environment are difficult enough within a nation; they become extremely complex in international situations. International law is different from domestic law in basic concept because there is no world government with enforcement authority over nations. As a result, international law must depend on the agreement of the parties to bind themselves to behavior that many residents of a particular nation may oppose. Certain issues of multinational concern are addressed by a collection of policies, agreements, and treaties that are loosely called international environmental law. There have been encouraging developments in this area, such as agreements to reduce air pollutants that destroy stratospheric ozone (the Montreal Protocol of 1987 and subsequent discussion and agreements; see Chapter 21).

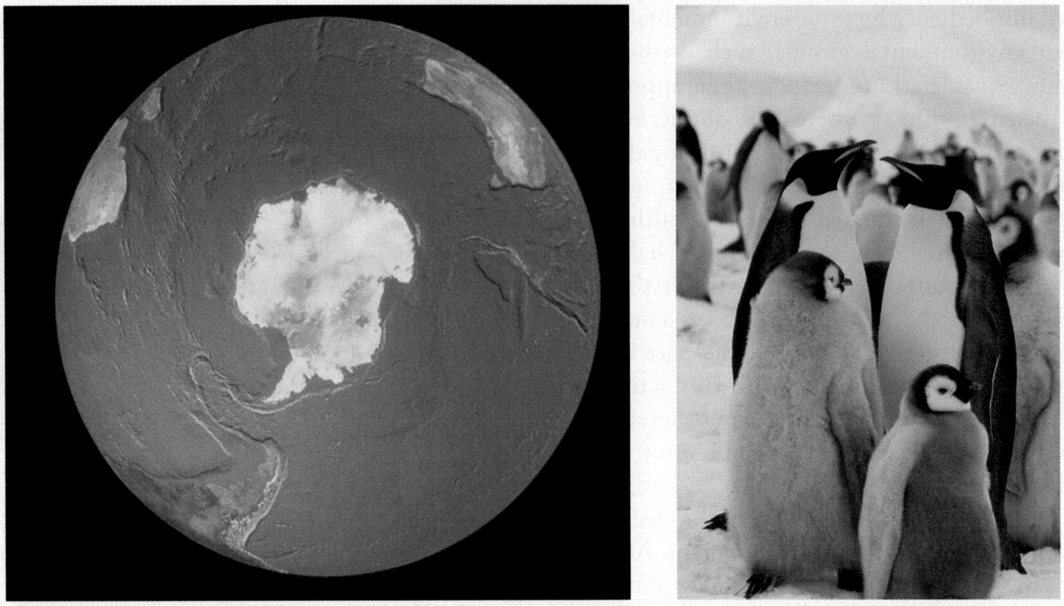

Tom Van Sant/Photo Researchers

David W. Hamilton/The Image Bank/Getty Images

(a) (b)

FIGURE 24.11 International agreements determine environmental practices in Antarctica. (a) Satellite image of Antarctica and surrounding southern oceans; **(b)** emperor penguins and chicks in Antarctica.

Antarctica provides a positive example of using international law to protect the environment. Antarctica, a continent of 14 million km^2, (5.4 million mi^2) was first visited by a Russian ship in 1820, and people soon recognized that the continent contained unique landscapes and lifeforms (Figure 24.11). By 1960, a number of countries had claimed parts of Antarctica to exploit mineral and fossil-fuel resources. Then, in 1961, an international treaty was established designating Antarctica a "scientific sanctuary." Thirty years later, in 1991, a major environmental agreement, the Protocol of Madrid, was reached, protecting Antarctica, including islands and seas south of 60° latitude. The continent was designated "nuclear-free," and access to its resources was restricted. This was the first step in conserving Antarctica from territorial claims and establishing the "White Continent" as a heritage for all people on Earth.

Other environmental problems addressed at the international level include persistent organic pollutants (POPs), such as dioxins, DDT, and other pesticides. After several years of negotiations in South Africa and Sweden, 127 nations adopted a treaty in May 2001 to greatly reduce or eliminate the use of toxic chemicals known to contribute to cancer and harm the environment.

24.8 Global Security and Environment

The terrorist attacks on New York City and Washington, DC, on September 11, 2001, brought the realization that the United States—in fact the world—is not as safe as we had assumed. The attacks led to a war on terrorists and their financial and political networks around the world. However, for every terrorist removed, another will fill the void unless the root causes are recognized and eliminated.

Achieving sustainability in the world today has strong political and economic components, but it also has an environmental component. Terrorism comes in part from poverty, overcrowding, disease, and conflicts that have environmental significance. Over 1 billion people on Earth today live in poverty with little hope for the future. In some large urban regions, tens of millions of people exist in crowded, unsanitary conditions, with unsafe drinking water and inadequate sewage disposal. In the countryside, rural people in many developing countries are being terrorized and displaced by armed conflicts over the control of valuable resources, such as oil, diamonds, and timber. Examples include oil in Nigeria, Sudan, and Colombia; diamonds in Sierra Leone, Angola, and the Democratic Republic of the Congo; and timber in Cambodia, Indonesia, and Borneo.[12]

The goal of the 1992 Rio Earth Summit on Sustainable Development was to address global environmental problems of both developed and developing countries, with an emphasis on solving conflicts between economic interests and environmental concerns. In many countries today, the gap between the rich and the poor is even wider than it was in the early 1990s. As a result, political, social, and economic security remains threatened, and serious environmental damage from overpopulation and resource exploitation continues. Environmental protection continues to be inadequately funded. Governments of the world are spending very little on the environment compared to what is spent on military purposes.[12]

24.9 Challenges to Students of the Environment

To end this book on an optimistic note—and there *are* reasons to be optimistic—we note that the Earth Summit on Sustainable Development, held in the summer of 2002 in Johannesburg, South Africa, had the following objectives:

- To continue to work toward environmental and social justice for all the people in the world.
- To enhance the development of sustainability.
- To minimize local, regional, and global environmental degradation resulting from overpopulation, deforestation, mining, agriculture, and pollution of the land, water, and air.
- To develop and support international agreements to control global warming and pollutants, and to foster environmental and social justice.

Solving our environmental problems will help build a more secure and sustainable future. This is becoming your charge and responsibility, as you, students of the environment and our future leaders, graduate from colleges and universities. This transfer of knowledge and leadership is a major reason why we wrote this book.

CRITICAL THINKING ISSUE

Is It Possible to Derive Some Quantitative Statements about Thresholds beyond Which Unacceptable Environmental Change Will Occur?

A *threshold* is a condition or level that, if exceeded, will cause a system to change, often from one mode of operation to another, in terms of actual processes or rates of processes. In the environmental literature, thresholds are sometimes spoken of as tipping points, beyond which adverse consequences are likely to occur. Other definitions of a tipping point are: a point when a system (say the global climate) changes from one stable state to another stable state (this is a threshold); and a point where slow, small changes over time result in a sudden large change (also a threshold). However, thresholds are not tipping points where change becomes catastrophic and may be irreversible. For example, some believe that if global warming continues past a particular point, say a two degree Celsius rise of temperature, then changes will become more rapid and the consequences of those changes more severe. The purpose of this critical thinking issue is to examine some of these hypotheses in more detail.[13]

In previous chapters, we discussed the major environmental problems related to human population, water, energy, and climate. In discussing human population, we introduced the concept of what Earth's carrying capacity might be. In answering that question, we posed another: "What would we *like* it to be?" It is acknowledged that human population growth is the environmental problem, but at what population level would the degree of environmental degradation become unacceptable to us? Similar limits or thresholds might be introduced for biological productivity; loss of biological diversity; use of nutrients, such as nitrogen and phosphorus; transformation of the land; and our use of freshwater resources. For this list, some scientists have tried to pinpoint thresholds beyond which environmental degradation is unacceptable (a value judgment).

Table 24.1 is based on a paper published in 2009 in the major scientific journal *Nature* and entitled "A Safe Operating Space for Humanity." You should treat these ideas as proposals for discussion, not as truths or facts. The table lists these systems in terms of parameters that may be measured, along with suggested thresholds, which are compared to the present status and also to pre-industrial levels. For example, for human population, a suggested threshold might be 5 billion people—fewer than are on Earth today and 4 billion more than the preindustrial level of about 1 billion. This 5 billion threshold might be based on the fact that biological productivity, when it was more in balance with human needs, peaked around 1985, when the population was 5 billion people. The arbitrary choice of 5 billion is obviously linked to other factors shown in the table, as they are interrelated. Any specific number for the optimum carrying capacity of the planet will be controversial, but your evaluation will depend on the knowledge you bring to bear and your values.

With respect to climate change, Table 24.1 lists a hypothetical threshold of 350 parts per million for carbon dioxide concentration in the atmosphere, versus the 2009 level of 390 parts per million and the preindustrial level 280 parts per million. Setting the threshold at 350 parts per million was based on examination of the geologic record, the possible effects of previous climate change, and the likely levels of carbon dioxide in the atmosphere. This table is intended solely for the sake of our discussion here. Similarly, the amount of land transformation or water use is also related to our present scientific knowledge.

Looking at Table 24.1, we can see that some of the suggested thresholds have already been exceeded and others have not. However, whether they actually have been exceeded will depend on how much we know about the particular system, whether the consequences are unacceptable, and whether this can be shown with some degree of certainty.

Critical Thinking Questions

1. Do you think it is a valid argument that some sorts of thresholds, or tipping points, exist beyond which unacceptable environmental degradation will occur?
2. Has science satisfactorily answered whether or not these thresholds, or tipping points, can in fact be established?
3. From your reading of *Environmental Science*, can you make other suggestions as to where thresholds or tipping points might be placed?
4. If you are not able to set thresholds, what sorts of studies might be necessary to establish them in the future? Of course, this assumes that the whole concept of thresholds, or tipping points, is a valid approach in environmental science.

TABLE 24.1 GLOBAL THRESHOLDS THAT, TRANSGRESSED, COULD CAUSE UNACCEPTABLE ENVIRONMENTAL CHANGE [FOR DISCUSSION PURPOSES ONLY; NOT TO BE TAKEN AS FACTS]

SYSTEM	PARAMETER	SUGGESTED THRESHOLD	PRESENT STATUS	PRE-INDUSTRIAL LEVEL
Human population	Billions of people	5.0	6.8	1.0
Climate	Carbon dioxide concentration (parts per million)	350	390	280
Biological productivity	Portion used by humans	0.6	1.2	<0.2
Biodiversity loss (extinction)	Extinction rate (number of species per million species per year)	10	>100	0.1–1.0
Nitrogen use	Amount removed from the air for human use (millions of tons per year)	35	120	0
Phosphorus use	Quantity flowing into the ocean (millions of tons per year)	11	9	−1.0
Land transformation	% of land converted to agriculture	15	12	Low
Global freshwater use	km^3/yr	4,000	2,600	415
Air pollution	Metric tons per year	To be determined	To be determined	To be determined
Water pollution	Metric tons per year	To be determined	To be determined	To be determined

Source: Modified from J. Rockström et al., 2009. "A Safe Operating Space for Humanity," *Nature* 461: 472–475. doi:10.1038/461472a.

SUMMARY

- A fundamental question, continuously debated in a democracy, is the extent to which human societies and their environment can function as self-organizing systems, and how much formal planning—laws and so on—is necessary.
- Both natural ecosystems and human societies are complex systems. The big question is how the interaction among these systems can lead to the long-term persistence of both, and perhaps even improvements.

- Mistakes are always likely; advance planning, including rapid response, is essential to maintaining the best environment.
- Planning for large, infrequently occurring natural disasters such as earthquakes, tsunami floods, and hurricanes is becoming more important as human populations increase and more people live in regions where these events are more likely to occur. A key to minimizing the effects of disasters is to be proactive (planning) rather than reactive (only responding after an event occurs).
- Our environmental laws have grown out of a combination of the English common law—derived from custom and judgment, rather than legislation—and American perspectives on freedom and planning.
- In the 19th and 20th centuries, America experimented with a variety of approaches to conserving nature, some involving laws, some new kinds of plans and organizations. The best combination is yet to be determined.
- International environmental law is proving useful in addressing several important environmental problems, including preservation of resources and pollution abatement.
- Global security, sustainability, and environment are linked in complex ways. Solving environmental problems will improve both sustainability and security.

STUDY QUESTIONS

1. Based on what you have learned in this book and in your studies about environment, what would an "ecotopia" include, in addition to what is mentioned in this chapter? Which of these items, if any, do you think could be achieved during your lifetime?
2. Just how big should a wilderness be?
3. The famous ecologist Garrett Hardin argued that designated wilderness areas should not have provisions for people with handicaps, even though he himself was confined to a wheelchair. He believed that wilderness should be truly natural in the ultimate sense—that is, without any trace of civilization. Argue for or against Garrett Hardin's position. In your argument, consider the "people and nature" theme of this book.
4. How can we balance freedom of individual action with the need to sustain our environment?
5. Visit a local natural or naturalistic place, even a city park, and write down what is necessary for that area to be sustainable in its present uses.
6. Should trees—and other nonhuman organisms—have legal standing? Explain your position on this topic.
7. What disasters are likely to occur in the region you live in? How does your list compare to other regions? What steps has your community taken to be prepared for future disasters? Finally, what more could be done to be more proactive?
8. Since there are no international laws that are binding in the same way that laws govern people within a nation, what can be done to achieve a sustainable environment for world fisheries or other international resources?
9. Do you think the Gulf oil spill could have been prevented? If so, how?
10. Do you think Garrett Hardin is right—that there are some technologies (such as drilling in deep water) that humans are not prepared to adequately address and that there will thus be continued accidents due to human error?

NOTES

1. Jusserand, J. 1897. *English Wayfaring Life in the Middle Ages (XIVth Century)*. London, UK: T. Fisher Unwin.
2. Cohen, B.S. 1970. The Constitution, the public trust doctrine and the environment. *Utah Law Review* 388.
3. Macintosh, B. 1999. *A Brief History of the National Park Service*. From http://www.cr.nps.gov/history/hisnps/NPSHistory/briefhistory.htm.
4. Bureau of Land Management Facts. http://www.blm.gov/wo/st/en/res/blm_jobs/blm_facts.html.
5. Steiner, F. 1983. Regional planning: Historic and contemporary examples. *Landscape Planning* 10:297–315.
6. Section 26 of the TVA Act. http://www.tva.com/abouttva/pdf/TVA_Act.pdf.
7. Berkes, H. 2003 (August 26). The vision of John Wesley Powell. National Public Radio. http://www.npr.org/programs/atc/features/2003/aug/water/part1.html.
8. Forman, D. 2004. *Rewilding North America: A Vision of Conservation for the 21st Century*. Washington, DC: Island Press.
9. Dettinger, M.D. and Ingram, L.B. 2013. The comming mega floods. *Scientific American*. January: 64-71
10. Yannacone, V.J., Jr., B.S. Cohen, and S.G. Davison. 1972. Environmental rights and remedies. *Lawyers Co-operative Pub.*, pp. 39–46.
11. Bacow, L.S., and M. Wheeler. 1984. *Environmental Dispute Resolution*. New York, NY: Plenum Press.
12. Renner, M. 2002. Breaking the link between resources and repression. In *Worldwatch Institute State of the World 2002*. New York, NY: W.W. Norton.
13. Rockström, J., et al. 2009. A safe operating space for humanity. *Nature* 461:472–475.

Appendix

A. Special Feature: Electromagnetic Radiation (EMR) Laws

Properties of Waves

- Direction of wave propagation

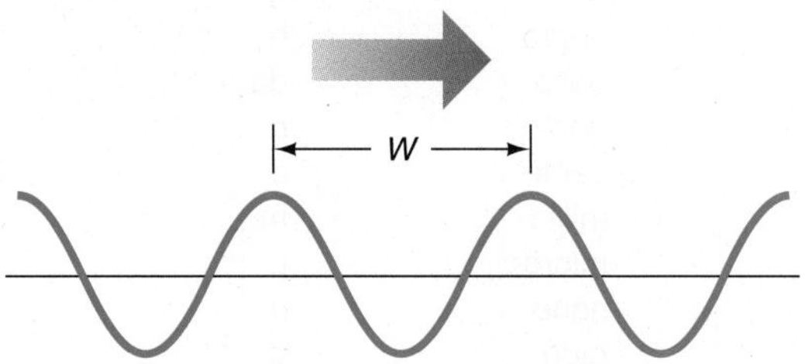

- W = wavelength (distance from one wave crest to the next)
- An EMR wave travels at the speed of light (C) in a vacuum, or about 300,000 km/s (3×10^8 m/s).
- The period T of a wave is the time it takes for a wave to travel a distance of one wavelength W. Then since distance is the product of speed (velocity) and time, W = CT.
- The frequency f of a wave is the number of cycles (each wavelength that passes a point is a cycle) of a wave that pass a particular point per unit time. Frequency f is measured in cycles per second (hertz). The frequency f is the inverse of T: f = 1/T and W = C/f. For example, the period T of a 6000 hertz EMF wave is: 6000 hertz = 1/T and T = 1/6000 S, or 1.7×10^{-4} S. The wavelength W = CT is 3×10^8m/s times 1.7×10^{-4} S, or 5.1×10^4 m, which according to Figure 20.18 is a long radio wave.

Absolute Temperature Scale (kelvin, K)

- Zero is really zero; there are no negative values of K
- Temperature in K = temperature in °C + 273

$$K = {}^\circ C + 273$$

- Example: water freezes at °C = 0 = 273 K
  water boils at °C = 100 = 373 K

Stefan–Boltzmann Law

- All bodies with a temperature greater than absolute zero radiate EMR. These bodies are called thermal radiators. The amount of energy per second radiated from thermal radiators is called *intensity* and is given by the Stefan–Boltzmann law

$$E = aT^4$$

where E is the energy per second (intensity); T is the absolute temperature; and a is a constant (the nature of this constant involves physical ideas beyond the scope of this text).

- The Stefan–Boltzmann law states that the intensity of EMR coming from a thermal radiator is directly proportional to the fourth power of its absolute temperature.

Wien's Law

$$W_P = a/T$$

where WP is the wavelength of the peak intensity of a thermal radiator; T is temperature in K; and a is a constant. For example, Figure 20.19 shows that WP for the Earth is about 10 μm. Wien's law states in a general way that the hotter a substance is, the shorter the wavelength of the emitted predominant electromagnetic radiation. That is, wavelength is inversely proportional to temperature.

B. Prefix and Multiplication Factors

Number	10×, Power of 10	Prefix	Symbol
1,000,000,000,000,000,000	10^{18}	exa	E
1,000,000,000,000,000	10^{15}	peta	P
1,000,000,000,000	10^{12}	tera	T
1,000,000,000	10^{9}	giga	G
1,000,000	10^{6}	mega	M
10,000	10^{4}	myria	
1,000	10^{3}	kilo	k
100	10^{2}	hecto	h
10	10^{1}	deca	da
0.1	10^{-1}	deci	d
0.01	10^{-2}	centi	c
0.001	10^{-3}	milli	m
0.000 001	10^{-6}	micro	μ
0.000 000 001	10^{-9}	nano	n
0.000 000 000 001	10^{-12}	pico	p
0.000 000 000 000 001	10^{-15}	femto	f
0.000 000 000 000 000 001	10^{-18}	atto	a

C. Common Conversion Gactors

LENGTH

1 yard = 3 ft, 1 fathom = 6 ft

	in	ft	mi	cm	m	km
1 inch (in) =	1	0.083	1.58×10^{-5}	2.54	0.0254	2.54×10^{-5}
1 foot (ft) =	12	1	1.89×10^{-4}	30.48	0.3048	
1 mile (mi) =	63,360	5,280	1	160,934	1,609	1.609
1 centimeter (cm) =	0.394	0.0328	6.2×10^{-6}	1	0.01	1.0×10^{-5}
1 meter (m) =	39.37	3.281	6.2×10^{-4}	100	1	0.001
1 kilometer (km) =	39,370	3,281	0.6214	100,000	1,000	1

AREA

1 square mi = 640 acres, 1 acre = 43,560 ft^2 = 4046.86 m^2 = 0.4047 ha
1 ha = 10,000 m^2 = 2.471 acres

	in^2	ft^2	mi^2	cm^2	m^2	km^2
1 in^2 =	1	—	—	6.4516	—	—
1 ft^2 =	144	1	—	929	0.0929	—
1 mi^2 =	—	27,878,400	1	—	—	2.590
1 cm^2 =	0.155	—	—	1	—	—
1 m^2 =	1,550	10.764	—	10,000	1	—
1 km^2 =	—	—	0.3861	—	1,000,000	1

VOLUME	in³	ft³	yd³	m³	qt	liter	barrel	gal (U.S.)
1 in^3 =	1	—	—	—	—	0.02	—	—
1 ft^3 =	1,728	1	—	0.0283	—	28.3	—	7.480
1 yd^3 =	—	27	1	0.76	—	—	—	—
1 m^3 =	61,020	35.315	1.307	1	—	1,000	—	—
1 quart (qt) =	—	—	—	—	1	0.95	—	0.25
1 liter (l) =	61.02	—	—	—	1.06	1	—	0.2642
1 barrel (oil) =	—	—	—	—	168	159.6	1	42
1 gallon (U.S.) =	231	0.13	—	—	4	3.785	0.02	1

Mass and Weight

1 pound = 453.6 grams = 0.4536 kilogram = 16 ounces

1 gram = 0.0353 ounce = 0.0022 pound

1 short ton = 2000 pounds = 907.2 kilograms

1 long ton = 2240 pounds = 1008 kilograms

1 metric ton = 2205 pounds = 1000 kilograms

1 kilogram = 2.205 pounds

Energy and Power[a]

1 kilowatt-hour = 3413 Btus = 860,421 calories

2 Btu = 0.000293 kilowatt-hour = 252 calories = 1055 joules

1 watt = 3.413 Btu/hr = 14.34 calorie/min

1 calorie = the amount of heat necessary to raise the temperature of 1 gram (1 cm^3) of water 1 degree Celsius

1 quadrillion Btu = (approximately) 1 exajoule

1 horsepower = 7.457×10^2 watts

1 joule = 9.481×10^{-4} Btu = 0.239 cal = 2.778×10^{-7} kilowatt-hour

[a]Values from Lange, N. A., 1967, *Handbook of Chemistry*, New York: McGraw-Hill.

Temperature

$F = \frac{9}{5}C + 32$

F is degrees Fahrenheit.

C is degrees Celsius (centigrade).

Fahrenheit		Celsius
32	Freezing of H_2O (Atmospheric Pressure)	0
50		10
68		20
86		30
104		40
122		50
140		60
158		70
176		80
194		90
212	Boiling of H_2O (Atmospheric Pressure)	100

Other Conversion Factors

1 ft^3/sec = 0.0283 m^3/sec = 7.48 gal/sec = 28.32 liter/sec

1 acre-foot = 43,560 ft^3 = 1233 m^3 = 325,829 gal

1 m^3/sec = 35.32 ft^3/sec

1 ft^3/sec for one day = 1.98 acre-feet

1 m/sec = 3.6 km/hr = 2.24 mi/hr

1 ft/sec = 0.682 mi/hr = 1.097 km/hr

1 atmosphere = 14.7 lb($in.^{-2}$) = 2116 lb(ft^{-2}) = 1.013×10^5 N(m^{-2})

D. Geologic Time Scale and Biologic Evolution

Era	Approximate Age in Millions of Years Before Present	Period	Epoch	Life Form
	Less than 0.01		Recent (Holocene)	
	0.01–2	Quaternary	Pleistocene	Humans
	2			
	2–5		Pliocene	
	5–23		Miocene	
Cenozoic	23–35	Tertiary	Oligocene	
	35–56		Eocene	Mammals
	56–65		Paleocene	
	65			
	65–146	Cretaceous		
Mesozoic	146–208	Jurassic		Flying reptiles, birds
	208–245	Triassic		Dinosaurs
	245			
	245–290	Permian		Reptiles
	290–363	Carboniferous		Insects
Paleozoic	363–417	Devonian		Amphibians
	417–443	Silurian		Land plants
	443–495	Ordovician		Fish
	495–545	Cambrian		
	545			
	700			Multicelled organisms
	3,400			One-celled organisms
	4,000	Approximate age of oldest rocks discovered on Earth		
Precambrian				
	4,600	Approximate age of Earth and meteorites		

Glossary

Abundance The size of a population.

Acid mine drainage Acidic water that drains from mining areas (mostly coal but also metal mines). The acidic water may enter surface water resources, causing environmental damage.

Acid rain Rain made acid by pollutants, particularly oxides of sulfur and nitrogen. (Natural rainwater is slightly acid owing to the effect of carbon dioxide dissolved in the water.)

Active solar energy systems Direct use of solar energy that requires mechanical power; usually consists of pumps and other machinery to circulate air, water, or other fluids from solar collectors to a heat sink where the heat may be stored.

Acute disease A disease that appears rapidly in the population, affects a comparatively large percentage of it, and then declines or almost disappears for a while, only to reappear later.

Adaptive management The application of science to the management process.

Adaptive radiation The process that occurs when a species enters a new habitat that has unoccupied niches and evolves into a group of new species, each adapted to one of these niches.

Advanced wastewater treatment Treatment of wastewater beyond primary and secondary procedures. May include sand filters, carbon filters, or application of chemicals to assist in removing potential pollutants such as nutrients from the wastewater stream.

Aesthetic justification for the conservation of nature An argument for the conservation of nature on the grounds that nature is beautiful and that beauty is important and valuable to people.

Age structure (of a population) A population divided into groups by age. Sometimes the groups represent the actual number of each age in the population; sometimes the groups represent the percentage or proportion of the population of each age.

Agroecosystem An ecosystem created by agriculture. Typically it has low genetic, species, and habitat diversity.

Air toxics Those air pollutants known or suspected to cause cancer and other serious health problems from either longor short-term exposure.

Allowance trading An approach to managing coal resources and reducing pollution through buying, selling, and trading of allowances to emit pollutants from burning coal. The idea is to control pollution by controlling the number of allowances issued.

Alternative energy Renewable and nonrenewable energy resources that are alternatives to the fossil fuels.

Aquaculture Production of food from aquatic habitats.

Area sources Sometimes also called nonpoint sources. These are diffuse sources of pollution such as urban runoff or automobile exhaust. These sources include emissions that may be found over a broad area or even over an entire region. They are often difficult to isolate and correct because of the widely dispersed nature of the emissions.

Asbestos A term for several minerals that have the form of small elongated particles. Some types of particles are believed to be carcinogenic or to carry with them carcinogenic materials.

Atmosphere Layer of gases surrounding Earth.

Atmospheric inversion A condition in which warmer air is found above cooler air, restricting air circulation; often associated with a pollution event in urban areas.

Autotrophs An organism that produces its own food from inorganic compounds and a source of energy. There are photoautotrophs and chemical autotrophs.

Average residence time A measure of the time it takes for a given part of the total pool or reservoir of a particular material in a system to be cycled through the system. When the size of the pool and rate of throughput are constant, average residence time is the ratio of the total size of the pool or reservoir to the average rate of transfer through the pool.

Bacillus thuringiensis A bacteria which causes a disease that affects caterpillars and the larvae of other insect pests.

Balance of nature An environmental myth that the natural environment, when not influenced by human activity, will reach a constant status, unchanging over time, referred to as an equilibrium state.

Barometric pressure Atmospheric pressure given in units that are the height to which a column of mercury is raised by that pressure.

Bioengineering A surface-water cleanup technique that uses plants and soil in an engineered landscape (especially urban lands) to minimize water pollution.

Biofuel Energy recovered from biomass.

Biogeochemical cycle The cycling of a chemical element through the biosphere; its pathways, storage locations, and chemical forms in living things, the atmosphere, oceans, sediments, and lithosphere.

Biogeography The large-scale geographic pattern in the distribution of species, and the causes and history of this distribution.

Biological control A set of methods to control pest organisms by using natural ecological interactions, including predation, parasitism, and competition. Part of integrated pest management.

Biological diversity Used loosely to mean the variety of life on Earth, but scientifically typically used asto consisting of three components: (1) genetic diversity—the total number of genetic characteristics; (2) species diversity; and (3) habitat or ecosystem diversity—the number of kinds of habitats or ecosystems in a given unit area. Species diversity in turn includes three concepts: *species richness*, *evenness*, and *dominance*.

Biological evolution The change in inherited characteristics of a population from generation to generation, which can result in new species.

Biological or biochemical oxygen demand (BOD) A measure of the amount of oxygen necessary to decompose organic material in a unit volume of water. As the amount of organic waste in water increases, more oxygen is used, resulting in a higher BOD.

Biological production The capture of usable energy from the environment to produce organic compounds in which that energy is stored.

Biomagnification Also called *biological concentration.* The tendency for some substances to concentrate with each trophic level. Organisms preferentially store certain chemicals and excrete others. When this occurs consistently among organisms, the stored chemicals increase as a percentage of the body weight as the material is transferred along a food chain or trophic level. For example, the concentration of DDT is greater in herbivores than in plants and greater in plants than in the nonliving environment.

Biomass The amount of living material, or the amount of organic material contained in living organisms, both as live and dead material, as in the leaves (live) and stem wood (dead) of trees.

Biome A kind of ecosystem. The rain forest is an example of a biome; rain forests occur in many parts of the world but are not all connected to each other.

Bioremediation A method of treating groundwater pollution problems that utilizes microorganisms in the ground to consume or break down pollutants.

Biosphere Has several meanings. One is that part of a planet where life exists. On Earth it extends from the depths of the oceans to the summits of mountains, but most life exists within a few meters of the surface. A second meaning is the planetary system that includes and sustains life, and therefore is made up of the atmosphere, oceans, soils, upper bedrock, and all life.

Biota All the organisms of all species living in an area or region up to and including the biosphere, as in "the biota of the Mojave Desert" or "the biota in that aquarium."

Biotic province A geographic region inhabited by life-forms (species, families, orders) of common ancestry, bounded by barriers that prevent the spread of the distinctive kinds of life to other regions and the immigration of foreign species into that region.

Birth rate The rate at which births occur in a population, measured either as the number of individuals born per unit of time or as the percentage of births per unit of time compared with the total population.

Black lung disease Often called coal miner disease because it is caused by years of inhaling coal dust, resulting in damage to the lungs.

Bog An open body of water with surface inlets—usually small streams—but no surface outlet.

Breeder reactors A type of nuclear reactor that utilizes between 40% and 70% of its nuclear fuel and converts fertile nuclei to fissile nuclei faster than the rate of fission. Thus breeder reactors actually produce nuclear fuels.

Burner reactors A type of nuclear reactor that consumes more fissionable material than it produces.

Carbon cycle Biogeochemical cycle of carbon. Carbon combines with and is chemically and biologically linked with the cycles of oxygen and hydrogen that form the major compounds of life.

Carbon-silicate cycle A complex biogeochemical cycle over time scales as long as one-half billion years. Included in this cycle are major geologic processes, such as weathering, transport by ground and surface waters, erosion, and deposition of crustal rocks. The carbonate-silicate cycle is believed to provide important negative feedback mechanisms that control the temperature of the atmosphere.

Carcinogen Any material that is known to produce cancer in humans or other animals.

Carnivores Organisms that feed on other live organisms; usually applied to animals that eat other animals.

Carrying capacity The maximum abundance of a population or species that can be maintained by a habitat or ecosystem without degrading the ability of that habitat or ecosystem to maintain that abundance in the future.

Catch per unit effort The number of animals caught per unit of effort, such as the number of fish caught by a fishing ship per day. It is used to estimate the population abundance of a species.

Certification of forestry The process of comparing the actual practices of specific corporations or government agencies with practices that are believed to be consistent with sustainability.

Chemical reaction The process in which compounds and elements undergo a chemical change to become a new substance or substances.

Chemoautotrophs Autotrophic bacteria that can derive energy from chemical reactions of simple inorganic compounds.

Chimney effect (or stack effect) A process whereby warmer air rises in buildings to upper levels and is replaced in the lower portion of the building by outdoor air drawn through a variety of openings, such as windows, doors, or cracks in the foundations and walls.

Chlorofluorocarbons (CFCs) Highly stable compounds that have been or are being used in spray cans as aerosol propellants and in refrigeration units (the gas that is compressed and expanded in a cooling unit). Emissions of chlorofluorocarbons have been associated with potential global warming and stratospheric ozone depletion.

Chronic disease A disease that is persistent in a population, typically occurring in a relatively small but constant percentage of the population.

Chronic patchiness A situation where ecological succession does not occur. One species may replace another, or an individual of the first species may replace it, but no overall general temporal pattern is established. Characteristic of harsh environments such as deserts.

City planning Formal, conscious planning for new cities in modern Western civilization.

Classical stability A system characterized by constant conditions that, if disturbed from those conditions, will return to it once the factor that disturbed the system has been removed.

Clear-cutting In timber harvesting, the practice of cutting all trees in a stand at the same time.

Climate The representative or characteristic conditions of the atmosphere at particular places on Earth. Climate refers to the average or expected conditions over long periods; weather refers to the particular conditions at one time in one place.

Climate forcing An imposed perturbation of Earth energy that balance major climatic forcings associated with global warming. Includes: greenhouse gases, such as carbon dioxide and methane; reflective aerosols in the atmosphere, black carbon. Forcing of the climate system also includes solar activity and Milankovitch Cycles.

Closed system A system in which there are definite boundaries to mass and energy and thus exchange of these factors with other systems does not occur.

Coal Solid, brittle carbonaceous rock that is one of the world's most abundant fossil fuels. It is classified according to energy content as well as carbon and sulfur content.

Codominants In forestry, fairly common trees that share the canopy or top part of the forest.

Cogeneration The capture and use of waste heat; for example, using waste heat from a power plant to heat adjacent factories and other buildings.

Commons Land that belongs to the public, not to individuals. Historically a part of old English and New England towns where all the farmers could graze their cattle.

Community effect (community-level effect) When the interaction between two species leads to changes in the presence or absence of other species or to a large change in abundance of other species, then a community effect is said to have occurred.

Competitive exclusion principle The idea that two populations of different species with exactly the same requirements cannot persist indefinitely in the same habitat—one will always win out and the other will become extinct.

Composting Biochemical process in which organic materials, such as lawn clippings and kitchen scraps, are decomposed to a rich, soil-like material.

Conservation With respect to resources such as energy, refers to changing our patterns of use or simply getting by with less. In a pragmatic sense the term means adjusting our needs to minimize the use of a particular resource, such as energy.

Consumptive use A type of off-stream water use. This water is consumed by plants and animals or in industrial processes or evaporates during use. It is not returned to its source.

Contamination The presence of undesirable material that makes something unfit for a particular use.

Continuous oil resources Resources that are regional in extent, occurring in broad geologic basins.

Controlled experiment A controlled experiment is designed to test the effects of independent variables on a dependent variable by changing only one independent variable at a time. For each variable tested, there are two setups (an experiment and a control) that are identical except for the independent variable being tested. Any difference in the outcome (dependent variable) between the experiment and the control can then be attributed to the effects of the independent variable tested.

Conventional oil resources Located in discrete oil fields where the oil and gas are trapped geologically.

Convergent evolution The process by which species evolve in different places or different times and, although they have different genetic heritages, develop similar external forms and structures as a result of adaptation to similar environments. The similarity in the shapes of sharks and porpoises is an example of convergent evolution.

Criteria pollutants Are the sixth most common air pollutants: sulfur dioxide, nitrogen oxides, carbon monoxide, ozone and other photochemical oxidants, particulate matter,and lead.

Crude oil Naturally occurring petroleum, normally pumped from wells in oil fields. Refinement of crude oil produces most of the petroleum products we use today.

Cultural eutrophication Human-induced eutrophication that involves nutrients such as nitrates or phosphates that cause a rapid increase in the rate of plant growth in ponds, lakes, rivers, or the ocean.

Cultural justification With respect to environmental values refers to the fact that different cultures have many of the same values but differ in others.

Death rate The rate at which deaths occur in a population, measured either as the number of individuals dying per unit of time or as the percentage of a population dying per unit of time.

Decomposers Organisms that feed on dead organic matter.

Deductive reasoning Drawing a conclusion from initial definitions and assumptions by means of logical reasoning.

Demographic transition The pattern of change in birth and death rates as a country is transformed from undeveloped to developed. There are three stages: (1) in an undeveloped country, birth and death rates are high and the growth rate low; (2) the death rate decreases, but the birth rate remains high and the growth rate is high; (3) the birth rate drops toward the death rate and the growth rate therefore also decreases.

Demography The study of populations, especially their patterns in space and time.

Denitrification The conversion of nitrate to molecular nitrogen by the action of bacteria—an important step in the nitrogen cycle.

Dependent variable A variable that changes in response to changes in an independent variable; a variable taken as the outcome of one or more other variables.

Desalination The removal of salts from seawater or brackish water so that the water can be used for purposes such as agriculture, industrial processes, or human consumption.

Direct costs Costs borne by the producer in obtaining, processing, and distributing a product.

Disease An impairment of an individual's health.

Disprovability The idea that a statement can be said to be scientific if someone can clearly state a method or test by which it might be disproved.

Divergent evolution Organisms with the same ancestral genetic heritage migrate to different habitats and evolve into species with different external forms and structures, but typically continue to use the same kind of habitats. The ostrich and the emu are believed to be examples of divergent evolution.

Dominant species Generally, the species that are most abundant in an area, ecological community, or ecosystem.

Dominants In forestry, the tallest, most numerous, and most vigorous trees in a forest community

Dose response The principle that the effect of a certain chemical on an individual depends on the dose or concentration of that chemical.

Doubling time The time necessary for a quantity of whatever is being measured to double.

Dynamic equilibrium A steady state of a system that with negative feedback will return to a quasi-equilibrium state following disturbance.

Dynamic system Characterized by a system that changes often and continually over time.

Early-successional species Species that occur only or primarily during early stages of succession.

Ecological community This term has two meanings. (1) A conceptual or functional meaning: a set of interacting species that occur in the same place (sometimes extended to mean a set that interacts in a way to sustain life). (2) An operational meaning: a set of species found in an area, whether or not they are interacting.

Ecological engineering The design of ecosystems for the mutual benefit of humans and nature.

Ecological gradient A change in the relative abundance of a species or group of species along a line or over an area.

Ecological island An area that is biologically isolated so that a species occurring within the area cannot mix (or only rarely mixes) with any other population of the same species.

Ecological justification for the conservation of nature An argument for the conservation of nature on the grounds that a species, an ecological community, an ecosystem, or Earth's biosphere provides specific functions necessary to the persistence of our life or of benefit to life. The ability of trees in forests to remove carbon dioxide produced in burning fossil fuels is such a public benefit and an argument for maintaining large areas of forests.

Ecological niche The general concept is that the niche is a species' "profession"—what it does to make a living. The term is also used to refer to a set of environmental conditions within which a species is able to persist.

Ecological restoration The actual activity of restoring ecosystems.

Ecological succession The process of the development of an ecological community or ecosystem, usually viewed as a series of stages—early, middle, late, mature (or climax), and sometimes postclimax. Primary succession is an original establishment; secondary succession is a reestablishment.

Ecosystem An ecological community and its local, nonbiological community. An ecosystem is the minimum system that includes and sustains life. It must include at least an autotroph, a decomposer, a liquid medium, a source and sink of energy, and all the chemical elements required by the autotroph and the decomposer.

Ecosystem effect Effects that result from interactions among different species, effects of species on chemical elements in their environment, and conditions of the environment.

Ecosystem energy flow The flow of energy through an ecosystem—from the external environment through a series of organisms and back to the external environment.

ED-50 The effective dose, or dose that causes an effect in 50% of the population of exposure to a particular toxicant. It is related to the onset of specific symptoms, such as loss of hearing, nausea, or slurred speech.

Effluent stream Type of stream where flow is maintained during the dry season by groundwater seepage into the channel.

Electromagnetic fields (EMFs) Magnetic and electrical fields produced naturally by our planet and also by appliances such as toasters, electric blankets, and computers. There currently is controversy concerning potential adverse health effects related to exposure to EMFs in the workplace and home from such artificial sources as power lines and appliances.

Energy efficiency Refers to both first-law efficiency and second-law efficiency, where first-law efficiency is the ratio of the actual amount of energy delivered to the amount of energy supplied to meet a particular need, and secondlaw efficiency is the ratio of the maximum available work needed to perform a particular task to the actual work used to perform that task.

Entropy A measure in a system of the amount of energy that is unavailable for useful work. As the disorder of a system increases, the entropy in a system also increases.

Environment health Human health and disease that is determined by or related to environmental factors such as toxic chemicals, toxic biological agents, or radiation.

Environmental audit A process of determining the past history of a particular site, with special reference to the existence of toxic materials or waste.

Environmental economics Economic effects of the environment and how economic processes affect that environment, including its living resources.

Environmental justice The principle of dealing with environmental problems in such a way as to not discriminate against people based upon socioeconomic status, race, or ethnic group.

Environmental law A field of law concerning the conservation and use of natural resources and the control of pollution.

Environmental tobacco smoke (ETS) Commonly called second-hand smoke from people smoking tobacco.

Environmental unity A principle of environmental sciences that states that everything affects everything else, meaning that a particular course of action could lead to a string of events. Another way of stating this idea is that you can't do only one thing.

Epidemic disease A disease that appears occasionally in the population, affects a large percentage of it, and declines or almost disappears for a while only to reappear later.

Equilibrium A point of rest. At equilibrium, a system remains in a single, fixed condition and is said to be in equilibrium. Compare with **Steady state**.

Eukaryote An organism whose cells have nuclei and organelles. The eukaryotes include animals, fungi, vegetation, and many single-cell organisms.

Eutrophication Increase in the concentration of chemical elements required for living things (for example, phosphorus).Increased nutrient loading may lead to a population explosion of photosynthetic algae and blue-green bacteria that become so thick that light cannot penetrate the water. Bacteria deprived of light beneath the surface die; as they decompose, dissolved oxygen in the lake is lowered and eventually a fish kill may result. Eutrophication of lakes caused by human-induced processes, such as nutrient-rich sewage water entering a body of water, is called cultural eutrophication.

E-waste Waste from electronic products such as televisions and computers.

Experimental errors There are two kinds of experimental errors, random and systematic. Random errors are those due to chance events, such as air currents pushing on a scale and altering a measurement of weight. In contrast, a miscalibration of an instrument would lead to a systematic error. Human errors can be either random or systematic.

Exponential growth Growth in which the rate of increase is a constant percentage of the current size; that is, the growth occurs at a constant rate per time period.

Exponential rate The annual growth rate is a constant percentage of the population.

Externality In economics, an effect not normally accounted for in the cost–revenue analysis.

Fact Something that is known based on actual experience and observation.

Fall line The point on a river where there is an abrupt drop in elevation of the land and where numerous waterfallsoccur. The line in the eastern United States is located where streams pass from harder to softer rocks.

Fecal coliform bacteria Bacteria that occur naturally in human intestines and are used as a standard measure of microbial pollution and an indicator of disease potential for a water source.

Feedback A kind of system response that occurs when output of the system also serves as input leading to changes in the system.

First law of thermodynamics The principle that energy may not be created or destroyed but is always conserved.

First-law efficiency The ratio of the actual amount of energy delivered where it is needed to the amount of energy supplied in order to meet that need; expressed as a percentage.

Fission The splitting of an atom into smaller fragments with the release of energy.

Flow The amount of material transferred.

Flux The rate of transfer of material within a system per unit of time.

Food chains The linkage of who feeds on whom.

Food webs A network of who feeds on whom or a diagram showing who feeds on whom. It is synonymous with **food chain**.

Fossil fuels Forms of stored solar energy created from incomplete biological decomposition of dead organic matter. Include coal, crude oil, and natural gas.

Founder effect Occurs when a small number of individuals are isolated from a larger population; they may have much less genetic variation than the original species (and usually do), and the characteristics that the isolated population has will be affected by chance.

Fuel cells Highly efficient power-generating systems that produce electricity by combining fuel and oxygen in an electrochemical reaction.

Fusion The fusing, or combining, of atomic nuclei with the release of energy.

Gaia hypothesis The Gaia hypothesis states (1) that life has greatly altered the Earth's environment globally for more than 3 billion years and continues to do so; and (2) that these changes benefit life (increase its persistence). Some extend this, nonscientifically, to assert that life did it on purpose.

Garden city Land planning that considers a city and countryside together.

Gene A single unit of genetic information comprising of a complex segment of the four DNA base-paircompounds.

General circulation models (GCMs) Consists of a group of computer models that focus on climate change using a series of equations, often based on conservation of mass and energy.

Genetic drift Changes in the frequency of a gene in a population as a result of chance rather than of mutation, selection, or migration.

Genetically modified crops Crop species modified by genetic engineering to produce higher crop yields and increase resistance to drought, cold, heat, toxins, plant pests, and disease.

Genotype The genetic makeup that is characteristic of an individual or a group.

Geologic cycle The formation and destruction of earth materials and the processes responsible for these events. The geologic cycle includes the following subcycles: hydrologic, tectonic, rock, and geochemical.

Geothermal energy The useful conversion of natural heat from the interior of Earth.

Global dimming The reduction of incoming solar radiation by reflection from suspended particles in the atmosphere and their interaction with water vapor (especially clouds).

Global extinction A species can no longer be found anywhere.

Green building Designing buildings that have a healthy interior environment and landscape that benefits the local external environment as well.

Green revolution Name attached to post-World War II agricultural programs that have led to the development of new strains of crops with higher yield, better resistance to disease, or better ability to grow under poor conditions.

Greenbelt A belt of recreational parks, farmland, or uncultivated land surrounding or connecting urban communities, forming a system of countryside and urban landscapes.

Greenhouse effect Occurs when water vapor and several other gases warm the Earth's atmosphere by trapping some of the heat radiating from the Earth's atmospheric system.

Greenhouse gases The suite of gases that produce a greenhouse effect, such as carbon dioxide, methane, and water vapor.

Gross production (biology) Production before respiration losses are subtracted.

Groundwater Water found beneath the Earth's surface within the zone of saturation, below the water table.

Growth rate The net increase in some factor per unit of time. In ecology, the growth rate of a population, sometimes measured as the increase in numbers of individuals or biomass per unit of time and sometimes as a percentage increase in numbers or biomass per unit of time.

Hazardous waste Waste that is classified as definitely or potentially hazardous to the health of people. Examples include toxic or flammable liquids and a variety of heavy metals, pesticides, and solvents.

Heavy metals Refers to a number of metals, including lead, mercury, arsenic, and silver (among others) that have a relatively high atomic number (the number of protons in the nucleus of an atom). They are often toxic even at relatively low concentrations, causing a variety of environmental problems.

Herbivores An organism that feeds on an autotroph.

Heterotrophs Organisms that cannot make their own food from inorganic chemicals and a source of energy and therefore live by feeding on other organisms.

High-level radioactive waste Extremely toxic nuclear waste, such as spent fuel elements from commercial reactors.

Historical range of variation The known range of an environmental variable, such as the abundance of a species or the depth of a lake, over some past time interval.

Hormonally active agents (HAAs) Chemicals in the environment able to cause reproductive and developmental abnormalities in animals, including humans.

Human carrying capacity Theoretical estimates of the number of humans who could inhabit Earth at the same time.

Hydrologic cycle Circulation of water from the oceans to the atmosphere and back to the oceans by way of evaporation, runoff from streams and rivers, and groundwater flow.

Hydrologic fracturing (fracking) The process of extracting methane gas from deep wells by using water and other chemicals to fracture the rocks.

Hypothesis In science, an explanation set forth in a manner that can be tested and disproved. A tested hypothesis is accepted until and unless it has been disproved.

Incineration Combustion of waste at high temperature, consuming materials and leaving only ash and noncombustibles to dispose of in a landfill.

Independent variable In an experiment, the variable that is manipulated by the investigator. In an observational study, the variable that is believed by the investigator to affect an outcome, or dependent, variable.

Indirect cost See **Externality.**

Inductive reasoning Drawing a general conclusion from a limited set of specific observations.

Industrial ecology The process of designing industrial systems to behave more like ecosystems where waste from one part of the system is a resource for another part.

Inferences 1) A conclusion derived by logical reasoning from premises and/or evidence (observations or facts), or (2) a conclusion, based on evidence, arrived at by insight or analogy, rather than derived solely by logical processes.

Inflection point The point where a graphed curve or an equation representing that curve changes from convex to concave, or from concave to convex.

Influent stream Type of stream that is everywhere above the groundwater table and flows in direct response to precipitation. Water from the channel moves down to the water table, forming a recharge mound.

In-stream use A type of water use that includes navigation, generation of hydroelectric power, fish and wildlife habitat, and recreation.

Input With respect to basic concepts of systems, refers to material or energy that enters a system.

Intangible factor In economics, an intangible factor is one you can't touch directly, but you value it.

Integrated energy management Use of a range of energy options that vary from region to region, including a mix of technology and sources of energy.

Integrated pest management Control of agricultural pests using several methods together, including biological and chemical agents. A goal is to minimize the use of artificial chemicals; another goal is to prevent or slow the buildup of resistance by pests to chemical pesticides.

Integrated waste management (IWM) Set of management alternatives including reuse, source reduction, recycling, composting, landfill, and incineration.

Intermediate In forestry, trees that form a layer of growth below dominants.

Keystone species A species, such as the sea otter, that has a large effect on its community or ecosystem so that its removal or addition to the community leads to major changes in the abundances of many or all other species.

Lag time The delay in time between the cause and appearance of an effect in a system.

Late-successional species Species that occur only or primarily in, or are dominant in, late stages in succession.

LD-50 A crude approximation of a chemical toxicity defined as the dose at which 50% of the population dies on exposure.

Leachate Noxious, mineralized liquid capable of transporting bacterial pollutants. Produced when water infiltrates through waste material and becomes contaminated and polluted.

Liebig's law of the minimum The concept that the growth or survival of a population is directly related to the single life requirement that is in least supply (rather than due to a combination of factors).

Life expectancy The estimated average number of years (or other time period used as a measure) that an individual of a specific age can expect to live.

Limiting factor The single requirement for growth available in the least supply in comparison to the need of an organism. Originally applied to crops but now often applied to any species.

Linear process With respect to systems, refers to the addition or subtraction of anything to a compartment in a system where the amount will always be the same, no matter how much you have added before and what else has changed about the system and the environment. For example, if you collect stones from a particular site and place them in a basket and place one stone per hour, you will have placed 6 stones in 6 hours and 24 in 24 hours, and the change is linear with time.

Local extinction The disappearance of a species from part of its range but continued persistence elsewhere.

Logistic carrying capacity In terms of the logistic curve, the population size at which births equal deaths and there is no net change in the population.

Logistic growth curve The S-shaped growth curve that is generated by the logistic growth equation. In the logistic, a small population grows rapidly, but the growth rate slows down, and the population eventually reaches a constant size.

Low-level radioactive waste Waste materials that contain sufficiently low concentrations or quantities of radioactivity so as not to present a significant environmental hazard if properly handled.

Macronutrients Elements required in large amounts by living things. These include the big six—carbon, hydrogen, oxygen, nitrogen, phosphorus, and sulfur.

Made lands Man-made areas created artificially with fill, sometimes as waste dumps of all kinds and sometimes to make more land available for construction.

Malnourishment The lack of specific components of food, such as proteins, vitamins, or essential chemical elements.

Manipulated variable See **Variable, independent.**

Mariculture Production of food from marine habitats.

Materially closed system Characterized by a system in which no matter moves in and out of the system, although energy and information may move across the system's boundaries. For example, Earth is a materially closed system for all practical purposes.

Materials management In waste management, methods consistent with the ideal of industrial ecology, making better use of materials and leading to more sustainable use of resources.

Maximum lifetime Genetically determined maximum possible age to which an individual of a species can live.

Maximum sustainable yield The maximum usable production of a biological resource that can be obtained in a specified time period without decreasing the ability of the resource to sustain that level of production.

Maximum sustainable-yield population The largest population size that can be sustained indefinitely.

Medieval Warm Period (MWP) A period of approximately 300 years from A.D. 950 to 1250 when Earth's surface was considerably warmer than the normal that we experience today. The warming was particularly relevant and important in Western Europe and the Atlantic Ocean where the MWP was a time of flourishing culture and activity, as well as expansion of population.

Megacities Urban areas with at least 8 million inhabitants.

Mega disaster A very large disaster/catastrophe. Hurricane Katrina in 2005 (Gulf Coast), the Indonesian tsunami of 2004 and the 1861-62 floods in California are examples.

Meltdown A nuclear accident in which the nuclear fuel forms a molten mass that breaches the containment of the reactor, contaminating the outside environment with radioactivity.

Methane hydrate A white icelike compound made up of molecules of methane gas trapped in "cages" of frozen water in the sediments of the deep seafloor.

Micronutrients Chemical elements required in very small amounts by at least some forms of life. Boron, copper, and molybdenum are examples of micronutrients.

Middle-successional species Species that occur in between early or late-successional species.

Migration The movement of an individual, population, or species from one habitat to another or more simply from one geographic area to another.

Minimum viable population The minimum number of individuals that have a reasonable chance of persisting for a specified time period.

Mobile sources Sources of air pollutants that move from place to place; for example, automobiles, trucks, buses, and trains.

Model A deliberately simplified explanation, often physical, mathematical, pictorial, or computer-simulated, of complex phenomena or processes.

Monoculture The planting of large areas with a single species or even a single strain or subspecies in farming.

Moral justification for the conservation of nature An argument for the conservation of nature on the grounds that aspects of the environment have a right to exist, independent of human desires, and that it is our moral obligation to allow them to continue or to help them persist.

Mutation Stated most simply, a chemical change in a DNA molecule. It means that the DNA carries a different message than it did before, and this change can affect the expressed characteristics when cells or individual organisms reproduce.

Nanotechnology A surface-water cleanup technique that uses extremely small material particles (10–9m size, about 100,000 times thinner than human hair) designed for a number of purposes.

Natural capital Ecological systems that provide public service benefits.

Natural gas Naturally occurring gaseous hydrocarbon (predominantly methane) generally produced in association with crude oil or from gas wells; an important efficient and clean-burning fuel commonly used in homes and industry.

Natural selection A process by which organisms whose biological characteristics better fit them to the environment are represented by more descendants in future generations than those whose characteristics are less fit for the environment.

Naturalization The process by which a species, introduced into a new habitat or ecosystem, adjusts to the environmental and biological conditions, and is able to successfully reproduce and populations are able to persist.

Negative feedback A type of feedback that occurs when the system's response is in the opposite direction of the output. Thus negative feedback is self-regulating.

Net production (biology) The production that remains after utilization. In a population, net production is sometimes measured as the net change in the numbers of individuals. It is also measured as the net change in biomass or in stored energy. In terms of energy, it is equal to the gross production minus the energy used in respiration.

Nitrogen cycle A complex biogeochemical cycle responsible for moving important nitrogen components through the biosphere and other Earth systems. This is an extremely important cycle because nitrogen is required by all living things.

Nitrogen fixation The process of converting inorganic, molecular nitrogen in the atmosphere to ammonia. In nature it is carried out only by a few species of bacteria, on which all life depends.

Noise pollution A type of pollution characterized by unwanted or potentially damaging sound.

Nonlinear process Characterized by system operation in which the effect of adding a specific amount of something changes, depending upon how much has been added before.

Nonpoint sources Pollution sources that are diffused and intermittent and are influenced by factors such as land use, climate, hydrology, topography, native vegetation, and geology.

Nonrenewable alternative energy Alternative energy sources, such as deep-earth geothermal energy, where output exceeds input.

Nonrenewable resources A resource that is cycled so slowly by natural Earth processes that once used, it is essentially not going to be made available within any useful time framework.

Nuclear energy The energy of the atomic nucleus that, when released, may be used to do work. Controlled nuclear fission reactions take place within commercial nuclear reactors to produce energy.

Nuclear fuel cycle Processes involved with producing nuclear power from the mining and processing of uranium to control fission, reprocessing of spent nuclear fuel, decommissioning of power plants, and disposal of radioactive waste.

Nuclear reactors Devices that produce controlled nuclear fission, generally for the production of electric energy.

Obligate symbionts A symbiotic relationship between two organisms in which neither by themselves can exist without the other.

Observations Information obtained through one or more of the five senses or through instruments that extend the senses. For example, some remote sensing instruments measure infrared intensity, which we do not see, and convert the measurement into colors, which we do see.

Ocean acidification An increase in the acidity of seawater.

Off-stream use Type of water use where water is removed from its source for a particular use.

Oil shale A fine-grained sedimentary rock containing organic material known as kerogen. On distillation, it yields significant amounts of hydrocarbons, including oil.

Old-growth forest A nontechnical term often used to mean a virgin forest (one never cut), but also used to mean a forest that has been undisturbed for a long, but usually unspecified, time.

Omnivores Organisms that eat both plants and animals.

Open system A type of system in which exchanges of mass or energy occur with other systems.

Operational definitions Definitions that tell you what you need to look for or do in order to carry out an operation, such as measuring, constructing, or manipulating.

Optimum sustainable population The population size that is in some way best for the population, its ecological community, its ecosystem, or the biosphere.

Ore deposits Earth materials in which metals exist in high concentrations, sufficient to be mined.

Organelle Specialized parts of cells that function like organs in multi-celled organisms.

Organic compounds Carbon compounds produced naturally by living organisms or synthetically by industrial processes.

Organic farming Farming that is more "natural" in the sense that it does not involve the use of artificial pesticides and, more recently, genetically modified crops. In recent years governments have begun to set up legal criteria for what constitutes organic farming.

Outbreaks Sudden occurrences of waterborne disease.

Output With respect to basic operation of system, refers to material or energy that leaves a particular storage compartment.

Overdraft Groundwater withdrawal when the amount pumped from wells exceeds the natural rate of replenishment.

Overshoot and collapse Occurs when growth in one part of a system over time exceeds carrying capacity, resulting in a sudden decline in one or both parts of the system.

Ozone (O_3) A form of oxygen in which three atoms of oxygen occur together. It is chemically active and has a short average lifetime in the atmosphere. Forms a natural layer high in the atmosphere (stratosphere) that protects us from harmful ultraviolet radiation from the sun, is an air pollutant when present in the lower atmosphere above the National Air Quality Standards.

Ozone shield Stratospheric ozone layer that absorbs ultraviolet radiation.

Paleocene-Eocene Thermal Maximum (PETM) A period 50 million years ago when the concentration of atmospheric carbon dioxide was higher than it is today and the Earth was free (or nearly so of polar glaciers). It was a time of natural global warming.

Pandemic A worldwide disease outbreak.

Parasitism When one organism (the parasite) lives on or within another (the host) and depends on it for existence but makes no useful contribution to it and may in fact harm it.

Particulates Small particles of dust (including soot and asbestos fibers) released into the atmosphere by many natural processes and human activities.

Passive solar energy systems Direct use of solar energy through architectural design to enhance or take advantage of natural changes in solar energy that occur throughout the year without requiring mechanical power.

Pasture Land plowed and planted to provide forage for domestic herbivorous animals.

Peak oil Refers to the time in the future when one-half of Earth's oil has been exploited. Peak oil is expected to occur sometime between 2020 and 2050.

Pelagic ecosystem An ecosystem that occurs in the floating part of an ocean or sea, without any physical connections to the bottom of the ocean or sea.

Persistent organic pollutants (POPs) Synthetic carbon-based compounds, often containing chlorine, that do not easily break down in the environment. Many were introduced decades before their harmful effects were fully understood and are now banned or restricted.

Phosphorus cycle A major biogeochemical cycle involving the movement of phosphorus throughout the biosphere and lithosphere. This cycle is important because phosphorus is an essential element for life and often is a limiting nutrient for plant growth.

Photochemical smog Sometimes called L.A.-type smog or brown air. Directly related to automobile use and solar radiation. Reactions that occur in the development of the smog are complex and involve both nitrogen oxides and hydrocarbons in the presence of sunlight.

Photosynthesis Synthesis of sugars from carbon dioxide and water by living organisms using light as energy. Oxygen is given off as a by-product.

Photovoltaics Technology that converts sunlight directly into electricity using a solid semiconductor material.

Plantation In forestry, managed forests, in which a single species is planted in straight rows and harvested at regular intervals.

Plate tectonics A model of global tectonics that suggests that the outer layer of Earth, known as the lithosphere, is composed of several large plates that move relative to one another. Continents and ocean basins are passive riders on these plates.

Point sources Sources of pollution such as smokestacks, pipes, or accidental spills that are readily identified and stationary. They are often thought to be easier to recognize and control than are area sources. This is true only in a general sense, as some very large point sources emit tremendous amounts of pollutants into the environment.

Polar amplification Processes in which global warming causes greater temperature increases at polar regions.

Policy instruments The means to implement a society's policies. Such instruments include moral suasion (jawboning—persuading people by talk, publicity, and social pressure); direct controls, including regulations; and market processes affecting the price of goods, subsidies, licenses, and deposits.

Pollution The process by which something becomes impure, defiled, dirty, or otherwise unclean.

Pollution prevention Identifying ways to avoid the generation of waste rather than finding ways to dispose of it.

Population A group of individuals of the same species living in the same area or interbreeding and sharing genetic information.

Population dynamics The causes of changes in population size.

Positive feedback A type of feedback that occurs when an increase in output leads to a further increase in output. This is sometimes known as a vicious cycle, since the more you have, the more you get.

Precautionary Principle The idea that even full scientific certainty is not available to prove cause and effect, we should still take cost-effective precautions to solve environmental problems when it appears to be a threat of potentially serious and irreversible environmental damage.

Predation When an organism (a predator) feeds on other live organisms (prey), usually of another species.

Primary pollutants Air pollutants emitted directly into the atmosphere. Included are particulates, sulfur oxides, carbon monoxide, nitrogen oxides, and hydrocarbons.

Primary production See **Production, primary**.

Primary succession The initial establishment and development of an ecosystem.

Primary treatment (of wastewater) Removal of large particles and organic materials from wastewater through screening.

Prokaryote A kind of organism that lacks a true cell nucleus and has other cellular characteristics that distinguish it from the *eukaryotes*. Bacteria are prokaryotes.

Public-service functions Functions performed by ecosystems that benefit other forms of life in other ecosystems. Examples include the cleansing of the air by trees and removal of pollutants from water by infiltration through the soil.

Qualitative data Data distinguished by qualities or attributes that cannot be or are not expressed as quantities. For example, blue and red are qualitative data about the electromagnetic spectrum.

Quantitative data Data expressed as numbers or numerical measurements. For example, the wavelengths of specific colors of blue and red light (460 and 650 nanometers, respectively) are quantitative data about the electromagnetic spectrum.

Radioactive decay A process of decay of radioisotopes that change from one isotope to another and emit one or more forms of radiation.

Radioisotope A form of a chemical element that spontaneously undergoes radioactive decay.

Radionuclides Atoms with unstable nuclei that undergo radioactive decay.

Radon A naturally occurring radioactive gas. Radon is colorless, odorless, and tasteless and must be identified through proper testing.

Rangeland Land used for grazing.

Reclamation The restoration of land degraded by mining.

Recreational justification for the conservation of nature An argument for the conservation of nature on the grounds that direct experience of nature is inherently enjoyable and that the benefits derived from it are important and valuable to people.

Renewable energy Alternative energy sources, such as solar, water, wind, and biomass, that are more or less continuously available in a time framework useful to people.

Renewable resources A resource, such as timber, water, or air, that is naturally recycled or recycled by artificial processes within a time frame useful for people.

Reserves Known and identified deposits of earth materials from which useful materials can be extracted profitably with existing technology and under present economic and legal conditions.

Residence time The average time that an atom is stored in a compartment.

Resources Reserves plus other deposits of useful earth materials that may eventually become available.

Responding variable See **Variable, independent.**

Restoration ecology The field within the science of ecology whose goal is to return damaged ecosystems to ones that are functional, sustainable, and more natural in some meaning of this word.

Risk assessment The process of determining potentially adverse environmental health effects to people following exposure to pollutants and other toxic materials. It generally includes four steps: identification of the hazard, dose-response assessment, exposure assessment, and risk characterization.

Risk-benefit analysis In environmental economics, weighing the riskiness of the future against the value we place on things in the present.

Rock cycle A group of processes that produce igneous, metamorphic, and sedimentary rocks.

Rotation time Time between cuts of a stand or area of forest.

R-to-C ratio A measure of the time available for finding the solutions to depletion of nonrenewable reserves, where R is the known reserves (for example, hundreds of thousands of tons of a metal) and C is the rate of consumption (for example, thousands of tons per year used by people).

Sanitary landfill A method of disposal of solid waste without creating a nuisance or hazard to public health or safety. Sanitary landfills are highly engineered structures with multiple barriers and collection systems to minimize environmental problems.

Scientific method A set of systematic methods by which scientists investigate natural phenomena, including gathering data, formulating and testing hypotheses, and developing scientific theories and laws.

Scientific theory A grand scheme that relates and explains many observations and is supported by a great deal of evidence, in contrast to a guess, a hypothesis, a prediction, a notion, or a belief.

Scrubbing A process of removing sulfur from gases emitted from power plants burning coal. The gases are treated with a slurry of lime and limestone, and the sulfur oxides react with the calcium to form insoluble calcium sulfides and sulfates that are collected and disposed of.

Second law of thermodynamics The law of thermodynamics which states that *no use of energy in the real (not theoretical) world can ever be 100% efficient.*

Secondary pollutants Air pollutants produced through reactions between primary pollutants and normal atmospheric compounds. An example is ozone that forms over urban areas through reactions of primary pollutants, sunlight, and natural atmospheric gases.

Secondary production See **Production, secondary.**

Secondary succession The reestablishment of an ecosystem where there are remnants of a previous biological community.

Secondary treatment (of wastewater) Use of biological processes to degrade wastewater in a treatment facility.

Second-growth forest A forest that has been cut and has regrown.

Second-law efficiency The ratio of the minimum available work needed to perform a particular task to the actual work used to perform that task. Reported as a percentage.

Secure landfill A type of landfill designed specifically for hazardous waste. Similar to a modern sanitary landfill in that it includes multiple barriers and collection systems to ensure that leachate does not contaminate soil and other resources.

Seed-tree cutting A logging method in which mature trees with good genetic characteristics and high seed production are preserved to promote regeneration of the forest. It is an alternative to clear-cutting.

Selective cutting In timber harvesting, the practice of cutting some, but not all, trees, leaving some on the site. There are many kinds of selective cutting. Sometimes the biggest trees with the largest market value are cut, and smaller trees are left to be cut later. Sometimes the best trees are left to provide seed for future generations. Sometimes trees are left for wildlife habitat and recreation.

Shale gas A natural gas within tiny openings of shale rock.

Shelterwood cutting A logging method in which dead and less desirable trees are cut first; mature trees are cut later. This ensures that young, vigorous trees will always be left in the forest. It is an alternative to clear-cutting.

Sick building syndrome (SBS) A condition associated with a particular indoor environment that appears to be unhealthy for the human occupants.

Silviculture The practice of growing trees and managing forests, traditionally with an emphasis on the production of timber for commercial sale.

Single-stream recycling Recycling process in which paper, plastic, glass, and metals are not separated prior to collection.

Sink With respect to systems operation, refers to a component or storage cell within a system that is receiving a material, such as a chemical. The donating compartment is called a source, and there generally is a flux or rate of transfer between the source and sink.

Site (in relation to cities) Environmental features of a location that influence the placement of a city. For example,

New Orleans is built on low-lying muds, which form a poor site, while New York City's Manhattan is built on an island of strong bedrock, an excellent site.

Site quality Used by foresters to mean an estimator of the maximum timber crop the land can produce in a given time.

Situation (in relation to cities) The relative geographic location of a site that makes it a good location for a city. For example, New Orleans has a good situation because it is located at the mouth of the Mississippi River and is therefore a natural transportation junction.

Smog A term first used in 1905 for a mixture of smoke and fog that produced unhealthy urban air. There are several types of fog, including photochemical smog and sulfurous smog.

Solar collectors Usually flat, glass-covered plates over a black background where a heat-absorbing fluid (water or some other liquid) is circulated through tubes.

Source With respect to storage compartments within a system, such as the atmosphere or land, refers to a compartment that donates to another compartment. The donating compartment is the source; the receiving compartment the sink.

Species A group of individuals capable of interbreeding.

Species evenness The relative abundance of species.

Species richness The total number of species in an area.

Srban-runoff naturalization An emerging bioengineering technology with the objective to treat urban runoff before it reaches streams, lakes, or the ocean.

Stand An informal term used by foresters to refer to a group of trees.

Static system A fixed condition that tends to remain in that exact position.

Stationary sources Air pollution sources that have a relatively fixed location, including point sources, fugitive sources, and area sources.

Steady-state system The inputs (of anything of interest) are equal to the outputs, so the amount stored within the system is constant.

Stratosphere Overlies the troposphere and the atmosphere-from approximately 20-70 kilometers above the Earth. The stratosphere contains the higher concentrations of ozone at about 25 kilometers above the Earth known as the ozone layer.

Strip cutting In timber harvesting, the practice of cutting narrow rows of forest, leaving wooded corridors.

Succession The process of establishment and development of an ecosystem.

Sulfurous smog Produced primarily by burning coal or oil at large power plants. Sulfur oxides and particulates combine under certain meteorological conditions to produce a concentrated form of this smog.

Suppressed In forestry, describes tree species growing in the understory, beneath the dominant and intermediate species.

Sustainability Management of natural resources and the environment with the goals of allowing the harvest of resources to remain at or above some specified level, and the ecosystem to retain its functions and structure.

Sustainable energy development A type of energy management that provides for reliable sources of energy while not causing environmental degradation and while ensuring that future generations will have a fair share of the Earth's resources.

Sustainable water use Use of water resources that does not harm the environment and provides for the existence of high-quality water for future generations.

Symbiont Each partner in symbiosis.

Symbiosis An interaction between individuals of two different species that benefits both. For example, lichens contain an alga and a fungus that require each other to persist. Sometimes this term is used broadly, so that domestic corn and people could be said to have a symbiotic relationship—domestic corn cannot reproduce without the aid of people, and some people survive because they have corn to eat.

Synergism Cooperative action of different substances such that the combined effect is greater than the sum of the effects taken separately.

Synergistic effect When the change in availability of one resource affects the response of an organism to some other resource.

Synfuels Synthetic fuels, which may be liquid or gaseous, derived from solid fuels, such as oil from kerogen in oil shale, or oil and gas from coal.

Synthetic organic compounds Compounds of carbon produced synthetically by human industrial processes, as for example pesticides and herbicides.

System A set of components that are linked and interact to produce a whole. For example, the river as a system is composed of sediment, water, bank, vegetation, fish, and other living things that all together produce the river.

Systematic errors Errors that occur consistently in scientific experiments, such as those resulting from incorrectly calibrated instruments.

Tangible factor In economics, something you can touch, buy, and sell.

Tar sands Sedimentary rocks or sands impregnated with tar oil, asphalt, or bitumen.

TD-50 The toxic dose defined as the dose that is toxic to 50% of a population exposed to the toxin.

Tectonic cycle The processes that change Earth's crust, producing external forms such as ocean basins, continents, and mountains.

Terminator gene A genetically modified crop that has a gene to cause the plant to become sterile after the first year.

Thermal pollution A type of pollution that occurs when heat is released into water or air and produces undesirable effects on the environment.

Thermodynamic equilibrium With respect to systems, is a physical concept of equilibrium where everything is at the lowest energy level in the system, and matter and energy are dispersed randomly.

Thinning The timber-harvesting practice of selectively removing only smaller or poorly formed trees.

Threshold A point in the operation of a system at which a change occurs. With respect to toxicology, it is a level below which effects are not observable and above which effects become apparent.

Tidal power Energy generated by ocean tides in places where favorable topography allows for construction of a power plant.

Tight gas A natural gas that is produced from continuous deposits (resevoirs) of dense sandstone or limestone.

Time series The set of estimates of some variable over a number of years.

Tolerance The ability to withstand stress resulting from exposure to a pollutant or other harmful condition.

Toxicology The science concerned with the study of poisons (or toxins) and their effects on living organisms. The subject also includes the clinical, industrial, economic, and legal problems associated with toxic materials.

Toxin Substances (pollutants) that are poisonous to living things.

Transuranic waste Radioactive waste consisting of human-made radioactive elements heavier than uranium. Includesclothing, rags, tools, and equipment that has been contaminated.

Trophic level In an ecological community, all the organisms that are the same number of food-chain steps from the primary source of energy. For example, in a grassland the green grasses are on the first trophic level, grasshoppers are on the second, birds that feed on grasshoppers are on the third, and so forth.

Troposphere The atmospheric zone from the surface of the Earth to an altitude of approximately 20 kilometers above the Earth. The troposphere is the zone of the atmosphere we are most familiar with because we spend most of our lives in it.

Ultrafine particles Particles that are smaller than 0.18 micrometers in diameter, released into the air by vehicles on streets and freeways.

Ultraviolet (UV) Index An index based on the exposure to ultraviolet radiation to humans. Varies from low to ex treme and is useful for people wishing recommendation of how much exposure to the sun they should incur and how much sun block to use.

Undernourishment The lack of sufficient calories in available food, so that one has little or no ability to move or work.

Uniformitarianism The principle stating that processes that operate today operated in the past. Therefore, observations of processes today can explain events that occurred in the past and leave evidence, for example, in the fossil record or in geologic formations.

Utilitarian justification for the conservation of nature An argument for the conservation of nature on the grounds that the environment, an ecosystem, habitat, or species provides individuals with direct economic benefit or is directly necessary to their survival.

Variables A variable is any factor that can be controlled or changed in an experiment.

Virtual water The amount of water necessary to produce a product, such as rice or, in industry, an automobile.

Wastewater treatment The process of treating wastewater (primarily sewage) in specially designed plants that accept municipal wastewater. Generally divided into three categories: primary treatment, secondary treatment, and advanced wastewater treatment.

Water budget Inputs and outputs of water for a particular system (a drainage basin, region, continent, or the entire Earth).

Water conservation Practices designed to reduce the amount of water we use.

Water power An alternative energy source derived from flowing water. One of the world's oldest and most common energy sources. Sources vary in size from microhydropower systems to large reservoirs and dams.

Water reuse The use of wastewater following some sort of treatment. Water reuse may be inadvertent, indirect, or direct.

Watershed An area of land that forms the drainage of a stream or river. If a drop of rain falls anywhere within a watershed, it can flow out only through that same stream or river.

Weather What is happening in the atmosphere over a short time period or what may be happening now in terms of temperature, pressure, cloudiness, precipitation, and winds. The average of weather over longer periods and regions refers to the climate.

Wetlands A comprehensive term for landforms such as salt marshes, swamps, bogs, prairie potholes, and vernal pools. Their common feature is that they are wet at least part of the year and as a result have a particular type of vegetation and soil. Wetlands form important habitats for many species of plants and animals, while serving a variety of natural service functions for other ecosystems and people.

Wilderness An area unaffected now or in the past by human activities and without a noticeable presence of human beings.

Wind power Alternative energy source that has been used by people for centuries. More recently, thousands of windmills have been installed to produce electric energy.

Work (physics) Force times the distance through which it acts. When work is done we say energy is expended.

Zero population growth Results when the number of births equals the number of deaths so that there is no net change in the size of the population.

Index

Entries followed by an f refer to figures; those followed by a t refer to tables.

D

Q

R

S

X

Y

Z